D1350798

McGraw-Hill Yearbook of Science and Technology 1978 REVIEW

1979 PREVIEW

1978

COMPREHENSIVE COVERAGE
OF THE IMPORTANT EVENTS
OF THE YEAR AS COMPILED
BY THE STAFF OF THE
McGRAW-HILL
ENCYCLOPEDIA OF
SCIENCE AND TECHNOLOGY

McGraw-Hill Yearbook o

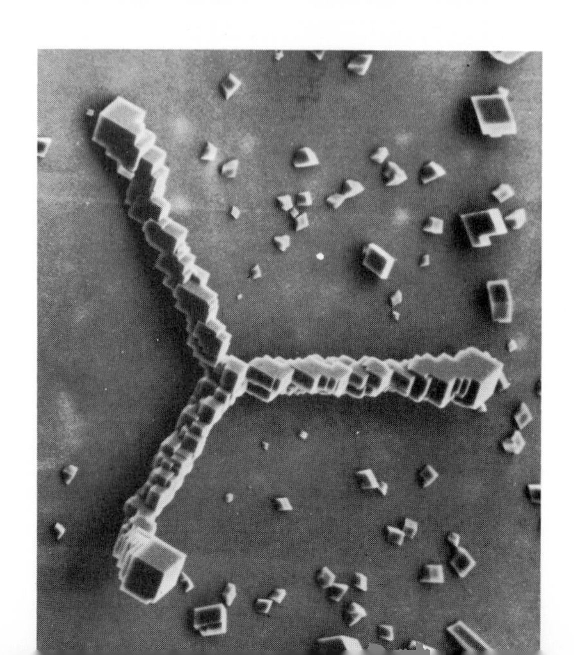

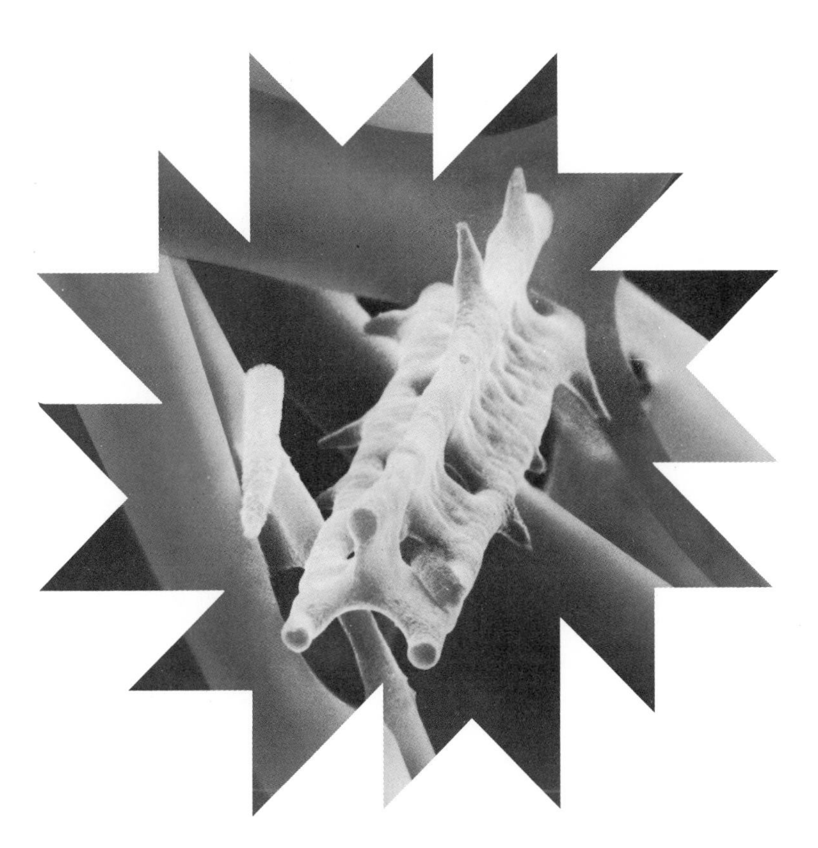

Science and Technology

McGRAW-HILL BOOK COMPANY

NEW YORK ST. LOUIS SAN FRANCISCO
AUCKLAND BOGOTA DUSSELDORF JOHANNESBURG LONDON MADRID MEXICO MONTREAL NEW DELHI PANAMA PARIS SAO PAULO SINGAPORE SYDNEY TOKYO TORONTO

On preceding pages:

Left. Calcite biocrystals decorating a triradiate spicule from a sea urchin embryo. The spicule, incubated in a solution of calcium chloride and sodium bicarbonate, acted as a seed for crystal growth.

Right. Scanning electron micrograph of skeletal spicule from Arbacia.

(From S. Inoué and K. Okazaki, Biocrystals, Sci. Amer., 236(4):82–95; copyright © 1977 by Scientific American, Inc.; all rights reserved)

Library of Congress Catalog Card Number: 62-12028

International Standard Book Number: 0-07-045349-7

The Library of Congress cataloged the original printing of this title as follows:

McGraw-Hill yearbook of science and technology. 1962–
New York, McGraw-Hill Book Co.

v. illus. 26 cm.
Vols. for 1962– compiled by the staff of the McGraw-Hill encyclopedia of science and technology.

1. Science–Yearbooks. 2. Technology–Yearbooks. I. McGraw-Hill encyclopedia of science and technology.
Q1.M13 505.8 62-12028
Library of Congress (10)

Table of Contents

Editorial Advisory Boards

1960 – 1978

1968 – 1978

Editorial Staff

Consulting Editors

Consulting Editors *(continued)*

Contributors

A list of contributors, their affiliations, and the articles they wrote will be found on page 421.

Preface

The 1979 *McGraw-Hill Yearbook of Science and Technology*, eighteenth in the series of annual supplements to the *McGraw-Hill Encyclopedia of Science and Technology*, presents the outstanding scientific and technological achievements of 1978. Like its predecessors, this Yearbook contains articles reporting on those developments judged to be most significant by the 66 consulting editors and the editorial staff. Thus, the Yearbook serves not only as a ready reference to the progress in science and technology but as a means of updating the basic information in the fourth edition (1977) of the Encyclopedia.

The organization of the 1979 Yearbook follows the traditional three-section format. The first section includes seven feature articles on topics chosen for their broad interest and future significance. The second section, photographic highlights, presents a collection of photographs considered noteworthy because of the nature of the subject or the photographic technique. The third section consists of 150 alphabetically arranged articles on advances and discoveries of the past year in many disciplines, including archeology, virology, energy technology, microcomputers, and atomic physics.

For the past 18 years the Yearbook has served as an information source for the general public and the scientific community on developments in science and technology. While the ideas and efforts of the 66 consulting editors and the editorial staff have guided this effort, the real creators of the Yearbook are the eminent specialists who wrote the articles.

DANIEL N. LAPEDES
Editor in Chief

McGraw-Hill Yearbook of Science and Technology 1978 **REVIEW**

1979 **PREVIEW**

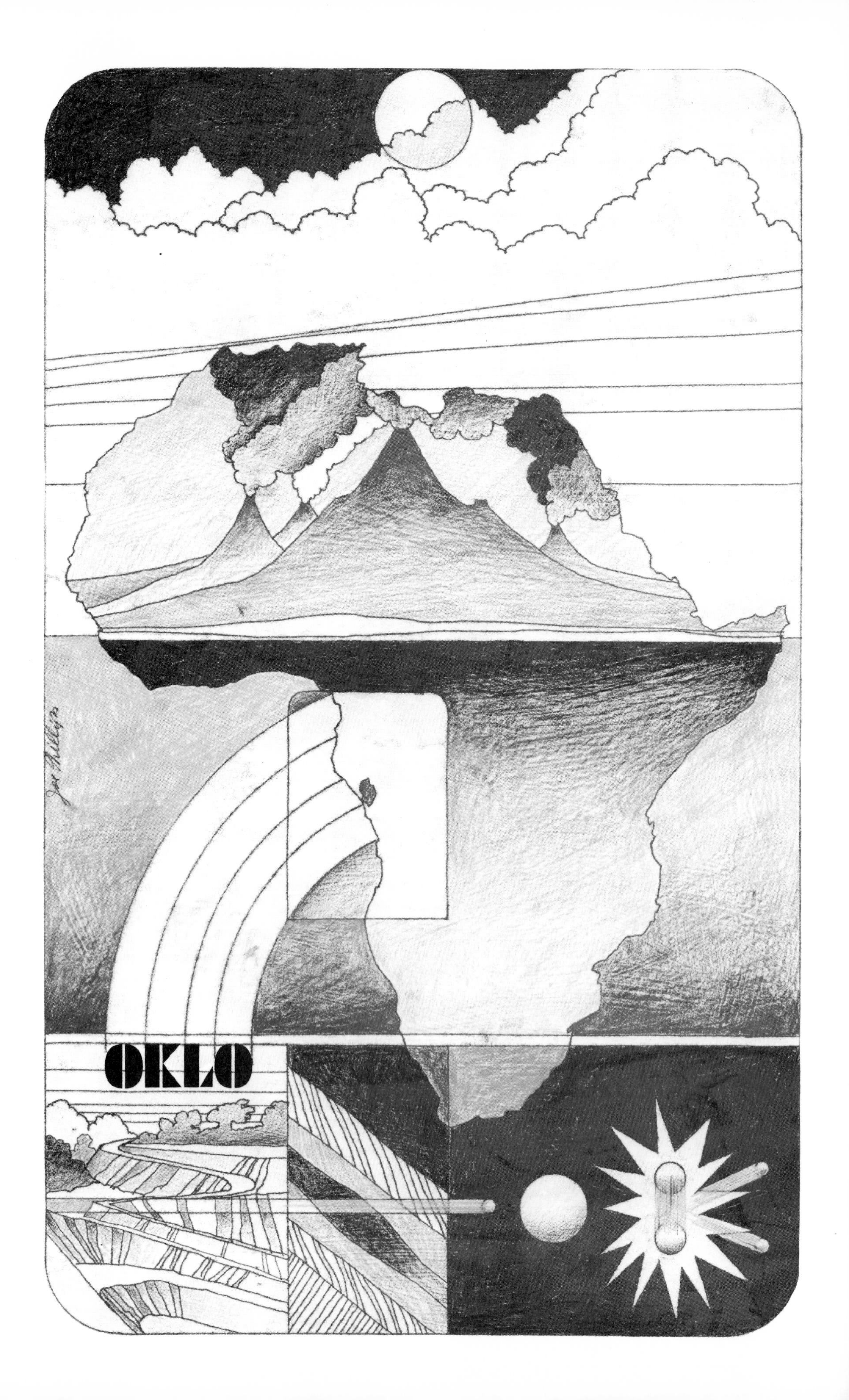
OKLO

Naturally Occurring Nuclear Reactor

Paul K. Kuroda is professor of nuclear chemistry at the University of Arkansas. He started his career as assistant professor of inorganic chemistry at the University of Tokyo, in his native country, and moved to the United States in 1949. He has been active in nuclear chemistry, radiochemistry, analytical chemistry, geochemistry, and cosmochemistry.

On Sept. 25, 1972, the French Atomic Energy Commission reported to the world scientific community the discovery of an anomaly in isotopic composition in a uranium deposit from Oklo (now in the Republic of Gabon). This evidence supported the theory that self-sustaining nuclear chain reactions had occurred on the Earth approximately 2×10^9 years ago. Prior to the discovery of the "Oklo phenomenon," it was believed that the world's first nuclear chain reaction was artificial and had occurred at the University of Chicago on Dec. 2, 1942. The discovery of the Oklo phenomenon thus contradicted this view, since it contained experimental evidence that nature, not humans, had produced the first chain reaction. However, the occurrence of chain reactions in nature had been theoretically predicted by P. K. Kuroda in 1956, 16 years prior to the discovery of the Oklo phenomenon.

EARLY THEORY AND INVESTIGATIONS

The concept that some unknown large-scale nuclear reactions may have occurred in the Earth's interior was first proposed in the early 1900s. Francis W. Aston alluded to the possibility of the release of an enormous amount of energy by the conversion of matter in such a reaction in his Nobel lecture of Dec. 12, 1922: "Should the research workers of the future discover some means of releasing this energy in a form which could be employed, the human race will have at its command powers beyond the dreams of scientific fiction; but the remote possibility must always be considered that the energy once liberated will be completely uncontrollable and by its intense violence detonate all neighboring substances. In this event the whole of the hydrogen on the earth might be transformed at once and the success of the experiment published at large to the Universe as a new star."

The possibility of the occurrence of large-scale nuclear reactions in the interior of the Earth's crust was discussed by J. Noetzlin in France in 1939, M. Odagiri in Japan in 1940, and N. Efremov in Germany in 1946. In 1949 and 1950 J. B. Orr in the United States investigated the ^{235}U content of thucholite from the Besner mine in Ontario for vestiges of a natural nuclear chain reaction. The results of his investigations, however, did not validate his hypothesis that a chain reaction might have taken place in the Canadian thucholite in the past if its mass was sufficiently great and the carbon contained in the mineral acted as a neutron moderator. In 1951 R. R. Edwards initiated a research project to investigate the contents of radioactive fission products (such as ^{89}Sr and ^{90}Sr) in the water of Hot Springs National Park in Arkansas. He theorized that a uranium chain reaction might be occurring in the deep interior of the Earth's crust in the Hot Springs region.

The concept of naturally occurring self-sustaining nuclear chain reactions, however, became quite unpopular with scientists in the 1950s for two reasons: (1) the $^{238}U/^{235}U$ ratio in uranium minerals and ores from various localities of the world appeared to be constant within ±0.1%; and (2) it appeared that Enrico Fermi's pile theory, when applied to uranium ore deposits, led to the unequivocal conclusion that a natural chain reaction could not have become self-sustaining.

Uranium isotopic composition. Table 1 lists a set of uranium isotope data compiled by F. E. Senf-

Table 1. $^{238}U/^{235}U$ ratios in various uranium minerals*

Minerals and ores	Location	$^{238}U/^{235}U$
Sample G	Mineral Joe Mine, Colorado	137.1±0.35
Sample S	Mineral Joe Mine, Colorado	137.7±0.64
Sample J	Mineral Joe Mine, Colorado	137.8±0.31
Sample N	Mineral Joe Mine, Colorado	137.7±0.31
Uraninite	Happy Jack Mine, Utah	137.8±0.26
Oxidized ore	Happy Jack Mine, Utah	137.8±0.31
Carbonaceous ore	Temple Mountain, Utah	137.8±0.31
Uraninite	Mi Vida Mine, Utah	137.7±0.31
Coffinite	Woodrow Pipe Mine, New Mexico	137.6±0.24
Coffinite	Poison Canyon Mine, New Mexico	137.8±0.31
Black oxidized ore	J. J. Mine, Paradox Valley, Colorado	137.8±0.31
Uraninite	Joachimsthal, Czechoslovakia	137.8±0.28
Uraninite	Great Bear Lake, Canada	137.8±0.25

*Determined at the Mass Assay Laboratory of Union Carbide Company, as reported by F. E. Senftle et al., Comparison of the isotopic abundance of U^{235} and U^{238} and the radium activity ratios in Colorado Plateau uranium ores, *Geochim. Cosmochim. Acta*, 11:189–193, 1957.

Table 2. Constancy of $^{238}U/^{235}U$ ratio presented as deviations from the standard*

Ore deposit	Location	Maximum possible percentage of deviation in $^{238}U/^{235}U$
Wheal Edward	Cornwall, England	0.025
Magnesia uranium concentrate	Portugal	0.041
Shinkolobwe pitchblende	Belgian Congo (Zaire)	0.046
Shinkolobwe "ionex"	Belgian Congo (Zaire)	0.046
Mindola, lot 1	Northern Rhodesia	0.028
Mindola, lot 4 (high copper content)	Northern Rhodesia	0.024
South African concentrate	The Rand, South Africa	0.033
Rum Jungle	Northern Australia	0.038
Mary Kathleen	Queensland, Australia	0.043
Radium Hill	South Australia	0.033
Blind River	Ontario, Canada	0.028
Beaver Lodge	Alberta, Canada	0.028

*From A. N. Hamer and E. J. Robbins, A search for variations in the natural abundance of uranium-235, *Geochim. Cosmochim. Acta*, 19:143–145, 1960.

Table 3. Constancy of $^{238}U/^{235}U$ ratio compared with the standard*

Sample number	Mineral formation	Mine location	Ratio†	95% confidence limits
542	Autunite in granite	Marysvale, UT	1.0002	± 0.0004
640–651	Arsenical in sandstone	Camp Bird Mine, Temple Mountain, UT	0.9999 (0.9997	± 0.0002 ± 0.0002)
1235	Uraninite-fluorite-tyuyamunite in Todilto limestone	Haystack Mountain, McKinley County, NM	0.9993 (0.9994	± 0.0002 ± 0.0003)
1487	Uraninite in sandstone	Big Indian Wash, UT	1.0001 (1.0000	± 0.0001 ± 0.0002)
1740	Low-vanadium oxide ore from sandstone, Petrified Forest, Chinle Formation	Ramco No. 17 Mine, Cameron, AZ	0.9997	± 0.0002
1757	Uraninite-copper-vanadium in sandstone Shinarump, Chinle Formation	C-3 Mine, Monument Valley, Utah-Arizona	0.9994 (0.9995	± 0.0002 ± 0.0002)
1762	Partly oxidized coffinite in sandstone, Brushy Basin, Morrison Formation	Jackpile Mine, Valencia County, NM	0.9993 (0.9993	± 0.0002 ± 0.0002)
2294	Arsenical uraninite-coffinite in volcanic rocks	White King Mine Lakeview, OR	1.0002	± 0.0002
3033	Uraninite-copper in sandstone	White Canyon, UT	0.9995 (0.9992	± 0.0002 ± 0.0002)
3034	Uraninite in sandstone, Upper Wind River Formation	Aljob Claims, Natrona County, WY	0.9996	± 0.0003
3035	Uranium-vanadium in Salt Wash Sandstone, Morrison Formation	Paradox D Mine, Montrose County, CO	0.9998	± 0.0003
3036	Uraninite (some coffinite) in metamorphic rocks	Schwartzwalder and Mena mines, Denver, CO	1.0001 (0.9999	± 0.0002 ± 0.0002
3039	Schroeckingerite in mudstone, Lost Creek deposit	Sweetwater County, WY	0.9995 (0.9995	± 0.0002 ± 0.0002)
Lot 19	Orange oxide	Belgian Congo (Zaire)	1.0000	± 0.0001
Lot 305	Orange oxide	Port Hope, Canada	0.9998	± 0.0001

*From L. A. Smith, *U.S. At. Energy Comm. Rep. no. K-1462*, Jan. 19, 1961.
†Value of the $^{235}U/(1 - ^{235}U)$ ratio of the sample divided by the $^{235}U/(1 - ^{235}U)$ ratio of the standard.

tle and associates. In 1960 A. N. Hamer and E. J. Robbins reported on the results of a comparison of uranium isotopic compositions in 12 ore concentrates by high-precision gas-source mass spectrometry (Table 2). Measurements of the $^{238}U/^{235}U$ ratio in various uranium ores were also carried out by L. A. Smith in 1961 (Table 3). The results from these and other studies appeared to be sufficient evidence to rule out the possibility that self-sustaining uranium chain reactions had occurred in the uranium ore deposits during the history of the Earth. It was reasoned that the occurrence of natural nuclear reactors would have caused a variation in $^{238}U/^{235}U$ ratios of the uranium ores from various localities of the world. The fact that these ratios were constant thus appeared to be decisive evidence against the hypothesis.

Xenon isotopic composition. During the early 1950s H. G. Thode and coworkers at McMaster University in Canada and G. W. Wetherill and M. G. Inghram at the University of Chicago investigated the isotopic compositions of xenon and krypton found in uranium minerals. They noted that both spontaneous fission of ^{238}U and neutron-induced fission of ^{235}U contribute to the production of stable fissiogenic xenon isotopes in uranium ores. The contribution from ^{238}U was found to be predominant in the uranium ores containing large amounts of rare-earth elements, which act as neutron-absorbing "poison." In uranium ores with relatively low rare-earth contents, for example, in pitchblende in the Belgian Congo (now Zaire), the contribution from thermal neutron-induced fission of ^{235}U was found to be as high as 25–30%.

In 1953 Wetherill and Inghram estimated the degree to which the pitchblende deposit was an operating reactor. They found that 10% of the neutrons produced in the deposit were absorbed to produce fission. Thus, the deposit was 25% of the way to becoming a reactor. Such a deposit would have had conditions closer to an operating reactor 2×10^9 years ago, when ^{235}U abundance was 6% instead of 0.7%.

The results from these studies thus appeared to suggest the possibility that some of the ore deposits might have been natural reactors billions of years ago. However, while Wetherill and Inghram concluded that such deposits were closer to being operating reactors 2×10^9 years ago, they did not actually state that these deposits were operating piles.

OBJECTIONS TO THE THEORY

As mentioned above, one of the reasons why the theory of natural reactors was unpopular during the 1950s was the fact that Fermi's pile theory, when applied to uranium ore deposits, appeared to negate the possibility that a chain reaction could have become self-sustaining. The fact that this is not necessarily the case was pointed out by Kuroda in 1956. Kuroda stated that the infinite multiplication constant k_∞ is an indicator of the stability of uranium minerals, which are natural assemblages of uranium, moderator, and impurities. Such a system is considered to be stable when k_∞ is much less than unity; the system is unstable when k_∞ is greater than unity.

Results of Fermi's pile theory. According to Fermi's pile theory, k_∞ is given by Eq. (1), where ϵ

$$k_\infty = \epsilon p f \eta \tag{1}$$

is the fast fission factor, p is the resonance escape probability, f is the thermal utilization factor, and η is the number of fast neutrons available per neutron absorbed by uranium. The major sources of neutrons in minerals are the spontaneous fission of ^{238}U and the (α,n) reactions.

In order to understand the operation of a nuclear reactor, one must trace the life history of n fast neutrons produced by thermal fission of ^{235}U. Some of these fast neutrons have high enough energies to produce fission in ^{238}U, and there is a small amount of fast fission of ^{235}U, so that to allow for this increase in the total number of neutrons one must introduce the fast fission factor ϵ, which is defined as the ratio of the total number of fast neutrons produced by fissions due to neutrons of all energies compared with the number resulting from thermal neutron fissions. For a natural uranium-graphite reactor, ϵ has a value of 1.029. In the case of natural reactors, it can be assumed that ϵ is close to 1 and hence ϵ can be neglected in the following calculations.

The $n\epsilon$ fast neutrons are then rapidly slowed down to thermal energies by collisions with moderator nuclei, such as hydrogen, deuterium, or carbon, but during this thermalizing process they are susceptible to resonance capture by ^{238}U until their energy is reduced below a value of about 5 eV. The probability that a fast neutron will reach thermal energies without capture by ^{238}U is given by the resonance escape probability p. The value of p can be obtained from Eq. (2), where N_0 is the number

$$p = \exp\left[-\frac{N_0}{\xi \Sigma_S}\int_E^{E_0} (\sigma_a)_{\text{eff}}\, dE/E\right] \tag{2}$$

of atoms of ^{238}U per cubic centimeter; ξ is the average loss in logarithm of the neutron energy from an elastic collision; Σ_S is the macroscopic scattering cross section of the moderator; σ_a is the microscopic thermal neutron absorption cross section for the material in question (in this case, ^{238}U); E is the energy of the neutron; and $\int_E^{E_0} (\sigma_a)_{\text{eff}}\, dE/E$ is the effective resonance integral. The value of the effective resonance integral for ^{238}U is determined by the ratio Σ_S/N_0, which represents the scattering cross section associated with each atom of absorber. The mass or properties of the particular moderator used does not affect the value of this integral, so that there is no essential difference between resonance absorption in graphite- and water-moderated reactors.

There are thus $n\epsilon p$ neutrons reaching thermal energies, and some of these neutrons are absorbed by moderator, coolant, and other poisons present, while a fraction, denoted by the thermal utilization factor f, is absorbed by the fuel to produce fission; f is defined as the ratio of the number of thermal neutrons absorbed in the fuel to the total number of thermal neutrons absorbed by any process.

The factor f can be expressed in the form of Eq. (3), where N_u and N_m are, respectively, the

$$f = \frac{N_u(\sigma_a)_u}{N_u(\sigma_a)_u + N_m(\sigma_a)_m + N_1(\sigma_a)_1 + N_2(\sigma_a)_2 + \cdots} \tag{3}$$

number of atoms of uranium and the number of atoms of moderator per cubic centimeter; N_1, N_2, . . . are the number of atoms of impurities per cubic centimeter; and σ_a is the microscopic thermal neutron absorption cross section for the material in question.

If, on the average, η fast neutrons are produced per thermal neutron capture by uranium, the number of fast neutrons produced must be $n\epsilon p f \eta$. Here η can be written as Eq. (4), where ν is the

$$\eta = \frac{{}^{235}N(\sigma_f)_{235}}{{}^{235}N(\sigma_a)_{235} + {}^{238}N(\sigma_a)_{238}} \cdot \nu \tag{4}$$

average number (equal to 2.5) of fast neutrons emitted per thermal fission of ^{235}U.

The value of k_∞ is that for a reactor of infinite size since any loss of neutrons by leakage from the system has not been included in the calculations. Since the multiplication factor is the ratio of the number of neutrons in any generation to the corresponding number in the previous one, the value of k_∞ can be expressed as Eq. (5). This

$$k_\infty = \frac{n\epsilon p f \eta}{n} = \epsilon p f \eta \tag{5}$$

equation is sometimes called the four-factor formula, and specifies the conditions under which a chain reaction can take place in a reactor of infinite size.

The values of p and f can be calculated if the chemical composition of the mineral is given: ϵ is always close to unity, and p, f, and η as a function of the uranium enrichment can be obtained from Eqs. (2), (3), and (4). Hence the value of k_∞ of a mineral at any geological time can be calculated.

A very accurate chemical analysis of Johanngeorgenstadt (Saxony) pitchblende was carried out by W. F. Hillebrand at the end of the 19th century, as shown in Table 4. Table 5 shows the present-day calculated values of p, f, and k_∞ for pitchblendes and uraninites found at various localities. As can be seen in Table 5, a sample of pitchblende from Johanngeorgenstadt has the largest value of k_∞. This means that this ore was the most likely to have become unstable during the history of the Earth. This pitchblende was thus chosen for the calculation of the values of k_∞ as a function of geological time. The results from such calculations were rather disappointing, however, as shown in Table 6: the ore appears to have been always stable during the past, even as far back as 2.8×10^9 years ago.

Moving back in time, the values of f and η steadily increase as the ^{235}U abundance increases. The value of p depends on the ratio of the number of atoms of ^{238}U to the number of atoms of moderator, which in this case is the hydrogen in the water molecule. The value of the resonance escape prob-

Table 4. Analysis of Johanngeorgenstadt (Saxony) pitchblende*

Component	Percentage
UO_3	22.33
UO_2	59.30
ThO_2	None
CeO_2	None
ZrO_2	None
$(La,Di)_2O_3$	None
$(Yt,Er)_2O_3$	None
Al_2O_3	0.20
Fe_2O_3	0.21
PbO	6.39
CuO	0.17
MnO	0.09
CaO	1.00
MgO	0.17
Bi_2O_3	0.75
V_2O_5, MoO_3, WO_3	0.75
Alkalies	0.31
SO_3	0.19
P_2O_5	0.06
As_2O_5	2.34
He	Trace
H_2O	3.17
SiO_2	0.50
Total	97.93

*Data from W. F. Hillebrand, *Bull. U.S. Geol. Surv.*, no. 78, p. 43, 1891; no. 90, p. 23, 1892; no. 220, pp. 111–114, 1903. See also, F. W. Clarke, *The Data of Geochemistry*, U.S. Government Printing Office, p. 725, 1924.

ability in Eq. (2) increases as Σ_S/N_0 increases. Since Σ_S is proportional to the number of atoms of hydrogen per cubic centimeter, and since N_0 would have been greater in the past than it is today, the values of Σ_S/N_0 and p decrease as one moves back in time. Increases in f and η are thus counterbalanced by the decrease in p, so that k_∞ also decreases after reaching a maximum value approximately 1.4×10^9 years ago. The results of these calculations thus seemed to indicate that the chain reaction could never have become self-sustaining.

Effect of increased water content. It was discovered, however, that the above calculations were based on an overly simplified model, in which it was assumed that an ore deposit was somehow instantaneously created in nature in essentially its present chemical state. The above calculations indicate that these uranium ores were stable during the past 2.8×10^9 years, provided that water content of the ores remained unchanged. However, a slight increase in the water-to-uranium ratio could have resulted in a sharp rise of p, without affecting f considerably, causing the system to become unstable.

However, it is probable that uranium ore deposits were formed by a process in which small quantities of uranium were extracted from the rocks and

Table 5. Values of p, f, and k_∞ for various pitchblendes and uraninites*

Locality	p	f	k_∞
Placer de Guadalupe, Mexico	0.08	0.03	0.003
Black Hawk, Colorado	0.30	0.90	0.36
Kirk Mine, Colorado	0.15	0.19	0.038
Iizaka, Japan (clevite)	0.10	0.006	0.001
Hale's Quarry, Connecticut	0.12	0.30	0.048
Blanchville, Connecticut	0.13	0.91	0.16
Boqueirão, Brazil	0.20	0.21	0.06
Johanngeorgenstadt, Saxony	0.47	0.93	0.58
Shinkolobwe, Katanga	0.08	0.20	0.021
Morogoro, East Africa	(0.22)†	0.08	(0.023)
Xique-Xique, Brazil	0.40	0.21	0.11
Gustay's Mine, Norway (bröggerite)	0.16	0.09	0.018
Lac Pied des Monts, Quebec	(0.22)†	0.14	(0.040)
Wilberforce, Ontario	0.16	0.02	0.004
Baringer Hill, Texas (nivenite)	0.33	0.007	0.003
Arendal, Norway (clevite)	(0.22)†	0.007	(0.002)
Great Bear Lake, Canada	0.37	0.08	0.041
Ingersoll Mine, South Dakota	0.09	0.08	0.010
Winnipeg River, Manitoba	(0.22)†	0.07	(0.020)
Sinyaya Pala, Karelia, Soviet Union	0.29	0.024	0.009

*From P. K. Kuroda, On the multiplication constant and the age of uranium minerals, *J. Chem. Phys.*, 25:1295–1296, 1956.

†The values of p cannot be calculated, since water contents of the minerals are unknown. An assumed value of $p = 0.22$ (an average value of p of the 16 samples of minerals) has been used for the calculation of k_∞.

Table 6. Values p, f, n, and k_∞ for a Johanngeorgenstadt pitchblende as a function of geological time*

Geological time, 10^6 years before present	^{235}U enrichment, %	p	f	η	k_∞
0 (present)	0.7	0.47	0.93	1.32	0.58
700	1.3	0.45	0.95	1.57	0.67
1000	1.6	0.43	0.96	1.66	0.69
1400	2.3	0.42	0.97	1.77	0.72
2100	4.0	0.38	0.98	1.91	0.71
2800	7.0	0.34	0.99	1.98	0.67

*From P. K. Kuroda, On the nuclear physical stability of the uranium minerals, *J. Chem. Phys.*, 25:781–782, 1956.

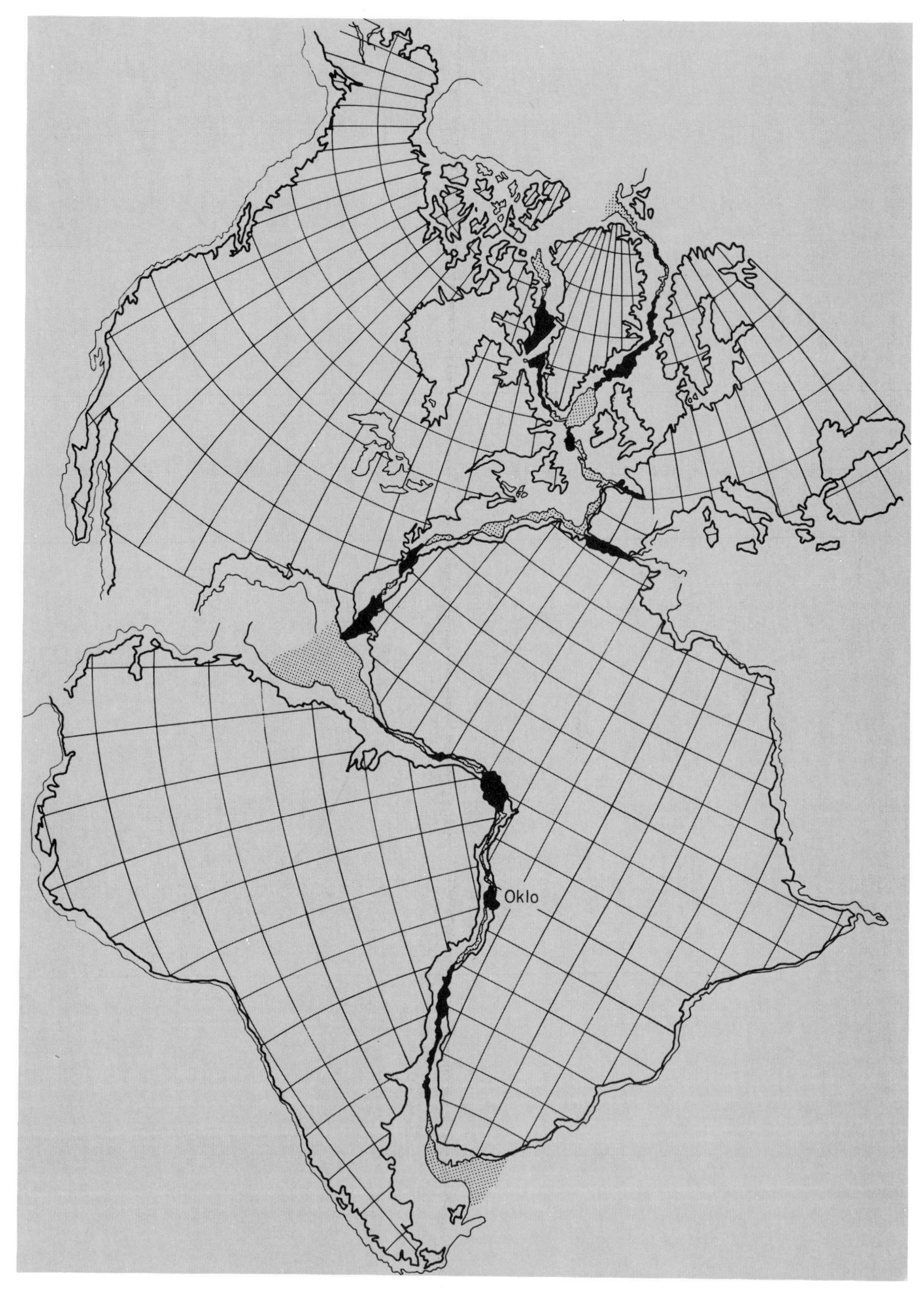

Fig. 1. Location of Oklo in the Republic of Gabon, Africa. The continents are shown in the positions which they occupied 2×10^9 years ago, when the Oklo reactor was operating.

transported by water. Hence it was decided to carry out the calculations based on a new model in which it was assumed that the formation of uranium ores comprised the following sequence of events: an aqueous solution of uranium (^{235}U-enriched) is gradually converted to an assemblage of 1 mole of uranium plus *m* moles of H_2O, where *m* is a variable, and finally is converted to an almost water-free uranium deposit.

Table 7 shows the result of calculations based on

this model, which is considered to be more realistic than the earlier one. Table 7 shows that the assemblages of the Johanngeorgenstadt pitchblende plus water were unstable 2.1×10^9 years ago, and the critical uranium chain reactions could have taken place if the size of the assemblages was greater than, say, a thickness of a few feet.

DISCOVERY OF THE OKLO PHENOMENON

A few months after Kuroda proposed the theory of naturally occurring nuclear reactors in 1956, research workers of the French Atomic Energy Commission discovered a new uranium ore deposit at Oklo which contained the experimental evidence needed for the verification of the theory. If representative uranium specimens from Oklo had been carefully examined mass-spectrometrically at the time, an anomaly in the isotopic composition of uranium indicating a depletion of ^{235}U in the ores would have been noted. However, it was not until 16 years later that such an anomaly was actually detected.

On June 7, 1972, the anomaly in the isotopic content of a sample of natural uranium was observed by H. Bouzigue and colleagues at the French Atomic Energy Establishment at Pierrelatte: the ^{235}U abundance was 0.7171 at. %, as opposed to 0.7202 ± 0.0010 at. % for normal natural uranium. A group of scientists was then sent by the French government to the U.S. National Bureau of Standards in Washington, DC, to make a comparison of the sample with the American uranium standard; the results of this comparison were identical to the original results.

During June–August 1972 the following new facts emerged: (1) the anomalous ore was traced to Oklo, and it was found that the ore which originated from Oklo between December 1970 and May 1972 was deficient by a total of 200 kg of ^{235}U; (2) uranium with an extremely low isotopic abundance of 0.440% ^{235}U was discovered; and (3) fission-produced neodymium and samarium isotopes were found in the ore.

Description of reaction site. The reaction site consisted of several bodies of very rich uranium ore, and it was estimated that more than 500 tons of uranium had been involved in the reactions, with a quantity of energy released equal to about 10^{11} kWh. The integrated neutron flux at certain points exceeded 1.5×10^{21} n/cm^2, and samples have been found in which the concentration of ^{235}U is as low as 0.29%, as compared with 0.72% in natural uranium.

Table 7. Values of water-uranium ratio (m), p, f, η, and k_∞ for a Johanngeorgenstadt pitchblende; 2.1×10^9 years ago*

m	p	f	η	k_∞
1/4	0.29	0.99	1.91	0.55
1/2	0.47	0.98	1.91	0.88
1	0.62	0.97	1.91	1.15
2	0.74	0.95	1.91	1.34
3	0.79	0.93	1.91	1.40
4	0.82	0.91	1.91	1.42
5	0.84	0.89	1.91	1.43
10	0.86	0.81	1.91	1.33

*From P. K. Kuroda, On the nuclear physical stability of the uranium minerals, *J. Chem. Phys.*, 25:781–782, 1956.

Figure 1 shows how the continents of Africa and America were joined together 2×10^9 years ago when the Oklo reactor was operating. The structure of Africa was established at the end of the Precambrian Era ($5-6 \times 10^8$ years ago). Figure 2 shows a geological map of Africa prepared by P. Molina and J. C. Besombes. They distinguish four major orogeneses: Precambrian A ($5-6 \times 10^8$ to $0.9-1.2 \times 10^9$ years ago), Precambrian B ($0.9-1.2$ to $1.8-2 \times 10^9$ years ago), Precambrian C ($1.8-2.0 \times 10^9$ to 2.5×10^9 years ago), and Precambrian D (before 2.5×10^9 years ago). Africa is made up of four consolidated and granitized cratons: the West African, Congolese, Kalahari, and nilotic cratons. Gabon lies in the northwest part of the Congolese craton. The Franceville basin (labeled 22 in Fig. 2) is one of the intracratonic basins of the Congolese craton. The age of its sediments has been estimat-

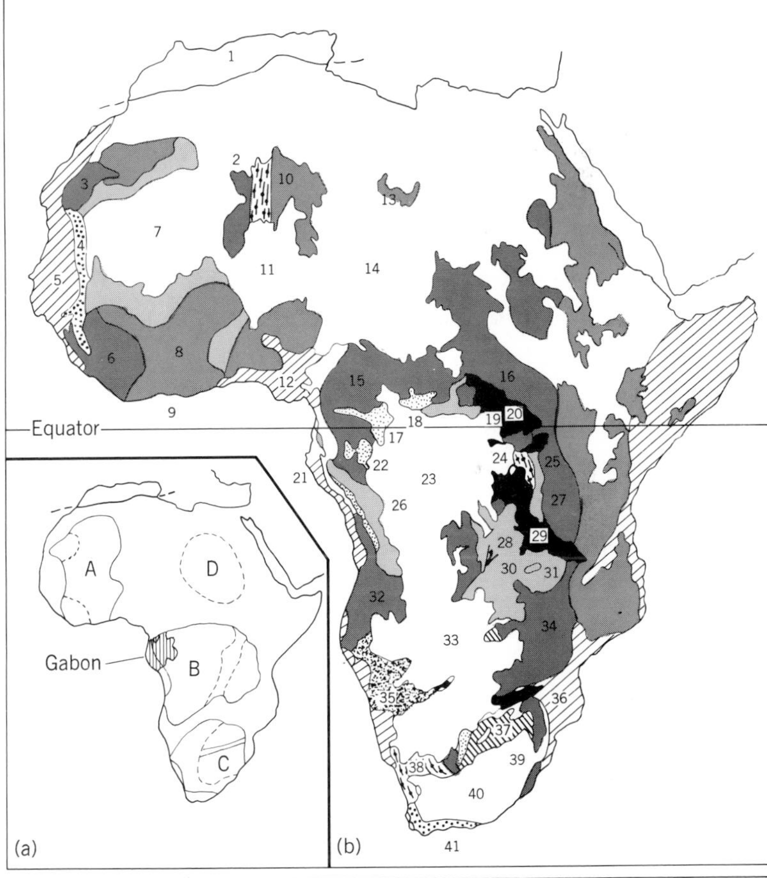

chronological divisions (× 10⁶ years)		cratons	mobile zones	basins: intracratonic	basins: coastal
0					
500–600					
900–1200	Precambrian A				
1800–2000	Precambrian B				
2500–3000	Precambrian C				
	Precambrian D				

Fig. 2. Geological map of Africa. (*a*) Cratons. A = West African craton, B = Congolese craton, C = Kalahari craton, D = nilotic craton. (*b*) Principal structural units, indicated by numbers. *(From P. Molina and J. C. Besombes, The Oklo Phenomenon, International Atomic Energy Agency, Vienna, 1975)*

Fig. 3. Oklo uranium deposit showing preservation of the fossil reactor core (center). *(From Activités Scientifiques et Techniques, Commissariat à l'Énergie Atomique, Paris, 1976)*

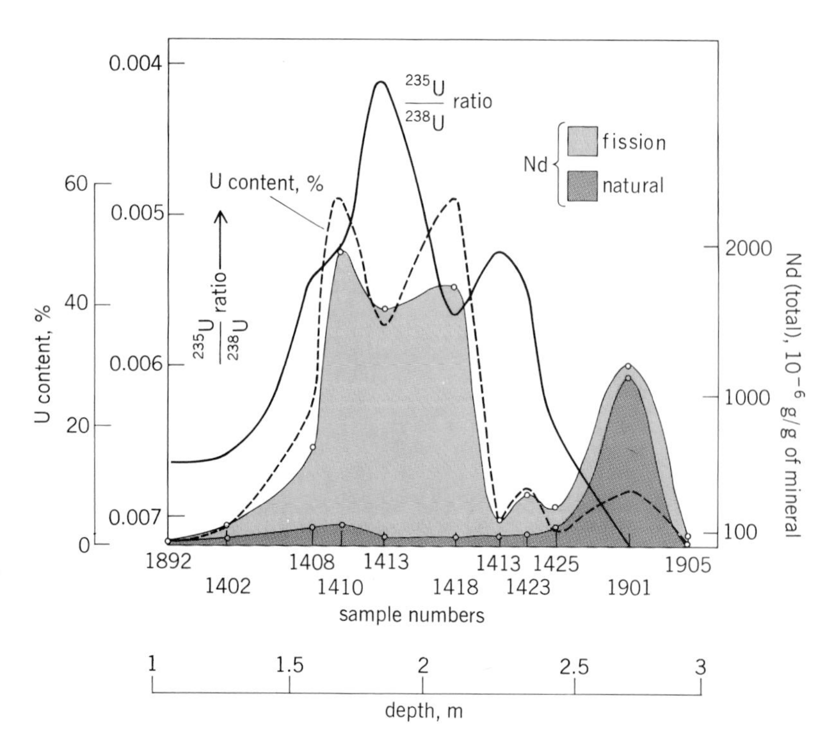

Fig. 4. Uranium contents, $^{235}U/^{238}U$ ratios, and neodymium isotopes in the Oklo reactor site. *(From Activités Scientifiques et Techniques, Commissariat à l'Énergie Atomique, Paris, 1976)*

ed at $1.74 \pm 0.02 \times 10^9$ years. The Franceville basin can thus be assigned to the Precambrian B orogenesis. Figure 3 shows a photograph of the Oklo mine, in which the reactor core is being preserved.

Experimental data. The experimental data obtained were reported in the fall of 1972 by R. Bodu and coworkers, M. Neuilly and coworkers, and G. Baudin and coworkers. Table 8 shows part of the data reported by Neuilly and coworkers. The isotopic compositions of uranium and neodymium in the samples Oklo-M and Oklo-310 are compared with those in ordinary rocks. It should be noted that ^{235}U is markedly depleted in these samples relative to the normal abundance of 0.7202% observed in uranium ores from other localities, while the isotopic compositions of neodymium in these samples resemble those of fissiogenic neodymium isotopes. Since ^{142}Nd is not formed by fission, it is possible to correct for the part due to the natural element, and the corrected isotopic compositions are compared with the known yields from the thermal neutron-induced fission of ^{235}U. Table 8 indicates that there is good agreement between these values, strongly suggesting that the neodymium isotopes were produced by fission.

Figure 4 shows the distributions of the uranium contents, $^{235}U/^{238}U$ ratios, and the contents of fission-produced and natural neodymium in the vicinity of one of the reactor cores at Oklo. Note that the assemblage has a thickness of a few feet, as envisioned by Kuroda in 1956. It is also interesting that the fission-produced neodymium isotopes appear to have been well preserved within the reactor core for nearly 2×10^9 years. Figure 4 shows that only small amounts of isotopes have migrated out of the reactor core and were found in the samples taken only a short distance away.

OPERATING CONDITIONS OF THE OKLO REACTOR

Results from the investigations of the Oklo reactor generally confirmed that the theory of a naturally occurring nuclear reactor is essentially correct as originally proposed in 1956. However, there were a few unexpected results. First, the reactor seemed to have operated for a period as long as 0.6 to 1.5×10^6 years without destroying itself. In 1956 Kuroda theorized that the critical uranium chain reactions occurring in nature would have caused a sudden elevation in temperature, resulting in the complete destruction of the critical assemblage. He further hypothesized that such destructive processes might account for the fact that the ages of the large uranium deposits never exceed 2×10^9 years, or might explain the marked discrepancies between the $^{206}Pb/^{238}U$ and $^{207}Pb/^{206}Pb$ ages of uranium minerals.

Another unexpected discovery was the fact that the formation of natural reactors appeared to be closely related to the appearance of life on Earth. Michel Maurette noted that the high uranium overconcentrations found at Oklo resulted from a long series of repetitive fractionation processes, in which oxygen played a dominant role as an oxidizing agent. It is generally assumed that oxygen was injected into the Earth's atmosphere only 2×10^9 years ago by the new generation of living organisms capable of carrying out photosynthesis. Thus, the high uranium overconcentration (greater than about 20%) needed to trigger a nuclear chain reac-

Table 8. Isotopic anomalies observed in uranium ores from Oklo*

Isotope	Oklo-M	Oklo-310	Natural abundance	Oklo-M†	Oklo-310†	Expected in ^{235}U fission products
^{235}U	0.4400 ±0.0005	0.592 ±0.001	0.7202			
^{142}Nd	1.38	5.49	27.11	0	0	0
^{143}Nd	22.1	23.0	12.17	22.6	25.7	28.8
^{144}Nd	32.0	28.2	23.85	32.4	29.3	26.5
^{145}Nd	17.5	16.3	8.30	18.05	18.4	18.9
^{146}Nd	15.6	15.4	17.22	15.55	14.9	14.4
^{148}Nd	8.01	7.70	5.73	8.13	8.20	8.26
^{150}Nd	3.40	3.90	5.62	3.28	3.46	3.12

*From M. Neuilly et al., Sur l'existence dans un passé reculi d'une réaction en chaîne naturelle de fissions, dans le gisement d'uranium d'Oklo, *C. R. Acad. Sci. Paris 275 D*, pp. 1847–1849, 1972.
†Corrected for the part due to the natural element.

Table 9. Reactor operating conditions*

Sample	α, %	β, %	τ, n/cm^2	C†	Δt, 10^5 years
KN50-3548	2.5 ± 1	4 ± 1	1.23×10^{21}	0.47	6.4
KN50-323	3 ± 1	3 ± 1	1.02×10^{21}	0.42	6.3
SC36-1413-3	3 ± 1	4 ± 1	1.32×10^{21}	0.37	5.4
SC36-1418	3.5 ± 1	4 ± 1	0.81×10^{21}	0.58	5.8

*From R. Hagemann et al., *The Oklo Phenomenon*, International Atomic Energy Agency, Vienna, pp. 415–423, 1975.
†Proportion of ^{235}U atoms resulting from the α-decay of ^{239}Pu.

tion was probably never achieved in uranium ore deposits until 2×10^9 years ago, and the occurrence of fossil reactors was probably limited to a relatively brief period of time ranging from 1 to 2×10^9 years ago.

Formation of Oklo deposit. Figure 5 is a schematic representation of a plausible scenario for the formation of the Oklo uranium deposit according to Maurette, who envisioned the following sequence of events: "(1) uranium was dissolved as a result of surface leaching, which acted either on volcanic ashes from the Francevillien series or an older igneous rock from the basement; (2) the ore-forming fluid infiltrated layers of sediments already covered by the sea, and organic matter then played a fundamental role in reducing and precipitating uranium as pitchblende in favorable geological traps; (3) the sediments were probably buried at shallow depths (perhaps <1000 m) at the time of the primary uranium mineralization, but as the relative level of the sea increased, the ore bodies were gradually buried at great depths (≳5000 m) by further sedimentation; and (4) during the subsequent geological evolution the relative level of the sea decreased and the U-rich sediments became exposed to erosional processes acting at an estimated rate of ~10 m/10^6 years, which brought the Oklo ores into a near-surface position quite recently."

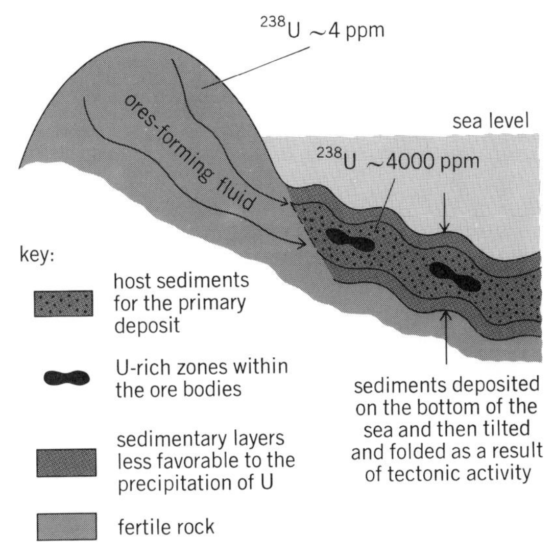

Fig. 5. Schematic representation of a plausible scenario for the formation of the Oklo uranium deposit. *(From Michel Maurette, Fossil nuclear reactors, Annu. Rev. Nucl. Sci., 26:319–320, 1976; used with permission; © 1976 by Annual Reviews Inc.)*

Duration of chain reaction. R. Hagemann and coworkers have estimated the duration of the nuclear reaction by measuring several isotopes and assuming competition between two simultaneous reactions: a neutron absorption event leading to fission or a neutron capture event resulting in radioactive decay. ^{239}Pu, which has a half-life of 24,000 years, can be used to illustrate these processes. It is formed by reaction (5).

$$^{238}U + \text{neutron} \rightarrow {}^{239}U \xrightarrow[23.5\ \text{min}]{\beta^-} {}^{239}Np \xrightarrow[2.35\ \text{days}]{\beta^-} {}^{239}Pu \quad (5)$$

^{239}Pu decays to ^{235}U by emitting an alpha particle, as shown in reaction (6).

$$^{239}Pu \xrightarrow[24{,}000\ \text{years}]{\alpha} {}^{235}U \quad (6)$$

If the duration of the natural reactor is comparable to or much longer than the half-life of ^{239}Pu, much of the ^{239}Pu synthesized in the reactor undergoes thermal neutron-induced fission and produces its own fission products according to reaction (7), while reaction (8) is occurring in the reactor. Thus

$$^{239}Pu + \text{neutron} \rightarrow \text{fission products} + \text{neutrons} \quad (7)$$

$$^{235}U + \text{neutron} \rightarrow \text{fission products} + \text{neutrons} \quad (8)$$

the fission products found in the remains of natural reactors are expected to be mixtures of the fission products of ^{235}U and ^{239}Pu.

The yields of the fission products from ^{235}U and ^{239}Pu are different and are accurately known. Thus one should be able to estimate the contribution from the ^{239}Pu fission by carrying out isotopic analysis of elements for which the plutonium gives fission yields different from those of uranium. Actually, ^{238}U fission also contributes to the production of the fission products.

According to Hagemann and coworkers, Eq. (9)

$$dN_{235}/dt = -N_{235}\sigma_{235}(1-C)\phi = -N_{235}\sigma_{235}\,\phi + N_{239}\lambda_{239} \quad (9)$$

holds, where C is a conversion factor; ϕ is the neutron flux; λ_{239} is the decay constant of ^{239}Pu; and $N_{239} = CN_{235}\sigma_{235}\tau/\lambda_{239}\Delta t$, where Δt is the duration of the reaction and τ is the fluence.

The percentage of fissions due to ^{239}Pu can be written as Eq. (10), where α and β are the percent-

$$\beta = N_{239}\sigma_{f239}\left[\frac{1-\alpha-\beta}{N_{235}\sigma_{f235}}\right] \quad (10)$$

ages of fissions due to ^{238}U and ^{239}Pu, respectively. The rest of the reaction can be obtained from the above equations according to Eq. (11). In Eq. (11) τ

$$\Delta t = \frac{\sigma_{f239}\sigma_{235}}{\sigma_{f235}} \cdot \frac{1-\alpha-\beta}{\beta} \cdot \frac{C\tau}{\lambda_{239}} \quad (11)$$

and C can be obtained relatively easily from the isotopic analysis of uranium and rare earths, especially neodymium and samarium.

Hagemann and coworkers determined the values of α and β from the observed ratios R of several fission products found in the ore samples. The following ratios were used for the calculations: $^{150}Nd/(^{143}Nd + ^{144}Nd + ^{145}Nd + ^{146}Nd)$, $^{154}Sm/(^{147}Sm + ^{148}Sm)$, $(^{157}Gd + ^{158}Gd)/(^{155}Gd + ^{156}Gd)$, $^{110}Pd/^{105}Pd$, $^{110}Pd/^{106}Pd$, $^{104}Ru/^{101}Ru$, and $^{104}Ru/^{102}Ru$.

$$R = \frac{\rho'_{235}(1-\alpha-\beta) + \rho'_{238}\,\alpha + \rho'_{239}\,\beta}{\rho''_{235}(1-\alpha-\beta) + \rho''_{238}\,\alpha + \rho''_{239}\,\beta} \quad (12)$$

Equation (12) shows the relationship between R and the fission yields (ρ' and ρ'') from ^{235}U, ^{238}U, and ^{239}Pu. The results of the calculations are shown in Table 9.

Competition between the radioactive decay of ^{99}Tc and neutron capture in ^{99}Tc and ^{99}Ru leading to ^{100}Ru also enables the determination of the duration of the nuclear reactor operation:

$$\begin{array}{ccc} & t_{1/2} = 2.13 \times 10^5 \text{ years} & \\ ^{99}Tc & \longrightarrow & ^{99}Ru \\ +n \downarrow & & \downarrow +n \\ ^{100}Tc & \longrightarrow & ^{100}Ru \\ & t_{1/2} = 17 \text{ s} & \end{array}$$

Unfortunately, however, the cross sections for the neutron capture reactions are not accurately known. The ^{99}Tc method yields higher values (0.74 to 3.5×10^6 years) for the duration of the chain reaction, which are considered as being less reliable than the ^{239}Pu method.

According to Hagemann and coworkers, competition between the radioactive decay of ^{129}I to ^{129}Xe, with a half-life of 1.6×10^7 years, and neutron capture in these two isotopes leading to ^{130}Xe cannot be used for the determination of the duration of the chain reaction, due mostly to the migration of the iodine and the xenon during the reaction.

It is important to note the conclusion reached from radiation damage studies on the rocks and minerals found in the reactor site. According to Maurette—in striking contrast to expectations based on the high neutron fluences shown in Table 9—no high concentration of radiation damage defects has thus far been observed in insulator grains extracted from the reactor cores. These results showed that the Oklo ores had to be buried most of the time at great depths (probably more than 5000 m) in order to achieve the relatively high temperatures (>300°C) that are required for track annealing.

Static and dynamic models. The concept of a static model for the Oklo reactors was first developed by R. Naudet and coworkers. In this model, with the exception of water which is considered an adjustable parameter, the present-day characteristics of the Oklo ore bodies are used to evaluate the values of the neutron multiplication factor k expected in the uranium-rich zones at the time of the chain reactions (1.8×10^9 years ago). The calculations show that the criticality was indeed easily achieved under the following conditions: $^{235}U/(^{235}U + ^{238}U) = 3\%$; ^{238}U concentration = 30%; the thickness ϕ of the uranium-rich lenses = 1 m; and $H_2O/U = 0.15$.

This set of values is very similar to those observed for artificial reactors which use natural water as a moderator. The very efficient control of the chain reactions in the natural reactor is attributed to the removal of water during boiling, which causes a sudden drop in k. Then, as the reactor cools down, liquid water is reinjected into the core and the chain reaction starts again.

G. A. Cowan and coworkers have developed a dynamic model for the Oklo reactor based on a simmering water reactor, where a water reflux action reconcentrates the uranium from the borders to the center of the reactor. In this model, the variations of n isotope ratios are computed by using as variables the neutron fluences (thermal, epithermal, and fast), the uranium distribution prevailing during the chain reactions, and the duration of the reactions.

SEARCH FOR ADDITIONAL NATURAL REACTORS

Prior to the development of the theory of natural reactors by Kuroda, the probability of discovering a natural reactor seemed very small. The same can be said today about the prospect of finding another natural reactor. If it were not for the careful observations made by Bouzigues and coworkers, the Oklo reactor might not have been discovered. It is therefore quite possible that similar reactors have already been mined out without being discovered.

Maurette believes that neither the Oklo phenomenon nor the characteristics of the deposit are particularly unique. What is unique is the careful analysis of uranium performed by Bouzigues and associates. The relative mass of uranium cycled through the chain reactions at Oklo was evaluated at about 10^{-3}. To detect fissile reactors in other uranium ores would therefore necessitate the frequent control of the isotopic composition of uranium with an accuracy greater than about 0.1%, and there is no certainty that such accurate analyses have been or will be conducted in other deposits.

G. A. Cowan has proposed that not all natural reactors would necessarily create depleted ores. If a natural reactor formed as late as 8×10^8 years ago, when the relative abundance of ^{235}U was about 1%, it might actually have become a breeder reactor, and the ^{235}U consumed in the reaction would have been more than replaced by new ^{235}U created by the decay of ^{239}Pu. Although the pitchblende deposits in the former Belgian Congo have now been mined out, precise isotopic analyses are available for a few samples of the ore, and they appear to be slightly enriched in ^{235}U. Cowan also noted that uranium from the Colorado Plateau has a ^{235}U content slightly smaller than the world average, and it is possible that a natural reactor once operated in this region.

SUMMARY

Until recently, scientists believed that large-scale transmutations of the elements occur only in stars, but the discovery of the Oklo phenomenon revealed the fact that a nuclear chain reaction existed on the Earth billions of years ago. Further investigation of the remains of such a reaction may hold the key to the solution of one of the most critical problems confronting the human race today: how to deal with the dangerous waste products from the nuclear reactions created by 20th-century humans.

[PAUL K. KURODA]

Micro-encapsulation

Curt Thies is professor of chemical engineering at Washington University in St. Louis. He joined the faculty in 1973 after 8 years in microencapsulation research at National Cash Register.

Microencapsulation is a unique way to package solids, liquids, dispersions, and gases inside plastic or waxlike coatings. Current microencapsulation technology is very diversified and capable of handling different types of materials. Figure 1 shows what typical microcapsules look like. Capsule sizes range from well below 1 μm in diameter to over 2000 μm. The amount of active material contained in a capsule may vary from a small fraction of total capsule weight to well over 90%. Although microcapsules have yet to achieve the large-scale usage originally envisioned for them, the field is steadily developing. The following discussion provides an overview of various technologies being used to make microcapsules and summarizes selected capsule applications.

MICROENCAPSULATION TECHNOLOGY

The patent literature describes an overwhelming number of different processes for manufacturing microcapsules. It is difficult to catalog them into a few simple categories. Nevertheless, this must be done in order to present a reasonably clear picture of the cur-

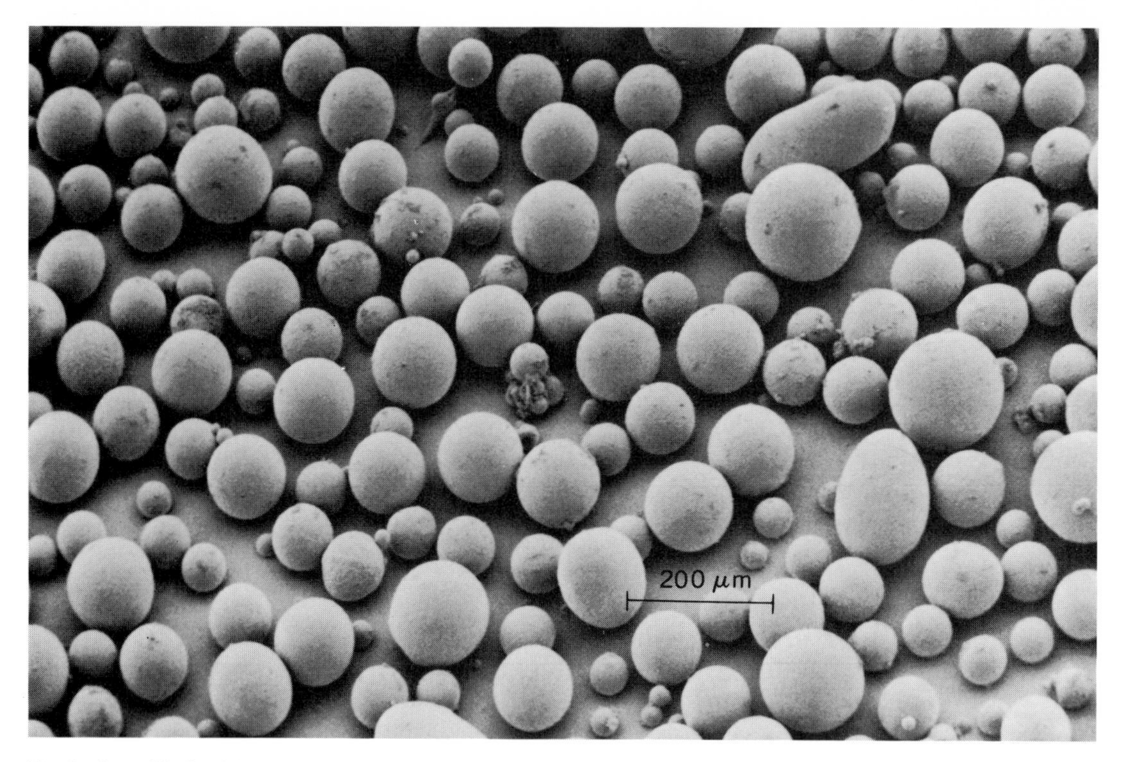

Fig. 1. Drug-filled poly(lactic acid) microcapsules prepared by a polymer phase separation process.

rent status of encapsulation technology. The approach taken in this article is to arbitrarily define seven categories that will accommodate most encapsulation processes and to outline briefly how a typical process in each category operates.

Polymer phase separation. Complex coacervation and polymer-polymer incompatibility are two well-known polymer phase separation phenomena used to make microcapsules. Figure 2 is a flow diagram of an encapsulation process based on complex coacervation of gelatin and gum arabic. This is the process that was originally used to make capsules for carbonless copy paper. Because the encapsulation process is run in water, it is suitable for encapsulating only water-immiscible materials. In principle, any pair of oppositely charged polyelectrolytes is capable of forming a complex coacervate suitable for use in a microencapsulation process. Gelatin-based systems predominate because they consistently yield capsules suitable for many uses.

Gelatin capsule walls are typically insolubilized by cross-linking with glutaraldehyde or formaldehyde. Capsule walls treated this way remain hydrophilic and swell measurably in water. Walls that do not visibly swell in water are obtained by cross-linking gelatin-based capsules with urea-formaldehyde polymers. There are microencapsulation processes based exclusively on this type of condensation reaction. In such processes, a water-insoluble material to be encapsulated is dispersed in an aqueous solution of urea-formaldehyde prepolymer. Polycondensation is initiated and a highly cross-linked capsule wall is formed.

Figure 3 is a flow diagram of an encapsulation process based on polymer-polymer incompatibility. The capsule wall material, a partially hydrolyzed ethylene–vinyl acetate copolymer, is dissolved in

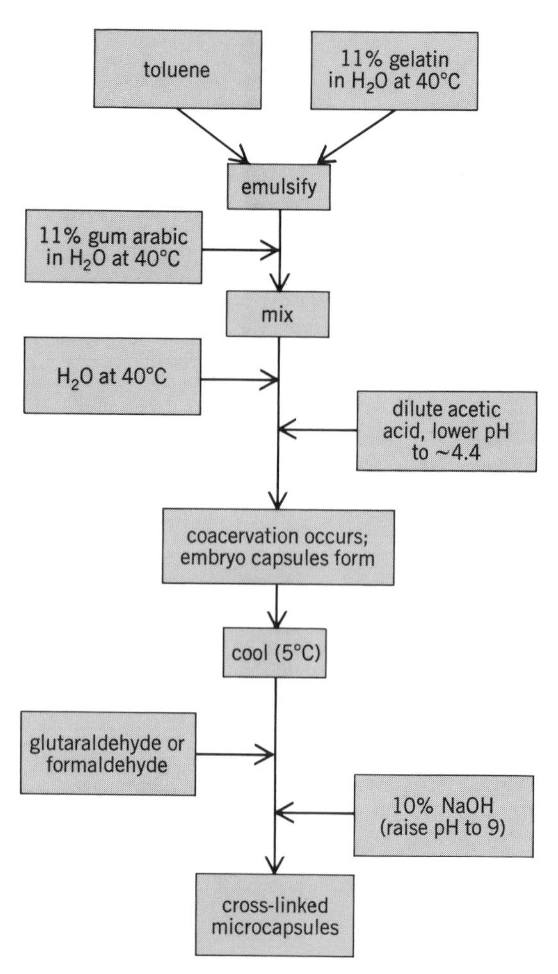

Fig. 2. Flow diagram for preparation of microcapsules by a process based on complex coacervation.

toluene and phase-separated by cottonseed oil. A copolymer-rich phase is formed. This engulfs the droplets of water to form embryo capsules which are then cross-linked with tolylene diisocyanate.

Although water-soluble and organic solvent–soluble polymers exhibit polymer-polymer incompatibility, most encapsulation processes based on this phenomenon are carried out in organic solvents. Defining optimum conditions for the many operating parameters involved is a major task. For this reason, the number of polymers used successfully to prepare capsules by processes based on polymer-polymer incompatibility is small relative to the number of candidate coating polymers available.

Some polymers will spontaneously precipitate at an interface to form microcapsules. This phenomenon is used to encapsulate aqueous solutions of enzymes with nitrocellulose. The capsule walls formed are very thin, but this is desirable when high rates of solute transport across the capsule wall are needed.

Interfacial polymerization. The rapid polymerization of reactive monomers at liquid-liquid interfaces is a convenient way to form microcapsules (Fig. 4). As shown, a multifunctional acid chloride is dissolved in the organic phase, and the organic solution is then emulsified with water. A polyfunctional amine and sodium hydroxide are added to the water phase. The amine reacts at the liquid-liquid interface with the acid chloride to form a cross-linked capsule coating. The HCl reaction by-product is neutralized by the NaOH. As long as the organic solvent is the dispersed phase, the

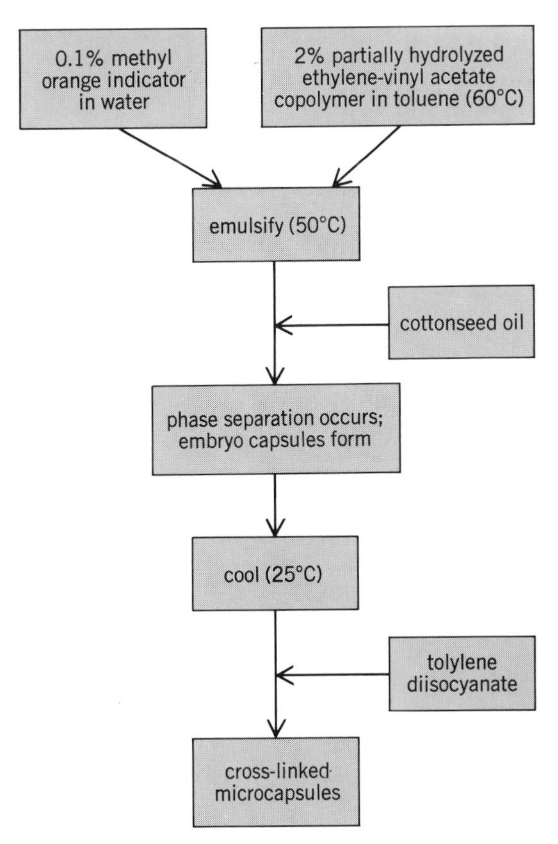

Fig. 3. Flow diagram for preparation of microcapsules by a process based on polymer-polymer incompatibility.

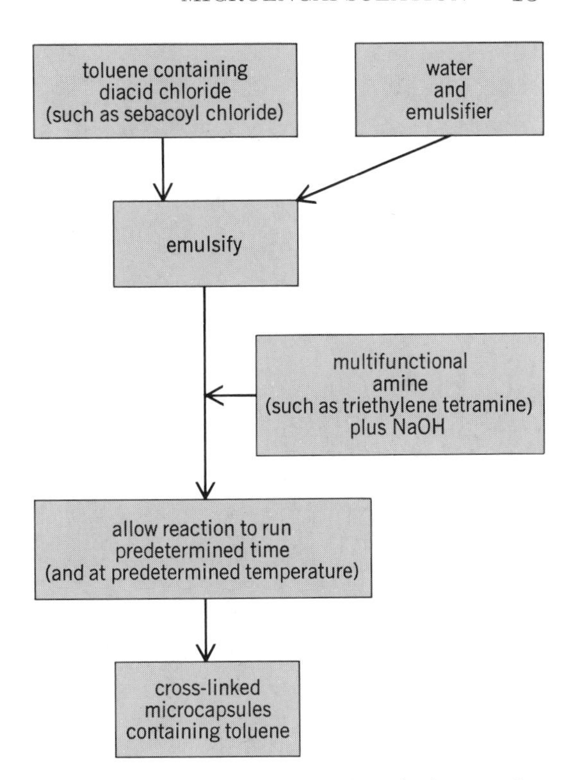

Fig. 4. Flow diagram for preparation of microcapsules by a process based on interfacial polymerization.

capsules will contain the organic solvent. If water is to be encapsulated, the amine-water solution is emulsified in the organic solvent. The diacid chloride is added to the solvent, thereby causing a capsule wall to form around the dispersed water droplets.

Isocyanates, like acid chlorides, undergo interfacial polymerization and are used to form microcapsules. By varying the isocyanates, acid chlorides, and amines used in an interfacial polymerization process, a variety of capsule walls can be deposited. Interfacial polymerization is a good way to encapsulate liquids and solid-liquid dispersions. It can yield small capsules (with diameters below 5 μm), and has been commercialized. A range of polyester, polyamide, polyurea, and urethane coatings can be deposited in this way. The number of combinations of reactants that should form microcapsules by interfacial polymerization is virtually infinite. However, the process is not problem-free; for example, it cannot handle solids, except as solid-liquid dispersions. The types of solutes that can be encapsulated often are limited because of limited solubilities in the liquid internal or continuous phases. Finally, the reactants used to form the capsule coatings are subject to side reactions with the material being encapsulated or water.

Spray drying. Spraying an emulsion or dispersion into a stream of hot, inert gas is used to encapsulate a variety of substances. Spray-dry encapsulation of water-immiscible perfumes, flavors, and vitamin oils has received considerable attention. Typical coatings for these materials include various carbohydrate polymers such as gum arabic, modified starches, and dextran, as well as water-soluble cellulose gums. Figure 5 shows a cocurrent-type spray dryer used by one manufacturer.

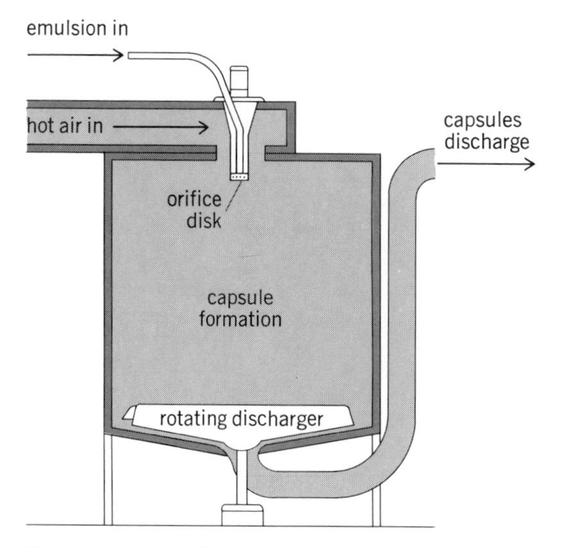

Fig. 5. Schematic diagram of cocurrent-type spray dryer used to form microcapsules. *(PFW, Middletown, NY)*

Spray-dry encapsulation processes for volatile flavors and fragrances typically involve dissolving the coating material in water and then emulsifying in this solution the oil being encapsulated. Oil droplet size in the final emulsion is approximately 1 μm. The emulsion is fed into the spray dryer, which causes the water to evaporate, thereby forming solid particles 10 to 250 μm in diameter that contain entrapped oil. Many small oil droplets are uniformly dispersed throughout each particle. If the coating material is not chemically cross-linked during the spray-drying process, the isolated capsules dissolve in water to release their oil payload. Such capsules are difficult to prepare by coacervation or interfacial polymerization processes. Gum arabic was formerly the favored material for spray drying from water, but modified water-soluble starches are also used now. Spray drying from organic solvents can also be done to form capsules containing dispersed solids or aqueous solutions. Furthermore, the coating material used in a spray-drying process can be a reactive monomer or prepolymer that polymerizes to a high-molecular-weight material under the heated conditions that exist in a spray dryer.

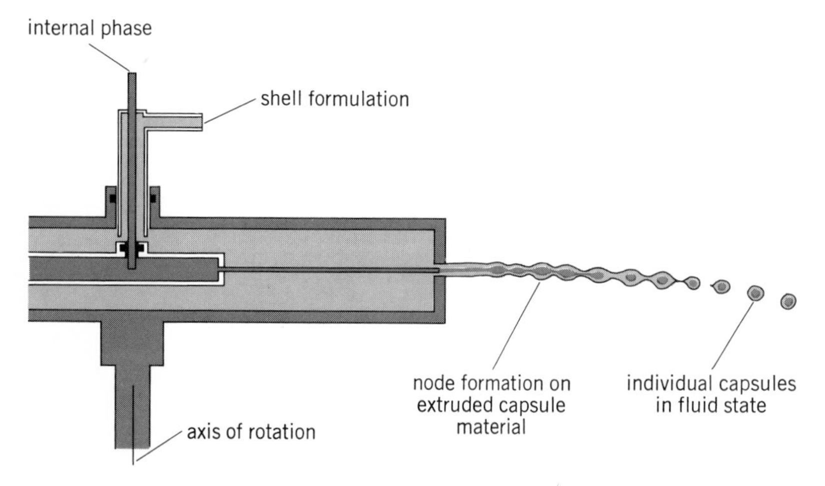

Fig. 6. Schematic diagram of a multiorifice centrifugal extrusion head used to form capsules by an engulfing process. *(Southwest Research Institute, San Antonio, TX)*

Spray drying is a well-established industrial operation and suitable for high volume throughput. Thus, it is reasonable to suggest that encapsulation technology based on this approach will continue to develop.

Air suspension encapsulation. This technique, the Wurster process, is well suited for encapsulating various solid particles. The particles are held suspended by a vertical current of air while they are sprayed with a solution of the coating material. The particles are then moved by the airstream into a drying section of the unit. The particles are recycled through the unit so as to receive a series of alternating spray and drying treatments. Coating thickness is controlled by the number of times the particles recycle. The Wurster process is best suited for solids that have a particle size of several hundred micrometers or more. Particles down to 40 μm can be encapsulated, but smaller particles agglomerate. A broad range of natural and synthetic coatings have been successfully deposited by this process.

Engulfing processes. Microcapsules can be made by mechanical or simple melt engulfing processes. In a mechanical engulfing process, droplets of core material (that is, the material being encapsulated) are forced through a film of liquid coating material, which completely engulfs the core material to form an embryo microcapsule. Figure 6 is a schematic diagram of a centrifugal extrusion device used to make capsules in this way. The core and coating materials form a biliquid column which breaks up into embryo capsules as it leaves the centrifugal head. Molten waxes, polymers, and wax-polymer blends are typical coating materials used. Capsule walls formed from these materials are solidified by air cooling. Aqueous polymer solutions represent another class of coating materials. In such cases, wall solidification is achieved by ejecting the embryo capsules into a hardening bath. Capsules formed from aqueous sodium alginate hardened in a calcium chloride bath are a specific example. Such capsules can carry a higher payload than those formed from molten coating materials because significant coating shrinkage occurs when the capsules are dried. Mechanical engulfing processes are well suited for encapsulating a range of solutions, especially aqueous solutions. Although such processes can be used to make small microcapsules, most capsules seem to be larger than 500 μm.

Simple melt engulfing processes consist of dispersing the core material in a molten wax or polymer and then dispersing this mixture in cool water. The molten coating solidifies to yield microcapsules. Alternately, the core and coating materials are codispersed in hot water, which is then cooled to give solid microcapsules. The rates of cooling and agitation determine capsule size and shape.

Drying-in-liquid process. Simple evaporation of solvent from an agitated polymer solution can be used to make microcapsules. The material being encapsulated is dispersed, dissolved, or emulsified in a volatile, water-immiscible solvent containing dissolved polymer. Water containing a protective colloid such as polyvinyl alcohol or gelatin is then added. The resulting mixture is agitated, and the organic solvent is evaporated (or extracted), leaving a solidified microcapsule that can be isolated.

Capsules with nonrigid walls. The encapsulation procedures discussed thus far involve forming well-defined particles with rigid walls. Liquid membranes, liposomes, and loaded erythrocytes have capsulelike structures with walls that differ greatly from those of classical microcapsules. They are designed to serve many of the same functions as microcapsules, and hence warrant a brief discussion.

Liquid membrane capsules use specifically formulated liquids to form small-diameter spherical shells separating two phases. They are generally formed by first making an emulsion of two immiscible phases (A and B). This emulsion is dispersed in a third phase (C) called the continuous phase. Phases A and C are usually miscible but separated from each other by phase B, the liquid membrane. If phases A and C are water, phase B is an "oil." This oil usually contains surfactants, additives, and a solvent for these ingredients. The surfactants and additives serve to control the stability, permeability, and selectivity of the membrane.

Liposomes are lipid spherules that range in size from 20 nm up to 1 μm or more. Unilamellar and multilamellar forms exist. Figure 7 illustrates a procedure to prepare negatively charged liposomes. The lipid coating materials are codissolved in chloroform-methanol. This solvent mixture is then evaporated under nitrogen to leave a lipid film. An aqueous solution of the substances to be encapsulated is added and the system sonicated. Sonication entraps the compound being loaded inside the liposomes. Encapsulated and unencapsulated material are separated on a Sepharose 6B column.

Erythrocytes, or red blood cells, can be loaded with substances by hypotonic exchange. Figure 8 illustrates schematically the loading of enzyme into two erythrocyte cells by this process. The cells are immersed in buffered salt solution (pH 6.5) containing an enzyme to be loaded. Hypotonic exchange is induced by rapid addition of excess distilled water. This causes the hemoglobin in the cells to escape and be replaced by enzyme. After a specific amount of time, enough NaCl is added back to the system to restore isotonicity. Unentrapped enzyme is washed away, leaving enzyme-loaded erythrocytes.

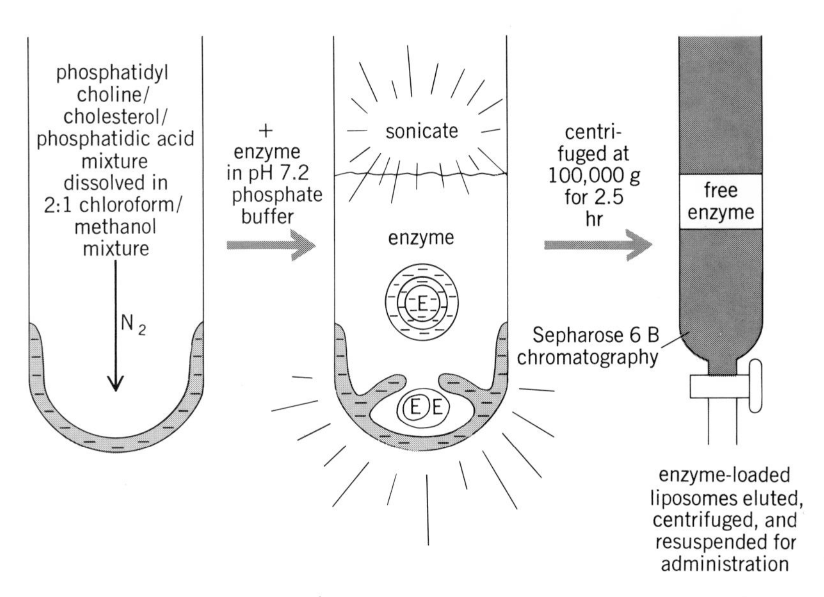

Fig. 7. Diagram of the process by which negatively charged liposomes containing an enzyme (E) are formed. *(From R. J. Desnick et al., in T. M. S. Chang, ed., Biomedical Applications of Immobilized Enzymes and Proteins, vol. 1, Plenum Press, 1977)*

APPLICATIONS OF MICROCAPSULES

The number of suggested microcapsule applications appears to be endless, and new ideas are constantly being generated. Major potential capsule applications exist in every branch of science and industry. Although many suggested capsule applications will never develop beyond the conceptual stage, the probability that some will is high. Indeed, the virtually inexhaustible supply of ways to use microcapsules is a rationale for the belief that microcapsule use and technology have a viable future. In order to provide a current picture of capsule uses, a few applications in a number of different fields will be discussed. Many of the applications mentioned have been commercialized, but some are still at the developmental stage.

Graphic arts. Carbonless copy paper is the primary application of microcapsules. Carbonless copy papers are formed by coating the back of a sheet of paper with a layer of small capsules that contain a dye in a colorless form dissolved in a high-boiling hydrocarbon solvent. The top of the paper sheet is coated with a clay or phenolic resin. When writing or printing pressure is applied to a sheet, the capsules on the back of the paper are broken, thereby releasing the dye solution. The dye is transferred to the top of a second sheet of coated paper located directly below the first sheet. A visible color develops in this receiving sheet due to a reaction between the dyes and clay-phenolic coating. From 7 to 10 copies can be made from a single impression.

The first carbonless copy papers were based on small aggregates (10–20 μm in diameter) of gelatin-based capsules formed by complex coacervation. The color-forming system was a mixture of crystal violet lactone and *N*-benzoylleukomethylene blue. These dyes give a blue print. Throughout the years carbonless copy papers have steadily improved in quality. Current papers often use capsules formed by interfacial polymerization or polycondensation of urea-formaldehyde prepolymers. Both of these latter encapsulation pro-

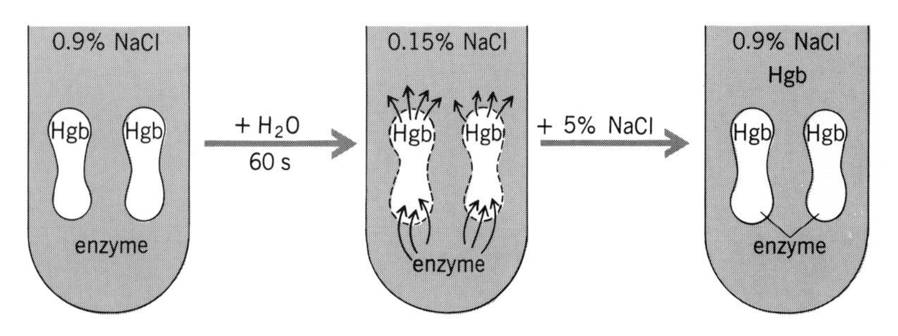

Fig. 8. Diagram of the process by which erythrocytes are loaded with enzymes. *(From R. J. Desnik et al., in T. M. S. Chang, ed., Biomedical Applications of Immobilized Enzymes and Proteins, vol. 1, Plenum Press, 1977)*

cedures yield individual capsules as small as 2 to 5 μm in diameter. The small capsules cause sharper reproduction of multiple copies. Smudge problems caused by premature capsule rupture during normal handling are minimized by incorporating starch granules into the coating. These are somewhat larger than the capsules and act as inert spaces. A movement toward carbonless copy paper with black print has begun.

At present, a preponderance of the carbonless copy paper sold is coated at the mill and then converted into business forms. The capsules are manufactured, pumped as a slurry to the coating operation, and coated onto the paper. It has been suggested that spot printing of capsules onto forms at the printshop may be a better approach. In such cases, dry powder capsule formulations would be shipped from the capsule producer to the consumer. Whether or not this concept will be adopted remains to be seen.

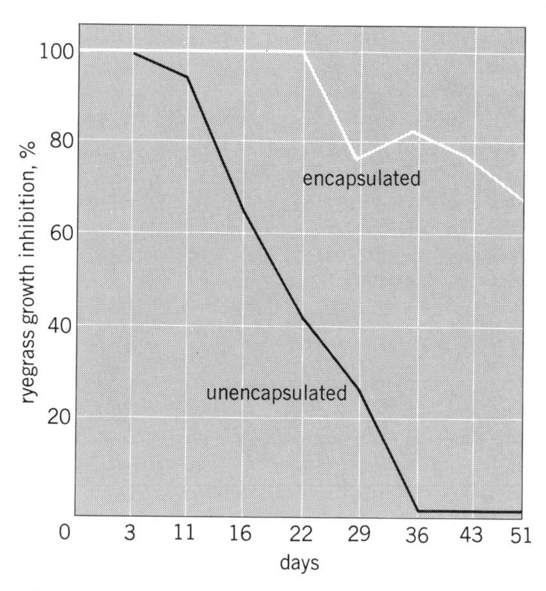

Fig. 10. Herbicidal performance of encapsulated and unencapsulated chlorpropham, as determined from degree of inhibition of ryegrass growth. *(From W. A. Gentner and L. L. Danielson, in N. F. Cardarelli, ed., Proceedings of the 1976 Controlled Release Pesticide Symposium, University of Akron, 1976)*

Pharmaceuticals. Microencapsulated drugs can be administered orally, topically, or by injection. Capsules given orally are designed to provide sustained drug release, taste-mask bitter drugs, stabilize oxygen or hydrolytically unstable drugs, or separate incompatible materials. Encapsulated drug formulations that accomplish a number of these tasks are already being marketed. Time-release encapsulated aspirin has been sold for some time. In addition to providing sustained release, encapsulated aspirin reportedly is tolerated in the gastrointestinal tract much better than raw aspirin. Similar benefits are attributed to encapsulated KCl given orally. Encapsulated antibiotics designed to reduce the bitter taste for children are also available. A microcapsule formulation containing both vitamin B complex and calcium pantothenate provides taste-masking and separation of incompatibles. Spray-dried microcapsules containing oxygen-sensitive vitamin A or vitamin A derivatives reportedly are stable for prolonged periods at room temperature in air. Such capsules have carbohydrate coatings that are very effective barriers to oxygen transport.

Topical applications of encapsulated drugs have received some consideration. Bandages containing drug-filled capsules that release their payload slowly may be suitable for the treatment of various skin disorders. Insertion of drug-containing capsules into the eye as an ointment is a possible means of delivering drugs for the treatment of a variety of ophthalmic disorders.

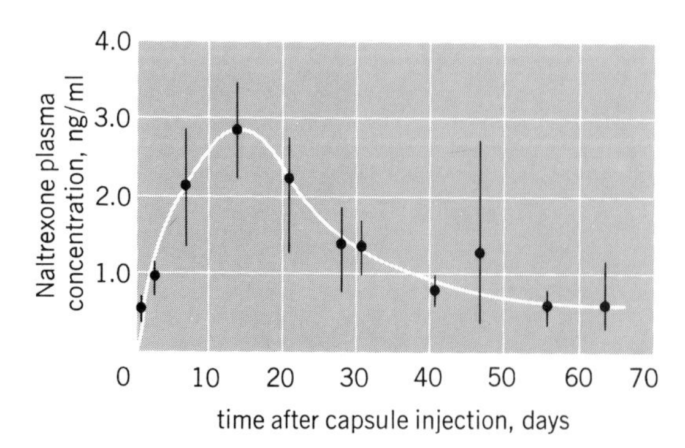

Fig. 9. Naltrexone plasma levels in monkeys receiving an intramuscular injection of polylactic acid capsules (diameter less than 106 μm) containing naltrexone pamoate, a narcotic antagonist. *(From S. Harrigan et al., in R. Kostelnik, ed., Polymeric Delivery Systems, Gordon and Breach Publishers, 1978)*

Some of the potentially most important applications of microcapsules involve their use as drug carriers in injectable drug formulations. This is an active area of current research. Studies with living animals indicate that a single intramuscular or subcutaneous injection of drug-filled microcapsules is capable of providing slow drug release for prolonged periods. Figure 9 contains results of drug plasma assays in monkeys that demonstrate this point. Injectable capsules are fabricated from polylactic acid or lactic acid – glycolic acid copolymers because these materials are biocompatible and biodegradable. Intravenous injection of small capsules is also being explored; a specific example is the incorporation of sodium stibogluconate into liposomes. This drug-carrier system has shown promise as an effective way to treat leishmaniasis.

Liposomes and drug-filled erythrocytes are being evaluated as drug carriers for the treatment of cancer and enzyme deficiency diseases. The concept of building into these carriers a homing mechanism that guides them to a specific target is being pursued vigorously. Such carriers would be particularly valuable for the treatment of cancer. Although a degree of targeting has been achieved, significant improvements in the homing mechanism(s) are needed before the full benefits of selectively targeted drug carriers are realized.

In summary, microencapsulated drugs have enormous potential that is just beginning to be tapped. Most current microencapsulated drug formulations are specialty products administered orally. However, this situation could change dramatically as the technology needed to manufacture usable capsules at a reasonable cost becomes better developed.

Agricultural applications. Conceptually, microcapsules can release pesticides at a controlled rate for prolonged periods and thereby greatly alter how pesticides are delivered to the environment. Although agricultural applications of microcapsules have been known for some time and are well publicized, the number of commercial pesticide formulations containing microcapsules is limited. One spray concentrate being marketed contains encapsulated methyl parathion. The capsules are formed by interfacial polymerization, and average 30–50 μm in diameter. The capsule wall is a polyamide cross-linked polyurea. Advantages claimed for encapsulated methyl parathion include reduced handling toxicity, longer persistence in the field, and reduced amounts of active material needed for effective pest control. Recently, an unexpected problem with the capsules was discovered. Because the capsules resemble pollen grains, foraging bees carry them from the fields to the hives. There the methyl parathion is released and kills the bees. This problem, now overcome, illustrates the complexity of developing a microencapsulated pesticide formulation that functions as intended and does not generate new environmental problems.

Microcapsules containing pesticide should completely release their payload to the environment at as uniform a rate as possible. Variations in temperature, wind velocity, humidity, and sunlight can produce significant fluctuations in release rate which affect capsule performance. The rate and extent of pesticide release under field conditions have not been well studied. These types of data are needed in order to better assess the type of microcapsules best suited for use as pesticide delivery devices. Some efforts are being made to gather such data. For example, it was found that disparlure, the sex attractant phenomone of the gypsy moth, is incompletely released from experimental gelatin-based microcapsules formed by coacervation. The disparlure that is released comes out of the capsules at a steadily declining rate. Whether this type of behavior occurs with a range of pesticides or different types of capsules is unknown. However, it is very undesirable. Once the rate of pesticide release from a capsule sample drops below that needed to cause a meaningful toxic effect, the capsule formulation will cease to function as planned. The degree to which measurable but marginally lethal levels of pesticide released over prolonged periods from a microcapsule affect development of pest resistance to the pesticide is unknown. Some entomologists fear that if this type of release were to occur, it would greatly accelerate development of resistance to pesticides. The potential impact of this possibility on the widespread use of long-acting pesticide delivery devices cannot be underestimated. There admittedly is much to be gained by developing suitable pesticide delivery devices such as microcapsules. However, successful development is a very complex process that must consider a number of environmental factors. This undoubtedly is a contributing factor to the limited number of encapsulated pesticide formulations on the market. The rate of development of microcapsules for many agricultural applications will be controlled by the envi-

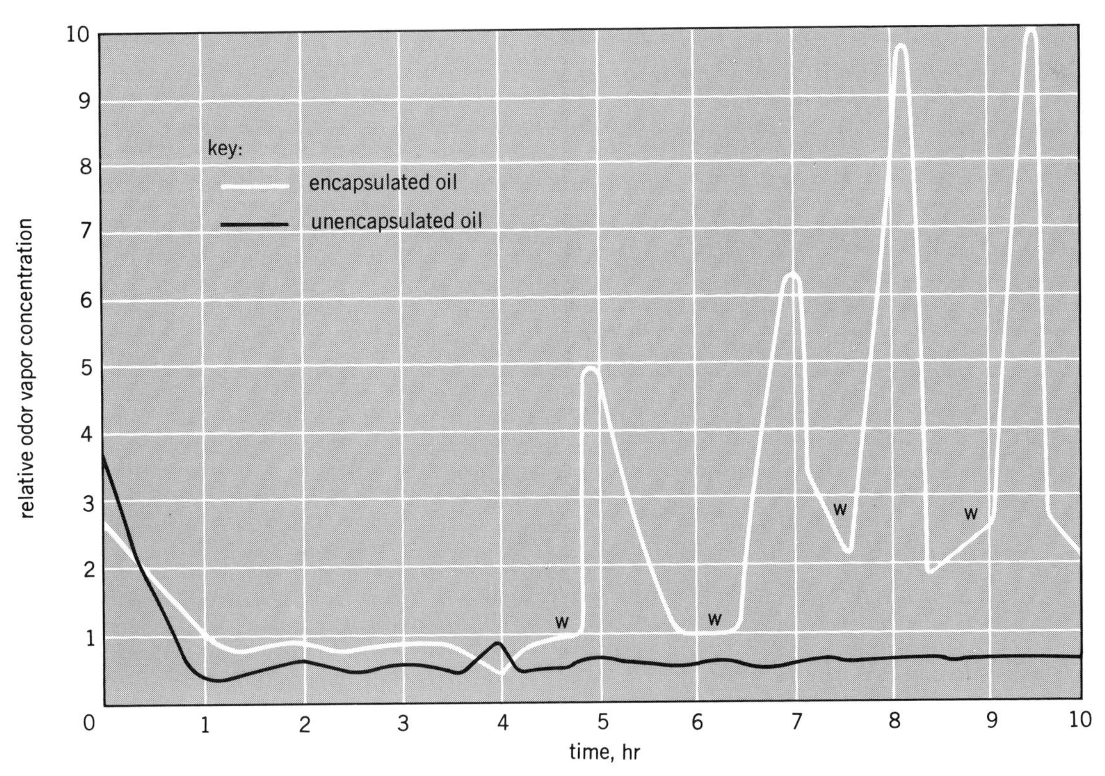

Fig. 11. Release of fragrance from spray-dried oil-loaded capsules subjected to repeated applications of moisture, compared with release from unencapsulated oil. The W's indicate times at which moisture was applied. (*From J. M. Miles et al., Encapsulated perfumes in aerosol products, J. Soc. Cosmet. Chem., 22:655–666, 1971*)

ronmental regulatory agencies and by economics.

Microcapsules may make their greatest contribution to pest control by stabilizing the so-called second- and third-generation pesticides which have low mammalian and wildlife toxicities. These pesticides tend to be relatively unstable in the field and need some type of controlled-release device in order to be effective. Long-lasting encapsulated herbicide formulations may also be environmentally acceptable. Figure 10 shows that the encapsulated herbicide chlorpropham is effective in the field significantly longer than the unencapsulated herbicide. The effectiveness of the herbicide was determined by measuring the degree of inhibition of ryegrass growth it caused.

Other areas where microcapsules can contribute quickly with minimal environmental impact include long-lasting fertilizers and coated seeds. At present these are specialty applications oriented largely toward the home gardener.

Fragrances and flavors. Encapsulated fragrances and flavors have been an active area of product development. Products obtained thus far fall into two broad categories: encapsulated flavors-fragrances with water-soluble walls for food and cosmetic applications; and encapsulated flavors-fragrances with water-insoluble walls for advertising and promotional purposes.

Capsules in the first category can be made by spray drying. The dry powder products remain stable for a prolonged period and protect flavors and fragrances from oxidation or volatilization. Spray-dried capsules with modified starch walls have the ability to encapsulate measurably higher flavor-fragrance payloads than formerly was possible with gum arabic, the traditional coating material. Spray-dried capsules are used to incorporate flavors into a variety of prepared food products. They also have been used in an antiperspirant formulation. This latter application is based on the release of fragrance induced by moisture. Figure 11 illustrates how repeated applications of a water mist onto a surface coated with spray-dried fragrance capsules induces fragrance release. The use of spray-dried capsules in various food and cosmetic products should grow steadily.

Fragrance capsules with water-insoluble walls have been used in a variety of advertising and promotional applications. The capsules are coated onto a paper or plastic film substrate which contains an advertisement or product description. When the capsule-coated area is rubbed or scratched, the fragrance is released, thereby stimulating the consumer's sense of smell. A variety of magazine advertisements, labels, and children's books have utilized fragrance-filled capsules. Whether or not this is a novelty application with a limited lifetime remains to be seen.

Enzyme immobilization. A number of enzymes have been immobilized by microencapsulation. Enzymes encapsulated by an interfacial polymerization process usually have a polyamide wall, whereas capsules formed by interfacial precipitation generally have a cellulose nitrate wall. As in any encapsulation process, various parameters must be controlled if useful capsules are to be obtained. Which parameters are most important will be determined by the specific enzyme system being encapsulated and the encapsulation process used.

Microcapsules containing enzymes have a large surface area which maximizes enzyme-substrate contact. This is regarded as a major advantage of encapsulated enzymes. Another significant feature is the ability of the capsule wall to act as a screen that excludes large molecules from the capsules while freely passing small molecules. Of course, this also means that the activity of encapsulated enzymes is limited to substrate molecules small enough to pass across the capsule wall. A number of different enzymes can be placed inside the same capsule, thereby allowing sequential reactions to proceed. Even intact cells can be encapsulated.

Encapsulated enzymes retain 10 to 100% of the activity they had before encapsulation. Interfacial polymerization tends to cause a higher loss in activity than interfacial precipitation. A soluble enzyme enclosed inside a capsule should have the same properties it had before encapsulation. This is not always true, however, perhaps due to changes in the environment inside a capsule or some form of enzyme – capsule wall interaction.

Most studies of microencapsulated enzymes have focused on biomedical applications such as tumor suppression, replacement of congenital enzyme deficiencies, and use in an artificial kidney. The potential of enzymes encapsulated in liquid membranes for wastewater treatment has also been discussed. No microencapsulated enzymes have been commercialized as yet.

Enzymes can be immobilized in many different ways, so microencapsulated enzymes must in effect compete with alternate technologies capable of immobilization. A critical assessment of the properties of enzymes immobilized in several different ways would be worthwhile. Likewise, enzyme-containing microcapsules prepared from a broader range of wall materials by a wider variety of encapsulation processes merit consideration.

Adsorption and extraction. Encapsulated solid and liquid adsorbents have a number of applications. For example, encapsulated charcoal particles are used to treat patients suffering from chronic renal failure and acute drug or chemical intoxication. The blood from such patients is passed through a column packed with coated carbon particles. The polymer coating freely passes the toxic substances in the blood, so that they are adsorbed by the charcoal. The coating prevents charcoal fines from getting into the blood, thereby preventing emboli formation.

Microencapsulated charcoal packed in columns has also been examined for a number of biochemical applications. A specific example is the adsorption of kanamycin from broth filtrate without the adsorption of other components of the medium, such as glucose, polypepton, and melanoid pigment. A fermentation broth filtrate containing alkaline protease is decolorized when passed through a packed column of encapsulated charcoal. Finally, lysozyme from egg white has been purified by passing diluted fresh egg white through a column of coated charcoal. These examples establish that a broad range of separations is possible using encapsulated charcoal.

Liquid membranes have been considered for purifying wastewater. They can be designed to remove one or more of a broad number of water contaminants such as phenol, mercury ions, and cyanide ions. Preliminary economic studies suggest that liquid membranes will be cost-effective, particularly for advanced water treatment applications not easily handled by conventional techniques.

In summary, microcapsules in various forms can be used to effect a broad range of separations. Initial steps to commercially utilize capsules for this purpose have been taken, and the potential for large-scale growth exists.

SUMMARY

The numerous, large-scale commercial applications originally envisioned for microcapsules have yet to be realized. Nevertheless, microencapsulation is steadily evolving into a major coating technology. The number of encapsulated specialty products is steadily broadening, and it is reasonable to suggest that an increasing number of these will develop into major products. Nevertheless, successful encapsulation still remains more an art than a science. Much of the technical knowledge needed to prepare usable microcapsules remains tightly held as proprietary information. This situation should change in the next few years, thereby contributing to an accelerated rate of development of encapsulation technology. [CURT THIES]

Arecibo
Arizona
New Mexico

Star Formation

Gillian R. Knapp received her B.Sc. in physics from the University of Edinburgh in 1966 and her Ph.D. in astronomy from the University of Maryland in 1972. Her research work has focused on radio-astronomical spectral line studies of the gas content of galaxies, the large-scale structure of galaxies, the interstellar medium, dust clouds, and star formation regions.

On a dark night several thousand individual stars are visible in the sky. But also crossing the sky can be seen a faint band of light, the Milky Way. The light comes from billions of stars, too distant and faint to be seen individually. How and where do stars form? How do they produce their light? What happens when their source of fuel runs out, and how do they die? Intimately related to these questions are the dark patches and stripes which can be seen against the bright background of the Milky Way. These are not holes in the distribution of stars, but are huge clouds of gas and dust, many light-years in extent and so opaque that they completely block the light of the stars behind them. It is in these clouds that new stars are born; and these clouds capture gas flung into space by the explosions of dying stars. This captured material becomes part of the substance from which new stars are born, as the cycle of stellar birth and death continues.

This article discusses recent discoveries about the formation of stars in the Galaxy. Newly developed techniques of infrared and radio astronomy, in particular the discovery and measurement of signals from numerous species of interstellar molecules, have allowed astronomers to penetrate the interiors of the dark clouds and to study the actual process of star formation. A description of the life cycle of stars, the structure of the Galaxy and of the interstellar gas, the physical processes involved in the emission of radiation from interstellar molecules and the emission of infrared radiation, and the astronomical telescopes used in this work is given as background information. The rest of the article focuses on the interstellar environment, the interstellar gas, and the processes of star formation.

EVOLUTION OF A STAR

A star, such as the Sun, is held together by the attractive force of gravity (hence its spherical shape). It does not collapse, however, because it is "held up" by thermal motion in the nuclei and electrons in the enormously hot gas of which it is composed. Heat is produced in the star's center, where densities and temperatures are sufficiently high for the nuclear fusion of hydrogen into helium to take place. The star is thus in equilibrium between the inward force of gravity and the outward pressure force due to the production of energy in its center.

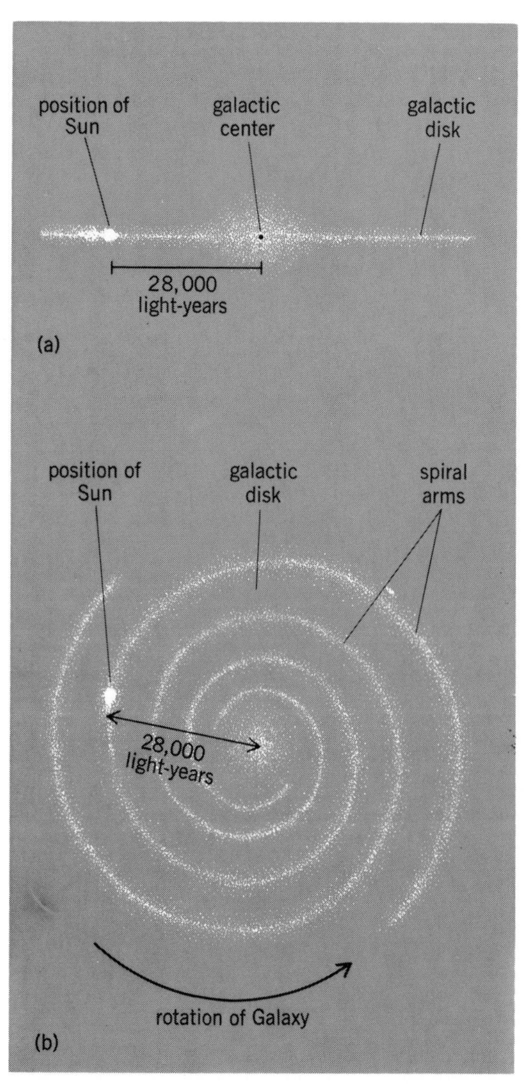

Fig. 1. The Galaxy, as seen from the outside: *(a)* side view and *(b)* top view.

The star is also in equilibrium in another way (a second balance); it is radiating energy from its surface, and thus continually losing energy, which must be replaced by energy produced in the star's center. Thus the rate of energy production in the center of the star equals the rate of energy loss from the surface of a star.

Newborn stars. The hotter the surface of a star, the bluer is the light emitted from it. Hotter stars are more massive, since they must produce more energy to balance the force of gravity. The luminosity L (total energy radiated) of a star is related to its surface temperature T by Eq. (1), where σ is

$$L = \sigma A T^4 \quad (1)$$

Stefan's constant and A is the surface area of the star. Thus a star which is twice as hot as the Sun (whose surface temperature is about 6000°C) but the same size radiates 16 times as much energy as the Sun, and thus its internal energy production is 16 times as high. Since the hotter stars are more massive, the lifetime of a massive star before it runs out of fuel is much shorter than that of a lower-mass star. The lifetime of the Sun is estimated to be about 9×10^9 years; that of a star 20 times as massive as the Sun (designated as 20 $M_\odot$ where $M_\odot$ is the mass of the Sun, about 2×10^{33} g) is only about 10^6 years. Thus (on an astronomical time scale) bright blue hot stars must be relatively newborn.

At the beginning of their life stars reflect the composition of the cloud from which they formed, typically, by number of atoms, about 90% hydrogen, 10% helium, and 1% everything else (the "heavy elements," including such species as carbon, sulfur, silicon, oxygen, iron, and calcium). Thus a star begins its life with a great deal of hydrogen.

Red giant. However, eventually the hydrogen in the center of the star is all consumed. At this point, the star begins to burn the hydrogen in a shell around the helium core; added to this is heat released from the collapsing core. The increased energy output causes the outer envelope of the star to expand; as it moves farther from the source of energy, the envelope cools down and glows red; the star has become a red giant. (When the Sun reaches this stage, it will expand until it extends past the orbit of Mars.)

Supernova. When the collapsing helium core becomes hot and dense enough, nuclear reactions converting helium to carbon begin. In some stars, the burst of energy produced in the "helium flash" causes the outer envelope (containing perhaps 0.2 $M_\odot$ of gas) to expand away from the star. What happens next depends on the mass of the star; for one as massive as the Sun, the remnant core will slowly, over billions of years, cool down and "go out"; for a more massive star, however, the core material will become so compressed that more complex nuclear reactions involving heavier nuclei can take place. A runaway reaction occurs which ends with the violent explosion of the whole star. Such an explosion is called a supernova; the light from a supernova is so bright that it can briefly outshine a whole galaxy. The remnant star quickly fades; but in the violence of the explosion the nuclei of heavy elements are synthesized. The gaseous material from the explosion is flung into interstellar space, where it mingles with the gas already there. In this way, the interstellar gas, which was presumably originally composed of only hydrogen and helium, is progressively enriched by the addition of heavy elements with the passage of time.

STRUCTURE OF THE GALAXY, AND THE INTERSTELLAR MEDIUM

The Galaxy is a spiral galaxy containing several hundred billion stars. As seen from the edge (Fig. 1*a*), it would have a flat platelike appearance, with a large central bulge. As seen from the top (Fig. 1*b*), the galactic disk has a spiral pattern. The Sun is a typical star in the Galaxy and is about 28,000 light-years (1 light-year = 9.46×10^{15} m) from the galactic center, around which it revolves once every 240,000,000 years.

In the neighborhood of the Sun, the average distance between the stars is about 5–10 light-years. Compared with the sizes of the stars, these distances are truly enormous. However, the space between the stars is not empty, but is filled by very tenuous gas. The average density of this gas, which again is mostly hydrogen, is only about 1 atom/cm^3. For comparison, the density of the Earth's atmosphere at sea level is about 2×10^{19} molecules/cm^3. The density of the gas in interstellar space is very low, but because of the great size of the Galaxy the total mass of gas is very large. In the Galaxy, the total amount of gas is about 5×10^9 $M_\odot$, or about 5% of the total known mass of the Galaxy. This is enough gas to form about 5×10^9 new stars of the same brightness as the Sun. Mixed in with the gas, fairly uniformly as far as can be determined, is a small amount of interstellar dust. The total mass of the dust is about 1% of that of the gas, and as far as is known at present, the dust is composed mostly of heavy elements.

INTERSTELLAR RADIATION AND MAGNETIC FIELDS

As well as being filled by gas, the space between the stars is flooded by various types of radiation, which react with the gas, as a source of energy input to the gas.

Source of radiation. The most obvious source of radiation is starlight. Of particular importance is the ultraviolet light (emitted at wavelengths shorter than 3000 A (300 nm), for this radiation can destroy molecules and ionize atoms. A second type of radiation is that from the microwave blackbody background. This radiation permeates all of space and is the remnant of the original "big bang" with which the universe began some $1-2 \times 10^{10}$ years ago.

Interstellar magnetic field. Also of great importance is the interstellar magnetic field. Large-scale, low-strength magnetic fields are present in most astronomical bodies. The Galaxy likewise has a weak magnetic field which pervades the interstellar medium. Space is also permeated by cosmic rays and x-rays, both of which are thought to originate in the violent explosions which are novas and supernovas, and in other energetic events, for example, the falling of matter into a black hole.

OBSERVATIONAL TECHNIQUES

Information concerning the nature of the interstellar medium can be obtained by using the techniques of radio and infrared astronomy.

Radio astronomy. The energy of a photon of electromagnetic radiation E is expressed by Eq. (2),

$$E = \frac{hc}{\lambda} \qquad (2)$$

where h is Planck's constant and c is the speed of

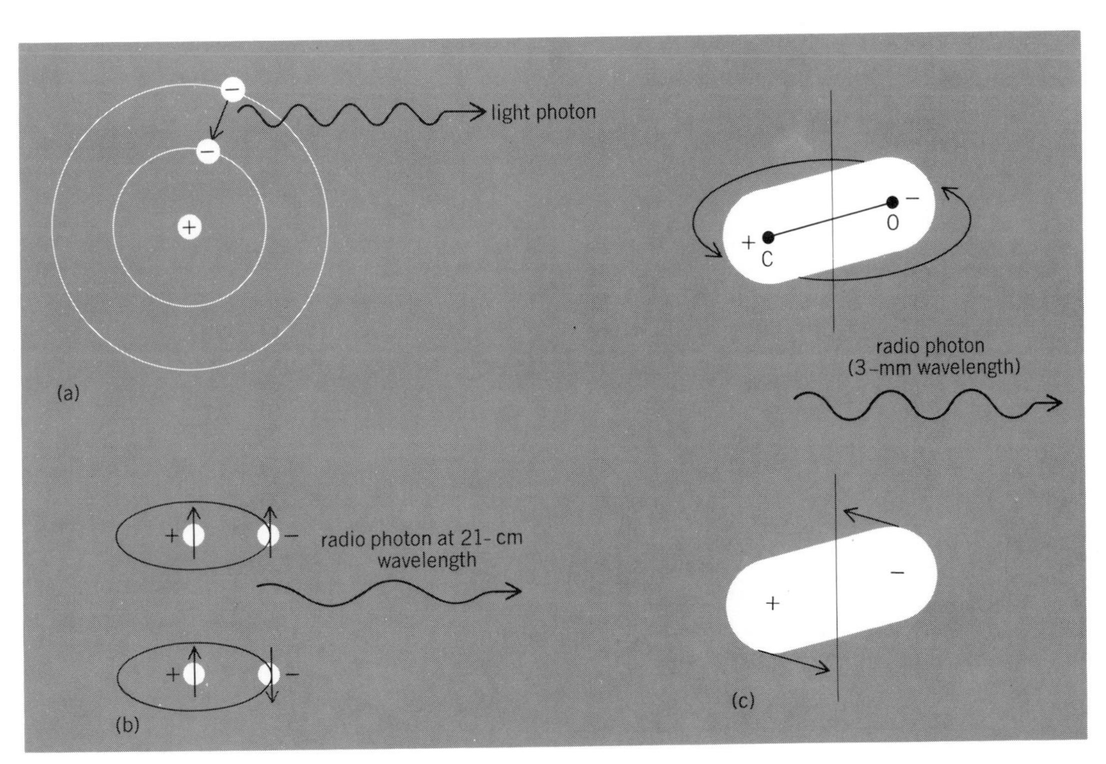

Fig. 2. Emission mechanisms for spectral lines used in studying interstellar gas. (*a*) Electronic transition (hydrogen). (*b*) Spin-flip transition (hydrogen). (*c*) Rotational transition (carbon monoxide).

light. Since, for blackbody radiation, the wavelength λ_p at which the source is brightest is related to the temperature T of the body by notation (3), it

$$\lambda_p \sim \frac{1}{T} \quad (3)$$

can be seen that for very low temperatures most of the radiation is emitted at low energies, or long wavelengths. For temperatures a few degrees above absolute zero (the temperature of the interstellar gas), most of the radiation is emitted at wavelengths of a few millimeters to a few centimeters, the microwave radio range.

Spectral line emission. Most of the information about interstellar gas comes from spectral line radiation. Atoms and molecules are made up of individual particles bound together by electromagnetic forces. The strength of the electromagnetic force depends both on the intrinsic strength of the electric charge on particles and on the distance between them. A system of bound particles thus loses or gains energy as its particles move closer together or farther apart. In the atomic regime, energy loss or gain occurs in small packets or quanta, where the energy is a multiple of Planck's constant h. As shown by Eq. (2), this energy loss or gain corresponds to the absorption or emission of a quantum of radiation (a photon) at a particular wavelength determined by the properties of the atom or molecule emitting the radiation.

Perhaps the most familiar example of the absorption or emission of radiation occurs in electronic transitions in atoms (Fig. 2*a*). The electron in the atom can occupy only discrete energy levels with respect to the nucleus, and thus the jump of the electron from one level to a lower one causes the emission of radiation at a particular wavelength which depends on the properties of the atom. This is termed spectral line emission. Thus the observation of spectral line emission at specific wavelengths shows which atoms are present. In this way the chemical composition of distant astronomical bodies can be found by analyzing their light.

Electronic transitions usually cause radiation to be emitted at optical wavelengths, which are much shorter than radio wavelengths. There are various other types of interaction between molecules or atoms and the electromagnetic field, however. The hydrogen atom, consisting only of a proton with an orbiting electron, undergoes what is known as a spin-flip transition (Fig. 2*b*). When both the proton and electron are "spinning" in the same direction, the atom has slightly more energy than when they are spinning in opposite directions. A spontaneous "somersault" by the electron thus causes the atom to emit radiation. When the atom is in its lowest energy state (which happens when the atom is in a very cold environment such as interstellar space), radiation is emitted at a wavelength of 21.1 cm.

Another type of transition is the rotational transition, emitted by many molecules. A molecule such as CO (Fig. 2*c*) is shaped like a dumbbell, with two atoms bound together by a cloud of electrons. Since the atoms are not identical, the molecule has a net positive and negative charge at opposite ends of the dumbbell. The molecule can rotate, but again only at certain quantized speeds. Molecules can be set rotating by colliding with another molecule. The change from one state of rotation to another again involves the emission or absorption of radiation; typical wavelengths are a few millimeters. The lowest-level transition of the CO molecule occurs at 2.6 mm. Unfortunately, molecules made up of two identical atoms (such as H_2) do not have a net charge and so do not emit radiation from rotational transitions.

Thus, the observation of molecular spectral line emission from a region of interstellar space can

Known interstellar molecules*

Molecule†	Chemical symbol	Molecule†	Chemical symbol
Methyladyne	CH	Hydrogen sulfide	H_2S
Cyanogen radical	CN	Methanimine	H_2CNH
Methyladyne ion	CH^+	Sulfur monoxide	SO
Hydroxyl radical	OH	‡	N_2H^+
Ammonia	NH_3	Ethynyl radical	C_2H
Water	H_2O	Methylamine	CH_3NH_2
Formaldehyde	H_2CO	Dimethyl ether	$(CH_3)_2O$
Carbon monoxide	CO	Ethyl alcohol	CH_3CH_2OH
Hydrogen cyanide	HCN	Sulfur dioxide	SO_2
Cyanoacetylene	HC_3N	Silicon sulfide	SiS
Hydrogen	H_2	Acrylonitrile	H_2CCHCN
Methyl alcohol	CH_3OH	Methyl formate	$HCOOCH_3$
Formic acid	$HCOOH$	Nitrogen sulfide radical	NS
‡	HCO^+	Cyanamide	NH_2CN
Formamide	NH_2CHO	Cyanodiacetylene	HC_5N
Carbon monosulfide	CS	Formyl radical	HCO
Silicon monoxide	SiO	Acetylene	C_2H_2
Carbonyl sulfide	OCS	Cyanoethynyl radical	C_3N
Methyl cyanide	CH_3CN	Ketene	H_2C_2O
Isocyanic acid	$HNCO$	Cyanotriacetylene	HC_7N
Methylacetylene	CH_3C_2H	Nitroxyl	HNO
Acetaldehyde	CH_3CHO	Cyanotetracetylene	HC_9N
Thioformaldehyde	H_2CS	Nitric oxide	NO
Hydrogen isocyanide	HNC	Butadynyl radical	C_4H

*Courtesy of B. E. Turner, National Radio Astronomy Observatory.
†Molecules are listed in order of detection in the interstellar medium. Complete as of June 1978.
‡Molecular ions with no chemical names.

Fig. 3. The 10-m telescope at the Owens Valley Radio Observatory, California Institute of Technology.

yield a great deal of information, including: which molecules are present (by which wavelengths are detected); the number of each type of molecule (the more molecules of a given type, the stronger the signal); the temperature of the region (the hotter the region, the faster the molecules spin); the total density of the region (the denser the gas, the more often collisions between molecules occur and so the more molecules are spinning); and the size of the region (from the area over which molecular emission can be detected). An additional, very important piece of information can be obtained: the wavelength of the radiation is changed slightly by the Doppler effect (the change in wavelength of waves emitted from a moving object), so that the motions of the gas in the region can also be measured. Finally, because their wavelengths are much longer than the sizes of the dust grains (most are less than about 1 μm in diameter), radio waves can pass unaffected through dusty regions.

In recent years many different types of molecules have been discovered in interstellar space. A list of these, complete as of June 1978, is given in the table. Studies of the radiation from these molecules leads to an understanding of interstellar chemistry, the physical properties of the interstellar medium, and the events preceding star formation.

Radio telescope. Radiation from interstellar molecules is observed and measured with radio telescopes. A new 10-m-diameter radio telescope built by R. B. Leighton at the Owens Valley Radio Observatory of the California Institute of Technology is shown in Fig. 3. The surface of the telescope is accurate to a small fraction of the shortest wavelength to be observed with the telescope, about 0.8 mm. (This means that the surface accuracy is about 0.02 mm.) This requirement of high surface accuracy limits the size of the telescope to about 10 m; radio telescopes built to observe longer-wavelength radiation do not have such stringent requirements on surface accuracy and so can be much larger.

Although it may not look much like a conventional optical telescope, the radio telescope is in fact very similar. Radio waves are reflected off the main bowl of the telescope onto the secondary mirror held above the bowl by support members. The waves are then reflected back down to the center of the dish, where they are detected by a radio receiver. The signals are then amplified and fed through cables to instruments located near the base of the telescope, for analysis. For each particular molecule the radio receiver is tuned to the desired wavelength in a manner similar to tuning in various radio stations on a home tuner. Signals from space are extremely weak, however, and the receiving equipment must be very sensitive. To achieve this sensitivity, radio receivers are often cooled to the temperature of liquid nitrogen or liquid helium to reduce their own contribution to radio "noise," which one hears in a radio receiver as a hissing sound which can mask the faint signals from distant broadcasting stations.

Infrared astronomy. When an object is warm (about 100 to 1000 K) most of its radiation is emitted at wavelengths longer than that of visible light [Eq. (3)]. These infrared waves are felt on the skin as heat.

Infrared radiation is often emitted from hot spots in interstellar clouds, regions in which a hot object (that is, a star) is heating the dust grains around it. Again, the infrared waves are long enough in wavelength to pass right through an interstellar dust cloud. Thus not only hot spots in dark clouds but also the stars themselves, or even stars behind the clouds, can be detected by their infrared radiation.

The existence of infrared radiation was first demonstrated by William Herschel in 1800 when he passed the light of the Sun through a prism, displaying its spectrum from violet to red. A thermometer placed next to the red band registered the presence of heat, or infrared radiation. A very sensitive version of the same technique is used in modern infrared detectors. The infrared radiation from the object, gathered by a telescope, is focused on a detector whose resistance increases slightly as its temperature rises, so that the measurement of the resistance measures the amount of radiation falling on it. Again, the signals from space are very weak, so that to increase the sensitivity of the system the detectors are cooled to the temperature of liquid helium.

PHYSICAL STATE OF THE INTERSTELLAR MEDIUM

By means of the observational techniques described above, the physical state of the interstellar medium can be studied.

Gas clouds. As mentioned earlier, most of space is filled with very-low-density (approximately 1 atom/cm^3) gas. This gas is almost entirely in atomic form, is quite cold (approximately 50 K), and is detected by its emission in the 21-cm wavelength line of atomic hydrogen. Because the Galaxy is rotating slowly like a giant whirlpool, the different gas clouds are moving at different speeds and the wavelength of their radiation is shifted by different amounts. In this way, the motions of these gas clouds can be measured, and the structure of the Galaxy deduced. The gas is concentrated into a flat, thin, platelike disk, and the disk is further compressed into great spiral ridges. Mixed in with the gas are the dust grains, whose cumulative blocking effect means that the distant parts of the Galaxy (beyond about 6000 light-years from the Sun) cannot be seen.

Dark clouds. Embedded in the diffuse interstellar medium are numerous large, dark clouds. Altogether, these clouds concentrate more than 50% of the interstellar gas within their boundaries. The gas in these clouds is almost entirely in molecular form and is measured by emission at the frequency of the CO molecule. These clouds often have densities of about 10^4–10^5 molecules/cm^3. The molecules in these clouds are formed by two processes. First, they form on the surfaces of the dust grains; if two atoms stick side by side to a dust grain, they can combine to form a molecule. The energy released by combination often causes the molecule to jump off the surface of the grain. A second method of forming molecules in dark clouds is by gas-phase interactions, wherein the simple molecules combine in the gas phase to form more complex ones. The energy input to this process appears to be provided indirectly by ionization caused by low-energy cosmic rays passing through the clouds.

In addition to providing surfaces on which molecules can form, dust grains shield the interior of the cloud from ultraviolet starlight which would destroy the molecules, and does so in lower-density regions of space.

Dark clouds are quite cold (about 10 K), and a study of their motions using observations of molecular emission lines shows that they are fairly quiescent, that is, they are not expanding, collapsing, or rotating to any great extent. Furthermore, the internal motions of the clouds (due to heat and to turbulence, that is, swirling and eddying within the clouds) are roughly the amount needed to balance the gravitational forces.

Herbig-Haro objects and T Tauri stars. Many of the dark clouds, however, show circumstantial evidence that star formation is taking place in them, or has done so recently. On the surfaces of some clouds are small bright patches, known as Herbig-Haro objects, after astronomers G. Herbig and G. Haro who first discovered them. In addition, slightly embedded in many clouds are stars of a particular type called the T Tauri stars. These stars, which are named after the prototypical member of the class, appear to be very young. They are of roughly the same mass as the Sun and are surrounded by an envelope of hot gas in turbulent motion which, along with irregular variations in their brightness, marks the stars as having only recently (within the past few hundred thousand years or so) begun their life. The similarity of the optical spectra of the T Tauri stars and of the Herbig-Haro objects has led to the suggestion by S. Strom and K. Strom that the Herbig-Haro objects are reflection nebulosity due to newborn stars embedded in the dark clouds. In other words, the

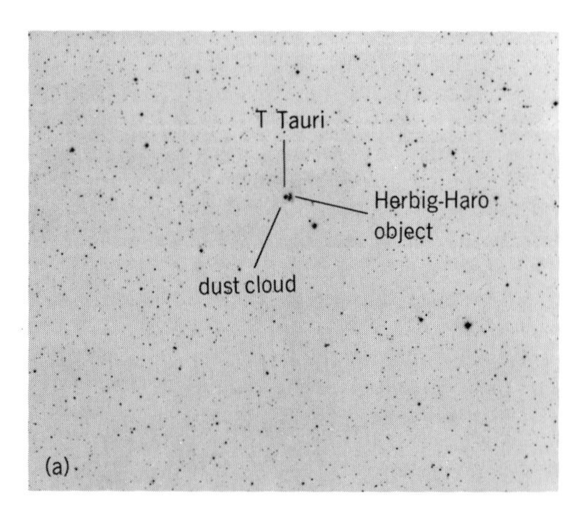

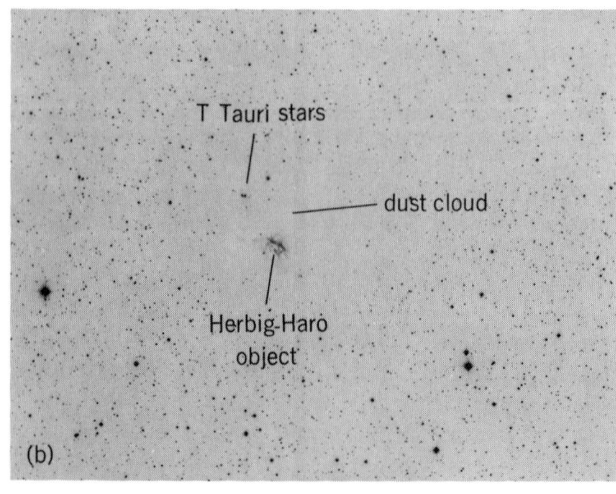

Fig. 4. Two small dust clouds in the constellation Taurus, containing newborn stars and Herbig-Haro objects; parts *a* and *b* are explained within the text. *(Hale Observatories)*

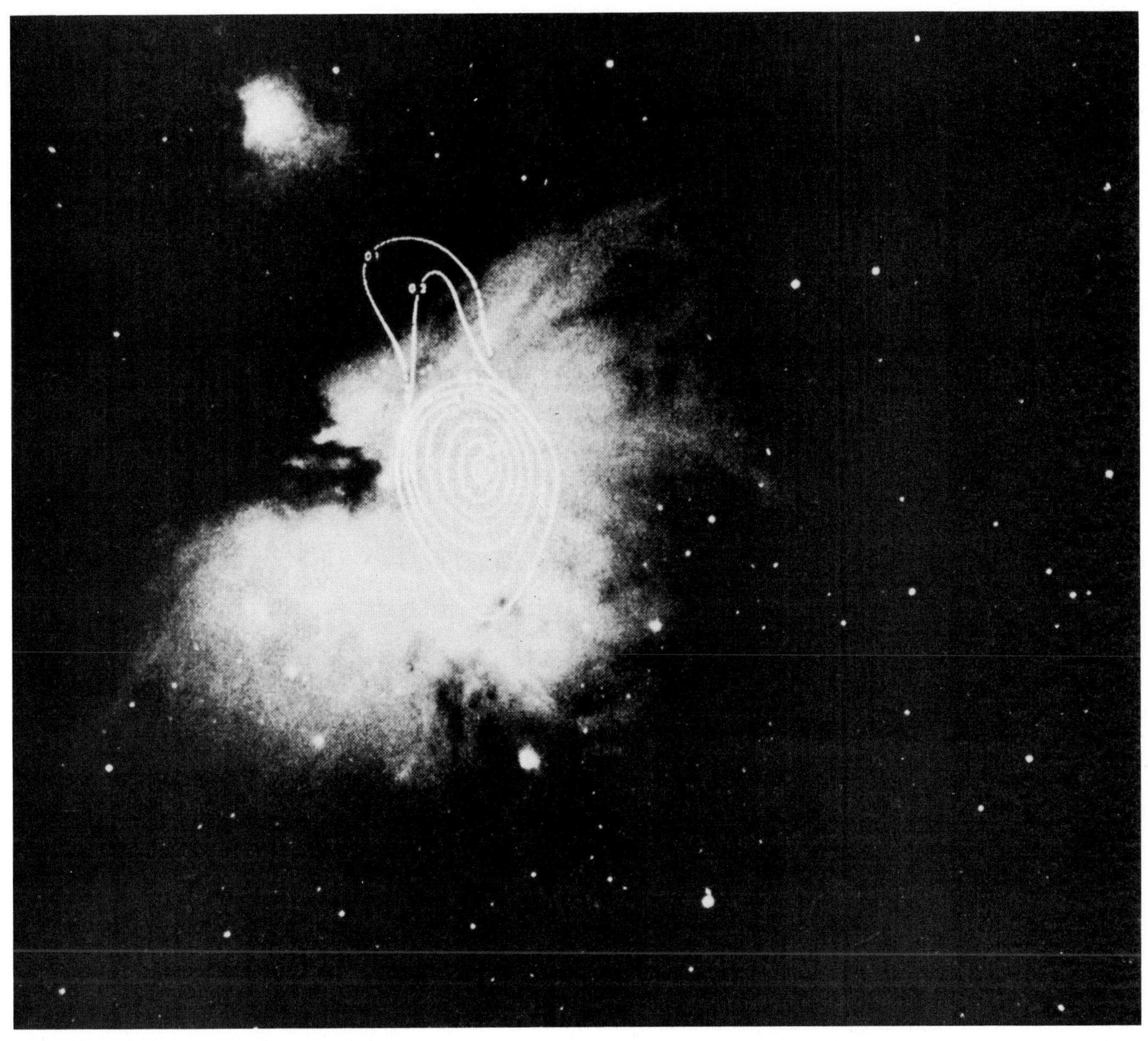

Fig. 5. The great nebula in Orion. The contours show the maxima in emission observed in the radiation of the formaldehyde molecule. (*From N. J. Evans et al., Interstellar H_2CO: II. 2-cm emission from the Orion molecular cloud, Astrophys. J., 199:383–397, 1975*)

light of the embedded star is scattered and reflected from the dust grains until a small amount reaches the surface, even though the density of dust is so great that the star itself cannot be seen directly. The effect is similar to that seen on a very overcast day, when bright patches on the clouds can be seen as the sunlight diffuses through slightly less opaque pathways in the cloud, while in the actual direction to the Sun the cloud is so thick that the Sun's disk cannot be seen. Figure 4 shows two small dark clouds in the constellation Taurus; Fig. 4*a* shows the cloud containing T Tauri itself; to one side of the star is a Herbig-Haro object, produced by the reflection of starlight from the dust grains; Fig. 4*b* shows a nearby small cloud, with two T Tauri stars (also surrounded by nebulosity) and a large Herbig-Haro object.

The proposed association of Herbig-Haro objects with embedded stars has been confirmed by the detection of sources of infrared radiation within the clouds. The light from the hidden stars may not be able to emerge out of the dust clouds, but the infrared radiation can, and the strength of the radiation measures the brightness of the stars, which appear also to be T Tauri stars. But another interesting thing is hidden in these little clouds; radiation from many molecules is seen. In addition to the ubiquitous CO molecules, radiation from HCN, H_2CO, CS, and other molecules is detected. These last molecules must be in regions of fairly high density (about 10,000 molecules/cm^3) for their rotational transitions to be excited; and the measurement of this radiation allows one to map small regions of gas of much higher density than their surroundings deep inside the dark clouds. Further, the molecular emission lines show that some parts of the gas are in relatively rapid motion, as might happen, for example, if some of the gas were fall-

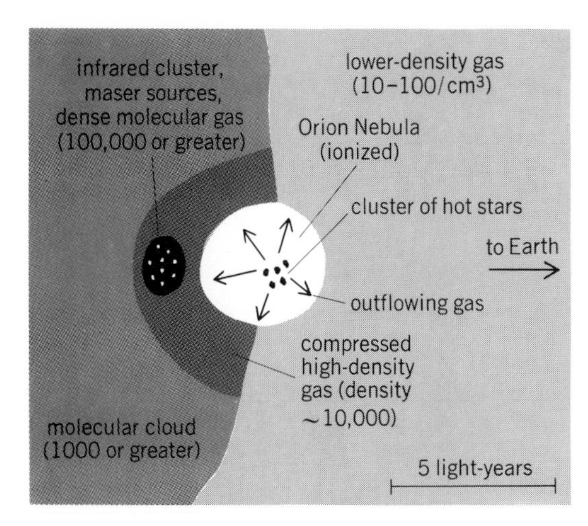

Fig. 6. Cross section through the Orion Nebula, showing the location of the compact cluster of newborn stars behind it.

ing inward to a central point under the force of gravity.

Ionized hydrogen region. Another state of the interstellar matter is the ionized hydrogen region. These are huge (several light-years in extent) glowing clouds of gas, of which surely the best known is the Orion Nebula (Fig. 5), which are heated to temperatures of about 10,000 K by extremely hot stars embedded in the gas. The hydrogen in these regions is completely ionized by ultraviolet radiation from the stars and glows with fluorescent light. The gas is so hot that it is flowing away from the hot stars, and a nebula such as the Orion Nebula will dissipate in much less than a million years. The fact that these hot young stars are embedded in very dense gas is yet another piece of circumstantial evidence that stars formed in large regions of dense gas. But, interesting as the Orion Nebula is, perhaps more fascinating is what lies behind it. The Orion Nebula is only a small part of a much larger, mostly dark dense cloud, with its gas in the form of molecules. This great cloud, which extends across the whole constellation of Orion, was paid very little attention by astronomers until quite recently. The cloud appears on photographs of the sky as large, dark patches, which are certainly less noticeable than spectacular bright objects such as the Orion Nebula and its smaller neighbors. Further, since the hydrogen in the cloud is essentially all in molecular form, the cloud was not noticeable on the maps of the emission from atomic hydrogen. And yet, this dark cloud is dotted by many regions containing newborn stars, embedded infrared sources, small dense regions, and so on. Perhaps the most active region of star formation is located behind the nebula itself. In this region, a bright, hot star, seen only by its infrared radiation, which began to shine perhaps only a thousand years or so ago, was discovered by E. Becklin and G. Neugebauer. There are several other clusters of newborn stars embedded in the large cloud; each is surrounded by hot gas and dust which glows in infrared radiation. The gas surrounding these newborn clusters is particularly dense. The gas between the dark cloud and the ionized hydrogen region is also very dense, and appears to be compressed by the expansion of the Orion Nebula.

Also behind the nebula are small regions emitting intense radiation in the spectral lines of H_2O and OH. The arrangements of the atoms in these molecules and the environment in which they are found are such that their radiation is amplified throughout the region emitting it. The process is the same as that giving rise to intense beams of laser light, and these regions are known as interstellar masers. The presence of a maser indicates that the gas in the region must be very dense (perhaps 10^7 to 10^8 molecules/cm^3) and that there is a great deal of infrared radiation present.

In summary, there is much of the sort of activity associated with star formation going on in the region behind the Orion Nebula. This is shown schematically in Fig. 6. Is this coincidence? Not so, according to a picture most recently described by B. Elmegreen and C. Lada. The expansion of the Orion Nebula causes compression of the cloud, so that parts of it become denser and collapse to form

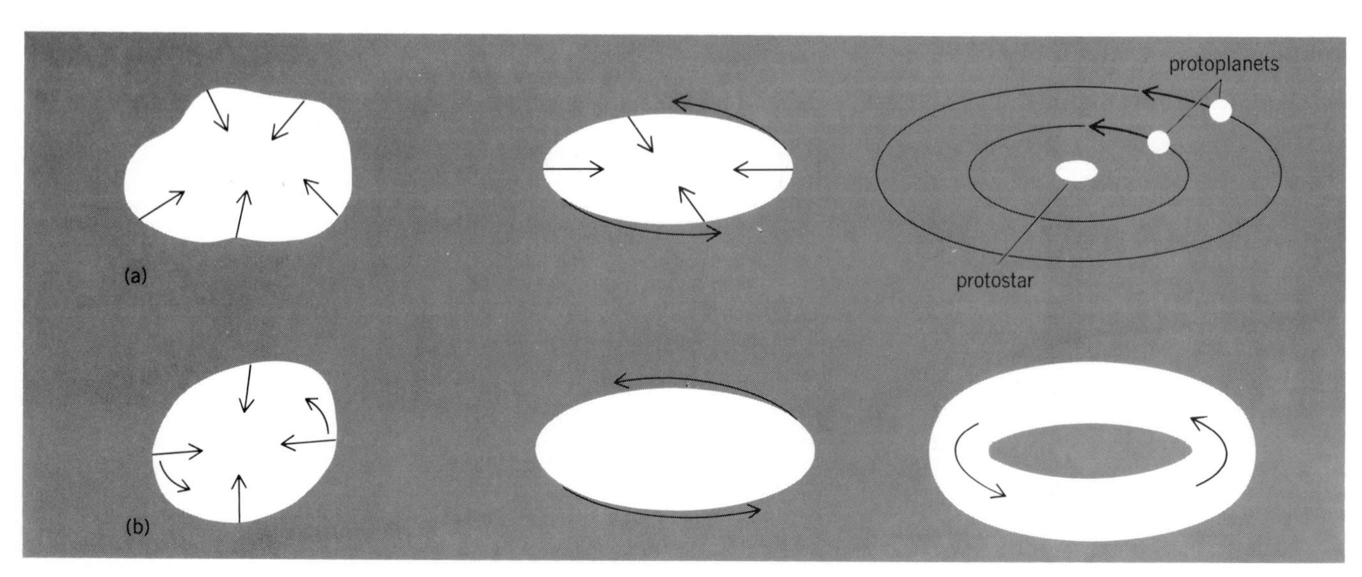

Fig. 7. Collapse of a rotating cloud into *(a)* a disk or *(b)* a "doughnut."

new stars. According to B. Zuckerman, the Orion Nebula itself appears like a "blister" on the surface of the dark cloud, with the expanding gas moving away from the cloud. This picture has recently been shown by F. Israel to apply to large numbers of ionized regions near molecular clouds.

What sort of picture of star formation emerges from all of the above somewhat disparate information? Like most events in astronomy, the formation of stars takes place so slowly that it cannot be watched as it happens, and so it must be pieced together from the "still" pictures taken of some events at various stages in the process.

SCENARIO FOR STAR FORMATION

About 95% of the known mass of the Galaxy is in stars, and most of this is locked up in small stars which age so slowly that those which were born at the formation of the Galaxy are still with us. About 5% is in the form of gas, which comes from two sources. Some of it is leftover gas from the initial formation of the Galaxy; the rest has been dispersed by dying stars. The complex interplay between the stars and the interstellar gas results in the Galaxy as it is, even to such "small details" as the formation of planets and the formation of life. Perhaps nowhere is this interplay more vividly illustrated than in the formation of interstellar dust. Infrared observations of red giant stars show that these aging stars are forming the dust grains in their cool, expanding outer envelopes, where the grains condense out as solid particles in the cooling gas.

Gas compression and molecule clouds. The interstellar gas and dust are, on the average, of very low density. Presumably density concentrations continuously form and disperse in the swirling gas. But as the gas passes through the density concentration which is a spiral arm of the Galaxy, it is compressed. This can be seen clearly in detailed maps made of emission from the hydrogen atom in other galaxies.

As a cloud of gas is compressed, it becomes more opaque to the interstellar radiation field because of its dust. Also, because the density is increased, the atoms collide with each other and with the dust grains much more often. Both of these circumstances allow molecules to form. First the atomic hydrogen forms into molecules, and at the same time or very soon after, CO molecules form. When this happens, the cloud is able to cool itself by radiation. The higher the temperature of the cloud, the faster the molecules move around. If a molecule collides with a dust grain, the grain is heated up and so radiates infrared energy into space. If a molecule collides with a CO molecule, the latter is set spinning more rapidly and radiates energy which is detected as spectral line radiation from the cloud. Thus the cloud is able to radiate heat and reduce the motion of its molecules, and can continue to contract under its internal force of gravity.

The molecule clouds vary greatly in mass, from a few hundred $M_\odot$ on up. Some are very large indeed, containing about 10^7 $M_\odot$ of gas—with this mass, they are the largest single objects in the Galaxy. Some of the clouds appear to be fairly quiescent, having settled down at a size where their internal energy balances the inward force of gravity. Their internal temperature of about 10 K is maintained by energy input to the clouds by cosmic rays passing through them.

Cloud collapse. In the next stage, some clouds start to fragment internally. Some of these pieces do not have enough internal energy to prevent further collapse, in which case the fragment contin-

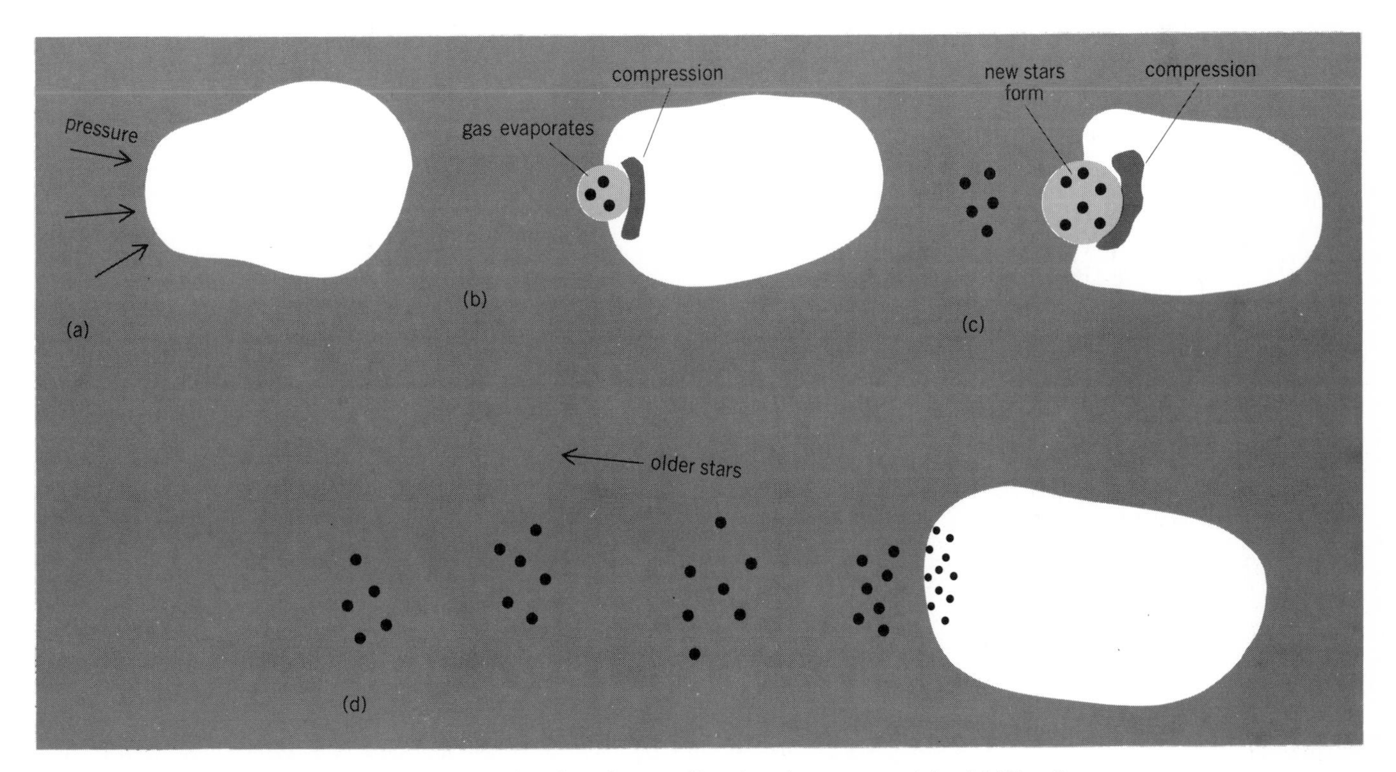

Fig. 8. Sequential star formation. *(a)* Molecular cloud. *(b)* Stars form. *(c)* Stars form in compressed cloud. *(d)* Result.

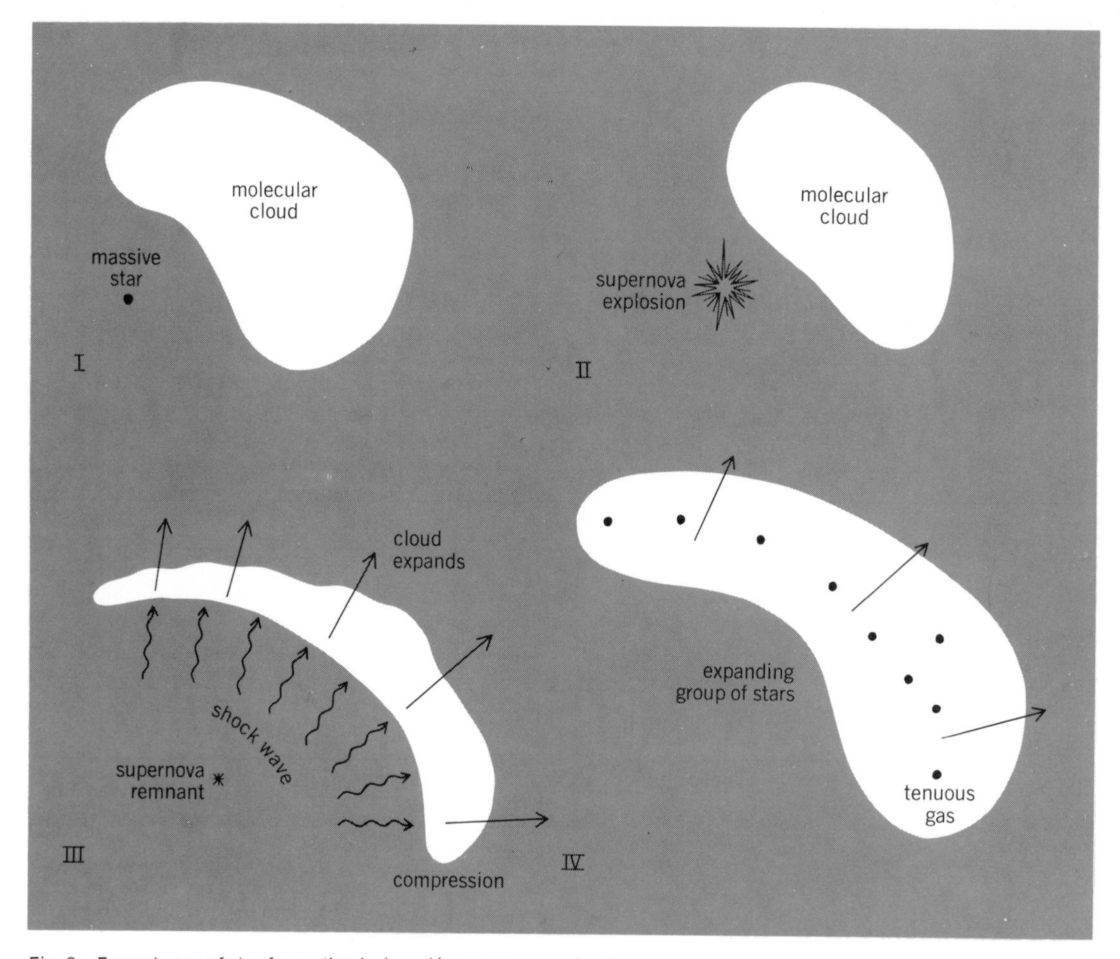

Fig. 9. Four stages of star formation induced by supernova shock.

ues to collapse, becoming denser. At some point, the cloud fragments into still smaller pieces, some of which will become stars. The so-called protostar continues to collapse, becoming ever more centrally condensed. The gas falling to the center of the cloud heats it up, and the center begins to glow in the infrared. Eventually, in about 10^6 to 10^7 years, depending on the mass of the star (the larger protostars collapse more rapidly), the center of the object is so hot and dense that the nuclei of the hydrogen atoms are completely stripped of electrons, and as they are pressed closer together, nuclear reactions begin. The resulting energy outflow halts the collapse of the outer envelope, the object settles into equilibrium, and a star is born (Fig. 7).

Galactic magnetic field. As well as the internal motions of the cloud, there are two other forces which can inhibit the cloud collapse. The first of these is the galactic magnetic field. As a cloud of gas contracts, it pulls the magnetic field with it. The magnetic field resists this pull and thus produces a "pressure" which resists the collapse of a cloud. The neutral particles (atoms and molecules) slip through the field. The ionized particles, however, are caught by the field lines and communicate this to the other particles via collisions. In dense clouds the ionized particles recombine to form a neutral atom (or molecule). If enough recombination takes place, as appears to occur in dark clouds of sufficient age, the gas slips almost unencumbered through the magnetic field. In this case the remaining ions are caught by the field and sieved out of the cloud; the friction between them and the collapsing cloud can be a significant source of heat input to the cloud.

Angular momentum. The other inhibitor of cloud collapse is angular momentum. The large cloud inherits a very small amount of rotation from the rotation of the Galaxy, often much too small to measure. However, as a cloud fragment contracts, it spins faster and faster, to the point where centrifugal forces can inhibit further collapse.

It has long been known that the Sun (and essentially all other stars of similar type, that is, similar mass) is spinning very much more slowly for its mass than the very massive stars. However, while the planets in the solar system contain negligible mass, relative to the Sun, they contain most of the angular momentum of the solar system. If this angular momentum were all put into the Sun, then it would have as much angular momentum per unit mass as the larger stars. This fact raises the exciting possibility that at least the lower-mass stars slow their rotation by forming planetary systems. Promising as this idea is, it is not yet understood theoretically; for example, calculations by P. Bodenheimer and D. Black suggest that a rotating cloud first collapses into a disk; however, it then evolves into a "doughnut," with no central concentration which could form into a star! These possibilities are depicted in Fig. 8.

Stable stars. Once a star has formed, it spends the next period of between a few thousand and a million years settling down into a stable star in the equilibrium described earlier. This settling down often involves much turbulence and activity on the surface of the star, resulting in irregular flickering of the star's brightness. In the process, hot gas flows from the surface of the star, dispersing the last remaining traces of gas and dust. At this point the star may still be embedded deep in the dust cloud and surrounded by darkness.

A small star such as the Sun will drift slowly through and perhaps out of the cloud, or may become liberated when some other agency dissipates the cloud. The star then moves freely in its orbit around the center of the galaxy. A more massive and luminous star, however, heats, ionizes, and expands the gas around it. Stars are born in clusters rather than singly, and the flowing away of much of the material from the cluster means that the cluster is no longer bound together by the force of gravity, so that the stars escape into galactic space.

Life and death cycle. It can be seen from the above that star formation is a highly inefficient process; perhaps only a few percent of the whole amount of gas in a cloud forms into stars before the cloud is ionized, heated, and dispersed back into the interstellar medium. However, the formation of hot massive stars at one edge of a large cloud can set off star formation in successive waves throughout the cloud, as the expansion of an ionized hydrogen region around clusters of newborn stars compresses the gas and begins the process of star formation in an adjacent part of the cloud. This description of the successive formation of stars (Fig. 8) due to Elmegreen and Lada explains an old observation by A. Blaauw, who showed that in a cluster of stars the stars increase or decrease in age across the cluster.

Compression appears to be important in initiating the collapse of part of the cloud. One type of pressure, due to expanding gas around hot stars, is described above. A second type is encountered by a cloud when it passes through the density enhancement of a spiral arm (most of the star formation in the Galaxy takes place in the spiral arms). A third type which appears to be important is due to the blast wave of a supernova explosion. If the blast wave hits an interstellar cloud, the pressure can cause parts of the cloud to begin to collapse (Fig. 9). This process has been described by R. Sancisi, G. Assousa, and W. Herbst, who noted that some associations of newborn stars occur in expanding rings of gas. Thus, the death throes of one star may result in the birth of others.

In the Galaxy, approximately 1 $M_\odot$ of gas per year is shed by dying stars, while the same amount is formed into new stars. Some overall equilibrium is thus achieved; but some of the stars forming are low-mass objects which will last for a very long time and return nothing to the interstellar gas. The time must come, then, when no new stars can form in the Galaxy, which will slowly, over many tens of billions of years, become cooler, redder, and dimmer, and will eventually fade out.

[GILLIAN R. KNAPP]

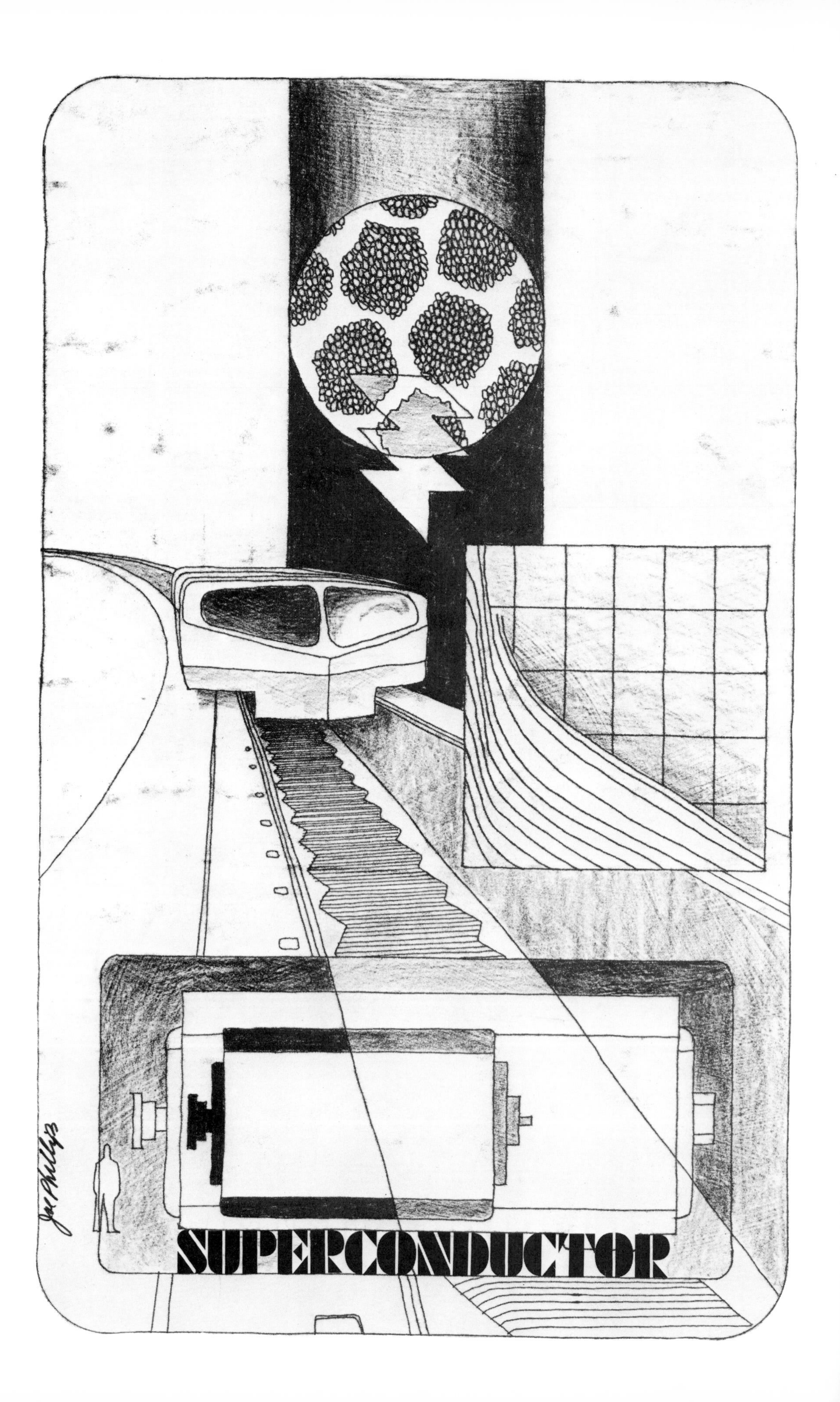
SUPERCONDUCTOR

Applications of Superconductivity

Guy Deutscher received a doctorate in physics from the University of Paris in 1966. He is a professor of physics at Tel Aviv University and has been a visiting professor at Rutgers University and the University of California, Los Angeles. He has worked on the properties of composite superconducting films and their application to switching devices.

In 1911 H. Kamerlingh Onnes discovered the phenomenon of superconductivity when he observed that the electrical resistance of a sample of mercury dropped to an immeasurably small value when it was cooled below a critical temperature of −269°C. (or about 4 K). Subsequent, more accurate measurements confirmed the complete absence of ohmic losses in the superconducting state. This simple observation clearly posed a challenge to physicists, and it took, in fact, almost 50 years until J. Bardeen, L. N. Cooper, and J. R. Schrieffer were able to develop a microscopic theory of superconductivity. From the outset, however, it was obvious to Kamerlingh Onnes that the absence of ohmic losses should have important practical applications. In fact, in his original article he mentioned the possibility of obtaining very high magnetic fields with superconducting coils, since the absence of ohmic losses permitted the use of very high current densities. Although his original attempts to achieve this failed, because mercury was not an appropriate superconducting material for that purpose, Kamer-

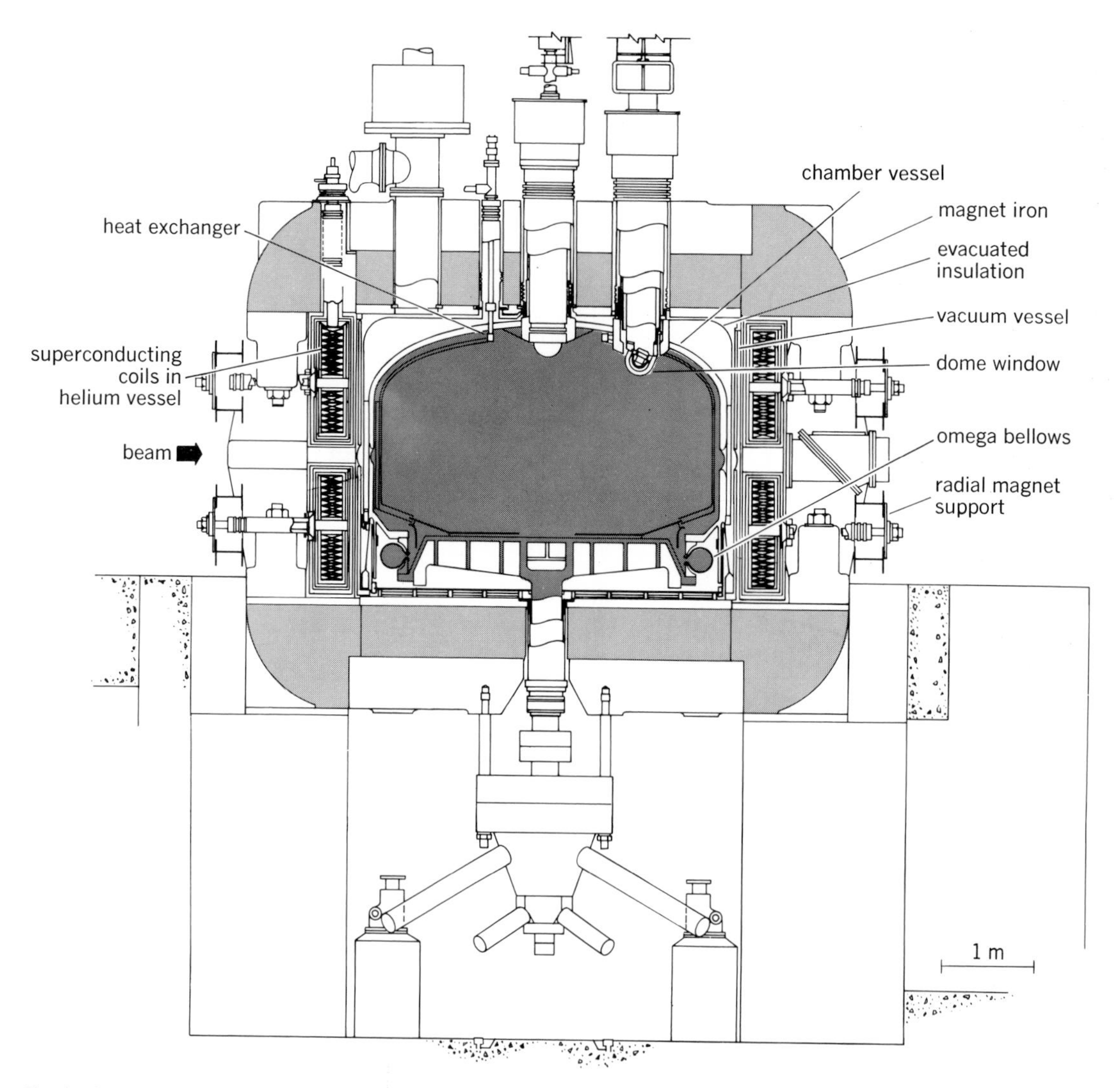

Fig. 1. Hydrogen bubble chamber at CERN. *(Proceedings of the 1967 CERN Conference on Bubble Chambers)*

lingh Onnes's vision was later to be fulfilled, probably beyond his own expectations. The superconducting coil of the bubble chamber at the European Commission for Nuclear Research (CERN) shown in Fig. 1 gives an idea of present large-scale applications of superconductivity.

This article reviews the scientific background and large-scale applications of superconductivity and then discusses detector and computer applications of Josephson devices.

SCIENTIFIC BACKGROUND

The reason why mercury was not appropriate for the creation of high magnetic fields was that a low magnetic field of a few hundred gauss (1 gauss = 0.0001 tesla) was sufficient to quench superconductivity in this material. Although many more superconducting elements were subsequently discovered—in fact, most metals are superconducting at low enough temperatures with the exception of the alkalies, the monovalent metals (copper, silver, and gold), and the magnetic metals—they all were found to have the same limitation of a low critical field. Moreover, they all had relatively low critical temperatures. The strongest superconducting element is niobium, which has a critical temperature of 9.2 K and a critical field of 0.2 T.

Type II superconductivity. The application of superconductivity to the construction of high-field coils for high-energy physics, motors, generators, levitated trains, and other applications was made possible by the discovery of type II superconductivity by L. V. Shubnikov (before World War II) in the Soviet Union. He observed that superconducting alloys composed of a superconducting element with impurities in solution (such as bismuth in lead) have a much higher critical field than that of the pure superconductor. Shubnikov's work, however, remained unknown until the late 1950s when type II superconductivity was rediscovered and investigated in detail. In the mid-1960s practical superconducting wires became commercially available with critical fields of 5 to 10 T, and only then did projects for large-scale applications of superconductivity receive serious attention.

It can be said that the problem of manufacturing superconducting wire with high enough critical fields and current-carrying capability has been essentially solved. Progress can certainly still be made, but the present state of the art, as described below, is already quite satisfactory.

Higher critical temperatures. Unfortunately, progress in finding alloys with higher critical temperatures has been much slower, and superconductivity remains more or less a low-temperature phenomenon; the highest critical temperature known, that of Nb_3Ge, is equal to 23 K, and present commercial wires have to be operated below 10 K, thus requiring the use of expensive refrigerators. Whether this is a fundamental limitation of superconductivity, or whether a superconducting alloy that can be operated at liquid hydrogen temperature (21 K) or liquid nitrogen temperature (77 K) will be discovered eventually, is a question that remains open to a certain extent. The conservative opinion is that superconductors will remain essentially in their present form, and that consequently researchers should focus on building cheaper, more reliable helium refrigerators, rather than hope for the discovery of a better superconductor that will allow the use of simpler refrigeration techniques. On the other hand, one cannot rule out the possibility of the discovery of an alloy that will superconduct at, say, 30 K and thus could be used at liquid hydrogen temperatures. Other significant breakthroughs are possible, if unlikely: recently a group in the Soviet Union claimed to have observed signs of superconductivity in CuCl at liquid nitrogen temperatures; organic one-dimensional conductors with chainlike structures have been under intense examination in recent years, and although no high-temperature superconductivity has been found yet in these compounds, there may still be some hope.

Josephson junctions. Besides the absence of electrical resistance, another aspect of superconductivity is useful in some applications and is directly linked to the quantum nature of the superconducting state. This phenomenon, which was discovered by B. D. Josephson in the early 1960s, can be briefly described as follows: When two superconducting electrodes are separated by a very thin oxide barrier, or by a geometrical constriction of small dimensions, a supercurrent of small intensity can flow between the two electrodes without any electrical resistance. In the superconducting state, the electron gas condenses in a lower energy state which consists of electron pairs of opposite spin and momentum. The supercurrent that flows through the thin oxide barrier is actually a tunneling current of such electron pairs. Unlike the supercurrent that flows in a macroscopic wire, this tunneling or Josephson current is a quasiperiodic function of an applied magnetic field and has the general aspect of an interference pattern. The superconducting state can be characterized by a wave function which has both a phase and an amplitude. The phase is sensitive to the presence of a magnetic field, or more specifically to the vector potential from which the field derives. For certain well-defined values of the field, the vector potential at opposite sides of the junction is such that the phases of the superconducting wave functions in the two electrodes are exactly opposite to each other, so that no net supercurrent can flow between them.

A similar interference effect occurs when two Josephson junctions are placed in parallel in a superconducting loop. The net supercurrent that can be carried by the loop is then a periodic function of the magnetic flux contained within the loop. The periodicity is equal to the flux quantum, whose value is 2×10^{-15} weber, which makes such loops very sensitive detectors of weak magnetic fields. Superconducting quantum interference devices (SQUIDs) use this high sensitivity for the detection of the weak magnetic signals produced by the human body or geological anomalies, for instance. In addition to being very sensitive, Josephson devices are also extremely fast, with switching times of the order of picoseconds. This property may lead to a completely new generation of computers.

MATERIALS FOR LARGE-SCALE APPLICATIONS

Normal conductors such as copper or aluminum have a current-carrying capability limited to 2×10^7 A/m^2 even with forced water cooling. Although it is possible to manufacture normal coils that produce very high fields in small volumes (up to 25 T in cores a few centimeters in diameter), their large size and huge power consumption (several megawatts) rule them out for practical applications.

On the contrary, superconducting coils can be used to produce high fields in large volumes, the only energy cost being that required to refrigerate the coil. The field intensity that can be reached is determined by the current-carrying capability of the superconducting wire in the presence of the field generated by the coil. This current-carrying capability is expressed in terms of a critical current density, which is the maximum current density that can be reached before a finite resistance appears in the wire. Critical current density is a decreasing function of the temperature and the applied magnetic field. Typical values of current densities used in superconducting coils are 10^9 A/m^2 or more—about two orders of magnitude

Fig. 2. Cross section of NbTi wire consisting of NbTi filaments (the smaller components) embedded in copper matrix. Current capacity is 20 kA at 5 T and 4.2 K. *(From G. Bogner, Large-scale applications of superconductivity, in B. B. Schwartz and S. Foner, eds., Superconductor Applications: SQUIDS and Machines, chap. 20, pp. 547–719, 1977)*

higher than values used in normal conductors, resulting in a large reduction in size and weight as well as power consumption.

Niobium-titanium alloy. The most commonly used superconducting alloy is NbTi. This consists of Ti impurities in solution in Nb, the effect of the Ti impurities being to transform the pure Nb into a strong type II superconductor, as discovered by Shubnikov. Commercial NbTi wires are composed of thin NbTi filaments, typically 25 μm in diameter, embedded in a copper matrix. The purpose of the copper matrix is to stabilize the superconducting wire: it provides a parallel low-resistance conducting path for the case where a normal spot would appear accidentally along one of the NbTi filaments. Current is then diverted momentarily into the copper, giving time for the hot spot to cool down and return to the superconducting state, thus avoiding an avalanche effect that could quench the whole superconducting coil. The cross section of a typical NbTi-Cu wire is shown in Fig. 2, and its superconducting characteristics are given in Fig. 3.

Niobium-tin alloy. Another superconducting alloy that has found commercial application is Nb_3Sn. It belongs to a different family of alloys, called the A15 compounds, after their crystallographic classification (Nb_3Ge belongs to that same family). Discovered by B. T. Matthias, they have the highest known critical temperatures and critical fields. Manufacture of A15 wire is more difficult than that of NbTi because A15 compounds are very brittle. However, filamentary Nb_3Sn in a copper-tin matrix is used for very-high-field applications. Nb_3Sn can also be used at much higher temperatures than NbTi due to its high critical temperature (18 K). The current-carrying capability of Nb_3Sn at 10 K is about the same as that of NbTi at 4K. This advantage may be very important for applications where the cost of refrigeration is crucial, such as superconducting cables, and for applications where a large margin of safety is required.

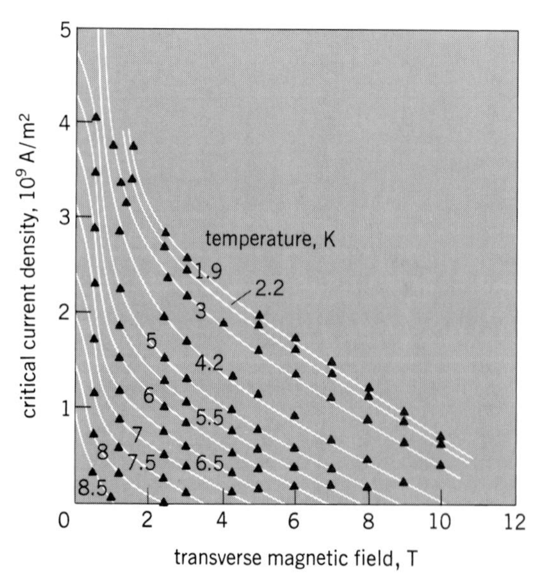

Fig. 3. Critical current density of NbTi as function of transverse magnetic field in the temperature range of 1.9–8.5 K. *(From G. Bogner, Large-scale applications of superconductivity, in B. B. Schwartz and S. Foner, eds., Superconductor Applications: SQUIDS and Machines, chap. 20, pp. 547–719, 1977)*

Hybrid magnets. NbTi and Nb_3Sn wires are often used in combination in hybrid magnets consisting of a large-bore NbTi external coil, which provides a background field of about 5 T and an inner Nb_3Sn coil, which provides additional field strength. Such combinations are presently used to obtain fields in excess of 10 T, and can reach fields of up to 18 T in small bores.

LARGE-SCALE MAGNETS

Magnetic fields are used extensively in high-energy physics to accelerate, bend, focus, and store particle beams, and to detect and identify elementary particles. The introduction of superconducting magnets in high-energy physics is due to the fact that they allow the utilization of higher fields in larger volumes at a lower cost in capital investment and energy expenses. This economic aspect may be the main reason why all major high-energy facilities are already using, or are planning to use in the near future, large-scale superconducting magnets. As an example of the orders of magnitude involved, the bubble-chamber magnet at CERN, the largest in operation at present, requires a power of 70 MW with conventional coils, while the superconducting version consumes less than 1 MW. The CERN superconducting magnet gives a field of 35,000 G (3.5 T) inside a coil about 4.7 m in diameter and 4.4 m in height which is cooled in a liquid helium bath. The energy stored in the coil approaches 1 gigajoule. Thus, high-energy physics has provided the first large-scale applications of superconductivity.

Fermilab accelerator. At present the most ambitious superconductivity program in the United States is the construction of a 1000-GeV proton accelerator at the Fermi National Laboratory (Fermilab) in Batavia, IL. The ring with superconducting magnets is to be built in the tunnel of the existing 400-GeV conventional accelerator; the idea is that a higher energy can be achieved in the same geometry due to the higher field provided by the superconducting magnets. This accelerator illustrates the economies in size, and therefore in building cost, that can be realized with the superconducting solution. More than 40,000 kg of superconducting wire is involved in this program. Superconducting transmission lines and a superconducting energy storage system will complement the accelerator in order to eliminate the large power pulses drawn from the utility's system.

Colliding-beam facilities. There is a growing interest in colliding-beam facilities because they can produce high center-of-mass energies with modest laboratory energies. It is likely that all future facilities will use superconducting magnets. The facility planned at Brookhaven National Laboratory is to have two superimposed rings each 2700 m long. One of the prototype dipole magnets to be used in this proton storage accelerator is shown in Fig. 4.

An interesting aspect of colliding-beam experiments is a new type of detector magnet. Fields of the order of 1.5 T are required in volumes of several cubic meters for the detection and analysis of

Fig. 4. Beam-guiding dc dipole superconducting magnet to be used in proton storage facility, with bore of 250 mm, length of 2.5 m, and operating field of 4 T. (*Brookhaven National Laboratory*)

the particles created in the collision. The coil winding must be very thin to permit the study of the particles outside the magnet. This implies very high current densities (10^9 A/m^2) which can be achieved only with superconducting wire; prototype coils of this type have already been built and tested.

World trend. The programs mentioned above are only a few examples of a worldwide trend: All major high-energy physics centers are involved in the planning, construction, and utilization of superconducting magnets of many different types, including CERN in Switzerland, Saclay in France, Karlsruhe in Germany (in conjunction with Deutsches Elektronen-Synchrotron), Rutherford Laboratory in England, and Fermilab, Brookhaven, and Stanford in the United States. Superconductivity has permitted new high-energy physics experiments by making them either more affordable (through savings in energy and construction costs) or more technically feasible (such as in the detection of short-lived particles). On the other hand, these experiments are an invaluable proving ground for large-scale superconducting magnets and cryogenic installations. In the future, it may be beneficial to have more installations built in order to spread the acquired knowledge to other areas of large-scale applied superconductivity.

POWER GENERATION AND TRANSPORT

During the last few years, electric power generation has emerged as one of the most promising areas of application of superconductivity.

Superconducting ac generators. There is a very good chance that superconducting alternating-current (ac) generators will be in use before the year 2000, and that they will have become established as the standard generators of the future. This, however, is by no means a certainty for several reasons: (1) there has been continuous progress in conventional machines. (2) The reliability of superconducting machines will be established only after large-scale machines have been built and tested. (3) The energy crisis has produced consid-

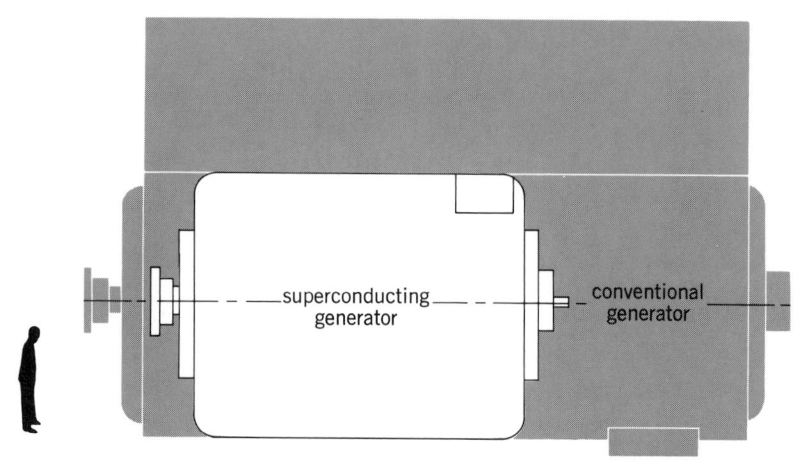

Fig. 5. Schematic representations of superconducting and conventional 1200 MVA generators, showing relative sizes. *(From M. Rabinowitz, Cryogenic power generation, Cryogenics, 17(6):319–330, 1977)*

erable uncertainty about the future evolution of energy consumption, and this uncertainty necessarily affects all large-scale funding programs on new technologies.

Conventional ac generators are composed of a rotating multipole electromagnet (the rotor) and a fixed armature (the stator). Power is produced in the stator windings as the magnetic flux lines produced by the rotor cut them periodically. As the machine rotates at a fixed speed, the power output is determined by the amplitude of the magnetic field produced by the rotor.

Superconducting ac generator models already built or currently being developed have a superconducting rotor and a normal armature. The reason for this combination is the poor behavior of superconductors in high-intensity ac fields, the losses becoming quickly prohibitive. Superconducting generators did not originally attract much attention (as compared, for instance, with direct-current, or dc, motors) because a cryostat rotating at high speed seemed to be an object of discouraging complexity. However, in 1971 a group of engineers at the Massachusetts Institute of Technology (MIT) demonstrated with their 45-kVA machine that a superconducting rotor is actually a practical solution for power generation.

Advantages. The advantages of the superconducting machine over conventional generators derive from the higher magnetic field intensity provided by the superconducting coils of the rotor and from the absence of electrical losses in the rotor. The higher field intensity results in a smaller volume and weight (due in part to total or partial suppression of the iron), while the absence of losses in the rotor results in a higher conversion efficiency.

Most of the original interest in superconducting generators stemmed from their smaller size and weight (by a factor of about 2) rather than from their increased efficiency (by approximately 0.5%). A schematic representation of conventional and superconducting generators is shown in Fig. 5, and the table gives a somewhat more detailed description of the comparison between the two. Recently, due to higher fuel costs, more emphasis has been placed on the gain in efficiency. Although 0.5% seems to be a modest improvement, it has been calculated that the resulting savings would pay for the generator in 25 years—the normal lifetime of this type of machine, assuming that the reliability of the cryogenic generator is the same as that of the conventional machine.

In the late 1960s and early 1970s, the interest in a reduced-size and -weight generator came from the realization that conventional generators were approaching their maximum size. With ever-expanding energy demand and production, there was a need for larger and larger generators; hence the interest in the superconducting generators, which could be used when generator ratings of 2000 to 3000 MVA were required.

Present-day reasoning is quite different. Energy

Fig. 6. Model of a three-phase superconducting power cable. *(Linde Division, Union Carbide)*

Comparison of superconducting and conventional 1200-MVA generators*

Characteristic	Superconducting	Conventional
Phase to phase voltage, kV	26–500	26
Line current, kA	26.6–1.4	26.6
Active length, m	2.5–3.5	6–7
Total length, m	10–12	17–20
Stator outer diameter, m	2.6	2.7
Rotor diameter, m	1	1
Rotor length, m	4	8–10
Synchronous reactance, pu	0.2–0.5	1.7–1.9
Transient reactance, pu	0.15–0.3	0.3–0.4
Subtransient reactance, pu	0.1–0.2	0.3
Field exciter power, kW, continuous	6	5000
Generator weight, tons†	160–300	600–700
Total losses, MW	5–7	10–15

*From M. Rabinowitz, Cryogenic power generation, *Cryogenics*, 17(6):319–330, 1977.
†1 short ton = 0.9 metric ton.

production is not expanding as fast as in the past, and the ultimate justification for the construction of ever larger power plants (such as breeder reactors) is being seriously questioned. In addition, there is an increasing emphasis on energy conservation. Any technology that can save energy, such as superconducting generators, is preferred. Savings are already sizable at current generator ratings of about 1000 MVA; therefore, superconducting generators are already of practical interest today, even if power plant size does not increase much in the near future, for much the same reason that superconducting magnets are attractive for high-energy physics centers: the superconducting solution is simply more economical.

Current development. The United States has assumed leadership in the development of superconducting generators, dating from the construction of the early MIT machine to present-day large-scale programs sponsored by the Electric Power Research Institute (EPRI) in Palo Alto, CA, and developed mostly at General Electric and Westinghouse. Westinghouse has already built and tested a 5-MVA machine and is currently testing a 20-MVA machine that has just been built. Current programs are aimed at the detailed engineering design and eventual construction of a 300-MVA machine. This machine would have a 5-T rotor field, as compared with the maximum 2-T field of the conventional machine, whose limit is due to iron saturation. The 300-MVA target was selected by EPRI so that design and construction of a commercial-size 1000-MVA machine would be the next stage in development.

Superconducting power cables. Practically all developed countries have programs on superconducting generators along similar lines, and interest has been accentuated due to the energy crisis. On the contrary, programs on superconducting power cables have been adversely affected by the energy crisis in every country except the United States. Cutbacks in cable programs in Japan and western Europe have been made after several years of considerable design and experimental work. It is not that this research was unsuccessful or that it has revealed unsurmountable problems; indeed, a number of dc and ac projects have reached an advanced stage. For example, short-length prototype cables have been built and tested with success. A model of a three-phase cable is shown in Fig. 6.

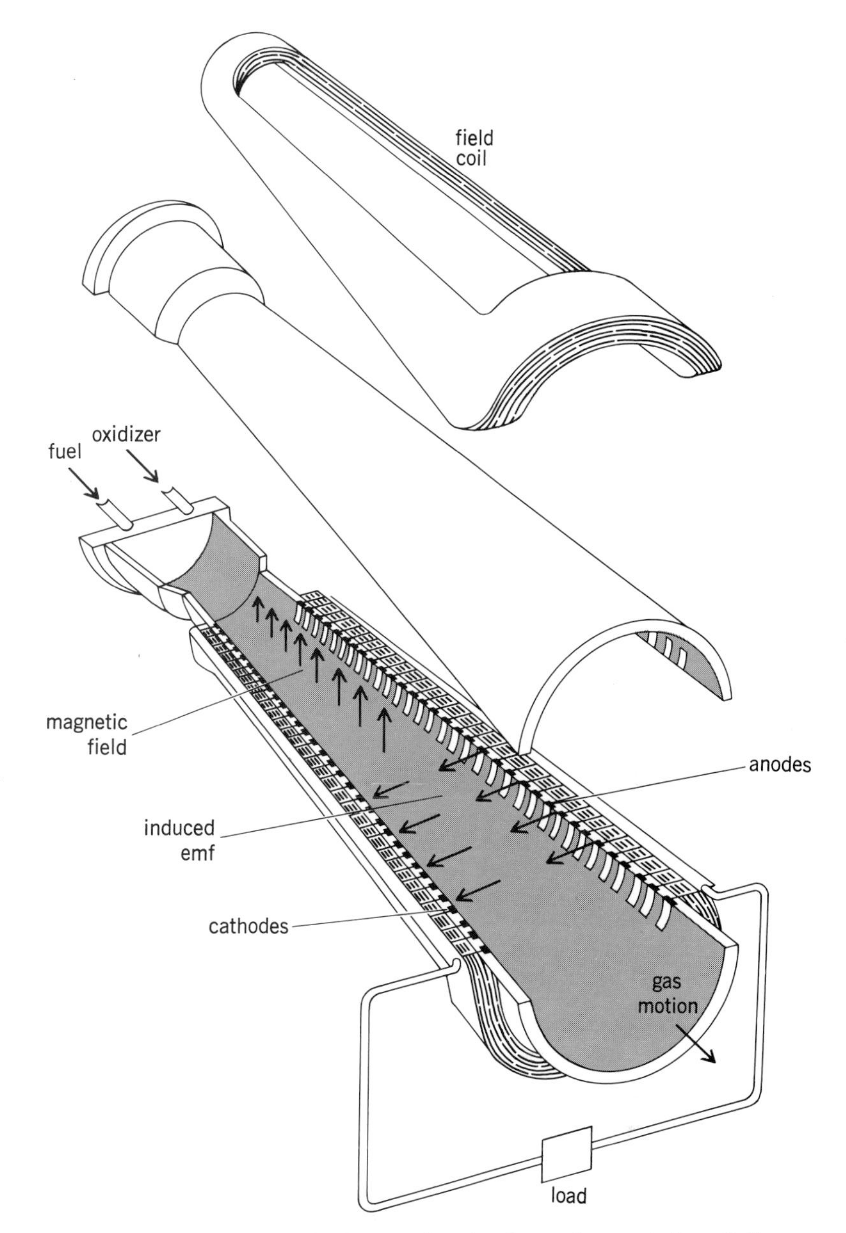

Fig. 7. Exploded view of an MHD expansion chamber and magnet. *(From G. Bogner, Large-scale applications of superconductivity, in B. B. Schwartz and S. Foner, eds., Superconductor Applications: SQUIDS and Machines, chap. 20, pp. 547–719, 1977)*

Loss of interest in power cables in Europe and Japan is due to economic studies which have shown that up to a rating of the order of 5 GVA the conventional forced cooled solution is less expensive than the superconducting one. Further, no need for cables above 2–3 GVA is foreseen until the year 2000—again due in part to the energy crisis and the slower rate of growth of electrical energy demand. On the other hand, in the United States, where there is more electrical energy consumed per capita than in Europe and Japan, there will be a need for such high-power cables before the end of the 20th century.

In any case, there is no question that superconducting cables become economical at a higher rating than ac generators: the refrigeration cost of the cable is simply higher due to its larger surface-to-volume ratio. However, the crossover rating of 5 GVA has been calculated (assuming an operational temperature of about 5 K) based on the use of Nb or NbTi as the superconductor (with a critical temperature of less than 10 K). The use of Nb_3Ge, which has a critical temperature of 23 K and can be operated between 15 and 20 K with subcooled liquid hydrogen, would radically change present prospects. Of all the large-scale applications of superconductivity, power cables are probably the one whose economic value is most sensitive to the critical temperature of the superconductor used. In any event, superconducting power cables will be used only in highly populated areas where space is at a premium; wherever free space is available at a low cost, overhead lines will remain cheaper for the foreseeable future.

VERY-LARGE-SCALE SUPERCONDUCTING MAGNETS

Very large superconducting magnets with stored energies of the order of 10 gigajoules or more have several areas of application. A stored energy of the order of 10 GJ is required for magnetohydrodynamics (MHD) electricity generation and for fusion. Superconducting magnets for energy storage in utility systems become economically attractive at stored energies of the order of 10^4 GJ.

As mentioned above, the largest magnet built to date, the bubble-chamber magnet at CERN, has a stored energy of less than 1 GJ. There is no practical experience in the many problems (mechanical, stability, and so forth) posed by larger magnets, which will also have to function under more severe operating conditions. However, the incentive to continue in this direction is extremely strong, because MHD, fusion, and large-scale energy storage will all be needed in the future, and only superconductivity is practical for these applications.

MHD power plants. In an MHD power plant an electrically conducting expanding fluid moves across magnetic flux lines and electricity is produced in much the same way as in a conventional generator, where conducting wires move in a magnetic field. This principle has been known for over a century, but progress has been very slow for two main reasons: (1) the corrosion problems of the electrodes and insulating material in the hot expansion chamber; and (2) the high intensity of the magnetic field (about 5 T) required in the large volume (approaching 100 m^3) of the expansion chamber necessary to make the MHD plant economical. In recent years renewed interest in MHD plants has derived from the realization that they can burn coal (even with high sulfur content) with very high efficiency (about 50%, as compared with a maximum of 40% in a conventional coal-fired plant). It now appears that MHD plants could be one of the most efficient and ecologically acceptable ways to make use of the large coal reserves in the United States and elsewhere (especially in the Soviet Union). Many of the corrosion problems have been solved, and it is thought that the large-scale superconducting magnets necessary for economical operation can be built (Fig. 7).

Fig. 8. A 1000-hp (746-kW) superconducting homopolar motor for naval propulsion. *(From B. W. Birmingham and C. N. Smith, Survey of large-scale applications of superconducting in the U. S., Cryogenics, 16(2):59–71, 1976)*

A large-scale cooperative program between the United States and the Soviet Union in the field of MHD generators underlines its worldwide significance. The first 25-MW MHD plant was built in Moscow. A superconducting magnet built by the Argonne National Laboratory was recently shipped to Moscow for testing in this plant. The next step in MHD development is the construction of a 500-MW commercial-size plant, which may take place at the end of the 1980s.

Controlled fusion. A controlled thermonuclear reaction can in principle be achieved in two ways: magnetic confinement of the plasma and laser ignition. In magnetic confinement, large-scale coils are needed to produce the field; in laser ignition, large energies have to be stored to fire the lasers. In both cases, superconductivity provides the most economical (and probably the only practical) way to achieve the very strong fields needed for these applications.

The United States fusion-confinement program calls for the construction of a 100-MW power reactor in 1985, a 500-MW reactor in 1990, and a 5000-MW demonstration commercial power plant by the end of the 20th century. The Japanese program also provides for a 2000-MW plant by the year 2000. A similar tokamak program has been agreed upon in Europe [called the JET (Joint European Toroidal Magnet System) program]; the reactor will be built at Culham, in Great Britain. All these programs will use superconducting coils ranging in diameter from about 1 to 20 m, as larger plants are built. The Soviet tokamak program follows similar lines as far as is known. If all goes according to plan, controlled fusion should be carried out by the mid-1980s and commercial fusion plants should be constructed by the beginning of the 21st century. Economic evaluations show that these plants will have energy costs comparable to that of present nuclear fission plants.

Energy storage. No matter which energy solutions are implemented (breeder, fusion, solar, and so forth), there will be a need for large-scale energy storage systems as a means of power leveling. Since sites for pumped hydroelectric plants are limited, utility companies are looking for other solutions. A 100-MJ superconducting storage system will be built at Los Alamos, NM, to study the problems associated with this type of storage, which is expected to become economical only at much larger sizes.

Motors for marine propulsion. Among the most promising applications for superconducting magnets for the short term are motors for marine propulsion. The much smaller weight of the superconducting motor makes electrical propulsion practical. The complete propulsion unit is then composed of a prime mover (and a superconducting generator located where convenient) and a small superconducting motor next to the propeller. A 1000-hp (746-kW) motor has already been tested by the U.S. Navy (Fig. 8), and larger units are under construction.

Magnetic separation. Another attractive application of superconducting magnets is for magnetic separation, already widely used for removal of iron from fluids in industrial processes and ore beneficiation. Higher field gradients achievable with superconducting magnets would permit higher enrichment ratios and separation of weakly magnetic particles.

Levitated trains. One of the most spectacular applications of superconducting magnets is for levitated trains for high-speed transportation. Their principle of operation is quite simple: When a magnet moves over an electrically conducting medium (such as a metallic sheet), the excited eddy currents are equivalent to an image magnet of opposite polarity located beneath the conducting sheet. Repulsion between the magnet and its image results in levitation of the magnet, provided the field intensity and the speed of motion of the magnet are high enough.

At present, very advanced projects are being developed in West Germany by a consortium formed by Allgemein Elektrizitäts Gesellschaft–Telefunken, Brown Boverie, and Siemens Corporation, and in Japan by the Japanese National Railway in collaboration with the Fuji, Hitachi, Mitsubishi, and Toshiba electrical companies. Due to their high-density population, both countries have an interest in fast ground transportation with a minimum of environmental impact.

The German and Japanese projects are quite similar. The speed of the levitated vehicles should reach 500 km/h. The superconducting magnets, working in the persistent mode, will be refrigerated by an on-board cooling system. The track is the active element of the propulsion system, which does not require any physical (mechanical) contact with the vehicle. A traveling electromagnetic wave is excited in the track, and the vehicle essentially rides along with the wave in a sort of electromagnetic "surf" mode. Only a portion of the track is excited at a time. Clearance between the vehicle and the track is comfortable—on the order of 100 mm or more at speeds exceeding 120 km/h.

Test facilities have been constructed in both Germany and Japan, and experimental vehicles with superconducting magnets on board have been tested successfully. The German vehicle and track are shown in Fig. 9. The track is a ring 280 m in diameter. By the early 1980s a commercial vehicle (120 tons, or 108 metric tons; 200 seats) should be ready for testing as well as a 60-km track. Similar progress toward commercialization is reported in Japan.

APPLICATIONS OF JOSEPHSON JUNCTIONS

The applications of Josephson devices are also becoming increasingly important.

Magnetic field detection. The very high sensitivity of SQUIDs to magnetic fields is finding a

Fig. 9. Superconducting test vehicle and track at Erlangen, West Germany. *(Siemens Research Laboratory)*

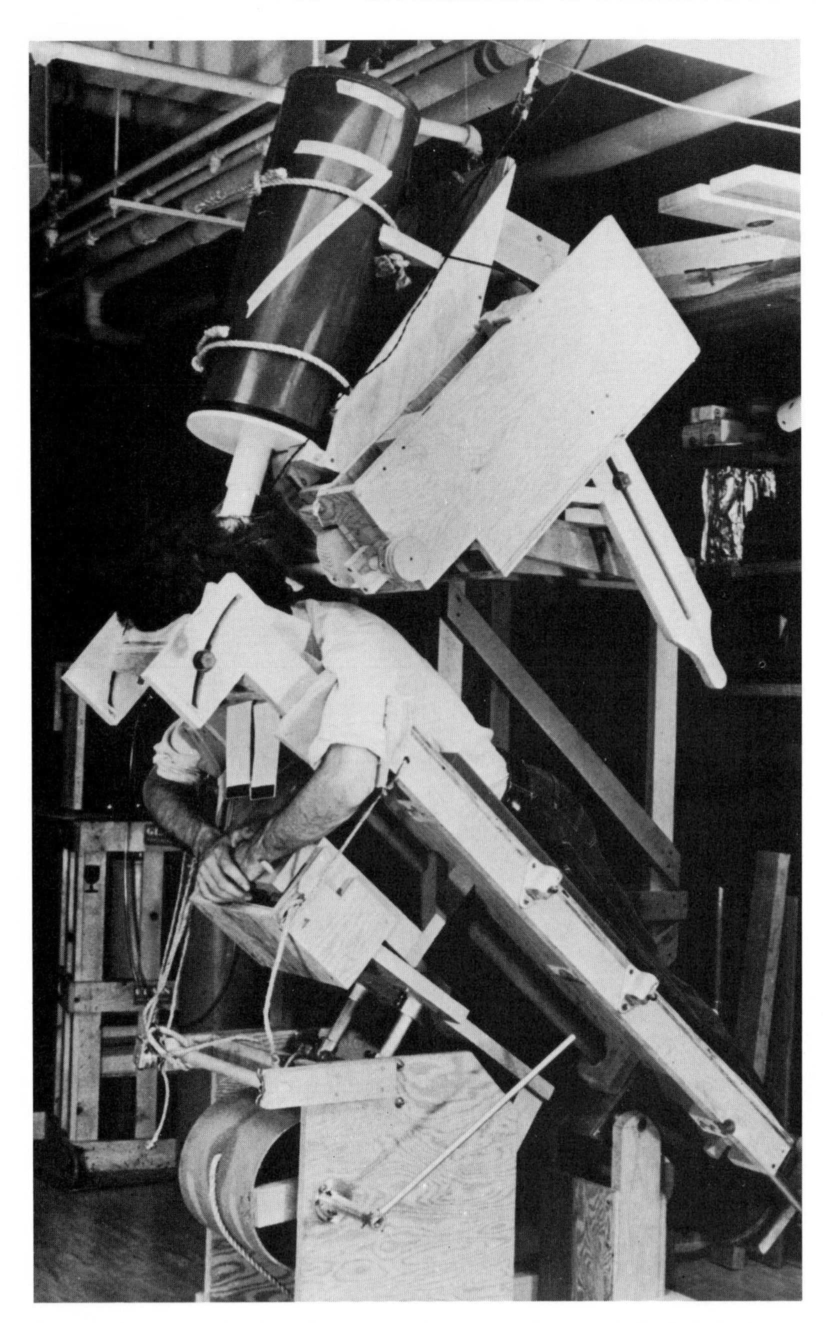

Fig. 10. Apparatus for detection of somatically evoked magnetic fields in the brain. *(From S. J. Williamson, L. Kaufman, and D. Brenner, Biomagnetism, in B. B. Schwartz and S. Foner, eds., Superconductor Applications: SQUIDS and Machines, chap. 8, pp. 355–402, 1977)*

number of applications in geophysics and medicine. Magnetic anomalies may be due to ore deposits, but more generally, any formation (including oil and water) that has an electrical conductivity different from that of the surrounding medium can be detected by a skin-depth measurement performed as a function of frequency. Local measurement of the magnetic and electrical fields gives a determination of the surface impedance of the Earth at the selected location, and a proper Fourier analysis of the signals can be used to determine the skin depth. An anomaly of the skin depth around a certain frequency indicates the presence of a formation at a certain depth.

SQUIDs have been used to study the magnetic signals produced by the human heart and by lung particles, and more recently by the brain (Fig. 10), both spontaneously and in response to visual excitation. The latter measurement was the first to show the existence of somatically evoked magnetic fields in the human brain. The field of biomagnetism is still in its infancy; study of other organs and of muscle activity, for instance, will undoubtedly lead to new developments. Various relevant magnetic field strengths are given in Fig. 11.

Computers. The use of Josephson junctions in computers was mentioned above. Their very short intrinsic switching time and very low power dissipation (less than 1 μW) can in principle be used to build computers one or two orders of magnitude faster than those in use today. The speed of present-day computers is in fact limited by the power dissipation of the semiconducting switching elements. This power dissipation leads to stringent

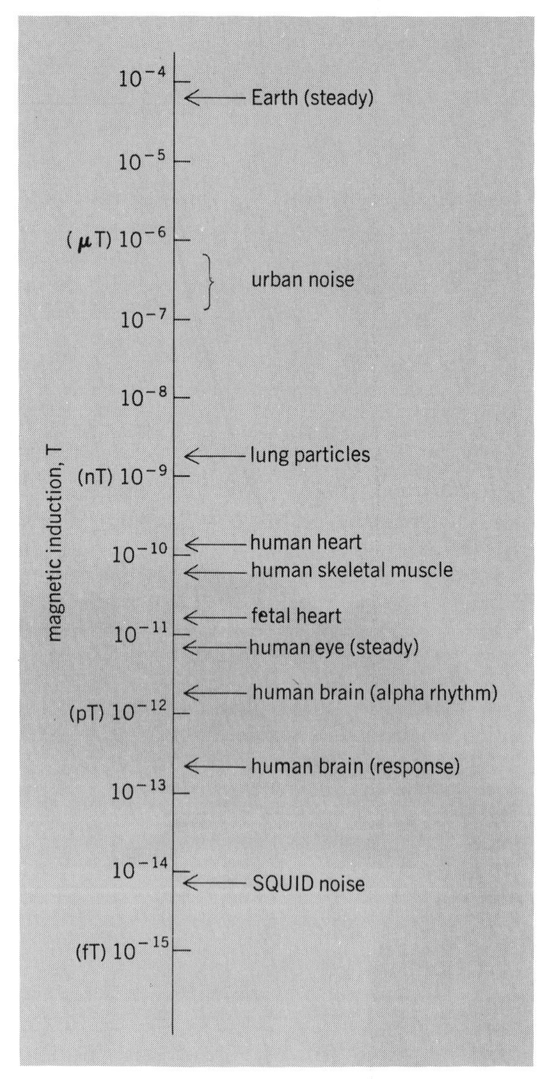

Fig. 11. Chart of typical biomagnetic field strengths compared with background (urban) noise and the intrinsic noise level within a 1-Hz bandwidth for present SQUID detectors. *(From S. J. Wilkinson and L. Kaufman, Application of SQUID detectors in biomagnetism, in J. E. Zimmerman and T. M. Flynn, eds., Applications of Closed-Cycle Cryocoolers to Small Superconducting Devices, NBS Spec. Publ. 508, pp. 177–204, 1978)*

cooling requirements and eventually prevents very close packing of the elements, thus producing relatively long transit and cycle times. A major effort at IBM Corporation is being directed toward the construction of a Josephson computer which would use very densely packed, micrometer-size junctions. Problems posed by the very thin (20-nm) oxide layers of the junctions are of course formidable, but IBM is optimistic about the chances of success. Should the oxide barriers prove to be an insurmountable problem, there would still be the possibility of using instead weak links of the constriction type with granular superconductors whose structure is precisely that of superconducting regions separated by very thin barriers.

CONCLUSION

As more powerful machines are built, whether they are power generators or computers, power dissipation becomes more of a problem. Superconductivity can then provide a logical solution. This is probably the underlying reason for the very broad spectrum of applications of superconductivity. It is likely that superconductivity will become an important branch of technology by the end of the 20th century. [GUY DEUTSCHER]

Human Ecology in the Tropics

Denis F. Owen received his M.A. degree from the University of Oxford and his Ph.D. from the University of Michigan. He has published six books and over 140 scientific papers on ecology and genetics, and is especially interested in tropical ecology. In 1974 he was appointed Distinguished Visiting Professor at the University of Massachusetts. He now lives and works in Oxford, England.

Ecology is the study of the relationships between plants and animals and their environment. The human being is an animal, and despite the unique characteristics of humans, their needs are fundamentally the same as those of other species. Human ecology is therefore an attempt to examine humans in relation to the environment. It is concerned with the balance (or lack of balance) between birthrates and death rates; how the environment is exploited for food and shelter; how the carrying capacity of the land can be increased or decreased by human activities; the nature and importance of diseases; and indeed, all aspects of human interaction with the physical and biotic surroundings. In a diffuse and as yet far from satisfactory way, human ecology links the natural and the social sciences. From an ecologist's viewpoint the human is a difficult species to study. No other species relies so much on tradition, belief, and knowledge for survival and reproduction, and no other species is as varied in its ecological relationships with the environment.

The tropics—the region between the Tropics of Cancer and Capricorn—support the poorest people in the world. For hundreds of years this region has been exploited for raw materials by the wealthy nations to the north, and in many areas this exploitation continues today. Population growth rates are far higher in the tropics than in most temperate regions, yet the land is less productive in food and much more easily ruined by overuse. Natural ecosystems are more complex (there are more species of plants and animals and hence more biotic interactions), but they are more fragile and less able to resist perturbations than temperate ecosystems.

TROPICAL ENVIRONMENT

Especially on or near the Equator and at low altitudes, year-round warmth and lack of frost are the most obvious characteristics of the tropical environment. Day length does not change much as the seasons progress, and provided there is rain it is possible to grow crops all year. The amount and seasonal distribution of rainfall are the most important determinants of human settlement and survival. Within the tropics are some of the wettest and also some of the driest spots on Earth. The Amazon Basin, coastal West Africa, and much of Southeast Asia are extremely wet, with annual rainfall totaling more than 200 cm. At the other extreme, the Peruvian desert, the southern Sahara, the Arabian peninsula, parts of the Indian subcontinent, and much of central Australia are arid, and some areas may receive no rain at all for several consecutive years. Even in the wettest places rainfall is seasonal; in some, notably East Africa, there are two wet and two dry seasons each year. Throughout the tropics cultivation and the raising of livestock are intimately related to the alternation of wet and dry seasons; at any given locality the inhabitants know what to expect, and adapt their activities accordingly.

Human impact on arid, semiarid, and some subhumid areas has been to reduce the vegetational cover, and this, in turn, has decreased the biological effectiveness of rainfall by increasing the evaporation rate. Large areas of Africa and India are much drier now than they were a few hundred years ago, and in the absence of evidence for climatic change it must be assumed that people are responsible. In the southern Sudan, land that produced crops and supported livestock as recently as 20 years ago has been converted to unproductive desert by removal of trees for firewood and by overcultivation and overgrazing by livestock.

Vegetation. The vegetation map of Africa shows an area of evergreen lowland forest extending from Sierra Leone in the west to Uganda in the east, with isolated patches throughout East Africa on the lower slopes of mountains. To the north, east, and south of this area there is savanna dominated by grass with or without scattered trees. In the north and southwest the savanna merges into desert. There are extensive areas of swamp, especially in the east and center, and some of the largest lakes and highest mountains in the world. In some small African countries, such as Uganda, variations in altitude and the proximity of large bodies of fresh water produce an almost complete spectrum of world vegetation types: semidesert in the north; evergreen forest in the south; montane vegetation, including bamboo forest, on the slopes of mountains; rich grasslands and extensive swamps in the center; and permanent snow and ice on the tops of high mountains such as Ruwenzori. However, this vegetation map is misleading because it shows what the vegetation would be like if there were no people. Much of the lowland forest has been cleared and is now secondary bush—a complex mosaic of original forest species, introduced crops and their weeds, and elements of the savanna vegetation that have infiltrated after the forest has been partially or completely felled. Similar changes have occurred in the surrounding savanna, which is almost completely cultivated or grazed by livestock except where tracts have been set aside as national parks for the conservation of large mammals. Throughout Africa the savanna grasslands burn at least once a year in the dry season; the fires may extend over vast areas, and most are deliberately started to clear the land of dead vegetation and to promote the growth of new grass to feed cattle (Fig. 1).

Fig. 1. Result of slash-burn process of clearing land. Vegetation is burned off to allow erosion. *(Courtesy of U. P. Garrigus, University of Illinois)*

Soils. The soil cover of the tropics is variable. Extensive areas are covered with laterite, a hard rocky substance formed when mineral-rich earth is exposed to sun and rain. This process, called laterization, is increasing as more of the natural vegetation is removed. There are now enormous areas of hard laterite that are impossible to cultivate. All soils develop primarily from the weathering of rocks, but also include quantities of humus derived from decomposing vegetation. The rate of decomposition is high in warm and humid climates, which means that the soil contains little humus and that most of the nutrients in an ecosystem are locked up in living plants. If a forest is cut down and the timber removed, most of the nutrients are lost. This is one reason why cultivation for more than a few consecutive seasons is difficult and why so many peasant cultivators practice shifting cultivation, a technique in which worked areas are abandoned and new areas constantly sought. Removal of vegetation not only negates the possibility of humus formation, but also encourages erosion by wind and rain which leaves the land barren and unfit for growing crops. In arid regions erosion is especially important and is one reason why the use of tractors is discouraged: they break the soil up too much, and their use can lead to the formation of dust bowls.

Pests and weeds. Tropical ecosystems, especially in high-rainfall areas, are extremely rich in species of plants and animals; indeed there is a latitudinal gradient of decreasing species diversity away from the Equator. One consequence of high species diversity is that there are more potential pests and weeds. Whenever an area of natural vegetation is cleared and cultivated, the ecosystem is simplified; that is, there are fewer species of plants and animals, which in turn means that any species may become so common as to constitute a weed or pest. In the tropics an increasing number of species are being designated as troublesome. For example, the grasshopper, *Zonocerus variegatus*, is beginning to be considered a pest in West Africa as a spoiler of cassava and other crops.

The hazards to tropical agriculture and livestock rearing are only just beginning to be understood. That understanding is dependent in part on an awareness of the dynamics of natural ecosystems. From past mistakes, ecologists know that the agricultural technology of the temperate regions cannot be transposed piecemeal to the tropics.

POPULATION TRENDS

Almost everywhere birthrates are high, but the average life expectation varies from place to place and is as low as 35 years in some African countries. Death rates among children are often particularly high, largely the result of diseases such as malaria and an inadequate diet. Nevertheless, throughout the tropics populations are increasing at rates of 2–3% per year; in the temperate countries growth rates are considerably lower, often under 1% per year, and in some countries the population is stable. A growth rate of 3% per year means that the population will double in as little as 24 years; if 1%, the population will double in 70 years.

Nigeria. Population census figures for many tropical countries are notoriously unreliable; one example is the widely differing census results for Nigeria. According to the latest figures, Nigeria's population is about 80,000,000, which is 25,000,000 (or 45%) higher than the 1963 census which indicated a modest growth rate of 2% annually. The recent figure of 80,000,000 gives an annual growth rate (over a period of 10 years) of 5%. This is higher than any other country in the world and if true probably represents the maximum possible for a human population. It is likely, however, that the 1963 census was a serious underestimate.

India. India has long been considered as a nation where the pressures of overpopulation are likely to be felt sooner than anywhere else. In 1971 there were about 548,000,000 people in India, and the population was increasing at a rate of 2.5% a year, which means the addition of about 13,000,000 people annually. The present (1978) population must be nearly 700,000,000, and there may be over 1,000,000,000 people by the year 2000.

Hundreds of thousands of Indians have adopted birth-control methods or accepted sterilization, but with little impact on India's population growth rate; in fact, the present estimated growth rate is slightly higher than it was in the 1950s, and contrasts markedly with the early 1920s when births and deaths were balanced at about 49 per 1000 people. The application of medical technology has reduced the death rate to about 14 per 1000, while the birthrate has been only slightly reduced and at times has been higher than 49 per 1000. Medical and social workers in India are often well satisfied with local successes of programs for limiting population growth, but the total situation is fast becoming unmanageable. An additional and also (it is claimed) unexpected problem is that conflicting religious interests generate suspicion of the motives of those wishing to introduce birth-control measures. Thus the Hindus, who make up 80% of the population, believe that the Muslims are increasing in numbers more rapidly than themselves and are suspicious of government intentions.

India's position is by no means unique; similar conditions may be found in other tropical countries, but because the population of India is so large and because a crisis seems imminent, the situation there is cause for special concern. Countries such as El Salvador in Central America and Burundi in Africa are just as overcrowded, but because they are so small their population problems are rarely publicized.

Urbanization. Most people in the tropics live in what might be described as rural areas. However, recently there have been unprecedented movements to urban centers, triggered, it is believed, by dissatisfaction and a feeling that cities offer more economic opportunity. In 1957 the population of Kinshasha in Zaire was about 380,000; 10 years later it was over 900,000, and by 1978 it must be well over 1,000,000. Similar increases have occurred in virtually all tropical cities and have created social problems and unemployment on a scale that would be utterly unacceptable in developed countries.

CROPS AND CULTIVATION

Peasant cultivators growing food for their families know their own patch. They can estimate fertility and suitability for a particular crop by the natural vegetation and soil characteristics; they can assess the number of seasons during which a crop can be farmed with worthwhile results, and the number of seasons during which the land must be rested. Judgment of initial fertility is based on the natural climax vegetation of the patch, and judgment of continuing fertility is based on the species composition of successional vegetation, including weeds, that develops following cultivation. Peasant cultivators have a vocabulary of hundreds of words for indicator plants and can identify particular vegetational associations by specific terms. This knowledge has been derived from trial and error and from oral tradition. If these peasants are moved to another environment, they will probably fail in their efforts at cultivation, and possibly even starve.

Multiple cropping. Perhaps the most striking feature of peasant cultivation is the extraordinary mixture of species grown at a single site. Multiple

Fig. 2. Examples of multiple cropping. (*a*) Banana trees in the foreground, a mixture of sorghum and cowpeas in the background. (*b*) Cowpeas in the foreground; maize is intermixed with cowpeas until in the background maize predominates. *(Courtesy of J. deWit, University of Illinois)*

cropping is a form of cultivation by which productivity is sustained by growing several species of crops simultaneously (Fig. 2*a* and *b*). Monocultures are favored in developed countries, where sowing and harvesting are done mechanically; polycultures are much more labor-demanding, but are quite feasible where there is an abundance of field laborers and a dearth of machines and fuel.

The reasons for multiple cropping can be deduced from the results of scientific investigation and from a great deal of circumstantial evidence. Multiple cropping is reliable: if one crop fails there are always others, and it can therefore be viewed as a form of insurance against annual fluctuations in yield. However, there are also ecological reasons which help to explain its widespread occurrence in the tropics. Most cultivators know intuitively that no two crop species have exactly the same cultivation requirements. Different crops react to variations in the season in different ways; they have different water and nutrient requirements and different responses to shade and sunlight; and they are susceptible to different species of pests and diseases. All plants, crops included, possess an array of chemical defenses against insect attack. Each species or variety produces chemical compounds which play no part in the growth and development of the plant but which deter the laying of eggs by plant-feeding insects. Aromatic compounds also act as attractants to plant feeders. Extensive monocultures attract a few species in large numbers so that they become pests. If crops are mixed, there is a confusing array of aromas whose effect is to reduce overall insect damage. This may well be the most important reason for multiple cropping in the tropics, where the number of potential pests is high.

Seasonal cultivation. The problems of cultivation in an area of highly seasonal rainfall, where diseases and ill-health are a constant threat to work efficiency, are well illustrated by rural life in the Gambia in West Africa. Here the people are mainly of the Mandinka tribe, and they subsist by cultivating land around their own villages. The people live in mud huts with thatch or corrugated-iron roofs. Numerous pathways develop to cultivated fields, but these are periodically abandoned as the pattern of land use changes, especially when plots are left to rest.

The long dry season lasts from November to May; there is virtually no rain during this period and the humidity is low; night temperatures may drop to 15°C, while the day may become extremely hot and unpleasant, especially just before the onset of the rains when temperatures as high as 45°C are common. The rains begin with violent thunderstorms, and the heaviest rainfall is in August and September. The timing of the rains and the amount of precipitation that falls vary from year to year and determine the success of cultivation.

Most cultivation is concentrated during the wet season. The work is done by hand, mainly with locally made tools, the men being responsible for groundnuts which are sold, and for grains such as millet, sorghum, and maize which are consumed locally (Fig. 3). The women help in the preparation of the soil for these crops, but their main responsibility is the growing of rice in swamps some distance away. The rice fields are cleared of weeds at the onset of the rains, planting and transplanting are done in August and September, and the rice ripens and is harvested in November and December. The productivity of the rice fields is low, and there are few facilities for and little interest in controlling the numerous pests, including grain-eating birds, that attack the crop.

Energy budget. The concentration of heavy work into a restricted period of the year, together with seasonal changes in the availability of food, has a pronounced effect on the energy budget of the people. From March to May there is little or no work to do, the body weights of the people remain constant, and their intake and expenditure of energy is low. With the onset of the rains heavy work commences and there is an increase in energy expended, which continues to increase as the rains develop and more heavy work becomes necessary.

Fig. 3. Africans working in the fields with hand tools. *(Courtesy of J. Harlan, University of Illinois)*

At this time of the year the store of food from the previous harvest is almost exhausted, and a seasonal low of available energy is reached at the time when energy expenditure is greatest. This deficit results in a decrease in body weight as the tissues of the body are drawn upon. Weight is regained at the beginning of the dry season, when food from the harvest becomes available.

Food supply and disease. A Mandinka family produces all its own food. Rice and millet are the staples; groundnuts and wild leaves and fruits are used as sauces, and the women and children collect snails and small fish which provide some variation in the diet. Domestic animals such as cattle, goats, and chickens are regarded as status symbols or as an insurance against real shortage, and are killed only on special occasions or when absolutely necessary. On the whole, food is plentiful in the dry season, but during the wet season, when the need to work is at its peak, food is scarce.

Diseases are prevalent throughout the year, but especially during the wet season when insect vectors are abundant. Most children suffer from malaria and many die from it. There are periodic measles epidemics which can kill up to 40% of the children in a village.

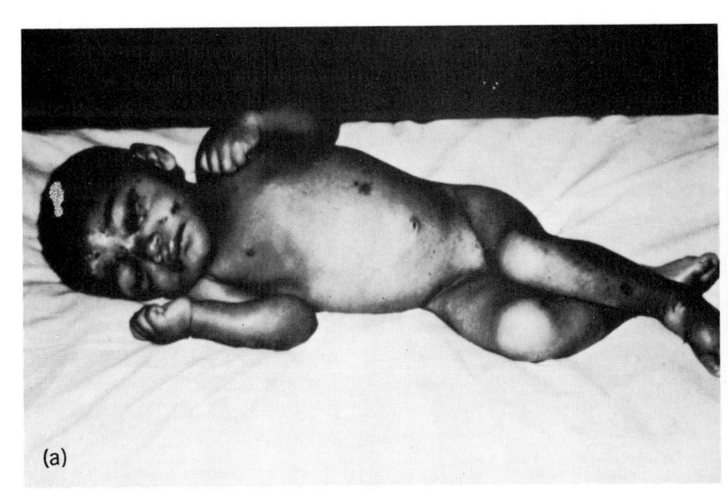

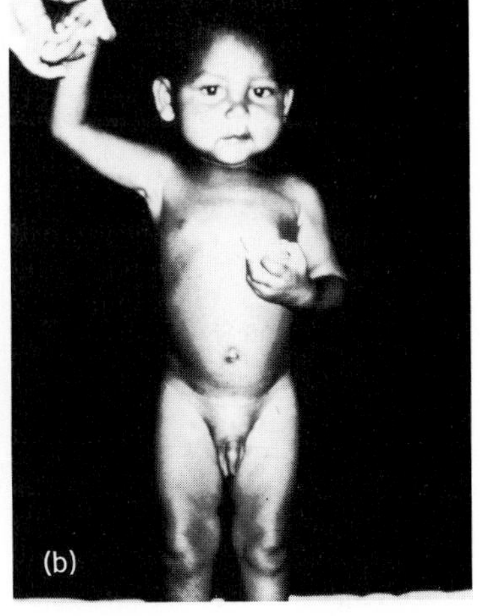

Fig. 4. Effect of malnutrition. (*a*) Child suffering from kwashiorkor. (*b*) Same child 6 weeks later after adequate diet and other therapy has been applied. *(Courtesy of D. E. Alexander, University of Illinois)*

Malnutrition. In this part of Africa, as in many other areas of subsistence cultivation, there is a deficiency of protein in the diet. Malnutrition is common in young children and arises as a result of a chronic shortage of protein and a relative excess of carbohydrate. Some tropical staples, such as cassava and cooking bananas, contain very little protein and are the main cause of this form of malnutrition, often called kwashiorkor (Fig. 4*a* and *b*). Several effects of kwashiorkor have been identified: deficiencies in enzymes and proteins in the body, low lipoprotein level, a reduction in bone marrow activity, and a failure to produce enough hemoglobin. In extreme cases the hair turns golden, the skin becomes pale, lesions appear on the body, the eyesight deteriorates, and the child refuses to eat. Even when cured, there may be long-term effects including heart disease in later life and mental retardation.

Cash crops. The European colonial powers introduced to the tropics the concept of cash crops and the idea that crop products could be exported for profit. Some of these crops—cocoa is the best example—can be viewed as inessential luxuries. The cocoa tree is native to tropical South America. It was cultivated by the indigenous people for several centuries before being discovered by the Spaniards and brought to Europe. As the drink made from cocoa became popular, the tree was introduced to other parts of the tropics. It was first planted in West Africa (now the main cocoa-producing area) toward the end of the 19th century.

The conditions under which cocoa may be grown are well known, and much research has been devoted to the crop, particularly on how to control its numerous pests and diseases. Cocoa is a tree crop and must be grown under essentially forest conditions (Fig. 5). Because of this, much planning and substantial investment are required for the establishment of plantations and small holdings, which means that once a piece of land has been developed for cocoa production it is difficult to revert to an alternative crop. Ghana is now the world's largest producer, and about 60% of the country's foreign exchange comes from the export of cocoa. The most fertile areas of the country, which might otherwise be devoted to food production, are committed to cocoa forest. Ghana now has to import substantial quantities of food, chiefly for its expanding urban population, and this food must be paid for with earnings from the export of cocoa. The variety of cocoa-based foods and drinks now available is truly astounding, but from every point of view cocoa is an unnecessary luxury: it is produced by the poor and consumed by the rich and is a likely casualty in any future world recession. A country like Ghana is thus in a precarious position; nevertheless, the aim is to increase cocoa production and to make the cocoa farms more efficient.

Fig. 5. Branches and fruit of cocoa tree. *(Courtesy of J. Harlan, University of Illinois)*

PASTORALISM

Cattle, sheep, goats, and camels are typically kept in the drier regions of the tropics. The forest region of Africa is unsuitable for cattle because of the presence of trypanosomiasis (sleeping sickness), which is transmitted by flies of the genus *Glossina*. Forest areas are more suitable for cultivation and a sedentary way of life; it is in arid and semiarid areas, where cultivation is difficult or impossible, that there are high concentrations of pastoral people. Some of these people are nomadic, moving their herds over considerable distances as grazing opportunities change, while others are sedentary (Fig. 6). In Africa sedentary cultivators are preferred by national governments to nomadic pastoralists. Cultivators are easier to tax, govern, and count, and it is the policy of most governments to try to settle nomads in order to control their activities.

Risky ecosystems. Nomadic pastoralists dependent on livestock live in "risky" ecosystems. They obtain higher yields from arid marginal lands than do hunters and gatherers; they manipulate vegetation (especially by burning) in order to provide fresh grass for their herds; and they live off their animals in a subtle and efficient way. Their way of life has persisted for thousands of years and, like peasant cultivators, they have accumulated a vast fund of knowledge which is difficult to replace with modern technological innovation. During the 1968–1973 drought in the Sahel region of West Africa, outside aid organizations provided the means for drilling deep boreholes to provide water for livestock. The consequences were disastrous, as the soil was so trampled by the concentration of livestock around the waterholes that the vegetation disappeared: there was water but no food, and the animals starved.

The Kel Adrar Tuareg tribe of northern Mali subsist in what by any standards is a hazardous and risky environment. There wide annual variations in the always scarce rainfall and therefore in available pasture. Diseases can wipe out entire herds in a matter of weeks, and dry-season sandstorms disperse and kill animals. In a subsistence pastoral society loss of part of a herd leads to immediate destitution, a risk that pastoralists in this region face several times during their lives. The Kel Adrar Tuareg tend camels, cattle, sheep, and goats, as well as donkeys to carry water. Cattle and sheep are grazers, camels graze and browse, and goats browse, which means that by diversifying the herds the ecosystem is exploited more effectively. The different species have different water requirements: cattle and sheep need water constantly so must be kept near a water source; camels require water less often and so can travel further afield; goats drink small amounts often and so must remain near the camp. Each species also requires a different amount of attention for herding, milking, and watering: sheep are labor-intensive, necessitating full-time shepherds; camels and cattle can be left for long periods without attention; goats need constant attention and are tended by children. Cows and camels produce young in the wet season when fresh pasture is available. Goats do not require green leaves and can therefore breed throughout the year, but both goats and sheep are prevented from breeding too often unless conditions are exceptionally favorable. Nomads possess a thorough knowledge of the requirements of these animals and the carrying capacity of the range.

Strategy for movement. Movement is the most conspicuous ecological strategy of nomads, and although they may move because of social or political difficulties, the main stimulus is to provide fresh resources for their herds. The decision when and where to move involves predicting rainfall patterns over large areas where rainfall is unreliable. Cultivators can store food in the dry season, but for nomads food storage in the conventional sense is impossible as milk and meat simply do not keep. Storage is in the form of live animals, and herds are thus allowed to build up beyond subsistence needs. This strategy has been misinterpreted as an accumulation of animals for status and prestige. In reality, it is an insurance against disaster: animals are traded for grain and other crops when conditions are bad. The Kel Adrar Tuareg also have an elaborate system of lending animals to fellow pastoralists who are in need. The loans (and gifts) have restrictions on them, but the possibility that others can become destitute is acknowledged and help is freely given. All these practices suggest a complex and well-tried social system, a considerable degree of generosity and understanding of the plight of others, and a thorough adjustment to a hostile environment which few outsiders could survive in for more than a short time.

Outside pressures. However, this delicate ecological balance is being eroded from the outside. The demand for meat and skins is increasing and has led to overstocking and conversion of pasture to desert. The nomads have been forced into becoming more restricted in their movement, and have received "help," ranging from the creation of water holes to introduction of antibiotics, which has upset the balance to such an extent that the ecosystem may never recover. The people of the arid tropics are faced with a series of development options that they neither understand nor want; the problem is that outsiders are interfering, and little can be done to arrest the inevitable changes that are taking place.

Fig. 6. African nomads with their cattle. *(Courtesy of J. Harlan, University of Illinois)*

PROSPECTS FOR THE TROPICS

Cultivators and pastoralists producing food for themselves are well adjusted to the tropical environment. They are poor and are not significant consumers of the world's natural resources, yet despite widespread illiteracy they know how the rest of the world lives and have understandably developed expectations for a better life. The rich, temperate countries have two different and to some extent conflicting tendencies in dealing with the tropics: the desire to exploit and the desire to help.

In South America, Africa, and Southeast Asia there are vast mineral resources to be exploited. Because of outside intervention, cash-crop economies aided by technological innovations and demands from the rich nations will replace peasant cultivation. Meanwhile, as populations grow, food will become scarcer. With changes in land use, diseases spread: almost everywhere in tropical Africa and in many parts of Southeast Asia, schistosomiasis is on the increase, partly the result of new irrigation projects, canals, and dams which provide bodies of fresh water in which live the aquatic snails which are the intermediate hosts of the microorganisms causing the disease.

Tropical peoples, it is urged, must be helped to achieve a better standard of living. Outside aid continues to finance large-scale development projects, but for most tropical countries the money received for social improvements is less than is paid out for armaments.

Tropical peoples do not need large-scale projects, armaments, and an array of often conflicting advice; instead they need help with small-scale projects, economic independence, an understanding of and sympathy for existing ways of life, and above all a chance to find their own solutions to problems. Human ecology in the tropics is a story of sudden disruption of life-styles evolved over thousands of years; the future is precarious, but there is still time for constructive change.

[DENIS F. OWEN]

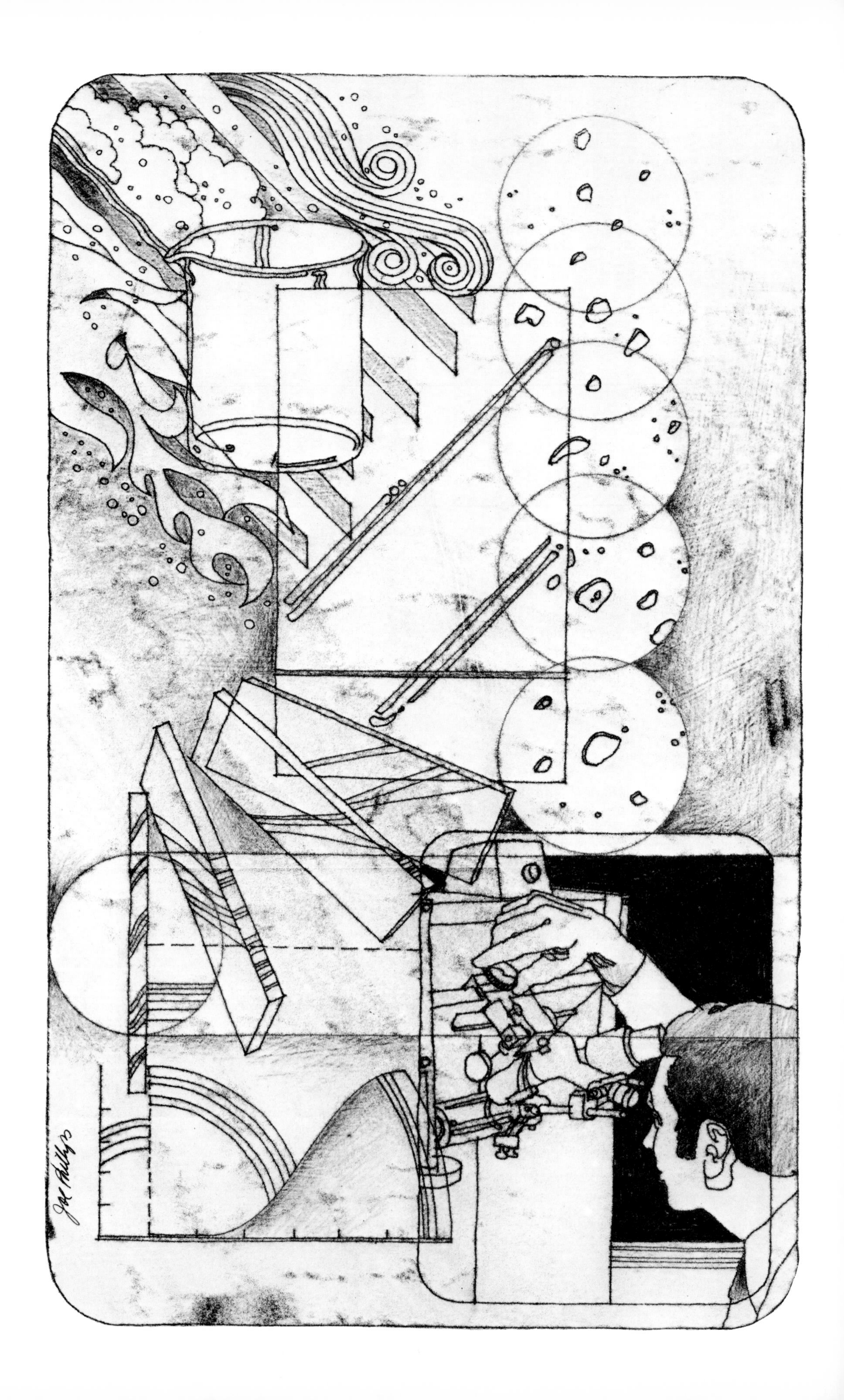

Chemical Durability of Glass

Robert H. Doremus is the New York State Professor of Glass and Ceramics in the Materials Engineering Department, School of Engineering, Rensselaer Polytechnic Institute. With Ph.D. degrees from the University of Illinois and Cambridge University, he was previously associated with the General Electric Research and Development Center as a physical chemist.

Glass is one of the most chemically durable materials in common use. It is highly resistant to corrosion by water and acids and is stable at high temperatures. Thus it is especially useful as a container for corrosive fluids, as a liner for chemical reactors, and as an electrical insulator. Nevertheless, glass reacts slowly with various chemical environments. These reactions are important in determining the limits of usefulness of glass in severe environments. The most important reactants are water (liquid and vapor) and aqueous solution.

In the last few years, new techniques have been developed for studying the compositions of thin reacted layers on glass surfaces, and these results give fresh insight into the mechanisms of the reaction of water with glass. These techniques allow measurement of the profiles of concentrations of various constituents of a glass close to its surface. One fascinating result of these measurements has been the dating of glass objects as divergent as natural obsidian and a recent facsimile of a 19th-century vase.

In this article, understanding of the chemical durability of glasses of commercial importance is emphasized. By far the most important commercial glass is soda-lime, used for containers, windows, lamps, architectural materials, and a host of other applications. In most uses the chemical durability of the glass is critical. The raw materials for soda-lime glass are cheap and readily available, and this glass can be melted at temperatures lower than other, more durable glasses.

Another important class of commercial glasses comprises the borosilicates. Pyrex borosilicate is more chemically durable than soda-lime glass, and has a lower coefficient of thermal expansion, so that it is used in laboratory ware, cooking utensils, automobile headlights, and telescope mirrors. Because Pyrex glass must be melted at a higher temperature and because it contains more expensive components than soda-lime glass, Pyrex glass costs more. Other sodium borosilicate glasses are important for sealing, and can be matched to the expansion coefficients of a variety of other solids.

Lead-containing silicate glasses are important in decorative glass objects and as low-melting sealing and solder glasses. Questions have arisen about the toxicity of lead leached from lead glasses used as glazes in dishes and as goblets. Understanding this leaching behavior is essential for predicting the amount of lead released.

In the following, the stages in the reaction of water with glass are described, and some experimental measurements of the reaction are reviewed; these results lead to a model for the reaction of water with glass, which is then compared with experimental results. The remaining sections of the article include discussions of the influence of glass composition, weathering due to water vapor, and solution pH as it affects glass durability.

REACTION OF WATER WITH GLASS

At least three steps are involved in the reaction of water with a silicate glass containing alkali ions: (1) ion exchange of hydronium (H_3O^+) ions from the water with alkali ions in the glass, (2) the formation of a gel layer on the glass surface, and (3) dissolution of the glass into the contacting solution. Either or both of the first two steps can be absent, depending upon glass composition or solution pH (acidity), and the third is absent when the glass reacts with water vapor instead of liquid water or a solution.

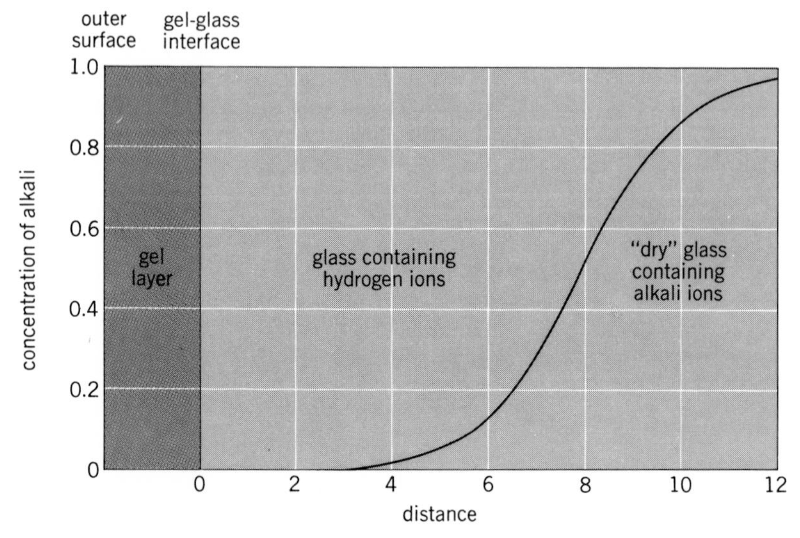

Fig. 1. Schematic diagram of the hydrated layers on the surface of an alkali silicate glass.

A schematic diagram of the surface structure of an alkali silicate glass after reaction with water is shown in Fig. 1. The "dry" glass contains alkali (usually sodium) ions at their original concentration; as one proceeds outward to the glass surface, there is a decrease in the concentration of alkali ions as a result of their replacement with hydronium ions. In this layer of partial exchange the network structure of the glass is intact, and the only change is replacement of one ion for another. Closer to the glass surface, the network can become partially hydrated by reaction of silicon-oxygen bonds with water, Eq. (1). This partial hydration

$$-\overset{|}{\underset{|}{Si}}-O-\overset{|}{\underset{|}{Si}}+H_2O = -\overset{|}{\underset{|}{Si}}OH\ HO\overset{|}{\underset{|}{Si}}- \qquad (1)$$

leads to a more open structure than in the original glass; ions from solution and water molecules can penetrate through this partially hydrated or gel layer with mobilities much higher than in the glass network that has not been broken up by the reaction in Eq. (1).

Experimental studies. This schematic model of the surface of glass after its reaction with water is substantiated by a number of experimental studies. Profiles of sodium-, lithium-, potassium-, and hydrogen-containing ions in a number of different glasses have been measured by a variety of techniques, and all showed the S-shaped concentration dependence, as in Fig. 1. These techniques include chemical analysis of layers etched off the glass, nuclear resonance reactions with specific elements to give gamma rays, alpha-particle backscattering measurements, luminescence of the glass surface during ion sputtering, and Auger analysis of surfaces as they are removed by sputtering. The electrical resistance of hydrated glass bulbs as they are etched by hydrofluoric acid gives information about ionic mobilities in different layers at the glass surface. The resistance is nearly constant initially, which is evidence for rapid ionic motion in a hydrated gel layer. Next there is a sharp drop in resistance as the partially exchanged layer is etched. This sharp drop corresponds to the part of the glass with almost complete replacement of sodium ions by hydronium ions and shows that the hydronium ions have low mobilities compared with sodium ions. Finally, the slope of the resistance-versus-time plot reaches an intermediate value, corresponding to the dry glass, in which sodium ions have mobilities between the high mobilities of ions in the gel layer and the low mobility of hydronium ions in the unhydrated glass.

The exchange of alkali ions in the glass and hydronium ions from water can be described by Eq. (2), where g is glass and s is solution. In most glass-

$$Na^+(g) + H_3O^+(s) = Na^+(s) + H_3O^+(g) \qquad (2)$$

es this exchange appears to be complete at the glass-solution interface. Complete exchange results because of the strong affinity of the SiO^- group for hydrogen (or hydronium) ions, as demonstrated by the weak-acid character of this group in aqueous solution.

Static solutions. To get reliable experimental results on the time dependence of the reaction of water with glass, some care must be taken with experimental conditions. Many experiments measuring this rate have been done in "static" solutions, that is, solutions in which the products of the reaction accumulate. As the reaction shown in Eq. (2) proceeds, the solution becomes more basic and the rate of dissolution of the silicon-oxygen network becomes more rapid. Thus it is essential to keep the pH of the solution constant to study reaction mechanisms. Furthermore, products of the dissolution of the glass can accumulate at the glass surface, changing processes there. Experiments are sometimes performed in a static solution because this arrangement simulates practical conditions, for example, the reaction of a bottle with its contents. Such experiments are unsuitable for deducing mechanisms of reactions because of the changing conditions in the solution.

Interdiffusion of ions. In one method, fresh hot water is continuously passed over a glass powder in a Soxhelt extractor, thus keeping the solution conditions constant and removing products of the reaction. With this experimental technique, the amount of alkali ion that appeared in the washings, as measured by chemical analysis, was initially proportional to the square root of time for alkali silicate glasses, suggesting a diffusion-controlled process. This process is the interdiffusion in the glass of hydronium ions from solution with alkali ions in the glass. The anionic oxygen sites for these exchanging cations are immobile, so that this interdiffusion process is similar to that in an organic ion exchanger with fixed anionic sites.

Dissolution of glass. At long times of reaction the amount of sodium appearing in solution becomes proportional to time, rather than to its square root. Furthermore, silicon and other glass constituents are found in the solution. These results suggest that the glass dissolves into the solution by reactions of types (3)–(5). These reactions

$$2H_2O + SiO_2 = Si(OH)_4 \quad (3)$$

$$H_2O + CaO = Ca(OH)_2 \quad (4)$$

$$3H_2O + Al_2O_3 = 2Al(OH)_3 \quad (5)$$

show the result of dissolution, not its mechanism. The breakdown of the glass presumably starts with reactions such as (1); then second, third, and fourth silicon-oxygen-silicon bridges react, finally giving silicic acid in solution, as in reaction (3). There is some evidence in experiments on dissolution of silica that "islands" of silica break into solution and that monomeric silicic acid results only after further hydration of these polymeric islands in solution. The simplest assumption about the rate of dissolution of glass in a fixed solution (unchanging pH and ionic concentrations) is that the rate is linear.

Time and age measurements. As mentioned, the thickness to which the surface of glass is hydrated can be used to determine the time that it has been exposed to water vapor in the ambient atmosphere. At least two different measurements of this kind have been reported. The thicknesses of hydrated layers on the surfaces of obsidian, natural glasses of approximate composition 75 wt % SiO_2, 14% Al_2O_3, 4% Na_2O, 4% K_2O, 2% Fe_2O_3, and 1% CaO, were measured in a microscope and found to be proportional to the square root of time of exposure. From these calibrating measurements the geological ages of various natural obsidians were deduced.

Hydration profiles on a number of glasses of known ages were measured by using the ^{15}N nuclear resonance technique for depth profiling, and again the hydration thicknesses were proportional to the square root of age. With this method it was shown that a cut-glass vase purportedly made in the 19th century was actually a recent imitation.

Diffusion model. The two steps of interdiffusion of ions and dissolution of the glass can be combined in a diffusion model in which the surface of the glass is progressively removed, giving a "moving boundary." When two ions of different mobilities interdiffuse on fixed sites, the fast ion tends to outrun the slow one, setting up an electrical field (the diffusion potential). This field speeds up the slow ion and slows the fast one, so that their fluxes become equal and electroneutrality is preserved. The resultant effective diffusion coefficient depends on the concentration ratio of the two ions. Profiles calculated for this kind of interdiffusion have the S-shape of the profile in Fig. 1.

The best way to calculate the diffusion coefficients and dissolution rates during the reaction of water with glass is from profiles of the type shown in Fig. 1. The shape of the curve gives a measure of the ratio of the diffusion coefficients, and the diffusion coefficients themselves can be calculated from time of reaction and distance of penetration.

The rate of dissolution of the glass can be examined with the model of surface structure of alkali silicate glasses shown in Fig. 1. Reaction (1) occurs at the gel-glass interface and determines the rate of movement of the diffusion interface. Water and ions can readily diffuse through the porous gel layer to the gel-glass interface. The gel itself dissolves more slowly into the solution, so that the thickness of the gel layer increases as reaction of the glass with water proceeds. Sodium and hydroxyl ions are formed at the gel-glass interface, so that their concentration in the gel layer is probably higher than in solution.

INFLUENCE OF COMPOSITION

Many experiments show the effect of composition on chemical durability of glass. Most of these were done with static solutions, and thus their pH could change as reaction with water proceeded. Such results are of qualitative or at best semiquantitative value. Nevertheless, certain trends of composition effects are clear, and will be discussed in terms of the model previously described. These trends should be valid in solutions that are quite acidic ranging to those that are mildly alkaline (of pH from about 1 to 9); effects of stronger acid and base are considered in a later section.

As the amount of alkali silicate glass is increased, with the remaining constituents held in the same ratios, the rate of reaction with water increases. In static solutions this trend is primarily related to the increase in total quantity of alkali that builds up in the solution surrounding the glass, which gives a progressively more alkaline solution and more rapid attack of the glass. In binary alkali silicate glasses the diffusion coefficient of sodium increases as the amount of sodium in the glass increases; concurrently the rate of reaction with water increases as the amount of alkali increases. There appear to be no reliable measurements of the influence of total alkali concentration on the rate of dissolution of these glasses.

Soda-lime and sodium silicate glass. Even the most ancient glass samples are not binary alkali silicates, but invariably contain some calcium and magnesium oxides. These alkaline-earth oxides may have been part of the starting materials, but it is also possible that ancient glassmakers realized that the binary alkali silicates were easily attacked by water, and the addition of lime or magnesia improved durability. Even today the vast majority of commercially made glass has a soda-lime composition.

Diffusion measurements show why the addition of calcium oxide improves durability. The diffusion coefficient of sodium ions in glasses containing 5–10% CaO is up to a factor of 50 times lower than in a binary sodium silicate glass with the same soda concentration. The rate of dissolution may also be reduced by addition of lime. The lower diffusion coefficient of sodium in the lime glass can be understood as resulting from a blocking of alkali ion motion by the doubly charged calcium ions that are bound tightly in the silicate network.

A number of other oxides of divalent metals give a similar enhancement of durability, although calcium and magnesium oxides are the ones usually used commercially. There is a similar rate of extraction of sodium for glasses of the composition 80% SiO_2, 15% Na_2O, and 5% RO, with R being magnesium, calcium, strontium, barium, or cadmium. Glasses with lead or zinc as R usually have significantly lower extraction rates, although lead does not always give better durability, perhaps because some of it can be present in the Pb^{4+} state. The lower extraction rate in lead glass may result from the large radius of the Pb^{2+} ion, but the position of zinc appears anomalous.

In all of this discussion it has been assumed that changes in sodium ion mobility are accompanied by similar changes in hydronium ion mobility; the hydronium ion is actually the more important one in controlling the rate of ion exchange, because it is much less mobile than sodium. From measurements of mobilities of sodium, hydronium, and potassium ions in soda-lime and binary sodium silicate glasses, it appears that addition of calcium to the glass lowers the mobilities of larger ions such as potassium and hydronium even more than of sodium. Thus the addition of divalent oxides is probably even more effective in increasing durability than would be deduced from sodium ion mobilities alone.

Addition of oxides of the higher-valent ions titanium, zirconium, and aluminum to sodium silicate glasses gives a substantial increase in chemical durability. Presumably these ions also decrease the mobility of hydronium ions in the glass, although there are apparently no measurements of ionic mobilities in these glasses except for those containing aluminum, as discussed below. One might predict that oxides of any tervalent or tetravalent ion added to sodium silicate glasses would increase their durability; confirmation of this supposition awaits measurements on such glasses.

Commercial soda-lime glasses usually contain some alumina, and alumina has long been known to increase the chemical durability of these glasses. The maximum increase in the durability of soda-lime glasses was found to occur with an alumina addition of about 2%. The addition of alumina to a sodium silicate glass reduces the diffusion coefficient of sodium much less than divalent oxides such as calcium. Nevertheless, the rate of the early diffusion-controlled part of the reaction of sodium aluminosilicate glass with water is much slower than for binary sodium silicate glasses and even for soda-lime glasses. Thus one must conclude that addition of aluminum strongly reduces the mobility of hydronium ions in silicate glasses, even though it does not have much influence on sodium mobility.

If a second alkali oxide such as potassium is added to a sodium silicate glass, the durability of the glass is increased. The increase is greatest when the molar ratio of alkali ions is about equal. This increase is a result of the "mixed-alkali" effect, in which the mobility of an alkali ion is reduced when another alkali ion is added. The mechanism of this effect is being hotly debated.

Binary potassium silicate glasses have poorer durability than binary sodium silicate glasses of the same molar alkali composition. The most likely reason is that the hydronium ion has a higher mobility in potassium than in sodium silicate glasses because potassium and hydronium ions have about the same effective radius (1.3 A or 0.13 nm).

Pyrex borosilicate glass. Corning Glass Company developed Pyrex borosilicate glass as a highly durable glass that can be melted at temperatures much less than those needed for fused silica.

Weight loss of different silicate glasses after 6 hr in 5% sodium hydroxide at 100°C*

Corning number	Glass	Weight loss, mg/cm²	Approximate compositions, wt %							
			SiO_2	B_2O_3	Al_2O_3	CaO	MgO	PbO	Na_2O	K_2O
7900	96% silica	0.9	96	4						
7740	Pyrex borosilicate	1.4	81	13	2				4	
0080	Soda-lime	1.1	72.6	0.8	1.7	4.6	3.6		15.2	
0010	Lead glass	1.6	77			1		8	9	5
7050	Sodium borosilicate	3.9	68	24	1				7	0.7
8870	High-lead	3.6								
1710	Aluminosilicate–low alkali	0.35	64	4.5	10.4	8.9	10.2		1.3	
7280	Alkali-resistant (with ZrO_2)	0.09								

*From A. M. Filbert and M. L. Hair, Surface chemistry and corrosion of glass, in M. N. Fontana and R. W. Staehle (eds.), *Advances in Corrosion Science and Technology*, Plenum Press, 1976.

The composition of this glass is approximately 81% mole SiO_2, 13% B_2O_3, 4% Na_2O, and 2% Al_2O_3. The low sodium concentration is one reason for its high durability; in addition, the diffusion coefficient of sodium in this glass is low compared with the value in other silicate glasses, being even lower than in fused silica.

The presence of phase separation in borosilicate glasses also strongly influences their chemical durability. As Pyrex borosilicate is heated at 600°C or above, its durability progressively deteriorates as a sodium borosilicate phase separates from a silica-rich phase. The sodium borosilicate glass separates into a disconnected sodium borosilicate phase in a silica-rich matrix on a scale of 20 A (2 nm) or less, as shown in Fig. 2. This separation may also contribute to chemical durability of the Pyrex glass, since the silica matrix contains little alkali and is itself very durable.

Thirsty glass. A variety of sodium borosilicate glasses containing up to about 10% Na_2O and from about 10 to 70% B_2O_3, with the balance SiO_2, can readily separate into two amorphous phases. The two phases often have an interconnected or "wormy" morphology, making it possible to etch out the soluble sodium borosilicate phase completely (Fig. 3). The resulting porous glass, known as "thirsty" glass, is about 96% silica, and can be sintered to a dense fused silica known as Vycor.

Phase separation in binary and ternary alkali silicate glasses can also reduce their durability, especially when a matrix phase high in alkali is formed. The lime and especially the alumina in commercial soda-lime glasses reduce their tendency to phase separation and therefore protect their good durability.

Lead glass. The corrosion of lead glasses containing no alkali apparently proceeds by exchange of hydronium ions from solution with lead ions in the glass, much the same as for the alkali silicate glasses. The profile of lead ions in the surface of binary lead silicate glasses is similar to the profiles found for alkali silicate glasses. The rate of reaction of this lead glass with water is much more rapid than the reaction of soda-lime glasses with water at the same temperature. Apparently the binary lead glass has a more open structure than the soda-lime glasses, leading to higher mobility of hydronium ions.

Quenched glass. Water reacts more rapidly with quenched glass than with annealed glass. This result can be understood from the higher ionic mobility in the quenched glass, which has a lower density and a more open structure, and the greater reactivity of strained Si-O bonds.

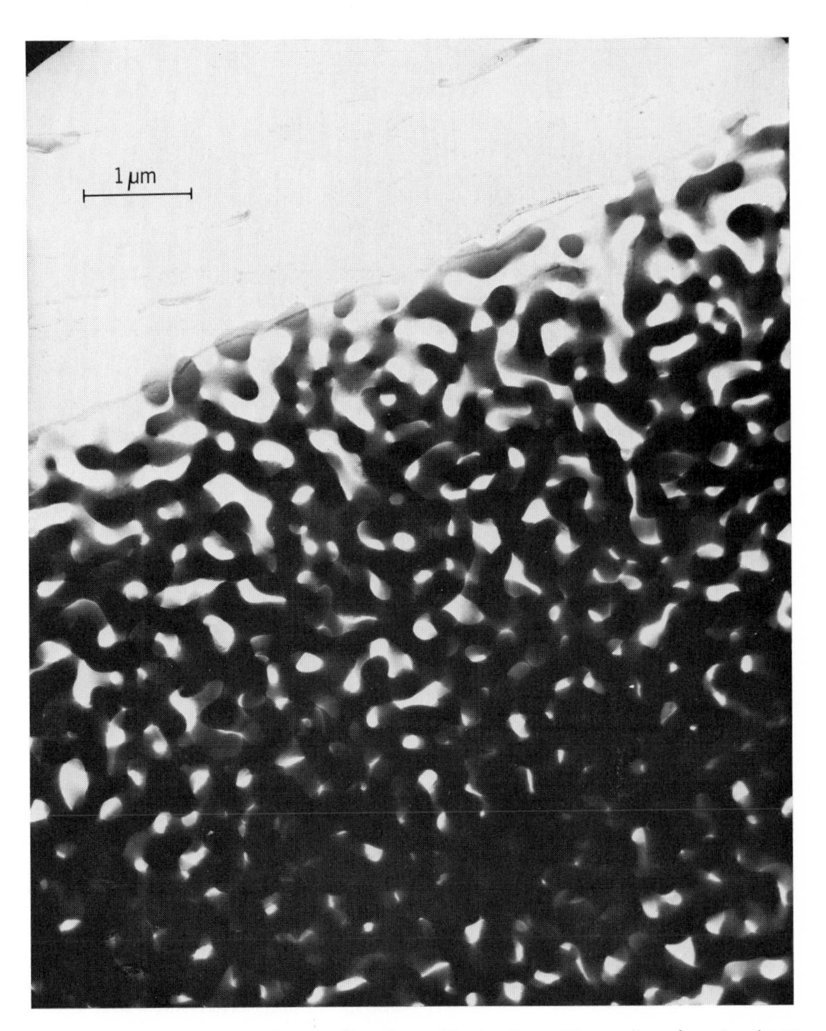

Fig. 3. Phase separation in a sodium borosilicate glass. The sodium borate phase was etched out with HCl. *(Transmission electron micrograph by A. M. Turkalo)*

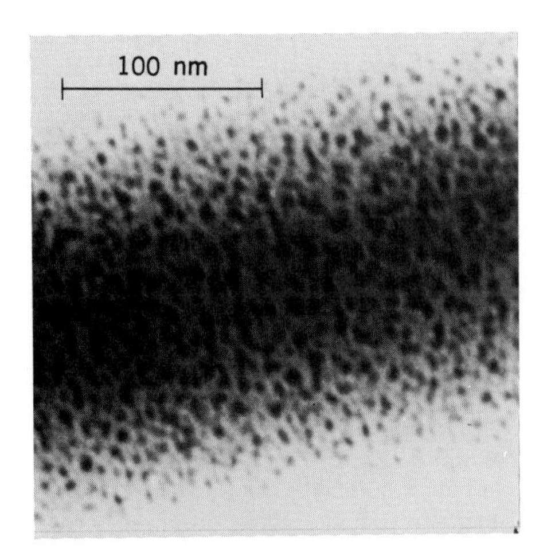

Fig. 2. Transmission electron photograph of Pyrex borosilicate glass in which silver ions have been exchanged for sodium ions to enhance contrast between the phases. *(From R. H. Doremus and A. M. Turkalo, Phase separation in Pyrex glass, Science, 164:418–419; copyright 1969 by the American Association for the Advancement of Science)*

WEATHERING OF GLASS

The reaction of glass with water vapor in the ambient air, as contrasted with liquid water, is called weathering. The mechanism of weathering involves the ion exchange of Eq. (2), but no dissolution of the glass takes place. Furthermore, the reaction products (sodium hydroxide or sodium carbonate) crystallize on the surface of the glass. Since these crystals are strongly alkaline and absorb water from the air (they are hygroscopic), they can attack the glass in the local areas where they have formed, leading to a pitted surface. The rate of weathering is directly related to the interdiffusion coefficients of hydronium and sodium ions,

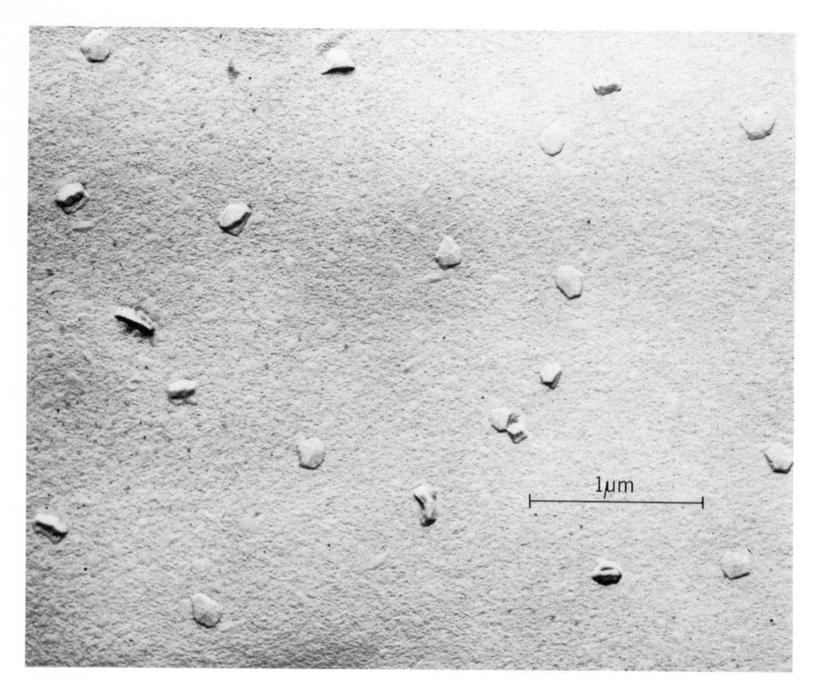

Fig. 4. Crystals on a soda-lime silicate glass surface, shown in an electronmicrograph of a replica of a fractured surface.

just as is the diffusion-controlled stage of reaction with liquid water, so that resistance to weathering and resistance to aqueous corrosion are closely related.

An example of weathering is shown in Fig. 4. A sample of soda-lime glass was fractured, and a replica of the fracture surface was examined in the electron microscope. Although the replica was taken only a short time after fracture, crystals formed on the glass surface as a result of the weathering reaction during this time.

Fig. 5. Scanning electron micrograph of solid material on the inside of a tiny glass microsphere as a result of reaction with water in the sphere. *(KMS Fusion, Inc., Ann Arbor, MI)*

The surface of a tiny glass shell shows another example of weathering in Fig. 5. Both the inside and the outside of this shell show surface deposits resulting from reaction of the glass with water.

The reaction of glass with furnace atmospheres containing sulfur improves the chemical durability of the glass. The furnace gases must also contain oxygen and water. The resulting ion exchange process is shown in Eq. (6). The sodium sulfate

$$2Na^+(g) + SO_2 + \tfrac{1}{2}O_2 + 3H_2O = 2H_3O^+ + Na_2SO_4 \quad (6)$$

crystallizes on the glass surface, but is not as alkaline as sodium carbonate or hydroxide and so does not attack the glass. It can be washed off at lower temperatures. Since the ion exchange of Eq. (6) involves no dissolution of the glass, a relatively thick layer containing H_3O^+ (or perhaps H^+) ions is formed on the glass surface, and the glass is quite durable at lower temperatures where diffusion is much slower.

pH AND DURABILITY

The previous discussion shows that the chemical durability of alkali silicate glasses in water can be understood well in terms of the model of ionic interdiffusion and dissolution of the glass. In the pH range from about 1 to 9, these are the dominant processes, with diffusion being the most important factor since it controls the rate of the initial states of reaction of the glass with water. In highly acidic and mildly alkaline solutions, dissolution becomes the important process, and the chemical stabilities of the various oxide components of a glass influence its durability at extremes of pH.

In neutral or acidic solutions the rate of dissolution of the silicate lattice in either fused silica or an alkali silicate glass is not much influenced by pH, because the dissolving species, un-ionized silicic acid, has nearly constant solubility. However, as the pH increases above about 9, the formation of silicate ions leads to higher effective solubility and a higher rate of dissolution of the silicate lattice.

Alkaline solutions. In the table the weight loss of several commercial glasses in sodium hydroxide is compared. The first four glasses have about the same rate of loss, in spite of their widely different compositions and much different rates of reaction at lower pH. The similarity shows that the rate of attack of these glasses is controlled by the dissolution process, not ion exchange and diffusion, and that the rate of dissolution is not much affected by the different constituents in these glasses, especially alkali and alkaline-earth oxides. The weight loss is linear with time, consistent with dissolution control.

Certain oxide components of silicate glasses have a specific influence on the rate of dissolution of the glass in alkali, as shown in the table. Boron oxide in Pyrex borosilicate glass does not much influence the rate of dissolution of the silicate lattice, but in the 7050 glass the higher amount of B_2O_3 leads to reduced alkali durability. Similarly, in 0010 glass a small amount of lead does not much influence durability, but the much larger amount in 8870 reduces alkaline durability.

Certain multivalent ions lead to precipitation of a solid silicate when added to a silicate solution. Among the most effective precipitating ions are zirconium, thorium, aluminum, beryllium, and zinc; transition-metal ions are somewhat less effective, but still can induce precipitation. Some of these same ions in solution have been reported to reduce the rate of alkali corrosion of silicate glasses. Beryllium seems to be particularly effective, with zinc less so. Zirconium in silicate glasses imparts substantial durability to alkaline attack. This effectiveness presumably is related to the strong resistance of zirconium oxide (ZrO_2) to alkaline attack, as compared with almost all other oxides. As shown in the table, aluminum also appears to reduce alkaline attack somewhat.

Addition of lanthanum oxide to silicate glasses increases resistance to alkaline attack. Lanthanum oxide has a high melting point (about 2300°C), and its hydroxide is precipitated from alkaline solutions. Perhaps these properties are related to the enhancement of alkaline durability by lanthanum oxide. Tin and chromium oxides have also been found to impart better alkaline resistance to silicate glasses.

Acid solutions. Concentrated acid solutions (above 1 mole/liter) attack soda-lime glass more rapidly than dilute acid or neutral solutions. There is an initial rapid attack and subsequent reduction in rate of attack to zero. The rate of this attack depends upon the acid concentration and the anion, with hydrochloric acid attacking more rapidly than sulfuric.

Just as for alkaline solutions, certain ions in the glass can lead to preferential attack by acid solutions. Glasses containing substantial amounts of boron, aluminum, or lead, such as 1720, 7050, and 8870 in the table, are much more rapidly attacked by acid (1 *N*) than soda-lime glass (0080). Pyrex borosilicate and vitreous silica glasses retain their improved durability in this concentration of acid, especially vitreous silica. The high solubilities of boron, aluminum, and lead oxides in acids apparently lead to their deleterious influence on acid durabilities of the glasses containing them.

Hydrofluoric acid dissolves silicate glasses rapidly, probably because of the formation of silicofluoride complexes in solution. It is also possible that fluoride ions react directly with the silicon-oxygen bonds in the silicate lattice.

Phosphoric acid attacks silicate glasses at temperatures above 200°C. After long treatment in the acid, the glass surface becomes coated with silicon phosphate crystals which protect the glass from hydrofluoric attack.

[ROBERT H. DOREMUS]

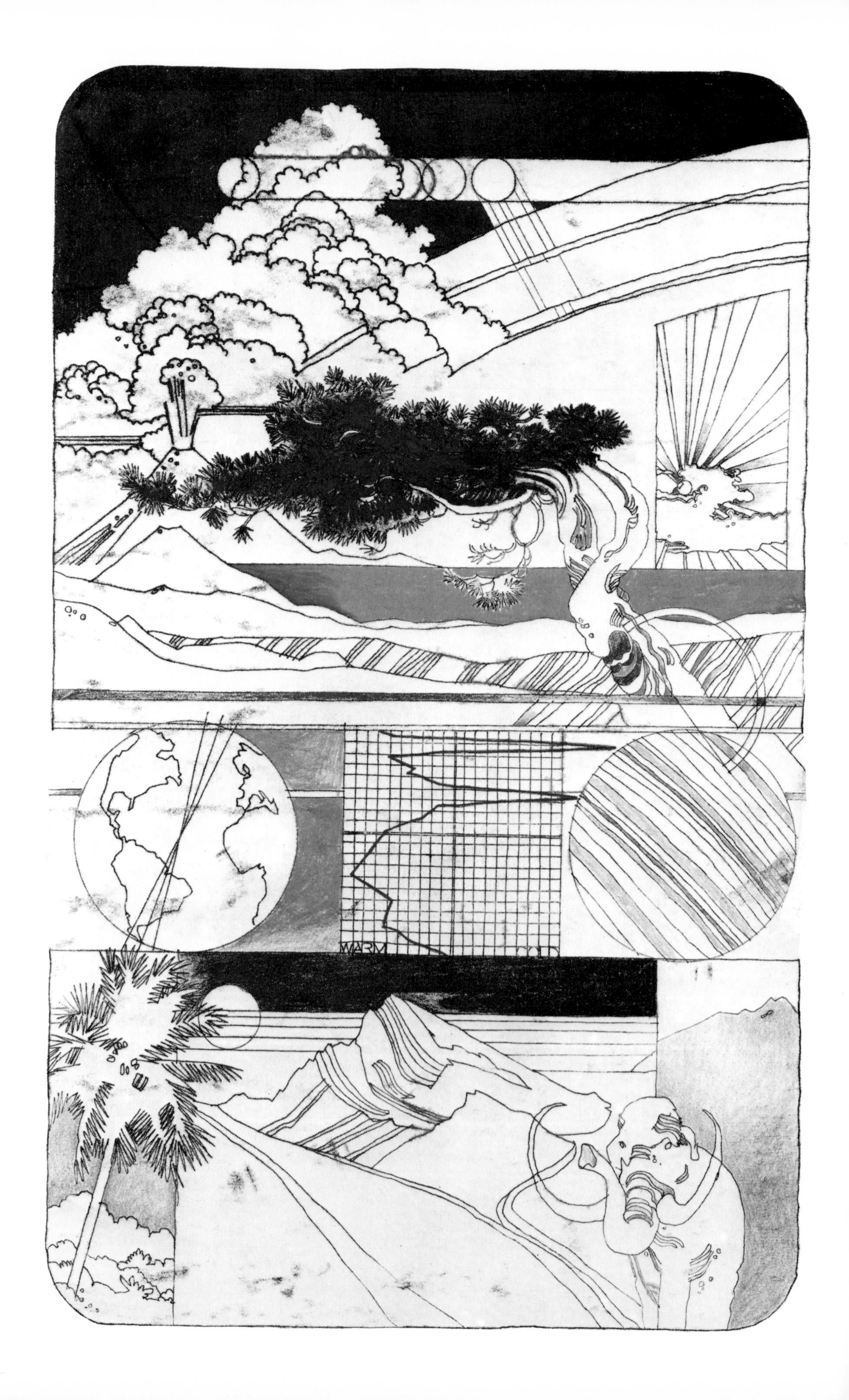
WARM

Climatic Fluctuations

Helmut E. Landsberg received his Ph.D. in geophysics and meteorology from the University of Frankfurt. In addition to his academic career, he held important positions in the U.S. Weather Bureau, the Environmental Science Services Administration, and the World Meteorological Organization. He is president of the American Institute of Medical Climatology and a member of the National Advisory Committee on Oceans and Atmosphere.

Climate is not a constant factor on either the global or local level. It fluctuates with time, and is necessarily a composite of the unending sequence of weather conditions. However, for convenience, climate is often viewed within a fixed time frame, usually several decades in duration. There is no scientific reason for choosing one particular time interval over another, although a period of 30 years is customary.

Within a 30-year period weather observations at a given location are summarized by statistical measures. Most common among these are the arithmetic mean, the median (the value which divides a series of observations so that half are smaller and half are larger), the range (the difference between the highest and lowest values), and the standard deviation (a statistical measure of dispersion of the values). These statistics can be applied to the observations of various meteorological variables, including temperature, precipitation, solar radiation, cloudiness, visibility, and wind direction and speed. As a rule, most attention is focused on temperature and

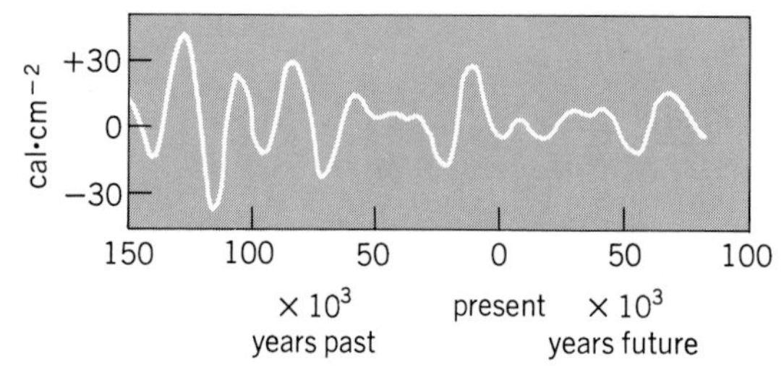

Fig. 1. Insolation changes at 45° latitude, in cal · cm^{-2} (cal/cm/cm; 1 cal = 4.1855 joules) due to cyclical elements of the Earth's position with respect to the Sun reconstructed for the past 150,000 years and projected for 75,000 years into the future. *(Adapted from W. S. Broecker and Jan van Donk, Insolation changes, ice volumes, and the O^{18} record in deep-sea cores, Rev. Geophys. Space Phys., 8:169–198, 1970)*

precipitation, which have the greatest impact on human activities. The mean (or median) values of these variables are often used immediately not only to describe a local climate but also to compare climates at different locations.

TEMPERATURE VARIATION

It makes little sense to discuss precipitation on a global scale, but temperature can indeed be considered an astronomical characteristic of the Earth in its exchange of energy with space.

Earth-space energy exchange. This energy exchange is the primary cause for climatic variations. The Sun is the principal energy source for the Earth. At the mean distance of the Earth from the Sun, at right angles to the solar rays, this energy is calculated at 1360 W/m² outside the Earth's atmosphere. It is not clear whether this has always been the Earth's energy income, or whether this level will remain constant in the future. Observations at the Earth's surface indicate that the equilibrium temperature value is now around 12°C (285 K). Calculations have shown that this value will rise about 5°C for each 1% increase in the solar energy income and drop a proportionate amount for each 1% decrease in solar radiation. A 5° increase in temperature would result in the spread of tropical conditions far polewards; a 5° decrease would precipitate an ice age with glaciation far into the middle latitudes.

Energy variation effect. The geological record indicates that both these phenomena have occurred in the past. There is no firm proof of long-term variation in solar output, although variable energy output of stars is the rule rather than the exception. However, there are examples of short-term changes in solar energy output, the magnitude of which will have to be quantified by measurements in space. There are many terrestrial phenomena which point to differences in climate from one era to another.

Splitting of continental plate. One such occurrence is the splitting of a single original continental plate into several smaller plates and their subsequent drifting apart. The position of these plates during their drift relative to the poles has had a profound influence on their respective climates; for example, those plates closer to the poles experienced glaciations at earlier ages.

Raising and erosion of mountains. Raising and subsequent erosion of tall mountain chains were often accompanied by significant volcanic eruptions. The impact of these events on climate is subject to debate. On the one hand, eruptions cause large masses of dust to rise to stratospheric heights. These are likely to scatter some of the incoming solar radiation back to space and hence may cause cooling. On the other hand, volcanic exhalations include carbon dioxide, which is an interceptor of outgoing long-wave radiation from the Earth, and therefore may result in warming.

Global temperature variations. There are some quantitative estimates of global temperatures for the past half-million years of the Earth's history. Much evidence has been obtained from the ratio of two isotopes of oxygen, $^{18}O/^{16}O$. Oxygen is incorporated into the shells of small one-celled sea animals, and the ratio of ^{18}O to ^{16}O reflects the sea-surface temperature at the time the animals lived, with warm eras indicated by less ^{18}O. After the animals die, the shells sink to the bottom to become part of the sediment. Thus deep-sea cores permit estimates to be made of the sequence of past climates. The time scales are relatively crude, and resolution is in hundreds or thousands of years. A plot of a time series obtained from Caribbean deep-sea cores shows fairly regular variations in temperature. The major dips can be identified with geologically established ice ages in the Northern Hemisphere, the rises with interglacials.

Earth-Sun angle position effect. The notable apparent periodicities in these records have been explained by the Yugoslavian astronomer M. Milankovich. He showed that three basic periodic elements of the motion and position of the Earth relative to the Sun (causing changes in exposure to solar radiation, especially at high latitudes) are involved: (1) the periodic change in obliquity of the Earth's axis with respect to the ecliptic (40,000 years); (2) the change in obliquity of the Earth's path around the Sun (92,000 years); and (3) the precession of the spring point (21,000 years). The precession of the spring point governs the seasonal variation in the distance of the Earth from the Sun. At present the Earth is closest to the Sun (perihelion) during the Northern Hemisphere winter. The calculated radiation values from these path elements over the past few hundred thousand years show time sequences similar to the data recovered from sea-sediment cores (Fig. 1).

The likelihood that these periodic changes in radiation exposure of the Earth will continue in the

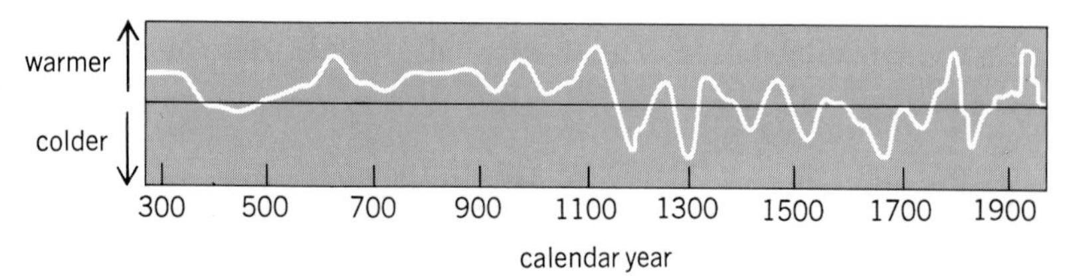

Fig. 2. Oxygen isotope ratio from ice cores of the Greenland ice cap, reflecting temperature changes over the past 1700 years. *(Adapted from W. Dansgaard et al., One thousand centuries record from Camp Century on the Greenland Ice Sheet, U.S. Army Cold Regions Eng. Lab. Tech. Note, 1969)*

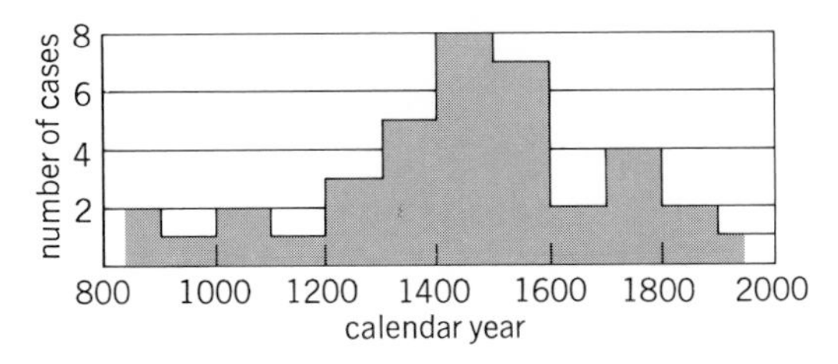

Fig. 3. Number of cases in which upper Lake Constance was frozen in the past 1100 years, in each century. *(From data published by V. W. Steinjans, A stochastic point-process model for the occurrence of major freezes in Lake Constance, Appl. Stat., 25:58–61, 1976)*

future has led many climatologists to assume that the present interglacial will end and that another ice age is imminent. The most recent calculations by John Imbrie predict that this will occur thousands of years from now. Extrapolating these path elements into the immediate future indicates less radical changes in the next 50,000 years than have occurred in the last 150,000 (Fig. 1).

Interstellar cloud effect. Other mechanisms for interference with the radiation budget of the Earth have been suggested. One hypothesis suggests that the Earth on occasion encounters interstellar clouds with a high density of hydrogen molecules ($>10^3$ cm^3). This would expose the Earth to a high influx of this interstellar material for several thousand years. Photochemical interactions in the higher atmosphere would effect an increase in water vapor at the 80-km level, leading to cloud formation. The clouds would increase the planetary albedo (reflectivity) and cause cooling, because less solar radiation would traverse the atmosphere. Such a sequence of events is thought to initiate an ice age. It has been estimated that such encounters with interstellar hydrogen clouds have occurred a half-dozen times in the Earth's history due to the movement of the solar system through the spiral arms of the Galaxy.

Change in last 11,000 years. Since the end of the last ice age about 11,000 years ago, there have been many climatic variations of shorter durations. These have been determined from analyses of pollen deposits in lake sediments. These analyses show whether heat- or cold-adapted plants were present at a particular time period. Approximate dating of the warm and cool episodes is possible by ^{14}C analysis. This radioactive isotope of carbon, incorporated into the organic substances by its decay (half-life of 5730 years) indicates the time interval since their deposition. The pollen analysis shows that there was a very warm era, the hypsithermal, about 5000 years before present, the warmest interval since the last ice age.

More specific information about climate since that time has been derived from historical records and tree rings. Tree rings are widely used as so-called proxy data to extend the instrumental record. In cold regions tree rings reflect variations in temperatures, while in dry regions tree growth responds more to precipitation fluctuations. Values for earlier years are derived by correlating temperature or precipitation values observed at the site with the annual growth increments shown by the tree rings. Dating by means of this procedure is highly reliable.

All these data indicate that there have been only relatively small climatic variations since the beginning of the Roman Empire. Yet the time of the Viking explorations of the North Atlantic must have been relatively warm because of the favorable conditions for settlement in Iceland and Greenland. There are also indications for cooler conditions in the 16th century, especially in western Europe.

North Atlantic area. Some of the conditions in Greenland are reflected by the $^{18}O/^{16}O$ ratios obtained from ice cores of the Greenland ice cap (Fig. 2). These isotope ratios represent the temperatures of the precipitated water that became incorporated into the ice. (The dating is most accurate for whole decades.) The warmer and colder intervals are probably representative of the North Atlantic area. These ratios indicate the occurrence of cold stages in the 15th and 17th centuries. The 15th-century stage coincides with freezes of central European lakes (Fig. 3), while the 17th-century stage is coincident with a decrease in solar activity (Maunder minimum of sunspots). Yet tree ring analyses for northeast Canada by Harold Fritts show no unusually prolonged cold spells in the second half of the 17th century (Fig. 4). On the other hand, the Greenland cores do not reflect cold conditions in the 19th century, which are clearly shown in the northeast Canadian tree rings and by the instrumental records at many points in the Northern Hemisphere.

Instrumental observations. Systematic meteorological observations yielding coherent data series did not become possible until the end of the 17th century. One such long-term study is the homogeneous time series derived for central England temperatures of 1659–1973 by Gordon Manley (Fig. 5). A noteworthy feature of this and similar graphic representations is their unrest, which is termed a climatic "noise" pattern. The variability between adjacent years is high. Equally striking in

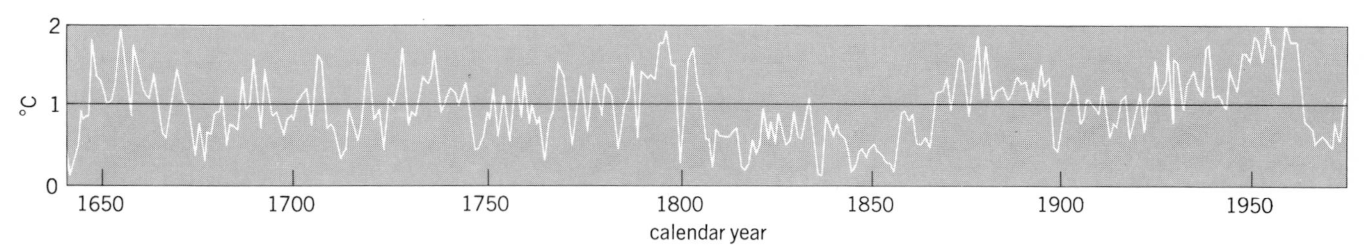

Fig. 4. Estimate of temperature fluctuations based on ring measurements of trees growing near the tree line at Fort Chimo, Quebec, Canada, from 1640 to 1973. *(Adapted from H. C. Fritts, A record of climate past, Environ. Data Serv., pp. 3–10, July 1977)*

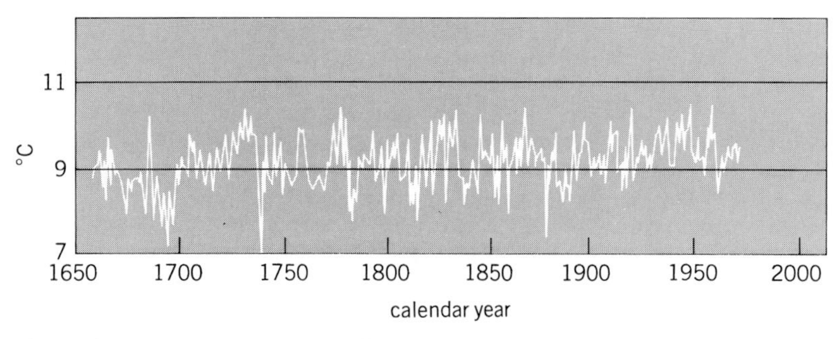

Fig. 5. Annual mean temperatures for central England, 1659–1973. *(From data published by G. Manley, Central England temperatures, 1959–1973, Quart. J. Roy. Meteorol. Soc., 100:389–405, 1974)*

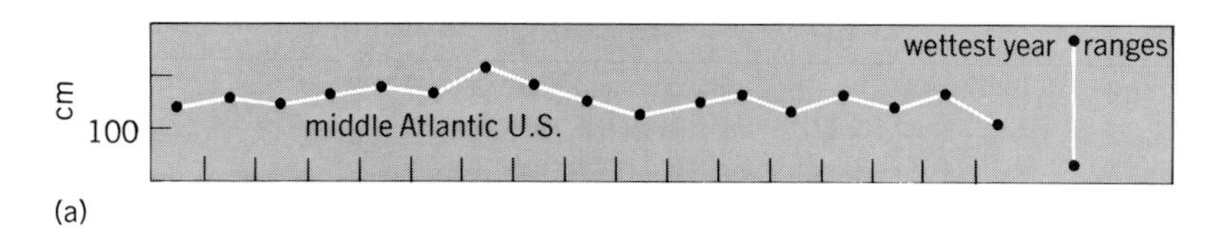

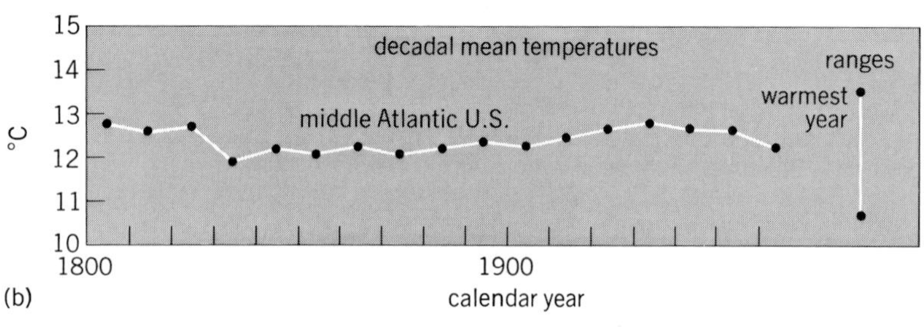

Fig. 6. Graph of (*a*) average total precipitation and (*b*) average temperatures for 10-year intervals, 1800–1970; for the eastern seaboard of the United States, centered at Philadelphia.

all of the long observational series is the absence of sustained trends. If the temperature observations are averaged for 10- and 30-year intervals, the noise is suppressed and one can discern measurable but apparently relatively small differences between the decadal and longer mean values (Fig. 6).

All such instrumental observations from the 17th to the 20th century indicate the same broad trends. (However, in the earlier years they are markedly biased in favor of observations from central and western Europe and eastern North America.) After the cool period in the 16th century there was a gradual but discontinuous temperature rise to the end of the 18th century, followed by a cool period in the 19th century. The first half of the 20th century was a time of rising temperatures. This was followed by a cooling trend to the end of the 1960s, which leveled off in the 1970s, at least in the Northern Hemisphere.

Little Ice Age. A number of scientists have given the name "Little Ice Age" to the entire period from the 16th through the 19th century. However, it might be better to restrict this term to the middle of the 19th century during which, in addition to relatively low temperatures, advances of glaciers in the central European mountains are well documented. This cold period is also clearly indicated in the northeast Canadian tree rings (Fig. 4).

Standardized networks. The first standardized network of meteorological observations was begun in 1781 by J. J. Hemmer of Mannheim; some of the stations established at that time are still operative. Meteorological networks have grown rapidly since the middle of the 19th century; the greatest impetus came from the founding in 1873 of the International Meteorological Organization (now the World Meteorological Organization), which established standards of observational practice.

There are now many locations from which a century of observations is available. Several scientists have used these data to derive a composite estimate of global temperatures. However, the reliability of such estimates is still limited by the lack of information from oceanic areas. This is particularly true for the Southern Hemisphere, where oceans predominate. In addition, for much of the polar regions less than 30 years of reliable data are available. Figure 7 gives estimates of climatic fluctuations, as departures from a mean, for most of the Northern Hemisphere from 1881 to 1975, as compiled by a team of Soviet scientists. Perhaps most remarkable is the indication that, on a hemispheric scale, the total range of temperature fluctuations is only 0.8°C.

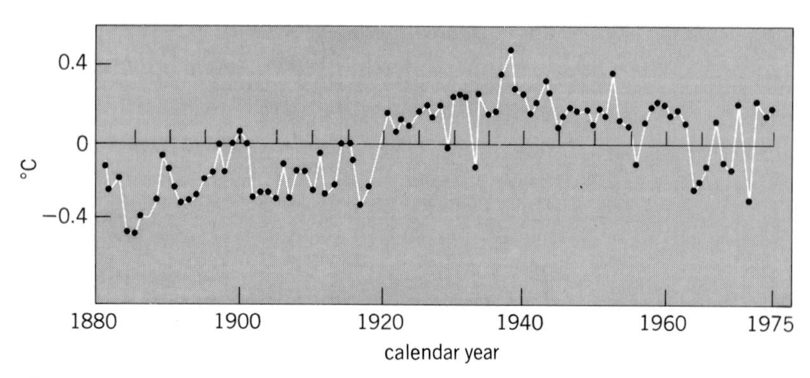

Fig. 7. Departures from average of annual temperatures for the Northern Hemisphere, latitudes 17.5 to 87.5°, 1881 to 1975. *(After I. I. Borsenkova et al., Izmenenie temperatury vosducha Severnogo Polyshariya za period 1881–1975 gg., Meteorol. Gidrol., no. 7, pp. 27–35, 1976)*

Sea-level changes. Another useful indicator of climatic fluctuations is a change in sea level. During periods of glaciation vast amounts of terrestrial water are trapped in ice masses; during periods of melting, sea level rises. Determination of sea levels for the ice ages can be made by examining animal and plant material that originated in a coastal zone, now submerged. Periods when the sea is receding, indicating increases in ice masses, can be determined from beach terraces. Uncertainty is introduced by independent vertical tectonic motions of the coastal crusts in various localities. For the last glacial period it has been inferred that there was a general sea-level depression of close to 125 m, about 15,000 years ago. At present there is still a little over 2% of all terrestrial water, about 2.9×10^7 km^3, locked up as ice.

As the climate fluctuates, some of this ice melts or more is added. This is reflected in year-to-year sea-level fluctuations. These have been watched by many tide-recording stations for decades. A recent evaluation by Steacy Hicks for the United States coastlines (except Alaska and Hawaii) shows a rise of 64 mm from 1940 to 1975 (Fig. 8).

By using diverse coastal sectors the influence of tectonic coastal movements is minimized. Sea-level fluctuations are large and show the integrated climatic effects of temperatures and of snowfall in the glaciated areas. On the whole, these observations point to a continuing decline in ice volume.

Snow cover. For the past dozen years surveillance of hemispheric snow covers has been possible. Over North America the areas covered by snow have been fairly constant, but over Eurasia fluctuations in areas covered by winter snow have been quite large (Fig. 9). However, there is no indication of a one-sided trend.

Latitude and temperature change. Temperature fluctuations are very small in the tropics, in contrast to the polar regions in which fluctuations are great. This is of great practical importance because in the warmer periods previously marginal lands can be brought under cultivation; alternatively, when a cooling trend occurs productivity is radically reduced. For example, in the 1960s forage crops in Iceland and Finland became inadequate for herds of sheep and reindeer, respectively. In fact, fluctuations near the Arctic Circle were four to ten times greater than in the subtropics.

PRECIPITATION

Fluctuations in precipitation are even more unpredictable than variations in temperature.

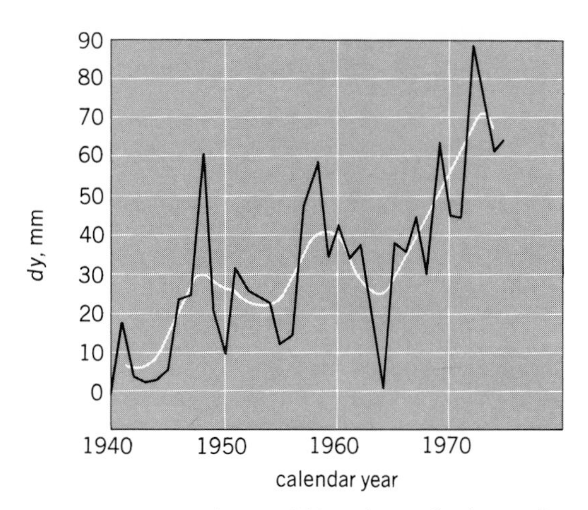

Fig. 8. Sea-level changes (*dy*) and smoothed curve for the United States coasts (except Alaska and Hawaii), 1940–1975. *(From S. D. Hicks, An average geopotential sea level series for the United States, J. Geophys. Res., 83:1377–1379, 1978)*

Estimates of variation. Sampling methods for catching rain and snow in cans are subject to far greater errors than temperature measurements. Variations in time and place are also considerable, so that longer intervals of time are necessary to permit comparisons. Mean values are generally

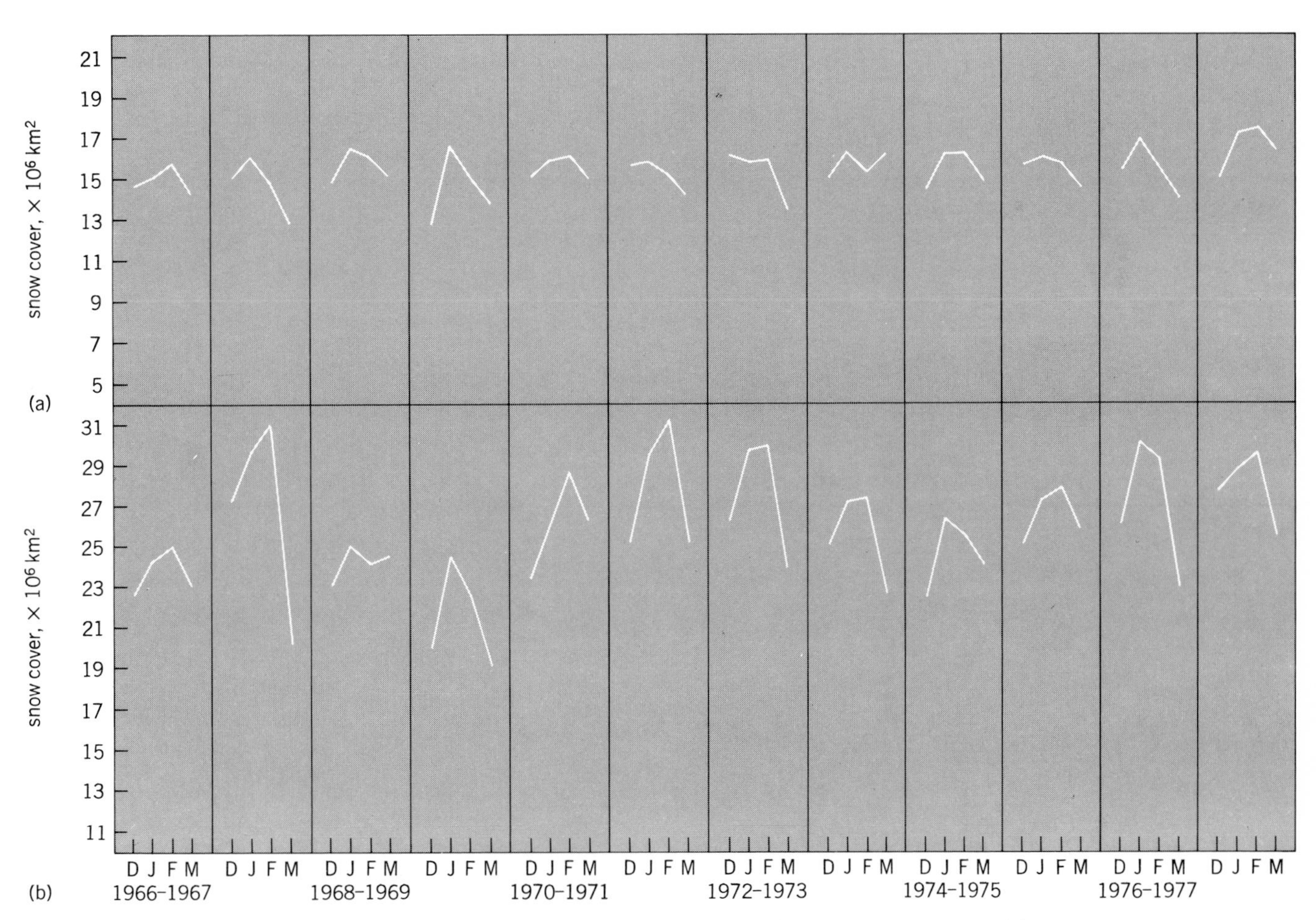

Fig. 9. Winter (December–March) snow cover over land areas of the Northern Hemisphere, as obtained from satellite observations. *(a)* North America. *(b)* Eurasia. *(Courtesy of Michael Matson, National Environmental Satellite Service, National Oceanic and Atmospheric Agency)*

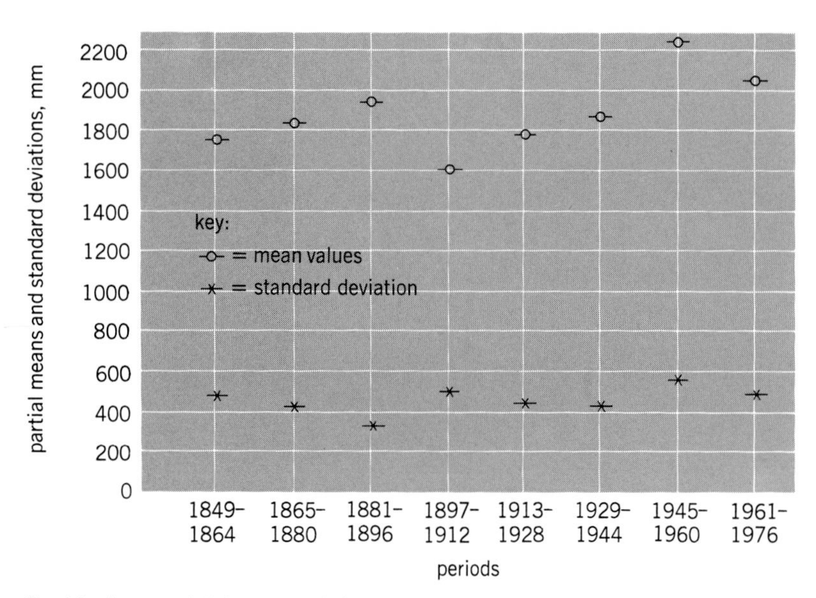

Fig. 10. Average total annual rainfall values and their standard deviations for eight consecutive 16-year intervals, 1849–1976, for Bombay, India.

not the best measure for comparing precipitation observations. Because the frequency distributions of rainfall are often skewed, the median is a better value for comparison. In numerous regions of the world many years with small rainfall amounts are balanced by a few years with large amounts. Standard deviations of the time series of precipitation observations are usually large. The variability from year to year generally increases as the amount of annual rainfall decreases. Thus the fringe areas of deserts show the highest variability of rainfall. In these semiarid regions there are occasionally years of appreciable precipitation. This often leads to more intense agricultural development than is warranted by the climate, resulting in severe problems when the wetter period ends. The drought years of 1973–1975 in the area just south of the Sahara, the Sahel, were a typical case.

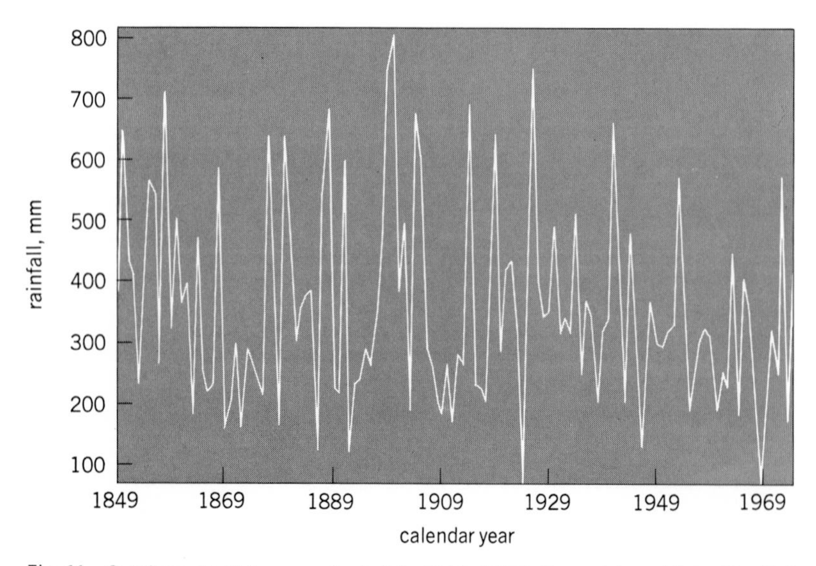

Fig. 11. Santiago de Chile annual rainfall, 1849–1975. *(From data published by E. P. Heilmaier, Periodizitäten in den jährlichen Niederschlagshöhen Mittel-Chiles und deren Beziehung zur magnetischen Sonnenaktivität, Wetter und Leben, 28:243–251, 1976)*

Drought. Drought is more common in areas of high variability of precipitation, but occurs occasionally even in regions with usually reliable precipitation. There is no specific definition of drought; the term usually refers to an inadequacy of water for the customary activities of a region, such as farming, hydroelectric power production, or other consumption. The ramifications of drought often depend on regional patterns of precipitation.

West Coast drought. In the drought of 1976–1977 on the West Coast, the usual winter precipitation was lacking and the snow pack was sparse. The reduction in runoff affected both power production and farm irrigation. This placed heavy burdens on groundwater supplies and required drastic measures of curtailment of domestic water use. This drought, as is not uncommon, was relieved by a season of excessive rainfall in 1977–78.

Southern and eastern Asia. Focal areas in terms of precipitation are the densely populated regions of southern and eastern Asia. The success or failure of crops depends on moisture-laden winds (monsoons) from the Indian and East Pacific oceans which deposit copious rains on the continent. They are a feature of the general air circulation of the atmosphere. As such, they are intimately tied to a characteristic of the tropical airflow which results in a clash between the trade winds of both hemispheres. There are well over 100 years of data at a number of localities in India. The statistical analysis of the rainfall history reveals no striking long-term anomalies in the rainfall regimes, and standard deviations are fairly constant (Fig. 10). There have been years of monsoonal failure in both the 19th and 20th centuries, followed by a return of the rains in an as yet unexplained sequence.

Great Plains. At the present time much attention is also being focused on the few countries with potential crop surpluses; that is, Australia and Argentina in the Southern Hemisphere, and the United States and Canada in the Northern Hemisphere. The Great Plains, east of the Rocky Mountains, and the Midwest comprise the grain center of the American continent. In the relatively drier western portion wheat cultivation dominates; to the south and east maize (corn) prevails. This area has seen disastrous droughts. The worst to date occurred in 1934–1936 when, aided by land abuse, whole states were turned into semideserts. Only strenuous rehabilitation measures reclaimed much of the land for production. Irrigation using accumulated pond and ground water has sustained the area during occasional rain deficiencies. Yet the climatic history of the region indicates that major droughts are recurrent features. In addition, tree ring analysis by J. Murray Mitchell, Jr., and Charles Stockton have also revealed periodic droughts in the area. Their data and the instrumental observations have revealed a 22-year cycle for these droughts.

Hale cycle. The determination of this 22-year drought cycle has spurred climatologists to determine the causes for this and other climatic fluctuations. The length of the drought rhythm corresponds to the double sunspot cycle (Hale cycle), in which there is a reversal of the magnetic field of the sunspots. Other climatic elements reveal rhythmic elements corresponding to the sunspot cycle of about 11 years in length. The term "cycle" is really misapplied because there is a variable length of the interval between maxima and minima of sunspots. However, sunspots are not dominant in the fluctuations of climate.

This is best illustrated by mathematical analysis of meteorological time series. The analysis is designed to resolve the series into a number of harmonic components or to correlate segments of equal time intervals with each other. Such analyses can be accomplished by computers. They reveal the contribution of any chosen time interval to the total variance in such a time series. The intervals are arbitrarily fixed but, taken together, they account for all the departures or variances from the average value of the series.

An example will reveal the usefulness of this procedure. Figure 11 shows the annual rainfall in Santiago de Chile for the interval from 1849 to 1974. The variability from 800 mm in 1899 to zero in 1923 is impressive. The series also exhibits a lack of trend. When the data are subjected to analysis for hidden periodicities, one finds that most of the variance is contributed by short-term fluctuations between 2 and 5 years in length (Fig. 12*a*). The causes for these fluctuations have not yet been explained. In many meteorological elements at the surface and even in the upper winds, a fluctuation just slightly over 2 years in length predominates. This "quasibiennial" oscillation can be seen in tree rings and lake sedimentation. It is responsible for the often dissimilar weather character of two successive years. The Santiago rainfall accounts for about 8% of the total variability in the 9- to 13-year interval of the sunspot rhythm. But if the coherence to the sunspots is tested, one finds a very large apparent influence (Fig. 12*b*).

Satellite data. These solar-terrestrial relations are just beginning to be explored by satellites. The influence of changing corpuscular particle flux from the Sun on the outer reaches of the atmosphere is great, but the influence on lower levels of the atmosphere has yet to be determined. Two possibilities are suggested: (1) that there is a difference in total energy output from the Sun between the active periods of high sunspot numbers and the quiescent intervals with few or no sunspots; and (2) that the solar particle injections directly or indirectly affect cloud formation and perhaps stimulate precipitation.

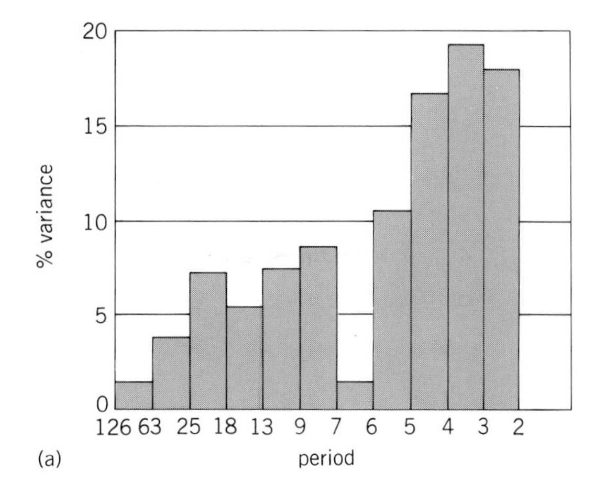

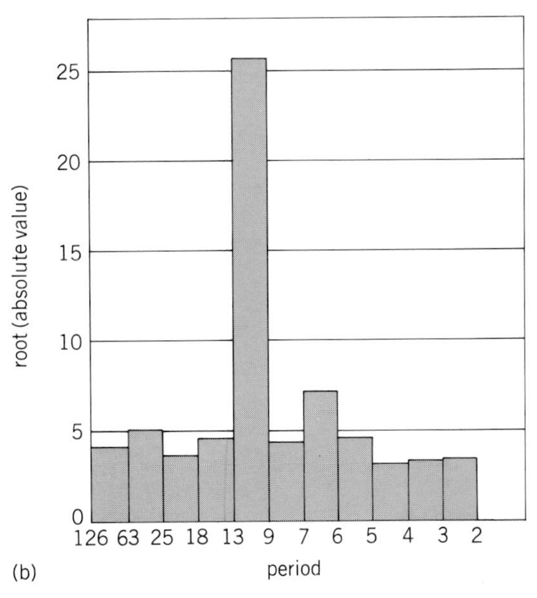

Fig. 12. Analysis of Santiago de Chile annual rainfall data shown in Fig. 11. (*a*) Contributions to the variance of various time intervals (power spectrum) by spectral analysis using fast Fourier transform. Note the large contribution of the relatively short time intervals. (*b*) Coherence of the annual rainfall with sunspot numbers (sunspot co-spectrum). Note the strong "signal" in the 9- to 13-year interval indicating a hidden periodicity of the same length in the rainfall data, coincident with the sunspot rhythm.

EFFECT OF HUMAN ACTIVITY

Possible alterations of climate due to human activity are well established locally.

Urban heat island. One result of human influence is the so-called urban heat island which, because of the modification of the surface and the heat rejection from human activities, raises urban temperatures, on an average, by 1–2°C. On the evenings of calm, clear days temperature differences between urban centers and the surrounding rural areas can reach from 6 to 10°C. In a few very densely settled areas, such as New York City, heat rejection can reach or exceed the energy locally received from the Sun, especially in winter. These effects, however, are strictly local; on a continental or global scale they are a negligible influence on climate.

The urban heat island is also responsible for a local increase in precipitation, generally somewhat downwind from the urban center. This increase is most noticeable in summer, when convective cloud systems and local thunderstorms are affected, and can reach 10 to 15% annually.

Pollutants. Other changes in climate have been ascribed to the effects of pollutants. Aerosol or human-caused dust content have been cited as primary among these. Their thermal effects appear to be minimal, and their effect on precipitation has not been fully established except in a few isolated cases. Natural dust sources still appear to have a far greater effect on climate.

Among pollutants carbon dioxide (CO_2) has been singled out as having a potentially disruptive influence on climate. This gas has been steadily increasing in the Earth's atmosphere, in proportion to the increase in large-scale use of fossil fuels. Carbon dioxide is an interceptor of outgoing terrestrial long-wave radiation. Its effect is being viewed as a cause for warming, because the Earth loses less heat. Since 1860 there has been about a 10% increase in atmospheric CO_2, from 290 to 320 parts per million. According to a crude model calculation, this would lead to a 0.2°C global temperature increase. Such an increase would readily disappear in the large natural fluctuations. However, continually accelerated use of fossil fuels might lead to a more pronounced global warming if no other factors in the climatic energy balance interfere. [HELMUT E. LANDSBERG]

Photographic Highlights

These photographs have been chosen for their scientific value and current relevance. Many result from advances in photographic and optical techniques as humans extend their sensory awareness with the aid of the machine, and others are records of important natural phenomena and recent scientific discoveries.

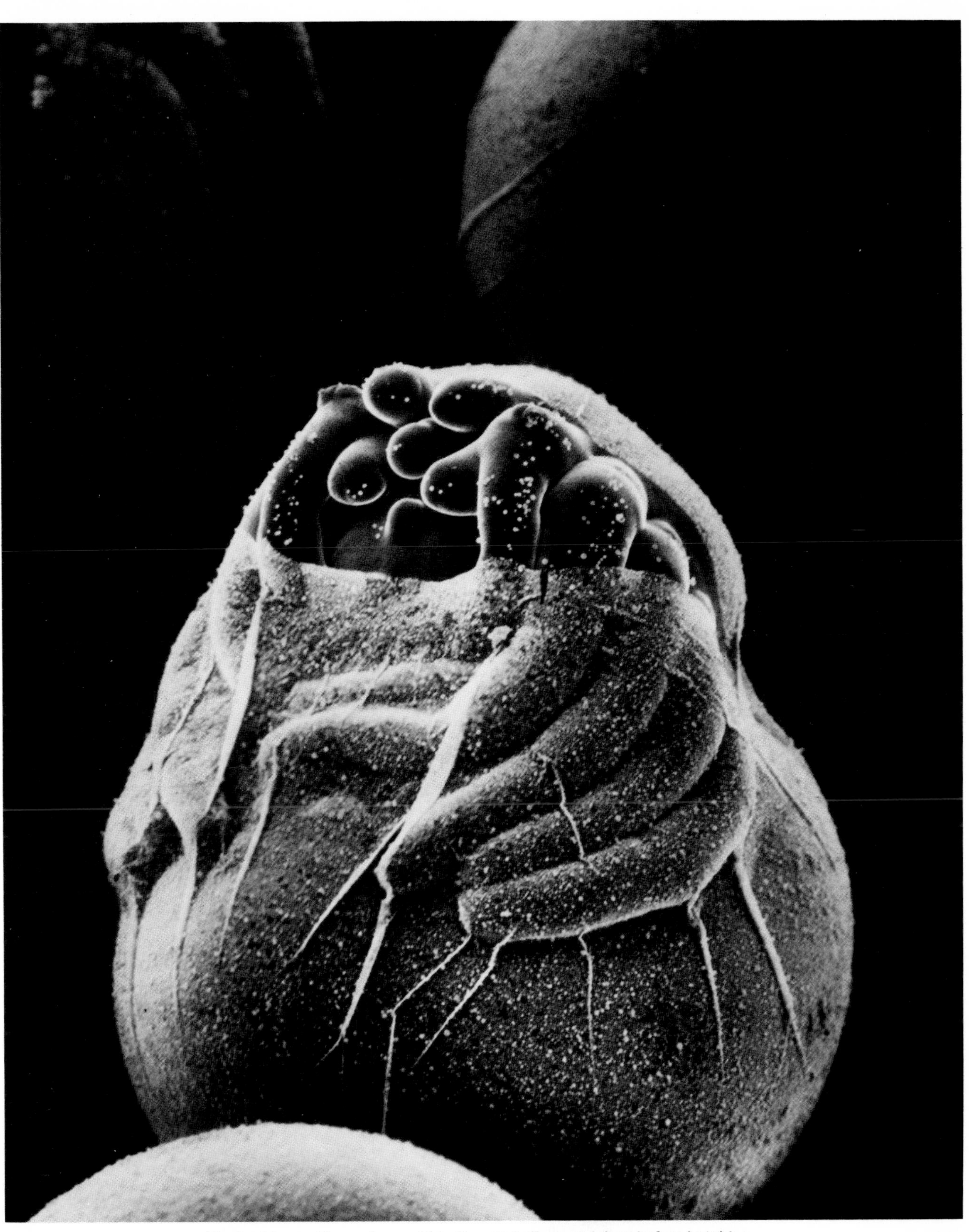

Black widow spider embryo within its eggshell several days before hatching. The four left legs are apparent as well as primitive mouthparts (top center). The shell typically cracks at this stage due to uneven body growth.
(Scanning electron microscope view by J. Norman Grim, Biology Department, Northern Arizona University)

Ancient Metallurgical Skills

Iron implements from ruins in the Near East dating from late in the 2d millennium B.C. to late in the 1st millennium B.C.

From R. Maddin, J. D. Muhly, and T. S. Wheeler,
How the Iron Age began,
Sci. Amer., 237(4):122-131;

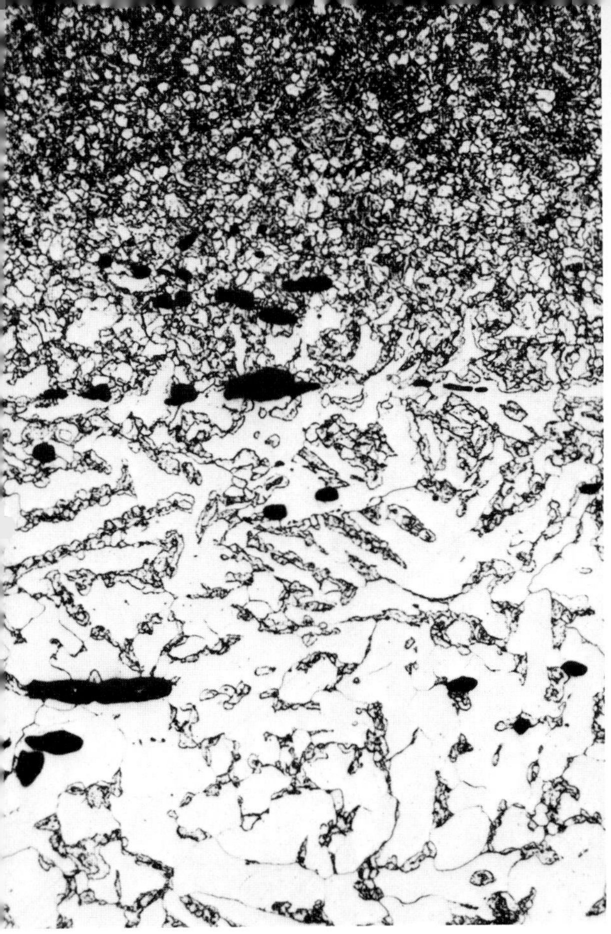

Section of an adze blade (4th century B.C.) from Al Mina on the coast of Turkey shows masterful skill in using fine-grained carburized iron for one face and coarse-grained uncarburized iron for the other.

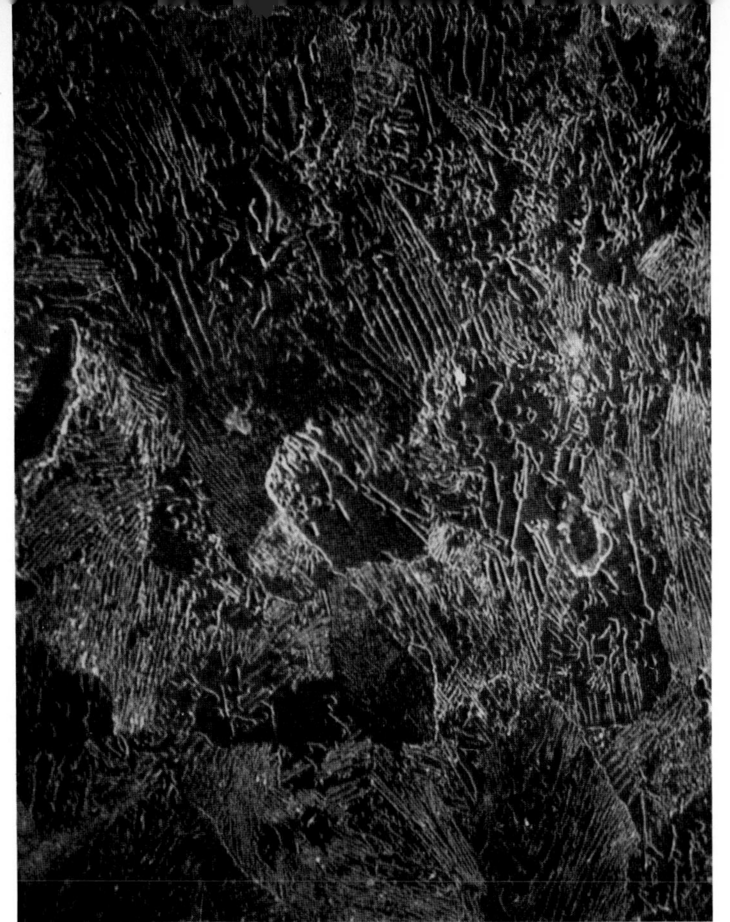

Pearlite microstructure characteristic of carbon steels is shown in this micrograph of a steeled-iron blade from Tel Fara South in Israel.

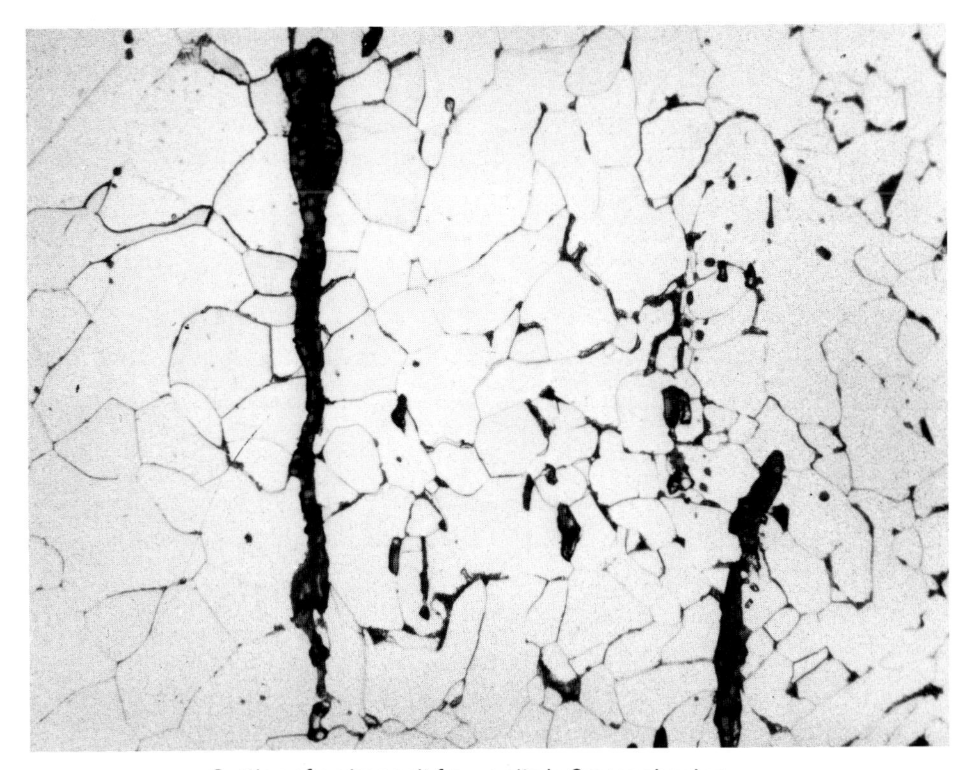

Section of an iron spit from a site in Greece showing stringers (narrow dark areas), which are slag fragments not expelled during hammering before forging.

Crustacean Digestive Tracts Devoid of Microorganisms

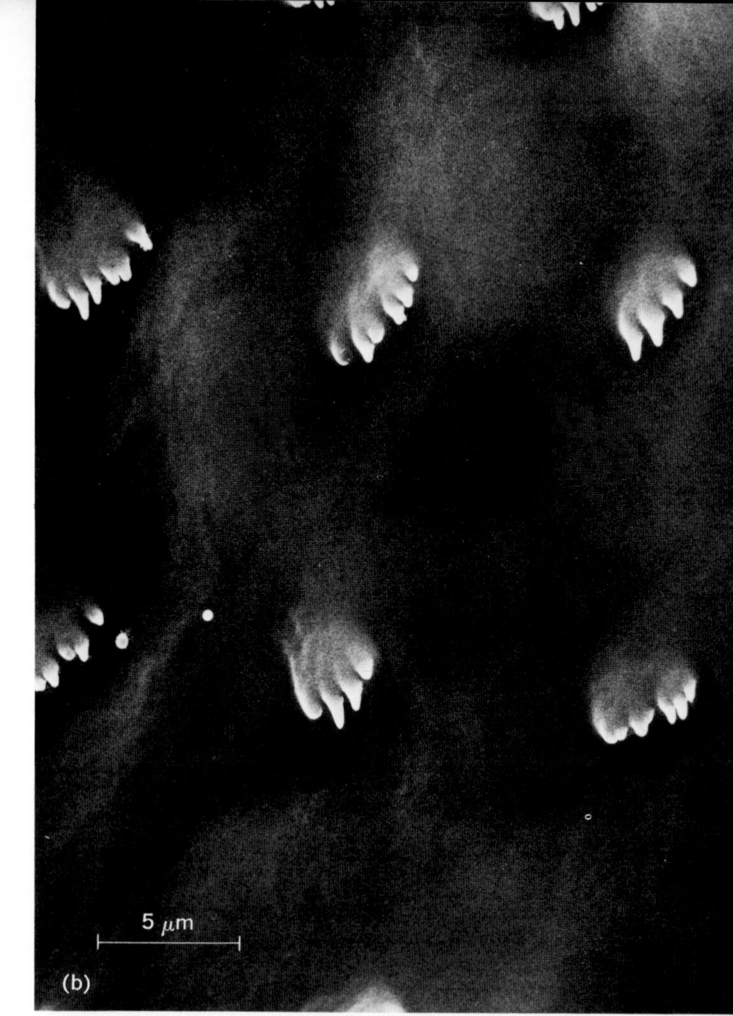

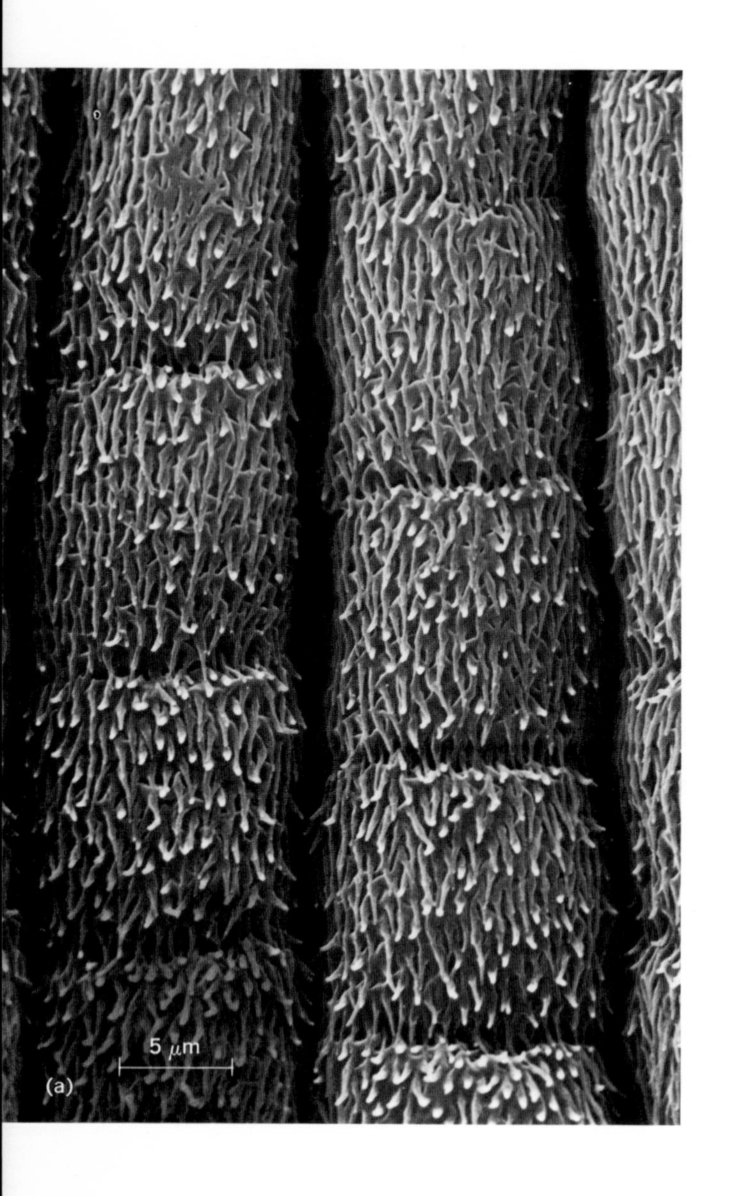

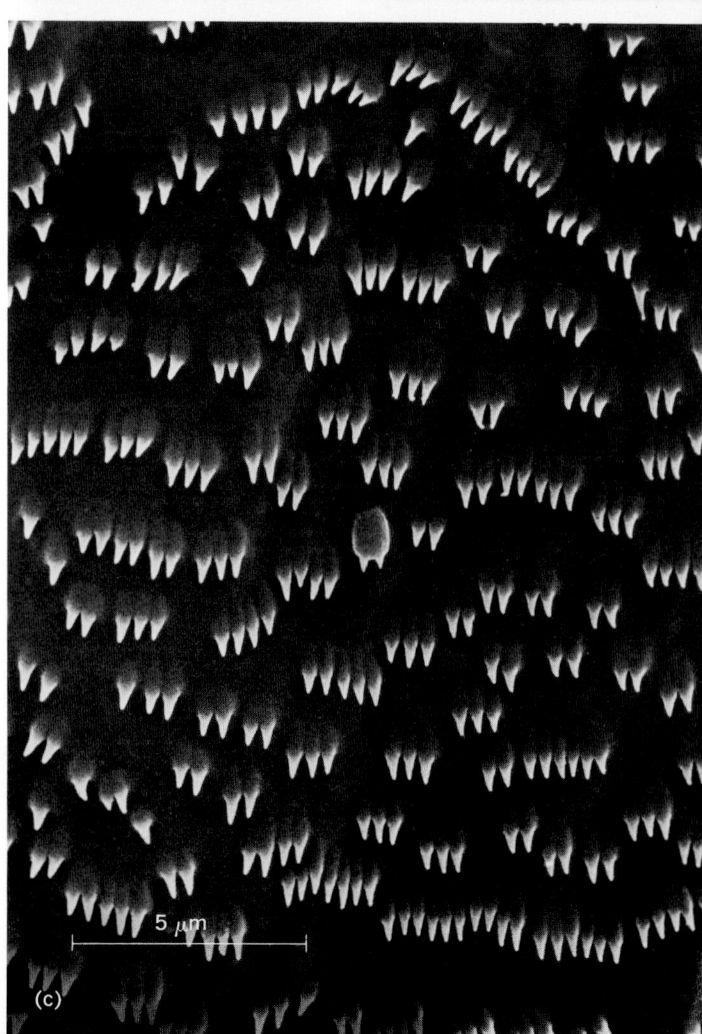

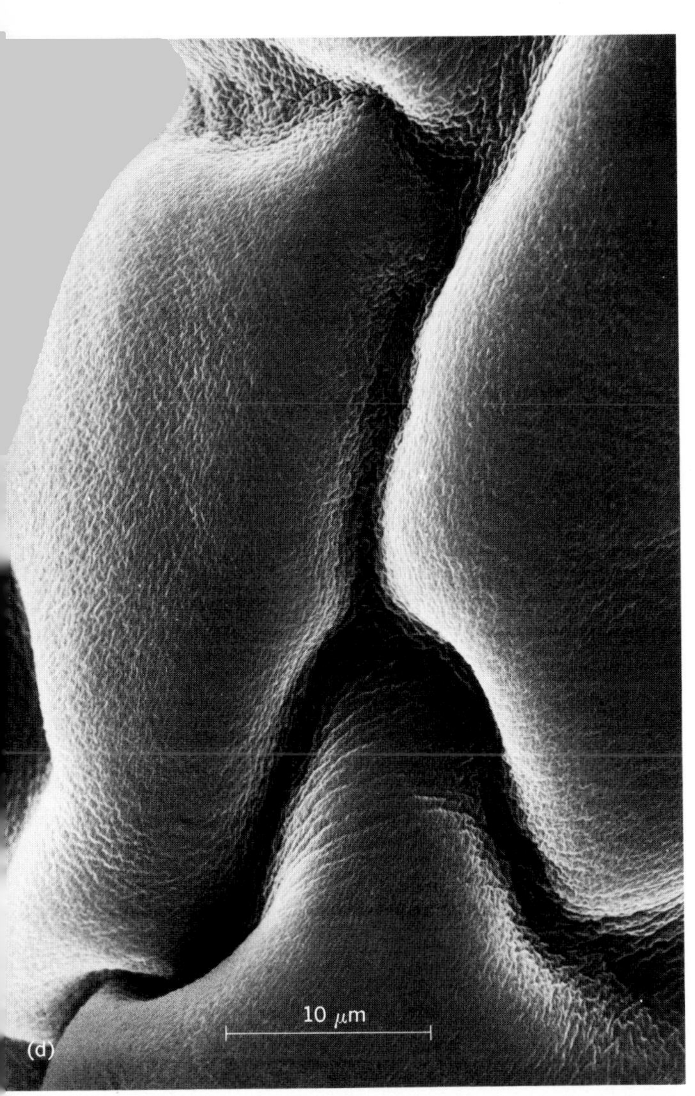

Absence of indigenous microflora is demonstrated in the gut of *(a) Limnoria tripunctata, (b) L. lignorum, (c) Chelura terebrans,* and *(d) Oniscus asellus.* This is a clear exception to the dense bacteria shown on *(e)* the exoskeleton of the crustacean *L. lignorum* and *(f)* the hindgut of the termite *Reticulitermes flavipes.*

From P. J. Boyle and R. Mitchell, Absence of microorganisms in crustacean digestive tracts, Science, 200(4346):1157-1159;

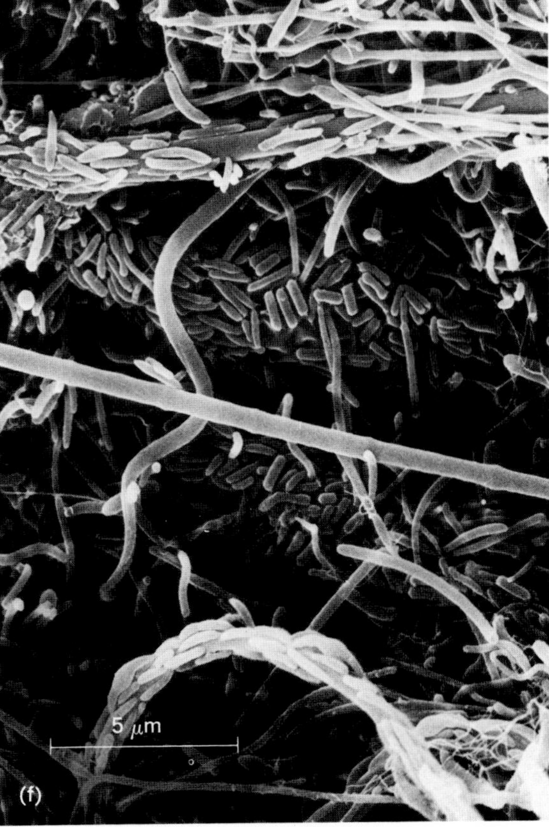

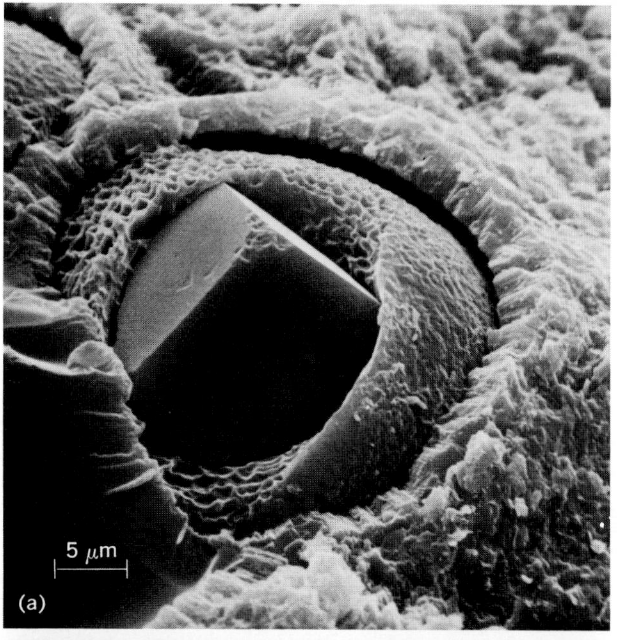

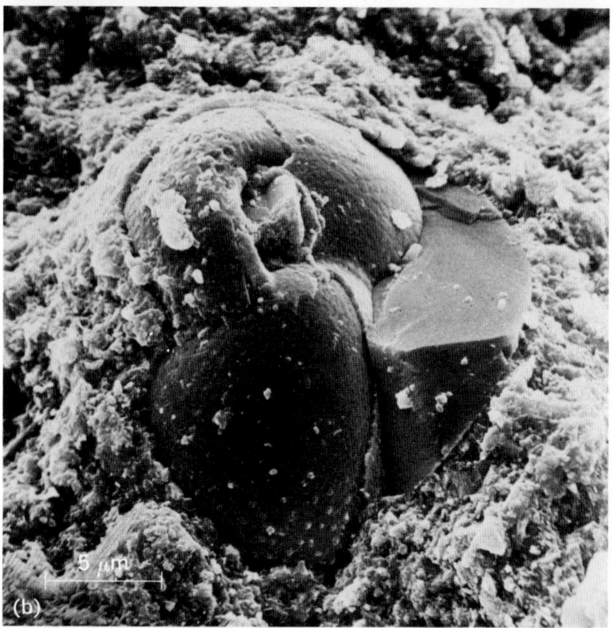

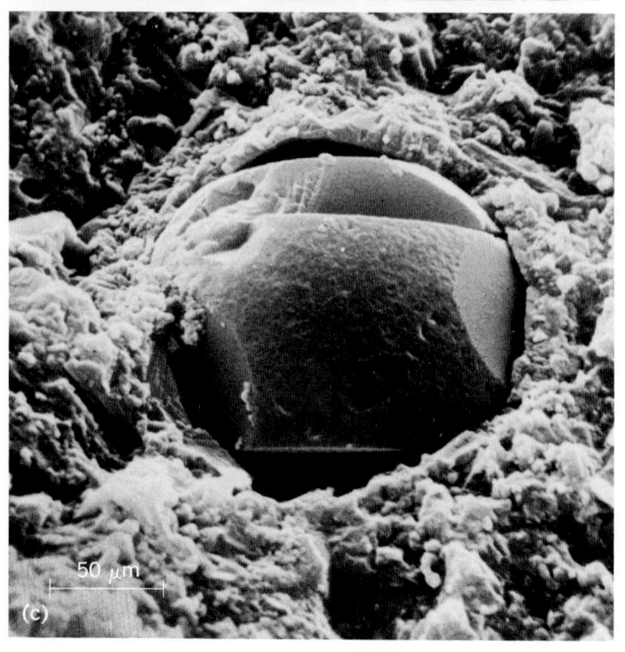

Stages of calcite replacement of gypsum in tests of foraminifera reflecting a set of environmental conditions during the formation of organic-rich sediments in a hypersaline lagoon in Maastrichtian times (about 70,000,000 years ago). *(a)* Calcite crystal in center of chamber filled with gypsum in state of dissolution. *(b)* Chamber partially filled with crystal. *(c)* Test completely filled by single crystal. *(From B. Spiro, Bacterial sulphate reduction and calcite precipitation in hypersaline deposition of bituminous shales, Nature, 269(5625):235-237, 1977)*

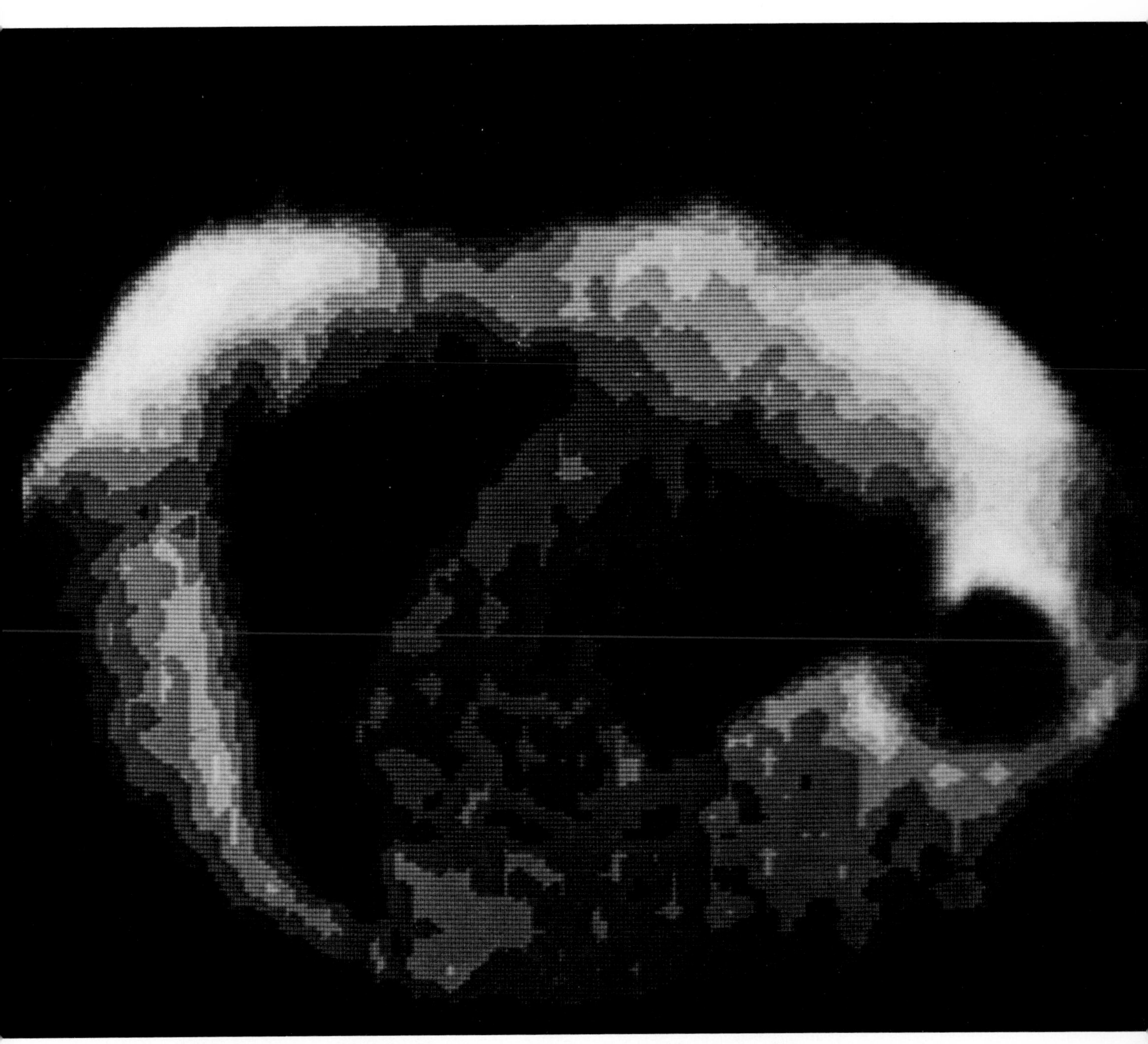

Nuclear magnetic resonance image of a slice of calf heart 18 mm thick. The image was produced by Paul Lauterbur and his collaborators (State University of New York, Stony Brook) from 65 different field-gradient orientations. The experiment was done at a frequency of 4 MHz.
(From NMR imaging technique provides high resolution, Phys. Today, p. 17, May 1978; copyright © 1978 by American Institute of Physics)

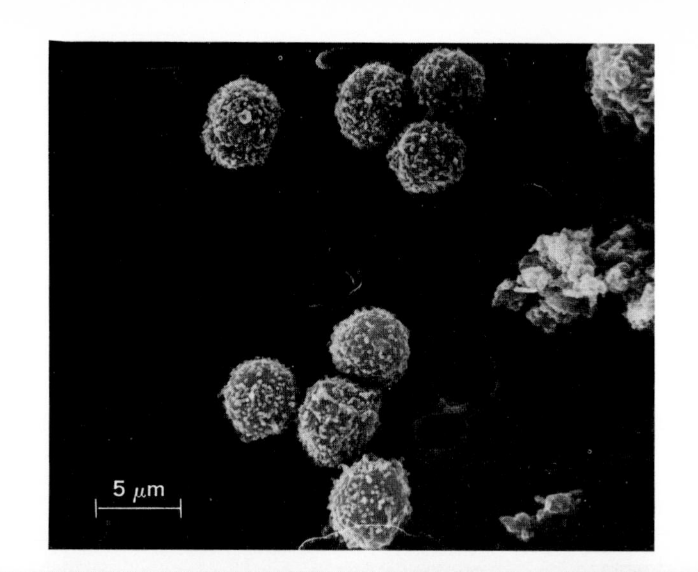
5 μm

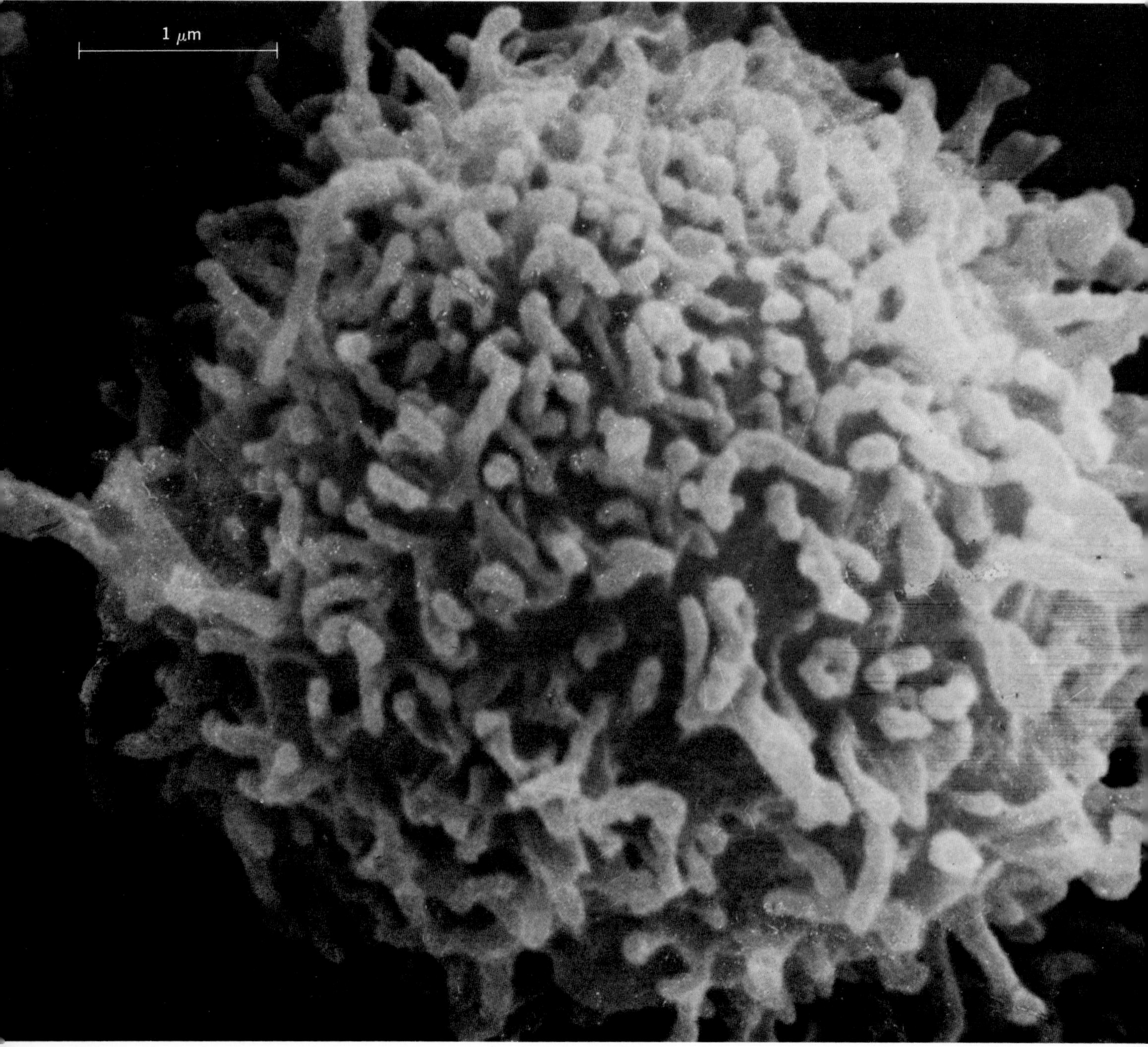
1 μm

Normal human peripheral blood lymphocytes showing covering of microvilli on the surface. As shown in the highly magnified views, microvilli may vary in length from long (opposite page, bottom) to stublike (below). *(From S. Roath et al., Scanning electron microscopy and the surface morphology of human lymphocytes, Nature, 273:15-17, May 4, 1978)*

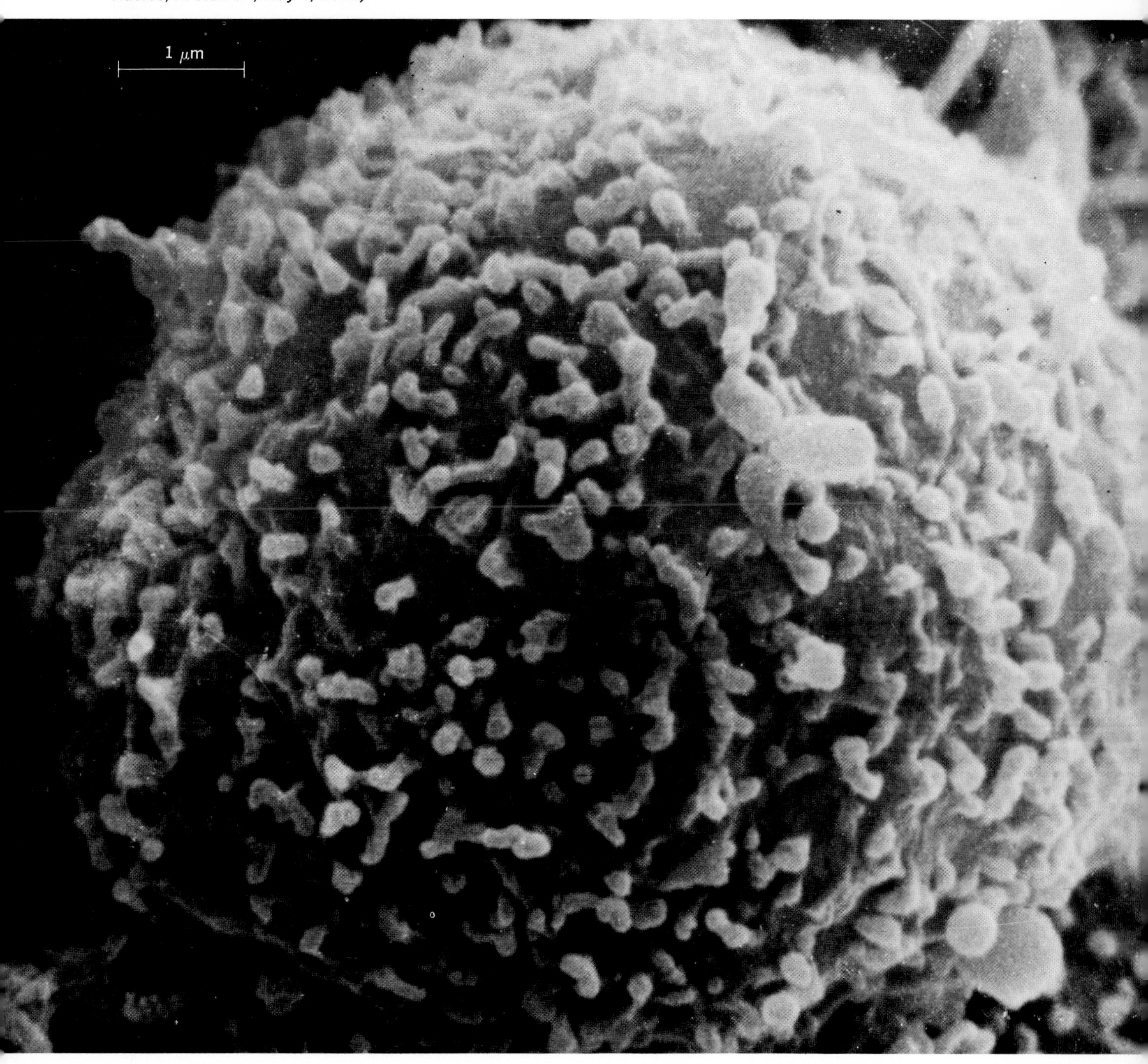

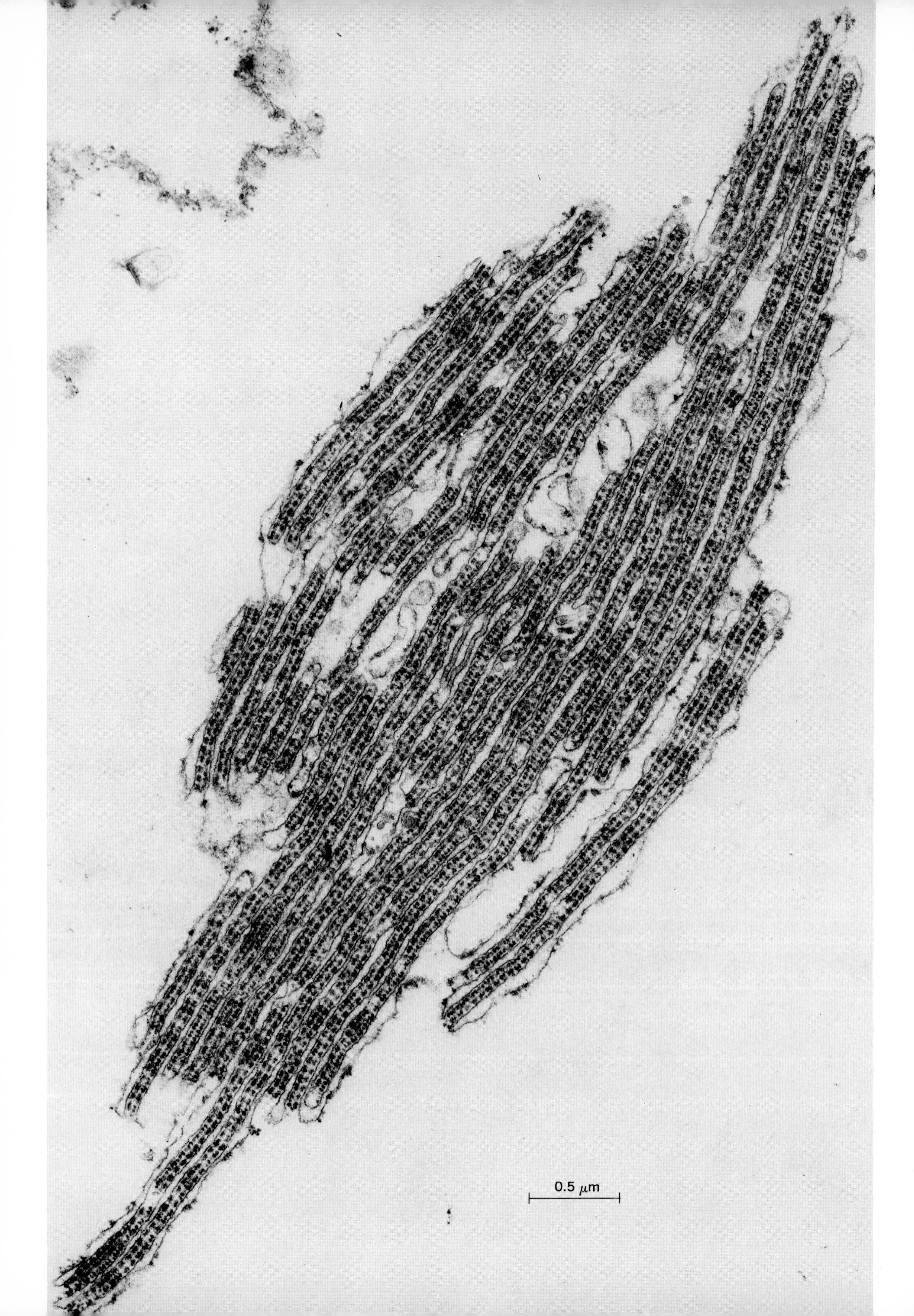
0.5 μm

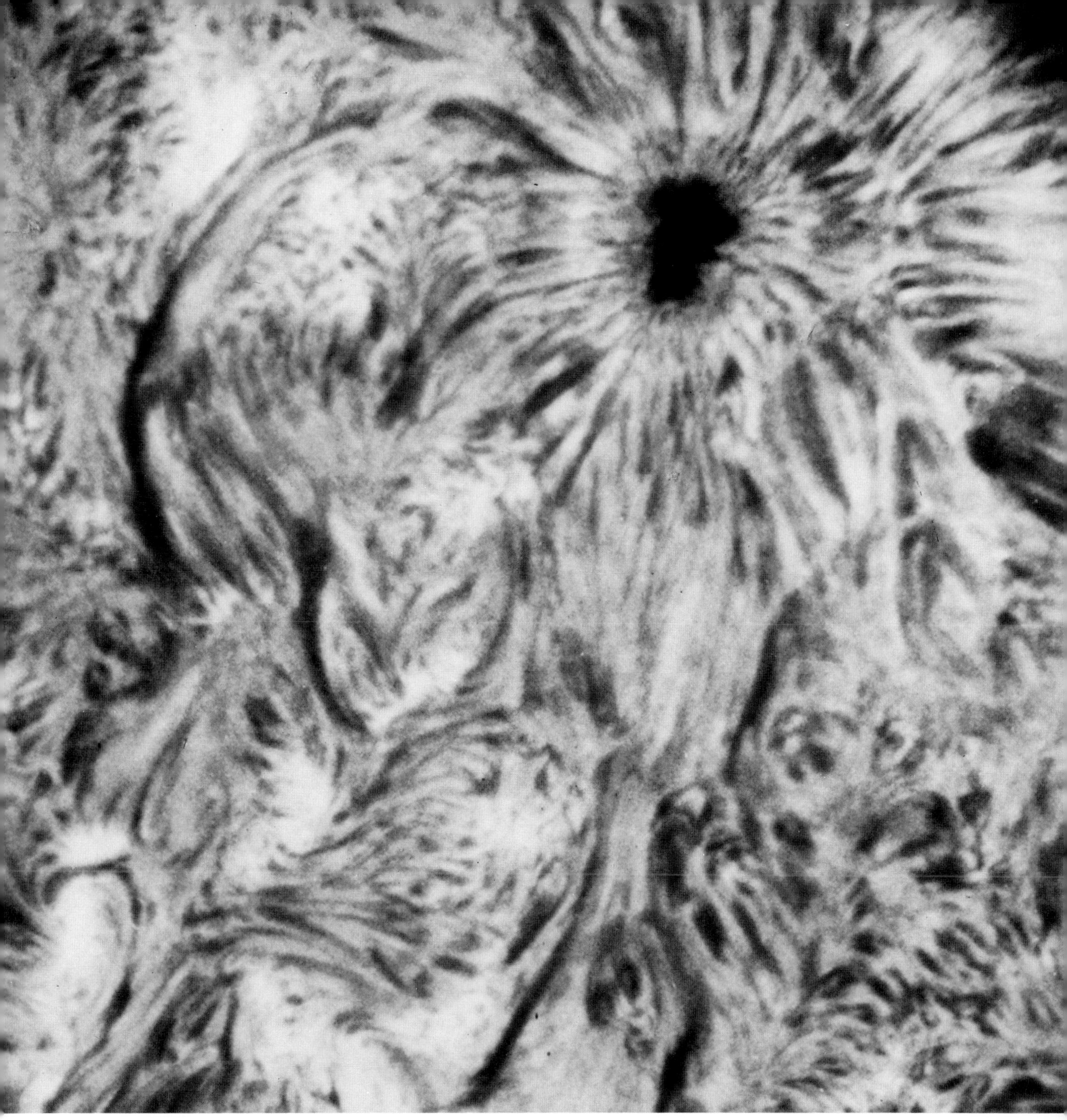

Closeup view of the surface of the Sun viewed with a 12-in. (30-cm) telescope. *(From The shivering Sun, New Sci., 78(1104):509, 1978)*

Opposite page. Crystalline ribosome-membrane complex as found in the oocytes of the lizard *Lacerta sicula* during winter. *(From cover photograph, courtesy of P. N. T. Unwin, Nature, vol. 269, Sept. 8, 1977)*

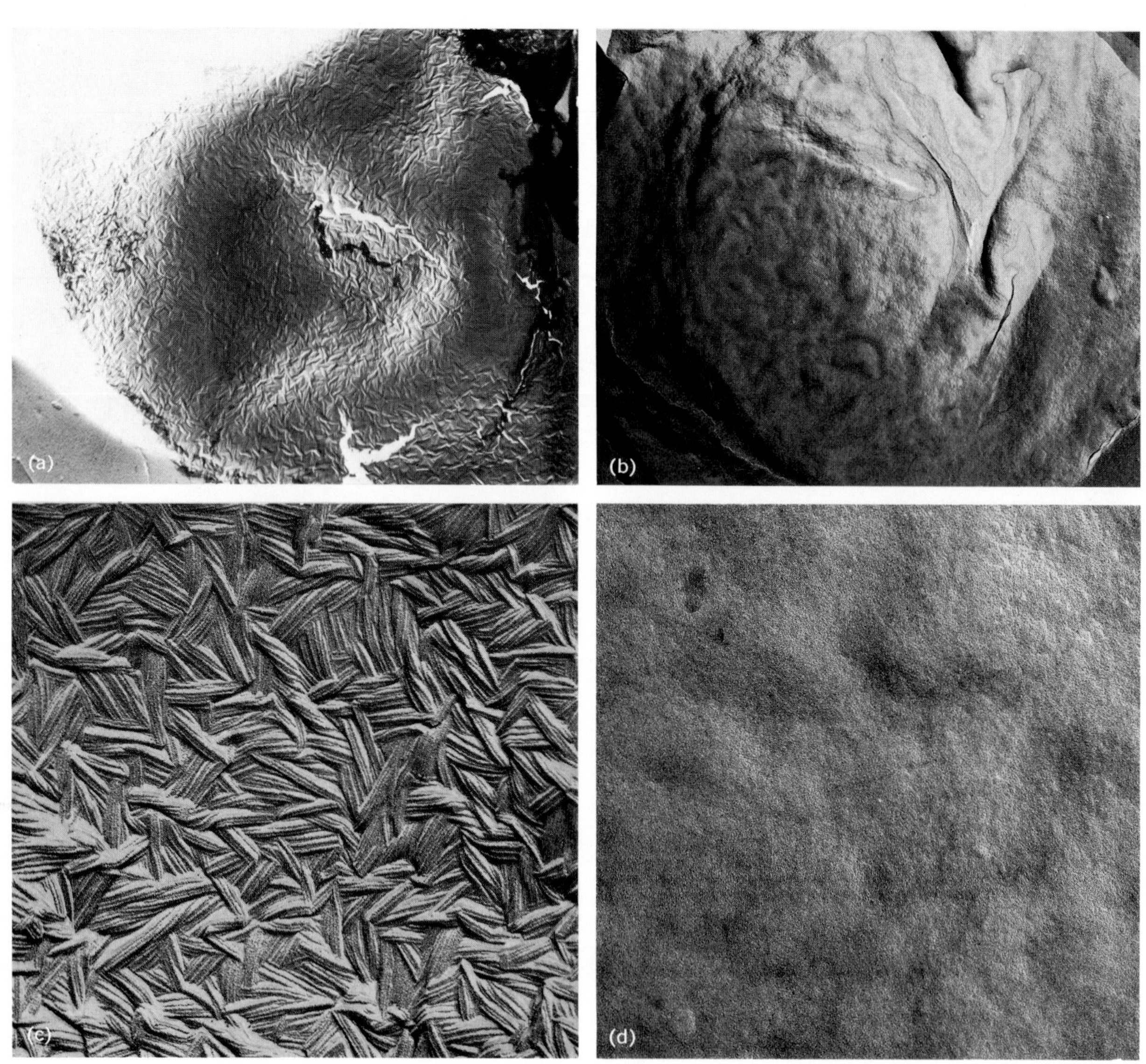

Replicas of the surface of conidia of *Neurospora crassa*, *(a, c)* wild type and *(b, d)* the *eas* mutant. Characteristic rodlets appear in the wild type and are absent in the mutant. *(From R. E. Beever and G. P. Dempsey, Function of rodlets on the surface of fungal spores, Nature, 272(5654)608-610, 1978)*

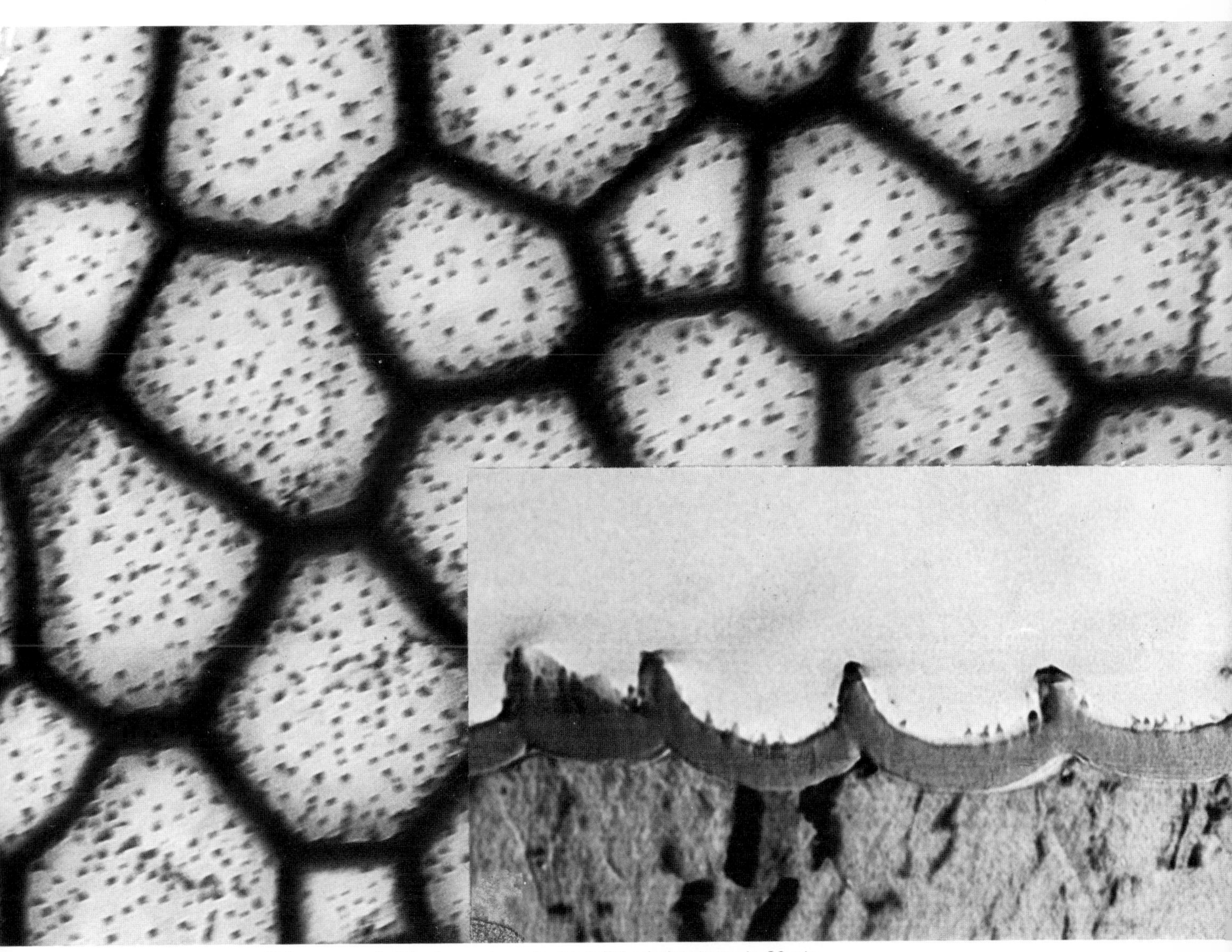

Stripped anodic film formed on electropolished aluminum in 20 s in phosphoric acid. Ultramicrotomed section (inset) shows scalloped nature of the metal film interface. *(From G. E. Thompson et al., Nucleation and growth of porous anodic films on aluminum, Nature, 272(5652):433-435, 1978)*

Particle contamination of microcircuit; particle wedged under wire bond contacts metallized strip on semiconductor and shorts the circuit. *(Hughes Aircraft Company)*

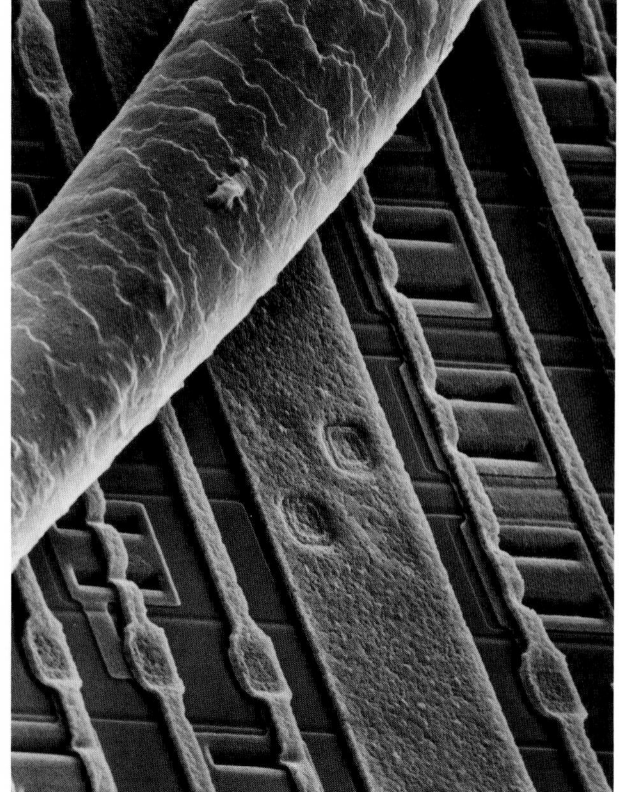

Opposite page. Crinoid *Comantheria briareus*, photographed from ventral side showing arms from their point of origin. *(Courtesy of D. B. Macurda, Jr.)*

Electron micrograph of computer memory transistors of such reduced dimensions as to be dwarfed by a human hair. *(From 3-D ICs reduce memory chip size, Ind. Res. News, 19(8):17, 1977)*

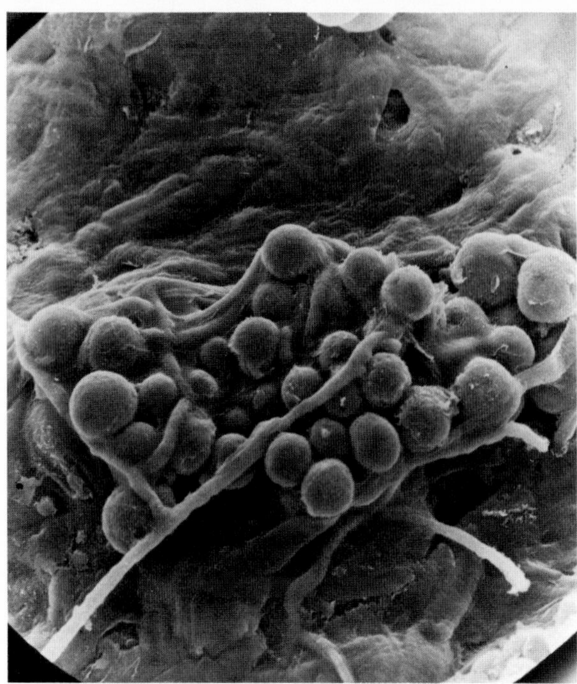

Early stages of synthesis between the fungal and algal symbionts of the lichen *Lecidea albocaerulescens* as seen with the scanning electron microscope. *(From V. Ahmadjian, J. B. Jacobs, and L. A. Russell, Scanning electron microscope study of early lichen synthesis, Science, 200(4345): 1062-1064; copyright © 1978 by the American Association for the Advancement of Science)*

Tracks left by a fibroblast moving across a glass plate covered with a fine layer of gold particles. *(From G. Albrecht-Buehler, Cell, 11:395-404, 1977)*

ground preparation and cultivation. However, this savings is offset to some extent by the additional herbicides and pesticides required. The net reduction from conventional to no-tillage systems may be as high as 5 gal of fuel per acre (47.5 liters/ha). However, no-tillage systems are adaptable only in certain areas, and the use of these systems can also result in decreased yields that may more than offset the fuel savings.

There are a variety of new technologies under development to decrease drying fuel requirements, ranging from solar drying and cob gasification to modifications of existing systems such as dryeration (a high-heat system of drying in which heat is shut off before complete drying occurs; a high volume of airflow is used to complete the process). They can provide substantial fuel savings at varying economic cost. All of these technologies necessitate a new capital investment—which is an investment in new embodied energy as well—and require some increase in management requirements. In some cases, the risk of spoilage is also increased.

Since fertilizer and irrigation are the largest energy users, they provide the best opportunities for reducing energy needs. One promising development that reduces nitrogen fertilizer use is the use of nitrification inhibitors to prevent nitrogen loss to the soil system before the plant can fully utilize it. This has the additional benefit of reducing the pollution that occurs in streams or water systems from nitrogen fertilizer. This new technology can be especially useful to producers who want to increase the efficiency of their operation by applying nitrogen in the fall for the following year's crop. This technique enables the producer to plant the crop early in the spring so as to take advantage of maximum solar radiation at the summer solstice.

Many new approaches are being explored to reduce the energy required to provide sufficient

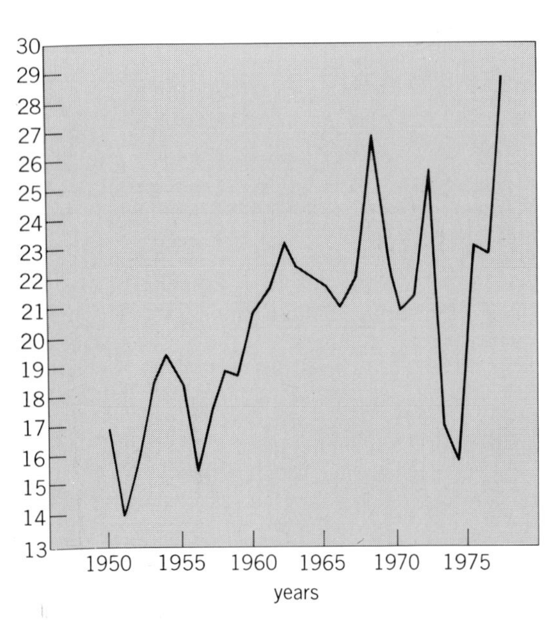

Fig. 3. Ratio of alfalfa hay (dollars per ton) to corn (dollars per bushel) prices in 1950–1977.

moisture for plant growth, for example, improv ment in the efficiency of water distribution ar utilization with drip irrigation or improvement i irrigation pump efficiency. Price increases in nat ural gas, electricity, and diesel fuel have already made many of these improvements economic necessities. In addition, there is now economic pressure against growing low-value crops in arid regions. A change in crop mix is taking place as well as technical adaptation. In some areas, such as the Great Plains, improved moisture control through retention of moisture in the soil is occurring as a result of minimum-tillage practices. In this case, the energy savings from reduced irrigation requirements may be greater than those from reduced tillage. Improved moisture control has also boosted production and thus increased the productivity of other energy inputs in areas of low to moderate rainfall without irrigation.

Low-energy cropping systems. These are defined as land-extensive systems in which an increased land resource base is required to maintain a level of production equivalent to that obtained under high-energy systems. Most low-energy systems being investigated today are based on the replacement of chemical nitrogen energy with rotations of legumes which fix nitrogen in the soil, with animal manure, or with a combination of the two. Such operations can be economically viable. Mixed systems of corn, soybeans, wheat, and alfalfa can be as profitable as continuous corn or corn and soybean systems if the alfalfa has a value of $65 or more per ton ($72 per metric ton). The problem for producers is that alfafa is a perennial (3- to 5-year) crop, but like other crops its price may change substantially from year to year. Producers of corn or soybeans are able to make cropping decisions annually and further reduce this annual risk through forward selling. The profitability of a system that includes an alfalfa rotation as compared with continuous corn thus depends upon the relative prices of corn and alfalfa as well as the price of the chemical nitrogen which the alfalfa partially replaces. Not only have alfalfa prices in the central Corn Belt been below $65 a ton in recent years, but the relationship between corn and alfalfa prices has been so variable (Fig. 3) that producers are discouraged from making the long-term commitment that alfalfa requires.

Mixed enterprises that include cropping and animal operations and utilize manure to replace fertilizer are another form of land-extensive system that can reduce fossil energy use per unit of output. However, animal waste is not usually as rich in nitrogen as it is in phosphorus and potassium, and it is nitrogen that accounts for the bulk of the energy input in corn production. Many low-energy systems combine an animal operation with legumes such as alfalfa, which provide forage as well as contribute nitrogen to the soil. It then becomes very difficult to distinguish specific costs and benefits among the cropping system, forage production, and animal operation. Unless the livestock operation is physically integrated with the cropping operation, the energy cost of transporting manure and the losses of nitrogen that occur in transit decrease the benefits derived from such system. With current prices, it is usually econor

Adiabatic demagnetization

A new method of adiabatic demagnetization has recently been developed that bridges a gap which existed between techniques using electronic magnetic moments and nuclear magnetic moments for cooling. The new technique is known as hyperfine enhanced nuclear cooling, and utilizes the large hyperfine magnetic fields which are present at the nuclei of magnetic ions.

Nuclear versus electronic methods. The use of the small nuclear magnetic moments in metals such as copper, aluminum, or indium for adiabatic magnetic cooling has the advantage that lower end temperatures can be reached than with the use of electronic magnetic moments (due to the much weaker interaction between nuclear magnetic moments) and that thermal contact to other metallic samples is easily established. The disadvantage, however, is that much larger magnetic fields or lower starting temperatures are needed, because nuclear magnetic moments are on the average about 1500 times smaller than the electronic magnetic moments used in conventional paramagnetic salts. In such salts, one can remove nearly the maximum possible amount of magnetization heat ΔQ_m in an applied field of 30 kilooersteds (1 Oe = 79.6 A/m) at a temperature of 1 K. That is, one removes nearly the maximum possible amount of magnetic entropy $\Delta S_m = \Delta Q_m/T \cong nR \ln(2J+1)$, where T is the temperature, R is the gas constant, n is the number of moles of magnetic ions that are used, and $2J+1$ is the number of possible orientations that the magnetic moment can assume according to the laws of quantum mechanics.

In copper, on the other hand, the amount of magnetization heat that one can extract is much smaller. Even if the starting temperature is as low as 15 millikelvins, one removes in 30 kOe only 0.55% of the maximum possible amount of heat or entropy. Although one can use superconducting solenoids capable of generating fields of 80 kOe or more, the amount of usable cooling entropy is still a small fraction (about 4%) of the maximum possible value.

Principles of the new method. By the technique of hyperfine enhanced nuclear cooling, one can bridge the gap that exists between the techniques of using electronic magnetic moments and nuclear magnetic moments for cooling. This new technique, which was theoretically proposed by the Soviet physicist S. A. Al'tshuler in 1966, makes use of the large hyperfine magnetic fields that exist at the nuclei of magnetic ions. These fields are largest for the magnetic ions of the rare-earth series (cerium to ytterbium) and are caused by the electronic magnetic moments of the spinning electrons in the $4f$ shell. In free and unperturbed rare-earth ions, this hyperfine field can reach a magnitude of several million oersteds. Since it is constant, however, it cannot be used for adiabatic demagnetization of their nuclei. When the ions are embedded in suitable crystals, on the other hand, the crystalline electric field can completely quench the electronic magnetic $4f$ moment. This is possible when the number of $4f$ electrons in the ion is even (that is, for praseodymium, europium, terbium, holmium, and thulium, the so-called non-Kramers ions). No hyperfine field then exists at the nucleus of such ions in their ground state.

By applying a moderate external magnetic field, however, a fraction of the original electronic moment (and with it a fraction of the original hyperfine field) can be reinduced by virtue of the so-called Van Vleck susceptibility χ_{VV}. The total magnetic field at the nucleus is then the sum of the external applied field H and the induced hyperfine field H_{hf} and is given by the equation below, where

$$H_{tot} = H + H_{hf} = H(1 + h_f\chi_{VV}) = H(1+K)$$

h_f is a constant characteristic of the ion. The factor $1+K$ is called the hyperfine enhancement factor. Nonmagnetic ground states are most often found in praseodymium and thulium compounds, and the magnitude of the enhancement factor $1+K$ is typically between 5 and 20 for the former and of order 100 for the latter. Especially attractive is the use of metallic Van Vleck paramagnetic praseodymium or thulium compounds, since their better thermal conductivity makes it easier to absorb heat during the nuclear cooling process.

Historically, the first experiments on hyperfine enhanced nuclear cooling were carried out in the compounds TmSb and PrSb at Bell Laboratories, Murray Hill, NJ, in 1967. For reasons which are still not yet understood, the thermodynamic reversibility of the magnetic cooling process is always better in praseodymium compounds. A number of such compounds have been experimentally investigated in the past decade, and their main properties are summarized in the table.

Advantages and limitations. The two main advantages of the hyperfine enhanced nuclear cooling method are the fact that the cooling power is greatly enhanced [by the factor $(1+K)^2$ in low fields] and that the cooling compounds are metals. This latter fact is essential both for good heat conductivity and for good thermal contact between the conduction electrons and the praseodymium nuclei. The enhancement factor makes it possible to reach the 1 mK temperature range, starting from about 20 mK with a magnetic field of only 20 kOe. As an example, when using a $PrNi_5$ cooling pill of 110 g weight (of approximately 13 cm³ volume), the end temperature after such a demagnetization is about 1 mK and the temperature can easily be kept below 2 mK for 24 hr.

Disadvantages are, first, that the lowest attainable temperature is always higher than what one can reach with a "bare" (or brute force) nuclear demagnetization (that is, without hyperfine enhancement and using higher magnetic fields). This is due to remaining "effective" interactions both between the nuclei and (in noncubic crystals) between the nucleus and the 4f electron shell. Actual end temperatures, obtained by demagnetizing from starting temperatures between 20 and 30 mK and fields of less than 30 kOe, are listed in the table. Somewhat lower temperatures could be obtained by using larger magnetic fields

Some properties of compounds used for hyperfine enhanced nuclear cooling

Material	Crystal structure	Enhancement factor (1+K)	End temperatures reached, mK
$PrCu_6$	Orthorhombic	15.3	2.5
$PrPt_5$	Hexagonal	23.5	3
$PrNi_5$	Hexagonal	14.1	0.8
$PrTl_3$	Cubic	20	1.5
$PrBe_{13}$	Cubic	8.7	0.85

or lower starting temperatures. Second, the compounds listed have to be of sufficient purity in order to minimize irreversible heat production during demagnetization. Such effects are observed, for example, in $PrNi_5$ samples when traces of nickel metal are present.

A promising application of hyperfine enhanced nuclear cooling is a two-stage nuclear demagnetization cryostat, in which a conventional nuclear cooling stage is precooled to millikelvin temperatures with a hyperfine enhanced stage. Demagnetization of the second stage could then yield end temperatures in the microkelvin range.

For background information *see* ADIABATIC DEMAGNETIZATION; CRYOGENICS; LOW-TEMPERATURE THERMOMETRY in the McGraw-Hill Encyclopedia of Science and Technology.

[KLAUS ANDRES]

Bibliography: S. A. Al'tshuler, *JETP Lett.*, 3:112, 1966; K. Andres and E. Bucher, *J. Appl. Phys.*, 42: 1522, 1971; K. Andres and S. Darack, *Physica*, 86–88B + C:1071, 1977.

Agriculture

There has been increasing concern about the amount of energy used in intensive cropping systems in the United States, even though agricultural production accounts for only 2–3% of the nation's energy demand. This concern has resulted in the use of energy input/output accounting based on heat units (British thermal units or kilocalories) to calculate the energy efficiency of agriculture. This type of accounting assumes that society places a value on a Btu of coal equal to that it places on a Btu of human labor or corn, which is clearly not the case. A better perspective on energy use in agriculture can be obtained through an analysis of the basis for the choice of resources used in cropping systems and through an understanding of the nature of these cropping systems.

Energy use and resource choices. Energy has been increasingly applied to agricultural production to replace human and animal labor, to increase production, and to reduce risk. These uses are not mutually exclusive, but each forms the basis for cost-benefit decisions that are made continuously by individual farmers. The producer must determine the cost of fertilizer relative to the cost of other inputs that might substitute for fertilizer, and relative to the price of the additional product that fertilizer makes possible. The corn/nitrogen price ratio given in Fig. 1 shows that nitrogen has become more expensive relative to corn in recent years. However, purchases of nitrogen have continued at high levels except in 1975, because relative to the price of the final product, nitrogen has remained an economically productive input.

Nature of cropping systems. Cropping systems are tightly integrated systems which may be subject to substantial change if one component of the system is altered or withdrawn. In addition, a major environmental factor, weather, is beyond the control of the producer, and production is determined by weather as much as by management and the level of inputs. For example, a corn crop may be planted, fertilized, and then ruined by drought, resulting in no return on a substantial energy expenditure.

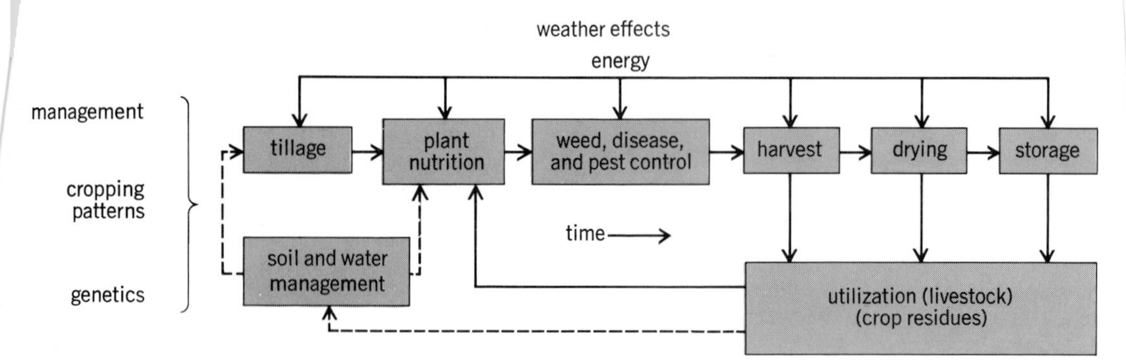

Fig. 2. Dynamic effects of various components on the crop production system.

Figure 2 demonstrates that components of a cropping system are interrelated. The manipulation of one component, such as tillage, to improve the energy cost-benefit ratio for that component may result in either a gain or a loss for the total system. For example, changing to a less fuel-intensive tillage system under unfavorable weather and soil conditions may result in lower yields due to weed and pest problems, which would offset any gain in fuel savings. Only by considering the total cropping system in its weather environment can the total costs and benefits of modifying such a system be analyzed.

The analysis of cropping systems is made more difficult by what amounts to a synergistic effect from the combination of energy inputs to cropping systems. In the case where a crop is already receiving chemical fertilizer and irrigation, the addition of a chemical herbicide or introduction of mechanical cultivation would add substantially to yield. In fact, the contribution of weed control once other productivity-boosting energy inputs are in place may be proportionately greater than if energy-intensive weed control is added where other productivity-boosting energy inputs are not available. The critical problem for researchers is how to separate out the contribution of individual energy-based inputs to productivity when a portion of their contribution depends upon the interaction of their individual effects with the effects of other energy-based inputs; a further complication

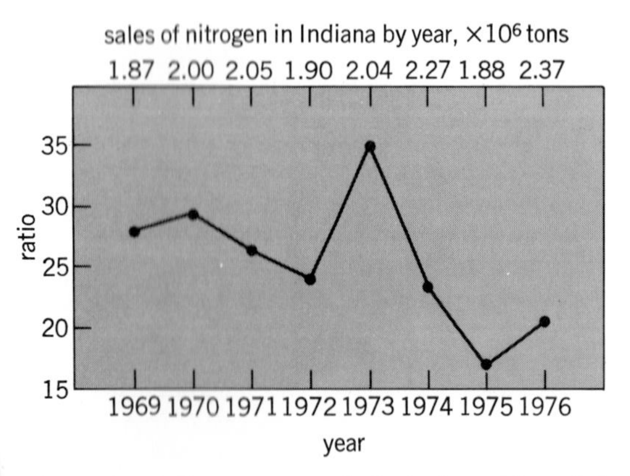

Fig. 1. Corn/nitrogen price ratio for Indiana; 1 ton = 0.9 metric ton.

involves accounting for the effects of weather.

Timing is another factor involved. Production inputs result in much greater productivity if they are introduced at the right moment in the plant's growth cycle. Here again it is difficult to measure precisely the contribution of a particular input and separate it from management practices and weather, both of which are overriding considerations, as shown in Fig. 2.

High-energy cropping systems. High-energy cropping systems use large amounts of fossil-based energy to increase production and reduce risk. The substitution of energy for human and animal labor is often a decision by society to improve the quality of rural life and reduce the amount of human labor, rather than a decision relating to energy intensity in agriculture. Societies that have already decided to move away from human and animal labor in agricultural production are not likely to reverse this trend for both social and economic reasons.

Fertilizers, drying fuels, pesticides, and sometimes fuels for irrigation are the large energy inputs in energy-intensive corn production. Considering a corn crop with supplemental irrigation (12 in., or 30 cm, of water), the proportional energy use stated in terms of gallons of liquefied petroleum gas (LPG) energy equivalents for 1 acre is as follows: 5 gal (47.5 liters/ha) for the energy in insecticides and herbicides, 10 gal (95 liters/ha) for all machinery fuel, 20 gal (190 liters/ha) for drying fuel, 40 gal (380 liters/ha) for the energy in fertilizer, and 60 gal (570 liters/ha) for irrigation pumping.

Over the last several years the benefits from such energy inputs have justified their use at the prevailing price levels in the United States, with exceptions in some instances for irrigation. Even though the United States is indirectly subsidizing domestic energy prices, this cost-benefit relationship holds for many other countries which are not keeping energy prices below prevailing world levels.

Adaptations of high-energy systems. There have been increasing attempts to develop adaptations to high-energy systems that would maintain the same level of production at lower energy levels on the same land base. Integrated pest management systems have achieved adequate levels of protection against insects with about half the chemical energy requirement. Minimum- or no-tillage systems can reduce the fuels required for ically beneficial to utilize animal waste as fertilizer in a well-integrated system. However, labor requirements are high for such systems, and this acts as an obstacle to the effective utilization of animal wastes even in cases where a cropping system is integrated with animal enterprise.

Future prospects. The alternatives of energy-intensive and less energy-intensive systems are not mutually exclusive. There is a continuum of more to less energy-intensive practices or system components which can be integrated into different cropping systems in an almost infinite variety of ways. If farmers change to less energy-intensive cropping systems, there will have to be more extensive use of land in order to maintain production. On the other hand, an increased land base is not necessary if successful adaptations can be made to high-energy systems, which allow the same level of output with less fossil energy input. The land use question is critical because a portion of the most energy-efficient land base, fertile soil with adequate moisture and drainage, continues to be irrevocably lost to urbanization every year.

Energy reductions within the context of high-energy systems are feasible, yet most adaptations require increased levels of management as well as increased dollar and energy capital investment. In some cases, the energy capital investment required may offset the savings in day-to-day energy use. The requirements for capital replacement and increased management time and expertise will likely place small or part-time farming units at a disadvantage in terms of improving their energy efficiency.

Petroleum supplies are now predicted to be more than adequate on a worldwide basis until the mid-1980s. Based on this assumption, the cost of energy inputs for agricultural production should not increase very much relative to the cost of other inputs. Continued inflation could result in land prices increasing faster than energy prices. This would make land-extensive, low-energy production systems even less economically attractive. There will be little incentive for producers to invest management and capital in more energy-efficient adaptations of current high-energy systems unless there are increases in relative energy costs along with sufficient profits to finance the investment.

One promising research area is plant genetics. This has the potential to make the plant a more efficient producer through better utilization of solar energy and soil nutrients. Research investment in this area offers tremendous benefits in terms of higher productivity without necessarily requiring any greater direct energy input.

For background information *see* AGRICULTURE in the McGraw-Hill Encyclopedia of Science and Technology.

[OTTO C. DOERING, III]

Bibliography: T. J. Considine et al., in R. C. Loehr (ed.), *Food, Fertilizer and Animal Residues*, 1977; O. C. Doering et al., *An Energy Based Analysis of Alternative Production Methods and Cropping Systems in the Corn Belt*, Agricultural Experiment Station, Purdue University, Lafayette, IN, NSF/RA-770125, June 1977; G. H. Heichel, *Amer. Sci.*, 64:64–72, 1976; R. Klepper et al., *Amer. J. Agr. Econ.*, 59(1)1–12, February 1977; C. B. Richey, D. R. Griffith, and S. D. Parsons, in *Adv. Agron.*, 29:141–182, 1977; M. D. Skold, *Farmer Adjustments to Higher Energy Prices: The Case of Pump Irrigation*, USDA Economic Research Service, Publ. no. 663, November 1977.

Amaranth

The amaranth is any one of a large number of plant species in the genus *Amaranthus* (family Amaranthaceae) which are annuals (seldom perennials) and are distributed worldwide in warm and humid regions. Amaranths are botanically distinct for their small and chaffy flowers, arranged in dense green or red, monoecious or dioecious inflorescences (spikes), with zero to five perianth segments and two or three styles and stigmata, and for their dry membranous, indehiscent, one-seeded fruit. Amaranths are of great scientific interest at present as highly productive food plants.

Recent physiological, genetic, and nutritional studies have discovered their great potential economic value and have attracted the attention of researchers and laypeople concerned with world food problems. Of particular interest are the high productivity as a rapidly growing summer crop, large amounts of protein in both seed and leaf with high lysine, overall high nutritional value, and water use efficiency for the C photosynthetic pathway, as is the important place of amaranths in the culture, diet, and agricultural economy of the people of Mexico, Central and South America, Africa, and northern India.

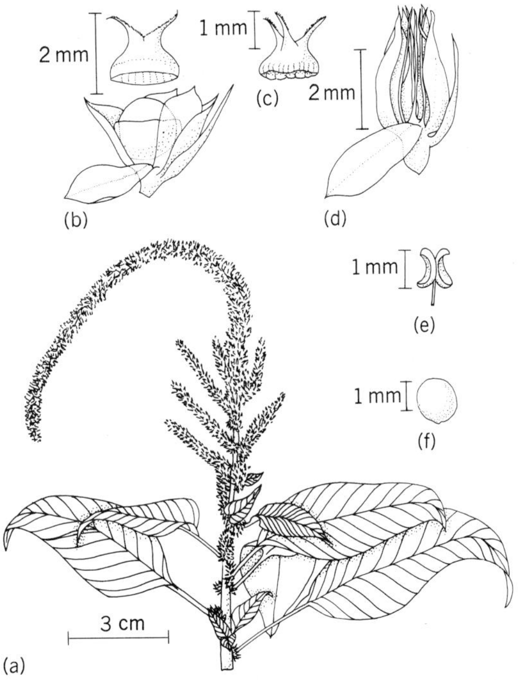

Fig. 1. *Amaranthus caudatus* L.: (*a*) branch with inflorescence; (*b*) adult female flower with dehiscent cap; (*c*) cap with three stigmata; (*d*) male flower; (*e*) anther; and (*f*) seed, lateral view. (*From G. J. H. Grubben, The Cultivation of Amaranth as a Tropical Leaf Vegetable, Roy. Trop. Inst. Commun. no. 67, 1976; reprinted with permission; Laboratory for Plant Taxonomy and Plant Geography, Agricultural University of Wageningen, and Royal Tropical Institute, Amsterdam*)

Classification and species relationships. The genus *Amaranthus* has been described by botanists in terms of up to 200 or more species, of which nearly 50 are distributed in the New World. However, removal of a great deal of confusion and synonymy will probably result in a much smaller number of distinct species. Some of the better-known and more widespread weedy species are red amaranth or pigweed *(A. hybridus)*, thorny amaranth *(A. spinosus)*, tumbleweed *(A. albus)*, wild beet *(A. retroflexus)*, and *A. powelli*. The common ornamental forms such as Prince's Feather *(A. hybridus* ssp. *hypochondriacus)*, Joseph's Coat *(A. tricolor)*, and Love-Lies-Bleeding *(A. caudatus)* are recognized by the form of their spikes and foliage colors.

Two sections of the genus are *Amaranthotypus*, with large terminal inflorescence, five tepals and stamens and fruit opening circularly; and *Blitopsis*, with axillary cymes, often two to four tepals and stamens, and fruits opening irregularly. The cultivated grain species *(A. cruentus, A. caudatus,* and *A. hypochondriacus = A. leucocarpus)* with ornamental and crop forms belong to the section *Amaranthotypus*, whereas two important cultivated vegetable species *(A. tricolor* and *A. lividus)* belong to the section *Blitopsis* (Fig. 1). Several Asian natives, such as *A. tricolor, A. gangeticus,* and *A. melancholicus*, have been grown as potherbs or ornamentals but never as a grain crop. The cultivated grain species are largely pale-seeded, whereas the ornamentals and weedy forms are black-seeded, presumably an important trait used by early peoples in keeping them apart. Interspecific hybridization, reported by several botanists in various regions, is considered to be a primary source of evolutionary diversity. The grain crop forms and their weedy relatives include both 32- and 34-chromosome species; much of the preliminary biosystematic and cytogenetic research has not yet provided conclusive species relationships. However, the weedy and cultivated species of *Amaranthotypus* are separable into two distinct groups on the basis of recent taxonomic and protein variation studies. Use of electrophoretic assays of genetic variation for many enzymes seems promising in identifying species and their hybrids. *See* BREEDING (PLANT).

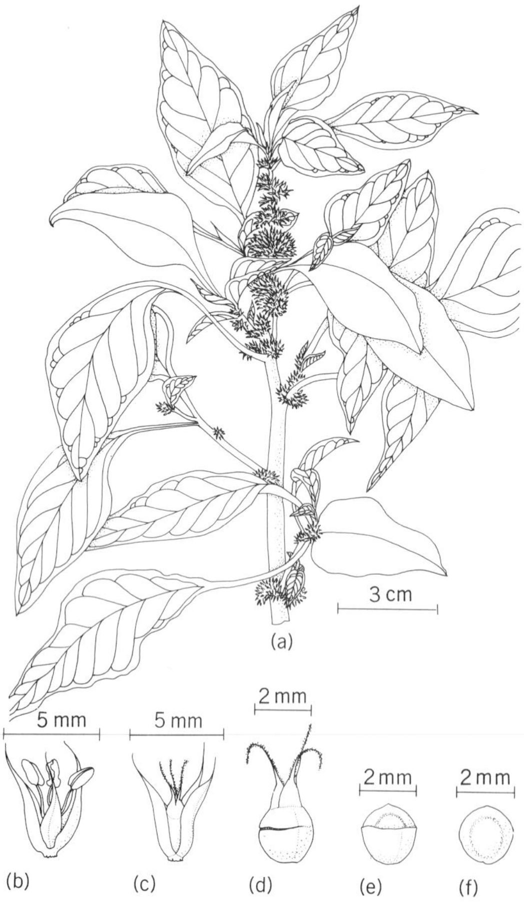

Fig. 2. *Amaranthus tricolor* L.: (*a*) branch with flowers; (*b*) male flower; (*c*) female flower; (*d*) fruit with dehiscing cap; (*e*) fruit without cap; and (*f*) seed, lateral view. *(From G. J. H. Grubben, The Cultivation of Amaranth as a Tropical Leaf Vegetable, Roy. Trop. Inst. Commun. no. 67, 1976; reprinted with permission; Laboratory for Plant Taxonomy and Plant Geography, Agricultural University of Wageningen, and Royal Tropical Institute, Amsterdam)*

Origin and history of domestication. In domesticated forms some of the main changes from the wild species are development of short and weak bracts and perfectly dehiscent and white seed with good popping quality and flavor in many cultivars. Ornamental forms are highly pigmented. The leafy vegetable forms permit many harvests of leaves, and some even have delayed flowering. According to Jonathan Sauer of the University of California, Los Angeles, three different wild or weedy forms *(A. hybridus, A. powelli,* and *A. quitensis)* growing in Mexico, Central America, Andean South America, and Bolivia-Argentina regions gave rise to the three grain species *(A. hypochondriacus, A. cruentus,* and *A. caudatus*, respectively) through domestication within each region (Fig. 2). This hypothesis is based on careful surveys of systematic, geographic, and other sources of evidence. Alternatively, some authors believe that *A. hybridus* might be a widely distributed wild progenitor, which in the north gave rise to two of the grain species, and which moving southward and hybridizing with a wild form, *A. quitensis*, gave rise to *A. caudatus* (including *A. edulis*) forms.

Amaranths are of ancient New World origin and were important as grains in the Aztec, Incan, and Mayan civilizations where wheat and rice were not grown. Archaeological and ethnobotanical evidence suggests that the earliest domestication took place in South America at least 4000 years ago. About A.D. 500 grain amaranths were a widely cultivated food crop in Mexico. Some of the ornamental forms moved from the New World to Africa and Asia in recent times (early 1800s), and may have spread from India to the East Indies, China, Manchuria, and Polynesia as a leaf vegetable in recent times. Ceremonial sacrifice and use in rituals drew attention of the Spaniards, who suppressed the Aztecs and the amaranth cultivation.

Cultivation. Amaranths are a very widely grown vegetable in tropical regions, including southern India, West and Equatorial Africa, and New Guinea (Fig. 3). G. J. H. Grubben of the Royal Tropical Institute in Amsterdam found amaranths to be the most popular vegetable in South Dahomey, and conducted extensive work there on various crop development aspects. Varieties such as Chinese

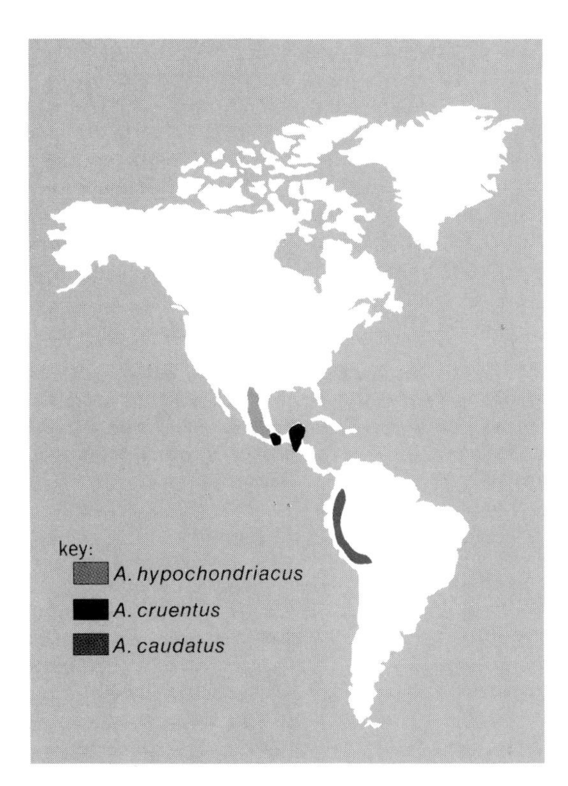

Fig. 3. Distribution of amaranths and areas of origin according to Sauer. *(From G. J. H. Grubben, The Cultivation of Amaranth as a Tropical Leaf Vegetable, Roy. Trop. Inst. Commun. no. 67, 1976; reprinted with permission; Royal Tropical Institute, Amsterdam)*

spinach, Ceylon spinach, and African spinach suggest its use as a salad ingredient; however, somewhat high oxalic acid content of leaves makes it unpleasant for this purpose.

The crop may be raised by transplanting seedlings in narrowly spaced rows and harvesting by uprooting, or by sowing seeds in wider rows, where repeated cuttings are used for 35–60 days before flowering. Amaranth production in swampy "chinampas" (pre-Hispanic floating gardens) to raise seedlings is practiced in Tulychualco, the "amaranth candy capital" of Mexico. Amaranths raised as a grain crop may also be mixed with millet, maize, or a legume, or even pepper or brinjal, as is done in northern India. Although seeds are small, a large number of seeds produced per plant allows optimistic yields to be as high as 3–6 metric tons/ha in Ethiopia and commonly 0.5–1.0 ton/ha. In Dahomey, a leaf form yielded 2.7 tons of dry matter per hectare in 6 weeks (6.4 g/m^2/day).

Food value. Leaf amaranths are rich in provitamin A, vitamin C, iron, calcium, and protein, with lysine as high as 5.9% of protein (equal to soymeal, and more than some of the best maize strains) and 10.8% glutamic acid in *A. edulis*. Amaranth grains, with very high starch quality and quantity, high protein (up to 17%), and high digestibility, rate better in nutritive value than all cereals. Primary modern uses in Mexico include making of alegrias (cakes of popped seed bound with sugar syrup), flour made of popped seed (with sugar) to prepare pinole, paste to make chuale, and amaranth milk (atole) from the black-seeded and variegated forms. In India, amaranth seed is used in alegrias and confectionaries (flour is cooked with rice and mustard) and as a cereal (popped seed).

Genetic, ethnobotanical, and agronomic researches are under way to develop amaranths as an important food plant in modern agriculture.

For background information *see* LYSINE; PHOTOSYNTHESIS; PROTEIN in the McGraw-Hill Encyclopedia of Science and Technology.

[SUBODH K. JAIN]

Bibliography: G. J. H. Grubben, *The Cultivation of Amaranth as a Tropical Leaf Vegetable*, Roy. Trop. Inst. Commun. no. 67, 1976; J. D. Sauer, *Ann. Missouri Bot. Gard.*, 54:103–137, 1967; *Proceedings of the 1st Amaranth Seminar*, 1977; H. B. Singh, *Grain Amaranths, Buckwheat and Chenopods*, Indian Council of Agricultural Research, 1961.

Analytical chemistry

The analysis of air, water, soil, food, body fluids, fuels, medicines, and a wide variety of consumer and industrial products for chemical components such as carcinogenic hydrocarbons, drugs, toxic and nutritional elements, naturally occurring organic species such as glucose and bilirubin, and enzymes is vitally important to society. Analytical spectroscopy represents one of the most powerful tools for these analyses. One of the significant developments in analytical spectroscopy in recent years is the use of television camera tubes (called imaging detectors) for a variety of applications.

An imaging detector is a collection of several hundred to several thousand individual photodetectors. One- and two-dimensional arrangements of photodetectors are available. When illuminated with light of different intensities, different detectors produce separate electrical signals with magnitudes that depend upon light intensity. Light transmitted through or emitted from a sample is separated into different wavelengths by a prism or grating device and directed to the television camera tube, which then measures several different wavelengths (colors) simultaneously. Imaging detectors, which were designed primarily for television applications initially, are useful in analytical spectroscopy because of this ability to simultane-

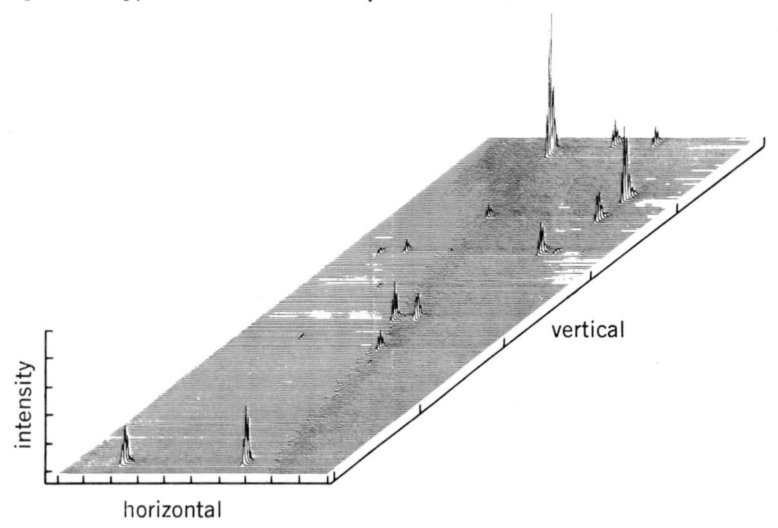

Fig. 1. Emission spectrum for mercury recorded with an image dissector interfaced to an echelle spectrometer.

ously monitor different wavelengths and hence to identify and quantify chemical components in complex samples.

Elemental analysis. Figure 1 shows a three-dimensional display of spectral peaks recorded with an imaging detector for mercury excited by an electrical discharge. Each position represents a different point along the horizontal and vertical axes of the two-dimensional imaging detector. Each peak represents the intensity of light emitted at a different wavelength (displayed along the horizontal axis) for different regions of the ultraviolet-visible spectrum (displayed along the vertical axis). The positions of the different peaks in the display are unique for this element, and the heights of the peaks depend upon the amount of mercury present. Thus, a chemist can use this information both to confirm the presence of mercury and to determine how much of it is present. Another element, such as cadmium, lead, or arsenic, gives an entirely different spectrum. If several elements are present in the same sample, then peaks for the elements appear at different points in the two-dimensional display, and these peaks can be used to identify and quantify the different elements present. The instrumental system used to produce Fig. 1 has been utilized to determine 10 elements in the same sample without a separation, and is capable of determining 20 or more elements simultaneously. These types of analyses are useful to physicians trying to determine which of several possible elements is responsible for a patient's illness, to scientists studying the relationship between trace elements and occupational diseases such as black lung, and to environmentalists, food scientists, and agronomists.

Simultaneous determinations of multiple trace metals have been carried out in substances such as serum, water samples, and alloys. Some investigators have used systems that produce two-dimensional spectra such as that shown in Fig. 1, while others have used simpler systems that produce one-dimensional spectra. These simpler systems are less expensive, but give less information. Methods have included atomic absorption, atomic emission, and atomic fluorescence spectroscopy.

Drug determinations. Another type of application involves the absorption of ultraviolet energy (Fig. 2). This figure shows different amounts of energy absorbed by a drug, procainamide (PA), and a metabolic product of the drug, *N*-acetylprocainamide (NAPA). PA and NAPA are used in the control of erratic heartbeats in some patients. Determining the amount of drug and metabolite in a patient's serum is important for dosage control and for analysis of the drug's effectiveness. The different spectral properties measured with an imaging detector can be used with specialized computer data-processing methods to determine these drugs in serum without having to separate them.

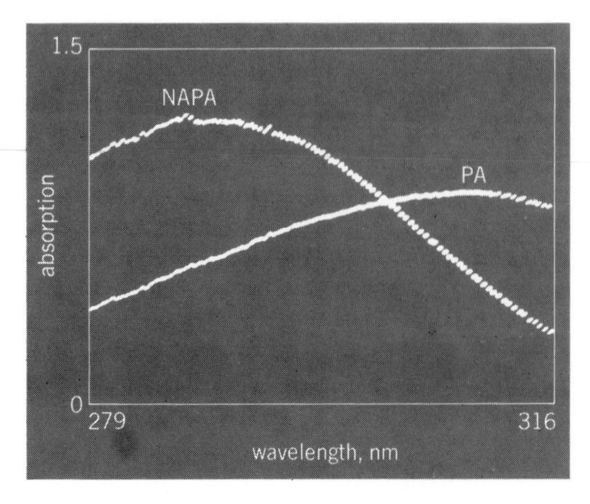

Fig. 2. Absorption spectra for procainamide (PA) and *N*-acetylprocainamide (NAPA) recorded with vidicon spectrometer.

The procedure illustrated in Fig. 2 for two components has been extended successfully to samples containing up to seven different drugs.

A second approach uses liquid chromatographic procedures to separate the drugs and an imaging detector to record the spectrum of each component. Both absorption and fluorescence methods have been used to identify and quantify components. In a third approach similar to that in Fig. 2, a two-dimensional imaging detector is used to record fluorescence excitation and emission spectra simultaneously for drugs. These data are much more informative than conventional fluorescence spectra, and the method is being developed for simultaneous multicomponent determinations.

Kinetic studies. Still another type of application involves the measurement of the visible light absorbed at different points in time when bilirubin in serum reacts with a particular organic compound (Fig. 3). Bilirubin is a product in the breakdown of hemoglobin in the blood, and gives the yellow color to the skin in jaundice. It is important in many types of illnesses, and is particularly significant in newborn infants because too much bilirubin can cause brain damage resulting in mental retardation or even death in some severe cases if corrective action is not taken. The chemical reaction represented by data in Fig. 3 is very fast, taking place in less than 1 s. The imaging detector is very useful in this area because it permits scientists to monitor these very fast spectral changes and to understand the chemical processes involved. Information gained in this type of study has been used to develop a new approach for the simultaneous determination of the different forms of bilirubin that exist in body fluids.

More generally, the processes of controlling industrial chemical reactions so that they will produce needed products at economical costs, of understanding disease processes so that they can be diagnosed and treated, or of evaluating the effectiveness of drugs and other medicines often involve detailed studies of very complex chemical reactions. Many of these reactions involve several steps. Because imaging detectors can monitor several species simultaneously (as shown in Fig. 3), they have been used successfully in a variety of basic studies by D. W. Margerum and coworkers. One example involves the very complex behavior of so-called chloramine compounds that are formed between chlorine and ammonia. These studies are important because of both the desirable effects of chlorine when added to swimming pools and the undesirable toxic effects of chloramine compounds on plant and fish life in natural waters.

Hemoglobin. Carbon monoxide poisoning results from an interaction of the gas with hemoglo-

A-Z

Adiabatic demagnetization

A new method of adiabatic demagnetization has recently been developed that bridges a gap which existed between techniques using electronic magnetic moments and nuclear magnetic moments for cooling. The new technique is known as hyperfine enhanced nuclear cooling, and utilizes the large hyperfine magnetic fields which are present at the nuclei of magnetic ions.

Nuclear versus electronic methods. The use of the small nuclear magnetic moments in metals such as copper, aluminum, or indium for adiabatic magnetic cooling has the advantage that lower end temperatures can be reached than with the use of electronic magnetic moments (due to the much weaker interaction between nuclear magnetic moments) and that thermal contact to other metallic samples is easily established. The disadvantage, however, is that much larger magnetic fields or lower starting temperatures are needed, because nuclear magnetic moments are on the average about 1500 times smaller than the electronic magnetic moments used in conventional paramagnetic salts. In such salts, one can remove nearly the maximum possible amount of magnetization heat ΔQ_m in an applied field of 30 kilooersteds (1 Oe = 79.6 A/m) at a temperature of 1 K. That is, one removes nearly the maximum possible amount of magnetic entropy $\Delta S_m = \Delta Q_m/T \cong nR \ln(2J+1)$, where T is the temperature, R is the gas constant, n is the number of moles of magnetic ions that are used, and $2J+1$ is the number of possible orientations that the magnetic moment can assume according to the laws of quantum mechanics.

In copper, on the other hand, the amount of magnetization heat that one can extract is much smaller. Even if the starting temperature is as low as 15 millikelvins, one removes in 30 kOe only 0.55% of the maximum possible amount of heat or entropy. Although one can use superconducting solenoids capable of generating fields of 80 kOe or more, the amount of usable cooling entropy is still a small fraction (about 4%) of the maximum possible value.

Principles of the new method. By the technique of hyperfine enhanced nuclear cooling, one can bridge the gap that exists between the techniques of using electronic magnetic moments and nuclear magnetic moments for cooling. This new technique, which was theoretically proposed by the Soviet physicist S. A. Al'tshuler in 1966, makes use of the large hyperfine magnetic fields that exist at the nuclei of magnetic ions. These fields are largest for the magnetic ions of the rare-earth series (cerium to ytterbium) and are caused by the electronic magnetic moments of the spinning electrons in the $4f$ shell. In free and unperturbed rare-earth ions, this hyperfine field can reach a magnitude of several million oersteds. Since it is constant, however, it cannot be used for adiabatic demagnetization of their nuclei. When the ions are embedded in suitable crystals, on the other hand, the crystalline electric field can completely quench the electronic magnetic $4f$ moment. This is possible when the number of $4f$ electrons in the ion is even (that is, for praseodymium, europium, terbium, holmium, and thulium, the so-called non-Kramers ions). No hyperfine field then exists at the nucleus of such ions in their ground state.

By applying a moderate external magnetic field, however, a fraction of the original electronic moment (and with it a fraction of the original hyperfine field) can be reinduced by virtue of the so-called Van Vleck susceptibility χ_{VV}. The total magnetic field at the nucleus is then the sum of the external applied field H and the induced hyperfine field H_{hf} and is given by the equation below, where

$$H_{tot} = H + H_{hf} = H(1 + h_f\chi_{VV}) = H(1+K)$$

h_f is a constant characteristic of the ion. The factor $1+K$ is called the hyperfine enhancement factor. Nonmagnetic ground states are most often found in praseodymium and thulium compounds, and the magnitude of the enhancement factor $1+K$ is typically between 5 and 20 for the former and of order 100 for the latter. Especially attractive is the use of metallic Van Vleck paramagnetic praseodymium or thulium compounds, since their better thermal conductivity makes it easier to absorb heat during the nuclear cooling process.

Historically, the first experiments on hyperfine enhanced nuclear cooling were carried out in the compounds TmSb and PrSb at Bell Laboratories, Murray Hill, NJ, in 1967. For reasons which are still not yet understood, the thermodynamic reversibility of the magnetic cooling process is always better in praseodymium compounds. A number of such compounds have been experimentally investigated in the past decade, and their main properties are summarized in the table.

Advantages and limitations. The two main advantages of the hyperfine enhanced nuclear cooling method are the fact that the cooling power is greatly enhanced [by the factor $(1+K)^2$ in low fields] and that the cooling compounds are metals. This latter fact is essential both for good heat conductivity and for good thermal contact between the conduction electrons and the praseodymium nuclei. The enhancement factor makes it possible to reach the 1 mK temperature range, starting from about 20 mK with a magnetic field of only 20 kOe. As an example, when using a $PrNi_5$ cooling pill of 110 g weight (of approximately 13 cm^3 volume), the end temperature after such a demagnetization is about 1 mK and the temperature can easily be kept below 2 mK for 24 hr.

Disadvantages are, first, that the lowest attainable temperature is always higher than what one can reach with a "bare" (or brute force) nuclear demagnetization (that is, without hyperfine enhancement and using higher magnetic fields). This is due to remaining "effective" interactions both between the nuclei and (in noncubic crystals) between the nucleus and the $4f$ electron shell. Actual end temperatures, obtained by demagnetizing from starting temperatures between 20 and 30 mK and fields of less than 30 kOe, are listed in the table. Somewhat lower temperatures could be obtained by using larger magnetic fields or lower starting temperatures. Second, the compounds listed have to be of sufficient purity in order to minimize irreversible heat production during demagnetization. Such effects are observed, for example, in $PrNi_5$ samples when traces of nickel metal are present.

Some properties of compounds used for hyperfine enhanced nuclear cooling

Material	Crystal structure	Enhancement factor (1+K)	End temperatures reached, mK
$PrCu_6$	Orthorhombic	15.3	2.5
$PrPt_5$	Hexagonal	23.5	3
$PrNi_5$	Hexagonal	14.1	0.8
$PrTl_3$	Cubic	20	1.5
$PrBe_{13}$	Cubic	8.7	0.85

A promising application of hyperfine enhanced nuclear cooling is a two-stage nuclear demagnetization cryostat, in which a conventional nuclear cooling stage is precooled to millikelvin temperatures with a hyperfine enhanced stage. Demagnetization of the second stage could then yield end temperatures in the microkelvin range.

For background information *see* ADIABATIC DEMAGNETIZATION; CRYOGENICS; LOW-TEMPERATURE THERMOMETRY in the McGraw-Hill Encyclopedia of Science and Technology.

[KLAUS ANDRES]

Bibliography: S. A. Al'tshuler, *JETP Lett.*, 3:112, 1966; K. Andres and E. Bucher, *J. Appl. Phys.*, 42: 1522, 1971; K. Andres and S. Darack, *Physica*, 86–88B + C:1071, 1977.

Agriculture

There has been increasing concern about the amount of energy used in intensive cropping systems in the United States, even though agricultural production accounts for only 2–3% of the nation's energy demand. This concern has resulted in the use of energy input/output accounting based on heat units (British thermal units or kilocalories) to calculate the energy efficiency of agriculture. This type of accounting assumes that society places a value on a Btu of coal equal to that it places on a Btu of human labor or corn, which is clearly not the case. A better perspective on energy use in agriculture can be obtained through an analysis of the basis for the choice of resources used in cropping systems and through an understanding of the nature of these cropping systems.

Energy use and resource choices. Energy has been increasingly applied to agricultural production to replace human and animal labor, to increase production, and to reduce risk. These uses are not mutually exclusive, but each forms the basis for cost-benefit decisions that are made continuously by individual farmers. The producer must determine the cost of fertilizer relative to the cost of other inputs that might substitute for fertilizer, and relative to the price of the additional product that fertilizer makes possible. The corn/nitrogen price ratio given in Fig. 1 shows that nitrogen has become more expensive relative to corn in recent years. However, purchases of nitrogen have continued at high levels except in 1975, because relative to the price of the final product, nitrogen has remained an economically productive input.

Nature of cropping systems. Cropping systems are tightly integrated systems which may be subject to substantial change if one component of the system is altered or withdrawn. In addition, a major environmental factor, weather, is beyond the control of the producer, and production is determined by weather as much as by management and the level of inputs. For example, a corn crop may be planted, fertilized, and then ruined by drought, resulting in no return on a substantial energy expenditure.

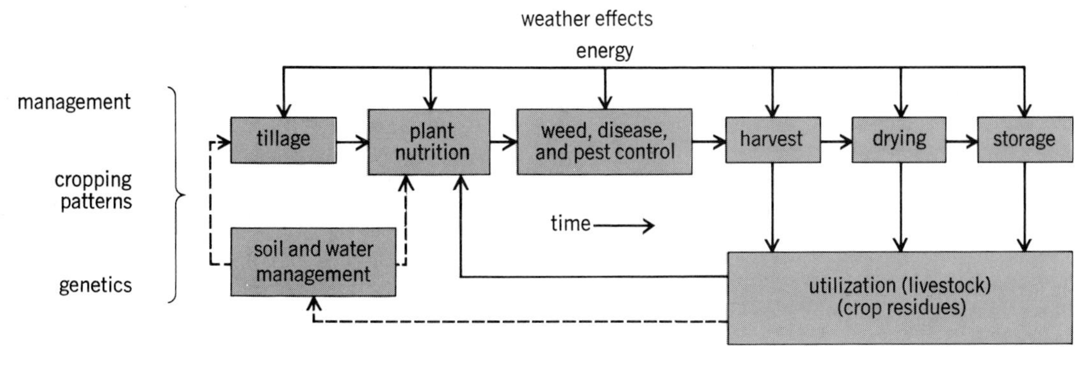

Fig. 2. Dynamic effects of various components on the crop production system.

Figure 2 demonstrates that components of a cropping system are interrelated. The manipulation of one component, such as tillage, to improve the energy cost-benefit ratio for that component may result in either a gain or a loss for the total system. For example, changing to a less fuel-intensive tillage system under unfavorable weather and soil conditions may result in lower yields due to weed and pest problems, which would offset any gain in fuel savings. Only by considering the total cropping system in its weather environment can the total costs and benefits of modifying such a system be analyzed.

The analysis of cropping systems is made more difficult by what amounts to a synergistic effect from the combination of energy inputs to cropping systems. In the case where a crop is already receiving chemical fertilizer and irrigation, the addition of a chemical herbicide or introduction of mechanical cultivation would add substantially to yield. In fact, the contribution of weed control once other productivity-boosting energy inputs are in place may be proportionately greater than if energy-intensive weed control is added where other productivity-boosting energy inputs are not available. The critical problem for researchers is how to separate out the contribution of individual energy-based inputs to productivity when a portion of their contribution depends upon the interaction of their individual effects with the effects of other energy-based inputs; a further complication involves accounting for the effects of weather.

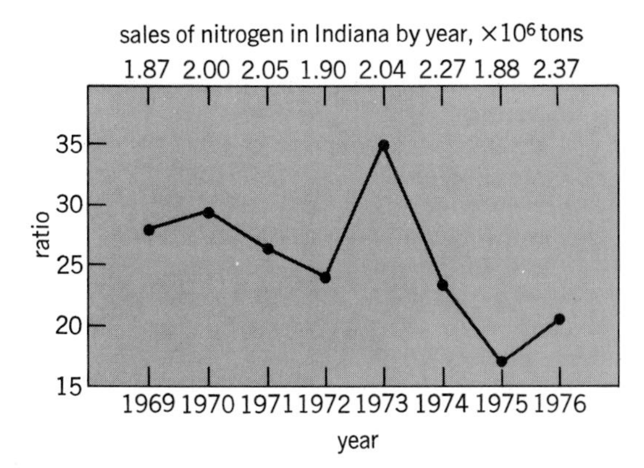

Fig. 1. Corn/nitrogen price ratio for Indiana; 1 ton = 0.9 metric ton.

Timing is another factor involved. Production inputs result in much greater productivity if they are introduced at the right moment in the plant's growth cycle. Here again it is difficult to measure precisely the contribution of a particular input and separate it from management practices and weather, both of which are overriding considerations, as shown in Fig. 2.

High-energy cropping systems. High-energy cropping systems use large amounts of fossil-based energy to increase production and reduce risk. The substitution of energy for human and animal labor is often a decision by society to improve the quality of rural life and reduce the amount of human labor, rather than a decision relating to energy intensity in agriculture. Societies that have already decided to move away from human and animal labor in agricultural production are not likely to reverse this trend for both social and economic reasons.

Fertilizers, drying fuels, pesticides, and sometimes fuels for irrigation are the large energy inputs in energy-intensive corn production. Considering a corn crop with supplemental irrigation (12 in., or 30 cm, of water), the proportional energy use stated in terms of gallons of liquefied petroleum gas (LPG) energy equivalents for 1 acre is as follows: 5 gal (47.5 liters/ha) for the energy in insecticides and herbicides, 10 gal (95 liters/ha) for all machinery fuel, 20 gal (190 liters/ha) for drying fuel, 40 gal (380 liters/ha) for the energy in fertilizer, and 60 gal (570 liters/ha) for irrigation pumping.

Over the last several years the benefits from such energy inputs have justified their use at the prevailing price levels in the United States, with exceptions in some instances for irrigation. Even though the United States is indirectly subsidizing domestic energy prices, this cost-benefit relationship holds for many other countries which are not keeping energy prices below prevailing world levels.

Adaptations of high-energy systems. There have been increasing attempts to develop adaptations to high-energy systems that would maintain the same level of production at lower energy levels on the same land base. Integrated pest management systems have achieved adequate levels of protection against insects with about half the chemical energy requirement. Minimum- or no-tillage systems can reduce the fuels required for ground preparation and cultivation. However, this savings is offset to some extent by the additional herbicides and pesticides required. The net reduction from conventional to no-tillage systems may be as high as 5 gal of fuel per acre (47.5 liters/ha). However, no-tillage systems are adaptable only in certain areas, and the use of these systems can also result in decreased yields that may more than offset the fuel savings.

There are a variety of new technologies under development to decrease drying fuel requirements, ranging from solar drying and cob gasification to modifications of existing systems such as dryeration (a high-heat system of drying in which heat is shut off before complete drying occurs; a high volume of airflow is used to complete the process). They can provide substantial fuel savings at varying economic cost. All of these technologies necessitate a new capital investment—which is an investment in new embodied energy as well—and require some increase in management requirements. In some cases, the risk of spoilage is also increased.

Since fertilizer and irrigation are the largest energy users, they provide the best opportunities for reducing energy needs. One promising development that reduces nitrogen fertilizer use is the use of nitrification inhibitors to prevent nitrogen loss to the soil system before the plant can fully utilize it. This has the additional benefit of reducing the pollution that occurs in streams or water systems from nitrogen fertilizer. This new technology can be especially useful to producers who want to increase the efficiency of their operation by applying nitrogen in the fall for the following year's crop. This technique enables the producer to plant the crop early in the spring so as to take advantage of maximum solar radiation at the summer solstice.

Many new approaches are being explored to reduce the energy required to provide sufficient

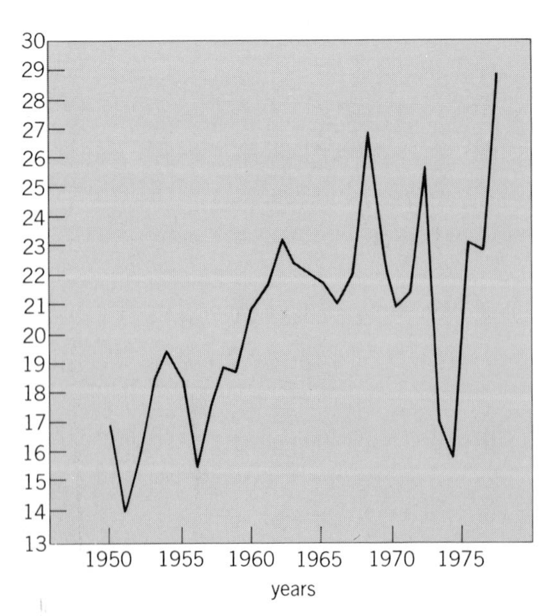

Fig. 3. Ratio of alfalfa hay (dollars per ton) to corn (dollars per bushel) prices in 1950–1977.

moisture for plant growth, for example, improvement in the efficiency of water distribution and utilization with drip irrigation or improvement in irrigation pump efficiency. Price increases in natural gas, electricity, and diesel fuel have already made many of these improvements economic necessities. In addition, there is now economic pressure against growing low-value crops in arid regions. A change in crop mix is taking place as well as technical adaptation. In some areas, such as the Great Plains, improved moisture control through retention of moisture in the soil is occurring as a result of minimum-tillage practices. In this case, the energy savings from reduced irrigation requirements may be greater than those from reduced tillage. Improved moisture control has also boosted production and thus increased the productivity of other energy inputs in areas of low to moderate rainfall without irrigation.

Low-energy cropping systems. These are defined as land-extensive systems in which an increased land resource base is required to maintain a level of production equivalent to that obtained under high-energy systems. Most low-energy systems being investigated today are based on the replacement of chemical nitrogen energy with rotations of legumes which fix nitrogen in the soil, with animal manure, or with a combination of the two. Such operations can be economically viable. Mixed systems of corn, soybeans, wheat, and alfalfa can be as profitable as continuous corn or corn and soybean systems if the alfalfa has a value of $65 or more per ton ($72 per metric ton). The problem for producers is that alfafa is a perennial (3- to 5-year) crop, but like other crops its price may change substantially from year to year. Producers of corn or soybeans are able to make cropping decisions annually and further reduce this annual risk through forward selling. The profitability of a system that includes an alfalfa rotation as compared with continuous corn thus depends upon the relative prices of corn and alfalfa as well as the price of the chemical nitrogen which the alfalfa partially replaces. Not only have alfalfa prices in the central Corn Belt been below $65 a ton in recent years, but the relationship between corn and alfalfa prices has been so variable (Fig. 3) that producers are discouraged from making the long-term commitment that alfalfa requires.

Mixed enterprises that include cropping and animal operations and utilize manure to replace fertilizer are another form of land-extensive system that can reduce fossil energy use per unit of output. However, animal waste is not usually as rich in nitrogen as it is in phosphorus and potassium, and it is nitrogen that accounts for the bulk of the energy input in corn production. Many low-energy systems combine an animal operation with legumes such as alfalfa, which provide forage as well as contribute nitrogen to the soil. It then becomes very difficult to distinguish specific costs and benefits among the cropping system, forage production, and animal operation. Unless the livestock operation is physically integrated with the cropping operation, the energy cost of transporting manure and the losses of nitrogen that occur in transit decrease the benefits derived from such a system. With current prices, it is usually economically beneficial to utilize animal waste as fertilizer in a well-integrated system. However, labor requirements are high for such systems, and this acts as an obstacle to the effective utilization of animal wastes even in cases where a cropping system is integrated with animal enterprise.

Future prospects. The alternatives of energy-intensive and less energy-intensive systems are not mutually exclusive. There is a continuum of more to less energy-intensive practices or system components which can be integrated into different cropping systems in an almost infinite variety of ways. If farmers change to less energy-intensive cropping systems, there will have to be more extensive use of land in order to maintain production. On the other hand, an increased land base is not necessary if successful adaptations can be made to high-energy systems, which allow the same level of output with less fossil energy input. The land use question is critical because a portion of the most energy-efficient land base, fertile soil with adequate moisture and drainage, continues to be irrevocably lost to urbanization every year.

Energy reductions within the context of high-energy systems are feasible, yet most adaptations require increased levels of management as well as increased dollar and energy capital investment. In some cases, the energy capital investment required may offset the savings in day-to-day energy use. The requirements for capital replacement and increased management time and expertise will likely place small or part-time farming units at a disadvantage in terms of improving their energy efficiency.

Petroleum supplies are now predicted to be more than adequate on a worldwide basis until the mid-1980s. Based on this assumption, the cost of energy inputs for agricultural production should not increase very much relative to the cost of other inputs. Continued inflation could result in land prices increasing faster than energy prices. This would make land-extensive, low-energy production systems even less economically attractive. There will be little incentive for producers to invest management and capital in more energy-efficient adaptations of current high-energy systems unless there are increases in relative energy costs along with sufficient profits to finance the investment.

One promising research area is plant genetics. This has the potential to make the plant a more efficient producer through better utilization of solar energy and soil nutrients. Research investment in this area offers tremendous benefits in terms of higher productivity without necessarily requiring any greater direct energy input.

For background information *see* AGRICULTURE in the McGraw-Hill Encyclopedia of Science and Technology.

[OTTO C. DOERING, III]

Bibliography: T. J. Considine et al., in R. C. Loehr (ed.), *Food, Fertilizer and Animal Residues*, 1977; O. C. Doering et al., *An Energy Based Analysis of Alternative Production Methods and Cropping Systems in the Corn Belt*, Agricultural Experiment Station, Purdue University, Lafayette, IN, NSF/RA-770125, June 1977; G. H. Heichel, *Amer. Sci.*, 64:64–72, 1976; R. Klepper et al., *Amer. J. Agr. Econ.*, 59(1)1–12, February 1977; C. B. Richey, D. R. Griffith, and S. D. Parsons, in *Adv. Agron.*, 29:141–182, 1977; M. D. Skold, *Farmer Adjustments to Higher Energy Prices: The Case of Pump Irrigation*, USDA Economic Research Service, Publ. no. 663, November 1977.

Amaranth

The amaranth is any one of a large number of plant species in the genus *Amaranthus* (family Amaranthaceae) which are annuals (seldom perennials) and are distributed worldwide in warm and humid regions. Amaranths are botanically distinct for their small and chaffy flowers, arranged in dense green or red, monoecious or dioecious inflorescences (spikes), with zero to five perianth segments and two or three styles and stigmata, and for their dry membranous, indehiscent, one-seeded fruit. Amaranths are of great scientific interest at present as highly productive food plants.

Recent physiological, genetic, and nutritional studies have discovered their great potential economic value and have attracted the attention of researchers and laypeople concerned with world food problems. Of particular interest are the high productivity as a rapidly growing summer crop, large amounts of protein in both seed and leaf with high lysine, overall high nutritional value, and water use efficiency for the C photosynthetic pathway, as is the important place of amaranths in the culture, diet, and agricultural economy of the people of Mexico, Central and South America, Africa, and northern India.

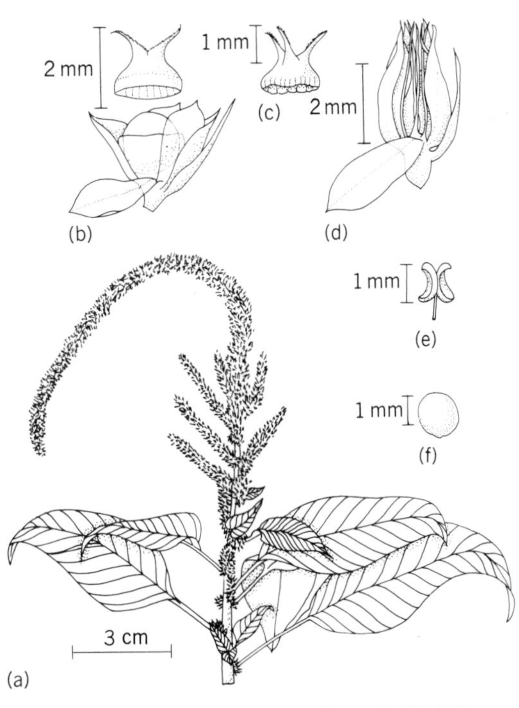

Fig. 1. *Amaranthus caudatus* L.: (*a*) branch with inflorescence; (*b*) adult female flower with dehiscent cap; (*c*) cap with three stigmata; (*d*) male flower; (*e*) anther; and (*f*) seed, lateral view. (*From G. J. H. Grubben, The Cultivation of Amaranth as a Tropical Leaf Vegetable, Roy. Trop. Inst. Commun. no. 67, 1976; reprinted with permission; Laboratory for Plant Taxonomy and Plant Geography, Agricultural University of Wageningen, and Royal Tropical Institute, Amsterdam*)

Classification and species relationships. The genus *Amaranthus* has been described by botanists in terms of up to 200 or more species, of which nearly 50 are distributed in the New World. However, removal of a great deal of confusion and synonymy will probably result in a much smaller number of distinct species. Some of the better-known and more widespread weedy species are red amaranth or pigweed *(A. hybridus)*, thorny amaranth *(A. spinosus)*, tumbleweed *(A. albus)*, wild beet *(A. retroflexus)*, and *A. powelli*. The common ornamental forms such as Prince's Feather *(A. hybridus* ssp. *hypochondriacus)*, Joseph's Coat *(A. tricolor)*, and Love-Lies-Bleeding *(A. caudatus)* are recognized by the form of their spikes and foliage colors.

Two sections of the genus are *Amaranthotypus*, with large terminal inflorescence, five tepals and stamens and fruit opening circularly; and *Blitopsis*, with axillary cymes, often two to four tepals and stamens, and fruits opening irregularly. The cultivated grain species *(A. cruentus, A. caudatus,* and *A. hypochondriacus = A. leucocarpus)* with ornamental and crop forms belong to the section *Amaranthotypus*, whereas two important cultivated vegetable species *(A. tricolor* and *A. lividus)* belong to the section *Blitopsis* (Fig. 1). Several Asian natives, such as *A. tricolor, A. gangeticus,* and *A. melancholicus*, have been grown as potherbs or ornamentals but never as a grain crop. The cultivated grain species are largely pale-seeded, whereas the ornamentals and weedy forms are black-seeded, presumably an important trait used by early peoples in keeping them apart. Interspecific hybridization, reported by several botanists in various regions, is considered to be a primary source of evolutionary diversity. The grain crop forms and their weedy relatives include both 32- and 34-chromosome species; much of the preliminary biosystematic and cytogenetic research has not yet provided conclusive species relationships. However, the weedy and cultivated species of *Amaranthotypus* are separable into two distinct groups on the basis of recent taxonomic and protein variation studies. Use of electrophoretic assays of genetic variation for many enzymes seems promising in identifying species and their hybrids. *See* BREEDING (PLANT).

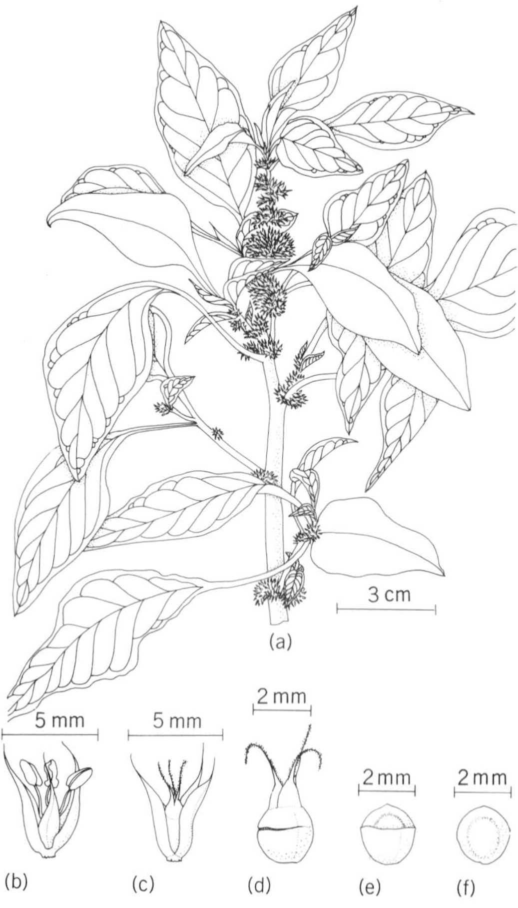

Fig. 2. *Amaranthus tricolor* L.: (*a*) branch with flowers; (*b*) male flower; (*c*) female flower; (*d*) fruit with dehiscing cap; (*e*) fruit without cap; and (*f*) seed, lateral view. *(From G. J. H. Grubben, The Cultivation of Amaranth as a Tropical Leaf Vegetable, Roy. Trop. Inst. Commun. no. 67, 1976; reprinted with permission; Laboratory for Plant Taxonomy and Plant Geography, Agricultural University of Wageningen, and Royal Tropical Institute, Amsterdam)*

Origin and history of domestication. In domesticated forms some of the main changes from the wild species are development of short and weak bracts and perfectly dehiscent and white seed with good popping quality and flavor in many cultivars. Ornamental forms are highly pigmented. The leafy vegetable forms permit many harvests of leaves, and some even have delayed flowering. According to Jonathan Sauer of the University of California, Los Angeles, three different wild or weedy forms *(A. hybridus, A. powelli,* and *A. quitensis)* growing in Mexico, Central America, Andean South America, and Bolivia-Argentina regions gave rise to the three grain species *(A. hypochondriacus, A. cruentus,* and *A. caudatus*, respectively) through domestication within each region (Fig. 2). This hypothesis is based on careful surveys of systematic, geographic, and other sources of evidence. Alternatively, some authors believe that *A. hybridus* might be a widely distributed wild progenitor, which in the north gave rise to two of the grain species, and which moving southward and hybridizing with a wild form, *A. quitensis*, gave rise to *A. caudatus* (including *A. edulis*) forms.

Amaranths are of ancient New World origin and were important as grains in the Aztec, Incan, and Mayan civilizations where wheat and rice were not grown. Archaeological and ethnobotanical evidence suggests that the earliest domestication took place in South America at least 4000 years ago. About A.D. 500 grain amaranths were a widely cultivated food crop in Mexico. Some of the ornamental forms moved from the New World to Africa and Asia in recent times (early 1800s), and may have spread from India to the East Indies, China, Manchuria, and Polynesia as a leaf vegetable in recent times. Ceremonial sacrifice and use in rituals drew attention of the Spaniards, who suppressed the Aztecs and the amaranth cultivation.

Cultivation. Amaranths are a very widely grown vegetable in tropical regions, including southern India, West and Equatorial Africa, and New Guinea (Fig. 3). G. J. H. Grubben of the Royal Tropical Institute in Amsterdam found amaranths to be the most popular vegetable in South Dahomey, and conducted extensive work there on various crop development aspects. Varieties such as Chinese

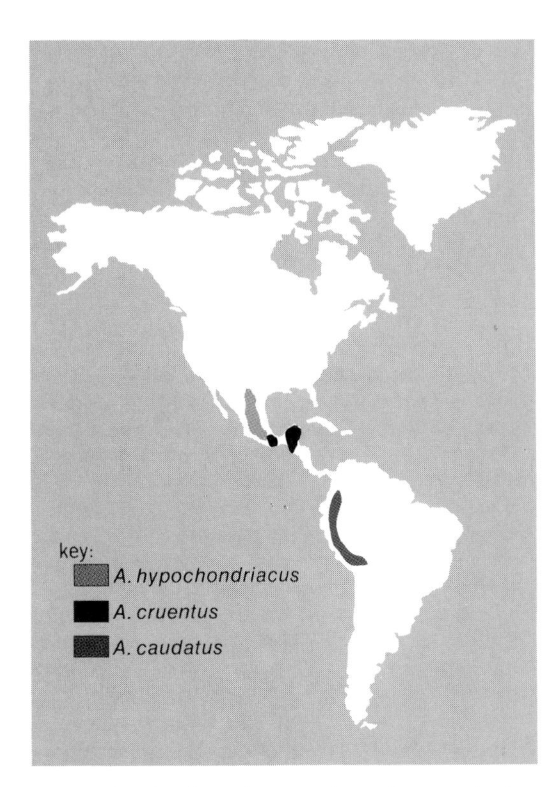

Fig. 3. Distribution of amaranths and areas of origin according to Sauer. *(From G. J. H. Grubben, The Cultivation of Amaranth as a Tropical Leaf Vegetable, Roy. Trop. Inst. Commun. no. 67, 1976; reprinted with permission; Royal Tropical Institute, Amsterdam)*

spinach, Ceylon spinach, and African spinach suggest its use as a salad ingredient; however, somewhat high oxalic acid content of leaves makes it unpleasant for this purpose.

The crop may be raised by transplanting seedlings in narrowly spaced rows and harvesting by uprooting, or by sowing seeds in wider rows, where repeated cuttings are used for 35–60 days before flowering. Amaranth production in swampy "chinampas" (pre-Hispanic floating gardens) to raise seedlings is practiced in Tulychualco, the "amaranth candy capital" of Mexico. Amaranths raised as a grain crop may also be mixed with millet, maize, or a legume, or even pepper or brinjal, as is done in northern India. Although seeds are small, a large number of seeds produced per plant allows optimistic yields to be as high as 3–6 metric tons/ha in Ethiopia and commonly 0.5–1.0 ton/ha. In Dahomey, a leaf form yielded 2.7 tons of dry matter per hectare in 6 weeks (6.4 g/m^2/day).

Food value. Leaf amaranths are rich in provitamin A, vitamin C, iron, calcium, and protein, with lysine as high as 5.9% of protein (equal to soymeal, and more than some of the best maize strains) and 10.8% glutamic acid in *A. edulis.* Amaranth grains, with very high starch quality and quantity, high protein (up to 17%), and high digestibility, rate better in nutritive value than all cereals. Primary modern uses in Mexico include making of alegrias (cakes of popped seed bound with sugar syrup), flour made of popped seed (with sugar) to prepare pinole, paste to make chuale, and amaranth milk (atole) from the black-seeded and variegated forms. In India, amaranth seed is used in alegrias and confectionaries (flour is cooked with rice and mustard) and as a cereal (popped seed).

Genetic, ethnobotanical, and agronomic researches are under way to develop amaranths as an important food plant in modern agriculture.

For background information *see* LYSINE; PHOTOSYNTHESIS; PROTEIN in the McGraw-Hill Encyclopedia of Science and Technology.

[SUBODH K. JAIN]

Bibliography: G. J. H. Grubben, *The Cultivation of Amaranth as a Tropical Leaf Vegetable*, Roy. Trop. Inst. Commun. no. 67, 1976; J. D. Sauer, *Ann. Missouri Bot. Gard.*, 54:103–137, 1967; *Proceedings of the 1st Amaranth Seminar*, 1977; H. B. Singh, *Grain Amaranths, Buckwheat and Chenopods*, Indian Council of Agricultural Research, 1961.

Analytical chemistry

The analysis of air, water, soil, food, body fluids, fuels, medicines, and a wide variety of consumer and industrial products for chemical components such as carcinogenic hydrocarbons, drugs, toxic and nutritional elements, naturally occurring organic species such as glucose and bilirubin, and enzymes is vitally important to society. Analytical spectroscopy represents one of the most powerful tools for these analyses. One of the significant developments in analytical spectroscopy in recent years is the use of television camera tubes (called imaging detectors) for a variety of applications.

An imaging detector is a collection of several hundred to several thousand individual photodetectors. One- and two-dimensional arrangements of photodetectors are available. When illuminated with light of different intensities, different detectors produce separate electrical signals with magnitudes that depend upon light intensity. Light transmitted through or emitted from a sample is separated into different wavelengths by a prism or grating device and directed to the television camera tube, which then measures several different wavelengths (colors) simultaneously. Imaging detectors, which were designed primarily for television applications initially, are useful in analytical spectroscopy because of this ability to simultane-

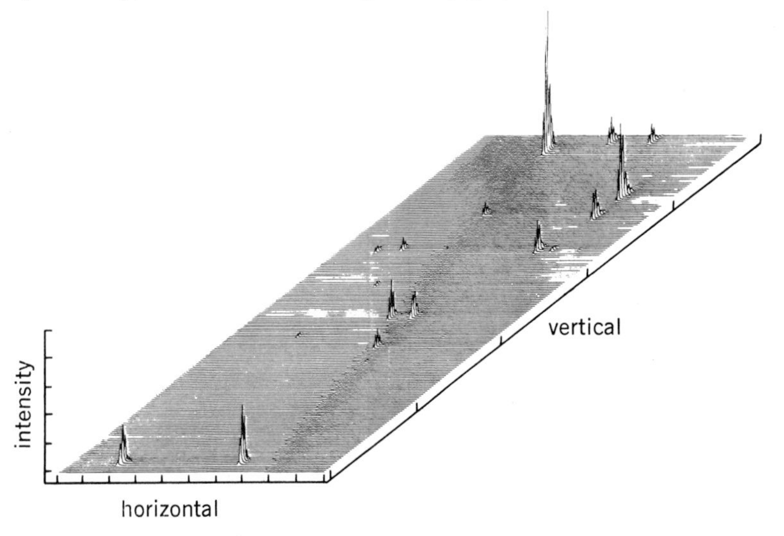

Fig. 1. Emission spectrum for mercury recorded with an image dissector interfaced to an echelle spectrometer.

ously monitor different wavelengths and hence to identify and quantify chemical components in complex samples.

Elemental analysis. Figure 1 shows a three-dimensional display of spectral peaks recorded with an imaging detector for mercury excited by an electrical discharge. Each position represents a different point along the horizontal and vertical axes of the two-dimensional imaging detector. Each peak represents the intensity of light emitted at a different wavelength (displayed along the horizontal axis) for different regions of the ultraviolet-visible spectrum (displayed along the vertical axis). The positions of the different peaks in the display are unique for this element, and the heights of the peaks depend upon the amount of mercury present. Thus, a chemist can use this information both to confirm the presence of mercury and to determine how much of it is present. Another element, such as cadmium, lead, or arsenic, gives an entirely different spectrum. If several elements are present in the same sample, then peaks for the elements appear at different points in the two-dimensional display, and these peaks can be used to identify and quantify the different elements present. The instrumental system used to produce Fig. 1 has been utilized to determine 10 elements in the same sample without a separation, and is capable of determining 20 or more elements simultaneously. These types of analyses are useful to physicians trying to determine which of several possible elements is responsible for a patient's illness, to scientists studying the relationship between trace elements and occupational diseases such as black lung, and to environmentalists, food scientists, and agronomists.

Simultaneous determinations of multiple trace metals have been carried out in substances such as serum, water samples, and alloys. Some investigators have used systems that produce two-dimensional spectra such as that shown in Fig. 1, while others have used simpler systems that produce one-dimensional spectra. These simpler systems are less expensive, but give less information. Methods have included atomic absorption, atomic emission, and atomic fluorescence spectroscopy.

Drug determinations. Another type of application involves the absorption of ultraviolet energy

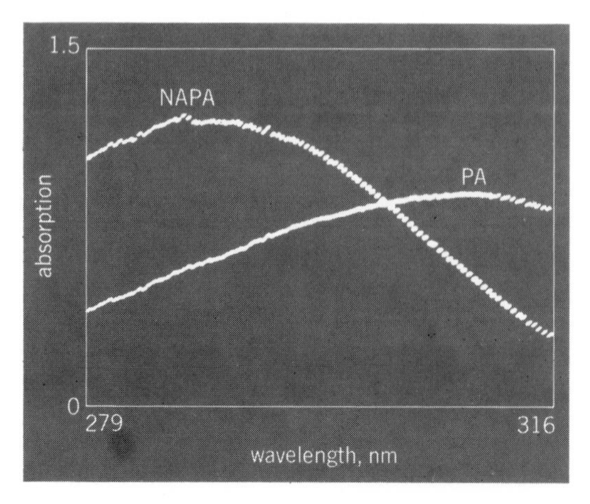

Fig. 2. Absorption spectra for procainamide (PA) and *N*-acetylprocainamide (NAPA) recorded with vidicon spectrometer.

(Fig. 2). This figure shows different amounts of energy absorbed by a drug, procainamide (PA), and a metabolic product of the drug, *N*-acetylprocainamide (NAPA). PA and NAPA are used in the control of erratic heartbeats in some patients. Determining the amount of drug and metabolite in a patient's serum is important for dosage control and for analysis of the drug's effectiveness. The different spectral properties measured with an imaging detector can be used with specialized computer data-processing methods to determine these drugs in serum without having to separate them.

The procedure illustrated in Fig. 2 for two components has been extended successfully to samples containing up to seven different drugs.

A second approach uses liquid chromatographic procedures to separate the drugs and an imaging detector to record the spectrum of each component. Both absorption and fluorescence methods have been used to identify and quantify components. In a third approach similar to that in Fig. 2, a two-dimensional imaging detector is used to record fluorescence excitation and emission spectra simultaneously for drugs. These data are much more informative than conventional fluorescence spectra, and the method is being developed for simultaneous multicomponent determinations.

Kinetic studies. Still another type of application involves the measurement of the visible light absorbed at different points in time when bilirubin in serum reacts with a particular organic compound (Fig. 3). Bilirubin is a product in the breakdown of hemoglobin in the blood, and gives the yellow color to the skin in jaundice. It is important in many types of illnesses, and is particularly significant in newborn infants because too much bilirubin can cause brain damage resulting in mental retardation or even death in some severe cases if corrective action is not taken. The chemical reaction represented by data in Fig. 3 is very fast, taking place in less than 1 s. The imaging detector is very useful in this area because it permits scientists to monitor these very fast spectral changes and to understand the chemical processes involved. Information gained in this type of study has been used to develop a new approach for the simultaneous determination of the different forms of bilirubin that exist in body fluids.

More generally, the processes of controlling industrial chemical reactions so that they will produce needed products at economical costs, of understanding disease processes so that they can be diagnosed and treated, or of evaluating the effectiveness of drugs and other medicines often involve detailed studies of very complex chemical reactions. Many of these reactions involve several steps. Because imaging detectors can monitor several species simultaneously (as shown in Fig. 3), they have been used successfully in a variety of basic studies by D. W. Margerum and coworkers. One example involves the very complex behavior of so-called chloramine compounds that are formed between chlorine and ammonia. These studies are important because of both the desirable effects of chlorine when added to swimming pools and the undesirable toxic effects of chloramine compounds on plant and fish life in natural waters.

Hemoglobin. Carbon monoxide poisoning results from an interaction of the gas with hemoglo-

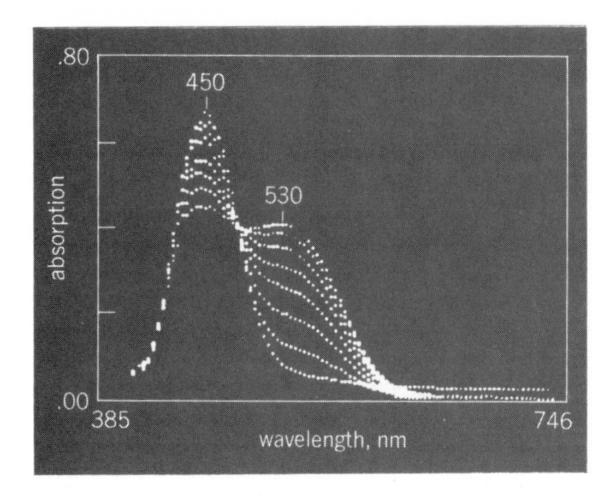

Fig. 3. Time-dependent absorption spectra for reaction of bilirubin with *para*-diazobenzene-sulfonic acid recorded with vidicon spectrometer.

bin so that the ability of the blood to transport oxygen to vital organs is reduced. This interaction results in changes in the absorption spectrum that can be detected. M. J. Milano at the State University of New York at Buffalo used a photodiode array in a unique device to monitor changes in hemoglobin spectra in the bloodstream. A miniature photometer cell fashioned with a light-pipe is placed in the tip of a hypodermic needle, so that when the needle is inserted into a vein, the system monitors the light absorption of the blood. A computer processes the data to obtain the first derivative spectrum, which is very sensitive to changes in hemoglobin. This technique permits the continuous monitoring of chemicals in living animals.

Clinical chemistry. Imaging detectors have numerous applications in clinical chemistry and the diagnosis of illness. These include the determination of drugs and trace metals mentioned earlier, as well as other naturally occurring chemical species such as glucose, uric acid, cholesterol, bilirubin, and enzymes such as lactate dehydrogenase and alkaline phosphatase.

Summary. The applications discussed above represent highlights of recent studies involving imaging detectors in analytical spectroscopy. These detectors offer real advantages compared with some conventional instrumentation because they can monitor many different wavelengths rápidly and repetitively. This area is in its infancy in terms of technical developments, and the applications reported to date should encourage manufacturers to develop better detectors designed explicitly for specific applications. It is not expected that imaging detectors will replace their more conventional counterparts, but rather that they will fulfill an important role, along with the conventional detectors, in spectrochemical instrumentation of the future.

For background information *see* ANALYTICAL CHEMISTRY; SPECTROSCOPY in the McGraw-Hill Encyclopedia of Science and Technology.

[HARRY L. PARDUE]

Bibliography: H. L. Felkel and H. L. Pardue, *Anal. Chem.* 50:602, 1978; G. Horlick, *Appl. Spectrosc.*, 30:113, 1976; D. W. Johnson, J. B. Callis, and G. D. Christian, *Anal. Chem.*, 49:747A, 1977; A. E. McDowell and H. L. Pardue, *J. Pharm. Sci.*, 67: 822, 1978.

Animal systematics

Paleontology is currently undergoing considerable reevaluation in the area of theory and methodology of phylogenetic analysis. Most thinking on the subject is traditional, the major premises having been established by Charles Darwin in 1859 in *On the Origin of Species*. Traditional thought is characterized by the following assumptions: (1) Species are nondiscrete entities through time, and their taxonomic delineation involves an arbitrary division of an evolutionary continuum. (2) Evolution is viewed as a slow, gradual transition through time, with occasional geographic segregation of populations. (3) Because evolution is a gradual process through time and space, the study of fossils offers the only avenue for reconstructing the phylogenetic and distributional history of plants and animals. (4) Phylogeny can be reconstructed by an empirical approach, namely, by linking organisms from successive strata on the basis of overall similarity, and in this way fossils provide "directly historical data" on patterns of evolutionary change.

In recent years many neontological and paleontological systematists have begun to challenge these traditional assumptions. They have pointed out that all observations are theory-laden and that the data of many classic paleontological studies can be interpreted differently under the guise of new theory. In addition, within neontology the emergence of new theoretical and methodological concepts proposed by the followers of Willi Hennig, constituting the school of phylogenetic systematics, is beginning to have a profound impact on traditional paleontological thinking relating to the reconstruction of phylogeny.

Species and the evolutionary process. The question of whether species are real and discrete or whether they are arbitrary segments of an evolutionary continuum created by humans has been under debate for many years. Recently, the debate has intensified, and the outcome has implications for the evaluation of methods to reconstruct phylogenetic history. Niles Eldredge and Steven Gould have proposed the application of the neontological allopatric speciation model to the paleontological record. Under this view, morphological gaps in the stratigraphic record are interpreted as real, not as the result of an imperfect record. Furthermore, morphological stasis is not to be ignored in documenting change through time, but rather is to be used as the norm for most of a species' history. Morphological change, then, is postulated to occur rapidly in a geological sense and to take place near the time of geographical isolation of species' populations.

One important implication of this model is that species can be viewed as distinct evolutionary units through time, the boundaries between species being demarcated by these punctuational speciation events. Methodologically, this suggests paleontology must first identify these unit taxa (species) and then develop a hypothesis about their genealogical (cladistic) relationships; it is not important if there are gaps in the fossil record, because hypotheses are created only for the taxa known to exist. In the more traditional view of phyletic gradualism, reliance on a continuous stratigraphic record was considered paramount; the phylogenetic history of groups with a poor fossil

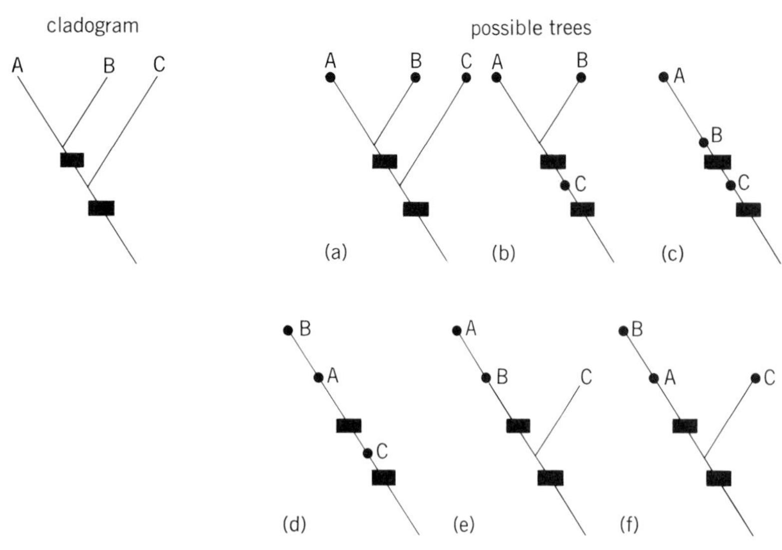

Fig. 1. A cladogram for three species with synapomorphies, symbolized by black rectangles. The six evolutionary trees (*a-f*) are all consistent with the cladogram.

record was typically considered incapable of analysis.

Phylogenetic reconstruction. Paleontologists are beginning to recognize that fossils, in themselves, are not a necessity for developing hypotheses about phylogenetic history. This view does not deny the importance of fossils, but the phylogenetic status of fossil taxa must be evaluated within the framework of conceptions of the relationships of the Recent biota. The process of identifying a fossil involves a series of statements about its hypothetical relationships to Recent taxa, that is, one classifies the fossil as a vertebrate, mammal, perissodactyl, and so on.

Cladograms and trees. If it is accepted that phylogeny involves branching and divergence, then the pattern of relationships among taxa is hierarchical, and the first goal of phylogenetic analysis is to formulate a hypothesis about that hierarchical pattern. This hypothesis may be represented by a cladogram, or branching diagram. In this sense the word does not necessarily infer the types of relationships—common ancestry or ancestral-descendant—that might obtain among the taxa; it is simply a hypothesis about the nested hierarchical pattern of similarities. Trees, on the other hand, are based on a cladogram, but contain hypotheses about the types of relationships among the taxa (Fig. 1).

What types of similarities might be expected to show a nested pattern? Following a speciation event, it is expected that the resultant lineages will differentiate morphologically in one or more characters. Methodologically, if such characters cannot be identified, it is not possible to document the existence of the lineage or the speciation event; that is, ancestors and descendants are indistinguishable. Thus, each lineage is characterized by features inherited from its ancestor (primitive or plesiomorphic) and is defined by its own unique, derived (apomorphic) characters. It is these derived characters, at each level of the hierarchy, that form a nested pattern among the taxa; characters derived at one level of the hierarchy will be primitive at all inclusive levels. A derived feature shared by the taxa of a lineage is termed a synapomorphy, and thus cladograms are hypotheses about nested synapomorphy.

Cladogram analysis. There are three types of criteria that can be used to formulate hypotheses about directionality of character-state transformation. First, the character state occurring earlier in the stratigraphic record can be hypothesized to be primitive relative to subsequent character states. This may have some validity as a general statement, but the inadequate fossil record of many groups and discontinuities in the stratigraphic column dictate extreme caution when this criterion is applied. Second, the character state occurring earlier in a developmental sequence can be hypothesized to be primitive relative to later character states of the transformation sequence. Given the availability of such data—and they would be restricted to well-studied Recent forms—this criterion potentially offers an important tool for postulating derived characters. Third, the character state occurring widely in presumably closely related groups is considered to be primitive. This is the most frequently applied and generally useful criterion.

Cladogram construction and testing. A cladogram is a branching diagram on which the pattern of synapomorphous similarity can be superimposed. Of the numerous alternative cladograms generally possible for a group of taxa, the cladogram that is preferred is the one that best depicts the nested pattern of synapomorphy. Each postulated synapomorphy serves as a test of each cladogram in that it either is or is not congruent with other synapomorphies defining nested sets of taxa. A synapomorphy not defining a nested set within a cladogram is said to be inconsistent with that cladogram. Applying a parsimony criterion, the preferred cladogram is the one in which such inconsistencies (retrospectively defined as convergences) are minimized.

Hypotheses about trees. Given the acceptance of a particular cladogram for a group of species taxa, there may be a number of evolutionary trees consistent with that cladogram. In formulating hypotheses about trees, a decision is made to recognize branch points of a cladogram as speciation events or to recognize one of the coordinate species of each branch point as the ancestor of the other. (Clearly, if the taxa are supraspecific in rank, then only common ancestry can be postulated, because only species-level taxa can evolve and be ancestors.)

Without any precisely stated criteria, paleontologists have traditionally used earlier stratigraphic occurrence to identify ancestral species. However,

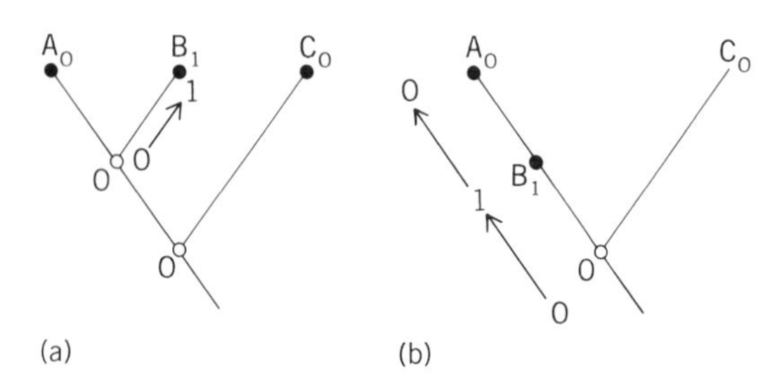

Fig. 2. Testing the hypothesis of whether species B is an ancestor of species A. Explanations of *a* and *b* are found in the text.

recent cladistic formulations have provided more stringent morphological criteria for choosing among tree hypotheses (Fig. 2). A given species can be rejected as an ancestor if it can be shown to possess a derived character not shared with other species (an autapomorphy). Species B in Fig. 2 has such a state [indicated by (1)]; species A and C have the primitive character state (0). The species is rejected as an ancestor on the basis of parsimony (Fig. 2), since it involves fewer specific assumptions to postulate the single origin of a derived character state (Fig. 2*a*) than its origin and subsequent loss (Fig. 2*b*). Thus, the presence of an autapomorphy rejects a hypothesis of ancestry but remains consistent with a common-ancestry hypothesis. If an autapomorphy is not found, either because it is not present or is undetected, an ancestor-descendant hypothesis might then be postulated (Fig. 2*b*). However, the tree postulating all possible speciation events (Fig. 2*a*) is still consistent with the available data. In conclusion, it seems impossible to choose only a tree involving direct ancestry because the alternative tree, in which none of the species are ancestors, can itself never be rejected.

For background information *see* ANIMAL SYSTEMATICS; PALEONTOLOGY; PHYLOGENY in the McGraw-Hill Encyclopedia of Science and Technology.

[JOEL CRACRAFT]

Bibliography: J. Cracraft and N. Eldredge (eds.), *Phylogenetic Analysis and Paleontology*, in press; S. J. Gould and N. Eldredge, *Paleobiology*, 3:115–151, 1977; N. I. Platnick, *Syst. Zool.*, 26: 438–442, 1977; E. O. Wiley, *Syst. Zool.*, 23: 233–243, 1975.

Archeology

Coastal shell middens have long been used as evidence of human occupation, but detailed investigations have traditionally focused on associated artifacts and burials. It now appears that the shells themselves can provide useful information on the patterns of occupation and utilization of the sites.

This potential is becoming realized through growth line analysis. Shell growth varies with the seasons, and once the pattern is understood it is possible to determine the season in which an individual shell was harvested. Analysis of a representative sample of shells from a single horizon permits the interpretation of the season and duration of the harvest, as well as any seasonal patterns in the harvesting of various shellfish species. The extension of this approach to other horizons can identify long-term variations in site occupation and utilization which might otherwise escape notice or remain speculative. Such information can be of immense value in studying prehistoric cultures.

Growth lines. Many mineralized tissues, such as mollusk shells and coral skeletons, are continually grown or accreted during the life of the organism. Because the physiology of cold-blooded organisms is strongly influenced by temperature and other environmental variables, new material added during different seasons will commonly exhibit characteristic variations. Such periodic variations are called growth lines. Growth lines can be more precisely defined as abrupt or repetitive changes in the character of an accreting tissue. A growth increment is the interval between two successive growth lines. A shell may exhibit more than one type of growth line; some bivalves, for example, appear to have both daily and annual growth lines. Growth lines can also be separated into external lines, which can be seen on the outer surface, and internal lines, which are best seen by sectioning the shell (Fig. 1). Some skeletal tissues, such as pearls, have internal growth lines only, but most have both internal and external lines.

Although these organisms are influenced by the external environment, many other factors are involved in their shell growth. Thus growth line patterns are not simply reflections of variations in the environment. A few shell characteristics may be expected to show extremes which coincide with, in this case, the times of warmest and coldest water temperatures. Other characteristics exhibit limit or threshold effects; organic content, for example, may increase slowly with water temperature until a certain temperature is reached and then show no further change until the water temperature drops below that point. In another species growth rate may initially increase with water temperature but suddenly drop when the water is warm enough for spawning; in each of these instances the growth line pattern is annual, but the maximum temperature is not indicated.

Many mollusks occupy environments whose seasonal extremes put them under severe stress. During such times of stress shell growth is nearly if not completely halted, and resumption of normal growth leaves a very prominent growth line called an annual ring. Under some circumstances this growth halt can continue for several months; in such cases not only is there no indication of the time of most extreme temperatures, but there is no way of distinguishing between an individual which was harvested at the beginning of the growth halt and one which was harvested at the end. Equally prominent growth lines, called disturbance rings,

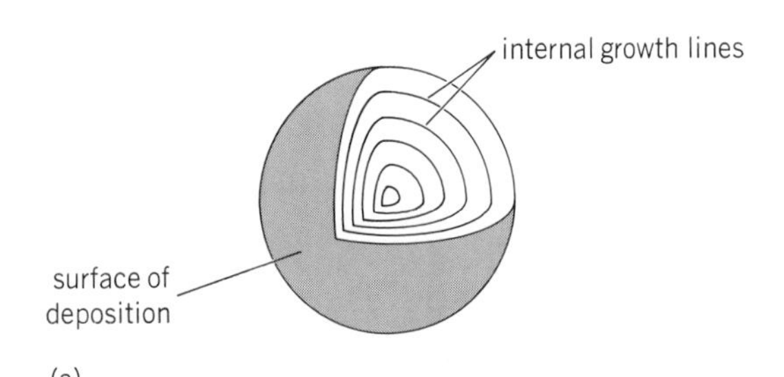

external growth lines
internal growth lines
surface of accretion
surface of deposition
(b)

Fig. 1. Two examples of the accretion of skeletal tissue and the development of growth lines. (*a*) Pearl. (*b*) Limpet. *(From G. R. Clark, II, Growth lines in invertebrate skeletons, Annu. Rev. Earth Planet. Sci., 2:77–79, 1974; used with permission; © 1974 by Annual Reviews Inc.)*

can be caused by much shorter growth halts related to random events such as storms or unsuccessful attacks by predators. These can closely resemble the annual rings when examined on the external surface only.

An additional complication is that many shellfish grow rapidly for a few years and then dramatically reduce their growth rate. This has the effect of crowding the annual growth lines so closely together on the older shells that they cannot be distinguished by external examination alone. Moreover, the shift from the mature, relatively fast-growing phase to the senile, very slow-growing phase is quite sudden. This makes it especially difficult in a mature shell to decide, from external examination alone, whether a narrow strip of new growth beyond an annual ring represents a small part of another fast-growth year or a large part of the first slow-growth year. These observations lead to two conclusions about annual growth lines: the most prominent lines are the least accurate, and examination of external lines alone is a hazardous approach.

Fig. 2. Growth lines in the bivalve *Mercenaria mercenaria*. *(a)* External lines on left valve of individual in senile stage. *(b)* Cut and polished section of the margin of the same shell. Note relationships between internal and external lines. *(c)* Thin section of a portion of the same shell. *(From G. R. Clark, II, Seasonal growth variations in the shells of recent and prehistoric specimens of Mercenaria mercenaria from St. Catherines Island, Georgia, Anthropol. Pap. Amer. Mus. Nat. Hist., in press)*

Many short-period growth lines are also present in mollusk shells. These appear to be variously related to day and night, the rise and fall of the tides, long-term tidal cycles, and combinations of these effects. Such lines seem most useful as indicators of growth rate; during periods of slow growth they are tightly spaced, whereas during periods of fast growth they are far apart. Some attempts have been made to determine the time of year by counting the lines, but this requires both a meaningful starting point and an accurate knowledge of the periodicity of the lines.

Methods of study. Growth lines are expressed in many ways. The most common lines visible on the outer surface of mollusk shells are grooves or ridges, although color variations are found on a few species. Because of its nature as a surface of deposition, the inner surface of a shell lacks growth lines. It may, however, exhibit a wide variety of growth lines when properly sectioned and prepared. Variations in transparency, color, mineralogy, abundance of organic matter, and size, shape, and orientation of crystals may all vary enough to produce noticeable growth lines. More subtle variations, such as trace-element chemistry and oxygen isotope ratios, are generally studied only to confirm the significance of the obvious growth lines.

External lines are usually studied by simple examination (Fig. 2*a*). Because of the crowding mentioned earlier, only shells still in a relatively mature stage are suitable for this approach. Disturbance lines may confuse the interpretation, but can often be determined to be narrower or fainter than annual lines.

Internal lines are far more dependable, but require additional preparation. An exception to this is the crude but effective technique of "candling"—holding the shell in front of a strong light to reveal the zones of relatively transparent shell. These usually indicate long periods of stress and thus correlate with annual growth lines; this is one of the best ways to distinguish annual lines from disturbance lines when using external examination only.

The usual first step in the examination of internal lines is to section the shell. For certain studies of crystal form and orientation a fracture section is preferred, but for most purposes the shell is cut on a diamond saw and polished to remove scratches. This preparation may suffice if variations in color or transparency are prominent (Fig. 2*b*), but if more detail is desired either of two additional steps may be taken. For many purposes, particularly if there are variations in transparency, the best preparation is a thin section. For this technique the polished surface is cemented to a glass slide, and the remaining shell is sliced off just above this surface. With further polishing the section is reduced to a highly transparent film whose details can be readily studied under a microscope (Fig. 2*c*). An alternative preparation method is to etch the polished surface with a weak acid or chelating agent to emphasize differences in crystal size, organic content, or mineralogy. This etched surface can be studied directly by reflected light or scanning electron microscopy, or an acetate peel replica may be made for examination by transmitted light microscopy. Both thin sections and peels can be used as negatives to make photographic enlargements for easier study.

Determination of season. The identification of annual growth lines on a shell is of little value, in itself, to archeology. It is essential to be able to determine the time of year in which every part of the annual increment has formed and, if pertinent, the time and duration of halts in growth. Fortunately, nearly all species of mollusks found in middens are also living today in nearby areas, and it is relatively simple to make collections of these specimens during the various seasons. (A collection made during a single season is of little value.) This comparative material can then be examined by the same techniques as the archeological specimens, and the precise seasonal relationships determined.

Occasionally the recent shells do not display growth patterns like those in the archeological shells. This may be due to differences in collecting areas; for example, clams living on wave-swept beaches in the high intertidal experience different stresses from those living in tidal channels in the low intertidal or subtidal. Every effort should be made to collect recent specimens from all types of environments; even so, changing sea levels in the past few thousand years may have eliminated the particular local environment from which the archeological mollusks were harvested.

Another factor contributing to differences between modern and ancient shell growth patterns is climatic change. Although such changes are not known to have been extreme in the past few thousand years, relatively small changes can bring about dramatic differences in those growth lines related to environmental stress. For example, the common Atlantic bivalve *Mercenaria mercenaria* experiences environmental stress in the winter in areas north of the Carolinas, but is stressed in the summer further south. Thus the annual growth line indicates winter in cold climates and summer in warm climates, and might indicate either or both, or be absent, in the intermediate zone. A collection of this species from middens many thousands of years old—in, for example, South Carolina—may not have the same seasonal growth patterns as specimens living there today. Independent checks by paleotemperature methods, such as oxygen isotopes, are desirable to avoid misinterpretation in such cases.

Although the prominent annual rings are the easiest to use, most shells have many other kinds of seasonal growth lines. A thorough understanding of these will help to avoid some of the problems associated with the interpretation of annual rings, and in any case will increase general confidence in interpretations.

Applications and potential. The application of seasonal information of this type to the interpretation of patterns of migration and occupation must be done with some caution. It should be noted, for example, that while shells whose growth has ended in September indicate harvesting and therefore occupation of the site in September, they do not prove that the site was not occupied in July and August. Other foods may have been preferred, or the shellfish may have been toxic (due to red tides) during that period.

Occasional shells are to be expected with death indicated some months apart from the rest of the collection. These can be explained as those harvested as empty shells, and upon close examination they will often be found to contain worm tubes or other evidence of death before harvesting.

Shells found in burial mounds must be interpreted with particular care. If evidence is lacking of the ceremonial function of layers of shells, it must be noted that either empty shells or shells containing meats could have been used to serve the purpose. If clean empty shells are preferred, either beach debris or older middens are a potential source, and neither is likely to be restricted to shells with death occurring in the same month as the burial. On the other hand, shells found articulated (both valves touching) are a good indication that death occurred shortly before (or at the time of) burial.

Although dried shellfish meats are known to have been transported considerable distances in primitive cultures, it is unlikely that the shells in middens were transported to a significant degree. Ornamental and ceremonial shells are a different matter, however, and this is another reason for caution when dealing with burials.

The sensitivity of growth line patterns in some species might prove helpful in situations where climatic change seems involved. Thus, what adds to the confusion during initial interpretation may prove a benefit later.

Growth lines in other organisms are also being studied and applied to seasonal determinations. Mammal teeth and fish remains seem particularly useful in this regard. As mammals, fish, and mollusks make up a substantial part of the diet of coastal cultures, a combined investigation of all types of growth lines may eventually permit interpretations of year-round migrations and food-gathering strategies.

For background information *see* ARCHEOLOGY; MOLLUSCA in the McGraw-Hill Encyclopedia of Science and Technology. [GEORGE R. CLARK, II]

Bibliography: G. R. Clark, II, *Annu. Rev. Earth Planet. Sci.*, 2:77–99, 1974; G. R. Clark, II, *Anthropol. Pap. Amer. Mus. Nat. Hist.*, in press; L. C. Ham and M. Irvine, *Syesis*, 8:363–373, 1975; M. J. Kennish and R. K. Olsson, *Environ. Geol.*, 1: 41–64, 1975; R. A. Lutz and D. C. Rhoads, *Science*, 198:1222–1227, 1977.

Asteroid

A new, unclassifiable object has been discovered between the orbits of Saturn and Uranus. This object seems too small to be a planet, too large to be a comet, and too distant to be related to the asteroids. The discovery was made on Nov. 1, 1977, by astronomer Charles T. Kowal of the Hale Observatories. Meanwhile, at the other extreme of distance, additional "close-passing" asteroids were discovered during 1977.

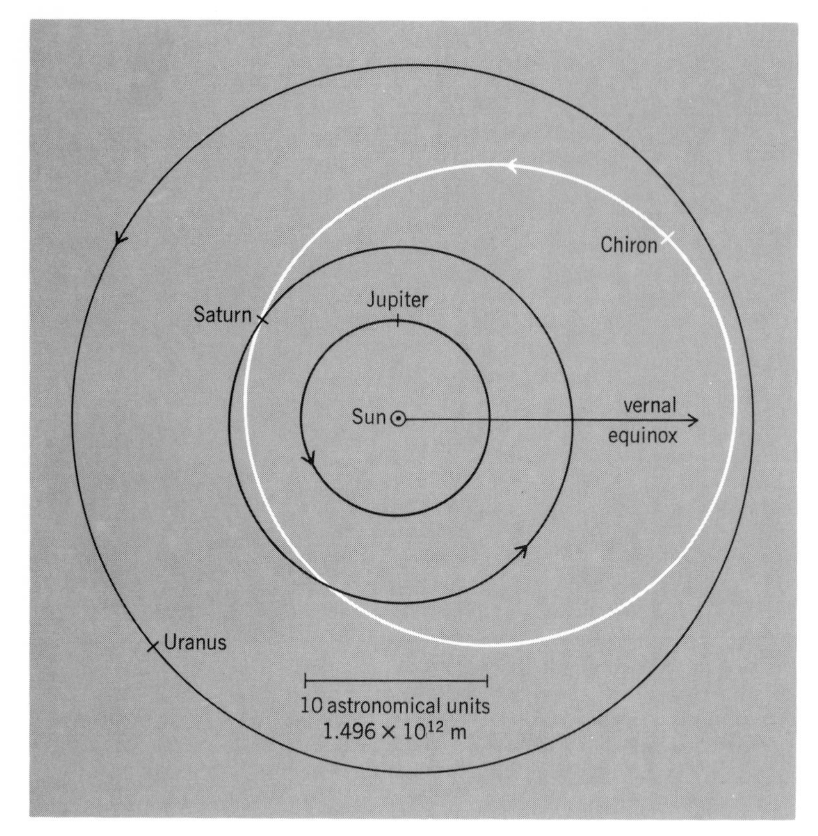

Fig. 1. Orbits of Chiron and of the planets Jupiter, Saturn, and Uranus. Tick marks show location of each planet on Jan. 1, 1978. *(From J. Ashbrook, Kowal's strange slow-moving object, Sky Telesc., 55:4–5, 1978)*

Chiron. The new object has been named Chiron by its discoverer. Chiron was the most prominent of the Centaurs, those creatures of Greek mythology who were half man, half horse. Kowal suggests that if more such objects are found in the future, they could be named after other Centaurs, and the objects as a group could be called the Centaurian Objects.

Chiron travels around the Sun in an eccentric orbit which takes it from inside the orbit of Saturn out to the orbit of Uranus (Fig. 1). It takes 50.7 years to make one revolution around the Sun.

The only objects previously known in that part of the solar system were the outer planets and the comets. The new object is estimated to be between 75 and 300 mi (120 and 480 km) in diameter. This is roughly 20 times smaller than the smallest planet, but about 100 times larger than a typical comet nucleus. Chiron is comparable in size to the largest asteroids, but it never goes near the asteroid belt.

Discovery and orbit computation. Kowal obtained his first photographs of Chiron on Oct. 18 and 19, 1977 (Fig. 2), but the actual discovery was made during a microscopic examination of the photographs on November 1. The object was noticed because of its unusually slow motion—much slower than the motion of the asteroids. This implied that the object was at a very great distance from the Earth.

After more observations were obtained, it became possible to compute a preliminary orbit for the new object. This was done by B. G. Marsden of the Smithsonian Astrophysical Observatory. The preliminary orbit was used to compute the position of Chiron in past years. In this way, images of Chiron were found on photographs taken in 1969 and 1952. These old positions were then used to improve the orbit computations, after which it became possible to find the object on still older photos. Finally, Chiron was traced all the way back to photographs taken in 1895—practically the dawn of astronomical photography! Chiron had not been discovered on any of those old photographs, simply because no one was looking for such an object.

Origin and nature. Since Chiron crosses the orbit of Saturn, it can occasionally come quite close to that planet. It is now known that Chiron passed within 1.6×10^7 km of Saturn in 1665 B.C. This is as far back as the orbit can be reliably computed. Such close approaches to Saturn could change the orbit of Chiron considerably. It is still not known, however, if Chiron was formed near its present location or in an entirely different part of the solar system.

One theory holds that Chiron is a giant comet that was captured into its present orbit by the gravitational pull of Uranus and Saturn. If this is true, Chiron is the biggest comet ever seen. It does not look like a comet, simply because the gases which give comets their characteristic appearance are all frozen into solid ice.

It is also conceivable that Chiron originated in the asteroid belt and was "kicked out" by a collision or some other process. If such processes can actually occur, there should be many other asteroids outside the asteroid belt. No such asteroids have been found. There is one asteroid, named Hidalgo, that does travel out past the orbit of Saturn, but it returns to the asteroid belt during part of its orbit. Only Chiron stays far beyond the orbit of Jupiter at all parts of its orbit, and only Chiron reaches the orbit of Uranus.

Another possibility is that Chiron was formed in approximately its present location and somehow avoided collision with Saturn during the past 4 or 5×10^9 years. If this is true, then Chiron is a remnant of the building blocks that formed the outer planets. Chiron may then give much information about the formation of the solar system.

Chiron was farthest from the Sun in 1970. It is now coming closer and getting brighter. By the time Chiron reaches its smallest distance from the

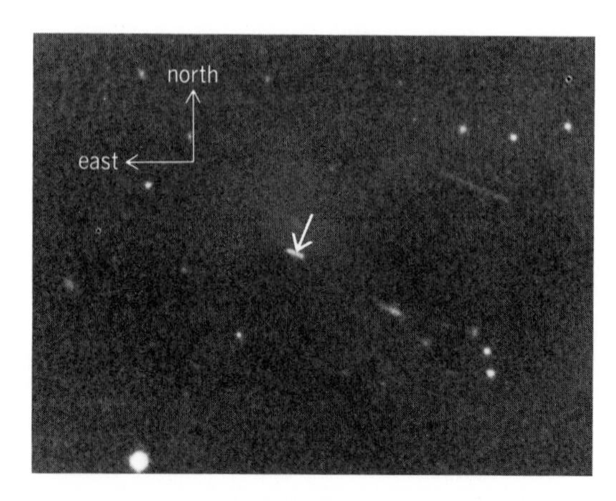

Fig. 2. Photograph of Chiron taken on Oct. 19, 1977. The 75-min exposure was taken with the Palomar 48-in. (1.2-m) Schmidt telescope. *(Hale Observatories)*

Sun, in 1996, it will become bright enough to be studied in considerable detail. It should then be possible to learn the composition of Chiron and, hopefully, how the object originated.

Close-passing asteroids. While Chiron is the most distant "asteroid" known, asteroids at the other extreme of distance are also being discovered. In April 1977 Kowal discovered asteroid *1977 HB*, which passed within 1.8×10^7 km of the Earth. In November 1977 E. Helin and E. Shoemaker of the California Institute of Technology discovered asteroid *1977 VA*, which was 2.1×10^7 km away. These asteroids are important because they are among the easiest asteroids to reach with a space probe. Since they pass close to the Earth at a relatively slow speed, they could be reached with a fairly small expenditure of energy and fuel. Both of these asteroids are less than 2 km in diameter.

These close-approach asteroids may be of enormous practical importance in the future. There has been considerable discussion of actually mining such asteroids for their raw materials, such as iron and nickel. These materials could be transported to the Earth, to provide an inexhaustible supply of metals, or they could be used in space to construct gigantic satellites and space colonies.

If astronomers continue to discover more of these asteroids, it is quite likely that some will be found which can be reached more easily than the Moon. When that happens, it will open up new possibilities for the commercial utilization of space.

For background information *see* ASTEROID; COMET in the McGraw-Hill Encyclopedia of Science and Technology. [CHARLES T. KOWAL]

Bibliography: J. Ashbrook, *Sky Telesc.*, 55:4–5, 1978; B. O'Leary, *Science*, 197:363–366, 1977.

Atom

In the last 3 years several groups have developed methods for the detection and identification of small amounts of matter, including single atoms. Such methods permit the study of several rare events which provide important clues for the understanding of elementary particles, cosmology, nuclear matter, and quantum electrodynamics. Moreover, these new sensitive methods can be used to study atomic collisions and chemical processes. The first demonstration of the detection and identification of single atoms was carried out in 1976 by Samuel Hurst, M. H. Nayfeh, and Jack Young. Shortly after this demonstration, successful single atom detection was reported by J. Gellbach and colleagues, D. Lewis and colleagues, and V. Letokhov and colleagues, and various applications were carried out or suggested.

Resonance fluorescence. Although the concept of counting and identifying individual atoms predates Democritus (400 B.C.), sufficiently sensitive instrumentation was not available until the advent of narrow-band tunable lasers. The first major advance came in 1975 when W. Fairbank, Jr., T. Hänsch, and A. Schawlow at Stanford University detected sodium at a concentration of 100 atoms/cm³ and thus improved the existing sensitivity by seven orders of magnitude. This was achieved by exciting sodium atoms with a continuous collimated laser beam resonant with the D-line transition. The presence and density of sodium were then inferred from the amount of light scattered from the forward direction of the exciting beam. This method is known as resonance fluorescence.

Resonance ionization spectroscopy. As the reduction in the number of atoms to be detected continues, resonance fluorescence loses some of its sensitivity and becomes quite complicated. The method of resonance ionization spectroscopy (RIS), however, is very sensitive in this region. RIS involves saturated absorption from the ground state followed by one or more additional selective saturated absorption steps that eventually result in selective ionization of the atom. Figure 1 shows the atomic energy levels and transitions involved in this process for two-photon ionization. An atom in its ground state is excited to a bound state when a photon is absorbed from a laser beam at a very well-controlled wavelength. Only if this resonance photoabsorption occurs will a second photon have sufficient energy to remove an electron entirely from the atom.

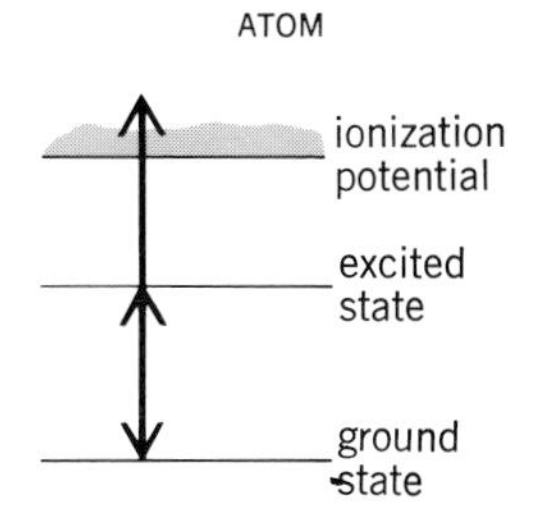

Fig. 1. Atomic energy levels involved in resonance ionization spectroscopy in the case of two-photon ionization. Arrows indicate transitions involved in the process.

In addition to the extra selectivity achieved through the multistep absorption process, resonance ionization differs from resonance fluorescence in a basic way. In the ionization method, detection involves massive particles (electrons or ions) which can be easily controlled, while it is very hard to control scattered photons. This property makes elimination of wall events possible, resulting in considerable reduction of background. In the resonance fluorescence method, the detector must observe a solid angle of less than 4π steradians in order to avoid the main unscattered beam, while 100% collection efficiency is achieved in the ionization method. More importantly, the measurement of a small number of electrons or ions is much easier and more quantitative than is the measurement of a small number of photons.

Demonstration of a single-atom detection. The detection of single atoms depends on the fulfillment of two conditions: (1) the probing laser must selectively eject one electron from the atom being detected with 100% efficiency, and (2) the ejected electron must be detected with 100% efficiency. The amount of laser power needed to satisfy the first condition depends on the properties of the atom, namely, the absorption and photoionization cross sections and lifetimes of the levels being considered.

Two-photon ionization. The first successful demonstration of single-atom detection used the cesium transition, $6^2S_{1/2} - 7^2P_{3/2}$ (455.5 nm), which is conveniently in the range of pulsed-dye laser sources. Since the photons at 455.5 nm also ionize the $7P$ level, only one laser source is required for the two-photon ionization process of the cesium ground state. The power dependence of the two-photon ionization yield begins as a quadratic function, becomes linear, and finally becomes independent of power level as this level is increased. It is only in the high-power regime that one achieves ejection of an electron from every atom probed.

Electron detection. Detection of single electrons is achieved by using a proportional counter as a gaseous amplifier. The counter is filled with P-10 gas (90% Ar + 10% CH_4), and its central wire is maintained at 1000 V. The detector is shown in Fig. 2. The principles of operation of the counter are given in Fig. 3. The selectively ejected elec-

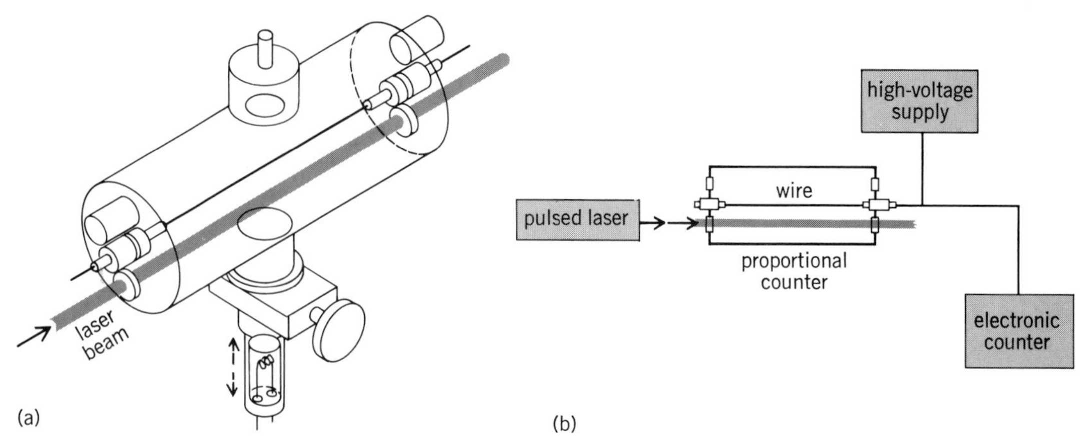

Fig. 2. Single-atom detector. *(a)* Proportional counter. *(b)* Counter and associated electronic equipment.

tron from the atom being detected is accelerated by the nonlinear field of the counter's central wire. The moving electron produces more charges due to the presence of high-pressure gas in the counter, and thus amplification takes place. Low concentrations of cesium in the laser beam are controlled by regulating the temperature of the sample, pressure of the counting gas, and distance between the sample and the laser beam.

Saturation of ionization process. The amount of laser power needed to saturate the two proton ionizations was measured experimentally. With the condition that the laser always saturate the ionization, pulse-height distributions were made for a wide range in the number of cesium atoms (10^2-10^6). From known data on the number of electrons that are produced by radioactive sources in P-10 gas, the number of cesium atoms in each distribution was measured.

Density fluctuations. Direct density fluctuation of atoms, when the number of atoms n is so small that $1/\sqrt{n}$ becomes appreciable, can also be studied by using the two-photon ionization process. Figure 4 shows a pulse-height distribution for an average number of atoms equal to 67 but with considerable fluctuations. The laser sampling of the atoms represents successive emptying and refilling of a small volume surrounded by an effectively infinite source of free atoms. These fluctuations were not previously observable, but were deduced from such experimental observations as Brownian motion and light scattering.

Detection of extremely low concentrations. The number of cesium atoms reaching the laser beam can be further reduced so that one ionization signal is produced for every 20 or more laser shots. The resulting pulse-height distribution at a counting gas pressure of 200 torr (27 kPa) is given in Fig. 5, and is compared with the distribution due to single photoelectric electrons produced by photons from a very weak incoherent ultraviolet light source (mercury lamp) as they strike the inner walls of the counter. The latter distribution does not change any further even if the ultraviolet light intensity is reduced, which indicates that it is a distribution of single electrons. The distribution from the two-photon ionization of cesium is identical to this single-electron distribution, and is therefore interpreted as being due to one atom in the laser beam.

In Fig. 5 the fluctuations in the distributions reflect the counting statistics of the proportional counter as predicted, and they do not reflect uncertainty in the total number of atoms counted. By integrating the total number of pulses produced above the normal electronic and laser transient noise, 95% of the single-atom pulses were counted. At all population levels, including the single atom, signals vanished when the laser was detuned from the cesium transition, which indicates the identification of the cesium atoms, even in the presence of 10^{19} other atoms.

Rare events. A method capable of detecting and identifying single atoms in the presence of 10^{19} or more of other types of atoms is uniquely suited to the spectroscopy of rare events.

Quarks. A number of important physical processes in nature, which contain important clues for

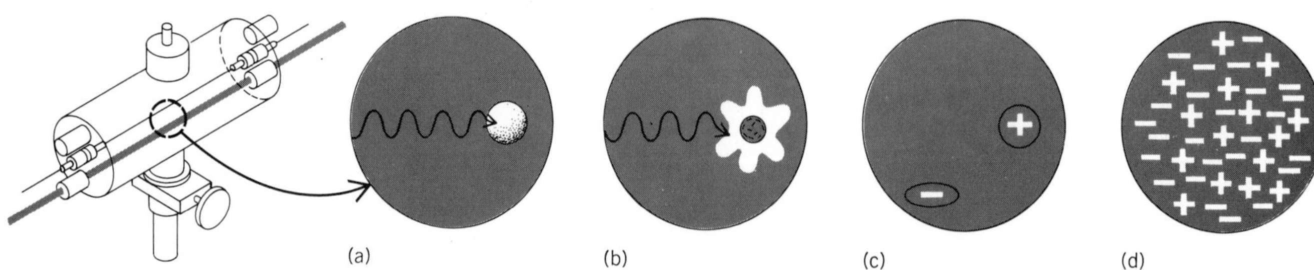

Fig. 3. Steps in process of detecting a single atom. *(a)* Photon is absorbed by atom in its ground state. *(b)* Excited atom absorbs second photon. *(c)* Free electron (negative charge) is removed from atom, leaving atom with positive charge. *(d)* In the process of colliding with molecules of the counting gas, free electron creates additional ionization. *(From M. H. Nayfeh, Laser detection of single atoms and applications, Amer. Sci., vol. 67, no. 1, January–February 1979)*

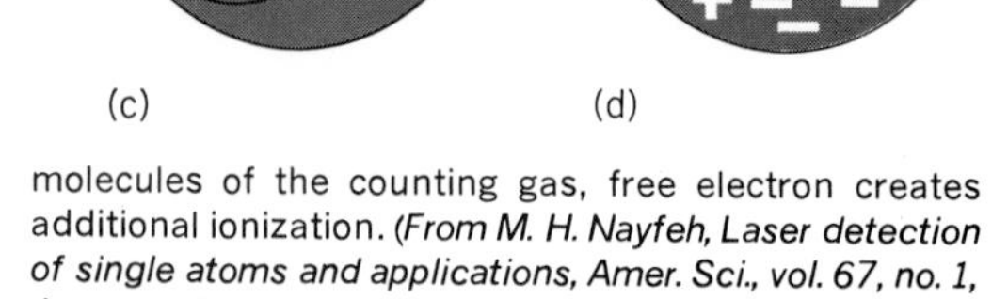

the understanding of cosmology and the laws of nature, could, in principle, manifest themselves in the formation of rare and quite different forms of atoms. One such case involves the quark, an elusive postulated elementary particle with fractional charge, first suggested in 1960 by M. Gell-Mann and G. Zweig. In spite of the many methods used in searching for them, there is as yet no convincing proof of the physical existence of quarks. If a free charged quark were attached to an atomic nucleus, the quark's fractional ($-\frac{1}{3}e$) unshielded charge would certainly change the atom's spectrum. Guided by quantum-electrodynamic calculations of the level shifts, experiments are under consideration which would employ single-atom detection to search for such quark atoms among numerous "ordinary" atoms. *See* QUARKS.

Solar neutrinos. Another rare event which might be studied is the interaction of solar neutrinos with matter. Solar neutrinos are emitted in nuclear interactions occurring in the solar core, and quantitative measurements of their flux striking the Earth are important in establishing the correct model of the solar interior. Since neutrinos are weakly interacting, zero-mass particles, they will convert, for example, a thallium-205 nucleus to lead-205, with a yield on the order of 1 in 10^{17} or less, and single-atom detection might possibly serve to detect these rare lead nuclei.

Study of atomic collisions. The saturated resonance ionization method has been extended by Nayfeh for use in absolute measurements of collisional line broadening at very low gas densities where absorption and fluorescence become extremely small. Traditionally, distortion of atomic transitions because of collision of the atoms with atoms of foreign gases, from which the interatomic forces can be measured, was studied by photon absorption and fluorescence. In these studies, high-temperature absorption tubes are required in order to induce appreciable absorption or fluorescence; but dense samples introduce not only self-broadening but also dimer absorption. Moreover, high-pressure buffer gases can result in three-body collisions which mask the satellite structure otherwise observable in the two-body collision. In the new method every absorption event in a colliding system is converted to an ion pair by absorption of another photon. Consequently, the measurement of a small number of absorbed or emitted photons is replaced by more sensitive detection of free electrons; this extra sensitivity allows studies of optically thin samples. A reduction of seven orders of magnitude in the required density was achieved with a very simple capacitor-plate charge collector. This reduction in density permitted the resolution of a new satellite structure in the Cs($7p$)-Ar interaction which was not resolved in previous absorption measurements at higher densities. Densities of even a few atoms per cubic centimeter or very close collisional distances at which only single absorption events occur can be studied with a proportional counter since single absorption events are detectable.

Detection of molecules. A further application of the resonance ionization method involves the combination of selective photoionization and mass spectrometry. This technique was developed by Letokhov and colleagues for the detection of complex molecules. The selective ionization that is achieved in this method and that is lacking in the usual nonselective ionization by an electron beam or by continuous vacuum ultraviolet radiation can be used to record the optical (infrared or visible) absorption spectrum by tuning the excitation wavelength as well as the mass spectrum.

Study of reactive atoms. Still another application involves the study of reactive atoms. The photodissociation of a spatially defined population of molecules by a pulsed laser serves as a source of ground-state atoms at a well-defined time. This time- and space-defined preparation of atoms fol-

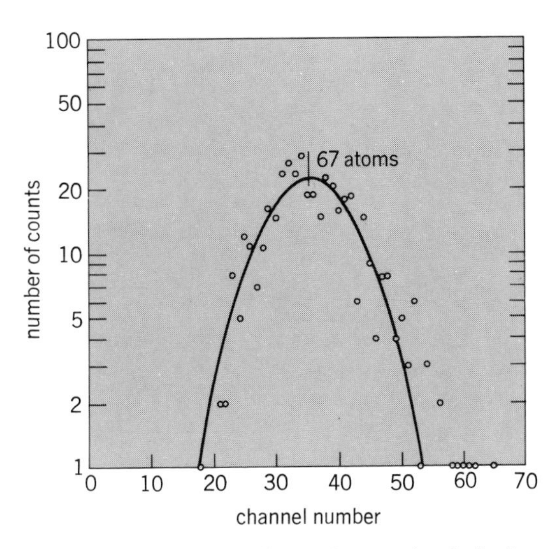

Fig. 4. Pulse-height distribution from cesium ionization showing the direct measurement of density fluctuations of an average number (67) of cesium atoms. (*From G. S. Hurst, M. H. Nayfeh, and J. P. Young, One-atom detection using resonance ionization spectroscopy, Phys. Rev. A, 15: 2283–2292, 1977*)

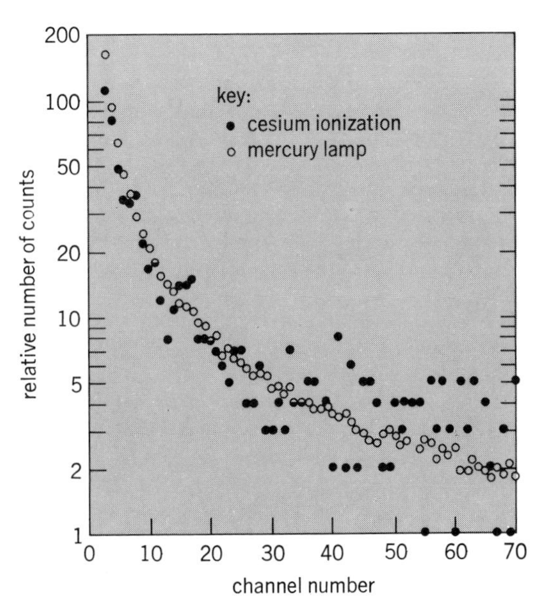

Fig. 5. Pulse-height distribution from cesium ionization for case where less than one cesium atom was counted per 20 laser probing pulses. Distribution is compared with that due to one electron produced by a very weak incoherent light source (mercury lamp). (*From G. S. Hurst, M. H. Nayfeh, and J. P. Young, A demonstration of one-atom detection, Appl. Phys. Lett., 30:229–231, 1977*)

lowed by multistep resonance ionization was used by L. W. Grossman and colleagues to study diffusion, chemical reactions with controlled species, and photodissociation cross sections. Because of the extra sensitivity achieved in the resonance ionization, lower-temperature samples can be used, resulting in a reduction of the vibrational populations of the molecules. Moreover, RIS is crucial in determining the population of reactive atoms, because pulsed probing can be concluded within a few nanoseconds following the preparation of the atoms with essentially no interference from the environment. This freedom from environmental interference is required for many investigations. One case arises in the study of sodium contamination of silicon, which is important in the manufacture of microelectronics.

For background information *see* ATOMIC STRUCTURE AND SPECTRA; IONIZATION CHAMBER; LASER in the McGraw-Hill Encyclopedia of Science and Technology.

[MUNIR HASAN NAYFEH]

Bibliography: W. M. Fairbank, Jr., T. W. Hänsch, and A. L. Schawlow, *JOSA*, 65:199–204, 1975; G. S. Hurst, M. H. Nayfeh, and J. P. Young, *Appl. Phys. Lett.*, 30:229–231, 1977; G. S. Hurst, M. H. Nayfeh, and J. P. Young, *Phys. Rev. A*, 15: 2283–2292, 1977; V. S. Letokhov, in A. Mooradian, T. Jaeger, and P. Stokseth (eds.), *Tunable Lasers and Applications*, pp. 122–139, 1976; M. H. Nayfeh, *Amer. Sci.*, vol. 67, no. 1, January–February 1979.

Atomic structure and spectra

A new technique has been developed for measuring the lifetimes of low-lying atomic energy levels in refractory metals. It is based on direct measurement of the decay in optical emission from sputtered atoms, following bombardment of a solid surface by a beam of energetic heavy ions.

Direct measurements of atomic lifetimes in refractory and rare-earth metals are of great importance to astrophysics. A recent study by E. Biémont and N. Grevesse reveals that many of the solar abundances for these elements are over- or underestimated. In the last 10 years, there have been successful efforts to determine the solar abundances of the various elements. However, accurate lifetimes for the refractory metals are still lacking, and their solar abundances, whose values are based on these lifetimes, are therefore poorly determined.

Difficulties in measuring lifetimes. The measurements of the lifetimes are complicated by several factors. First, the melting and boiling points of these refractory metals are rather high (about 2000°C), and it is difficult to produce atomic vapors of these elements. Many of them tend to be highly reactive in the molten state. Second, the atomic spectra of these metals are very complex. For example, over 92,000 lines in neutral and singly ionized uranium have been reported between 300 and 600 nm with an average density of 160 lines per nanometer. The reason is that numerous electron configurations exist in these atoms, the terms of which exhibit considerable mixing. Recently developed laser techniques have been of great help in resolving these spectra. Third, beam-foil spectroscopy, which has yielded several lifetimes of astrophysical interest in the past few years, is not feasible for these measurements because of problems of foil breakage and because it has been extremely difficult to produce intense ionic beams of refractory metals. Furthermore, lasers are not available to populate the excited states of the different refractory metals. Therefore, a universal method of measuring lifetimes in these metals seemed desirable.

Optical emission by sputtered atoms. Optical emission of radiation from solids under heavy-ion impact has been known for a long time and can be classified into three categories: optical emission from sputtered atoms or molecules, from backscattered ions, and from the solids themselves. The optical emission from sputtered atoms and backscattered ions consists of discrete lines, and normally the spectrum results from the lowest-lying levels of the atoms. Sometimes, especially for elements with open *d* shells, a strong continuum is also seen. This has been attributed variously to the emission from metal-oxygen molecules, metal-atom clusters, or the solid itself. However, the mechanism for this continuum emission has not been satisfactorily explained so far.

A sample spectrum of uranium bombarded by 400-keV argon ions is shown in Fig. 1. The salient features of the spectrum are: (1) All the spectral lines belong to neutral atoms of the sputtered species (in this case, neutral uranium atoms, symbolized UI). (2) Instead of a very complicated spectrum consisting of thousands of lines, only very few discrete lines are seen. (3) There is an overall continuous background. The reason for the appearance of such a simple spectrum consisting of only a few lines can be understood qualitatively. The typical distance between atoms in a metal is of the order of 0.1 nm, whereas the classical Bohr radii of highly excited orbits correspond to about 1 nm. Therefore, these highly excited states are quenched inside the solid due to electron loss. Furthermore, if ions are created inside the solid, they will capture electrons. Thus, optical emission results only from the low-lying excited states of the sputtered atoms.

There is also a region close to the surface of the solid where radiationless decay can take place. Only those atoms or ions with sufficient velocity to pass through this barrier region contribute to the emission and, consequently, the velocity of the atoms which contribute to light emission far away from the solid is fairly high (typically 4000 to 5000 m/s).

Measurement of atomic lifetimes. Once the atom is outside the solid, it is free to radiate. The most straightforward method of measuring the lifetime of the excited states is to pulse the beam and to look for the intensity of emission as a function of the time following excitation by the pulse. The beam is pulsed by sweeping it across a narrow slit, typically 1 mm in width, kept just before the solid; and by a proper choice of the experimental setup, pulses with half width as small as 8 ns can be obtained. The light from the sputtered particles is detected via a monochromator and a photomultiplier in delayed coincidence with the beam pulse (Fig. 2). The repetition rate of the pulsing system is adjusted so that a decay over several decades can

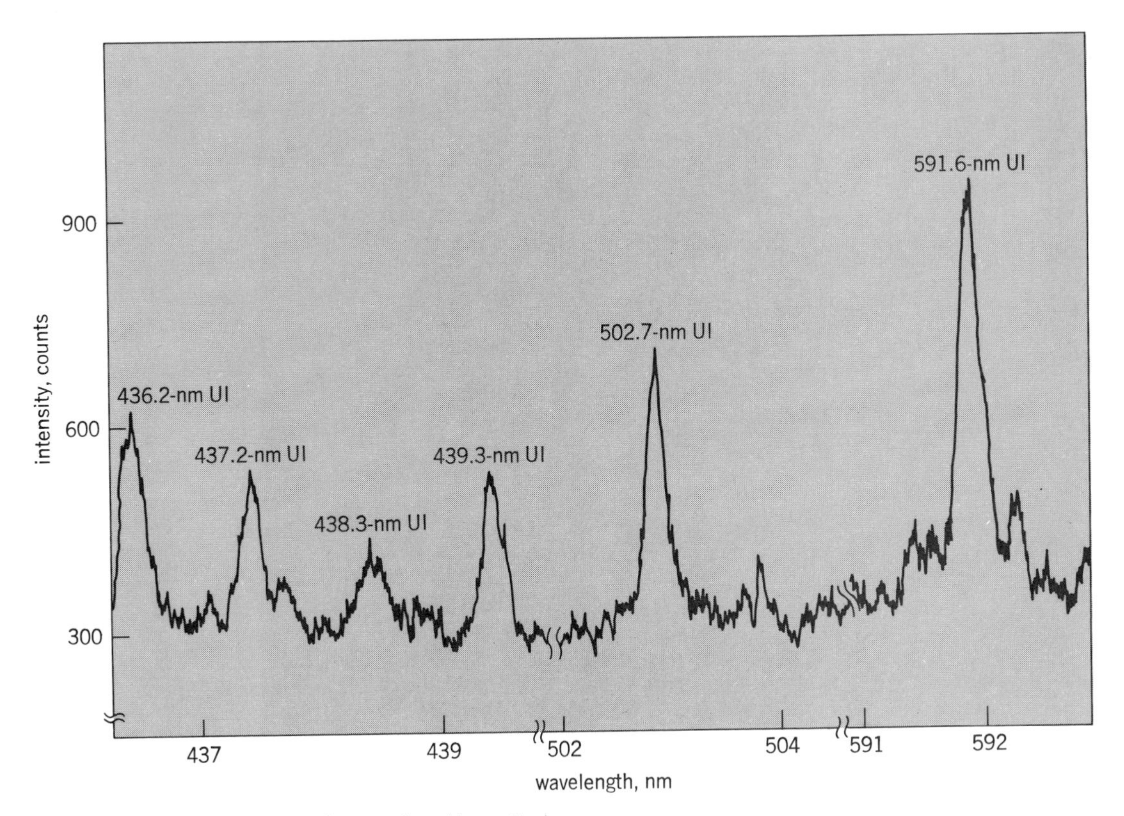

Fig. 1. Spectrum of atomic uranium produced in sputtering.

be observed before the next beam pulse starts. The decay consists typically of one or two exponentials and a constant background, from which the lifetimes are extracted by a computer fit.

Typical decays from the 26,875-cm^{-1} level of neutral iron and from the 16,900-cm^{-1} level of the neutral uranium are shown in Fig. 3. Intensity is shown on a logarithmic scale, so that a pure exponential decay would follow a straight line. The neutral iron decay (Fig. 3*a*) has been fitted as the sum of an exponential decay, indicated by the sloping straight line, with a lifetime τ of 67.7 ± 3.0 ns, and a constant background, indicated by a horizontal line. Because of the presence of the continuum (which also undergoes decay), it was necessary to fit the neutral uranium decay (Fig. 3*b*) with two

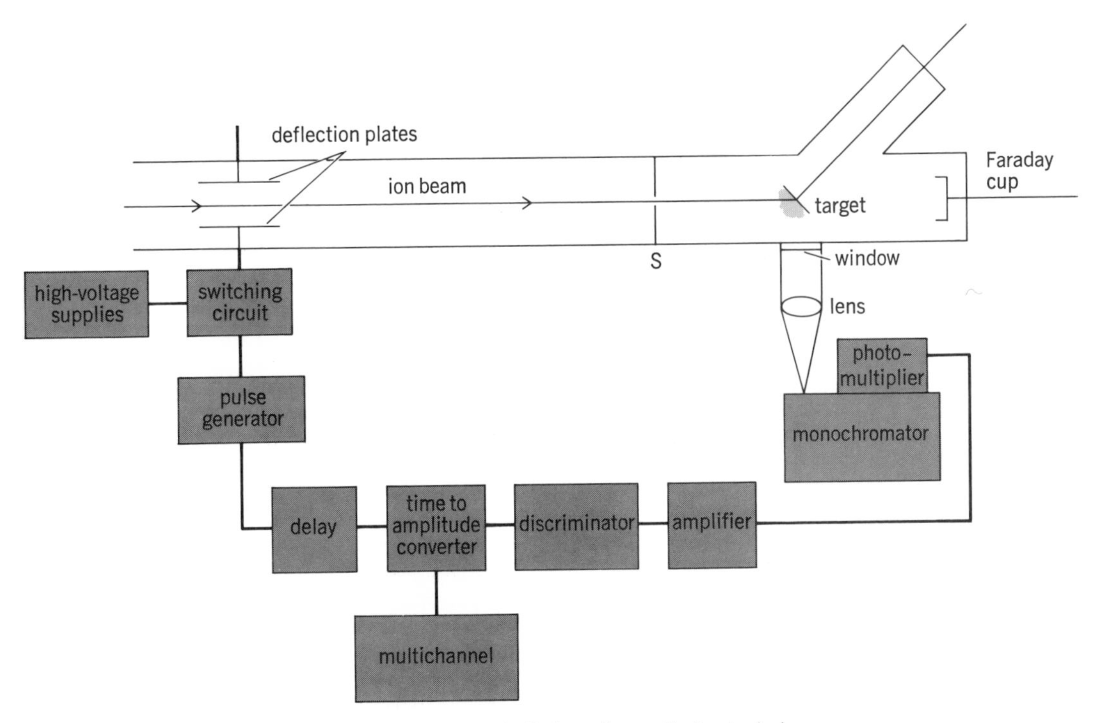

Fig. 2. Experimental setup for measurement of atomic lifetimes by sputtering technique.

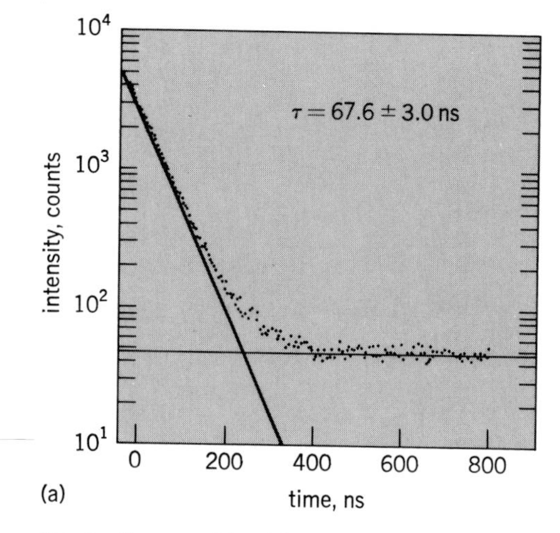

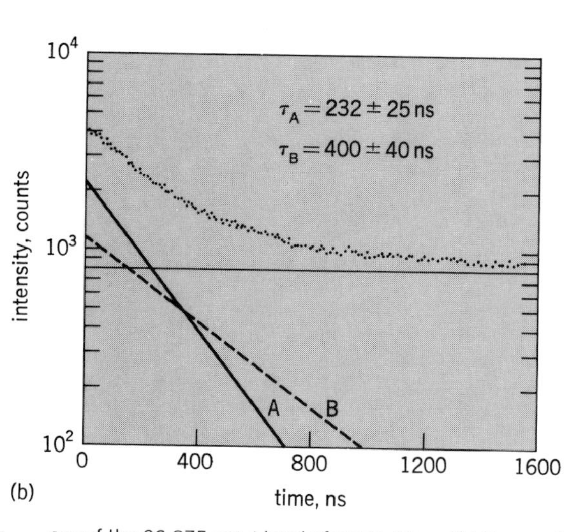

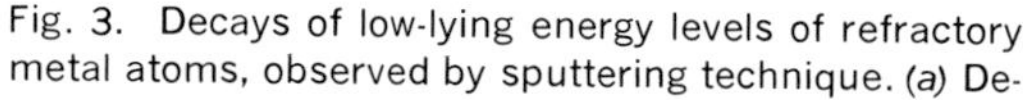
Fig. 3. Decays of low-lying energy levels of refractory metal atoms, observed by sputtering technique. (*a*) Decay of the 26,875-cm^{-1} level of neutral iron. (b) Decay of the 16,900-cm^{-1} level of neutral uranium.

exponentials (in addition to a constant background). The continuum lifetime was measured just outside the uranium line and used as a parameter in the fit. In Fig. 3*b* line A indicates the contribution to the total intensity from decay of the uranium line, with a lifetime of 232 ± 25 ns; likewise, line B indicates decay of the continuum with a lifetime of 400 ± 40 ns, and the horizontal line indicates the constant background. The lifetimes obtained by this technique generally agree with those determined by different techniques, such as the Hanle effect and phase-shift method.

The advantages of the method are immediately obvious. Since the number of ions per pulse is rather small (typically 10^5-10^6 ions), no perturbation of the excited atomic-state lifetimes is expected due to Stark effects and so forth. Since the light emitted comes predominantly from the low states, no cascades from high-lying levels are seen. This is in contradistinction to the beam-foil technique, where several high-lying levels are excited which decay to the state of interest, distorting the lifetime observed.

The lifetimes measurable with this technique depend on the velocity of the sputtered particles. If the velocity is too large, depending on the lifetime, some of the particles could escape the observation region, giving a shorter lifetime. Thus, it is imperative to have an optical system collecting light from all the sputtered particles.

Significance of results. As mentioned earlier, determination of the lifetimes in refractory metals allows one to measure their solar abundances. The solar abundance values are based upon intensity measurements in the solar spectrum and the oscillator strengths of the observed transitions, and the latter are derived from the lifetime of the decaying level and the branching ratios. The usual technique for measuring solar abundances relies on the measurement of oscillator strengths in an arc discharge, as was done by C. H. Corliss and W. R. Bozman. Since the numbers of atoms in excited states are assumed to follow the Boltzmann distribution, even a small inaccuracy in the measurement of the arc temperature can lead to incorrect abundances. The measurements of lifetimes in yttrium, zirconium, iridium, and platinum by the sputtering technique yield values for the solar abundance of zirconium and platinum that support the accepted values; however, for yttrium the solar abundance has to be increased by a factor of 2, and for iridium by a factor of 4. The new abundance values are in agreement with the values obtained by chemical analysis of meteorites. It might even be possible to determine molecular and free-radical abundances by using the sputtering technique.

Furthermore, a measurement of lifetimes of low-lying states is also of interest to laser isotope-separation research. In this technique, an isotope of the material is excited by a laser to one of the low-lying levels, from which it is ionized by absorption of a second photon from a second laser. The sputtering value for the lifetime of the intermediate level in uranium which has been used for isotope separation is in excellent agreement with that measured by the photoionization technique.

For background information *see* ATOMIC STRUCTURE AND SPECTRA; ELEMENTS, COSMIC ABUNDANCE OF; ISOTOPE (STABLE) SEPARATION; SPUTTERING in the McGraw-Hill Encyclopedia of Science and Technology. [P. S. RAMANUJAM]

Bibliography: E. Biémont and N. Grevesse, *Phys. Scripta*, 16:39–47, 1977; C. H. Corliss and W. R. Bozman, *NBA Monogr. 53*, 1962; P. S. Ramanujam, *Phys. Rev. Lett.*, 39:1192–1194, 1977.

Behavioral genetics

Behavioral genetics is an interdisciplinary field which applies genetic techniques to a large variety of animal and human behaviors in an effort to extend the understanding of the origins and maintenance of these behaviors. In recent years most standard genetic techniques, including selection studies, pedigree analyses, classical breeding programs, and quantitative, physiological, molecular, and cytological genetics, have been used by psychologists, ethologists, and anthropologists to study behaviors. Although the discipline of behavioral genetics has been in existence for less than 20 years, speculation about genetic influences on behavior extend back to early recorded history.

History. Historical references and folklore abound with references to the concept that "like begets like" in terms of both physical characteristics and behavior. It was not until the latter half of the 19th century, however, that the foundations for the systematic study of the inheritance of behavior began to be established. In *On the Origin of Species* and later works, Charles Darwin proposed that behavioral as well as physical characteristics were subject to the process of natural selection. During the period of intense debate following the introduction of Darwin's theory, Francis Galton's work on the inheritance of mental characteristics polarized a scientific community already divided on the role of evolution and inheritance in determining human behavior. Despite Galton's pioneering statistical and methodological developments, few behavioral scientists were willing to accept his view that behavioral characters are inherited. Their reluctance stemmed not only from theological objections to Darwinian theory but also from the philosophical orientation of the time, which emphasized the role of experience in altering behavior. Although Gregor Mendel had already published his research on plant hybridization, which provided the basis for the laws of inheritance, his work was unknown to Darwin and Galton, and was not rediscovered until 1900. During the following 20 years, the field of genetics grew rapidly, and the relationship of mendelian genetics to natural selection was clarified by Ronald Fisher and Sewall Wright.

Early research. Despite rapid developments in the field of genetics, applications of techniques and genetic principles to behavioral characters were limited during the period 1900–1950. In 1907 R. M. Yerkes described a genetic anomaly which produces "waltzing" mice, and in the 1920s and 1930s R. C. Tryon published his famous studies on selection for "bright" and "dull" rats based on maze-learning ability. Early research on genetic influences in human behavior largely focused on studies of "eminent" or "degenerate" families and on pedigree studies on mental defects. During the 1920s several studies examining the possible genetic basis for schizophrenia and for general intelligence were completed using twins, siblings, and other relatives.

With the possible exception of some of the selection studies with animals, the research on genetic influences on behavior had relatively little impact on the behavioral sciences at the time. Family studies were unable to rule out environmental influences, and other studies often lacked rigorous control procedures. During the 1930s instinct theory was attacked by behavioral psychologists, and an extreme environmentalist position was popular in the United States and Great Britain. In addition, although Sigmund Freud and other leaders of "dynamic" psychiatry were trained in biology, the field had rapidly moved away from its biological foundations, and data implicating genetic factors in the etiology of psychopathology were often ignored. Finally, the association of many early investigators with the eugenics movement, which hoped to improve the human race by limiting the breeding of "unfit" individuals, had created hostility to genetic analysis of behaviors as the eugenics movement was rejected on moral grounds. Research implicating genetic factors in behavior had to wait for acceptance until the 1960s, when a biological point of view reemerged as a major approach to the study of behavior.

Animal research. Although many species of insects, fish, birds, rodents, and canines have been used in behavioral genetic research, most work has been carried out on two species: *Drosophila melanogaster*, a species of fruit fly, and *Mus musculus domesticus*, the common house mouse. A great amount of genetic data is available on each species, and both have numerous characteristics which make them ideally suited to genetic research on behavior.

Most contemporary behavioral genetic research on animals involves the use of selected lines or various types of inbred strains and their crosses.

Selected lines. These are derived from a randomly bred population which is tested on a behavioral character and then divided into high- and low-scoring groups. In each succeeding generation the highest-scoring offspring of the high line and the lowest-scoring offspring of the low line are selected for further breeding. Although the amount of genetic information obtainable from selection studies tends to be limited, their use has continued in behavioral genetics since the early work of Tryon and others. After many generations of selection the highly divergent lines are often maintained for further research. Bright and dull, high- and low-emotional, active and inactive, and large- and small-brained lines of mice and rats have, for example, been established and used in numerous studies. Results generally show that selected lines differ considerably less than was expected or not at all on characteristics somewhat different from those originally selected for, suggesting that selection is limited to a highly specific character and does not involve a general capacity related to a large number of behaviors.

Inbred strains. These are derived by crossing closely related individuals, usually siblings, for many successive generations. After approximately 30 generations all offspring are genetically identical, and the inbred strain is established. If a difference is found among two or more inbred strains on a behavioral measure, genetic factors are implicated as a source of variation in the behavior. During the 1950s and 1960s a large number of strain-differences studies were carried out on a wide range of behaviors; it became apparent that strain differences occurred more often than not, suggesting that there were varying degrees of genetic influence on nearly every sensory, learning, activity, sexual, maternal, and temperament character studied.

Most contemporary behavioral genetic research on animals involves more elaborate breeding designs which help provide a better picture of the genetic architecture of the behavioral characters of interest. The classical breeding technique, in which two inbred strains are crossed to produce F_1 and F_2 populations and backcrosses to the original parent strain, and more extensive breeding procedures such as the diallel cross, where a large number of inbred strains are crossed in all possible combinations, are now used in conjunction with modern biometrical genetic analysis techniques. Not only can the relative contributions of genetic and environmental influences be assessed by using these methods, but environmental influences can often be subdivided into maternal effects and variation between and within families. Genetic variation can likewise be subdivided into the relative influences of additive genetic effects, dominance, epistasis, and sex linkage.

Investigators have recently begun to focus more on the nature of the genetic variation itself than on estimating relative genetic and environmental effects, since the genetic architecture of a character can provide important clues to its adaptive significance, in an evolutionary sense, to the organism being studied. Elaborate breeding designs are now often carried out in conjunction with environmental manipulation or developmental studies to help determine the interaction of genetic influences with age and environment. Although approximately 200 inbred strains are available for laboratory research, concern about the unknown origin of many of these strains and the possibility that they were derived from populations which developed under radically different selection pressures is encouraging the development of new inbred strains derived from known base populations.

A recent technique in animal behavior genetics involves the use of recombinant inbred strains, in which an F_2 population generated from two inbred strains is used as a base population to produce a number of new inbred strains. A comparison of scores on a given behavior in these new strains can often be matched with one previously established for distinctive loci, which indicates possible identity or close linkage of the two genes. The chromosomal segments carrying the genes which influence the behavior being studied can thus be identified.

Occasionally a mutation is identified within an inbred strain. If the mutant can be successfully bred through backcrossing with its original strain, a coisogenic strain may be developed which is genetically identical to the original parent strain except for a gene at a single locus. If a pair of coisogenic strains differ on a behavior, the single gene in which the lines differ must necessarily be implicated. Discovering behavioral characters which are under the control of a single gene is very useful, because such systems are potentially easier to trace from the initial gene alteration to the final behavior than systems involving a large number of genes.

Because investigators of laboratory animal behavior are becoming increasingly concerned about the biological relevance of their behavioral measures, and because of the growing interest in the emerging field of sociobiology, the application of complex behavioral genetic methods to animal behavior is receiving increasing attention among psychologists and ethologists.

Human research. Many studies of genetic influences on human personality, cognitive factors, sensory processes, and abnormal behavior are now available from many countries.

Twin study method. The most frequently used technique in human behavioral genetics is the twin study method. In its simplest form, the similarity of identical or monozygotic (MZ) twins is compared with the similarity of fraternal or like-sex dizygotic (DZ) twins. Since MZ twins are genetically identical and DZ twins share only half their genes, one would expect MZ twins to be more similar than DZ twins on any character influenced by genetic factors. When the character of interest is distinctive, such as a sensory deficit or a diagnosed illness, similarity of twins is usually assessed by calculating the proportion of twin pairs in which both members are affected. If the trait is a continuous one, such as the score on a written test or a rating on a personality scale, similarity is usually indicated by correlation coefficients between pairs. In most cases, MZ-DZ differences are used to estimate broad heritability, the fraction of the observed variance in behavior which can be attributed to genetic effects.

Several criticisms have been raised concerning the twin study method. MZ twins may be more similar to each other on certain behavioral characteristics simply because they exert a strong psychological influence on one another or because they were reared more alike than DZ twins. Although this does not seem to be the case with respect to effects on many behaviors, a desirable extension of the twin method involves the inclusion of a group of MZ twins who were separated at an early age and reared apart. Another extension is to compare individuals of all degrees of genetic similarity ranging from identical twins, through siblings and cousins, to unrelated persons. If such studies also include pairs of individuals reared together and apart, reliable estimates of genetic influences on behavior can be made. The illustration summarizes correlation coefficients from a variety of intelligence test scores between individuals of various degrees of genetic and environmental relationships. The average correlation between pairs of individuals varies systematically with the degree of relationship that exists between individuals. There is also a consistently greater similarity between individuals reared together than those reared apart, indicating the importance of environmental as well as genetic influences on intelligence measures.

One of the limitations of twin and family studies is the difficulty in locating MZ twins reared apart. Recent mathematical developments have uncovered two alternative designs which provide the same genetic information as the classical twin design, but do not require MZ twins reared apart. Future research in human behavior genetics should thus be considerably easier to carry out.

Alternatives to twin study method. A powerful alternative to twin and family studies are studies where adopted offspring are compared to both foster and biological parents. If environmental influences prevail, foster children should resemble more closely their foster parents; whereas if biogenetic influences prevail, these children should be more similar to their biological parents. Other procedures often used in human behavioral genetic research include pedigree analysis, in which complete family pedigrees for three or more generations are obtained, studies of children of consanguineous marriages (marriage among first cousins), and dual mating studies, where both marital partners have a trait or disorder of interest.

Intelligence tests. Human intelligence testing has been the subject of controversy for many years. Interpretation of results of intelligence tests given to individuals in different social conditions is highly complex and sometimes misleading. However, the results of studies of genetic influences on intellectual characters so overwhelmingly point toward the presence of a strong genetic influence on intellectual characters that it is difficult to deny that such influences exist. The illustration, for

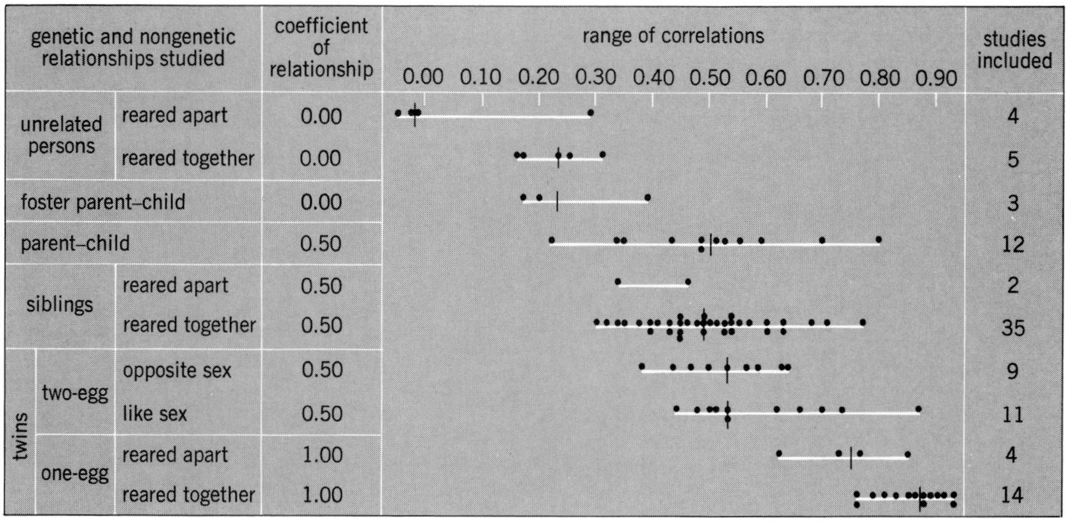

genetic and nongenetic relationships studied			coefficient of relationship	range of correlations 0.00 0.10 0.20 0.30 0.40 0.50 0.60 0.70 0.80 0.90	studies included
unrelated persons		reared apart	0.00		4
		reared together	0.00		5
foster parent–child			0.00		3
parent–child			0.50		12
siblings		reared apart	0.50		2
		reared together	0.50		35
twins	two-egg	opposite sex	0.50		9
		like sex	0.50		11
	one-egg	reared apart	1.00		4
		reared together	1.00		14

Summary of correlation coefficients compiled from various sources. Horizontal lines indicate range of correlation coefficients in intelligence between individuals of various degrees of genetic and environmental relationship. Vertical lines indicate the medians. *(After L. Erlenmeyer-Kimling and L. F. Jarvik, Genetics and intelligence: A review, Science, 142:1477–1479, 1963)*

example, shows the consistent relationship between the similarity in intelligence and the degree of relationship of members of families. Adoption studies have been equally consistent, showing much higher correlations between the intelligence of adopted children and their biological parents than their foster parents. Present evidence suggests that a little over half the variance in general intelligence may be attributed to genetic factors, a small proportion to interactions between genotype and environment and correlations between genotype and environment, and the remainder to environmental influences.

When general intelligence is broken down into more specific factors such as word fluency, perceptual speed, and verbal, spatial, memory, number, and reasoning abilities, it appears that the degree of genetic influence varies considerably among the factors. In several studies where these abilities were assessed, reasoning ability showed less genetic influence than verbal and spatial abilities. From a genetic point of view these individual factors appear to be somewhat independent of each other, suggesting that general intelligence test scores involve an aggregate of abilities which are influenced differentially and somewhat independently of genotype.

Personality tests. Personality tests typically measure a number of personality traits which can be analyzed separately by using standard behavioral genetic methods. Results suggest that genetic influences on various personality traits are smaller than those found on intellectual characters. Although environment carries substantial weight in determining all personality dimensions, a recent large-scale study showed that the effect of environment was uniform across widely different traits even though parental and social pressures most likely vary for these traits and between sexes. Differences in the treatment of twins during childhood were not found to be predictive of adolescent personality differences. These results suggest that the consistent directional factors in personality development are the genes and that important environmental influences consist of highly variable situational experiences.

Genetic influences on psychopathology. Extensive work on genetic influences on psychopathology has focused heavily on schizophrenia, where strong evidence for genetic influence continues to accumulate. Despite the fact that genetic factors are implicated, present data do not provide clear support for any single genetic theory. Research is increasing on the study of manic-depressive psychosis, psychoneurosis, psychopathy, alcoholism, and criminality. Unfortunately, genetic research on psychopathology continues to be hampered by problems of diagnosis and classification. Until more reliable classification procedures and further understanding of the physiological, toxological, and biochemical bases of schizophrenia and other abnormal behaviors are provided, an understanding of the genetic components of these disorders is likely to remain superficial.

Heritability estimates. A misunderstanding of the meaning of heritability estimates has frequently led to inappropriate statements concerning both the effectiveness of environmental intervention in altering behavior shown to be highly heritable and the genetic basis of racial or ethnic differences in behavior. All genetic research estimates of the relative contributions of genetic and environmental factors to variability of different individuals are limited to the population studied. Heritability estimates can differ considerably in a new population with different gene frequencies or in a genetically similar population in a different environment. The fact that a behavior is highly heritable does not necessarily imply that a change in environmental conditions will have little effect on the behavior. In addition, high heritabilities within populations are insufficient evidence that differences between populations are influenced by genetic factors.

For background information *see* BEHAVIOR AND HEREDITY; GENETICS; HUMAN GENETICS in the McGraw-Hill Encyclopedia of Science and Technology. [NORMAN D. HENDERSON]

Bibliography: J. DeFries and R. Plomin, *Annu.*

Rev. Psychol., 29:473–515, 1978; L. Ehrman and P. Parsons, *The Genetics of Behavior*, 1976; J. Loehlin, G. Lindzey, and J. Spuhler, *Race Differences in Intelligence*, 1975; J. Loehlin and R. Nichols, *Heredity, Environment and Personality*, 1976.

Biophysics

The traditional means for measuring the velocities of the motions of living organisms has been visual observation of the distance traversed in a measured period of time. Recently a new technique has been developed which utilizes the special properties of optical laser light to permit the direct, continuous measurement of the velocities of moving objects and particles. Like radar, this technique, which is called laser Doppler velocimetry (LDV), operates on the principle of the Doppler effect. Laser light which is reflected or scattered from moving objects has a measured frequency which is slightly shifted by the motion of the object. LDV now has numerous applications to the measurement of the flow of fluids in a variety of engineering problems. This technique can also be applied to the study of motion in such diverse biological systems as bacteria, plants, animals, and humans.

The laser is an ideal source for a velocimetry measurement because its output beam has a low divergence and because the light is highly monochromatic (single frequency) and spatially coherent (all waves in phase). Moreover, optical light has intrinsic advantages over microwave radiation because it can be focused to observe much smaller objects. In addition, due to the much higher frequency of optical light, the Doppler shift is much greater, thereby permitting the measurement of much lower velocities. For very low velocities, the maximum Doppler shift is a very small fraction of the optical frequency. (The actual fraction is equal to the observed velocity divided by the speed of light.) Direct observation of this slight difference in frequency is not possible, so the technique of beating, used in radar and radio technology, must be employed. Optical beating is accomplished by mixing the light from the moving object with an unshifted light beam from the same laser. The unshifted beam is usually called the local oscillator. The Doppler-shifted light from the moving object and the unshifted light from the local oscillator are mixed at the surface of a photodetector to produce a beat whose frequency is exactly the Doppler shift magnitude. The frequency of this oscillating photocurrent is measured by an electronic spectrum analyzer. Many particles can be observed simultaneously, and if they have different velocities, a spectrum of Doppler frequencies is produced from which the distribution of velocities can usually be determined. A diagrammatic representation of an LDV is shown in the illustration. By using such an instrument, it is possible to determine the complete distribution of velocities in a complex sample in a few seconds. This great savings in time over conventional methods is a primary reason for the growing impact of LDV technology on biological and medical research.

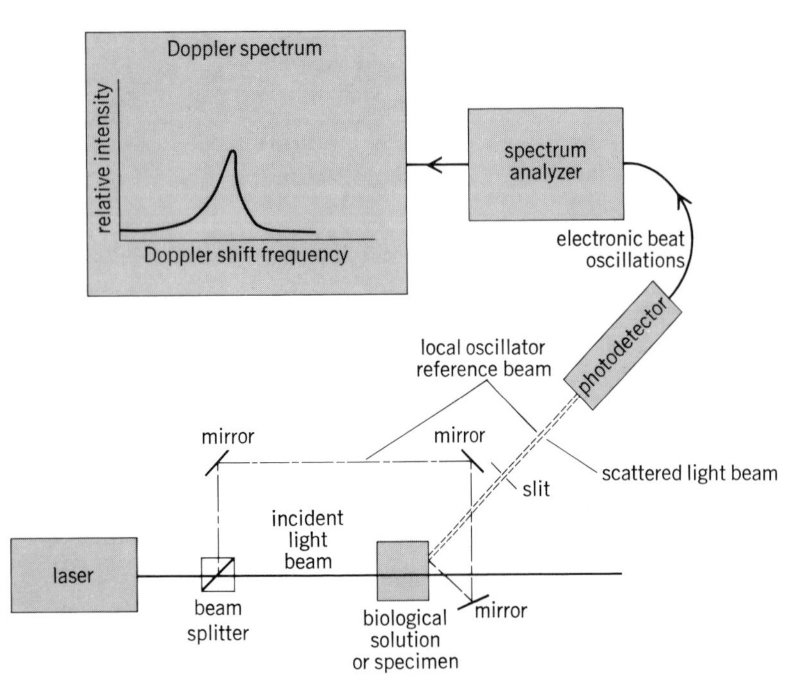

Schematic diagram of laser Doppler velocimeter and typical Doppler spectrum.

Motile cells. The first biological application of LDV was to the study of the swimming behavior of motile cells, particularly spermatozoa and bacteria. The swimming behavior of spermatozoa is an important index of the probable fertility of a particular sample. Rapid LDV analyses of sperm swimming-speed distributions are important for the diagnosis of marginal fertility in human males and for the characterization of commercial samples of bull semen. In addition, the LDV spectrum can be used to infer more subtle details of the motion of the spermatozoa, such as the rotations and undulations of the sperm head as the cell swims. Similar applications of this technique to the study of bacteria are capable of characterizing both the swimming-speed distributions and some of the finer details of the motile behavior of these microorganisms. Among the practical applications of these studies is the detection of very low levels of water pollutants by observing their effect on the swimming behavior of particularly sensitive bacteria.

Protoplasmic streaming. A different type of motility which has been effectively studied by LDV methods is protoplasmic streaming, the active transport of protoplasm inside living cells. A highly attenuated, finely focused laser beam can selectively illuminate tiny portions of the protoplasm inside a living cell without damaging the cell or interfering with its normal activity. The light scattered from moving particles in the region of illumination is then collected, and the Doppler spectrum of the light is determined. In this way the distribution of velocities at various points in the cell can be determined, and the experiment can be repeated as a function of temperature, light level, nutrient availability, or the concentration of an added chemical or drug. From these data a much clearer concept of the mechanism which produces the motive force may be inferred. Although this type of experiment has been used thus far to study motion in lower organisms such as algae and fungi, the mechanism is believed to be of significance to the active processes of all living cells, since component proteins almost identical to the muscle proteins actin and myosin are known to be involved.

Blood flow. The flow of blood in higher animals is a phenomenon of obvious biological and medical significance for which there has never been an adequate means of quantitation. Recent applications of LDV instrumentation in this field show great promise for potential clinical application. The first LDV study of human blood flow was a characterization of retinal vessels made by workers at the Massachusetts Institute of Technology and the Massachusetts Eye and Ear Institute. (Retinal vessels are the only vessels which are commonly accessible to optical examination, and visual observation of the state of flow in the retinal vessels is a common procedure used by physicians to detect general effects of atherosclerosis as well as retinal disorders.) By using a highly attenuated laser beam to avoid retinal damage, these workers were able to illuminate single vessels in the retina and to infer the flow velocities from the Doppler shifts of the light emanating from the eye. This technique is now in practical clinical use, and research on the expansion of information available from this method continues.

Unfortunately, most blood vessels are not accessible to optical probes because they lie under opaque layers of skin and other tissue. It has been demonstrated that LDV methods may be used to measure the velocities of flow in these vessels by means of the insertion of a fiber optic catheter into the vessel. The catheter functions as a light pipe in carrying the incident laser light to the blood vessel and returning the Doppler-shifted light to the instrument for analysis. However, the use of fiber optic catheters may induce blood clotting and thus involves significant risks to the subject. As a result, experimentation of this type has not been extended to humans. However, blood flow in any area of the body may be manifest near the skin by the flow rate in the arterioles, venules, and capillaries. It may therefore be possible to characterize the blood flow in a particular region by illumination of the skin with a laser and measurement of the Doppler spectrum of the light scattered back from the skin.

The first study by LDV of blood flow in the surface epithelial tissue of a human subject was reported in 1974. The light from a safe, low-power laser was scattered from the surface tissue with a wide range of Doppler frequencies which were related to blood flow. Studies by a number of groups in 1978 have demonstrated a quantitative relationship between the Doppler spectrum and known indicators of blood flow. Several groups have initiated efforts to develop clinical instruments for LDV characterization of surface blood flow. Potential uses of such a device include measuring the extent of cauterization in severe burn victims, evaluating the return of circulation in tissue grafts, and quantifying the state of circulation in patients with circulatory disorders such as atherosclerosis, hypertension, and sickle cell anemia. The device could also be utilized by dentists for detection of the onset of gingivitis (gum disease), and could be used in hospitals as a simple, painless, nontactile means of monitoring the state of blood flow in critically ill patients.

Fluid flow around organisms. LDV has been used for several years to measure the flow of fluids in and around airplanes and ships. This technique has recently been extended to a study of water flow around certain types of marine life, for example, water flow through the olfactory organs of fishes. For these experiments the fish is generally immobilized and the laser is focused on the entrance or exit of the olfactory organ. The geometry of the laser apparatus can be manipulated to measure any desired component of the velocity. The LDV approach has provided the first quantitative determinations of the velocities of fluid flow in these organs, and further studies of this type are anticipated.

Electrophoretic light scattering. In addition to its applications to the characterization of natural biological motions, LDV has been applied to the design of new types of experiments on biological specimens. One example is a new technique called electrophoretic light scattering (ELS), in which the laser Doppler principle is utilized for measuring the velocities of particles migrating in an electric field. Since biological particles generally have an electrical charge, they migrate in an electric field toward the electrode of opposite polarity, a phenomenon called electrophoresis. The velocity with which they migrate divided by the strength of the electric field is a characteristic number for each particle called its electrophoretic mobility. This number can be used both to characterize the particle and to distinguish it from other particles. Electrophoresis has been used in biochemical and clinical laboratories for decades. By using the Doppler shifts of scattered laser light, it is now possible to detect electrophoretic velocities spectroscopically and thus to perform precise measurements in a small fraction of the time required by the more tedious and less objective classical techniques.

ELS was invented at the University of Illinois in 1971. The first experiments were conducted on solutions of proteins, but the more recent applications have been primarily in the study of the surface charge of the membranes of living cells and membranous subcellular particles. Cell membranes are complex combinations of proteins, glycoproteins (proteins attached to carbohydrates), and carbohydrates anchored in or on a bilayer of lipid (fat) molecules. Some of these components are free to move in the plane of the bilayer, and interactions of the membrane components in response to external stimuli and extracellular reagents are fundamental elements of the biological function and response of many cells and organelles. Of the intrinsic physical parameters which may be used to characterize a living cell, the one most closely related to the cell surface is electrophoretic mobility, which may be related to the surface density of electrical charge at the hydrodynamic surface of the cell. The characterization of cell surface charge by electrophoresis has been performed for nearly a century by microelectrophoresis, in which the motion of individual cells is clocked in a microscope. The introduction of ELS has drastically reduced the amount of time necessary to perform such measurements while also improving precision, accuracy, and objectivity. The use of a rapid automatic technique is important for the characterization of cells, because it permits the monitoring of dynamic cellular processes in time intervals short enough that the cells can remain viable and stable. ELS instrumentation is also more appropriate to routine clinical analysis of numerous samples in the event that the biological research leads to significant clinical applications.

T and B human lymphocytes. A great deal of the ELS research in the past year has been focused on the study of human lymphocytes. Two major types of lymphocytes are recognized by immunologists: T cells (so named because they mature in the thymus gland) are responsible primarily for direct cellular immunity; and B cells (directly derived from bone marrow) are responsible for the synthesis of antibodies. These two types of cells are not morphologically differentiable, and rather complex immunological tests are required to distinguish them. Recently two independent studies have shown that T and B cells may be distinguished in seconds by the ELS spectrum. T cells have higher electrophoretic mobility and therefore move faster and produce a higher Doppler shift than B cells. The determination of the relative proportions of T and B cells in both humans and animals is important both for research in immunology and for the characterization of certain diseases. The ELS method is by far the fastest method of making this determination, and it is currently being introduced in both research and clinical laboratories for this purpose.

Lymphocyte pathology. ELS has also been used to detect and characterize pathological states of lymphocytes. Much of the effort in this area has been focused on human leukemia. In leukemia a particular type of bone marrow cell becomes malignant and produces large numbers of malignant daughter cells which resemble normal white blood cells but are not generally immunocompetent. When the malignant cells resemble lymphocytes, the disease is known as lymphocytic leukemia, and there are two distinct types, acute and chronic lymphocytic leukemias. Both diseases are fatal if untreated, although treatments are becoming increasingly effective. Drugs which are given to the patients selectively kill rapidly dividing cells. These drugs kill many healthy cells, but they have a higher toxicity for the malignant leukemia cells. If after treatment the leukemia cells can no longer be detected, the patient is said to be in remission. Remission is often followed after a period of months or years by a relapse, in which the leukemia cells return and predominate. Several successive remissions may be achieved in a given patient, but the prognosis becomes progressively poorer with each relapse.

It has recently been shown by ELS and by classical methods that the leukemia cells produced in both acute and chronic lymphocytic leukemias are electrophoretically distinguishable from normal lymphocytes. By using selective reagents, it has also been shown that the chemical groups responsible for the electric charge on the surface of the leukemic cells are different from those of either T or B cells. The peak electrophoretic mobility of leukemic cells shows more variability among different patients than the variability of normal lymphocytes among different normal donors, and the question of whether the mobility of the leukemic cells from a particular patient may be a useful prognostic indicator is currently being pursued. Perhaps the most exciting prospect is the possibility of monitoring patients throughout the course of treatment and during remission to evaluate their progress. Preliminary findings have indicated that the change in the electrophoretic mobility distribution may be the earliest indication of relapse. Early detection of relapse is extremely important in the treatment of leukemia patients, since the severity of the treatment may be mitigated considerably if therapy is administered earlier.

Cell surface studies. ELS has been applied to a number of other kinds of cell surface studies. For example, a combination of velocity sedimentation separation with ELS has been used to define subpopulations of thymus lymphocytes from mice. The goal of such studies is to provide physical characterization of different cell types for identification and separation, much as the early work in physical biochemistry permitted identification and separation of proteins. ELS has also been used to probe the fundamental forces responsible for cell-cell adhesion. In one study the effects of various agents on the rates of adhesion of mouse fibroblasts (dividing connective tissue cells) were correlated with the change in surface charge as determined by ELS. It was found that electrostatic repulsion between the cells was not a major force in determining their propensity to aggregate. The effect of stimulating agents on cell surfaces have been measured to determine the degree to which the changes induced in the cell surfaces are manifest in the electrophoretic mobility.

Secretion studies. Another new application of ELS has been the study of the secretion process. The release of secreted chemicals from various glands and nerve endings is mediated by secretory granules, spherical membrane packets in which the hormones and neurotransmitters are stored. Secretion is believed to occur when the membranes of these granules adhere to and then fuse with the membranes of the cells in which the granules are stored. This process is stimulated by an influx of calcium ions. The reactions at the surfaces of the granule and cell membranes are not well characterized, nor is it known what physical forces induce adhesion and fusion of the two membranes. ELS studies have been used to probe the reduction in charge which may be accomplished by controlled concentrations of calcium ions and other agents. In addition, specific chemical reactions at the surfaces of these particles may be monitored by ELS through their effect on the cell surface charge.

Other laser uses. ELS and other forms of laser Doppler velocimetry have been used to probe various physical phenomena which may be important under certain circumstances for understanding biological systems, particularly in the interpretation of experiments in culture. LDV methods have been useful in the study of thermal and mechanical fluid instabilities and in the study of the rheology of liquids, liquid crystals, and solutions of biological interest. ELS and other light-scattering methods have been used for the study of the hydrodynamics of macroions in an attempt to determine the molecular size, charge, and conformation and to understand the physical forces which dictate the structure and dynamics of ionic suspensions and solutions.

For background information *see* BIOPHYSICS;

DOPPLER EFFECT in the McGraw-Hill Encyclopedia of Science and Technology. [B. R. WARE]

Bibliography: B. J. Berne and R. Pecora, *Dynamic Light Scattering*, 1976; B. A. Smith, B. R. Ware, and R. A. Yankee, *J. Immunol.*, 120(3): 921–925, 1978; B. R. Ware, Applications of laser velocimetry in biology and medicine, in C. B. Moore (ed.), *Chemical and Biochemical Applications of Lasers*, vol. 2, chap. 5, 1977; D. Watkins and G. A. Holloway, *IEEE Trans. Biomed. Eng.*, BME-25:28–33, 1978.

Blood

Heparin is a natural substance found throughout the animal kingdom. It is not a homogeneous "pure" polysaccharide, but rather a group of polymers within an extremely heterogeneous family of polysaccharides. Thus a chemically homogeneous heparin molecule with a well-defined structure, where relationships between the structure and biological properties could be easily established, has not been isolated. Nevertheless, many recent and older studies have provided evidence that heparin(s) has important intravascular functions. In this regard, a demonstration of the presence of heparin in blood would be decisive, but since chemically defined heparin has not been isolated from normal blood, some investigators doubt its presence there. However, an extract of normal human blood in which heparin activity is present has been obtained by several researchers. Endogenous heparin has also been identified in rat blood by chemical degradation methods. The suggestion that the human blood extract might contain heparan sulfate (heparitin)—closely related to heparin but having only about 10% of its anticoagulant potency—is academic in view of the recent finding by x-ray fiber diffraction pattern analysis that the heparan molecule contains a large heparin moiety. Thus functionally it matters little whether blood contains heparan, heparin, or both. What is important is that heparin activity is normally present in the bloodstream. In human plasma the amount of heparin activity varied from 0.1 to 0.24 unit per milliliter, as determined by anticoagulant activity of the extract in sheep plasma. (This is not a negligibly small quantity.)

Fluidity of blood. Heparin almost instantaneously accelerates the action of plasma antithrombin in neutralizing the activity of the series of enzymes involved in the formation of thrombin, as well as accelerating the neutralization of thrombin itself. The formation of activated factor X, Xa, is a key prerequisite to the formation of thrombin. It has been shown that as little as 0.01 unit of heparin in 0.8 ml of normal plasma markedly accelerates the neutralization of factor Xa by antithrombin. Thus there is 10 to 20 times more endogenous heparin activity present in human plasma than the amount required, together with antithrombin, to inhibit the production of factor Xa. When one considers the large amount of thrombin potentially available in the blood, and its explosive generation following the injection of a few units of factor Xa, it is apparent that the maintenance of blood fluidity depends primarily on the prevention of the formation of thrombin. This, in turn, depends on the action of antithrombin plus heparin, with heparin probably playing an essential catalytic role.

When tissue is cut, platelets aggregate at the cut surface and release platelet factor 4. This substance has a binding affinity for heparin and neutralizes its activity, thus permitting the formation of thrombin and the production of a fibrin clot which prevents further bleeding. In rat blood 50% of the heparin activity was associated with platelets. This link has not been determined in humans because heparin levels were obtained from platelet-poor plasma. However, it is likely that some heparin is also associated with platelets in humans, as there is more circulating lipemia-clearing activity (lipoprotein lipase) in platelet-rich plasma. Since platelet function is essential to coagulation, and since platelets contain a potent heparin-neutralizing factor, it is probable that a platelet-heparin and antithrombin balance system functions in blood. There is recent evidence that excessive platelet activity may disturb that balance. Such excessive platelet activity in relation to heparin has been found in men with severe coronary heart disease.

Antithrombin plus heparin may also be involved in the regulation of plasmin activity in the fibrinolytic system, and perhaps in the kinin-generating system. However, this theory requires further investigation.

Complement. The complement system of blood is involved in the inflammatory response. Portions of the clotting sequence, the fibrinolytic system, and the kinin-generating system interact with the complement system. Activation of the complement system results in the sequential interaction of the serum proteins or components of the system, with the development of a series of active enzymes from their precursors. It has long been known that heparin inhibits complement. Recent work has probed the nature of this inhibition in greater detail. Heparin, in concentrations comparable to those that facilitate the action of antithrombin, inhibits the activation of complement at various sites. The action of heparin is very rapid and, at least at some points, requires the cofactor activity of a serum protein. Although the mechanisms have not been fully elucidated, it is apparent that the kinetics of the heparin action are strikingly similar to its effect, together with antithrombin, on the coagulation sequence.

Fat removal. Lipoprotein lipase activity is the major pathway for the removal of ingested fats (triglycerides), or of fats synthesized by the liver, from the blood. Lipoprotein lipase splits the triglyceride component of lipoproteins into free fatty acids and glycerol. It is generally accepted that this enzyme functions at the surface of the vascular endothelium. However, the occurrence of this activity to a lesser degree has been demonstrated in the blood of most people when platelet-rich plasma was studied. In any event, enzyme activity at the surface of the cells lining the inner walls of blood vessels and capillaries is essentially intravascular, since the fat-containing particles upon which the enzyme acts are circulating in the blood. Injected heparin releases the enzyme from the capillary wall and markedly increases lipoprotein lipase activity in the blood. Heparin also increases the activity of lipoprotein lipase obtained by tissue extraction. Heparinase decreases lipoprotein lipase activity. In humans, lower average circulating

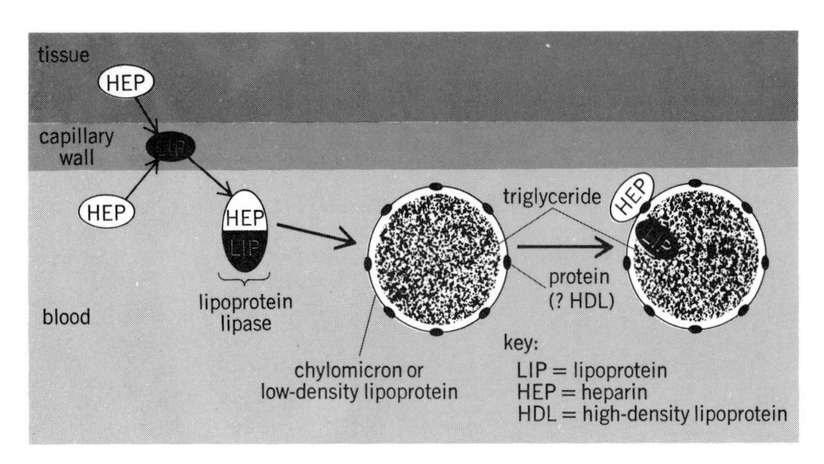

Schema representing formation and action of lipoprotein lipase. *(From H. Engelberg, Probable physiologic functions of heparin, Fed. Proc., 36:70–72, 1977)*

heparin levels are found in individuals with higher concentrations of serum triglycerides. These various observations indicate that heparin is normally involved in triglyceride lipolysis via lipoprotein lipase activity. A proposed mechanism of action is shown in the illustration. The true protein enzyme (lipoprotein lipase) is present in the capillary wall. Heparin acts together with it at the endothelial cell surface and perhaps in the adjacent bloodstream. Heparin facilitates contact with the fat-containing particles in the blood (chylomicrons and very-low-density lipoproteins) by interaction with surface proteins (high-density lipoproteins?) of the fat particles. This brings the enzyme into contact with the triglyceride core of the chylomicrons, and lipolysis occurs. It is probable that at the same time heparin binds to the surface protein of the fat particle, the bond between heparin and lipoprotein lipase is at least partially being broken as further enzyme activity decreases, unless additional heparin is added.

Mode of action. Heparin functions in a similar manner in the three areas discussed above. It is involved in a series of interactions which facilitate the action of various proteins. This implies some type of catalytic role. Heparin is a highly charged anionic substance which electrostatically interacts with cationic molecules such as proteins. It is capable of convolutional changes in solution. Its bonding with proteins enhances or inhibits their activity. It changes its attachment to proteins where conditions change, and the original activity of the protein is restored when its bond with heparin is broken. These considerations suggest that heparin may play a catalytic or regulatory role in those physiological processes involving a series of protein-protein interactions.

For background information *see* BLOOD; COMPLEMENT, SERUM; LIPID METABOLISM in the McGraw-Hill Encyclopedia of Science and Technology. [HYMAN ENGELBERG]

Bibliography: E. D. T. Atkins and I. A. Nieduszynski, *Fed. Proc.*, 36:78–82, 1977; H. Engelberg, *Fed. Proc.*, 36:70–72, 1977; R. D. Rosenberg, *Fed. Proc.*, 36:10–18, 1977; J. M. Weiler et al., *J. Exp. Med.*, 147:409–421, 1978.

Bond angle and distance

Certain quadruple bonds, that is, bonds with a multiplicity of four, between transition-metal atoms may be the shortest and strongest of all chemical bonds, provided the term "shortest" is used in a relative sense, based on the intrinsic sizes of the atoms involved, rather than in an absolute sense.

The multiplicity of a chemical bond is determined by the number of electron pairs that are shared between the bonded pair of atoms. Single, double, and triple bonds, where one, two, and three electron pairs, respectively, are shared, have long been known, but quadruple bonds were recognized only in 1964, when the nature of the Re-Re bond in the $Re_2Cl_8^{2-}$ ion (Fig. 1) was elucidated. Since then, scores of other examples have been found, and the quadruple bond has taken its place among the basic concepts in molecular structure and chemical bonding.

Quadruple bonds are formed only by transition metals, because the formation of a quadruple bond requires the use of *d* orbitals, and only the transition metals have *d* orbitals available for use in forming bonds. Figure 2 shows, at the left, the shapes of the five *d* orbitals that make up a complete set on any one metal atom, and the entire figure shows how pairs of *d* orbitals on adjacent metal atoms can overlap to form bonds.

Types of bonds. The overlap of two d_{z^2} orbitals (Fig. 2*a*) allows the formation of a very strong bond when two electrons occupy the spatial region defined by the resulting two-center orbital. This type of bond, which is symmetrical around the internuclear axis, is called a sigma (σ) bond. Similarly, in Fig. 2*b* and *c*, the overlap of other pairs of *d* orbitals which leads to the formation of pi (π) bonds is shown. These two π-bonds, which exist in mutually perpendicular planes, are equivalent to each other.

In Fig. 2*d*, a third type of overlap of *d* orbitals which gives rise to a delta (δ) bond is shown. While the remaining pair of *d* orbitals (Fig. 2*e*) could, if free to do so, also overlap to form a δ-bond, this does not occur in the molecules to be discussed in this article. Their lobes point toward the other atoms that are attached to the metal atoms, and they are therefore used primarily to form bonds to

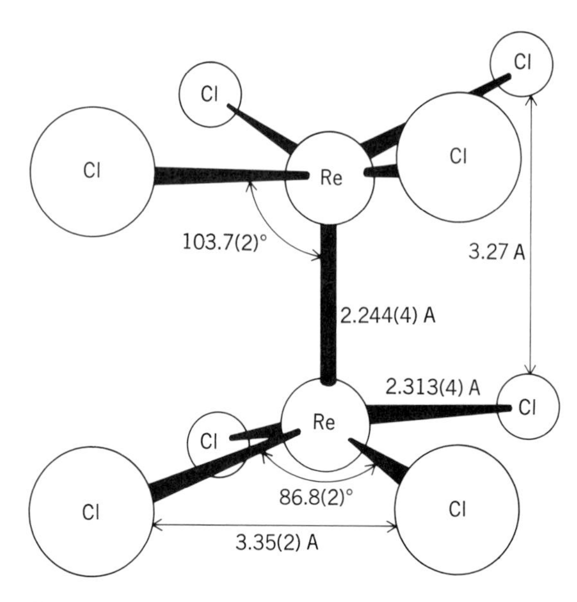

Fig. 1. Structure of $Re_2Cl_8^{2-}$ ion. 1 A = 0.1 nm. *(From F. A. Cotton and C. B. Harris, The crystal and molecular structure of dipotassium octachlorodirhenate(III) dihydrate, $K_2[Re_2Cl_8]\cdot 2H_2O$, Inorg. Chem., 4:330, 1965)*

these so-called ligand atoms, for example, the chlorine atoms in $Re_2Cl_8^{2-}$.

To summarize, then, there is a total of four important overlaps between pairs of *d* orbitals on adjacent transition-metal atoms that can occur in the molecules being considered, namely, one of σ type, two of π type, and one of δ type. When all four occur simultaneously, and when there is a pair of electrons available to occupy each of the resulting two-center bonding orbitals (four electron pairs in all), a quadruple bond is formed.

Quadruple bond formation requirements. The formation of quadruple bonds depends primarily upon the suitability of the metal atoms, with the following characteristics being essential: (1) Each of the two atoms must have precisely four electrons available for metal-to-metal bonding after all metal-ligand bonds are formed. (2) The pertinent *d* orbitals on the two metal atoms (which need not be, but almost always are, identical) must be of such energy and size that they can overlap well. The first requirement means that for homonuclear quadruple bonds there is the following relationship between the group of the periodic table, N, and the oxidation number n: $N - n = 4$. Thus, for the first few groups in the periodic table, where the elements are listed in parentheses following N, the required oxidation numbers would be: group 4 (Ti,Zr,Hf), 0; group 5 (V,Nb,Ta), +1; group 6 (Cr,Mo,W), +2; group 7 (Mn,Tc,Re), +3; group 8 (Fe,Ru,Os), +4; and so on. The second requirement, that the energy and spatial extent of the *d* orbitals allow good overlap, eliminates some of these formal possibilities: For group 4, with oxidation number 0, the *d* orbitals are too high in energy and too diffuse to overlap well. Besides, under normal conditions diatomic molecules, such as Zr_2, would revert to their normal metallic state. For group 5 also, orbital overlap may be inadequate, although this case has not been adequately studied as yet. With groups 6 and 7 conditions are optimum, and it is these elements which consistently form quadruple bonds. For group 8 the high charge causes the *d* orbitals to contract so much that they cannot overlap well and, in addition, close approach of the metal atoms is opposed by repulsion between the highly charged ions.

There are at least two other factors that also affect the capability of a given metal atom (or ion) M^{n+} to form a quadruply bonded species M_2^{2n+}: the row of the periodic table in which the metal occurs and the electronic properties of the associated ligands. In group 6, the ability of the metal atoms to form quadruple bonds appears to be Mo>Cr>W, while in group 7 it is Re>Tc≫Mn. The theoretical basis for these orders is purely speculative at present. The influence of the associated ligands is pronounced but again not well understood. Thus, $M_2X_8^{2-}$ species, where X = CH_3, have been prepared and studied for Cr, Mo, and W, but for X = Cl or Br, only the $Mo_2X_8^{2-}$ ions appear to exist.

Ultrashort bonds. With any given M_2^{2n+} unit, the length of the quadruple bond, which is related to the bond strength, varies by about 0.1 A (0.01 nm) with the ligand used, but when M^{n+} is Cr^{2+} this variation is very much larger. Cr-Cr distances, in compounds with quadruple bonds, range from as much as 2.54 A (0.254 nm) to as little as 1.83 A (0.183 nm). The 1.83-A distance and a few others below 1.9 A (0.19 nm) are characterized as ultrashort. The compounds in which these ultrashort distances have been found contain ligands of a certain general type, as shown in Fig. 3. In these ligands the X and Y atoms or groups are the donor atoms. The general structure of these compounds is shown in Fig. 4 for the case where X, Y, and Z are C, OCH_3, and OCH_3, respectively.

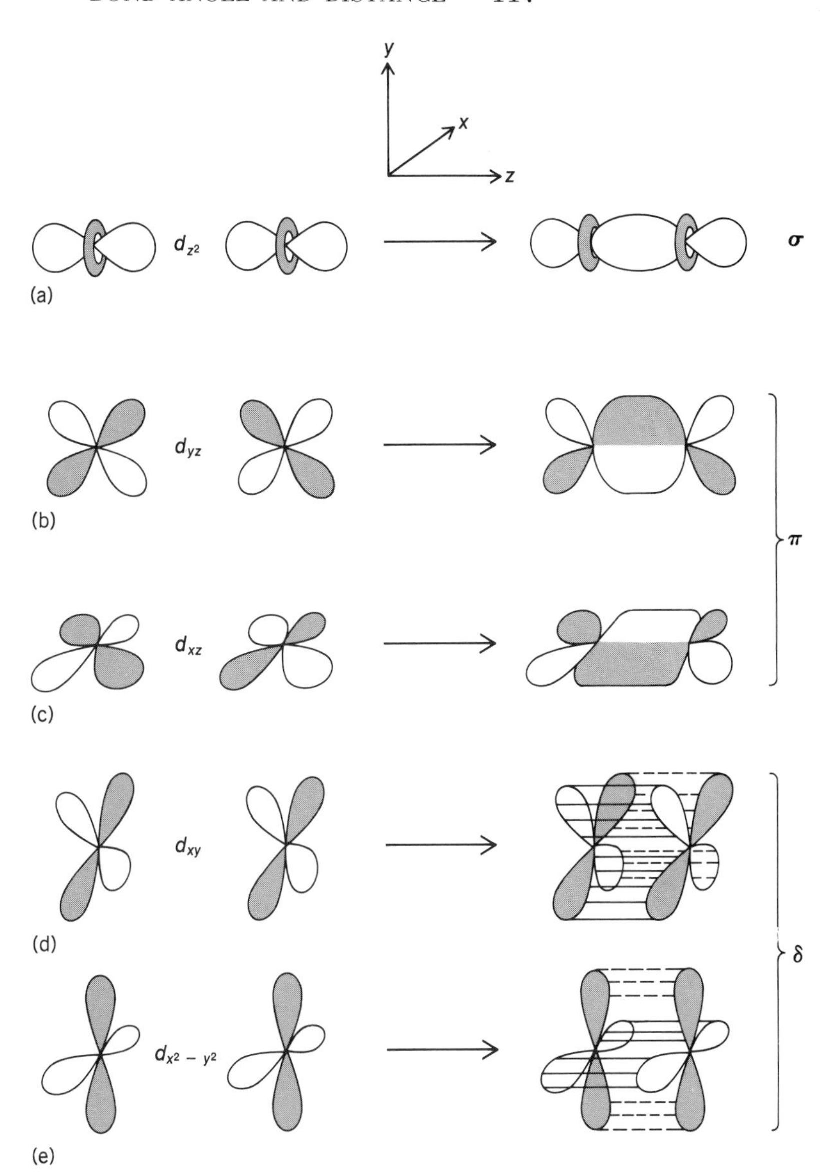

Fig. 2. Diagrams showing the five *d* orbitals that make up a complete set with the same principal quantum number, indicating how matched pairs of these orbitals on adjacent metal atoms can overlap to form *(a)* σ, *(b, c)* π, and *(d, e)* δ bonds.

ligand type (ring with substituent Z, ring atom X, substituent Y)

X	Y	Z	Cr—Cr, A
N	O	CH_3	1.889
C	OCH_3	OCH_3	1.847
N	NH	CH_3	1.860
C	O	H	1.830

Fig. 3. Compounds in which ultra-metal-to-metal bonds are found and the type of ligand they contain. 1 A = 0.1 nm.

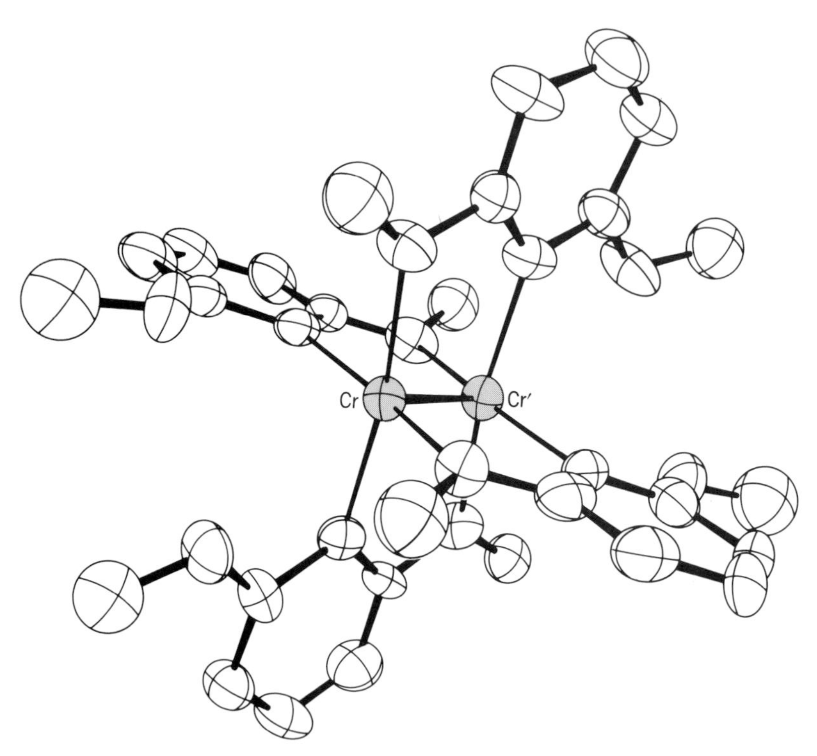

Fig. 4. The molecular structure of the dichromium complex with four 2, 6-dimethoxyphenyl ligands, in which there is an ultrashort (1.847-A, or 0.1847-nm) Cr-Cr quadruple bond. *From F. A. Cotton, S. Koch, and M. Millar, Exceedingly short metal-to-metal bonds, J. Amer. Chem. Soc., 99:7372, 1977)*

If the normal bond radius R_X of an atom X is used as a gage of its inherent size in chemical bonding situations, then the relative shortness of any given M-M bond can be expressed as the ratio of the actual internuclear distance, r_{MM}, to $2R_M$. For a Cr-Cr bond with $r_{CrCr} = 1.830$ A (0.183 nm), this ratio is 0.772. The only other M-M distances for which this ratio is less than 0.80 are found in a few quadruply bonded Mo-Mo compounds with the same types of ligands. The lowest such Mo-Mo ratio is 0.796. It is notable that even the strongest chemical bonds for which bond energies are known, namely N≡N and C≡C bonds, have higher ratios, namely 0.786 and 0.783, respectively.

The reason why ligands of this particular type (and no other, so far as is known) cause metal atoms to form such short bonds is unknown. From the variety of combinations of coordinated atoms (C,O; N,O; N,N) one concludes that this factor is not critical. Indeed, other kinds of ligands employing the same donor atoms have longer bonds. The presence of the aromatic ring seems to be the key feature, and it is possible to speculate as to how the π-orbital system of the aromatic ring might interact with the metal orbitals to promote stronger M-M bonding. However, a detailed thoretical study of this problem is only in its early stages.

For backgound information *see* BOND ANGLE AND DISTANCE in the McGraw-Hill Encyclopedia of Science and Technology.

[F. ALBERT COTTON; GEORGE G. STANLEY]

Bibliography: F. A. Cotton, *Account. Chem. Res.*, 11:224–231, 1978; F. A. Cotton, *Chem. Soc. Rev.*, 4:27–53, 1975; F. A. Cotton, *Inorg. Chem.*, 4: 334–336, 1965; F. A. Cotton, S. Koch, and M. Millar, *J. Amer. Chem. Soc.*, 99:7372–7374, 1977.

Breeding (plant)

Although the ability to regenerate whole flowering plants from protoplasts (plant cells from which the cell wall has been removed by enzymatic digestion) is still restricted to approximately eight genera, and has not yet proved possible for many crop species, significant progress in the production of somatic hybrid plants from protoplast fusion has been made with those species more amenable to cell culture.

Interspecies somatic hybrid plants have been produced by protoplast fusion of sexually compatible species in the *Nicotiana* and *Petunia* genera, and in that of *Daucus*. These studies have emphasized the importance of adequate cultural capability of the species being fused and the need for stringent methods for the selection of hybrids from parental material. The usefulness of this method of nonsexual plant hybridization in plant breeding will depend largely on whether genetically novel plants can be produced; the somatic hybridization methodology that has proved successful for sexually compatible species is providing an approach to the hybridization of sexually incompatible species. In most higher plants a genetic analysis of the role played by nonchromosomal genes has been impeded by the inability to obtain, by sexual crosses, plants that are heterozygous for extranuclear genes. Protoplast fusion provides this opportunity, since uniparental maternal inheritance is eliminated. Even when species can be hybridized sexually, the resultant diploid sexual hybrid is sometimes highly infertile, and increased fertility results only if the tetraploid can be obtained; in these instances tetraploid somatic hybrid plants produced as a direct result of the fusion of diploid species may be very useful. In the case of crop species that are usually vegetatively propagated, the use of protoplast fusion to produce tetraploids may eliminate the need for sexual hybridization and associated segregation problems.

Intrageneric hybrid. The induced fusion of protoplasts either at the intra- or interspecies level is now readily achievable by using either polyethylene glycol, alkaline conditions in the presence of calcium ions, or a suitable combination of both. While interfamilial heterokaryons, and occasionally somatic hybrid cells, can be produced, these are primarily of cytological interest, and presently available technology has not produced somatic hybrid plants from such combinations. Since many of the hybrid combinations required by the plant breeder are intrageneric, most researchers are presently concentrating on the extension of the somatic hybridization methodology to species within the same genus which are either unidirectionally or completely sexually incompatible.

Following the induced fusion of protoplasts of different species, either a somatic hybrid is produced, or the complete elimination of the chromosome of one of the species results in the formation of a somatic cybrid (Fig. 1). This cybrid would contain the cytoplasms of both species, but the chromosomes of only one.

Somatic hybrid selection. Selection of somatic hybrids from the parental species is best achieved by using complementation selection procedures. Theoretically, this is best carried out by using recessive auxotrophic, or dominant drug-resistant,

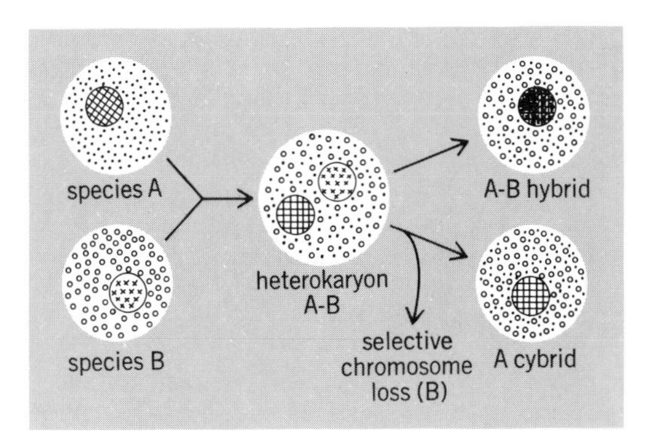

Fig. 1. Some consequences of the fusion of protoplasts from different species. (*From E. C. Cocking, Microbial and Plant Protoplasts, copyright 1976 by Academic Press, Inc. (London), Ltd.*)

mutants. Unfortunately, such mutants are not readily available and are often difficult to isolate, and therefore need to be specially obtained. Moreover, such mutant cells frequently have lost the ability to develop into whole plants, and this greatly detracts from their use. Albino mutant seedlings which occur naturally, or which can be readily obtained by mutagen treatments of seeds, are particularly useful as source material for the isolation of albino protoplasts (either directly from the plant or from suspension cultures) which can be used in complementation selection procedures (Fig. 2) as also successfully done by G. Melchers and his coworkers. They also have the great advantage that whole plant regeneration from protoplasts is not impaired. Naturally occurring differences between species in their sensitivity to drugs and culture media can also be utilized without loss of regenerative capability.

Somatic cybrid. As can be seen from Fig. 1, if fusion of nuclei does not occur in heterokaryons, then cybrids can be produced if one of the nuclei persists. A stable heterokaryon condition is not found in higher plants and, as has been suggested from studies of sexually incompatible interfamilial fusions, failure of nuclear fusion in heterokaryons may result in selective nuclear loss and cybrid formation. This may represent a level of heterospecific incompatibility. The use of plastome (cytoplasmic) and genome (nuclear) chlorophyll deficiencies in tobacco has enabled a fuller understanding of these interactions, since it is possible to distinguish somatic hybrids from cybrids.

Recent work on the transfer of male sterility from one variety of tobacco (cytoplasmic male sterile) to another utilized protoplast fusion to enable cytoplasmic mixing to take place, as well as any nuclear fusion that might also result. Products of fusion and the parental varieties were cultured as a whole until the plants had regenerated. Several nuclear somatic hybrids were readily detected by their leaf morphology, and also many plants with altered flower morphology and different degrees of expression of the male sterility character. These had clearly resulted from the biparental transmission of cytoplasmic genes to form a cybrid comparable to that in Fig. 1.

Chromosome selective elimination. Interspecies sexual hybridization in plants often results in an instability in the embryos during their subsequent development. For *Hordeum vulgare* and *H. bulbosum* and *N. tabacum* and *N. plumbaginofolia*, the production of interspecific hybrid embryos is followed by the selective elimination of the chromosomes of one of the species. Apart from the developments already discussed, the extent to which such events could occur in interspecies somatic hybrids, and the extent to which selective chromosome elimination could be advantageously used by the plant breeder, is not yet known. The further improvement of the karyotyping of plants and the ability to distinguish between chromosomes in interspecies somatic hybrids would be significant developments. Biochemical cytoplasmic markers such as fraction-1 large subunits have already been used to determine the stability of mixed cytoplasms in somatic hybrids, but sometimes species being hybridized (even though sexually incompatible) have identical profiles, and more refined methods such as chloroplast or mitochondrial deoxyribonucleic acid restriction nuclease profiles may be required.

For background information *see* BREEDING

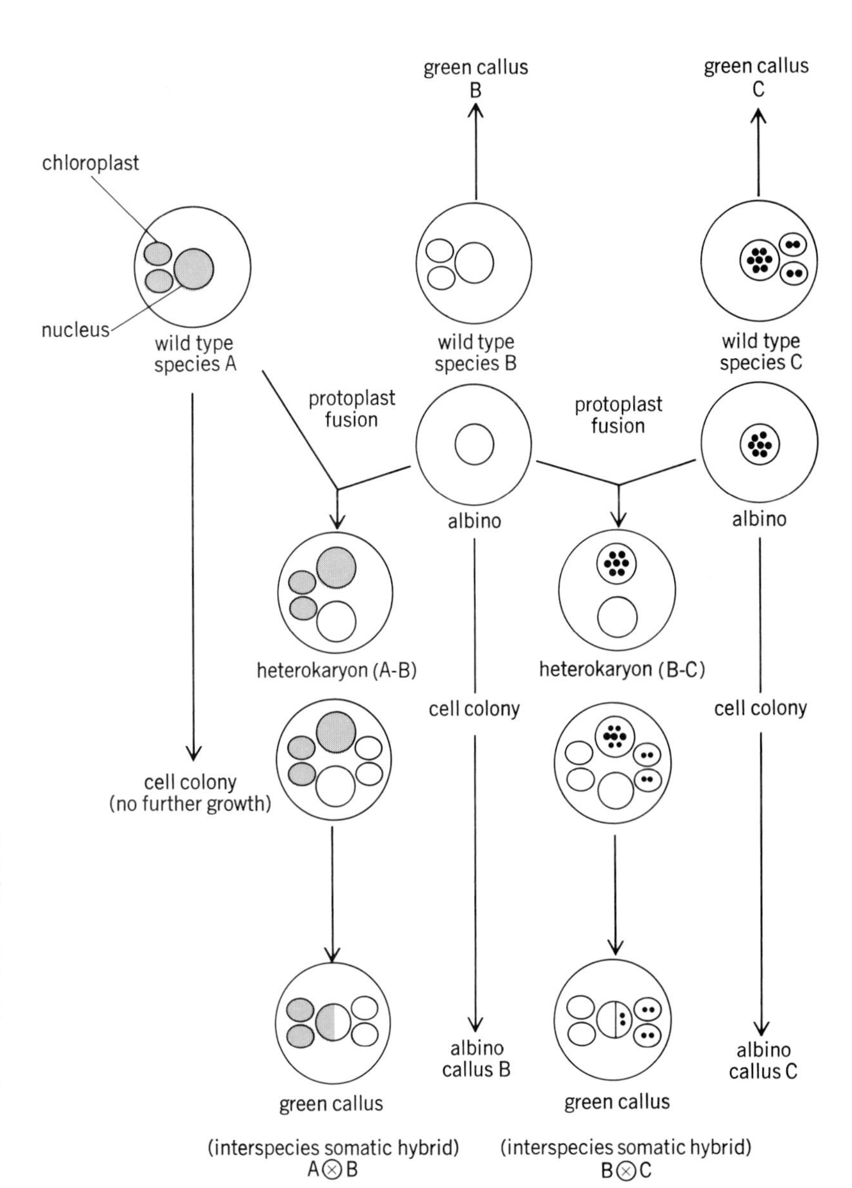

Fig. 2. Schemes for albino complementation selection of interspecies somatic hybrids. (*From E. C. Cocking et al., Plant Sci. Lett., 10:7–12, 1977*)

(PLANT); CHROMOSOME; HYBRIDIZATION, SOMATIC CELL in the McGraw-Hill Encyclopedia of Science and Technology. [E. C. COCKING]

Bibliography: E. C. Cocking, Fusion and somatic hybridization of higher plant protoplasts, in J. F. Peberdy et al. (eds.), *Microbial and Plant Protoplasts*, 1976; E. C. Cocking et al., *Plant Sci. Lett.*, 10:7–12, 1977; D. Dudits et al., *Theor. Appl. Genet.*, 51:127–132, 1977; Y. Y. Gleba, Non-chromosomal inheritance in higher plants as studied by somatic cell hybridization, in *4th Biological Sciences Colloquium*, Columbus, OH, 1978.

Cancer (biology)

Biochemical abnormalities caused by the synthesis and release of tumor products in cancer patients can be a valuable aid in tumor detection and diagnosis. Much progress has been made in the use of antigens, enzymes, peptide hormones, and other proteins as biochemical markers. The more promising tumor markers are discussed below.

Carcinoembryonic antigen. One of the most extensively investigated biochemical markers for cancer is the carcinoembryonic antigen (CEA) of the human digestive system. CEA is an acid glycoprotein with a molecular weight of approximately 200,000, consisting of 14 amino acid residues and 6 carbohydrate moieties. The sedimentation constant is between 6.9 and 8.0. The peptide portion is about 60% of the molecule and has been shown to be a straight chain. CEA migrates at the β-globulin region in electrophoresis. The antigen was first demonstrated in embryonic and fetal gut, liver, and pancreas in the first 6 months of gestation, and was thought to be specific for entodermally derived digestive system cancers, but later was found to be common to other malignancies. The CEA molecule has been localized to the glycocalyx of the tumor cell surface as well as to the cytoplasm. Currently a number of sensitive and reliable radioimmunoassays for CEA are available which are capable of measuring nanogram quantities of circulating CEA. The incidence of elevated CEA in the blood of patients with gastrointestinal neoplasm in early stages is about 20–50%.

CEA assay has been shown to be of value as an adjuvant tool in conjunction with conventional cancer diagnostic procedures. It has been reported that CEA determination is useful in preoperative prediction of tumor stage and postoperative prognostication of patients' chances for survival and disease recurrence, as well as postoperative management of patients with curative surgery. However, it is generally believed that CEA assay in its present form cannot be employed as a screening test for cancer in the general population. This is due to cross-reaction of various proteins of glycoprotein nature with anti-CEA antiserum in the radioimmunoassay system. Also, small tumors have been known to be associated with normal plasma levels of CEA; however, high levels of CEA in patients with an obvious but undiagnosed illness are highly suggestive of hidden carcinoma. Recently, a species of CEA, known as CEA-S, has been reported to be a specific antigen for human gastrointestinal malignancies.

α-Fetoprotein. This is a normal embryonic serum protein, biochemically very similar to human albumin, which migrates at the α-globulin region in electrophoresis. α-Fetoprotein levels decline after birth, and the protein is present only in trace amounts in the bloodstream of the healthy child and adult. α-Fetoprotein is generally considered to be a marker for primary hepatocellular carcinoma and embryonal cell carcinomas of the testes and ovary, as a significant number of patients with these tumors have showed an elevation of serum α-fetoprotein. α-Fetoprotein has been used for large-scale screening for hepatoma in Senegal and China, since Africans and Orientals have a very high incidence of liver cancer. The results from these screening tests are encouraging. Almost half of the patients with hepatoma were detected prior to clinical evidence of the disease.

It should be mentioned that α-fetoprotein measurement also plays another important diagnostic role. Elevation of α-fetoprotein in maternal serum has been associated with fetal distress or death. Further, a relationship between elevated α-fetoprotein levels in amniotic fluid and neural tube defects in the fetus has been reported. A large-scale screening test for this defect is being conducted in Great Britain.

Prostatic acid phosphatase. Acid phosphatase was first reported 40 years ago to be associated with cancer of the prostate. The activity of serum acid phosphatase, as measured by conventional spectrophotometric techniques, is elevated in 60–90% of patients having prostate cancer with bone metastasis, 30% of patients without roentgenologic evidence of bone metastasis, and only 5–10% of patients without demonstrable metastasis or with early stages of prostate cancer. However, serum acid phosphatase can originate from tissues other than the prostate and from red blood cells, leukocytes, and platelets. Therefore, serum acid phosphatase can be elevated in diseases other than prostate cancer.

Conventional and biochemical approaches (heat inactivation and "specific" substrates or inhibitors) to assay the specific prostatic fraction of acid phosphatase have been developed and found to be unsatisfactory. Recently, acid phosphatase of prostate origin has been shown to exhibit immunologic specificity, that is, antiserum produced from prostatic acid phosphatase does not react with acid phosphatases of other tissues. Immunoassays (radioimmunoassay, counterimmunoelectrophoresis, and immunofluorometric assay) have been developed for the measurement of serum prostate–specific acid phosphatase.

By using these newly developed specific and sensitive immunochemical techniques, a substantial number of patients (30–78%) with early stages of prostate cancer can now be detected. Since tumors in these patients are usually small and treatable, aggressive therapies can potentially provide a high cure rate. This new finding is particularly important since, at present, 90% of prostate cancers are first detected when tumors have already metastasized to distant organs, a condition which presents a uniformly poor prognosis. Prostate cancer is the third leading cause of death due to cancer in men under age 55 and the leading cause in men older than 70. The effect of this early detection on the survival of the patients remains to be determined. A national trial on the immunochemical detection of early prostate cancer is under way, which may be one of the most significant developments in cancer detection and diagnosis in recent years.

Thyrocalcitonin. Thyrocalcitonin is a peptide of 32 amino acids with a molecular weight of about 2500. Calcitonin is a native hormone which lowers the blood calcium and is normally produced in the C cells of the thyroid gland. Medullary carcinoma of the thyroid produces much larger amounts of calcitonin than in the normal thyroid gland. The recent development of radioimmunoassay for the measurement of serum calcitonin level thus provides a valuable tool for the early detection of this thyroid cancer. Almost all of the patients with medullary carcinoma of the thyroid have exhibited an elevated level of serum thyrocalcitonin, which has been used as a biochemical marker for diagnosis. Moreover, since this tumor may be inherited within the family as an autosomal dominant trait, measurement of serum thyrocalcitonin in family members of the patient has been an effective screening technique. Although some patients with C-cell-hyperplasia thyroiditis also demonstrate an elevation of serum calcitonin, as do some patients with bronchogenic carcinoma and other tumors, infusion of calcium or pentagastrin has greatly improved the diagnostic specificity for members of kindreds in early detection of this cancer.

Fibrinogen degradation products. A biochemical marker may play a role in the early detection of bladder cancer. Elevation of fibrinogen degradation products has been reported in the urine of patients with cancer of the bladder, as measured by a hemagglutination inhibition technique. A high degree of accuracy (90%) was found in correlating urinary fibrinogen degradation products and cytology with the activity of bladder cancer. No correlation between fibrinogen degradation products in serum and activity of disease was observed. It is possible that more specific assay methods may detect minimal quantities of fibrinogen degradation products in the urine and improve the accuracy of detecting clinically occult disease states, including carcinoma in place.

Peptide hormones. Human chorionic gonadotropin is a normal placental product present in the serum of pregnant women. The presence of elevated levels of human chorionic gonadotropin has been reported in cancer, particularly in those patients with trophoblastic neoplasms and cancer of the testes and breast. Other peptide hormone markers, ectopically produced by tumors, such as pro-ACTH, β-lipotropin, and arginine vasopressin-vasotocin, have been found to be useful in the early diagnosis of lung cancer.

Ferritin. Ferritin is the major iron storage protein in tissue. Elevated levels of circulating ferritin were first found in leukemia and Hodgkin's disease. The analysis of ferritin from normal tissue by isoelectric focusing has revealed considerable heterogeneity. Ferritin isolated from breast and pancreatic tumors as well as from the placenta contained acidic isoferritins not found in adult liver. The "carcinofetal" isoferritin has been studied extensively as a marker for breast tumors. The concentrations of circulating ferritin are generally higher in breast cancer patients than in healthy women. Patients with an elevated level of circulating ferritin have been shown to have a higher tumor recurrence rate. The reason why patients with early breast cancer have serum ferritin levels somewhat higher than normal is still unclear, although it may be due to the nonspecific effect of malignancy or reticuloendothelial iron metabolism.

Other biochemical markers. The list of substances receiving attention as biochemical markers for cancer is expanding. Aryl hydrocarbon hydroxylase has been proposed as an enzyme marker to identify individuals at risk for carcinoma of the lung. Ovarian cystadenocarcinoma antigen has provided initially promising results in the detection of ovarian cancer in selected asymptomatic patients. Tissue polypeptide antigen is a marker for tissue proliferation which may also serve as an aid in cancer diagnosis. Preliminary findings on "B-protein" indicate that this acetyl coenzyme A system–related compound may be a tumor marker. A pilot study from a "Tennegen" assay (an antigen assay that has been developed by a group working in Tennessee) has given a set of results similar to those obtained from CEA. Terminal deoxynucleotidyl transferase, deoxyribonuclease, and protease have been shown to be effective tools for the differential diagnosis of leukemia. Further studies are necessary to determine whether these markers will ultimately become additional biochemical markers for the early diagnosis of cancer.

For background information *see* ANTIGEN; CANCER (BIOLOGY); PROSTATE GLAND (HUMAN); RADIOIMMUNOASSAY; THYROCALCITONIN in the McGraw-Hill Encyclopedia of Science and Technology.

[T. MING CHU]

Bibliography: T. M. Chu et al., *Invest. Urol.*, 15: 319, 1978; R. B. Herberman, *Amer. J. Clin. Pathol.*, 68:688, 1977; L. H. Lees et al., *Ann. N.Y. Acad. Sci.*, 297:603, 1977; S. Sell and F. F. Becker, *J. Nat. Can. Inst.*, 60:19, 1978.

Cartilage

The role of cartilage tissues in the evolution of the vertebrate skeleton has been of such importance in understanding vertebrate skeletal biology that the existence of mesenchymally or mesodermally derived endoskeletal cartilages in invertebrate animals has been almost completely overshadowed. Indeed, many present-day biologists are unaware that invertebrates have endoskeletal cartilages, although such structures have been known for more than 100 years. In 1969, after the study of invertebrate cartilages had been almost completely neglected for nearly 50 years, P. Person and D. E. Philpott revived interest in the tissues by summarizing the then extant knowledge and established their true cartilaginous nature by correlated microscopic and chemical studies.

Invertebrate phyla in which cartilage tissues have been reported are Coelenterata, Annelida, Mollusca, and Arthropoda. Interesting features of invertebrate cartilages are (1) the wide diversity of tissue types and chemical components; (2) the fact that, insofar as is known, these tissues never mineralize in nature; and (3) their possible contributions to the understanding of the origins of bone and cartilage, and of the evolution of primitive vertebrates.

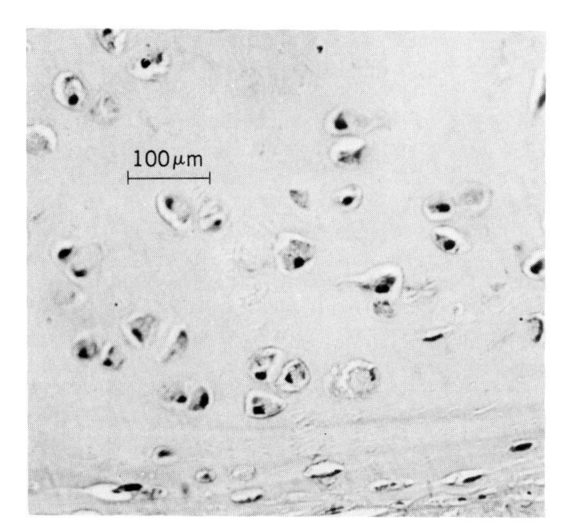

Fig. 1. Photomicrograph of a tissue section of cranial cartilage from the squid *Loligo pealii*. Irregularly shaped cells are distributed within a relatively abundant, fairly homogeneous matrix. *(From P. Person and D. E. Philpott, On the occurrence and the biologic significance of cartilage tissues in invertebrates, Clin. Orthop. Related Res., 54: 185–212, 1967)*

Morphology. Because of the wide variety of histologic tissue types and chemical components, it is impossible to characterize a "typical" invertebrate cartilage. The histologic variability of tissues derives in large part from different ratios of cells to intercellular matrix in different tissues. At one extreme is the head cartilage of *Loligo* (squid), shown in Fig. 1, which resembles hyaline cartilage of vertebrates in being composed of relatively few cells enclosed in an abundant (metachromatic) matrix. At the other extreme are the cartilage tissues of the feather-duster marine worm *Eudistylia*, shown in Fig. 2. In *Eudistylia* there is a central supporting rod or core of predominantly rectangular cartilage cells with deeply staining cell wall–like matrix between the cells. The central core gives rise to linear arrays of almost completely vacuolated cells which are almost at right angles to the core

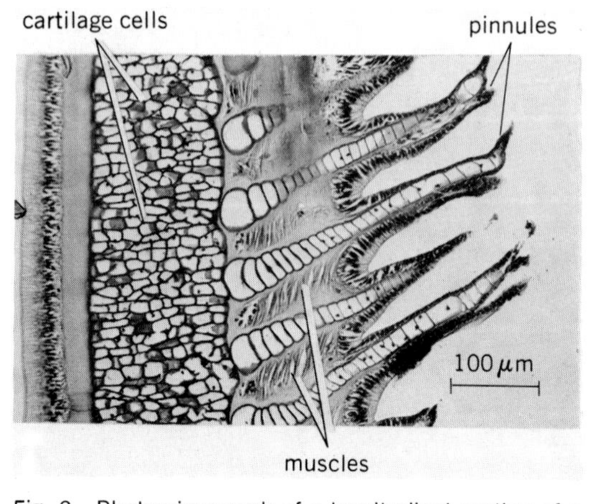

Fig. 2. Photomicrograph of a longitudinal section of a portion of a tentacle from the crown of the marine feather-duster worm *Eudistylia polymorpha*. *(From P. Person and M. B. Mathews, Endoskeletal cartilage in a marine polychaete, Eudistylia polymorpha, Biol. Bull., 132:244–252, 1967)*

and which support feathery appendages known as pinnules. As the tentacles move in the water, muscles set the pinnules in motion, creating water currents which sweep food and organisms toward the animal's mouth.

The cells in Fig. 2 are far more numerous than in Fig. 1 and separated only by relatively thin seams of matrix. The bioarchitectural regularity (of cell shape and arrangement) of the *Eudistylia* cartilage makes it very similar to plant tissue. Indeed, in this instance the cartilage matrix might be considered a cell wall, although the term is almost never used in reference to animal tissues.

There is a broad spectrum of invertebrate cartilage tissues in which cell-to-matrix ratios in between those of *Loligo* and *Eudistylia* are found. In addition, in some tissues, for example, odontophore cartilage of the mollusk *Stagnicola* (the oyster drill), there is an intimate admixture and interweaving of muscle cells among cartilage cells.

To existing light and electron microscopic characterizations of invertebrate cartilage tissues H. Kryvi has now added correlated scanning electron microscope observations. He has shown that the cartilage tissues of *Sabella penicillum*, another marine feather-duster worm, vary depending on their location within the animal. For example, cells of the basal cartilaginous complex, which supports a radial array of tentacles about the oral end of the organism, are spherical in shape and are surrounded by narrow seams of matrix containing a dense pile of collagenlike fibrils. These fibrils are oriented parallel to the long axis of the worm. Close to the cell periphery, the matrix is thickened to form a wall-like support structure. On the other hand, the cells of internal support for the tentacles and their pinnules vary from cuboidal to cylindrical in shape, and their matrix surround is more loosely organized than that of the basal complex. Cells of both areas contain large vacuoles which occupy most of the cell volume, and which are surrounded by a very thin cytoplasmic layer (barely visible in the light microscope) on the order of 0.1 μm thick which contains typical cytoplasmic organelles including granular endoplasmic reticulum, Golgi bodies, free ribosomes, and scattered mitochondria. The matrix fibrils resemble vertebrate embryonic precollagen in their striation pattern, while the *Sabella* cartilage itself is similar to vertebrate elastic cartilage and also bears some resemblances to vertebrate embryonic notochord and cartilage.

No definite evolutionary pattern or progression of cartilage tissue types can be discerned as one surveys the various invertebrate phyla. It would appear, rather, that as is the case with vertebrate cartilages, the histology, chemistry, and physicochemical properties of tissues correlate more closely with the function of the cartilages and the ecology of the organisms in which they are found.

Chemistry and mineralization. Although the chemical study of invertebrate cartilages is only beginning, the data available to date also illustrate a great diversity of chemical composition in tissues of different organisms. M. B. Mathews has summarized data for the acidic glycosaminoglycan composition of various invertebrate cartilages: the molar ratios of hexuronic acid to hexosamine are 0.99–1.18, and of sulfate to hexosamine 0.07–1.88 for cartilage tissues from annelid, mollusk, and

Chemical components of invertebrate cartilage tissues*

Tissue	Ribonucleic acid	Deoxyribonucleic acid	Collagen	Noncollagenous protein	Lipid
Limulus (gill)	0.60 ± 0.10	0.09 ± 0.03	4.32 ± 1.46	19.52 ± 3.74	0.11 ± 0.08
Eudistylia (crown and tentacle)	1.31 ± 0.09	0.16 ± 0.03	19.05 ± 2.17	18.73 ± 2.45	0.39 ± 0.17
Loligo (cranial)	0.36 ± 0.07	0.09 ± 0.01	6.35 ± 0.43	10.12 ± 1.94	0.22 ± 0.12
Busycon (odontophore)	0.09 ± 0.01	0.06 ± 0.02	1.01 ± 0.23	3.68 ± 1.85	0.80 ± 0.34

*Percentages as calculated on a net weight basis by using the average of three analyses plus or minus scanning electron microscopy.
SOURCE: From J. L. Rabinowitz et al., Lipid components and in vitro mineralization of some invertebrate cartilages, *Biol. Bull.*, 150:69–79, 1976.

arthropod specimens. The acidic glycosaminoglycans are important characteristic compounds found in all cartilage tissues, and the above values may be compared with reference compound molar ratios for hexuronic acid and sulfate to hexosamine, of 1.15 and 0.97, respectively. J. L. Rabinowitz and coworkers have analyzed similar cartilages for other essential tissue components (see the table), with similar variations in results. Such differences in composition undoubtedly reflect, in great part, the pleomorphic nature of the tissues studied.

Person and Philpott pointed to a unique characteristic of invertebrate cartilages in which they differ from most vertebrate cartilages, namely, that none of the currently known invertebrate tissues form a mineral phase in the course of their natural history. Why this should be true is not understood, particularly since in some animals other tissues do have the ability to mineralize. Thus, calcium carbonate will form in the outer shell of mollusks, while odontophore cartilages remain nonmineralized. For this reason it may well be that the invertebrate tissues hold important clues for the understanding of biological mineralizations. The lack of mineralization may be explainable on the basis of the presence (or absence) of unique chemical constituents in the invertebrate tissues. Alternatively, there may be specific repressor substances which inhibit mineralization of the cartilages, or the lack of mineralization may be genetically determined.

In an attempt to investigate these hypotheses, Rabinowitz and coworkers studied the mineralizability in cell cultures of cartilage tissues from *Limulus*, *Loligo*, and *Busycon*. In these studies, cartilages from all of these organisms were shown to be capable of mineralizing when placed in metastable solutions. Of greater interest, however, was the fact that in each instance the mineral phase formed was hydroxyapatite, which is characteristic of vertebrate skeletal mineralization and only rarely encountered in invertebrates! To date, it has not been possible to induce invertebrate cartilage mineralization by calcium carbonate, which is the characteristic invertebrate tissue-mineralizing phase.

For background information *see* ANNELIDA; ARTHROPODA; CARTILAGE; COELENTERATA; MOLLUSCA in the McGraw-Hill Encyclopedia of Science and Technology.

[PHILIP PERSON]

Bibliography: H. Kryvi, *Protoplasma*, 91:191–200, 1977; M. B. Mathews, *Connective Tissues: Macromolecular Structure and Evolution*, 1975; P. Person and D. E. Philpott, *Biol. Rev.*, 43: 1–16, 1969; J. L. Rabinowitz et al., *Biol. Bull.*, 150:69–79, 1976.

Cell division

During cell division the formation of the mitotic spindle provides the polarity and force-generating mechanisms for the complex movements of the chromosomes during mitosis. Attempts to understand the structure and function of the mitotic spindle have long been frustrated by its transitory, dynamic nature. Each time cell division occurs (Fig. 1), a mitotic spindle is assembled anew from subunits available in the cell, and upon completion of cell division the spindle disappears. Also, even while the spindle exists, its configuration and dimensions are changing. To further complicate the problem, the mitotic spindle, particularly its changing microtubular structure, is extremely difficult to observe by most microscopy techniques—partly because of its inherent composition and partly because it is often obscured by other organelles and cellular inclusions. Scientists are still trying to identify the molecular components of the mitotic spindle and to ascertain how the molecules interact to produce the forces for chromosome movements.

Recent developments in experimental approaches to these problems are providing new information about spindle composition and operation. Among these approaches are the production of fluorescent-tagged antibodies to known and suspected spindle molecules, and the isolation from their cellular environs of mitotic spindles that retain their structural integrity as well as a number of important functional capabilities. Results from these approaches are showing the presence in the mitotic spindle of molecules, such as actin and dynein (an enzyme that catalyzes the breakdown of adenosinetriphosphate, or ATP, to adenosinediphosphate, or ADP, and inorganic phosphate in order to release energy), which are known to be involved in other cellular motile activities, but evidence for their role in generating chromosome movement is still incomplete.

Spindle microtubules. It is well documented that the principal structural components of the mitotic spindle are the microtubules—250-A-diameter (25-nm) hollow cylinders of variable length polymerized from a cellular pool of tubulin, a globular protein. Clusters of microtubules, the spindle fibers, extend from the kinetochore of each chromosome and the two spindle poles (Fig. 2). It ap-

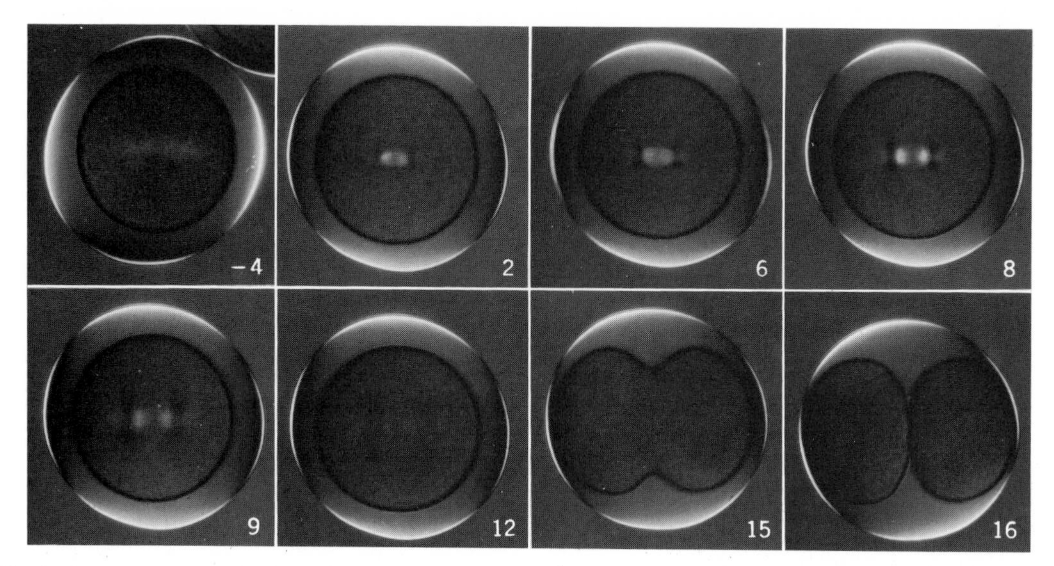

Fig. 1. Polarization micrographs of cell division in living egg of sea urchin *Lytechinus variegatus* showing density and distribution of bundles of parallel microtubules. Birefringent spindle fibers appear light or dark depending on their orientation in the field. Time in minutes from nuclear membrane breakdown is given on each frame. Temperature is 25°C. Egg diameter (not including fertilization membrane) is about 100 μm.

pears that most microtubules have one end anchored at either a pole or a kinetochore, but how many, if any, have both ends anchored (at a kinetochore and a pole or at both poles) is still uncertain. Chromosome movement is produced by forces directed along the microtubules of the mitotic spindle. Microtubules are required both for generating and for orienting chromosome movements. The microtubules are not static structures, but appear to be in a dynamic state of flux with the pool of presynthesized tubulin subunits. Assembly of the spindle fiber microtubules is somehow controlled by the organizing or orienting centers at the kinetochores and the poles. Experiments have indicated that local and selective growth and shortening of spindle fiber microtubules are directly related to the production of forces. As a result, recent efforts have been directed toward discovering molecular mechanisms of force generation along the spindle fiber microtubules, as well as molecular mechanisms controlling microtubule assembly and disassembly.

Chromosome movement theories. Theoretical models for the role of the spindle in chromosome movement abound, but three are currently considered most plausible. One theory, proposed by S. Inoué, is that the selective depolymerization and shortening of the spindle fiber microtubules produces pulling forces, while polymerization and elongation produces pushing forces. While agreeing that depolymerization and polymerization of microtubules are essential for regulating the rate of chromosome movement, other researchers feel that an additional mechanism for generating force is necessary. Some, such as J. R. McIntosh, believe that between interdigitating microtubules there is an active sliding produced by cross bridges such as those of dynein in cilia and flagella. Other workers, such as A. Forer, suggest that there is an actin-myosin contractile system associated with the microtubules. Evidence for either of the latter two models depends on identifying the relevant molecules in the spindle and demonstrating their activity in generating chromosome movement.

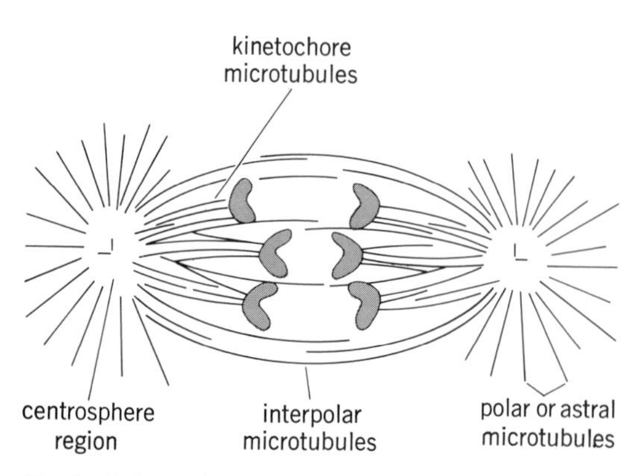

Fig. 2. Schematic diagram of the probable distribution of microtubules in the early anaphase spindle.

Mitotic spindle structure. Mitotic spindles treated with fluorescent-tagged antibodies to tubulin show dramatically the spindle's structure in fluorescence microscopy. Of more theoretical interest, however, are immunofluorescence experiments with antibodies to actin, myosin, and α-actinin, proteins involved in generating and transmitting forces for muscle contraction. When it was found that fluorescent-tagged antibodies to actin bind to spindles in fixed cells in the regions of the kinetochore fibers, speculation increased that the forces for chromosome-to-pole movement are produced by the interaction of actin and myosin molecules. Further experiments with antibodies to myosin and α-actinin, however, were inconclusive. While these antibodies do bind to molecules in the mitotic spindle, they produce only diffuse fluorescence, which indicates a low concentration. In contrast, the same antibodies produce intense fluorescence in the cell's cortex at the site of the cleavage furrow, indicating a very high concentration. When antibodies to myosin were injected into

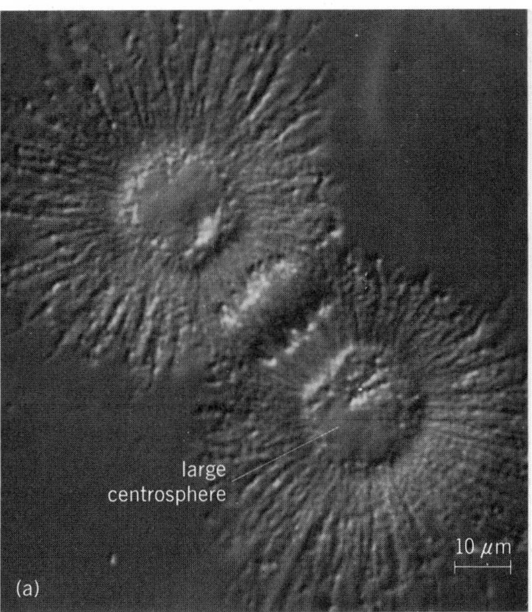

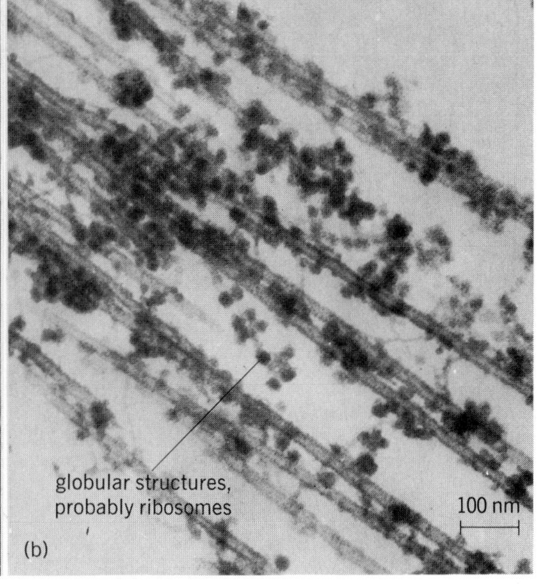

Fig. 3. Anaphase mitotic spindles isolated in calcium-free detergent buffer from eggs of the sea urchin *Stronglyocentrotus droebachiensis*. (*a*) Whole spindle seen with differential interference microscopy. (*b*) Thin section through a kinetochore fiber viewed with transmission electron microscopy.

living cells by D. Kiehart, division of the cytoplasm ceased, whereas nuclear division with normal chromosome movements continued. Presumably, the antibodies bound to molecules of myosin, thus preventing an interaction between myosin and actin. It is generally accepted that actin-myosin interactions generate the constricting forces in the cleavage furrow for division of the cytoplasm, but it now seems less likely that such interactions play a significant role in chromosome movement.

While fluorescent antibodies appear to be effective in deducing the molecular structure of the mitotic spindle, other approaches are needed to answer questions about its function. To facilitate various experimental manipulations, scientists need to isolate an intact mitotic spindle with normal functional capabilities from the cellular environment. A major problem in obtaining an isolated spindle is that when a cell is lysed, and the membranes and cytoplasm are washed away, the spindle usually dissolves rapidly. It has been possible for many years to isolate intact spindles if they were first stabilized or fixed with substances such as hexylene glycol. Unfortunately, after isolation these spindles are no longer functional, and they do not exhibit the same assembly-disassembly characteristics as untreated spindles inside cells.

Calcium ion effect. Recently it has been discovered that if, at the time a dividing cell is lysed, calcium ions are removed by a strong chelator such as EGTA [ethylene glycol bis(β-aminoethyl ether) N,N,N',N'-tetraacetic acid], the spindle's microtubular structure is preserved (Fig. 3) while remaining responsive to changes in composition of the surrounding medium. This discovery is important for two reasons. First, it has focused attention on the possibility that the cell itself uses local changes in calcium ion concentration to regulate the assembly and disassembly of spindle microtubules. Second, it is hoped that spindles isolated in this manner may retain sufficient functional capacity that they can be induced to move chromosomes by proper manipulation of experimental conditions. There has been significant progress in research along both these lines.

E. D. Salmon and R. Jenkins have shown that spindles which are rapidly isolated, and detergent-extracted, into a calcium-free medium can be dissolved by adding calcium ions in very low (micromolar) concentrations. Microtubules polymerized in a test tube from microtubule protein purified from brain neurons are also sensitive to slight elevations in calcium ion concentration if they are assembled in the presence of certain proteins called calcium-dependent regulatory proteins (CDRPs) as discovered recently by J. M. Marcum and coworkers. Microtubules associated with CDRPs in a test tube disassemble in the presence of micromolar concentrations of calcium ions and reassemble when the ions are washed away. Observations by M. J. Welsh and coworkers with fluorescence microscopy of mitotic spindles treated with fluorescent-tagged antibodies to CDRP have shown CDRPs distributed along the mitotic spindle fibers. When D. Kiehart and Inoué microinjected calcium ions into specific spindle regions, the spindle microtubule birefringence was temporarily abolished in the immediate vicinity of the injection, indicating a very local, and reversible, depolymerization of microtubules in response to increased calcium ion concentration. Apparently, calcium does not diffuse freely within the spindle region and is rapidly chelated. An adenosinetriphosphatase (ATPase) activated by calcium ions, found by D. Mazia and C. Petzelt to be concentrated in the mitotic apparatus, may be involved in the sequestering and release of calcium ions. L. Rebhun and coworkers have recently shown that this Ca^{++}-ATPase activity varies with the reduced state of its sulfhydryls which, in turn, may be controlled by cyclic changes in the concentration of reduced glutathione during cell division.

Scientists are now searching for structural components of the mitotic spindle that might be involved in regulating calcium ion concentration. Research in this area is being focused particularly on the numerous membrane vesicles interspersed among the spindle fibers.

Dynein effect on chromosome movement. It is still too early to specify the conditions under which spindles isolated into calcium-free media can be induced to move chromosomes, but the outlook is promising. Recently, E. D. Salmon has seen that kinetochore microtubules shorten and chromosomes start to move poleward when isolated spindles (Fig. 3) are bathed in solutions with ATP and micromolar concentrations of calcium. Refining the conditions for activating chromosome movement in an isolated spindle would help greatly to distinguish the actual mechanism by which the cell's mitotic spindle produces forces for chromosome movement. Analyses of the protein composition of calcium-free spindle isolates have shown the presence of actin, as well as dynein. Speculations that forces for chromosome movement might be generated by dynein-like cross bridges have had only tentative support from some electron micrographs that suggest the presence of cross bridges between adjacent microtubules. However, it now appears that the calcium-free spindle isolates contain several proteins with the same molecular weight as ciliary dyneins and that fluorescent-tagged antibodies to dynein bind to the mitotic spindle.

Experiments with "partially isolated" spindles are adding to evidence for the possible role of dynein in moving chromosomes. W. Z. Cande has recently demonstrated that dividing cells can be treated with low concentrations of detergent to partially solubilize their membranes so that although the spindle remains in its cytoplasmic matrix, various molecules of experimental interest can penetrate to the spindle region. In such cells, chromosome movement continues if ATP is present in the surrounding medium. Sodium vanadate, which specifically inhibits the activity of dynein, inhibits the movement of chromosomes, and norepinephrine, which blocks the inhibition of dynein activity by sodium vanadate, starts it again.

For background information *see* ADENOSINEDIPHOSPHATE (ADP); ADENOSINETRIPHOSPHATE (ATP); CELL DIVISION in the McGraw-Hill Encyclopedia of Science and Technology.

[E. D. SALMON]

Bibliography: W. Z. Cande, *J. Cell Biol.*, in press; E. R. Dirksen and D. Prescott (eds.), *J. Supramol. Struct.*, 29(suppl. 2):282–338, 1978; R. Goldman, T. Pollard, and J. Rosenbaum (eds.), *Cold Spring Harbor Conferences on Cell Proliferation*, vol. 3: *Cell Motility*, 1976; E. D. Salmon and R. Jenkins, *J. Cell Biol.*, 75:295a, 1977.

Chlorophyll

Laser excitation has been increasingly used in the study of the light-induced photochemical and photophysical processes involved in plant photosynthesis. The general objective of these studies has been to determine the rates and mechanisms of the processes involved in the absorption of light by antenna chlorophyll molecules and the subsequent transfer of this excitation energy to the photoactive centers (special pairs) where the energy is trapped. The experimental results, particularly where high-powered lasers are used, are not always consistent or readily interpreted. Part of the problem arises from the fact that the photophysical processes appear to be a function of the photon fluxes used in the experiments. At high light intensities a significant fraction of the chlorophyll molecules in a photosynthetic unit may be excited, and processes that do not occur in low light intensities may become important. Examples are the nonexponential fluorescence decay times observed in photosynthetic units excited with high-intensity picosecond laser pulses and the intensity-dependent fluorescence quantum yields observed in both picosecond and nanosecond excitation of photosynthetic units. One explanation for these phenomena is that once the traps are closed as a result of the energy transfer from the antenna, further excitation in the antenna leads to the buildup of chlorophyll excited states. These are then deactivated by energy transfer to yield heat and chlorophyll molecules in the ground state (singlet-singlet annihilation). Research has been directed toward determining whether there are alternative photophysical processes that can result in similar phenomena. Recent studies on laser-excited chlorophyll molecules outside of living plants have shown that the phenomena of lifetime shortening and decrease in fluorescence quantum yield can take place under conditions where singlet-singlet annihilation does not occur.

There is also the more fundamental question of the fate of energy transferred to the photoactive centers. The preparation of the linked covalent dimers of chlorophyll, bis(chlorophyllide *a*) ethylene glycol diester, has made it possible to study the effect of light on species that have many of the properties of the photoactive (special-pair) centers. It is found that laser excitation results in significantly different effects when the dimers are in open and folded (reaction-center) configuration. When the dimer is in folded form, its behavior is very similar to monomer chlorophyll in terms of intensity-dependent phenomena; for example, laser output is obtained and lifetime-shortening with increasing pump power is observed. In the open configuration there is still an unidentified mechanism that prevents the buildup of significant excited-state concentrations. These observations suggest that configurational factors may play an important part in determining the fate of energy

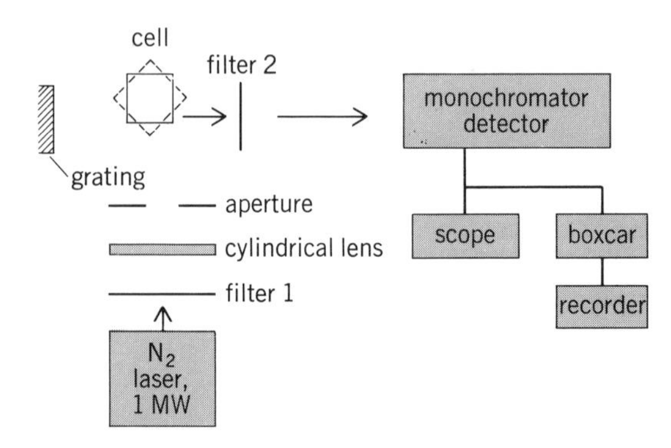

Fig. 1. Experimental setup for observing lasing in solutions of chlorophyll. For simple lasing experiments, the cell is set at right angles to the nitrogen pump beam.

transferred to chlorophyll molecules in reaction centers.

Stimulated light emission. Experiments on solutions of monomeric chlorophyll *a* and related molecules outside of living plants have shown that optical pumping with an ultraviolet nitrogen laser can lead to significant population of the first excited singlet state of these molecules. At sufficiently high light intensities the population of excited molecules becomes greater than the number of molecules in the ground state. If the solution is in a simple resonator consisting of a 1-cm-square quartz fluorescence cell, stimulated emission (lasing) can be observed. In the experimental setup shown in Fig. 1, the grating is used to demonstrate tunability of the chlorophyll laser, the cylindrical lens focuses the nitrogen laser beam on the cell, and the filters are used to reduce the beam intensities. Under these experimental conditions, the laser light emitted has a relatively narrow bandwidth (relative to normal fluorescence), with its maximum intensity at a wavelength near the maximum in the fluorescence curve (Fig. 2). For a nitrogen laser with a pulse width of approximately 9 ns, the photon density required to produce population inversion is of the order of 5×10^{16} photons/cm^2.

It is possible to obtain laser emission not only from chlorophyll *a* but also from other chlorophylls and related compounds in a variety of solvents. In fact, material which exhibits lasing behavior similar to chlorophyll *a* can be obtained by simple extraction with organic solvents from grass clippings, leaves, and so on.

Chlorophyll lasers. The fact that chlorophylls and related compounds can be made to emit laser light is of importance for several reasons: First, whether a compound will function as a laser dye depends on specific properties of the molecule in its ground and excited states. It is therefore possible to derive information about these properties from an analysis of the conditions required to induce laser action. For example, high photon densities relative to those required for pumping dyes such as rhodamine 6G have been found to be necessary to initiate laser action in the chlorophylls. This finding can be directly related to the small wavelength shift and strong overlap between the fluorescence and absorption bands in the chlorophylls and related compounds. It is also observed that the lasing wavelength is near the maximum of the fluorescence band. This is rather unusual behavior, and suggests that there may possibly be an excited-state absorption which tends to shift the laser wavelength from its expected frequency. A third reason that the emission of laser light is of importance is related to the high concentration of molecules in excited states that are produced in the excitation process.

Nonlinear optical phenomena. There are a number of effects associated with stimulated emission and high concentrations of molecules in excited states, including excited-state lifetime shortening, nonequivalent rates of photon and excited-state decay, fluorescence spectral narrowing, and a nonisotropic distribution of fluorescence intensity around a nonspherical excited volume. There is also the phenomenon of fluorescence quenching that results from absorption of either pump or fluorescence photons by the excited-state molecules. All of these phenomena are well

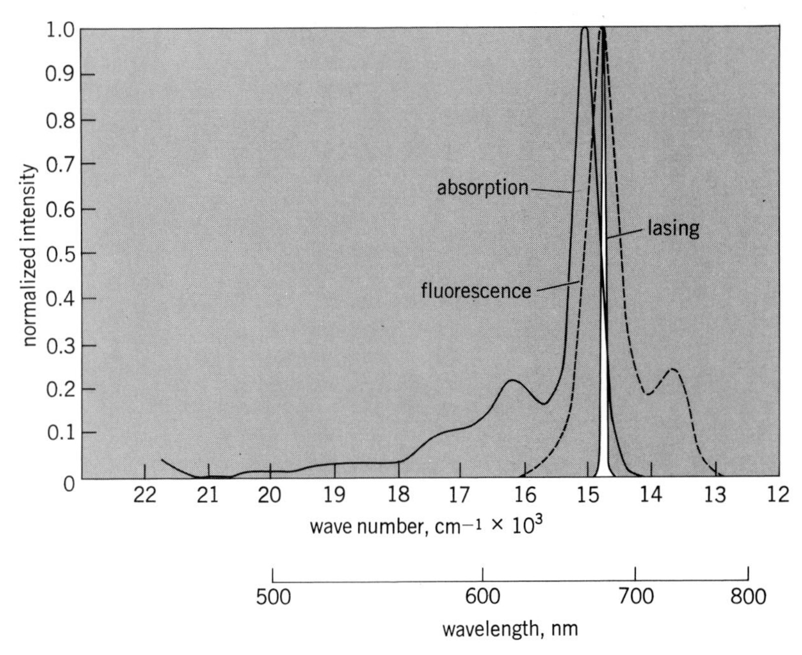

Fig. 2. The absorption, fluorescence, and lasing spectrum of a millimolar solution of chlorophyll *a* in ethanol.

known in laser dyes.

The effect of stimulated emission on the fluorescence lifetime and the effect of stimulated emission and excited-state absorption on the fluorescence quantum yield are the two phenomena of particular interest. The photophysical consequences of stimulated emission on the fluorescence lifetimes are formally the same as those ascribed to exciton annihilation processes. In high photon flux experiments the fluorescence lifetime becomes nonexponential and shortens as the concentration of excited states increases. Similarly, the probability of light absorption by the excited states increases as the population increases, with a consequent drop in the apparent fluorescence quantum yield. In both cases, however, the quantitative interpretation of the observations is complicated by the fact that the fluorescence properties are affected by the nonisotropic distribution around the nonspherical volumes excited by the nitrogen laser. Further experiments are necessary to determine whether or not the changes in fluorescence properties with increasing photon densities within and outside living plants are consequences of the same or different photophysical processes.

Linked chlorophyll dimers. Chlorophyll species that have optical and redox behavior very similar to that of special-pair chlorophyll in the photoreaction centers of green plants have been prepared in the laboratory. These species are formed from covalently linked bis(chlorophyllide *a*) ethylene glycol diester by the addition of a small amount of water to a benzene solution or of alcohol to a toluene solution. The photophysical processes that occur in these linked dimers under laser excitations are of interest not only because of the possible relationship between the folded linked dimer and the chlorophyll special pair in photosynthetic organisms but also because they may elucidate the types of interactions which take place between two chlorophyll molecules in close proximity. As an example, the chlorophyll dimers formed by keto C=O . . . Mg interactions in a nonpolar solvent such as toluene are essentially nonfluorescent.

Significant differences are found in the laser behavior of the open and folded forms of the covalent linked chlorophyll dimers when excited by a nitrogen laser. The folded form behaves as a single chromophore, and laser output is observed at 733 nm in a wet benzene solution. On the other hand, in the open form, the observed optical properties are very similar to those of monomeric chlorophyll molecules, except that it is not possible to obtain laser output. This failure to obtain laser emission results from the fact that no significant concentration of fluorescent excited molecules is produced even at very high excitation powers. This difference in the properties of the two configurations of the dimer is also reflected in the effect of increasing pump laser power on the fluorescence lifetimes. In the case of the folded dimer, increasing the pump power decreases the fluorescence lifetime as a result of stimulated emission. However, no such effect is observed for the open dimer. There is obviously a mechanism for energy loss (nonradiative process) in the open form that prevents a significant population of the fluorescing species from being produced. There are a number of possible explanations for the difference in behavior of the open and folded forms. As an example, excitation of the open dimer may lead to a conformational change in the excited-state species, yielding a nonfluorescent dimer similar to the chlorophyll *a* dimer present in a nonpolar solvent such as CCl_4 or toluene.

An interesting aspect of the experimental results is the observation that simply bringing two chlorophyll molecules into close proximity does not necessarily result in the formation of a nonfluorescent species or species having short fluorescence lifetimes. Further studies on these linked dimers may shed light on other, perhaps steric, factors that will have to be considered in arriving at a suitable model for energy transfer for systems within living plants.

For background information *see* CHLOROPHYLL; PHOTOSYNTHESIS in the McGraw-Hill Encyclopedia of Science and Technology.

[JAMES C. HINDMAN]

Bibliography: Chem. Phys. Lett., 53:197, 1978; J. C. Hindman et al., *Proc. Nat. Acad. Sci. U.S.A.*, 74:5–9, 1977; D. Leupold et al., *Chem. Phys. Lett.*, 45:567–571, 1977; *Proc. Nat. Acad. Sci. U.S.A.*, 75:2076–2079, 1978.

Chromatography

Gas chromatography (GC) is a separation method in which a sample (mixture of compounds) is made volatile, passed through a tube with a sorption medium (chromatographic column), and detected by appropriate means. During the migration of the sample through the chromatographic column, sample components (ideally, individual compounds) are separated from each other due to their different interaction with the column packing and emerge as molecular bands at the column end. The GC detector attached to the column records these molecular bands as chromatographic peaks; the peak position within a given chromatographic spectrum (chromatogram) bears certain qualitative indication, while the area under the peak can be related to the amount of substance. GC is applicable only to those compounds that have sufficient vapor pressure at temperatures up to about 300°C, or to substances that can be easily made volatile.

Frequently, there is a need to analyze extremely complex mixtures both qualitatively and quantitatively. The success of such determinations is strongly dependent on the quality of resolution of individual mixture components, or chromatographic efficiency. Capillary GC provides resolving power superior to any other technique for this purpose.

Capillary gas chromatography. In conventional GC, packed columns are used that consist of tubes with typically 2-mm internal diameters and 2–4-m lengths, which are filled with a granular packing material. The sorption agent is most often a nonvolatile liquid embedded on the surface of an inert solid. Such columns have at best a separation efficiency of several thousand theoretical plates (hypothetical volume elements of a chromatographic column). The number of theoretical plates is related to the width of a given chromatographic peak; that is, the wider the peak, the lower the number of theoretical plates.

In contrast, capillary columns have efficiencies in the range of 10^4–10^6 theoretical plates. Consequently, capillary GC bands are narrower and more components can be spaced and resolved within a given chromatogram. Typical capillary columns (sometimes referred to as open tubular columns) may have lengths up to 100 m, with internal diameters between 0.2 and 0.5 mm. The chromatographic (retentive) medium is spread on the inner wall of the capillary tubes.

The high separating performance of capillary GC is a combined result of optimized diffusion phenomena inside the columns with respect to gas flow, and of the great coaxial permeability through a length of such tubes to gas. Their separation efficiency is primarily the result of a significant difference in column length as compared with packed columns. The resolving power of capillary GC is demonstrated in Fig. 1. Numerous components are separated in this method, whereas a typical packed column would hardly yield more than 30–40 peaks from the same mixture. However, packed columns can handle samples of much greater size; capillary GC is typically restricted to samples of less than 10^{-6} g per component.

Column types. The two main column types in capillary GC are wall-coated capillary columns and support-coated open tubular columns. Whereas the wall-coated column is characterized by the layer of stationary liquid coated directly on the inner wall of a capillary coiled tube, the support-coated column utilizes fine granular solid support to disperse the stationary liquid. Thus, in the latter case, liquid can be encountered on the surfaces of both solid support and the tubing material (Fig. 2). The support-coated column can tolerate larger samples, but its typical efficiency is significantly lower.

For both column types, the liquid phase is distributed on the respective surfaces as a very thin film, typically around a few tenths of a micrometer. It is essential that the film be highly uniform along the entire column length. The solid supports used

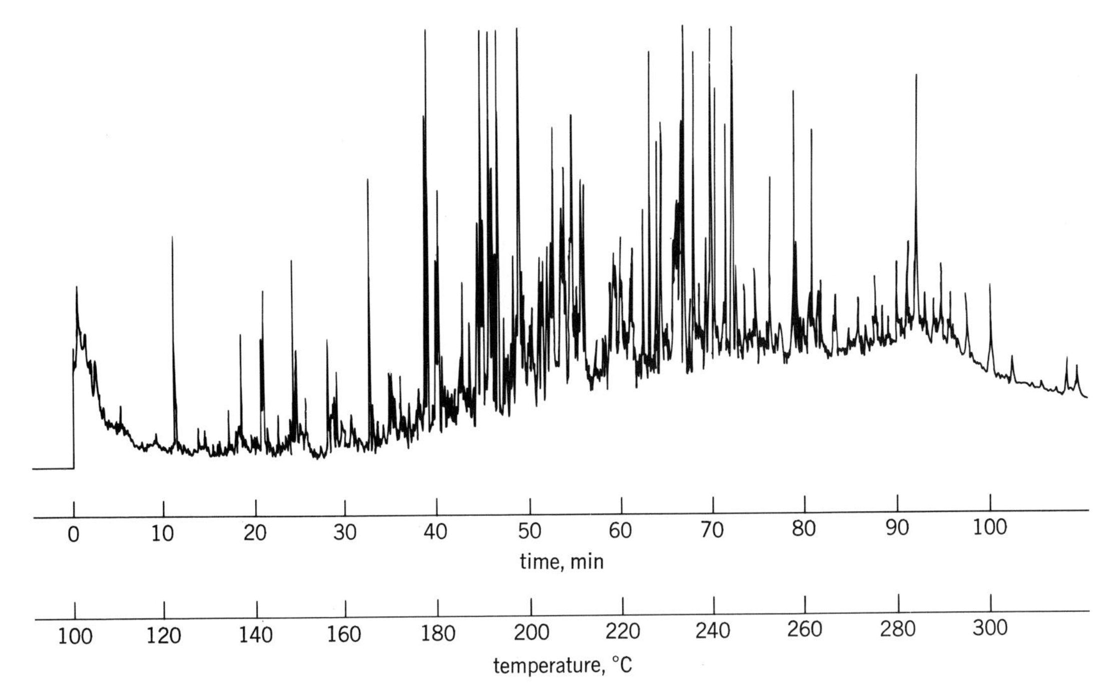

Fig. 1. Chromatogram of nitromethane extract of an engine oil as detected by flame ionization.

in the column preparation are conventional chromatographic materials. For the currently less popular GC alternative, gas-solid chromatography, capillary tubing can also be provided with a retentive adsorbent layer.

The chemical nature of the tubing material is frequently crucial to the quality of separations. Although capillary columns have been prepared from various tubing materials in the past, glass is currently the most popular because of its high inertness and defined surface characteristics. Long helixes of glass capillary tubes can be prepared with a special apparatus. However, the inner surface of glass capillaries has to be treated chemically in order to promote the highly uniform and stable deposition of the nonvolatile stationary liquids. There are two methods for the controlled deposition of submicrometer-size organic films: dynamic and static coating procedures.

The efficiency of a capillary column is largely dependent on its length and internal diameter; the columns of greater lengths and smaller diameters tend to be more efficient. However, many practical considerations necessitate columns with inner diameters of 0.2–0.3 mm and lengths below 100 m. Certain sample characterization studies require capillary columns of a larger bore.

Analytical and instrumental aspects. Capillary GC is a departure from conventional GC techniques in the following respects: (1) The volumes and gas flows are only a fraction of those encountered in typical work with packed columns; sampling techniques and the problems of "dead volumes" are more critical in capillary GC. (2) Smaller quantities of samples are handled with capillary GC; consequently, inertness of the used analytical system is emphasized. A lack of inertness may result in compound losses, irreversible adsorption of samples, peak tailing, and so on. (3) Information obtained on multicomponent samples is often extremely complex.

The design of a capillary gas chromatograph is somewhat critical. Since small enough samples cannot be introduced into a capillary column in a conventional manner, two alternative approaches are used: (1) The sample is injected into the stream of carrier gas in a hot space, is rapidly volatilized, and is split into two unequal parts. The larger fraction is discarded, and a small sample fraction (typically 0.1%) is introduced into the column. (2) A dilute sample (in either a convenient solvent or a gaseous medium) is introduced into the system, with the capillary column held at sufficiently low temperature. The bulk of the sample is permitted to pass through the column, and the components of interest are effectively retained in a short column section for analysis.

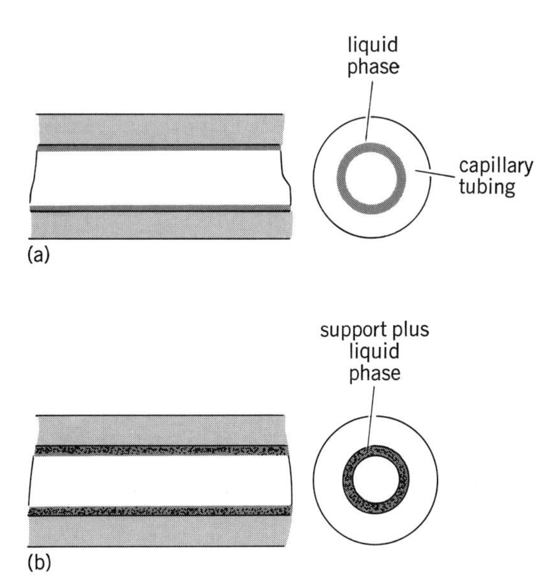

Fig. 2. Difference in *(a)* wall-coated and *(b)* support-coated open tubular columns.

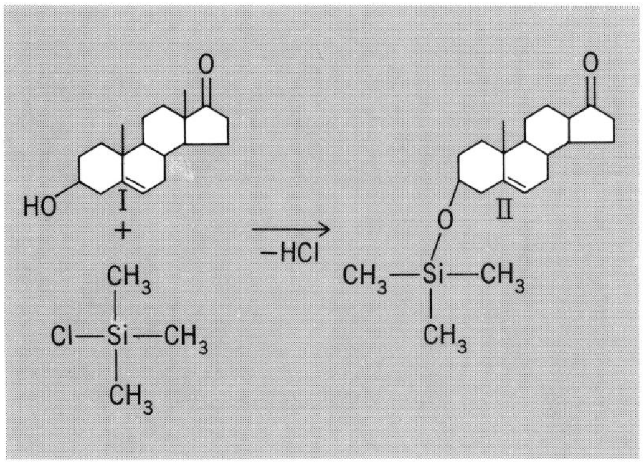

Fig. 3. Conversion of dehydroepiandrosterone (I) to its trimethylsilyl derivative (II).

Detectors used in capillary GC must have small effective volume, high sensitivity, and rapid response. Several useful detectors fulfill these criteria, including flame-ionization, electron-capture, flame-photometric, thermionic, and optical spectroscopic detectors. Selective detectors can be used for identification purposes. Capillary gas chromatographs can also be attached to mass and infrared spectrometers.

Applications. Capillary GC has found its most important application in the analyses of complex mixtures. Frequently, hundreds of components can be resolved, simultaneously quantitated, or even identified. The method has been applied to complex petrochemical samples, combustion mixtures, natural products, physiological fluids and tissue constituents, mixtures of environmental importance, and so on.

The second most important use of capillary GC is in resolving substances with very similar structural characteristics. This method has been used to resolve cis-trans isomers, positional isomers, compounds with different optical activity, various isotopic species, and the spin isomers of hydrogen or deuterium.

The significance of capillary columns in numerous practical problems also lies in the speed of analysis. Capillary columns can provide separations comparable to or better than packed columns in a considerably shorter analysis time.

Reaction gas chromatography. A combination of GC techniques with chemical reaction principles can frequently provide information on the analyzed sample that is otherwise difficult to obtain. Chemical approaches are often a valuable addition to spectroscopic identification techniques such as mass spectrometry, nuclear magnetic resonance, or infrared and ultraviolet optical spectroscopy.

Means of structural identification. Sample alteration either before or after the chromatographic column can be of substantial help in the elucidation of structure. For example, ozonolysis of an unsaturated molecule results in reaction products which are characteristic of the site of unsaturation within a given molecule; such products are indicated by their retention times in GC.

Microreactors used prior to GC can selectively subtract certain types of chemicals or convert them into characteristic products. Reagents known from general techniques of organic analysis are typically used. Alternatively, if the fractions emerging from the GC column are trapped, their small quantities can be reacted with selective agents to determine the compound class (such as characteristic precipitation and change of color).

Sample pyrolysis (controlled thermal decomposition) can be used in conjunction with GC to investigate large, nonvolatile molecules, such as biological or synthetic polymers. Controlled pyrolysis yields molecular fragments that can be related to the structure of a large molecule under investigation. Such fragments are separated and characterized through GC.

Sample derivatization. When dealing with substances that possess reactive functional groups, it is frequently advantageous to convert them into various derivatives prior to GC. Such techniques are particularly necessary to increase sample volatility and thermal stability. For example, it is easier to analyze dehydroepiandrosterone (structure I in Fig. 3) as its trimethylsilyl derivative (structure II in Fig. 3) prepared quantitatively by a simple organic reaction. Even very polar and nonvolatile compounds such as carbohydrates, amino acids, carboxylic acids, and alkaloids can be converted into relatively volatile derivatives and determined by GC. Additional advantages of easier separation or higher sensitivity of determination can often be derived.

For background information *see* GAS CHROMATOGRAPHY in the McGraw-Hill Encyclopedia of Science and Technology.

[MILOS V. NOVOTNY]

Bibliography: L. S. Ettre, *Open Tubular Columns in Gas Chromatography*, 1965; L. S. Ettre and W. H. McFadden (eds.), *Ancillary Techniques of Gas Chromatography*, 1969; M. Novotny, *Anal. Chem.*, 50:16A–25A, 1978; M. Novotny and A. Zlatkis, *Chromatogr. Rev.*, 14:1–44, 1971.

Clear-air turbulence (CAT)

Clear-air turbulence ahead of an aircraft can be detected in real-time by an infrared (IR) radiometer. The alert time and reliability depend upon the passband of the IR filter used and the altitude of the aircraft. Preliminary results indicate that a passband of 26 to 33 μm appears to be optimal to alert the aircraft crew to CAT from 1.5 to 6.0 min before the encounter. Alert time increases with altitude as the atmospheric absorption, determining the horizontal weighting, is reduced.

Characteristics. CAT is frequently associated with shear-induced Kelvin-Helmholtz atmospheric waves, as described by P. C. Patnaik, F. S. Sherman, and G. M. Corcos in 1976. These waves, frequently invisible, produce a fluctuating burden of water vapor as a result of entrainment. Thus the output of a moving radiometer, sensing in the rotational water vapor band and scanning slightly upward into such waves, is proportional to the varying emission of water vapor in the line of sight.

Initial experiments. Initial flight experiments in a forward-looking mode (Fig. 1) have demonstrated that a broad-passband (19–37 μm) radiometer can predict CAT. Of 194 in-flight encounters analyzed, the IR radiometer system incorporating a standard deviation of the radiometer signal has correctly alerted 155 times (of 165 total alerts, 6% were false alarms). The alerts averaged 1.5 to 6.0 min advance warning.

Flight testing of filters. In January 1978 flight research with a three-filter system was begun to find the filter with the combination of best range and best intensity of alert. Data were acquired at altitudes from 15,000 to 45,000 ft (4.5 to 14 km). The filter selected is the middle-range filter (SrF_2 in Fig. 2), providing a range of radiant emission acceptance out to 70 km horizontally.

In the high-altitude dry atmosphere, an alert time of 5 to 6 min is generally observed, corresponding to a physical range of about 70 km. The alerts decrease to less than 1 min at altitudes of approximately 20,000 ft (6100 m) and lower. The shortening of alert times at lower altitudes reflects the stronger atmospheric absorption due to the greater optical mass and pressure.

Preliminary analyses have indicated the physical principles necessary to refine the concept for application to commercial aviation. A filter passband of approximately 26 to 33 μm appears to be optimal for a single-channel CAT radiometer at normal jet altitudes. The use of a multichannel radiometer may be desirable for lower altitudes.

In-flight performance record. To determine the efficiency of the IR water vapor radiometer to detect CAT, three parameters are required: time, radiometer output voltage, and data indicating the onset of turbulence. Samples of time and radiometer output voltage are recorded on a nine-track tape every 2 s during flight on the C-141-A data system. Vertical acclerometer output is recorded on an analog chart during flight.

It is also necessary to ensure that all evaluations are based on the same type of data sets. Therefore the following restrictions were imposed on the C-141-A flights: (1) Only data from 41,000 ft (12,505 m) were used. (It may be noted here that when data of other altitudes were evaluated, the results were unchanged.) (2) The gain setting of the radiometer system was the same in all cases. (Again, when other settings were used, results were unchanged but the threshold of alert varied with radiometer gain.) (3) The elevation angle of the radiometer was 10° in all cases. (4) CAT was defined as equal to or exceeding an accelerometer value of 0.1 *g*, peak to peak.

Algorithms—for arc length, standard deviation, and second difference—were employed to process the radiometer output signal to determine an alert or no-alert. Several different data time intervals were used to determine the following: (1) the individual algorithm which achieves the highest optimum number of alerts followed by real encounters and which reduces the number of false alarms; (2) the number of data points in each algorithm which produces the best score; (3) the radiometer numerical threshold value (or alert value) which produces the best score; and (4) the correlation, if any, between the numerical intensity of the alert and the intensity of the encounter.

CAT radiometer results

Total encounters	194
Total alerts	165
Alerts followed by encounter	155
Alerts not followed by encounter	10
Percent of true alerts	80
Percent of false alerts	6
Percent of no-alerts	20

Fig. 1. NASA Learjet 701 used in testing IR radiometric warning system for subsequent CAT encounters. Note radiometer assembly on top of nose radar cover.

The CAT radiometer was operational for 13 C-141 flights from Sept. 30, 1977, to Jan. 27, 1978. This includes one flight, calendar day 327 of 1977, where cirrus produced many false alarms. One hundred ninety-four encounters separated by at least 3-min intervals were examined. The results for one algorithm (standard deviation of the radiometer signal over 12 s) are given in the table.

The significance of the score to date is that one can expect the existing system incorporating the standard deviation to produce valid alerts prior to CAT encounters 80±2% and false alerts 6±1% of the time. The intensity of the threshold varies with the gain of the radiometer and the type of algorithm used. Using a large threshold may eliminate false alarms, but success is also diminished.

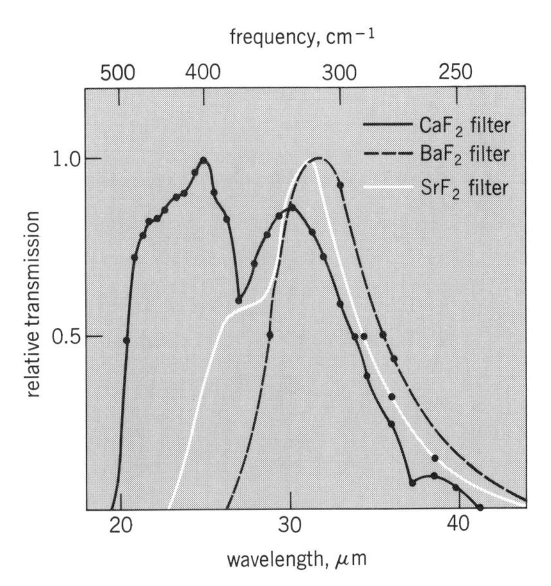

Fig. 2. Relative transmission versus wavelength and frequency for three crystal filters: calcium fluoride (CaF_2), strontium fluoride (SrF_2), and barium fluoride (BaF_2).

Correlations were made between the intensity of signal and intensity of the encounter, yielding the following results: arc length=0.1; standard deviation = 0.1; and second difference = −0.02. Such a small correlation is not significant.

For background information *see* CLEAR-AIR TURBULENCE (CAT); ELECTROMAGNETIC RADIATION; INFRARED RADIATION in the McGraw-Hill Encyclopedia of Science and Technology.

[PETER M. KUHN]

Bibliography: P. Kuhn, F. Caracena, and C. M. Gillespie, Jr., *Science*, 196:1099–1100, 1977; P. C. Patnaik, F. S. Sherman, and G. M. Corcos, *J. Fluid Mech.*, 73(2):215, 1976.

Climatology

Design climatology, which is included under the general heading of applied climatology, involves the scientific analysis of climatic data for the purpose of improving the design of equipment and structures intended to operate in, or withstand, extremes of climate. For centuries people have built bridges, planned battles, embarked on voyages, and conducted other activities with some understanding of the weather risks involved. This understanding was based on a knowledge of historical events and the judgment of individuals. However, with the development of the statistical theory of extreme values during the past 40 years, risks associated with extreme conditions can now be calculated.

This article discusses the ways in which design climatology uses standards, statistics, and meteorology to study climatic extremes. It focuses on some of the work of the following individuals: (1) Norman Sissenwine and Rene Cormier in establishing a standard of climatic extremes for military equipment; (2) E. J. Gumbel in modeling the statistics of extremes; and (3) Arnold Court and Irving Gringorten in developing procedures for reducing calculated risk by considering a return period or design life.

Military standard. Sissenwine spearheaded a coordinated effort by the U.S. Department of Defense to establish a design standard of calculated risk and to determine the thresholds of climatic extremes. Because it would often be prohibitive in cost or technologically impossible to design military equipment to operate anywhere in the world under the most extreme environmental conditions, military planners take a calculated risk and accept equipment designed to operate under environmental stresses except for a certain small percent of time. A Department of Defense task group agreed on the following standards for the determination of thresholds of weather elements: (1) For each weather element the 1% extreme in the month of the year with the most severe weather has been selected as the design limit during operations, when life is not endangered. (2) Extremes that have a near zero chance of occurring are to be used as the criteria for design when operational failure due to an environmental extreme would endanger life. (3) Equipment under constant exposure should be designed to withstand more severe conditions than the design criteria for operations, so that the equipment does not incur irreversible damage and become permanently inoperable. Unlike the criteria for operations where the risk is expressed in terms of the percent of time (for example, in hours) that the equipment is expected to be inoperable in the most extreme area during the most extreme month, the criteria for withstanding is expressed in terms of the probability that an extreme weather event will occur at least once in the expected lifetime of the equipment. A 10% risk, for equipment expected to be exposed 2, 5, 10, or 25 years, is used for withstanding. For each item of equipment, the likely duration (number of years) of exposure must be considered separately with respect to each climatic element. For example, intended durations of exposure to tropical conditions may be 10 or 25 years, whereas, for the same item, exposure to the cold extremes of the Arctic might not be planned for more than 2 years. A 10% risk in 2, 5, 10, or 25 years implies that such extremes will recur in approximately 20, 50, 100, and 250 years, respectively; hence these numbers are called return periods. Design criteria for withstanding sometimes exceed the extremes of record, especially when the latter are found in short periods of actual observation.

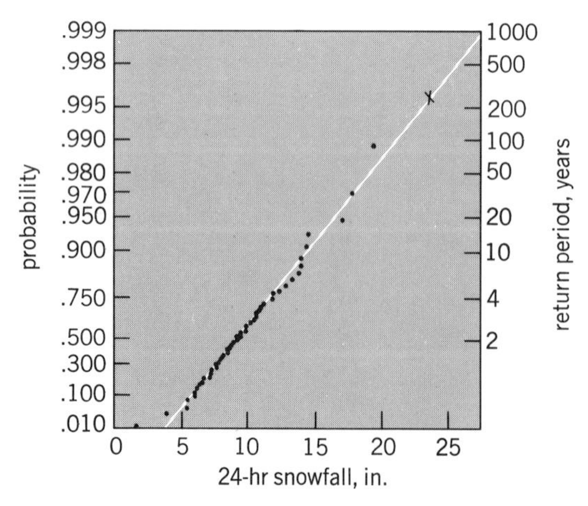

Fig. 1. Heaviest 24-hr snowfall, in inches, observed in Boston during each of the 50 winters from 1927–1928 through 1976–1977 plotted on extreme probability graph and fitted by linear least squares. The X marks the heaviest 24-hr snowfall observed in the 1977–1978 winter season. 1 in.= 2.54 cm.

Statistical theory of extremes. Weather events such as extremely strong winds or heavy snowfalls, which are very important in design, are not always well documented. Even the more readily available longer records of daily maximum temperature and precipitation amount are often too short to provide reliable estimates of future weather extremes without the aid of a statistical model. The search for a satisfactory statistical theory has resulted in the publication of hundreds of papers on the subject in the last 50 years. Gumbel's book includes more than 600 references on the statistics of extremes. The most commonly used model for estimating extremes is generally known as the Gumbel distribution, shown in Eq. (1), where $P(x)$ is the cumulative probability of the variable x, which may be, for example, the heaviest snowfall of the winter in a single storm, and y is a reduced variate, in a one-to-one correspondence with x.

$$P(x) = \exp(-e^{-y}) \tag{1}$$

The heaviest 24-hr snowfall in Boston during each of the 50 winter seasons, from the winter of 1927–1928 through the winter of 1976–1977, is plotted in Fig. 1. It can be seen that the points are approximately linear. A least-squares linear fit to the points is shown by the line on the graph. This line can be used to estimate the probabilities of exceeding 24-hr snowfall amounts.

The heaviest 24-hr snowfall, occurring in the storm of Feb. 6 and 7, 1978, was 23.6 in. (59.9 cm). According to Fig. 1, storms of this intensity have a return period of 245 years, as shown by the X on the graph. The table gives snowfall estimates for Boston obtained from Fig. 1.

Return period. Court's development of the principle of reducing calculated risk by considering a return period can be illustrated with the snowfall example.

Consider a structure that is intended to last for a period of n years, after which it is to be declared obsolete. Assume that the concern is the snow load on the roof, and that this can be represented by the 24-hr snowfall. For any specific 24-hourly snowfall threshold, the return period is the reciprocal of its probability: an event with probability p of happening in any one year will, on the average, occur once in every $1/p = n$ years. This is the average interval between occurrences, but the average must be taken over many years. Actually, the probability that the event will not occur during n years is as shown in Eq. (2). As p becomes very small,

$$P\text{ (no success in } n \text{ years)} = (1-p)^n = (1-p)^{1/p} \quad (2)$$

P approaches $1/e = 0.36788$. Actually, the approach is so rapid that this exponential value is sufficiently exact for all practical purposes when $P < 0.02$, corresponding to return periods that are greater than 50 years. The probability of at least one occurrence before the nth year is $1 - (1/e) = 0.63212$.

If a structure is not designed to withstand a load greater than the extreme event having probability $P = 1/n$, it has a 63% chance of failing before its desired lifetime n has expired. Therefore, a calculated risk of 63% is taken in designing an n-year structure for only the "n-year event." To reduce the risk of failure in n, a structure should be designed to withstand an extreme with probability less than $1/n = p$. This design probability may be taken as kp, with $k < 1$, making the design return period $1/kp$. The corresponding probability of nonoccurrence during n years of the event having probability kp of occurrence in any one year is given in Eq. (3). Hence k can be adjusted to provide

$$P\text{ (no success in } n \text{ years)} = (1-kp)^{1/p} \xrightarrow[p \to 0]{} e^{-k} \quad (3)$$

any designed calculated risk r, because $r = 1 - e^{-k}$. Thus $k = -\ln(1-r)$. Values of k for various r are:

r:	0.50	0.33	0.25	0.20	0.10	0.05	0.01	0.001
k:	0.69	0.40	0.29	0.22	0.11	0.05	0.01	0.001

To have a calculated risk of 5% during a desired lifetime of 50 years, a structure should be designed to withstand the event having the probability $0.05 \times 1/50 = 0.001$ of occurring in any one year; that is, it should be designed for the event with 1000-year return period. In the Boston snowfall example, this corresponds to a 24-hr snowfall amount of 27.6 in. (70.1 cm).

This procedure implies that the series of observation (years) from which the extreme is estimated is stationary, or unchanging. In other words, the probability of occurrence of a given extreme is assumed to be the same throughout the 100 or 1000 years. However, after developing this straightforward approach for calculating risk, Court goes on to write: "But few things in the world are truly constant, least of all climate. Return periods greater than 20 or 30 years are fictional, because climatic conditions change more frequently than that. The observations of the past 30 or 40 years will provide a good estimate of the event having probability of 4 or 5%, but offer no reliable basis for estimating the magnitude of the event with probability of 0.01 or 0.001. Estimates of such events must be used with caution in any application of the principles of calculated risks." However, other workers find the assumption of stationarity more acceptable than Court does.

Example. The severe New England storms of January and February 1978 resulted in millions of dollars of damage through the failure of structures

Snowfall estimates for Boston obtained from Fig. 1

Heaviest 24-hr snowfall, in. (cm)	Probability of exceeding heaviest snowfall	Return period, years
9.1 (23.1)	0.50	2
11.6 (29.5)	0.25	4
14.4 (36.6)	0.10	10
16.5 (41.9)	0.05	20
19.1 (48.5)	0.02	50
21.1 (53.6)	0.01	100
23.0 (58.4)	0.005	200
25.6 (65.0)	0.002	500
27.6 (70.1)	0.001	1000

Fig. 2. Collapse of Hartford (CT) Civic Center Coliseum. (*The Hartford Courant*)

to withstand the extremes of weather. The collapse of the Hartford (CT) Civic Center Coliseum roof, shown in Fig. 2, is one example. Hundreds of tons of wet snow caused the roof to collapse only 6 hr after a college basketball game attended by 5000 people. The property damage and potential loss of life is a reminder of the importance of establishing low-risk withstanding design criteria and of meeting these specifications in construction.

For background information *see* CLIMATOLOGY; PROBABILITY; STATISTICS in the McGraw-Hill Encyclopedia of Science and Technology.

[IVER A. LUND]

Bibliography: A. Court, *Advances in Geophysics*, pp. 45–85, 1952; I. I. Gringorten, *J. Appl. Meteorol.*, 2:82–89, 1963; E. J. Gumbel, *Statistics of Extremes*, 1958.

Cocoa powder and chocolate

Major cocoa bean growers are Brazil and the developing West African countries of Ghana, Nigeria, the Ivory Coast, and Camerouns. Ghana, the largest producer, contributes approximately one-third of the world's supply. The major importers of cocoa beans have been the United States and West European countries. Recently, Eastern bloc countries have been increasing their purchases. For a long time cocoa powder had been considered a by-product of the manufacture of cocoa butter and chocolate liquor from cocoa beans. However, the sharp increases in demand for cocoa powder and poor cocoa bean crops have raised the price of cocoa powder and chocolate liquor so high that food manufacturers are looking for natural and artificial substitutes.

Products and bean production. Three basic products of cocoa beans are cocoa butter, cocoa powder, and chocolate liquor. Until recently, cocoa butter and chocolate liquor, used in chocolate coatings, were the higher-priced derivatives while cocoa powder was the least costly. This relationship has been reversed, probably due to the expanded role of cocoa powder as a principal flavoring for beverages, confectionery, baked goods, ice cream, and desserts. These products now compete for a limited supply of cocoa powder. Furthermore, cocoa butter is increasingly being replaced in compound coatings with hardened vegetable fats, decreasing its demand and, in turn, its price.

Cocoa bean production is sensitive to environmental and economic changes. Worldwide production has rarely exceeded 1,500,000 metric tons. Poor crop yields, resulting from drought conditions, coupled with the above-mentioned demand factors, caused an upward surge in cocoa prices from a 1971 low of 26¢/lb (57.2¢/kg) to a 1977 high of $2.44/lb ($5.37/kg). These higher prices, coupled with dwindling inventories, market volatility, and strong consumer demands, are forcing many food manufacturers to turn to natural and artificial substitutes for cocoa and chocolate.

Replacing cocoa attributes. Cocoa is valued for its unique combination of flavor, aroma, bitterness, astringency, texture, and fatty mouth feel. In addition, cocoa has desirable color and bulking properties. Simultaneous replacement of all these attributes has not yet been achieved. Food manufacturers are in the process of determining which of these attributes are essential to their products, and are looking to their suppliers for help in finding suitable replacements. The food industry has taken various approaches to solving this problem.

Reformulation. As much as a 30% cutback in cocoa has been achieved in many formulations simply by replacing cocoa with other indigenous ingredients, such as starch and sugar.

Substitutes of natural origin. Further reductions of cocoa can be effected by simulating the flavor aspects of cocoa with other natural ingredients. Carob powder is made from the dried pods of the carob tree which are broken into pieces, roasted, and ground. Its flavor and particulate texture are similar though not identical to cocoa. Its usage is expanding rapidly. In some products cocoa, for example, is completely eliminated and replaced by carob in the "health foods" industry. Normally, carob powder can successfully replace as much as 25% of cocoa. Malt, caramelized sugar, and brown sugar function in a similar manner in replacing cocoa.

Molasses is being touted as a substitute for up to 10% of formulation cocoa. Besides supplying comparable color and flavoring, it provides additional sweetness. In products where only the aromatic quality of cocoa is desired, vanilla is frequently used.

Artificial flavors. These are generally used where the manufacturer's policy permits the word "artificial" on the label. Artificial cocoa and chocolate flavors have greater flexibility in price, quality, and functionality than their natural counterparts. In a recent count, more than 400 identifiable components have been found in cocoa, providing the flavor chemist with an insight into the synthesis of cocoa flavor. Artificial flavors are further improved when coupled with products of the Maillard reaction (induced thermally from selected amino acids and reducing sugars). At present, many flavor manufacturers offer artificial cocoa and chocolate flavors of excellent quality.

Complete cocoa substitutes. Several products offered today are reputed to be complete substitutes for cocoa, supplying flavor, mouth feel, texture, and color, as well as bulking properties. The following list includes the major types of complete substitutes:

1. Thermally reacted whey. A relatively inexpensive, pound-for-pound cocoa substitute is composed primarily of whey and flavorings. Water is added to whey solids, and this slurry is processed at high temperatures forming Maillard-reaction and melanoidin condensation products. The dark brown slurry is then dutched, supplemented with flavor and fat, and spray-dried to yield a powder similar to cocoa in flavor, texture, appearance, and functional properties.
2. Grains. Another type of cocoa substitute is made up of refined vegetable protein, vegetable fats, flavoring, and color. Its color can be adjusted to simulate tan natural, red dutch, and red double dutch types of cocoa powder.
3. Compounded mixtures. Extenders for cocoa powder have been created exclusively from noncocoa sources. Ingredients of such extenders may include modified food starch, dextrins, maltodextrins, salt, tapioca starch, soy flour, vegetable oils (soybean, palm, and cottonseed), color (such as caramel), and flavoring.

4. Yeast products. By means of a patented process, treated brewer's yeast has successfully replaced cocoa powder in several products. The flavor and aroma are similar to cocoa, and the treated brewer's yeast is free of yeastlike characteristics. By tailoring the process, colors ranging from natural tan through dark dutched brown may be produced.

Complete cocoa butter substitutes. These "hard butter" replacers may be derived from palm, coconut, cottonseed, soybean, and sunflower seed oils, shea butter, and beef tallow, or combinations of these. Methods of manufacture include hydrogenation, rearrangement, and fractionation of the fats and oils. These substitutes are compatible with cocoa butter and have flavor and texture characteristics similar to pure chocolate.

For background information *see* COCOA POWDER AND CHOCOLATE in the McGraw-Hill Encyclopedia of Science and Technology.

[IRA LITMAN; SCHELLY NUMRYCH]

Bibliography: L. Cook, *The Manufacturing Confectioner*, p. 106, June 1977; S. Crocco, *Food Eng.*, 3:93, 1977; A. Fincke, *CCB Rev. Choc. Confectionery Bakery*, 2(1):8, 1977; *Food Eng.*, 49(9):47, 49(9):ef30–33, 49(9):31, 1977; *Food Eng. Int.*, 1(7): 8, 1976; P. Keeney, *Amer. Oil Chem. Soc.*, 49(10): 567, 1972; N. Post, *Candy Snack Ind.* 2:26, 1973.

Connective tissue

Connective tissue functions principally as a means of attachment for cells, as a rigid or flexible framework for muscles, and as support for the body. Other functions include mechanical protection for cells, organs, and parts of or even whole bodies; thermal insulation and antifreeze protection; and the locus of energy depots in the body. Connective tissues comprise primarily extracellular materials, both organic and inorganic, together with the cells that secrete them. The organic substances include a wide range of proteins, carbohydrates, and protein-carbohydrate complexes, while the inorganic material is mostly calcium carbonate, although silica and salts of several other metals are also found.

It is difficult to strictly delineate which materials are connective tissues. For example, mucous secretions of epithelia are not considered connective tissues; yet the same or related glycosaminoglycan molecules alone, in ground substance of interstitial fluid, or when incorporated in proteoglycans and ensheathed in collagen may, in a major way, determine the unique properties of short-term rigidity and long-term viscoelasticity of cartilage. These secretions are found not only in vertebrates but also in the horseshoe crab *Limulus* and some other arthropods, in the cephalopod mollusks, and in other phyla. Many recent advances are related to an appreciation of the properties of composite materials.

Innervated echinoderm connective tissue. A recent study by I. C. Wilkie focuses on intervertebral ligaments in the arms of ophiuroids (brittle starfish) and their role in autotomy (that is, voluntary shedding of an appendage, with minimal loss to the animal, in response to a noxious stimulation). The response involves a nervously mediated, rapid, dramatic change in the mechanical properties of the connective tissues. Wilkie states that the five classes of the echinoderms are unique in that the collagenous tissues—various ligaments, dermis, tube feet, and spine capsules—are capable of undergoing rapid, nervously mediated changes, and his study provides some of the best evidence to date on the mechanisms involved. Most observations were made on *Ophiocomina nigra*, in which the breakage plane occurs one or two segments proximal to an applied noxious stimulus. That muscles are not involved in effecting autotomy was shown by observations that the muscles detach before arm separation (parting at a tendon characteristic for each muscle) and that a segment can be stimulated to autotomize when gravity is the only force acting. A direct influence of nerves upon the connective tissue is indicated by the very fast response time (mean of 1.6 s) after stimulus and by the fact that anesthetics inhibit the response. Noxious stimuli result in a dramatic loss of ligament tensile strength, and the microscopic appearance of such autotomized ligaments is quite distinct from the appearance of an intact arm ruptured by a sudden pull.

The intervertebral ligament, which structurally is a pliant composite of overlapping collagen fibers in a viscous matrix of ground substance, behaves like a simple viscous fluid during mechanical studies of slow creep (Fig. 1*a*). There is no elastic recoil when the distending force is removed. Application of excess K^+ initiates a marked decrease in

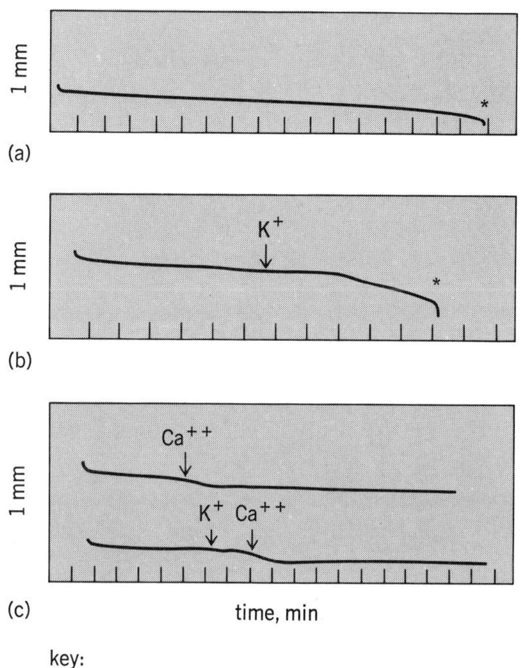

Fig. 1. Tracings of oscillograph records showing extension curves from ligaments under a 30-g load. (*a*) Ligament in sea water. (*b*) Addition of isosmotic KCl to elongating ligament is followed by abrupt increase in rate of extension and premature rupture. (*c*) After addition of isosmotic $CaCl_2$ to an elongating ligament, rate of extension is reduced until no further elongation occurs (upper curve), and a similar effect is observed on a ligament which had begun to respond to previous stimulation by KCl. (*Adapted from I. C. Wilkie, Nervously mediated change in the mechanical properties of a brittlestar ligament, Mar. Behav. Physiol., 5:289–306, 1978*)

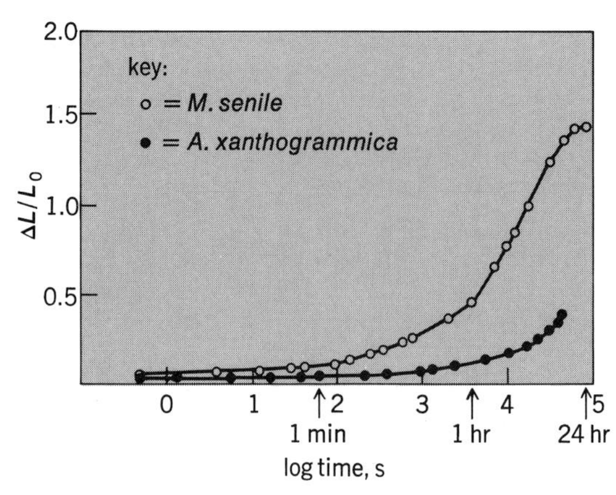

Fig. 2. Typical graphs of the extension ($\Delta L/L_0$) against log time (in seconds) for mesogloea from *Metridium senile* and from *Anthopleura xanthogrammica*. Specimens were tested along their longitudinal axes at 6°C with stresses on the order of 3000 N·m^{-2} (N/m/m). (*From M. A. R. Koehl, Mechanical diversity of connective tissue of the body wall of sea anemones, J. Exp. Biol., 69:107–125, 1977*)

viscosity, while excess Ca^{++} renders the ligament inextensible (Fig. 1*b* and *c*). Histochemical evidence indicates the presence of calcium associated with glycoprotein in the processes of neurosecretory cells within the ligament, and it is hypothesized that these cells are involved in control of the availability of Ca^{++} to the connective tissue ground substance.

Mechanical design in organisms. In recent years it has become appreciated that the mechanical properties of many connective tissues are characteristic not only of component substances but also of composite materials. Two examples will be discussed to illustrate this point: anthozoan mesogloea and molluscan shells.

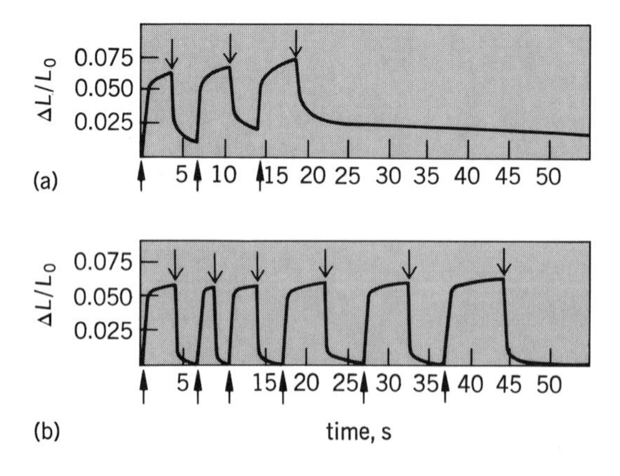

Fig. 3. Short-term creep and recovery tests (at 6°C). Graphs of extension against time for (*a*) *Metridium senile* mesogloea pulled longitudinally with a stress of 29,946 N·m^{-2} (N/m/m). (*b*) *Anthopleura xanthogrammica* mesogloea pulled longitudinally with a stress of 52,238 N·m^{-2}. (*From M. A. R. Koehl, Mechanical diversity of connective tissue of the body wall of sea anemones, J. Exp. Biol., 69:107–125, 1977*)

Anthozoan mesogloea. Mesogloea of anemones is a composite material of collagenous fibers in a hydrated, amorphous matrix of neutral polysaccharides. The more rapidly a stress is applied, the more rigid the material appears, whereas the response to a slowly applied force is a slow creep or extension (by as much as 200% in some species) from which the viscoelastic matrix slowly recovers its original dimension upon removal of the stress. In other words, the properties are time-dependent. This is illustrated for two anemones, *Metridium senile* and *Anthopleura xanthogrammica* in Fig. 2. *Metridium senile*, which lives in quiet waters, is relatively rigid when stressed for periods of only 1 or 2 min, but is highly extensible over much longer periods, as might be expected for an animal which lives in a slow tidal current. Conversely, the mesogloea of *A. xanthogrammica*, which is exposed to strong wave action, is relatively inextensible over several hours. Short-term stress properties show a similar difference. The viscoelastic response of *M. senile* body wall shows an incomplete elastic recoil in Fig. 3*a* over the test time, while that of *A. xanthogrammica* (Fig. 3*b*) has a much more complete recovery (fast enough, as observed in the natural state, to allow it to return to its original shape between each wave stress). The mechanical properties of the connective tissues are thus finely adapted to the specific requirements of the animal species.

Molluscan shells. In a study by J. D. Currey and J. D. Taylor, the structure and properties of molluscan shells have been described as small-grain ceramics in an organic matrix. Of eight described composites of crystallites and organic matrix, nacre and calcite prisms showed superior ultimate tensile strength and rupture modulus, as well as having an organic content (approximately 2–8% by volume) which is one to two orders of magnitude greater than that of the weakest material, crossed lamellar shell. Also, due to their organic content, nacreous shells show considerable plastic deformation. The small, uniform crystals undoubtedly help in limiting the spread of cracks. Probably significantly, many of the weaker crossed lamellar shells were found in burrowing bivalves, which also have thinner shells and are not exposed to the same stresses as epifaunal mollusks. S. A. Wainwright and coworkers argue that the form of these and other shell types—small, uniform ceramic in organic matrix—has evolved because it makes the shells much stronger than composites with larger or uneven-sized crystallites.

For background information *see* CONNECTIVE TISSUE in the McGraw-Hill Encyclopedia of Science and Technology. [HUGH Y. ELDER]

Bibliography: C. H. Brown, *Structural Materials in Animals*, 1975; J. D. Currey and J. D. Taylor, *J. Zool., Lond.*, 173:395–406, 1974; S. A. Wainwright et al., *Mechanical Design in Organisms*, 1976; I. C. Wilkie, *J. Zool.*, in press.

Construction methods

The conventional house of today has evolved from its predecessors through trial and error rather than applied engineering. Although experience has shown that present construction techniques produce sound and serviceable structures, more material is being used than is essential to structural safety. No single area offers greater potential savings of timber resources than house construction. The application of new engineering principles can lead to material savings without sacrificing the structural integrity in the house.

The development and use of prefabricated wood roof and floor truss systems have led to more economical utilization of natural resources and faster and safer construction. The lightweight truss-framed house extends these concepts and offers an improved framing system directly applicable to residential and light-frame construction. It consists of an open-web floor system, trussed rafters, and wall studs. These components are tied together by rigid or semirigid joints into an assembly forming a unitized framework capable of supporting loads over long spans without intermediate support (Fig. 1). The joints can be made with conventional truss plates, plywood gusset plates, or any fasteners capable of transmitting moment, shear, and axial forces between members (Fig. 2). The resultant structure provides considerable in-plane rigidity in that any loaded member shares its load with other elements within the system.

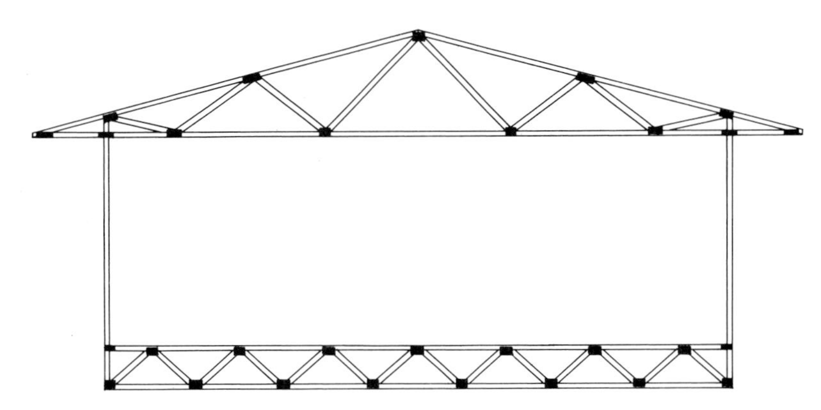

Fig. 1. Section of lightweight truss-framed house. This system extends the engineering principles used in roof and floor trusses to the entire framing system.

Material requirements. The material savings and ease of erection with trussed rafters over conventional site-built rafters have been well demonstrated. These savings will be even more pronounced with the lightweight truss-framed house system. Estimates of material requirements show at least a 30% saving in exterior framing lumber over a conventional house with the same floor plan. Support beams and columns in the basement can be eliminated because of the span capabilities of this system.

A further advantage is that only one lumber size, nominal 2- by 4-in. (5- by 10-cm) lumber, is required for fabrication. This eliminates the need to stockpile the several sizes normally required for conventional construction. Also, the availability of the larger-sized lumber is decreasing with the harvesting of second-growth and plantation timber.

Another area involving serious waste and economic hardship to individuals is damage to houses caused by hurricanes and other severe natural forces. Major weaknesses in conventional construction are in the connections between foundations, floors, walls, and roofs. The conventional house is not positively tied together, and roofs are often blown off or entire houses are displaced from their foundations. Laboratory tests have confirmed these field observations. The lightweight truss-framed system provides positive connections with far greater resistance to natural disasters.

Site construction. Conventional light-frame construction involves the piece-by-piece assembly of members at the site. This involves a high degree of skill in layout and cutting and a knowledge of construction details, nail sizes, and spacing. Work progresses in stages from the foundation up to the roof system. Interior framing and bearing partitions are constructed simultaneously with exterior framing. Finally, the roof is constructed and sheathed. Considerable time and field labor have been expended before the house is finally enclosed so that work can continue unhindered by wind, snow, or rain.

With lightweight trussed framing, all of the structural connections are made in a plant under controlled conditions. Each assembly, being relatively light in weight, approximately 250 lb (115 kg), can be erected using tilt-up construction methods without the need for lifting equipment. The floor, wall, and roof framing goes up as a unit, and the house can be quickly enclosed. Floor, wall, and roof coverings are applied in a conventional manner. A less skilled labor is required with this system, and protection from the elements is achieved in a short time.

Disaster-resistant construction. Property losses from natural disasters have averaged approximately $1,000,000,000 per year in the United States. Losses in life, property, and resources are one of the nation's most serious problems. The three major natural phenomena which have the

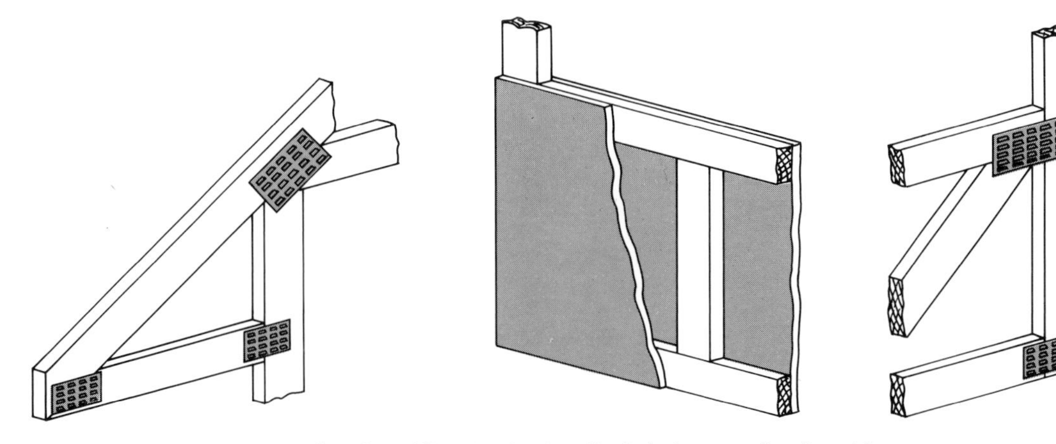

Fig. 2. Typical joint details capable of providing structural continuity between roof, wall, and floor.

Fig. 3. Full-scale sections of truss-framed house being tested in Structures Laboratory at West Virginia University.

greatest impact are earthquake, wind, and flood. As indicated earlier, major damage in house construction is due to inadequate fastening of the components.

Tornadoes produce the greatest wind forces, with velocities up to 300 mph (480 km/hr). However, Fujita found that 91% of tornadoes had wind velocities of less than 158 mph (250 km/hr). Hurricanes and severe local windstorms normally generate maximum wind speeds under 150 mph (240 km/hr).

It would not be practical or even possible to design houses to withstand a direct hit by a 300-mph tornado. Thus the economics of added cost of construction versus the added safety gained must be carefully weighed. Standards that are set too high would price houses out of the reach of most homebuyers.

A good compromise might be to shoot for a wind-resistance rating of about 150 mph. This level of performance should provide safety against tornado destruction about 90% of the time; it should also be sufficient to withstand local windstorm hazards and most hurricane wind damage. Such a design would also rectify many of the structural deficiencies responsible for earthquake damage. The truss-framed house is positively connected from the roof down to the foundation and can readily be designed to achieve that level of performance.

Structural analysis. The truss frames are highly engineered units, and a complete structural evaluation of all the elements was required. Since the frames are composed of a multitude of members, the solutions are not amenable to simple hand calculations, and the analyses were accomplished with a computer program.

Truss frames can be designed to cover a great number of applications. Specific requirements and designs are controlled by many factors, such as roof loads (geographical considerations), frame spacing, desired span, member size, and lumber grades. The system is extremely flexible and can readily be designed for frame spacings of 2 to 4 ft (0.6 to 1.2 m) with clear spans up to 32 ft (9.8 m). Many of these possibilities were investigated, but the initial design was established for 2-ft (0.6-m) frame centers to be compatible with commercially available sheet products such as plywood and gypsum board.

For commercial application, truss fabricators generally have access to computers and provide the engineering services, and with a repetitive market the fabricators can standardize designs so that the need for recurring engineering can be considerably minimized.

Full-scale tests. To verify the computer solutions, the U.S. Forest Products Laboratory (FPL) entered into a cooperative research program with West Virginia University to test full-sized frames. This work also included additional computer solutions for various frame configurations, strength and stiffness tests of typical joints, and tests of frames and assemblies.

Three frames, spaced 2 ft (0.6 m) apart and constructed entirely of nominal 2- by 4-in. (5- by 10-cm) lumber, were erected over a 28-ft (8.5-m) span. Floors were loaded with two rows of 50-lb (23-kg) concrete blocks stacked three high from wall to wall (Fig. 3). The roof was loaded with sandbags.

The lumber was not of particularly high quality—spruce-pine-fir construction grade with a fairly low allowable unit bending stress. The grade of material was not crucial for these tests since material properties are part of the program input. Even with the quality of material used, stiffness was adequate to meet deflection limitations, and the floor carried more than twice the normal design load.

Prototype house. The final step in the development of a new concept is the construction of an actual house. FPL entered into a cooperative agreement with the University of Wisconsin to erect a prototype house at the Arlington Experimental Farms near Madison, with the house to be occupied by a family.

The prototype house should disclose any problems that might be inherent in a new system, such as fabrication, transportation, erection, and occupant reaction. It was felt that the best way to obtain this information was to let a contract for public bids. One contract was awarded to a truss manufacturer to fabricate, deliver, and erect the frames. A second was awarded to a local builder to finish the house by using standard methods. The house was bid two ways, by conventional construction and by using the truss frames. A direct savings of $2300 was realized with the truss-framed system.

The demonstration house was successfully erected on July 27, 1978. The speed and ease of erection for this first house were very encouraging. The prefabricated components for the complete house were easily transported on a single truck to the construction site from the fabrication plant. A crew of three plus light crane erected the 48- by 26-ft (14.4- by 7.8-m) house in only 6 hr. There were some minor procedural problems but, in general, the demonstration was highly successful. Improvements in the fabrication and erection process would obviously come with experience.

For background information *see* BUILDINGS; LUMBER MANUFACTURE in the McGraw-Hill Encyclopedia of Science and Technology.

[ROGER L. TUOMI]

Bibliography: T. T. Fujita, *Wind-Resistant Design Concepts for Residences*, Civil Defense Preparedness Agency, TR-83, p. 13, July 1975; S. K. Suddarth, *A Computerized Wood Engineering System: Purdue Plane Structures Analyzer*, U.S. Dep. Agr. For. Serv. Res. Pap. FPL 168, 1972; R. L. Tuomi and W. J. McCutcheon, *Testing of a Full-Scale House under Simulated Snowloads and Windloads*, U.S. Dep. Agr. For. Serv. Res. Pap. FPL 234, 1974.

Corn

Within the past several years, two major developments in corn *(Zea mays)* virus diseases have occurred in the continental United States. The first development was the discovery of six new viruses and a mycoplasmalike organism (MLO) infecting United States corn (see the table). (Mycoplasma are prokaryotic microorganisms lacking a cell wall and belong to the class Mollicutes.) The second development was the expansion of the geographical distribution of the maize dwarf mosaic virus (MDMV) into several northern states and one western state where sweet corn production is a major agricultural industry. This expansion has reduced yields in late-planted sweet corn and has created a major problem for the sweet corn canning and processing industries in these states. The occurrences of the new viruses and MLO have created no major problems in corn production, except in the case of the newly discovered maize chlorotic mottle virus (MCMV) in Kansas. However, at least one other of the newly discovered pathogens has the potential to cause serious damage to United States corn.

Infection of corn by virus frequently causes leaf discoloration that is seen as mosaic or mottle patterns or, in other cases, as reddish, purplish, or yellowish blotches. Also, virus-infected plants frequently do not elongate as much as healthy plants, and thus appear stunted. Because of the similarities in symptoms, individual diseases frequently cannot be distinguished consistently on the basis of symptoms alone. Thus, the characteristics or identity of the pathogen must be determined in order to differentiate the diseases. This approach will be followed in this article.

Recent pathogen discoveries. Of the seven pathogens new to United States corn, only one has been involved in an economically important disease. This disease, corn lethal necrosis, is caused by infections of MCMV plus MDMV, the Johnson grass or A strain (MDMV-A). Prior to its appearance in the United States in 1976, MCMV was known to occur only in Peru, where it causes a major disease problem. In contrast, MDMV-A has been known in the United States since its discovery in 1963. As the name "corn lethal necrosis" suggests, susceptible corn plants die following infection by both viruses. Infection by either of the viruses alone does not cause plant death. In the United States, MCMV has been found in Kansas, Nebraska, and Texas.

Two other new viruses, maize rayado fino virus (MRFV) and maize stripe virus (MSV), were also known to occur previously in Latin America. MSV also occurs in Africa. Both viruses cause major disease problems in these areas. In the United States, MRFV has been found in Texas and Florida and MSV in Florida.

MLO, recently named maize bushy stunt mycoplasma (MBSM), causes symptoms similar to those of the Mesa Central strain of the corn stunt disease. This disease has been known in Latin America for several decades. MBSM has been found only in Texas.

The fifth disease is associated with a rhabdovirus, or bullet-shaped particle. The maize mosaic virus (MMV), which causes extensive damage to corn in Hawaii and in some South American countries, has a similar shape, but further evidence of a relationship between the newly discovered rhabdovirus and MMV has not been presented. Rhabdoviruses have been identified in corn in Alabama, Mississippi, Louisiana, Texas, and Iowa.

The sixth disease is caused by the barley yellow dwarf virus. This virus is a major pathogen of barley *(Hordeum vulgare)*, oats *(Avena sativa)*, and wheat *(Triticum aestivum)* in the United States but not of corn.

The seventh new disease is associated with a virus which has a long filamentous particle that resembles the particles of MDMV and wheat streak mosaic virus. (The latter virus was first identified in United States corn in the mid-1950s and has been of only minor importance as a corn pathogen since then.) The relationship of this new, unnamed corn virus to other filamentous viruses is as yet unknown. It has been found only in Texas.

Among the newly discovered pathogens, only MCMV and possibly MRFV appear to pose an

First reported occurrences of virus and viruslike pathogens of corn in the continental United States

Pathogen	Sites of first reported occurrence 1942–1961	1962–1968	1969–1972	1973–1974	1975	1976	1977
Barley yellow dwarf virus	...	...	...	...	...	...	South Dakota
Corn stunt pathogen	Texas	Arizona, Florida, Louisiana, Mississippi	...	...	...	...	...
Maize bushy stunt mycoplasma	...	...	...	...	...	Texas	...
Maize chlorotic dwarf virus	...	...	Ohio plus 5 southern states	10 additional southern and adjacent states	...	...	Pennsylvania
Maize chlorotic mottle virus	...	...	...	...	...	Kansas	Nebraska, Texas
Maize dwarf mosaic virus	...	25 states within the region from 40°N latitude to the southern border and from Atlantic to Pacific coasts	Massachusetts, New York	...	...	Northern Ohio	Idaho, Minnesota, North Dakota, Wisconsin
Maize rayado fino virus	...	...	...	...	...	Texas	Florida
Maize rhabdovirus	...	...	...	Texas, Alabama, Louisiana	Iowa	...	...
Maize stripe virus	...	...	...	Florida	...	...	...
Unidentified corn virus	...	...	...	...	...	Texas	...
Wheat streak mosaic virus	Idaho	Iowa, Nebraska, Ohio	South Dakota	Oklahoma	...	...	...

immediate danger to United States corn production. MCMV is highly infectious and appears to survive winters in corn fields where it has caused damage. It is transmitted by several beetle species which are abundant throughout the Corn Belt. Preliminary tests indicate that only a few corn inbreds or hybrids are resistant to MCMV. Also, the damage caused by MCMV in Kansas in 1976 and 1977 is a further reason for concern.

MRFV is potentially dangerous because the leafhopper, *Graminella nigrifrons*, which transmits it is widely distributed throughout the Corn Belt and because no immune corn varieties are presently known. On the other hand, MRFV has a very limited number of susceptible plant species, and no species is known that could harbor the virus between corn crops in the Corn Belt.

Maize dwarf mosaic outbreaks. The expansion of maize dwarf mosaic was primarily into northern sweet corn–growing areas where the disease had not been previously found. With few exceptions prior to 1976, maize dwarf mosaic, as caused by MDMV-A, occurred only where Johnson grass *(Sorghum halepense)*, the overwintering host of the virus, survives as a perennial weed grass. This area extends from about 40°N latitude to the southern United States border and from the Atlantic to Pacific coasts. (This had been the distribution of MDMV-A in the United States from about the mid-1960s to the mid-1970s. Notable exceptions were outbreaks in late-planted sweet corn in New York and Massachusetts in the early 1970s.) It appears that a major change of distribution began in 1976 when a maize dwarf mosaic epiphytotic (a widespread occurrence of an infectious plant disease) occurred in late-planted sweet corn in northern Ohio, north of where Johnson grass is abundant. In 1977 epiphytotics occurred in late-planted sweet corn in northern Ohio, Minnesota, Idaho, and probably Wisconsin. In North Dakota the disease was observed in late-planted dent as well as sweet corn. These were the first major occurrences of the disease in these states, except for Ohio where the disease has occurred in the southern half since about 1962. The other three states are far north of where Johnson grass is a perennial weed.

The outbreaks in Ohio, Minnesota, Wisconsin, Idaho, and the northeastern states involved both strains A and B (non-Johnson-grass strain) of MDMV. Only MDMV-A was reported in North Dakota. For northern Ohio the incidences of MDMV-A and MDMV-B were about equal for the 2 years. This MDMV-B incidence was several-fold greater than that found in southern Ohio. The reason for the dramatic increase in the MDMV-B incidence is unknown. Also unknown are the sources for initial infections of MDMV-A and MDMV-B in these northern areas. Control of maize dwarf mosaic in dent corn has been accomplished by planting tolerant or resistant hybrids. Unfortunately, presently available sweet corn hybrids do not possess satisfactory tolerance or resistance.

It is important to note that maize chlorotic dwarf, the other major United States corn virus disease, has not been observed in these northern states where maize dwarf mosaic has been epidemic. Maize chlorotic dwarf remains confined to the southeastern United States and states bordering this region. This disease, along with maize dwarf mosaic, has been a major limiting factor in corn production in this region since the mid-1960s.

For background information *see* CORN; PLANT VIRUS in the McGraw-Hill Encyclopedia of Science and Technology. [DONALD T. GORDON]

Bibliography: D. T. Gordon et al., *Proc. Amer. Phytopathol. Soc.*, 4:92, 1977; D. T. Gordon and L. R. Nault, *Phytopathology*, 67:27–36, 1977; L. R. Nault and O. E. Bradfute, in K. Maramorosch and K. Harris (eds.), *Leafhopper Vectors and Plant Disease Agents*, 1978; C. L. Niblett and L. E. Claflin, *Plant Dis. Rep.*, 62:15–19, 1978.

Cosmology

There have been several recent developments in tests designed to predict whether the universe will expand forever or eventually collapse. The question arises from two discoveries made in the 1920s and 1930s that enormously widened astronomers' understanding of the universe, namely, that the Milky Way Galaxy is but one of countless billions of galaxies, and that those galaxies are moving away from each other, so that the whole universe is expanding.

Expansion of the universe. A key property of the expansion, discovered by E. P. Hubble, is that each galaxy's speed of recession from any other is proportional to the distance between them. This means that the scale of the entire universe is expanding: 5×10^9 years from now, for example, the relative spacings of the galaxies will be the same, but the spacings will all be about 30% greater. Hubble's discovery also implies that if the motions are extrapolated backwards, with each galaxy supposedly maintaining a constant speed, one surmises that the galaxies were all crushed together at a particular instant in the past. The time interval between that instant and the present is called the Hubble time, and its value is simply the ratio of distance apart to recession speed for any pair of galaxies. (In practice, one must exclude very close neighboring galaxies, because their relative motions may be orbits about each other, so that they move together as a single element in the expanding universe. The Milky Way has several neighbors that are bound to it in this way.) The value of the Hubble time is currently estimated to be between 10 and 20×10^9 years, the uncertainty arising from the difficulty of determining distances to galaxies. Although the picture of galaxies moving apart at constant speeds is an oversimplification, observations do indicate that the universe has been expanding for about 10 or 20×10^9 years, starting from an extremely dense, hot state known as the big bang.

Fate of the universe. The future of the universe depends to a large extent on whether or not it will expand forever. Figure 1 shows the possibilities allowed by the simplest models derivable from Einstein's general theory of relativity, which is usually assumed to be the applicable theory. (Other theories have been suggested, including some in which the universe is not seen as expanding because Hubble's observations are given a different explanation.) One possibility, known as the open universe, theorizes that the expansion will go on forever: each galaxy will become darker and darker as its stars die out, and all but its closest neighbors will move away out of sight. The other possibility suggests a closed universe, in which the expansion will slow to a halt billions of years from now, at which time the universe will contract, becoming progressively denser until the galaxies and stars are crushed out of existence and everything ends in a fireball similar to the big bang from which it emerged. The theory of relativity predicts that various observable quantities would differ in the two cases; thus astronomers approach the question of whether a hot or cold fate lies ahead by trying to determine the relevant quantities.

Deceleration of the expansion. The best-known cosmological test is designed to measure the past expansion rate of the universe to determine whether it is slowing down enough to reverse itself eventually. This test was first performed by M. L. Humason, N. U. Mayall, and A. R. Sandage in the early 1950s, and has been carried out subsequently by Sandage and others with increasing penetration into the past. A quantity called q_0 is used to describe the slowing down: if $q_0 = 0$, there is no deceleration and the expansion will go on forever at the same rate; if q_0 is less than ½, the expansion is slowing down but will still go on forever; and if q_0 exceeds ½, there is enough deceleration for an eventual halt and contraction.

The method involves studying galaxies so far away that their light left them billions of years ago. Shifts in the wavelengths of the light can be used to determine how fast each galaxy was receding at that time. Because the expansion rate should be faster in the past, depending on the amount of deceleration, q_0 should be obtainable from the excess recession speeds observed over the speeds expected via the usual proportionality to distance. Despite decades of effort, this test has not yielded firm results. The problem is that the relative distances of galaxies must be known within a few percent if the change in the expansion rate is to be measured accurately. There is no direct way of measuring distances of galaxies. However, objects of a given intrinsic luminosity appear dimmer with distance in a known way. Thus, if the intrinsic

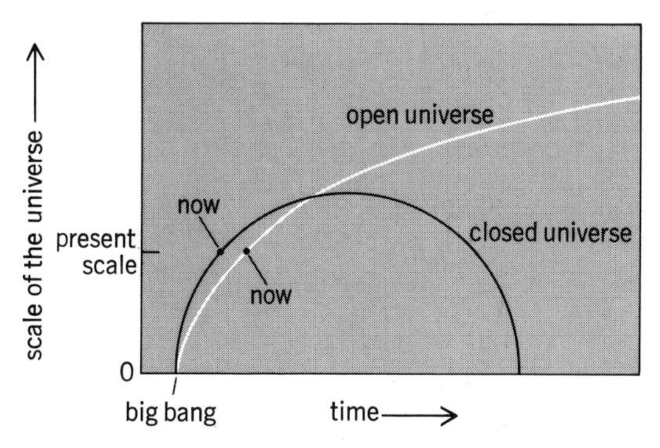

Fig. 1. History of the scale of the universe in the simplest types of cosmological models allowed by the general theory of relativity.

luminosities of a sample of galaxies are known, the apparent brightnesses of the galaxies would give their distances. Unfortunately, galaxies have a wide range of intrinsic luminosities, and there is no way of choosing a sample with exactly known values, so distances are correspondingly uncertain. Moreover, the luminosity of each galaxy changes with time because its individual stars evolve, a fact which must be taken into consideration when comparing nearby galaxies with those seen as they were billions of years ago.

Some recent results are shown in Fig. 2, which is a plot of recession speed versus apparent brightness for a sample of galaxies studied by J. E. Gunn and J. B. Oke. An approximate distance scale (roughly the average for a given brightness) is given at the top of the figure. The data points do not define a narrow line because at each distance the galaxies observed have a range of intrinsic luminosities, but the general trend is for galaxies with greater speeds to be fainter, illustrating Hubble's discovery that speeds increase with distance. Two theoretical lines are drawn: the line labeled $q_0 = 0$ shows the average relation expected if there is no deceleration, and that labeled $q_0 = 1$ shows more than enough deceleration to reverse the expansion. It appears that the scatter in the points is too great for a clear choice to be made as to which type of model fits the data best, an impression that is confirmed by careful statistical analysis. Moreover, scatter is not the only problem.

Evolution of galaxies. Galaxies grow dimmer as their brighter stars die, so that the most distant galaxies in Fig. 2 are overluminous relative to nearby ones. To correct for this, the points near the top of the diagram should be shifted to fainter positions, as indicated by the short arrow on one point. Clearly, if all the points for distant galaxies were shifted by that amount, their positions would correspond to a significantly smaller value of q_0. Until recently, data with this type of correction for evolution tended toward values of q_0 less than ½, corresponding to an open, ever-expanding universe.

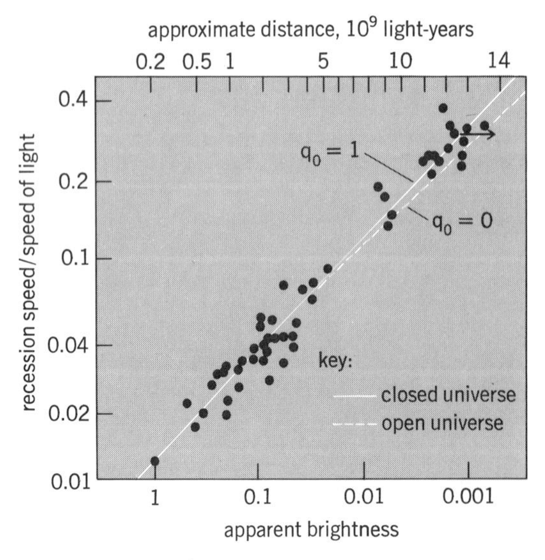

Fig. 2. Relation between the recession speeds and apparent brightnesses of galaxies. Apparent brightness is relative to the brightest galaxy. 1 light-year = 9.46055 × 10^{15} m.

However, in 1975 J. P. Ostriker and S. D. Tremaine discovered a process that compensates for dying stars: large galaxies sporadically gain more stars by cannibalizing their smaller neighbors. Thus the distant galaxies, seen as they were billions of years ago, are underpopulated relative to nearby ones, and the previous correction for dying stars has exaggerated the extent to which they are overluminous. As a result, the arrow in Fig. 2 should be shorter or possibly even reversed! There is not yet a good estimate of this correction, which may be the deciding factor as to whether the value of q_0 is less than or greater than the critical value of ½. At present, this test cannot predict the future of the universe until more is understood about the past of galaxies.

Other tests. Other approaches to this problem seem to be more promising. The general relativistic models give a simple relation between the future of the universe and the average density of matter in it: if the density exceeds a certain critical value, the gravitational attraction between all galaxies is sufficient to stop and reverse the expansion; however, if the density is less than critical, gravity slows the expansion, but by such a small amount that it can continue indefinitely. The density of matter in the universe can be estimated by counting galaxies in a given volume of space and multiplying by an average mass per galaxy. First estimates yielded a value that was less than 1% of the critical density, clearly favoring the hypothesis of an open universe. However, it has been suggested in recent years, initially by astronomers at Princeton University and in the Estonian S.S.R., that galaxies are much more massive than was previously thought, being surrounded by heavy "halos" of invisible matter. However, after allowing generously for possible heavy halos around all the galaxies counted, the density is still less than critical by a factor of 3 to 10. Apparently there is not enough matter in the universe for gravity to stop the expansion.

Yet another approach is to compare the actual age of the universe with the time that would have elapsed since the big bang if the expansion rate had remained constant. The latter value is equal to the Hubble time, mentioned above. With deceleration, the expansion would have been faster in the past, so that the actual time since the big bang must be shorter than the Hubble time; the critical test is whether the actual age is less than two-thirds of the Hubble time, in which case the deceleration was sufficient in the past to lead to an end and reversal of expansion in the future. The oldest bodies known in the universe are star clusters and radioactive elements whose ages are estimated as between 1.2 and 2 × 10^{10} years; the actual age of the universe must therefore be at least 1.2 × 10^{10} and possibly 2 × 10^{10} years. For comparison, the Hubble time is between 1 and 2 × 10^{10} years. There are unsatisfactory uncertainties in these numbers, but it seems most likely that the universe is older than two-thirds of the Hubble time. If so, the deceleration has been small enough for expansion to continue forever.

Considering all these results, one finds that the firmest evidence points to an open universe. However, astronomers are by no means convinced that a final answer has been reached. Observers and theorists will continue to work on the past expansion rate and the density and age of the universe, as well as devising other cosmological tests, to determine whether the universe will expand forever and whether the general relativistic models provide an adequate theoretical framework for describing the universe.

For background information *see* COSMOLOGY; GALAXY, EXTERNAL; HUBBLE CONSTANT in the McGraw-Hill Encyclopedia of Science and Technology. [BEATRICE M. TINSLEY]

Bibliography: J. R. Gott et al., *Sci. Amer.*, 234(3): 62–79, March 1976; J. E. Gunn and J. B. Oke, *Astrophys. J.*, 195:255–267, 1975; B. M. Tinsley, *Phys. Today*, 30(6):32–38, June 1977; S. Weinberg, *The First Three Minutes*, 1977.

Cracking

Recent advances in cracking have been concerned with the development of cracking catalysts with specific performance characteristics.

New catalysts. Energy conservation, environmental restrictions, the need for higher octanes for unleaded gasoline, and the continued pressure for higher performance from existing cracking units have provided the incentives for the development of a wide variety of new cracking catalysts with specific performance properties in addition to the conventional yield and product quality characteristics. Recent developments are concentrated in four areas: the promotion of carbon monoxide combustion in regenerators, the mitigation of the deleterious effects of metals deposited on the catalyst from heavy feeds, the reduction of sulfur (SO_x) emissions from regenerators, and improved attrition resistance.

Catalysts for improved octane numbers are an active but not yet clearly defined area of development. Catalysts with significantly higher octanes, for example, the Houdry-Engelhard HFZ and HEZ series, appeared on the market several years ago, and several new catalysts with enhanced octane capabilities are undergoing commercial tests by other manufacturers.

CO combustion promoters. Prior to about 1972, typical catalytic cracking regenerators burned coke from the catalyst with a flue gas CO_2/CO ratio of approximately unity. The advantages of complete combustion to CO_2—the release of about 50% more heat and the reduction of CO emissions—have long been recognized. These benefits have been obtained through the use of external CO boilers or by operating the regenerator at increased temperatures, a procedure referred to as high-temperature regeneration and first introduced by the American Oil Company in 1972. Catalytic promotion of CO to CO_2 is obtained with a small amount of noble metal, either on the cracking catalyst or present as a separate particle, referred to as promoted catalyst and solid promoter, respectively. In either case, the concentration of noble metal is extremely small, estimated to be less than 100 parts per million (ppm) and possibly less than 10 ppm.

Promoted catalysts are available with various levels of CO promotion: full promotion with essentially complete oxidation of CO and generally less than 1000 ppm CO in the flue gas, and partial promotion with, for example, 5% CO (50,000 ppm) in the flue gas. Independent control of cracking activity and the CO promotion effect can be achieved with solid promoters; these are noble metal–containing materials with a particle size distribution similar to that of cracking catalyst and an extremely high level of CO oxidation activity. On most units, the use of 1–10 lb of solid promoter per ton (0.5–5 kg per metric ton) of makeup catalyst is sufficient for full oxidation of CO to CO_2. Solid promoters are currently available from the Houdry Division of Air Products and Chemicals, Inc., the Davison Chemical Division of W. R. Grace & Company, Filtrol Corporation, Engelhard's Minerals and Chemicals Division, and Akzo NV (Ketjen) in Holland. Prices range from $4 to $17 per pound ($8.80 to $37.40 per kilogram) and at least one manufacturer offers two grades. Use of the solid promoter is simple, and most refiners inject small batches, 5–10 lb (2.3–4.5 kg), into the regenerator through a small blowcase once a day or periodically. In addition, a liquid promoter is offered by Universal Oil Products.

The use of CO promoters is now widespread. Their principal advantages are elimination of the need for a CO boiler, reduction of stack CO content below pollution control levels, enhanced stability of the regenerator system, reduction of afterburning and the possibility of flameout, and increased liberation of heat, resulting in higher temperatures in the regenerator. In any given situation, the incentives for CO promotion depend on how that particular unit can accommodate the excess heat provided by enhanced CO oxidation. In order to keep the unit in heat balance, more air must be added or carbon production must be reduced. This reduction can be achieved by eliminating auxiliary fuels such as torch oil (if used), reducing reactor holdup, lowering reactor temperature, or shifting to a lighter feed. Liquid yield credits are generally obtained through higher regenerator temperature which results in lower carbon on regenerated catalyst and lower catalyst-to-oil ratio and thus lower coke make.

Metals passivators. The Phillips Petroleum Company technique diminishes the effects of metals on cracking catalysts and consists of injecting an antimony-containing additive, Phil-Ad CA, along with the feed. In a manner which has not been elucidated, the antimony reacts with the metals on the catalyst, reduces their dehydrogenation potential, and significantly reduces coke and hydrogen yields. The Phil-Ad CA material is an oil-soluble composition containing antimony and has a low pour point and a relatively high flash point. The cost of the injection system, primarily a small metering pump, is low. Addition rates can be changed to reflect variation in metals content of the feeds, and are believed to be roughly one atom of antimony per atom of nickel plus vanadium in the feed. Additive costs are also low, reported to be about 4¢ per barrel converted (25¢ per cubic meter), for a West Texas long-residuum feed containing 7 ppm Ni, 13 ppm V, and 20 ppm Fe.

Since most commercial catalytic crackers are limited by coke-burning capacity or dry gas–handling capacity, a reduction in coke and hydrogen yields by use of a metals passivator not only increases selectivity to useful liquid product, but may also allow a significant increase in unit throughput or an increase in conversion level. In a commercial-scale test in the heavy-oil (residuum) cracker at the Phillips refinery at Borger, TX, metals passivation reduced the hydrogen yield by 46.6%, dry gas by 27.5%, and coke yield by 15%. After passivation, the unit throughput was increased from 26,000 to 30,000 barrels (4134 to 4770 m^3) per day and conversion was increased 3.9% with a 6% increase in the gasoline yield. Passivation produced no apparent change in octane number, and the yields of C_3 and C_4 olefins were essentially constant. No safety hazards or environmental or metallurgical problems have resulted with the proper use of Phil-Ad CA metals passivating agent.

Reduction of sulfur emissions. As early as 1949 it was reported that the use of silica/magnesia catalysts resulted in as much as an 85% reduction in SO_x emissions. More recently, Amoco successfully developed catalyst technology which reduces SO_x emissions by 60–75%. Among other firms active in this area are Davison Chemical Division of Grace, Chevron Research Company, Akzo NV (Arnhem, Netherlands) and, jointly, Atlantic Richfield Company and Engelhard's Minerals and Chemicals Division.

The basic chemistry involved in SO_x reduction is illustrated in the reactions below for MgO; how-

Regenerator: $SO_2 + \frac{1}{2}O_2 \rightarrow SO_3$
$MgO + SO_3 \rightarrow MgSO_4$
Reactor/stripper: $MgSO_4 + 4H_2 \rightarrow MgS + 4H_2O$
$MgS + H_2O \rightarrow MgO + H_2S$

ever, CaO, other oxides, and mixtures have been used. Thus, sulfur carried into the regenerator as part of the coke deposit on the catalyst is burned to form SO_2 and SO_3; the SO_3 reacts with MgO to form $MgSO_4$ and is returned to the reactor rather than emitted with the regenerator gases. In the reactor/stripper system, the $MgSO_4$ is reduced by H_2 and steam to regenerate the MgO and release the sulfur to the normal gas-handling facilities as H_2S. Proper control of regenerator conditions to effect oxidation of SO_2 and reaction with MgO must be achieved without upsetting the other functions and limitations of the regenerator. SO_x control catalysts are purported to be most effective with CO burning, and in some cases require it.

In May 1978 Amoco stated that SO_x reductions in excess of 80% could be achieved commercially and stressed the importance of proper regenerator and process control with SO_x reduction catalysts. In all of their pilot plant and commercial unit trials with several alternate SO_x reduction catalyst formulations, there were no significant yield differences when comparisons were made with standard catalysts. In earlier pilot plant studies, with a feed containing 2.5% sulfur, various Amoco catalysts reduced the SO_x emission from 2350 wppm to 280–890 wppm (weight parts per million), a reduction of 62–88%, which is sufficient for most state air-pollution standards for existing or new SO_x sources.

Attrition-resistant catalysts. Attrition-resistant catalysts, for example, Davison's AGZ series, have been available since the late 1960s; however, they did not capture a major portion of the market. More recently several new series of attrition-resistant catalysts have been introduced and are becoming widely used to comply with increasingly more stringent particulate emission standards.

In 1970 Houdry introduced HFZ-20, a catalyst with greatly increased attrition resistance. Since that time, Houdry-Englehard has developed the HFZ and HEZ series of catalysts, covering a range of activity, stability, and selectivity, but all with the same attrition resistance as HFZ-20. Laboratory attrition rates of fresh HFZ catalysts have been reported to be 12–30% that of other commercial catalysts. In a series of commercial plant comparisons, the use of HFZ catalysts reduced catalyst makeup rate by up to 50% and decreased stack losses by as much as 70%.

In 1976 Davison introduced the Super-D line of attrition-resistant catalysts. Six grades of varying activity are available. Laboratory attrition rates of the Super-D and Super-D Extra grades of this series are 7 and 13, respectively (expressed as the Davison index), compared with values of 23–27 for other current Davison catalysts. The improved properties of the Super-D series are attributed to a new low-surface-area, low-activity-matrix material which not only reduces attrition but also reportedly results in improved selectivity due to more zeolite-type cracking, increased thermal stability, improved stripping of hydrocarbons from the coked catalyst, and higher metals tolerance.

For background information *see* CRACKING; PETROLEUM PROCESSING in the McGraw-Hill Encyclopedia of Science and Technology.

[ALFRED D. REICHLE]

Bibliography: G. H. Dale et al., *43d Midyear Meeting*, Toronto, API Preprint no. 60–78, May 11, 1978; W. D. Ford et al., *43d Midyear Meeting*, Toronto, API Preprint no. 59–78, May 11, 1978; T. A. Montgomery, *Catalgram*, no. 56, Davison Chemical Company, 1978; L. L. Upson, E. E. Winfree, and E. L. Leuenberger, *N.P.R.A. Annual Meeting*, AM-78-50, Mar. 19–21, 1978; W. H. Wallendorf, *Catalgram*, no. 53, Davison Chemical Company, 1977.

Crustacea

The celebrated Middle Cambrian Burgess Shale (about 540,000,000 years old) of British Columbia, discovered by C. D. Walcott in 1910, is of great importance as one of the oldest deposits in which the soft parts of organisms are exceptionally well preserved. It has yielded a great variety of invertebrate fossils, over one-third of which are arthropods, and it is therefore an obvious place to look for early Crustacea. Most of the arthropods in the Burgess Shale, however, apart from the trilobites, cannot be assigned to any of the major groups. It appears that a radiation in late Precambrian and Early Cambrian time resulted in a variety of "experimental" and subsequently unsuccessful forms which enjoyed a relatively brief heyday in the absence of severe competition or predation. D. E. G. Briggs demonstrated in 1978 that the Crustacea are represented in the Burgess Shale by specimens of *Canadaspis perfecta*.

Canadaspis perfecta is one of the most abundant of the arthropods which occur in the Burgess Shale; some 5000 specimens have been collected. The fauna was buried almost in place by successive clouds of sediment. The fossils are largely complete, indicating that transport was neither far nor very turbulent, but specimens nevertheless came to rest in a variety of orientations to the bedding. This facilitates their reconstruction in three dimensions (Fig. 1), even though they were subsequently compacted into the plane of bedding of the shale; it is almost like working from a large number of photographs taken from different angles. During burial, sediment was trapped beneath the carapace and between the appendages of the arthropods, and this matrix separates the layers of the compacted fossil and allows them to be flaked off so that a "paleodissection" can be carried out. The carapace, for example, can be removed to reveal the trunk and appendages beneath.

Morphology of C. perfecta. The cephalon (head), thorax, and all but the extremities of the appendages of *C. perfecta* were covered in life by the carapace (Fig. 2), which consisted of two suboval valves joined along the hinge line by a strip of flexible cuticle. Valves of larger specimens average about 40 mm in length, but examples less than 10 mm are known. The cephalon bore a pair of eyes on short stalks which flanked a pair of small projections interpreted as the first antennae (antennules). A median spine projected forward just below these. It was not part of the carapace but may have served a function similar to the rostrum of other Crustacea in protecting the sensory organs. An additional pair of sensory appendages, the second antennae, lay below and behind the cephalic spine. Food particles were broken up mainly by the mandibles, which consisted of a series of large spines just behind the mouth forming a molar and incisor process (Fig. 3). The posterior two pairs of cephalic appendages characteristic of the Crustacea, the first and second maxillae, are very similar in *C. perfecta* to the limbs of the thorax. Only the first maxilla differs in bearing a series of bunches of spines along its inner margin.

The thorax consisted of eight somites (divisions of the body), each bearing a pair of biramous (two-branched) appendages. The inner branch was made up of 14 segments and terminated in a number of hooked spines or "claws." It was presumably used in walking and in raking the sediment into suspension during feeding. The outer branch, which is thought to have functioned in respiration, was flaplike and consisted of a number of rays attached to a basal lobe. The appendages probably moved in a rhythmic (metachronal) fashion characteristic of the arthropods. The outer gill branches would have created currents ensuring a constant stream of fresh water for respiration, and also perhaps carrying suspended material into the space between the thoracic limbs. *Canadaspis perfecta* did not, however, feed by filtering small particles out of these currents. Each appendage bore a bunch of inwardly facing spines near the base of the segmented branch which were used to transport relatively large food particles forward from one appendage to the next until they reached the mandible, where they were broken up before being passed to the mouth. The mouth was situated on the ventral side of the head and was covered by a lip (labrum) which prevented food being lost.

The abdomen of *C. perfecta*, which extended beyond the protection of the carapace, consisted of seven somites and terminated in a telson bearing the anus. Traces of the alimentary canal are commonly preserved, often filled with sediment (Fig. 2). The first six somites lacked appendages, but the seventh bore a pair of spinose projections which extended beneath and beyond the small telson. These projections, unlike the posterior appendages of some other crustaceans, were unsuitable for swimming; they may have assisted the flexible abdomen in pushing the arthropod out of soft sediment.

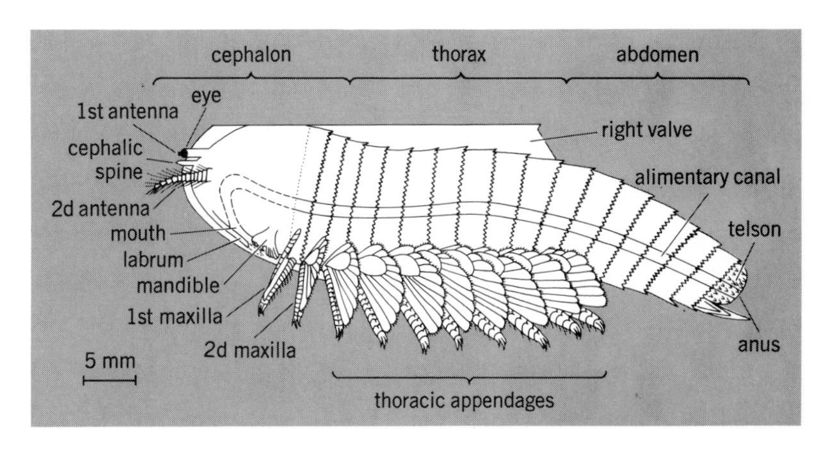

Fig. 1. Diagrammatic reconstruction of *Canadaspis perfecta* viewed from the left side with the left valve removed and only the left appendages shown. *(From D. E. G. Briggs, The morphology, mode of life, and affinities of Canadaspis perfecta (Crustacea: Phyllocarida), Middle Cambrian, Burgess Shale, British Columbia, Phil. Trans. Roy. Soc. Lond. B, 281:439–487, 1978)*

Evolutionary significance of C. perfecta. The present-day Crustacea to which *C. perfecta* shows the greatest similarity are the leptostracans (of which *Nebalia* is perhaps the best-known example). There are, however, a number of important differences. The Leptostraca bear limbs on the abdomen and a pair of usually leaf-shaped appendages (the caudal furca or "tail fork") on the telson.

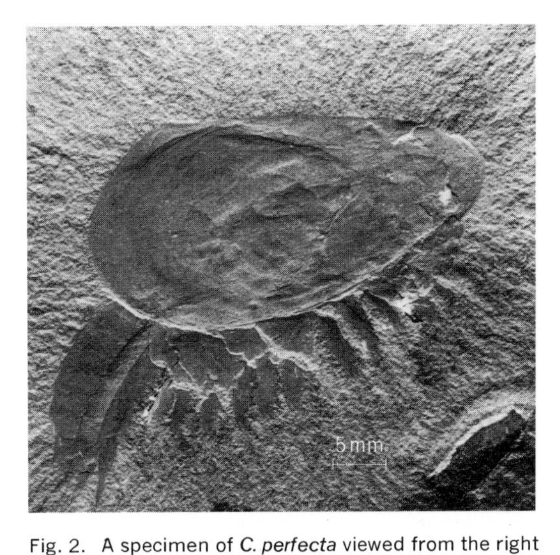

Fig. 2. A specimen of *C. perfecta* viewed from the right side showing the appendages and abdomen extending beyond the carapace. Note the trace of the alimentary canal which was originally sediment-filled. *(From D. E. G. Briggs, The morphology, mode of life, and affinities of Canadaspis perfecta (Crustacea: Phyllocarida), Middle Cambrian, Burgess Shale, British Columbia, Phil. Trans. Roy. Soc. Lond. B, 281:439–487, 1978)*

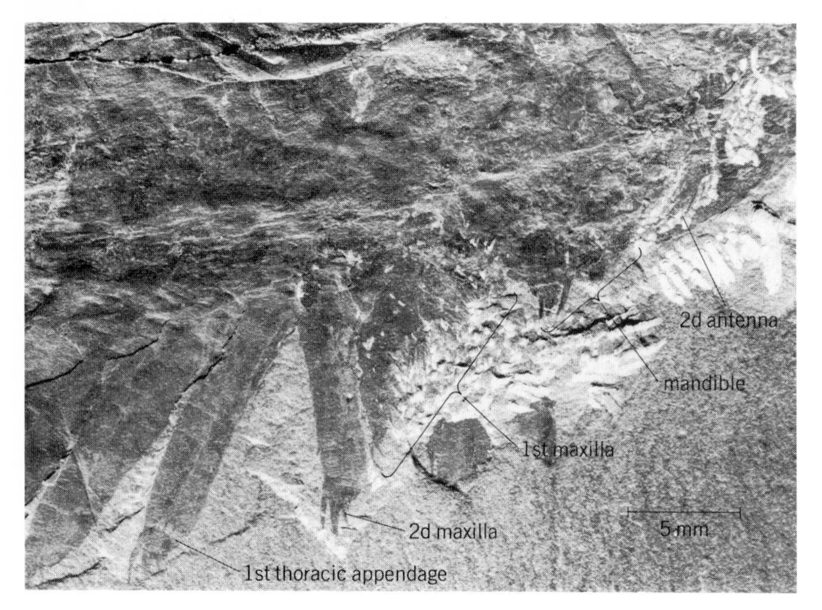

Fig. 3. The appendages near the boundary between the cephalon and thorax of *C. perfecta*. This specimen is viewed from the right side, but the right appendages have been removed and those of the left side are displayed. *(From D. E. G. Briggs, The morphology, mode of life, and affinities of Canadaspis perfecta (Crustacea: Phyllocarida), Middle Cambrian, Burgess Shale, British Columbia, Phil. Trans. Roy. Soc. Lond. B, 281: 439–487, 1978)*

These limbs are adaptations for swimming and were absent in *C. perfecta*, which presumably used walking as its normal mode of locomotion. The thoracic appendages included a much larger number of segments than is found in present-day Crustacea, and this condition, which also occurs in some other fossil arthropods, may be primitive. The large number was perhaps necessary to give the appendage comparable flexibility to that found in other arthropods which have evolved adaptations such as a variability in the length of segments and specialization of the articulations between them.

A more striking difference between *C. perfecta* and present-day Crustacea is the lack of a clearly defined boundary between the cephalon and thorax. The posterior appendages of the cephalon (the first and second maxillae) are very similar to those of the thorax immediately succeeding them (Fig. 3); an equal similarity occurs only in the Cephalocarida (such as *Hutchinsoniella*) of present-day Crustacea. Considerations of early arthropod evolution suggest that the differentiation of various sections of the body of a serially segmented ancestral form for particular functions (the process of tagmosis) started at the anterior end and progressed backward. The evidence of embryology shows that the crustacean second maxilla is functionally part of the trunk in the earliest juvenile stage and is a late addition to the cephalon. *Canadaspis perfecta* may therefore represent a stage in crustacean evolution before the complete differentiation of the posterior divisions of the cephalon from the trunk.

If the appendages interpreted as the first and second maxillae are assigned to the cephalon of *C. perfecta*, the thorax and abdomen include the number of somites (eight and seven, respectively) characteristic of the malacostracan Crustacea. *Canadaspis* is therefore classified with the Leptostraca in a subclass of the Malacostraca, the Phyllocarida. A closely related bivalved form, *Perspicaris*, was described by Briggs from the Burgess Shale in 1977, but it is much rarer and not as well preserved. *Perspicaris* appears to have employed swimming, as opposed to walking, as its primary mode of locomotion. Although *C. perfecta* has barely acquired the characteristic morphology of the crustacean cephalon, it would be foolhardy to suggest that it was the first arthropod to do so. It is, however, the oldest fossil with appendages preserved in sufficient detail to identify it as a crustacean. Its association with a number of arthropods with similar bivalved carapaces which are evidently not Crustacea suggests that, in the absence of evidence of soft parts, other early records of Crustacea should be viewed with caution.

For background information *see* CRUSTACEA in the McGraw-Hill Encyclopedia of Science and Technology. [DEREK E. G. BRIGGS]

Bibliography: D. E. G. Briggs, *Palaeontology*, 20:595–621, 1977; D. E. G. Briggs, *Phil. Trans. Roy. Soc. Lond. B*, 281:439–487, 1978.

Cyclone

Several recent observational and theoretical studies have expanded knowledge of the processes which cause tropical cyclone development. Vorticity advection (import of rotating air), strong large-scale horizontal wind shear, and weak vertical wind shear at the system center are now thought to be crucial in storm formation. Modest improvements in the quality of tropical cyclone motion forecasts may be possible based on the results of observational research and new or updated forecast models. It has long been hypothesized that mature storms can be modified to reduce their intense central winds. This theory is now ready for extensive experimental field testing and may become an operational reality within the near future.

Storm formation. Tropical storms form only in selected oceanic regions. By analyzing the seasonal climates of these and other regions it is possible to deduce large-scale conditions which are favorable for the transformation of a common cloud cluster or easterly wave into an intense tropical cyclone. In the mid-1970s W. Gray showed that storms tend to form in regions where the following mean conditions prevail: sea surface temperature $\geq 26°C$ to a depth of 60 m, above-average middle-level relative humidities, strong low-level relative vorticity, weak vertical wind shear, and latitudes poleward of 5°N or 5°S.

The first two criteria usually indicate a conditionally unstable atmosphere favorable to the growth of deep convective clouds, and are satisfied most of the time over large regions of the tropical oceans. Tropical cyclones, however, are rather rare—only about 80 storms per year develop into hurricane-strength circulations. The key genesis parameters appear to be the vorticity and vertical wind shear patterns. This has been substantiated by comparisons of developing and nondeveloping tropical disturbances. Systems which intensify exhibit much larger values of positive relative vorticity near the surface and of negative vorticity in the upper troposphere than nondeveloping disturbances. Recently, J. McBride has noted that individual disturbances tend to develop when located in an area with strong vertical shears of different sign on opposite sides of the circulation center (see illustration). This is analogous to the above-mentioned vorticity relationships. Since the developing storm's radial circulation is inward at the lower levels and outward in the upper troposphere, the divergent winds act to increase vorticity in the center of the system spinning up the tangential winds. It appears that tropical cyclones form only when the large-scale environment has sufficiently large gradients of vorticity to permit inward vorticity advection to overcome natural dissipative processes.

L. Shapiro developed a genesis prediction scheme based on the assumption that an easterly wave disturbance will intensify into a tropical depression only if it moves into a region in which vorticity advection becomes significant. He tested his scheme during the 1975–1977 hurricane seasons by making daily analyses of the winds in the West Indies region to forecast favorable regions for storm genesis. When a wave moved into such an area, a forecast of intensification was made. This method has been quite accurate in forecasting storm development and has thus become the first objective technique to be effective in this area. Applications of this method to other regions are anticipated. Forecast accuracy should improve with the addition of modifications to eliminate development in regions with cold sea surface temperatures, large vertical wind shears, and so on.

Effect of vertical shear. Observational studies show that while strong vertical shear north and south of the disturbance are favorable for storm genesis, weak vertical shears are found at the center (see illustration). To intensify the circulation the system must concentrate heat near its center. This is difficult to accomplish in the tropical atmosphere where any changes in horizontal pressure gradients result in rapid convergence/divergence with compensating vertical motions due to the slow geostrophic adjustment times. Thus a gradual heat buildup must occur. Since the warming occurs as a result of strong latent heat release, it is also necessary to develop high relative humidities in the system to enhance convection. Strong shears of the vertical wind would result in blow-through or ventilation of the system, dissipating the heat and moisture in the core.

Convection and storm intensity. Since the energetics of tropical cyclones are largely dependent upon the release of latent heat near the center of the circulation, one might expect a correlation between the total amount of deep precipitating convection and the storm intensity or intensification rate. A few researchers have noted some weak statistical relationships between total latent heat release and the system's intensity. However, tropical cyclones are extremely variable with respect to convective amount, size, intensity, and so on. The standard deviation of the total latent heat release associated with storms with similar intensities or intensity tendencies is much greater than the mean differences between different classes of storms. Many loosely organized cloud clusters have far more convection than certain very intense hurricanes.

The formation of tropical cyclones is better correlated with the organization of convection than with the amounts of rainfall. As the storms develop, there is an early tendency to concentrate the deep clouds in the central core region without changing the overall cloud amount. Forecast schemes based on convective pattern recognition utilizing satellite data have had some success in determining current storm intensity and forecasting intensity changes.

Cloud pattern forecast techniques require the classification of a system into a category so that its life cycle may be predicted by using statistical analog techniques. Therefore, this approach can be highly unreliable when an atypical system is encountered. If satellite observations of tropical cyclones are ever to be relied upon exclusively, it will be necessary to improve measurements of other types of parameters such as wind, temperature, and pressure fields. A great deal of research activity is under way in these areas.

Storm motion. Models used in tropical cyclone motion prediction are continually becoming more sophisticated. There are two basic approaches: The first approach is to forecast motion with analog models emphasizing sophisticated statistical treatments of storm climatology. These models

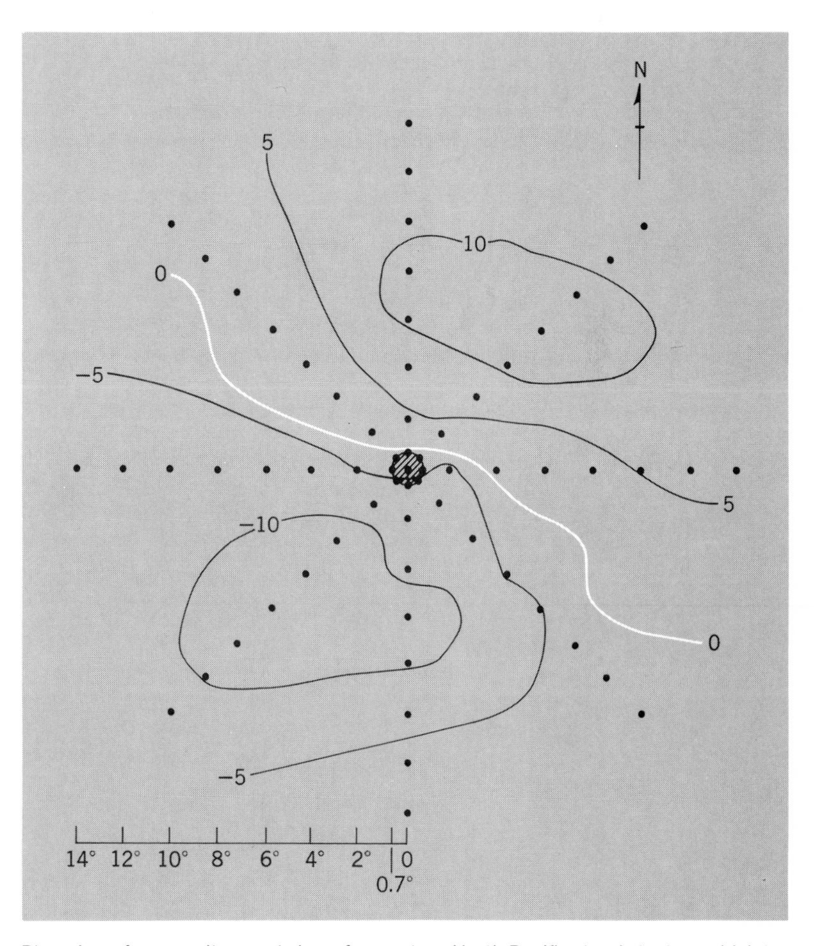

Plan view of composite zonal shear for western North Pacific cloud clusters which later become typhoons. Zonal shear is defined as the easterly wind at the 200-mb (20-kN/m²) level minus the easterly wind at 900 mb (90 kN/m²). Units are meters/second. Note that the intensifying disturbance is located on a zero vertical shear line between regions of large positive and negative shears. The scale at the bottom shows the size of the circulation in degrees of latitude (1° latitude = 111.1 km). *(From J. McBride, 11th Technical Conference on Hurricanes and Tropical Meteorology, 1977)*

work fairly well for the majority of storms which follow normal tracks, but they tend to fail spectacularly in the other cases. While modest improvements have been made in analog forecast models due to increasing data samples, this approach may be nearing its theoretical limit of accuracy.

The second approach, involving numerical models or dynamic forecast models, seems to offer more hope. In particular, recent advances have been made in modeling the motion of a point vortex embedded in a steering current. Other numerical forecast techniques simply predict synoptic steering currents at one or more levels. The biggest problems with these techniques are in determining the proper steering levels and in accurately predicting large-scale wind fields in the tropics. Solution of the prediction problem is not yet in sight, but there has been good progress in relating storm motion to the large-scale wind field.

Recent observational studies of composite typhoon and hurricane data by J. George and W. Gray indicate that tropical cyclones are steered by the 500–700-millibar (mb; 50–70-kilopascal) winds averaged from about 100- to 800-km radius around the center. The storms tend to move about 15° to the left of the winds at 500 mb at a speed approximately 1.15 times the mean 700-mb wind speed. Recurvature of storms, the leading cause of forecast error, is best determined from analysis of the 200-mb (20-kPa) wind fields poleward of the center.

Tropical cyclone modification. Due to the awesome destructive potential of tropical cyclones, attempts to modify them have received high priority. Several hypotheses have been advanced, but only one is currently being considered for field testing and operational use.

The intense cyclonic surface winds near the center of tropical cyclones result from a spinning up of the low-level inflowing air—the "piano stool" effect. Frictional processes only partially retard the tangential acceleration. Based on angular momentum considerations, the maximum wind speed attained by the air increases as the radius of innermost penetration (the eyewall) decreases. It is hypothesized that the central core of the storms can be weakened by causing more of the air to rise to the outflow level before it reaches the eyewall. Therefore, the proposed method of modification is to seed clouds just outside the eyewall with silver iodide to enhance convection, thereby "short-circuiting" part of the mass inflow. A great deal of research has gone into assessing the feasibility of this technique, including numerical modeling simulations, cloud physics measurements, and observational studies using radar, aircraft, and rawinsonde data.

Any modification hypothesis must ultimately be proved in the field. Based on the results of years of theoretical studies, Project Stormfury has been authorized to begin experimental seeding of selected tropical cyclones in the western North Atlantic and eastern North Pacific regions. Possible extensions of the program to the east and west coasts of Australia are also being considered. It is hoped that the results of these experiments will be conclusive enough to assess the potential for using this cloud-seeding technique operationally to protect populated areas. Although any weather modification technique must be carefully scrutinized for potential detrimental side effects, the most recent studies indicate that the Stormfury seeding should not have any measurable effect on large-scale storm circulation or rainfall.

For background information *see* CYCLONE; HURRICANE; METEOROLOGY; STORM; WEATHER MODIFICATION in the McGraw-Hill Encyclopedia of Science and Technology. [WILLIAM M. FRANK]

Bibliography: J. George and W. Gray, *J. Appl. Meteorol.*, 15:1252–1264, 1976; W. Gray, Tropical cyclone genesis, CSU Pap. no. 234, 1975; J. McBride, *11th Technical Conference on Hurricanes and Tropical Meteorology, AMS*, 1977; L. Shapiro, *J. Atmos. Sci.*, 34:1007–1021, 1977.

Cytoplasmic inheritance

The discovery in 1963 that mitochondria contain deoxyribonucleic acid (DNA) added new impetus to the field of cytoplasmic genetics. The study of the coding functions of this mitochondrial DNA (mtDNA) has advanced far more rapidly than other aspects of cytoplasmic genetics, due mainly to the combination of formal genetic analysis with the biochemical techniques of molecular biology. The genetic and biochemical data correspond very closely, and at present about 50% of the genes on mtDNA in yeast have been identified and located. Comparative data have recently emerged from other organisms, for example, the fungi *Aspergillus* and *Neurospora*, the single-celled *Paramecium* and *Tetrahymena*, *Xenopus*, *Drosophila*, many animals, and humans.

Yeast mitochondrial mutants. Normal haploid yeast cells have 50–100 mtDNA molecules (several per mitochondrion), which exist as circles about 25 μm in circumference and 5×10^7 daltons in molecular weight. They are thus approximately five times larger than animal mtDNAs. All mitochondria have their own protein synthetic apparatus, including ribosomes, ribosomal ribonucleic acids (rRNAs), transfer RNAs (tRNAs), and charging enzymes. Yeast mitochondria synthesize three of the seven subunits of cytochrome oxidase, four of the nine subunits of the adenosinetriphosphatase (ATPase) complex, and one of the seven subunits of the cytochrome bc_1 complex. These are all enzymes involved in respiration and energy or adenosinetriphosphate (ATP) production, and the remaining subunits are synthesized in the cytoplasm.

Baker's yeast *(Saccharomyces cerevisiae)* can grow without oxygen if glucose is present. However, for growth on glycerol it requires oxygen and must use its mitochondria. Thus any impairment in respiratory (mitochondrial) function can be detected by the inability to grow on glycerol whereas growth on glucose is unaffected. Three broad classes of mitochondrial mutants in yeast have been described. All three were selected by using this fundamental property.

Petite mutants. The first class are the respiratory-deficient "petite" mutants, which have no detectable mitochondrial protein synthesis. The petite mutations apparently involve a fragmentation of normal mtDNA, followed by replication of some of the mtDNA pieces. Thus petites have lost some mtDNA genes, while other genes are retained and greatly amplified.

Antibiotic-resistant mutants. The second class are mutants resistant to antibiotics which inhibit mitochondrial functions. This class includes mutants resistant to antibiotics such as chloramphenicol, erythromycin, and paromomycin which are inhibitors of mitochondrial protein synthesis; to antibiotics such as oligomycin which are inhibitors of mitochondrial ATPase; and to antibiotics such as antimycin A, funiculosin, mucidin, and diuron which are inhibitors of coenzyme QH_2-cytochrome *c* reductase and which block electron flow between cytochromes *b* and c_1.

mit⁻ Mutants. The third group are the *mit⁻* mutants which cannot grow on glycerol but which have retained mitochondrial protein synthetic activity. Their inability to grow on glycerol is caused by specific deficiencies in ATPase, cytochrome oxidase, or cytochrome *b* and reductase.

All these mutations can be shown to be cytoplasmically inherited by their pattern of inheritance after meiosis and mitosis, which differs from that of nuclear genes. Loss of the mutant characteristic in cytoplasmic petites indicates that the mutant gene is on mtDNA.

Mapping. The position of the loci controlling these mutations on the mtDNA can be determined by genetic and biochemical procedures.

Genetic procedures. There are several genetic methods of mapping the mutations on mtDNA. Recombination analysis involves the crossing of different mutants, and the measurement of the frequency in the progeny of the cross of cells (recombinants) with combined mutant characteristics. It is assumed that the farther apart are two genes on DNA, the greater the rate of recombination occurring between them. Similarly, deletion analysis is based on the assumption that the greater the distance between two genes on mtDNA, the greater the chance that they will not be retained together when that cell is mutated to a petite.

Both types of analysis give similar results in yeast. There is a gene on mtDNA, ω, with two naturally occurring alleles ω^+ and ω^-, which influences the proportion of the various types of recombinants in $\omega^+ \times \omega^-$ crosses. Cells which are ω^+ have an insertion of about 1000 base pairs into their mtDNA. It also appears that different populations of mtDNA molecules can undergo many recombination events, and that this results in the mtDNA becoming "homogenized."

Biochemical procedures. Molecular RNA/DNA hybridization involves the binding of an RNA sequence to a region on DNA by base pairing, that is, to the gene which codes for that RNA. If the RNA molecules are labeled with ferritin, the RNA/DNA hybrids can be seen on the mtDNA with electron microscopy. Thus far, one gene for each of the two mitochondrial rRNAs and 26 mitochondrial tRNA genes have been located on yeast mtDNA.

Hybridization has also been used to map mutations on mtDNA. For example, yeast strains containing several antibiotic-resistance mutations are further mutated to petites, a process which results in the loss of many normal mtDNA genes (see above). Crosses between these petites and antibiotic-sensitive normal cells show the appearance of antibiotic-resistant progeny, indicating that some petites have retained one or more of the original antibiotic-resistance genes. Mitochondrial DNA from these petites is then fragmented with restriction enzymes, and the pieces are hybridized with single-stranded normal mtDNA. The hybridized regions are clearly visible with the electron microscope, and indicate the positions on normal mtDNA of the antibiotic-resistance genes retained in the petite. Similar experiments have been done with pieces of mtDNA that have been cleaved at specific sites with restriction enzymes, and a map of these restriction sites has been made.

Although some areas are in dispute, there is good overall correlation between the results with different methods and from different laboratories. A composite map of yeast mtDNA is shown in the illustration. The *mit⁻* genes code for cytochrome oxidase (OXI 1, 2, and 3), ATPase (two loci), and reductase. CAP, ERY, OLI 1 and OLI 2, ANT, FUN, MUC, DIU, and PAR denote the genes controlling resistance to, respectively, chloramphenicol, erythromycin, oligomycin, antimycin A, funiculosin, mucidin, diuron, and paromomycin. Mutants at the var 1 locus produce polymorphic proteins of different molecular weights. Genes involved in protein synthesis have been described above. The location of these genes coincides with the sites of hybridization of mitochondrial messenger RNAs (mRNAs).

Other organisms. Genetic analysis in organisms other than yeast has been limited by a relative lack of mutants. Nevertheless, several antibiotic-resistant mutants have been isolated in obligate aerobic yeasts, *Aspergillus*, *Paramecium*, and mouse and human cultured cells. Genetic recombination has been demonstrated in the yeasts and *Aspergillus*. Chloramphenicol resistance is cytoplasmically inherited in all organisms examined thus far.

Mitochondrial DNA has been examined with biochemical techniques in a wide variety of organisms. There is one gene for each mitochondrial rRNA and at least 15 tRNA genes in all species. The rRNA genes are adjacent, and the tRNA genes are fairly randomly distributed, in all organisms except yeast. The base sequence of mtDNA can vary considerably between closely related animals such as donkeys and horses, sheep and

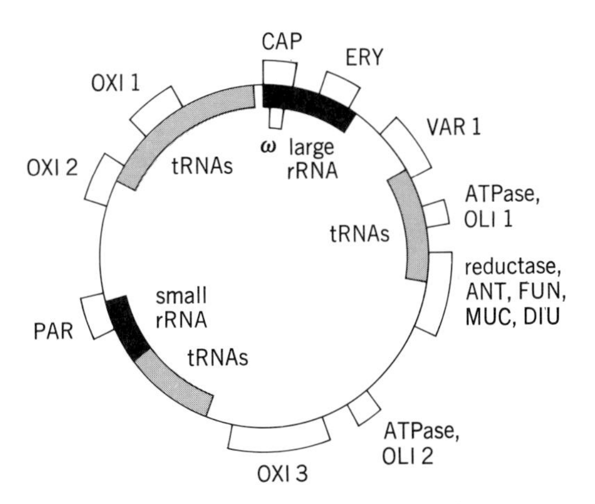

A composite map of the known genes in mtDNA in yeast. *(From B. Dujon, A. M. Colson, and P. P. Slonimski, The mitochondrial map of Saccharomyces cerevisiae: Compilation of mutations, genes, genetic and physical maps, in W. Bandlow et al., eds., Mitochondria 1977, Walter de Gruyter, Berlin, pp. 579–669, 1977)*

goats, or humans and chimpanzees. Even within horses, humans, and rats, different individuals can have mtDNAs differing in sequence.

Thus throughout evolution mtDNA has conserved certain coding functions, while it has evolved rapidly in other, at present unknown, coding functions. The elucidation of these remaining unknown functions represents the next phase of mitochondrial genetics.

For background information *see* CYTOPLASMIC INHERITANCE; GENETICS; MITOCHONDRIA; MOLECULAR BIOLOGY in the McGraw-Hill Encyclopedia of Science and Technology. [CLIVE L. BUNN]

Bibliography: W. Bandlow et al. (eds.), *Mitochondria 1977*, 1977; C. Saccone and A. M. Kroon (eds.), *The Genetic Function of Mitochondrial DNA*, 1976; W. B. Upholt and I. B. Dawid, *Cell II*, pp. 571–584, 1977.

Dam

In the United States the most significant recent developments relating to dams have been actions by the Federal government to improve dam safety. Tragic failures have focused public and congressional attention on dam safety problems. President Jimmy Carter has asked all Federal agencies concerned with dams to review dam safety practices, has established ad hoc committees to survey dam safety problems, and has directed the U.S. Army Corps of Engineers to inspect all non-Federal dams that represent high hazard potentials because they are situated upstream from nearby developments.

Background. On June 5, 1976, the Teton Dam on the Teton River in southeastern Idaho failed and released about 250,000 acre-ft (3.08×10^8 m^3) of stored water. The failure resulted in 11 deaths and property damages amounting to about $500,000,000. On Nov. 6, 1977, the Kelly Barnes Dam on Toccoa Creek at Toccoa Falls in northern Georgia failed. This failure released about 630 acre-ft (777,000 m^3) of water over a 180-ft-high (54.86-m) waterfall into a community of mobile homes, resulting in 39 deaths.

The Teton and Toccoa Falls failures represent a number of contrasts in addition to the differences in volume of water involved. Teton was a new 300-ft-high (91.4-m) earth dam that was being filled for the first time. It was designed by and constructed under the direction of the Bureau of Reclamation of the U.S. Department of the Interior, an agency known throughout the world for its work in dam engineering. The Kelly Barnes Dam had been constructed in 1899 as a rock-filled timber crib and subsequently raised to a height of 42 ft (12.8 m) by earth fills, apparently with no professional engineering supervision.

Failure investigations. A number of investigations were launched after the Teton Dam failure. An independent panel of nine experts in dam engineering, headed by W. L. Chadwick, former president of the American Society of Civil Engineers, undertook an intensive investigation of the failure itself. A parallel investigation was made by an Interior Review Group, composed of engineers lent by Federal agencies and having expertise in dam engineering. The group was chaired by D. N. Sachs, Deputy Assistant Secretary of the Interior, and later F. W. Eikenberry of the Office of the Secretary of Interior. Although the flood flows through the dam break had removed any possible physical evidence of the cause of failure, the consensus of both investigating groups was that the Teton failure resulted from a process termed piping. By this process, reservoir water seeping through or along the boundaries of the impervious earth core of the dam causes erosion of the materials, thus opening up a continually enlarging hole or "pipe" through the core. Both groups agreed that the design of the dam was deficient in that in some areas it did not incorporate the usual provisions to ensure that the core materials would not be eroded by seepage.

Committees of both the Senate and the House of Representatives held hearings on the Teton failure. Other investigations have focused on the organization and procedures of the Bureau of Reclamation relating to dam safety. Several Bureau projects nearing completion were intensively reviewed by an outside consultant firm. As a result of these investigations, the organization and procedures of the Bureau have been revised and some Bureau projects are being modified.

Following the failure of the Kelly Barnes Dam, Governor George Busbee of Georgia requested Federal assistance in investigating this tragedy. A Federal Investigative Board was formed with R. L. Crisp, Jr., of the Corps of Engineers as chairman and with representatives of the Departments of Agriculture, Commerce, and Interior as members. The Board could not isolate one factor as sole cause of the failure because of the lack of physical evidence. However, the Board noted that several types of failure of the dam embankment were possible causes, considering the evident poor design and construction and lack of maintenance. For example, it was established that the dam had not been overtopped.

Actions regarding safety of Federal dams. By a memorandum dated Apr. 23, 1977, addressed to the Federal agencies having major responsibilities for dams, President Carter issued the following directives:

1. A thorough review of practices affecting the safety and integrity of dams, to be undertaken by the head of each Federal agency responsible for these structures.
2. The convening of an ad hoc interagency committee by the chairman of the Federal Coordinating Council for Science Engineering and Technology (FCCSET) to coordinate dam safety programs and provide recommendations for improving the government-wide dam safety effort. Duties of this FCCSET committee included the preparation of proposed Federal dam safety guidelines.
3. Establishment by the director of the Office of Science and Technology Policy (Office of the President) of a panel of recognized experts to review agency practices and proposed dam safety guidelines and to report by Oct. 1, 1978.

The reviews of practices by Federal agencies were carried out by September 1977. The methods used for these reviews varied with the agency: some were accomplished by contract with engineering firms; others were carried out "in house," sometimes utilizing data developed by previous independent-of-agency investigations. Reports of such reviews were made to the FCCSET intra-agency committee by the Department of Agriculture (covered three operating agencies), Corps of Engineers, Department of the Interior (covered six operating agencies), U.S. Nuclear Regulatory Commission, and the Tennessee Valley Authority. Based on the reports of these agencies, the FCCSET interagency committee issued a report, *Improving Federal Dam Safety*, dated Nov. 15, 1977. It presented proposed Federal dam safety guidelines including: independent reviews throughout the development of a dam; staffing for activities related to dam safety; research and testing; coordination of design, construction, and operating forces; documentation of design, construction, and operation processes; instrumentation of dams; periodic inspections; site investigations; construction inspection and quality control; and operation and maintenance.

The FCCSET report also proposed the following future actions to improve dam safety: inclusion of downstream hazard study and recommendations for practical measures to reduce dam failure potential in planning studies for dams; assessment of research needs related to dam safety; development of a comprehensive policy for dealing with problems of aging dams; revision of individual agency policies and practices to conform to the committee's guidelines; recommendations for security measures against sabotage; and Federal interagency cooperation in regard to safety of non-Federal dams.

The Office of Science and Technology Policy has established the panel of experts as directed by the President. F. E. Perkins, head of the Department of Civil Engineering, Massachusetts Institute of Technology, is chairing that panel.

Federal inspection of non-Federal dams. The unprecedented current program for inspection of non-Federal dams is based on legal authority enacted in 1972. At that time the United States had experienced a series of serious dam failures. (One such failure involved a coal mine tailings embankment on Buffalo Creek in West Virginia that claimed about 125 lives.) Congress reacted by passing Public Law 92-367, which has the following provisions:

1. The Chief of Engineers of the U.S. Army is to undertake a national program for inspection of all dams except those owned or regulated by other Federal agencies.
2. The governors of the respective states are to be informed of the results of the inspections.
3. A report is to be furnished the Congress including: an inventory of all dams in the United States; a review of each inspection, with recommendations for action, to be sent to the governor of the state in which the dam is located; recommendations for a comprehensive national program for the inspection and regulation of dams, including the respective responsibilities to be assumed by Federal, state, and local governments and by public and private interests.
4. For application in this act a dam is defined as an artificial barrier across a stream, for the purpose of impounding or directing water, which is 25 ft (7.62 m) or more in height or which is capable of impounding 50 acre-ft (61,700 m^3).

At the time he signed P.L. 92-367 on Aug. 8, 1972, President Richard Nixon stated that he felt the Federal government should not inspect non-Federal dams. Both the Nixon and Ford administrations implemented that policy statement, and the Corps of Engineers was not provided funds to make dam inspections. However, funds were provided to permit the Corps to make an inventory of all dams 25 ft or more in height or impounding 50 acre-ft; conduct a survey of state safety practices relating to inspection and regulation of dams; develop (with the assistance of other Federal and state agencies and technical societies) recommended guidelines for the safety inspection of dams; and prepare a report to the Congress.

The inventory of dams provided the first reliable estimate of the total magnitude of the dam safety problem in the United States. The Corps found that 49,500 dams meet the criteria established by P.L. 92-367. Based on a sampling technique, it was estimated that about 20,000 dams were so located that their failure could cause serious losses to life and property. Of these, 9000 were estimated to represent a high hazard potential.

The 1974 Corps survey found that, with few exceptions, the states did not have effective programs to inspect and regulate dams. Ten states had no legislation to regulate dam building and operation, while many others had inadequate legislation or inadequate implementation of dam safety laws. In line with policy directives of P.L. 92-367, the Chief of Engineers report to Congress recommended that Federal agencies be authorized, funded, and staffed to provide continuing periodic inspections and surveillance of Federal dams, and to make such repairs and modifications as needed to ensure dam safety. The report also recommended that the state governments be encouraged to adopt effective inspection and regulation systems for non-Federal dams within their jurisdictions.

The next development in Federal activity relating to non-Federal dams came in August 1977, when Congress appropriated $15,000,000 for the Corps of Engineers to inspect such dams. This appropriation posed a policy problem to the Carter administration, since a position in regard to Federal inspection of non-Federal dams had not been determined. However, the Kelly Barnes Dam failure focused attention on this problem, and on Nov. 28, 1977, President Carter directed the Secretary of the Army to proceed with a program to inspect the estimated 9000 high-hazard dams. The President indicated that the Federal government would use this initiative to establish a partnership with the states in developing state programs. He further stated that the Federal effort would be limited to initial inspections only, would involve no assumptions of Federal liability, and would be completed in 4 years. The cost of this program was tentatively estimated at $70,000,000.

Within 2 weeks of the President's directive the Corps of Engineers had begun some dam inspection activity in all states, although severe winter weather hampered inspections in the northern states for several months. The Corps has attempted to secure active state participation in the program, as directed by the President, and 20 states have taken over management responsibilities for the inspections within their respective areas. Other states are participating in lesser degrees. The actual dam inspections are being made by Corps personnel, state employees, or private engineering firms under contract with the Corps or with individual states. It is anticipated that most of the inspections will be made by private firms.

After the first half-year of inspection activities, 549 dams had been inspected and 37 dams had been reported as unsafe to the respective state governors. Seven of the dams were considered to be in such unsafe condition that emergency measures, such as breaching the dam or draining the reservoirs, were recommended and carried out.

Future programs. Activities now under way in regard to safety of Federally owned dams should result in much more emphasis being placed on dam safety in the operations of a number of Federal agencies, and should effectively meet national needs in this sector. However, the national program for inspection of non-Federal dams has an extremely limited objective and will not meet the needs for effective inspection and supervision of the thousands of dams involved. A few states have met these needs with their own programs, but most states are still lacking in this respect. Legislation has been introduced to provide Federal assistance to the states in establishing and maintaining effective supervision programs. Until such programs are in effect in all states, dam safety will continue to be a perplexing public policy question.

For background information *see* DAM in the McGraw-Hill Encyclopedia of Science and Technology. [HOMER B. WILLIS]

Bibliography: Department of the Army, *National Program of Inspection of Dams*, vol. 1, May 1975; Federal Coordinating Council for Science, Engineering, and Technology, *Improving Federal Dam Safety*, Nov. 15, 1977; Federal Investigative Board, *Report of Failure of Kelly Barnes Dam, Toccoa, Georgia*, Dec. 21, 1977; Independent Panel to Review Cause of Teton Dam Failure, *Failure of Teton Dam*, Dec. 21, 1976; U.S. Department of the Interior Teton Dam Failure Review Group, *Failure of Teton Dam: A Report of Findings*, April 1977.

Dating methods

A new technique for determining gas residence times in sea water and lakes has been developed. The tritium-helium (^3H-^3He) dating method enables investigators to study the rates of physical, chemical, and biological processes in oceans and lakes from time scales as short as a few days to as long as decades.

Tritium-helium method. This dating method is based on the measurement of tritium, ^3H, and its stable daughter product ^3He in a water sample. Tritium, the heaviest isotope of hydrogen, is radioactively unstable and decays to ^3He with a half-life of 12.3 years. Tritium occurs in nature predominantly as part of a water molecule, so that it is almost ideal as a tag or label for water movements. Although it occurs naturally, most of the present-day ^3H was produced by the thermonuclear fusion reactions of the nuclear weapons testing of the 1950s and 1960s. As a result, environmental levels are quite variable and difficult to predict, so that one cannot perform simple ^3H dating, as is possible with ^{14}C dating.

The simultaneous measurement of ^3He not only solves the problem of variable ^3H levels, but enhances the sensitivity of the technique. Consider a water parcel in contact with the atmosphere. Although ^3H in this parcel continually decays and produces ^3He, no excess ^3He can build up because of gaseous exchange with the atmosphere. Once the parcel loses contact with the atmosphere, however, excess ^3He begins to accumulate, and the "clock" is then running. At some later time, the measurement of the ^3H and excess ^3He contents yields a tritium-helium age according to $\tau = 1/\lambda \log(1 + [^3He]/[^3H])$, where $1/\lambda$ is the mean life of ^3H (18.8 years).

The minimum "isolation" time which can be detected by this technique is limited by how small a ^3He excess can be measured and by the ^3H content of the sample. Obviously, the higher the ^3H concentrations, the greater the rate at which ^3He is produced and the shorter the minimum detectable time. For example, for ^3H-^3He dating in North Atlantic surface water ($[^3H] \sim 10$ tritium units, 1 T.U. $\equiv 1\ ^3H/10^{18}$ H), this minimum time is about 1 month. In the Great Lakes ($[^3H] \sim 100$ T.U.), times as short as a few days can be detected.

The maximum detectable time depends on more subtle factors. Primarily, the upper limit to the technique is governed by the transient nature of ^3H distribution. This limits the useful range of the ^3H-^3He method to the order of a decade or so. In addition, the interpretation of these older ages is more complex, since mixing nonlinearities come into play. That is to say, if two water parcels of differing ^3H concentrations were mixed together, the measured ^3H-^3He age would not reflect the true "averaged" age, but rather would be weighted toward the component of higher ^3H content. For mixing time scales less than, say, 5 years in the North Atlantic, these effects are negligible, but to extend the technique to decadal time scales requires caution in interpretation.

Analytically, the excess ^3He is determined by comparing the isotopic ratio of the dissolved He to atmospheric He by using a dual-collection, statically operated mass spectrometer. In this manner, the background of dissolved atmospheric ^3He (except for a small, predictable isotopic shift due to solubility effects) is "blanked out," and the excess ^3He is seen as an "isotope ratio anomaly." By this technique, excesses as small as 10^{-16} cm^3 (standard temperature and pressure), or about 3000 atoms, of ^3He per gram of water can be seen.

The ^3H can be measured by either of two techniques. The more conventional technique involves electrolytic enrichment of the water, conversion to a "counting gas," and counting in a shielded, low-level proportional counter for several hours. A more recent technique, developed by W. B. Clarke, W. J. Jenkins, and Z. Top, lends itself to the ^3He measurement technique. The water sample is degassed (that is, all the helium is removed), and the sample is stored in a vacuum for a period of time (usually a few months or more) to allow more ^3He to "grow in" by ^3H decay. This grown-in ^3He is subsequently reextracted and analyzed, thereby allowing a determination of the ^3H. The advantages of this recent technique are that it is inherently less susceptible to contamination and potentially more sensitive than the older techniques.

Application to oceanography. One of the major difficulties in descriptive physical and chemical oceanography is to establish the rates of processes. In other words, it is possible to measure and describe the spatial distribution chemical properties, but it is very difficult to ascertain the relative magnitude (that is, rates) of those physical, biological, and chemical processes which control the distribution. This difficulty arises from the fact that the observed distributions are in the steady state because the processes or "forces" all balance out. This is somewhat analogous to trying to judge the tone and quality of a bell just by looking at it.

The ^{3}H transient, that is, the bulge of ^{3}H produced by the nuclear testing of the 1950s and 1960s, can be of use in this regard. The ^{3}H generated was largely injected into the upper atmosphere, from which it "rained out" into the oceanic-hydrologic system. One can trace this ^{3}H "dye" as it first washes into the surface layers of the oceans and later bleeds down into the deeper waters. By this process, one can learn about the physical transport processes which affect the distribution of chemicals in the sea.

However, although this concept is simple, the actual procedure of mathematically extracting the processes from what is observed is rather complex, and often the results are too qualitative and ambiguous to be of great practical use. In addition, this approach requires a large data base and long sampling periods. Use of the ^{3}H-^{3}He dating method lessens many of these problems, and is accurate for much shorter time scales. Whereas studying the ^{3}H transient gives time scales of the order of decades, the ^{3}H-^{3}He method is applicable to time scales from a few days (in lakes) to a few months (in oceanic surface waters) to several years.

Tritium-helium results from the Geochemical Ocean Sections Study (GEOSECS) Expedition of 1972 were recently reported by Jenkins and Clarke, and revealed many interesting features about the North Atlantic. This represented the first substantial development of the technique, although some preliminary work was reported earlier by Jenkins and associates. However, the full potential of the technique was not demonstrated until Jenkins reported a study in the Sargasso Sea which used ^{3}H-^{3}He dating to determine the rate of oxygen utilization.

The oxygen clock is very similar to the ^{3}H-^{3}He clock. The concentration of oxygen in the upper layer of the sea is fixed by photosynthesis and gas exchange with the atmosphere. Once a water parcel leaves the sea surface (and sunlight), the oxygen begins to decrease due to biological consumption (that is, respiration). By correlating the ^{3}H-^{3}He age with the apparent oxygen utilization (namely, the deficit in oxygen relative to its original surface concentration), it is possible to obtain the oxygen utilization rate (see illustration).

While these results are of biological interest, they also have a much broader scope. If it is assumed that this consumption rate is constant for a given water type (that is, water of a given temperature and salinity from a particular area), then the oxygen clock, which has been calibrated by the ^{3}H-^{3}He clock, may be used. Dissolved oxygen measurements are easily taken and, in fact, are made routinely on most oceanographic cruises. Consequently, a large data base exists which can be used on a much broader scale.

Tritium-helium dating will eventually prove very useful in studying those physical processes which transport or redistribute substances in the oceans. This kind of information will prove vital in assessing the growing impact of humans on the oceans. For example, ^{3}H-^{3}He dating may help determine how rapidly the oceans can take up the enormous amounts of carbon dioxide produced by the burning of coal and oil, for it is the oceans which will ultimately absorb this waste product.

Application to limnology. In general, lakes evolve on a much faster time scale than the oceans. Consequently, one would expect that the limitations of the minimum isolation time would be more severe. Fortunately, because of their shorter

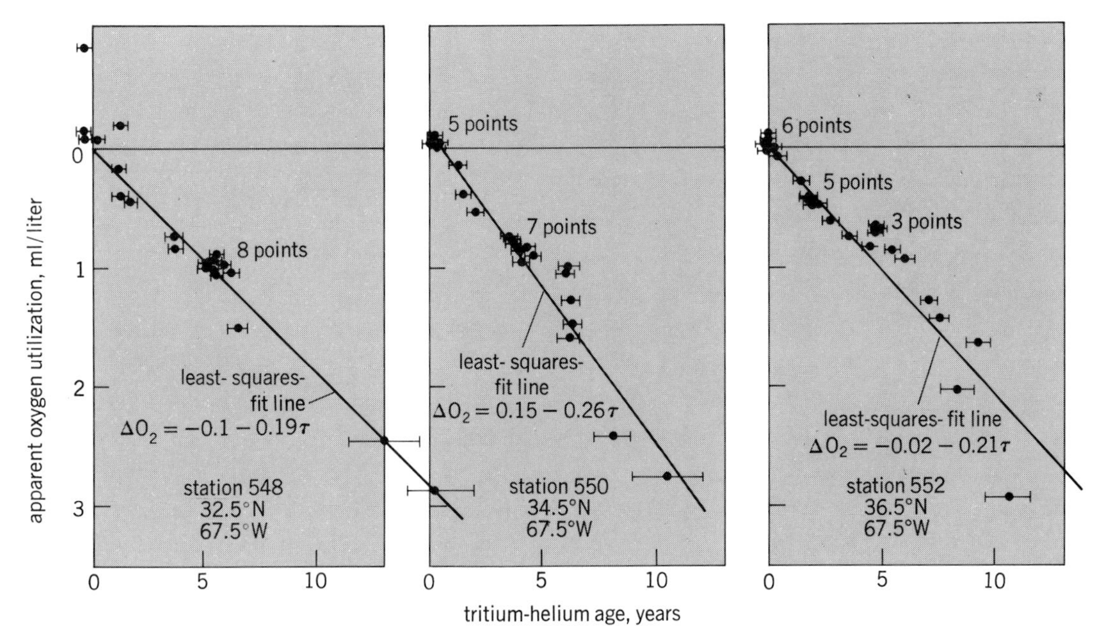

The ^{3}H-^{3}He age plotted versus the apparent O_2 utilization. *(From W. J. Jenkins, Tritium-helium dating in the Sargasso Sea: A measurement of oxygen utilization rates, Science, 196:291–292, Apr. 15, 1977, copyright © 1977 by the American Association for the Advancement of Science)*

residence times, lakes have felt the "bomb" ^{3}H transient more strongly, and exhibit much higher ^{3}H concentrations than the oceans. This reduces the minimum isolation time to the order of days, which is more than adequate for useful limnological studies.

The major advances in dating with regard to limnology have been initiated by T. Torgersen, who studied ^{3}H-^{3}He in a variety of lakes, including the Great Lakes. Because of the short minimum isolation time, the efficiency of gas exchange during lake turnover (destratification) and the rates of vertical diffusion could be examined.

For background information *see* DATING METHODS; HELIUM; TRITIUM in the McGraw-Hill Encyclopedia of Science and Technology.

[W. J. JENKINS]

Bibliography: W. J. Jenkins, *Science*, 196:291, 1977; W. J. Jenkins et al., *Earth Planet. Sci. Lett.*, 16:122, 1972; W. J. Jenkins and W. B. Clarke, *Deep-Sea Res.*, 23:481, 1976; T. Torgersen et al., *Limnol. Oceanogr.*, 22:181, 1977.

Deoxyribonucleic acid (DNA)

Recently the complete nucleotide sequence of a DNA virus, ϕX174, was reported. This virus is a small bacteriophage that infects *Escherichia coli* and contains single-stranded circular DNA as its genetic material, which is made up of 5386 nucleotides (1.7×10^6 daltons as single-stranded and 3.4×10^6 daltons as double-stranded DNA). From previous genetic and biochemical studies, it was known that ϕX174 coded for 10 proteins (Fig. 1). These viral proteins are labeled proteins A–H plus A′ and J. The largest protein, protein A, which is coded by the gene A coding region, has a molecular weight of 56,000 daltons and is an endonuclease that initiates viral DNA synthesis by cutting one of the strands of the viral DNA duplex at a specific site called the origin of DNA replication. Proteins F (48,000 daltons), G (19,000 daltons), and H (37,000 daltons) are the structural proteins that make up the mature virion. There are 60 copies of F, the capsid protein, 60 copies of G, and 12 copies of H, which together make up the 12 virion spikes of the icosahedral particle. Protein J (5000 daltons) is a small, highly basic protein which is present in the virion and appears to be involved in condensation and packaging of the DNA. Proteins B (19,000 daltons) and D (14,500 daltons) are morphogenetic proteins concerned with the assembly of the virion. Protein E (9000 daltons) is the lysis protein that breaks down the cell membrane, prior to release of the virus. The functions of proteins C (7000 daltons) and K (5000 daltons) are not known.

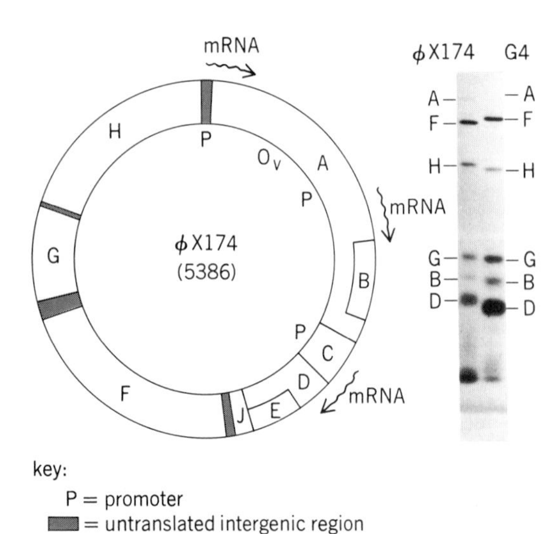

Fig. 1. Genetic map of bacteriophage ϕX174 and the proteins that it synthesizes. *(Adapted from B. G. Barrell, G. M. Air, and C. A. Hutchinson III, Overlapping genes in bacteriophage ϕX174, Nature, 264:34–41, Nov. 4, 1976)*

Discovery of overlapping genes. These protein molecular weights were derived from SDS polyacrylamide gel electrophoresis, which is accurate only to ±5%, but when the exact sizes of the proteins were determined from the nucleotide sequence (that is, from the number of nucleotides coding for each of the proteins), it became obvious that the sizes of the proteins added up to more than the total coding capacity of the DNA. This problem was resolved by the significant discovery of B. G. Barrell and coworkers that protein E was encoded entirely within the protein D coding region. This means that to code for protein E the nucleotide sequence of the gene D coding region is being read in different reading frames. Both the start and termination of protein E are within gene D. The discovery of this overlapping coding sequence was quickly followed by the recognition that coding sequence B is also overlapping and is entirely within the A coding region. This arrangement reduced the number of nucleotides required to code for all of the ϕX174 proteins to within the observed number of the total genome.

Evidence for overlapping genes comes from two types of experiments, genetic marker rescue experiments and nucleotide sequencing experiments. In the case of the gene E overlap, it was found that the restriction endonuclease DNA fragment Hae III-7, which was known from DNA sequence studies to be completely contained within the D coding region, could rescue genetic markers of not only gene D mutants but gene E mutants as well. This showed that this piece of DNA contained sequences that coded for both proteins D and E. This sort of evidence for overlapping genes was substantiated by examining the nucleotide sequence of the mutant DNA directly. In the case of the gene A/B overlap, for instance, it was found that a gene A mutant, amber 18 (am18), is caused by C → T transition which produces an amber termination codon in the reading frame of the A gene (Fig. 2). In the B gene reading frame of the A gene coding region, however, this same C → T change produces a change in the B amino acid sequence of an alanine to a valine, which apparently has no phenotypic effect. However, a revertant of am18, [caused by a G → C transversion which changes the phase A amber chain termination codon to a tyrosine codon and, in the B phase, converts a glutamic acid (GAA) to a glutamine codon (CAA)] causes, in addition to the alanine to valine change, the protein B product to be temperature-sensitive. These changes are shown in detail in Fig. 2.

Evolution of overlapping genes. When the ϕX174 overlapping genes were first described, they were greeted with skepticism because it seemed unlikely that evolutionary change could

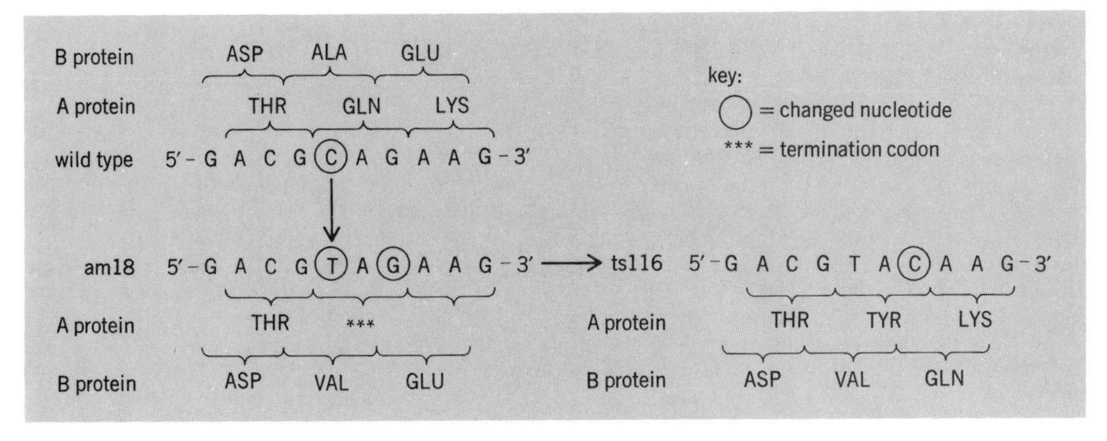

Fig. 2. Changes in the ϕX174 nucleotide sequence caused by am18 and its effect on gene A and B proteins.

proceed at a reasonable rate if two proteins were changed by each mutation rather than one protein. However, recently the complete nucleotide sequence of a bacteriophage called G4, which is closely related to ϕX174, has been established. This provides an opportunity to examine the evolutionary changes that are possible in overlapping gene regions of the DNA of two bacteriophages. A comparative nucleotide and amino acid sequence of the ϕX174 and G4 overlapping gene E coding regions is shown in Fig. 3. In the first two lines (40 codons) there is considerable conservation of both the nucleotide and amino acid sequence, but in the last three lines (52 codons) there are considerable differences. Out of a total of 273 nucleotides, 51, that is, approximately one in five, are different. This is somewhat below the average found in nonoverlapping gene regions, which is approximately one nucleotide in three, but when the top two lines of the gene E coding regions are examined, it will be seen that the conserved amino acids are all hydrophobic amino acids and, in particular, there are runs of leucine residues located in such a way that the protein has a strongly hydrophobic N-terminus. This is consistent with the function of the protein in lysing the cell membrane, and it is the very part of protein E that one would expect to be conserved between the two phages. The conserva-

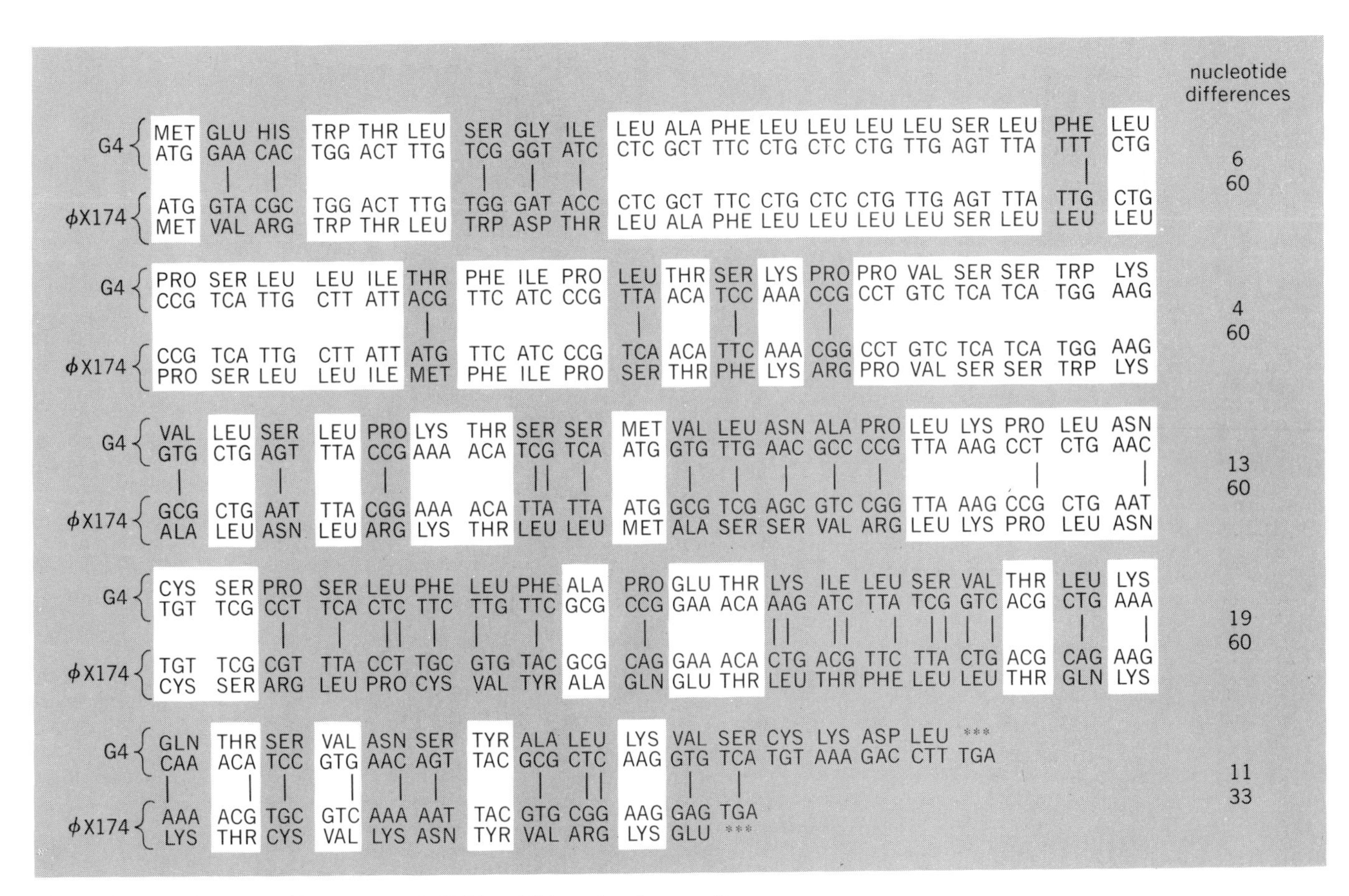

Fig. 3. Comparative nucleotide sequence of ϕX174 and G4 overlapping gene E.

tion of this part of the gene is therefore a result of conservation of function and not conservation because it codes for two proteins.

In the overlapping gene A/B regions, the nucleotide sequence differences between the bacteriophages φX174 and G4 are hardly different than in the surrounding coding regions which code only for a protein in one of these possible phases. It is reasonable to conclude, therefore, that the simultaneous use of the nucleotide sequence in the two reading frames is not as severe a block on evolutionary change as might be expected.

Overlapping gene prediction. The question as to why overlapping genes exist in only some parts of the genome was posed by Barrell, who defined the gene as an AUG (adenylic, uridylic, and guanylic acids) initiation codon (AUG coding for methionine is always the first triplet of a protein coding region) preceded by a nucleotide sequence complementary to some part of the 3′ end of the 16S ribosomal ribonucleic acid (rRNA) (that is, a ribosome binding site) and followed by an open phase of at least 20 codons. Twenty such sequences exist in φX174 which could code for new proteins of at least 20 amino acids in length. Thirteen such sequences exist in G4 and some of these are the same as in φX174. All of these new possible proteins would be small, of less than 100 amino acids, but such small proteins had hitherto been neglected. This led to a search for new small viral coded proteins, and a new overlapping gene was uncovered. This is now called protein K, which is coded as a second reading frame of part of genes A and C and starts where the overlapping gene B terminates. Gene K is unique in that not only is most of it coded as a second phase of preexisting genes but five nucleotides of its coding region are read as a third phase of preexisting overlapping genes. This is the first demonstration that the DNA nucleotide sequence can be used in all three reading frames simultaneously. Whether there are other overlapping genes in φX174 and G4 remains to be seen, and only now are very small proteins, less than 50 amino acids long, being examined. However, the detection of such small proteins is complicated by the fact that they may be regulatory proteins which are produced only in small amounts.

Untranslated noncoding intergenic regions. The main difference between small viral DNA genomes and large chromosomes is the presence in the large chromosome of large untranslated nucleotide sequences. These sequences are called repetitive DNA sequences, and are believed to have a regulatory role in chromosome organization and gene expression. The intergenic regions in φX174 and G4 are useful models for examining the function of these spaces. In φX174 there are 110 untranslated nucleotides between genes F and G, 63 between genes H and A, 36 between genes J and F, and 10 between genes G and H (Fig. 1). All of these intergenic spaces have nucleotide sequences which can form stable secondary-structure hairpin loops, with which control DNA sequences are often associated. For instance, the intergenic regions between φX174 or gene H and A contains a promoter (RNA polymerase binding sites) and the start of messenger RNA (mRNA), the main mRNA termination site and a ribosome binding site for the gene A. The hairpin loop in the intergenic sequence between genes J and F is an mRNA termination site, but the function of the two loops present in the large intergenic space between genes F and G is not yet known. G4 also has intergenic untranslated regions between these genes but, interestingly, they are different in size and very different in nucleotide sequence. However, they too can form stable secondary-structure hairpin loops which have functional control sequences associated with them. This suggests that it is the secondary structure of these regions that is important rather than the specific nucleotide sequence.

The untranslated intergenic regions, therefore, differ in size and vary considerably in nucleotide sequence. It is only a short step from this type of untranslated region to the structure of a large eukaryote DNA which has extensive intergenic untranslated regions. In the φX174 and G4 viral systems, however, untranslated regions can be genetically engineered and their relations to gene expression can be studied in a way that cannot yet be accomplished with higher-organism untranslated sequences.

For background information *see* CHROMOSOME; DEOXYRIBONUCLEIC ACID (DNA) in the McGraw-Hill Encyclopedia of Science and Technology.

[G. NIGEL GODSON]

Bibliography: B. G. Barrell, G. M. Air, and C. A. Hutchison, III, *Nature*, 264:34–41, 1976; G. N. Godson, *Virology*, 58:272–289, 1974; G. N. Godson, B. G. Barrell, and J. F. Fiddes, in preparation; F. Sanger et al., *Nature*, 265:687–695, 1977; M. Smith et al., *Nature*, 265:702–705, 1977.

Desertification

Desertification is the spread of desertlike conditions in arid and semiarid areas, due to human influence or climatic change. Natural vegetation is being lost over increasingly larger areas because of overgrazing, especially in Africa and southwestern Asia. Where the original state of vegetation has been retained, for example, in military compounds, the contrast is dramatic. In such a case, near Nefta in southern Tunisia, the coverage of vegetation inside an area fenced 60 years ago is 85%, in contrast to 5% outside the area; here the original dry steppe has changed into a semidesert, without any appreciable variation of precipitation (about 80 mm per year). Another dramatic example of this type is revealed in a NASA Earth Resources and Technology Satellite (ERTS) photograph of the Sinai-Negev desert region. The political boundary established in the 1948 armistice between these two regions is clearly visible, with the lighter Sinai region to the west of the boundary and the darker Negev region to the east. The sharp demarcation is due to the fact that Bedouin Arabs' goats in the Sinai have defoliated enough land on the Egyptian side to make the boundary visible from space. Similar conditions have been observed in India.

In many areas the deserts seem to be spreading, with an apparent speed of about one or more kilometers per year (the Sahara Desert appeared to be advancing into the Sahel—consisting of Mauritania, Senegal, Mali, Upper Volta, Niger, and Chad—during the 1968–1973 drought at the rate of 50 km

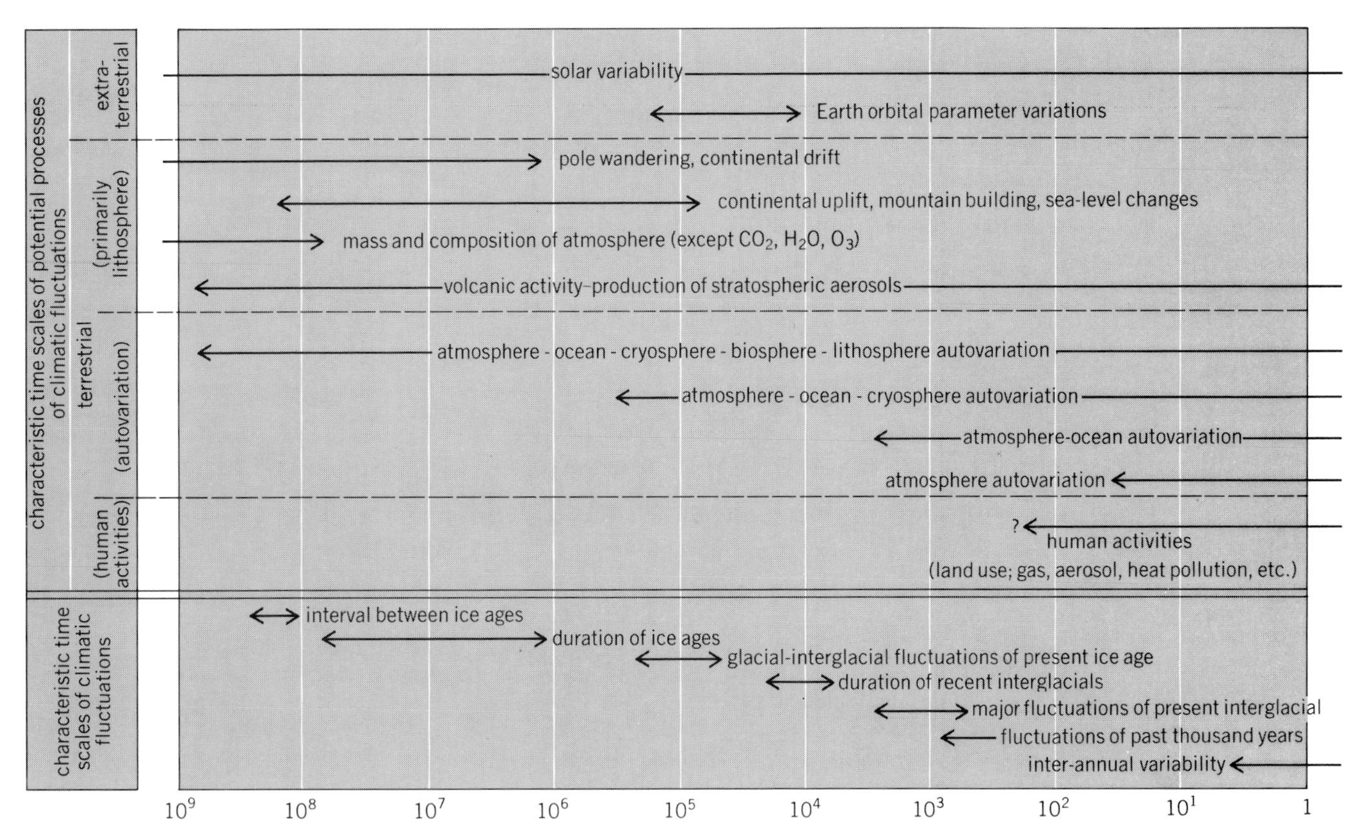

Processes that may cause climatic variation, plotted against their characteristic time scale. *(From United States Committee for the Global Atmospheric Research Program, Understanding Climatic Change: A Program for Action, National Academy of Sciences, Washington, DC, 1975)*

per year), depending on the density of the population, as a consequence of grazing animals (especially goats). However, recent desertification is the result of the interaction of naturally recurring drought with unwise land-use practices. In order to understand to what extent each of these factors plays a part in the total desertification process, and to consider whether the process can be slowed down or reversed, it is helpful to review natural and human causes of deserts.

Natural causes of deserts. The world distribution of dry climates depends mainly on the subsidence associated with the subtropical high-pressure belts, which migrate poleward in summer and equatorward in winter. These migrations are connected with the atmospheric general circulation. There results a threefold structure in the arid zone: a Mediterranean fringe, with rains occurring only in winter; a desert core (in about 20–30° latitude), with little or no rain; and a tropical fringe, with rains mainly in the high sun season. Throughout the arid zone, rainfall variability is high. Aridity arises from persistent widespread subsidence or from more localized subsidence in the lee of mountains. In some regions, such as the northern Negev, the combination of widespread and local subsidence is clearly visible as clouds on the windward side of mountains form and immediately dissipate on the leeward side. Aridity may also be caused by the absence of humid airstreams and of rain-inducing disturbances. Clear skies and low humidities in most regions give the dry climates very high solar radiation (averaging over 200 W/m^2), which leads to high soil temperatures. The light color and high reflectivity (albedo) of many dry surfaces cause large reflectional losses, and long-wave cooling is also severe. Hence net radiation incomes are relatively low—of the order of 80–90 W/m^2.

Climatic variation takes place on many time scales. The processes that may cause climatic variation, plotted against their characteristic time scale, are shown in the illustration, in which autovariation means internal behavior. The world's deserts and semideserts are very old, although they have shifted in latitude and varied in extent during geological history. The modern phase of climate began with a major change about 10,000 years before present, when a rapid warming trend removed most of the continental ice sheets. The Sahara and Indus valleys were at first moist, but since about 4000 years before present, when severe natural desiccation took place, aridity has been profound.

Recent climatic variations, such as the Sahelian drought, are natural in origin, and are not without precedent. Statistical analysis of rainfall shows a distinct tendency for abnormal wetness or drought to persist from year to year, especially in the Sahel. Prolonged desiccation, lasting a decade and more, is common and often ends abruptly with excessive rainfall. This persistence suggests that feedback mechanisms may be operating, whereby drought feeds drought and rain feeds rain.

Human causes of deserts. Recent desertification has resulted from the spatially uneven pressure exerted by humans on soil and vegetation, especially at times of drought or excessive rainfall.

The Dust Bowl in the Great Plains of the United States and the degeneration in the Sahel, the Ethiopian plateaus, and the Mendoza Province of Argentina are all manifestations of human misuse of the land at times of climatic stress.

A common mechanism of desertification includes the following steps: (1) expansion and intensification of land use in marginal dry lands during wet years, including increased grazing, plowing, and cultivation of new lands, and wood collection around new camps or settlements; and (2) wind erosion during the next dry year, or water erosion during the next maximum rainstorm.

This desertification process has the following climatological implications: (1) Increased grazing during wet years tends to compact the soil near water holes, and increased livestock numbers cause pressure on perennial plants during dry seasons. The result is to expose surface soil to erosion by wind. Runoff may be increased, and so may albedo (that is, reflectivity of the ground with respect to solar radiation). (2) Increased cultivation during wet years greatly improves the chances of wind erosion of fine soil materials during the dry season, and possibly increases evapotranspiration (in the case of water-demanding crops). (3) Removal of wood increases direct solar heating and may considerably decrease evapotranspiration. (4) In the ensuing dry years, the acceleration of wind erosion by the above processes further reduces water storage capacity by removal of topsoil. The loss of some or all perennial plant cover lowers infiltration rates and hence the potential percolation. During subsequent rains, surface runoff is increased, with an attendant loss of water for subsequent use by shrubs, herbage, or crops.

In order to test hypotheses of both natural and human-caused effects on the climate, one must have effective mathematical models of the dry climate.

Modeling attempts. In the mathematical model, the particular physical phenomenon to be studied is described in mathematical terms, and the equations are then solved by means of high-speed computers. In this way, predictions may be made of the effects of changes in external forcing, including the inadvertent or intentional changes caused by humans. Experiments thus far comprise:

1. General circulation models (GCMs) applied to the specific problems of the causes of climatic fluctuations in the dry climates. These models simulate in three dimensions the general circulation of the hemisphere or globe, and can be subjected to chosen perturbations, such as sea surface temperature anomalies, to determine the effects.

2. Investigations of specific feedback processes, such as changed albedo or variations of soil moisture storage. The models seek to predict the effect of feedbacks on circulation and precipitation over the arid zone. One recent experiment of this type was carried out in the United Kingdom. A model investigation tested the speculation that rainfall variations over north and central Africa were related to sea surface temperature anomalies over the tropical Atlantic. The experiment showed that introduction of an extensive sea surface temperature anomaly over the tropical Atlantic was related to higher precipitation amounts over North Africa. While the experiment did not actually prove the causal relationship, it changed the speculation into a credible hypothesis.

Most recent modeling attempts have been related to specific feedback processes that may augment or retard naturally induced climatic variations along the desert margin, especially the effect of changed albedo and other consequences of the degradation of vegetation cover.

The albedo feedback hypothesis argues that the destruction of vegetation and exposure of soil increase albedo and hence lower surface temperatures and suppress convective shower formation. This hypothesis has been considered to be a mechanism for desertification. Since certain regions of the central and northern Sahara, eastern Saudi Arabia, and southern Iraq have a negative radiation balance at the top of the atmosphere on hot summer days, in spite of the intense input of solar radiation through the cloudless atmosphere, the following argument has been advanced: Since the ground stores little heat, it is the air that loses heat radiatively. In order to maintain thermal equilibrium, the air must descend and compress adiabatically. Since the relative humidity then decreases, the desert increases its own dryness. A biogeophysical feedback mechanism of this type could lead to instabilities or metastabilities in borders themselves, which might conceivably be set off or maintained by anthropogenic influences.

This hypothesis has been tested by means of a dynamical model, in which the surface albedo was increased over large desert regions. Sharp reductions of cloud and rain followed the increase in albedo. Possible mechanisms by which albedo is changed are removal of vegetation by drought, overstocking, cultivation, or all of these, or by desiccation of the soil itself, soil albedo being related to soil water content. In practice, the three mechanisms are likely to occur simultaneously, so that soil moisture content may itself serve as a positive feedback for drought, wet soil favoring renewed rainfall derived from local evapotranspiration.

All of the above hypotheses work by influencing the overall dynamics of the desert margin climates, essentially via their effect on rates of subsidence and hence stability. There is also the possibility that cloud microphysics may be affected by surface conditions. It has been suggested that cumulus and cumulonimbus clouds of the Sahelian and Sudanian belts of western Africa are "seeded" by organic ice nuclei raised from the vegetated surface below. Removal of the vegetation destroys the local source of organic nuclei. Hence, well-formed clouds remain unseeded and rainless, thereby accelerating the decay of surface vegetation.

In summary, modeling experiments suggest that there may well be a positive feedback process along the desert margin, operating via the increase of surface albedo attendant upon the destruction of the surface vegetation layer, and possibly also upon the decrease of soil moisture and surface organic litter. It thus appears possible that widespread destruction of the vegetation cover of the dry world may tend to further reduce rainfall over these areas.

Possible solutions. Suggestions for solution of the desertification problem include using techniques for weather and climate modification, control of surface cover, and maximum application of modern technology.

Weather and climate modification. Cloud seeding, establishing green belts, and flooding of desert basins are a few of the weather and climate modification techniques under consideration.

Special conditions are required for rainfall augmentation by cloud seeding. If the proper type of seedable cloud exists, it may be possible to increase the rainfall locally by this method.

Establishment of green belts along the northern and southern margins of the Sahara is considered to be of dubious value climatologically, as desertification does not spread outward from the desert. Thus the green belt would not serve as a "shelter belt." However, this hypothesis should be tested by means of a model.

Precipitation depends largely on water vapor which has traveled great distances, along with upward motion of the air. The experience with artificial oases to date has shown little or no change of climate in their vicinity. Modeling experiments carried out by simulating the "flooding" of Lake Sahara predicted no significant change of rainfall around the lake, although rainfall was increased over an isolated mountain region 900 km from the shore. However, these results do indicate the possibility of creating artificial bodies of water judiciously with respect to the areas where rainfall augmentation is desired. This hypothesis may also be tested by model experiments.

Control of surface cover. The key to the control of the desertification process is the control of surface cover. If the relatively secure wind-stable surfaces of some desert areas, or a reasonably complete vegetation cover (even if dead) can be maintained, soil drifting and deflation are minimized. Overstocking, unwise cultivation, and the use of overland vehicles weaken and ultimately destroy these protective covers. The proposed green belt is of value because of the added protection it affords the soil.

The usefulness of land depends on the surface microclimate. The ability to conserve this microclimate rests on transformation of desert technology, rather than on transformation of climate. The ability of the surface to respond quickly and generously to renewed abundance of rainfall depends on the soil's capacity to retain nutrients, organic substances, and fine materials; on high infiltration capacity; and, of course, on viable seeds as well as a surviving root system. A surface litter of organic debris may also be important for precipitation mechanisms, and has some effect on surface radiation balance.

Use of modern technology. Satellite data could be used in tracking the major rainstorms of the rainy season to study their habits and, if possible, to predict their displacement. Other important satellite data, for example, radiation, could be collected and used in connection with feedback studies.

Future research. More research is needed into the relationship between climate and the desertification process. There should be a concerted international research effort, with major attention focused on deserts as well as oceans, and on surface physical and bioclimatic processes as well as atmospheric dynamics.

Further experimentation is needed that specifically examines the problems of the general circulation of the Earth's atmosphere, and even of the smaller-scale circulation, over the dry land areas of the subtropics. There is also a need for more detailed study of the physical climatology and bioclimatology of dry land surfaces, and particularly for a closer synthesis of climatology with geomorphology, soil science, hydrology, and ecology.

For background information *see* CLIMATE, MAN'S INFLUENCE ON; CLIMATIC CHANGE; DESERT EROSION FEATURES; DESERT VEGETATION; DROUGHT in the McGraw-Hill Encyclopedia of Science and Technology. [LOUIS BERKOFSKY]

Bibliography: M. H. Glantz (ed.), *The Politics of Natural Disaster*, 1976; F. K. Hare, T. Gal-Chen, and K. Hendrie, *Climate and Desertification*, Institute for Environmental Studies, University of Toronto, 1976; Massachusetts Institute of Technology, *Inadvertent Climate Modification: Report of the Study of Man's Impact on Climate (SMIC)*, 1971; S. H. Schneider and L. E. Mesirow, *The Genesis Strategy*, 1976.

Devonian

During earliest Devonian time a worldwide lowering of sea level produced widespread marine regression on all continents, which were in a state of structural (orogenic) quiescence. Marine animals evolved rapidly in separate areas, forming distinct biogeographic realms. Renewed sea-level rises, beginning before Middle Devonian time, resulted in transgression of epicontinental seas in two pulses, only one of which was felt in the interior of Gondwanaland. Orogenic deformation on two sides of North America accompanied transgression of continental interiors in Middle and Late Devonian time. The continued spread of epicontinental seas, except in Gondwanaland, allowed unrestricted migration and competition which eventually caused elimination of biogeographic realms. Many animal groups died out near the end of Middle Devonian time for this reason. Other animal groups successfully colonized vast regions of the epicontinental seas only to become extinct within the Late Devonian, perhaps because of a rapid drop in sea level that altered all marine habitats.

Boundaries and nomenclature. The Silurian-Devonian boundary has been fixed, by international agreement, at an actual outcrop of sedimentary strata in Czechoslovakia, where it corresponds to the base of the *Monograptus uniformis* graptolite zone. The Devonian-Carboniferous boundary has not yet been similarly fixed, but many stratigraphers employ the base of the *Siphonodella sulcata* conodont zone as the base of the Carboniferous and the top of the Devonian. The Lower, Middle, and Upper Devonian series boundaries have not yet been precisely defined, but are under study by an international stratigraphic subcommission.

For many years German and Belgian stage names have been widely employed as Devonian time-stratigraphic units except in eastern North America, where stage names based on the New York rock succession have been in use. Defects relating to utility of the German Lower Devonian stages, combined with growing acceptance of Czechoslovakian stage names, has led to a recom-

	stages	conodont zones
Devonian	Famennian	*praesulcata*
		costatus
		styriacus
		velifer
		marginifera
		rhomboidea
		crepida
		Pal. triangularis
	Frasnian	*gigas*
		Ancy. triangularis
		asymmetricus
	Givetian	*herm.-cristatus*
		varcus
		ensensis
	Couvinian	*kockelianus*
		costatus costatus
		patulus
	Dalejan-Zlichovian	*serotinus*
		inversus
		gronbergi
	Pragian	*dehiscens*
		sulcatus n. subsp.
		sulcatus
	Lochkovian	*pesavis*
		Ozark. n. sp. D
		eurekaensis
		hesperius

Devonian stages and conodont zones.

mendation that the Czech names Lochkovian, Pragian, Zlichovian, and Dalejan represent the Lower Devonian and that the Belgian names Couvinian, Givetian, Frasnian, and Famennian represent the Middle and Upper Devonian (see illustration).

Although evolution of ammonoid cephalopods has long provided a zonal reference standard for most of the Devonian, taking up when the graptoloid graptolites disappeared, the rapid growth of a precise, time-significant conodont zonation of worldwide utility has shown that conodonts are the most useful Devonian biostratigraphic tools.

Paleogeography. The Devonian world comprised a newly sutured low-latitude Euramerica (consisting of Europe and North America), an already-old Gondwanaland (consisting of South America, Africa, Antarctica, and possibly Australia), situated mostly in southern high-latitude and polar regions, and a still-disjunct group of Asian plates.

The North American and Russian continental platforms, comparable in age, were sites of broad epicontinental seas which evolved through harmonic transgressive-regressive cycles to produce marine sedimentary rock sequences. These cycles were probably of eustatic origin. They include widespread regression, inherited from the Silurian and ending in mid–Early Devonian time, above the Lochkovian stage (see illustration), followed by transgression in two phases (Pragian to early Givetian and mid-Givetian to mid-Famennian) which ended in high eustatic standstill, with geographically stable shorelines. In North America, deposition patterns of sedimentary rocks and the resultant marine environments that controlled most Devonian life were enormously affected by orogenic highlands of the Antler-Acadian orogeny.

The interior of Gondwanaland had undergone structural consolidation at an earlier time, perhaps as long as 250,000,000–300,000,000 years prior to the time that Euramerica reached the same stage of development. Consequently, by Devonian time much of Gondwanaland had attained paleoelevations which were too high to be flooded by widespread epicontinental seas controlled only by rising sea level. Paleogeographic maps for the South American continent strikingly illustrate the dearth of any broad epicontinental seas there. Apparently, the marine seaways that did cross parts of Gondwanaland during the Devonian followed narrow downwarpings caused by structural (epeirogenic) changes.

Euramerica. Biogeographic realms, set up first as provinces based on the distribution of brachiopod genera, have not been significantly revised by recent work. These include an Eastern Americas Realm (EAR, formerly the Appalachian Province) in eastern and central North America and northern South America (Colombia and Venezuela) and an Old World Realm in western and arctic North America, Europe, Asia, and North Africa. Recent work on ostracods, trilobites, and corals essentially adds to the data base which describes these animal distributions.

It seems clear that the origin of the EAR is due to isolation afforded by adjacent land barriers. Provinciality of brachiopods, ostracods, trilobites, and corals ended with the Taghanic onlap, during mid-Givetian to early Frasnian time (see illustration), which allowed easy faunal migration across the transcontinental arch dividing North America.

Analysis of the changing environmental parameters that existed during the Early Devonian in Nevada—while that region was changing its faunal affinity from Old World to Eastern Americas and back to Old World—reveals that EAR faunas entered the faunally rich central Nevada area during the major regression at the end of the Lochkovian, and were supplanted by Old World faunas again during the first Devonian transgression, beginning in Pragian time. This suggests that, in addition to physical barriers, Devonian realm boundaries were primarily biofacial (environmental) in nature, with EAR and Old World faunas representing epicontinental and marginal-continental biofacies, respectively. Biofacies, physical barriers, and climate, acting together, account for the major biogeographic marine units of the Devonian, which are therefore similar to the trilobite biofacies of the Cambrian.

Long-held ideas relating regression of epicontinental seas to extinction of marine invertebrates have recently been developed as workable models to help explain environmental stress (during regression) and adaptive radiation (during transgression). Several transgressive-regressive or transgressive-standstill episodes are now evident during the Devonian. The transgression inceptions are as follows (roman numerals refer to depositional phases):

Ia. McColley-Oriskany (dated as *sulcatus* conodont zone in Nevada; see illustration).

Ib. Denay-Simonson-Onondaga-Couvinian (*patulus* or *costatus costatus* conodont zones).

Ic. *Castanea* Zone–Centerfield (dated as lower *varcus* conodont zone).

IIa. Taghanic-Frasnian (begins in middle *varcus* conodont zone).

Because of Taghanic onlap, competition between animals in analogous communities of the once-separate realms probably was responsible for many end-of-Givetian extinctions. This is reflected in much lower census totals for Frasnian brachiopod genera. End-of-Frasnian extinctions that followed had an even larger effect on brachiopods and corals, such that Famennian and early Mississippian coralline buildups are very rare.

Gondwanaland. The Gondwanaland supercontinent was fringed in the Devonian by epicontinental seas in North Africa and northern South America (Colombia and Venezuela) that contained Euramerican marine faunas. At its other extremity, and at similar low latitudes, Gondwanaland also harbored Euramerican faunas in Australia, particularly in the Middle and Upper Devonian basins of Western Australia and in the Tasman geosyncline along Australia's present east coast. At higher latitudes, and even near the south Devonian pole in southern Africa, marine faunas of Gondwanaland were established as a cold-water cratonic biofacies of monotonous and low-diversity composition. These faunas, constituting the Malvinokaffric Realm, inhabited shallow seaways, none of which reached far into the interior of Gondwanaland.

The greatest record of marine sedimentation in the Malvinokaffric Realm is of late Early and early Middle Devonian age. The Givetian-Frasnian record is much less extensive, although this was the time of the greatest transgression on the Euramerican platforms, that is, by the Taghanic onlap. This anharmonic relationship tends to confirm the suggestion that the Malvinokaffric faunas inhabited seaways that formed primarily due to structural rather than eustatic causes.

Lower Devonian brachiopods of North America belong to one of two major biofacies: a *Gypidula-Atrypa-Schizophoria* biofacies and an acrospiriferid-leptocoeliid biofacies. These represent offshore and onshore environments, respectively, and it is significant that the Malvinokaffric brachiopods compose a fauna which is a remarkable homolog of the acrospiriferid-leptocoeliid biofacies. All three of the nominal taxa of the offshore *Gypidula-Atrypa-Schizophoria* biofacies are absent from the Malvinokaffric fauna. One hypothesis explaining the absence of these taxa assumes them to have inhabited only warm waters, and that the Late Devonian extinctions of all atrypid and gypidulid brachiopods, as well as other animal groups, was due to the influence of cold waters periodically invading Devonian equatorial belts. However, the *Gypidula-Atrypa-Schizophoria* biofacies represents a deeper water environment which must have been colder than the shallow, onshore acrospiriferid-leptocoeliid biofacies of the EAR, which the Malvinokaffric fauna most closely resembles. Therefore, although the Malvinokaffric faunas inhabited cold waters, temperature was evidently not a primary factor controlling their makeup. It is probable, however, that the large size of many Malvinokaffric invertebrate animals was a response to low temperature.

Recent paleoecological studies bear out the nearshore, relatively shallow-water characteristics of almost all Malvinokaffric benthic marine animal associations. A different view would distribute the known Malvinokaffric benthic marine associations out to maximum depths of the most offshore communities known elsewhere in the Silurian and Devonian. Apparently, benthic marine animals of the Devonian had not yet successfully colonized the deeper oceanic environments. Instead, they were confined to the bottoms of epicontinental seas and upper regions of the continental slopes. *See* PALEOCEANOGRAPHY.

For background information *see* ANIMAL EVOLUTION; BRACHIOPODA; CONODONT; DEVONIAN; EXTINCTION (BIOLOGY); PALEOECOLOGY; PALEOGEOGRAPHY in the McGraw-Hill Encyclopedia of Science and Technology. [J. G. JOHNSON]

Bibliography: M. R. House, *Proceedings of the Yorkshire Geological Society*, vol. 40, pt. 2, pp. 233–288, 1975; J. G. Johnson, *Geol. Soc. Amer. Bull.*, 82:3263–3298, 1971; J. G. Johnson, N. L. Penrose, and M. T. Wise, *J. Paleontol.*, 52: 793–806, 1978; W. A. Oliver, Jr., *Palaeogeogr. Palaeoclimatol. Palaeoecol.*, 22:85–135, 1977.

Dog

Dogs *(Canis familiaris)* are a unique domesticated animal in many respects. It is generally agreed that dogs were domesticated from wolves *(C. lupus)*. This domestication occurred earlier than with any other animal, and recent finds extend the date of fully domestic dogs back about 12,000 years. Another characteristic that distinguishes dogs from other domestic animals is the great variety of roles that dogs, in a wide assortment of breeds, have performed in human culture. The social behavior and genetic plasticity of wolves made possible this early alliance and continued association of domesticated dogs with humans.

Early remains. Remains of domestic dogs have been identified from widely separated locations as early as the 7th millennium B.C. The earliest of these finds, dating about 10,000 B.C., is from Palegawra Cave in northeastern Iraq. Other remains of domestic dogs were found in Jaguar Cave in Idaho, and these are associated with deposits that date to 10,000 years or more before present. Dog remains that date at about 9000 years have been found in such widely separated places as Illinois, Japan, England, Greece, Crete, and Turkey. These finds indicate that not only have dogs been associated with people for at least 12,000 years (and probably a great deal longer), but they have also accompanied people during their dispersal throughout the world.

It is generally believed that the wild ancestor of the domestic dog was the relatively small Asian

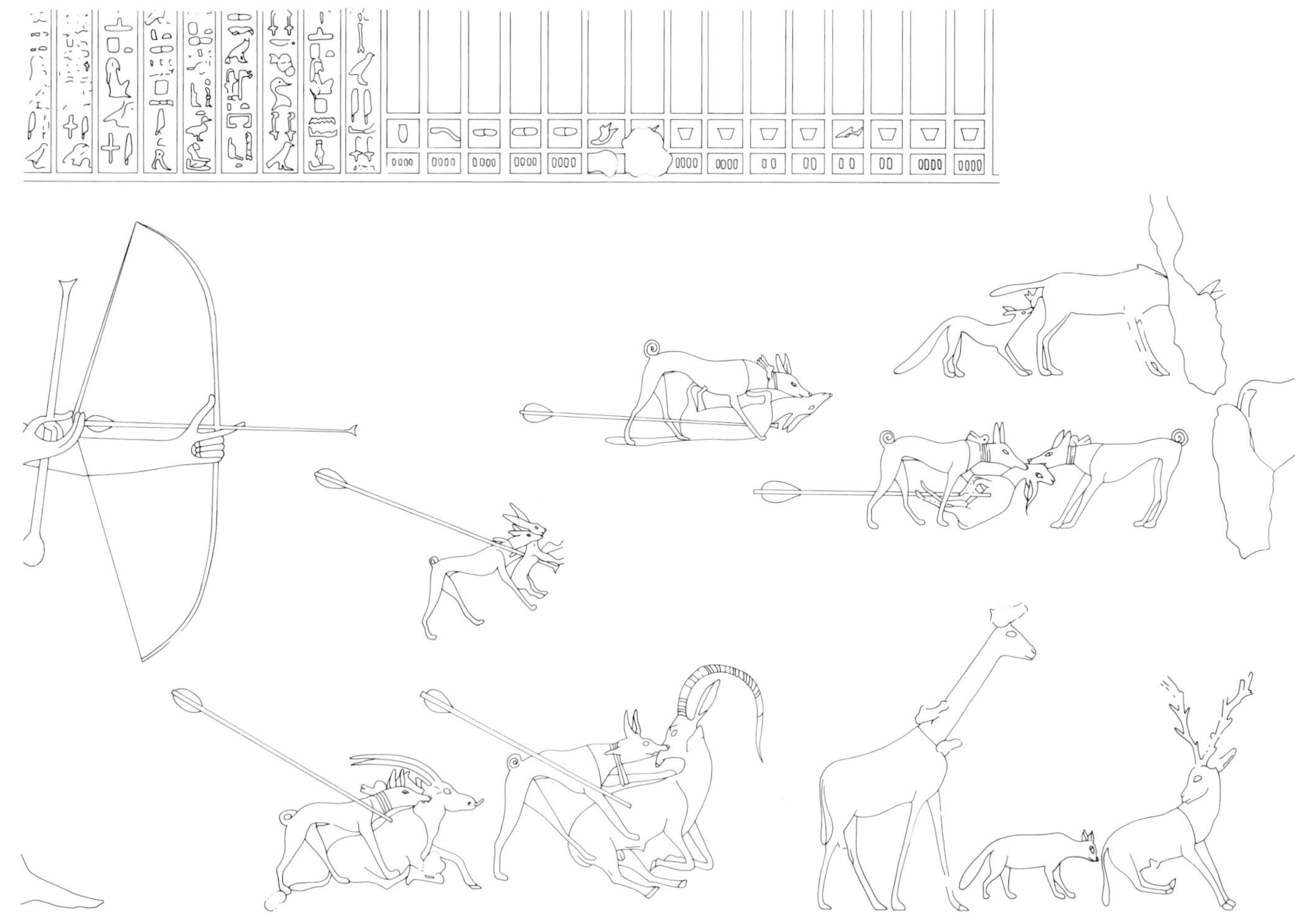

Fig. 1. Hunting scene from an Egyptian tomb, with dogs participating in the capture of a variety of game. Note that the dogs with collars all have curly tails. *(From W. J. Darby, P. Ghalioungui, and L. Grivetti, Food: The Gift of Osiris, Academic Press, vol. 1, 1977)*

wolf. The close similarities between the morphology and behavior of dogs and wolves support this belief. Dogs are able to interbreed and produce fertile offspring with wolves as well as with other members of the genus *Canis* such as coyotes *(C. latrans)* and jackals *(C. aureus)*. Such crosses have probably occurred repeatedly throughout the existence of the domesticated dog and have contributed to the enormous variability of the animal. It has also been suggested that wolves were domesticated more than once, which would also contribute to the variability seen in dogs. Evidence for such hybridization during prehistoric times comes from animal remains excavated from the neolithic Vlasac Site in Hungary. Measurement of teeth and jaws of the canids from this site are interpreted as including wolves, dogs, and transitional individuals in the Vlasac fauna.

A convincing case has been made for the domestication of another canid, the Falkland Island wolf *(Dusicyon australis)*, or Aguara dog as it is also called. Differences between this now extinct wolf and other members of the genus *Dusicyon* are comparable to those seen in dogs. The characteristics associated with both animals are white markings on the coat, a wide muzzle with somewhat crowded teeth, and a prominent forehead caused by the expanded frontal bones. An early account, referring to these animals as the indigenous dogs of the American Indians, states that the animals were gradually replaced by domestic European dogs and finally died out about 1880.

Selection. There are so many varieties of dogs that it is difficult to distinguish the characteristics which set them apart from other animals except for their continued association with humans. Physical characteristics described above—shortening of the muzzle accompanied by crowding of the premolars and the prominent forehead—as well as a decrease in overall size, are frequently seen in the remains of prehistoric dogs. Behavioral characteristics for which there is scant evidence, but which were probably eliminated early in the domestication process, are those associated with displays of dominant or aggressive behavior in wolves. Aggressive behavior is signaled by pricking the ears, erecting the hair of the shoulders, and raising the tail. Many modern breeds of dogs cannot display these aggressive signals because they have long droopy ears, do not have a shoulder ruff, and have curled tails. It seems likely that early in the association between people and tamed wolves the less dominant animals, that is, those less apt to give strong aggressive signals, were favored. The only evidence available to support the hypothesis that these characteristics occur in early dogs is that one or more of these traits are seen in the early human representations of dogs (Fig. 1). White markings or spots are frequently seen in domesticated animals, including dogs. A spotted coat pattern is evident on some of the early drawings of dogs.

Modern dog breeds are the result of selection over generations for structural and functional characteristics that satisfy certain breed standards. This selective breeding brings out innate patterns that are deemed desirable and suppresses others; in extreme cases genetic anomalies such as dwarfism and gigantism are maintained through selection. Breed standards are established to achieve an animal best suited for a particular role, such as a lap dog as opposed to a guard dog, or to create a particular appearance, such as breeding the Pekingese to resemble the spirit-lion of Buddha. Although modern selective breeding is carried out with a knowledge of genetics, selection, either purposeful or accidental, also occurred in prehistoric times, resulting in distinct types of dogs. During Archaic times in the New World and the Neolithic in Europe, dogs were largely uniform in size. From the millennium prior to the Christian Era there is an increase in the variety of dogs with respect to size and proportions of the skull and postcranial skeleton. It is likely that these variations in stature were accompanied by other differences such as coat color, hair length and texture, and ear length.

This observable variation in types of dogs may have resulted from the diversity in the roles dogs played in human society. For example, at the time of conquest of the New World, Indians had dogs that were fattened for feasts, and some whose hair was plucked for use in textiles. Based on archeological remains alone, it is often difficult to interpret what their roles might have been. However, two different types of deposits of dog remains suggest different values placed on dogs: deposits indicating intentional burial, and others indicating use as food.

Dog burial. Dog burials are found in most parts of the world, and date from several millennia before Christ to the present. It is not known precisely what motivated different people to bury their dogs during prehistoric times. Burial was afforded few other animals, and for no other animal is the practice so widespread in time and place.

The earliest dog burials in the New World (from 7000 to 7400 years ago) are found in the Rodgers Shelter in Missouri and the Koster Site in Illinois. Numerous other Archaic dog burials have been found in Kentucky, Alabama, and Florida. Four dog burials, dating between 2500 and 1750 B.C., were recently found in a site on the coast of Ecuador. Contemporary with these burials are several others found in England. Approximately 3000 years later the burial of dogs by the European colonists in the southeastern United States was found. Three burials of Spanish and English dogs in southern Georgia and St. Augustine, FL, are of uniformly large, robust animals, lending support to the theory that large dogs were brought to the Americas to subdue the Indians.

A distinctive series of dog burials has been excavated from burial mounds on the Pacific Coast of Mexico in the states of Sinaloa and Nayarit. The association of these dog burials with human burials makes them unique. Farther south in the state of Colima pottery representations of dogs are associated with human burials (Fig. 2). There are accounts documenting the belief, at the time of the Spanish conquest of Mexico, in life after death and the role of dogs in ensuring that their masters reach a peaceful resting place. Many such beliefs endured for a long time, so it is conceivable that dogs associated with human burials dating from about A.D. 500 to 1000 were sacrificed for use by their masters in the afterlife.

Association with food remains. At the same time, in some parts of Mexico, particularly the Gulf Coast, dogs were raised for food. Fray Diego Dur-

Fig. 2. A reproduction of a Colima dog holding an ear of corn in its mouth.

an, writing in the 16th century, describes a market near Mexico City in which 400 dogs were being offered for sale as food. This is not a unique use of dogs. Dog meat has been eaten by people in a number of places in the New World as well as in Africa, China, and Southeast Asia. Evidence for the consumption of dog meat in antiquity comes from finds of fragmentary dog bones associated with the remains of other animals that were used for food. Some of these remains, excavated from archeological sites, have butchering marks or charred portions, lending further support to the interpretation of these remnants as food.

Dogs are carnivores, that is, they are grouped taxonomically with meat-eating mammals; however, they can subsist on a diet composed largely of plants. A number of new techniques, such as analysis of the trace-mineral strontium in the bone apatite crystal and analysis of the ratio of the isotopes of carbon-12 and carbon-13, have shed light on the amount of meat and plant foods in the dogs' diets. All of the Mexican dogs studied by using these new techniques indicate that their diet was composed almost entirely of plant foods, primarily corn. It thus appears that rather than being carnivorous, competing with their owners for scarce meat resources, dogs were herbivorous and had a diet similar to many other domesticated animals.

The implications of the use of corn-fed dog meat are particularly important in the context of prehistoric Mexico. Mexico is one area in which a complex society developed. Turkey and dog were the only two important domestic animals used by that society. The subsistence use of dogs by the Olmec, Maya, and Aztec Indians may be more important than has been realized.

Summary. During the long history of association of dogs and humans many different breeds have evolved and disappeared, just as present-day breeds are ephemeral and will disappear as soon as they are not actively maintained. The uses to which dogs have been put are likewise varied, and reflect the importance of this animal to human society.

For background information *see* DOGS AND ALLIES in the McGraw-Hill Encyclopedia of Science and Technology. [ELIZABETH S. WING]

Bibliography: S. Bökönyi, Vlasac: An early site of dog domestication, in A. T. Clason (ed.), *Archaeozoological Studies*, pp. 167–178, 1975; R. Burleigh et al., *J. Archaeol. Sci.*, 4(4):353–366, 1977; J. Clutton-Brock, *Science*, 197:1340–1342, 1977; W. J. Darby, P. Ghalioungui, and L. Grivetti, *Food: The Gift of Osiris*, vol. 1, 1977; H. Epstein, *The Origin of the Domestic Animals of Africa*, vol. 1, 1971; Fray Bernardino de Sahagun, *General History of the Things of New Spain* (transl. by A. J. Anderson and C. E. Dibble), 1950; P. F. Turnbull and C. A. Reed, *Fieldiana Anthropol.*, 63(3): 81–146, 1974.

Earth, heat flow in

The importance of groundwater circulation in transporting heat and minerals has only recently been recognized. Sea-floor studies have shown that hydrothermal circulations of sea water near the mountainous mid-ocean ridges have a number of important effects, including the chemical modifications of the crustal rocks which can create new minerals and the concentration of already existing minerals into deposits. A similar process occurs on the continents, where thermally driven groundwater circulations are responsible for concentrating many deposits of minerals. Without this process, most mining would be impractical. Finally, groundwater circulation is vital in creating steam reservoirs that may be tapped for geothermal power plants.

Hot springs. The near-surface circulation of groundwater is common to many areas. These circulations are usually driven by variations in the altitude of the water table. When the water table reaches the surface, springs result. A common occurrence is for rainwater to percolate into the soil along a ridge, migrate under the influence of gravity through the layer of permeable rock such as limestone or along a fault, and appear as a spring in a nearby valley.

In some cases these groundwater circulations in stable continental areas penetrate to depths of several kilometers, so that the water is significantly heated. When the water emerges, it creates hot springs (water temperatures greater than 38°C) which may range from the relatively mild, 38°C springs of the Virginias and Georgia to the boiling springs of the Great Artesian Basin in Australia.

However, the vast majority of hot springs occur not in stable continental areas but in regions of active mountain building and volcanism, such as the western United States. Unlike the hot springs previously mentioned, these springs are believed to be associated with cooling magma bodies at depth. With this type of hot spring, the motion of the circulating water is not caused as much by the altitude of the water table as it is by a process called hydrothermal convection. This occurs when the magma body heats the surrounding rock and the groundwater circulating through it. Because heated water is less dense, it rises to the surface. If the heat given off by the magma body is sufficient, steam is generated which forms geysers if it reaches the surface. The steam geysers in Yellowstone National Park are good examples of this process. The Yellowstone area was the site of massive volcanic activity about 600,000 years ago, and smaller eruptions have probably continued up to the present. Seismic and gravity-measurement evidence indicate a large magma body several kilometers below the park which is large and hot enough to stimulate hydrothermal convection.

These groundwater circulation systems above cooling magma bodies are primary sites for geothermal power developments. Such a plant has been built in the Geysers area north of San Francisco. There, wells reaching depths of only a few thousand feet tap steam reservoirs, and the steam is fed directly into turbines which generate the electricity. The plant has an installed generating capacity of over 500 MW.

Mineral deposits. The important role of hydrothermal circulations in generating continental economic mineral deposits (ones that can be profitably mined) is a recent discovery. Studies by Hugh Taylor of the concentrations of oxygen isotopes in the older rocks adjacent to newer, cooled magma bodies show that large quantities of groundwater circulated through the older rocks during the cooling of the magma that intruded them. Studies of the fluid inclusions in rocks associated with many mineral deposits show that boiling of the groundwater occurred during the mineral deposition. As groundwater was heated near a magma body, the solubility of many minerals increased and the minerals were leached from the surrounding rocks. When boiling occurred, the minerals were precipitated, since they had near zero solubility in steam. This resulted in a local concentration of minerals and deposits that can be economically mined. Examples of minerals deposited in this manner include copper, zinc, tin, molybdenum, gold, and silver.

Hydrothermal circulations in the sea floor. The mountainous mid-ocean ridges which lie on the sea floor are areas of extensive volcanic activity. At these volcanic ridges new oceanic crust is created from solidified magma. Since most of the ridge system lies below sea level, the ocean water rapidly cools the hot magma. This rapid cooling of the brittle magmatic rock creates large-scale fracture systems, which allow sea water to percolate through the oceanic crust to considerable depths. There is evidence that sea water not only seeps through the crust but circulates back into the ocean again; hot springs have been discovered on the sea floor near the Galapagos Islands, and similar studies along the volcanic ridge that bisects the Red Sea have indicated the presence of heated, mineral-rich brines which circulate through the ocean crust, creating mineral deposits. The circulation of sea water continues until a cover of slowly accumulating sediment closes off the fractures in the ocean crust, leaving behind mineral deposits.

This process is illustrated by a copper-bearing outcrop in Cyprus. This outcrop is composed of old sea floor which has been uplifted and exposed by erosion. Studies indicate that the extensive hydration of the rock in the outcrop was caused by circulating sea water. It is believed that the copper deposit in the outcrop was emplaced by the hydrothermal circulation of sea water near an oceanic ridge. Studies of the distribution of the near-surface temperature gradient in the oceanic crust near the Galapagos Islands have shown that similar mineral deposition may be occurring there.

For background information *see* EARTH, HEAT FLOW IN; GEOTHERMAL POWER; GROUNDWATER; ORE AND MINERAL DEPOSITS in the McGraw-Hill Encyclopedia of Science and Technology.

[DONALD L. TURCOTTE]

Bibliography: E. Bonatti, *Sci. Amer.*, 238(2): 54–61, 1978; C. H. Sondergeld and D. L. Turcotte, *J. Geophys. Res.*, 82:2045–2053, 1977; H. P. Taylor, *J. Geol. Soc. London*, 133:509–558, 1977.

Earthquake

Of the thousands of earthquakes that occur in a year, only a small number are large enough and located so as to cause engineering damage. Two such earthquakes occurred recently: one on June 12, 1978, offshore from Sendai, Japan, of magnitude 7.5; and the other on June 20, 1978, near Salonika, Greece, of magnitude 6.4. The Sendai earthquake resulted in 21 fatalities and 400 injuries as well as engineering damage estimated in the tens of millions of dollars. The Salonika earthquake caused 50 deaths and over 100 injuries. There is no known way to stop such earthquakes from occurring, but steps perhaps can be taken to prevent the loss of life and mitigate destructive effects.

Aftermath studies. The destructive potential of earthquakes is not very well understood. Scientists are seldom fortunate enough to have the right combination of instruments at the proper place and sufficient time to study these earthquakes. Even when the right setup occurs, the complexities of the Earth inhibit scientists' understanding of ground motion. The usual method of studying destructive earthquakes is to send a field reconnaissance team to perform an aftermath study. These studies are important in that they provide first-hand information on structural failure and often indicate simple, effective measures which could be used in the construction of new structures to ensure their safety. This experience is useful for studying the effects of earthquakes on a particular class of structures which are still being built. However, this is not the case for unique structures such as power plants, dams, and port facilities. There are not enough situations where such structures have survived earthquakes to permit design decisions to be made based on aftermath studies.

Design decisions. Lacking such practical experience, design decisions must be based on predictions of ground motion during an earthquake, of the response of a structure to these ground motions, and of the consequences in terms of damage and injuries. The response of real structures to an input ground motion can be modeled by civil and structural engineers using modern techniques, such as finite element analysis, which takes into account nonlinear properties of materials. However, engineers and seismologists readily acknowledge the inability of present methods to accurately predict earthquake ground motions.

Ground-motion time studies. Ground-motion time history is very complicated. It consists of compressional- and shear-wave energy arriving at the earthquake site after being reflected, refracted, converted, and scattered. The time history may involve peak accelerations ranging from a few percent to over 100% of the acceleration of gravity. The time duration of the strong ground motion may vary from a few seconds to a few tens of seconds. Finally, time history is a function of both the nature and physics of the earthquake source, as well as the nature of the medium in which the waves propagate.

Until recently, the estimation of earthquake ground motion was based upon a correlation of particular characteristics of the strong-motion time history—such as peak acceleration, peak velocity, peak displacement, and duration or response spectra—with some measure of the size of the earthquake, usually magnitude. The geometrical spreading of energy from the earthquake source was also included in the empirical model. The main problem with this approach is that it is a simplification. Its success can be gaged by the fact that a standard deviation in the estimate is a factor of 2 or 3. This scatter makes the decision of input motion to the structure both difficult and cautious.

Recent work has taken another approach, namely, an attempt to better understand the physics of the earthquake source and the propagation of seismic waves. Theory already has been proposed for the generation of elastic waves in a plane-layered earth model due to a point earthquake dislocation source. An assumption of lateral homogeneity is also made in the model. The time history to be computed is in the form of the double Fourier-Bessel integral equation, shown below. Here, r, z,

$$u(r,z,\phi,t) = \int_{-\infty}^{\infty} \int_{0}^{\infty} \sum U_n(k,z,\phi,f) J_n(kr)\, dk \; \exp(j2\pi f)\, df$$

and ϕ are the radial, axial, and azimuthal coordinates in a cylindrical coordinate system; k represents the wave number; f represents the frequency; J_n is the Bessel function of the first kind of order n; and the U_n are integral transforms of u. The three-dimensional vector of ground displacements u is a function of the source time history, the earth model, the orientation of the dislocation source, and the depth of the source. The evaluation of the Fourier-Bessel integral in the equation is complicated by the presence of branch points and poles in the complex k plane. One recent approach is to introduce anelastic attenuation into the earth model. This results in complex singularities being removed from the real k axis, and integration is straightforward. Another approach being used is to assume a nonattenuating earth model, in which case the complex singularities lie along the real k axis. The wave-number integration is then performed by using the methods of contour integration. A third approach used for simple earth models is to expand U_n into individual generalized ray contributions. Laplace transform techniques are then used.

The results of the numerical evaluation of the equation are quite satisfying to date. The displacement time histories of the strong ground motions of several earthquakes (including the Apr. 9, 1968, Borrego Mountain earthquake in southern California of magnitude 6.4) were successfully matched by using realistic source and transmission models. Velocity and acceleration time histories are harder to fit because the simple assumptions made break down at high frequencies at which the Earth acts in a heterogeneous manner with respect to the transmission of seismic energy.

Nature of seismic source. Given improved models for the transmission of seismic waves through a realistic earth model, research on the nature of the seismic source is also undergoing new developments. The effect of the medium upon the strong ground motion observed can now be decoded, yielding a better insight into the source processes. Studies are being performed on the mechanics of failure, the coherency or incoherency of the rupture process, and a verification of the source theories on the basis of detailed studies of actual faults and strong-motion time histories.

Recent numerical applications of elastic-wave theory offer some exciting prospects for better estimation of strong ground motion generated by earthquakes as well as insights into the physics of an earthquake itself.

For background information *see* EARTHQUAKE; FOURIER SERIES AND INTEGRALS in the McGraw-Hill Encyclopedia of Science and Technology.

[ROBERT B. HERRMANN]

Bibliography: D. M. Boore, *Sci. Amer.*, 237: 68–78, December 1977; D. J. Leeds (ed.), *Newsletter of the Earthquake Engineering Research Institute*, vol. 12, July 1978.

Ecological interactions

All members of an ecological community ultimately interact with and influence each other, resulting in mutual selection between the species and evolutionary change, among which is coevolution—the process by which species evolve together. Many types of interactions within a community are well known and easily detected. Predation, competition, parasitism, and symbiosis occur frequently among members of a community. On the other hand, some types of ecological interactions are extremely subtle and, therefore, are difficult to decipher without detailed studies. Perhaps the most interesting, but often the most obscure, associations involve cases of mutualism.

Definition of mutualism. Two species exhibit mutualism when their close association is of benefit to both. Usually, the mutualistic relationship is the result of a long period of coevolution during which each member of the mutualistic pair has accommodated its basic life history and ecology so as to facilitate the interrelationship with the other species. Many examples of coevolution which resulted in mutualism are well known. Root nodule bacteria which live in the roots of leguminous plants and convert atmospheric nitrogen into compounds that can be used in the nitrogen metabolism of the plants are a classic example. Flower morphology and color and the animal pollinator are another.

Obligatory mutualism. Occasionally, the species in a mutualistic relationship not only benefit from the association, but actually become totally dependent on their mutualistic partners. Coevolution has, in such cases, reached the point where one, or perhaps all, of the members in a mutualistic relationship cannot survive without their partners. In these instances of obligatory mutualism, the extinction of one species can cause the demise of other dependent species.

Although ecological chain-reaction extinctions of this type are clearly possible, ecologists have thus far discovered only a few specific instances in nature. Unless the nature of a mutualistic relationship is understood before the extinction of one of the involved species, the subsequent extinction of a dependent species may be difficult to explain, and may in fact be attributed to other factors.

Dodo–tambalocoque tree mutualism. At least one case of obligatory plant-animal mutualism is known in which the extinction of one species almost caused the extinction of its mutualistic partner. The two species involved were the now extinct dodo *(Raphus cucullatus)* and the now nearly extinct tambalocoque tree *(Calvaria major)*, which were both endemic to the small Indian Ocean island of Mauritius.

The dodo was a large, flightless bird found only on Mauritius, which remained uncolonized by humans until the end of the 16th century. Slaughtered by sailors and attacked by introduced mammalian predators which also destroyed its eggs, the dodo became extinct in about 1681. Little is known about the dodo's ecology, except for information pieced together from the accounts of early travelers. What is known is that the dodo weighed over 12 kg; it fed on fallen fruits and seeds, and its gizzard contained large stones which were probably used to crush large seeds.

The tambalocoque tree is a large, tropical-forest species that reaches heights of over 30 m. The tambalocoque tree was formerly rather abundant in the forests of Mauritius; early forestry records show that it was frequently exploited because of its fine wood. However, by 1973 only 13 old and overmature trees survived in the remaining native forests of the island. Each of these surviving trees was estimated by foresters to be over 300 years old. Despite the fact that each of these trees produced well-formed, apparently fertile seeds each year, no young trees could be found. Apparently none of the seeds germinated naturally, and even when planted under nursery conditions the seeds remained dormant. It seemed that the tambalocoque tree would soon become extinct as old trees died with no new trees to replace them. Various hypotheses have been proposed to explain the tambalocoque's predicament, the most tenable of which is that the plants have some lethal genetic condition which renders them sterile.

It now appears that all these hypotheses were incorrect; the tambalocoque was actually becoming extinct because there were no more dodos. The fruit of the tambalocoque contains a hard pit about the size of a walnut. The walls of this pit are so thick (up to 15 mm) and hard that the embryos within are apparently unable to break through the wall and sprout. Upon examining a tambalocoque seed, it occurred to S. A. Temple that this type of seed probably needed to be worn thin by abrasion before it could germinate successfully. One possible way for the seed to be abraded would be in the gizzard of a large bird. The fact that no tambalocoque seeds had apparently germinated in 300 years suggested a link with the dodo, which became extinct 300 years ago.

It seems likely that dodos ate the fallen fruits of the tambalocoque tree, which even today are consumed by other fruit-eating animals. However, unlike these smaller animals which eat only the fruit and leave the large pit untouched, the dodo was large enough to have swallowed the fruit whole. If the dodo was an efficient consumer of tambalocoque seeds, the tree would have had to evolve some sort of protection for its seeds so they would not be crushed in the dodo's powerful gizzard. To prevent this, the tambalocoque apparently evolved an extremely tough seed. The seeds were hard enough to withstand ingestion by the dodo, but they were so tough that they were unable to germinate without first being abraded in the dodo's gizzard.

To test his hypothesis, Temple force-fed fresh tambalocoque seeds to a domestic turkey, which served as a surrogate dodo. Some of the seeds were crushed by the turkey's gizzard, but other seeds were either regurgitated or defecated after being abraded and reduced in size. When these seeds were later planted under nursery conditions, they germinated, becoming perhaps the first tambalocoque seeds to sprout in 300 years. It is possible, therefore, to save the tambalocoque from extinction by propagating artificially abraded seeds.

Implications for other communities. This case of mutualism between the tambalocoque and the dodo demonstrates that chain-reaction extinctions can actually occur in ecological communities and that predicting all the consequences of a species' extinction can be difficult. Some communities are likely to include many examples of obligatory mutualism. For example, tropical forests, which have many plants with specific insect pollinators, could be seriously affected by the indiscriminate application of insecticides. When one considers that many, or perhaps most, cases of mutualism are as yet undetected, the possible impact of unsuspected chain-reaction extinctions may be even more significant than was previously supposed.

For background information *see* ECOLOGICAL INTERACTIONS in the McGraw-Hill Encyclopedia of Science and Technology. [STANLEY A. TEMPLE]

Bibliography: R. L. Dressler, *Evolution*, 22:202–210, 1968; M. S. Henry, *Symbiosis*, 1966; D. H. Janzen, *Evolution*, 20:249–275, 1966; M. Proctor and P. Yeo, *The Pollination of Flowers*, 1972; C. C. Smith, *Ecol. Monogr.*, 40:349–371, 1970; N. G. Smith, *Nature*, 219:690–694, 1968; S. A. Temple, *Science*, 197:885–886, 1977.

Electric power systems

The application of electric load management techniques which alter the electric usage patterns of consumers to level the burden placed on electric power systems recently moved forward on a wide front. Moreover, in an urgent effort to hold down the rising cost of electricity, researchers are presently studying the feasibility of managing additional electrical appliances within the home and the possibility of extending the application to the commercial, industrial, and agricultural sectors as well.

Background. The use of electricity rises and falls throughout the day. Since electric utilities must install sufficient electrical capacity, power plants, and transmission lines to serve all customers at any time, these wide variations in demand for electric energy increase overall cost. Leveling these peaks and valleys not only would result in a more efficient use of existing equipment but would also help delay the need to construct expensive facilities in the future.

If all consumers of an electric utility turned on all of their appliances at once, the utility's equipment would be strained beyond the breaking point. The likelihood of this happening is remote because of the natural diversity in personal habits and the operation of electrical appliances. In planning the electrical supply system serving its consumers, the utility takes advantage of this natural diversity. Load management techniques attempt to optimize this diversity by controlling the usage patterns of certain appliances to reduce the demand for electricity during certain periods of peak consumption.

Direct control. The electricity usage patterns of consumers may be affected by either direct or indirect means. In the direct method, control devices, which are under the direct control of the utility, are installed in the home to disconnect certain electrical appliances such as water heaters, space heaters, or air conditioners from the electrical supply lines for short periods of time.

In one recent permanent application, radio-controlled switches were placed on water heaters with the permission of the consumers in over 30,000 homes served by a northern electric utility. The objective is to reduce the demand for electricity during the peak hour of the year, which usually occurs during the coldest winter day. On those days when a new system peak demand is possible, the utility disconnects all water heaters from the electrical supply lines by sending a radio signal through the air. The water heaters are usually disconnected for periods of time ranging up to 2 hr. Since the natural diversity of the total group of water heaters is disturbed during this process, they are restored to service in smaller groups, allowing ample time to permit natural diversity to be reestablished. In this process some individual water heaters may be disconnected for time periods up to 5 hr.

In this successful application of load management, the hourly diversified demand as seen by the utility's supply lines at its peak hour ran about 1.1 kW per unit. Since each individual water heating unit contains heating elements rated at about 4.5 kW, the results indicate that on the average about one out of every four units is operating at any given moment during the hour.

In another application of load management, a southern utility installed radio-controlled switches to control central air conditioners in 5000 homes. However, in this case the air conditioners are disconnected from the electrical supply lines for only short periods of time, averaging about 7½ minutes off out of every 30 min. On an hourly basis, this represents a 25% load shed (see illustration). Theoretical calculations indicate that during heat spells the hourly diversified demand of air conditioners ranges from 4 to 5 kW per house or more. Therefore, the 25% load shed strategy reduces demand on the utility by approximately 1 to 1¼ kW per home.

As with the water heaters, the air conditioners are controlled only rarely by the utilities and only during those times when an annual system peak demand is imminent. The feasibility of extending the control of air conditioners and also space heaters, on a cyclical basis, to longer periods of time, up to 16 hr a day, that correspond to the peak load duration time of some utilities is being studied.

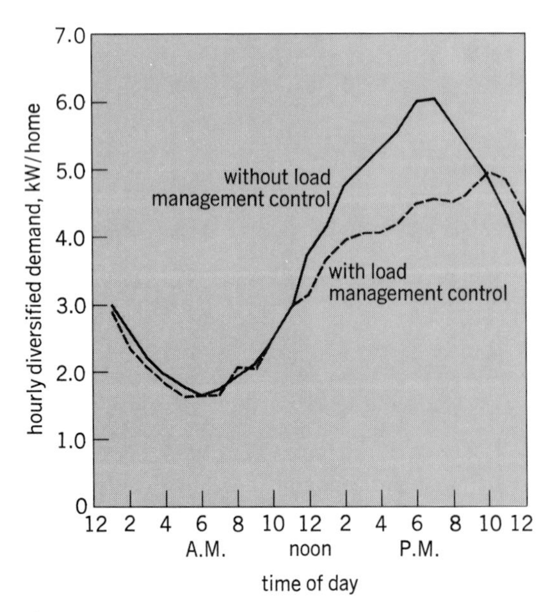

Comparison of the hourly diversified demand of homes with and without load management control of central air conditioners.

In addition to load management control by radio signal, other direct control schemes are being investigated. These include a ripple system, successfully utilized in Europe for many years, in which an audio frequency is superimposed on the existing power supply conductors of the electric utility as the transmission medium for control purposes. The use of high-frequency carriers injected directly on the power system conductors or as an alternate on the telephone wires serving consumers to disconnect appliances is also being studied. Systems that combine features of these schemes are also available.

These control systems communicate in only one direction, from the electric utility to the control device. However, two-way systems in which communication is possible in both directions have also been utilized recently. With the two-way control systems, additional functions, such as automatic meter reading and the verification of the operation of various devices, are possible.

Indirect control. Electricity usage patterns may also be affected by influencing the consumer indirectly through price signals built into the design of applicable rates. Time-differentiated rates, in which prices for electricity vary according to the time of day or season, may influence the amount of consumption in each time period. With time-of-day rates, each hour of the day is classified as either a peak or off-peak hour, depending upon the overall load characteristics of the utility. Depending upon the cost allocation, prices charged for electricity consumed during the peak load hours may be 10 or more times higher than electricity consumed during off-peak hours.

The Electric Power Research Institute and other sponsoring organizations working on behalf of the National Association of Regulatory Utility Commissions reported during the last year on their continuing study of rate design and load control. This comprehensive study is divided into 10 topics which are being researched by separate task forces, namely: peak load pricing, elasticity of demand, load management experiments, costing for peak load pricing, rate design, cost benefit analyses of load management, metering, equipment for using off-peak energy, load controls and penalty pricing, customer acceptance of load management.

In addition, a number of time-of-day rate demonstration projects and experiments, some funded by the former Federal Energy Administration, are in various stages of completion. These programs are generally designed to test the response of a sample group of consumers to the stimulus of time-of-day rates in real-life situations.

Storage devices. As already mentioned, the hourly diversified demand of central air conditioners may reach as high as 6 kW or more per unit. Independent studies indicate that similar demands for space heaters are in the range of 6 to 9 kW per home, depending upon local weather conditions. However, consumer reaction during some recent load management tests indicates that disconnecting these appliances for periods greater than 20–25% of the time may not be tolerable.

In an attempt to utilize the full load management potential of space heaters and air conditioners, a number of demonstration projects are in progress utilizing thermal storage devices for both heating and cooling. For example, in one program a core of ceramic bricks is heated electrically to very high temperatures during the utility's off-peak period. During periods of high consumption, the storage device is disconnected from the electrical supply lines and the entire heating requirement of the house is supplied by the heat stored within the ceramic material. Other, similar cooling storage device demonstration projects are in progress.

For background information *see* ELECTRIC POWER SYSTEMS in the McGraw-Hill Encyclopedia of Science and Technology.

[WILLIAM E. MEKOLITES]

Bibliography: R. G. Uhler, *Rate Design and Load Control*, November 1977.

Electrical utility industry

It became clear as 1978 drew to a close that the electrical utility industry had undergone changes that were possibly both radical and permanent. Growth of both peak demand and kilowatt-hour usage had historically oscillated about an annual level of 7.2% for several decades preceding the oil embargo of 1973. Since that time, yearly growth has been extraordinarily erratic, influenced by various forces of economic recession, conservation, drought, and other extremes of regional weather. In 1978, however, conditions overtly appeared to be normal. Against this background, the 1.4% growth in peak demand and the 2.4% increase in overall kilowatt-hour usage came as a surprising indication that basic structural changes may have occurred in the patterns of electric power use, having been masked previously by the other changing factors.

In other important areas, in 1978 the Environmental Protection Agency passed new air quality regulations that, in requiring environmental review and public hearings for coal-fired generating stations at the state level, effectively added an additional 2 years to the present 6-year overall lead time to construct such units. Further, other rulings mean that all future-coal-fired units would have to add flue gas scrubbers to reduce SO_2 emissions, regardless of the sulfur content of the coal burned. Nuclear power programs also sustained additional setbacks when several states, led by California, imposed what amounts to a moratorium on nuclear plants by requiring proof that high-level radioactive waste could be stored safely, as a prerequisite to permitting contruction. This proof depends on Federal action and is still years away.

General. The United States electrical utility industry has a pluralistic ownership which is unique in the world. Ownership is shared by private investors, cooperatives (such as the Rural Electrification Agency), Federal agencies, and various municipal, district, and state political bodies. Investor-owned utilities now serve 77.6% of all customers, cooperatives serve 10.0%, and public bodies serve 12.4%. In terms of capacity installed, investor-owned utilities own 78.3% of all operating capacity, publicly owned bodies 10.2%, cooperatives 1.8%, and Federal agencies 9.7%.

Capacity additions. Utilities had a total generating capability of 556,818 MW at the end of 1977. An additional 35,350 MW was added during 1978, of which 16,400 MW was fossil-fueled, 8300 was nuclear, 6400 was conventional hydroelectric, 2000 was pumped storage hydroelectric, 2200 was combustion turbines, and 20 was internal combustion engines. This additional capacity, coming into service at a time when overall reserve margins were already well in excess of need, further exaggerated the apparent overbuilding of generation plant. National reserve jumped from 31 to 38% in 1978, while the lowest permissible reserve is generally only 18%. However, the long lead times for major plants—currently 6 years for a fossil-fueled unit—means that the capacity now coming into service was started before the period of reduced growth began. In spite of massive, continuing cancellations and deferrals, it has not been possible to cut the construction program quickly enough to prevent continuing buildup of excess capacity. In 1979, as deferrals and cancellations begin to be more effective, sharply reduced capacity will come into service, amounting to 28,800 MW.

The additions in 1978 resulted in the following overall composition of the present generating plant: fossil-fueled, 69.2%; nuclear, 9.0%; conventional hydroelectric, 10.5%; pumped storage hydroelectric, 1.8%; combustion turbines, 8.6%; and internal combustion engines, 0.9%.

Fossil-fueled capacity. Fossil-fuel units constitute 46.5% of the total new capacity added in 1978. Thirty-one individual units went into service, of which 87% were coal-fired, 20% were oil-fired, and 3% were fueled by gas. The composite average cost per kilowatt of capacity was $277. Coal-fired units averaged $300; oil-fired units averaged $230/kW, and gas-fired units averaged $100/kW to construct. Nuclear units averaged $470.

Operating costs presented a different picture. Nuclear units produced power at 6.0 mills/net kWh, coal units at 8.6 mills/net kWh, and oil units at 20.7 mills/net kWh. Because of the wide variation in gas prices throughout the United States, operating costs for gas-fired units are indeterminate on a national average basis.

The $10,550,000,000 expended in 1978 for fossil-fired construction was up substantially from the $9,060,000,000 expended in 1977, an increase of 9.5% in real dollars.

Nuclear power. Utilities added 7 more nuclear units in 1978 to bring the total of reactors now operating in the United States to 72. Total capacity of the plants brought into service during the year was 8260 MW, raising the total now operating to 58,150 MW. Of the units added, 1 was a boiling water reactor (BWR), 5 were pressurized water reactors (PWR), and the other was a high-temperature gas-cooled unit. There are now 41 PWRs and 26 BWRs operating in the United States; the remaining 5 units use other technologies. Nuclear units now already planned or in construction have a total capacity of 141,000 MW which, if current plans hold, will bring nuclear capacity to 22% of all installed capacity by 1985. During 1977, nuclear units generated a total of 2.5×10^{11} kWh, accounting for 13% of the year's total.

Combustion turbines. Combustion turbines, because of their quick-start capabilities, that is, the ability to go from cold to full load in 2–3 min, and their low initial capital cost of $175/kW, have served admirably as peak-load units. Utilities keep an average of 8% of their peak demand in gas turbine capacity, using them for several hundred hours a year to meet annual peak loads, or to go on line quickly to supply load in emergency situations. However, because of the uncertainty of the future supply of the distillate oil or gas that these machines burn, and the uncertainty of national policies concerning the permissible use of petroleum fuels and the permissible levels of the nitrogen oxides that these machines produce in their exhaust gases, this percentage will undoubtedly decline substantially.

Utilities brought 1600 and 2200 MW into service in 1977 and 1978, respectively. Of this total, however, 535 MW represented the combustion turbine portion of combined cycle installation, in which the 900–1000°F exhaust gas of the turbines is used to generate steam for a conventional steam turbine. The 2200 MW brought into service in 1978 was made up of 40 individual units. The total installed combustion turbine capacity for the entire industry is now 49,940 MW, or 8.4% of total capacity. Utilities spent $155,700,000 on combustion turbine construction in 1978.

Hydroelectric installations. Installation of conventional hydroelectric capacity continued, with 6365 MW coming on line in 1978, and a total of about 19,500 MW additional capacity planned for the future. Hydroelectric capacity amounting to 58,271 MW provided 10.5% of total industry capacity. However, as sites become increasingly more difficult to find and develop, this percentage is expected to decline to 8% by 1985. During 1978, the industry spent $908,600 on hydroelectric projects, the bulk of which, $739,537,000, was by Federal agencies. The future of hydroelectric power will depend on economical development of low-head, or run-of-river, turbines, since environmentally acceptable high dam sites are almost nonexistent.

Pumped storage. Pumped storage represents one of the few methods by which electrical energy can be stored. In this mechanical analog, water is pumped to an elevated reservoir during off-peak periods and is released through hydraulic turbines during the subsequent peak period, recovering 65% of the original fuel energy. Although the 10,041 MW installed at the end of 1977 was only 2% of the industry's total capacity, another 10,600 MW is planned for 1979–1980, and utilities continue to spend about $375,000,000 annually on such projects. The world's largest pumped-storage installation is Ludington Station on Lake Michigan, rated at 970 MW.

Growth rate. Rate of growth of peak demand in 1978 was very substantially below the 1977 forecast of 6.4%. A stronger economy, continued weakening of the conservation thrust engendered by the 1978 oil embargo, breaking of the drought in the western states, and increasing electrification were expected to produce strong growth. Instead, peak demand rose only 1.4%, for reasons which are not yet clear.

A negative growth in the industrial Midwest, Northeast, and Mid-Atlantic regions was the primary contribution to the lower growth. These regions did experience lower than normal summer temperatures, which generally has the effect of reducing growth by 2–3%, and had a high growth the previous year, which also tends to make the current year's growth relatively lower. However, the magnitude of the decrease indicates that other forces are at work. For instance, higher than normal summer temperatures in the Southwest and the high economic activity in that area should have produced a 10–12% growth, rather than the 6% actually achieved.

Comparison of industrial kilowatt-hour use to residential and commercial use suggests that the major cause of the low peak growth is to be found in the industrial sector. This in turn suggests that industry has achieved significant electrical energy savings in production and that load management techniques have been more effective than expected. Load management, in which controls are used to limit total peak demand by monitoring and coordinating all power-consuming devices in a plant, is becoming increasingly popular, and should act to flatten and broaden future peaks. Long-term growth of overall peak will probably reflect this new situation and will average 5.0% in the 1978–1980 period, 4.8% between 1980 and 1990, and stabilize at 4.5% thereafter.

Future decline in rate of growth is expected to result from a balance between the upward-acting forces of increasing electrification brought about by unavailability and price of primary petroleum fuels, the need to increase productivity at point of use, and industrial use of electricity to avoid pollution, versus the downward pressures generated by a declining rate of population growth, a slower growth rate for the gross national product, conservation, and load management.

Usage. Sales of electricity rose at a rate almost twice as great as peak demand, reaching 2.4% for 1978. This was, however, still less than half of the forecast rate. Total usage for 1978 was 2×10^{12} kWh. The first quarter was dragged down by a long coal strike which caused industry to cut usage to avoid forced shutdown. However, the rest of the year saw little recovery. Total industrial sales for the year were 7.642×10^{11} kWh, compared with 7.572×10^{11} in 1977, for an increase of only 0.92%.

Residential and commercial growth were much stronger than that of the industrial sector. Residential usage rose to 6.79×10^{11} kWh for 1978, and commercial use to 4.836×10^{11}, increases of 4.1% and 3.1%, respectively, over 1977. Electric heating continued to contribute strongly to residential sales, accounting for 1.316×10^{11} kWh, or 19.6% of the total kilowatt-hours used in residences. Residential revenues also rose from $26,-317,000,000 in 1977 to $27,635,000,000 in 1978, a 5% rise in constant 1978 dollars. These sales reflect an average annual use per residential customer of 8876 kWh and an average annual bill of $361.25.

In the industrial sector, generation by industrial plants contributed 8.42×10^{10} kWh, up slightly from the 8.29×10^{10} of 1977. This increased industrial generation's share of the total very slightly to 9.9%. The long-term trend of industrial generation as percent of total is consistently downward, however, and by 1988 is expected to decline to only 6.5% of the total.

Fuels. The energy embargo brought tremendous Federal pressures on utilities to convert from petroleum fuels to coal. Subsequently, all new planned stations have been designed to fire coal.

The use of coal in 1977 rose from 4.484×10^8 tons (4.036×10^6 metric tons) in 1976 to 4.771×10^8 tons (4.294×10^8 metric tons) in 1978, a rise of 6.4%. Oil, however, also rose from 5.559×10^8 (8.839×10^7 m^3) bbl in 1976 to 6.235×10^8 bbl (9.914×10^7 m^3) in 1977, a rise of 12%. Although real quantity of gas used rose from 3.0806×10^{12} ft^3 (8.7181×10^{10} m^3) in 1976 to 3.1942×10^{12} ft^3 (9.0396×10^{10} m^3) in 1977, the rate of increase fell by 3.7%.

Energy generated by each of these major fuels and their percentage share of total generation were as follows for 1977: coal, 9.853×10^{11} kWh (51.7%); oil, 3.579×10^{11} kWh (18.8%); gas, 3.055×10^{11} kWh (16.0%); and nuclear, 2.497×10^{11} kWh (13%). Converting these quantities to coal equivalent gives an equivalent total fuel consumption of 9.212×10^8 tons (8.291×10^8 metric tons) of coal for the entire industry in 1977. During the year, it took only 0.968 lb (436 g) of coal to produce 1 kWh.

Transmission. Utilities spent $3,476,000,000 on transmission construction in 1978. This included $813,000,000 for overhead lines below 345 kV, $815,000,000 for overhead lines at 345 kV and above, $60,200,000 for underground construction, and $1,013,000,000 for substations. These costs bought 7015 km of overhead lines at 345 kV and above, and 15,555 km at lower voltages. Only 307 mi (491 km) of underground cables were installed, primarily because of the 8:1 ratio of underground to overhead costs. Utilities also installed a total of 81.31 GVA of substation capacity during the year. Maintaining existing lines cost utilities $406,-000,000 in 1977, and the utilities planned on spending $412,000,000 in 1978 for such maintenance.

Distribution. Distribution facilities required the expenditure of $5,134,000,000 in 1978. Of this, $1,444,000,000 was spent to build 61,115 km of three-phase equivalent overhead primary lines ranging from 5 to 69 kV, with the majority at 15 kV. Of all overhead lines constructed in 1978, 3% was rated at 4 kV, 80% at 15 kV, 10% at 25 kV, and 7% at 34.5 kV. Expenditures for underground primary distribution lines amounted to $661,500,000 in 1978. In underground construction, of the 22,985 km of three-phase equivalent lines built, 3% were rated at 5 kV, 78% at 15 kV, 11% at 25 kV, and 8% at 34.5 kV. Utilities energized 27,251 MVA of substation distribution capacity in 1977 at a total cost of $629,800,000. Maintenance costs for distribution were $1,561,000,000 in 1978.

Capital expenditures. Utilities increased their capital expenditures in 1978 to $30,700,000,000, up 4.07% in constant 1978 dollars from 1977. Of this total, $21,400,000,000 was for generation, $3,-170,000,000 for transmission, $4,860,000,000 for distribution, and $1,320,000,000 for miscellaneous uses, such as headquarters buildings, services, and vehicles, which cannot be directly posted to the other categories. Total assets for the investor-owned segment of the industry rose from $171,-900,000,000 in 1976 to $190,000,000,000 in 1978. The equivalent 1978 figure for cooperatives is $11,700,000,000.

For background information *see* ELECTRIC POWER GENERATION; ELECTRIC POWER SYSTEMS; ENERGY SOURCES; TRANSMISSION LINES in the McGraw-Hill Encyclopedia of Science and Technology. [WILLIAM C. HAYES]

Bibliography: Edison Electric Institute, *Statistical Yearbook of the Electric Utility Industry*, 1978; 1978 annual statistical report, *Elec. World*, 189(6): 75–106, Mar. 15, 1978; 20th steam station cost survey, *Elec. World*, 188(10):43–58, Nov. 15, 1977; 29th annual electrical industry forecast, *Elec. World*, 190(6):61–76, Sept. 15, 1978.

Electrochemistry

The advent of the chemically modified electrode may offer a new dimension in electrochemistry. By a variety of recently developed techniques, molecular species may be attached to electrode surfaces. In the event that these surface-bound molecules retain the chemical, electrochemical, or physical characteristics they exhibit in the free state, the opportunity exists for tailoring an electrode surface to a particular application. Hence, in contrast to conventional electrodes, chemically modified electrodes may perform preselected chemical or electrocatalytic reactions at the same time as the electron transfer process, thereby affecting rates or product distributions in electrochemical reactions. Such is the case for carbon electrodes modified by covalently attaching optically active amino acid molecules. These electrodes have been shown to display selectivity in electrochemical reactions where chiral products are generated.

Chemically modified electrodes have great possibilities as electron transfer catalysts. When the attached molecules are themselves electroactive, they may be oxidized or reduced by varying the electrode potential and yet remain immobilized at

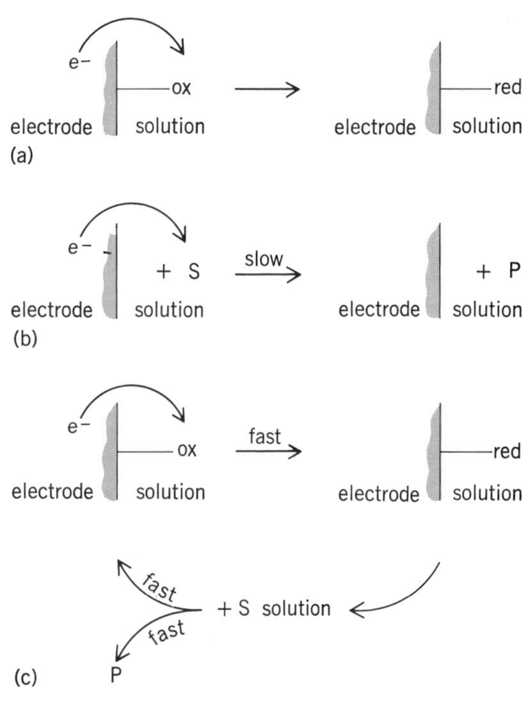

Fig. 1. Schematic of electrochemical reactions at a chemically modified electrode. *(a)* Reduction of surface-attached molecule (ox) by electron transfer from electrode, yielding attached molecule in its reduced state (red). Reaction at bare electrode *(b)* is compared with *(c)* electrocatalysis using chemically modified electrode. S = solution component in oxidized state; P = solution component in reduced state.

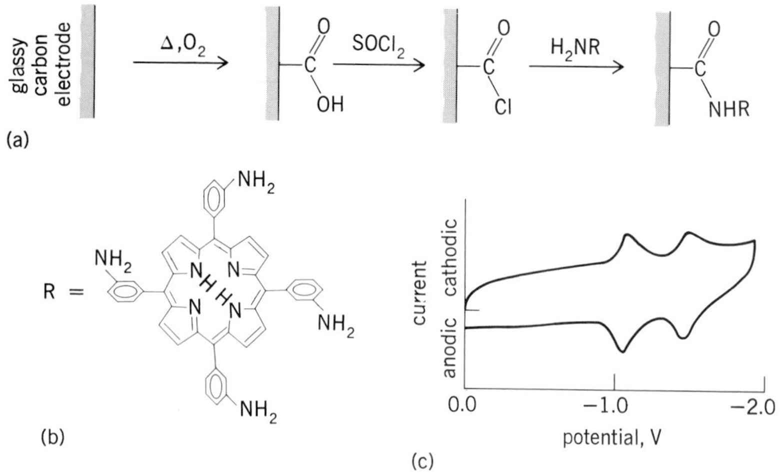

Fig. 2. Chemically modified carbon electrode. *(a)* Surface modification chemistry utilized for attachment of *(b)* tetra(amino phenyl) porphyrin molecule (R). *(c)* Cyclic voltammogram for the modified carbon electrode; potentials are referred to the saturated calomel electrode.

the electrode surface (Fig. 1*a*). These charge transfer surface states can in principle act as electron transfer mediators to achieve fast electron transfer from the electrode to electroactive species in solution. A given species S, which undergoes electron transfer reactions with bare (unmodified) electrodes very slowly (being converted to its reduced form P; Fig. 1*b*), reacts at greatly accelerated rates with an electrode bearing electroactive molecule "red" (Fig. 1*c*). By attachment of the appropriate electron transfer mediator, an array of electrocatalytic schemes can be foreseen analogous to known homogeneous solution chemistry. Chemically modified electrodes have implications not only for electrocatalysis but for electroanalysis, electrosynthesis, and energy conversion schemes as well. Research efforts thus far, however, have emphasized the development of reagent immobilization techniques and the determination of the fundamental properties of the chemically modified electrode. Their implementation in applied areas is expected in the future.

Methods of attachment. A variety of approaches have been developed for immobilizing molecules of interest on electrode surfaces. These approaches can be classified into three broad categories: chemisorption, deposition techniques, or covalent bonding. Electrodes modified by irreversible adsorption of the desired molecule onto the surface represent the simplest method; however, such attachments are typically nonpermanent. Alternatively, the desired molecules may be deposited on an electrode surface by polymerization processes, by coating techniques or by inducing some type of insolubility of the molecule such that it precipitates at the electrode surface. The most popular method is covalent bonding of the molecular species to the electrode surface. This can be accomplished either directly by covalent reactions with electrode surface functional groups or by initial bonding of a reagent to the surface with subsequent coupling to the molecule of interest. These methods are somewhat more predictable and versatile than the others, in that known solution chemistry may be employed for controlled synthesis of surface structures. Furthermore, the covalent bonds are significantly more durable than adsorptive or insolubility interactions. Examples which illustrate covalent bonding of electroactive species to carbon and metal oxide electrodes are discussed below.

Carbon electrodes. Methods employed for attaching molecules to the surfaces of carbon electrodes by covalent bond formation have typically involved oxygen-containing groups such as quinones, phenols, or carboxylates present on the carbon surface. Thermally oxidized carbon electrodes are purported to have an abundance of carboxylic acid groups. After treatment with $SOCl_2$ to form the acid chloride, the active surface can be reacted with amine-bearing molecules to form an amide bond. Figure 2*a* shows the use of the amidization reaction to couple an electroactive porphyrin molecule (Fig. 2*b*) to a glassy carbon electrode. The porphyrin ring system is an interesting R group because it displays well-behaved nonaqueous electrochemistry and can be metalated to yield an array of surface redox systems. Perhaps more significant is the fact that porphyrin systems are known to exhibit catalytic effects.

The surface-bound porphyrin exhibits reversible electrochemistry similar to that of the amino porphyrin in dissolved form. Figure 2*c* shows the current potential behavior (cyclic voltammogram) for the modified carbon electrode in an electrolyte which is devoid of electroactive species (other than

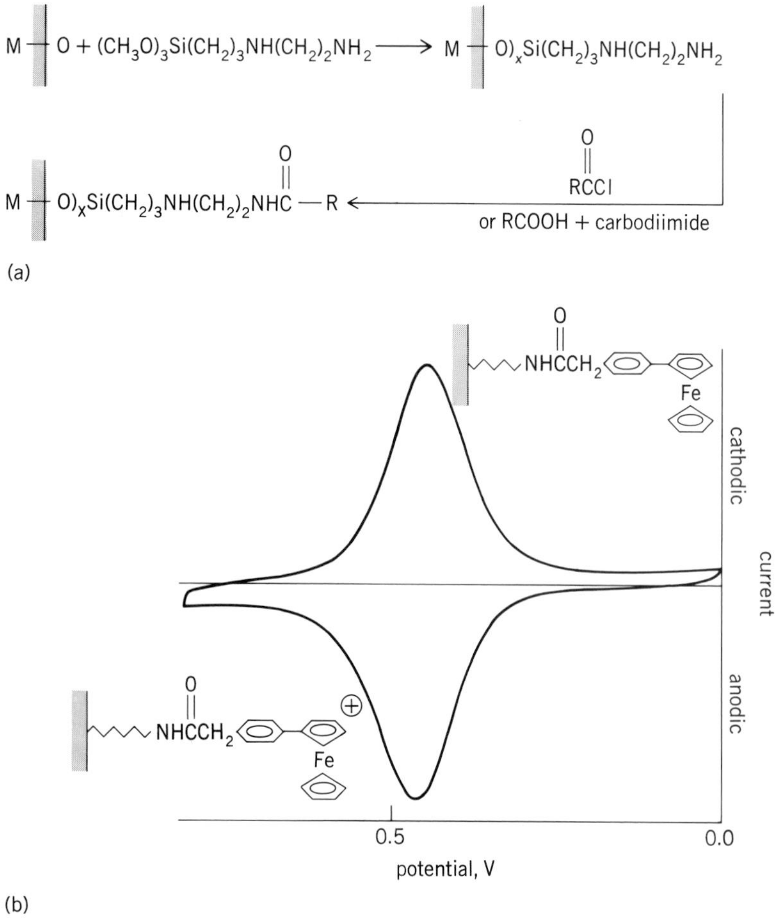

Fig. 3. Chemically modified metal oxide electrode. *(a)* Two-step covalent attachment of molecule R to electrode. *(b)* Surface structure and cyclic voltammogram of chemically modified platinum electrode, where R is 4-(ferrocenyl)phenyl acetic acid.

surface-attached molecules). The pair of electrochemical waves observed on the cathodic portion (upper half) of Fig. 2*c* occur when the electrode potential is swept from 0 to −1.8 V (versus a saturated calomel electrode). These correspond to two reduction steps forming the immobilized porphyrin anion radical and dianion. Scanning the electrode potential in the opposite direction yields the corresponding reoxidation waves in the anodic portion. The symmetric shape and nearly coincidental peak potential of the cathodic- and anodic-directed waves are indicative of rapid electron transfer between electrode and porphyrin system. Chemical insertion of Co(II) into the immobilized porphyrin ring yields a new electrochemical wave (at −0.86 V), which is associated with the Co(II) to Co(I) redox process. Hence, the chemical and electrochemical properties of the bound porphyrin closely parallel those for the unbound species.

Metal oxide electrodes. The most versatile and most widely used chemical method involves the reaction of organosilane reagents with metal oxide electrode surfaces. The silanes are similar to those employed for modification of silica and alumina surfaces used in liquid chromatography, affinity chromatography, catalyst immobilization, and adhesion technology. Typically, binding of silanes to metal oxide electrode surfaces (SnO_2, RuO_2, and Pt/PtO) is a first step toward attaching various molecules to the electrode. Figure 3*a* shows the reaction sequence involved for attaching molecules which bear a carboxylic acid (or acid chloride) functionality. The surface oxides or hydroxides are silylated with 3-aminoethylaminopropyltrimethoxy silane to produce stable M-O-Si bonds.

Subsequent amidization of the surface-bound amine yields covalent attachment of the desired molecule to the electrode surface. Figure 3*b* shows the surface structure and cyclic voltammogram for a silanized platinum/platinum oxide electrode which was amidized (carbodiimide-assisted) with 4-(ferrocenyl)phenyl acetic acid. The symmetrically shaped wave can be ascribed to the electrochemical oxidation of the surface-bound ferrocene (anodic portion) to the ferricinum cation and subsequent re-reduction to ferrocene (cathodic portion). An outstanding feature of these modified platinum electrodes is their chemical and electrochemical stability. Some specimens have been continuously cycled between the oxidized and reduced forms up to 15,000 times before total depletion of ferrocene molecules from the electrode surface occurs due to side reactions.

Prospects. While a substantial number of methods have been developed for immobilization of a variety of organic and inorganic redox systems, demonstration of synthetic or catalytic capabilities has been minimal. The "ideal" surface-bound redox site for catalytic purposes is one which is both chemically and electrochemically reversible, that is, one in which the redox site undergoes a rapid electron exchange reaction with the electrode surface to which it is bound, and is stable in all oxidation states involved in the catalytic cycle. Efforts thus far appear to be hampered by the modest chemical lifetimes of surface redox molecules. There remains a great deal of uncertainty associated with the chemical properties of immobilized redox systems and the surface structure of modified electrodes. Little is known about the stereochemical requirements placed on surface-bound molecules for optimum electrode and electrocatalytic reactions. Whether the chemically modified electrode is applied to technology depends on its stability and predictability, both of which remain to be adequately tested.

For background information *see* ELECTROCHEMICAL TECHNIQUES; ELECTROCHEMISTRY in the McGraw-Hill Encyclopedia of Science and Technology.

[JERRY R. LENHARD]

Bibliography: J. R. Lenhard and R. W. Murray, *J. Electroanal. Chem.*, 78:195, 1977; J. C. Lennox and R. W. Murray, *J. Electroanal. Chem.*, 78:395–401, 1977; P. R. Moses, L. Wier, and R. W. Murray, *Anal. Chem.*, 47:1882–1886, 1975; B. F. Watkins et al., *J. Amer. Chem. Soc.*, 97:3549–3550, 1975.

Electrode

Single crystals of anisotropic polymeric sulfur nitride, $(SN)_x$, are currently being studied as a new electrode material, with the ultimate goal of producing practical chemically modified electrodes. This substance has been classified as a metal even though there are no metal atoms present in its structure. The $(SN)_x$ exhibits electrical and chemical anisotropy.

A great deal of research has recently been initiated in the study and development of chemically modified electrodes which involve primarily the immobilization of oxidation-reduction systems on the surfaces of conducting materials. The conducting substrate serves as an infinite source or sink of electrons for regeneration of the active valence state of the immobilized molecule as it reacts with another redox system in the solution phase. The immobilized redox system, which would have fast electron transfer properties with respect to reaction with the solution-phase system, thus acts as regenerative catalyst for the solution-phase system, which usually undergoes slow rates of electrolysis at the normal surface of the electrode material. In general, the object is to bond the redox catalyst covalently to the electrode material in order to have a long lifetime without loss or change of its electron transfer properties. The interest in using $(SN)_x$ as the conducting substrate for modified electrodes arises from the tremendous variety of reactions of sulfur and nitrogen possible for covalently immobilizing the desired redox system to the surface. Thus, it is felt that $(SN)_x$ may have more potential for the production of selective redox catalysts than graphite, metal oxide, or noble metal electrodes.

Because single crystals of $(SN)_x$ are easily grown, the chemistry and electrochemistry of the different crystal faces can be studied. Electrodes which have the axis of the polymer chain parallel to the crystal face (designated parallel electrodes below) and electrodes which have the chain axis perpendicular to the crystal face (designated perpendicular electrodes below) exhibit different chemical and redox behavior in similar circumstances, as might be expected.

General electrochemical behavior. Recent studies using standard electroanalytical techniques have shown that $(SN)_x$ can be used as an electrode material in aqueous media under a variety of conditions, including temperature, pH, and nature of the supporting electrolyte. Observation of background currents for both parallel and perpendicular electrodes, prior to cathodic or anodic breakdown, indicates that surface functional groups undergo oxidation and reduction reactions as the potential is varied. The background currents decrease in magnitude and the breakdown current limits increase as the cation of the supporting electrolyte is changed in the following order: cesium, potassium, sodium, lithium, and calcium. The nature of the anion of the supporting electrolyte has no effect. The cathodic breakdown potential becomes negative with decreasing pH, which is exactly opposite to what is observed on all other electrode materials. These results indicate that cations interact strongly with the $(SN)_x$ surface, which is also the opposite of what occurs with metal electrodes.

The ferro-ferricyanide couple which exhibits reversible diffusion-controlled redox properties at metal, metal oxide, and carbon electrodes was found to behave exactly the same way at both the parallel and perpendicular $(SN)_x$ electrodes. Thus, this redox couple does not interact with the $(SN)_x$ surface and undergoes a fast outer-sphere (no bond disruption of the complex) electron transfer just as it does in all other known electrode materials (metal, metal oxide, and carbon).

Surface-modified electrodes. Because the $(SN)_x$ surface was expected to have a Lewis base nature, the first attempts to modify the surface chemically involved immersing the electrode in solutions containing transition metal and other cations prior to immersing the electrode into an electrolysis solution. As expected, the cations are bound irreversibly to the surface, presumably through the formation of coordinate bonds with the sulfur and nitrogen atoms or functional groups containing these elements on the surface. Still undetermined are what the valence state of the surface coordinated ions is and whether the valence state varies with potential of the electrode; however, it is evident that the presence of the coordinated ion drastically affects the rates of electron transfer at the modified surface. For example, the illustration shows the effect of surface modification of a parallel $(SN)_x$ electrode on the electrochemical kinetics of the reduction of the iodate ion (IO_3^-) to iodide (I^-). Curve A1 shows the behavior of an untreated electrode, which provides virtually no evidence for the reduction of the iodate ion. The rate of this electrode reaction is extremely slow at an $(SN)_x$ surface compared with that at a platinum electrode (curve B). However, after dipping the same electrode into a chromic ion (Cr^{3+}) solution, washing it, returning it to the iodate solution, and repeating the experiment, a well-defined iodate reduction wave indicating a fast electrode reaction is observed (curve A2). As the catalytic effect of the Cr^{3+} is permanent after one treatment, it is felt that some form of chromium has been attached irreversibly to the surface. Other metal ion pretreatments, such as palladium(II), mercury(I), and silver(I), also catalyze the IO_3^- reduction. Other examples are silver(I)- and palladium(II)-pretreated electrodes, which catalyze the oxidation of iodide ion to iodine, and Hg(I), which catalyzes the electrode deposition of lead on parallel $(SN)_x$ electrodes. (The kinetics of the electrode deposition of lead is slow on untreated parallel electrodes and fast on perpendicular electrodes.)

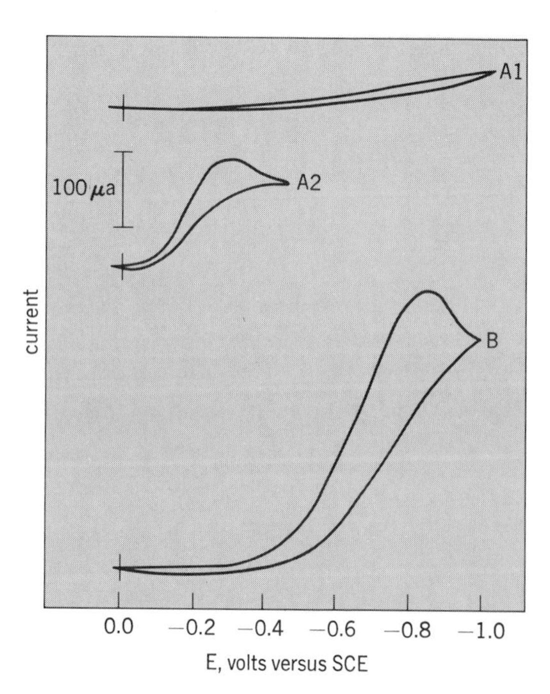

Influence of chromium(III) pretreatment on the cyclic voltammetric behavior at a parallel $(SN)_x$ electrode of a 0.04 *M* KIO_3, 0.5 *M* acetate buffer solution (pH = 4.75). Curve A1 represents the untreated electrode; curve A2, pretreatment by dipping in 1.0 *M* $Cr(ClO_4)_3$; and curve B, the same solution at a platinum disk electrode. SCE = saturated calomel electrode.

A more complex molecule which exhibits electrocatalytic effects has recently been immobilized on an $(SN)_x$ surface. A perpendicular electrode that has been dipped into a pentaamminechlororuthenium(IV) chloride $[Ru(NH_3)_5Cl \cdot Cl_2]$ solution exhibits a significant increase in the rate of oxidation of iodide to iodine. Again, the catalytic effect is permanent, and presumably the sulfur or nitrogen of the $(SN)_x$ substitutes for the coordinated chloride and perhaps one or more of the ammonia ligands of the ruthenium(IV) to form a coordinate bond(s) holding the complex to the surface. The fact that the parallel $(SN)_x$ surface is not affected by pretreatment with the $Ru(NH_3)_5Cl \cdot Cl_2$ solution, and also that the substitutionally inert $Ru(NH_3)_6 \cdot Cl_3$ shows no catalytic effects on $(SN)_x$ of either orientation, also indicate that a mixed ruthenium coordination compound is formed on the perpendicular surface.

Research in this area will continue to receive attention because of the great potential for bonding biological and other redox catalysts covalently to the $(SN)_x$ surface to produce low-voltage electrosynthetic surfaces of practical importance.

For background information *see* ELECTROCHEMICAL TECHNIQUES; ELECTROCHEMISTRY; ELECTRODE in the McGraw-Hill Encyclopedia of Science and Technology. [HARRY B. MARK, JR.]

Bibliography: C. M. Mikulski et al., *J. Amer. Chem. Soc.*, 97:6358–6370, 1975; R. J. Nowak et al., *J. Chem. Soc. Commun.*, pp. 9–11, 1977; R. J. Nowak et al., *J. Electrochem. Soc.*, 125:232–240, 1978; A. N. Voulgaropoulos et al., *J. Chem. Soc. Commun.*, pp. 244–245, 1978.

Energy sources

The oil embargo by the Organization of Petroleum Exporting Countries (OPEC) in the early 1970s brought home to many nations their increasing dependence on foreign oil as the primary energy source. After the embargo the United States instituted programs to lessen its dependence on imported oil. The results of two of these programs are discussed in this article. The first program focuses on the use of hydrogen as an energy source; the second on the use of energy recovered from solid waste.

Hydrogen. The Hydrogen Homestead, recently completed in Provo, UT, is the first step of a multiphase project directed toward the commercial utilization of hydrogen energy. During the initial phase, the feasibility of using hydrogen in domestic applications is to be evaluated. In the Homestead, natural gas appliances have been converted for hydrogen operation. Data from the conversions will be used to determine the technical feasibility of expanding these applications to other residences. During a later phase, the project will be expanded to a 38-home Hydrogen Village. Several of the expanded village homes are presently under construction, and will operate on natural gas for 2 years to provide base-line data.

Energy storage shed. A separate garage-size building was constructed to house experimental prototypes used in connection with the Homestead (Fig. 1). Solar collectors are mounted on the south-facing roof and side wall. There are 10 roof-mounted collectors to heat a water-antifreeze mixture for use with the metal hydride vessel. Two 275-gal (1000-liter) insulated holding tanks are provided for hot-water storage, one for each solar collector array. Experimental equipment currently located in the shed includes a metal hydride vessel, an

Fig. 1. Garage-type building with solar collectors on roof and wall.

Fig. 2. Inside view of Fig. 1 showing computer monitoring system, metal hydride vessel, and hot-water storage tanks.

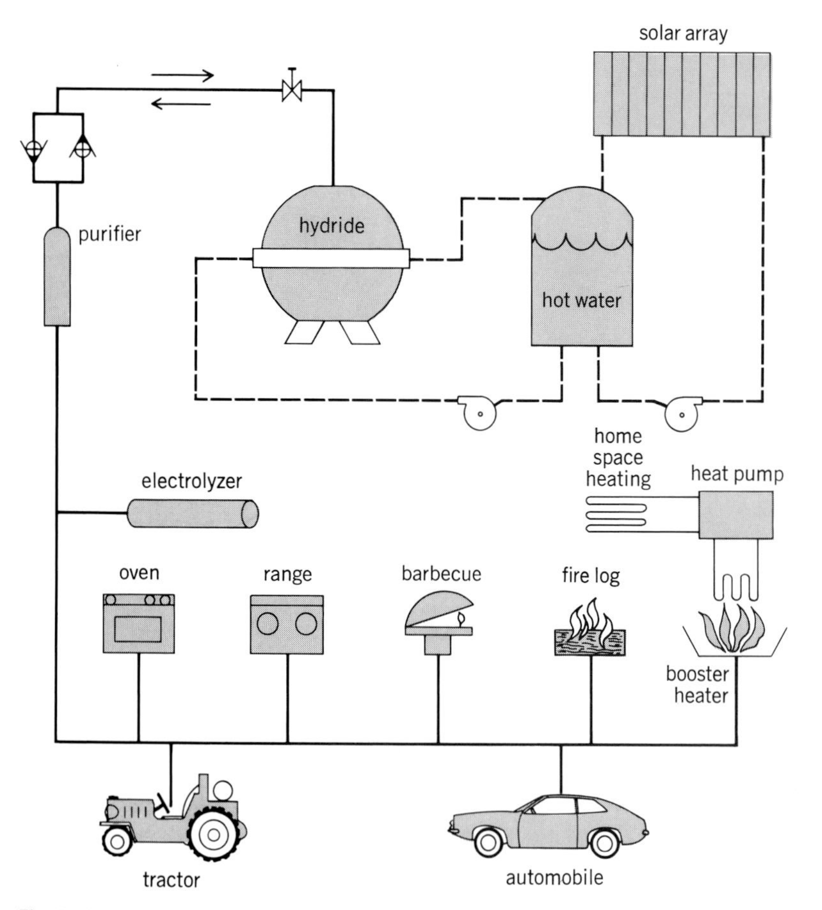

Fig. 3. Hydrogen Homestead energy system.

electrolyzer, hot-water storage tanks, and the computer monitoring system (Fig. 2).

Hydrogen system. The hydrogen system shown schematically in Fig. 3 has been installed in the Homestead. Hydrogen is produced by electrolysis of water using a Billings solid polymer electrolyzer. After passing through a water trap in the electrolyzer unit, hydrogen flows either to the Homestead or to the metal hydride storage vessel, according to demand. Before entering the hydride, the hydrogen is further purified by passing it through a catalyst to convert the oxygen to water vapor. The gas then passes through a molecular sieve dryer. Hydrogen pressure for flow into the hydride is 500 psig (3.4 MPa), the pressure for recharge equilibrium with the hydride and the design output pressure of the electrolyzer.

Hydride equilibrium temperature for the design recharge is approximately 55°C (131°F). The entire hydride bed and vessel are maintained at this temperature under equilibrium conditions by heat liberated by the hydriding reaction as hydrogen enters and by heat exchange with solar-heated water. When Homestead demand exceeds electrolyzer output, the hydrogen flows from the hydride vessel and the pressure drops. The temperature of the hydride also drops as heat is used to release the hydrogen. Hydrogen flows back through the purifier at reduced pressure because of the use of large lines and few restrictions.

In the Homestead, hydrogen is used for gas appliances, to boost the heat input to the heat pump used for space heating, and to fuel two vehicles. The operation of each of these experiments is described below.

Hydrogen production. In the first phase of operation, hydrogen is generated by electrolysis using power from a commercial hydroelectric source. Later, the electrolyzer will be interfaced with a wind machine and with several types of solar-electric generators. An industrial park south of the Homestead is available for future installation and testing of a prototype coal gasifier for hydrogen production to supply the proposed Hydrogen Village.

The electrolyzer installed in the energy storage shed at the Homestead is designed to deliver 3 lb (1.36 kg) of hydrogen per day at a pressure of 500 psig (3.4 MPa). Hydrogen from the electrolyzer recharges the hydride vessel for the Homestead and the vehicle hydride tanks (Fig. 4).

The BEC electrolyzer uses a DuPont membrane material known as Nafion for its electrolyte. This plastic absorbs water and conducts hydrogen ions between electrodes. The membrane, which replaces acid or caustic electrolytes, allows the use of deionized tap water in the cell. The Nafion membrane also acts as a separator to prevent mixing of hydrogen and oxygen within the cell. Disk-shaped electrodes 3.5 in. (8.9 cm) in diameter are pressed against the membrane to obtain good electrical contact. Twenty cells are fitted within housing to make up a 1-lb-per-day (0.45-kg) module.

Hydrogen evolves at a pressure sufficient to charge hydride vessels without the use of a compressor. Oxygen is generated at atmospheric pressure. The pressure difference across the membrane forces the Nafion membrane against the anode for better electrical contact. Water consumption is 3.2 gal (12 liters) per day and electrical consumption is 6 kW. This electrolyzer operates at a high current density of 4000 A/m² to minimize capital cost and size. Lower-current-density units may be constructed for higher electrical efficiency.

Fig. 4. Photograph of Billings electrolyzer.

Hydride vessel. Hydrogen storage is a key feature of the Homestead system. Many of the energy sources proposed for use in small-scale appliances, such as in the Homestead, are intermittent (that is, they use solar and wind electrical generation for electrolysis). In addition, hydrogen demand is high when a vehicle is being refueled or meals are being cooked, but can be near zero for extended periods. A storage system thus decreases the required size of the hydrogen production unit.

Testing and evaluation of a large metal hydride vessel in the Homestead application is being conducted. The objective of this experiment is to evaluate the operational characteristics of a large continuous bed of hydride and the integrity of the containment vessel.

Because the flow rate requirements are quite small compared with vehicle specifications, heat exchange with the vessel is simplified. The storage vessel is composed of two mild steel hemispherical ends welded to a cylindrical center section. The specifications and composition of the vessel are given in Table 1.

A steel vessel of certain compositions may be subject to hydrogen embrittlement under select loading conditions. The storage vessel was examined by Sandia Laboratories, and they determined that the tank was suitable for hydrogen containment. The vessel is constructed of a low-carbon low-manganese steel of moderately fine grain size and low strength. Chemical composition is similar though not identical to materials which have been utilized for hydrogen and hydride storage service. Its compatibility should be similar to or slightly less than that of alloy A515 Grade 70. A proof test at 1000 psi (6.9 MPa) was conducted to clear the vessel for 500-psi (3.4 MPa) service. The hydriding alloy was specified to be $Ti_{0.51}Fe_{0.44}Mn_{0.5}$. A certified analysis of the material indicated a weight composition of 46.8% Fe, 5.2% Mn, 0.006% N, 0.059% O_2, and the balance Ti.

Heat exchange between the vessel and the hot-water storage tank is intended to maintain the hydride at 55°C over slowly varying charge and discharge cycles. The dissociation isotherm for this hydride is shown in Fig. 5. During operation the hydrogen pressure will not exceed 500 psig (3.4 MPa), the set point for electrolyzer shutoff. This pressure is adequate for recharging vehicle hydride tanks, which are much smaller than the Homestead hydride vessel. Line pressure drop to

Table 1. Hydride vessel specifications and composition

Height	48.5 in. (123.2 cm)
Diameter	38.3 in. (97.3 cm)
Wall thickness	0.937 in. (2.38 cm)
Vessel material	Mild steel (0.15 wt % C, 1.1 wt % Mn, 0.09 wt % Cr, 0.005 wt % N)
Internal volume	21.08 ft^3 (0.5969 m^3)
Service pressure	500 psig (3.4 MPa)
Test pressure	1000 psig (6.9 MPa)
Hydride composition	$Ti_{0.51}Fe_{0.44}Mn_{0.05}$
Hydride mass	3950 lb (1791 kg)
Service temperature	131°F (55°C)
Pressure excursion	1 – 500 psig (0.007 – 3.4 MPa)
Usable wt %	1.72%
Stored hydrogen	67.94 lb (30.81 kg)
Stored energy (higher heating value)	4.14 × 10^6 Btu (4.37 GJ)

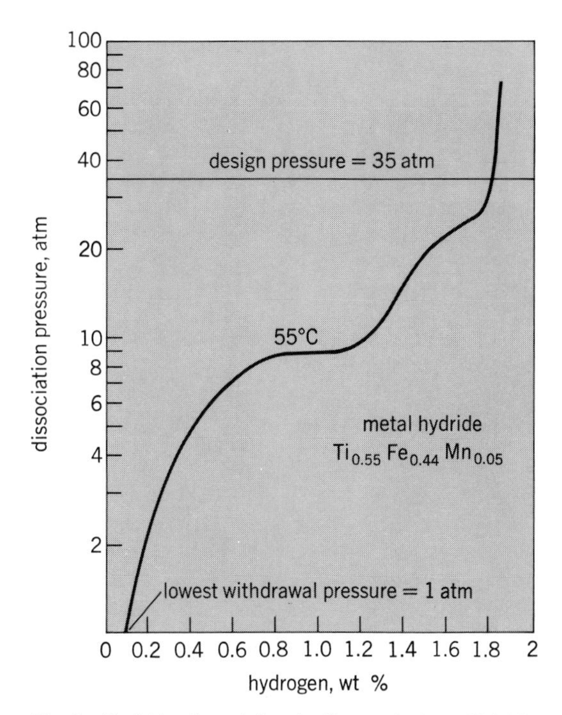

Fig. 5. Hydride dissociation isotherm. 1 atm = 101,325 Pa.

the range is very small, providing a usable pressure excursion from atmospheric to service pressure. Assuming that the discharge takes place over several days and the hydride is maintained at 55°C, the usable weight percent of stored hydrogen is 1.74%. The energy storage for this system is thus 4.14 × 10^6 Btu (4.37 GJ) based on the higher heating value of hydrogen.

Thermistor sensors are installed in the hydride bed and on the vessel surface. Pressure transducers are connected to read hydrogen charge-discharge pressure, bed center-line pressure, wall pressure at the midsection, and wall pressure at the bottom center line. There is also a view port and a hydride sample port. Two wires encircle the vessel girth with linear transducers to measure vessel expansion. A vernier scale measuring tap accompanies the lower wire for redundant measurement of girth changes.

All of the transducers, flow controllers, water circulation pumps, and switches are interfaced with a Billings computer monitoring system. The computer records pertinent data periodically on magnetic disks. It also controls water flow in the heat exchanger and solar collectors and performs hydrogen mass flow of integrations and other data reduction.

As a result of the swelling of the alloy as it hydrides (combines with hydrogen), it is possible for the hydride to become "locked up," preventing expansion to the free surface. Brookhaven National Laboratory tests on small containment vessels have shown that the vessel in that condition is strained beyond the elastic limit. It is believed that the hemispherical shape of the Homestead vessel will be less conducive to lockup than other shapes. However, to ensure against lockup, a loosening jet was installed in the bottom of the tank for the purpose of lifting and loosening the hydride bed with blasts of hydrogen. The same port will also be used

ENERGY SOURCES

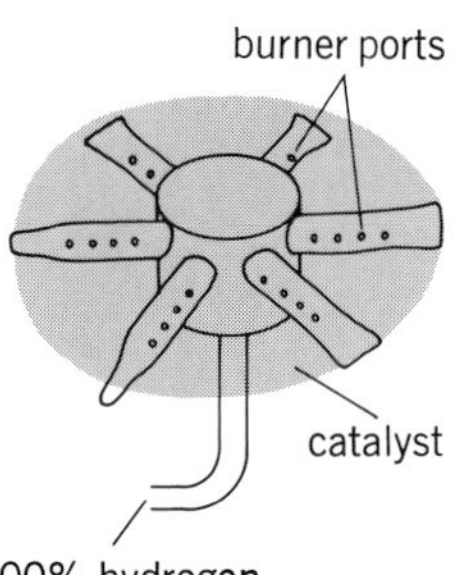

Fig. 6. Hydrogen stove-top burner.

to evaluate methods of hydride heat transfer with hydrogen recirculation.

Tests planned for the system include the following: (1) total hydrogen capacity at equilibrium; (2) discharge capacity at constant flow rate; (3) change of hydride particle size and activity with extended use; (4) pressure drop through the hydride bed; (5) pressure drop across the delivery line filter; (6) temperature profile within the hydride bed; (7) temperature gradient on the vessel surface; (8) heat transfer to the circulation water; (9) filter clogging; (10) alloy mobilization in the outlet gas; (11) change in tank girth with pressure, temperature, and hydride packing; (12) observation of hydride motion and particle size change; (13) effectiveness of loosening jets in breakup of hydride packing and lockup; and (14) effectiveness of circulation of heated hydrogen in discharging the vessel.

These tests will provide engineering scale information relative to the use of hydriding alloys in pressure vessels. The storage vessel represents a practical application of new technology as an important part of an integrated energy system using hydrogen.

Appliances. Five natural gas appliances—an oven, range, barbeque, fireplace log, and booster heater for the heat pump system—were converted to hydrogen to determine their practicality and investigate advanced burner designs.

It has been demonstrated during previous experiments that the nitric oxide (NO_x) formation from an uncontrolled hydrogen burner can exceed levels from conventional natural gas devices by as much as a factor of 10 (25 to 35 parts per million is typical for natural gas appliances while levels in excess of 250 ppm NO_x have been observed for hydrogen devices). The NO_x are formed when nitrogen and oxygen, the two major constituents of air, are heated above a threshold temperature of about 1315°C (2400°F). The higher concentrations for hydrogen are a result of the fact that the laminar flame speed of a stoichiometric mixture of hydrogen and air is 10.7 ft/s (3.26 m/s), compared with 1.5 ft/s (0.46 m/s) for natural gas–air mixtures. Higher hydrogen flame speed results in a larger fraction of hydrogen being burned in a smaller area and with a higher peak temperature; the result is an increase in NO_x formation. Although the combustion temperatures vary only modestly, natural gas combustion takes place at the lower end of the NO_x formation threshold, and consequently only a small increase in temperature is needed to cause a significant increase in NO_x formation.

Appliance conversions were accomplished by using a technique developed under a contract from the Mountain Fuel Supply Company of Salt Lake City. In this technique, NO_x formation is controlled by the application of two interacting phenomena. Hydrogen-air mixing is inhibited by (1) blocking all primary air and (2) placing a stainless steel wire mesh around the burner ports. The placement of the wire mesh in a proper configuration allows the gradual mixing of hydrogen and air throughout the operating flow rates of the burner design. A hydrogen-rich atmosphere exists close to the burner, with the oxygen concentration increasing with distance from the burner head. If the stainless steel material is properly configured, a region of inflammable hydrogen concentration immediately surrounding the burner openings will exist. This region, referred to as the inflammable

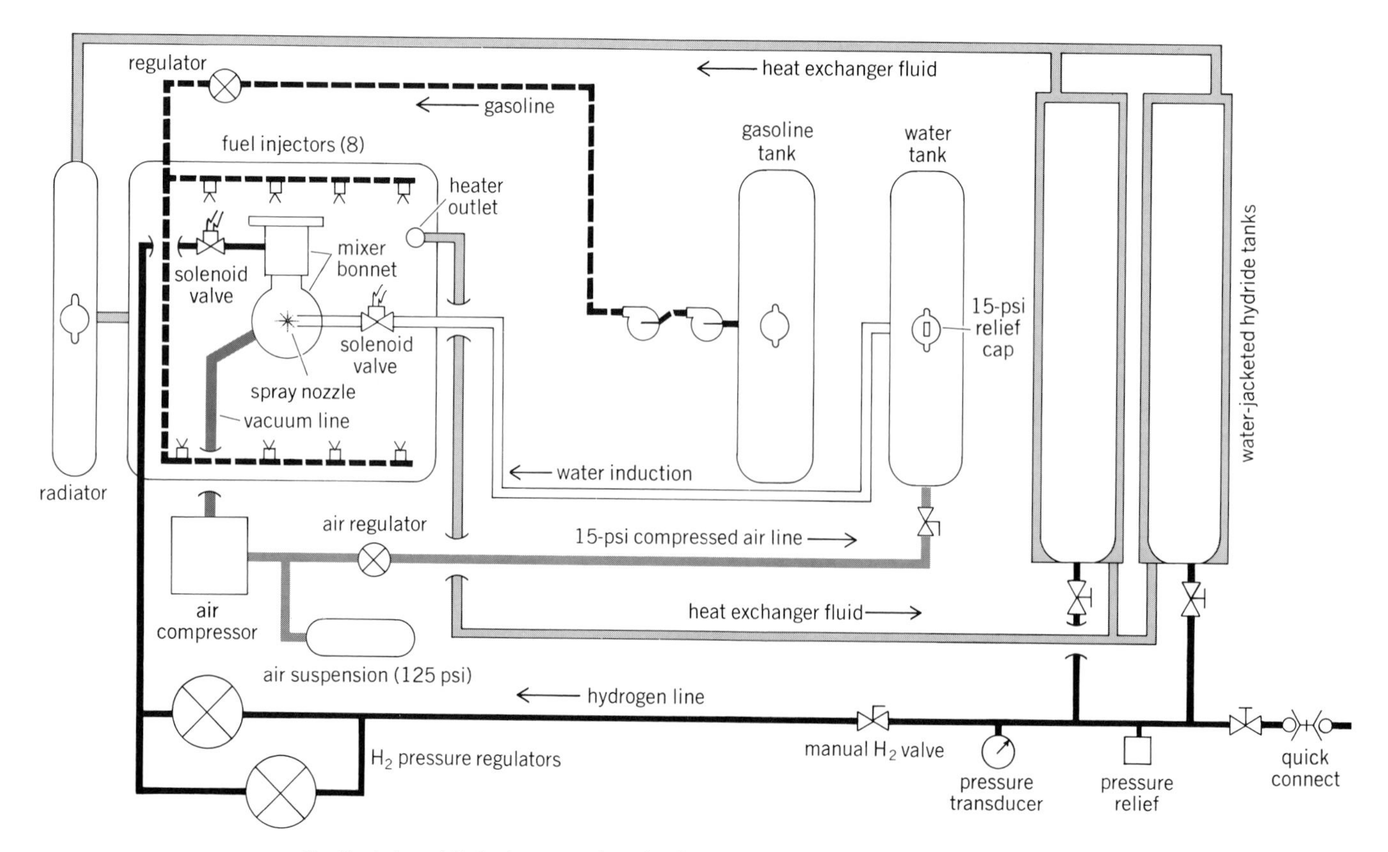

Fig. 7. Automobile hydrogen system. 1 psi = 6895 Pa.

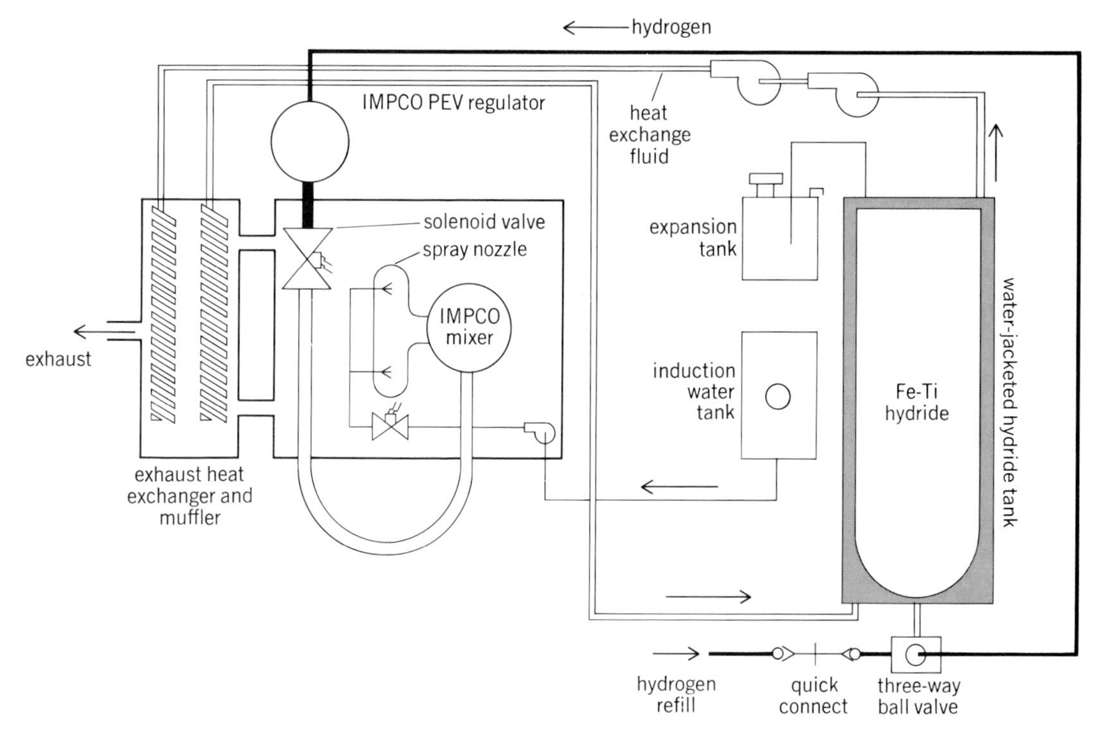

Fig. 8. Tractor hydrogen system that uses an air-cooled engine.

zone, moves in toward and out from the burner head, depending on hydrogen flow rate. A proper burner design incorporates sufficient stainless steel material to ensure that the inflammable limit zone is always located within the outside perimeter of the stainless steel material, as is illustrated in Fig. 6.

The stainless steel also provides a very important secondary function in addition to its mixing-inhibiting function. At high temperatures, stainless steel is an excellent catalyst for hydrogen combusting. Shortly after ignition, the temperature of the stainless steel is raised sufficiently by hydrogen combustion so that catalysis begins. At this point, the hydrogen in the inflammable zone begins to react with the dilute quantities of oxygen present through the action and on the surface of the stainless steel catalyst. In this manner, a controlled reaction occurs in the inflammable zone where mixture limitations will not permit the rapid combustion of hydrogen which would occur under normal conditions. Consequently, peak combustion temperatures are maintained below the threshold level for NO_x formation. NO_x data of 1 to 5 ppm are thus achieved.

Vehicles. A passenger car and a garden tractor have been built as part of the Hydrogen Homestead to demonstrate the use of hydrogen as a substitute for gasoline. The range of a car of this size is limited by the weight of the installed hydride tank. Research into improved hydride materials continues. In the meantime, the practicality of a dual fuel system has been demonstrated by the hydrogen automobile. Since a large portion of total road miles accrue from short trips, for example, commuting to work and shopping errands, a small hydrogen storage system that is switchable to gasoline service while the vehicle is being driven is satisfactory at present. Garage recharge of the hydride tank from the electrolyzer permits predominant hydrogen utilization in the vehicle for a typical driving cycle. For longer trips gasoline is used. The hydrogen automobile dual fuel system is shown in Fig. 7.

A garden tractor powered exclusively by hydrogen can operate for 1.5 hr on one charge of hydrogen. The hydrogen system is shown schematically in Fig. 8. Hydrogen is stored in an Fe-Ti hydride within a water-jacketed aluminum vessel. Pressure is regulated by an Impco model PEV-2 regulator feeding an Impco model CA50 gaseous mixer. A solenoid valve is used to shut off hydrogen supply to the engine. The ignition system was modified by elimination of centrifugal spark advance and by decreasing plug gap. No changes in compression ratio were made on this air-cooled Kohler engine. Water induction into the engine-intake gas stream is accomplished with two spray nozzles mounted above the intake valves. The induction pump and solenoid valve are under manual control. Heat exchange with the hydride is provided by circulation of water-antifreeze solution through an exhaust gas–heat exchanger replacing the original equipment muffler. Two small centrifugal pumps circulate fluid between the heat exchanger and hydride tank as shown in Fig. 8.

Decreased vehicle power and performance were not noticeable on the tractor, probably because of the low-gear drive train and hydrostatic drive. The tractor will be used to mow the Homestead lawn and to cultivate the garden. It demonstrates the practicality of hydrogen fuel for use in small work vehicles (such as tractors, fork-lift trucks, front-end loaders, and sweepers) operated in proximity to a source of hydrogen.

Conclusions. The Hydrogen Homestead demonstrates the possibility of near-term application of the hydrogen-electricity-solar design concept for residential use. The key elements are the production, storage, and utilization of hydrogen in ways that complement both solar and electrical systems.

In this example, production of hydrogen is accomplished by electrolysis of water using hydroelectric energy. However, electricity could also be generated from other renewable energy resources such as the Sun or wind. Future plans call for hydrogen supply by pipeline from direct conversion of coal. The hydrogen would be stored in a vessel containing Fe-Ti hydriding alloy. Operational characteristics of the vessel and hydride are now being studied.

The Homestead serves as a test facility for evaluation of hydrogen systems and the interface with electric and solar systems. It is believed that test operation of available hardware will promote interest in hydrogen applications and encourage the development of hydrogen-related technology.

[ROGER E. BILLINGS]

Energy recovery from solid waste. Although the concept of energy recovery from municipal solid waste is not new, it has only recently become of major interest in the United States. Conditions favorable to the use of municipal refuse as an energy source include the higher cost of conventional (oil, gas, and coal) and nuclear fuels; the limited availability of land for use as landfills in urban areas; the improvements of energy recovery technologies; and the enforcement of more stringent environmental regulations by Federal and state governments.

Energy has been reclaimed from refuse in Europe for over 20 years and, more recently, in Japan, Canada, and Australia. As energy conservation and improvement of the quality of the environment become key policy goals in the United States, the concept of refuse-to-energy has gained increasing importance. Domestic refuse-to-energy approaches generally include the application of improved combustion practices as demonstrated in Europe, as well as the development of new techniques for producing solid, liquid, and gaseous combustible products from municipal solid waste.

Table 2. Summary of energy recovery facilities implementation[a]

Location[b]	Type[c]	Capacity, tons per day[d]	Products (markets)	Startup date
Operational facilities				
Ames, IA	RDF	400	RDF, Fe, Al	September 1975
Blytheville, AR	MCU	50	Steam (process)	November 1975
Braintree, MA	WWC	240	Steam (process)	1971
Southwest Chicago	RWI	1200	Steam	1963
Northwest Chicago	WWC	1600	Steam (no market)	1970
Bridgewater, MA (N, E)	RDF	160	RDF (utility)	1974
Groveton, NH	MCU	30	Steam (process)	1975
Harrisburg, PA	WWC	720	Steam (no market)	1972
Merrick, NY	RWI	600	Electricity	1952
Miami	RWI	900	Steam	1956
Nashville	WWC	720	Steam (heating and cooling)	July 1974
Norfolk, VA	WWC	360	Steam (Navy base)	1967
Oceanside, NY	RWI/WWC	750	Steam	1965–1974
Palos Verdes, CA	Methane recovery		Gas (utility) and Fe	June 1975
St. Louis (D)[e]	RDF	300	RDF (coal-fired utility)	1972
Saugus, MA	WWC	1200	Steam (process)	October 1975
Siloam Springs, AR	MCU	20	Steam	September 1975
South Charleston, WV (N)	Pyrolysis	200	Gas, Fe	1974
Washington, DC (N)	RDF	80	RDF, Fe, Al, glass	1974
Facilities under construction				
Baltimore (D)	Pyrolysis	1000	Steam (heating and cooling), Fe, glass	June 1975
Baltimore County (G)	RDF	550	RDF, Fe, Al, glass	April 1976
Chicago (Crawford)	RDF	1000	RDF (utility)	March 1977
Hempstead, NY	WRDF/WWC	2000	Electricity, Fe, Al, glass	N.A.[f]
Milwaukee	RDF	1000	RDF, corrugated, Fe	1977
Mountain View, CA	Methane recovery		Gas (utility)	June 1977
New Orleans (N)	RDF[g]	650	Nonferrous, Fe, glass, paper	November 1976
Portsmouth, VA (shipyard)	WWC	160	Steam loop	December 1976
San Diego County (D)	Pyrolysis	200	Liquid fuel (utility)	April 1977
St. Louis	RDF	6000	RDF (utility), Fe, glass, Al	N.A.[f]

[a]From *Resource Recovery and Waste Reduction: Fourth Report to Congress (3)*, August 1977.

[b]D = EPA demonstration grant; G = EPA implementation grant; N = non-EPA pilot or demonstration facility; E = Energy Research and Development Administration grant.

[c]RDF = refuse-derived fuel; WRDF = wet-pulped refuse-derived fuel; WWC = waterwall combustion; RWI = refractory wall incinerator with waste-heat boiler; MCU = modular combustion unit.

[d]1 short ton = 0.9 metric ton.

[e]Plant closed down in 1976.

[f]N.A. = not applicable.

[g]Uses RDF technology, but current plan is to landfill the light fraction because of lack of market.

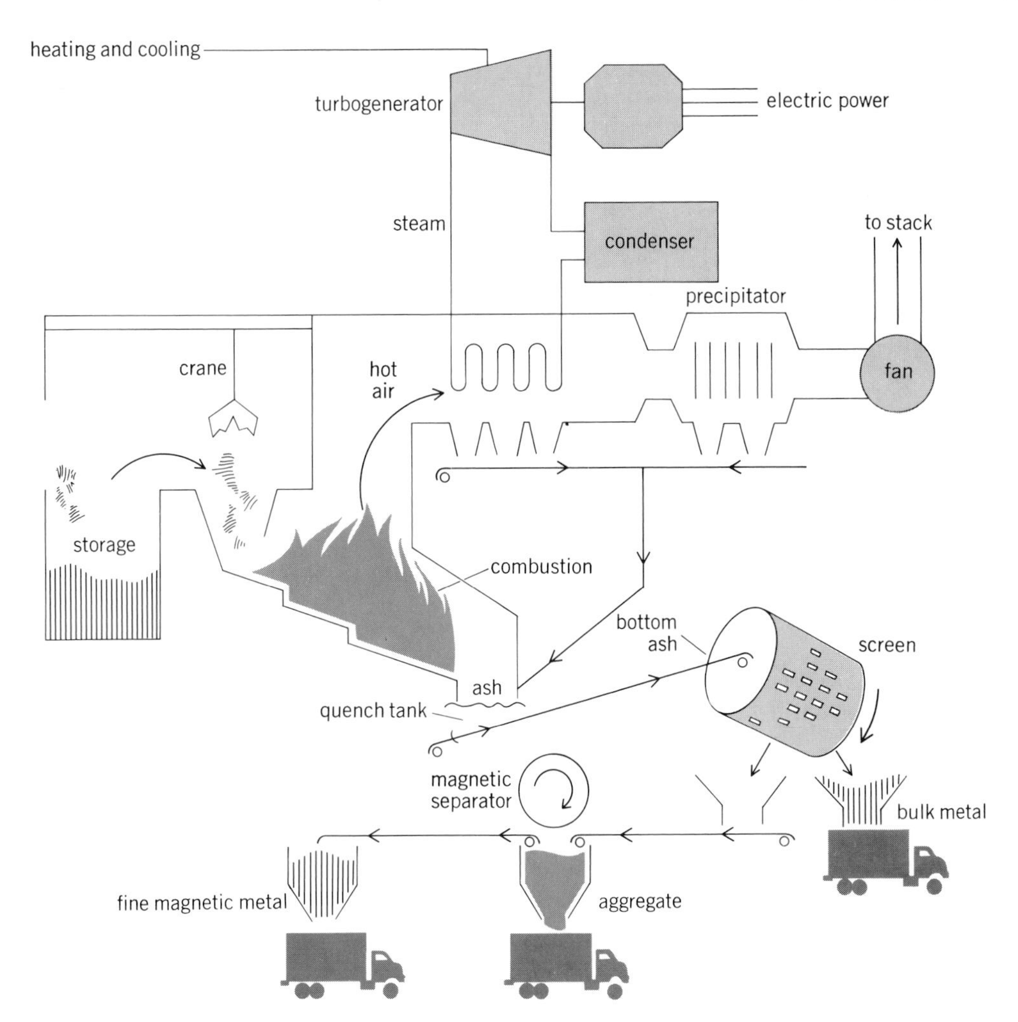

Fig. 9. Schematic of a refuse-to-energy process.

To date, only one technology has been fully developed for commercial use in the United States, the refuse-fired steam-generating system, in which refuse is burned "as received" in large, specially designed heat recovery furnaces. A second emerging technology is the preparation of shredded solid waste combustibles, sometimes referred to as refuse-derived fuel (RDF), for use either as a fuel supplement in electric utility or industrial boilers, or as a feedstock for conversion into liquid or gaseous derivatives, usually through a pyrolytic process. Other experimental techniques include the biological conversion of solid waste through bacterial decomposition into a combustible gas and burning solid wastes to recover hot gases which are used to drive gas turbines.

Table 2 shows the status of energy recovery from refuse in the United States. The listed capacities do not necessarily imply that these levels have been achieved on a regular basis. Many of these facilities are of a pilot or demonstration nature, so that the start-up date does not imply continuous operation from that date. [The Environmental Protection Agency (EPA) may be consulted for more detailed information on the status and operating data of individual projects.]

Refuse-fired steam-generating systems. According to the EPA, the "combustion of solid waste on mechanical grates in waterwall furnaces to recover steam is the most thoroughly proven resource recovery technology."

In the refuse-fired steam-generating system, municipal refuse trucks unload waste into a large storage pit. An overhead crane transfers waste from the pit and discharges it into a feed hopper from which it flows into the furnace, where it is burned on inclined reciprocating grates (Fig. 9). The waste discharged into the furnace feed hopper is in the form "as received" from the delivery vehicles.

From the feed hopper, waste is fed onto the furnace grates where combustion takes place in specially designed waterwall refractory furnaces under controlled conditions. Heat of combustion is absorbed by the water tubes in the furnace itself, and by a high-temperature boiler capable of producing turbine-quality steam. (There is no processing by means of shredding, shearing, or flailing mechanisms as is done to produce the various forms of shredded waste combustibles described below.)

Hot combustion gases within the furnace range in temperature from 800 to 1000°C. As the combustion gases pass through the boiler, they are cooled. The cooled gases pass through air-pollution-control devices (usually electrostatic precipitators) in order to remove dust and smoke in compliance with environmental standards. The cleaned combustion gases are then discharged through a stack.

The noncombustibles in the waste, such as bottles, cans, bicycles, or bed springs, are cooled in a trough of water and conveyed to a glass and metal reclamation area. Magnetic metals, nonferrous metals, glass, and residue can then be sorted and processed for resale.

Refuse-fired steam-generating plants are operating in several parts of the country. Of these, two of the largest facilities, in Nashville and in Saugus, MA, produce and sell steam commercially. The Boston North Shore Project in Saugus is an excellent example of the viability of this technology. Since late 1975, it has processed over 850,000 tons (771,000 metric tons) of municipal waste and has delivered over 4.4×10^9 lb (2.0×10^9 kg) of steam for sale, without turning away or diverting waste from a contract community.

Early refuse-to-energy facilities were designed to produce low-pressure steam and hot water for central heating or industrial processing. This is still a widespread application. However, as a result of several design improvements during recent years to reduce and control corrosion, production of steam at higher temperature and pressure is now possible, enabling efficient electric power generation.

With the production of steam of a quality suitable for efficient electric power generation, a refuse-fired steam-generating facility can be sited independent of a specific industrial or commercial steam user. Power can be introduced into a transmission grid from one of many locations. Thereby, the plant's vulnerability to steam load fluctuation is reduced, flexibility in plant location is broadened, and the sale of highly marketable electric power is possible. The additional power supplied to an electric utility company has minimal impact on its grid network and adds to its ability to meet demand requirements.

Preprocessed waste combustibles. In this approach, refuse "as received" is initially processed through one or more sorting devices and shredders where it is reduced in size. At this point, magnetic metals can be removed for sale if markets exist. The remaining waste passes through an air-gravity separator where it is separated into lighter particles (primarily paper, plastics, and varying amounts of noncombustibles) and heavier materials (mainly glass, other metals, and some combustibles). The lighter, shredded waste combustibles can then be introduced into a utility boiler as a fuel supplement to coal, generally at a level of 5–15% of the boiler's heat output. The noncombustible portion of the waste undergoes further processing to reclaim metals and glass, leaving a residue suitable for landfill.

There are several limitations to the production and use of shredded waste combustibles. According to the EPA, design and operating parameters have not been well defined for the most cost-effective approach to the production, storage, transport, and firing of shredded waste combustibles. Also, many potential users, primarily electric utility companies, are reluctant to use this material in their power plants because of potential adverse effects on boiler operation and internal pollution-control requirements.

Pyrolysis of shredded waste combustibles. Pyrolysis occurs when organic materials are destructively distilled or "baked" by intense heating in the absence of oxygen. When subjected to pyrolysis, solid waste is reduced to a gas (containing primarily hydrogen, carbon monoxide, carbon dioxide, and methane); a liquid (containing water and a variety of organic chemicals); and a carbonlike char or a glassy molten slag material. The products of pyrolysis are either gaseous or liquid, depending on such factors as temperature and pressure in the pyrolysis reactor, retention time, feed particle size, and use of auxiliary fuels or catalysts.

Before pyrolysis, refuse undergoes the same sorting, shredding, magnetic separation and, in some cases, air-gravity separation performed to obtain shredded waste combustibles. This primary combustible mixture is then injected into a furnacelike reactor where it is subjected to high-temperature destructive distillation in an oxygen-deficient atmosphere. The gas or liquid products are subsequently tapped from the reactor while the remaining residue is ejected from the bottom. In some reactors, temperatures are high enough to melt bottles and cans into a molten slag material which, upon cooling, has the appearance of a black, glassy aggregate.

There are a small number of developmental pyrolysis systems and numerous experimental systems in existence in the United States. Two of these produce a low-Btu gas which can be burned to produce steam; another produces a medium-Btu gas planned for marketing as a fuel. A fourth produces an oillike liquid product.

Because pyrolysis systems are still experimental, full-scale applications must be proved before becoming commercially available. At the same time, additional testing and experimentation are necessary before potential users of pyrolytic products will enter into long-term purchase agreements.

Experimental facilities. Several facilities are being constructed as pilot plants to experiment with new approaches for extracting energy from municipal solid waste.

One such experimental approach is methane recovery from landfills. In this process, gas generated from decomposing waste deep in the ground is reclaimed through a deep well and a gas collection system on the surface. Prototype systems are currently in operation near Los Angeles. A second prototype is being developed to recover methane from decomposing solid waste mixed with sewage sludge in Pompano Beach, FL. In a gas turbine pilot project in Menlo Park, CA, combustion gases from burning solid waste drive a gas turbine. Research and development programs with this system have been going on for many years with inconclusive results.

Future developments. According to the EPA, the next few years should improve the yet uncertain technical reliability of many emerging systems. However, the longer-term questions concerning economic feasibility still remain. Thus far, favorable competitive economics are found in those areas where municipal solid waste disposal costs are high and a great value is placed on substituting energy from refuse for existing energy sources.

EPA survey data indicate that there are 21 pilot, demonstration, and operating resource recovery projects, 10 under construction, and 33 in the advance planning category. In addition, 54 localities have energy recovery projects in the early planning stages. Of the operational plants, 19 recover energy, 13 of which do so by direct combustion of waste. Six of these employ integrated refuse-fired boilers, while the remaining seven recover energy with waste heat boilers.

Energy recovery from waste is becoming an increasingly attractive alternative to waste disposal from several standpoints: It can be integrated into existing waste collection and transfer methods, particularly with newer transfer compactor substations, permitting more efficient haulage of compacted waste loads. It can reduce land disposal requirements substantially by decreasing as-received refuse volume to as little as 5% in the case of refuse-fired steam generators. Finally, in the case of proved refuse-to-energy systems, it can economically conserve valuable energy in an environmentally safe and acceptable manner, thereby improving the quality of life.

For background information *see* ENERGY SOURCES in the McGraw-Hill Encyclopedia of Science and Technology.

[JOHN M. KEHOE, JR.; JOSEPH FERRANTE, JR.; CHRIS G. GANOTIS]

Bibliography: N. R. Baker, *Oxides of Nitrogen Control Techniques for Appliance Conversion to Hydrogen Fuel*, Billings Energy Corp., Provo, UT, 1974; A. U. Blackham, *The Flameless Catalytic Oxidation of Hydrogen Odorants for Hydrogen Illuminants for Hydrogen*, Mountain Fuel Supply Company Project for Billings Energy Corp., 1974; B. C. Campbell, *Development of Billings SPE Electrolyzer*, 2d World Hydrogen Energy Conference, Zurich, August 1978; W. G. F. Grot, G. E. Munn, and P. N. Walmsley, *Perfluorinated Ion Exchange Membranes*, E. I. DuPont de Nemours & Co., Wilmington, DE, 1972; W. K. MacAdam, *IEEE Spect.*, 12(11):46–50, 1975; S. L. Robinson, *Sandia Laboratories Rep. SAND77-8293*, January 1978; G. Strickland, J. Milan, and M. J. Rosso, *BNL23130*, Brookhaven National Laboratory, August 1977; U.S. Environmental Protection Agency, *Resource Recovery Plant Implementation: Technologies*, Environ. Protect. Publ. SW-157.2, 1976; U.S. Environmental Protection Agency, *Resource Recovery and Waste Reduction: Fourth Report to Congress*, Environ. Protect. Publ. SW-600, pp. 45–51, 1977.

Energy storage

Electrical energy is not easily stored and is usually generated when needed. Satisfying instantaneous demand thus requires utilities to install generation capacity to meet peak demand load, and hence to have excess capacity under off-peak conditions. Because average generation requirements for utilities are less than 60% of peak load, electricity generated during off-peak times can decrease peak generation input energy requirements (Fig. 1) if energy storage costs are favorable.

Parameters important in determining which electrical storage method to use are given in Table 1. Electricity has been stored in various energy forms by use of natural phenomena as follows: (1) in a chemical compound by charging an electric storage battery; (2) in an electric field by electrostatic charge separation in a capacitor; (3) in a magnetic field by an electric current flowing in a coil; (4) in a gravity field by pumping water to an elevated reservoir; (5) in a mechanical system by spinning a flywheel with electricity; (6) in a pressurized gas by compressing air into a reservoir; (7) in a nuclear isotope by inducing radioactivity in the isotope; (8) in a thermal system by heating or changing the phase of a material.

Battery. A national battery energy storage test (BEST) facility for utility application is scheduled for operation in 1980 with 5000–10,000-kWh battery unit tests planned. Today's lead-acid battery is not economical for utility electricity storage due to its high cost ($100/kWh) and limited life cycles (300 cycles, 80% discharge; 10,000 cycles, 25% discharge). A utility lead-acid battery system, including ac to dc conversion and cooling, would cost about $1000 to store enough energy to provide 1 kW for 10 hr ($100/kWh). This is not economical based on existing and predicted cost ratios of peak to off-peak power. Utility target costs are $10–50/kWh for 10 hr discharge based on 10 to 20 years' battery life. Advanced lead-acid, sodium-sulfur, zinc-chlorine, and sodium-antimony batteries are being developed to meet these targets.

Lead-acid batteries used in the two electric cars sold in the United States store 6.25 kWh (0.025 kWh/kg) and cost about $600, but are not adequate. Batteries being developed such as zinc–nickel oxide improve performance, but cost about five times as much.

In comparison to battery storage efficiency of 80%, electrical energy storage by electrolytic production of hydrogen, storage, and later conversion to electricity in a fuel cell has an overall efficiency of less than 25%.

Capacitor. Electrostatic energy E_e stored in a capacitor is equal to $E_e = CV^2/2$, where E_e is in joules, C is the capacitance in farads, and V is the voltage in volts.

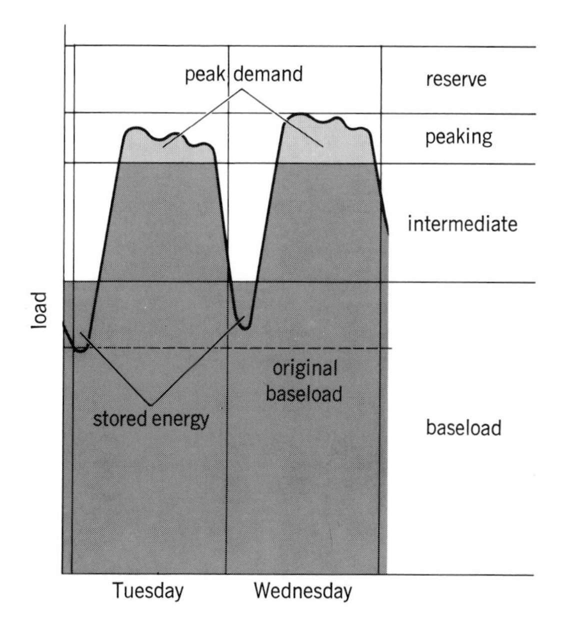

Fig. 1. Load graph showing peak demand reduced by stored energy. *(From R. Whitaker and J. Birk, Storage batteries: The case and the candidates, EPRI J., 1(8):6–13, October 1976)*

Table 1. Energy storage comparison

Energy form	Device	Date commercially available	Efficiency, %	Unit energy, kWh†	Cost, $/kWh†	Weight‡, kg/kWh
Chemical compound	Battery	1985*	60–80	1–10	50–100	40
Electrical field	Capacitor	*	90	0.01	500,000	–
Magnetic field	Superconducting magnet	After 2000*	70–85	$0.1–10^7$	50–300	–
Gravity field	Pumped storage	Now	70–75	$10^5–10^7$	25–100	4000
Mechanical system	Flywheel	1985*	70–85	$1–10^5$	100–1000	25–1000
Pressurized gas	Compressor and turbine	Now	50–70	$10^5–10^7$	100–200	60
Nuclear isotope	SNAP	*	<1	$1–10^4$	$10^4–10^6$	0.33
Heat	Thermal storage device	1985	25–75	$10–10^6$	200–300	23–200

*Units now available are not economic for large energy storage.
†kWh = 3.6×10^6 J.
‡Kilograms of energy storage material not including conversion equipment.

At Princeton Plasma Physics Laboratory a 0.01-F capacitor bank is charged to 15,000 V to store 10^6 J (0.28 kWh) of electrical energy for use with a Tokamak fusion device. Similar capacitor banks have stored five times as much energy. Such capacitors cost about $700,000/kWh ($0.20/J), and are used for storing small amounts of electricity with discharge times of about RC seconds, where R is the resistance of the circuit in ohms and C the capacitance in farads.

Inductor. Electricity stored in a magnetic field is equal to $E_L = i^2L/2$, where E_L is joules of energy stored and i is amperes of current flowing in the magnet coil with inductance L in henries. The coil wire must be superconducting or the energy would be dissipated in a time of about L/R seconds, where R is the resistance of the coil in ohms. Electricity can circulate in superconducting coils without significant loss.

Superconducting coils which store 300,000 J (0.08 kWh) are in use experimentally at Los Alamos and are about the size of an oil barrel. A bubble-chamber superconducting magnet in Europe stores 8×10^8 J (225 kWh) in a 3.5-tesla magnetic field.

A University of Wisconsin group has analyzed superconducting coils 100 m in diameter (2920 H) for storing 10^7 kWh (3.6×10^{13} J) for utilities. This energy would be stored by operating a 1000-MW nuclear plant 10 hr at night when excess low-cost power is available, and used to reduce the peak generation requirements the next day. The cost of such a system, including ac to dc conversion and cryogenic cooling, has been estimated to be about $500,000,000 ($50/kWh). Whether such large devices are ever built depends upon resolution of technical and cost uncertainties.

Pumped storage. The most economical method used to store electricity for utility application is the pumping of water up into an elevated reservoir and later allowing the water to run back through a water turbine generator to produce electric power. Pumped-hydro capacity installed in the United States as of 1978 was about 12,000 MW. Costs depend upon the availability of a suitable site for an elevated lake, although studies of digging underground storage caverns for this purpose have been made.

The Ludington Pumped-Storage facility on Lake Michigan, operating since April 1974, has generated 2076 MW and is the largest facility of this type in the world. Overall generating-to-pumping efficiency is 72%, which is 3% above the guarantee. It can store 1.5×10^7 kWh of electrical energy with 54,000 acre-ft (6.4×10^{10} liters) usable volume of water at a height of 90 m in a 4.5-km² reservoir. The final cost was $350,000,000 ($175/kW) for the plant and its associated transmission requirements, which amounts to $23/kWh for energy storage. It can be operated and maintained by a crew of 30 people, which is less than one-fifth that required for a steam plant of comparable size. Availability of power is about 300 MW in less than 10 min, and full output in about 30 mins from standstill.

Flywheel. Electricity can be stored by using an electric motor to spin a flywheel and later using the electric motor as a generator to convert the rotating mechanical energy back into electrical energy. Charge and discharge times are set by motor-generator ratings, while energy storage capability E_f is given by $E_f = 2 \times 10^{-6} N^2 I$, where E_f is in joules, I is the rotational inertia in g-cm², and N is the speed of rotation in revolutions per second.

As an example, a system to supply 1860 kWh (6.7×10^9 J) of electrical energy to a fusion experiment was designed as two vertical-shaft motor-generators and 818-metric-ton flywheels each with 9.5×10^{13} g cm² of rotational inertia. Each stores 1250 kWh (4.5×10^9 J) at 290 revolutions per minute (rpm) with rotation of the generator-flywheel de-

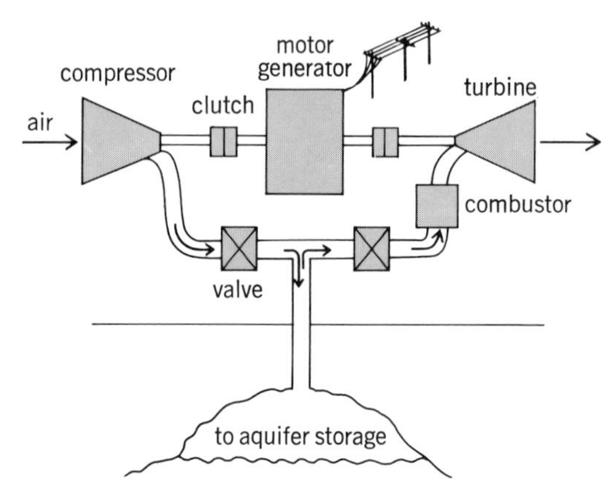

Fig. 2. Diagram of compressed-air-storage gas turbine power station. *(From W. R. Lang, Air stored for peaking power, Elec. World, 187(1):30–31, 1977)*

Table 2. SNAP units

SNAP number	Use	Power, W	Weight, kg	Isotope	Half-life	Generation life
3	Demonstration device	2.5	1.8	Polonium-210	138 days	90 days
7B	Fixed navigational light	60	2100	Strontium-90	28 years	3–5 years
9A	Satellite power	25	12	Plutonium-238	89.6 years	5 years

creasing to 145 rpm during the pulsed operation to supply 3.35×10^9 J per generator. At 290 rpm the flywheels each store 0.0007 kWh/lb (5500 J/kg) and are made of steel laminations bolted together and conservatively designed for use, as in a water-wheel application, for 50 years.

In contrast, a flywheel made of kevlar fibers held together with an epoxy binder is being tested for an electric car in which it will store energy during deceleration and generate electricity during acceleration to smooth battery requirements. It can operate at 25,000 rpm and store 1 kWh (3.6×10^6 J) in a 26-kg rotor, or 1.4×10^5 J/kg, which is 25 times as much energy per unit weight as the low-speed flywheel.

Compressed air storage. In a compressed-air-storage combustion turbine power station (Fig. 2), an electric motor compresses air into a cave. The energy is reclaimed by using the compressed air as input to the combustor of a combustion turbine where it reacts with fuel to produce hot gas which spins a combustion turbine and generator.

The world's first compressed-air-storage station is the Huntorf 290-MW plant in Germany, with commercial operation planned by September 1978. For 8 hr its motor uses 58 MW to compress air (mode 1). For 2 hr 290 MW of electricity is generated (mode 2) by using the compressed air as input to the combustor, with a heat rate of 5500 Btu (1.4×10^6 cal or 5.8×10^6 J) of fuel per kilowatt-hour of electrical output. Because in conventional compressor-turbine operation (mode 3) about two-thirds of the turbine gross output power is used to operate the compressor, about three times as much electricity can be generated when stored compressed air is used.

Efficiency of electrical energy storage and regeneration is complicated by introduction of energy in the form of fuel for the gas turbine. One method for calculating efficiency e is given by the equation below. "Energy fuel could produce" is

$$e = \frac{\text{Energy out} - \text{energy fuel could produce}}{\text{Compressor energy input}}$$

assumed to be in a gas turbine (11,000 Btu/kWh).

Huntorf's 63% efficiency could be increased to over 70% by using exhaust gas from the turbine to heat air input from the reservoir. Total cost of the plant is approximately $200/kW. This includes the underground storage facility of two 150,000-m^3 caverns (with less than 0.001% leakage per day) for operation at a pressure of 1050 psi (70 bars, or 7 megapascals) which were solution-mined out of a geological salt deposit. About 4.6×10^5 kWh (1.7×10^{12} J) of electricity can be stored at a cost of about $125/kWh.

Radioisotopes. Energy can be stored by producing a radioactive isotope which will provide energy when it decays. Charged particles emitted can be collected on an electrode to produce electrical energy directly or, alternatively, the energy released as alpha, beta, and gamma radiation can be absorbed to produce heat which is then converted to electricity by thermocouples. The system for nuclear auxiliary power (SNAP-3) first produced 2.5 W of electricity from polonium in January 1959. Other facts about SNAP units are summarized in Table 2.

The overall energy storage and regeneration efficiency of a SNAP-3 unit is about 0.1% if it is assumed that a gram of polonium-210 with initial power density of 140 W/g is required to produce 2.5 watts-electric for 90 days. As a comparison, the same neutrons used to produce polonium by irradiating bismuth could have produced a gram of plutonium, which in turn could fuel about 5000 kWh of electrical output from a nuclear fission reactor.

Thermal storage. Conversion of electrical energy to thermal energy for later regeneration is not economical because the efficiency of such a system would be limited to less than 40% by heat to electrical conversion efficiency. A more efficient scheme is to store thermal energy at the generation station before it is converted to electrical energy. For example, a boiler could heat water or oil and later use the stored thermal energy to produce steam and generate electricity with about the same efficiency as immediate generation. However, the thermal input to a 1000-MW electric plant for 1 min would raise the temperature of 10^6 gals (3600 metric tons) of oil 20°F (11°C). Alternatively, an hour's input would melt 3000 tons of salt (NaCl). Hence thermal methods are more applicable to smaller plants where economics is not the main consideration.

For background information *see* BATTERY, ELECTRIC; CAPACITANCE; FLYWHEEL; INDUCTANCE; NUCLEAR BATTERY; WATERPOWER in the McGraw-Hill Encyclopedia of Science and Technology. [ROBERT J. CREAGAN]

Bibliography: H. L. Forgey, C. T. McCreedy, and R. L. Seguin, in *Proceedings of the American Power Conference*, vol. 36, pp. 797–841, 1974; H. C. Herbst and Z. S. Stys, in *Proceedings of the American Power Conference*, vol. 40, April 1978; H. A. Peterson et al., *Wisconsin Superconductive Energy Storage Project*, University of Wisconsin, July 1974; R. Whitaker and J. Birk, *EPRI J.*, 1(8): 6–13, October 1976.

Epstein-Barr virus

Epstein-Barr virus (EBV), first detected in cultures of African Burkitt's lymphoma cells, is the etiologic agent of heterophile antibody-positive and occasionally heterophile antibody-negative infectious mononucleosis. EBV has the unique property of inducing human (and primate) cell transformation, and is closely associated with human neoplasms, in particular, Burkitt's lymphoma and nasopharyngeal carcinoma.

Seroepidemiologic studies have shown that the

acquisition of antibodies to EBV commonly occurs in early childhood and that, generally, the prevalence increases to 80–100% in adulthood. The vast majority of these initial or primary EBV infections in children are asymptomatic or manifested by mild signs common to other upper respiratory viral infections. If the initial infection does not occur until young adulthood, it is more like to be manifested by infectious mononucleosis. The persistent carrier state established after a primary infection with EBV in young children has recently been found to be reactivated as such individuals become older.

Viral carrier state. Various investigators have demonstrated that a persistent viral carrier state becomes regularly established after a primary infection with EBV. This phenomenon occurs regardless of whether the EBV infection is manifested by infectious mononucleosis.

The repository site or sites for the latent form of EBV is not clearly defined. It has been shown in cell culture that the genome of EBV can be uniformly detected in peripheral blood lymphocytes of individuals with a prior EBV infection. In addition, EBV (initially described as leukocyte-transforming factor) has been detected in oropharyngeal secretions from infectious mononucleosis patients and to a lesser degree in healthy individuals for prolonged and intermittent periods. (Leukocyte-transforming factor has the property of transforming susceptible lymphoid cells to lymphoblastoid forms with the ability to multiply indefinitely.) One investigator has found EBV genome in pharyngeal tonsillar tissue. These findings and the known predilection of EBV for lymphoid cells has led to the theory that lymphoidal tissue of the pharyngeal area may be the primary site for EBV replication and latency. There are other findings, however, which suggest that pharyngeal tonsillar tissue may play a minor or negligible role in EBV infection.

Serologic evidence for reactivation. A number of virus-related antigens of EBV have been differentiated. The detection of antibodies directed against these antigens is useful in determining whether the onset of infection is recent or old. In infectious mononucleosis, the prototype of a primary EBV infection, antibodies to capsid antigen (CA) and early antigen (EA) of EBV and specific immunoglobulin (IgM) antibodies against this virus are present early in the disease course. Antibodies to EBV-CA remain for life, while specific IgM antibodies and antibodies to EBV-EA are usually detectable for only 2 to 3 months after clinical onset of the infection. Antibodies to nuclear antigen (NA) of EBV appear 3 or more months after clinical onset and also probably are lifelong. A similar pattern of antibody responses appears to develop after EBV infections which are not manifested by infectious mononucleosis.

In a large seroepidemiologic study of an American semirural community, it was demonstrated that increased antibody titers to EBV-CA are characteristically found in two different age brackets, young children and individuals past the age of 50. It was not surprising that the high antibody titers to EBV-CA were observed in early childhood, since primary EBV infections commonly occur at this age. However, the presence of high titers in the older age group was unexpected. This finding was tentatively interpreted to mean that persistent infection with EBV reactivates with advanced age.

Further serologic investigations on the older age group corroborated this speculation. Sera of 14% of the adults sampled in this study contained antibodies to EBV-EA in association with elevated antibody titers to EBV-CA and the presence of specific IgM. These findings were serologic evidence that this group was experiencing a current or recent EBV infection. However, the sera of the sample group uniformly contained antibodies to EBV-NA, a late-onset antibody found in long-standing EBV infections. The data, therefore, suggested that the majority, if not all, of this group of adults were manifesting a host-immune response to reinfection with EBV. The acute response to EBV was most likely caused by endogenous reactivation of EBV. It is true, on the other hand, that exogenous reinfection with EBV would probably result in a similar serologic response. However, the well-established feature of EBV latency acquired after initial infection makes exogenous reinfection a less satisfactory explanation. The stimulus that prompted the reactivation is unknown. Perhaps the decline in immunologic responsiveness of older individuals permitted reactivation of latent virus.

None of the individuals with presumed EBV reactivation experienced a prior or current infectious mononucleosis–like illness. Their rate of other mild signs and symptoms was similar to the control group not experiencing viral reactivation. Although reactivation of other herpesviruses is known to be associated with characteristic signs and symptoms (for example, recurrent vesicular disease with herpes simplex virus and zosteriform lesions from reinfections with varicella-zoster virus), the clinical implications of EBV latency and reactivation are obscure.

IgM antibody response. It is of interest that specific IgM antibody to EBV, but not heterophile antibody, was elicited by EBV reactivation. Specific IgM antibody responses have likewise been documented with other viral reinfections. An IgM antibody response, therefore, does not necessarily signify an acute primary infection, as was commonly thought.

For background information *see* EPSTEIN-BARR VIRUS in the McGraw-Hill Encyclopedia of Science and Technology. [CIRO V. SUMAYA]

Bibliography: J. C. Niederman et al., *New Engl. J. Med.*, 294:1355–1359, 1976; K. Nilsson et al., *Int. J. Cancer*, 8:443–450, 1971; C. V. Sumaya et al., *J. Infect. Dis.*, 131:403–408, 1975; C. V. Sumaya, *J. Infect. Dis.*, 135:374–379, 1977.

Eucalyptus

In the coastal states of the southeastern United States, research is expanding in an effort to determine the potential climatic range in which the fast-growing exotic *Eucalyptus* tree will survive. The eucalypts have a worldwide reputation for high-volume production in short time periods. The distinctive foliage on some species makes them desirable for ornamental purposes in flower arrangements and for landscaping.

Because several pulp and paper companies presently need a tree that will grow rapidly on low-fertility soils and because *Eucalyptus* trees are grown in many countries for fuel wood and can perhaps contribute to the United States energy supply, these trees are receiving increased attention. At least 15 pulp and paper companies, the U.S. Forest Service, and several private industries are conducting research on *Eucalyptus*.

Habitat. Australia is the native habitat of these evergreen, broad-leaved, hardwood trees that are capable of growing year-round where cold temperatures do not restrict their growth. In tropic and semitropic countries they are harvested for pulpwood on a 6- to 10-year cutting cycle. Since most species sprout from the stump when cut, several rotations of trees can be harvested without replanting.

Freeze-susceptibility screening. There are more than 550 species of eucalypts, most of which are too freeze-susceptible to survive in the United States except in the most southern areas. Some freeze-susceptible species are being planted on a limited scale in southern Florida, but these species will not survive farther north along the Gulf and Atlantic coastal plains, where most of the pulp and paper industry is located. Considerable effort has been spent, since 1972, to screen for species of *Eucalyptus* that will survive the freezes in the geographic area where the hardwood fiber is required by industry.

Screening of the introductions was done by planting healthy eucalypt seedlings, allowing the environment to select those with the desired freeze-resistance, and assessing the growth rate and form of the survivors. Since several companies are cooperating by planting the same species and sources on a variety of soils and in different geographic areas of the Southeast, it has been possible to select the promising species based on more than one outplanting.

Results of the earliest screening trials of 24 species planted in 1972 and 1973 showed that several could survive the freezes and grow much faster than any of the native hardwood species. Even after a severe freeze of 10°F (−12.2°C) in February 1977 at one outplanting in southwest Georgia, the hardiest trees were still growing in excess of 1 ft/month (4 m/year). The tallest of these, *E. viminalis*, were 54 ft (16.5 m) tall and 8.6 in. (21.8 cm) in diameter at 4 years of age (see illustration). These early successes, when testing only one seed provenance of a species, were promising. Some of the species occur over very large areas in Australia, and it was expected that if the seed source tested in the 1972 and 1973 trials survived and grew well, there was a good chance that other seed sources

Eucalyptus viminalis, planted at Bainbridge, GA. Tree is 4½ years old, 56 ft (17.1 m) tall, and 11 in. (28 cm) in diameter.

might prove even better adapted to the new environment.

In 1975, in order to better test the genetic variation between seed provenances of a given species, five pulp and paper companies, under the guidance of the North Carolina State University Hardwood Cooperative, planted several hundred trees of each of 80 new seed sources. Measurements of height growth and freeze damage showed that there was large variation among species and within seed provenances of a given species.

The table shows the range of variation in average height between seed sources of the surviving species and an indication of their relative freeze-hardiness. *Eucalyptus viminalis* has the best growth rate, but there was terminal damage on many seed sources due to the freezes (some sources were almost frozen to the groundline) while other sources were undamaged. *Eucalyptus nova-anglica*, although shorter on the average, has some sources growing as well as *E. viminalis*, and

Average heights, height range, and relative freeze-hardiness of surviving species at Bainbridge, GA, after 35 months

Species	Number of seed sources tested	Range of average height of sources, m	Average height of all sources, m	Relative hardiness
Eucalyptus viminalis	44	2.5–6.9	4.3	Variable by source
E. camphora	9	3.0–5.8	3.8	Most sources hardy
E. macarthurii	8	3.0–5.2	3.6	Most sources hardy
E. nova-anglica	8	2.2–5.0	3.8	All sources very hardy

is much hardier. As observed by the intraspecific variation in growth rate and freeze-hardiness among seed sources, it is apparent that the evaluation of a species cannot be made based on only one or two seed sources. This is especially true of a species such as *E. viminalis* which has a very large geographic and altitudinal range. Based on results from the 100 seed sources of *E. viminalis* being tested, it is apparent that unless the hardiest sources are planted, failure may result.

From 1972 through 1976 the winter temperatures were not severe enough to truly test the plantings, since colder temperatures could eventually be expected. However, in February 1977 low temperatures were recorded across most of the southeastern coastal plain that were of a severity experienced only once in 30–50 years. Large variation was noted; some species were more heavily damaged than was expected and some were unharmed. It is apparent that some species are hardy enough to withstand extremely cold temperatures in the area and that there are numerous hardy seed sources of less cold-hardy species.

The best-formed cold-hardy individuals are being consolidated at one location so that a genetic program can begin to improve hardiness and growth.

Seed introductions are continuing in the Southeast, and there are now over 450 different seed provenances of 90 *Eucalyptus* species being tested. Most of these species and seed sources are being planted by several companies to better ascertain the geographic area in which they can survive. Perhaps other species will be found which have a higher potential than any found to date. Already it is apparent that some recently planted species are hardier than *E. nova-anglica*, the hardiest species tested before 1977. Tests are also being conducted to determine whether the young seedlings can be artificially screened in a cold chamber.

Genetic improvements. Most eucalypts produce flowers at an early age; therefore, genetic improvement of growth rate, form, and freeze-resistance can be rapidly initiated by combining the best individuals from the populations already established. In other countries *Eucalyptus* has been very responsive to genetic manipulation, and gains are expected to be greater than in most tree-breeding activities. This has been the case with the *Eucalyptus* species in southern Florida.

Summary. It is now known that there are some species of *Eucalyptus* which have sufficient freeze-hardiness to grow rapidly along the Gulf and southern Atlantic coastal plains. These trees are capable of outgrowing native hardwood species when planted on upland sites. Indications are that marketable-size trees can be produced in only one-third of the time required for pine.

Most of the 550 species cannot survive the rapid intense fluctuations in temperature experienced in the inland areas of the Southeast, but through a determined research effort in the use of fertilizers, herbicides, and intensive cultural treatments, and by developing genetically improved seed better adapted to the environment, the probability of success is very high.

Some species may be found that can be moved northward into central Alabama and Georgia, while other species will be restricted to the frost-free areas of southern Florida. Even the freeze-susceptible species that sprout rapidly when cut may have a potential as an annual wood fiber crop for energy use.

For background information *see* EUCALYPTUS in the McGraw-Hill Encyclopedia of Science and Technology. [RONALD HUNT]

Bibliography: R. Hunt and B. Zobel, *S. J. Appl. For.*, vol. 1, no. 1, February 1978.

Fast chemical reactions

The most frequently used approach to the analysis of fast chemical processes involves a study of relaxation phenomena in which chemical equilibria are displaced by a pulse, following which relaxation toward equilibrium is monitored. For the fastest relaxation processes, light pulses, and in particular laser pulses, are used. This not only permits the study of molecular dynamics but offers opportunities for the synthesis and analysis of new species having lifetimes from about 10^{-14} s to stable photoproducts.

Relaxation processes. After excitation by light, chemical systems undergo a sequence of relaxation processes spanning many decades of time until equilibrium is restored. The first discernible relaxation processes occur on a time scale of 10^{-14} to 10^{-13} s, involving relaxation of highly excited electronic states of aromatic molecules through a combination of processes including vibrational energy redistribution, internal conversion between different electronic energy levels and, in condensed media, thermalization of excess vibrational energy. Recent pulsed laser work by M. R. Topp and coworkers has shown that the different stages in these complex relaxation pathways can be sampled, since the different levels involved exhibit weak but measurable fluorescence. It has been shown that the fluorescence spectra of states directly excited by a narrow-band laser are much sharper than, and displaced to the blue from, the spectra of states populated by radiationless processes.

Eventually, as 10^{-12} s is approached, direct time resolution becomes possible through the use of mode-locked laser techniques. In this time region a variety of unimolecular and motional radiationless relaxation processes are encountered, such as intersystem crossing, molecular rotation, solvent rearrangement (including macromolecular environments, for example, around biological chromophores), microscopic diffusion, cis-trans isomerization, electron solvation following photoionization, vibrational dephasing, exciton and phonon migration, molecular dissociation (solvent-cage—restricted), proton transfer, and hydrogen bond formation and hydrogen atom transfer.

Processes longer than about 10^{-9} s in duration at room temperature result from the inhibition of reaction rates by activation energy factors and by longer-range communication such as Forster energy transfer and molecular diffusion.

Pulsed laser applications. Pulsed lasers have been used extensively to study fast chemical reactions, and refinements in experimental techniques over the past couple of years have produced a number of versatile instruments. Major developments have been in the following areas: shorter

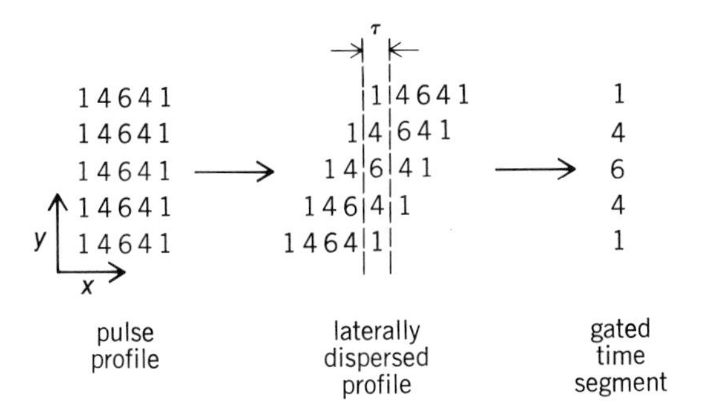

Fig. 1. Schematic of time-to-space conversion principle for analysis of picosecond events.

pulse durations; higher repetition rates for intense pulses; development of reliable fluorescence monitoring techniques; development of techniques for fast absorption spectroscopy; increased wavelength tunability and improved wavelength resolution; and development of spectroscopies useful where time resolution is insufficient for direct lifetime measurements.

Short pulse durations. In the short time limit, the duration of an optical pulse is determined by the available frequency bandwidth, such that $\delta\bar{\nu}\sim10\ cm^{-1}$ is necessary for picosecond pulses. Although several types of laser satisfy these requirements (such as Nd^{3+}-glass dye lasers), inherent instabilities, optical dispersion, and nonlinearities associated with high pulse intensities prevented the generation of reliably subpicosecond pulses until the work of E. P Ippen and C. V. Shank and, most recently, J. P. Heritage and R. K. Jain.

Subpicosecond pulses are generated in a continuously oscillating mode-locked dye laser, using an argon-ion laser as an excitation source. The pulses, 0.2–0.8 picosecond in duration, are available at 10^5–10^8 pulses per second (pps) at low power, or they may be amplified in a Nd^{3+}–yttrium-aluminum-garnet (YAG) pumped-dye-laser amplifier chain at ~10 pps. Otherwise, for slightly lower resolution, Nd^{3+}-glass (>5 ps), flashlamp pumped dye (>2 ps), ruby (>15 ps), and Nd^{3+}-YAG (25–30 ps) lasers are used. These lasers have various advantages, principally higher power, which facilitates the generation of photochemical intermediates and is essential for nonlinear optical sampling applications.

Absorption spectroscopy. Fast reactions can be followed through the absorption spectra of transient intermediates. Above 10^{-9} s, photoelectric detection provides time resolution following excitation by a laser or other source. For full spectral measurements, especially for high resolution, a pulsed spectral continuum is used. The continua are derived from laser pulses, and the sampling time is controlled by interposing an optical path difference between the excitation and probing pulses. In the nanosecond region, laser-excited broad-band fluorescence is the most convenient probe, whereas for picosecond applications the spectral continuum is generated by focusing a laser pulse into a low-dispersion liquid, such as H_2O. After correction for group dispersion, these continua are seen to have the duration of the exciting pulses, even below a picosecond.

One of the features of picosecond reactions is that the time scale is comparable to the time necessary for a pulse of light to travel through a sample a few millimeters long. Since physical displacements now appear on the time axis, experimental geometry is an essential consideration, and is in fact used to provide time resolution since electronic scanning is usually too slow.

Fluorescence spectroscopy. Short-lived fluorescent states may be directly time-resolved by photoelectric methods involving oscilloscope display or, in the picosecond region, by ultrafast streak cameras. Alternatively, high sensitivity can be achieved by signal averaging of repetitive pulses using gating techniques. In the nanosecond region, various electronically gated signal averagers can be used to time-resolve weak signals or to increase the signal-to-noise ratio of pulse-limited signals. Below 10^{-9} s, optical gating is necessary, in which a laser pulse and a fluorescence signal are combined in a nonlinear medium. The state of polarization of a small time increment of the fluorescence signal may be changed or, for higher sensitivity, the optical frequency can be shifted into a noise-free region of the spectrum. Again, signal averaging can be used with a high-repetition-rate laser, to increase the signal-to-noise ratio.

Picosecond time resolution is achieved by optical time-of-flight methods, to which there are several approaches. An integrated optical delay line introduces a lateral time dispersion across the spatial profile of a propagating optical pulse, such as in laser-excited molecular fluorescence. The most

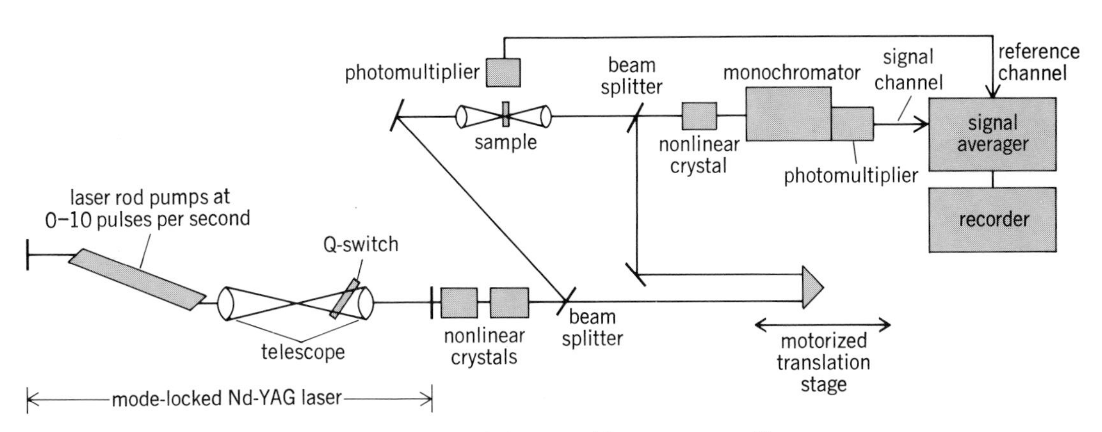

Fig. 2. Schematic of apparatus for measurement of picosecond fluorescence profiles.

accurate way of accomplishing this is to use a refracting prism or a diffraction grating in which pulse coherence is maintained. In a prism the time dispersion results from the difference between the group and phase velocities of light. On the other hand, the diffraction-grating time dispersion is due to the optical path difference for a coherent Bragg reflection multiplied by the number of grooves on the grating. Thus, a 100,000-groove grating gives a dispersion in first order of 166 ps at 500 nanometers.

In principle, a luminescence profile is sampled by a combination of an integrated delay line and an optical gate, as shown in Fig. 1. The overall operation involves the conversion of a variable along the time axis to a spatial image compatible with video digitization and computer analysis. This conversion also provides a ready method for direct oscilloscope display of picosecond laser pulses for accurate calibration purposes. With a standard diffraction grating, the potential time resolution is <0.005 ps.

An alternative method, having a higher signal-to-noise ratio, is to use a sequence of laser shots, scanning the optical delay between shots. If a high-repetition-rate laser is available, the delay line can be motorized, and the output coupled directly to a signal averager, as shown in Fig. 2. Using an optical frequency-conversion gate, this method is compatible with virtually any picosecond laser.

An example of the use of fluorescence spectroscopy to study a fast chemical reaction is given in Fig. 3. The time scale of 10^{-11} s is sufficiently long for the effects of motional relaxation to appear in a fluorescence spectrum. Figure 3 shows time profiles of the fluorescence of 2-amino,7-nitrofluorene in benzene–2-propanol solution following picosecond-pulsed excitation. The profiles are clearly different, and the falling portions at the two wavelengths represent two distinct processes. Short-rage diffusion of polar molecules, driven by a selective dipolar interaction, causes a relaxation of the fluorescent dipole, quenching the fluorescence at <500 nm and enhancing the fluorescence at longer wavelengths. Subsequently, an activation-controlled hydrogen-bonded interaction results in the removal of the molecules from the fluorescent state, as recorded at 680 nm.

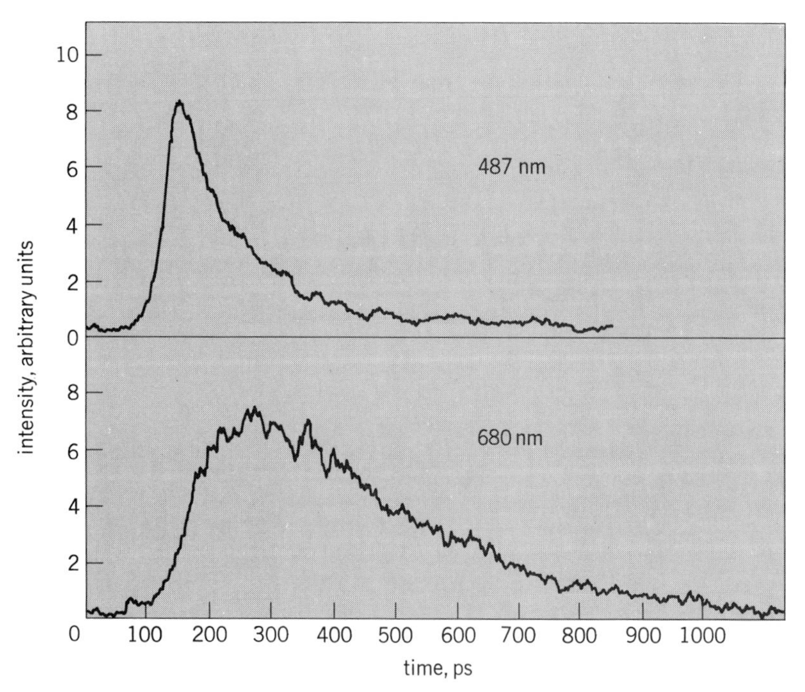

Fig. 3. Fluorescence time profiles of 2-amino,7-nitrofluorene in benzene–2-propanol solution, showing dynamic aspects of selective solvation and fluorescence quenching.

Thus, relatively straightforward measurements in this time region can yield important information about the primary processes in liquids which govern the rate and outcome of photochemical reactions.

For background information *see* FAST CHEMICAL REACTIONS; LASER; SPECTROSCOPY in the McGraw-Hill Encyclopedia of Science and Technology. [MICHAEL TOPP]

Bibliography: J. P. Heritage and R. K. Jain, *Appl. Phys. Lett.*, 32:101, 1978; R. M. Hochstrasser and A. C. Nelson, in J. Joussot-Dubien (ed.), *Lasers in Physical Chemistry and Biophysics*, 1975; S. L. Shapiro (ed.), *Ultrashort Light Pulses*, 1977; M. R. Topp, *Appl. Spectrosc. Rev.*, 14:1, 1978; E. B. Treacy, *J. Appl. Phys.*, 42:3848, 1971.

Fertilizer

The presence of nitrates in plants is a natural consequence of plant uptake from the soil in excess of that necessary for assimilation into nitrogen-containing compounds. Nitrate accumulation in plants is under genetic control, and is also influenced by a number of environmental and cultural variables. Fertilization with nitrogen-rich materials is foremost of the cultural variables in its effect on the nitrate content of plants. In addition to plant sources, nitrates are sometimes added to certain prepared or smoked meats, and may occur in drinking water from wells. Accordingly, ingestion of nitrates is an everyday occurrence and is harmless, except in very rare instances.

Nitrate has a rather low toxicity in humans, ranging between 15 and 70 mg nitrate-nitrogen per kilogram body weight. On the other hand, nitrite has somewhat greater toxicity, about 20 mg nitrite-nitrogen per kilogram. Nitrite may result from nitrate reduction by microorganisms when foods are improperly stored prior to consumption, or may be formed after ingestion, particularly in those suffering from gastrointestinal disturbances. Nitrate reduction is most likely to occur in infants because they are often prone to digestive disturbances and because of lower acidity in the gastrointestinal tract.

Nitrate and health. Ingestion of nitrite-containing foods or reduction of nitrate to nitrite after ingestion has been implicated in two potential hazards to human health. Methemoglobin, which lacks the ability to transport oxygen, is formed in the presence of nitrite, which oxidizes the ferrous iron of hemoglobin to the ferric form. Again, infants are much more susceptible since fetal hemoglobin is more likely to be converted to methemoglobin than the hemoglobin of older children and adults.

The other potential hazard associated with foods containing nitrate or nitrite and secondary amines is the formation of nitrosamines. These compounds have been shown to be mutagenic, teratogenic, and carcinogenic in tests with laboratory animals. Nitrosamines have been detected in a number of foods and in human blood following consumption of such foods. High temperatures enhance nitrosamine formation, so that food preparation methods have a great effect on the occurrence of these compounds. Based on these findings, governmental regulation of nitrate and nitrite addition to foods is currently under intensive study.

Nitrates in plants. Nitrate present in the harvested plant represents the difference between the nitrate absorbed by the plant's roots and that which is assimilated into the myriad of nitrogen-containing compounds found in plants. About 80% of the human daily ingestion of nitrates is from vegetables.

Variation among genotypes with respect to nitrate accumulation has been demonstrated in lettuce and spinach. Some crisphead lettuce cultivars consistently contain higher nitrate concentrations than other varieties. Likewise, spinach cultivars, notably those with small, ruffled, dark green leaves, are rich in nitrates, while cultivars having large, smooth, light green leaves are lower in nitrates. These differences have been attributed to a more efficient nitrate-reducing system in the smooth-leaved cultivars. Unfortunately, preliminary genetic studies suggest that efficient nitrate reduction has a rather low degree of heritability. Research is continuing, however, to develop vegetable cultivars which have efficient reducing systems.

Vegetables contain variable, sometimes appreciable, nitrate concentrations. The term "nitrate accumulator" is used to describe vegetables such as beets, collards, kale, lettuce, mustard greens, radishes, spinach, and turnips which typically have high nitrate concentrations (see the table). The edible portion of these vegetables is invariably higher in nitrates than vegetables which are nonaccumulators. Nitrates accumulate to highest concentrations in stems, leaf petioles, and leaf blades, while fruit and flower parts are usually low in nitrates. Roots, tubers, and bulbs are variable in nitrate content, with some, such as the turnip and radish, being quite high and others, such as the potato and onion, being low. The amount of nitrate ingested is thus dependent to some degree on the part of the vegetable being consumed.

Nitrate concentrations in fresh vegetables*

Plant part	NO_3-N, mg per 100 g fresh weight			
	<20	20–50	>50–100	100
Fruits and flowers	Cauliflower Sweet corn Cucumber Melons Peas Peppers Squash Tomato	Broccoli		
Roots, tubers, and bulbs	Onion Potato	Carrot	Beets	Radish
Leaves	Cabbage Onion	Lettuce	Beets Celery Parsley Spinach Turnip	Kale
Petioles		Celery		Beets Kale Spinach Turnip

*From D. R. Nielson and J. G. MacDonald (eds.), *Nitrogen in the Environment*, vol. 2, p. 208, Academic Press, 1978.

External factors affecting nitrate concentration. Use of commercial fertilizers or organic materials containing nitrogen is foremost among the cultural variables affecting nitrate content. Regardless of the source of nitrogen, heavy fertilization will result in abnormally high nitrate concentrations, since the pathways of microbial activity in aerated cultivated soils ultimately lead to the formation of nitrate in the soil solution. The nitrate ion is the primary form in which plants obtain nitrogen from the soil solution, and acquisition from a nitrogen-rich medium may far exceed plant requirements.

Judicious use of fertilizers precludes high nitrate accumulation without sacrificing high yields or product quality. Extensive surveys of vegetables purchased in local markets have shown that nitrate concentrations are at about the level required for maximum yields. Furthermore, periodic surveys made in the 20th century do not show any appreciable change in nitrate concentrations despite increased use of nitrogenous fertilizers.

Climatic variables do not influence nitrate accumulation to as great a degree as nitrogen fertilization; nonetheless, they do have a considerable effect. The nitrate-reducing system in plants requires light; therefore, nitrates accumulate in darkness and under low light intensities. Some measurements made in beets show that nitrate concentrations are more than twice as high at 4 A.M. than at 4 P.M. Likewise, spinach contains about twice as much nitrate when grown at low light intensities.

Temperature and water effects on nitrate accumulation are somewhat variable, but it appears that nitrate accumulation is enhanced by high temperature, high atmospheric humidity, and drought.

Regulation of nitrate accumulation. Without a doubt, fertilizer management is the most important factor in controlling nitrate accumulation. Crop requirements for fertilizer have been determined and fertilization guidelines have been developed to meet these needs in the major vegetable-growing areas of the United States. Timed applications based on the results of plant and soil analyses ensure adequate nutrition without high nitrate accumulation. Similar control of nitrate accumulation can be effected by the use of newly developed controlled-release fertilizers which meter nitrogen to the crop throughout the growth cycle.

Greater regulation of the form of nitrogen available to the plant may be possible in the future through the use of temporary chemical inhibitors of the microbial population which mediates the first step in nitrification, shown in the reaction below.

$$NH_3 \xrightarrow{\textit{Nitrosomonas}\ \text{spp.}} NO_2 \xrightarrow{\textit{Nitrobacter}\ \text{spp.}} NO_3$$

Nitrate accumulation has been greatly restricted in lettuce, radishes, and spinach by addition of a nitrification inhibitor to ammonium nitrogen fertilizers under experimental conditions in greenhouse and field studies.

The light requirement for nitrate reduction can also be exploited to provide vegetables of lower nitrate content. Harvesting crops in the late afternoon rather than in the morning and on bright, sunny days rather than on dark, cloudy days results in markedly lower nitrate concentrations in the harvested vegetable.

Preparation of vegetables prior to consumption greatly influences the amount of nitrate ingested. For example, removal of spinach petioles lowers nitrate ingestion by more than 50%. Similarly, discarding the outer petioles from a celery stalk and outer leaves of a head of lettuce will significantly lower nitrate intake. Proper storage to prevent microbial activity and timely use of vegetables eliminate microbial conversions of nitrate to nitrite before consumption. Therefore, vegetables should be rapidly cooled after harvest and remain refrigerated until prepared for consumption.

The foregoing discussion applies to fresh vegetables, but many vegetables are consumed from canned or frozen products. Generally, processed vegetables are considerably lower in nitrates than fresh vegetables because of the washing, preparation, and cooking involved in processing. Once opened, however, processed products must be refrigerated in closed containers to prevent the possibility of nitrite formation by contaminating microorganisms.

Although vegetables constitute the principal source of nitrates in foods and excessive nitrates have been implicated with impaired human health, no deaths have been recorded in the United States from this cause. Research results indicate that judicious use of nitrogen-containing fertilizers, improved crop management, and proper handling of harvested vegetables will ensure vegetables low in nitrates.

For background information *see* FERTILIZER in the McGraw-Hill Encyclopedia of Science and Technology. [DONALD N. MAYNARD]

Bibliography: D. N. Maynard et al., *Adv. Agron.*, 28:71–118, 1976; D. R. Nielson and J. G. MacDonald (eds.), *Nitrogen in the Environment*, vol. 2, 1978.

Fission, nuclear

The advent of the double-hump fission barrier was a major breakthrough in the history of nuclear fission. This barrier shape was obtained for actinide nuclei by using a macroscopic-microscopic method proposed by V. M. Strutinsky, in which the total potential energy of a nucleus, for a given deformation, is calculated by adding a small shell-energy correction to the energy derived from a macroscopic model (the liquid-drop model, for example). It was then possible to explain in a unified and coherent manner many aspects of fission, such as fission isomerism and various types of structure in fission cross sections. Recent results substantiate the existence of the double-hump fission barrier and suggest even more complicated shapes for this barrier, at least for light actinides.

Fission isomers. With the double-hump fission barrier, fission isomers could be immediately interpreted as being largely deformed nuclear states in the second well of the barrier. That fission isomers are indeed shape isomers was verified by measurement of the moment of inertia of the ^{240}Pu fission isomer with a lifetime of 4 ns. Recent experiments confirmed that fission isomers have a large deformation by means of measurements of either their moment of inertia or, even better, their quadrupole moment. Some information about the spin or the nuclear g factor of fission isomers can be deduced from other experiments, but the results are still preliminary.

Moment of inertia. The moment of inertia $\mathcal{I}$ of a deformed nuclear state can be determined experimentally from the spacings between adjacent levels of the rotational spectrum associated with this state. The energy spectrum $E_K(J)$ for a rotational band having spin projection number K, as a function of spin $J (J \geq K)$, is given approximately by the equation below, where E_K^0 is the energy of

$$E_K(J) = E_K^0 + \frac{\hbar^2}{2\mathcal{I}}\left\{J(J+1) - K(K+1)\right\} \text{ for } K \neq 1/2$$

the band head (K=J), and $\hbar$ is Planck's constant divided by 2π. For K=0, J has only even values (0, 2, 4, and so forth).

This spectrum is observed, with the ground state as a band head, for deformed nuclei. However, it was also observed with the 4-ns $^{240\text{m}}$Pu isomer as a band head by H. J. Specht and colleagues in 1972 through detection of the conversion electrons following the E2 transition in-band cascade γ-rays. A value of $\hbar^2/2\mathcal{I} = 3.343\pm0.003$ keV was obtained, as compared with 7.16 keV for the ground state, thus demonstrating for the first time that fission isomers are shape isomers.

A similar experiment was carried out by J. Borggreen and colleagues in 1977 for the 110-ns $^{236\text{m}}$U fission isomer, populated through the ^{235}U(d,p) reaction, for which a value $\hbar^2/2\mathcal{I} = 3.36 \pm 0.01$ keV could be deduced, which is almost identical to the value found for $^{240\text{m}}$Pu. Quantitatively, these two values of the moment of inertia are in good agreement with that predicted by microscopic calculations based on the "cranking model."

Quadrupole moments. Quadrupole moments Q of fission isomers can be determined by measuring the lifetime of rotational levels associated with these isomeric states. This lifetime depends, among other factors, on Q^{-2} but lies in the picosecond range and therefore cannot be measured directly by fast electronic timing. Two indirect methods were used to measure the half-life $T_{1/2}$ of rotational levels built on the 8-μs $^{239\text{m}}$Pu and the 37-ps $^{236\text{m}}$Pu shape isomers.

For $^{239\text{m}}$Pu, $T_{1/2}$ was obtained by D. Habs and colleagues in 1977 with the charge-plunger technique, taking into account the charge properties of the $^{239\text{m}}$Pu recoiling ion formed in the ^{238}U $(\alpha,3n)$ reaction. The in-band transitions between adjacent rotational levels are highly converted. As a result, many Auger electrons are ejected from the atomic shell, thus causing the charge state of the recoiling ion to be high and to vary in flight depending on the lifetime of the rotational levels. This high charge state is reduced to a few electronic charge units after passage of the ion through a thin carbon foil. The lifetime of the rotational levels can then be determined by analysis of the charge distribution of the recoiling ions after their passage through the carbon foil as a function of the distance between the foil and the ^{238}U target. By making further assumptions about the spin of this fission isomer, a

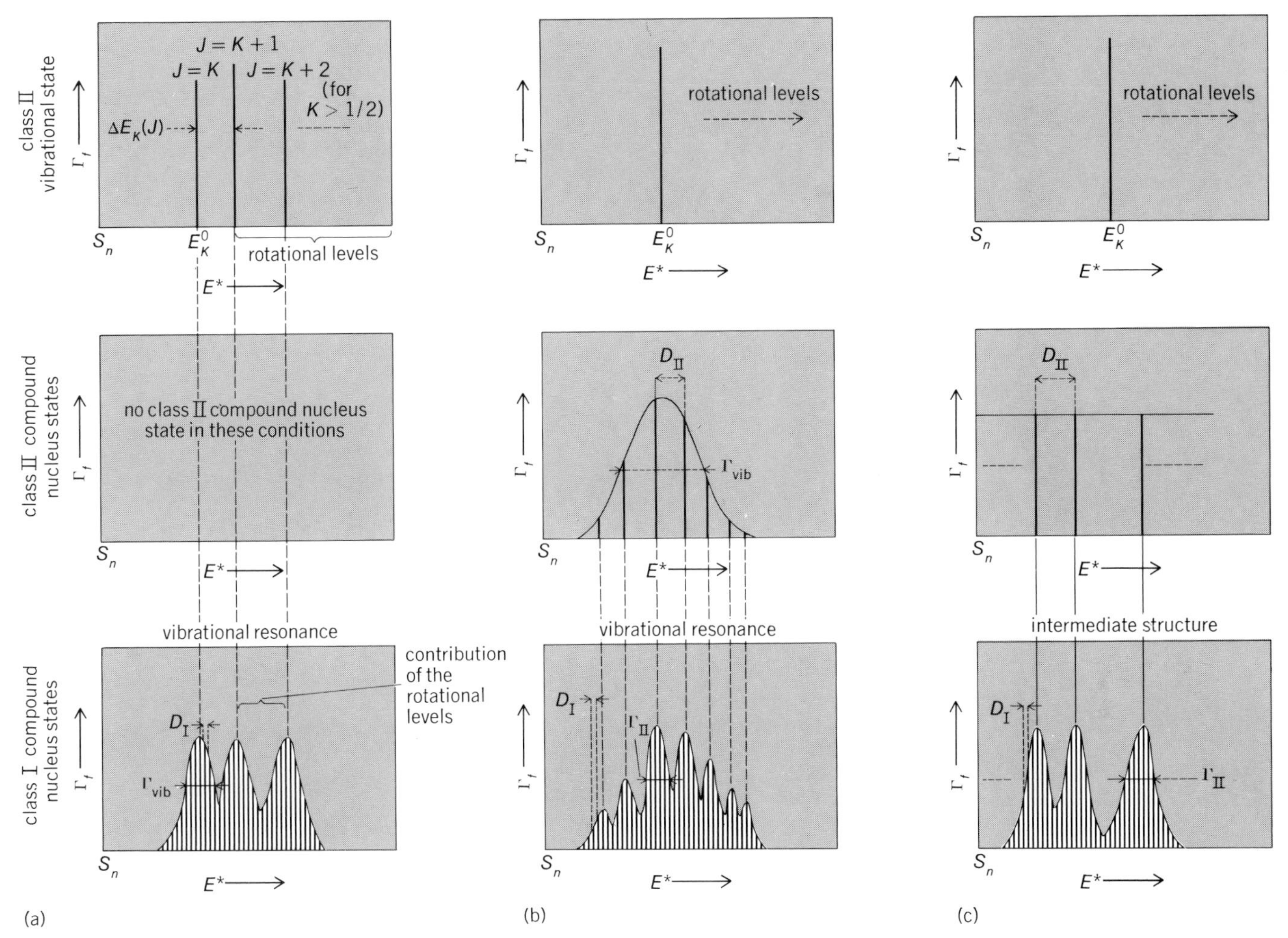

Fig. 1. Structure effects that can appear in neutron-induced fission cross sections, caused by class II states in the second well of the double-hump fission barrier. *(a)* No damping of class II vibrational states. *(b)* Moderate damping. *(c)* Full damping.

quadrupole moment $Q = (36.0 \pm 4.4)\ 10^{-28}\ m^2$ was deduced. This method can be applied only to relatively long-lived fission isomers.

For the ^{236m}Pu case, the rotational-level lifetime was derived by V. Metag and G. Sletten in 1977 from the angular distribution of the delayed fission fragments following the $^{234}U(\alpha, 2n)$ reaction. If the delayed fission decay originates primarily from the ground state (having spin and parity $J^\pi = 0^+$) in the second well, the angular distribution of the fission fragments is isotropic. However, if this ground state has a short fission lifetime comparable to the E2 rotational lifetimes, then delayed fission can also originate from the rotational levels for which the angular distribution of the fragments relative to the incident particle beam is anisotropic; the larger the spin of the fissioning state, the larger the anisotropy. Therefore, the measurement of the anisotropy yields the values of the E2 rotational lifetimes relative to that of the fission isomer. A quadrupole moment $Q = (37^{+14}_{-8})\ 10^{-28}\ m^2$ was obtained in this manner.

These two high Q values for ^{239m}Pu and ^{236m}Pu are very close to each other and much larger than the value of Q for the ground states of these two nuclei. Furthermore, they are in good agreement with theoretical predictions. Thus, experimental results on the moment of inertia and the quadrupole moments for a few fission isomers demonstrate convincingly that they are shape isomers.

Neutron-induced fission cross sections. The existence of nuclear states in the second well of the double-hump fission barrier can cause various types of structure in neutron-induced fission cross sections (designated σ_{nf}). Usually, these largely deformed states are called class II states, as compared with class I states located in the first well of the barrier. The type of structure that is caused by class II states in some σ_{nf} depends, among other factors, on the nature of these states, that is, whether they are vibrational, rotational, or compound nucleus states; on their coupling to the initial class I compound nucleus states formed by the capture of the incident neutrons by the target nucleus; and on the damping of vibrational class II states into intrinsic excitations. The damping of vibrational states in the second well is not well known but, as in the first well, increases with excitation energy. Also, for a given excitation energy, the density of intrinsic excitations, and consequently the damping of vibrational states, is larger for odd than for even nuclei.

Three types of structures with possible fine-structure components can be observed in σ_{nf} in the vicinity of a class II vibrational state, depending on the damping conditions of this state. The mecha-

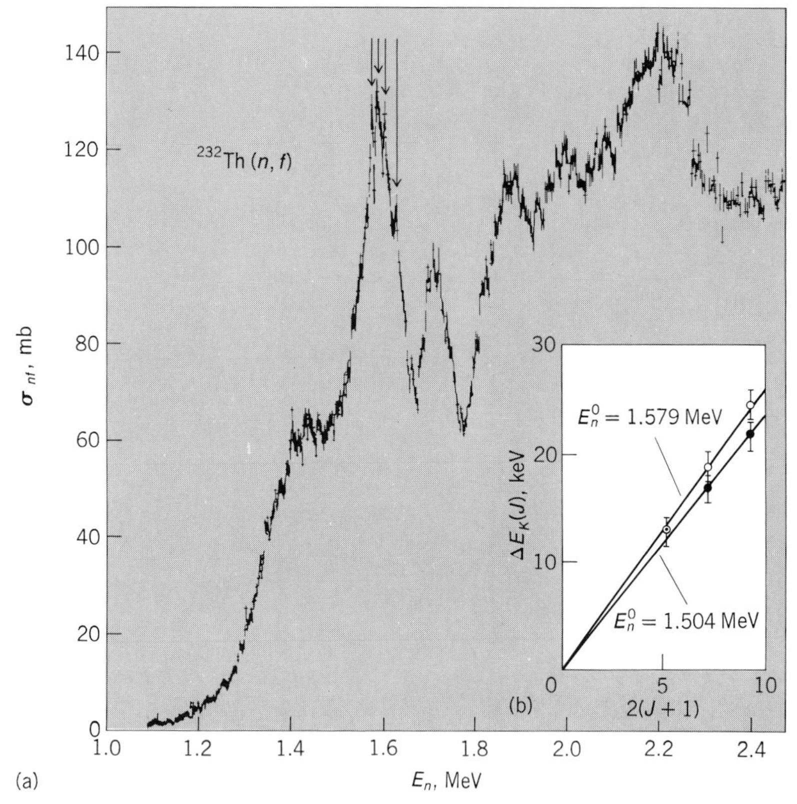

Fig. 2. Neutron-induced fission cross section of ^{232}Th. *(a)* Cross section as function of neutron energy E_n, showing vibrational resonances with fine-structure components. Fine-structure peaks are indicated by arrows at the vibrational resonance at 1.6 MeV. *(b)* Fits to the energy spacings $\Delta E_K(J)$ between the sharp peaks of the vibrational resonances at 1.5 and 1.6 MeV. The neutron energy of the band head in each case is called E_n^0. *(From J. Blons et al., Evidence for rotational bands near the ^{232}Th (n,f) fission threshold, Phys. Rev. Lett., 35:1749–1751, 1975)*

nisms for this are shown in Fig. 1, where the fission widths Γ_f of various states are plotted versus excitation energy E^* in the vicinity of a class II vibrational state. (Porter-Thomas fluctuations of the widths for compound nucleus states are ignored here.) For neutron-induced fission, E^* is greater than S_n, the neutron separation energy in the compound nucleus (typically 4.5–6.5 MeV for actinide nuclei, depending on their neutron number N).

In the absence of damping (Fig. 1*a*), the width Γ_{vib} of the vibrational class II state is very small and leads to a big peak in the cross section, known as a vibrational resonance. Such peaks are observed in the 100 keV–1 MeV incident neutron energy range for some nuclei (for example, ^{230}Th and ^{232}Th). Furthermore, the rotational band built on this vibrational state gives rise to a fine structure which can be detected in high-resolution measurements (as in the case of ^{232}Th discussed below). A still finer structure due to class I compound nucleus states, with spacing D_I between adjacent states, may also exist, but cannot be detected because of the large level overlap and because the experimental resolution is too broad in this energy range.

For moderate damping (Fig. 1*b*), the class II vibrational level is coupled to class II compound nucleus states over an energy interval comparable to its total width Γ_{vib}. The vibrational resonance is still the same as in Fig. 1*a*, but with a possible fine structure caused by class II compound nucleus states with width Γ_{II} and spacing D_{II} between adjacent states. For the same reasons as in Fig. 1*a*, the finer structure corresponding to class I compound nucleus states cannot be observed. This type of structure is not found frequently, but may have been observed for ^{234}U.

For full damping (Fig. 1*c*) the width Γ_{vib} exceeds the spacing of class II vibrational levels and the vibrational resonances disappear. Instead, an intermediate structure effect caused by class II compound nucleus states can be observed with a fine structure coming from the class I compound nucleus states. Since the intermediate structure effect is independent of the positions of the fully damped class II vibrational levels, it can appear at a much lower energy than the vibrational resonances discussed above. Therefore, the fine-structure components can be resolved in a low-energy class II fission cluster (electronvolt to kiloelectronvolt range) where the experimental resolution is excellent. This intermediate structure effect was first observed by the Saclay group in the subthreshold fission cross section of ^{237}Np, and was subsequently confirmed for other nuclei such as ^{240}Pu. The fission mechanism shown in Fig. 1 was substantiated later (1973) for ^{237}Np by G. A. Keyworth and colleagues, who demonstrated by polarization measurements that all the large fission resonances in a given class II cluster have the same spin.

Third well possibility. Interesting results have been obtained recently in the detailed study of vibrational resonances in some σ_{nf}, especially for even thorium isotopes. Vibrational resonances were known to exist in σ_{nf} for ^{230}Th and ^{232}Th and could be interpreted, at least qualitatively, in terms of double-hump barrier shapes. But a quantitative analysis of the data could not be achieved with calculated fission barriers since, in all realistic calculations, the second minimum was too low compared with the one deduced from the experiments. This situation, known as the thorium anomaly, seems to have been clarified by new high-resolution measurements of σ_{nf} for ^{232}Th together

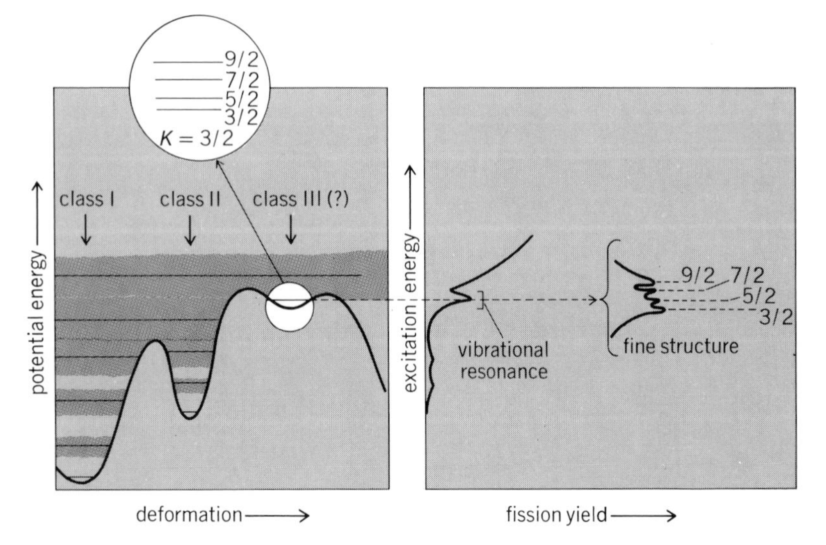

Fig. 3. A possible explanation of the thorium anomaly in terms of a third well in the fission barrier. The mechanism is illustrated from a $K = 3/2$ rotational band, as measured in the broad peaks at 1.5- and 1.6-MeV incident neutron energy for the ^{232}Th fission cross section.

with more refined fission barrier calculations.

The recent high-resolution measurements of σ_{nf} for ^{232}Th carried out by the Saclay group are shown in Fig. 2*a*. The well-known vibrational resonances appear as broad peaks at 1.4-, 1.5-, 1.6-, and 1.7-MeV incident neutron energies but, in addition, the fine-structure components in these broad peaks are resolved experimentally and are observed clearly in the data. The regularity in the spacings between the sharp peaks suggests the presence of a rotational spectrum as shown in Fig. 1*a*. This suggestion is quantitatively substantiated by the plot of Fig. 2*b*, where the spacings $\Delta E_K(J) = E_K(J+1) - E_K(J)$ show a linear variation with $(J+1)$, as implied by the equation for the energy spectrum given earlier, at least for the broad peaks at 1.5 and 1.6 MeV, for which $K = 3/2$. (The fine-structure pattern of the other peaks for which $K = 1/2$ is more complicated.) For the peaks at 1.5 and 1.6 MeV, the parameter $\hbar^2/2\mathcal{I}$ has values as low as 2.46 and 2.73 keV, respectively. These results are definitely lower than the already low value of about 3.3 keV obtained for the ^{240m}Pu and ^{236m}U fission isomers. In the absence of accurate calculations for the moment of inertia of largely deformed nuclei, the aforementioned results for ^{232}Th suggest, however, that the vibrational resonances for this nucleus are caused by states which have a deformation larger than that of the second well and which are likely to be in the vicinity of the second maximum. This suggestion is substantiated by potential-energy calculations in the region of the second maximum. By using the droplet model and the folded-Yukawa single-particle potential for the calculation of the macroscopic energy and the shell-energy correction, respectively, a shallow well (the third well) appears in the region of the second maximum if asymmetric deformations are taken into account. This third well provides the possibility for the existence of a third category of states (called class III states) which have an even greater deformation.

The existence of class III states can, in principle, explain the ^{232}Th data according to the mechanism shown in Fig. 3. At excitation energies which are reached by fast neutron capture, class I and class II vibrational states are above the inner barrier top, are completely mixed and fully damped, and are therefore incapable of causing any structure in σ_{nf}. However, the third well is at the right energy for class III vibrational states to cause vibrational resonances. Since this well is shallow, these vibrational levels are weakly damped, and the rotational levels add a fine structure which has been observed recently in the experiments.

Measurements with still higher resolution are being carried out to more accurately determine properties of the fine-structure components in the vibrational resonances of ^{230}Th and ^{232}Th.

For background information *see* FISSION, NUCLEAR; NUCLEAR STRUCTURE; REACTOR, NUCLEAR in the McGraw-Hill Encyclopedia of Science and Technology. [ANDRÉ MICHAUDON]

Bibliography: A. Michaudon, *Adv. Nucl. Phys.*, 6:1–217, 1973; *Phys. Today*, 31(1):23-30; *Proceedings of the International Conference on the Interactions of Neutrons with Nuclei*, CONF-760715-P, vol. 1, pp. 641–722; R. Vandenbosch, *Annu. Rev. Nucl. Sci.*, 27:1–35.

Flash smelting

Flash smelting is a continuous pyrometallurgical process developed by Outokumpu Oy of Finland in the mid-1940s. Although primarily used in the processing of nickel and copper sulfide concentrates, flash smelting has also been successfully applied in the treatment of pyrite concentrates, and has possible applications to other metalliferous sulfides. This article will discuss only the application of the flash smelting process to copper sulfides. Also, the process description will be limited to that developed by Outokumpu Oy, although other processes (notably INCO and Mitsubishi) have been successfully demonstrated on a commercial basis.

Flash smelting process. In the flash smelting process, metalliferous sulfides are partially oxidized. This oxidation liberates heat and a relatively high-strength SO_2 gas. The heat is utilized as an energy source for the smelting of the sulfides, thereby lowering the overall energy consumption of the system. The high-strength SO_2 gas may be treated in a number of alternative processes to remove the SO_2 to comply with various environmental regulations.

Flash smelting may be likened to the burning of pulverized coal. Copper sulfide concentrates replace the pulverized coal and are introduced into the flash furnace through a "burner" where the concentrates are intimately mixed with oxygen, as shown in the illustration. The oxygen source may range from air (ambient or preheated) to almost pure oxygen. At this point the copper concentrates "flash," as various reactions occur depending on the composition of the concentrates. Assuming a typical chalcopyrite ($CuFeS_2$) copper concentrate, general reactions (1)–(4) occur:

$$2CuFeS_2 \rightarrow Cu_2S + {}^1\!/_2S_2 + 2FeS \quad (1)$$
$$2FeS + 3O_2 \rightarrow 2FeO + 2SO_2 \quad (2)$$
$${}^1\!/_2S_2 + O_2 \rightarrow SO_2 \quad (3)$$
$$FeO + SiO_2 \rightarrow FeO \cdot SiO_2\ (\text{slag}) \quad (4)$$

The combination of these reactions results in a

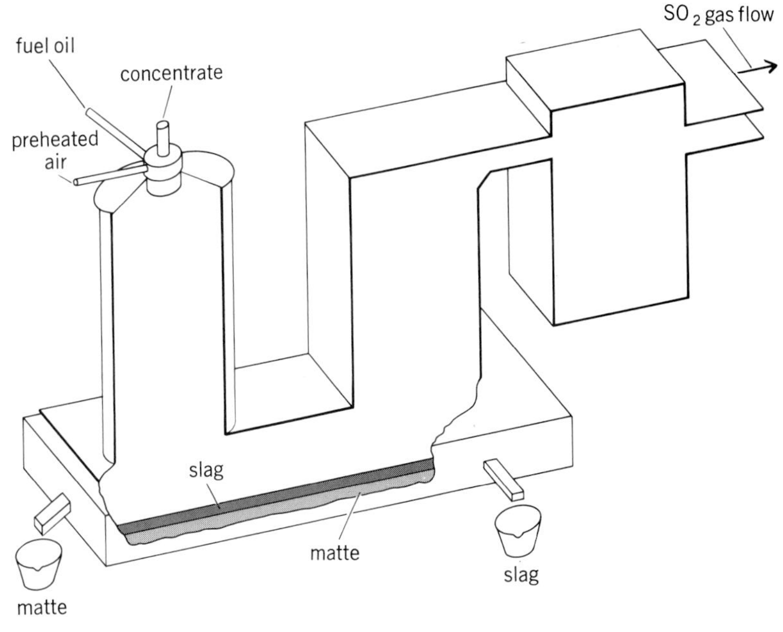

Cutaway of typical Outokumpu Oy flash furnace.

highly exothermic system, and the solid particulates become molten in 1 to 3 s. The degree of "burning" is controlled by adjusting the ratio of air or oxygen to concentrate. In this manner, a predetermined amount of sulfur may be left unburned to become part of the melt. Copper, having a high affinity for sulfur, settles in the furnace as molten copper sulfide which reacts with iron sulfides to form a molten matte ($Cu_2S \cdot FeS$). The nonsulfide particles (such as iron oxide, silica, alumina, magnesia, and lime) combine as a molten slag. This lower-density slag floats on the heavier molten matte, and matte and slag are separated by tapping the respective materials from the furnace through holes at different elevations, as shown in the illustration.

As might be expected, the molten slag and matte fractions are in an agitated state, which results in a relatively high copper content (1–2% Cu) in the flash furnace slag. Therefore, this slag must be retreated either by slag flotation or in an electric furnace to recover as much copper as possible before final disposal of the slag.

Comparison of smelting processes. The copper smelting industry in the United States historically has utilized reverberatory furnaces as primary smelting devices. A typical reverberatory furnace is rectangularly shaped and fired at one end with fossiliferous fuel, while sulfide concentrates are distributed along the sidewalls of the furnace. The heat liberated from the combustion of the fuel is transferred to the sulfide material via convection and radiation (the "reverberation" effect from the furnace roof). Even though some 30% of the total sulfur in the copper concentrate is liberated into the gas stream of the furnace, the SO_2 concentration of the off gas is very low due to the large amount of nitrogen introduced into the furnace via the combustion air.

The major advantages and disadvantages of flash smelting as compared with reverberatory smelting may be interpreted from Table 1 as follows:

1. The flash furnace is a high-capacity unit which is able to treat almost twice as much sulfide concentrates per unit volume as the reverberatory furnace. This is a result of the rapid rates of reaction which occur during the oxidation of the sulfide concentrates.

2. A greater portion of the sulfur is liberated in the flash furnace than in the reverberatory furnace, resulting in higher matte grades from the flash furnace. The lower matte volumes produced by the flash furnace normally mean that fewer converters are required to treat the matte.

3. The low copper content of the reverberatory slag allows it to be discarded. However, the flash furnace slag must be retreated for recovery of additional copper, which is a disadvantage. The overall copper recovery rates of the reverberatory and flash furnace systems are approximately the same.

4. The flash furnace produces lower volumes of higher-strength SO_2 off gases than the reverberatory furnace; this offers distinct advantages which will be discussed in detail below.

5. The lower energy consumption of the flash furnace is an important advantage which will also be discussed below.

Table 1. Comparison of reverberatory and flash smelting parameters*

Parameter	Reverberatory furnace	Flash furnace
Capacity, tons concentrate/day	915	2000
Internal furnace volume, ft^3	52,000	62,000
Specific capacity, tons concentrate/day/ft^3	0.017	0.032
Matte, % copper	39.0	55.0
Slag, % copper	0.4	1.5
Off gas, % SO_2	1.5	12.0
Energy required for smelting, 10^6 Btu/ton concentrate	6.82	3.17

*1 ton = 0.9 metric ton; 1 ft^3 = 2.83×10^{-2} m^3; 1 Btu = 1055 J.

Energy considerations. The advantages of the flash smelter system over conventional reverberatory furnace smelting with respect to unit energy consumption are shown by the comparative data in Table 2. As indicated above, for smelting alone without any pollution control, the reverberatory smelting system consumes more than twice the energy of the flash smelting system in treating 1 ton (0.9 metric ton) of concentrate. The energy advantages are even more dramatic if at least 90% of the input sulfur must be captured ("fixed"); this is the requirement for new smelters in at least one copper-producing state. In this case the reverberatory system consumes 2.7 times as much energy (1.57×10^7 Btus/ton versus 5.86×10^6 Btus/ton; 1 Btu = 1055 J) to smelt 1 ton of concentrate and fix 90% of the input sulfur. Assuming that the concentrate contains 25% Cu and also assuming a unit energy cost of \$2.50/10^6 Btus, the energy cost per pound (454 g) of copper is:

	Reverberatory	*Flash*
Smelting only	3.4¢	1.6¢
Pollution control	4.4	1.3
Total	7.8¢	2.9¢

From this comparison, it may appear advantageous to replace existing reverberatory smelting systems with flash smelting systems. However, building a flash smelter in an area where nothing has existed before is an extremely capital-intensive undertaking, with capital costs ranging from \$1900 to \$2500 (1976 dollars) per annual ton of copper, depending upon capacity and location of the facility. Approximately 35–45% of the total capital requirement is generally committed to pollution abatement facilities (for control of dust, sulfur oxides, fugitive gas, and other pollutants). Accordingly, a facility with a capability of treating 500,000 tons per year (450,000 metric tons) of concentrate containing 25% copper would require capital expenditures of approximately \$237,000,000–312,000,000. Amortization costs alone over a 20-year period would be 5–6¢ per pound of copper produced.

Considering both capital and energy costs, unit energy costs would have to be over \$20/10^6 Btus to justify the replacement of a reverberatory installation (smelting only) with a flash smelting system. Thus the smelting industry did not consider replacing reverberatory installations with flash smelters until the advent of stricter Federal and state pollution control regulations in the early 1970s.

Table 2. Comparison of total energy consumption of reverberatory and flash furnaces*

Process area	Reverberatory furnace: Energy consumption, 10^6 Btu/ton concentrate†	Reverberatory furnace: Percent of total	Flash furnace: Energy consumption, 10^6 Btu/ton concentrate†	Flash furnace: Percent of total
Smelting				
Materials handling and miscellaneous	0.30	2	0.30	5
Drying	–	–	0.70	12
Smelting	6.22	40	1.55	26
Converting	0.30	2	0.30	5
Slag processing	–	–	0.32	5
Total smelting	6.82	43‡	3.17	54‡
Pollution control				
Acid plant	1.81	12	1.63	28
Lime scrubber	4.80	30	–	–
Fugitive gas handling	2.27	14	1.06	18
Total pollution control	8.88	57‡	2.69	46
TOTAL	15.70	100	5.86	100

*Includes fuel and electric power.
†1 Btu = 1055 J; 1 ton = 0.9 metric ton.
‡The numbers involved in this sum have been rounded off.
SOURCE: S. N. Sharma and W. L. Davis, Jr., *Energy Conservation: A New Challenge*, Society of Mining Engineers, pp. 38–41, May 1977.

Environmental considerations. As noted earlier, a major advantage of flash smelting over reverberatory furnace smelting is the comparatively high SO_2 content of the flash furnace off gas.

Any new smelter in the United States must be capable of limiting SO_2 concentration in off gases exhausted to the atmosphere to not more than 0.065% (650 parts per million). This environmental constraint virtually eliminates the reverberatory furnace as a processing alternative, since a large amount of the total sulfur input to the furnace is liberated as SO_2 in the weak reverberatory furnace off gases, which cannot be treated effectively in sulfur-removal facilities.

The relatively high SO_2 content of the flash furnace off gas affords several advantages: (1) smaller acid plant facilities with associated lower capital and operating cost requirements; (2) greater flexibility in the converter operation; and (3) possible opportunities to produce a waste product other than sulfuric acid, such as liquid SO_2 or elemental sulfur. (The term "waste product" is used rather than "by-product." A by-product connotes something produced in expectation of economic gain. Almost without exception, no smelter operation has realized any economic gain in fixing the input sulfur as sulfuric acid to avoid exhausting the SO_2 to the atmosphere.)

Most existing reverberatory smelters have been forced to treat their converter off gases in sulfuric acid plants to comply with state and Federal environmental regulations. Since converting is essentially a batch-type process, the smelter operators attempt to keep at least one converter "blowing" at all times to generate sufficient gas to maintain the acid plant operation. During the periods when no converters are operating, supplemental energy must be consumed to maintain the overall heat balance of the acid plant. Air leakage into gas ducts further aggravates this problem. When a flash furnace is operated in conjunction with the converters, the furnace provides a steady base load of high-strength SO_2 gas to the acid plant, thereby significantly simplifying converter operation.

Since sulfuric acid is two-thirds air and water, the high cost of shipping acid from remote smelting locations to metropolitan markets has forced smelter managers to consider alternatives to producing sulfuric acid. One such alternative is to convert the SO_2 in the furnace off gas to elemental sulfur. The lower shipping costs for elemental sulfur greatly increase the potential marketing area for an elemental sulfur waste product as compared with a sulfuric acid waste product. Furthermore, if the waste product cannot be sold, it can be stockpiled as solid elemental sulfur, whereas sulfuric acid cannot be stored indefinitely. However, these marketing and stockpiling advantages generally do not outweigh the disadvantages of high energy consumption for reduction of SO_2 to elemental sulfur. For example, an additional 6×10^6 Btus/ton of concentrate would be consumed as fossiliferous fuels in the reduction process and tail gas treatment facilities associated with elemental sulfur production. Since the electrical energy consumption for the elemental sulfur facility is approximately equivalent to that for a sulfuric acid plant, this increased energy consumption would be a net addition to the energy consumed in the smelting system.

Conclusion. Since any new smelting facility in the United States is required to meet stringent Federal standards for SO_2 emissions, it is unlikely that reverberatory furnaces will be used in any new smelting facilities. Rapidly escalating energy costs coupled with environmental considerations make flash smelting an attractive alternative. Unfortunately, a flash smelter equipped with SO_2 recovery facilities is an extremely capital-intensive undertaking, and therefore a general industry changeover from existing reverberatory operations to new flash smelter installations appears unlikely at this time.

For background information *see* COPPER; PYROMETALLURGY, NONFERROUS in the McGraw-Hill Encyclopedia of Science and Technology.

[L. R. JUDD]

Bibliography: S. N. Sharma and W. L. Davis, Jr., *Soc. Min. Eng.*, 29(5):38–41, May 1977.

Flower

Plants, like animals, live and reproduce in a potentially hostile environment and have developed recognition systems that enable them to respond to external challenges and select appropriate gametes. In vascular plants, pollination leads to fertilization and seed set. The pollen grain, carrying the male gametes, interacts with the pistil, which bears within it the ovule containing the female gamete. Until recently, the processes by which compatible pollen was discriminated from foreign pollen could not be explained. Today, this discrimination of self from nonself is seen as part of the wider process of biological recognition which permits interactions between whole organisms, and between their component tissues and cells.

Sexual reproduction provides an experimental system for the exploration of male-female recognition: it is genetically well defined, and information on the nature of the interacting surfaces has implicated macromolecules which are capable of acting as messengers or receptors for recognition and cooperative interaction. In order to carry out this function, the molecules not only must be located on the surface, but also must possess some type of information-carrying capacity. This potential resides in the sequences of amino acids in proteins or sugar residues in carbohydrates, and glycoproteins with covalently linked protein and carbohydrate chains have dual potentialities as informational molecules. Proteins and glycoproteins are located in plant cells. However, the plasma membrane is usually surrounded by a thick polysaccharide wall. In some cases, this membrane may be a barrier to intercellular communication; in others, it may provide a matrix for cell-cell interactions. Intercellular cooperation is considered to be initiated through cell recognition, entailing molecular interaction at cell surfaces, which leads to defined responses in the cells involved.

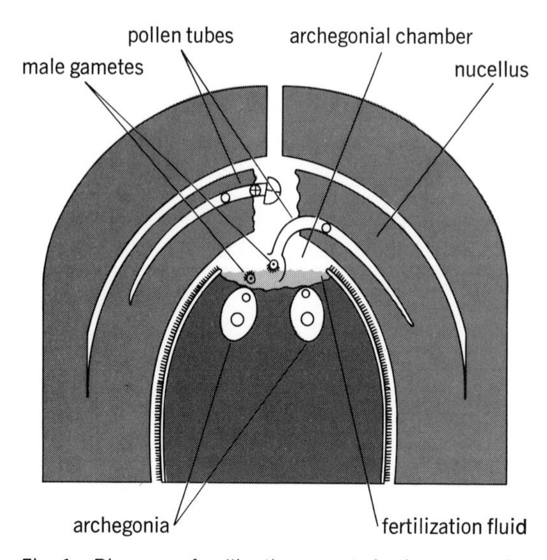

Fig. 1. Diagram of pollination events in the ovule of a cycad showing germinated pollen grains and their pollen tubes which penetrated the nucellus. Structures not drawn to scale. *(From J. M. Pettitt, Detection in primitive gymnosperms of proteins and glycoproteins of possible significance in reproduction, Nature, 266:530, 1977)*

Male-female recognition in cycads. Cycads, the remnants of an ancient group of gymnosperms, are dioecious, (have separate male and female trees) and possess motile sperm cells. Higher gymnosperms such as conifers and all angiosperms are siphonogamous, that is, the sperm are nonmotile and are transported to the egg within the pollen tube. In cycads, the pollen tube is produced when the pollen germinates within the archegonial chamber (Fig. 1). It acts simply as an anchor for the sperm sac, which ruptures and releases the sperm into a pool of fluid bathing the egg-containing archegonia. Recognition of compatible male gametes occurs within the fluid pool at the surface of the archegonia. J. Pettitt has shown that the fluid contains informational molecules, including proteins and glycoproteins derived from surrounding parental cells.

Pollination systems in flowering plants. After capture and acceptance at the stigma surface, the pollen of flowering plants germinates, and the tube bearing the sperm cells must usually penetrate the stigmatic cuticle and enter the wall system of the transmitting tissue, which provides a route through the style and supports the tube nutritionally in its growth to the ovule. The recognition events involved can be considered as a sequential series, each having an option in which the mutual interaction can be terminated (Fig. 2).

In plants which are self-fertile, for example *Gladiolus gandavensis* or *Crocus chrysanthus*, the first indication of acceptance after capture is the hydration and swelling of the pollen grains, which occur within a few minutes after arrival. In contrast, foreign pollen is often ignored and may fail to hydrate, or it may successfully pass this barrier only to be stopped at a later step. Pollen from closely related species usually germinates, but the tubes fail to penetrate the stigma surface. Compatible pollen tubes usually grow rapidly toward the ovule and, in plants such as *Helianthus annuus* and *Cosmos bipinnatus*, may travel distances of 7 mm within 40 min, while in trees such as *Pinus* and *Eucalyptus* the process may take several months.

Response to self. About two-thirds of the families of flowering plants exhibit self-incompatibility, in which self-pollen germination and growth are inhibited as a mechanism to favor outbreeding among the species. In most cases, self-pollen is recognized at the stigma surface, and may germinate and grow to a limited extent before arrest (Fig. 2). Self-incompatibility is controlled by the *S* gene, a multigene family similar to genes controlling immunoglobulin or ribosome biosynthesis.

Sporophytic self-incompatibility, in which pollen behavior is determined by the phenotype of the parent rather than the pollen grain itself, is characteristic of species with a dry stigma, and arrest usually occurs at the stigma surface. Rejection of self-pollen is correlated with the deposition of the polysaccharide callose, which may occlude both pollen tubes and adjacent stigmatic papillae. This has provided a cytochemically detectable marker for the rejection response in various Cruciferae and Compositae, and in *H. annuus* it is manifested following both self-pollination and cross-pollination with pollen of other Compositae genera. In compatible matings (in which the *S* alleles of pollen and stigma are different), the callose is confined to the germinating pollen and forms plugs which seal off the growing pollen tube tip containing the living cytoplasm. H. Dickinson and J. Heslop-Harrison and coworkers demonstrated that the isolated pollen surface proteins elicited the callose rejection response in self-stigmas of radish and kale, and recent experiments have implicated protein fractions in the 10,000–25,000 molecular weight range as the active components.

In gametophytic systems of self-incompatibility, expression of the *S* gene is determined by the genotype of the individual pollen grain at the stigma. Arrest of self-pollen tubes can occur at a variety of steps (Fig. 2). In the evening primrose, *Oenothera organensis*, the fate of the pollen is determined after capture in the surface mucilage of the wet stigmas, while in several Solanaceae, including *Petunia hybrida* and *Nicotiana peruvianum*, arrest occurs in the stylar transmitting tissue and is accompanied by marked thickening and swelling of the tube tips which become occluded by callose and often appear to burst as if in premature discharge of the gametes.

Pollen-stigma interface. Successful pollination requires a fine degree of reciprocity and coadaptation between the pollen and female organs. The interactions are initiated by contact between the pollen and stigma surface, and both the nature of these surfaces and their role in pollination have recently been explored.

The pollen-wall surface is intricately patterned, often with species- or group-specific features (Fig. 3). It consists of the outer patterned exine composed of sporopollenin, a polymer which is remarkably resistant to biodegradation, and an inner smooth layer, the pectocellulosic intine. The cavities in the exine are filled with proteins, glycoproteins, and lipids from the surrounding parental tapetal cells of the anther just prior to pollen maturation. In contrast, the intine proteins are products of the haploid pollen grain and are laid down within the polysaccharide matrix of the intine during its period of synthesis. They are especially concentrated at the germinal apertures where the intine is conspicuously thickened. The proteins include a variety of enzymes, some of which are antigenic in rabbits. The major macromolecular components of *Gladiolus* pollen contain protein, carbohydrate, and lipid in the ratio 10:6:0.2. Airborne pollen such as from ragweed and grasses contains wall-borne proteins and glycoproteins in both exine and intine sites that are allergens in humans.

A characteristic of pollen-wall proteins is that they are rapidly released when the pollen is moistened, with the exine-located components being released first and the intine proteins diffusing out more slowly.

The receptive stigma surface of flowering plants has recently been classified into two distinct types: wet stigmas (Fig. 3*a*), which bear a copious free-flowing secretion, and dry stigmas (Fig. 3*b*), which are coated with extracuticular hydrated secretion.

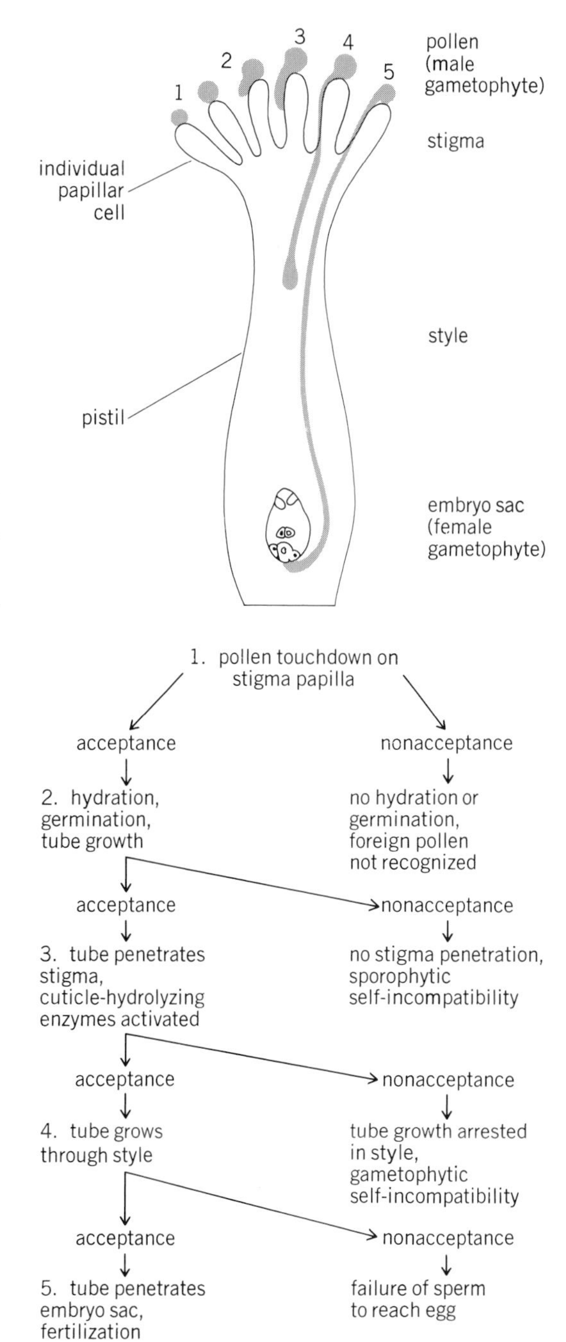

Fig. 2. Diagram representing behavior of pollen in terms of the events leading to a compatible pollination and the various incompatibility options. *(From A. E. Clarke and R. B. Knox, Cell recognition in flowering plants, Quart. Rev. Biol., 53:6, 1978)*

Further subdivision depends on the nature of the receptive cells—whether they protrude as papillae or form part of a rounded smooth surface. Elongate papillae are analogous to root hairs in their mode of origin from epidermal cells and may occur as filaments of cells, or single cells, in different species. The surface secretion of dry stigmas acts both as an adhesive and for pollen-stigma recognition. Although it is a true secretion and does not resemble a bilayered membrane in its ultrastructural appearance, this surface secretion can be compared with a membrane in its involvement in defined recognition reactions.

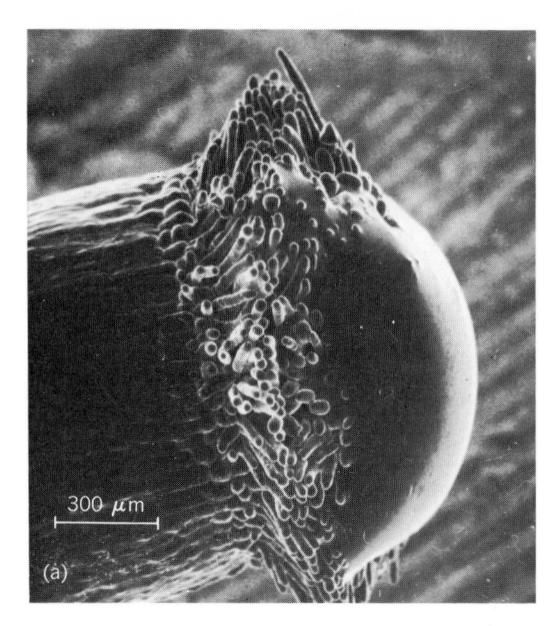

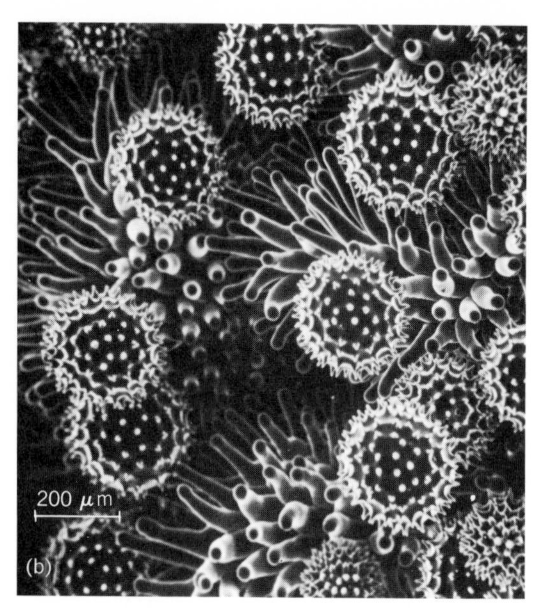

Fig. 3. Scanning electron micrographs of living pollinated stigmas of *(a) Eucalyptus* wet stigma with pool of free-flowing secretion; triangular-shaped pollen grains can be seen immersed in the fluid at left-hand side (upper part of micrograph) of stigma; *(b)* dry stigma of *Ipomoea* showing spine-covered pollen grains adhering to the stigmatic papillae.

In the dry stigma of *Gladiolus*, the papillae are coated when receptive with a surface secretion containing equal quantities of protein, carbohydrate, and trace amounts of lipid. Glycoproteins binding the lectin concanavalin A are present, together with a high-molecular-weight arabinogalactan protein with remarkable adhesive properties. The stigmas of most flowering plants also contain a surface component with nonspecific esterase activity, and the cytochemical detection of this enzyme provides a useful guide both to the receptive site of stigmas and to the onset of receptivity.

Attempts to demonstrate binding of isolated stigma and pollen surface components have been unsuccessful because of the remarkable general binding capacity of the stigma surface preparations. Pollen surface preparations show the greatest number of antigens expressed in any plant cell type, which may be indicative of their recognition function.

Search for recognition gene product. During the past 50 years, since E. East proposed that the *S*-gene product in pollen might resemble animal antigen and interact with antibody in the stigma, there has been continuing speculation as to its nature. D. Lewis advanced two widely accepted alternative hypotheses: (1) that the interactions depend on the binding of mutually complementary receptors on pollen and stigma, or (2) that identical molecules from the two partners interact to produce an inhibitory complex. Common antigens between pollen and stigma had been demonstrated in *Petunia* by H. Linskens, and in *Gladiolus* and *Prunus avium* by A. Clarke and coworkers, but whether or not these antigens would combine during pollination, and whether the combination would initiate rejection, has not been resolved.

The *S*-specific antigens of *Brassica* stigmas have been detected by immunoabsorption and shown to have a high isoelectric point, to be susceptible to periodate oxidation, and to bind the lectin concanavalin A. The use of affinity chromatography and immunological techniques should enable the isolation and characterization of the *S*-gene product and the elucidation of the recognition mechanism.

For background information *see* ANTIGEN; BREEDING (PLANT); FLOWER; POLLEN in the McGraw-Hill Encyclopedia of Science and Technology. [R. BRUCE KNOX]

Bibliography: A. E. Clarke and R. B. Knox, *Quart. Rev. Biol.*, 53:3–28, 1978; Y. Heslop-Harrison and K. R. Shivanna, *Ann. Bot.*, 41:1233–1258, 1977; R. B. Knox et al., *Proc. Nat. Acad. Sci. U.S.A.*, 73:2788–2792, 1976; J. Pettitt, *Nature*, 266:530–532, 1977.

Food engineering

Research and development in food engineering is concerned with the operations involved in food manufacturing from quality control of raw and processed material to storage of the processed food. This article will discuss (1) the use of lasers in particle-size analysis for quality control throughout the food manufacturing process; (2) the use of a retort pouch in packaging food; (3) the use of the collapsible metal tube as a food container; and (4) irradiation as a method of food preservation.

Laser particle-size analysis. Since the demonstration of the first operational laser in 1960, many applications have evolved. One recent advancement is the application to particle-size measurements of a variety of foods, chemicals, metal powders, ceramics, and pigments. The particle-size distribution, as well as mean diameter, surface area, and median diameter, provides a measure of quality of the material for the supplier or producer. Similarly, the consuming industry can monitor received product prior to use to avoid production delays caused by a low-quality product. In this way the producer and consumer are assured that the ground or milled product will properly disperse, dissolve, or coat a product; that the product should advance to another production step; that acceptable food texture will be obtained; and that production costs are minimized through energy conservation.

Application of the laser, microelectronics, and special optical features to long-established optical principles permits accumulation of particle-size data in the range of 2–300 μm (50 to 5000 mesh, Standard U.S. Sieve). The laser source is a mixture of helium and neon gases whose emitted light is directed on particles in a cloud constantly flowing across the path of the laser beam (Fig. 1). The particles scatter the light at specific angles and intensity depending on their size: large particles scatter light at smaller angles and higher intensity than smaller particles (Fig. 2).

The scattering effect by the particles is highly defined due to the specific wavelength emitted by the laser source. Since the scattered light is of highly defined angle and intensity, it contains information on particle size. The optical components are employed to analyze the information from the particulate cloud. The first component is a lens which focuses the scattered light on a patented optical mask which is a combination fixed disk and rotating transmission filter. The combination disk and filter selectively admit light from the entire scatter pattern of the particulate cloud, depending on the angle of scatter. A second lens collects the selected transmitted light and directs it to a photodetector which measures light intensity. The electronic signal from the photodetector is relayed to a microcomputer where the signals are mathematically manipulated to provide various types of particle-size data.

Process control. Of special interest to plant managers and process engineers is the capability of instrumentation to generate analytical data directly and continuously on process streams in real time as opposed to waiting for distant, lengthy laboratory analysis. The advantage of such a system is the ability to monitor a process at a critical location for product not complying with specifications. Data obtained from the on-line instrument are electronically transmitted to central control rooms to alert engineers of possible processing errors so that corrective action may be initiated.

The relatively recent spiraling cost of energy has stimulated interest in on-line measurements such as particle-size instruments. It has been estimated that 16% of energy in the United States is consumed in milling and grinding operations. Laser particle-size analysis with on-line capability aids in reducing production costs by monitoring for grinding errors in many operations, including cocoa grinding, corn wet milling, starch production, and wheat milling. Monitoring a grinding or milling operation ensures that specifications are maintained while preventing excessive grinding and unnecessary expenditure of costly energy.

Dispersal and dissolution of solids. It is generally desirable for solids to disperse or dissolve evenly in various liquid media. Settling of suspended materials may be counteracted by proper selection of particle size and emulsifying ingredients such as those naturally found in dairy products or dessert mixes. Suspension of cocoa in such products partly depends on particle size, for particles of excessive size will sediment following mixing.

Dissolution characteristics of a solid are also a function of process variables and particle size. The rate of dissolution of encapsulated materials has been determined kinetically, employing laser particle-size analysis. By using such data, the ingredient manufacturer can determine optimum particle size for rapid dissolution in liquids and can allow for process control through routine quality control. Other products of interest in dispersal and solution include canned and instant soup mixes, starch suspensions for paper coatings, food colorants, and powdered drinks.

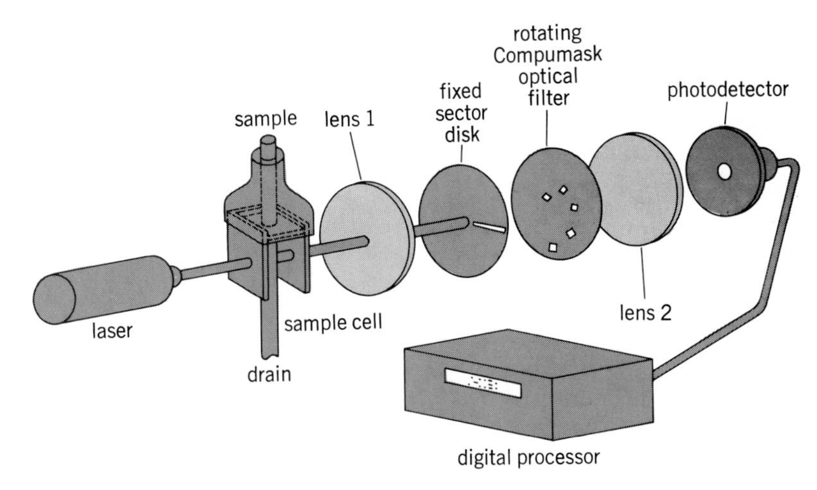

Fig. 1. System used for laser particle-size measurement.

Milling, grinding, and air classification. Particle-size measurement of milled cereals is carried out to ensure compliance with Federal standards and bakery specifications. Analysis of wheat flour is performed by dry-sample analysis using the laser system to control product quality. At appropriate process locations, particle-size analysis of hard wheat flour destined for fractionation may be performed. One such location or control point (Fig. 3) is the point of entry to the system to ensure uniformity of the flour for further fractionation, since variations in flour particle size occur, depending on wheat crop or age. Other control points analyze the particle size of various finished products to be shipped to bakeries. As an example, hard wheat flour is milled and separated on the basis of baking acceptability as a function of particle size. Once an acceptable particle size for baking is determined, the separators are set and monitoring of the three fractions (less than 17 μm, 17–35 μm, and greater than 35 μm) is performed.

Another use of the laser particle-size system in solving food industry problems has been to characterize the interaction of flour with water as a func-

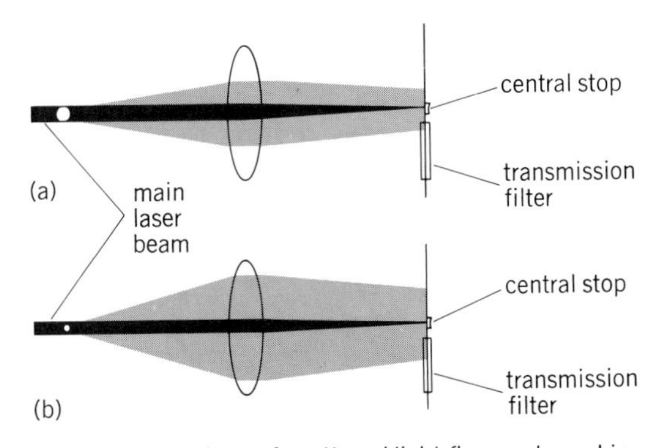

Fig. 2. Comparison of scattered light flux angle and intensity for *(a)* large and *(b)* small particles.

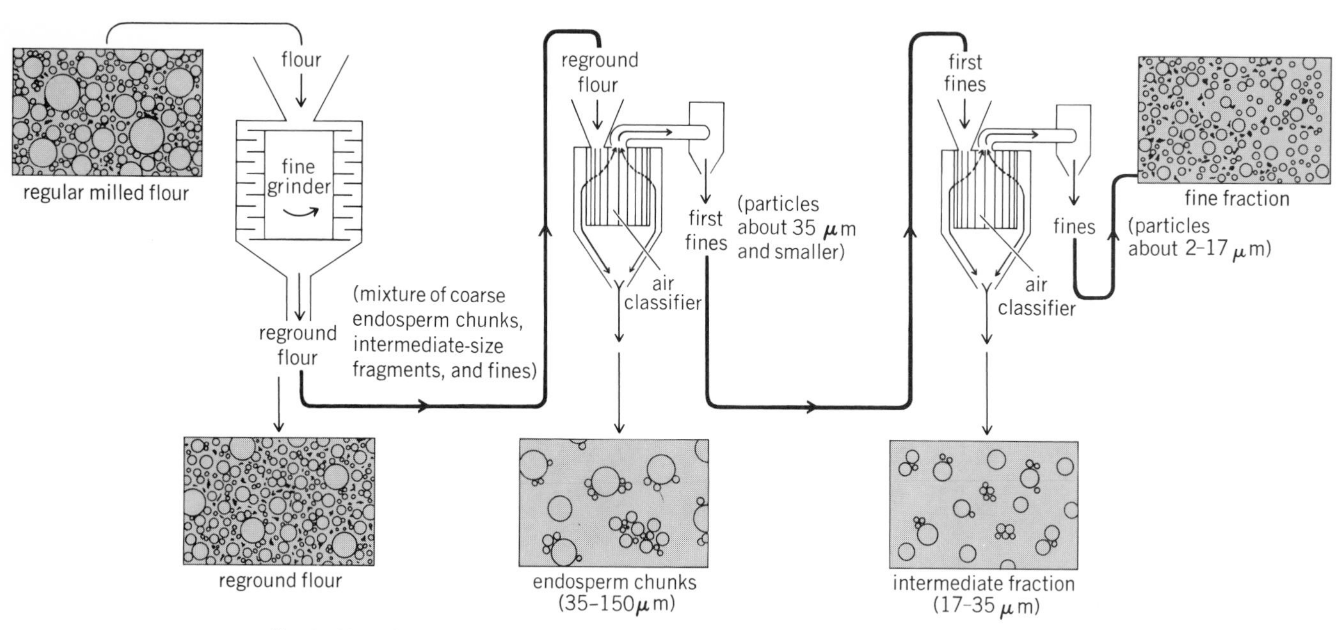

Fig. 3. Flour fractionation process.

tion of time. A portion of flour is supplied to the instrument, and the ability to measure volume, a particle-size parameter, as a function of time is utilized to observe relative volume changes. Such data show whether flour mixes sufficiently well with water for efficient dough production operations. From such studies, it has been shown that durum (macaroni), hard wheat (bread), and soft wheat (cake) flours interact quite differently and characteristically during the 15-min experimental period.

Texture. The flavor of foods encompasses several parameters, one of which is texture or mouthfeel. Various rheological instruments are available to assist food scientists in measuring chewability or break-off, but smoothness of texture is difficult to measure in products such as chocolate or powdered sugar due to the small particle sizes. Particles of sizes ranging from submicrometer to approximately 300 μm are important to palatability. Properly ground cocoa from press cakes will exhibit fineness, having particles nearly 100% smaller than 75 μm, as determined by wet-sieving procedures. Quality-control procedures are improved with concomitantly increased speed by employing the advanced laser particle-sizing technique. The instrument quickly measures the particle-size distribution of cocoa in water or the dry state to provide timely information for evaluation of production control parameters. Chocolate would be expected to exhibit larger particle size due to the presence of sugars and other components. Where grinding is insufficient or components are not small enough, unacceptable grittiness is imparted to the flavor profile of the product. Milk chocolate has related sampling problems in obtaining particle-size information; however, with milk chocolate the laser particle-size characterizing instrument performs the analysis simply by suspending the product in warm vegetable oil and transferring it to a specially designed sample container. Particle-size distribution may be obtained for similar products, such as fondants or sugars used in the manufacture of other confectioneries.

[PHILIP E. PLANTZ]

Retort pouch. One of the most important methods of food preservation is the heat treatment of foods within hermetically sealed containers. Although the process was carried out in the past in rigid containers such as metal cans or glass jars, the retortable pouch, governmentally approved and commercially available in the United States since May 1977, offers a new and most viable package opportunity for shelf-stable foods.

The term "retort pouch" is used to describe a flexible package, composed of a lamination of polyester, aluminum foil, and a polyolefin sealing ply, into which a food product is placed; the pouch is then sealed and sterilized at temperatures in excess of 212°F (100°C; generally the process temperature is 240–250°F or 116–121°C). The finished product is a commercially sterile, shelf-stable unit of food requiring no refrigeration. Additional modifications of pouch structure may be made, depending on the size of the pouch, pH level of the product, and process temperatures involved.

During the late 1950s Continental Flexible Packaging, a member of Continental Group, Inc., and the U.S. Army independently undertook the development of a retortable flexible pouch. Prior to that time there were few commercial, thermal plastic, heat-sealable films that could withstand the 250°F cooking temperature required in the commercial retort sterilization process. However, during the next 5 years polyurethane adhesives were adapted, permitting the use of high-density polyethylene and polypropylene films in retorting. While the United States government did not approve the retort pouch structure until May 1977, other parts of the world began to commercialize the use of the retortable pouch for processed food. The first commercialization of this three-ply pouch was for the packaging of meat products sold in England in 1967, and in 1968 Japan initiated a retort pouch process which grew rapidly and now exceeds 500,000,000 units per year.

Advantages. Due to the fact that the retortable pouch is flexible, it conforms to the shape of the product. This results in a thin cross section and large surface area/volume ratio, ensuring rapid heat penetration during processing. As much as a 60% reduction in total time required for commercial sterility can be obtained over an equivalent rigid package containing the equivalent amount of product. This reduction in process time minimizes overcooking of foods near the surface of the container, and the addition of liquid or brine, which is necessary for most vegetable products, is not required as a heat-transfer medium in retort pouches. In general terms, studies have indicated that because of the shorter processing time, foods undergo less thermal deterioration, which in turn results in less caramelization of sugars and starches, less breakdown of protein, less destruction of heat-sensitive vitamins, and less hydrolysis of flavor components. The most dramatic improvements in product quality have been demonstrated in multicomponent entree dishes where delicate flavor must be preserved, in sauces which are heat-sensitive, or particularly in the case of shellfish, such as shrimp, lobster, scallops, and crab.

Continental Flexible Packaging and its licensees have packaged and evaluated over 75 food products in retort pouches under the categories of beef, pork, seafood, poultry, pasta, vegetables, fruit, sauces, soups, and bakery items. Consumers today in Japan buy over 1,500,000 retort pouch products per day, including beef stew, curry dishes, hamburgers, sauces, and desserts. In Europe and Canada various marketed items include beef burgundy, ravioli, french fries, parsley sauce, chicken a la king, and hot dogs.

Process methods. Once pouches have been formed, either by the packaging convertor or online by the food processor, they are filled, the air is evacuated by means of mechanical vacuum or steam, and the pouches are top-sealed. Filled pouches are placed into retort racks, which aid in maintaining the cross section of the filled pouch for uniform heat penetration and which restrict the pouch from excess movement during processing. Overriding air pressure is required within the retort pouch so that pressure differentials are minimized between the inside and outside of the pouch during the heating and cooling phases of the retort cycle. This retort technology is commonly used by food companies processing glass jars. To prevent excessive pressure buildup in the pouch during retorting, the residual air content of the pouch after sealing should be less than 10 cm^3. Following retorting, the pouches are inspected and then cartoned for shipment.

Current filling speeds of retortable pouches are in the range of 40 to 60 pouches per minute, significantly below the higher-speed lines now used for filling metal cans. The prime factors affecting this speed include the type of product being filled, method of filling, and method of evacuation.

Equipment manufacturers have stated that by 1980 they expect filling and sealing equipment to operate at speeds in excess of 200 pouches per minute, a factor which will significantly improve the economics of retort pouch packaging.

Market status. Since the commercial introduction of the retort pouch in the United States in mid-1977, a major food company went into test market with seven entree products ranging from chicken a la king to veal scallopini in 8-oz (227-g) retort pouches, and a variety of pouched meat items are being supplied to the recreational/camping market by another food processor. A number of other companies are in the process of finalizing product formulations and consumer research for eventual market introduction of additional retort pouch items.

In more specialized market areas, retort pouch foods have been utilized in the Apollo program for feeding the astronauts. The National Aeronautics and Space Administration is spearheading a program to utilize retort pouch technology to feed the elderly, and the U.S. Army will be replacing the C-ration can with an annual requirement of 40,000,000 retort pouches comprising main entree items, cakes, and fruit products.

Companies are exploring retort pouch applications within four basic areas: retail distribution, institutional food-service applications, new product/new package applications for snack foods, and development of food items for rural areas in foreign countries. Within the retail sector, considerations include the multipacking of complete meals in one carton or providing single-service entrees. Institutional applications extend into specialized programs such as modified salt-free diet items for hospitals, vending machine products, and airline meals. Additional research is being conducted in utilizing larger pouches for up to 3 lb (1.36 kg) of product for distribution through restaurants and food-service outlets. Vegetable products now packed in institutional size number 10 cans require up to 40% of the total net weight to be liquid or brine for the heat-transfer medium. Studies have indicated that many of these products can be packed without these liquids in a 60-oz (1.70-kg) pouch, thereby improving color and texture of the product. Within the area of snack foods, shelf-stable meat items such as ham sticks with pineapple glaze or individual sausage-type items can be eaten directly from the pouch. Internationally, the retortable pouch offers a new packaging system for remote areas where distribution of raw product prior to processing may result in a high percentage of spoilage. Modular production systems can be transported by rail, truck, or ship directly to the site where crops, seafood, or livestock are raised. The 90% reduction in weight and 85% reduction in inventory space for empty pouches versus metal cans, coupled with shortened process times, affords tremendous savings in energy and increases the mobility of the packaging and processing system.

Outlook. Market expansion in retort pouch foods will undoubtedly be enhanced as pouch filling speeds are improved. Utilization of nonfoil pouch laminates and thermal or mechanical forming of web structures into dishlike containers will provide additional applications of processable laminations. Most importantly, the retort pouch offers a dynamic package medium for new products.

[RICHARD C. ABBOTT]

Collapsible metal tube. Ever since the collapsible metal tube was invented by the artist J. G. Rand, who used it to keep his paints from drying up, this unique package has grown in popularity for use in many different product lines. One relatively new use for the metal tube is as a food container. Recent upward trends in European consumption of food in collapsible metal tubes and the interest shown by several American food manufacturers indicate that the American consumer may soon be offered a variety of food products in tubes.

The use of metal tubes for food in the United States began with the G. Washington Company, which manufactured a coffee and milk concentrate in a metal tube about 25 years ago. In 1958 a Pennsylvania bakery included 6-oz (180-ml) tubes of jelly for its bread and pastry routes. A few years later, American Home Foods packaged Gulden's "Diablo" mustard in tubes to accompany the introduction of a new brand of frankfurters. In 1964 Otto Seidner, Inc., marketed 5-oz (150-ml) tubes of mayonnaise called "Picnic Packs."

The United States armed forces began using food in tubes for U-2 pilots in the early 1960s. Astronauts in the Gemini and Apollo programs also carried food tubes on space flights. Although these tubes were successfully used by the astronauts, food in tubes was not employed on a large scale by the Army.

In more recent years food in tubes has become increasingly popular in Europe. According to statistics of the European Tube Association, Europeans in 1976 consumed over 420,000,000 tubes of cheese spreads, mustard, catsup, butter, honey, and other food products. In some countries food-in-tube sales constitute two-thirds of the total collapsible metal tube market.

Although only one food product (anchovy paste) is currently manufactured in collapsible metal tubes in the United States, many food scientists believe that the metal tube is ideal for handling most foods. The package is lightproof and practically airtight, thus minimizing the possibility of contamination. Since the tube is virtually unbreakable and easily portable, it would be ideal as a convenience package for foods. The shelf life of food-in-tube products is generally rated at better than 90 days and, in most cases, these products do not require refrigeration after opening.

Tube sizes and linings. The size of a collapsible metal tube used for food products can vary from ½ to 2 in. (1.3 to 5 cm) in diameter and from 1½ to 7½ in. (3.8 to 19 cm) in length, with European tubes as large as 10 in. (25 cm) in length. Aluminum is the material most commonly used for food products. The tubes are composed of over 97% elemental aluminum and 3% various trace elements. Tin tubes are also used, and are usually formed from pure (99%) tin or from various tin alloys which contain a small percentage of copper.

The tubes are lined to prevent incompatibility with the food product. These linings must have good adhesion properties and should be highly flexible. Organic resins are the best-suited materials for inside linings, and either an epoxy, phenolic, or vinyl resin can be applied. Any of the inside coatings can be combined, although an epoxy base lining is generally employed as a base coat for a phenolic or vinyl cover. This method may be used for such high-acidity foods as fruit sauces, tomato juice, mustard, horseradish, or other volatile products. Any of these resins can be applied at high speeds by machinery integrated into the production line.

Most European food-in-tube products are pasteurized, although sterilization is not impossible. Steam sterilization is the most practical technique for foods in tubes, and is carried out at about 250°F (121°C).

Manufacturing costs. Energy costs have risen over the last few years, due largely to inflation and the increase in production costs of raw aluminum, which have risen by about 25% since 1974. Light, heat, and power costs have increased about 50% during the same time period. Manufacturing costs vary widely, since each plant has different equipment. The estimated selling price for a typical lined aluminum tube—1 in. × 5½ in., about 2 oz; 2.5 cm × 14 cm, about 60 ml—is roughly $75 per thousand.

Since the tubes are lightweight and can be shipped empty, there is considerable savings for the food manufacturer through a reduction in shipping and handling costs. An additional advantage of the metal tube is its versatility. Almost any type of food or food concentrate can be packaged in collapsible metal tubes, thus enabling the food manufacturer to tailor products to the needs of special markets.

With the current increase in consumer awareness, many food manufacturers are considering food in tubes. The overall costs for the production of metal tubes for food makes the package competitive with jars, cans, and other food packages. The unique features of the collapsible metal tube may lead to the introduction of many foods in tubes in the United States by 1980. [PETER KAUFMAN]

Food irradiation. Irradiation of food by high-energy electrons, x-rays, and gamma rays (from cobalt-60 or cesium-137) is a new preservation method. Radappertization, that is, sterilization with high doses of radiation [40 kilograys (kGy)], results in highly acceptable meats, poultry, and fish products, which are stable at room temperature and free of pathogenic and food-spoilage organisms. Radpasteurization, that is, pasteurization with low doses (0.5–5.0 kGy) significantly reduces or completely eliminates pathogens such as *Salmonella* spp. and *Escherichia coli.* Still lower doses (0.05–0.5 kGy) disinfect fruits, vegetables, and grain products, inhibit sprouting, and retard the ripening of fruits.

High-dose application. Recent developments have resulted in many highly acceptable ready-to-eat meat items that are stable over extended periods of time (several years) at room temperature. Contributing most to recent product improvements are the following advances: heat inactivation of proteolytic enzymes, elimination of oxygen, and irradiation at low temperature. Earlier food irradiation techniques primarily concentrated on eliminating microbes and placed less emphasis on these factors.

Enzyme inactivation. High doses of radiation completely eliminate microorganisms, but only slightly reduce the activity of most enzymes. Thus, food free of microorganisms but still containing active enzymes will enzymatically decompose. Such enzymes can be inactivated by heating the meats to about 68–75°C. Small amounts of salt (0.75% of sodium chloride) and phosphate (0.375% of sodium tripolyphosphate) are added to the meats to increase their water-holding capacity and to retain their juiciness and cohesiveness. This heat treatment also makes the meat "ready-to-eat."

Packaging. Prior to enzyme inactivation the meat can be prepared in different forms, depending on consumer preference. Beef can be made into rolls, steaks, roasts, hamburger, or frankfurters. Turkey can be processed as whole ready-to-eat turkey, turkey rolls, or smoked turkey slices. Before the product is irradiated, it must be hermetically sealed. This can be done in tin cans, aluminum cans, tray packs, or flexible retort pouches of plastic-aluminum laminates. Since cobalt-60 gamma rays can be used to irradiate samples of any size (up to 25 cm or 10 in.), the irradiation process does not restrict the form or size of the product. In the case of irradiating with 10,000,000-eV electrons from an accelerator, product thickness is limited to 3.3 cm (1⅓ in.). Because the heat treatment required for enzyme inactivation does not involve packaging, it places less restriction on the size and form of the product than the heat required for thermal sterilization.

Elimination of oxygen. Oxygen in the package will react with the lipids in the meat during irradiation to produce rancidity similar to that produced when meat is exposed to air. Therefore, it is important to reduce the oxygen content to less than ½ cm^3 air (equivalent to 0.1 cm^3 O_2) per 100 g. This level is highly recommended and is obtainable industrially.

Low-temperature irradiation. Approximately 60% of the enzyme-inactivated meat is water. If meat is irradiated while unfrozen, the radiolytic products formed in the water (such as $O\dot{H}$, H_2O_2, e^-, and H) will react with the organic components of the food. In irradiated frozen food most of these radiolytic products will be constrained and their chemical reaction limited. The ejected electrons may react at their site of origin, and the other radiolytic species in the ice may react with each other rather than with the organic solutes. Higher-order reactions such as $\dot{H} + O\dot{H} \rightarrow H_2O$; $e^- + H_3^+O \rightarrow H_2O + \dot{H}$; $e^- + OH \rightarrow OH^-$; and $\dot{H} + \dot{H} \rightarrow H_2$ tend to dominate.

The radiation effect on the most resistant microbial spore-formers, on the other hand, changes little from the liquid to the ice phase because the spores contain very little water. The quantities of radiolytic products produced in frozen meats are often 5 to 30 times smaller than in unfrozen meats.

Low-dose application. The number of *Salmonella* spp., *E. coli*, other pathogens, and food-spoilage organisms can be reduced and their growth delayed by the use of the low doses (about 2 kGy) of radiation. Low-dose irradiated meats can be processed and distributed at conventional refrigeration temperatures. Whole or cut-up chicken in carton boxes can pass from the processing floor through the irradiation facility directly to the warehouse or a waiting truck. The irradiation process is simple and requires no substantial change in existing processing and distribution methods. The cost of this process is about ½¢/kg (or ¼¢/lb).

Energy savings. The energy used for preservation of food by irradiation is small when compared with retorting, refrigeration, or frozen storage. However, one must consider not only the irradiation but also the other energy used in processing, storage, distribution, and home preparation. In all these areas significant energy savings would result from radiation processing. Frozen chicken rolls and radappertized enzyme-inactivated chicken rolls are comparable products. The total energy used for processing, storage (4 weeks), distribution, and home preparation of frozen chicken rolls is about 27,550 kJ/kg. For irradiated chicken rolls the comparable energy expenditure is about 14,260 kJ/kg. Radappertized enzyme-inactivated meat can be stored at room temperature (21°C) for several years. In many energy-poor developing countries where refrigerated distribution, marketing, and home storage are not highly advanced, radappertized foods offer the advantages of greater safety and reliability.

For background information *see* FOOD ENGINEERING in the McGraw-Hill Encyclopedia of Science and Technology.

[ARI BRYNJOLFSSON]

Bibliography: R. Bannar, *Food Eng.*, 49(7): 51–52, 1977; A. Brynjolfsson, *Energy and Food Irradiation*, U.S. Army Natick R&D Command Tech. Rep. TR-78/032, August 1978; A. Brynjolfsson, *The High Dose and the Low Dose Food Irradiation Programs in the United States*, U.S. Army Natick R&D Command Tech. Rep. TR-78/033, August 1978; E. P. Larkin, *Thermal Inactivation of Viruses*, U.S. Army Natick R&D Command Tech. Rep. TR-78/002, 1977; P. J. Mann, *Food Eng.*, 49: 85–88, 1977; A. Pinto, *Mod. Packag.*, 50(11): 46–48, 1977; W. Remington, *Mod. Packag. 1976–77 Encycl. Buyer's Guide*, 49(12):112–114, 1976; E. Robbins and B. Hagan, *Food Process.*, 39: 98–99, 1978; J. Stitley et al., *Food Prod. Develop.*, 10:72–74, 1976; A. L. Wertheimer et al., *SPIE*, 129:49–58, 1977; A. L. Wertheimer and W. L. Wilcock, *Appl. Opt.*, 15:1616–1620, 1976.

Foraminifera

Foraminifera are predominantly marine protozoa belonging to the class Sarcodina, which includes the amebas. Recent research on feeding patterns and prey apprehension has elucidated some of the adaptive mechanisms of planktonic foraminifera that enhance their survival in the uncertain existence imposed by a floating habit.

Although foraminifera are related to the amebas, their morphology and feeding behavior are markedly different. Amebas possess finger-shaped pseudopodia called lobopodia by which they move and engulf food. Foraminifera have fine, strandlike pseudopodia called rhizopodia that form a branching or netlike array. They secrete a calcium carbonate shell or gather fragments of debris around them to form a protective coat. The cytoplasm within the shell is coalesced to form a compact mass or is sometimes frothy due to the presence of many vacuoles and lacunae. One or more apertures in the shell permit continuity between cytoplasm inside and outside the shell. The rhizopodia emanate from a thin layer of cytoplasm surrounding the shell or sometimes from a frothy mass immediately inside each aperture. The pattern of rhizopodia is determined in part by the habitat and feeding behavior of the foraminifer. Benthic foraminifera, which dwell on the bottom of the ocean or on surfaces near the shoreline, possess a fan-shaped array of branching rhizopodia whereby they gather food. Planktonic foraminifera are floating organisms, and their rhizopodia are arranged as a corona radiating out in all directions in space around the shell.

As with many planktonic organisms, foraminif-

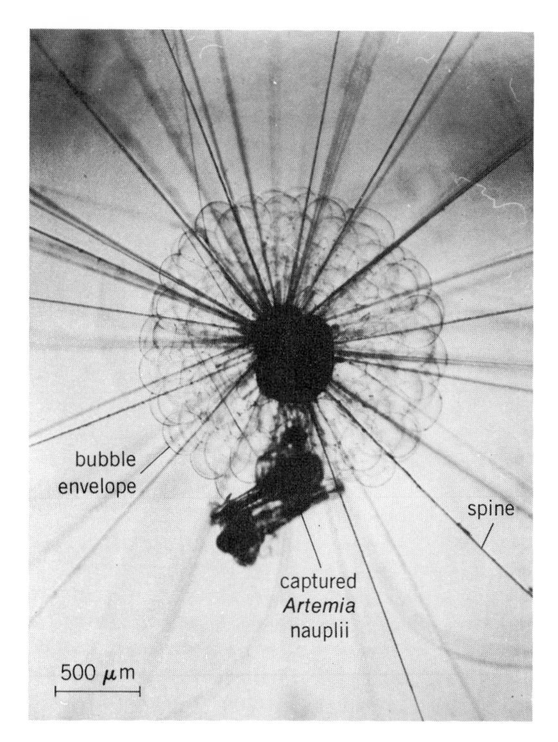

Fig. 1. Light micrograph of a living *Hastigerina pelagica*. Cytoplasmic bubble envelope surrounds opaque shell which bears numerous spines. Two captured *Artemia* nauplii are being drawn into bubble envelope. *(Courtesy of A. W. H. Bé)*

era float in ocean currents that carry them from one location to another. Therefore, their path is largely determined by the direction and magnitude of the currents. Food organisms are captured as they swim by or come into contact with the rhizopodia. Some prey, however, that are positively phototactic may be attracted to the glistening surface of the cytoplasm and shell. A general understanding of foraminiferal morphology is essential to an appreciation of the feeding behavior of these organisms.

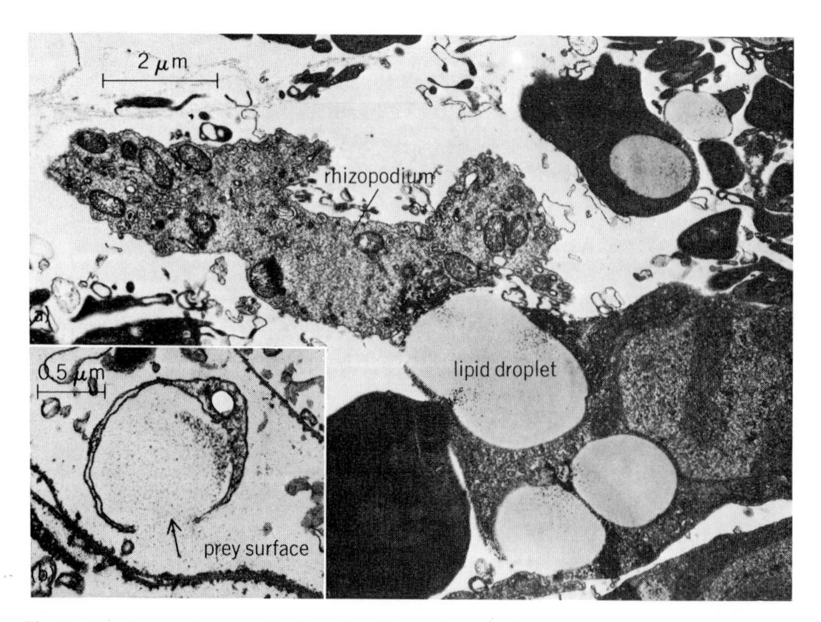

Fig. 2. Electron micrograph of section through prey tissue *(a)* invaded by a foraminiferal rhizopodium which is engulfing a lipid droplet extruded from a lysed prey cell; *(b)* rhizopodial adhesive vesicle has opened (arrow) and released adhesive substance near a prey surface. *(From O. R. Anderson and A. W. H. Bé, A cytochemical fine structure study of phagotrophy in a planktonic foraminifer, Hastigerina pelagica [d'Orbigny], Biol. Bull., 151:437–449, 1976)*

Morphology and fine structure. A light micrograph of a living *Hastigerina pelagica* (Fig. 1) illustrates the organization of the cytoplasm surrounding the spiral shell which has numerous spines. Other planktonic foraminifera, such as the Globorotaliidae, do not have spines and they are accordingly termed nonspinose species. The cytoplasm outside the shell of *H. pelagica* forms an envelope of bubblelike compartments, which undoubtedly aids buoyancy; if it is damaged or resorbed, the foraminifer sinks. Other planktonic foraminifera are surrounded by a mass of weblike rhizopodia. When spines are present, fine strands of rhizopodia stream along their surface and radiate from their tips as a fringe of sticky filaments. The large surface area produced by the rhizopodia increases the probability that prey will be captured and secured. The rhizopodia contain vesicles filled with an adhesive substance that is released to apprehend and hold prey (Fig. 2*b*). Fine-structure observations of the cytoplasm within the shell show that the adhesive-containing vesicles are secreted by the Golgi apparatus. Digestive vacuoles containing prey are found in the rhizopodia and in the intrashell cytoplasm.

Sources of nutrition. Pigmented dinoflagellates (flagellated algae) are commonly observed in vacuoles or are associated with the rhizopodia. It is not known whether they are symbionts, serving a beneficial role for the host, or merely commensals harbored by the host. They do not appear to be parasites, since they often occur in large numbers but produce no apparent deleterious effects. Fine-structure evidence indicates that dinoflagellates sequestered within cytoplasmic vacuoles in *Globigerinoides sacculifer* may nourish the host by secreting organic substances. However, much additional research needs to be done on the host-algal association.

Observations of feeding behavior in field-collected specimens and laboratory experiments on feeding and prey selectivity show that most planktonic foraminifera are omnivorous. They prey upon algae and small invertebrates, including crustacea (for example, copepods) and ciliated protozoa such as tintinnids. Among the algae consumed are pennate (boat-shaped) and centric (drum-shaped) diatoms and flagellates, including Coccolithophorida and dinoflagellates. Representative species of omnivorous planktonic foraminifera are *G. ruber*, *G. aequilateralis*, *G. menardii*, *G. truncatulinoides*, and *Pulleniatina obliquiloculata*. *Hastigerina pelagica*, however, is primarily carnivorous.

Feeding behavior. The feeding behavior and fate of ingested prey have been studied most thoroughly in *H. pelagica*. The following is a summary of events which occur during crustacean predation. When a crustacean comes into contact with *H. pelagica*'s spines, it is immediately snared by the rhizopodia (Fig. 1). The prey is gradually drawn into the bubble capsule, displacing some bubbles in the process. The rhizopodia exhibit a remarkable plasticity during prey engulfment. They constantly undergo transformations as they separate into numerous strands or coalesce to form sheet-like structures. The net result of this activity is to enshroud the prey and move it toward the shell. An adhesive substance is released from vesicles within the rhizopodia which covers the surface of the prey. The rhizopodia penetrate into crevices in the cuticle of the prey, split it open, invade the underlying soft tissue, and engulf lipid droplets and small particles of tissue that are released by the dying cells (Fig. 2*a*).

Engulfed particles of food are carried by rhizopodial streaming into the frothy layer of cytoplasm where food vacuoles are formed. The vacuoles containing engulfed prey tissue (Fig. 3*a*) become converted to digestive vacuoles by fusion with lysosomes containing digestive enzymes. Cytochemical staining for digestive enzyme activity confirms that the digestive vacuoles possess the lysosomal marker enzyme (Fig. 3*b*). Other digestive enzyme activity is found in the lacunae between the rhizopodia near the aperture. The origin of these enzymes is not clear. However, it is known that foraminifera purge their cytoplasm of nondigested material by excreting waste vacuoles containing residual enzymes at the aperture of the shell. Some of the residual enzymes may be used to begin digestion of freshly captured food. Digestive vacuoles are formed throughout the cytoplasm within the shell, and after digestion is completed, waste particles are centrifugally carried out of the aperture and discarded at some distance along the

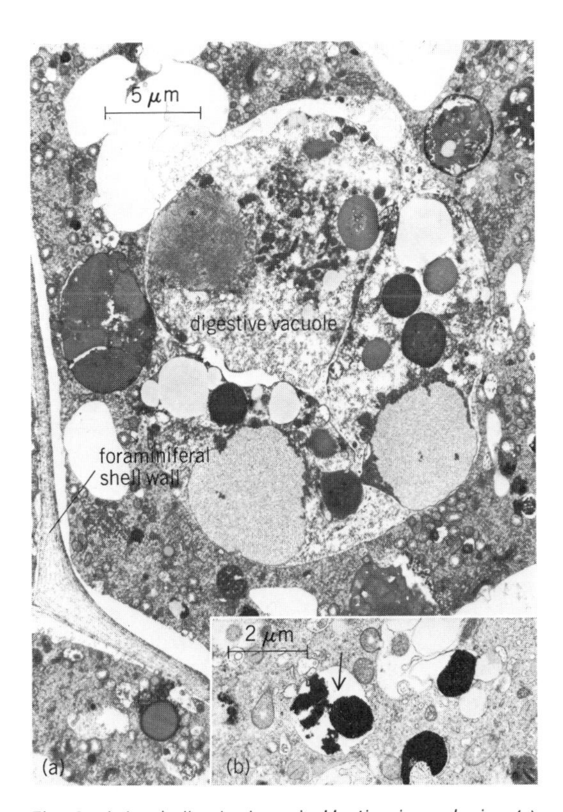

Fig. 3. Intrashell cytoplasm in *Hastigerina pelagica*: (*a*) Large digestive vacuole contains masses of partially digested prey tissue; segment of decalcified foraminiferal shell wall surrounds the cytoplasm. (*b*) Vacuoles contain dense stain (arrow) confirming presence of digestive enzymes. (*From O. R. Anderson and A. W. H. Bé, A cytochemical fine structure study of phagotrophy in a planktonic foraminifer, Hastigerina pelagica [d'Orbigny], Biol. Bull., 151:437–449, 1976*)

rhizopodia. Remaining pieces of prey exoskeleton or cuticle are also discharged.

Hastigerina pelagica has a large orange body in the proximal spiral region of the shell. Although its function and chemical composition are unknown, the body varies in size and color intensity depending on the nutritional state of the organism. When the organism is well fed, the body is large and nearly reddish in color. However, when the organism is starved, the body diminishes in size and eventually the total cytoplasm becomes white or colorless. Also, *H. pelagica* collected from the Sargasso Sea in the spring and early summer, when they feed most actively, frequently possess orange protoplasm, whereas those collected in winter have whitish protoplasm. It is possible, therefore, that the orange substance may be a food reserve.

Algal predation in omnivorous species begins when an acceptable algal prey contacts the rhizopodia and is snared within the adhesive substance. If the alga is unacceptable, it is promptly discarded or is not snared. After adhesion, the prey is engulfed in a rhizopodial food vacuole and carried by rhizopodial streaming into the apertural region where digestion begins. Some digestive vacuoles are also formed within the intrashell cytoplasm. The digested products of animal and algal prey pass through the membrane of the digestive vacuole and diffuse into the cytoplasm of the foraminifer.

Many species of planktonic foraminifera can be maintained in laboratory cultures if they are kept in a fresh supply of millepore-filtered open ocean water. They accept *Artemia* (brine shrimp) nauplii and laboratory-cultured algae as food. Preliminary studies with laboratory specimens indicate that a 3- to 6-day interval between feeding is optimum.

For background information *see* FEEDING MECHANISMS (INVERTEBRATE); FORAMINIFERIDA in the McGraw-Hill Encyclopedia of Science and Technology. [O. ROGER ANDERSON]

Bibliography: O. R. Anderson and A. W. H. Bé, *Biol. Bull.*, 151:437–449, 1976; O. R. Anderson and A. W. H. Bé, *J. Foraminiferal Res.*, 6:1–21, 1976; C. Febvre-Chevalier, *Protistologica*, 7: 311–324, 1971; J. J. Lee et al., *J. Protozool.*, 13: 659–670, 1966.

Fusion, nuclear

Nuclear fusion research has largely centered on magnetic confinement techniques and to a lesser extent on inertial confinement techniques using powerful lasers to implode tiny fuel pellets of deuterium and tritium. Recently, research has been undertaken on the implosion of fuel pellets by beams of electrons and both light and heavy ions. The article discusses electron-beam-induced fusion and the Alcator device, a highly promising approach to fusion by magnetic confinement.

Electron-beam-induced fusion. Scientific teams at Sandia Laboratories, Albuquerque, and at the Kurchatov Institute, Moscow, have begun to study the use of intense electron (and more recently, light-ion) beams to induce fusion. These beams are generated by relatively simple and thus inexpensive pulsed megavolt megampere generators (approximately 2% of the cost of a pulsed laser of comparable energy). Such devices, which were

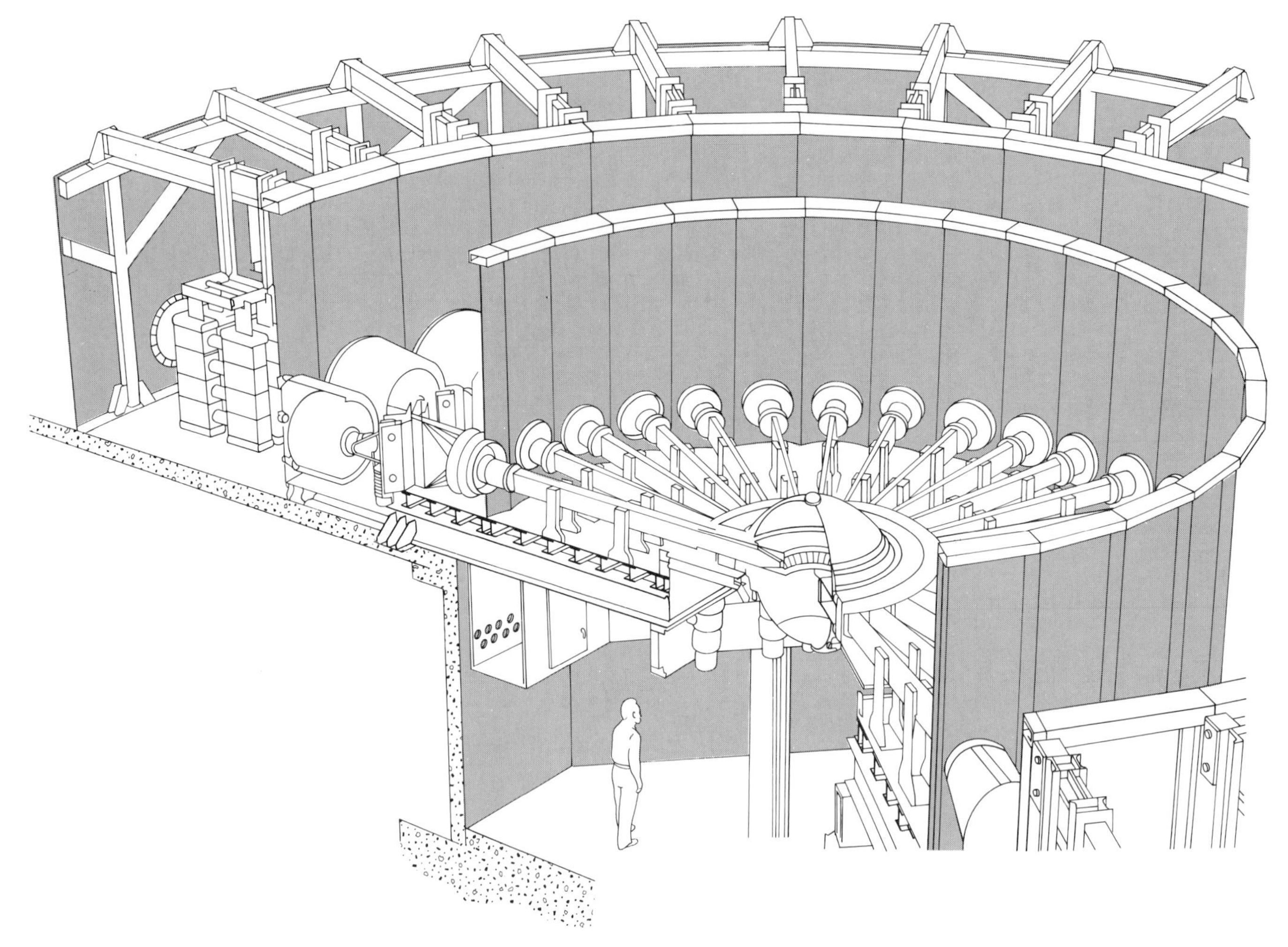

Fig. 1. Electron-beam fusion accelerator, under construction at Sandia Laboratories.

initially developed to provide powerful x-ray bursts for radiographic and nuclear weapons effects simulation, have evolved since their inception in 1965 to the 8×10^{12} W level, and $50-100 \times 10^{12}$ W generators are under development for use in fusion ignition experiments in the 1980s.

Technology. An intense particle-beam generator is basically a high-voltage source which stores energy in capacitors and then discharges it through switches into a liquid-dielectric-filled pulse-forming line and from there into a vacuum diode. Electrons are accelerated to the anode from a dense plasma which forms on the metallic cathode surface and, if a plasma exists on the anode, ions are accelerated in the opposite direction. In electron-beam applications, a 1-2-MeV approximately 0.5-MA electron beam is formed which self-constricts (pinches) due to its intrinsic magnetic field, and power densities of 10^{13} W/cm^2 are achieved on the anode. Single-beam experiments are carried out by placing a spherical target on a short stalk attached to the anode; uniform spherical irradiation is achieved because of the gaslike nature of the beam. Multiple-beam irradiation is to be achieved either by bringing many tapered diodes to the target surface (the Kurchatov approach) or by transporting the beams in preformed plasma discharge channels (the approach under study at Sandia). Such discharge channels could be initiated in a dense background by using laser beams, and thus there would be no mechanical connection with the pellet as required for postulated reactor applications.

At the Kurchatov Institute a 5-MJ 100-TW accelerator, Angara V, is being developed, and at Sandia a 1-MJ 30-TW accelerator, the electron-beam fusion accelerator (EBFA), is to be completed in 1980 (Fig. 1).

Magnetically insulated transmission lines. Both devices will rely on the application of a new power concentration technique called the magnetically insulated transmission line. This concept, which was developed at Kurchatov, Physics International Company, and Sandia, uses the self-electromagnetic field of a powerful pulse of electrons propagating in vacuum between two metal surfaces to prevent energy loss. Without the magnetic insulation provided by this self-field approach, electrons would cross the interelectrode gap and the propagating wave would lose energy, so that the power which could be transmitted in a reasonable-size transmission line would be 0.1 TW (Fig. 2*a*). However, at high enough power levels for self-insulation, the wave magnetic field turns electrons with an orbital radius smaller than the gap, so that power is transmitted efficiently and the transmitted power level exceeds 1 TW (Fig. 2*b*). Computer simulation of the process shows that electron loss occurs only at the head of the propagating pulse (Fig. 2c).

In EBFA, 36 rectangular transmission lines, each 7 m in length, are to be used to deliver 30 TW to within 50 cm of the pellet, and beam transport along plasma channels will deliver the beam to the target in a setup similar to spokes converging at the hub of a wheel.

Light-ion beams. In the light-ion approach, the self- or external magnetic fields are used to divert and thus suppress the flow of electrons across the diode, and the more massive ions from the anode plasma are extracted through apertures in the cathode. By shaping the diode properly in order to direct the ions to a common point, an intense focus should be formed. In one diode configuration being considered for EBFA, multiple ion beams are geometrically focused over a distance of less than 1 m and then transported in multiple plasma discharge channels to the target.

Space-time compression. EBFA will also be used to evaluate a beam concentration concept which is a standard technique used in conventional particle accelerators, namely, increasing the ion-beam power on target by space-time compression. Such compression is possible because ions in the energy range of a few megavolts are nonrelativistic, and by increasing the accelerating voltage in time, with a simple voltage ramp, the later ions in a pulse can be made to overtake slower-moving early ions, producing a shortened pulse and higher power at the target. A power gain of at least five may be achieved in this way. These concepts are also under investigation by R. N. Sudan at Cornell University and by a group directed by G. Cooperstein at the Naval Research Laboratory, Washington, DC.

Target physics. Of the many problems which are being faced in electron-beam fusion research, one of the most critical is the question of energy deposition. Efficient beam coupling and implosion of pellets by megavolt electrons require that the particles be stopped in a relatively thin outer region of the pellet. This can be done by using a dense pellet overcoat such as lead, but this increased mass which must be imploded leads to increased energy requirements. Fortunately, it appears that the self-magnetic field of an intense electron beam can lead to reduced penetration. In a simplified sense, this effect can be explained in terms of beam stagnation, which permits ordinary collisions to stop the beam in a thin layer. The most impressive demonstration of this effect was announced by L. I. Rudakov of the Kurchatov group. He reported heating of thin gold foils to a temperature of 80 eV, which amounts to a 10-fold electron deposition enhancement. The Kurchatov group also used this technique with a multishell pellet design to produce (in 1976) the first thermonuclear neutrons (over 10^6 were produced) from an electron-beam-driven pellet.

Electron-beam target experiments at Sandia in 1977 resulted in the production of a similar number of neutrons by using an approach adopted from magnetic fusion, namely, the use of a magnetic field to thermally insulate preheated fuel in the pellet from the surrounding high-density outer shell. In this way, the magnetically insulated pellet can be imploded more slowly without losing its energy prematurely by conduction. This approach permits the use of longer power pulses and places less stringent demands on beam-focusing.

Because of the complexity and as yet incompletely demonstrated scaling of enhanced electron deposition, interest has grown in the use of light ions instead of electrons because of the ion's short deposition length in matter. This shorter range eases both pellet design and beam generation requirements. For instance, calculations show that a magnetically insulated target should allow one to reach ignition conditions with about 10 TW absorbed as compared with about 50 TW with electrons. Experience with light ions is still insufficient, particularly in regard to focusing, to determine whether they will in fact be preferable to electrons. Electron- and ion-beam-driven target experiments with EBFA and Angara V projected for the early 1980s should resolve these questions and set the stage for the next probable step, a reactor experiment in the 1990s.

Reactors. In reactor system studies, Sandia is considering a small reactor chamber less than 3 m in radius. The reactor would operate at 100 MWe and would require a pellet gain as small as 30. Such a small pellet gain would still prove useful if the present low cost and efficiency of particle beam drivers can be extended to reactor drivers.

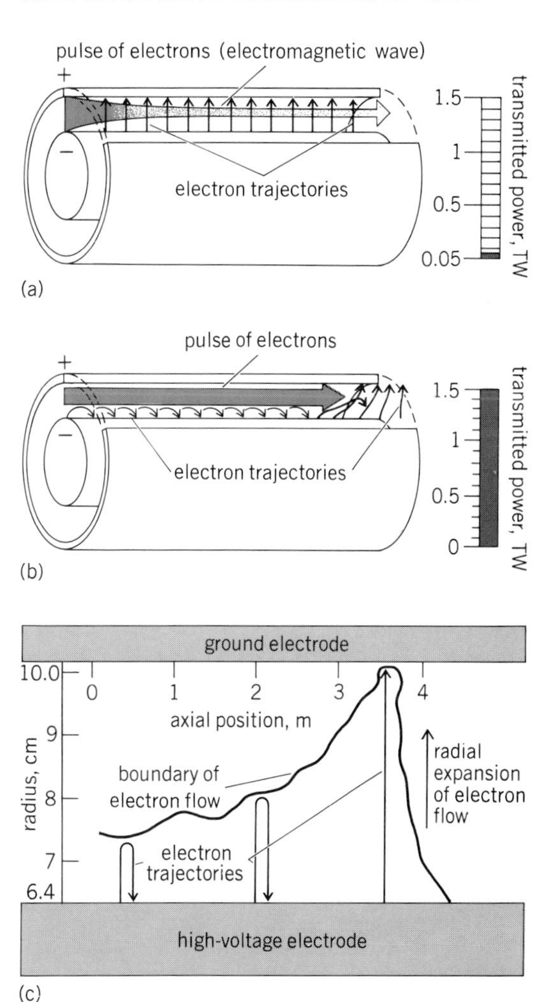

Fig. 2. Magnetically insulated transmission line. Operation *(a)* without and *(b)* with magnetic insulation. *(c)* Cross section of transmission line, showing computer simulation of electron flow.

This reactor design benefits from the dense background gas within which the discharge channels are formed. This buffer gas shields the wall of the reactor chamber from low-energy pellet-explosion by-products which would, in an evacuated chamber, lead to wall erosion. Instead, a "fireball" forms near the pellet, and this region of hot plasma expands and its pressure decays sufficiently so that a small radius chamber can accommodate this repetitive loading. The Kurchatov group is considering a different approach, namely, the use of much larger pellet gains (about 1000) and pellet chambers (8-m radius), and also a thorium blanket to produce ^{233}U in addition to useful power. Various systems studies of fusion-fission hybrids in the United States have shown that a fusion-driven source of fissionable material could economically supply fuel to a preestablished fission reactor economy. Whether such hybrids will prove desirable is the subject of considerable debate; regardless of the outcome, the high efficiency of particle accelerators makes it possible to consider the environmentally attractive approach of pure fusion with small pellet gains. The technology and physical understanding of intense electron and ion beams is rapidly evolving, and experiments now being designed should determine whether this approach will provide an avenue to practical fusion power. [GEROLD YONAS]

Alcator. Confinement of thermonuclear plasma in a tokamak is a highly promising approach to the realization of fusion reactors. The term "tokamak" is applied to an axially symmetric toroidal device in which the hot plasma is immersed in a strong toroidal magnetic field B_T, generated by an external coil. Current is induced in the plasma by transformer action from another external coil. This toroidal plasma current I_p generates a poloidal magnetic field B_Θ, which in conjunction with B_T gives the plasma column overall stability. I_p also serves to heat the plasma resistively.

Alcator is a high-toroidal-field high-density tokamak. Stable discharges with peak density $n_0 = 1.5 \times 10^{15}$ cm^{-3} have been obtained in deuterium at toroidal field $B_T = 87$ kilogauss (8.7 tesla). The best $n_0\tau_E$ value (product of peak deuteron density n_0, and the global energy confinement time τ_E) obtained in Alcator is 3×10^{13} cm$^{-3}\cdot$s. This is more than five times the best $n_0\tau_E$ value obtained in any other magnetic confinement device. It is only a factor of 3 less than the theoretical value required for fusion energy "break-even" at higher temperatures.

The main thrust of the Alcator program has been to study and elucidate tokamak operation in the high-density nearly collisional regimes, and to investigate the merits of this approach to thermonuclear fusion conditions. The physical significance of the ultra-high-density plasmas includes the following: (1) Nearly classical behavior, from the viewpoint of energy confinement, is observed. Values of τ_E and poloidal beta β_Θ (ratio of plasma kinetic pressure to magnetic pressure due to B_Θ) are close to neoclassical estimates. (2) Nearly complete thermal equilibration between electrons and ions is achieved, with temperatures close to 900 eV. (3) The flux of energetic charge-exchanged neutral particles with energy greater than 200 eV decreases by a factor of about 100 relative to the flux at low density. This reduces greatly the potential for sputtering of the vacuum wall by neutral particles. Consequently, the flux of heavy impurities into the plasma from the stainless steel wall is substantially reduced. (4) Impurities play a minor role in determining the power balance in the plasma core.

Energy confinement. A study of energy confinement and its dependence on B_T, I_p, $\bar{n}$ has been made ($\bar{n}$ is the interferometrically determined line-averaged plasma density). Such a study is facilitated by the wide range of parameters in which Alcator can operate, 30 kG (3 T) $\leq B_T \leq$ 100 kG (10 T), 80 kA $\leq I_p \leq$ 290 kA, and 5×10^{12} cm$^{-3} \leq \bar{n} \leq 7.5 \times 10^{14}$ cm^{-3}, combined with clean vacuum and wall conditions such that the measured plasma resistivity is nearly equal to the calculated classical value. When B_T and I_P are constant, the global energy confinement time τ_E is observed to increase approximately linearly with $\bar{n}$, as shown in Fig. 3. τ_E is defined as the ratio of the total plasma energy to the power put into the plasma. Total plasma energy is governed by various transport processes, a standard for which is the classical model in which dissipation is induced by Coulomb-collisional scattering. Neoclassical theory is the extension of the classical model to a magnetically confined plasma, specifically taking into account the effect of the topology of the confining magnetic fields on the particle trajectories. Neoclassical theory provides an irreducible minimum value for the various plasma transport coefficients. The curve labeled "neoclassical" in Fig. 3 gives the value of τ_E if neoclassical transport were the dominant energy loss mechanism, for plasma conditions identical to those in the experiment. Comparison between the two curves for τ_E shows that the plasma is afflicted by anomalous energy loss, and that this loss is reduced as density increases in Alcator. This excess energy loss is attributed to anomalous radial electron heat conduction.

A study of the energy balance in Alcator shows

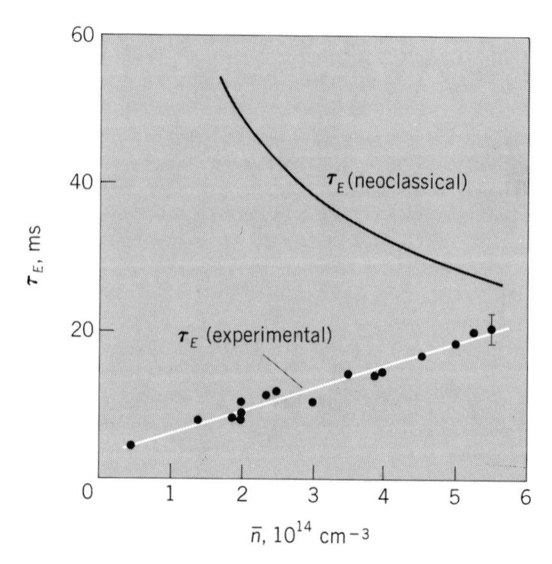

Fig. 3. Global energy confinement time τ_E as a function of average plasma density $\bar{n}$. Experiments were carried out in deuterium plasmas with $B_T = 60$ kG and $I_P = 150$ kA. Also shown are the corresponding neoclassical values for τ_E.

that at the highest densities the central region of the plasma column exhibits behavior characteristic of a regime dominated by neoclassical ion-thermal conductivity. The anomalous electron-thermal conductivity present at lower density is suppressed. Thus energy confinement at the plasma center does not continue to increase linearly with increasing density at fixed B_T and I_P, but approaches a value corresponding to neoclassical estimates for these plasma conditions. Radial energy fluxes for $r \leq 0.5a_L$ (where r is the distance from the center of the plasma column and a_L is the plasma radius) may be deduced from the experimental values of the electron temperature $T_e(r)$, the ion temperature $T_i(r)$, and the local plasma density $n(r)$, neglecting energy losses due to radiation and charge exchange. For $\bar{n} < 3 \times 10^{14}$ cm^{-3}, the ion-thermal conductivity deduced is in agreement with the neoclassical value, but the electron-thermal conductivity χ_e (experimental) exhibits a large anomaly relative to χ_e (neoclassical). At $\bar{n} = 1 \times 10^{14}$ cm^{-3}, χ_e (experimental) $\simeq 200\ \chi_e$ (neoclassical). In this regime, χ_e (experimental) $\propto 1/\bar{n}$. At $\bar{n} > 4 \times 10^{14}$ cm^{-3}, the dominant role in energy confinement can be attributed to ion-thermal conduction. At $\bar{n} = 5 \times 10^{14}$ cm^{-3}, χ_e (experimental) $\leq 10\ \chi_e$ (neoclassical). Thus as density increases in Alcator, a transition from the anomalous electron-thermal conduction regime to the neoclassical ion-thermal conduction regime is observed in the central core of the plasma.

Impurities. Impurities pose a serious obstacle to the production of reactor-grade plasmas, adversely affecting almost every aspect of tokamak operation. At present, vital questions about impurity origin, transport, and control remain unanswered. Theoretical models have predicted that impurity ions continuously accumulate in the plasma, rather than recycle along with the hydrogenic particles. However, there is no experimental confirmation of this hypothesis. The depressing prospect of impurity accumulation in the plasma is further alleviated by the recent observation of poloidally asymmetric impurity-ion emission in the high-density plasmas in Alcator. The sense of the asymmetry reverses when the direction of B_T is reversed, and all observed dependences of the strength of the asymmetry are consistent with the explanation that drift of the highly collisional impurity ions due to gradients and curvature of the magnetic field cause the asymmetry. The asymmetry implies that the drift could be a major process for transport of impurities out of a high-density tokamak plasma.

Reactor prospects. Since τ_E increases with $\bar{n}$, the Lawson parameter $n_0\tau_E$ increases in proportion to $\bar{n}^2$, as shown in Fig. 4. The Lawson parameter is a figure of merit for a confinement device and, together with T_i, determines when self-sustaining nuclear fusion reactions will occur. The strong dependence of $n_0\tau_E$ on $\bar{n}$ facilitates the conception of relatively compact and technologically simple high-power-density fusion reactors. Detailed tokamak reactor design calculations reveal that the high-field high-density tokamak approach offers great promise for rapid progress in the development of fusion power production.

For background information *see* FUSION, NUCLEAR; LAWSON CRITERION in the McGraw-Hill

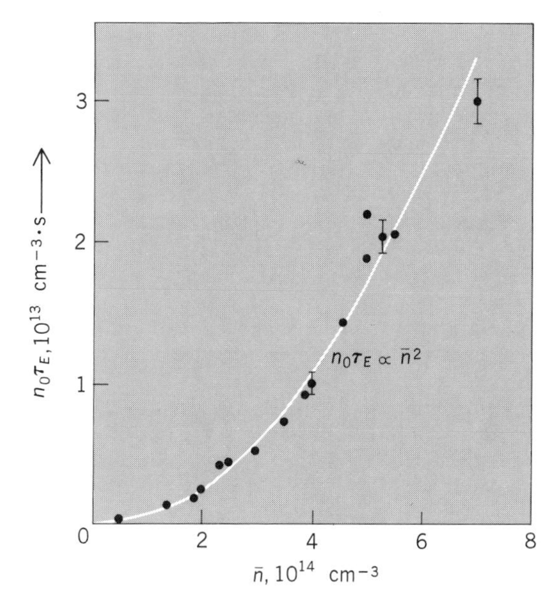

Fig. 4. Product of peak deuteron density n_0 and global energy confinement time τ_E as a function of average plasma density $\bar{n}$. Experiments were performed in deuterium plasmas.

Encyclopedia of Science and Technology.

[AWINASH GONDHALEKAR]

Bibliography: L. A. Artsimovich, *Nucl. Fus.*, 12: 215–252, 1972; D. R. Cohn, R. R. Parker, and D. L. Jassby, *Nucl. Fus.*, 16:31–35, 1045–1046, 1976; D. L. Cook and M. A. Sweeney, *Proceedings of the 3d ANS Topical Meeting*, Santa Fe, 1978; M. Gaudreau et al., *Phys. Rev. Lett.*, 39:1266–1270, 1977; A. Gondhalekar et al., *Proceedings of the 7th International Conference on Plasma Physics and Controlled Nuclear Fusion Research*, Innsbruck, 1978; F. L. Hinton and R. D. Hazeltine, *Rev. Mod. Phys.*, 48:239–308, 1976; J. A. Nation and R. N. Sudan (eds.), *Proceedings of the 2d International Topical Conference on High Power Electron and Ion Beam Research and Technology*, Ithaca, NY, 1977; J. L. Terry et al., *Phys. Rev. Lett.*, 39:1615–1618, 1977; G. Yonas (ed.), *Proceedings of the International Topical Conference on Electron Beam Research and Technology*, Albuquerque, 1975; G. Yonas (ed.), *Sandia Labs Report*, SAND78-0080, 1978.

Galaxy, external

Superclusters of galaxies (clusters of galaxy clusters) are the largest aggregates of matter currently known beyond doubt to exist. As such, they are as fundamental to understanding the nature of the universe as are galaxies, stars, and quarks. Unlike stars and galaxies—but like quarks—they are difficult to observe and so have been objects of controversy until recently. But the past year has seen several important observational studies of some of the nearer superclusters come to fruition, and an equal number of studies on other superclusters begun. *See* QUARKS.

Local Supercluster. In 1784 William Herschel noted that most of the "nebulae" he had discovered seemed to be concentrated in a wide belt across the northern sky, oriented roughly perpendicularly to the Milky Way. The Virgo Cluster, discovered previously by Charles Messier and

Pierre Méchain, formed a dense concentration near the pole of the Milky Way in Herschel's "stratum" of nebulae. This great band of nebulae was further explored and commented on during the next 150 years, but not until the work of Knut Lundmark, Eric Holmberg, and Anders Reiz at Uppsala and Lund observatories in the 1920s and 1930s was a reasonable explanation brought forth for this decidedly nonrandom distribution of the bright galaxies. It was the opinion of the Scandinavian astronomers that observers on Earth are simply seeing from the inside a great flattened system of groups and clusters of galaxies, with a few individual objects scattered between. Gérard de Vaucouleurs called this the Local Supercluster in 1958, though 5 years earlier he had revived the ideas of superclustering and had called attention to the work of Holmberg and Reiz.

The illustration shows the apparent positions of all galaxies in the northern galactic hemisphere with known radial velocities less than 2000 km/s. These objects are the Galaxy's nearest neighbors in extragalactic space and most are members of the Local Supercluster, which shows well as the band of objects running diagonally across the illustration. The Virgo Cluster is the concentration of points in this band to the upper left of center. The point toward which the Galaxy is moving is shown by a cross. If the Milky Way were shown here, it would be an irregular belt around the periphery of the illustration.

Rotation. The apparent flattening of the Local Supercluster naturally led de Vaucouleurs to test the idea that it might be rotating as well as expanding. The resulting model fit, and continues to fit, the totality of the data on the velocities and positions of the nearer galaxies very well. It also explains the apparent anisotropy of velocities of bright galaxies across the sky: at the tenth magnitude, galactian velocities in the northern hemisphere are smaller by nearly a factor of 2 than velocities in the south. Furthermore, the very marked inequality of the numbers of galaxies brighter than fourteenth magnitude between the two hemispheres (the southern sky has about half as many objects) arises because the Galaxy is not located in the center of the Local Supercluster.

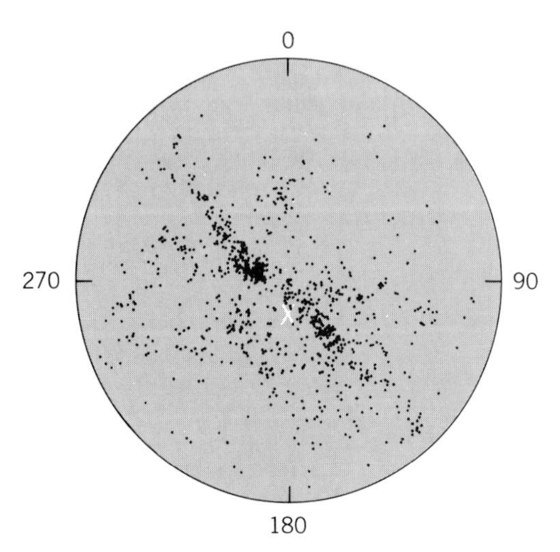

Apparent positions of all galaxies in the northern galactic hemisphere with known radial velocities less than 2000 km/s. Numbers are galactic longitudes. *(From R. B. Tully and J. R. Fisher, A picture of the Supercluster, Bull. Amer. Astron. Soc., 8:555, 1976)*

One consequence of this model should be an observable anisotropy in the 3 K microwave background radiation due to the Earth's rotational motion around the center of the Supercluster. Such anisotropies have been reported by several radio astronomers for the past 10 years. The observations are difficult, however, and the data are "noisy." Nevertheless, there is rough agreement with de Vaucouleurs's model, in both velocity and direction (350 km $\cdot$ s^{-1} toward galactic longitude 185°, galactic latitude +75°, a point in the sky near the supergalactic equator where it passes through southern Ursa Major).

Nature and origin. The foregoing has often been taken as evidence that the Local Supercluster is a physical gravitationally bound system. This is almost certainly not the case. Since the Supercluster is about 1.5 to 2 × 10^8 light-years (1.4 to 1.9 × 10^{24} m) across, its period of rotation should be comparable to or greater than the age of the universe. Moreover, all data show beyond doubt that it is expanding. So, the Local Supercluster is probably better thought of as simply a relic of a vast inhomogeneity in the distribution of matter in the early history of the universe. Similarly, its "rotation" may be a reflection of internal motions that may have played an important role in its creation. Several theoretical studies have been directed toward this viewpoint. Among the more successful at explaining the observations are the gravitational instability concept explored by P. J. E. Peebles and his coworkers at Princeton and by A. G. Doroshkevich and others in the Soviet Union; and the idea of primeval turbulence advanced by L. M. Ozernoy in the Soviet Union.

Other superclusters. Besides the Local Supercluster, there are perhaps 50 other superclusters that have been noted at one time or another. These have been found primarily through examination of the various surveys of faint galaxies and of clusters of galaxies. Most useful among these surveys are those conducted by Harlow Shapley at Harvard and by C. D. Shane at Lick Observatory—both concentrating on faint galaxies—and the two catalogs of rich clusters of galaxies by George Abell and by Fritz Zwicky, both working with the Palomar Sky Survey. The astronomer's view of the cosmos is decidedly confused, however, since galaxies at all distances are seen projected onto the plane of the sky. Distant, intrinsically bright galaxies look little different, at first glance, from nearby faint ones. This results in an apparent smearing out of the actual spatial distribution of galaxies, which has led in turn to general acceptance of the concept of the isotropy and homogeneity of the universe. In recent years, however, the painstaking collection of more and better data for some thousands of the brightest and nearest galaxies (for example, the work of Stephen Gregory and Laird Thompson), combined with new and more sophisticated methods of statistical data analysis applied to the surveys of the more distant objects, has given a drastically different view of the cosmos. Instead of a rather smooth distribution of galaxies where a variation of a factor of 2 or more in the counts could be passed over as a "random fluctuation," there now appears to exist a vast array of clusters and superclusters separated by equally vast stretches of amazingly empty space. Statistical studies of the subtle fluctuations in the surface density of galaxies and clusters also point to clustering on different orders—from single galaxies (zero-order clusters) to perhaps even clusters of superclusters (third-order clusters)—as a basic property of the distribution of matter.

Many problems remain. For example, the statistical studies suggest that there is no preferred size for clusters or superclusters. In other words, a single large cluster may have the same diameter as a small supercluster. Yet most observational studies of superclusters yield typical diameters on the order of 2×10^8 light-years (1.89×10^{24} m), just as clusters seem to have typical diameters of 10^7 light-years (9.46×10^{22} m). The Local Supercluster and the Virgo Cluster—often suggested as the "core" of the Local Supercluster—are good examples. The resolution of this apparent problem and others will most likely come with better, and more, data.

X-ray observations. There has also been recent discussion of the association of extragalactic x-ray sources with clusters and superclusters. The associations with clusters cannot be doubted: it seems likely that the x-rays originate by Compton scattering in clouds of hot gas dispersed through the clusters. There have been attempts to correlate the x-ray luminosity of clusters with their velocity dispersions; such a correlation would do much to remove the persistent virial mass discrepancy in clusters and would also provide important information on the formation and evolution of clusters.

In superclusters, though, the x-ray studies are hampered by data at the present limit of detectability. Several supercluster sources found by the *Uhuru* x-ray satellite observatory have not been "seen" by the similar *Ariel V* satellite. This implies that these sources do not exist, or are highly variable, in which case they must be very small compared with the sizes of superclusters. So, in spite of several reported associations of x-ray sources with superclusters, it is still not known if diffuse x-ray sources, independent of any included clusters, belong to superclusters. If such sources do exist, they would imply the presence of vast amounts of presently unseen matter in the universe. It is not known whether there is enough gas there to "close" the universe. *See* COSMOLOGY.

Extension of clustering hierarchy. Finally, the extension of the hierarchy of clustering beyond superclustering has been examined a few times. There is marginal evidence in the distribution of the most distant galaxies and clusters that clusters of superclusters (third-order clusters) may indeed exist. However, studies of the most distant radio sources do not confirm these suggestions. Again, the problem is most likely due to incomplete and unreliable data on the misty borderline between discovery and misinterpretation. Still, there are hints of grander structures in the universe than are presently seen, and the years ahead promise to be exciting ones for observational cosmology.

For background information *see* COSMOLOGY; GALAXY, EXTERNAL; X-RAY ASTRONOMY in the McGraw-Hill Encyclopedia of Science and Technology.

[HAROLD G. CORWIN, JR.]

Bibliography: G. O. Abell, in M. S. Longair (ed.), *Confrontation of Cosmological Theories with Observational Data* (IAU Symposium no. 63, pp. 79–92, 1973), 1974; S. A. Gregory and L. A. Thompson, *Astrophys. J.*, 222:784–799, 1978; M. S. Longair and J. Einasto (eds.), *The Large-Scale Structure of the Universe* (IAU Symposium no. 79, 1977), 1978; G. de Vaucouleurs, *Publ. Astron. Soc. Pac.*, 83:113–143, 1971.

Grass crops

Recent investigations have shown that factors associated with grass tetany such as mineral and organic acid concentrations are under genetic control in forage grasses. This provides an opportunity for forage grass breeders to manipulate these concentrations.

Mineral composition of forage grasses. Breeding forage grasses to prevent or reduce the incidence of grass tetany is a complex process since the disorder is not completely understood. Grass tetany is often related to a low level of blood serum magnesium in the animal, and it is therefore tempting to consider it as a simple mineral deficiency. A. Kemp established 0.20% on a dry-matter basis as the "safe" level for magnesium in the forage grass herbage. Herbage below this value would be considered tetany-prone. However, not all ruminants with low serum magnesium succumb to grass tetany. It is necessary to determine why the low levels of serum magnesium leading to tetany developed. Magnesium may not be available to the animal because it is either tied up in the chlorophyll molecule or coated by silica. Certain organic acids produced by the forage grass plant may act as chelators which withhold magnesium from ruminant tissues. In addition, the critical magnesium level may vary depending upon the concentration of other elements, especially sodium, potassium, and calcium, in the animal's diet. Kemp and M. L. t'Hart along with E. J. Butler concluded that with a K/(Ca + Mg) ratio (where K, Ca, and Mg are calculated as milliequivalent per kilogram of dry matter) of less than 2.2, the incidence of grass tetany was reduced. These and other environmental and physiological factors complicate the development of guidelines for forage grass breeding programs.

Mineral composition of forage grasses can be modified through fertilizer application. Direct supplementation of the animal's rations is often difficult and expensive. For example, magnesium supplements are relatively unpalatable; they can be mixed with salt, but they still may not be eaten in sufficient amounts by the animals to prevent grass tetany. In order to improve palatability, many mineral supplements fed to cattle have used grain as protein supplements or carriers, which in turn increases the cost. Diet supplements have been available for some time and have not eliminated grass tetany. Genetic variability for traits associated with the incidence of grass tetany is present in forage plants. It appears worthwhile to investigate the possibility of modifying mineral concentration in the forage grass through plant breeding.

Relation of diet to tetany. Grass tetany has been observed most frequently in ruminants grazing lush grasses in the spring or autumn. Initiation of lush growth in the spring occurs after periods of stress related to cold weather, and autumn growth occurs after the heat and drought stress of the summer months. D. L. Grunes and associates re-

Table 1. Mineral concentrations in herb, legume, and grass species*

Species	Magnesium†	Calcium†	Potassium†
Herbs			
Yarrow *(Achillea millefolium)*	0.99	1.3	3.7
Burnet *(Poterium sanquisorba)*	1.10	1.3	1.8
Plantain *(Plantago lanceolata)*	0.61	1.6	2.7
Chicory *(Cichorium intybus)*	0.64	1.4	4.6
Legumes			
Black medic *(Medicago lupulina)*	0.76	1.6	2.3
Alsike *(Trifolium hybridum)*	0.62	2.0	2.1
Alfalfa *(M. sativa)*	0.55	2.1	2.6
Sanfoin *(Onobrychis viciifolia)*	0.79	1.0	2.5
Grasses			
Perennial ryegrass *(Lolium perenne)*	0.21	0.46	2.0
Orchard grass *(Dactylis glomerata)*	0.22	0.42	2.3
Timothy *(Phleum pratense)*	0.25	0.41	2.2
Meadow fescue *(Festuca pratensis)*	0.26	0.44	2.1
Tall fescue *(F. arundinacea)*	0.32	0.30	2.0
Red fescue *(F. ruba)*	0.16	0.40	1.8

*From G. A. Fleming, Mineral composition of herbage, in G. W. Butler and R. W. Bailey (eds.), *Chemistry and Biochemistry of Herbage*, vol. 1, pp. 529–566, Academic Press, 1973.

†In percent of dry matter.

ported that grass tetany occurs on perennial grasses such as orchard grass (*Dactylis glomerata*), perennial ryegrass (*Lolium perenne*), timothy (*Phleum pratense*), tall fescue (*Festuca arundinacea*), crested wheatgrass (*Agropyron desertorum* and *Agropyron cistateum*), and smooth bromegrass (*Bromus inermis*). Grass tetany has also been reported on annual grasses such as soft chess (*Bromus mollis*) and mouse barley (*Hordeum leporinum*). Cereal grasses such as wheat (*Triticum aestivam*), rye (*Secale cereale*), oat (*Avena sativa*), and barley (*Hordeum vulgare*) have also been implicated in grass tetany when utilized by grazing animals as a green pasture.

Grass tetany is less frequent when animals consume legumes as a green pasture in pure stands or in combination wih grasses. It is also less frequent in animals which consume legume hay. When legumes and herbs are fed alone or in combination with forage grasses, they reduce the probability of grass tetany developing in the consuming ruminants since they contain higher levels of magnesium and calcium than forage grasses (Table 1). Although weedy herbs increase magnesium and calcium concentration in pastures, their presence is not welcomed since they reduce total forage production.

The occurrence of grass tetany in the tropics is minimal, with scattered incidences occurring only at higher elevations. Concentrations of magnesium in tropical grass species are generally low, with values commonly below 0.20%. The ratio of K/(Ca + Mg) in tropical grasses is often well below the 2.2 "safe" level established for temperate species. The mineral concentration in tropical grasses leads one to conclude that other factors such as differences in susceptibility of animals and temperature may be more important than plant composition.

Genetic control. Most of the discussion on the genetic control of minerals affecting grass tetany centers on continuous genetic variation. Characters that show continuous genetic variation are probably controlled by many genes. If the forage grass breeder is to modify mineral concentrations in the plant, it is important to know the magnitude of both the available genetic variation and its heritability.

Heritability is defined in a general sense as that proportion of the total variation in a character that is due to genetic causes. A heritability estimate refers only to the plant population or range of genetic material from which it came in a given environment. It is necessary to consider the environment in making heritability estimates so that the forage grass breeder can make an estimate of how stable a particular trait is under varying growth conditions.

Heritability estimates are expressed in percent and range from 0 to 100%. Heritability estimates of mineral elements important in grass tetany are given in Table 2 for ryegrass, tall fescue, and reed canary grass. These estimates appear to be adequate to breed plants for improved magnesium, calcium, and potassium concentrations in the three grass species. The heritability estimate of 81% for the ratio of K/(Ca + Mg) is adequate to manipulate this ratio in tall fescue. Reed canary grass values were obtained over 3 years and two locations, the tall fescue data were from 2 years and one location, and the number of locations and years in the ryegrass experiments is unknown.

Table 2. Narrow-sense heritability estimates for nitrogen and minerals in ryegrass, tall fescue, and reed canary grass

Element	Heritability, %		
	Ryegrass*	Tall fescue†	Reed canary grass‡
N	63	–	25
Mg	86	76	66
Ca	78	70	56
K	80	62	49
K/(Ca + Mg)	–	81	22

*Data from J. P. Cooper, Genetic variation in herbage constituents, in G. W. Butler and R. W. Bailey (eds.), *Chemistry and Biochemistry of Herbage*, vol. 2, pp. 379–417, Academic Press, 1973.

†Data from D. A. Sleper et al., Breeding for Mg, Ca, K, and P content in tall fescue, *Crop Sci.*, 17:433–438, 1977.

‡Data from A. W. Hovin, T. L. Tew, and R. E. Stucker, Genetic variability for mineral elements in reed canarygrass, *Crop Sci.*, 18:423–427, 1978.

Table 3. Average daily gains and intakes of guinea pigs consuming two selections of tall fescue*

		Average daily gain, g		Average intake/ pig/day, g	
Selection	Magnesium, % of dry matter	Without added magnesium	With added magnesium	Without added magnesium	With added magnesium
1†	0.13	5.7	7.1	22.9	19.8
2‡	0.15	7.5	7.7	31.3	22.3

*Unpublished data from G. B. Garner, University of Missouri.
†K/(Ca + Mg) = 3.3.
‡K/(Ca + Mg) = 2.0.

Organic acids such as *trans*-aconitate have been related to grass tetany. Little research information exists on genetic differences within forage grass species for organic acids. R. L. Boland and associates reported that in tall fescue total organic acid concentration was related to origin of the genotype. Genotypes selected from strains and varieties developed in the United States were higher in total organic acid content than genotypes from the Mediterranean region. There appears to be adequate genetic variability in organic acid content of tall fescue leaves to allow the acid content to be altered through breeding. Organic acid levels and their genetic control need to be researched in other cool-season forage grasses.

Effects of selection. Mineral elements are not inherited independently of each other. For example, it has been shown that breeding for improved levels of magnesium and calcium in certain forage grass species increases the concentration of phosphorus and potassium. The effects of selection for improved mineral concentrations on disease resistance, herbage yield, range of adaptation, and traits associated with forage quality are unknown. If the forage grass breeder modifies one or more components in the plant, care must be taken so that other desirable characteristics are not adversely affected.

The forage grass breeder needs to use a plant breeding procedure which will accurately characterize a genotype or a group of plant genotypes for its tetany potential. Breeding for improved mineral levels is both time-consuming and expensive, which makes individual plant selection impractical. Success in plant breeding often depends on having a high number of individuals evaluated. Therefore, the breeder needs to use a breeding method involving family selection. This would allow the bulking of all individuals in a particular family, thereby reducing considerably the total number of samples for chemical analysis.

The measure of success of any forage selection modified by plant breeding is whether it is significant in terms of animal performance. It is not entirely known whether small animal assays may be used to evaluate tetany-prone selections and predict the response in cattle or sheep. Small animals such as guinea pigs are more convenient and less expensive to use in research than, for example, cattle. G. B. Garner at the University of Missouri discovered that feeding two tall fescue genotypes which varied slightly in their magnesium content produced differing guinea pig responses (Table 3). The phosphorus, calcium, nitrogen, potassium, and fiber content was balanced in both genotypes so that only the magnesium content differed. A second group was given added magnesium to 0.3% of the diet so that it was not limiting. Better average daily gains and intakes were obtained with the low-cation genotype (selection 2). When ample magnesium was added, the average daily gain was similar for both plant genotypes. There appears to be a marked difference in either availability of the magnesium in these two genotypes or in palatability differences, which resulted in differences in magnesium intake. If guinea pigs can be used as a tool in the selection of non-tetany-prone genotypes, development of a nontetany forage through plant breeding is expected to be more rapid and less expensive.

For background information *see* GRASS CROPS in the McGraw-Hill Encyclopedia of Science and Technology.

[DAVID A. SLEPER]

Bibliography: R. L. Boland et al., *Crop Sci.*, 16: 677–679, 1976; J. P. Cooper, in G. W. Butler and R. W. Bailey (eds.), *Chemistry and Biochemistry of Herbage*, vol. 2, pp. 379–417, 1973; G. A. Fleming, in G. W. Butler and R. W. Bailey (eds.), *Chemistry and Biochemistry of Herbage*, vol. 1, pp. 529–566, 1973; D. L. Grunes, P. R. Stout, and J. R. Brownell, *Adv. Agron.*, 22:331–374, 1970; A. W. Hovin, T. L. Tew, and R. E. Stucker, *Crop Sci.*, 18: 423–427, 1978; D. A. Sleper et al., *Crop Sci.*, 17: 433–438, 1977.

Helium, liquid

There has been considerable investigation of the properties of thin films of liquid helium in the last few years. Recent advances in this area include (1) the development of methods for measuring persistent currents in thin films and the stability and decay of such currents, and (2) the development of a theory which predicts the existence of a sharp phase transition in thin films and successfully accounts for the transition temperature.

Persistent currents in thin films. The most extraordinary feature of superfluids is their ability to flow in currents which circulate around a perimeter with no apparent loss of energy. These are called persistent currents to emphasize their stability. However, there are conditions where these currents are observed to decay. In fact, the flow of helium in all persistent currents is believed to be metastable, with extremely long lifetimes in some cases. As the temperature is raised close to the superfluid transition, persistent currents decay increasingly rapidly to the point where any flow appears to be totally unstable. There is also a maximum persistent current velocity that can be created in any given situation, beyond which the flow becomes rapidly unstable. Curiously, this maximum velocity increases as the flow channel size is reduced. Maximum velocities of 200 cm/s are seen in thin films of only 8–10 atomic layers thickness, while less than 1 cm/s is observed in tubes of 1 mm diameter. For this reason thin-film geometries have become important in the investigation of the properties of superfluid helium.

Recently developments related to flow in thin films have been utilized to obtain new information about the stability of persistent currents. The first development is the introduction of a technique for generating persistent currents without rotation of the substrate. This greatly simplifies the experimental apparatus and permits the study of persistent currents under a wide variety of geometrical circumstances. In the past, rotation of the substrate failed to generate persistent currents on flat surfaces for reasons that are not yet understood. This problem is circumvented by using the trapping scheme described below. The second development is the utilization of acoustic techniques to measure directly the flow velocity of the persistent current. This is accomplished by performing Doppler shift measurements of the velocity of surface waves, known as third sound, propagated within the flowing film. This technique provides instantaneous measurements of the flow velocity and renders the decay directly observable with high precision. Some of the results obtained with the new trapping scheme and the third-sound probe are discussed below. *See* SOUND.

Trapping of circulation. A persistent current consists of fluid flowing around a closed path. As a measure of this current, a number equal to the product of the flow velocity and the path length is designated as the circulation. Since the path length is usually fixed, the flow velocity directly reflects the persistent current and the circulation. Superfluid helium is believed to flow only in states of certain prescribed circulation. The only flow states allowed are those where the circulation is an integral multiple of a certain fundamental unit (approximately equal to 0.001 cm^2/s); the reason for this is discussed below. The stability of the persistent current state derives from the lack of a mechanism for the entire flow to change its circulation by one of these units. This is in sharp contrast to ordinary fluids where the circulation can change continuously. The key to generating circulation in the helium film is to locally raise the temperature of the film above its transition temperature so that it becomes a normal fluid. If this process is carried out while a flow is under way, circulation is observed to be introduced into the film, and a persistent current is generated.

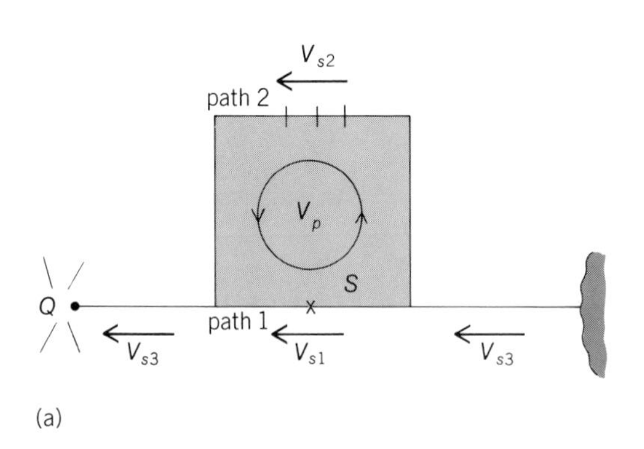

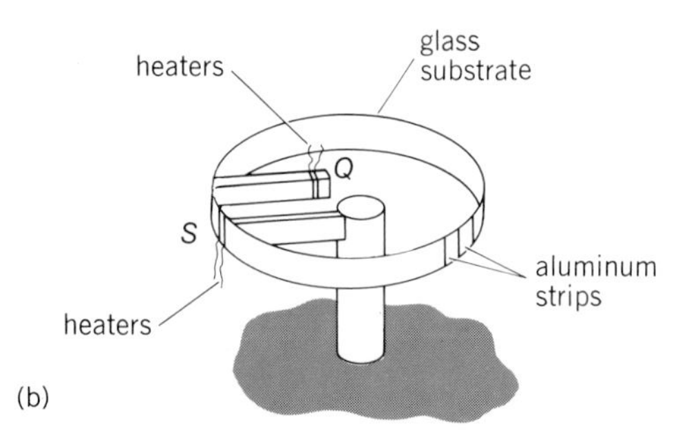

Fig. 1. Apparatus for generating persistent currents. *(a)* Schematic of the helium film flow path. *(b)* Glass ring substrate around which the persistent current flows. *(From K. L. Telschow and R. B. Hallock, Stability of persistent currents in unsaturated superfluid ^{4}He films, Phys. Rev. Lett., 37:1484–1487, 1976)*

Figure 1*a* illustrates the circulation trapping geometry. Helium film is caused to flow from a reservoir along two parallel flow paths of unequal length. On the short flow path 1, a heater *S* is placed which can completely stop all flow at that spot. When *S* is turned on, most of the flow is diverted around the long path 2, thus forcing circulation around the ring. After *S* is turned off, a persistent current V_p is observed to flow around the ring initiated by flow back through *S* as the short path is reopened. The amount of circulation remaining around the closed path (1 and 2) is directly related to the total amount of flow coming initially from the reservoir, V_{s3}. Therefore, any desired persistent current can be initiated by controlling V_{s3}. Also shown in Fig. 1*a* is a heater *Q* which drives the film from the reservoir by evaporating the fluid in contact with it. This is the simplest method for driving thin films of less than 20 atomic layers. The first experiments demonstrating the trapping technique utilized long thin glass tubes as the flow paths and puddles of liquid at both ends as reservoirs for the film. Gravity provided the driving force, with film flow initiated by upsetting the equilibrium in the two reservoirs.

Measurement with third sound. In order to utilize third sound to detect the persistent current velocity, the film flow is placed on the outside of a glass substrate. The flow geometry is still a double path consisting of different sections of a circular ring with a perimeter of 20 cm, as shown in Fig. 1*b*. *S* and *Q* are merely resistance elements of wire wound around the paths. The third-sound probe is placed on the long path to detect the flow velocity there, V_{s2}. If the helium film is "plucked" at some point by the application of a heat pulse, a dimple forms in the film, left over from the evaporating atoms. This dimple travels away from the source of heat as superfluid rushes in to fill the void. The result is the surface wave—third sound—which propagates with a well-defined wave velocity completely analogous to a water wave in a shallow pool. The wave is detected through the small temperature variations which accompany it by means of sensitive superconducting bolometers. These bolometers are in the form of thin (50-nm) strips of aluminum deposited on the long flow path directly opposite the two posts, and separated by a known distance. As the third-sound wave passes over these strips, the temperature of the strips changes by a small amount, easily detected through electrical measurements. The wave velocity is determined from the time of flight of these pulses between the strips. By directing the pulses upstream and downstream of the flowing film, V_{s2} is obtained from the net Doppler shift of these pulses.

Figure 2 shows the results of the third-sound measurements used in the trapping of circulation around the ring. V_{s3} is determined from the heat applied to Q, and the flow over the short path V_{s1} can be deduced from the others, assuming conservation of mass throughout the flow. Initially V_{s2} changes very little with V_{s3} when S is off, as shown in Fig. 2 by the path from A to A′. This is due to the fact that since the circulation around the ring was initially zero, it remains zero as long as the film is superfluid. Throughout this region most of the flow is through the short path. The trapping of circulation process is illustrated through the sequence A→B→C→D. If at point B heater S is turned on and off, point C results. The flow now consists of the zero circulation part and a persistent current circulating around the ring, as shown by point D. Figure 2 shows that various values of circulation can be created around the ring in this manner; the total circulation around the ring K is given in units of 10^5 times the fundamental unit.

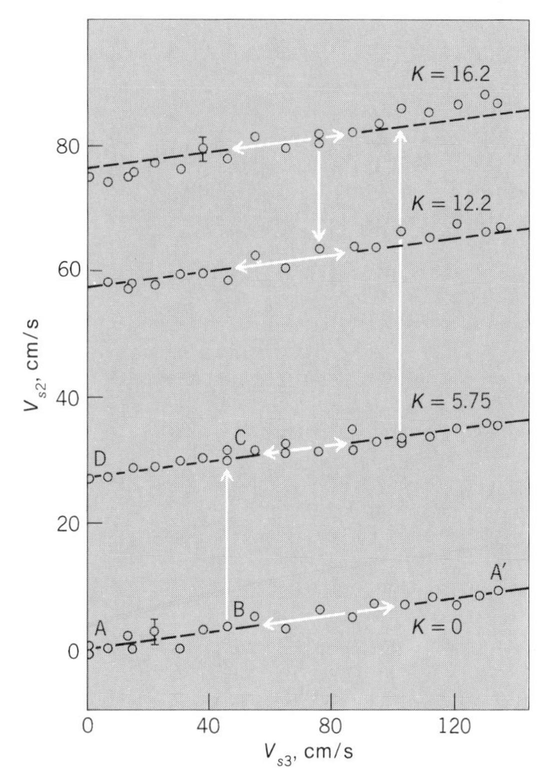

Fig. 2. Observed flow over the long path V_{s2} versus the flow from the reservoir V_{s3}. *(From K. L. Telschow and R. B. Hallock, Stability of persistent currents in unsaturated superfluid ⁴He films, Phys. Rev. Lett., 37:1484–1487, 1976)*

Decay of persistent currents. The Doppler shift of the third-sound pulses versus time yields the decay of the persistent flow from one circulation state to another. Generally the persistent currents are very stable, decaying almost immeasurably over laboratory time scales. However, as the temperature is raised or the film is thinned, a region is reached where rapid decay is seen. Figure 3 shows some typical decays observed for different film thicknesses. The rate of decay increases rapidly as the film is thinned, eventually reaching a point of complete instability beyond which no superflow is possible at any velocity. This is the transition point between superfluid and normal fluid. The form of the decay is actually logarithmic with time, and yields the levelling off of the curves in Fig. 3 as time increases. It appears that the net decrease in the persistent current velocity divided by the initial velocity changes by a fixed percentage for every logarithmic decade of time. These percentages are indicated next to the curves in Fig. 3. The curve shown in Fig. 3 for a film thickness of 7 atomic layers has a fractional decay per decade of 25%. Therefore, it would take 4 logarithmic decades of time for the velocity to decrease to zero. In contrast, the curve for 12 layers shows only a 2% decay per decade, requiring 50 logarithmic decades of time for completion. Surely, after the first decade or two of observation, this circulating current deserves the name "persistent."

Persistent currents are believed to be made up of large numbers of very small vortex lines, each of a single unit of circulation. The total circulation is simply the sum of these smaller units. Typical currents involve hundreds of thousands of these tiny vortexes all bound to the substrate. The trapping of circulation involves the generation of these vortex lines from the turbulent region near S and their distribution throughout the film. The persistent current decays as these vortex lines break loose from the substrate and destroy themselves through friction with the substrate. The entire array becomes more stable as fewer and fewer of the initial number of vortexes remain. Furthermore, starting from different initial velocities (that is, different numbers of vortexes), persistent currents are pro-

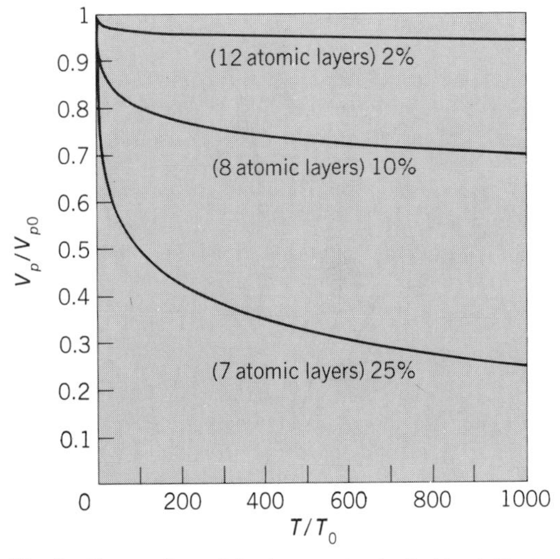

Fig. 3. Decay of persistent current velocity V_p relative to its initial value V_{p0} versus time T relative to the initial time T_0. Decays for the three film thicknesses are shown. T_0 was typically 1 min.

duced which decay at different rates even when they all pass through the same velocity state. Thus states of the system with the same temperature, film thickness, and persistent current velocity (number of vortexes) are not identical. To this list of parameters one must add the previous history of how that particular state was reached, since the flowing film somehow "remembers" its mode of origin.

Currently, experimental resolution is too low for direct observation of the discrete circulation states, as they are represented by velocity steps separated by only 5×10^{-5} cm/s. More precise acoustic measurement techniques exist with resolutions of better than 1 part per million. The feasibility of using third sound suggests that improved acoustic techniques may render the motion of individual vortexes directly observable. This would provide a unique opportunity to study these microscopic effects on a macroscopic scale.

[KENNETH L. TELSCHOW]

Phase transition in thin films. It has been understood for a long time that thin superfluid helium films differ in an essential way from bulk superfluid helium because they are two- rather than three-dimensional. There has been considerable theoretical controversy about the nature of these differences, and it is only recently that experimental measurements have been able to answer some of the questions raised. It is now clear that there is a sharp phase transition associated with the disappearance of superfluidity as the temperature is raised even in these two-dimensional systems and that the phase transition occurs when quantized vortices are formed spontaneously in the thin film.

Bose condensation. The properties of bulk superfluid ^{4}He have been elucidated in terms of Bose condensation, which means that a finite fraction of the helium atoms condenses into a single quantum state $\psi_0(\mathbf{r})$. In the equilibrium state, $\psi_0(\mathbf{r})$ is a constant except very close to the boundaries. In a state of superfluid flow, $\psi_0(\mathbf{r})$ is proportional to $e^{iS(\mathbf{r})}$, and the superfluid velocity is given by Eq. (1).

$$\mathbf{v} = \hbar \text{ grad } S/m \quad (1)$$

where m is the mass of a helium atom and $\hbar$ is Planck's constant divided by 2π. This formula, combined with the condition that the condensate wave function must be single-valued, leads to the quantization of circulation. If helium is flowing around the inside of a torus, S must change by an integral multiple of 2π around the torus, and so the integral of $\mathbf{v}$ around a closed loop is a multiple of $2\pi\hbar/m$. A quantum vortex is a line singularity within the superfluid around which the phase S changes by $\pm 2\pi$, so that there is a single quantum of circulation around the vortex. A straight vortex line produces a velocity in the tangential direction equal to $\pm \hbar/mr$, where r is the distance from the vortex line. According to the two-fluid model, only the superfluid component with superfluid density ρ_s need participate in this motion, while the normal component with density ρ_n behaves like an ordinary fluid; in a thin film the normal fluid is firmly attached to the substrate and follows its motion. Since the circulation around a torus can change only by multiples of the quantum of circulation, the only way in which circulating superfluid helium can come to rest is by an appropriate number of vortex lines crossing the torus.

In the two-dimensional case the question has been raised as to whether there can be any Bose condensation. It is known that even if there is a condensate its wave function must have thermal fluctuations which lead to incoherence in phase, so that its average value is zero. If such a fluctuating condensate wave function exists, it should be rather different from the nonfluctuating condensate of the bulk superfluid.

Theory of the phase transition. The existence of a condensate wave function is shown by the stability of superfluid flow and by the quantization of circulation. Since superfluid flow is changed by means of the movement of vortices, the question of whether or not there are vortices present in the fluid when it is in thermal equilibrium is important. If there are vortices present in the equilibrium state the superfluid flow can decay spontaneously, whereas if they are absent the flow should be stable.

In very thin films, the phase of the condensate wave function should depend primarily on the position in the surface and very little on depth, so that vortex lines should have one end on the free surface of the film and the other on the closest point of the substrate. Thus the depth dependence can be ignored and the problem treated as a two-dimensional one. The vortices are just point singularities around which the phase changes by $\pm 2\pi$, and the statistical mechanics of such vortices is quite simple. Each vortex has an energy E given by Eq. (2), where ρ_s is the superfluid density per unit

$$E = (\pi\hbar^2\rho_s/2m^2) \ln (A/a^2) \quad (2)$$

area, A is the area of the film, and a is a length of the order of the interatomic spacing. The entropy S of a single vortex is simply the Boltzmann constant k_B times the logarithm of the number of different places the vortex can occur, and so Eq. (3) holds, where T is the temperature.

$$TS = k_BT \ln (A/a^2) \quad (3)$$

At low temperature the free energy $E - TS$ is positive and tends to infinity in the thermodynamic limit in which A is infinite, so that no vortices occur spontaneously, while at high temperatures it is negative, so that vortices occur spontaneously and superfluid flow is destroyed. The condition for the existence of stable superfluid flow is therefore given by Eq. (4).

$$\rho_s/T > 2m^2k_B/\pi\hbar^2 \quad (4)$$

The phase transition to the nonsuperfluid state occurs when ρ_s/T is equal to its limiting value; thus, in contrast to bulk helium where ρ_s goes continuously to zero, the superfluid density per unit area has a nonzero value at the phase transition. These are the essential elements of the Kosterlitz-Thouless theory of this phase transition which was developed in 1972.

Confirmation of the theory. Two methods have been used to measure the superfluid density of a thin film and to determine how it varies with temperature. One involves the measurement of the velocity of third sound. The normal fluid participates only passively in third sound, as it is locked to the substrate by its viscosity, and so the velocity of third sound is inversely proportional to the square root of the superfluid density. A reanalysis of old experiments of this type has recently been carried out by I. Rudnick. The second method, developed by D. J. Bishop and J. D. Reppy, involves a study of the behavior of a helium film condensed onto a long strip of Mylar film rolled into a cylinder. The axis of the cylinder is attached to a torsion rod, and the oscillations of the cylinder are then studied. The normal fluid follows the motion of the cylinder exactly, but the superfluid slips; thus the moment of inertia of the cylinder is reduced by an amount proportional to the superfluid density. It has been found that the results can be fitted in detail by a theory based on this vortex model of the phase transition, and both types of experiment give a value for the ratio ρ_s/T at the transition which is in agreement with Eq. (4) over a range of film thicknesses in which the transition temperature varies by a factor of 10.

This confirmation of the predictions of the theory has implications for other branches of physics, since similar arguments have been applied to the melting of a solid, to magnetic transitions in two-dimensional systems, and to certain phase transitions of liquid crystals.

For background information *see* HELIUM, LIQUID; QUANTIZED VORTICES AND MAGNETIC FLUX in the McGraw-Hill Encyclopedia of Science and Technology. [D. J. THOULESS]

Bibliography: K. R. Atkins and I. Rudnick, *Progress in Low Temperature Physics*, vol. 6, 1970; D. J. Bishop and J. D. Reppy, *Phys. Rev. Lett.*, 40:1727–1730, 1978; B. I. Halperin and D. R. Nelson, *Phys. Rev. Lett.*, 41:121–124, 1978; J. M. Kosterlitz and D. J. Thouless, *Progress in Low Temperature Physics*, vol. 7B, 1978; J. S. Langer and J. D. Reppy, *Progress in Low Temperature Physics*, vol. 6, 1970; I. Rudnick, *Phys. Rev. Lett.*, 40: 1454–1455, 1978; S. B. Trickey, E. D. Adams, and J. W. Dufty (eds.), *Quantum Fluids and Solids*, 1977.

High-pressure phenomena

Calculation of the equilibrium temperatures and pressures of chemical reactions which take place at high pressures is of great interest to geologists in understanding processes taking place deep within the Earth. The most abundant volatile compounds in the Earth's crust and upper mantle are H_2O and CO_2, so that most reactions of geological interest involve dehydration and decarbonation with increasing temperature. The procedure for calculating the equilibrium temperature of dehydration and decarbonation reactions is the same at all pressures. However, until recently the lack of information regarding the thermodynamic properties of H_2O and CO_2 at high pressures had limited the pressure range where accurate predictions of reaction temperatures could be made to less than 10 kilobars (1 GPa). The accurate calculation of important H_2O and CO_2 reactions at pressures up to 100 kbars (10 GPa), equivalent to a depth of 300 km within the Earth, is now possible due to the recent introduction of a new equation of state for gases at high pressure. Experimental studies are very difficult at pressures greater than 30 kbars (3 GPa), so this advance allows the extrapolation of reactions which have been studied at low pressures into experimentally inaccessible pressure ranges, and the calculation of the positions of reactions for which necessary thermodynamic data are available.

Gas fugacities. At low pressures the pressure-temperature-volume relations of most gases are closely approximated by the ideal gas equation $PV = nRT$, where P is the pressure, V is the volume of the gas, n is the number of moles of gas, R is the gas constant, and T is the temperature. However, at higher pressures gases become more incompressible and no longer obey that relation. A variable called fugacity f, essentially a fictitious pressure, is used to indicate the true activity of nonideal gases at high pressures. It is defined by Eq. (1) at constant temperature. The ideal volume

$$RT \ln (f/P) = \int_{P_0}^{P} (V - V_{\text{ideal}})\, dP \tag{1}$$

V_{ideal} is that predicted by the ideal gas law. By measurement of the volume of a gas sample under a number of different pressures at constant temperature, the variation of fugacity with pressure may be determined. Accurate volumetric measurements have been made on H_2O, CO_2, and H_2O-CO_2 mixtures in recent years. This has led to a search for equations of state which can accurately describe the PVT relations observed and predict relations at higher pressures.

Modified Redlich-Kwong equation. The van der Waals equation has been used for many years to approximate the PVT relations of nonideal gases at relatively low pressures. It is useful because only four variables are used. A modification of the van der Waals equation, called the modified Redlich-Kwong equation (MRK equation), which uses only five variables, was shown by John Holloway in 1977 to be useful to very high pressures. The MRK equation is usually stated as Eq. (2), where $a(T) =$

$$P = RT/(V - b) - a(T)/[(V^2 + Vb)(T^{1/2})] \tag{2}$$

$a^0 + a_1(T) \cdot a(T)$ is a measure of molecular attraction as a function of temperature, and b is a measure of molecular volume. For H_2O, Holloway gives $a^0 = 35 \times 10^6$, $b = 14.6$, and $a(T) = (166.8 \times 10^6) - 193080T + 186.4T^2 - 0.071288T^3$. For CO_2, Holloway gives $a^0 = 46 \times 10^6$, $b = 29.7$, and $a(T) = (73.03 \times 10^6) - 71400T + 21.57T^2$. Here, the units for a are bar cm^2 $T^{1/2}$ mole^{-2}, and the units for b are cm^3/mole. These values were shown by Holloway to reproduce measured fugacity values with an error of only $\pm 5\%$ up to 10 kbars (1 GPa). Holloway suggested that the MRK equation could give useful estimates of fugacity at pressures up to 40 kbars (4 GPa) and temperatures up to 1800°C. More recent work determining CO_2 fugacities by phase equilibrium studies has shown that the fugacities predicted by the MRK equation are within $\pm 10\%$ of the true value at pressures up to 40 kbars, and may be reasonably accurate at much higher pressures because the MRK equation contains no pressure-dependent terms.

Fugacities of mixtures. The simplest way of estimating the fugacities of H_2O and CO_2 in mixtures of the two gases is to assume that there is no interaction between them. In this case, referred to as ideal mixing, the fugacity of a gas in the mixture is simply equal to the fugacity of the pure gas at the same pressure and temperature times the mole fraction of that gas in the mixture. This assumption has normally been made in the past simply because there were no data on the interaction of H_2O and CO_2. Experimental evidence now available indicates that H_2O and CO_2 interact significantly at high pressures by a reaction of the sort $H_2O + CO_2 \rightarrow$ complex. The extent of complex formation has been evaluated at pressures up to 1400 bars (140 GPa), and these data may be used to estimate the degree of complex formation at higher pressures. The departure from ideal mixing is predicted to increase as pressure increases, but approach an asymptotic limit at very high pressures. In H_2O-CO_2 mixtures it is assumed that $a = X^2_{H_2O}a_{H_2O} + X^2_{CO_2}a_{CO_2} + 2X_{H_2O}X_{CO_2}a_{H_2O-CO_2}$, and that $b = X_{H_2O}b_{H_2O} + X_{CO_2}b_{CO_2}$, where X_{H_2O} and X_{CO_2} are the mole fractions of H_2O and CO_2, respectively, in the gas mixture. The form of the complexing term used with the MRK equation is $a_{H_2O-CO_2} = (a_{H_2O}a_{CO_2})^{1/2} + \frac{1}{2}R^2T^{5/2}K$, where K is the observed equilibrium constant for complex formation and is given by $\ln K = -11.07 + 5953/T - (2746 \times 10^3/T^2) + (464.6 \times 10^6/T^3)$. Experimental studies have demonstrated that calculations of reaction temperature in H_2O-CO_2 mixtures are much more accurate when the nonideal mixing of H_2O and CO_2 in the vapor phase is considered.

Reaction equilibrium. The basic thermodynamic relation describing reaction equilibrium is Eq. (3), where ΔV is the volume of solid products mi-

$$RT \ln K - \Delta V(P-1) = -\Delta G = -\Delta H + T\Delta S \quad (3)$$

nus the volume of the solid reactants; $(P-1)$ is the pressure difference between the pressure of the calculation and the pressure of the standard state, 1 bar; ΔG is the change in Gibbs free energy of the reaction; ΔH is the enthalpy change of the reaction; and ΔS is the entropy change of the reaction. At equilibrium, $\Delta G = 0$, so by rearranging the equation one may derive the form most often used for the calculation of reaction equilibrium, Eq. (4).

$$\ln K = -\Delta H/RT + \Delta V(P-1)/RT + \Delta S/R \quad (4)$$

This general formulation is useful for calculations at all pressures. When this equation is used for calculating the positions of reactions, it is assumed that the changes in volume, enthalpy, and entropy remain constant at all temperatures and pressures. The error introduced by making this assumption is apparently too small to be detected by present experimental methods.

Reactions of pure compounds. The simplest case of a reaction whose position might be calculated using Eq. (4) and the fugacities predicted by the MRK equation is one in which all reactants and products are pure compounds. An example of such a reaction is $Mg(OH)_2 \rightarrow MgO + H_2O$. Because the solids are pure compounds, their activities are equal to 1.0 at all times. In this case the expression for the equilibrium constant, Eq. (5), reduces to

$$\ln K = \ln \left(\frac{a_{MgO} f_{H_2O}}{a_{Mg(OH)_2}} \right) \quad (5)$$

Eq. (6). [Here, a represents activity, and bears no

$$\ln K = \ln f_{H_2O} \quad (6)$$

relation to the quantity $a(T)$ introduced in Eq. (2).] In order to find the equilibrium temperature of this reaction at any given pressure, one simply determines at what temperature the ln K predicted by Eq. (4) is equal to the $\ln f_{H_2O}$ predicted by the MRK equation. By using this procedure at a number of pressures, one may easily calculate the pressure-temperature path of such a reaction.

If thermodynamic data for all compounds involved in a reaction are not available but experimental data on equilibrium temperatures and pressures are, the experimental data may be used to estimate enthalpy and entropy changes of the reaction. The fugacities at the temperature and pressure of equilibrium are known from the MRK equation. When the molar volumes of all solids involved in the reaction are known, the pressure correction term of Eq. (4), $\Delta V(P-1)/RT$, may be used to convert all the data to a single pressure. The pressure chosen is usually 1 bar (100 kPa), because that is the standard pressure to which most thermodynamic data are referred. When the pressure term is eliminated from the equilibrium equation, it becomes Eq. (7). This is an equation of

$$\ln K = -\Delta H/RT + \Delta S/R \quad (7)$$

the form $Y = aX + b$, so a plot of ln K versus $1/T$ yields estimates of both $\Delta H/R$ (the slope of the line plotted) and $\Delta S/R$ (the intercept when $1/T = 0.0$) for the reaction.

Reactions involving impure compounds. When one or more of the compounds participating in a reaction is not pure, the activity of that compound must also be considered in the equilibrium constant equation. For example, in reaction (8) the

$$CaCO_3 + SiO_2 + CaAl_2Si_2O_8 \rightarrow Ca_3Al_2Si_3O_{12} + CO_2 \quad (8)$$

$CaAl_2Si_2O_8$ is normally present in a solid solution with $NaAlSi_3O_8$, while $CaCO_3$, SiO_2, and $Ca_3Al_2Si_3O_{12}$ are relatively pure. The equilibrium constant equation for this reaction is therefore given by Eq. (9), which reduces to Eq. (10). Clearly the

$$\ln K = \ln \left(\frac{a_{Ca_3Al_2Si_3O_{12}} f_{CO_2}}{a_{CaCO_3} a_{SiO_2} a_{CaAl_2Si_2O_8}} \right) \quad (9)$$

$$\ln K = \ln \left(\frac{f_{CO_2}}{a_{CaAl_2Si_2O_8}} \right) \quad (10)$$

lower the concentration of $CaAl_2Si_2O_8$ in the solid solution, the lower the fugacity of CO_2 will be when the reaction is at equilibrium. This means that a reaction will occur at lower temperatures when it involves pure reactants than when one of the reactants is impure. The opposite is true if one of the products is impure.

Mixed volatile equilibria. Dehydration reactions are experimentally observed to occur at lower temperatures in the presence of mixed H_2O-CO_2 vapor than in the presence of pure H_2O, and an analogous relationship of reaction temperature to vapor composition is observed for decarbonation reactions. It is possible to make general thermodynamic predictions of the slopes of dehydration and decarbonation reactions on plots of temperature versus vapor composition. In order to simplify the study of mixed volatile equilibria, one variable, usually pressure, is held constant. Given a generalized reaction such as reaction (11), where A, B, and C are pure compounds, equilibrium is described by Eq. (12). In a vapor containing only H_2O and

$$A + B \rightarrow C + \nu_1 H_2O + \nu_2 CO_2 \quad (11)$$

$$\Delta G = -\Delta S\, dT + \nu_1 RT\, d\ln X_{H_2O} + \nu_2 RT\, d\ln X_{CO_2} = 0.0 \quad (12)$$

CO_2, $X_{H_2O} + X_{CO_2} = 1.0$, and $d \ln X_{H_2O} = -d \ln X_{CO_2}$. With these conditions, plus the fact that $d \ln X_{H_2O} = d X_{H_2O}/X_{H_2O}$, it is possible to rephrase Eq. (12) as Eq. (13).

$$dT/dX_{CO_2} = \frac{RT}{\Delta S}\left(\frac{\nu_{CO_2}}{X_{CO_2}} - \frac{\nu_{H_2O}}{1 - X_{CO_2}}\right) \quad (13)$$

Given that relationship, one can predict the slopes of reactions such as those illustrated in Fig. 1. Reactions such as that labeled (1) in Fig. 1, where both ν_{H_2O} and ν_{CO_2} are 0, are not affected by vapor composition and plot as lines of constant temperature on a T-X_{CO_2} diagram. For reactions which produce CO_2, such as that labeled (2) in Fig. (1), dT/dX_{CO_2} is finite and positive where $X_{CO_2} = 1.0$. As X_{CO_2} decreases, dT/dX_{CO_2} is predicted to become increasingly positive until it becomes infinite where $X_{CO_2} = 0.0$. Reactions which produce H_2O, such as that labeled (3) in Fig. (1), have the opposite relation of slope to X_{CO_2}.

When both ν_{H_2O} and ν_{CO_2} are positive, dT/dX_{CO_2} is equal to plus infinity when $X_{CO_2} = 0.0$ and minus infinity when $X_{CO_2} = 1.0$. These reactions, for example, that labeled (4) in Fig. 1, have a thermal maximum located at the value of X_{CO_2} given by Eq. (14). When H_2O and CO_2 are of opposite sign, as in

$$X_{CO_2} = \frac{\nu_{CO_2}}{\nu_{H_2O} + \nu_{CO_2}} \quad (14)$$

the reactions labeled (5′) and (5″) in Fig. 1, the sign of dT/dX_{CO_2} is constant, and is infinite at both $X_{CO_2} = 0.0$ and $X_{CO_2} = 1.0$. Such a reaction must have an inflection point in its slope located where Eq. (15)

$$\frac{d^2T}{dX^2_{CO_2}} = \frac{2RT}{\Delta H}\left(-\frac{\nu_{CO_2}}{X^2_{CO_2}} - \frac{\nu_{H_2O}}{(1 - X_{CO_2})^2}\right) = 0.0 \quad (15)$$

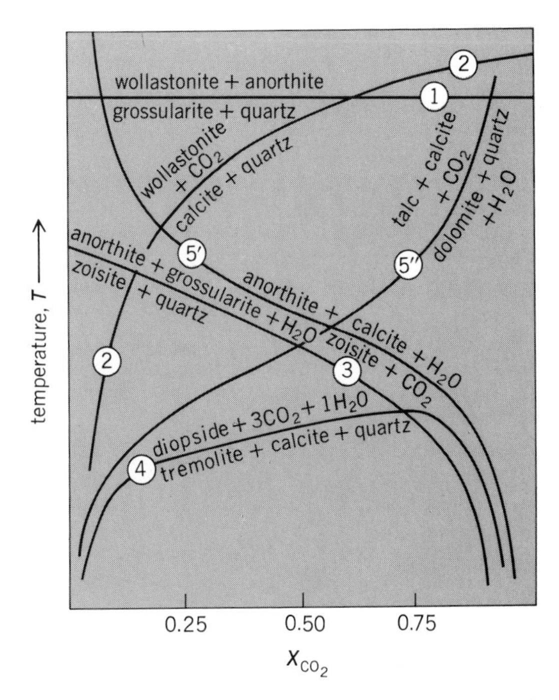

Fig. 1. Schematic illustration of various types of mixed-volatile equilibria on an isobaric T-X_{CO_2} diagram. *(From D. M. Kerrick, Review of metamorphic mixed-volatile equilibria, Amer. Mineral., 59:729–762, copyright 1974 by the Mineralogical Society of America)*

is valid, which is also where Eq. (16) holds. For a

$$\frac{\nu_{CO_2}}{X^2_{CO_2}} = -\frac{\nu_{H_2O}}{(1 - X_{CO_2})^2} \quad (16)$$

reaction where $\nu_{H_2O} = -\nu_{CO_2}$, the inflection point would be located at $X_{CO_2} = 0.5$.

To calculate the position of a reaction on an isobaric T-X_{CO_2} diagram, one calculates the equilibrium fugacity of H_2O or CO_2 for the reaction in question from the equilibrium equation at a number of temperatures. At each temperature, one then determines the mole fraction of H_2O or CO_2 with the correct fugacity by comparison with tables of fugacity versus mole fraction calculated using the MRK equation of state.

Application. The major application of reaction calculations at high pressures has been in geology. Such calculations are useful in modeling both crustal metamorphic rocks and the mineralogy of the Earth's upper mantle. Petrologists have been quite successful in explaining the mineralogy of metamorphic rocks by using such calculations in conjunction with experimental studies to construct petrogenetic grids (temperature-pressure diagrams showing reaction curves which limit the stability of specific minerals and mineral assemblages). Two varieties of petrogenetic grid are commonly used, those in which only one volatile is considered while pressure and temperature vary, and those in which pressure is held constant while temperature and vapor composition vary. Those with constant vapor composition are normally constructed with H_2O as the vapor, and are applicable to the metamorphism of carbonate-free rocks such as shales and sandstones. Many metamorphic rocks contain carbonates, so that petrogenetic grids showing temperature versus vapor composition are the most common. When constructing a petrogenetic grid, real rock compositions are usually approximated by simpler systems. For instance, siliceous dolomites are modeled by the system CaO-MgO-SiO_2-H_2O-CO_2, ignoring minor constituents such as K_2O, Na_2O, and Al_2O_3. Most metamorphic rocks equilibrate at pressures less than 5 kbars (500 MPa), so many petrogenetic grids are constructed for a total pressure of 2 kbars (200 MPa).

By comparing the mineral assemblages found in natural rocks to those indicated in petrogenetic grids, the conditions at which the rocks originated may be estimated, and the process of metamorphism may be studied. A schematic petrogenetic grid for the system MgO-SiO_2-H_2O-CO_2 for a pressure of 2 kbars is given in Fig. 2. Using this grid, one finds that any rock containing the minerals magnesite and talc originated in the region below the reactions involving that mineral assemblage. Metamorphic rocks frequently contain mineral assemblages which permit the petrologist to identify a specific reaction which was occurring in the rock. For instance, if a rock containing talc and magnesite also contained forsterite, a petrologist could check the grid given in Fig. 2 and see that the reaction magnesite + talc → forsterite + H_2O + CO_2 had been taking place in that rock.

Buffering of vapor-phase composition. An important concept pointed out by H. J. Greenwood in 1975 is that most metamorphic rocks contain sufficient quantities of hydrates and carbonates that the buffering capacity of the solids is much greater than the buffering capacity of the vapor. In this case, calculation of H_2O-CO_2 mixed-volatile reactions may be used to estimate the amounts of volatiles given off by a rock during metamorphism, which in some cases is unexpectedly small. For example, in a rock containing the mineral assemblage magnesite + quartz + talc, the composition of the vapor phase would be buffered by the reaction magnesite + quartz + $H_2O \rightarrow$ talc + CO_2, as shown in Fig. 2. As the temperature of the rock is raised, the vapor is required to become richer in CO_2. However, because the reaction consumes H_2O while producing CO_2, only a small amount of reaction is necessary to buffer the fluid composition. Only when the rock reaches a temperature where the mineral assemblage magnesite + quartz + talc is destroyed by another reaction will a significant amount of vapor be given off.

The same sort of buffering of vapor-phase composition is possible at much higher pressures in the Earth's upper mantle. The major constituents of upper-mantle peridotite appear to be forsterite, enstatite, and diopside. H_2O and CO_2 are known to be present in trace amounts, no more than 0.1 wt%. If hydrates or carbonates, or both, are formed by the reaction of mantle peridotite with vapor, they would be capable of buffering the composition of vapor and any magma which might be present. This has made the subject of H_2O-CO_2 mixed-volatile equilibria at very high pressures a topic of great interest to petrologists in the past several years.

Models for reactions at high pressures. D. E. Ellis and P. J. Wyllie in 1978 showed that the same calculation procedures applicable to H_2O-CO_2 reactions at low pressures, when used with the gas fugacities predicted by the MRK equation of state, could accurately model the positions of experimentally studied reactions at pressures as high as 33 kbars (3.3 GPa). They used the system MgO-H_2O-CO_2 to model buffering reactions at pressures as high as 100 kbars (10 GPa). The composition of vapor in equilibrium with carbonates rapidly becomes water-rich at pressures above that where carbonates first become stable. At pressures near 100 kbars the mole fraction of CO_2 in vapor in equilibrium with carbonate may be as low as 10^{-5}. At pressures equivalent to the upper mantle, carbonates are stable in the presence of all but the most H_2O-rich vapor.

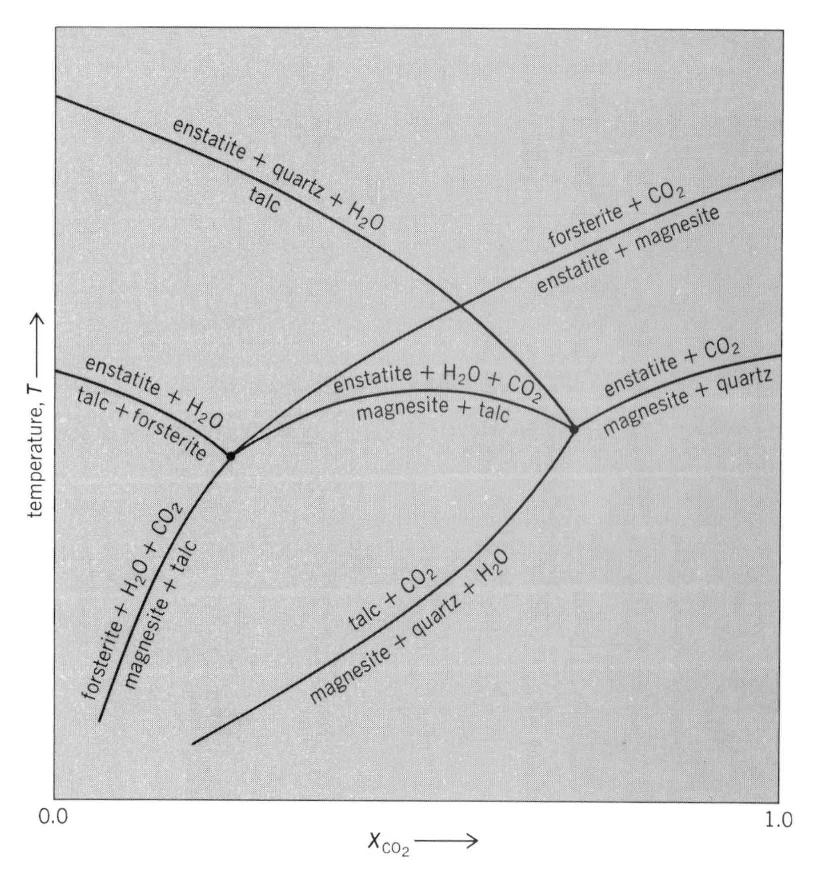

Fig. 2. Schematic isobaric petrogenetic grid for the system MgO-SiO_2-H_2O-CO_2.

Carbonates and hydrates in upper mantle. In 1978 Wyllie reviewed field and experimental evidence for the presence of carbonates, hydrates, and vapor in the upper mantle and the consequences of their presence. In the presence of pure CO_2, mantle peridotite would undergo a series of carbonation reactions with increasing pressure, producing dolomite at pressures greater than 27 kbars (2.7 GPa) and magnesite at pressures greater than 55 kbars (5.5 GPa). If water is also present in the vapor phase, the carbonation reactions take place at higher pressures, and the composition of the vapor phase is buffered in the presence of carbonate. Theoretical studies indicate that CO_2 reacts with forsterite at 42 kbars (4.2 GPa) to form enstatite + magnesite, and H_2O reacts with forsterite at about 90 kbars (9GPa) to produce enstatite + brucite. It is therefore impossible for a vapor phase to be present in the mantle at pressures greater than 90 kbars, equivalent to a depth of about 270 km. The presence of carbonates in the upper mantle buffers the composition of magma as well as the composition of vapor. In the absence of carbonate, the composition of magma formed in the upper mantle would be predicted to be basaltic. However, Wyllie has demonstrated that in the presence of carbonates, magma must have a very high CO_2 content, perhaps as much as 40 wt%, and very low silicate content. This sort of carbonate-rich magma must be the liquid which geophysicists have predicted, on the basis of seismic data, to be present in trace amounts in the asthenosphere.

For background information *see* ACTIVITY (THERMODYNAMICS); EQUILIBRIUM, CHEMICAL; FUGACITY; GAS; HIGH-PRESSURE PHENOMENA; PETROLOGY; THERMODYNAMICS, CHEMICAL in the McGraw-Hill Encyclopedia of Science and Technology. [DAVID E. ELLIS]

Bibliography: D. E. Ellis and P. J. Wyllie, *Amer. Mineral.*, in press; J. Greenwood, *Amer. J. Sci.*, 275:573–593, 1975; D. M. Kerrick, *Amer. Mineral.*, 59:729–762, 1974; P. J. Wyllie, in *Proceedings of the 2d International Kimberlite Conference*, in press.

Holography

Ever since the unveiling of laser holograms in the early 1960s, scientists have envisioned the creation of truly three-dimensional motion pictures. Success has recently been achieved in two independent formats.

The method developed in the United States, called multiplex holography, results in a rotating

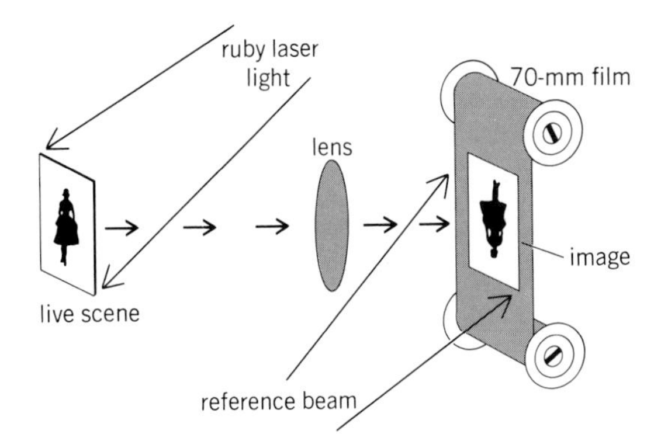

Fig. 1. Setup for recording a live scene onto 70-mm roll film in Soviet holographic motion picture process. *(From T. H. Jeong, Holography now in the USSR, Opt. Spectra, pp. 40–42, April 1978)*

cylindrical hologram illuminated by a tungsten-filament light bulb on the axis of the cylinder. As the cylinder rotates, or as the viewer moves around, images of objects or people in motion can be seen inside the cylinder in three dimensions.

The Soviet Union has developed a completely different system. It involves the projection of a sequence of 70-mm focused-image holograms onto a large holographic screen, in a format similar to that of a motion picture theater. The audience sits in fixed chairs and sees three-dimensional images in motion on the screen without having to wear any special spectacles.

Multiplex holography. The multiplex hologram is produced in two steps. The scene to be recorded is situated on a rotating platform, illuminated by common studio lights or natural light. A motion picture camera records the scene on black and white film. After the film is processed, its images are projected with red light from a helium-neon laser through a special cylindrical lens, forming real images in narrow vertical lines on a horizontal roll of holographic film. At the same time, a reference beam is directed at the location of the real image. A hologram is thus formed along this narrow strip. The successive frames of the motion picture are then recorded on the adjacent strips of the holographic film.

After the hologram is processed, it is mounted on a transparent plastic drum and illuminated with a filamental white-light source below the center of the drum. The light simultaneously illuminates all strip holograms on the drum. The viewer's left and right eyes thus are seeing through different strips corresponding to different views of the scene. If the scene involves people in motion, the holographic images correspondingly reproduce it.

Presently, multiplex holograms are made in small drums, typically 40 cm in diameter and 25 cm high. The drum is suitable as a small display in public, but not as a mass-audience device usable in theaters. It uses all the colors from the lamp that illuminates it and diffracts them out in different vertical positions. Thus, an image may appear blue to a tall person but red to a short one. Furthermore, this process sacrifices the vertical parallax.

Soviet motion picture system. The Soviet method uses a repeatedly pulsed ruby laser and makes a series of focused-image holograms on 70-mm film (Fig. 1). An imaging lens is used in a fashion similar to cinematography. However, a reference beam is used in each frame, thereby making each frame a hologram.

After processing, this film is projected by a collimated light from a mercury arc lamp in a direction precisely opposite to the original reference beam. The real image goes backward through the lens and forms a pseudoscopic image in the location of the original scene. This reconstructed image is projected onto each seat in the audience by a holographic screen which behaves as a multiple elliptical mirror (Fig. 2).

The screen, which can be made in an arbitrary size, is essentially a multiply exposed reflection hologram having an array of points as its object (Fig. 3). To produce this hologram, a converging reference beam is focused through the holographic plate to a point A, where the projected real image from the motion picture would be located. Beams arriving from points B_1, B_2, . . . , the locations of seats in the theater, interfere with the reference and form a white-light reflection hologram. Any point source emanating from A onto the processed holographic screen is thus focused simultaneously at the locations of the seats. Within the confinement of the viewer's seat, the viewer's head can move up and down or sideways and look around the image as if it were physically present.

The present system in Moscow presents a monochromatic image (yellow) and is limited to an audience of four. However, it is theoretically possible to construct a screen that can produce multiple-color images to an arbitrarily large audience.

White-light holograms. The recent emphasis on using common light sources for viewing has spurred a new generation of white-light reflection holograms of still images. New emulsions of silver halide and dichromated gelatin have been developed which create sparkling images.

The procedure for making these holograms is very simple when these emulsions are used. Typically, the object to be recorded is placed immediately next to the plate coated with the new emulsion. A single beam of laser light is directed through the plate from the opposite side of the ob-

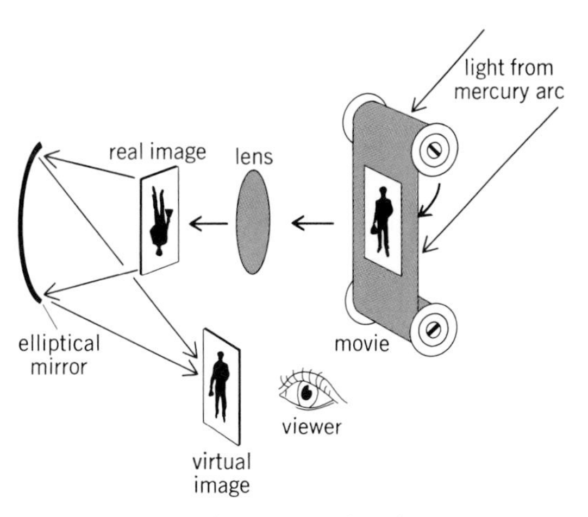

Fig. 2. Formation of image on an imaginary screen consisting of a large elliptical mirror in Soviet holographic motion picture process. *(From T. H. Jeong, Holography now in the USSR, Opt. Spectra, pp. 40–42, April 1978)*

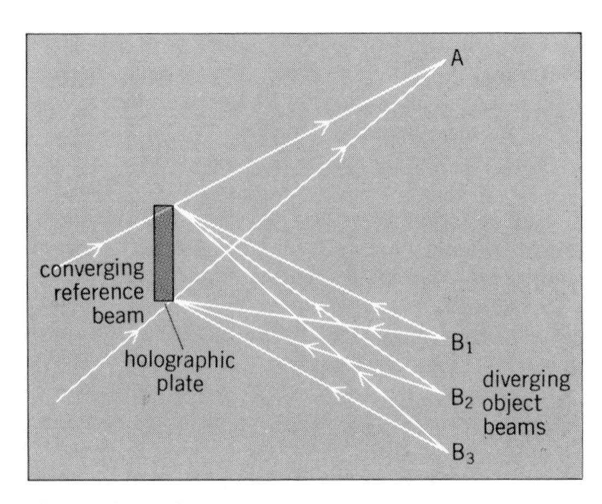

Fig. 3. Setup for a holographic screen that can be seen by more than one viewer. *(From T. H. Jeong, Holography now in the USSR, Opt. Spectra, pp. 40–42, April 1978)*

ject. The direct incidence provides a reference beam; the transmitted light is scattered by the object back onto the plate, providing the object beam. The processed hologram is viewed by shining any point source of white light on the hologram from the direction of the reference beam.

The Soviet Union is presently using this technique to recreate images of their national treasures, such as jewelry and other art objects from the Hermitage. In the United States, holographic jewelry is now sold, presenting images brighter than the original object.

For background information *see* HOLOGRAPHY in the McGraw-Hill Encyclopedia of Science and Technology.

[TUNG H. JEONG]

Bibliography: T. H. Jeong, *Encyclopedia of the Optical Industry and Systems*, E118–E136, 1978; T. H. Jeong, *Opt. Spectra*, pp. 40–42, April 1978; V. Komar, *Proceedings of the Society of Photo-Optical Instrumentation Engineers*, vol. 120, pp. 127–144, August 1977.

Holography, acoustical

In the new optical science of holography, the property which has aroused the most public interest has been its ability to generate an extremely realistic three-dimensional visual image. However, because the technique of holography involves the recording of a wave interference pattern, any process involving waves, such as sound waves, can benefit from the technique. Accordingly, numerous acoustical holography experiments have been made and have shown promise in the medical field, for locating cysts and tumors with x-ray-like acoustic pictures, and in underwater sound, for detecting underwater objects and as a supplement to present seismic procedures for locating offshore oil deposits.

Liquid surface holography. Since a hologram is a photographic record of the interference pattern generated between a set of waves of interest and a set of reference waves, it is obviously possible to make holograms by using sound waves, provided the wave interference pattern can be formed.

One procedure which permits the real-time viewing of acoustic holograms, bypassing the photographic recording process required in optical holography, uses two underwater transducer sound sources. The sound waves of interest (those originating at the object transducer and passing through the object) and the reference waves (those originating at the reference transducer) generate an interference pattern at the surface of the liquid, causing it to have extremely minute, stationary ripples (tiny ridges). These mechanical deformations of the liquid surface correspond exactly to the interference fringes of a hologram, so that reconstruction of the acoustic hologram image is accomplished by causing laser light to be diffracted by these ridges as it is reflected off the liquid surface.

Many parts of the human body react to high-frequency sound waves in a different manner than to x-rays. Accordingly, the application of hologram techniques as an alternative to x-rays has been examined. The most rapid development has been in the liquid surface technique just described, and equipment for this purpose has recently become available commercially. The (real-time) record obtained when a person's hand is placed between the liquid surface and the object transducer provides a differentiation between the various types of tissues; thus the acoustic holographic image of the hand clearly shows the tendons, the vascular structure, and the blood vessels, without the injection of dyes as is required for x-rays. The ability to acoustically delineate soft-tissue details has led to a fairly wide use of acoustic holography in the diagnosis of breast tumors, and extremely small tumors can now be detected.

Two organizations which have demonstrated significant progress in developing commercial through-transmission acoustical holography systems (as contrasted with the earlier reflection acoustic techniques, such as those embodied in the conventional B-scan systems) are Holosonics, Inc., and the Actron Division of McDonnell Douglas. Holography has also been applied, in the form of synthetic-aperture techniques, to the B-scan acoustic reflection systems to provide greater detail in the body areas located near the acoustic transducer. Also, a number of acoustic planigram B-scan records (sectional images called tomograms) have been used to record successively on the same hologram, planes at different depths (for example, in a series of different-depth mammograms). Reconstruction of these optical holograms provides a three-dimensional image of the body volume under examination.

Hologram sonar. The technique called sonar, employing acoustics for detecting underwater objects, has also benefited from acoustical holography developments. Recently a holography array of 400 acoustic transducers was used in such a sonar, and its success resulted in a succeeding version containing 5000 elements. The ability of hologram sonars to delineate nearby (near-field) objects is an important advantage.

Geophysical prospecting often employs acoustic waves which are sent into the earth, with the returning echoes then being analyzed to appraise the likelihood of oil or gas being present in the substructure. Here, too, acoustical holography has recently proved useful, permitting special tech-

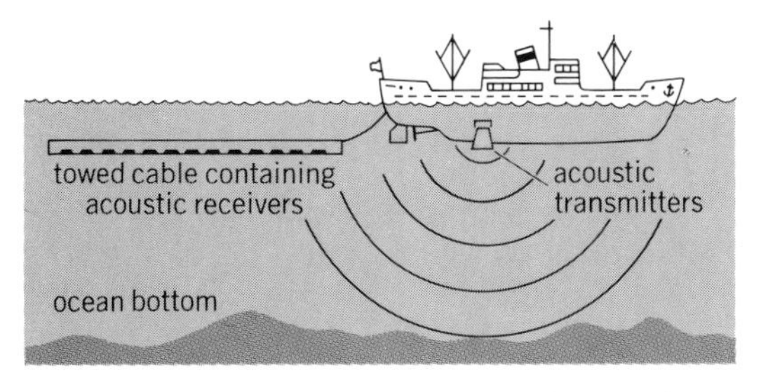

Underwater viewing by means of acoustic holography.

niques, such as spatial filtering, to suppress, in the records, artifacts which would otherwise obscure the desired information.

Use in offshore oil prospecting. In the technique for locating offshore oil deposits, a cable which in practice would be 100 wavelengths or more in length is towed behind a ship equipped with a high-power transmitter emitting low-frequency coherent acoustic energy (see the illustration). Signals reflected or scattered from the ocean bottom and from geological layers below are picked up by the cable array, with holographic processing of the seismic data obtained permitting the retrieval of useful information. One recent seismic holograph test was carried out in the Gulf of Mexico over a known and well-defined salt dome. The diameter of this salt dome is approximately 2 mi (3.2 km), and the top of the dome is at a depth of 1156 ft (352 m). The signal-processing procedure comprised mixing the received signal with a reference signal, the latter being the same signal used to drive the vibrators. Two signal values were recorded for every sampling plane, the second signal being the received signal mixed with the reference wave delayed by a quarter-wavelength. This procedure, developed by J. S. Keating of Bendix, provides a value which is in phase quadrature with the first, and by using both of these values in the holographic reconstruction, an unambiguous phase with respect to the reference was obtained. The exact outline of the dome in its expected position was observed in the hologram reconstruction.

Synthetic-aperture holography. The synthetic-aperture technique mentioned in connection with medical acoustics imaging has proved to be a very important addition to today's radar technology. Accordingly, extensive investigations are now under way to examine its possible application in various acoustical holography fields, including the medical one mentioned, and also in the sonar and the seismic technologies. Also there is a project under way to determine by acoustical holography whether a long-sought Leonardo da Vinci fresco is hidden under a more recent one painted by the Renaissance artist Giorgio Vasari on the walls of the Palazzo Vecchio in Florence.

For background information *see* HOLOGRAPHY; HOLOGRAPHY, ACOUSTICAL in the McGraw-Hill Encyclopedia of Science and Technology.

[WINSTON E. KOCK]

Bibliography: E. J. Barrakette et al., *Optical Processing*, vol. 2, 1978; W. E. Kock, *The Creative Engineer*, 1978; W. E. Kock, *Engineering Application of Lasers and Holography*, 2d ed., 1977.

Hydrography

It has recently become possible to make soundings, from an aircraft, of water depth from 2 to 30 m in coastal waters. In previous work a single beam of green laser light had been used, and depth was determined by the difference in time of arrival, at the aircraft, of reflections from the sea surface and bottom. A significant improvement in performance has been achieved by J. E. Clegg and M. F. Penny through the use of separate sensors to measure the height above the sea surface and bottom.

A neodymium yttrium-aluminum-garnet (Nd:YAG) laser supplies two pulses simultaneously to two sensors, one operating at 1060 nm and the other at 530 nm. This laser oscillates at 1060 nm, which is in the near-infrared part of the spectrum. It is convenient, though not essential, to use this wavelength for the surface sensor. Most of the energy of the laser operating at 530 nm is obtained by frequency doubling in a nonlinear optical crystal. This wavelength, which is in the green part of the spectrum, is used for the bottom sensor, and is very close to that which gives least attenuation in water. *See* OPTICS, NONLINEAR.

The Nd:YAG laser can readily generate very short pulses of 5 ns duration at a power level of 1 MW. As light travels through water at 0.22 m/ns, this pulse can be used to measure depths of less than 2 m with a resolution of 0.5 m.

Surface sensing. The aircraft carrying the depth sounder flies at a height of 500 m, which is measured accurately by the surface sensor by using the 1060-nm pulse. The sensor's beams are 1° (20 mrad) wide, providing good height resolution. The reflection at the water surface is largely specular, and thus the amplitude of the received pulse falls rapidly as the angle of incidence increases. The surface sensor is therefore mounted on a stabilized platform together with the laser and the bottom sensor.

Bottom sensing. About 95% of the 530-nm pulse enters the water and travels toward the bottom, and the amplitude of the received signal is not significantly affected by the angle of incidence at the

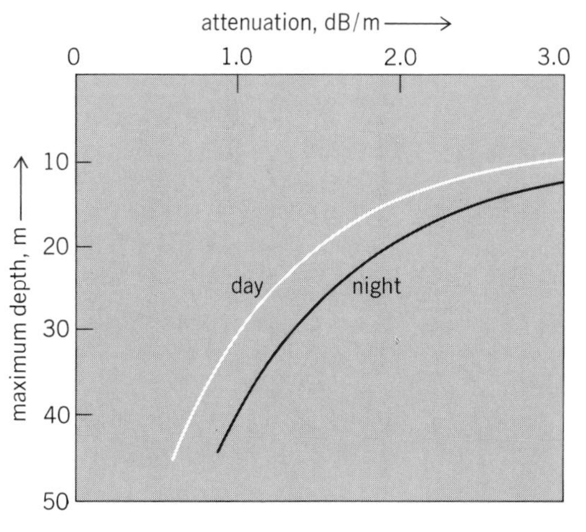

Fig. 1. Maximum depth which can be sounded by airborne laser system as a function of attenuation of light by water.

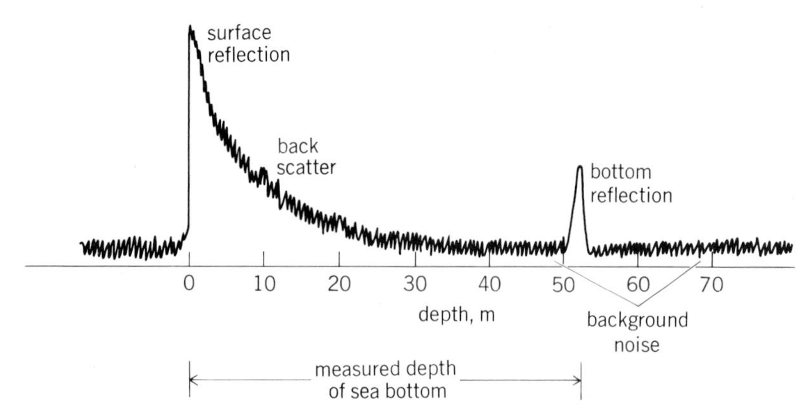

Fig. 2. Typical signal waveform received by bottom sensor.

water surface. It is therefore possible to scan this sensor across the track of the aircraft and to sound over an area instead of along a line. In an experimental installation the inclination of this sensor to the vertical was adjustable, and it was found that, over a wide range of sea states, the reduction in amplitude was acceptably small for inclinations up to 15° (250 mrad).

To provide good resolution in the horizontal plane and a good signal-to-noise ratio, the green transmitter pulse is concentrated into a beam 0.05° (1 mrad) wide which, at a flying height of 500 m, illuminates a patch of water 0.5 m in diameter. Ripples on the sea surface and forward scattering cause some broadening of the beam in the water. This broadening, along with diffuse reflection from the sea bottom, enables a comparatively broad receiver beam to be used. A small field stop can then be inserted in the receiver telescope to suppress the strong reflection from the sea surface, which occurs at small angles of incidence.

Maximum depth. The attenuation of light in water is the principal factor limiting the depth which can be sounded by this method. Many measurements of attenuation have been made from surface vessels and by divers off the coasts of South Australia and Queensland. In South Australia the attenuation is typically about 1 dB/m, but varies between 0.5 and 4 dB/m depending on location, depth, and sea state. In Queensland the attenuation is somewhat less, but is very high in the wet season when the rivers discharge colored water into the sea.

In daylight the maximum depth is reduced by background noise of solar origin, despite the insertion of an optical filter, 2 nm wide, in front of the receiver. For example, on a night flight over the Great Barrier Reef, it was possible to sound to 55 m, compared with 40 m in daylight. Figure 1 summarizes the effect of attenuation on maximum depth.

Signal components. The signal received by the bottom sensor has four components, as shown in Fig. 2: (1) reflection from the water surface, which is of variable amplitude and may be very large near vertical incidence; (2) backscatter, which decreases exponentially with depth from the surface; (3) background noise which, in daylight, is caused by sunlight; and (4) desired reflection from the bottom, if the depth is not too great.

System monitoring. The depth counter, which has been started by a pulse from the surface sensor, is stopped by the bottom reflection. The equipment operator has to adjust gain and range controls to prevent all but the bottom reflection from stopping the depth counter. Because the pulses are extremely short, and the time base less than 1 μs long, it has been difficult, until recently, to provide a satisfactory waveform monitor in the aircraft. Now, with the development of fast analog-to-digital converters, this has become feasible, and the bandwidth of the video waveform is effectively reduced from 300 MHz to 50 kHz. The use of analog-to-digital converters has the further advantage of making it possible to record the waveforms of every sounding and to store them on a modest amount of magnetic tape for subsequent analysis on the ground.

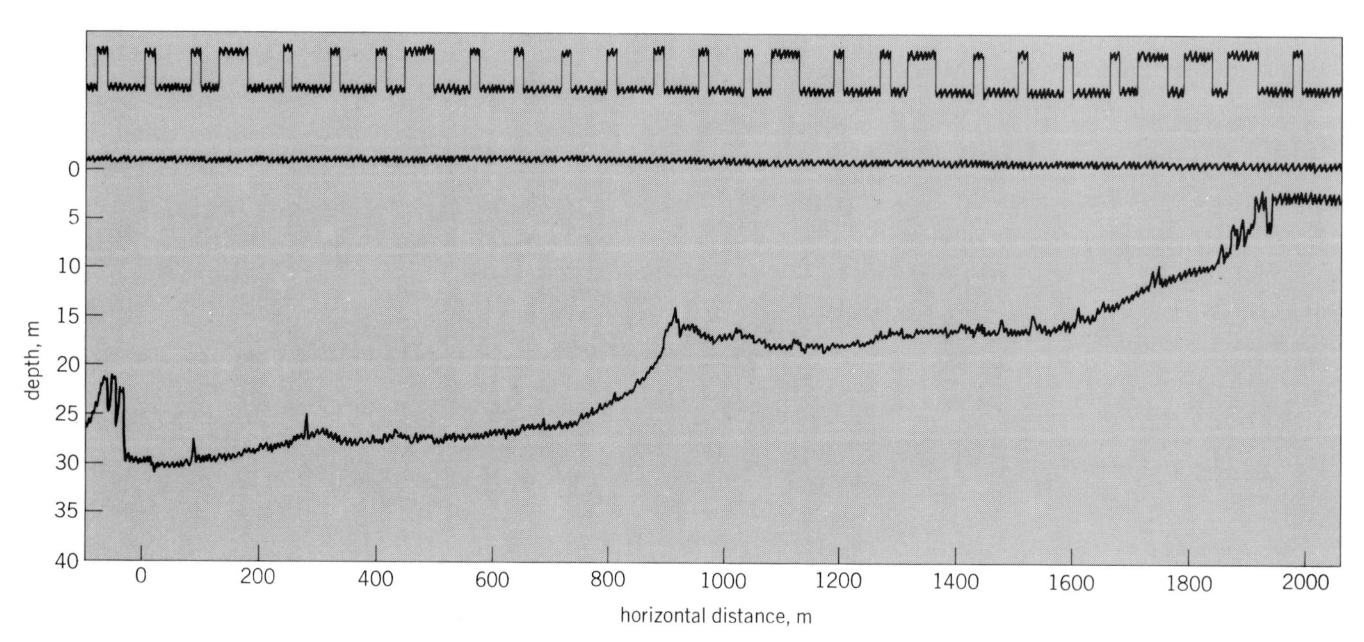

Fig. 3. Profile of sea bottom obtained by airborne laser depth-sounding system.

Pulse-repetition frequency. In the experimental systems, which are now being flight-tested, the pulse-repetition frequency of the lasers is about 25 pulses per second, giving soundings spaced about 3 m apart along the line of flight. A typical profile of the sea bottom obtained by this system is shown in Fig. 3. For an operational system, scanning a strip 240 m wide from a height of 500 m, the pulse-repetition frequency must be increased to about 200 pulses per second to provide a pattern of soundings 10 m apart.

Operational height. The conflicting requirements of coverage, minimum complexity, laser safety, and flight restrictions have resulted in the choice of a flying height of 500 m. A lower limit is placed on the size of the patch of sea illuminated by the laser by the need to keep the energy density below a safe level. The laser emits 5 mJ per pulse at 530 nm, which, for a uniformly illuminated area 0.5 m in diameter, gives an energy density of 20 mJ/m^2. Hence the amount of light which could enter the eye, with a night-adapted pupil 7 mm in diameter, is 1 μJ, which is below the danger level for this type of pulse. The laser is not switched on below this height.

Position fixing. When the sounding equipment is to be used operationally, it is necessary to integrate it with a precision position-fixing system so that the location of every sounding can be recorded with an accuracy consistent with the spacing of the soundings. With the aid of a minicomputer which forms part of the sounding equipment, this positional information can be processed to give navigational information which enables the aircraft to be piloted along prescribed tracks over the area to be charted.

For background information *see* HYDROGRAPHY; LASER in the McGraw-Hill Encyclopedia of Science and Technology. [JOHN E. CLEGG]

Bibliography: J. E. Clegg and M. F. Penny, *J. Nav.*, 31:52–61, January 1978; W. Rattman and T. Smith, *Hydrospace*, 5(1):57–58, February 1972.

Immunity

Macrophages constitute a class of heterogeneous cells known to be the main cell type constituting the reticuloendothelial system. These cells, both fixed and wandering, are thought to be derived mainly from bone marrow precursor cells. Recent studies have focused attention on the important role of these cells in many diverse biologic reactions, including those involved in immune responses.

Although macrophages are important for various metabolic activities, especially lipid and iron metabolism, it is now recognized that these cells constitute one of the most important classes of leukocytes involved in immune responses. It was widely known at the beginning of the 20th century that macrophages from many animal species, as well as humans, are "scavenger" cells and can rapidly interact with and digest many particulate or colloidal substances, but only during the last decade or so has a reawakening interest in the role of macrophages in host immune defense systems led to an understanding of their important functions. A number of recent conferences and meetings have been devoted to analyses of the role of macrophages in immunity, including their involvement in antibacterial responses and tumor resistance.

It is now thought that macrophages may be readily "activated" by various methods to show increased reactivity to viruses, bacteria, fungi, and other pathogenic microorganisms. Such activation can also be achieved merely by exposure to pathogens. In addition, activated macrophages stimulated either directly by various agents or following absorption of a specific antibody onto their surface acquire the ability to more efficiently recognize and interact with microorganisms or tumor cells. In addition as shown in the table, macrophages can be suppressed by a variety of procedures.

Macrophages are now known to play an important role in the early inductive stages of antibody formation, since they appear to be needed for the initial interaction and possibly for "processing" of particulate antigens. Furthermore, antimicrobial and antitumor immunity may be mediated not only by direct contact of these cells with antigen or reactive lymphocytes, but also by elaboration of biologically active soluble factors. Some of the recent advances in elucidating macrophage activities in these diverse reactions are described below.

General involvement of macrophages. Many recent studies, both in animals and in cell cultures, have documented the diverse role of macrophages in immunity. For example, macrophages are known to interact in animals with particulate or insoluble substances, antigen-antibody complexes, and so on. Macrophages also appear to take up and localize, as well as to degrade or process, antigen. They are also involved in microbicidal activity and bacteriastasis, and can protect animals from infection by intracellular parasites, although potentiation has also been noted. Similar reactions occur in cell cultures, including microbicidal and tumoricidal activity. Furthermore, macrophages are known to have "feeder" activity in cultures for other cells, especially lymphocytes. These cells release many enzymes in culture, including collagenase, lyso-

Agents shown to have activating or depressive effects on macrophages

Activating agents	Suppressing agents
Colloidal or particulate particles being phagocytized (including microorganisms)	High antigen concentration, debris (reticuloendothelial system "blockade")
Polynucleotides, nucleic acids	Antimacrophage serum
Microbial products	Radiation (high doses), radiominic agents
Oils, irritants	Silica, thorotrast
Interferon inducers	Carrageenin
Cytophilic antibody	Corticosteroids (high levels)
Antigen-antibody complexes	Antimetabolites, anti-inflammatory agents
Humoral factors (α_2-globulins, complement)	Microorganisms and their products
T cell products (lymphokines)	

zymes, and nucleases. They are also a source of interferon.

Specific involvement of macrophages. Many studies have shown that macrophages are necessary components for the immune responses in animals, since depression of macrophage activity by various methods often suppresses immune responsiveness to a wide variety of antigens. Recently such studies have been extended to cell cultures where various cell types can be readily separated and analyzed. Such studies have shown that macrophages may have a critical function in "presenting" antigen in a highly immunogenic form to responding antigen-specific B and T lymphocytes in the initiation of antibody information. Various experimental models have been used to demonstrate this phenomenon. One of the most recent experiments was conducted by C. W. Pierce. Earlier studies by him and others had shown that removal of macrophages from spleen cell suspensions immunized in cell cultures with sheep red blood cells impairs antibody formation, at least when macrophages are removed within the first 2 days of culture. Later removal of macrophages had little or no effect, indicating that macrophages are necessary during the initial phases of immune activation, at least to particulate or insoluble antigens.

More recent studies showed a genetic restriction regulating the efficient interaction among antigen-bearing macrophages and splenic lymphocytes in development of antibody responses to T cell–dependent antigens. Although such genetic restrictions seem quite important in many immune responses, recent cell-culture studies suggest that such restrictions are not applicable for primary antibody responses to a small polypeptide antigen which was first processed through macrophages. However, similar responses to complex antigens such as sheep red blood cells, as well as to tumor or normal tissue antigens, appear to involve genetic control mechanisms, especially in secondary responses. Thus, by using a synthetic polypeptide as antigen, Pierce and colleagues found that immune lymphoid cells develop secondary antibody responses preferentially when stimulated in cell culture with antigen on macrophages genetically identical to macrophages used to immunize the cells in animals. Such genetic restrictions involving macrophage-lymphocyte interactions in secondary immune responses appeared to be controlled by products associated with the histocompatibility complex of chromosomes. Furthermore, the genetic restrictions appeared to be antigen-specific, and operated at the level of immune T lymphocyte. During such studies it was also found that macrophages capable of "suppressing" immune responses may be involved in a regulatory mechanism.

Immunoregulation by macrophages. Macrophages have been found not only to enhance but also to depress immune responses. Many earlier studies had shown that excessive numbers of macrophages added to cell cultures actually inhibited the expected immune response to an optimum dose of antigen. In more recent studies it was found that when spleen cells from immunosuppressed animals were challenged in cell culture with an antigen, normal immune responses might develop upon removal of macrophages. Macrophages involved in such immunosuppression have been termed "suppressor" macrophages. For example, tumor virus–infected animals often show impaired lymphoid cell responses to a wide variety of immunologic stimuli in cell cultures. When such spleen cell suspensions are treated to remove macrophages, immune responsiveness often returns. Addition of the suppressor macrophage population to normal spleen cell suspensions results in depressed immune responsiveness. However, it should be noted that other cell types have also been found to be suppressive, especially T lymphocytes.

The nature and mechanisms of immunosuppression induced by suppressor macrophages, as well as by lymphocytes, are not clearly understood. However, many investigators feel that this phenomenon may be involved in the impairment of immunity often seen in tumor-bearing individuals, as well as in individuals infected with a wide variety of microorganisms which cause chronic systemic infections, including parasites and intracellular microbial pathogens. On the other hand, impairment of macrophage activity in certain infections, especially those induced by viruses, may itself result in immunosuppression.

Recent studies by Mauro Bendinelli and by S. Specter and H. Friedman have shown that immunosuppression induced by leukemia viruses may be reversed by "normal" macrophages. In this model system, spleen cells derived from leukemia virus–infected animals did not respond normally upon challenge immunization in cell culture with a wide variety of antigens. The impairment of the immune response appeared to be due to a defect of lymphoid cells rather than macrophages, as shown by appropriate cell separation techniques. Furthermore, infection of spleen cell suspensions with the tumor virus also induced immunosuppression; this appeared to be a direct effect of the virus on lymphocytes, especially B lymphocytes and their precursors involved in antibody formation. Addition of these suppressed lymphocytes to normal spleen cell suspensions transmitted the immunologic impairment. This appeared due, however, to a direct effect of the virus or virus-induced factor rather than to a suppressor lymphocyte.

Macrophages isolated from infected spleen cell populations had no suppressive effect on normal spleen cells. On the other hand, addition of relatively small numbers of normal macrophages to spleen cell suspensions from virus-infected animals restored the immune response. Addition of as little as 5% normal macrophages restored immune responsiveness to essentially normal levels. Thus macrophages can restore immunoresponsiveness of tumor virus–suppressed lymphocyte cultures. The mechanism of such restoration appeared to be due to functional activity of the macrophages themselves, including release of soluble mediators. For example, lipopolysaccharides (LPS), a known stimulator of immunity, may also restore some of the immunologic responsiveness of leukemia virus–infected spleen cell suspensions. Addition of LPS to normal spleen cell suspensions or purified macrophage populations resulted in activation of these cells so that fewer cells could restore immune responsiveness. Furthermore, cell-free culture supernatants derived from LPS-stimulated normal spleen cell suspensions could also restore responsiveness of virus-suppressed spleen cell cultures. Such supernatants were rich in immunostimulators known to be produced by activated macrophages or lymphocytes. Thus it appears that macrophages, at least during reversal of immunosuppression, may function by releasing various factors.

Macrophage secretory products. As indicated above, products released by macrophages have been shown to markedly influence immune responsiveness, as well as other metabolic activities of cells. Although macrophages are known to modulate the activity of T and B lymphocytes interacting with antigen, the phenomenon of antigen presentation is only one aspect of the complex mechanism whereby those cells are involved in such interactions. The secretory activity of macrophages is well known, especially the synthesis and release of many important molecules which may cause indirect activation of T and B lymphocytes. It has been known for some time that many molecules, including lysozymes, complement proteins, and interferon, are released by activated macrophages. Some of these factors, as well as others whose nature has not yet been elucidated, may have lymphostimulatory activities which modulate various immune functions. For example, recent studies by Emul Unanue have shown that cultured macrophages release a thymocyte-differentiating factor, so that these cells acquire a mature phenotype more rapidly. Mitogenic proteins released by macrophages also appear following interaction between these cells and immune T cells. Such interactions are thought to require full contact of the two cell types and may be regulated by the product of the major histocompatibility complex of chromosomes. Furthermore, soluble regulatory factors may be involved not only in activation but also in suppression of immune responses, especially to T cell–dependent antigens. Such factors may be induced not only by direct interaction with antigen-activated T cells, but also by LPS and other immunostimulatory substances from sources as diverse as microorganisms and serum factors.

The nature and characteristics of these factors appear amenable to physicochemical analyses not only of cell-free culture fluids from normal or activated macrophages, but also of cultured macrophage-derived tumor cells. Recent studies by J. Oppenheim and D. Rosenstreich have shown that lymphocyte-activating factor is released from cultures of a continuous cell line of macrophage-derived tumor cells, especially after stimulation of cell cultures with LPS. Thus the interaction between macrophages and lymphocytes has been studied not only with normal cells, but also with the readily cultured tumor cell line. It therefore seems likely that the complex interactions among macrophages and lymphocytes (including T and B cells), plus antigen, will be understood more fully because of the development of newer model systems.

For background information *see* IMMUNITY; IMMUNOLOGY in the McGraw-Hill Encyclopedia of Science and Technology.

[HERMAN FRIEDMAN]

Bibliography: P. Alexander, *Ann. Rev. Med.*, 27: 207, 1976; I. Kamo and H. Friedman, *Adv. Cancer Res.*, 25:271, 1977; D. S. Nelson (ed.), *Immunobiology of the Macrophage*, 1976; C. W. Pierce and J. B. Knapp, *Contemp. Top. Immunobiol.*, 5:91, 1976; A. S. Rosenthal and E. A. Shevach, *Contemp. Top. Immunobiol.*, 5:47, 1976; E. R. Unanue and H. Friedman, *Fed. Proc.*, 37:77–104, 1978.

Immunology

In the last few years perception of the role of the major histocompatibility complex (MHC) has changed from a passive role as a simple antigenic marker for tissue incompatibilities in transplantation, to a central role in the control of immune responses. This article focuses on three aspects: (1) the role of the MHC in the genetic control of immune response (Ir) genes; (2) the role of the MHC in macrophage-T_1 and T-B cell cooperation during the development of immune responses; and (3) the role of the MHC in recognition of virus-infected targets by cytotoxic T cells.

Background. The genetics of MHC has been studied in detail in two species, humans and mice. The illustration shows the current understanding of the map of human and mouse histocompatibility complex. The human histocompatibility (HLA) region is located on human chromosome 6 and the mouse histocompatibility region (H-2) is located on mouse chromosome 17. The genes *A*, *B*, and *C* each code for 45,000-dalton polypeptides which are extensively polymorphic within the human populations. Similarly, the mouse genes *K*, *D*, and *L* also code for polymorphic, 45,000-molecular-weight glycoproteins. Recently, Ia antigens of the mouse and Dr antigens of humans have been characterized. These are coded for by the I region in the mouse, and by the D region in humans. Ia and Dr antigens are polymorphic, two-chain glycoproteins of 28,000 and 30,000 molecular weight, respectively. Their determinants are responsible for the mixed lymphocyte response, a proliferative response of T lymphocytes to allogeneic cells. This response is unique in that it requires intact cells as the stimulating antigen and is strongest within a species. The mixed lymphocyte culture between species is much weaker.

Ir genes. H. McDevitt and B. Benacerraf have shown that in the branched polypeptide the genes in the murine I region may regulate the immune

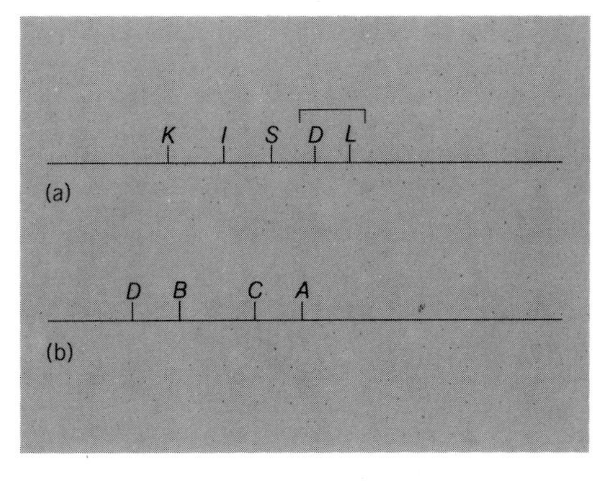

Genetic map of *(a)* the mouse major histocompatibility complex (H-2) and *(b)* the human major histocompatibility complex (HLA). The bracket over *D* and *L* indicates that the map order of the loci is unknown.

response to synthetic polypeptide antigens. The possession of specific responder alleles by an inbred mouse strain allows the members of that strain to respond with an antibody response, whereas other mice are unable to produce antibody. These results have been greatly extended and now cover a large series of antigens, including heterologous protein antigens, such as ovalbumin; synthetic polypeptides, such as mixed amino acid polymers; immunoglobulin allotypes; and cell-surface antigens such as H-4 and H-Y. Although it was originally suggested that Ir genes function solely in T cells, it appears that there may be Ir gene function in macrophages as well.

Two lines of evidence have suggested Ir function in the T cell. First, conjugation of the antigen to a highly immunogenic carrier, such as methylated bovine serum albumin, allows a productive antibody response, suggesting that a T-cell defect has been bypassed and B-cell function is intact and responsive. Second, the delayed-type hypersensitivity response mediated by T cells is governed by genetic controls identical to those governing the antibody response.

The results implicating Ir function in macrophages are based primarily on the work of A. Rosenthal and colleagues, who have shown that different strains of inbred guinea pigs respond to distinct portions of the insulin molecule and produce immune responses accordingly. Thus, in this particular case, guinea pig strain 2 recognized the C terminal portion of the molecule, while guinea pig strain 13 recognized the amino terminal portion. In effect, the two strains are responding to different antigens. They have shown that these differences depend on the macrophage which binds the antigen, and present the antigen to the lymphoid cells. This difference in binding would explain the appearance of Ir gene function in T cells in many systems, while the actual site was the macrophage.

H-2 in cell cooperation. The second major function of genes in the MHC is their role in cell-cell cooperation during the development of immune responses. Several years ago it was shown that antibodies directed against Ia could inhibit immune responses to non-Ir-controlled antigens such as sheep red blood cells. A number of groups demonstrated that I region compatibility, that is, a sharing of the identical allele in the I region, was necessary for both T-B cell cooperation and macrophage-T_1 cooperation. It was then demonstrated that both the soluble mediators of T- and B-cell cooperation, such as antigen-specific T-cell replacing factors and antigen-specific suppressor factors, had determinants which could be recognized by anti-Ia antibodies. Whether these molecules are identical to the Ia cell surface molecules is not known at this time, but it is thought that they are not structurally identical to the cell surface molecules. These results suggest that Ir genes are not the T-cell antigen receptor, as had been proposed earlier by McDevitt and Benacerraf. Clearly, one of the major mysteries in understanding how H-2 genes modulate immune responses is elucidating the mechanism by which specificity of the Ir genes is produced.

Role of K/D products. It has become clear that determinants in D and K are necessary for recognition of virus-infected cells by antigen-specific cytotoxic cells. In the mouse these cells provide the major protection against virus infection. Previously they were believed to recognize only the virus-induced glycoproteins on the cell surface as simple antigens. It has now been shown that in addition to the virus material on the cell membrane, the cytotoxic cell must share the D or K alloantigens with the virus-infected target. Thus, a cell from a mouse from strain A infected with vaccinia virus will be able to kill vaccinia-infected target cells from strain A, but not from strain C57 Bl/6.

Two hypotheses have been proposed to explain this phenomenon. One, called dual recognition, states that both the antigen—in this case the viral protein—and the histocompatibility antigens must be recognized by distinct receptors on the cytotoxic cell before a lytic interaction can occur and the infected cell can be destroyed. The second explanation is the altered-self hypothesis, which states that the cytotoxic effector cell recognizes the virus glycoprotein in a molecular complex on the cell surface with the H-2D or K antigens, and thus recognizes an altered-self K or D molecule.

While the altered-self hypothesis is attractive because of its requirement for one less receptor, it is difficult to envision such a model at the molecular level, since each virus glycoprotein must be able to associate with every K or D molecule in order to facilitate a cytotoxic event. Such a large number of different interactions in a fluid membrane seems unlikely. Dual recognition has the disadvantage of requiring two different receptors, only one of which need be antigen-specific. This hypothesis also presents some evolutionary problems in terms of coevolution of the K/D molecules and their receptors on cytotoxic cells, especially in a highly polymorphic system such as H-2 in the mouse. No definitive experiments have yet been performed to implicate one hypothesis over the other.

During 1978 R. Zinkernagel showed that T lymphocytes learn their self-recognition structures during maturation in the thymus, and that this is an active rather than a passive process. Further, the ability of cytotoxic cells to kill infected targets depends on the H-2 type of the thymus epithelium where it matured and the H-2 type of the virus-infected cell which sensitizes the cytotoxic effector. These studies will have a profound effect on the understanding of how cells learn to discriminate self from nonself.

Summary. It is clear that in the last 5 years great strides have been made in the understanding of the role of the MHC, but a great deal has yet to be learned. The MHC has been a productive system for study since the immunologic and genetic aspects of the study have continually reinforced each other. Thus the appearance of new genetic regions of H-2 have led to the definition of functional markers on lymphocyte subpopulation surfaces, and the examination of immune functions such as phenotypes has led to better definition of genetic regions.

For background information *see* IMMUNOLOGY in the McGraw-Hill Encyclopedia of Science and Technology.

[JEFFREY A. FRELINGER]

Bibliography: B. Benacerraf, *J. Immunol.*, 120: 1809–1812, 1978; A. S. Rosenthal et al., *J. Exp. Med.*, 147:882–896, 897–911, 1978; R. Zinkernagel et al., *Nature*, 254:78–79, 1977.

Infrared imaging devices

Infrared imaging devices convert an invisible infrared image into a visible image. Infrared radiation is usually considered to span the wavelengths from about 0.8 or 0.9 micrometer to several hundred micrometers; however, most infrared imaging devices are designed to operate within broad wavelength regions of atmospheric transparency, that is, the atmospheric windows. At sea level, for horizontal paths of a few kilometers' length, these are approximately at 8–14 μm, 3–5 μm, 2–2.5 μm, 1.5–1.9 μm, and wavelengths shorter than 1.4 μm. The radiation available for imaging may be emitted from objects in the scene of interest (usually at the longer wavelengths called thermal radiation) or reflected. Reflected radiation may be dominated by sunlight or may be from controlled sources such as lasers used specifically as illuminators for the imaging device. The latter systems are called active, while those relying largely on emitted radiation are called passive. In the past year striking advances have been made in active infrared systems which utilize the coherence available from lasers, while hybrid active-passive systems are being studied intensively.

Although developed largely for military purposes, infrared imaging devices have been valuable in industrial, commercial, and scientific applications. These range from nondestructive testing and quality control to earth resources surveys, pollution monitoring, and, more recently, energy conservation. Infrared images from aerial platforms are now used to accomplish "heat surveys," locating points of excessive heat loss. An example is shown in Fig. 1*a*. As discussed below, calibration allows association of photographic tones in this figure with values of apparent (that is, equivalent blackbody) temperatures. Dark areas in the figure are "colder" than light ones.

Scanning systems. Infrared imaging devices may be realized by electrooptical or optomechanical scanning systems. All have an optical head for receiving the infrared radiation and a display for the visible image. These are connected by electronics for the passage of electrical signals from the detector element(s) to the display input. Signal processing may be incorporated in the electronics to selectively enhance or reduce features in the produced visible image. For example, in Fig. 1*b* a "level-slice" technique presents in white all areas (mainly rooftops) with apparent temperatures between −7.9 and −8.9°C. (The ambient air temperature was −5°C.) Black regions in the figure correspond to apparent temperatures below or above the narrow "sliced" temperature range of the white regions.

Optomechanical methods such as rotating prisms or oscillating mirrors may be used to sample or scan the spatial distribution of infrared radiation in either the object or image plane. Electrooptical imaging devices may use an electron beam (for example, vidicons) or charge transport in solids (for example, charge-coupled devices, or CCDs) to scan the infrared image formed by the optics of the device. This image-plane scanning places more stringent requirements upon the optics for image quality off-axis than does use of mechanically moved optical elements placed before the entrance aperture of the system. Intensive development of pyroelectric vidicons, detector arrays, and infrared charge-coupled devices (IRCCDs) has taken place in recent years. This activity reflects the critical role played by the detector element in all infrared systems. The spectral, spatial, and temporal responses of detectors are the major factors in determining the wavelength regions, the spatial resolution, and the frequency response (that is, the time constant) of imaging devices. *See* TELEVISION CAMERA.

Detector arrays. Optomechanical scanning methods often stress the response time of the detector-electronics package. As a result, multiple detectors or detector arrays have begun to be incorporated in the focal planes, resulting in partially electronically scanned optomechanical systems. While the technology for use of a linear array of detector elements (most often lead selenide and indium antimonide detectors for the 3–5-μm region, and mercury-doped germanium and mercury cadmium telluride for the 8–14-μm window) is well developed, future advances are being sought by use of a two-dimensional array or matrix of detectors. Optomechanical imagers incorporating such arrays allow the use of time delay and integration (TDI) of the signals to improve the resulting signal-to-noise ratios.

Solid-state components such as CCDs afford the opportunity for implementation of signal processing directly at the focal plane. Two approaches are being taken to attain the focal-plane array technology of IRCCDs. In one, the development of a hybrid device, an infrared detector matrix of any suitable photodetector material, for example, indium antimonide, mercury cadmium telluride, and lead tin telluride, is mated with a conventional silicon CCD. Thus two solid-state wafers or "chips" are integrated to obtain an IRCCD. In the other, the goal is a monolithic chip, one incorporating the photodetection, charge generation, and charge transfer in a structure made of one material. Candidate materials include impurity-doped silicon, indium antimonide, and mercury cadmium tellu-

Fig. 1. Thermal imagery in the wavelength range 10.4–12.5 μm obtained during flights over Ypsilanti, MI, at 2400 hours, Nov. 23, 1975, by the Airborne Multispectral Scanner operated by the Environmental Research Institute of Michigan. *(a)* Calibrated thermal imagery. *(b)* Signal-processed thermal imagery of same scene. *(From F. Tanis, R. Sampson, and T. Wagner, Thermal imagery for evaluation of construction and insulation conditions of small buildings, Environmental Research Institute of Michigan, ERIM Rep. 116600-12-F, July 1976)*

ride. The hybrid device technology seems closer at hand than that needed for monolithic IRCCDs. The development of IRCCDs with the number of detecting elements in a sufficiently closely packed array required for high-performance infrared imaging devices is still further away.

Scanning motion. Some optomechanical imagers produce a two-dimensional scan entirely by movement of components of the device itself; others utilize the motion of a platform such as an aircraft or satellite. The first kind of system includes the forward looking infrared (FLIR) or framing imagers which usually scan in television-like raster patterns and display, synchronously if done in real time, a visible image corresponding to the spatial distribution of infrared radiation. These visible image outputs have been named thermographs and thermograms. Commercially available imaging devices of this type have used horizontally and vertically rotating germanium prisms, mirrors oscillated in orthogonal directions, two mirrors and a six-sided rotating prism, and other schemes to produce images at relatively slow rates from 16 per second to less than a quarter of a frame per second. Higher-performance systems have been produced for military purposes. The second class of imaging systems includes those often called line scanners or thermal mappers. One such system, the Airborne Multispectral Scanner operated by the Environmental Research Institute of Michigan (ERIM), is shown in Fig. 2. The ERIM 12-channel multispectral scanner included two thermal radiation channels, at 8.2–9.3 μm and 10.4–12.5 μm, whose magnetic-tape recorder output was processed to produce the thermal imagery in Fig. 1.

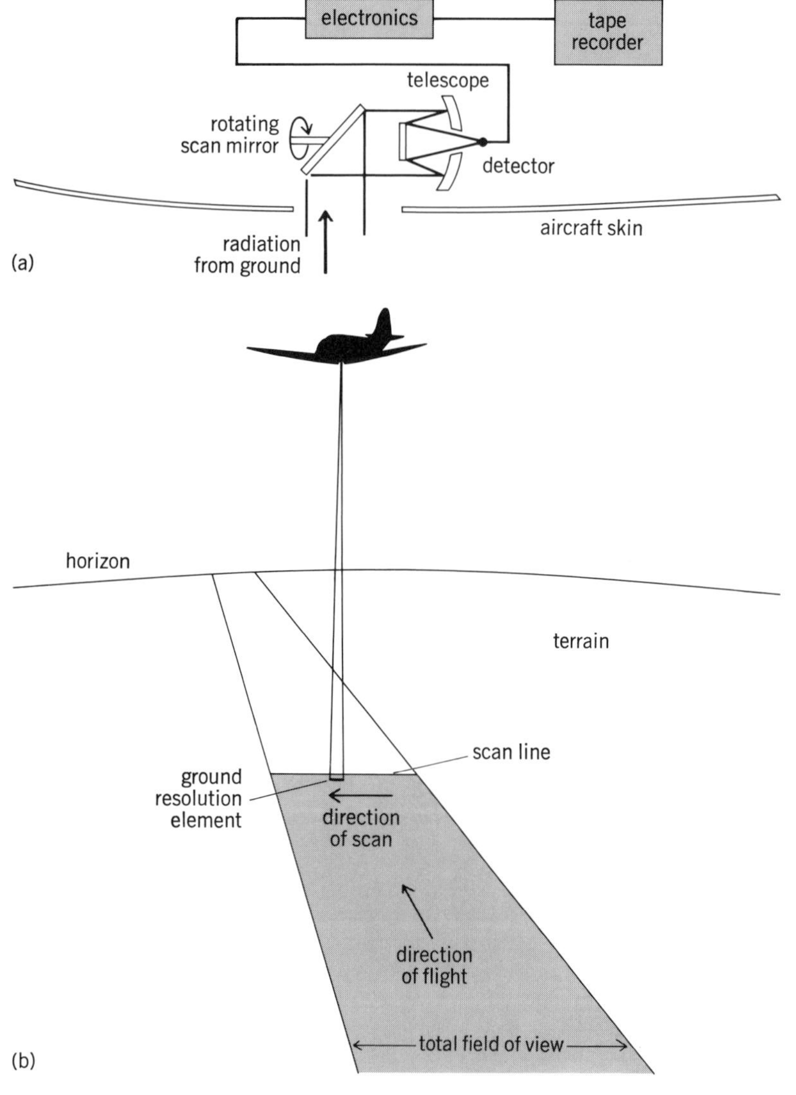

Fig. 2. Airborne Multispectral Scanner. *(a)* Schematic diagram of equipment. *(b)* Scanning operation, utilizing motion of aircraft. *(From F. Tanis, R. Sampson, and T. Wagner, Thermal imagery for evaluation of construction and insulation conditions of small buildings, Environmental Research Institute of Michigan, ERIM Rep. 116600-12-F, July 1976)*

Characterization of output. The instantaneous field of view (IFOV) or resolution element of imaging systems is always geometrically filled by the radiating source, so that the output of the device is a response to changes in amount of radiation from field of view to field of view. These changes are best characterized in terms of radiance L, the radiant flux per unit area per unit solid angle, usually in a selected spectral band of wavelengths, $\Delta\lambda$. Even in the infrared regions, the radiance variation may be ascribed to changes in reflectance, incident radiation, emissivity, or temperature. By restriction to wavelengths dominated by emission, the so-called thermal wavelengths longer than 3.5 μm, the radiance change can be described by the equation below, where T is the absolute tempera-

$$\Delta L = \frac{\partial L}{\partial T}\Delta T + \frac{\partial L}{\partial \epsilon}\Delta\epsilon$$

ture and ϵ is the emissivity. Contributions due to $\Delta\epsilon$ are usually treated as changes in an equivalent blackbody temperature by setting $\epsilon = 1$ and $\Delta\epsilon = 0$. Then T represents an equivalent blackbody temperature, and the radiance variation can be ascribed entirely to a value of ΔT. That value of ΔT corresponding to a radiance difference which will just produce a signal-to-noise ratio of 1 is called the noise equivalent temperature difference (NETD). One can also characterize the performance of an imaging system by a noise equivalent emissivity difference or even a noise equivalent reflectivity difference. The use of NETD as a figure of merit for thermal imagers is obviously more appropriate. For the higher-performance FLIRs, a useful figure of merit is the minimum resolvable temperature difference (MRTD), a figure of merit which includes the characteristics of the display and of the observer as well.

Display. The visible image which is the output of infrared imaging devices may be displayed in the same manner as a conventional television picture by means of a cathode-ray tube (CRT). Suitable CRT technology for most purposes is at hand now, but current research is directed toward creation of satisfactory flat panel displays using liquid crystal elements, light-emitting diodes, or plasma panels. Systems not requiring a real-time image display may utilize analog or digital data storage or transmission systems, which then are used to produce permanent visual records such as photographs. High-resolution "hard copy" images can be produced by sophisticated systems using electron-beam or laser recording on film. Complex signal-processing techniques are easily introduced before the final image recording is made.

For background information *see* EMISSIVITY; HEAT RADIATION; INFRARED IMAGE CONVERTER TUBE; INFRARED RADIATION in the McGraw-Hill Encyclopedia of Science and Technology.

[GEORGE J. ZISSIS]

Bibliography: J. M. Lloyd, *Thermal Imaging Systems*, 1975; *Proceedings of the IEEE: Special Issue on IR Technology for Remote Sensing*, vol. 63, pp. 1–208, January 1975; F. Tanis, R. Sampson, and T. Wagner, *Thermal Imagery for Evaluation of Construction and Insulation Conditions of Small Buildings*, Environmental Research Institute of Michigan, ERIM Rep. 116600-12-F, July 1976; W. Wolfe and G. Zissis (eds.), *The Infrared Handbook*, Infrared Information and Analysis Center, Environmental Research Institute of Michigan, 1978.

Integrated circuits

The vertical metal oxide–semiconductor field-effect transistor (MOSFET) has, during the past year, undergone several improvements which are expected to make it more attractive to larger segments of industry. Silicon-gate technology has in some cases supplanted the older metal-gate structure. Higher breakdown voltages exceeding 450 V now offer direct "off-line" operation without the need for either isolation or voltage-reducing transformers. Higher operating currents coupled with improved heat dissipation ratings allow direct logic control of power systems heretofore not easily implemented. Unquestionably, the most outstanding advance in vertical metal oxide–semiconductor (VMOS) technology has been the remarkable reduction in ON resistance due, in part, to the newer silicon gate, which now appears to place VMOS in direct competition with the power silicon bipolar transistor's low saturation voltage. *See* SEMICONDUCTOR DEVICES.

VMOS, which uses the principle of an anistropically etched V groove gate, was introduced by the Japanese in the late 1960s. However, American industry refined the technology, and by late 1976 both discrete power transistors and high-density semiconductor memories appeared on the market. Figure 1 compares the cross section of the discrete power transistor with that of the high-density memory, the subtle, yet fundamental differences in process technology.

Thermal and switching characteristics. In addition to the high packing density of VMOS for large-scale integration (LSI), its fundamental differences reside in its thermal and switching characteristics. Power bipolar transistors have a regenerative thermal effect; that is, heating tends to cause additional heating leading to catastrophic

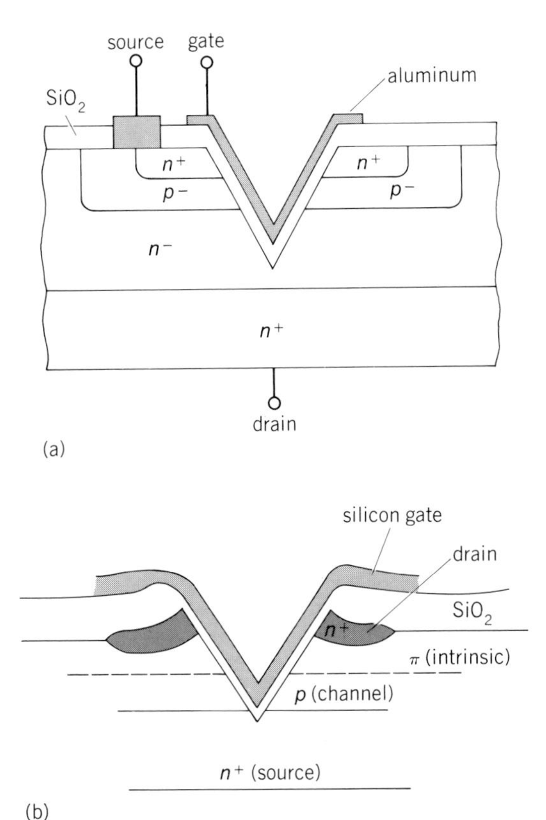

Fig. 1. Cross section of *(a)* discrete VMOS power transistor and *(b)* high-density VMOS memory.

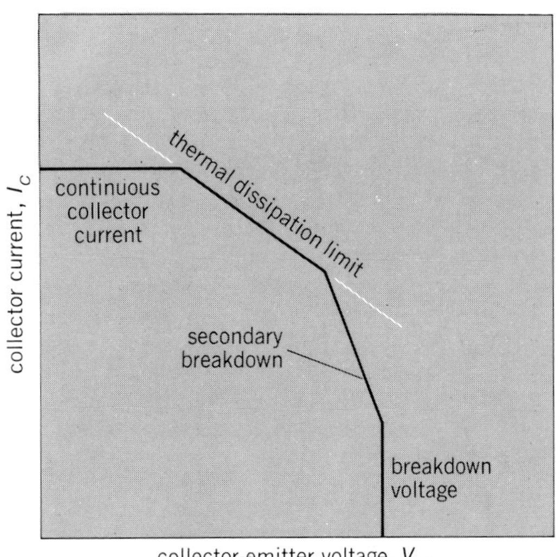

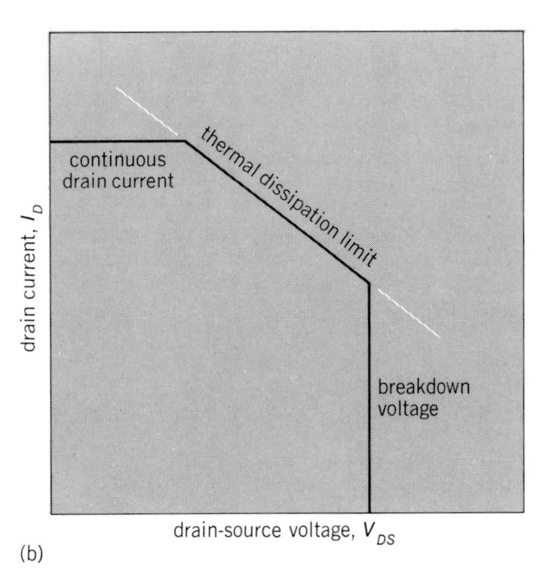

Fig. 2. Graphs showing safe operating area of *(a)* bipolar power transistor and *(b)* VMOS power transistor.

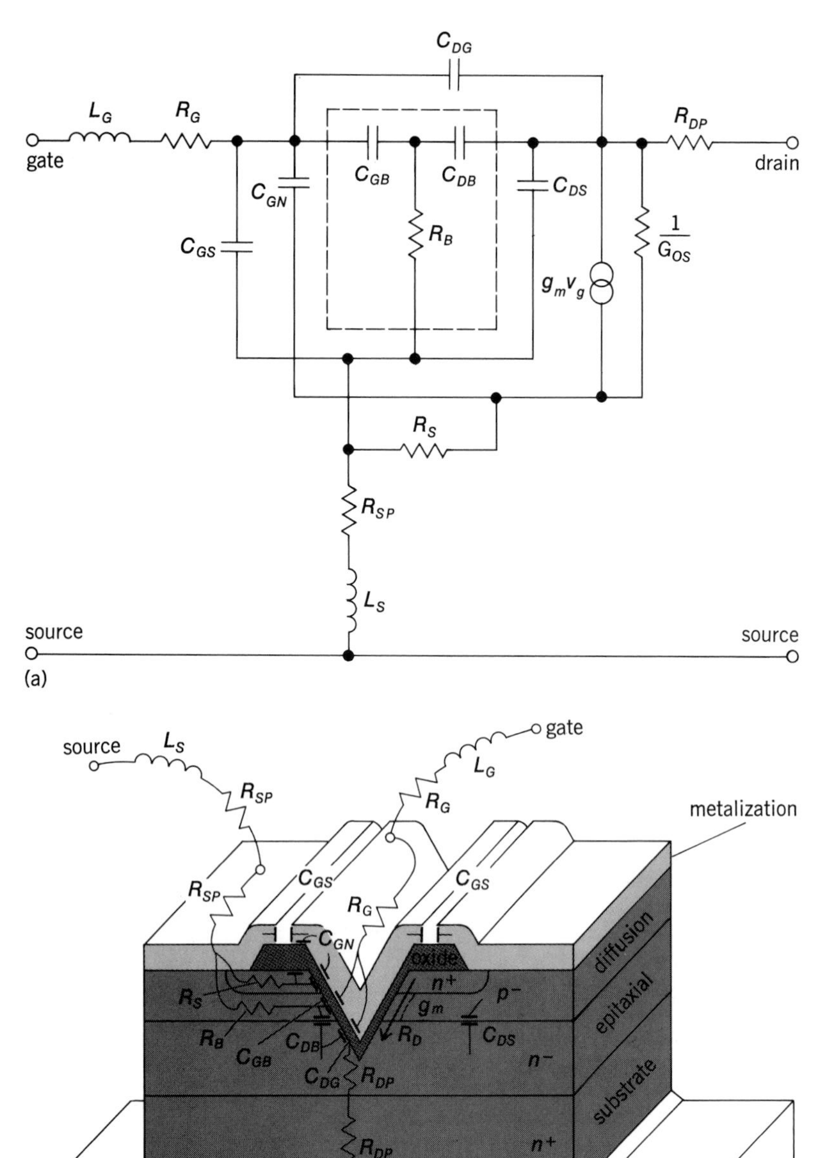

Fig. 3. VMOS power transistor, showing parasitic elements which affect high-frequency performance. *(a)* Equivalent circuit. *(b)* Sectional model. Here g_m is the forward transconductance of the VMOS; $g_m v_g$ is the forward transconductance times the analogy of the signal voltage impressed on the gate.

destruction. VMOS, on the other hand, exhibits a degenerative thermal characteristic, offering what appears to be vastly improved reliability.

VMOS, whether used as a discrete power transistor or in high-density LSI memories, relies upon the negligible storage time fundamental to all majority-carrier field-effect semiconductors. Switching can be executed in only a few nanoseconds ($3 \cdot 10^{-9}$ s). Another important characteristic of VMOS is its higher input impedance and very low drive current requirement that offers direct compatibility to many forms of logic drive circuitry such as transistor-transistor logic (TTL) and complementary MOS (CMOS). The β, or current gain, of VMOS far exceeds that of the best bipolar transistor and even surpasses the higher-performance Darlington pair. The β of a discrete VMOS transistor can exceed 900,000, while the β of the highest-performing Darlington can reach only 40,000. Logic compatibility gives VMOS the advantage of direct interface with the rapidly growing microprocessor market, providing, in effect, direct microprocessor control of power circuits. *See* MICROCOMPUTERS.

The outstanding operational reliability of VMOS, which stems from its degenerative thermal characteristic plus the exhibition of negligible storage time, has opened a great number of unique applications requiring high-speed switching and involving reactive loads. Those areas finding wide use of VMOS discrete power transistors include switching regulator power supplies, high-efficiency audio and radio-frequency amplifiers, as well as high-frequency amplifiers in general.

Audio amplifiers. VMOS provides several distinct advantages over the bipolar power transistor in both the more conventional high-fidelity audio amplifier and the high-efficiency switch-mode audio amplifier. Because of its degenerative thermal characteristic, there is no secondary breakdown phenomenon associated with VMOS, which allows VMOS power transistors to drive highly reactive loads such as loudspeakers without fear of catastrophic destruction of the transistors. Figure 2 compares the safe operating areas of bipolar and VMOS power transistors. For switch-mode (class D) audio power amplifiers, the improved safe operating area of the VMOS power transistor offers meaningful protection during the very brief cross-conduction period when, for an instant of time, both transistors in a push-pull configuration are conducting, causing, in effect, a short-circuit to the power supply. Such a condition is easily tolerated by VMOS power transistors, whereas elaborate protection circuitry is generally required for bipolar transistors to prevent catastrophic destruction. Bipolar power transistors generally exhibit a drive-dependent gain and frequency response which inhibit performance in a high-quality, high-fidelity audio amplifier, resulting in various forms of distortion. One of the several distinct advantages of the VMOS power transistor is that the β is not variable or drive-dependent, that is, the forward transconductance of VMOS is extremely constant above a minimum well-defined bias current. This singular advantage of a nonvariable β offers exceptional frequency response and excellent linearity at high frequencies when used in the power stages of audio amplifiers.

Another debilitating effect often observed in an all-solid-state high-fidelity audio amplifier is "hang-up," which is caused by the minority-carrier storage time characteristic of power bipolar transistors used in the output stages. VMOS, which has no minority-carrier storage time, exhibits no such effect and this property, coupled with the advantages mentioned above, enables it to produce a more pleasing acoustical effect for the critical listener.

Switching power supplies. For many of the same reasons that VMOS power transistors are advantageous to audio amplifiers, they also offer improvement in high-efficiency switching power supplies. The VMOS discrete power transistor has two fundamental properties that make it exceptionally suited to switching power supplies: the absence of both secondary breakdown and minority-carrier storage time. The absence of secondary breakdown allows the VMOS power transistor to perform reliably in highly reactive load environments as presented by the filtering networks; the absence of minority carrier storage time offers much higher switching speeds (that is, higher frequencies) which, in turn, reduce both the complexity of the filtering and its reactance, thus contributing to the overall reliability of both the VMOS power transistor and the power supply itself. The high gate impedance intrinsic to VMOS greatly simplifies the drive circuitry. With input drive currents in the low-nanoampere or picoampere range, direct CMOS drive is easily accomplished.

Radio-frequency amplifiers. The advantages of VMOS power transistors described above are also of significance in the design of reliable radio-frequency amplifiers. In limited situations, merely repackaging the VMOS semiconductor in an appropriate high-frequency package has provided unique performance generally unattainable with an equivalent-rated high-frequency bipolar transistor.

A high-frequency equivalent circuit and a sectional model (Fig. 3) show the important parasitic elements that affect high-frequency performance of the VMOS power transistor. The equivalent circuit (Fig. 3*a*) is basically similar to those previously modeled for VMOS, but with an important addition in body effect, shown in the dashed area. With the sectional model as a guide (Fig. 3*b*) one can identify the parasitic distributed drain-to-body capacity C_{DB}. This property, coupled with the gate-to-body capacity C_{GB} and the P^- body resistance R_B can cause a debilitating feedback, resulting in low forward power gain unless design precautions are taken in the initial VMOS transistor development.

Because of the absence of secondary breakdown, VMOS power transistors are far more tolerant of mismatched loads than are the typical high-frequency bipolar transistors. In addition, because the parasitic feedback capacitance of VMOS structure (C_{DG} in Fig. 3) has a low Q, spurious out-of-band responses are exceptionally low. As a result, high-frequency VMOS power transistors tend to offer improved performance.

Because they are not subject to minority-carrier storage, high-frequency VMOS power transistors allow the practical design of high-efficiency switch-mode radio-frequency amplifiers. Switch-mode amplifiers are generally identified as class D, E, and F, and have the capability of attaining a theoretical efficiency of 100%. This theoretical efficiency is not possible in practice—in bipolar transistors, because of their storage time and saturation voltage, and in VMOS, because of the finite ON resistance inherent in the structure. With the remarkable reductions in ON resistance of VMOS, the VMOS high-frequency power transistor will be the bulwark for future switch-mode designs, both for high-frequency applications and in any application where very high efficiencies are required.

Summary. The upper frequency limits of VMOS will probably settle in the low gigacycle area, development models having performed satisfactorily at 1.5 GHz. Power levels on the order of several hundreds of watts are clearly attainable. Because of the degenerative thermal characteristic of VMOS, device paralleling is easily accomplished to raise the current capabilities and power handling.

For background information *see* INTEGRATED CIRCUITS in the McGraw-Hill Encyclopedia of Science and Technology.

[EDWIN S. OXNER]

Bibliography: A. D. Evans et al., *Electronics*, vol. 51, pp. 105–112, June 22, 1978; E. Oxner, *EDN*, vol. 22, pp. 71–75, June 20, 1977.

Ion

In the relatively cold terrestrial environment, atoms and molecules are most often found in their electrically neutral, or at most singly charged (ionized), states. A far different situation prevails in stellar atmospheres, such as that of the Sun, and in similar artificial thermonuclear plasmas. There temperatures can reach 10^8 K, and highly charged ions (for example H^+, He^+, . . . , Fe^{24+}, . . .) in equilibrium with a compensating supply of free but hot electrons are the normal state of matter (sometimes called the fourth state).

Violent collisions. With the advent of heavy-ion accelerators that generate very highly ionized projectiles, and of other sources of multiply charged ions which can be used to produce slower projectiles (though in somewhat lower charge states), it is now possible to suddenly introduce into target atoms and molecules a high charge whose effects on the system electrons exceed those of the target atomic centers. Production of simultaneous many-electron transition phenomena is extremely probable under these conditions, requiring a new theoretical description wholly outside the framework of traditional single-electron perturbation theory. For example, in recent experiments carried out by a group of university scientists led by I. A. Sellin at the Oak Ridge National Laboratory, it was found that simultaneous 10-electron excitation and ionization of neon atoms struck by bare argon nuclei were probable events at distances of closest approach at which normal one-electron transitions would be expected to occur.

Analogy to solar system. Since atomic nuclei and electrons interact through the $1/r$ Coulomb potential (where r is the distance between the interacting particles), an analogy can be made to the solar system, whose Sun and planets also interact through a $1/r$ (but purely attractive) potential. A large passing comet might be expected to significantly perturb but not radically change the orbit of, say, Venus in a near-miss collision. By contrast, one can imagine what might happen if a very fast star appreciably more massive than the Sun were to pass through the solar system at a distance from the Sun of roughly one Earth orbital radius over a period of a few days. Violent and strongly correlated rearrangements of the planetary orbits might then be expected to occur.

Thus far, few experiments directly reveal the fate of any of the electrons that experience such violent interactions in the ion-atom collision case. The extent to which the electrons go into unbound states of the target atom (direct ionization) or of the projectile (capture into the continuum), or into higher-lying, previously unoccupied bound states of the target atom (excitation), or into bound states of the projectile (electron capture), has not been determined. The number of multiple-electron transitions produced is too great to be accounted for solely by any one of these mechanisms as described in conventional models.

New theoretical models. Very recently the rudiments of two new, nonperturbative models have been put forward. Both models are meant to apply when the projectile velocity v_p is greater than the target K (innermost) electron orbital velocity $v_K = Z_T v_B$, where Z_T is the atomic number of the target and v_B is the Bohr velocity ($v_B = \alpha c$, with α representing the fine-structure constant and c the speed of light).

In the model of C. Bottcher, the highly charged projectile ions can escape with most of the system electrons, which become long-term fellow travelers of the projectile as it moves outward from a collision, but are not attached to it in the sense of occupying bound orbits. Many-electron capture into continuum states of the projectile (that is, those ionization events for which the emitted electron's laboratory velocity $\vec{v}_e$ is approximately equal to $\vec{v}_p$) is an appropriate description of what Bottcher asserts to be the dominant process. The importance of this process for single electrons was realized a few years ago. While theory and experiment are sometimes in reasonable agreement for light projectiles, there are areas of significant disagreement for highly charged projectiles.

A second new approach has been used by J. Eichler, who has concentrated on a nonperturbative but otherwise more traditional description in terms of ionization of single electrons into the target continuum.

However, a recent experiment at Oak Ridge involving passage of highly ionized sulfur and chlorine ions through neon gas shows that multiple excitation of target electrons into higher, previously unoccupied bound orbits of the target atom is also extremely probable, indicating that multiple-target electron excitation provides an important if not dominant contribution to the multiple-electron rearrangement process, which may not be ignored compared to other processes.

In the solar system analogy, a composite description may well be required in which many planets end up in orbits larger than, say, that of Uranus, others depart in many different directions into outer space, still others drift off with the same speed and in the same direction as the departing projectile star, and some become its planets.

Unbound projectile-centered states. Perhaps the most controversial and incompletely considered element in the theories discussed thus far is the role of electron capture in the projectile-centered continuum. There is only moderate agreement between theory and experiment for the lightest projectiles, and there are striking discrepancies between theory and experiment for more highly charged, bare projectiles such as C^{6+}, O^{8+}, and Si^{14+}.

For example, theory predicts a Z^3 dependence on projectile atomic number Z for the yield of unbound projectile-centered electrons, for an incident bare nucleus at a given velocity. The Oak Ridge group did not find this dependence in the observed yield, but rather a weaker, $Z^{\sim 2.2}$ dependence. For incident projectiles possessing bound electrons just prior to a close collision, it is also easily possible to reach such projectile-centered unbound states by a simple excitation as opposed to capture process. Interesting comparisons of the rates for capture and loss into such states are just beginning.

Experiments with thin solid targets. It is also possible to study formation of unbound projectile-centered states when similar, highly ionized projectiles pass through thin solid targets a few hundred atoms thick, because of the much larger supply of electrons in such a macroscopic target. It has been of much recent interest whether electron "wakes" can form behind such projectiles. If so, crests and troughs might develop in the electron "sea" behind the projectiles. Whether a few electrons can ride along at the same speed as the projectile in the manner of a surfer riding directly in front of a wave crest is a hotly debated current question. At present there is inadequate agreement between speculative theoretical predictions of the properties of such wakes and corresponding experiments.

For background information *see* ATOMIC STRUCTURE AND SPECTRA; ION; ION SOURCES; SCATTERING EXPERIMENTS, ATOMIC AND MOLECULAR in the McGraw-Hill Encyclopedia of Science and Technology.

[I. A. SELLIN]

Bibliography: C. Bottcher, *J. Phys. B*, 10: L445–451, 1977; J. Eichler, *Phys. Rev. A*, 15: 1856–1861, 1977; J. R. Mowat et al., *Phys. Rev. Lett.*, 30:1789–1792, 1973; R. H. Ritchie, W. Brandt, and P. M. Echenique, *Phys. Rev. B*, 14: 4808–4812, 1976; I. A. Sellin and J. R. Mowat, *Comments on Atomic and Molecular Physics*, 1978.

Laser

A new process has been developed for the surface treatment of metals and alloys. Named Laserglaze, it involves the use of the high power densities characteristic of modern lasers to produce rapid surface melting which, when properly performed, is followed by extremely rapid solidification and subsequent solid-state cooling. Ideally, surface melting should be carried out rapidly, and at high power density, so as to have a maximum fraction of the absorbed energy used in melting (deposited within the liquid layer) and a minimum of the energy conducted into the solid substrate material. This condition produces a cold self-substrate and results in maximum thermal gradients at the solid-liquid interface, which in turn produce rapid solidification and solid-state cooling.

Laser processing techniques. The process of Laserglazing can be placed in perspective with regard to other techniques for laser processing of materials by referring to the graph of laser power density as a function of interaction time shown in Fig. 1. The various shaded regions represent the "operational regimes" for Laserglazing and other experimental and commercial laser materials processing techniques. The boundaries of the regions are necessarily approximate. The diagonal lines represent constant specific energy inputs in joules per square centimeter.

Typically, the highest power densities are applied for the shortest interaction periods. At lower power densities, longer interactions can be tolerated by the material. Typical transformation hardening of the surfaces of martensitic materials is usually accomplished in 10^{-2} to 1 s, at power densities of 10^3 to 5×10^4 W/cm². If the power density is raised to 10^5 to 10^7 W/cm² for periods of approximately 10^{-3} to 10^{-2} s, a vapor column (deep-penetration cavity) is created in the material. This cavity is surrounded by a liquid pool in equilibrium which, when translated through the work, has been used to produce high-quality fusion welds. At the higher range of welding power densities, approximately 10^6 to 10^8 W/cm², which is the limit of current-technology continuous high-power CO_2 lasers, shorter interaction times prevent the onset of deep penetration and produce shallow surface-localized fusion zones such as the featureless zones in Fig. 2. These zones are Laserglazed material. While the effects of the process are visually obvious in Fig. 2, the metallurgical implications of the Laserglaze technique will be discussed in more detail later.

Laserglazing apparatus. Figure 3 is a schematic diagram of the Laserglazing apparatus. The specimens to be surfaced are placed around the outer circumference of a rapidly rotating horizontal wheel and are adjusted in height so that the surface of the specimen is coincident with the focal point of a continuous high-power CO_2 laser beam which is held stationary. By rotating and translating the wheel beneath the beam, specimens can be surfaced at rates typically up to 1000 cm²/s at laser beam powers in the 3-kW range. Higher power will increase surfacing speed capability proportionately. The laser beam passes through a hole in an inert gas shield, also shown in Fig. 3. The purpose of this shield is to protect the molten layer from atmospheric contamination and to control the formation of energy-absorbing plasmas at the interaction point. If excessive plasma is allowed to form, it will diffuse the energy density and thus reduce the cooling rate.

Rapid solidification and cooling. The single most important problem in the primary fabrication of metallic alloys is phase separation and segregation that occurs in the slow cooling of multicomponent systems from the melt. The structures so produced are nonuniform, and therefore tend to be variable in properties. This segregation can be prevented by extremely rapid solidification and solid-state cooling, in which the near-perfect homogeneity of the liquid state is preserved in the solid. Current state-of-the-art techniques for rapid solidification and cooling, such as splat-quenching, melt-quenching, and atomization, all involve the quenching of liquid on a foreign substrate or chill, or into a gas. The interfaces present generally limit the controllability and reproducibility of the effect. The high available energy density and precise energy control which modern lasers provide has permitted the development of the Laserglaze technique, which has rapidly established itself as the most convenient and reproducible technique for producing and studying rapid chilling effects in metallic alloy systems. The high cooling rates obtained with Laserglaze are difficult to measure, but they can be accurately predicted by mathematical analyses, as described by E. M. Breinan and coworkers. By specifying the power density and interaction time or the total absorbed energy and absorption time, the melt depth can be accurately predicted if the thermal properties of the alloy are known. Once the melt depth and the power density are specified, the cooling rate is also calculable.

Structures produced by Laserglazing. As shown in Fig. 2, a variety of unusual microstructures have been produced by the Laserglaze process. The initial question of whether an amorphous (noncrystalline) metal surface layer can be pro-

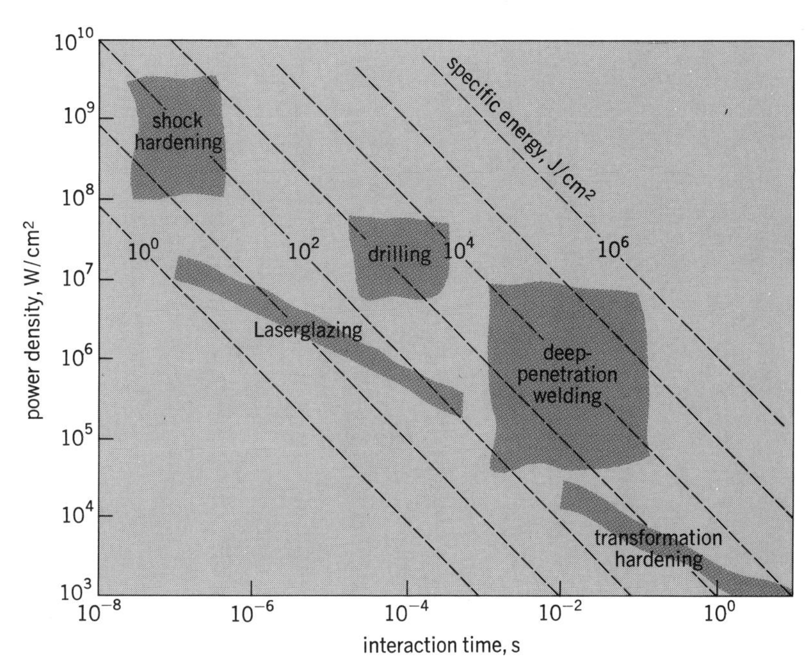

Fig. 1. Operational regimes for various laser materials processing techniques. Processes are represented in terms of the applied power density and interaction time necessary to achieve the desired effect.

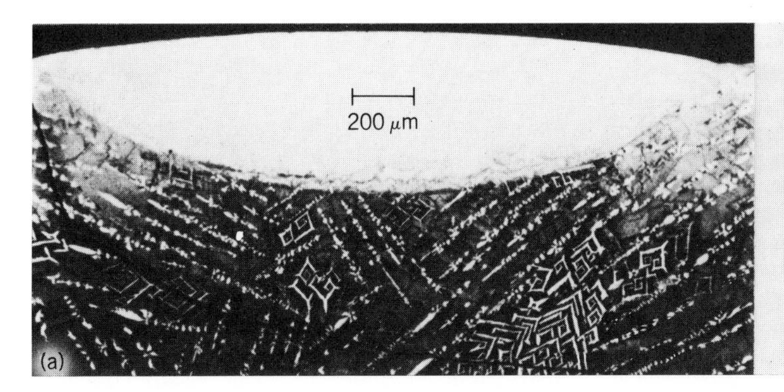

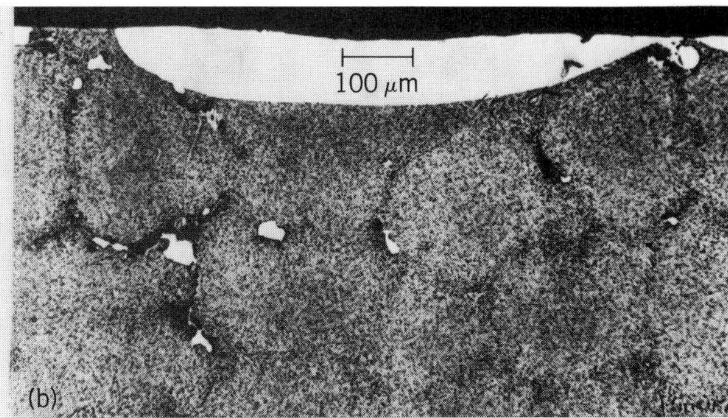

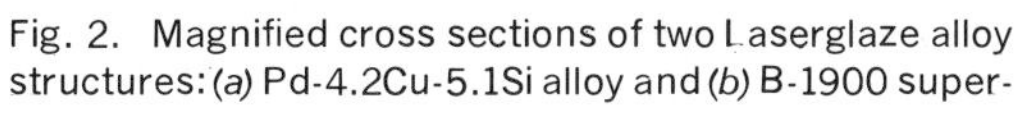

Fig. 2. Magnified cross sections of two Laserglaze alloy structures: *(a)* Pd-4.2Cu-5.1Si alloy and *(b)* B-1900 superalloy. The featureless zones are Laserglazed material. *(United Technologies Research Center)*

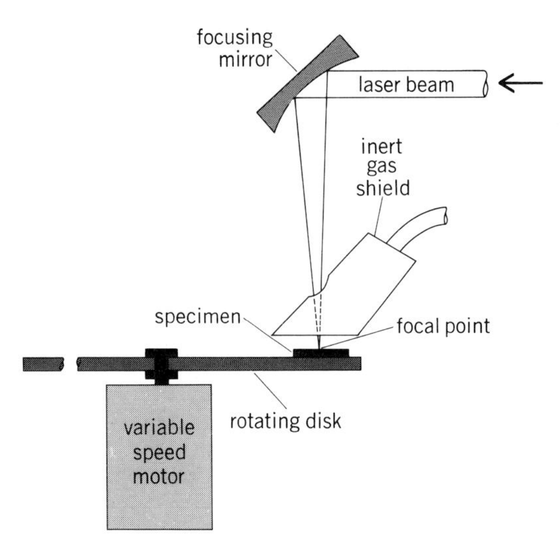

Fig. 3. Schematic diagram of Laserglaze processing apparatus.

duced on bulk crystalline substrates of the same composition, avoiding the tendency for homogeneous nucleation of the substrate, was answered with the production of a completely homogeneous and amorphous Laserglaze pass on a Pd-4.2Cu-5.1Si alloy (Fig. 2*a*). The Laserglaze pass was verified to be amorphous by electron diffraction as well as by fractography and observations of deformation mechanisms. The thorough homogenization of an advanced superalloy, B-1900, is shown in Fig. 2*b*. In this case, the structure is not amorphous, but is a homogeneous crystalline solid solution which is supersaturated, and therefore metastable. Other structures produced by Laserglazing include refined dendritic structures (typically solid solutions, though in some cases containing precipitates), ultramicrocrystalline structures, ultrafine and enriched eutectic structures, and changes in the normal morphologies of multiphase and eutectic microstructures.

Implications of the technology. Laserglazing has obvious advantages as a tool for modification and control of surface microstructures. Surface-controlled properties such as resistance to erosion, corrosion, wear, and fatigue can all be improved by Laserglazing. The ability to produce metastable, supersaturated solid solutions which can be aged to produce high-strength microstructures (strengthened with precipitates or dispersions) with good ductility has been demonstrated. Production of such structures will permit an increased efficiency of utilization of properties since smaller maximum flaw sizes can be used for design purposes. Finally, variation of the Laserglaze process, termed Layerglaze, is being developed to produce sequentially built-up bulk rapidly chilled structures. This promises to extend the ability to produce and control microstructure to bulk parts. In short, the metallurgy of rapid solidification, in conjunction with lasers as an advanced processing tool, has created new possibilities for the production and precise control of microstructures in metallic alloys.

For background information *see* LASER WELDING; SUPERSATURATION in the McGraw-Hill Encyclopedia of Science and Technology.

[EDWARD M. BREINAN]

Bibliography: E. M. Breinan et al., *A New Process for Production and Control of Rapidly-Chilled Metallurgical Microstructures*, SME Pap. no. MR76-867, 1976; E. M. Breinan, B. H. Kear, and C. M. Banas, *Phys. Today*, 29(11):44–50, November 1976.

Laser chemistry

Many chemical reactions do not occur spontaneously at room temperature. Before the development of lasers, the microscopic barrier to a reaction could be overcome either by heating the reactants or by irradiating them with visible and ultraviolet radiation. In the former case thermal energy is supplied randomly to the reactants, while in the latter method the reactants are excited electronically (that is, photochemically) and may yield products different from those of the thermal reaction. The new field of infrared laser-induced photochemistry offers a third means of inducing chemical reactions. When a molecule absorbs infrared radiation, it becomes vibrationally excited. Stretching and bending of the bonds can accelerate the reaction and lower the activation energy, while the reactants remain at room temperature in the ground electronic state.

One of the earliest examples of laser chemistry is the reaction of nitric oxide and ozone [reaction (1)].

$$NO + O_3 \rightarrow NO_2 + O_2 \qquad (1)$$

R. Gordon of the University of Illinois and M. C. Lin of the Naval Research Laboratory first discovered that excitation of ozone with a CO_2 laser can enhance the reaction rate by a factor of 10 at room temperature, while lowering the activation energy by 1.5 kcal/mole (6.3 kJ/mole). Since the infrared photon contains only 3.0 kcal/mole (12.6 kJ/mole) of energy, their results indicate that for this reaction vibrational energy is 50% effective in lowering the activation energy.

Types of lasers used. Because of its high efficiency and low cost, the CO_2 laser has been used for nearly all infrared laser-induced reactions reported to date. This laser emits radiation at wavelengths between 9.2 and 10.7 μm, corresponding to excitation energies between 3.1 and 2.7 kcal/mole (13.0 and 11.3 kJ/mole). In a few cases HF, HCl, and CO lasers have been used, with radiation at 2.4, 3.3, and 4.6 μm, or 11.8, 8.5, and 6.2 kcal/mole (49.4, 35.5, and 25.9 kJ/mole), respectively. Generally, pulsed lasers are used; however, in a number of cases continuously operating lasers have also been employed.

Multiphoton absorption. Since most chemical reactions have activation energies of several tens of kilocalories per mole, it would appear that infrared excitation would be of limited use. However, in 1973 R. V. Ambartzumian and V. S. Letokhov of the Institute of Spectroscopy in Moscow and J. L. Lyman and coworkers at the Los Alamos Scientific Laboratory discovered that if the laser beam is sufficiently intense (about 10^7 W/cm^2) some polyatomic molecules can absorb many photons in rapid succession. Such highly excited molecules are very reactive; in fact, if enough photons are absorbed, the molecule can rapidly dissociate into atoms and radicals. The most thoroughly studied example of multiphoton dissociation is the fragmentation of sulfur hexafluoride [reaction (2)],

$$SF_6 + 36h\nu \rightarrow SF_5 + F \qquad (2)$$

which can occur without any molecular lesions.

Selective excitation. One of the most useful properties of laser radiation is its sharp wavelength. This property makes it possible to selectively excite one species at a time in a reactant mixture. If the other molecules do not absorb the laser radiation, the temperature of the mixture need not rise appreciably, even though the excited compound may be in states that correspond to thermal excitations achieved only at temperatures of several thousand degrees. By conducting the reaction at room temperature, it is possible to avoid thermal decomposition of the products, thereby increasing the yield and suppressing unwanted side reactions. An interesting illustration recently reported by J. Steinfeld and coworkers at the Massachusetts Institute of Technology is the laser-induced reactions of chlorinated ethylenes. When *trans*-dichloroethylene is irradiated with a pulsed CO_2 laser, the only observed product is the cis isomer [reaction (3)]. The reverse reaction does

$$\text{(H)(Cl)C=C(Cl)(H)} \longrightarrow \text{(H)(Cl)C=C(H)(Cl)} \qquad (3)$$

not occur because the cis compound does not absorb CO_2 radiation. In contrast, when vinyl chloride is excited, HCl is eliminated in a concerted reaction producing acetylene [reaction (4)]. In

$$\text{(H)(Cl)C=C(H)(H)} \longrightarrow HC{\equiv}CH + HCl \qquad (4)$$

these and other examples the complicated mixtures of products produced in the thermal reactions are not observed.

Laser purification. A particularly promising application of laser chemistry is the purification of compounds. By selectively exciting an impurity, it is possible to destroy it without disturbing the compound of interest. Compounds which have been purified with lasers include SiH_4, BCl_3, and $AsCl_3$. *See* SOLAR CELL.

The ultimate example of laser purification is laser isotope separation. By tuning the laser to a wavelength absorbed by only one isotopic species, it is possible to remove that species selectively by either multiphoton dissociation or a vibrationally enhanced reaction. In the SF_6 example [reaction (2)], excitation of $^{32}SF_6$ results in the enrichment of $^{34}SF_6$, and vice versa. Elements which have been isotopically enriched by laser purification include H, B, C, N, O, Si, Cl, Br, S, U, Os, and many of the rare earths. The successful separation of uranium isotopes is of particular current interest.

Specificity of reaction mechanisms. One of the early expectations for laser-induced chemistry was the possibility of breaking specific bonds in complex molecules by selectively exciting the proper vibrational modes. However, because vibrational energy is rapidly redistributed in most molecules, the possibility of achieving such mode-selective chemistry in general has dimmed. Nevertheless, the specificity of mechanisms in laser-induced reactions [for example, reactions (3) and (4)] should be useful in synthetic chemistry. An interesting example reported by D. Brenner at Brookhaven National Laboratory is the multiphoton-induced dissociation of ethyl vinyl ether. When the compound was irradiated with a very short burst (2×10^{-7} s) of CO_2 laser radiation, a C-O bond was broken, producing radical products [reaction (5)]. On

$$CH_3CH_2{-}O{-}CH{=}CH_2 \longrightarrow \cdot O{-}CH{=}CH_2 + \cdot CH_2{-}CH_3 \longrightarrow \text{products} \qquad (5)$$

the other hand, when a longer laser pulse (2×10^{-6} s) was used, a concerted reaction took place, producing only acetaldehyde and ethylene [reaction (6)]. What is unusual is that the activation energy

$$CH_3CH_2{-}O{-}CH{=}CH_2 \longrightarrow O{=}CH{-}CH_3 + H_2C{=}CH_2 \qquad (6)$$

of reaction (5) is 20 kcal/mole (84 kJ/mole) higher than that of reaction (6). When the pulse duration is short (for the same total pulse energy), many photons are absorbed before molecular collisions can dissipate the absorbed energy, and the higher energy path is available.

Drawbacks. One drawback in laser-induced chemistry is that collisions between molecules compete with the reaction by converting internal energy to random thermal motion. Since the rates of such dissipative processes increase with the concentration of reactants, nearly all laser-induced reactions reported to date were carried out in the gas phase. An interesting exception reported by M. Poliakoff and coworkers at the University at Newcastle is the isomerization of $Fe(CO)_4$ carried out on low-temperature argon matrices, where molecular motion is restricted.

A second drawback is that only a limited number of molecules absorb radiation emitted by the intense infrared lasers currently available. A solution to this problem is to add an inert gas, which absorbs the laser radiation and can transfer its vibrational energy to the reactants through collisions. Examples of such sensitizing molecules are SF_6, SiF_4, and CH_3F. For example, E. Grunwald and coworkers at Brandeis University used SiF_4 to convert cyclopropane cleanly to propylene. In another instance, H. R. Bachmann and coworkers at Munich University used BCl_3 to trimerize tetrachloroethylene.

Use of visible and ultraviolet lasers. Thus far the discussion has been limited to the use of infrared lasers to stimulate chemical reactions. In addition, there have been a number of important developments utilizing visible and ultraviolet lasers. By placing the reaction vessel inside the cavity of a visible dye laser, M. Berry and coworkers at Allied Chemical Corporation were able to excite reactants to very high vibrational levels of the ground electronic state. They then used this technique to isomerize methyl isocyanide [reaction (7)]. In another example, P. Houston and C. B.

$$CH_3NC \rightarrow CH_3CN \qquad (7)$$

Moore of the University of California, Berkeley, used an ultraviolet laser to predissociate formaldehyde, producing carbon monoxide and hydrogen [reaction (8)]. By tuning the laser wavelength, it

$$CH_2O \rightarrow CO + H_2 \quad (8)$$

is possible to use this reaction to separate hydrogen, carbon, and oxygen isotopes.

The field of infrared laser-induced chemistry is still relatively new. The results thus far have shed much light on the kinetics and mechanisms of chemical reactions and promise many useful applications in the future.

For background information *see* ISOMERISM, MOLECULAR; LASER; MOLECULAR STRUCTURE AND SPECTRA in the McGraw-Hill Encyclopedia of Science and Technology. [ROBERT J. GORDON]

Bibliography: S. Kimel and S. Speiser, *Chem. Rev.*, 77:437–473, 1977; C. B. Moore (ed.), *Chemical and Biochemical Applications of Lasers*, vol. 1, 1976, vols. 2 and 3, 1977.

Lepton

In the past 3 years a great deal of experimental evidence has been accumulated to support the existence of a new, electrically charged elementary particle called the tau (τ) heavy lepton. The τ is called a heavy lepton because most of its properties are the same as the properties of the electron, the lightest lepton; however, the mass of the τ is 3500 times the mass of the electron. The τ decays spontaneously into other elementary particles such as electrons, muons, hadrons, and neutrinos. Its lifetime has been measured to be less than 10^{-11} s, but may be as short as 10^{-12} to 10^{-13} s. Since the τ is an unstable particle—unlike the electron, which is stable—it must be produced artificially. It was first discovered through reaction (1), in which a positron (e^+) and electron (e^-) anni-

$$e^+ + e^- \rightarrow \tau^+ + \tau^- \quad (1)$$

hilate and produce a pair of τ-leptons of opposite electrical charge. Reaction (1) is still the only method known for producing the τ in detectable quantities.

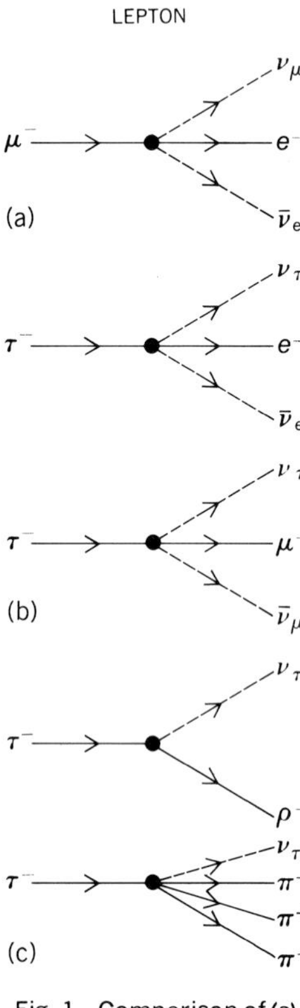

Fig. 1. Comparison of *(a)* the decay process for the muon, μ, and some of the *(b)* leptonic and *(c)* hadronic decay processes for the tau, τ. Solid lines indicate charged particles; dashed lines indicate neutral particles such as the neutrino, ν.

Leptons and hadrons. The significance of the discovery of a new, heavy lepton can be most easily understood by comparing the properties of leptons with the other family of elementary particles called the hadrons.

1. Size: Leptons are much smaller than hadrons. Hadrons have radii of about 10^{-13} cm, whereas the leptons are at least 100 times smaller.

2. Structure: According to present knowledge, a lepton is not composed of other particles, and it cannot be broken up or divided. Hence, the lepton is regarded as a true elementary particle. On the other hand, the hadrons have a complicated structure and appear to be composed of other particles called quarks. *See* QUARKS.

3. Forces: Closely related to the elementary nature of the leptons is the fact that they interact with themselves or other particles only through relatively mild forces or interactions: the electromagnetic force, the so-called weak interaction which leads to beta decay, and the gravitational force. The hadrons interact not only through these three mild forces but also through the nuclear force, also called the strong force. The strong force leads to complicated interactions between hadrons.

Since the hadrons are so complicated and since it is not known if quarks can be isolated, the leptons are the only truly elementary particles which can be isolated and studied directly.

Electrons, muons, and their neutrinos. The discovery of the τ is also important because the previously known leptons composed such a small family of particles (see the table). The previously known, electrically charged leptons are the electron *(e)* and muon (μ), and the previously known, electrically neutral leptons are the neutrinos associated with the e and μ. The masses of all these previously known leptons are less than the mass of the lightest hadron—the pion (π)—which is 140 MeV/c^2. It is these relatively low masses which led to the designation of these particles as leptons, which means "light ones" in Greek. Before the discovery of the τ one might have thought that all leptons had relatively small masses.

There is a unique neutrino-antineutrino pair (ν_e, $\bar{\nu}_e$) associated with the electron, and a unique neutrino-antineutrino pair (ν_μ, $\bar{\nu}_\mu$) associated with the muon. Furthermore, the electron subfamily of leptons (e^-, e^+, ν_e, $\bar{\nu}_e$) all possess a unique property called electron lepton number or, in common usage, "electronlikeness." This property is different from the analogous unique property of the muon subfamily of leptons (μ^-, μ^+, ν_μ, $\bar{\nu}_\mu$) called muon lepton number or "muonlikeness." These unique properties explain why the μ cannot change into an e through the simple electromagnetic reaction $\mu^\pm \rightarrow e^\pm$ + photon. In such a reaction, muonlikeness would have to change into electronlikeness. However, the $\mu^\pm$ does decay into the $e^\pm$ through the more complicated reactions (2), shown schematically in Fig. 1(*a*). In the decay

$$\mu^- \rightarrow \nu_\mu + e^- + \bar{\nu}_e \quad (2a)$$

$$\mu^+ \rightarrow \bar{\nu}_\mu + e^+ + \nu_e \quad (2b)$$

of the μ^- in reaction (2*a*), when the μ^- disappears, its muonlikeness property is transmitted to the ν_μ. In a similar manner, the creation of the e^- and its electronlikeness property is compensated for by the simultaneous creation of the $\bar{\nu}_e$ antineutrino and its anti-electronlikeness property. The μ^+ decay can be similarly understood. Therefore, in the rest of this article only the decays of negative leptons will be written explicitly.

Decays of the τ. Experimental studies of how the τ decays show that it behaves analogously to the μ. The electromagnetic decays $\tau^\pm \rightarrow \mu^\pm$ + photon or $\tau^\pm \rightarrow e^\pm$ + photon have not been found. However, the τ does decay through reactions (3) and (4), shown schematically in Fig. 1(*b*).

$$\tau^- \rightarrow \nu_\tau + \mu^- + \bar{\nu}_\mu \quad (3)$$

$$\tau^- \rightarrow \nu_\tau + e^- + \bar{\nu}_e \quad (4)$$

This leads to the conclusion that the τ carries a unique τ-lepton number or "taulikeness" property, and that there is a unique neutrino-antineutrino pair (ν_τ, $\bar{\nu}_\tau$) associated with the τ. At present the existence of unique neutrinos associated with the τ is not as well established as the existence of unique neutrinos associated with the electron or muon. Studies of reactions (3) and (4) have shown that the τ decays through the same weak interaction which causes the μ to decay [reaction (2)]. This is additional evidence that the τ is a lepton.

The mass of the τ is considerably greater than the mass of some of the lighter hadrons such as the π and the rho meson (ρ) which has a mass of 765 MeV/c^2. Hence, the τ should be able to decay into hadrons; indeed, several hadronic decay modes, such as reactions (5), (6), and (7), have been seen

$$\tau^- \rightarrow \nu_\tau + \rho^- \quad (5)$$

$$\tau^- \rightarrow \nu_\tau + \pi^- + \pi^+ + \pi^- \quad (6)$$

$$\tau^- \rightarrow \nu_\tau + \pi^- + \pi^+ + \pi^- + \pi^0 \quad (7)$$

(Fig. 1*c*). In each of these decays the taulikeness property of the decaying τ^- is carried off by the creation of the τ-neutrino (ν_τ).

Production of the τ. In reaction (1) the electron and positron annihilate each other and form a very small volume of electromagnetic energy technically called a virtual photon (Fig. 2). This virtual photon decays very rapidly into a large variety of elementary particles: several hadrons, a pair of muons, or an electron-positron pair, similar to the electron-positron pair which first annihilated into the virtual photon, may be produced. If the total energy of the initial electron-positron pair, E, is large enough, a pair of τ-leptons may be produced. Explicitly, E must be greater than the threshold energy $E_{\tau,\text{threshold}}$ given by Eq. (8). Here M_τ is the

$$E_{\tau,\text{threshold}} = 2M_\tau c^2 \quad (8)$$

mass of the τ, and c is the velocity of light. This is simply the Einstein energy-mass relationship multiplied by a factor of 2 because two τ's must be produced. Figure 3 shows how the rate of τ production, that is, the cross section for τ production, depends on E. Below $E_{\tau,\text{threshold}}$ there is no τ production: above $E_{\tau,\text{threshold}}$ the production rate follows one curve if the τ is a lepton with spin ½, and another curve if the τ is a hadron. In experiments the rate of τ production follows the lepton pair production curve, and this is one of the reasons why the τ is believed to be a lepton.

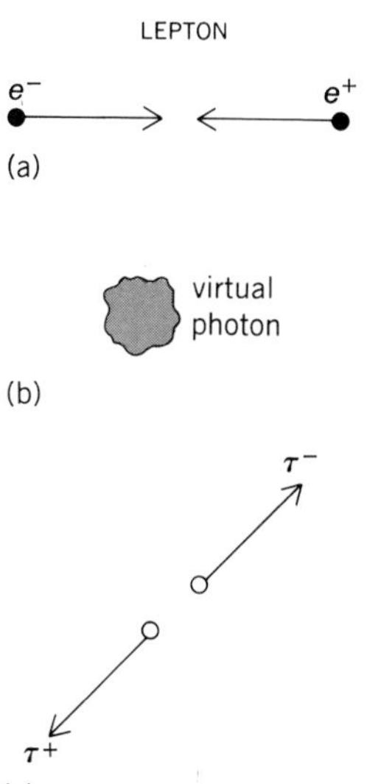

Fig. 2. Steps in the reaction $e^+ + e^- \rightarrow \tau^+ + \tau^-$. (*a*) The e^+ and e^- annihilate into (*b*) a virtual photon, which then decays into (*c*) a $\tau^+\tau^-$ pair.

In order to obtain enough total energy, the elec-

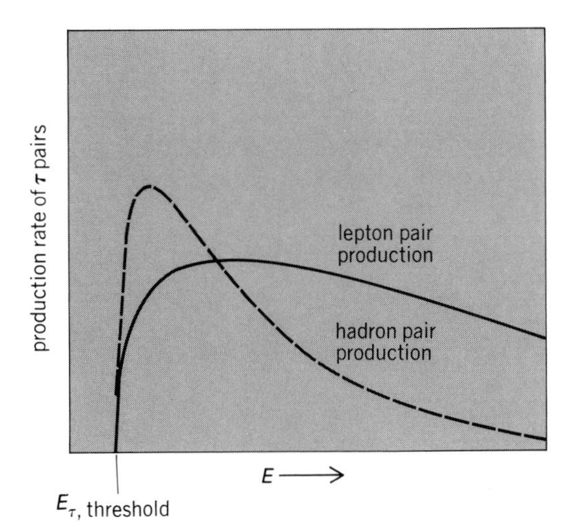

Fig. 3. Dependence of the production rate (cross section) of τ pairs on the total energy E of the electron-positron system. Curves show dependence of the actual lepton pair production rate on E and dependence which would be observed if the τ were a hadron.

Properties of leptons

Property	Electron	Muon	Tau
Symbol	$e^\pm$	$\mu^\pm$	$\tau^\pm$
Mass, MeV/c^2	0.51	105.7	1782^{+2}_{-7}
Lifetime, s	Stable	2.2×10^{-6}	Less than 10^{-11}
Spin	½	½	½
Does particle interact through electromagnetic and weak forces?	Yes	Yes	Yes
Does particle interact through strong force?	No	No	No
Associated neutrinos	$\nu_e, \bar{\nu}_e$	$\nu_\mu, \bar{\nu}_\mu$	$\nu_\tau, \bar{\nu}_\tau$
Mass of associated neutrinos, MeV/c^2	0	0	Less than 250, and could be 0
Are neutrinos associated uniquely only with corresponding charged lepton?	Yes	Yes	Probably

tron and positron annihilation must be accomplished in an electron-positron colliding-beams accelerator. This is a circular accelerator in which a beam of moving electrons collides with a beam of positrons moving in the opposite direction. These "head-on" collisions have enough energy to produce τ pairs. The first evidence for the existence of the τ was found at the SPEAR electron-positron colliding-beams accelerator at the Stanford Linear Accelerator Center in California. Additional evidence was found at a similar accelerator at Hamburg, Germany, called DORIS.

Heavy leptons in the future. No other charged leptons have been found in the mass range below 3000 MeV/c^2, and SPEAR and DORIS cannot search efficiently above that mass range. However, higher-energy electron-positron colliding-beams accelerators, PEP at Stanford and PETRA at Hamburg, will be able to search for masses up to at least 15,000 MeV/c^2. Such searches are important because there is no established theory of elementary particles which either predicts how many leptons exist or explains the masses of the known leptons (see the table).

One may also conceive of electrically neutral heavy leptons—particles similar to neutrinos but with masses greater than zero. No such particles have been found, although it is possible that the τ-neutrino has a mass greater than zero. Searches for neutral heavy leptons are more difficult than searches for charged heavy leptons because reaction (1) does not proceed nearly as easily for neutral leptons as it does for charged leptons.

For background information *see* ELEMENTARY PARTICLE; LEPTON; NEUTRINO; PARTICLE ACCELERATOR in the McGraw-Hill Encyclopedia of Science and Technology. [MARTIN L. PERL]

Bibliography: M. L. Perl, *Proceedings of the 1977 International Symposium on Lepton and Photon Interactions at High Energies*, Hamburg, pp. 145–164, 1977; M. L. Perl and W. T. Kirk, *Sci. Amer.*, 238(3):50–57, March 1978.

Linear accelerator

There has recently been interest in using linear accelerators in the production of nuclear power. Linear accelerators could be used to generate neutrons that would either produce fissile fuel for nuclear reactors or drive a subcritical nuclear reactor to produce power directly.

Power reactors and fuel regenerators. The linear accelerator breeder (LAB) or electronuclear breeder (ENB) had its origin in the Materials Testing Accelerator (MTA) Project which was carried out at the Lawrence Livermore Laboratory for the purpose of ensuring the production of fissile material for weapons. The LAB is a device which uses a powerful linear accelerator to produce an intense beam of protons or deuterons impinging on a target of a heavy element such as uranium, lead, or bismuth to produce spallation and cascade neutrons. These neutrons can in turn be absorbed in fertile uranium-238 or thorium-232 to produce fissile plutonium-239 or uranium-233.

A detailed design of such a system based on a 1-GeV proton linear accelerator delivering 65 MW of beam power to a rapidly flowing target of lead-bismuth eutectic alloy was made in the late 1960s at the Chalk River Nuclear Laboratories of Atomic Energy of Canada Limited as part of their Intense Neutron Generator (ING) program. Unfortunately, it was not possible to proceed in the construction of this system, but the design has been updated periodically and has been influential in this field.

Interest has recently been revived in the LAB as an alternative mode for the production of fissile fuels for the nuclear power program, where it could be a backup to the liquid-metal fast-breeder reactor (LMFBR) in case its development becomes delayed. Because of the advances in linear accelerator technology over the last 25 years, there is now considerable certainty that a high-current continuous-wave high-energy proton machine can be built to operate reliably at a reasonable cost.

LAB is also potentially competitive for the production of enriched uranium-235 in the head end of the fuel cycle, since the LAB would produce fissile fuel from natural fertile material. Enrichment, in conjunction with present-day nuclear burner power reactors, does not extend the nuclear fuel supply, whereas the linear accelerator fuel producer (LAFP) could extend the fuel resource.

Linear accelerator reactors could be used to produce fissile fuel either with or without fuel reprocessing. With fuel reprocessing, the fissile material would be separated from the target and refabricated into a fuel element for use in a burner power reactor. Without reprocessing, the fissile material would be produced in place either in a fresh fuel element or in a depleted or burned element after use in a power reactor. In the latter mode, the fissile material would be increased in concentration for reuse in a power reactor. This system is called a linear accelerator regenerator reactor (LARR; Fig. 1) or linear accelerator fuel regenerator (LAFR).

The linear accelerator reactor could also operate in a power production mode in which the spallation neutrons would be used to drive a subcritical assembly to produce power. This system is called a linear accelerator–driven reactor (LADR; Fig. 2).

Relation to United States policy. The United States administration's recent nuclear nonproliferation policy includes recommendations that plans for nuclear fuel reprocessing be stopped for the present, and that development of the Clinch River Breeder Reactor be halted.

A review is under way to determine the technical options available for reducing proliferation of nuclear weapons throughout the world. The following options are being explored: (1) eliminate reprocessing, (2) limit enrichment, (3) limit transport of fuel, (4) maintain low concentration of fissile material in power reactor fuel, and (5) approach a throwaway fuel cycle.

Many measures of this type would tend to reduce the total energy available from uranium resources. A number of alternatives are being explored to solve this problem, including the following: (1) finding additional natural uranium resources, (2) reducing enrichment plant tail concentrations, (3) using thorium as a nuclear fuel, and (4) exploring the use of new fuel cycles. *See* NUCLEAR POWER.

Both the LADR and the LARR appear to offer a high potential for extending the nuclear fuel supply while reducing the risk of proliferating weapons capability through misuse of the nuclear power fuel cycle.

Linear accelerator–driven reactor. The LADR comprises a subcritical reactor target containing either depleted, natural, or slightly enriched uranium, as well as thorium, producing a net amount of energy when supplied with neutrons generated by

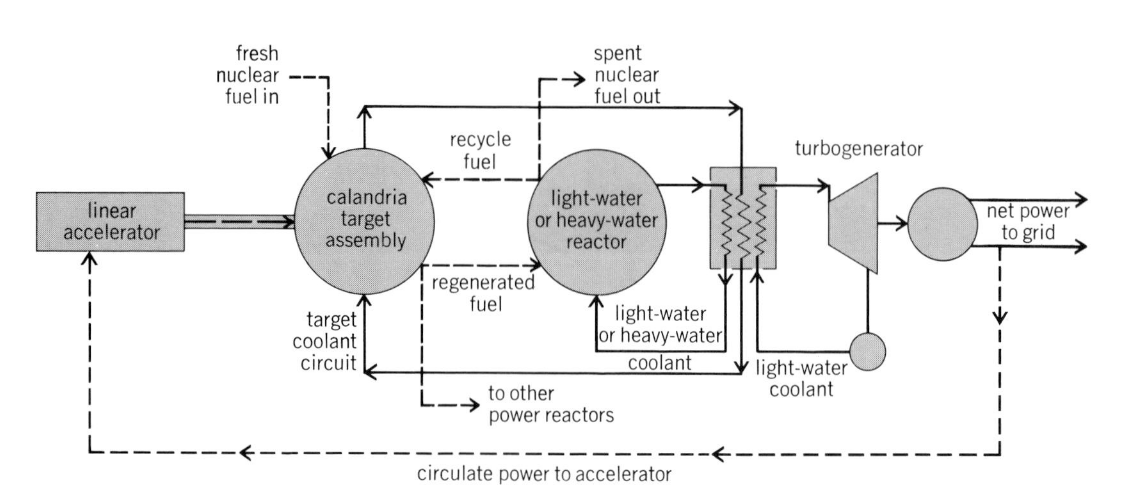

Fig. 1. Schematic diagram of linear accelerator fuel enricher and regenerator.

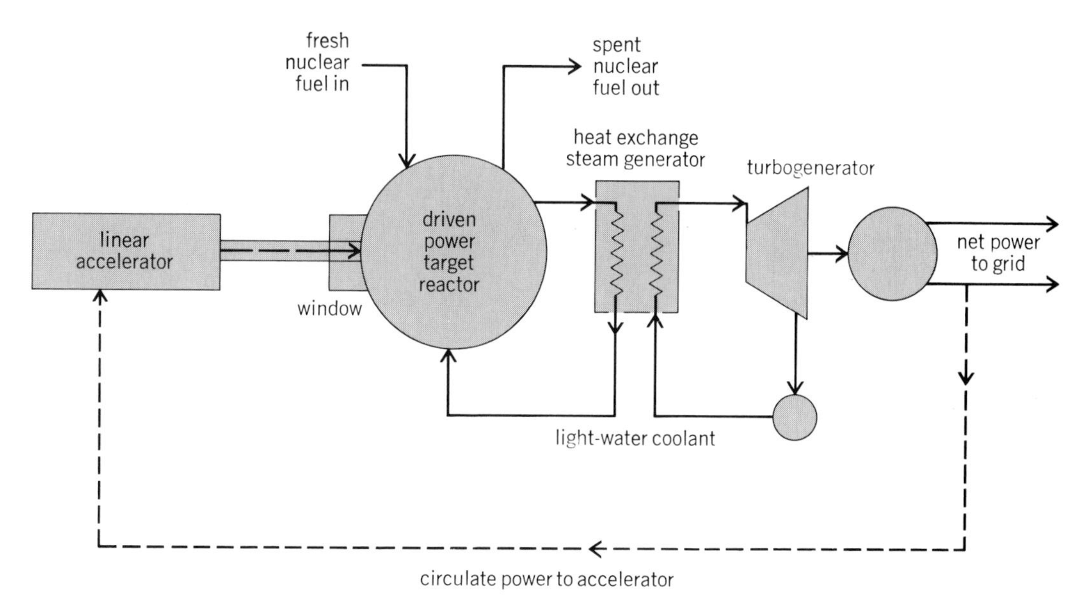

Fig. 2. Schematic diagram of linear accelerator–driven reactor.

stopping a high-energy proton beam from a linear accelerator. A critical economic parameter is the amount of electrical power needed to drive the accelerator, relative to the power output from the reactor target. As an economic rule of thumb, the power consumption of the accelerator should not be more than about 30% of the net electrical output of the target.

One mode of operation being explored is the following: (1) Fuel is initially burned in a conventional light-water reactor (LWR) with a conversion ratio of 0.6, to the usual burnup of about 30,000 megawatt-days (MWD) per ton. (2) The elements, which then contain about 1.8% fissile material (^{239}Pu and ^{235}U) are placed in the reactor target of an LWR-cooled LADR and burned to a value of k_∞ (multiplication factor) of about 0.9. With fuel shuffling in the LADR, additional burnup of an average of 30,000 MWD/ton would lead to burnup at an additional level of about 60,000 MWD/ton in fuel removed from the LADR. This would triple the energy available from uranium.

Another possibility is a natural uranium target with burnup to 30,000 MWD/ton, in which the average k_∞ is 0.85. The power circulated back to the accelerator would be 21.2% of reactor power (assuming 60 neutrons generated per accelerator proton), and the net power output would be 23,000 MWD/ton. This means that the utilization of fuel could be extended almost five times more than present LWR-enriched burner reactors. The target would be similar to an LWR reactor, and the elements would not be stressed to more than present LWR assemblies as far as burnup is concerned.

Linear accelerator regenerator reactor. The LARR would basically be a fuel producer, as opposed to the LADR, which would primarily be a power producer. The LARR would be used to irradiate fuel so that it could be produced in place to a level of reactivity for direct use in a normal power reactor. The LARR could be used to irradiate depleted, natural, or enriched uranium fuel, as well as thorium fuel.

For the limiting case of starting with unenriched ^{238}U, the relative capacity of power reactors which can be supported by the accelerator target reactor depends mainly on the power reactor conversion ratio. For a conventional LWR, the conversion ratio is 0.6, and an accelerator capacity of about 22% of the LWR capacity would be required to regenerate the fuel. It is desirable to produce a hard neutron spectrum in the regenerator to maximize fuel production, and therefore either light- or heavy-water steam coolant is preferred in the regenerator. By adopting a heavy-water-cooled power reactor such as the Canadian pressure-tube CANDU reactor, the conversion ratio can be increased and the regenerator made smaller.

The LARR could be located in an energy park, supplying regenerated fuel to a number of reactors at the same site. This would reduce the opportunity for theft of the nuclear material, since it would never be transported off-site and would be extremely radioactive in all cases where appreciable amounts of fissile material were involved. An advantage of the LARR over the LADR is that not every power reactor would require an accelerator, and thus the reliability for on-line power would be increased.

Preliminary economic tradeoff. Similar to critical breeder reactors, the linear accelerator reactors essentially increase the utilization of fertile fuel. The economics of critical breeder reactors is predicated on reducing fuel costs at the expense of increased capital cost. The net result is that breeder costs do not exceed present competitive LWR power costs. Likewise, LARs can trade off fuel cost for added investment in the accelerator-target assemblies. Estimates of accelerator costs indicate that direct unit capital costs are approximately equal to direct unit capital costs of LWRs, which at present are on the order of $600 per kilowatt of electrical power. Since the target assembly resembles an LWR, the direct capital cost is also approximately equal to an LWR. Based on the above assumption, k_∞ can be as low as 0.85, which allows for a number of LAR cases. The added value of a nonproliferation fuel cycle should make the system more desirable and thus economically reasonable.

For background information *see* NUCLEAR FUEL CYCLE; NUCLEAR POWER; REACTOR, NUCLEAR; REACTOR PHYSICS in the McGraw-Hill Encyclopedia of Science and Technology.

[MEYER STEINBERG]

Bibliography: H. J. C. Kouts and M. Steinberg, *Information Meeting on Linear Accelerator Breeders, Held at BNL on January 18 and 19, 1977*, ERDA Conf. Rep. 770107; F. R. Mynatt et al., *Preliminary Report on the Promise of Accelerator Breeding and Convertor Reactor Symbiosis (ABACS) as an Alternative Energy System*, Oak Ridge Nat. Lab. Rep. 5750, February 1977; M. Steinberg et al., *Electronuclear Fissile Fuel Production: Linear Accelerator Fuel Regenerator and Producer (LAFR and LAFP)*, Brookhaven Nat. Lab. Rep. 24356, April 1978; M. Steinberg et al., *Linear Accelerator Breeder (LAB)*, Brookhaven Nat. Lab. Rep. 50592, November 1976.

Magma

Studies of igneous processes in the 20th century have focused on magma generation and subsequent modification of these magmas by fractional crystallization. The importance of the additional complication of magma mixing is becoming widely recognized with the application of increasingly sophisticated analytical techniques. The probability that magma mixing will occur in any volcanic regime is likely to be related to the rate of magma production. Where generation rates are high, the magma stored at depth is unlikely to solidify completely prior to the influx of new melt, and will act as a trap by offering a path of least resistance. This process is expressed by the repeated eruption of fractionated lavas from the same vent system over a period of time such that a progressive, monotonic chemical evolution is not recorded in the volcanic stratigraphy. The tholeiitic shield volcanoes of ocean basins are excellent examples. Although the basalt-andesite-rhyolite associations of calc-alkalic volcanoes are less homogenized by magma mixing, there is abundant evidence that the process occurs and may also be responsible for triggering large, explosive eruptions.

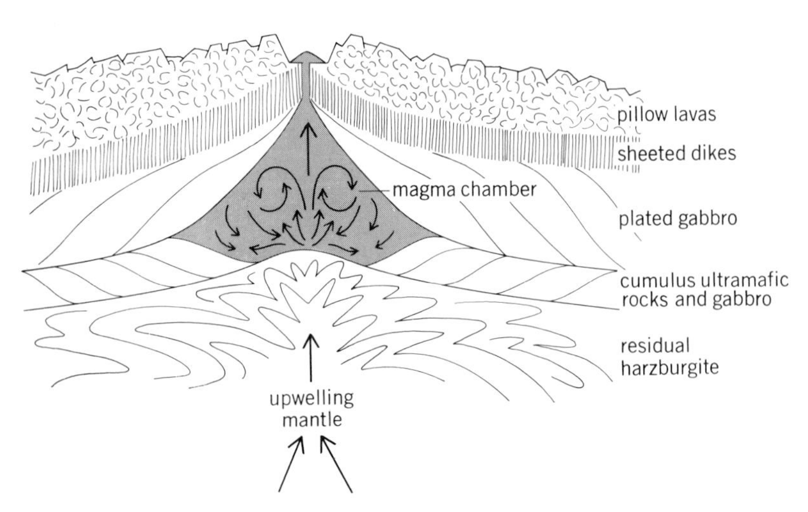

Fig. 1. Schematic cross-section through a mid-ocean ridge. The three basic components of the oceanic lithosphere are the eruptive basalts, the underlying plutonic dikes, gabbros, and ultramafic cumulates, and the residual mantle.

Ocean ridge setting. Divergent plate boundaries at which new oceanic lithosphere is being generated are highly probable sites for magma mixing (Fig. 1). A prime tenet of the conceptual model for the generation of new crust at mid-ocean ridges is that basalt eruptions and dike injection occur repeatedly at the axial singularity of the ridge and that the activity is localized by the inherent weakness of partially solidified magma.

The vast majority of volcanic rocks erupted at mid-ocean ridges are moderately fractionated tholeiites. Primitive magmas, unmodified during their ascent, and highly fractionated basalts are rare, and silicic derivatives (that is, rhyolites) produced by extreme fractionation are nearly absent. The uniformity of the lava compositions implies that there is something fundamental about the magmatic processes beneath mid-ocean ridges that prevents differentiation from proceeding unchecked. The mechanism involved appears to be physical buffering of the magmas in the subvolcanic holding chamber by periodic influx of unfractionated magma. The steady-state formation of new oceanic crust implies that plate divergence functions to maintain a magma chamber beneath the mid-ocean ridges of approximately constant dimensions.

Evidence for magma mixing. Recognition of magma mixing has traditionally been made on the basis of petrographic evidence for crystal-liquid disequilibrium in volcanic rocks (Fig. 2). Hybrid lavas characteristically contain suspiciously large numbers of different phenocryst phases, and some of these are likely to be resorbed, rimmed by a reaction zone, or reversely zoned. Although contamination of magmas by assimilation of country rocks and the presence of phenocrysts precipitated during high-pressure fractionation may also result in disequilibrium mineralogy, a systematic analysis may allow a distinction among these alternatives and magma mixing.

Disequilibrium mineralogy. A quantitative approach for determining whether or not a volcanic rock is a mixture of two liquids and their crystals is to apply known crystal-liquid elemental partitioning relations to the natural samples. Ocean floor basalts that are quenched against seawater during subaqueous eruption are particularly suited to this type of analysis because they preserve frozen samples of magmatic liquid in the glassy margins of pillows. Coexisting crystal-liquid pairs can therefore be analyzed with the microprobe. These data can be used to derive empirical partitioning relations for olivine and liquid. A complementary approach is to perform phase equilibria experiments on the samples under controlled conditions that simulate the natural conditions of crystallization. By using either one of these techniques it is possible to determine whether the most refractory phenocrysts in a sample crystallized from a liquid composition corresponding to the present bulk composition of the sample. If the phenocrysts are too refractory, they probably crystallized from a less fractionated magma.

Recognition of disequilibrium mineralogy generated by mixing of two basalts is facilitated if some crystals exhibiting reverse zoning can be found in addition to the excessively refractory phenocrysts. If not, the demonstration of disequilibrium may depend on careful analytical studies which indicate chemical incompatibilities.

In calc-alkalic systems, where the potential for mixing such diverse magmas as basalt and rhyolite is present, disequilibrium may be expressed more dramatically in thin section by reaction rims and the presence of mineral species, such as olivine and quartz, that clearly cannot have crystallized from the same magma. However, interpretation of less drastic crystal-liquid incompatibilities and zoning patterns in phenocrysts may be more difficult due to the increased probability that high-pressure crystallization, assimilation, or variation in vapor pressure has occurred.

Chemical modeling. A second, more indirect method applicable to testing for magma mixing is to model the chemical relationships among spatially and temporally associated magmas. This approach is particularly useful if a single parental magma can be identified. If evolved magmas can be derived from the parental liquid by subtraction of the observed phenocryst assemblage, an origin by fractional crystallization in a closed system is probably indicated; in other words, no magma mixing has occurred. Both major elements and trace elements can be utilized separately in the modeling, thereby providing a check on the solution. The difficult step in this scheme is the identification of the parental liquid.

Fortunately for ocean floor basaltic volcanism, there are some good estimates of this parental composition and evidence that it is relatively constant along the Mid-Atlantic Ridge. This probably results from large degrees of partial melting of a well-homogenized mantle below the ridge. The most primitive magmas sampled along the Mid-Atlantic Ridge are similar in composition and have chemical characteristics consistent with direct mantle derivation. Examples of these rocks are rare, however, and more compositional variation in primitive magmas may be recognized with additional sampling. In support of the relatively constant composition of ocean floor parental magmas are melt inclusions in highly refractory olivine phenocrysts, which also exhibit the distinctive chemical characteristics of primitive rock samples.

The chemical signature of the primitive ocean floor tholeiite is low TiO_2 and other lithophile elements in combination with high Cr and Ni abundances in the absence of accumulations of olivine phenocrysts. Of critical importance is an Mg/(Mg + Fe) ratio of about 0.70, which is an indication of compatibility with a residual mantle assemblage dominated by olivine with an Mg/(Mg + Fe) ratio of 0.9. The liquidus phase of these magmas is olivine with a similar composition. Other distinctive chemical characteristics are high Ca/Al and Ca/Na, which indicates that the first plagioclase to crystallize ought to be very calcic.

Figure 3 shows how crystal fractionation modeling, utilizing major elements in combination with both imcompatible and compatible trace elements, can be used to infer magma mixing. Given a parental magma and an evolved basalt that is potentially derived from it, an estimate of the percent solidified can be obtained by a least-squares fitting solution in which the phenocryst minerals are subtracted from the parent to give the derivative. By using the calculated proportions of crystals removed, the natural derivative can be compared to model values calculated from the parent magma in accordance with Rayleigh fractionation. Olivine fractionation from a magma will result, for example, in drastic reduction of the nickel content in the evolved derivative (C_l) relative to the starting concentration in the parent (C_i) because this element is strongly partitioned into the solid phase. Trace constituents not incorporated into fractionating mineral phases are gradually increased in the evolved derivatives until the late stages of crystallization when they are markedly enriched.

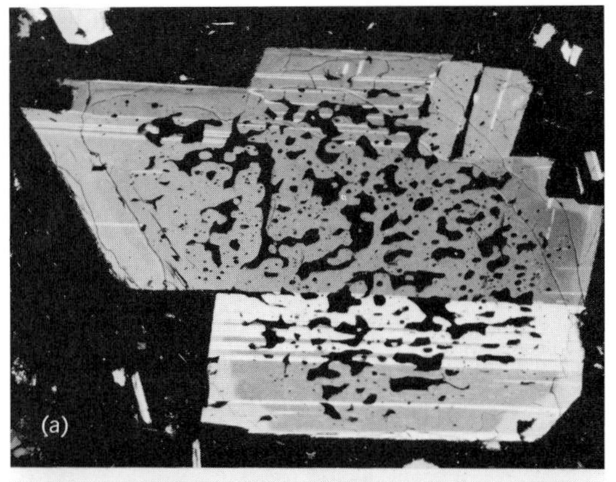

Fig. 2. Resorption textures in plagioclase phenocrysts indicative of crystal-liquid disequilibrium. *(a)* Crystal is normally zoned in that the rim is more sodic than the core. *(b)* An example of a relatively sodic core mantled by a more calcic rim with an intervening resorption zone.

As shown in Fig. 3, mixing primitive and highly evolved magmas results in apparently anomalously high concentrations of both incompatible and compatible trace elements in the hybrid magma. The maximum discrepancy occurs for approximately equal mixtures of the two end members. Significantly, a ubiquitous characteristic of moderately evolved ocean floor basalt magmas (that is, 20–50% fractionated) is their anomalously high content of trace elements, particularly the incompatible trace elements, compared with the values calculated in fractional crystallization models. Similarly, there is a lack of correspondence between the proportions of mineral phases that produce the best mathematical solutions in fractiona-

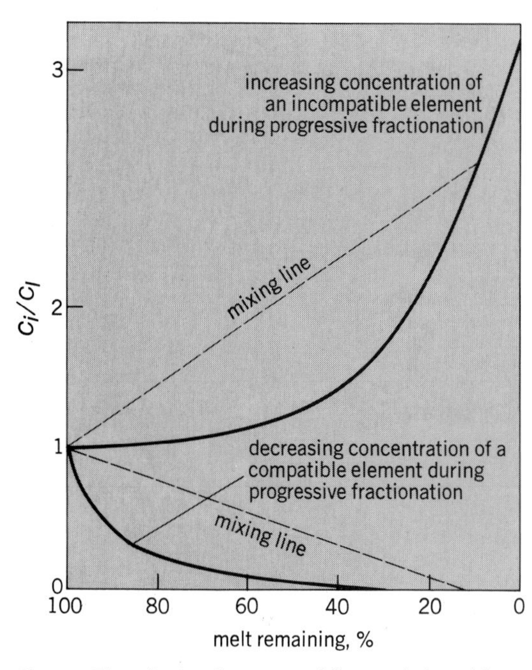

Fig. 3. The effects of magma mixing as deduced from trace-element fractionation. C_i = concentration in initial or parent liquid; C_l = concentration in derivative liquid.

tion calculations and the observed proportions of phenocryst phases. Specifically, clinopyroxene is disproportionately important in the calculated assemblage relative to olivine and plagioclase in comparison with its observed abundance. These two anomalies can be explained by a single mechanism if magma mixing is invoked. If the two end members consist of a primitive basalt crystallizing only olivine and plagioclase and a substantially evolved liquid which has fractionated clinopyroxene, both discrepancies will appear in the intermediate-composition hybrid. The removal of clinopyroxene from the evolved magma is probably facilitated by plating-on of gabbro (plagioclase + clinopyroxene) onto upper walls of the magma chamber as shown in Fig. 1.

For illustrative purposes the mixing process has been graphically shown as a one-stage mechanism between rather extreme compositions. In reality, moderately evolved ocean floor tholeiites are probably generated by repeated mixing of primitive magma with much less fractionated liquids. The excess lithophile element concentrations are probably generated incrementally by modest amounts of fractionation subsequent to periodic mixing events. As envisioned, this steady-state process is capable of producing repeated eruptions of the very similar basalts that are actually observed. Thus there is a mechanical buffering of the composition of magmas held in chambers beneath the mid-ocean ridges such that fractionation is prevented from proceeding unchecked.

[M. A. DUNGAN; J. M. RHODES]

Bibliography: A. T. Anderson, *J. Volcanol. Geotherm. Res.*, 1:3–33, 1976; J. F. Dewey and W. S. F. Kidd, *Bull. Geol. Soc. Amer.*, 88:960–968, 1977; M. J. O'Hara, *Nature*, 266:503–507, 1977; S. R. J. Sparks, H. Sigurdsson, and L. Wilson, *Nature*, 267:315–318, 1977; T. L. Wright and R. S. Fiske, *J. Petrol.*, 12:1–65, 1971.

Magnetic materials

The spin glass phenomenon in magnetism has been widely studied over the past several years. Spin glasses are magnetic solids in which the orientations of the atomic moments, or spins, are dispersed in a randomlike manner. Upon examining any two widely separated moments, one finds that there is no correlation in the relative orientation of the members of the pair. Instead, correlations exist only between members of a pair that happen to be near each other. The term "spin" here pertains to the spin on the electron, which produces the magnetic moment of certain atoms, and the moments themselves are often simply called spins. The term "glass" is used because of certain similarities to the properties of ordinary glass. The lack of a regular, or periodic, arrangement of spin orientations across the material is analogous to the absence of long-range order in the positions of the atoms in a noncrystalline, or glassy, solid.

Spin glass is thus distinguished from a ferromagnet, in which the spins have a preferred axis of orientation, and from various kinds of antiferromagnets, which display antiparallel ordering. The high-temperature disordered phase of a magnetic material is called the paramagnetic phase, in which the orientations of the spins are randomly fluctuating in time. The magnetically ordered phase is distinguished by a cessation of these fluctuations. In equilibrium each spin is aligned with the direction of the local magnetic field produced by the other spins in the material, so that each spin experiences no net torque. A distinction between the ordered spin glass and a hypothetical time-frozen paramagnet is due in part to short-range correlations among the spins.

Magnetic frustration. The conditions necessary for spin glass ordering, as opposed to ferromagnetic ordering, for instance, are not completely understood. A key feature seems to be that the interactions between the various pairs of spins compete with one another. The orientation of a given spin is dictated by its interactions with neighboring spins. The criterion for formation of a spin glass is that the alignments favored by the various interactions cannot all be satisfied simultaneously, and are thus said to be "frustrated." This can happen if the interaction between some pairs is positive, meaning it favors a parallel mutual alignment, while for others it is negative, favoring an antiparallel one. The result is that neither kind of alignment prevails on the average. An example of this type of interaction is given in the illustration. Here the spins are fixed at random positions in space. This is achieved in practice either by making the magnetic atoms part of an amorphous material, or by making the moments impurities in a nonmagnetic host crystal. In illustration *c* a situation is shown where certain pairs of spins are so closely spaced that they have comparatively strong interactions with one another and are mutually aligned. Otherwise orientations appear scattered in all directions. In theory, random positions are not really needed, only the presence of competing interactions. But in practice, spatial randomness is a direct way to achieve random interactions.

Examples of spin glasses. There is a wide variety of materials which display spin glass behavior and, in fact, this type of ordering phenomenon is not limited to magnetism. A few such examples are given below.

A nonmagnetic metal such as copper alloyed with a small concentration of manganese exhibits spin glass ordering below room temperature. The magnetic moments are localized on the manganese impurities, and interact with one another through a spatially oscillating spin polarization of itinerant conduction electrons in the copper host surrounding each moment. Because of the random positions of the impurity moments and the oscillating signs of the interactions, the conditions favoring spin glass ordering are present. Another example is the amorphous metal alloy composed of aluminum and the rare-earth metal gadolinium. Here the amorphous structure is the chief factor. In some cases one can start with a metal that orders ferromagnetically, such as a palladium-iron alloy, and then introduce a third impurity, such as manganese, which tends to create antiparallel interactions. Experiments on this ternary alloy system show that the ferromagnetism is supplanted by spin glass magnetism if the manganese concentration is large enough.

There are also ordinary glasses, which are insulating mixtures of glass-forming oxides, that display the same type of randomlike ordering. Insulating glasses composed of aluminum and silicon oxides and either a manganese or cobalt oxide display ordering among local groupings of many spins acting together as superlarge moments. The temperature-activated fluctuation of the moments in the paramagnetic phase is slow enough in these glasses to be measurable experimentally.

The ordering in spin glasses has similarities to the gelation of certain polymers. A notable breakthrough in the theory of spin glass ordering came about in 1975 when Sam Edwards and Philip Anderson recognized this analogy.

Certain molecules and chemical radicals having an electric dipole moment and present as impurities in solids show similar ordering effects because of the electric dipole interactions. Examples are the OH radical substituted for a halide ion in several alkali halides and polar molecules such as HCl trapped in organic solids. The ordering of the electric dipoles is possible because the trapped ions or molecules are free to reorient, even though they are held in a solid.

Nonzero entropy at zero temperature. There are a very great number of minimum-energy configurations of the spins, with each configuration having nearly the same total energy. Just how many such configurations exist is not known in general. As a result of this property, referred to as ground-state degeneracy, it is possible to cool a spin glass to low temperatures and yet not have the entropy of the system of spins approach zero. The third law of thermodynamics is not violated by this property since changes in entropy with temperature approach zero as the temperature approaches absolute zero. The entropy of any system is related by thermodynamics to the number of configurations available, and in ordinary materials it is nonzero at finite temperatures only because of thermally excited disorder. The ferromagnet, for example, has only one ground-state configuration, and its entropy approaches zero as the temperature is reduced toward absolute zero.

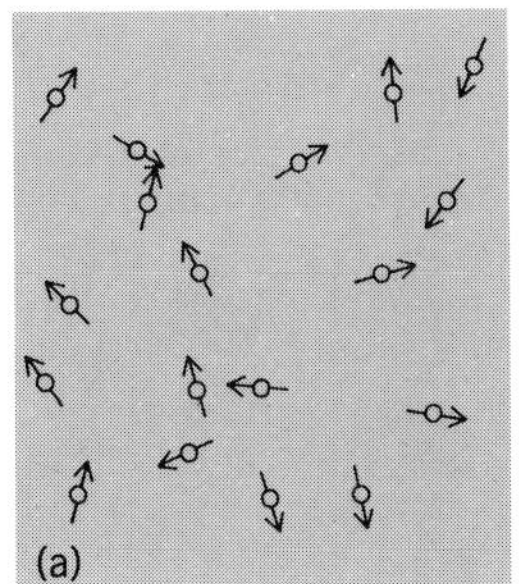

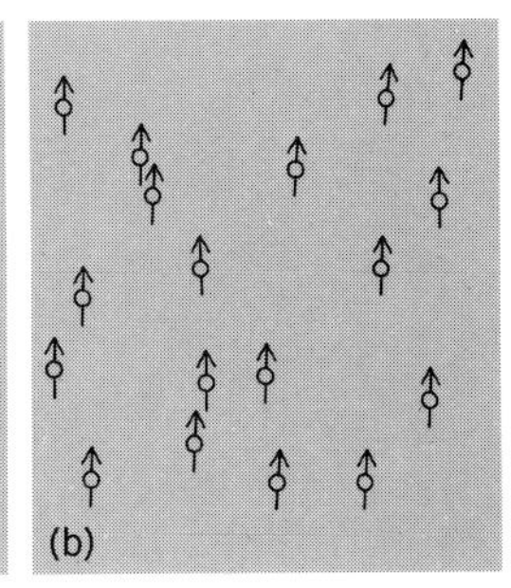

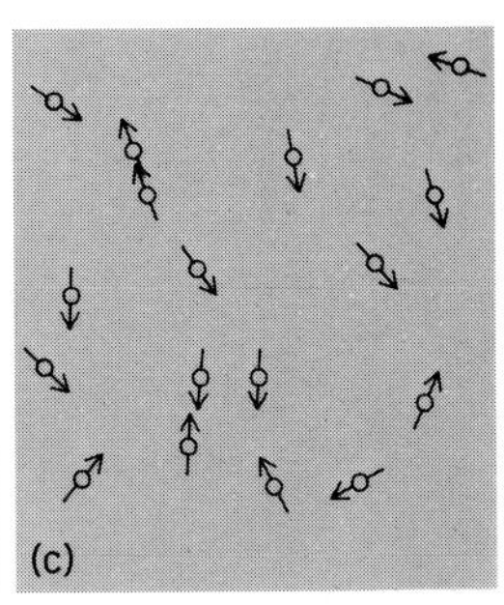

Two-dimensional representation of a random collection of magnetic spins: *(a)* random orientation in a paramagnet; *(b)* parallel arrangement in a ferromagnet; and *(c)* arrangement that might be found in a spin glass phase.

Consequences of this unusual property of spin glasses include the fact that there are numerous ways in which the system of spins can be perturbed at very little extra cost in energy. For an isotropic spin glass, one such perturbation can be visualized as a very gentle rotation in the orientations of the spins over large distances. The large number of low-energy excitations gives the material a large heat capacity at low temperatures, which is also a characteristic property of ordinary glasses.

Phase transition. The transition from the paramagnetic to spin glass phase occurs as the material is cooled through a certain temperature point. Magnetic susceptibility experiments on certain materials show that this temperature, called the spin freezing temperature, is well defined. Theoretically, spin freezing temperature marks the onset of a nonvanishing probability that spins in the system will remain frozen in some orientation. A significant unresolved difficulty with the theory of spin glasses has to do with the observation that the heat capacity varies smoothly with temperature at the spin-freezing point. The original formulation of the theory by Edwards and Anderson predicted a break in the slope of heat capacity versus temperature. The results of more recent theoretical calculations suggest that the absence of structure may be related to anisotropic interactions, which were not included in the original formulation. The anisotropy can arise from either the interactions between the spins or from random variations in the local environment of the host material. A phase transition is ordinarily associated with some kind of structure in the heat capacity. The spin glass transition is therefore unusual in that regard.

For background information *see* ENTROPY; FERROMAGNETISM; PARAMAGNETISM in the McGraw-Hill Encyclopedia of Science and Technology.

[ANTHONY T. FIORY]

Bibliography: J. A. Mydosh, in R. A. Levy and R. Hasegawa (eds.), *Proceedings of the 2d International Conference on Amorphous Magnetism*, pp. 73–83, 1977; P. W. Anderson, *J. Appl. Phys.*, 49: 1599–1603, 1978.

Mammalia

The Cenozoic Era (65 million years, or m.y., ago to the present) is often referred to as the Age of Mammals. However, the fossil record of mammals extends back almost 200 m.y. to the end of the Triassic Period. From this time until the end of the Mesozoic Era, approximately 65 m.y. ago, the fossil record of mammals is generally poor, and thus knowledge of this important interval which includes more than two-thirds of mammalian evolution remains extremely sketchy. During most of the Mesozoic, mammals were relatively insignificant compared with dinosaurs, the dominant land vertebrates. It is only near the end of the Mesozoic, in the Late Cretaceous, that fairly continuous sequences of mammal localities are known from the People's Republic of Mongolia and western North America. One sequence, in eastern Montana, is of particular interest because it straddles the Mesozoic-Cenozoic boundary. This is the time during which dinosaurs became extinct, relinquishing their title of dominant land vertebrates to the mammals.

Latest Cretaceous mammals. Late Cretaceous mammals have been known from western North America since the late 1800s. However, it was not until the mid-1900s that extensive collections of these mammals were made. The small size of these mammals, approximately between shrews and foxes, has hindered their collection. Most of the knowledge of these mammals comes from isolated teeth and jaw fragments. The use of underwater screening has made it possible to process large volumes of sediment, concentrating the relatively rare mammalian fossils. This technique has been applied in several regions in western North America to recover Late Cretaceous mammals. To date, a fairly continuous, although composite, sequence of mammal-bearing localities is known from about 77 m.y. ago (Campanian) through the end of the Cretaceous (Maestrichtian), 65 m.y. ago. The sediments containing most of the Late Cretaceous mammals in western North America were deposited on a low floodplain bordering a vast inland sea that bisected North America from north to south during much of the Cretaceous.

Until fairly recently most of the information concerning mammal evolution during the latest Cretaceous came from small collections made in various regions of western North America. In the late 1950s and 1960s these small samples were greatly expanded by large screen-washing collections from Wyoming and Alberta. In most aspects the composition of these mammalian faunas is very different from early Cenozoic (Paleocene) faunas. Unlike Paleocene faunas, latest Cretaceous faunas are dominated by marsupials and ptilodontoid multituberculates, with relatively few placental mammals. Most of the 12 or so marsupials assigned to three families superficially resemble the modern opossum. One of the families, the Didelphidae, which includes extant North and South American opossums, is also recognized in the latest Cretaceous of North America (Fig. 1*c*). The multituberculates are an extinct group of mammals that had a long evolutionary history separate from therians (marsupials and placentals). They probably filled an ecological role similar to modern rodents. The ptilodontoids are a suborder of multituberculates that are most common in the Late Cretaceous and early Cenozoic of North America (Fig. 1*a*). Unlike rodents and the taeniolabidoids, another group of multituberculates, the incisors of ptilodontoids were completely covered with enamel. Rodents and taeniolabidoids are convergent in restricting the enamel to a narrow band on the ventral side of at least the lower incisors (Fig. 1*b*). The rarer placental mammals are more difficult to classify, due in part to their fragmentary record and in part to their primitive dental morphology. Their molars are generally tall and sectorial (cutting), as seen in some extant insectivores. Along with a few early Cenozoic forms, seven species of Late Cretaceous placentals are usually included in the family Palaeoryctidae (Fig. 1*d*). One more unusual taxon, *Gypsonictops*, is commonly included with better-known Cenozoic forms in the family Leptictidae, although a recent study suggests it is a more distant relative. This rather distinctive mammalian fauna has been informally designated as being "typical latest Cretaceous."

Hell Creek Formation. In 1965 R. E. Sloan and L. Van Valen reported the discovery of new and distinctive latest Cretaceous mammals from the Hell Creek Formation, in McCone County, eastern Montana. They also noted the discovery of much smaller localities in the Hell Creek Formation that contained only typical latest Cretaceous mammals, and one locality in the overlying Tullock Formation that contained early Paleocene mammals. These mammals were unlike any previously reported from the latest Cretaceous in that they showed definite affinities with Paleocene mammals and, in fact, extended the record of some mammalian families back into the latest Cretaceous.

This group of taxa has been informally termed mammals of "Paleocene aspect." Sloan and Van Valen recognized three stratigraphically arranged localities containing Paleocene-aspect mammals. From lowest to highest, these are Bug Creek Anthills, Bug Creek West, and Harbicht Hill (Fig. 2). Bug Creek Anthills is by far the richest locality and occurs in the upper part of the Hell Creek Formation, 80 ft (24 m) below the Hell Creek–Tullock contact. Typical latest Cretaceous mammals, plus fragmentary dinosaurs (usually teeth) normally found with these mammals, are common at Bug Creek Anthills. However, a number of Paleocene-aspect mammals make their first appearance as common representatives, including *Protungulatum*, the earliest known member of the archaic ungulate order, Condylarthra (Fig. 1*e*); two taeniolabidoid multituberculates belonging to two separate families; and a new genus and species belonging to the family Palaeoryctidae. At each of the successively higher localities the relative abundance and taxonomic diversity of the typical latest Cretaceous mammals and dinosaurs decrease while those of Paleocene-aspect mammals increase. Most of the increased diversity is due to the invasion or origination of at least four additional species of condylarths representing at least two families. A single lower molar of the earliest primate *Purgatorius* was recorded from Harbicht Hill, but further attempts by others to recover additional specimens have not met with success. This sequence of localities clearly documents a fairly rapid turnover of a vertebrate as well as mammalian communities near the end of the Cretaceous.

Research being conducted just to the west of McCone County, in Garfield County, Montana, by a joint University of California, Berkeley–Los Angeles County Museum group has complemented the work of Van Valen and Sloan, and has added more complexity to the picture of mammalian evolution

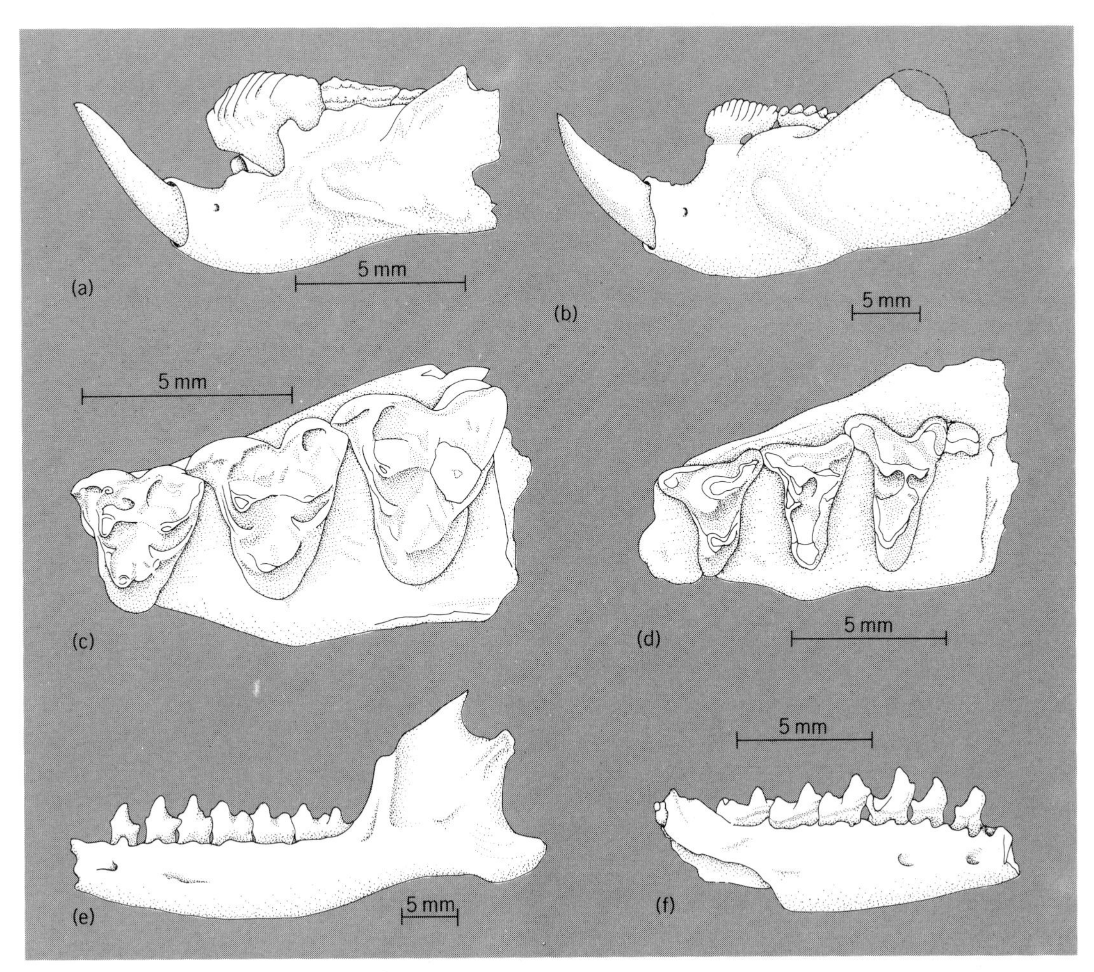

Fig. 1. Jaw fragments of some latest Cretaceous and early Paleocene mammals. Dentaries of (*a*) the ptilodontoid multituberculate *Mesodma formosa (modified from W. A. Clemens, Fossil mammals of the type Lance Formation, Wyoming, part I: Introduction and Multituberculata, Univ. Calif. Publ. Geol. Sci., 48:1–105, 1964)* and (*b*) the taeniolabidoid multituberculate *Stygimys kuszmauli (from R. E. Sloan and L. Van Valen, Cretaceous mammals from Montana, Science, 148:220–227, 1965)*; maxillae of (*c*) the didelphid marsupial *Alphadon rhaister (from W. A. Clemens, Fossil mammals of the type Lance Formation, Wyoming, part II: Marsupialia, Univ. Calif. Publ. Geol. Sci., 62:1–122, 1966)* and (*d*) the palaeoryctid placental *Cimolestes incisus (from W. A. Clemens, Fossil mammals of the type Lance Formation, Wyoming, part III: Eutheria and Summary, Univ. Calif. Publ. Geol. Sci., 94:1–102, 1973)*; and dentaries of (*e*) the earliest condylarth, *Protungulatum donnae (from R. E Sloan and L. Van Valen, Cretaceous mammals from Montana, Science, 148:220–227, 1965)*, and (*f*) the earliest primate, *Purgatorius unio (from W. A. Clemens, Purgatorius, an early paromomyid primate (Mammalia), Science, 184:903–905, 1974)*.

near the Cretaceous-Paleocene boundary.

In 1977 J. D. Archibald described several localities from the Hell Creek Formation that resemble the sequence in McCone County in possessing Paleocene-aspect mammals. The two richest localities, collectively known as the Hell's Hollow local fauna, were reported as being 32 ft (10 m) below the Hell Creek–Tullock contact. Field research conducted subsequent to that of J. D. Archibald suggests that these localities might be in the lowest portion of the Tullock Formation rather than the uppermost Hell Creek Formation. Although the exact stratigraphic position of these localities cannot presently be determined, it is clear that they are younger based on faunal comparisons, and to a lesser degree physical stratigraphy, than any of the Hell Creek Formation localities in McCone County that include Paleocene-aspect mammals. The Hell's Hollow local fauna localities are also older than any Paleocene localities now known from the Tullock Formation based on faunal comparisons. At the species level only Paleocene-aspect mammals are present in the Hell's Hollow local fauna, although one genus of typical latest Cretaceous multituberculate, *Mesodma*, is represented by a new species. Also of importance is the apparent lack of dinosaurs.

In addition to these sites, Archibald described a number of localities that included only typical latest Cretaceous mammals and abundant dinosaur remains. Of particular interest is the group of Flat Creek local fauna localities. Not only is this the richest of this type of mammalian occurrence in Montana, but it occurs only 17 ft (5 m) below the top of the Hell Creek Formation, which is stratigraphically higher than Harbicht Hill and possibly the two Hell's Hollow local fauna localities (Fig. 2). This is clear evidence that at least two different ecologies including mammals were coeval during the close of the Mesozoic in eastern Montana.

Bug Creek and Hell Creek faunal-facies. The environments of deposition of these various localities also suggest this relationship. Almost all of the localities containing Paleocene-aspect mammals occur in large, sandy channel-fill deposits within the upper 80 ft (24 m) of the Hell Creek Formation. Archibald has termed this the Bug Creek faunal-facies. Many localities including only typical latest Cretaceous mammals have been discovered throughout most of the stratigraphic extent of the Hell Creek Formation. Except for the Flat Creek local fauna, these localities are not very rich. They occur in small, silty channel-fill deposits or in paludal or overbank sediments. Archibald has termed this the Hell Creek faunal-facies. The limited evidence available suggests there was ecological and evolutionary stasis in this faunal-facies (that is, the faunal composition and individual species are similar) throughout the stratigraphic extent of the formation, in contrast to the rapid turnover that is clearly seen in the Bug Creek faunal-facies.

Late Cretaceous–Paleocene transition. The fates of the mammals in the two faunal-facies at the close of the Mesozoic are only now beginning to be understood. Early Paleocene mammals are being studied by various researchers. Early and middle Paleocene mammals from several rich localities in the Tullock Formation (Fig. 2) immediately overlying the Hell Creek Formation are currently under study by W. A. Clemens. In 1974 he reported on well-preserved material of the earliest primate, *Purgatorius* (Fig. 1*f*). Preliminary results of his work indicate the Paleocene faunas are a mixture of both the Bug Creek and Hell Creek faunal-facies.

Most of the Bug Creek faunal-facies mammals have close relatives or descendants in the Paleocene faunas. The condylarths that became increasingly diverse throughout the Bug Creek faunal-facies sequence continue to do so into the Paleocene. The two genera of taeniolabidoid multituberculates that made their first appearance at Bug Creek Anthills are represented by new species at the end of the Cretaceous that are also known from the early Paleocene. A third taeniolabidoid that first occurs in the Hell's Hollow local fauna also has close later Paleocene relatives. The only new palaeoryctid to occur at Bug Creek Anthills continued into the Paleocene, probably giving rise to the short-lived mammalian order Taeniodonta. As noted above, a single genus of Hell Creek faunal-facies mammal, *Mesodma*, occurs in the Hell's Hollow local fauna. This genus is known from the Paleocene, but the phylogenetic relationships of the Cretaceous to the Paleocene species are not well understood.

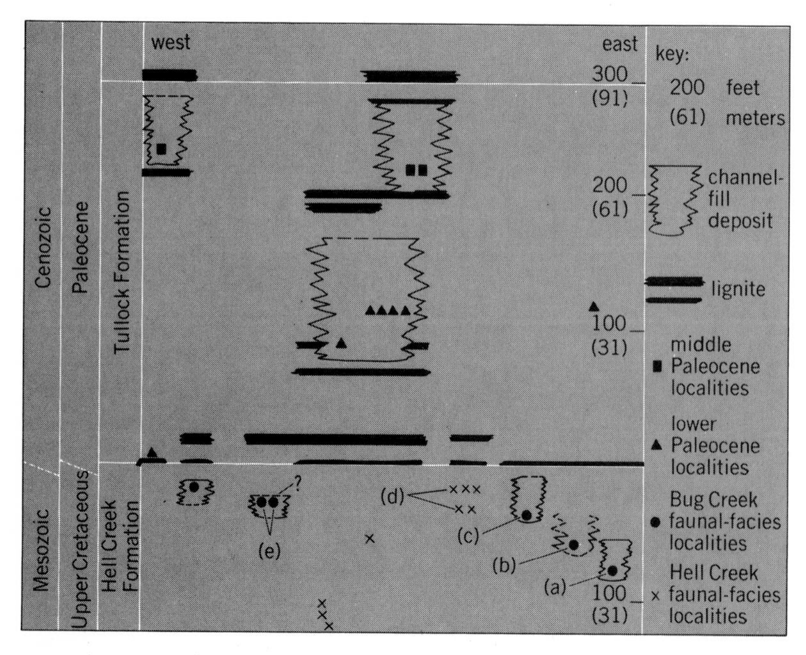

Fig. 2. Stratigraphic placement of important mammal-producing localities near the Cretaceous-Paleocene boundary in eastern Montana. Geographic placement is not to scale. *(a)* Bug Creek Anthills, *(b)* Bug Creek West, *(c)* Harbicht Hill, *(d)* Flat Creek local fauna, and *(e)* Hell's Hollow local fauna.

Fewer of the Hell Creek faunal-facies mammals appear to have left descendants. Of the approximately 12 species of marsupials, only one left a Paleocene descendant. The single Paleocene genus has also been recorded from the Hell's Hollow local fauna. The ptilodontoid multituberculates fared somewhat better. Of the four or five Cretaceous families, most left Paleocene descendants. The fates of the placental mammals in the Hell Creek faunal-facies are somewhat more difficult to determine. Most of the six or seven species of palaeoryctids have at one time or another been suggested as ancestral stock for various Cenozoic mammals. There is general agreement that the carnivores, true insectivores (for example, shrews), and a few extinct orders arose from this family. As noted above, the other placental, *Gypsonictops*, is at least distantly related to the Cenozoic family Leptictidae.

This well-documented sequence of latest Cretaceous and Paleocene mammal localities provides a reasonable indication of the events of mammalian evolution near the Mesozoic-Cenozoic boundary in eastern Montana. It remains to be seen whether this type of transition occurred in other parts of the world, although it is clear from the scant evidence available that distinctly different mammalian faunas were present in some other parts of the world at the Mesozoic-Cenozoic boundary. However, if the pattern of mammalian evolution at the Mesozoic-Cenozoic boundary in eastern Montana can be generalized to other faunas, several points are suggested. First, the origination or invasion and extinction of different mammalian lineages occurred at various times within the two faunal-facies, and were not geologically instantaneous events at the Mesozoic-Cenozoic boundary. Second, the invading Paleocene-aspect mammals may have gradually replaced the typical latest Cretaceous mammals (and possibly dinosaurs, as suggested by Van Valen and Sloan in 1977) within the Bug Creek faunal-facies through competition, but it is more difficult to extend this explanation to the Hell Creek faunal-facies. Third, although the mammals did replace dinosaurs as the dominant land vertebrates shortly after the Mesozoic-Cenozoic boundary, the differences between latest Cretaceous and early Paleocene mammalian faunas are not any greater than those of many of the other mammalian faunal transitions during the Cenozoic. Thus no extraordinary events or mechanisms are necessary to explain the mammalian faunal transition at the Mesozoic-Cenozoic boundary.

For background information *see* CENOZOIC; EVOLUTION, ORGANIC; MAMMALIA; MESOZOIC; PALEOCENE; PALEONTOLOGY in the McGraw-Hill Encyclopedia of Science and Technology.

[J. DAVID ARCHIBALD]

Bibliography: J. D. Archibald, unpublished doctoral dissertation, University of California, Berkeley, 1977; W. A. Clemens, *Science*, 184:903–905, 1974; R. E. Sloan and L. Van Valen, *Science*, 148: 220–227, 1965; L. Van Valen and R. E. Sloan, *Evol. Theory*, 2:37–64, 1977.

Mass spectroscopy

Nonvolatile materials are solids that have extremely low vapor pressures because the forces holding the atoms or molecules together are very strong. An analysis of these materials by mass spectroscopy has been difficult because it is necessary to produce molecular or atomic ions in the gas phase to obtain a mass spectrum. In the past 10 years, however, several new methods in mass spectrometry have been developed for analyzing nonvolatile molecules.

Nonvolatile, thermally unstable molecules. For materials that are thermally stable, vapor pressure can be increased by heating the material to high temperature. Nonvolatile molecules of biological origin, such as peptides, nucleotides, and many antibiotics, decompose when heated, so that it is not possible to obtain mass spectra by the methods employed for volatile molecules. These molecules can often be made more volatile by chemical modification (or derivitization) in a manner which decreases the strength of the forces between the molecules. An example is the attachment of trimethyl silyl ether groups at polar sites on the molecule. New techniques in the mass spectroscopy of nonvolatile biological molecules have recently been developed which produce gas-phase molecular ions without decomposing and without prior derivitization.

Field desorption. Field desorption (FD) is a method introduced by H. D. Beckey in 1969 for the analysis of nonvolatile biomolecules. A sample of the material is deposited on a thin tungsten wire containing sharp microneedles of carbon on the surface. When a voltage is applied to the wire, high electric field gradients are established at the sharp points on the wire. When the wire is then heated to a modest temperature (typically 300–500 K), molecular ions are desorbed from the surface of the wire and can be focused into a mass spectrometer. This method has had considerable impact on the application of mass spectrometry to nonvolatile biological molecules, and commercial FD mass spectrometers are now available.

Electrohydrodynamic ionization. A variation of the FD method called electrohydrodynamic ionization mass spectroscopy was developed in 1974 by D. S. Simons, B. N. Colby, and C. A. Evans, Jr. This method also utilizes high electric field gradients to induce ion emission, but from the surface of a droplet of a liquid solution rather than a solid film. The nonvolatile material is dissolved in a low-vapor-pressure solvent with a high dielectric constant such as glycerol. This reduces the forces between the nonvolatile molecules by surrounding them with solvent molecules.

Rapid heating. Another approach to the problem was introduced by R. J. Bleuhler and colleagues in 1974. They suggested that at high temperatures the rate of vaporization of a nonvolatile molecule from a solid surface may be enhanced relative to the rate of decomposition. Thus, if a sample is heated rapidly (for example, at 200 K/s) it may be vaporized without decomposing. Other techniques have evolved from this principle in which the energy deposition into the sample is so fast that the heating effect appears in the form of a shock wave. The means for transferring energy rapidly to a solid sample has involved the use of energetic positive ions and pulsed lasers.

Laser-induced mass spectroscopy. Laser-induced mass spectroscopy of nonvolatile biological molecules was reported in 1978 by M. A. Posthumus and associates. By using a transverse-electrical-atmospheric (TEA)–carbon dioxide pulsed laser with a 0.15-μs duration and a power density of 1 MW/cm^2, they were able to obtain mass spectra of amino acids, small peptides, sugars, and nucleotides.

Secondary-ion mass spectroscopy. The positive-ion-induced mass spectroscopy work in this field falls into two classes. Secondary-ion mass spectroscopy (SIMS), a method used routinely for surface analysis of inorganic materials, was shown by A. Benninghoven, D. Jaspers, and W. Sichtermann in 1976 to be applicable to the production of molecular ions of nonvolatile biomolecules from solid films. They exposed films to very weak beams of 2.5-keV Ar^+ ions. The ions strike the surface and transfer energy by collisions with molecules (nuclear scattering) to produce a localized hot spot. This can result in molecular excitation, chemical reactions producing molecular ion products, and volatilization in times on the order of 10^{-12} s.

Californium-252 plasma desorption. The other class of positive-ion-induced mass spectroscopy involves much higher-energy ions (in the range of 100 MeV). The amount of energy deposited in the sample by each ion is much larger than in SIMS, and the time scale may be as short as 10^{-16} s. The spontaneous nuclear fission of californium-252 produces positive ions in this energy range. A new method for mass analysis of nonvolatile biomolecules using californium-252 fission fragment excitation was reported by D. F. Torgerson, R. P. Skowronski, and R. D. Macfarlane in 1974. They named it californium-252 plasma desorption mass spectrometry (PDMS). Because of the high energy, these ions can penetrate thin films and deposit energy by electronic excitation in a cylindrical track having a diameter of about 20 nm. Fast chemical reactions can occur in the fission track, resulting in the formation of molecular ions. The ejection of molecular ions occurs from the surface of the film as a result of a shock wave which is produced by the energy released in the fission track. The PDMS method appears to cover a broader spectrum of types of nonvolatile biomolecules than the FD method. Its main application appears to be in the analysis of large biomolecules with molecular weights greater than 2000. Molecular ions up to 4000 daltons have been detected by this technique.

The PDMS method uses the time-of-flight method for obtaining mass spectra. In this method the ion mass is calculated by measuring the time it takes for an ion of known energy to travel a fixed distance (for example, 30–100 cm). The energy is set by accelerating the ions in an electric field. A diagram of the method applied to PDMS is shown in the illustration. The californium-252 source is mounted behind a film of the material mounted on a thin metal foil. A voltage of ± 9 kV (for positive or negative ions) is applied continuously to the foil. When a spontaneous fission occurs, two fission

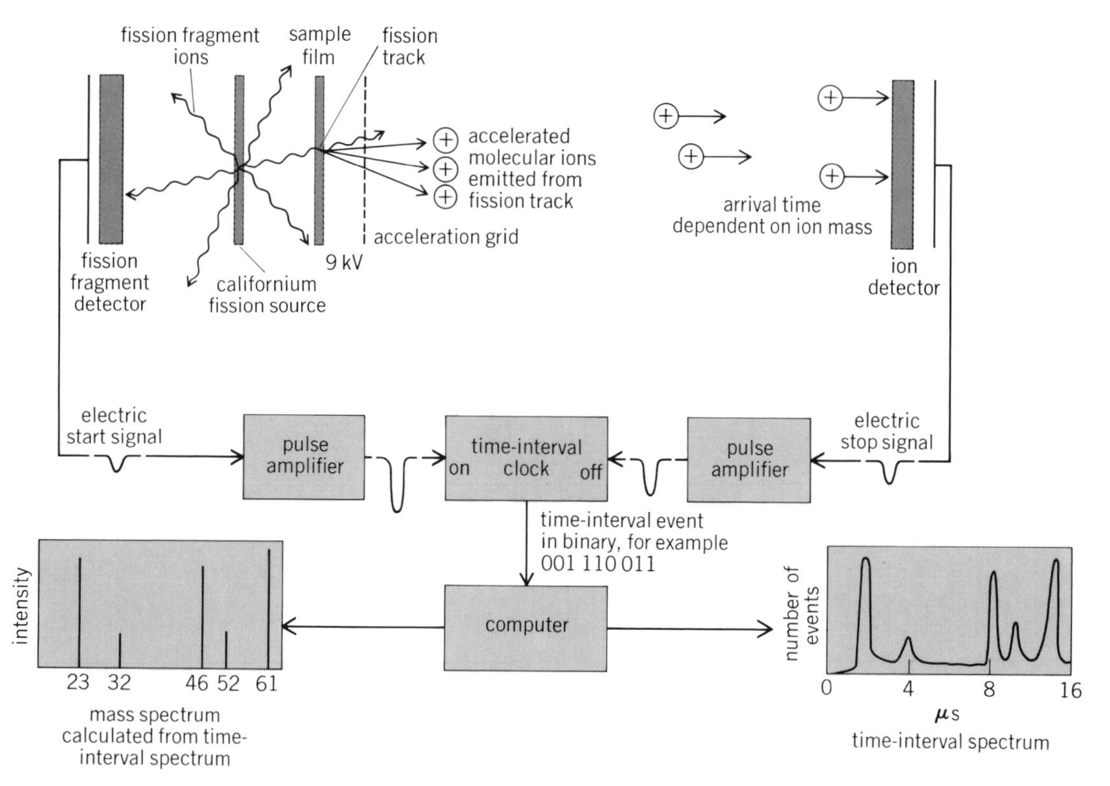

Flow diagram for method of californium-252 plasma desorption mass spectrometry.

fragments are emitted in opposite directions. If one fission fragment hits the foil, it passes through it, forming the fission track and ejecting molecular ions. The ions are accelerated to 9 keV and travel in vacuum down a cylindrical tube. To obtain the time-of-flight, the complementary fission fragment is detected and an electrical signal is generated which turns on a fast electronic clock. When an ion reaches the end of the tube, it strikes a detector which generates a second electrical signal. The signal is sent to the clock which turns it off. The accumulated time is transferred to a computer which stores the time information. A typical time is 20 μs. The measurement is repeated about 5000 times per second. The frequency of the measurement is dependent upon the californium-252 source intensity. The distribution of flight times can be translated into a mass spectrum when the kinetic energy of the ions is known. The computer sorts events after measurement, generates a time-interval spectrum, and calculates a mass spectrum.

For background information *see* MASS SPECTROSCOPE in the McGraw-Hill Encyclopedia of Science and Technology. [R. D. MACFARLANE]

Bibliography: R. D. Macfarlane and D. F. Torgerson, *Science*, 191:920–925, 1976; M. A. Posthumus et al., *Anal. Chem.*, 50:985–991, 1978; H. R. Schulten and H. M. Schiebel, *Naturwissenschaften*, 65:223–230, 1978.

Meristem, apical

The morphogenesis, that is, the development of form, of the primary plant body is governed by the activity of apical meristems located at the tips of the growing roots and shoots of the plant. Root apical meristems exhibit a wide variety of cell patterns; however, the simplest to analyze are those in roots that possess a single apical cell, as is the case in many ferns. The small water fern *Azolla* has been particularly useful for the study of the patterns of growth of root apical meristems in ferns, revealing new information on certain aspects of morphogenesis in and near plant meristems. By embedding roots in plastic and examining both transverse and longitudinal serial sections, it has been possible to build up a detailed knowledge of the cellular construction of the root. This has

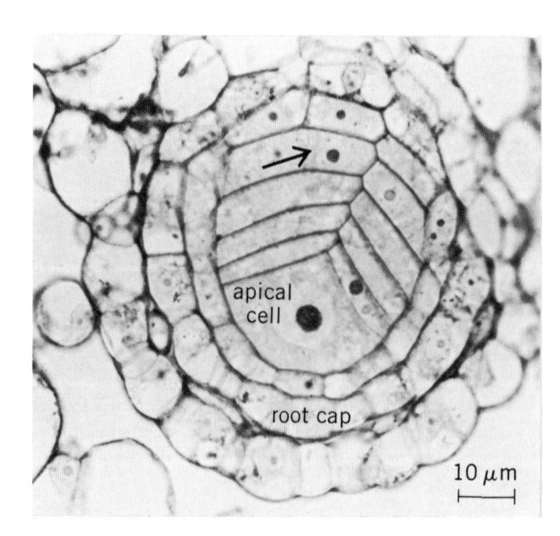

Fig. 1. Longitudinal section of a young *Azolla pinnata* root. Two of the three tiers of merophytes produced by division of the large tetrahedral apical cells can be seen, this root being composed of a total of 15 merophytes. First longitudinal formative division to occur is indicated by the arrow.

permitted the precise determination of the types, orientations, and locations of cell divisions; the cell lineages along which every cell in the root is formed; the patterns of cell differentiation; and the temporal changes that occur during root development.

Apical cell. The apical cell in the roots of *Azolla*, as in many ferns, is tetrahedral in shape. The curved distal face lies adjacent to the cells of the root cap, and the three proximal faces are flatter and approximately triangular in shape, tapering to the apex of the cell (Fig. 1). A single layer of root sheath cells surrounds the root cap in very young roots. Soon after the formation of the apical cell, a periclinal division parallel to the distal face produces the precursor cell of the root cap. All subsequent divisions of the apical cell are parallel to the three proximal faces occurring at each in turn, either in a clockwise or counterclockwise direction, and generate a succession of overlapping cells (Fig. 1). The root is thus built up of a helix of segments, or merophytes, each of which is the product of a division of the apical cell. In *Azolla* these merophytes can be recognized in all parts of the root even after extensive cell elongation, division, and differentiation have taken place within them (Fig. 2). The cell complement of the mature root arises by means of two classes of cell division. First, the initial cells of the cell files are generated by a sequence of formative divisions within each merophyte. Second, the initial cells that are formed by the formative divisions undergo proliferative divisions in the transverse plane.

Formative divisions. The merophytes, each of which occupies a 120° sector of the root, are subdivided by means of tangential longitudinal and radial longitudinal divisions, forming cell initials for every cell type in the root. The sequence of formative divisions within the merophytes in *Azolla* is very regular, and it is possible to assign arbitrary code numbers to each type of division that occurs. As an example, the divisions that complete the formation of the first cell initial, that of the outer cortex, will be outlined briefly. These and all other formative divisions are shown diagrammatically in Fig. 3. The first division within the merophyte is always a tangential longitudinal division, with the new cell wall being placed closer to the outside of the root than to the center. Two radial longitudinal divisions in the outer cell form three cells, which then divide in the tangential longitudinal plane. The division of the middle of the inner row of cells then completes the formation of four outer cortex cell initials in this merophyte. The same sequence of divisions in the other two 120° sectors gives initials for the total of 12 files of outer cortex cells seen in the mature root. The sequence of formative divisions continues in the other cells in the merophyte as shown in Fig. 3, progressively forming initials for each cell type in the root.

Proliferative divisions. Initial cells are continually being produced in young roots, thereby generating files of cells, as shown in Fig. 2. The initial cells then complete a number of rounds of transverse proliferative divisions before they begin to differentiate. The number of transverse proliferative division cycles completed by the cell initials is characteristic of each cell type, and is shown in the 11 columns of cells in Fig. 3. The simplest case is

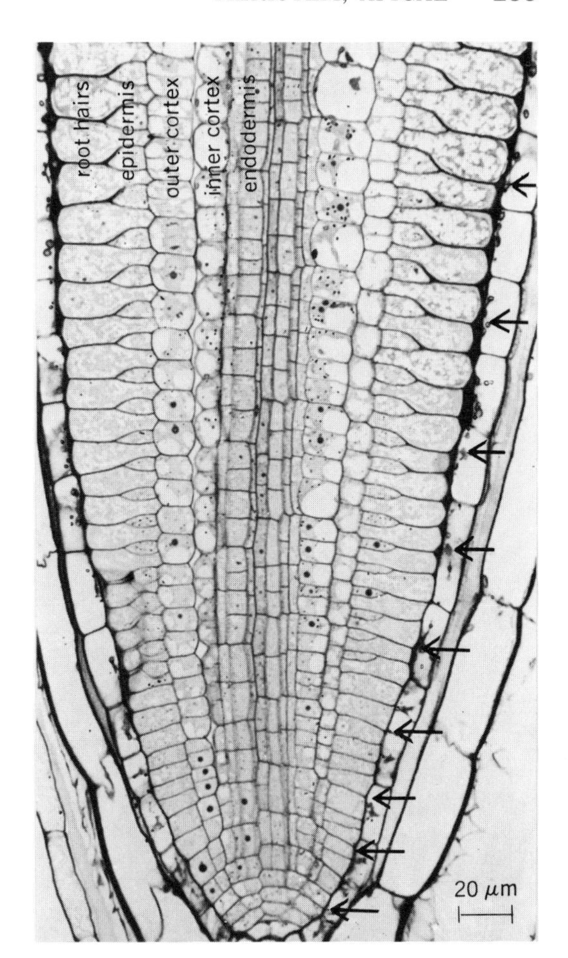

Fig. 2. Longitudinal section of an *A. pinnata* root, showing the files of cells that are generated by the sequences of formative divisions within the merophytes. Arrows indicate intermerophyte boundaries along the right side of the root.

that of the metaxylem cell initial, which begins its phase of differentiation without undergoing any proliferative divisions. On the other hand, the outer and inner sieve elements complete one and two rounds of transverse divisions, respectively, before they begin to differentiate.

Developmental changes. Active division of the apical cells in fern roots does not continue throughout the life of the root, but is confined to the early stages of root development. In young *Azolla* roots the apical cell divides every 3 to 5 hr until the root is about 1 mm in length. During subsequent development the apical cell becomes progressively more vacuolate and ceases to divide. The number of undivided merophytes adjacent to the apical cell and the sites of cell division and differentiation change during root development. As the root develops, there are fewer undivided merophytes adjacent to the apical cell. The different categories of formative divisions occur in merophytes closer to the apex, and the sites of the transverse proliferative divisions also shift toward the apex as the root ages (compare Figs. 1 and 2). The first round of proliferative divisions do not, however, invade the zone of formative divisions. When roots of different ages are compared, it is found that the length of the merophyte at which a given category of transverse divisions occurs is

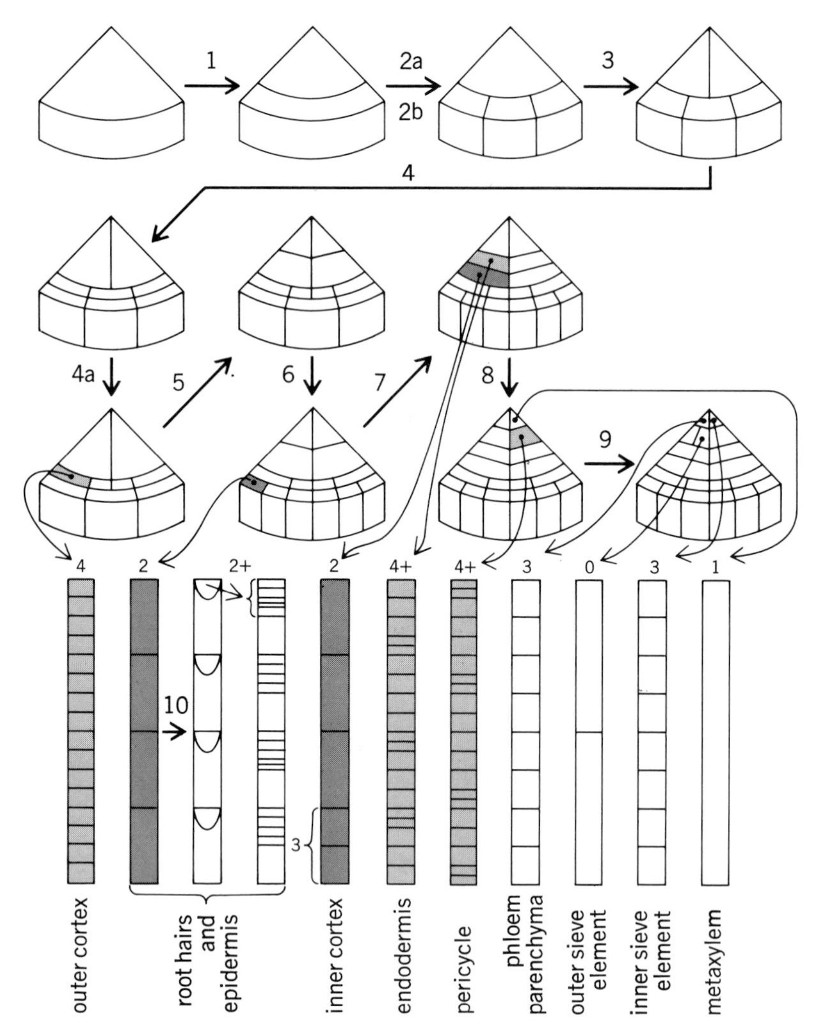

Fig. 3. Diagram summarizing formative divisions and final outcome of proliferative divisions in the *A. pinnata* root apex. Formative divisions that subdivide merophytes are numbered and shown at left and along top. Columns represent final stages of the cell files within merophytes after cell proliferation. Small numbers indicate the number of rounds of divisions in each case.

remarkably constant in roots at different stages of development. This suggests that cell dimensions may play a role in initiating this type of division. The differentiation of sieve and xylem elements also advances toward the apex of the root during development.

Number of cell generations. The study of the *Azolla* root has provided evidence that the number of successive cell generations may be limited. Mature *Azolla* roots contain about 50 merophytes, this number being attained by the time the root is about 1 mm in length. Thus the apical cell divides only about 50 times. When the number of cell cycles completed along the cell lineages of formative and proliferative divisions is calculated, again a maximum of about 50 cell generations is found. In contrast to the determinate growth shown by fern roots with a single apical cell, the multicellular apical meristems in roots of higher plants function actively throughout root development. In addition to the larger number of meristematic cells, higher plant root apexes also possess a quiescent center—a group of cells that are relatively immune to perturbations which damage more rapidly cycling cells. By acting as a very slowly cycling reservoir of replacement initials, each capable—like an apical cell—of contributing an alloted quota of cell generations, the quiescent center could effectively overcome the determinancy imposed by limitation of cell generations.

For background information *see* MERISTEM, APICAL in the McGraw-Hill Encyclopedia of Science and Technology.

[ADRIENNE R. HARDHAM]

Bibliography: B. E. S. Gunning, J. E. Hughes, and A. R. Hardham, *Planta*, 143:121–144, 1978; C. Hébant, R. Hébant-Mauri, and J. Barthonnet, *Planta*, 138:49–52, 1978; P. M. Lintilhac and P. B. Green, *Amer. J. Bot.*, 63:726–728, 1976.

Microcomputers

The successful entry of microcomputers into science and technology has become a widely accepted fact. Individual laboratories, advanced manufacturing plants, and even the designers of relatively inexpensive consumer products have found that the increased productivity and the higher level of sophistication possible with computer-based systems can often outweigh their additional cost.

Laboratory instrumentation. One area of science in which microcomputers have already made a tremendous impact is laboratory instrumentation. Before the invention of large-scale integrated (LSI) circuitry and the microcomputer, minicomputers were primarily employed only in the more complex and expensive laboratory instrumentation such as direct-reading emission spectrometers, inductively coupled plasmas, and mass, Fourier-transform infrared, and x-ray spectrometers.

The tremendous decrease in costs associated with microprocessors, semiconductor random-access memory (RAM), and a variety of types of read-only memory (ROM) has caused a virtual revolution in laboratory instrumentation. Today microcomputers are being found in an ever-increasing number of much less complicated laboratory instrumentation, such as ultraviolet/visible, infrared, and atomic absorption spectrometers, and even in support hardware such as the analytical balance.

Microcomputer applications. Microprocessors are used to do a wide variety of calculations for calibration, curve fitting, peak detection, peak area computation, scale expansion, data scaling, and statistical evaluation. In addition to providing the obvious capabilities for programming an instrument to reexecute a previous function, many manufacturers have introduced microprocessor-based systems capable of guiding inexperienced operators through complex setup procedures. A series of questions is presented which describes the next decision or step to be completed. The sequence moves from initially presenting the possible operating modes for a given instrument through the selection of a variety of pertinent parameters, followed by sample analysis and data reduction. Often certain parameters are set automatically, based on the operator's other selections. For instance, with one recently introduced infrared instrument, the operator chooses a given scan mode, and the instrument selects and sets the optimum speed, gain, and slit program, chart format, and wavelength range. External programming by the operator is generally accomplished by push-

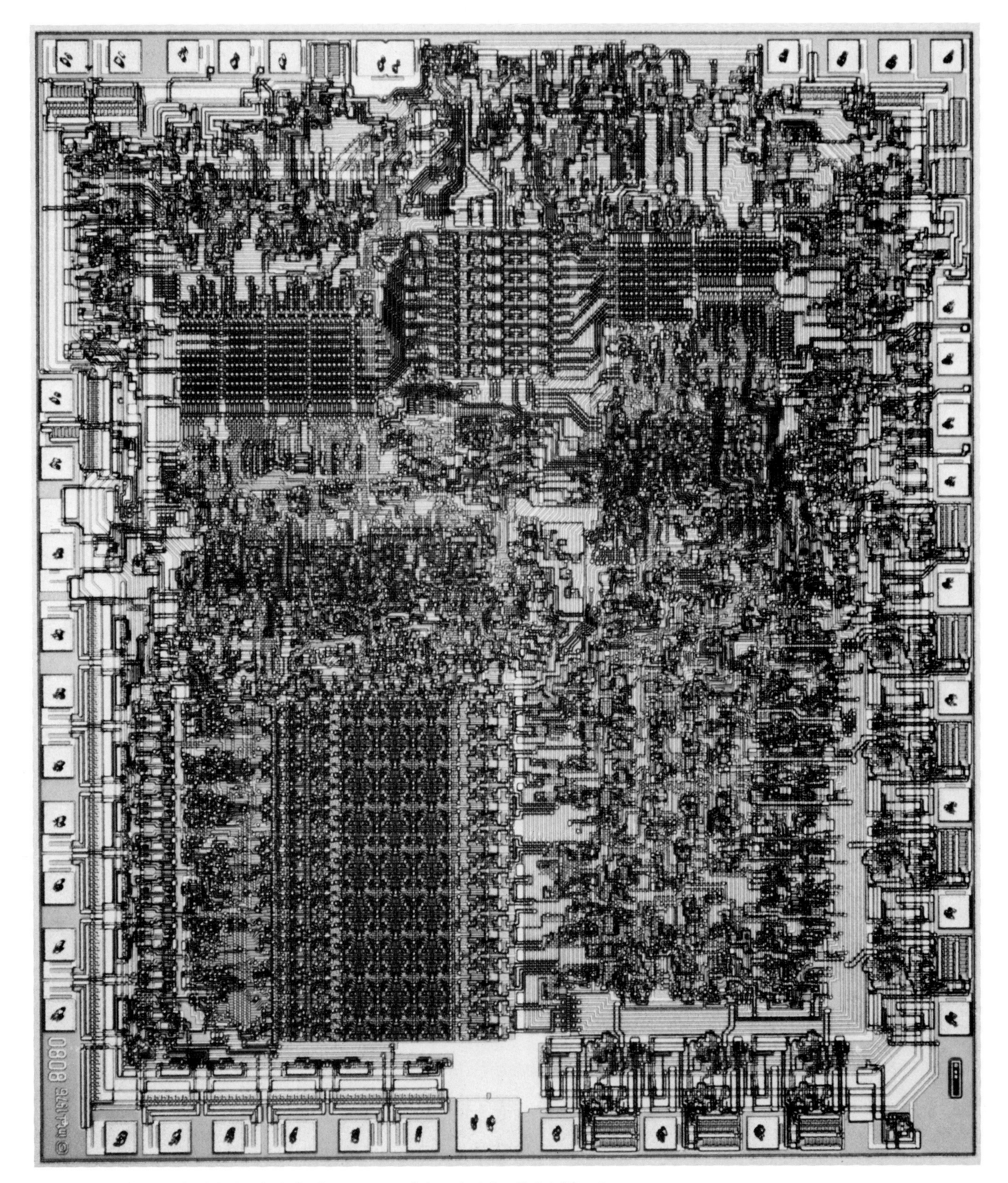

Fig. 1. Photomicrograph of the Intel 8080 microprocessor integrated circuit. *(Intel Corp.)*

button to select a desired function provided by the manufacturer in permanently programmed ROM.

Structure and programming. A microcomputer is typically composed of LSI circuit chips, each containing the equivalent of many thousand electrical components. The heart of any microcomputer is the central processor unit (CPU), normally referred to as a microprocessor. A photomicrograph of the Intel 8080 microprocessor is shown in Fig. 1. These highly complex but relatively inexpensive chips are combined to provide memory, input/output, and low-level computational functions, resulting in a microcomputer (Fig. 2). Most microcomputers require the appropriate combination of

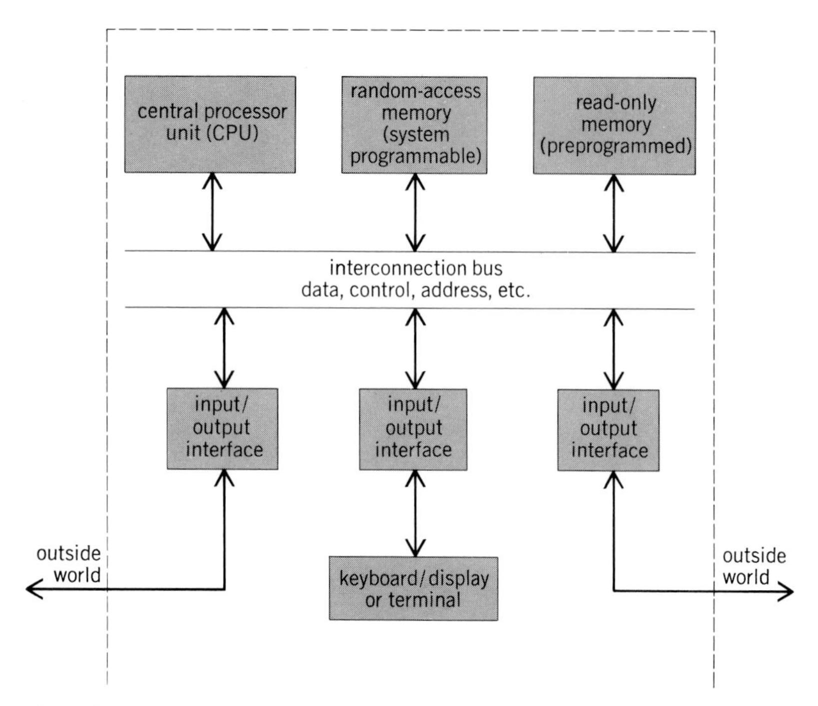

Fig. 2. Block diagram of a typical microcomputer.

a series of relatively low-level functions to implement a desired high-level task. To accomplish this, a series of binary programming instructions or software is required.

While such an approach allows tremendous flexibility and relatively low per-unit cost, it can also necessitate a substantial investment in programming. Initially, most instrument manufacturers utilized large computers which converted relatively high-level programming statements to reasonably effective binary machine code programs for use on a microcomputer. More recently, however, the trend has been toward software development using a microcomputer configured with a substantial amount of semiconductor memory, mass memory in the form of a cassette tape or a floppy disk, and an alphanumerics terminal, often referred to as a developmental system. After the desired algorithms have been written and debugged, they are transferred to permanent ROM in a much smaller and less expensive microcomputer within a given instrument or system.

Custom microprocessor applications. While the most widespread impact of microcomputers on science and technology has probably resulted from their extensive adaptation by instrument manufacturers, the number of custom microprocessor applications is growing rapidly. These systems range from such highly complex processor networks as the Smithsonian Institution's Multimirror Telescope and Lawrence Livermore Laboratory's SHIVA laser system (with a power of over 20 GW) to simple single-user laboratory data acquisition systems. As design engineers have become more familiar with microcomputers, their use has rapidly increased. One major manufacturer of aerial tramways is producing new systems as well as retrofitting old ones with microcomputers. The recent vast increase in availability of sophisticated, flexible, high-performance computer hardware, intended primarily for the "hobbyist" market, is also having a major effect. Large numbers of lay people as well as scientists have been purchasing microcomputers. Thus, a rapidly increasing percentage of the population is being educated in microcomputer capabilities. This increased public awareness will undoubtedly contribute to the proliferation of applications.

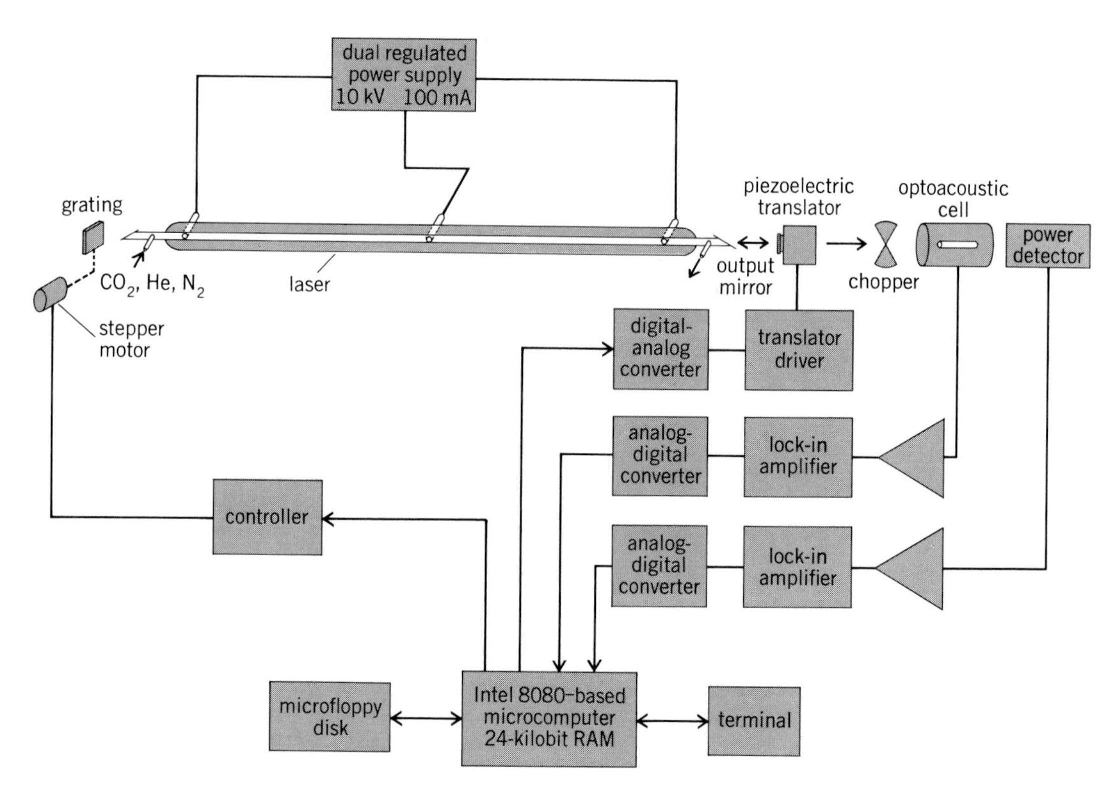

Fig. 3. Optoacoustic spectrometer employing a microcomputer.

High-level software. The vast individual market has also contributed to the level of software support. Most individuals do not want to program in binary or the slightly easier assembly level, where each single step of an algorithm must be tediously detailed. Due to the resulting demand, a number of high-level software packages have become available. Interpreters, such as a variety of BASIC operating systems, are relatively easy to use. These systems provide conversational interaction during programming, but may have drawbacks relating to memory requirements and execution speed. Compilers such as FORTRAN are also available, but are often difficult to program for use in interactive control of external devices. Continued advances can be expected in improved programming technology.

Optoacoustic spectrometer. A typical example of how a microcomputer can be employed to perform a variety of functions which would be difficult to implement without on-line computational capabilities is the optoacoustic spectrometer (Fig. 3). The overall principles behind this particular spectroscopic technique are simple. A reasonably high-intensity intermittent photon source is directed into a detection cell containing a gas mixture to be analyzed and a microphone (optoacoustic cell). When a component in the mixture absorbs energy, it heats up and causes a pressure wave or sound at the frequency of the intermittent excitation beam. This pressure wave is picked up by the microphone. If the excitation wavelength is then varied, it is possible to extract qualitative and quantitative information about the species in the cell. Unfortunately, implementation of such a system with a low-pressure carbon dioxide infrared laser presents a number of complex problems. This particular laser is not continuously tunable, and can be made to operate only at a number of discrete wavelengths near 10.6 μm. The observed signal is a function of the power delivered to the cell; however, the output of the laser varies from line to line. Slight changes in the distance between the grating and the output mirror (cavity spacing) are critical and must be optimized. Automatic wavelength calibration is also highly desirable. Finally, the absorption spectra of the species of interest can overlap, requiring simultaneous-equation data-reduction techniques.

The microcomputer can be employed to perform the functions of wavelength selection, automatic cavity spacing adjustment for maximum output power at a given line, correction for the variation of output power, and data-reduction computations. Data are permanently stored on the microfloppy disk or transferred to the terminal.

Programming in this particular application is accomplished by using an interpretive compiler operating language, CONVERS, which provides high-level user interaction with the experimental system by combining the most desirable features found in both interpreters and compilers.

Future prospects. Future applications of microcomputer technology will without question be closely linked with the rapid advance of available hardware technology and improvement in cost effectiveness. Current trends include continued reduction in memory cost, larger word size formats compatible with modern minicomputers, reduction in the total number of integrated circuit devices needed to implement a given capability, higher computational speed, and the direct availability of higher-level computational functions.

For background information *see* DIGITAL COMPUTER; INTEGRATED CIRCUITS in the McGraw-Hill Encyclopedia of Science and Technology.

[M. BONNER DENTON]

Microwave solid-state devices

A new class of microwave solid-state devices utilizes light to modulate or otherwise control the microwave energy output. The basic idea of such "optical control" is to use light to directly regulate the internal dynamics of a microwave source by the optical generation of charge carriers within that source. This is in marked contrast to the conventional approach whereby control is achieved indirectly through the interaction of some external device (for example, a varactor or PIN diode) with the microwave energy produced by the source.

Principle of optical control. When a semiconductor is illuminated with light having a photon energy slightly greater than the band gap, the incident photons are absorbed to a considerable depth in the material. The absorption of this light results in the generation of charge carriers (electron-hole pairs) within the semiconductor. Since the operation of a solid-state device depends on the dynamics of its internal carriers, the presence of such optically generated carriers alters the behavior of the device. For example, in an avalanche diode oscillator the rate of carrier buildup during an oscillation cycle depends on the number of carriers present at the beginning of the cycle. The additional carriers present when the device is illuminated with light will therefore result in a change in the carrier buildup rate which can produce changes in the frequency and amplitude of the oscillation.

Device configuration. A simple configuration for an optically controlled microwave source is shown in Fig. 1. A silicon avalanche diode is mounted in a coaxial microwave resonator which has a small opening in the outer wall near the diode. The diode is illuminated through the opening by a controllable light source, such as a miniature semiconductor laser. The structure of the avalanche diode, shown in Fig. 2, is essentially the same as that of conventional diodes, except that the top metal contact has been made small to allow light to reach the silicon surface. The depth to which this light is absorbed in the silicon depends

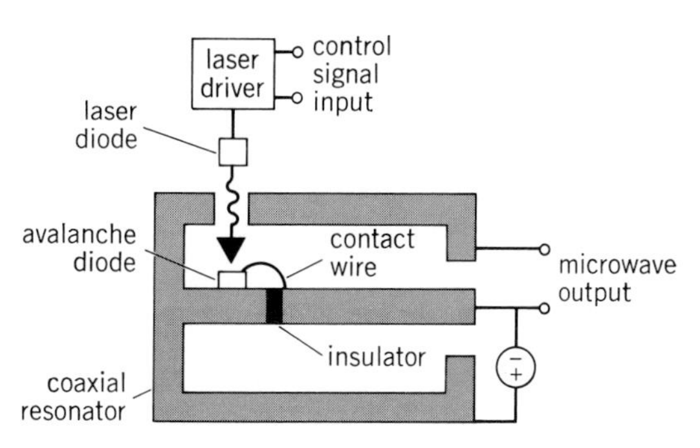

Fig. 1. Basic configuration of an optically controlled microwave oscillator.

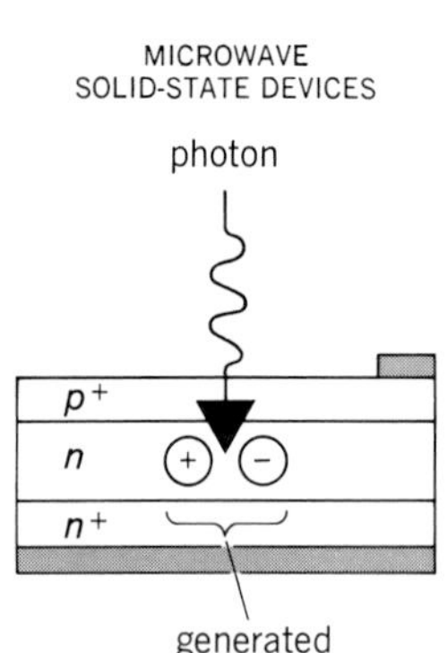

Fig. 2. Optical carrier generation in an avalanche diode.

on its wavelength. Hence, in order to produce optically generated carriers in the active *n* layer of the diode, the wavelength emitted by the light source must be tailored to the diode structure. For the dimensions of typical microwave devices (micrometers), this requirement is well met by the gallium-arsenide infrared lasers and light-emitting diodes recently developed for optical communication systems.

Optically modulated oscillators. Optical control is of particular interest for oscillators like the TRAPATT diode, for which high peak power levels severely hamper conventional control schemes. Amplitude modulation depths of 90% and frequency modulation of several percent have been demonstrated with optically controlled TRAPATTs operating at about 1 GHz with microwave output powers of nearly 100 W. Modulation at such high microwave power levels has been possible with optical power levels as low as a few milliwatts. The relative amounts of frequency and amplitude control can be adjusted by tuning the microwave circuit. Thus, it is possible to produce predominantly frequency or amplitude modulation in accordance with what is desired for a particular application. Similar results have also been obtained with IMPATT diodes operating at 10 GHz. For the IMPATT, however, frequency modulation has so far been limited to several tenths of 1%.

The capability for very-high-speed modulation of these devices has been demonstrated by illuminating the device with rapidly varying optical signals. Both frequency and amplitude shifts have been produced in nanosecond time intervals corresponding to the sharp rise and fall of the controlling optical signal itself. Figure 3 illustrates high-speed digital frequency modulation with an optically modulated TRAPATT diode. The top oscilloscope trace in this figure is proportional to the intensity of the optical control signal, and the bottom trace is proportional to the microwave oscillation frequency.

Optically induced phase control. Phase control of microwave oscillators by optical signals has also been achieved. In this case, carriers generated by a modulated optical signal are used to periodically disturb the normal operation of the oscillator. As with any oscillator, such a periodic disturbance causes the oscillation to lock in phase with the disrupting signal, provided that the signal frequency is close enough to the oscillation frequency or a subharmonic thereof. Such optically induced locking of transistor oscillators has been achieved at frequencies as high as 2 GHz with optical signals modulated at frequencies of about 100 MHz.

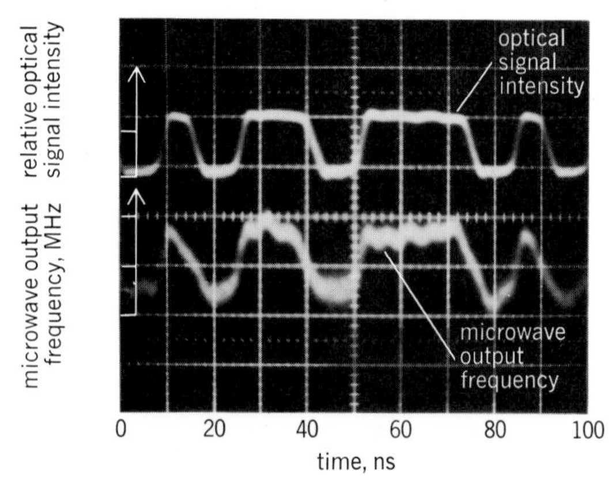

Fig. 3. Oscilloscope display showing frequency modulation with an optically modulated TRAPATT diode oscillator. (*From R. A. Kiehl and R. E. Hibray, IEEE Proc., 66: 708–709, June 1978*)

Applications. Optically modulated microwave devices are particularly attractive for application in high-speed data links and short-pulse radars. For such systems, the high-speed capabilities of optical techniques promise the realization of higher data rates and narrower microwave pulse widths than are possible with conventional schemes. Furthermore, the electrical isolation that exists between the microwave source and the control device (the light source) eliminates unwanted coupling of microwave energy into the modulator circuitry, a troublesome problem in conventional high-speed systems.

Optical phase control is attractive for active phased-array radars where it is necessary to control the phase of a great number of microwave sources spaced some distance apart. The use of a modulated optical signal rather than a microwave signal to achieve phase control allows the control signal distribution to be done with tiny optical fibers instead of coaxial or waveguide lines. This gives optical control an important advantage in lightweight airborne systems.

For background information *see* MICROWAVE SOLID-STATE DEVICES in the McGraw-Hill Encyclopedia of Science and Technology.

[RICHARD A. KIEHL]

Bibliography: R. A. Kiehl, *IEEE Trans. Electron Devices*, ED-25:703–710, June 1978; R. A. Kiehl and E. P. EerNisse, in *Digest of 1977 IEEE International Electron Devices Meeting*, pp. 103–106, December 1977; H. W. Yen and M. K. Barnoski, *Appl. Phys. Lett.*, 32:182–184, February 1978.

Mining

Recent developments in mining operations have been concerned with environmental protection in connection with surface and underground mines and with the safety and health aspects of mining and related operations.

Environmental protection. The protection of the environment in surface mining took a step forward when the U.S. Congress passed the Surface Mining Control and Reclamation Act (P.L. 95–87) in 1977.

Surface mining. Taking approximately 10 years in its progress through the Congress, the act applies primarily to coal mining, but will also have long-reaching potential applications to metallic mining, nonmetallic mining, and both deep and surface mining. The act applies interesting concepts to control of the environment and the environmental impacts of mining. It defines the operation of mining as a temporary land use, and states that after the mining operation is completed the land must be returned to its former use or to a higher and better land use. This new concept has wide-ranging meaning.

Many of the individual states had surface mining control legislation which varied in terms of technical complexity and the requirements placed on the mining industry. The Surface Mining Control and

Reclamation Act imposes regulations on the mining industry in two steps. The first set of regulations, identified as interim performance standards, went into effect for all new mines on Feb. 3, 1978, and for all active coal mines on May 3, 1978. This interim surface coal mining program regulates postmining land uses, backfilling and grading, topsoil removal, hydrologic balance, waste piles, explosives, revegetation, and steep-slope surface mining. In addition to the areas covered in the interim regulations, permanent performance standards are to be prepared and will go into the Code of Federal Regulations early in 1979. Additional areas included in the permanent regulations are conservation and recovery of coal, surface area stabilization, topsoil restoration, prime farmland, permanent water impoundments, augering, waste disposal, surface impacts of underground mining, fire hazards, air quality at mine sites, fish and wildlife protection, and other environmental concerns.

The act provides for the states to continue to be the primary agents for enforcing these regulations. Only in the event that a state fails to effectively enforce its own regulations—which are to parallel the Federal regulations—will the Federal government take over the administration of this program.

Some of the areas which are covered by the regulations deal with the preservation of prime farmland and certain tests and assurances that prime farmland must be restored to its original productivity following mining. Sedimentation ponds must be designed following a relatively sophisticated format, so that they will meet certain strict effluent guidelines that have been set forth by the Environmental Protection Agency. Arid land, principally in the West, must be revegetated with native, self-sustaining species, and the land must sustain this vegetation for at least 10 years. To ensure this revegetation in the West, the area will be kept under a performance bond for 10 years. In Appalachia and the central coal providence, revegetation must similarly be self-sustaining, but bonds must be retained only for a period of 5 years.

Provisions are included in the act dealing with a petition process, whereby "lands unsuitable for mining" can be identified and, following a rather detailed procedure, can be set aside by the Secretary of the Interior. Certain tests proving the existence of fragile flora or fauna, of unusual historical or cultural or scenic values, or of an unusual and irreplaceable water supply are some of the matters to be considered in determining an unsuitable area.

Another environmental control measure of great importance, especially in the Northern Great Plains coal fields, is the delineation of alluvial valley floors for special protection, as well as the restriction of mining on lands that had been set aside and identified as being flood-irrigated or subirrigated.

In addition, surface mining regulations cover the control of blasting, including the taking of a preblasting survey in homes within a short distance from the blast site. Furthermore, the frequency and magnitude of the blasting are carefully controlled.

Coal haul roads must be designed in such a way to have minimal adverse impact on the environment, and must be removed following the completion of the mining if they are not otherwise retained as part of an approved postmining land use.

A mining concept that had been growing in popularity in central Appalachia is mountain top removal. This procedure is still possible under the terms of the 1977 act, but the coal mine operator must show the regulatory agency that the postmining land use has been assured and that necessary facilities and utilities will be provided at the site. Certain mountain top areas that have been flattened through an approved mining process may be converted into farmland, and even into new town sites. However, the local planning agency must ensure that access roads, utilities, and other community facilities are made available on the mine site.

The environmental protection performance standards discussed above apply not only to surface mining of coal but also to the surface impacts of underground mining. The regulations deal with the problem of mine cave-ins or mine subsidence, as well as the disposal and covering of coal refuse and the discharge of mine drainage to receiving streams.

As part of the requirements of the act, Congress has required the Council on Environmental Quality to make a study on the possible applicability of this coal mining legislation to all surface mining, including the surface impact of deep mining for all metallic and nonmetallic minerals. This study, which was well under way in mid-1978, will first concentrate on oil shale development, then on sand, gravel, and crushed-rock production, and finally will cover all other metallic and nonmetallic mineral production. It is contemplated that certain philosophies may be applicable to these other minerals, for example, mining as a temporary and not a permanent land use. This landmark legislation also requires that a study be made of the applicability of the act to the mining problems in Alaska.

In order to aid the mining industry and government in finding solutions to newly raised problems, the Department of the Interior has set up mineral institutes throughout the United States not only to carry out research on mineral production but also to focus on the development of environmental protection in surface mining operations. Numerous other state and Federal agencies will be accelerating their development programs for environmental projects that will aid in refining and fine-tuning the technical requirements related to this surface mining act.

Underground mining. Environmental protection in underground mining operations is intertwined with health and safety developments. Principal improvements are related to equipment for dust control, methane dilution, and air quality evaluation. A new design for a mine stopping for controlling airflow was patented by the U.S. Bureau of Mines. A variety of new designs for cutter bit sprays, auxiliary fans, and electronic monitoring devices have been perfected and marketed. Much of the newest equipment is designed to improve safety and the environment for workers in the fast-growing mechanized longwall mining sections of flat or gently rolling coal beds.

[DAVID R. MANEVAL]

Safety and health aspects. The declining frequency of accidents during the past decade (see the illustration) is proof of the mining community's concern for safety. To ensure a continuing decline in the accident rate, Congress passed the Federal Mine Safety and Health Amendments Act of 1977, which drastically changed the previous coal mine health and safety act and repealed the earlier act for metal and nonmetallic mines. Basically, the 1977 act increases the role of the Federal government, strengthening its powers to develop and enforce regulations intended to militate against the hazards of mining. This report highlights important new technology intended to promote a safer, healthier, and more productive tomorrow.

Respirable dust. A comprehensive program to reduce hazardous respirable dust levels is being pursued in mills and in preparation plants as well as in mines. All studies to date show ventilation to be the most cost-effective means for dust control. Where the ventilation system alone is insufficient, notable reductions in respirable dust have been achieved by modifying auger-type miners and by redesigning bagging nozzles.

Auger-type miners are retrofitted by mounting two fans above the conveyor exhausting into an air duct discharging into the return. The cost to modify a machine approximates $15,000.

Partially surrounding each nozzle on a bagger with a sheet-metal shell forces air into the shell, entraining and exhausting the dust to a central collector. Modifying the bagger control circuit to delay bag release until a pulse of clean air is injected through the nozzle reduces spillage from two cups to one teaspoon per bag. These changes can be made for less than $2000.

Noise. Abatement of noise to acceptable levels is a difficult problem, for noise seriously affects productivity. The first and foremost effort is identifying sources of noise. An especially severe source is load-haul-dump machines. The Wagner Mining Machine and National Gypsum companies reduced noise from the transmission, engine, and cooling fan of an ST-5A Scooptram to below compliance levels. This noise abatement system, proved after a year under production conditions, costs $10,000–15,000 retrofitted on many load-haul-dump machines.

Noise from service vehicles affects the largest number of miners. A Getman Corporation dispatch vehicle was brought into compliance by a noise-quieting package that could be constructed and installed at a cost of $300 for materials and 10 hr of labor.

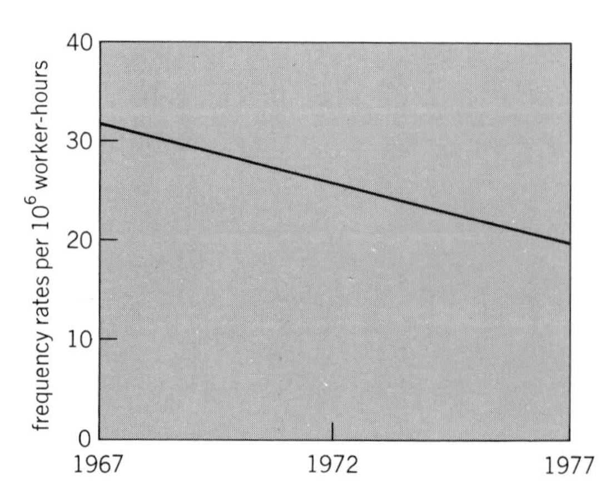

Fatal and nondisabling injuries in American mines and mills.

Resilient decking on vibrating screens, leaded-vinyl curtains around dryers and crushers, and impact-absorbent pads on chutes have reduced noise in coal preparation and ore processing plants to below compliance levels. Taconite producers who are members of the American Ore Association and the Consolidation Coal Company are conducting important studies in this area.

Toxic gases. Industrial hygiene studies focus on means for monitoring toxic gas concentrations in mines and surface installations. Teaching guides and handbooks, available through the U.S. Bureau of Mines and the Mining Safety and Health Administration, are important aids for companies when surveying inhalation contaminants and determining compliance with regulations on atmospheric pollutants.

Diesels operating in poorly ventilated areas produce carbon dioxide at a faster rate than other toxic exhaust gases. Monitors available from Beckman Instruments Company and Andros, Inc., can be installed on diesel-powered vehicles to give warning when the carbon dioxide concentration reaches an unsafe level, shutting off the engine if the concentration is not reduced to tolerable limits within 2 min.

Roof and rib control. Historically the greatest hazard to miners has been falls of the roof and the rib (the side of a pillar or wall of an entry). New technology is aimed at predicting hazardous geological conditions, which should allow preselection of safer and more productive mine design and roof control plans.

Seismic techniques, adapted from petroleum exploration, are being used for from-surface detection of channel sands over the Meigs No. 1 Mine (Ohio) and faults and rolls in the Lincoln Mine (Colorado). Ground radar is used to detect abandoned workings, fracture zones, and faults in metal mines. Satellite imagery and aerial photography techniques are being perfected to locate faults, linears, and zones of major earth fracturing.

On-site stress-measuring devices have improved greatly. A relatively low-cost gage when used with overcoring stress-relief and material property evaluation allows a reasonable characterization of the adequacy of roof stability before problems develop. Companies studying this method include Occidental, Anaconda, Amax, and Beckley Coal.

Full-column resin bolting has been effective in many areas where mechanical bolts have failed. Resin bolting is a roof and rib control technique in which steel or wood rods are secured in drilled holes by a polymer such as urethane, polyester, or epoxy. Among recent interesting studies were those in the main shaft of the Lucky Friday (Idaho) mine; resin bolting reduced closure to 3/4 ft (23 cm), instead of the 1½ feet (46 cm) previously experienced with jacket sets.

Where and when coal bumps and rock bursts are likely to occur is being determined by monitoring microseismic noise, tilt, and electromagnetic radiation. (Coal bumps and rock bursts are the result of the sudden yielding, sometimes with explosive-like violence, of strained coal or rock masses subjected to excessive stresses.) These observations fluctuate in a predeterminable manner prior to rock failure. Major efforts are centered in the Coeur d'Alene silver mining district and in coal mines in Colorado and Utah. The fact that three major mining companies have purchased and installed monitoring units illustrates the predictive practicality of microseismic noise generation.

Overstressed rock is being forced to burst deliberately during times when appropriate safety precautions can be taken. One destressing technique, being tried in the Star Mine (Idaho), involves drilling and lightly blasting long holes in the floor, sill, pillar, and first level of an undeveloped stope. This has been so promising that plans are being made to similarly destress an entire level.

A full-scale study of factors influencing slope stability is being conducted in Kennecott Copper Corporation's Kimberly pit (Nevada). Examination of the local structural geology, area tectonics, pit-wall stresses, strength of wall rock, and possible loading conditions led to the conclusion that the normal slope could be increased without slough benches for heights up to 550 ft (167.6 m). (Slough benches are relatively broad ledges, similar to steps or terraces, cut into one wall of an open-pit mine to reduce the hazard of rock slides.) After 18 months pit-wall stresses and deformations were as predicted, the wall remained stable, and mining safety and productivity were improved.

Microseismic monitoring of slopes has progressed to the point where failure surfaces can be located in advance of rock slides. At the Ruth copper mine (Nevada), monitors gave evidence of two slides beginning to form. There was enough advance warning to remove personnel to safety and to change and protect haulage roads. In the future, aerial photogrammetric techniques will be used to monitor surface movements.

Waste disposal impoundments. An engineering and design manual for coal refuse disposal facilities is now available from the U.S. Superintendent of Documents. This manual is an important adjunct to mine operators' efforts to determine the adequacy of their impoundments and for recognizing potential hazards.

Electrokinetic dewatering of coal sludge is being tested at the Henderson Mine (Colorado). Electrokinetic processing of mill slimes to make them suitable for backfill is being evaluated in the Los Torres Mine (Mexico), while safety practices are being developed by the U.S. Corps of Engineers.

Fire and explosion. Present concerns are the fire hazard of combustibles used in mines (such as timber, machinery, conveyors, plastics, fuels, and lubricants); fire and explosion hazards in oil shale mining; and the use of barriers to arrest explosion propagation. Among the studies of immediate interest are:

1. A mine ventilation computer program developed at Michigan Technological University based on 20 years of study and use in actual mine fire control. With this program, mine engineers can evaluate the ventilation network and also simulate fires and explosions, predicting their effect on escapeways and fire protection systems.

2. Water sprays for face ventilation developed by Foster-Miller Associates and being perfected by Consolidation Coal Company. Generally, only a small fraction of the main airflow reaches the face. Water sprays, strategically mounted on a miner, redirect the incoming air up the rib and sweep across the face. These sprays are quieter and more reliable than either diffuser fans or venturi sprays; they seem to be the most effective means yet developed for preventing frictional ignitions.

3. Underground fueling-area fire protection developed by Ansul Company. This system consists of rapid-response flame detectors that activate a dry chemical for fast fire quenching followed by a foam blanket over the area.

4. Shaft fire and smoke protection developed by Food Machinery Corporation. The system being tested in two deep mines consists of smoke, carbon monoxide, and temperature sensors installed at shaft stations, with signals transmitted to a main control for activating the appropriate smoke control doors and sprinklers. The system cost is estimated to be $100,000 for the main control and one level and $10,000–30,000 for each additional level.

Industrial hazards. Hazard and industrial analyses in underground and surface mines and of shaft sinking, haulage, and blasting practices show that there is a strong relationship among fatal accidents, task experience, and supervision. Management's philosophy and commitment to training are key factors in accident reduction.

Most industrial hazard studies involve the complex interaction among machines, people, and the mining environment. Few are near enough to completion to allow a meaningful evaluation. However, improved electrical circuitry is one area which has immediate application; for example, ground wire failure on many cables was traced to inappropriate manufacturing processes. A computer "model" of mine power systems, adopted by several companies, has improved voltage levels and short-circuit protection.

For background information *see* MINING; MINING, STRIP; MINING, UNDERGROUND in the McGraw-Hill Encyclopedia of Science and Technology. [DONALD W. MITCHELL]

Bibliography: *Coal Surface Mining Reclamation Costs*, Bur. Mines Inform. Circ. IC 8695, 1975; *Effectiveness of Surface Mine Sedimentation Ponds*, EPA 600/2.76.117, August 1976; *Environmental Protection in Surface Mining of Coal*, EPA 670/2.74.093, October 1974; *Mechanized Longwall Mining*, Bur. Mines Inform. Circ. IC 8740, 1977; Mine Ventilation Control Device, U.S. Patent no. 4,009,649; *Mining Research Review*, U.S. Bureau of Mines, August 1977; *Surface Mining and Reclamation Act of 1977—Public Law 95–87*, 91 Stat. 445, Aug. 3, 1977.

Molecular beams

Molecular beams are composed of molecules moving with nearly parallel velocities. In order to maintain this character, the molecules must move without undergoing any collisions in an evacuated apparatus. The most important feature of a molecular beam is that its molecules are isolated. This makes possible the study of the properties of individual molecules and of single collisions between molecules if one beam crosses a second beam.

This crossed-beam method has proved to be a very useful approach for studying the detailed dynamics of chemical reactions.

Experimental results. The main features of a molecular beam apparatus are shown in Fig. 1. Molecules emanate from an oven, are collimated into a beam by a slit, and may pass through one or more state selectors which will transmit molecules only in certain states (translational, rotational, and so forth) before they collide with the second beam. The entire apparatus is housed in a high-vacuum chamber to eliminate collisions. The density in each beam is sufficiently low that each beam is attenuated by the other by only a few percent or less. A detector can be rotated about the scattering center to detect any newly formed molecules as a function of scattering angle. A state selector can also be placed between the scattering center and the detector to analyze the states of the product molecules.

Study of reactions. In principle, one can study the reaction between any species, with each species in a selected quantum state, and analyze the quantum states of the product. Unfortunately, the reactive probability is very low (product intensities 6–10 orders of magnitude less than main beam intensities), and experiments have mainly been restricted to highly reactive systems such as the alkali, halogen, or hydrogen atoms reacting with halogen-containing molecules. A number of state selectors or analyzers have been used (but almost never at the same time because the reactively scattered intensity is too low), and two well-studied systems illustrate methods and results: For the $K + Br_2 \rightarrow KBr + Br$ reaction, measurement of the angular distribution of KBr shows that KBr is forward-scattered. The K atom rushes toward the Br_2 molecule and strips off a Br, and the KBr continues in roughly the same "forward" direction as the K atom was traveling. Measurement of the speed of the KBr shows that only a small fraction of the energy released in the reaction appears as translational energy, and direct measurement (with a rotational state analyzer) shows that the average rotational energy of KBr is very low. The excess energy must consequently appear in vibrational excitation of KBr.

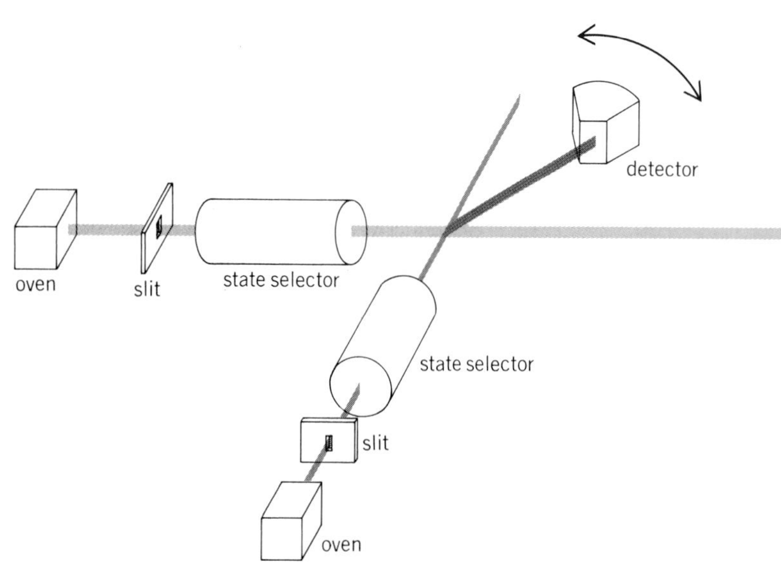

Fig. 1. Schematic of crossed-molecular-beam apparatus.

In contrast to this system, the $K + CH_3I \rightarrow KI + CH_3$ reaction is observed to scatter KI backwards (the K atom collides with the CH_3I, and the KI bounces back along the direction of the incident K), and the speed is high enough to account for most of the reaction energy, so that KI is not highly vibrationally excited. In addition, for $K + CH_3I$ the CH_3I speed has been varied and the reactivity found to rise to a maximum near 0.2 eV. The CH_3I orientation has also been specified and the reactivity at the I end found to be greatest.

Use of photons. A variety of other experiments rely upon the interaction of photons with molecules in the beam. Resonant chemical lasers have been used to produce reagent molecules in vibrationally excited states prior to intersecting the crossed beam, and tunable lasers have been used to identify the states of the product molecule by inducing fluorescence from those molecules. Photons have also been used as reagents. In these experiments one molecular beam crosses a laser beam instead of a second molecular beam, and unimolecular processes are observed. Photodissociation processes have been studied by K. R. Wilson, in which translational energy of the products and the dependence on laser polarization are measured to give information about unbound molecular states.

More recently, Y. T. Lee and collaborators have studied multiphoton dissociation processes. By measuring the angular distribution of the photofragments, they have unambiguously demonstrated that a single molecule absorbs 30–40 photons in the absence of collisions. The dissociation products can be identified and their speed can be measured. All beam results up to this point show that the highly excited molecule formed after absorption decays randomly without mode specificity.

Theoretical interpretation. Although the experimental results are frequently clearcut, their interpretation is not always as straightforward. Early chemical reaction theories were mainly based on "near-equilibrium" hypotheses, which treated a reaction as proceeding through an activated "complex" which decays randomly. Extensive averaging of the theoretical results was necessary to compare with experiment (rate constants), and much detail was obscured. While some of these theories, such as that of Rice-Ramsperger-Kassell-Marcus (RRKM), still find application, it is clear that all reactions cannot be described as proceeding through an activated complex. For example, both the $K + Br_2$ and $K + CH_3I$ reactions occur without formation of a long-lived complex. A variety of phenomenological models have been introduced to account for the experimental data. These include hard sphere interaction models, stripping models, and charge-transfer (or curve-crossing) models. Each is useful in interpreting the data, especially in giving some physical insight into the processes which are occurring. Unfortunately, however, these phenomenological models do not provide a general theoretical interpretation.

Born-Oppenheimer approximation. The most general theoretical treatment of a chemical reaction is to solve directly the Schrödinger equation for the entire system. While this is possible in principle, it is totally impractical and the results would be too detailed to digest. It is thus more meaningful to describe the system in terms of the interactions between the nuclei rather than between nuclei and electrons. This can be done in the context of the Born-Oppenheimer approximation, which treats the nuclei as so massive and slow relative to the electrons that they are able to instantaneously accommodate themselves to minute changes in the position of the nuclei. The nuclei may then be treated as moving under the influence of a potential which depends upon the positions of all of the nuclei. The theoretical treatment is now composed of two parts: first, the determination of the potential energy surface and, second, the description of the motion of the nuclei on that potential surface.

Treatment of collinear configuration. Even for the simplest possible chemical reaction, A + BC → AB + C, the potential energy will depend on three variables, R_{AB}, R_{BC}, and R_{AC}, the distances between the respective nuclei, and is not easily visualized. However, if A, B, and C are restricted to a collinear configuration, the potential may be displayed in the form of a contour plot; such a plot is displayed schematically in Fig. 2. Point 1 represents A at large distance from the BC molecule, R_{AB} is large, and R_{BC} is the equilibrium BC bond distance. An observer in this canyon would see a steep wall corresponding to compression of the BC bond and a much more gentle wall corresponding to dissociation of BC. The floor of the canyon (as drawn here) slopes up gently to a saddle point at 2, whereupon the floor then drops off sharply into the much deeper canyon characteristic of the product AB + C. The potential at point 3 is characteristic of the diatomic molecule AB far from atom C.

If the motion of the nuclei can be treated classically, the coordinate axes may be slightly changed so that a ball rolling on the surface $V(R_{AB}, R_{BC})$ will describe the motion of the system. A typical trajectory of this type is shown schematically by a curve in Fig. 2. If the system (ball) possesses enough energy to pass over the saddle point 2, it is accelerated into the product valley, but the curvature of the surface introduces an oscillatory motion which becomes vibration of the product AB. This type of potential surface thus qualitatively describes the K + Br_2 reaction in which the product is observed to be very highly vibrationally excited. If the downhill portion of the potential surface were around the "corner," the product would tend to slide down the floor of the exit valley without vibration and be more characteristic of the K + CH_3I reaction.

Construction of potential surfaces. The qualitative results are clearly very highly dependent upon the curvature of the potential surface near the corner. Potential surfaces have been constructed in various ways, ranging from the semiempirical LEPS (London, Eyring, Polyani, Sato) method to various methods based on first principles. The LEPS surfaces have been very useful in extracting qualitative information about how various surface bumps and hollows affect the reaction dynamics. Unfortunately, the dynamics are too sensitive to surface features to expect semiempirical methods to definitively determine the surface for a chemical reaction. The various methods based on first principles are time-consuming and expensive and are currently being investigated to determine how to best characterize the surface at the most sensitive points, and even to determine which points are most sensitive.

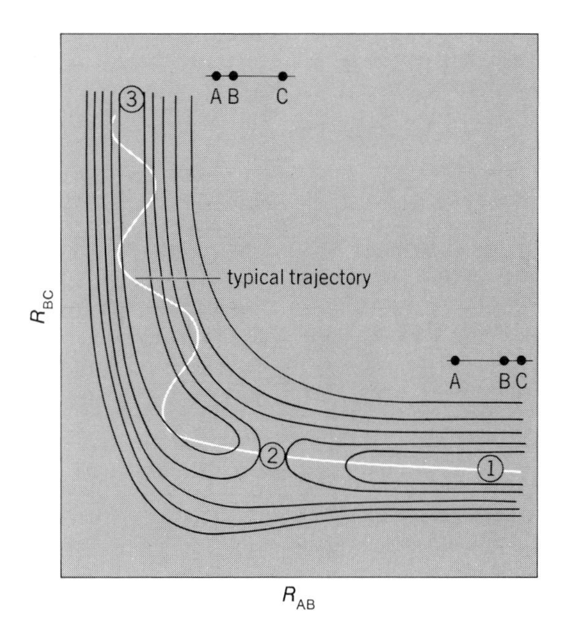

Fig. 2. Schematic of potential energy surface for the collinear exothermic reaction A + B → AB + C.

Quantum-mechanical description. The description of the motion of the nuclei on the potential surface has been almost entirely restricted to classical mechanics. A quantum-mechanical description must by necessity be correct, however, and recently R. E. Wyatt and associates and A. Kuppermann and associates have studied the quantum dynamics of the H + H_2 reaction on the semiempirical Porter-Karplus surface. As expected, quantum dynamics are most important at low energies, but some interference effects may be experimentally observable and may serve to characterize the potential energy surface in a manner similar to that used to invert elastic scattering data to give accurate atom-atom potential curves.

More complex reactions. Many reactions do not take place on a single potential surface and cannot be characterized by the Born-Oppenheimer approximation. These represent an additional complication, that of "hopping" from one surface to another, and are currently under investigation.

For background information *see* MOLECULAR BEAMS; MOLECULAR STRUCTURE AND SPECTRA; SCATTERING EXPERIMENTS, ATOMIC AND MOLECULAR in the McGraw-Hill Encyclopedia of Science and Technology.

[PHILIP R. BROOKS]

Bibliography: P. R. Brooks and E. F. Hayes (eds.), *State-to-State Chemistry*, ACS Symposium Series, vol. 56, 1977; R. D. Levine and R. B. Bernstein, *Molecular Reaction Dynamics*, 1974; Molecular scattering. *Advances in Chemical Physics*, vol. 30, 1975; Potential energy surfaces, *Faraday Discussions of the Chemical Society*, vol. 62, 1977.

Nobel prizes

The Swedish Royal Academy of Sciences announced eleven recipients of the Nobel prizes for 1978.

Physics. This prize was awarded to three men: Arno Penzias and Robert Wilson, of Bell Telephone Laboratories in Holmdel, NJ; and Pyotr Kapitsa, director of the S. I. Vavilov Institute of Physical Problems in Moscow. Penzias and Wilson were cited for their identification of the fossil heat remaining from the "big bang" that created the universe, according to one prevalent theory. Kapitsa was recognized for his fundamental work on magnetism and the physics of supercold temperatures, specifically for his discovery of the superfluid properties of supercold helium (helium II).

Chemistry. The British chemist Peter Mitchell received this prize for his work in the field of bioenergetics. In explaining how plants and animals convert nutrition into energy, Mitchell showed that the driving force behind this conversion process is the generation of photons in the oxidation process.

Medicine or physiology. Three microbiologists, Werner Arber, of the University of Basel, Switzerland, and Hamilton O. Smith and Daniel Nathans, both of the Johns Hopkins University, were awarded this prize. They were honored for their pioneer work in the field of genetic engineering. Arber's discovery of restriction enzymes permitted the analysis of the chemical structure of genes and the mapping of their sequence along deoxyribonucleic acid (DNA) strands. Smith discovered the restrictive enzyme produced by the bacterium *Hemophilus influenza*. Nathans applied Smith's enzyme to the monkey virus SV40, which was thus broken into 11 well-defined fragments.

Economics. Herbert A. Simon, professor at Carnegie-Mellon University, was awarded this prize for his economic theory of "satisficing" behavior. In his psychological approach, Simon contested the traditional view that businesses seek to maximize their profits and that individuals seek to maximize their satisfaction—and that society benefits when they behave in this way. Rather, they set goals that represent reasonable levels of achievement and try to fulfill them. This behavior does not guarantee that the economy will benefit.

Literature. Isaac Bashevis Singer, a native of Poland who emigrated to the United States in 1935, has written about the East European Hasidim in stories containing elements of the supernatural and folklore. He was praised by the Academy "for his impassioned narrative art which, with roots in Polish-Jewish cultural tradition, brings human conditions to life."

Peace. This prize was awarded jointly to Egyptian president Anwar Sadat and Israeli prime minister Menachem Begin. With the intention of spurring on Egyptian-Israeli negotiations by extending the honor to both men prior to realization of a treaty, the Academy asked them to make new efforts to "secure a future without war to the war-exhausted people of the Middle East."

Nuclear power

The years 1977 and 1978 were marked by the development of United States policies on nuclear power which discouraged the breeding and recycling of plutonium and promoted uranium enrichment and alternative nuclear fuel cycles, and by a large number of cancellations and delays in the construction of nuclear plants. Other important developments included the effecting of amendments to the Price-Anderson Act, and the resolution of a legal challenge to portions of this act.

United States policies. United States policies toward the development of nuclear power during 1977–1978 were strongly influenced by concern that the use of plutonium as a fuel in other parts of the world might encourage the proliferation of nuclear weapons.

National Energy Plan. The National Energy Plan, presented by President Jimmy Carter on Apr. 29, 1977, states that "access to plutonium, or even the capacity to recover or isolate it, can lead to the risk of diversion of material that could be used for nuclear explosive devices. The United States should develop advanced nuclear technologies that minimize the risk of nuclear proliferation, but with the knowledge that no advanced nuclear technology is entirely free from proliferation risks."

It therefore became the policy of the United States to seek an alternative approach to plutonium recycling and the plutonium breeder for the generation of nuclear power. Commercial reprocessing and recycling of plutonium as well as the commercial introduction of the plutonium breeder were to be deferred indefinitely, while priority would be given to light-water reactor safety, licensing, and waste management. The President proposed to cancel construction of the Clinch River Breeder Reactor Demonstration Project and to withhold Federal funding for the completion of the reprocessing facility at Barnswell, SC. It was hoped that these measures would encourage other nations to slow down their development of plutonium-based technology.

It was recognized that other nations which view nuclear power as the only feasible alternative to dependence on oil and gas imports could be persuaded to abandon plutonium technology only if they were assured supplies of the slightly enriched uranium required for light-water reactors. The United States, therefore, proposed to reopen uranium enrichment services (which had been closed since June 30, 1974, when the capacity of the government's three gaseous diffusion plants was fully committed); to guarantee by law the delivery of enrichment services to any country sharing nonproliferation objectives; and to expand the United States enrichment capacity (with the next plant to be based on the new gaseous centrifuge technology).

In order to develop alternative nuclear fuel cycles which do not involve direct access to materials that can be used in nuclear weapons, a proposal was made to redirect United States nuclear research and development and to establish an international nuclear fuel cycle evaluation program. Finally, in lieu of fuel reprocessing, it was announced that the waste management program had been expanded to include the development of techniques for the long-term storage of spent fuel, and that technologies, designs, and environmental criteria for waste repositories would be developed by 1978.

Nuclear Non-Proliferation Act. Congress expressed a similar concern about the availability of nuclear materials by passing the Nuclear Non-Proliferation Act of 1978. According to Senator John Glenn, the purpose of this act is to "establish strict export criteria or guidelines to assure that nuclear exports are used only for peaceful purposes." Criteria include the right of the United States to veto both the re-export and reprocessing of nuclear material originating in the United States. Agreements already in force must therefore be renegotiated. European countries are resisting these renegotiations and, as a result, further shipments of United States fuel to Euratom are threatened. The European nations argue that the act anticipates the judgments of the 2-year International Nuclear Fuel Cycle Evaluation, in which they had agreed to participate, and that renegotiations are therefore premature.

Foreign reprocessing plants and breeders. As a further indication of foreign resistance to the anti-plutonium policies of the United States, on Mar. 22, 1978, the British House of Commons endorsed the plans of British Nuclear Fuels, Ltd., to build a new oxide fuel–reprocessing plant at Windscale to serve overseas as well as domestic markets. It was argued that the best way to discourage the spread of reprocessing plants in countries which do not have nuclear weapons capability is to provide them with reprocessing services. A small commercial reprocessing plant for oxide fuel has been operating since May 1976 at La Hague, France, where Compagnie Générale des Matières Nucléaires (COGEMA) is now building a much larger facility. Both companies have already signed contracts for the transport and reprocessing of spent fuel from other countries, including Japan. Furthermore, France, West Germany, Belgium, Italy, and the Netherlands have signed agreements for joint research and development of fast breeders and the formation of a group to market them. Prototype breeders are operating in France, England, and the Soviet Union, and a full-scale (1250 megawatts electric power, or MWe) commercial breeder, the Super Phénix, is being built in France.

Congressional response to policies. Although Congress has supported the Carter administration's nonproliferation policies, it has shown a reluctance to accept the antireprocessing and anti-breeder policies by insisting on continued funding of the Barnswell project and the Clinch River Breeder Reactor ($80,000,000 in fiscal year 1978, as opposed to the $33,000,000 termination funding request by the administration). Plans for increasing enrichment capacity and for proceeding with spent fuel storage and waste repositories have been received more favorably. An interagency task force has been created by the President to make recommendations on waste management.

Power use and plant construction. The use of nuclear power during 1977 increased about 30% over 1976, with 64 operating commercial plants supplying almost 12% of total electrical production. On the other hand, since 1975 there have been 23 nuclear power plant cancellations with only nine new orders. Schedule slippages during 1976 were equivalent to a delay of about a year for each of the 103 plants under construction.

Cancellations and delays are reported by the U.S. Department of Energy to be due to a variety of problem areas. These include the belief by many electric utility executives that the administration is ambiguous toward nuclear power and that the impact of the nonproliferation initiatives on the domestic nuclear fuel cycle is uncertain. There are also concerns over long and uncertain lead times, financing difficulties, fuel supplies, lack of standardization, debates regarding the need for increased electrical capacity, the perceived failure of the Federal government to solve waste-disposal problems, state and local efforts to restrict or prohibit nuclear shipments, and the strength of intervener groups. [PHILIP N. POWERS]

Price-Anderson Act. On Dec. 31, 1975, a bill was signed into law extending the Price-Anderson Act for an additional 10 years. In renewing this act several amendments were incorporated into the existing legislation, the most significant being a requirement for nuclear power reactor licensees to assume, in the event of a significant nuclear incident, additional financial responsibility in excess of available private liability insurance. The primary objective of this amendment is to reflect the intent of Congress to phase out the need for continuing government indemnity as soon as practicable, but in no event later than 1985.

Another current development concerns a legal challenge to the Price-Anderson Act related to the constitutionality of the "limitation of liability" provision. In 1977 a U.S. district court ruled this section of the act to be unconstitutional. The decision was appealed to the U.S. Supreme Court, which heard oral arguments during March 1978 and rendered its decision on June 26, 1978, upholding the constitutionality of the act and thereby reversing the earlier district court ruling.

Background. The Atomic Energy Act of 1954 ended the government monopoly on atomic energy in the nonmilitary sector and encouraged private industry to develop peaceful applications of this energy source. It soon became evident that the risk of uninsurable liability for a nuclear incident could be a major deterrent to private industry in the development of this new technology. Thus, an extensive joint study by Federal and industry leaders resulted in the enactment in 1975 by Congress of the Price-Anderson Act, encompassing two major objectives: (1) to ensure the availability of funds to reimburse the public in the event of a major nuclear incident not compensated by available insurance; and (2) to protect the industry against the risk of unlimited liability for catastrophic accidents.

Protection of the public was afforded by requiring that licensees who operate nuclear power plants maintain "financial protection" in an amount set by the Atomic Energy Commission (AEC). The AEC has subsequently been disbanded, and its regulatory functions in administering the Price-Anderson Act now rest with the Nuclear Regulatory Commission (NRC).

For electric power reactors of substantial size (over 100 MWe) the financial protection amount was fixed at a level equal to the amount of private liability insurance available in 1957, that is, $60,-000,000. In addition, a provision for $500,000,000 of government indemnity would take effect above

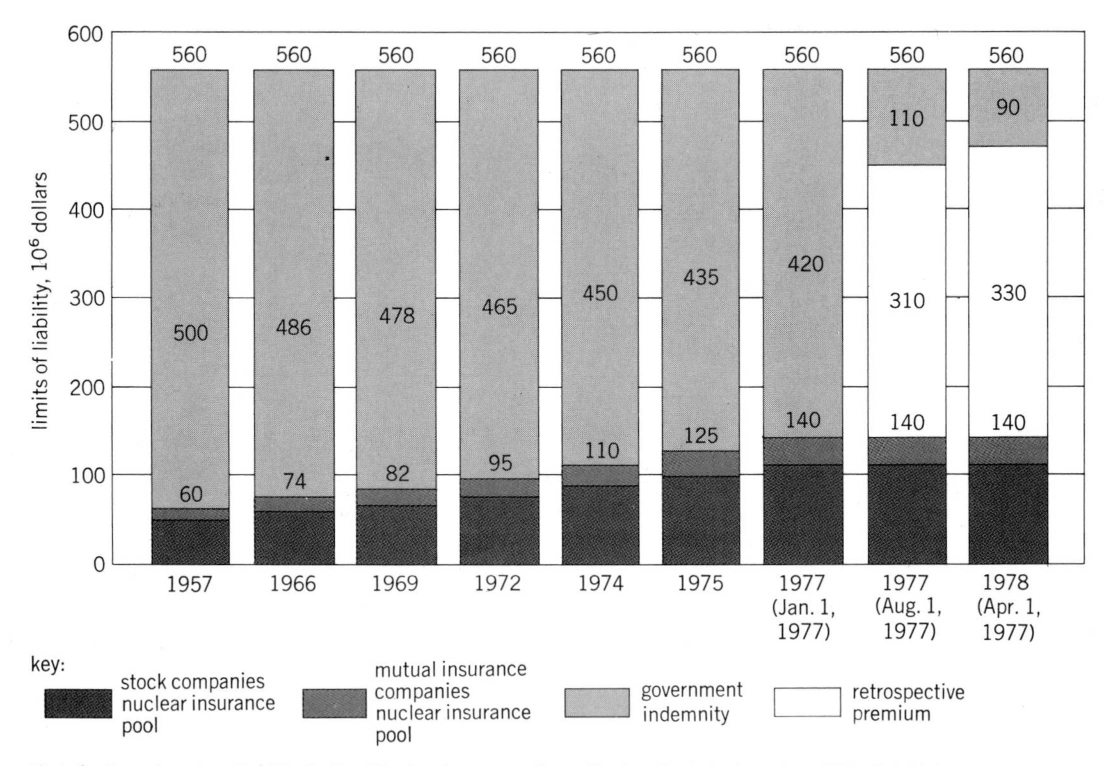

Distribution of nuclear liability limits. (*Nuclear Insurance Consultants, a technical service of Marsh & McLennan*)

the level of required financial protection. The act further provided that the total of $560,000,000 from private insurance and government indemnity would serve as a limitation of liability with respect to the licensee and its contractors and suppliers. Should additional limits of private insurance subsequently became available, the amount of government indemnity would decrease by that amount, and the overall limit of liability of $560,000,000 would be retained. In reality the amount of private nuclear liability insurance has increased from $60,000,000 in 1957 to $140,000,000 in 1977.

Latest amendments. The most important recent amendment to the act requires power plant licensees to assume additional financial protection. Prior to Aug. 1, 1977, each licensee of a nuclear power reactor was required to provide the primary financial protection discussed above. Effective August 1, this condition remains but, in addition, each licensee of an operating reactor larger than 100 MWe is liable for retrospective premium assessments (that is, if liability to the public from a nuclear reactor accident exceeds the total of the utility company's insurance) in an amount equal to $5,000,000 per reactor per incident. This assessment is over and above the primary financial protection insurance available from the insurance pools. Although there is no limit on the number of nuclear incidents in a given year for which a licensee may be assessed premium, the amendment limits the aggregate assessments per reactor for which payment is required in any one calendar year to $10,000,000. This provision in effect defers payment of amounts in excess of $10,000,000 to subsequent years. The illustration shows the financial protection makeup for the period 1957–1978. The retrospective premiums provided $310,000,000 (62 reactors × $5,000,000) of protection for a single nuclear incident effective Aug. 1, 1977, and $330,000,000 (66 reactors × $5,000,000) effective Apr. 1, 1978.

The recent amendment also provides that the future combination of available private insurance and assessments will serve as the total amount of financial protection available to the public once the amount exceeds $560,000,000. For example, current industry projections for 1985 indicate that 150 reactors would be licensed and the amount of primary insurance capacity would be $225,000,000. Based upon these projections, the combination of primary insurance and retrospective premiums of $750,000,000 (150 reactors × $5,000,000) would equal $975,000,000 of public liability protection for a single nuclear incident, as opposed to the $560,000,000 available in 1978.

Other amendments incorporated into the act include a financial protection requirement for certain plutonium licensees; revision of the fee schedule payable by reactor licensees who receive government indemnity; and a provision which clearly requires Congress to review any case in which a nuclear incident results in damages exceeding the limit of liability and to take appropriate action to protect the public.

Constitutionality of act. The "limitation of liability" section of the act was initially ruled unconstitutional in 1977 by a U.S. district court. The outcome of this case could have had a significant impact on the continuing development of nuclear power in the United States. This case involved a group of property owners known as the Carolina Environmental Study Group, who initiated legal action in 1973 against the Duke Power Company and the NRC. The study group contended that in the event of a nuclear incident occurring in a plant constructed in the vicinity of their properties they might not be adequately compensated for damages. In 1976 the district court ruled this section of the act to be unconstitutional, holding that it violated the due-process clause of the Fifth Amendment and was contrary to the Constitution's equal-protection guarantee. The Duke Power Company presented three points for Supreme Court consideration: (1) whether the "limit of liability" provision providing for an ascertainable sum is consistent with constitutional due process and equal protection; (2) whether the district court's decision improperly substitutes the policy judgment of a Federal court for that of Congress; and (3) whether the district court incorrectly found that the plaintiffs had standing and incorrectly took jurisdiction of the case in presenting an actual case or controversy. From the public viewpoint the invalidation of the limitation of liability would have created a system of financial recovery which was uncertain as to amount and timely settlement under tort liability procedures, as opposed to the existing system which essentially affords a "no-fault" indemnity cover with a guaranteed minimum of $560,000,000 available as public financial protection.

In a unanimous decision rendered on June 26, 1978, the U.S. Supreme Court ruled that the "limitation of liability" provision was constitutional and found that the limiting of liability is an acceptable method for Congress to utilize in encouraging the private development of electric energy by atomic power. The Court held that the congressional decision to fix a $560,000,000 ceiling was within permissible limits and did not violate due process. The Court further indicated its view that the congressional assurance of a $560,000,000 fund for recovery, accompanied by an express statutory commitment to take whatever action is deemed necessary and appropriate to protect the public from the consequences of a nuclear accident, was a fair and reasonable substitute for the uncertain recovery of damages of this magnitude from a utility or component manufacturer whose resources might well be exhausted at an early date.

Liability loss experience. Since 1957 the electric utility industry has constructed and received licenses to operate 68 nuclear power reactors. To date there has been no incident involving nuclear injury or damage to the public. Since 1957 the nuclear industry has paid premiums in excess of $107,000,000 for nuclear liability insurance and over $19,000,000 in fees for the excess government indemnity coverage.

For background information *see* NUCLEAR POWER; REACTOR, NUCLEAR in the McGraw-Hill Encyclopedia of Science and Technology.

[LAWRENCE G. CUMMINGS]

Bibliography: Joint Committee on Atomic Energy, U.S. Congress, *Selected Materials on Atomic Energy Indemnity and Insurance Legislation*, 93d Cong. 2d sess., March 1974; V. Meckoni, R. J. Catlin, and L. L. Bennett, *Energy Policy*, International Atomic Energy Agency, Vienna, 5(4):267–281, December 1977; Office of Energy Technology, Nuclear Power Development Division, U.S. Department of Energy, *A Study of Factors Inhibiting Effective Use of Domestic Nuclear Power*, Feb. 21, 1978; Office of Technology Assessment, U.S. Congress, *Nuclear Proliferation and Safeguards*, 1977; D. J. Rose and R. K. Lester, *Sci. Amer.*, 238(4): 45–56, April 1978; *United States Nuclear Regulatory Commission Rules and Regulations*, Code of Federal Regulations, Title 10, Chap. 1, Energy Pt. 140.

Nuclear reaction

The first measurements of pion double-charge exchange reactions to definite nuclear states have opened a new realm of nuclear studies. Information has been obtained which is applicable to a variety of areas of nuclear science. The importance of the absorption of mesons as they propagate in the nucleus has been illuminated. The relative strength of nonanalog transitions to double-analog transitions has proved to be surprisingly high. The details of nuclear structure of the target have been shown to be necessary to understand the reaction, and the reaction has been shown to be sensitive to nuclear correlations. All these new insights have made this field of research one which is quite exciting and one which has received international attention.

These reactions comprise two charge changes involving pions and nucleons. Pions, which are the lightest strongly interacting elementary particles, exist in three forms which are differentiated by their charge—positive, neutral, and negative. The nucleons, which make up the nucleus, come in two types—protons which are positively charged and neutrons which are neutral. The double-charge exchange measurements that have been made involve a positive pion as the incident particle and a negative pion as the detected particle. Since the nucleons come in only two types and total charge must be conserved, the pions can produce only a single-charge exchange by transmuting a proton into a neutron. Hence, this reaction must involve at least two separate nucleons interacting with the pions. Now, for the first time, reactions are being observed in which fundamental particles are interacting with two separate nucleons as the dominant process; this is giving a new understanding of nuclei.

Experimental observations. The probability for such reactions reaching definite final states, where the structures of the final products are reasonably well understood, is very small. Therefore, the experiments are complex, and it is only in the last 2 years that technology has advanced enough to carry them out. The major technological advance is the completion of a new breed of particle accelerators called meson factories. Two international facilities have participated in these studies to date. The reaction was first observed in 1976 at the Clinton P. Anderson Meson Physics Facility (LAMPF) in the United States, and it was confirmed shortly thereafter at the Swiss Institute for Nuclear Science (SIN) in Switzerland. Both of these facilities have the unique capability of producing sufficiently copious numbers of low-energy pions to be converted into beams for nuclear studies of rare processes.

Experimental signature. The experimental signature for pion double-charge exchange to definite final nuclear states is quite distinctive. Quantum mechanics restricts bound nuclei to be populated by nuclear reactions at definite energies of the outgoing particles. Thus, the first piece of evidence for the reaction is an enhancement of the number of outgoing pions of the proper momentum. Second, the outgoing pions are of the opposite sign with respect to the incident beam. The experiments at LAMPF and SIN have both made use of this signature in the same way: An incident beam of nearly monoenergetic positive pions illuminates a target. The outgoing pions are selected by momentum and charge by using a magnetic spectrometer.

Elimination of background. The primary experimental difficulty arises because the probability for observing the reaction (that is, the cross section) is so small that other processes can mock up the momentum and charge of the pions. The dominant source of this background is negatively charged electrons of the proper momentum produced by electromagnetic processes from positive electron contaminants in the incident beam. Backgrounds of this type are rejected by factors of about a million by using detectors which are highly efficient for distinguishing electrons from pions and by measuring the velocity of the pions, which is quite different from that of the electrons.

Initial experiments. Initial experiments have used the isotopes of oxygen as targets. Differential cross sections, $d\sigma/d\Omega$, for pions exiting an ^{18}O target at 0° with respect to the beam are shown as a function of incident energy in the illustration. The experimental measurements are indicated by the three data bars. Other published measurements include two data points for scattering from ^{18}O at angles other than 0° and one datum point at 0° on ^{16}O. In addition, there are preliminary results at 0° for a variety of nuclei from ^{9}Be to ^{58}Ni.

Interpretation of the data. Due to the limited number of experiments, the phenomenology of the reaction is not very extensive. However, enough data already exist to shed light on two important issues: the role of absorption of pions in the reaction and the relative strength of analog and nonanalog transitions.

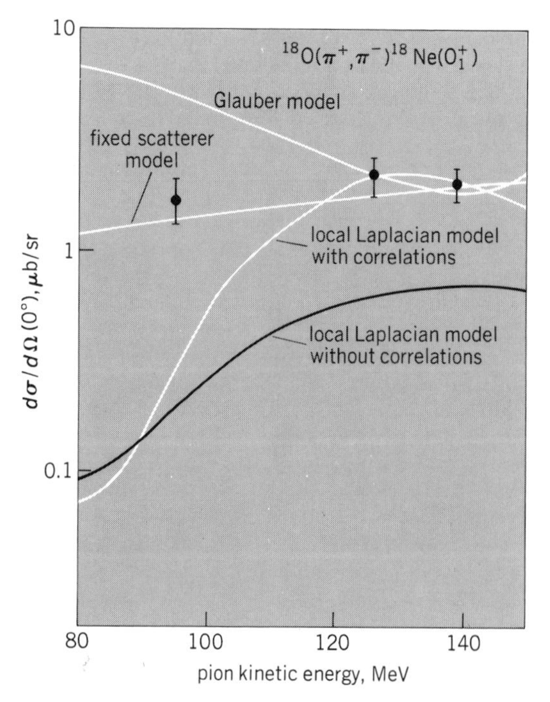

Comparison between the data and various theoretical approaches to calculating the energy dependence of the 0° differential cross section for the pion double-charge exchange reaction which converts ^{18}O into ^{18}Ne. *(From R. L. Burman et al., Pion double-charge exchange on ^{16}O and ^{18}O, Phys. Rev. C, 17:1774–1786, 1978)*

Absorption. Absorption plays a special role in double-charge exchange since the pion must propagate in the nuclear medium to go between the two nucleons on which it is charge-exchanging. If the details of the initial and final nuclear states make these nucleons far apart, then the propagation distance will be large. At energies at which the reaction has been studied, the absorption length is short, and the cross section is greatly reduced. No observation of the reaction has been made for larger nuclei, where all distances are greater. Silicon-28 is the largest nucleus in which double-charge exchange has been observed. The absence of any observation in heavy nuclei implies that absorption plays an important role in the reaction and must be understood in detail before a quantitative prediction of the measured cross sections can be made.

Analog and nonanalog transitions. A double-charge exchange reaction in which the nucleus is changed only by transmuting two protons into two neutrons is called a double-analog transition. If other changes to the nucleons take place, it is termed a nonanalog transition. On the basis of theoretical reasoning alone, it was suspected that the simplicity of the double-analog transition might make it so dominant that it would be the only reaction observed. The measurements reported that the ratio for double-charge exchange from ^{16}O and from ^{18}O is 1/3. The transition of ^{16}O to ^{16}Ne is nonanalog and ^{18}O to ^{18}Ne is double-analog. Higher-order corrections to the nuclear structure cannot explain the large ratio. Therefore, processes such as these, which change the angular momentum of the nucleons in addition to charge exchange, are now known to be important.

Detailed nuclear structure. In addition to the general phenomenology, there is considerable interest in the role played by the detailed nuclear structure of the initial and final nuclei. In the case of the oxygen isotopes, the data require not only nonanalog transitions but also more complicated nuclear configurations than predicted by the simplest models. Although the results are still preliminary, the data on a number of nuclei besides oxygen are highly dependent on the details of the nuclear structure since the cross sections vary substantially from target to target.

Nuclear correlations. After all the effects of the reaction mechanism and specific nuclear structure are understood, it should be possible to measure the effect of nuclear correlations. Nuclear correlations are all the deviations in the orbits of the nuclei not predicted by assuming that the orbits may be derived from a central potential. In particular, two nucleon correlations measure whether there are forces in the nuclei which tend to specially attract or repel two nucleons. Since double-charge exchange must take place on two nucleons, it should be particularly sensitive to the correlations.

The curves in the illustration reflect the status of most recent calculations. The discrepancy among various theoretical approaches (indicated by the curves labeled Glauber model, fixed scatterer model, and local Laplacian model with correlations) and the data reflects the uncertainties in the understanding of the reaction. Within a single approach, the local Laplacian model, the sensitivity to nuclear correlations is predicted by comparing the curves for the model without correlations to the corresponding curve for the model with correlations. The sensitivity appears to be quite large.

Future study. The future study of the pion double-charge exchange reaction promises to be quite exciting. New data are expected which will measure the complete angular dependence of the reaction. Efforts will be made to extend the range of nuclei measured, particularly into the unexplored region of heavy nuclei. Also, the question will be investigated of whether there are any differences between doing the reaction as it has been done or using incident negative pions and outgoing positive pions. Important complementary data on pion single-charge exchange, where only one charge exchange takes place, are expected in 1979. This wealth of new data will be supplemented by new theoretical efforts. The combination of new data and refined theoretical calculations promises to give a new understanding both of the interaction of pions with nuclei and of the structure of the nucleus.

For background information *see* ANALOG STATES, NUCLEAR; NUCLEAR REACTION; NUCLEAR STRUCTURE; PARTICLE ACCELERATOR in the McGraw-Hill Encyclopedia of Science and Technology.

[MARTIN D. COOPER]

Bibliography: R. L. Burman et al., *Phys. Rev. C*, 17:1774–1786, 1978.

Oceanic islands

Linear island chains, particularly the well-studied Hawaiian-Emperor chain in the North Pacific Ocean, have played an important role in the development of absolute plate motion models. In 1963, J. T. Wilson proposed that linear island chains, such as the Hawaiian Islands, marked the direction of motion of the oceanic crust. He proposed that the islands were sequentially formed as the oceanic crust passed over a magma source in the mantle. Wilson's idea lay dormant until 1971, when Jason Morgan greatly expanded the original proposal and constructed a global absolute plate motion model based to a large degree on the trend and age relations along linear island chains. Many aspects of the Wilson-Morgan hot spot hypothesis could be tested, and much effort has gone into testing the kinematic and dynamic aspects of the model. *See* PLATE TECTONICS.

Ages along the chain. One of the corollaries of the Wilson-Morgan hypothesis is that the ages of the volcanoes in the chain should increase away from the center of volcanic activity, presently located at Kilauea Volcano on the island of Hawaii. The available radiometric age data for the Hawaiian-Emperor chain are shown in the illustration, plotted as a function of distance from Kilauea. Several important additions have been made recently, including a redetermination of the age of Midway Atoll and an improved estimate of the age of the bend between the Hawaiian and Emperor chains. The revised age for Midway Atoll is considerably older than that previously reported, and eliminates the necessity of having a major change in the rate of volcanic migration between 40 million years (m.y.) ago and the present. The average rate of volcanic migration along the Hawaiian chain is 8.0 cm/year or 0.83° of angular rotation about a Pacific hot spot Euler pole located at 69°N, 68° W.

The age of the bend between the Hawaiian and Emperor chains is now well established at 42.0± 1.4 m.y. This substantiates the age of the bend proposed by Morgan in 1971, demonstrates the predictive value of Morgan's proposals, and indicates that a major change in Pacific plate motion occurred 42 m.y. ago. Within the Hawaiian Islands, the island of Lanai has been dated at 1.25± 0.04 m.y., identical to the age predicted by interpolated age data from adjacent volcanoes.

During the summer of 1977 the *Glomar Challenger* drilled into four of the central Emperor seamounts during Leg 55 of the Deep Sea Drilling Project. Work on the recovered core is still in the preliminary stages, but shipboard evaluation of the microfossils indicates that all four seamounts are at least Eocene in age. These paleontological ages appear to confirm that the Emperor Seamounts increase in age to the north, as predicted by the model. The paleontologic age of Suiko Seamount is in agreement with a single K-Ar age for a dredged mugearite reported by M. Ozima. The hot spot model of Wilson and Morgan has had great success in predicting the ages of volcanoes along the Hawaiian-Emperor chain.

Absolute reference frames. This aspect of the model has proved to be exceedingly difficult to test in a rigorous way, although three significant experiments were carried out in 1977–1978. In each of these studies it can be demonstrated that the central Emperor seamounts, now located at 40–45°N, formed far south of their present locations. Extinct reef complexes have been identified capping seamounts as far north as 45°30′. The water temperatures at this latitude are not presently warm enough to permit reef growth, nor do they appear to have been warm enough at any time in the last 60 m.y. The coral reefs capping these seamounts are therefore inferred to have formed when the seamounts were located farther south. The results of Leg 55 confirm that these seamounts are capped by warm-water reef structures.

Two different paleomagnetic studies have been carried out which demonstrate that Suiko Seamount formed far south of its present location. Paleomagnetic measurements on lava flows recovered during Leg 55 indicate a paleolatitude of 25±4°N, while a seamount paleolatitude based on shipboard geophysical data has been calculated at 17±5°N. Both paleolatitudes are close to the present latitude of Kilauea at 19°N, and both indicate significant northward motion since Suiko Seamount formed about 58 m.y. ago. These results are in agreement with the model proposed by Wilson and Morgan, and suggest that the Hawaiian hot spot has been fixed or has moved only very slowly during the last 50–70 m.y.

Composition of lavas. Many of the models advanced to explain the origin of the Hawaiian-Emperor chain predict that the volcanic rocks of the Emperor Seamounts and the Hawaiian Ridge should be chemically similar to Hawaiian volcanic rocks. In the volcanoes of the Hawaiian Islands, chemically distinct lavas occur during four erup-

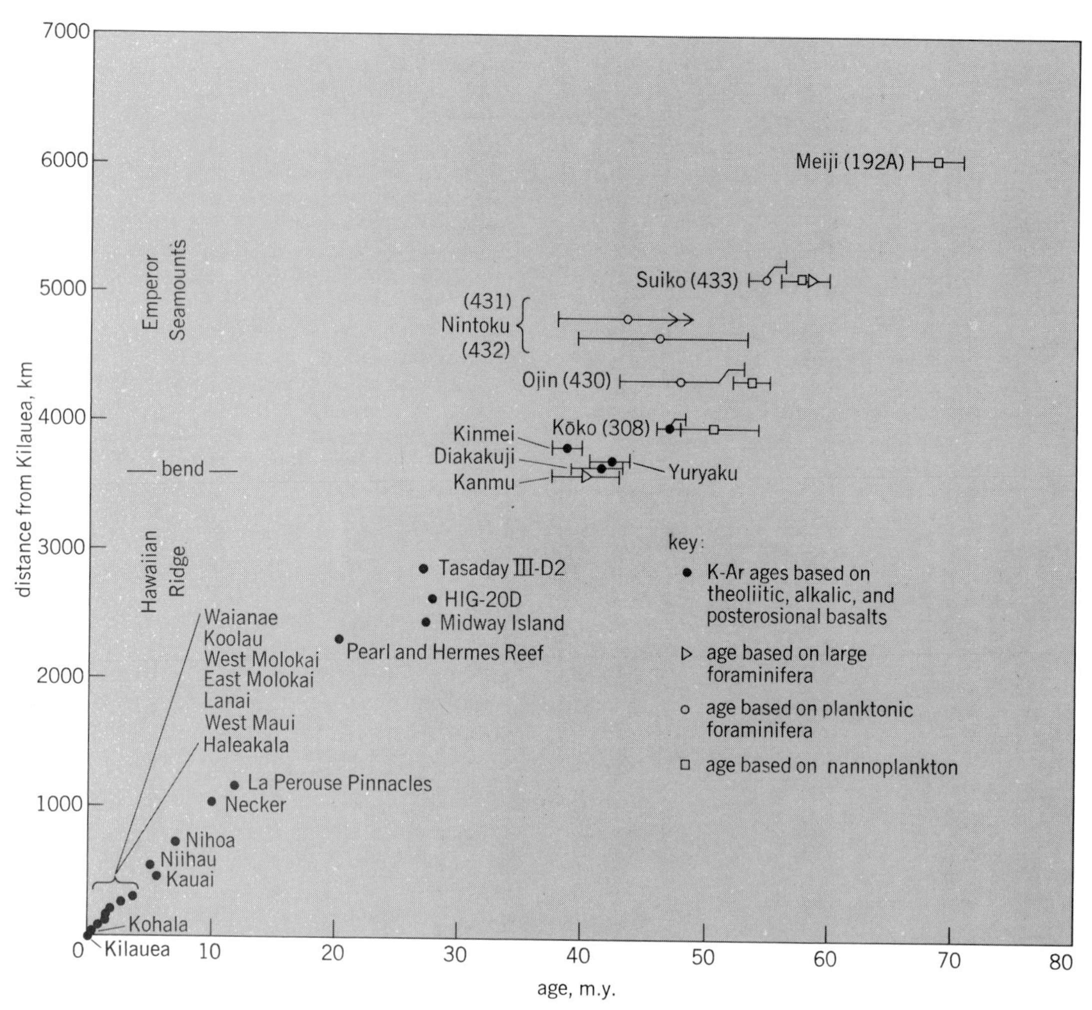

Radiometric age data for Hawaiian-Emperor chain, showing age relative to distance from Kilauea. Bars indicate the estimated standard deviations. Numbers in parentheses are Deep Sea Drilling Project site numbers.

tive stages: shield building, caldera filling, post-caldera, and posterosional. The lavas of the shield-building stage are tholeiitic basalts, which erupt rapidly and in great volume. This stage is quickly followed by caldera collapse and by the caldera-filling stage, during which the caldera is filled with tholeiitic and alkalic lavas. During the post-caldera stage a relatively thin veneer or cap of alkalic basalts and associated differentiated lavas is erupted, sometimes accompanied by minor eruptions of tholeiitic basalts. After a period of quiescence and erosion, lavas of the nephelinic suite, which include alkalic basalts as well as strongly undersaturated nepheline basalts, may erupt from satellite vents during the posterosional stage. Although many Hawaiian volcanoes develop through all four stages, individual volcanoes may become extinct before the cycle is complete.

The many samples dredged along the Hawaiian and Emperor chains have proved to be generally similar to the lavas found within the Hawaiian Islands, although some exceptions exist. Unfortunately, it is not possible to establish the sequence of eruption on these seamounts from the dredged samples. The majority of the dredged lavas are alkalic basalts and differentiated lavas. It is not surprising that the lavas recovered are generally alkalic lavas since, in Hawaii, these are the last lavas to erupt and thus cover the earlier tholeiitic flows. Leg 55 recovered flows from three of the central Emperor seamounts. In each case the top flows were alkalic basalts or differentiated alkalic basalts; at two of these sites, tholeiitic basalt was recovered from beneath the alkalic flows. The general model of volcanism observed in the Hawaiian Islands is also observed in older volcanoes, lending further support to the model that the Hawaiian-Emperor chain is a continuous feature formed by a single hot spot. The lavas from the Emperor Seamounts, both dredged and drilled, do not appear to be chemically identical to those in the Hawaiian Islands, and models in which the mantle source evolves through time are presently being evaluated.

Vertical motions. Recent studies in the Emperor Seamounts by both the U.S. Geological Survey and the Deep Sea Drilling Project demonstrate that many of the seamounts have experienced large vertical motions. The Emperor Seamounts all appear to have erupted subaerially and to have been eroded at sea level to form flat guyot tops. Following this erosional phase, the seamounts subsided and the coral reefs that capped them became extinct. There is evidence that at least some of these seamounts reemerged and that the coralline caps were cemented. Ascertaining the details of the up-

and-down motion of these seamounts will require considerably more work, but the major pattern is clear. The central Emperor seamounts have subsided as much as 2000 m since the seamounts were eroded at sea level. The observed subsidence is greater than was expected based on aging of the underlying sea floor or on lithospheric deformation caused by the loading of these large volcanoes onto the lithospheric plate. This observation has led to some new insights into the nature of hot spots.

Melting in hot spots. It has not been clear whether hot spots are truly hot or if the melting is caused by some other mechanism such as decompression of, or addition of water to, the mantle. This uncertainty led some workers to refer to hot spots as melting spots or melting anomalies. R. S. Detrick and S. T. Crough have presented an explanation of the excessive subsidence noted above that suggests that hot spots are indeed hot. They explain the subsidence of these volcanoes as being due to subsidence along an age-depth subsidence curve for lithosphere much thinner than oceanic crust as old as that underlying the chain. They propose that the Hawaiian swell, a broad region of elevated sea floor surrounding the island chain, is due to heating and thinning of the overlying lithosphere. After the lithosphere moves horizontally away from the hot spot, it cools and contracts, and the overlying swell and islands subside.

During the past few years, especially latter 1977 and early 1978, it has been possible to test the Wilson-Morgan hot spot hypothesis that linear island chains mark the paths of the lithospheric plates as they pass over hot spots in the mantle. Study of the Hawaiian-Emperor chain has largely confirmed the kinematic aspects of the hot spot hypothesis, while more recent studies have demonstrated that hot spots are truly hot.

Petrologic studies are continuing on lavas dredged and drilled from the submarine portion of the chain. These studies, combined with geochemical data from the Hawaiian Islands, may make it possible to evaluate the evolution of the Hawaiian hot spot during roughly the last 60 m.y. of activity.

For background information *see* OCEANIC ISLANDS; PLATE TECTONICS in the McGraw-Hill Encyclopedia of Science and Technology.

[DAVID A. CLAGUE]

Bibliography: G. B. Dalrymple and D. A. Clague, *Earth Planet. Sci. Lett.*, 31:313–329, 1976; R. S. Detrick and S. T. Crough, *J. Geophys. Res.*, 83:1236–1244, 1978; H. G. Greene, G. B. Dalrymple, and D. A. Clague, *Geology*, 6:70–74, 1978: Staff of Deep Sea Drilling Project 55, *Geotimes*, pp. 23–26, February 1978.

Oceans and seas

The promise of a renewable source of base-load electric power from ocean temperature gradients has led to United States government funding of research and development for the ocean thermal energy conversion (OTEC) process over the past several years. From a very modest beginning in 1972, the OTEC program has grown to a 1978 research and development funding level of about $36,000,000.

A recent analysis sponsored by the Congress has identified the following primary unresolved engineering problems in the OTEC concept: the heat exchangers (evaporators and condensers), including the working fluid, construction materials, biofouling, and corrosion; the ocean platform and cold water pipe; open-cycle versus closed-cycle processes; underwater electric power transmission lines; and the constructability and long-term reliability of an entire plant.

Types of OTEC plants. The conversion of energy from the temperature difference between warm surface water and cold deep-ocean currents was first proposed by the French physicist J. A. d'Arsonval in the 1880s, but a full-scale OTEC plant has not yet been built and operated successfully over a long period of time. Since a temperature difference of about 15°C is necessary to operate an OTEC system, there are a limited number of sites, primarily between 35°N and 35°S latitudes, where these plants can operate without excessively long cold water intake pipelines. Both coasts of Africa, the southeastern coast of the United States, the coast of Brazil, and many Caribbean and Pacific islands are located where the sea water decreases in temperature from 24 to 30°C at the surface to 4 to 7°C at a depth of about 800 m.

Open-cycle plants. An open-cycle OTEC pilot plant, in which the working fluid was low-pressure steam evaporated from the warm water flow, was built and operated in Cuba by Georges Claude in the late 1920s. The plant delivered only 22 kilowatts of electric power (kWe), and ceased to operate when the cold water pipe was destroyed. Claude later experimented with an 800-kWe floating plant, but was successful only in generally proving the overall OTEC concept.

As a result of these early attempts to generate power from the ocean's thermal gradients, the French formed a partly government-owned company called Énergie des Mers, subsidiary of Électricité de France, and construction of an open-cycle plant was planned for a site near Abidjan, Ivory Coast, during the 1940s and 1950s. This plant was never completed, although several of the subsystems were built, tested, and installed. Progress has recently been made toward overcoming the basic difficulties encountered by Claude, and more advanced concepts for the development of the open-cycle OTEC process have been proposed.

Closed-cycle plants. Power can also be generated by using the vapor of a secondary working fluid in a closed cycle instead of the open-cycle process. This approach actually was suggested by d'Arsonval during the late 1800s, and a 100-MWe closed-cycle floating plant with propane as the working fluid was proposed in 1966. A secondary working fluid with a relatively high vapor pressure results in a reduction in the size of the plant's power turbines from the large low-pressure units of the open cycle, but enormous heat transfer surface areas are required in the evaporators and condensers of the system.

OTEC process. Theoretically, work can be done by a heat engine when a temperature difference exists between two heat reservoirs, in this case the warm and cold sea water. The temperature difference is hundreds of degrees in nuclear and conventional power plants and gas turbines, but only about 20°C in the sea. The maximum theoretical or Carnot efficiency of an ocean thermal power plant is about 3 to 6%. When the typical operating losses

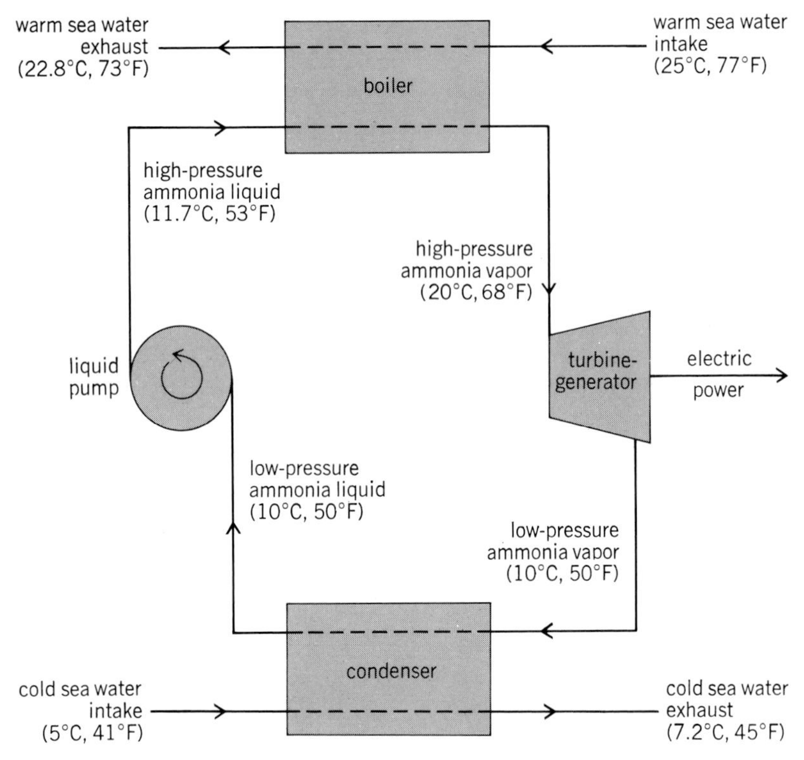

Fig. 1. Schematic diagram of a closed Rankine power cycle for the ocean thermal energy conversion (OTEC) process. *(From Owen M. Griffin, Power from the oceans' thermal gradients, in N. T. Monney, ed., Ocean Energy Resources, ASME, vol. 4, pp. 1–21, September 1977)*

in an actual system are taken into account, the efficiency is likely to be only about 2% or less. However, this low thermal efficiency is not a fatal drawback for such a device, because the boiler of an OTEC plant is designed to operate at relatively low pressure and temperature differentials and does not need the complex containment vessel required for operation at both high pressures and temperatures.

A closed-cycle OTEC process diagram is shown in Fig. 1. This is a Rankine power cycle consisting of a turbine, pump, condenser, and evaporator or boiler. The working fluid is transformed to a vapor in the evaporator, and power is generated by expansion of the vapor through the turbine. Upon exiting the turbine, the low-pressure vapor is condensed in a heat exchanger with water from the oceans' depths used as the cold sink fluid. The working fluid is then pumped back to the boiler pressure and recycled.

Typical operating temperatures for an ammonia cycle are shown in Fig. 1. Of particular note are the temperature drops in the evaporator and condenser which reduce the available temperature difference between the warm and cold water inlets from 20°C to a temperature difference of 10°C for the ammonia loop. For the conditions shown, the maximum thermal efficiency of the ammonia cycle is about 3%; the actual efficiencies will be reduced even further by the plant's operating losses.

There are some inherent advantages to the closed cycle. As mentioned previously, the power turbines are greatly reduced in size due to the higher operating pressures (typically several atmospheres; 1 atm = 10^2 kilopascals) and working fluid densities. It is also not necessary to remove the dissolved gases in the warm sea water as must be done in the open cycle. Some disadvantages in using a secondary working fluid are the temperature losses and presence of biofouling in the heat exchangers. Other disadvantages are that working fluids such as propane and ammonia require enormous heat exchanger surface areas and pose potential handling, materials-compatibility, and corrosion problems in the presence of sea water.

Recent design concepts. Both open- and closed-cycle OTEC plant designs have evolved over the past 3 decades, but by far the most attention in the last several years has been given to the closed-cycle process.

Open-cycle plant layouts have been proposed by D. F. Othmer and O. A. Roels for various land-based systems composed of an OTEC power plant in combination with desalination and mariculture facilities. Recent developments described by Othmer apparently have eliminated the major faults of Claude's original open-cycle design by utilizing the controlled flash evaporation (CFE) of steam for the production of power and fresh water. In this process the warm sea water enters a vacuum chamber and, as the pressure lowers, the water is evaporated. The low-pressure steam next passes to an expansion turbine to generate power, and then to a surface condenser for condensation by cold, deep sea water. The CFE process has minimized the problem of removing dissolved gases from the water and has reduced corrosion problems. It also reduces brine entrainment and temperature losses between the sea water and the vapor. Most open-cycle plant designs have been limited in size to about 10 MWe.

In 1966 J. H. Anderson and J. H. Anderson, Jr., of Sea Solar Power, Inc., proposed an ocean-based closed-cycle OTEC system. The plant was to have net power output of 100 MWe, and included submerged turbines, evaporators, and condensers in a floating platform anchored to the ocean bottom. A long vertical pipe was used to raise the cold ocean water from a depth of about 600 m. The working fluid for the proposed plant was propane (later Freon), and large areas of compact plate heat exchangers were required to operate the system. Based upon this and other recent studies, several overall features of the closed-cycle OTEC system have emerged. They are: a plant that is floating or submerged in deep water and anchored to the ocean bottom; modular evaporator, condenser, and turbo-generator systems; ammonia or propane as the most desirable working fluid (because of their high vapor pressure at the temperature of warm sea water, and their compatibility with sea water, as opposed to Freon-type fluids); and a single large-diameter cold water pipe.

Several industrial and laboratory teams have undertaken overall systems studies and three basic designs have emerged, as follows.

Lockheed design study. In the original concept (Fig. 2) there are four 40-MWe power modules mounted externally on a semisubmersible spar-type platform. The plant includes a 305-m long cold water pipe and a mooring system capable of deployment to depths of 6100 m. Each of the power

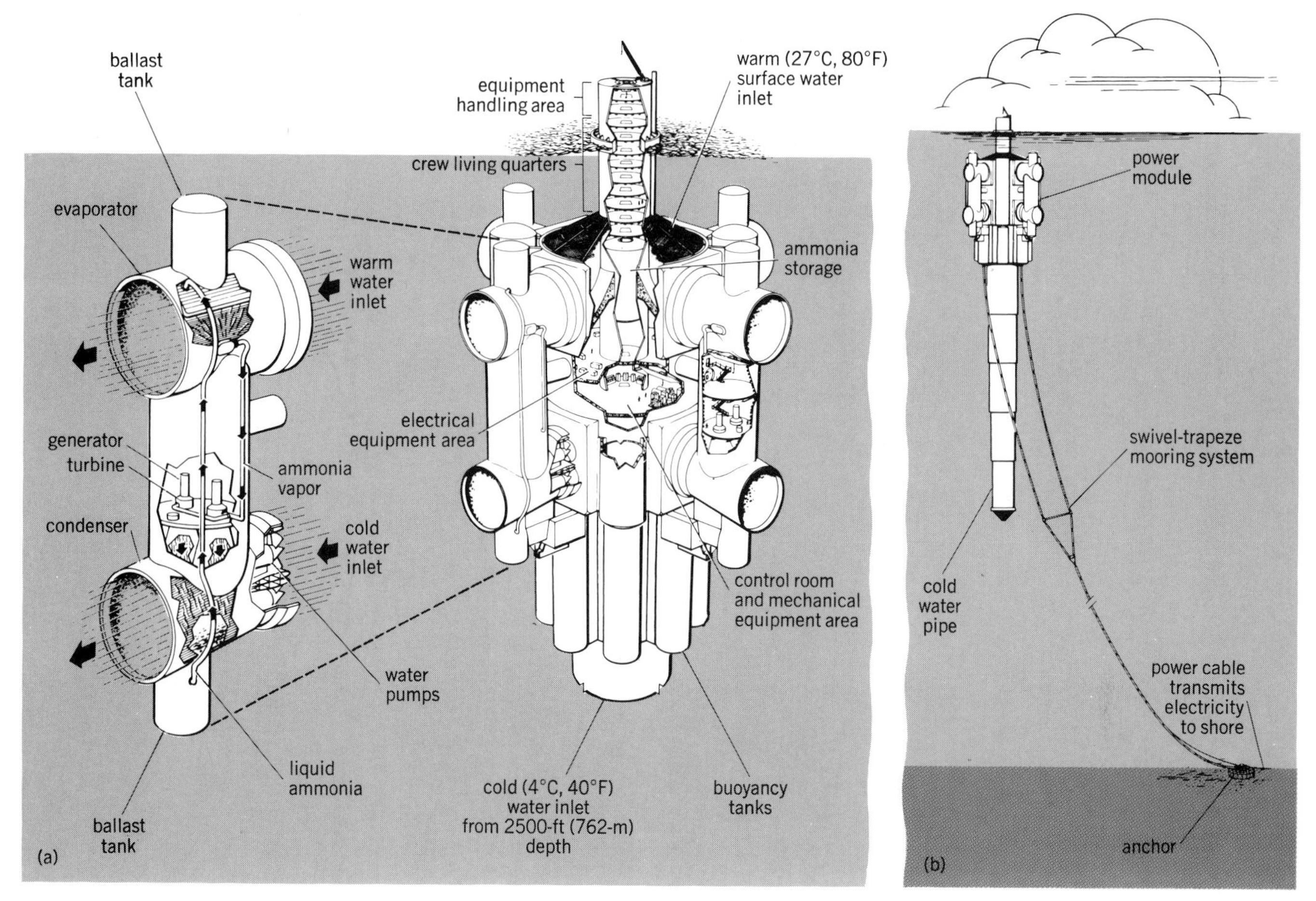

Fig. 2. OTEC system proposed by Lockheed. *(a)* General arrangement of components for OTEC plant. *(b)* Deployment of the OTEC platform by a single mooring line and anchor. *(Lockheed Missiles and Space Company)*

modules is an integral unit composed of sea-water and working fluid pumps, turbines, generators, evaporator, and condenser. A more recent Lockheed design study has resulted in an increase in the net plant power to 260 MWe and an extension in the length of the cold-water pipe to 776 m. Ammonia was chosen as the working fluid because of its superior thermodynamic characteristics and relatively wide history of industrial use. Shell-and-tube heat exchangers with titanium tubes were selected for the baseline configuration of the evaporators and condensers. The platform and the telescoping cold water pipe are of reinforced concrete construction, and the single-point mooring line is made up of steel cylindrical sections varying in diameter from 1.8 m at the platform to 0.3 m at the anchor. A swivel-trapeze mooring system prevents the pipe from becoming entangled as the power plant adjusts to changing currents and a power cable transmits electricity from the anchor to shore. Several design trade-offs were made before the final plant configuration was agreed upon; these included dynamic positioning versus fixed moorings and aluminum or stainless steel versus titanium for the heat exchanger tubes.

TRW design study. A second OTEC design study was made by TRW. In contrast to the Lockheed design, the TRW plant includes a surface-deployed cylindrical hull with four 25-MWe power modules mounted in the hull's interior. Reinforced concrete was chosen for the hull construction, while the cold water pipe, 1220 m long and 15 m in diameter, was proposed to be constructed in 18-m sections of stiffened, fiber-reinforced plastic. Ammonia was chosen over propane and Freon as the working fluid for the power turbines because of its superior thermodynamic properties and relatively wide history of use. Conventional shell and titanium tube heat exchangers were chosen as a baseline since titanium is compatible with ammonia and is corrosion-resistant over long periods.

Some perception of the size of the OTEC heat exchangers can be obtained from the TRW design parameters. About 93,000 m^2 of heat transfer surface area is required for each 25-MWe power module; approximately 65,000 titanium tubes, each 38 mm in diameter and 13.1 m long, are necessary to meet this requirement. The total flow rate of sea water through the evaporators and condensers of a 100-MWe OTEC plant is estimated to be 2½ times the flow of the Potomac River at Washington, DC!

A system of shrouded-pipe water jets was chosen as a dynamic positioning system for the TRW plant, since wire or chain deep-water mooring systems were considered to be unproved for water depths greater than 1000 m. The water jet positioning system is designed to make use of the kinetic

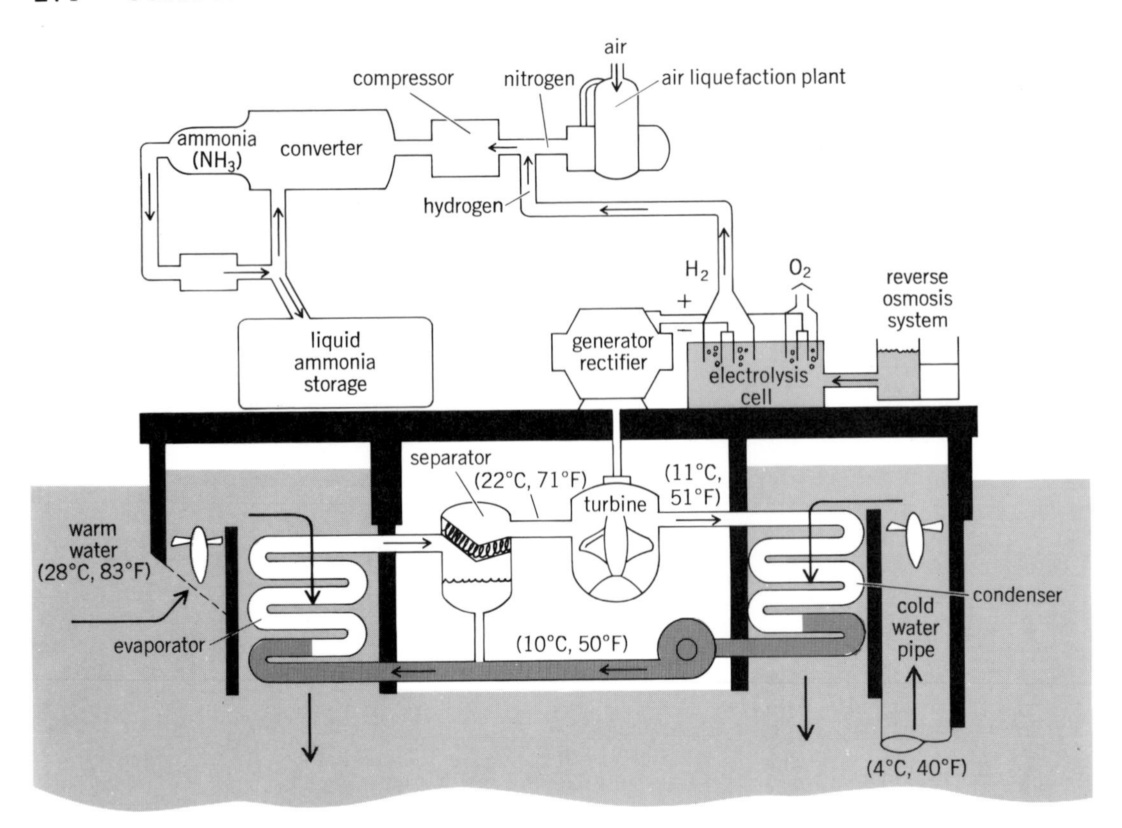

Fig. 3. Sectional view of an OTEC plant-ship proposed by the Applied Physics Laboratory of Johns Hopkins University, illustrating general arrangement of plant and ammonia production facilities. *(From W. H. Avery et al., Maritime and Construction Aspects of Ocean Thermal Energy Conversion Plant-Ships, Johns Hopkins Univ./Appl. Phys. Lab. Rep. SR 76-1A, April 1976)*

energy of the plant's enormous water throughput for station-keeping.

Applied Physics Laboratory proposal. A proposal was made by the Applied Physics Laboratory of Johns Hopkins University to construct OTEC plant-ships for the production of ammonia at sea, as well as electric power. The OTEC plant-ship is designed for grazing operation at a speed of about 1 knot (0.5 m/s) in tropical waters, and both a 100-MWe demonstration system and follow-on 500-MWe production plant-ships have been proposed.

The basic hull and cold water pipe are to be constructed of reinforced concrete, with the water side of the evaporator and condenser modules an integral part of the hull (Fig. 3). Ammonia was chosen again as the working fluid for the closed-cycle OTEC power modules, and aluminum heat exchanger tubes were chosen for the evaporators and condensers in order to minimize costs and weight.

Although a number of technical problems have been solved in recent years, the low thermal efficiency of the OTEC process is still an inherent drawback in the overall plant design. Many of the economic and environmental questions concerning OTEC plants await answers until the problems mentioned at the beginning of the article are further resolved.

For background information *see* ENERGY CONVERSION; ENERGY SOURCES; RANKINE CYCLE in the McGraw-Hill Encyclopedia of Science and Technology.

[OWEN M. GRIFFIN]

Bibliography: R. H. Douglass and P. J. Bakstad. *J. Hydronaut.*, 12(1):18–23, 1978; O. M. Griffin, *Ocean Energy Resources*, ASME, vol. 4, pp. 1–21, 1977; Office of Technology Assessment, U.S. Congress, *Renewable Ocean Energy Sources*, part 1: *Ocean Thermal Energy Conversion*, May 1978; D. F. Othmer, *Mech. Eng.*, 98(9):27–34, 1976.

Octopus

In the past 2 decades interest in the brain and behavior of *Octopus* has centered on the ready trainability of the animals, so that research has focused on discrimination experiments and the effect of brain lesions on learning. This phase appears to be ending, to be superseded by a broader interest in cephalopod physiology. In the past 3 years there have been only four studies on octopus learning.

Learning experiments. J. Z. Young's 1978 study of learning by the subesophageal brain comes closest to the traditional experimentation. It uses a familiar technique: reward-and-punishment training of blinded animals to make a rough/smooth tactile discrimination. This is combined with longitudinal sectioning of the supraesophageal lobes, and sectioning of one or of both of the brachiocerebral tracts joining the front parts of the sub- and supraesophageal brains. Previous work by M. J. Wells and Young had implied that it is impossible for an octopus to learn to discriminate by touch after removal of the inferior frontal system at the anterior end of the supraesophageal brain, and since sectioning the tracts cuts the sensory input to this part, one would expect a total failure to discriminate. However, when the inferior frontal system was removed, Young demonstrated that signs of discrimination can sometimes be seen, superimposed on large alterations in general response level. (There is no doubt that the animals can alter their responses, increasing the number of rejections or acceptances of objects presented to them; the question is whether they can learn to discriminate.) The differences in response to the rough and smooth objects are always small (only occasionally reaching significance at the $p = 0.01$ level) and always transient, disappearing between sessions, and do not appear in unrewarded tests run after training. Young's finding, therefore, is of interest in relation to the evolution of learning systems—since there is some evidence that the higher centers in the supraesophageal lobes are a comparatively late addition to the brain of cephalopods—but it does not significantly alter the picture of the octopus brain touch-learning mechanism.

Tactile training was also used as the basis for an attempt to demonstrate chemical transfer of learning by G. F. Domagk, W. R. Alexander, and K. H. Heermann. They made extracts from the supraesophageal lobes of animals trained to accept only smooth spheres and injected these into octopuses trained to accept both rough and smooth. Control animals received injections from the brains of untrained octopuses. No transfer was found; both control and experimental animals showed considerable disturbance of behavior, rejecting a much higher proportion of spheres, when injected with 0.5 ml (equivalent to one brain) of the supernatant from a crude homogenate.

In a more conclusive series of experiments, J. B. Messenger confirmed that *Octopus* is color-blind. His own previous optomotor and training experiments had shown this quite convincingly, but were open to the criticism that the training technique employed (successive presentation of colored plaques) was not the best means of extracting discrimination in difficult tasks. In the second set of experiments Messenger used simultaneous presentation (that is, the octopus had to discriminate between two alternatives, both visible to it) and an experimental design that included alternate trials with rectangles of different hue (matched for brightness) and rectangles differing either in brightness or in orientation. The animals always discriminated in the brightness and orientation tasks, but consistently failed to distinguish yellow or violet rectangles from gray shapes matched for subjective brightness. The spectral sensitivity of *Octopus* corresponds closely to that of the dark-adapted human eye.

The conclusion that *Octopus* cannot see colors is rather surprising in view of its impressive capacity for background-matching. The explanation appears to be careful tone control by adjustment of the expansion of the yellow-red-brown chromatophores, coupled with reflection by leucophores and iridiophores lying beneath these in the skin.

In a later series of tests *Octopus* was shown to prefer (that is, to be more ready to attack under aquarium conditions) shapes that contrast strongly with the background to shapes that are merely reflective. This finding is of interest in relation to the design of tanks for testing visual discrimination, since it has been sometimes suggested that the animals have an innate preference for dark or light figures; it now seems that this is simply a function of background lighting.

Brain function. The small peduncle lobes on the optic stalks continue to attract attention. Messenger showed that these were somehow involved in the fine adjustment of visually controlled movement, which is jerky and imprecise if they are removed. M. J. Hobbs and Young began to refer to the peduncle lobe as a "cephalopod cerebellum," after the demonstration (in decapods) that its input included both visual and positional (statocyst) information. Messenger and P. L. Woodhouse gave further details of the system in *Octopus*, and pointed out the existence of large numbers of small parallel fibers similar to those well known from the vertebrate cerebellum. J. Woodhams has followed this up with an ultrastructural study. He suggests a close functional and morphological analogy between the spine of the peduncle lobe and a folium of the vertebrate cerebellum. The cells are so small (most are less than 5 μm across) that study of them is hampered, although the system, like so much of the cephalopod central nervous system, plainly deserves the close attention of future electrophysiologists.

Endocrine systems. Other work on *Octopus* physiology has moved away from direct concern with brain and behavior toward brain-controlled endocrine systems. R. K. O'Dor and Wells, following their earlier studies of the effect of the optic glands on yolk production by the follicle cells, have shown that the same glands dominate somatic protein metabolism. J. Wodinsky, in the course of an investigation into the effect of the optic glands on brooding behavior, found that removal of the glands was followed by a resumption of feeding and growth. It has become clear that *Octopus* (and presumably all other cephalopods) function on a predominantly protein-energy economy, with muscle as the main store, and with the optic glands controlling the balance between reproductive and somatic growth.

There is still some doubt about the means by which the optic glands have their widespread effects. Cutting the nerve supply from the brain causes the glands to swell, and this is followed by an early onset of sexual maturity. Implanting the glands (including those derived from other species) has the same effect. Sexually mature females always seem to have enlarged glands, and immature females do not (the situation is less clear-cut in males, where the onset of sexual maturity is less dramatic). The tissue-culture medium in which glands have been incubated will cause additional amino acid uptake by the follicle cells. There is abundant circumstantial evidence for secretion of a gonadotropin. Yet the glands themselves show no such indications; ultrastructural examination shows that the yellowish (in *Octopus vulgaris*, but not, for example, in *O. macropus* where the glands are white) grana hitherto thought to include the gonadotropin in fact consist of masses of residual or lytic bodies; the yellow material is a lipofuscin, apparently a waste product. K. Mangold and D. Froesch reported that this product increases as the gland swells. Horse spleen ferritin, injected into the bloodstream, is taken up, accumulated, and then voided by the optic glands. Evidently the glands are somehow involved in the catabolism of materials taken up from the blood and, if this is the case, it is conceivable that their function is not so much secretion of a gonadotropin as removal of an inhibitory material which normally prevents the onset of sexual maturity.

Further work on endocrines has shown that a product of the neurosecretory nerves running from the subpedunculate lobe to the preorbital vein can be extracted and used to excite the systemic heart isolated from the body. However, a study of the effects of injection of the same material into free-moving intact octopuses has revealed that its physiological effect is vasoconstrictive rather than cardioacceleratory.

For background information *see* CEPHALOPODA; OCTOPUS in the McGraw-Hill Encyclopedia of Science and Technology. [MARTIN J. WELLS]

Bibliography: E. A. Bradley and J. B. Messenger, *Mar. Behav. Physiol.*, 4:243–251, 1977; G. F. Domagk, W. R. Alexander, and K. H. Heermann, *J. Biol. Psychol.*, 18:15–17, 1976; D. Froesch, K. Mangold, and W. Fritz, *Experientia*, 34:116–117, 1978; D. Froesch and J. B. Messenger, *J. Zool.*, p. 186, in press; K. Mangold and D. Froesch, *Symposium of the Zoological Society of London*, vol. 38, pp. 541–555, 1977; J. B. Messenger, *J. Zool.*, 174: 387–395, 1974; J. B. Messenger, *J. Exp. Biol.*, 70: 49–55, 1977; M. J. Wells, *Octopus: Physiology and Behaviour of an Advanced Invertebrate*, 1978; M. J. Wells and J. Wells, *Symposium of the Zoological Society of London*, vol. 38, pp. 525, 540, 1977; P. L. Woodhams, *J. Comp. Neurol.*, 174:329–346, 1977.

Optics, nonlinear

In the past few years coherent radiation has been generated at progressively shorter wavelengths, extending the spectral range in which such radiation is available into the extreme ultraviolet, that is, wavelengths below 100 nm, and almost to the soft x-ray range, which starts at about 30 nm. Such sources of coherent short-wavelength radiation have important potential applications in many areas of science and technology. For example, short-wavelength radiation can provide improvements in resolution in both photolithography and photographic imaging, since the size of the smallest object that can be resolved is limited by the wavelength of the illumination source. The coherent nature of a short-wavelength laser would allow holographic pictures to be made, eliminating the need for imaging elements, such as lenses or mirrors, which are of relatively poor quality at wavelengths below about 120 nm.

Applications in several areas of scientific research are also possible. The highly monochromatic nature of the short-wavelength laser sources should provide improvements in vacuum ultraviolet spectroscopy, studies of solid surfaces by photoelectron spectroscopy, and photochemical excitation studies. The penetrating nature of the short-wavelength radiation should also prove useful in studying the very dense plasmas that are created in experiments designed to investigate controlled thermonuclear fusion. *See* FUSION, NUCLEAR; HOLOGRAPHY; SPECTROSCOPY, LASER.

Optical harmonic generation. The extension of laser radiation to shorter wavelengths has been accomplished by a process known as optical harmonic generation, or more generally, optical frequency conversion. When an intense laser beam is passed through a transparent medium, such as a clear crystal or gas, new laser radiation can be produced at wavelengths corresponding to multiples, or harmonics, of the original laser frequency. As the frequency of the radiation increases, its wavelength decreases, since their product, which is equal to the speed of light, must remain constant. The laser radiation interacts with the electrons in the medium to produce a polarization which can radiate optical energy. When the incident, or pump, radiation is sufficiently intense, the interaction with the electrons becomes nonlinear, and the induced polarization can radiate at multiples of the original frequency. All of the properties of the incident radiation that characterize laser light, such as spectral purity, temporal and spatial coherence, and spatial collimation, are also present in the harmonic radiation.

Suitable media. In order for the process to be effective in producing new radiation, the medium must be transparent at both the pump wavelength and the generated wavelength. For wavelengths ranging from the infrared to the near-ultraviolet, harmonic generation can be accomplished by using suitable crystalline solids. However, suitable crystals become opaque at wavelengths below 200 nm, and transparent gases must be used for harmonic generation in the extreme ultraviolet. Because of the symmetry of a gaseous medium, the strongest interactions produce only odd harmonics, corresponding to wavelengths at $1/3$, $1/5$, $1/7$, . . . , of the original pump wavelength.

Conversion efficiency. Conversion efficiency, that is, the ratio of the power in the generated radiation to the power in the pump light, is a measure of the effectiveness of the harmonic conversion process. Although these nonlinear effects are exceedingly weak for optical intensities encountered in more usual light sources, such as incandescent lights or arc lamps, they can be quite strong at the optical intensities present in laser beams, and in principle at least, all of the energy present in the pump laser can be converted to harmonic radiation. With pulsed laser sources at visible wave-lengths, conversion efficiencies as high as 85% have been observed. Corresponding conversion efficiencies in the extreme ultraviolet using gases have been considerably lower (of the order of 0.001%), but sufficient radiation has already been produced for some applications.

Frequency mixing. If more than one optical frequency is present in the pump laser, radiation can also be generated at wavelengths corresponding to sum and difference combinations of the pump frequencies in a process known as frequency mixing. Both harmonic generation and frequency mixing have been used to produce laser radiation at several wavelengths in the extreme ultraviolet.

Extreme ultraviolet generation. The first technique used for producing coherent radiation in the extreme ultraviolet involved third-harmonic conversion of a pump wavelength in the ultraviolet.

This is the lowest-order process possible in gases, and it has also been used extensively for harmonic generation at longer wavelengths in the infrared and visible. Starting with the fourth harmonic of radiation from a neodymium:yttrium-aluminum-garnet (Nd:YAG) laser at 266.1 nm, researchers at Stanford University generated coherent radiation at 88.6 nm in argon.

Resonant enhancement. Further progress was made by using a pump laser at a shorter wavelength and by taking advantage of resonances between the pump frequency, or its multiples, and certain atomic transitions in the gas. These resonances result in increased conversion efficiency in a manner similar to the increase in linear absorption that occurs when the wavelength of a light source is tuned to match an atomic absorption line. Two-photon, or two-quantum, resonances are particularly important in harmonic generation because the absorption associated with these resonances is much weaker than the absorption associated with one- or three-photon resonances. In a two-photon resonance, the pump frequency is equal to one-half of the frequency of a suitable atomic transition. Only certain transitions may be used, namely those that are forbidden by symmetry to radiate a single quantum of energy but are allowed to radiate two quanta.

Occasionally a sufficiently close match occurs between a two-photon atomic transition and the wavelength of radiation produced in a fixed-frequency laser. More generally, a tunable laser must be used to take advantage of such resonant enhancements. Both these situations have been encountered experimentally in extreme ultraviolet generation.

Researchers at Imperial College in London have used radiation from a xenon excimer laser (the active medium in this laser is the short-lived Xe_2 molecule which exists only in an excited state) to generate radiation at 57 nm, again by third-harmonic generation in argon. The wavelength of the xenon laser can be tuned in a narrow band near 171 nm and can be made to coincide with a two-quantum resonance in argon (Fig. 1*a*). As shown in Fig. 1*b*, the strength of the third harmonic increases dramatically as the pump wavelength is brought into exact coincidence with the double-quantum resonance.

A similar two-photon resonance was used by researchers in the Soviet Union to generate radiation at 89.6 nm. They took advantage of a naturally occurring resonance between one of the transitions in mercury vapor and the fourth harmonic of a neodymium:glass laser operating at a wavelength of 1.0752 μm.

Higher-order harmonics. In addition to production by third-harmonic conversion, short-wavelength radiation can be generated by using fifth-, seventh-, or higher-order harmonics of a pump laser. The use of these higher-order harmonics allows a larger step to be taken along the wavelength scale in a single frequency-conversion process, extending the spectral range in which coherent radiation can be generated. This approach was followed by researchers at the U.S. Naval Research Laboratory (NRL) to extend the generation of short-wavelength radiation even further into the extreme ultraviolet. Starting with the fourth harmonic of a Nd:YAG laser at 266.1 nm, they produced radiation at 53.2 nm by a fifth-harmonic process in four of the rare gases (helium, neon, argon, and krypton).

A microdensitometer tracing of a photographic spectrum of the fifth-harmonic radiation generated in neon, showing a single spectral line, is shown in Fig. 2*a*. A spectrum from a helium arc lamp with three emission lines evident in the same spectral region is also shown (Fig. 2*b*). Using seventh-harmonic conversion of the same pump laser, the NRL group also generated coherent radiation at 38 nm in helium. This wavelength is very close to the soft x-ray range.

More recently researchers in Bulgaria have used

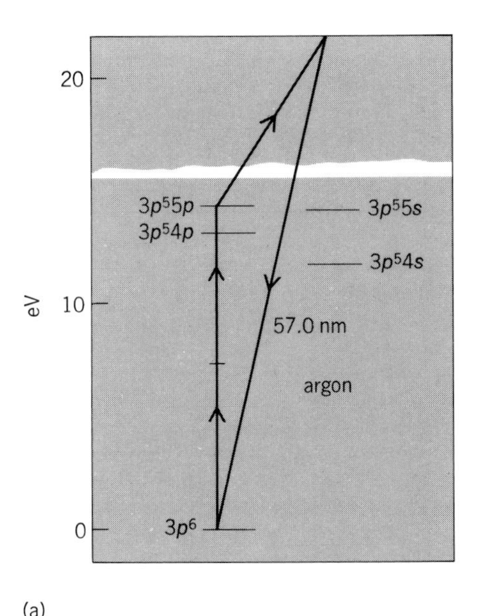

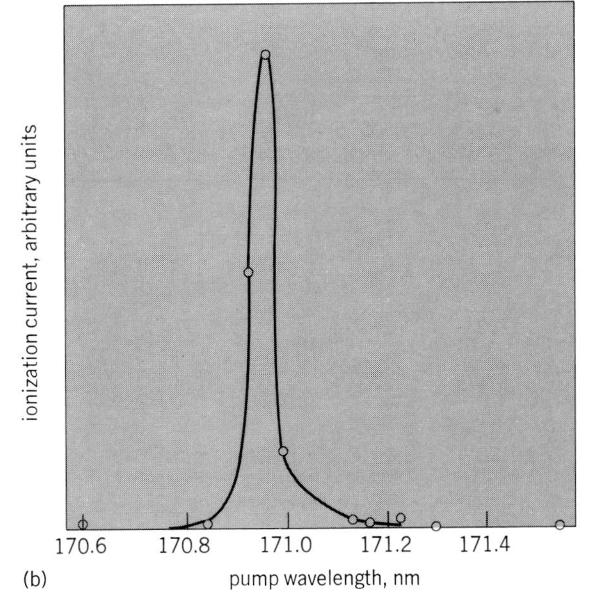

Fig. 1. Third-harmonic conversion in argon. (*a*) Energy level diagram showing two-quantum resonance. (*b*) Variation of third-harmonic signal as pump wavelength is tuned through the two-photon resonance. (*From M. H. R. Hutchinson et al., Generation of coherent radiation at 570 Å by frequency tripling, Opt. Commun., 18:203–204, 1976*)

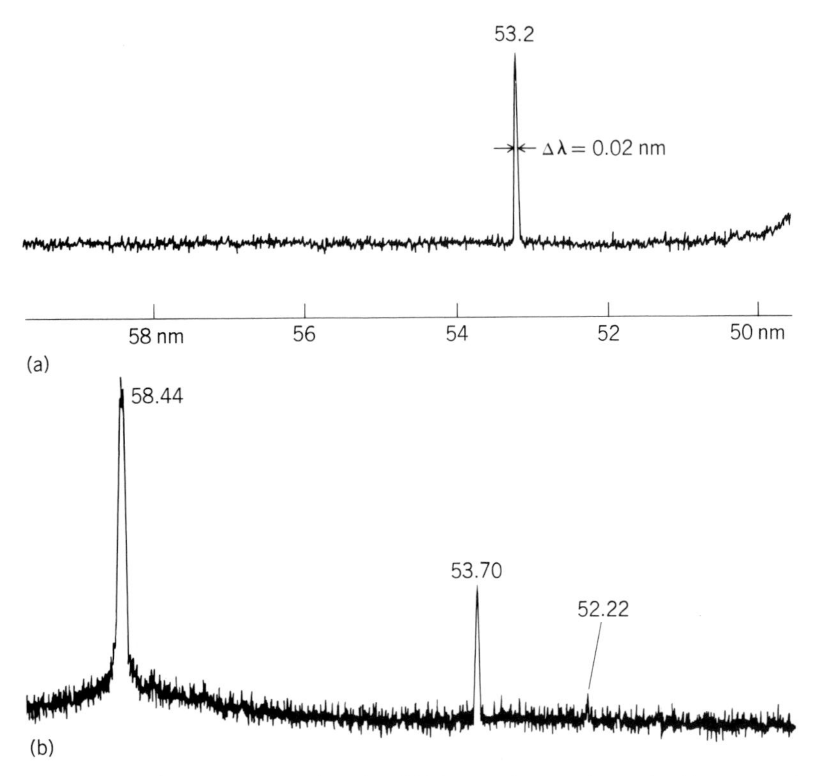

Fig. 2. Microdensitometer tracing of *(a)* fifth-harmonic signal generated in neon and *(b)* helium emission spectrum. *(From J. Reintjes et al., Generation of coherent radiation at 53.2 nm by fifth harmonic conversion, Phys. Rev. Lett., 37:1540–1543, 1976)*

ninth-harmonic conversion of radiation from a neodymium:glass laser to produce radiation at 118.2 nm. Such studies indicate that further progress in short-wavelength generation by using even higher-order harmonics may still be possible.

Fifth-order frequency mixing. The NRL group has also used frequency mixing to generate radiation at four additional wavelengths in the extreme ultraviolet: 59.1, 62.6, 70.9, and 76 nm. In these experiments, pump radiation was obtained from a Nd:YAG laser at 1.06 μm, its second harmonic at 532 nm, and its fourth harmonic at 266.1 nm. The extreme ultraviolet radiation was produced by combining four photons of the 266-nm radiation with one photon from either the 532-nm or the 1.06-nm light. The four additional wavelengths generated in this manner correspond to upper and lower sidebands (addition or subtraction of the final frequency) for each combination of pump wavelengths.

Substitution of continuously tunable radiation, such as that produced by dye lasers, for the fixed frequencies used in these experiments would result in continuously tunable, highly monochromatic radiation extending from about 38 to 100 nm. Such an advance would greatly increase the applicability of the technology described in this article to areas such as vacuum ultraviolet spectroscopy.

For background information *see* LASER; OPTICS, NONLINEAR in the McGraw-Hill Yearbook of Science and Technology.

[JOHN F. REINTJES]

Bibliography: M. G. Grozeva et al., *Opt. Commun.*, 23:77–79, 1977; S. E. Harris et al., in *Laser Applications to Optics and Spectroscopy, Physics of Quantum Electronics*, vol. 2, pp. 181–197, 1975; M. H. R. Hutchinson et al., *Opt. Commun.*, 18: 203–204, 1976; J. Reintjes, C. Y. She, and R. C. Eckardt, *IEEE J. Quant. Elec.*, QE-14:581–596, August 1978.

Paleoceanography

Paleoceanography is a new field among the earth sciences which has come of age only within the past 10 to 15 years. Earlier workers were largely hampered by the lack of material from the ocean basins and by the somewhat random distribution in space and time of material recovered. Knowledge of the major climatic and oceanographic events, as derived from land-based studies, was also in a very imperfect state—so much so that there was no general agreement as to the number of glacial events (ice ages) in the Pleistocene epoch or as to the ages of the major epoch boundaries. With the advent of the International Decade of Ocean Exploration, a program of the National Science Foundation, there has been an exponential increase in paleoceanographic research. In this article the methods used in this study will be surveyed and some of the major results will be summarized.

Methods. Most of the information concerning paleoceanographic events is derived from the study of microfossils preserved in ocean sediments. Several groups of unicellular plankton secrete skeletons of calcium carbonate or hydrous silicon dioxide, which settle to the ocean floor and may be preserved for millions of years, if the chemical composition of the bottom waters remains favorable. Two types of analysis based on these microfossils have produced valuable data from which paleoceanographic events may be deduced.

Microfossils. One method involves the use of individual microfossil species as ecologic indicators. Judging by the distribution of modern forms, many species appear to have definite preferences for waters of a certain temperature, salinity, or nutrient content. Thus, the species composition of any sample gives some indication of the conditions prevailing in the surface waters when the assemblage was alive. The use of statistical techniques can provide much detailed information on the significance of changes in the assemblages over a large area of space or time. It is also possible to relate the statistical assemblages of modern species to the present temperature of the surface waters and to use this relationship to estimate water temperatures in the past, based on fossil assemblages. Geographical distribution of extinct species can be used to gain insight into their general preferences, although exact temperature values cannot be assigned.

Stable isotope ratios. A second method involves the use of stable-isotope ratios in the calcareous skeletons. The most useful ratios are $^{18}O/^{16}O$ and $^{13}C/^{12}C$. The ratio of oxygen isotopes in the water changes in proportion to the amount of water locked in the polar ice caps, since isotopically "light" water (with ^{16}O) is preferentially included in the ice. Thus, during a period of ice growth, water becomes isotopically heavier. In addition, the calcareous organisms (foraminifera and coccolithophorids) tend to shift their isotopic-equilibrium relationship with the water as it becomes progres-

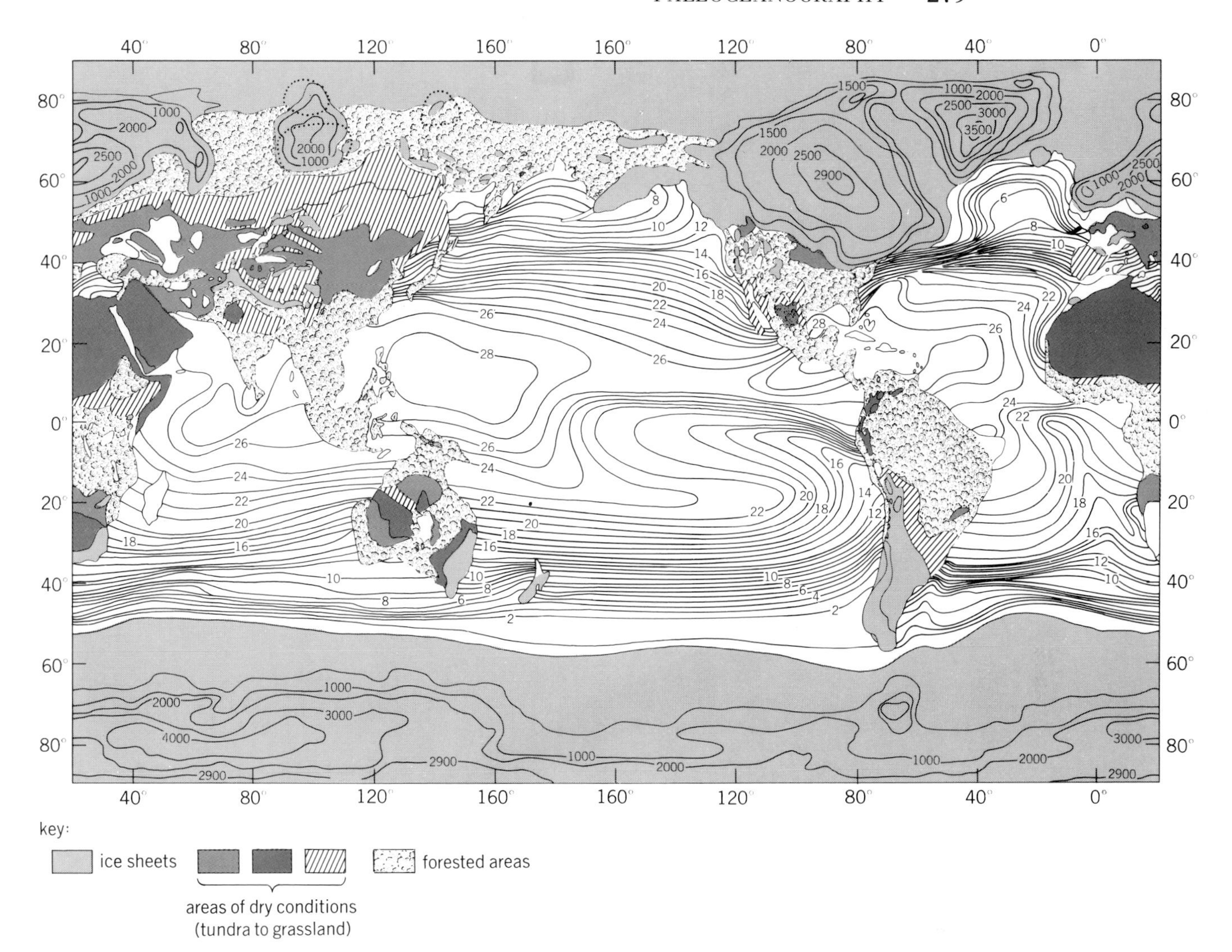

Fig. 1. Reconstruction of the Earth's climate during summer in the last ice age. Temperatures for isopleths in the oceans are in °C; numbers on ice sheets represent thickness of ice in meters. *(From CLIMAP project members, Science, 191:1131–1137; copyright © 1976 by the American Association for the Advancement of Science)*

sively colder. Quantitative changes in the oxygen-isotopic composition of a species can thus be interpreted as quantitative changes in the water temperature over time. The use of $^{13}C/^{12}C$ is very recent, and as yet the significance of many of the results is ambiguous. It appears, however, that a change in carbon isotopes is related to changing river input, which in turn is related to changing productivity of plants.

Other methods. These include the use of ancient reefs to determine paleo–sea level; of ash layers to infer wind- and water-current patterns; and of circulation models based on changes in ice cover, continental position, and wind patterns, along with sediment composition.

Pleistocene paleoceanography. Much detailed work has been carried out on the sediments of the past 700,000 years, due to the large number of sediment cores available and popular interest in this period, commonly called the Ice Ages. Work in the 1960s for the most part concentrated on stratigraphic analysis in a few cores and consisted essentially of plotting the changes in oxygen isotopes or "cold" versus "warm" species of microfossils downcore. For example, zones of high ^{18}O, or high proportions of cold species, were judged to represent glacial periods. Most of the research was concentrated on defining the number of glacial-to-interglacial cycles and determining the ages, based usually on a few ^{14}C dates or estimated sedimentation rates. A long record produced from ^{18}O analysis by C. Emiliani and N. J. Shackleton indicates the existence of 10 cycles of alternating cold-warm events—far more than the 4 that had been classically recognized. The curve plotting this record has a sawtoothed shape from past to present, indicating that in general the pattern is a period of gradual cooling for about 40,000 years, followed by a very abrupt return to warm conditions in less than 2000 years—in other words, ice ages advance slowly but end quickly. Statistical studies on such long curves have revealed a set of regular periodicities which correspond to the fluctuation of various parameters in the geometry of the Earth-Sun relationship.

More recently, scientists in the Climate Long-range Investigation and Mapping Project have undertaken to model one time plane during the peak of the last glacial event, about 18,000 years ago. By using various methods of time control, they

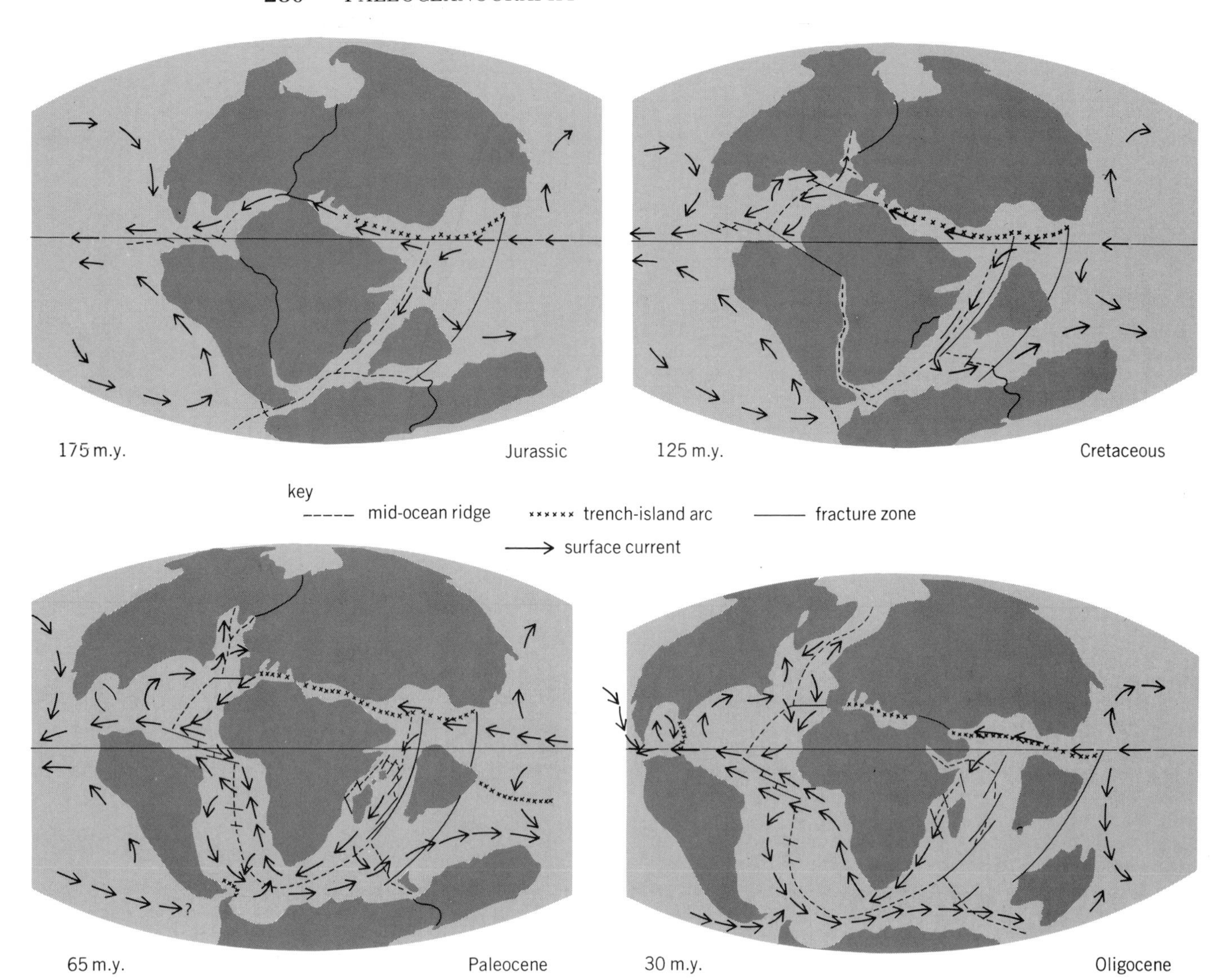

Fig. 2. Maps showing estimated continental positions, tectonic features, and surface ocean currents for four key periods during the pre-Pleistocene (m.y. = millions of years before present). *(From T. J. van Andel, An eclectic overview of plate tectonics, paleogeography and paleoceanography, 37th Biology Colloquium: Historical Biogeography, Plate Tectonics and the Changing Environment, in press)*

chose depth levels representing this age in a large number of deep-sea cores throughout the world ocean and examined the microfossil assemblages in them. By means of the relationship between the assemblage and water temperature mentioned above, they reconstructed a map of the world ocean at this time (Fig. 1). The data derived from this map were then used in a meteorological model to reconstruct the wind patterns, cloud cover, precipitation, and so on over the land. For the oceans, distribution of the isotherms allows interpretations concerning the relative strength and position of the major oceanic currents. The most significant features are the relative lack of change in the tropical regions, the extreme cooling of the high-latitude regions, and a compression of the transitional waters between the two. Particularly in the North Atlantic, there is an extremely sharp barrier (caused by the paleo–Gulf Stream) between the cold waters north of 40°N latitude and the main warm-water gyre to the south. There is also an expansion of the Antarctic ice sheet and an increase in strength of the Peru-Chile Current in the eastern Pacific. Studies of bottom-dwelling microfossil species at this time also indicate a change in the structure and thermal composition of the bottom water.

Pre-Pleistocene paleoceanography. Most work on older sediments has been based on long cores taken by the drilling vessel *Glomar Challenger.* Some of the sediments penetrated are well over 100 million years (m.y.) old, well into the Cretaceous period. There is no single record over this time range, but by piecing together results from sediment cores of different ages and from different regions, paleoceanographers are beginning to form a picture of the evolution of the oceans over this time. S. Savin and associates have produced a temperature scale for the Cenozoic – the last 65 m.y. – using the oxygen-isotope method. This indicates that while the tropics have remained fairly consistently warm (20 – 25°C) throughout the Cenozoic, the poles have experienced several progressive cooling events, most notably in the middle

Eocene and middle Miocene, during which polar temperatures dropped from 15°C to almost 0°C. This polar refrigeration is a direct result of continental drift, which has isolated Antarctica at high southern latitudes, as will be seen below.

While many questions remain, the broad outlines of the last 200 m.y. of ocean evolution have been pieced together. (W. Berggren and C. Hollister have worked out a useful summary, which provides more detail on the sequence given here.)

Late Triassic (200 m.y. ago). The ocean consists of one great basin, Panthalassa, which corresponds to a larger version of the Pacific. The continents are grouped into one continental mass, called Pangaea. The mass is solid in the western part (the Americas, Europe, and western Africa), but is separated in the east by a wide equatorial seaway, Tethys, which brings a warm tropical current from the main ocean westward as far as mideast Asia. The northern portion is called Laurasia, while the southern portion (South America, Africa, India, Australia, and Antarctica) is termed Gondwanaland. India and Australia are far south of their present position, impinging on Antarctica, so that the Indian Ocean is part of the Panthalassa Ocean.

Late Triassic to Middle Jurassic (200–150 m.y. ago). Sea-floor spreading results in an opening in Laurasia which is the beginning of the North Atlantic (Fig. 2). The opening is quite narrow, so that at most only a sluggish gyre is formed. Eurasia and Africa begin to separate, widening the Tethys and extending it westward into what is now the Mediterranean. The proto-Pacific circulation probably consists of a westward equatorial current which enters Tethys, and two large warm gyres on either side of the Equator.

Late Jurassic (150–130 m.y. ago). The North Atlantic continues to expand, forming a warm surface gyre, but with little movement of the bottom water. A branch of the gyre flows north as a proto–Gulf Stream. North and South America separate along the Straits of Panama, so that the Tethys seaway becomes circumglobal and creates a warm-water belt around the world, resulting in a widespread flora and fauna. In the Southern Hemisphere, eastern Gondwanaland (Australia and Antarctica) moves southward, while Africa and Australia separate, so that a southern Indian Ocean begins to form. The Arctic becomes compressed due to the northern movement of Laurasia.

Early Cretaceous (130–100 m.y. ago). The southern South Atlantic begins to open as Africa and South America separate (Fig. 2). The basin is shallow, however, and in both Atlantic basins circulation is quite slow and the deeper waters are stagnant, producing layers of carbon-rich sediments. India begins to move northward, and currents coming from the Pacific flow both northward and southward. Australia and Antarctica are still firmly connected.

Late Cretaceous (100–65 m.y. ago). Africa and South America separate completely, and the South Atlantic becomes a deep-water basin, so that vertical mixing occurs and anoxic sediments are no longer formed. Exchange of surface and, later, deep waters occurs between the two basins, and the Atlantic becomes one ocean, with marine deep-water sediments, mostly carbonates. The Tethys continues to funnel warm waters from the Pacific into the Atlantic. Eurasia and Labrador separate, and a warm proto–Gulf Stream flows northward to the Arctic and across into the Pacific through the ancient Bering Straits. Cool Atlantic faunas thus spread into the warmer North Pacific. New Zealand separates from Austral-Antarctica, and cooler faunas appear in this region also. Due to the cooler north and south waters, the Pacific's two large gyres are compressed toward the Equator and two high-latitude cool-water gyres form, so that the homogeneous Pacific is broken into several zones.

Paleocene to early Eocene (65–50 m.y. ago). In the North Atlantic, Greenland and Scandinavia separate, forming a full connection between the Arctic and the Atlantic. Later, North America and Greenland separate to create the Labrador Sea, so that cool Arctic water begins to flow south and temperature gradients are established. Extensive erosion occurs in the North Atlantic along the western margin, which may reflect the development of proto–North Atlantic Deep Water. In all ocean basins, along the equatorial belt, widespread formation of siliceous deposits occurs, indicating optimal conditions for organic productivity. Australia begins to separate from Antarctica and move north. This separation causes a cooling, and a southwest Pacific–Indian Ocean fauna forms. India is in a central position in the ocean.

Eocene to early Oligocene (50–38 m.y. ago). The Atlantic is relatively stable during this period, with high equatorial production of both siliceous and calcareous organisms. The slowly moving bottom water is still warm, due to sinking of the warm saline Tethys waters. India approaches Asia. Antarctica continues to cool, and mountain glaciers spread toward the sea. Antarctica and Australia separate, and the Drake Passage opens between South America and Antarctica, so that only the Tasman plateau forms a barrier to complete circum-Antarctic flow.

Eocene-Oligocene boundary (38 m.y. ago). Glaciers on Antarctica reach sea level, causing a sharp drop in water temperature. Antarctic Bottom Water is formed in the Ross and Weddell seas and spreads northward along the western margins of the Indian and southwest Pacific oceans, causing extensive hiatuses in the sediments due to erosion and dissolving of the calcium carbonate. The Tasman Sea opens, isolating Antarctica completely. Carbonate production becomes dominant in the equatorial Pacific, while silica dominates in the higher latitudes.

Oligocene (38–22 m.y. ago). The Atlantic remains stable, with circulation essentially the same as that of today. The Tethys still exists, although it is considerably narrowed, and flows across the Pacific and Indian oceans through the Mediterranean to the Atlantic (Fig. 2). Tasmania is completely separated from Antarctica, allowing the formation of the Antarctic Circumpolar Current, a cool, swift current which results in high production of siliceous organisms. Antarctic Bottom Water continues to spread, causing erosion in the South Atlantic. Upwelling of the nutrient-rich bottom water forms high carbonate production in the eastern equatorial Pacific.

Early to middle Miocene (22–12 m.y. ago). The Tethys becomes divided into two basins due to the suturing of Africa and Asia. The eastern Tethys becomes the Indo-Pacific connection, where free exchange of the waters is somewhat blocked by the uplift of numerous volcanic islands. The western Tethys becomes the Mediterranean-Atlantic exchange, which connects to the Pacific through the Caribbean and the Straits of Panama, so that equatorial waters are still warm and of high salinity. Higher-latitude waters, however, are cooler as North Atlantic Deep Water forms in the north due to cooling of Gulf Stream water in the Norwegian Sea and the continued spreading of the Antarctic ice cap.

Middle to late Miocene (12–5 m.y. ago). A worldwide lowering of sea level occurs as large amounts of ice are formed in Antarctica. The Mediterranean basin becomes cut off, and thick deposits of salt are formed in the basin. Climate deteriorates, and the Earth enters the glacial mode. Production in the North Pacific and around Antarctica increases, as the cooler polar currents create more vigorous circulation and mixing of nutrients.

Early Pliocene (5–3 m.y. ago). The Gibraltar sill opens, and the Mediterranean is flooded almost immediately. The Antarctic ice sheet continues to expand, extending the area of cold waters and compressing the equatorial system still more. Toward the end of this time the Straits of Panama close, separating the Pacific and Atlantic and destroying the last vestiges of the Tethys seaway. The Gulf Stream thus increases in volume, since the water is no longer drained into the Pacific, and transports more water northward, so that Arctic precipitation increases.

Late Pliocene to present (3–0 m.y. ago). The northern ice sheet begins to form, and the first glaciations begin in the Northern Hemisphere. Atlantic-Pacific exchange now occurs only through the Bering Strait, with the flow reversed in direction. Alternate cold and warm climatic cycles begin, with the intensity of cooling decreasing but the frequency increasing toward the present.

For background information *see* GEOLOGICAL TIME SCALE; GLACIAL EPOCH; MARINE SEDIMENTS; PALEOCLIMATOLOGY; PLEISTOCENE in the McGraw-Hill Encyclopedia of Science and Technology. [CONSTANCE SANCETTA]

Bibliography: W. A. Berggren and C. D. Hollister, *Tectonophysics*, 38:11–48, 1977; CLIMAP project members, *Science*, 191:1131–1137, 1976; C. Emiliani and N. J. Shackleton, *Science*, 183: 511–514, 1974; J. D. Hays, J. Imbrie, and N. J. Shackleton, *Science*, 194:1121–1132, 1976; S. M. Savin, R. C. Douglas, and F. G. Stehli, *Geol. Soc. Amer. Bull.*, 86:1499–1510, 1975.

Paleogeography

New developments in the fields of paleomagnetism, biogeography, and paleoclimatology have been inspired by plate tectonics theory and have made it possible, for the first time, to reconstruct the paleogeography of past geologic periods with considerable accuracy. The past 600 million years (m.y.) of Earth history is potentially accessible because of its rich fossil record, and it is now known that the lithospheric plates, with which the continents are associated, move characteristically at rates of several centimeters per year. This motion is indeed slow, but in geologic time it is possible for a continent in the polar region to move to the tropics, and in fact, Australia has moved in such manner in the past 50 m.y. The juxtaposition of the continents has thus changed radically in the past. This article explains how this complicated history is elucidated, and gives a brief outline of this history in terms of plate tectonics.

It is often assumed that the movement of lithospheric plates began with the breakup of the supercontinent Pangaea in the Early Jurassic approximately 200 m.y. ago. However, the hallmarks of plate tectonics, the arcuate volcanic chains of andesitic rock and the remnants of ocean floor basalts, are found in more ancient deposits, such as in the Paleozoic rock of the Ural Mountains, and some experts believe that tectonic processes observable today were in existence in some form at least 2,000,000,000 years ago. The driving forces are still poorly understood, but the continents have probably experienced several phases of breaking up and colliding.

Paleogeographic techniques. The last 200 m.y. of Earth history is relatively easy to decipher because the spreading history can be determined from the rocks of the ocean floor. It takes approximately this length of time for material to form at the mid-oceanic ridges and to migrate to the trenches, where it is consumed or deformed against the continental margins. The age of the ocean floor thus increases away from the spreading centers. Enough is now known about the ocean floor through the Deep-Sea Drilling Program that the more recent portions can be conceptually eliminated, and older portions refitted to determine accurately the geometrical relationships of the plates during earlier times (Fig. 1). The age of the ocean floor has been determined radiometrically where drill holes have penetrated the basaltic crust. Elsewhere, age is established on the basis of the magnetic polarity preserved in the rocks. Because the Earth's polarity has reversed in an irregular way in the past, and because the resultant magnetic anomalies can be measured on oceanographic vessels, the polarity reversal signature can be readily determined and correlated with drilled areas where absolute ages have been measured.

Ocean floor ages establish the relative positions of the continents during the last 200 m.y., but an additional step is necessary to determine the location of the pole, and hence the latitudinal framework of the continents. To accomplish this, oriented rock samples are collected and their magnetic moments are determined. It is assumed that magnetic minerals in sedimentary and volcanic rocks align themselves on deposition with the Earth's magnetic field and that the field has always been roughly coincidental with the axis of rotation. Minerals in rocks are thus similar to compasses that rotate in three-dimensional space, and yield information on both the direction of the pole and the declination of the magnetic field, and it is the declination that establishes the latitude of the continent. Paleomagnetic orientations are now being determined for rocks of all ages, and extensive catalogs of the results are available.

The relationships of the continents prior to the existence of Pangaea is more difficult to determine. Paleomagnetism is as effective in establish-

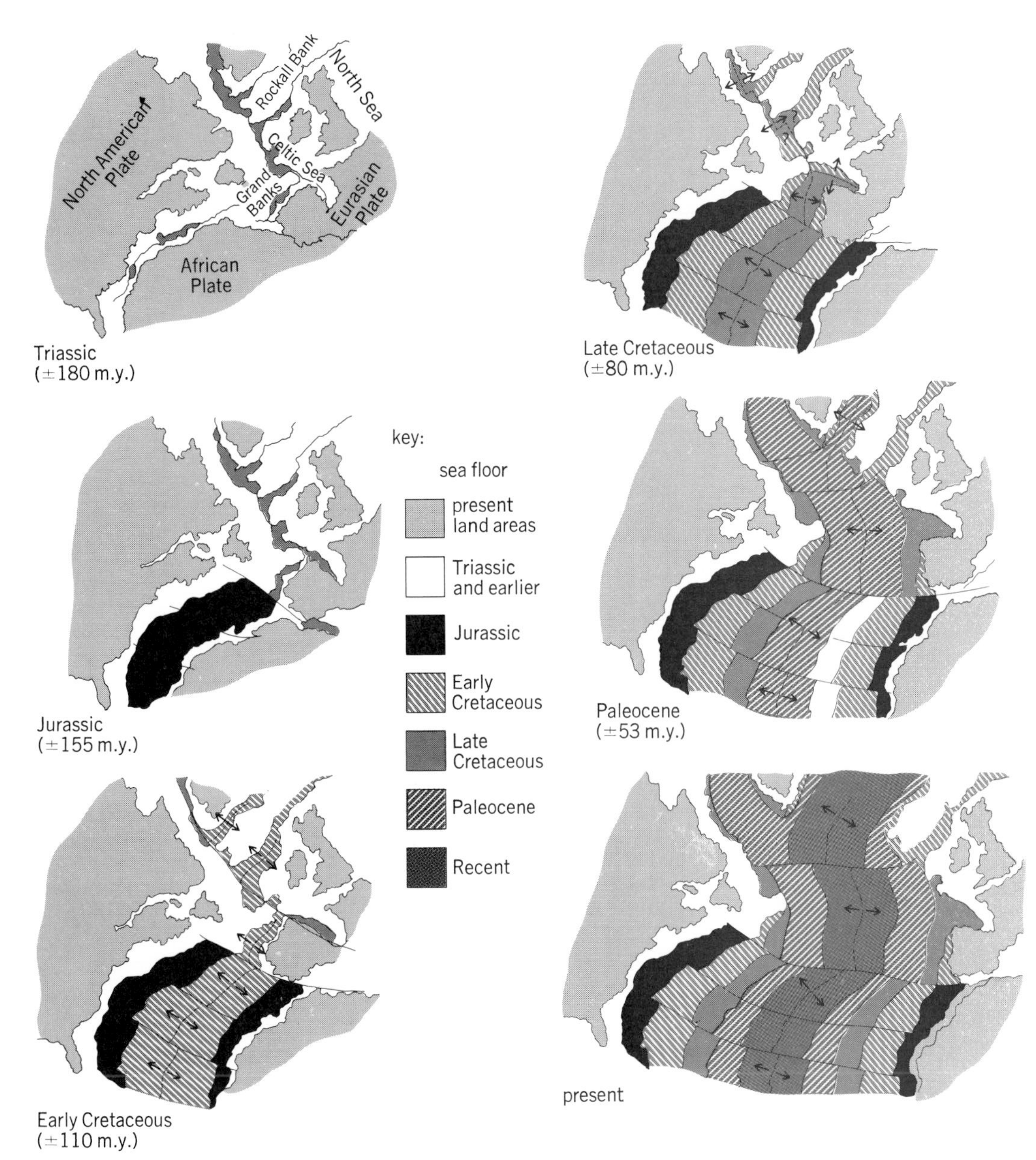

Fig. 1. History of the North Atlantic from the Mesozoic to the present showing the age of the ocean floor; m.y. = millions of years before present. *(From R. M. Pegrum and N. Mounteney, Rift basins flanking North America, Amer. Ass. Petrol. Geol. Bull., 62:419–441, 1978)*

ing latitude and orientation, but the older ocean floor has been severely deformed, so relative longitude must be determined by other methods. Moreover, paleomagnetism is fallible because it can be altered by thermal and chemical events, and even by bolts of lightning. The geologist must resort to paleoclimatic, biogeographic, and tectonic arguments to verify the paleomagnetic data and to try to gain some understanding of the relative longitude of continents during earlier periods.

Abundant evidence of past climates is preserved in the form of coral reefs, salt deposits, coal swamps, and glacial deposits. As continents drift from one latitudinal belt to another, the climatic indicators change accordingly. Striking examples of this effect are the Ordovician glacial deposits of the Sahara Desert, the Silurian coral reefs of northern Greenland, and the equatorial coal swamps of Pennsylvanian age of northern Europe. These climatic indicators provide a useful check on the paleomagnetic data as both are a direct test of latitude.

Biogeographic patterns yield information on both latitude and relative longitude, because both climatic gradients and geographic distance create migration barriers for animals and plants. Thus, two continents with distinctively different shallow tropical faunas would be thought to occupy the same latitude but to be separated by a wide ocean basin. This is the case with North America and Siberia in the Cambrian Period. However, in the succeeding Ordovician Period the faunas become increasingly similar, suggesting that these continents were moving together.

Another way of gaining some idea of relative distance is by timing the continental splits and collisions. These events give fixed points in geologic history when relationships of continental pairs are known, and then assumptions can be made, based on present-day movement, as to the rate of diver-

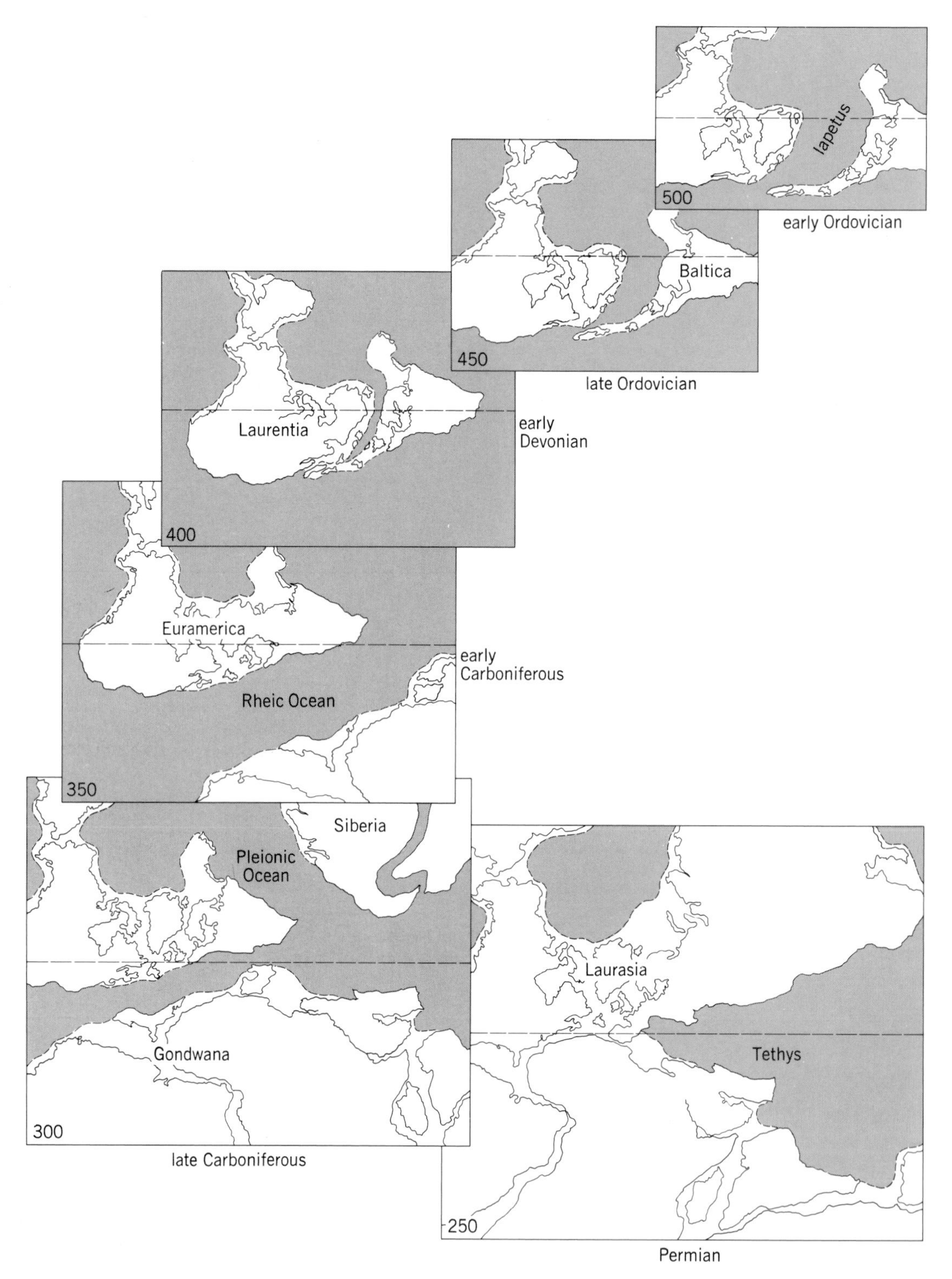

Fig. 2. History of the Atlantic-bordering continents (Laurentia, Baltica, Siberia, China, and Gondwana) during the Paleozoic Era. Numbers indicate millions of years before present. *(Courtesy of A. M. Ziegler)*

gence or convergence of the continents. Admittedly, none of the arguments concerning the relative position of the continents prior to Pangaea gives concrete information, and in practice the paleogeographer tries to produce a set of reconstructions that adequately accounts for all the evidence. No paleogeographer can produce a map that is accurate in detail; however, reconstructions are still the most useful way to depict geological evolution.

Paleogeographic development. Detailed paleogeographic knowledge of Earth begins about 600 m.y. ago due to the fact that animals first became abundant about that time, and their fossil remains provide geologists with the ability both to correlate rocks and to establish the conditions of the environment in which these organisms lived. Many continents show evidence of rifting at this time, suggesting the breakup of a super-continent about the end of the Precambrian. In any case, by the early Paleozoic the continents were widely disseminated (Fig. 2), and consisted of Laurentia (North America, Greenland, and the eastern Soviet Union), Baltica (northern Europe west of the Urals, and Nova Scotia), Siberia, China, and Gondwana (present southern continents with the addition of India, southern Europe, and Florida). Generally speaking, most of these land masses occupied low latitudes, with the exception of Gondwana, which was over the South Pole, while the North Pole was covered by an immense ocean.

About the middle of the Paleozoic, Laurentia and Baltica collided producing the northern Appalachians of New England and maritime Canada and the Caledonian Mountains of Great Britain and Norway. This produced a large stable land mass in low latitudes, and vascular plants and fresh-water fish developed in association with this "Old Red Continent" in the Devonian Period. Later in the Paleozoic and early Mesozoic, further collisions took place resulting in the formation of Pangaea at the end of the Triassic. The Appalachians, Urals, and central Asian mountain ranges are all part of this series of collisions. Climatic conditions were at first wet in the Carboniferous Period as indicated by the abundant coals, and then became generally dry in the Permian and Triassic periods as the amount of land area and the height of mountain ranges increased. The amphibian-reptile transition is correlated in time with this general climatic transition. Glaciations were especially widespread in the late Carboniferous and early Permian of Gondwana.

The breakup of Pangaea began about 200 m.y. ago, about the beginning of the Jurassic Period, when North America and Eurasia parted from Gondwana. However, the major split took place at about the beginning of the Cretaceous, when Gondwana split to form the South Atlantic and Indian oceans; the process of splitting continued into the Tertiary. The dinosaurs were thereby isolated on continental fragments that were drifting into higher latitudes, and became extinct at the end of the Cretaceous. The flowering plants, with their increased dispersal powers, developed in the Cretaceous and came to dominate the floras of the Earth.

At present the Earth is again in a period of continental collision. The Alpine and Himalayan chains are relatively recent, and Australia will become part of Eurasia. As in the late Paleozoic, the Earth has entered a major glacial period, which began just 2 m.y. ago. Both glacial periods are characterized by major oscillations; the present glacial cycle is now in a warm phase. However, the next few millennia will produce ice sheets, miles in thickness, which will advance on areas in North America and Europe that are presently heavily industrialized.

For background information *see* PALEOGEOGRAPHY; PALEOMAGNETICS; PLATE TECTONICS in the McGraw-Hill Encyclopedia of Science and Technology.

[ALFRED M. ZIEGLER]

Bibliography: M. W. McElhinny, *Paleogeography and Plate Tectonics*, 1973; R. M. Pegrum and N. Mounteney, *Amer. Ass. Petrol. Geol.*, 62:419–441, 1978; A. M. Ziegler et al., *Tectonophysics*, 40:13–51, 1977.

Perception

In the last few years two traditional areas of research on word perception in reading have been converging. One area concerns the perceptual processes used by adults to read familiar words, and has been carried on by academic experimental psychologists. The second area concerns the nature of reading disability in children, and has been carried on by clinical practitioners, neurologists, and educators. Experimentalists have tended to regard work in the second area as scientifically shoddy and inconclusive. Practitioners have tended to regard the theoretical work as divorced from reality and irrelevant to practical concerns. These criticisms have now begun to have their effect, and the work of both groups has improved as a result.

Perception of words. In the 19th century it was discovered that people did not move their eyes smoothly across the page as they read, but rather jumped from point to point, apparently taking in about three words per jump. Many investigators thought that they could analyze the reading process by learning what happened in a single jump. Thus the tachistoscope, a device for presenting visual stimuli for a fraction of a second, became widely used in reading research. Early experiments with the device showed that at short exposures adult subjects could report words much better than random strings of letters, or even, in some cases, single letters. It was thought that people had learned to see words as wholes, so that a single word was as easy to see as a single letter or a simple visual form. Such experiments were used by some investigators to justify whole-word reading instruction, but actually the conclusion that people perceived words as wholes did not follow from the evidence at the time, and subsequent research showed that the conclusion was wrong.

In particular, when people were shown words, they might have seen only a fragment of the word and guessed the rest. Further, the guess might have taken a much longer time than people would utilize when actually reading. These objections were answered in a series of experiments done in the last few years. In one experiment, subjects were asked to say as quickly as possible whether two words were the same or different, and the reaction time was measured. The pressure for prompt response prevented the subjects from thinking over what they saw, and the task was thus more like normal perception in reading. Subjects were faster at making the decision about words (such as *word-work*) than about nonwords, which were actually words with the letters rearranged (such as *owrd-owrk*). This showed that people had learned something about words, from reading, that helped them make these comparisons more quickly. But had they learned to perceive the words more quickly? Perhaps the subjects were actually comparing the sounds of the two words, or their meanings. They would be able to do this only when real words were presented, and this might explain why they were faster with words than nonwords. This would not count as a true perceptual difference between words and nonwords. To test this idea, another condition was included in the experiment: the words or nonwords differed only in the lower or upper case of a single letter (such as *word-*

worD or *owrd-owrD*). Then the subjects could not use the sound or meaning to make their decision, since in these respects *word* is the same as *worD*. However, this condition showed the same results as the first condition; words were still compared more quickly than nonwords. It thus seems that people do perceive words more quickly as a result of having learned to read.

In a similar experiment, besides comparing words and nonwords, subjects were asked to compare pseudowords such as *mord* and *mork*. Subjects could compare the pseudowords almost as quickly as the words. It seems that readers have learned something general about the properties of words. The superior perception of words is not due to people having memorized all the words "as wholes," as if they were familiar faces. Rather, people have learned what makes a possible word in English. Perhaps they have learned which letters are likely to occur in which parts of a word. Because this explanation of the original results was largely ignored, some workers falsely took the experimental data to support whole-word reading instruction.

Reading disability. It was noticed in the 19th century that certain kinds of brain damage led to total inability to read (alexia) or severe reading disability (dyslexia). In the early 20th century it was also noticed that many children seemed to have great difficulty learning to read, despite normal intelligence and instruction. These children were thought to have some sort of subtle neurological defect analogous to the defect in adult patients with brain damage. Hence, the term "developmental dyslexia" was coined to refer to reading disability in children without any observable brain damage. These dyslexic children made a large number of reversal errors in reading, mistaking *was* for *saw*, *big* for *dig*, or *form* for *from*. It was thought that these children saw the letters or words as backward on many occasions, and that errors in reading were due to misperceptions. Again, this conclusion did not follow from the evidence, for possibly the children perceived the words perfectly but had trouble recalling the "name" of the word when two words looked somewhat alike, just as a person might have trouble recalling which of two look-alike brothers is which. However, this did not stop many efforts to teach dyslexics to distinguish right from left in word perception.

Recent evidence has cast serious doubt on this whole theory. First, dyslexics do not make any more reversal errors than normal children who read just as poorly. In one study a group of sixth-grade poor readers got about the same average score on a reading test as a group of second-grade good readers. These two groups made the same number of reversal errors in reading. Another study shows a similar result for spelling. Thus, it seems that reversal errors are a sign that the children are reading at a certain low level, not that they have some subtle perceptual problem. Such errors are probably the result of poor reading rather than indicators of the cause.

Second, direct tests of the perceptual abilities of poor readers have shown no major deficits. In one study poor readers did just as well as normal readers at copying complex geometric designs after a brief presentation. This is an example of the sort of straightforward and well-done experiment that has resulted from a synthesis of experimental methods with practical concerns. In a less conclusive but more dramatic demonstration, inner-city children who were severely deficient in reading had little difficulty in learning the English words corresponding to Chinese characters. Chinese characters seem as perceptually complex as English words, but they lack correspondences between letters and sounds. It may be that these correspondences are the greatest source of difficulty for the beginning reader. Other evidence shows that poor readers have particular trouble identifying sounds within words. For example, they are less likely to recognize that *bee* and *bow* start with the same sound.

Perceptual learning. Poor readers seem not to have a deficit in perception as once thought. But recent evidence suggests that inept readers may have trouble with the kind of perceptual learning that accounts for the superior perception of words in normal readers. It has been found several times that when poor readers are asked to do perceptual tasks with words and nonwords, they are less likely to show the normal superior performance with words. But this may be a trivial finding; poor readers read less than normal readers, so they have not had as much of the experience required to show a word-superiority effect.

A recent study, however, set up an artificial analog of perceptual learning, so that good and poor readers could get equal experience with the particular materials used. Strings of six symbols each (from an IBM typewriter font ball—mostly Greek letters and mathematical symbols) were used. In the regular condition, certain symbols could occur only in certain positions in the stimulus (a string of letters), just as certain letters are especially likely to occur in certain positions in English words. In the random condition, each symbol occurred equally often in every possible position in the stimulus. The subjects, good and poor sixth-grade readers, were shown a symbol before each trial and asked to decide as quickly as possible whether the symbol was in the string of letters presented. This task thus measured their ability to find the symbol in the string or to decide that the symbol was not in the string, and thus probably measured some sort of perceptual ability. The good readers learned quickly to take advantage of the regularity; they were faster at finding (or not finding) the symbol in the string of letters in the regular condition than in the random condition. The poor readers did not learn to take advantage of the regularity. Thus, it seems that the poor readers may have a deficit in perceptual learning. They do not, however, seem to have a deficit in perception itself; in the random condition, the poor readers found (or did not find) the symbol as quickly as the good readers.

More work needs to be done, however, before this conclusion is definitive. For example, it might be that the good readers were doing the task differently; they may have learned just where to look for each letter in the regular condition, and decided that it was not in the string at all if it was not where it was supposed to be. What is clear, however, is that researchers are now on the path to getting answers to long-standing questions about perception in reading.

For background information *see* PERCEPTION in the McGraw-Hill Encyclopedia of Science and Technology. [JONATHAN BARON]

Bibliography: J. Baron, in *Estes' Handbook of Learning and Cognitive Processes*, vol. 4, 1978; M. Mason and L. Katz, *J. Experiment. Psychol. (Gen.)*, 105:338–348, 1976; A. S. Reber and D. Scarborough (eds.), *Toward a Psychology of Reading*, 1976; F. R. Vellutino et al., *Child Develop.*, 46: 487–493, 1975.

Periodicity in organisms

The pineal gland has repeatedly been associated with functions that involve changes in environmental lighting (in nature this means the daily light-dark cycle and the annual fluctuation in photoperiod). Most of the evidence for pineal function comes from three phenomena: body color changes in lower vertebrates, photoperiodic control of reproduction in mammals, and circadian rhythms in birds. Since body color changes have daily rhythms, and the photoperiodic control of reproduction involves circadian timing, a unifying concept for pineal function is the participation of the gland in circadian rhythms and the regulation of the rhythms by environmental light-dark cycles.

The pineal gland has well-defined daily rhythms of biochemical activity which can be affected by changes in environmental lighting. Pineal glands of chickens kept in a 24-hr light-dark cycle have 10-fold rhythms in the content of the pineal hormone, melatonin, and 27-fold rhythms in the content of an enzyme, *N*-acetyltransferase (NAT), partially responsible for melatonin synthesis. NAT converts serotonin to *N*-acetylserotonin in the pineal gland; a second enzyme, hydroxyindole-*O*-methyltransferase (HIOMT), converts the *N*-acetylserotonin into melatonin. While NAT activity is responsive to daily changes in lighting, HIOMT activity is affected only by weeks of treatment with constant light or dark.

Recent studies of the pineal gland have concentrated on the elucidation of its rhythms and their control. These studies include (1) work with sparrows to further explore pineal function in timing of circadian locomotor behavior; (2) comparative studies of rhythms in pineal gland biochemistry; (3) investigations of pineal regulation as an adrenergic nerve receptor; (4) explorations of input to the pineal gland from the hypothalamus; (5) pineal organ culture experiments to determine what timing ability resides within the gland itself; and (6) assays of the pineal hormone in blood and cerebrospinal fluid made with newly developed radioimmunoassay techniques. Thus, questions of input to and output from the pineal gland have been current areas of research in the context of periodicity.

Perch hopping in sparrows. S. Gaston and M. Menaker established that removal of the pineal gland in house sparrows eliminated the normal circadian rhythm of perch-hopping activity that was observed in constant dark. S. Binkley and coworkers verified the findings of Gaston and Menaker (see Fig. 1). N. Zimmerman and Menaker have extended this work by showing that replacement of the pineal gland (by transplantation into the anterior chamber of the eye) both restores circadian rhythmicity and sets the time for the recipient bird. The possibility that pineal function in circadian rhythms of sparrows is regulated by melatonin has been supported with experiments done by F. Turek and Menaker using implantations of Silastic capsules which release melatonin at an even rate over many weeks. The melatonin caused the sparrows to become arrhythmic in their perch-hopping activity or to change the period length of their circadian rhythms. The research on sparrows has firmly established the role of the pineal in circadian rhythm timing; however, this work is not generally applicable to other species. Since the circadian system can be considered an adaptation for environmental fitness, it seems reasonable to expect that different vertebrates would have specialized circadian systems in accordance with their individual adaptations.

Biochemical studies. The properties of pineal gland rhythms have been studied most fully in chickens and in rats (where they were discovered by D. Klein). In chickens the daily change in NAT activity (Fig. 2*a*) is established before hatching, and full dark-time activity is attained by about 18 days after hatching. The exact shape of the rhythm in a 24-hr light-dark cycle is affected by the light-dark ratio (or photoperiod), so that the peak of NAT activity is confined to the dark time; the activity achieved, however, is higher when the dark period is short (8 hr) than when it is long (16 hr). The daily rhythm in pineal gland NAT activity can be characterized as circadian because it persists in constant darkness. Constant light damps the rhythm to low values. The effect of unexpected lighting transitions on pineal NAT was studied, and it was found that (1) there is a refractory period which coincides with the light time during which dark will not result in high NAT values, and (2) light causes a rapid decline in NAT activity at any point during the dark time. The system that has been described applies to nocturnal rats as well as diurnal chickens; the daily change in NAT activity (or melatonin) with a nightly peak has now been observed in quail, hamsters, sparrows,

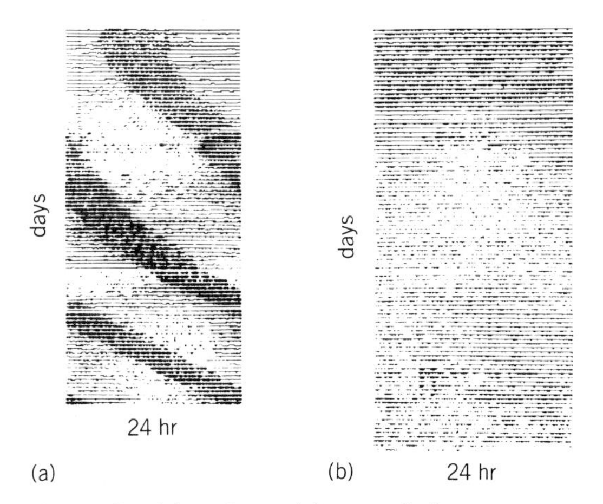

Fig. 1. Perch-hopping activity records from sparrows kept in constant dark. Each line represents 24 hr of data (a day). *(a)* Record showing persistent circadian rhythm from a control bird. *(b)* Record from a pinealectomized bird. *(From S. Binkley et al., Pineal and locomotor activity, J. Comp. Physiol., 77:163–169, 1972)*

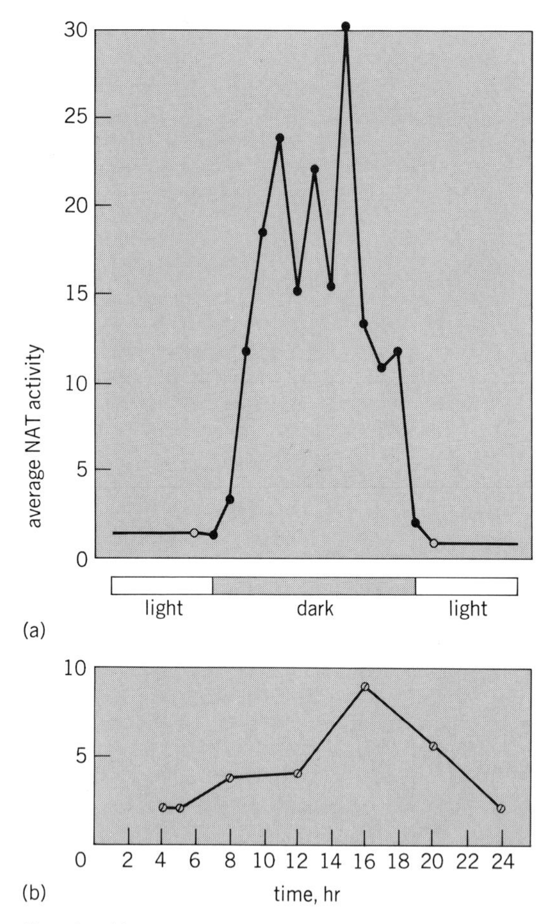

Fig. 2. *N*-acetyltransferase (NAT) activity in pineal glands of chickens over time. *(a)* Record of average NAT activity at different times relative to a cycle consisting of 12 hr of light alternating with 12 hr of dark *(from S. Binkley et al., Regulation of pineal rhythms in chickens, Gen. Comp. Endocrinol., 32:411–416, 1977)*. *(b)* Record of average NAT activity at different times in organ culture relative to the "projected" light-dark cycle to which the chickens would have been exposed had they not been killed for the experiment.

ground squirrels, gerbils, guinea pigs, sheep, weaver finches, and humans.

Neural mechanism. What input received by the pineal gland could account for the daily rhythm in NAT and melatonin synthesis? In some species, such as rats, there is a pathway by which the pineal gland receives timing, photic information, or both from the nervous system; it is thought that rhythmic signals (originating in the suprachiasmatic nucleus) and light information (from the eyes) are integrated in the suprachiasmatic nucleus of the hypothalamus located at the base of the brain over the optic chiasm. The integrated information is relayed by nervous connections through the superior cervical ganglion of the sympathetic chain to the pineal gland. In the pineal gland, nerve endings stimulate pineal cells by secretion of a neurotransmitter, norepinephrine. The pineal cells have receptors (beta-adrenergic receptors) which receive the norepinephrine signal and relay it to the pineal cell biochemical machinery by a "second messenger" system, cyclic adenosinemonophosphate. This pathway (for nervous input to the pineal gland from the hypothalamus) is well established for rats and hamsters. However, in some other species, such as chickens and sparrows, the nerve-input concept for the source of timing is insufficient to explain all the experimental results. In birds, especially, the pineal gland appears to have functional timing ability that resides within the gland itself.

Organ-culture experiments. The chicken pineal gland, unlike the rat pineal gland, was not stimulated by norepinephrine. The time course of NAT activity was studied in chicken pineal glands in organ culture, and it was found that NAT activity exhibited changes that were independent of neurotransmitters but were dependent on the chickens' prior light-dark cycle: (1) When pineal glands were obtained from chickens killed in the period of light, subsequent organ cultures began with low NAT levels and then showed a peak in NAT activity which coincided roughly with the projected period of dark (Fig. 2*b*). (2) When pineal glands were obtained from chickens killed in the period of dark, subsequent organ cultures began with high NAT levels and then showed a decline to low levels at a time which depended on the projected time of "lights on" and not on the time of the dark when the chickens were killed. These studies illustrate that the pineal gland, at least in chickens, has timing ability in the absence of external input.

Radioimmune assays. Determining what output of the pineal accounts for its function in daily rhythms has been facilitated by the recent development of sensitive radioimmunoassays for melatonin. By using the new assays, daily rhythms of melatonin have been studied in the blood and cerebrospinal fluid of a number of species, including sheep, monkeys, and humans. The results of these studies have mirrored those of melatonin synthesis (NAT and melatonin content) in the pineal gland itself. Although melatonin may be synthesized in other places in the body, the rhythm in circulating melatonin appears to originate in the pineal gland, since removal of the gland eliminates the daily change in circulating melatonin (in chickens).

Pineal physiology and rhythms in function. The information about the rhythms in pineal function matches well with the known physiology of the pineal gland. In those lower vertebrates where melatonin affects body coloration, melatonin acts as a blanching agent, and blanching occurs in the dark period when melatonin levels are usually high.

A pineal role in the photoperiodic control of reproduction in hamsters has been established, and it has been shown that melatonin may act as an antigonadal factor; this work shows good correlation with the biochemical studies demonstrating that light inhibits melatonin synthesis. The fact that the pineal gland is able to time synthesis of melatonin (in chickens) may account for the transplant restoration of rhythmicity and time setting in pinealectomized sparrows.

Conclusion. Caution must be exercised in elucidating pineal function because there appears to be variation associated with specialized environmental adaptations. The general agreement derived from recent studies, however, is that the pineal gland is part of the system responsible for circadian timekeeping in vertebrate organisms.

For background information *see* PERIODICITY IN

ORGANISMS in the McGraw-Hill Encyclopedia of Science and Technology. [SUE A. BINKLEY]

Bibliography: S. Binkley, J. B. Riebman, and K. B. Reilly, *Science*, 197:1181–1183, 1977; D. C. Klein, Circadian rhythms in indole metabolism in the rat pineal gland, in *The Neurosciences Third Study Program*, pp. 505–515, 1974; R. J. Reiter, *The Pineal—1977*, 1977; N. H. Zimmerman and M. Menaker, *Science*, 190:477–479, 1975.

Petroleum processing

Pollution control in petroleum processing has been concerned with fluid catalytic cracking units (FCCUs) as potential sources of both air and water pollutants. Potential air pollutants include carbon monoxide (CO), hydrocarbons (HC), ammonia (NH_3), hydrogen cyanide (HCN), and sulfur oxides (SO_x), which are by-products of coke combustion in the FCCU regenerator; and particulates, which are incompletely recovered by the regenerator cyclones. Water pollutants include NH_3, H_2S, and various organic compounds which result from steam stripping in both the FCCU reactor and the primary fractionator. In response to a number of regulatory pressures, several new technologies have been developed to control these pollutants. *See* CRACKING.

High-temperature regeneration. While not developed as a pollution control technique, high-temperature regeneration (HTR), which was introduced in the early 1970s, is an effective means of controlling combustible pollutants CO, HC, NH_3, and HCN. HTR involves operating the FCCU regenerator at 1300–1400°F (705–760°C), rather than at the approximately 1200°F (650°C) typical of earlier practice. If sufficient oxygen is supplied to the regenerator, HTR can reduce FCCU flue gas CO concentration to less than 0.05 vol %. Concentrations of HC, NH_3, and HCN can be reduced to less than 10 volume parts per million (vppm).

CO boiler. If HTR is not used, CO, HC, NH_3, and HCN are typically controlled by combustion in a CO boiler. Some auxiliary fuel is usually fired in the CO boiler for stability; more may be fired to provide additional energy. Flue gas composition exiting a CO boiler is similar to that produced by HTR except for one component; the nitrogen oxide (NO_x) content of flue gas exiting a CO boiler is typically 100–200 vppm, compared with less than 10 vppm from HTR. Neither NO_x level is currently the subject of regulatory concern.

CO combustion promoters. The CO level in FCCU flue gas can be lowered at constant regenerator temperature by the use of CO combustion promoters. These promoters, which were first introduced several years ago, are identified in the patent literature as noble metals. They can be incorporated directly into the catalyst or injected separately into the regenerator. CO promoters can be used either to lower the CO content of flue gas being fed to a CO boiler or to lower the temperature required to achieve complete combustion of CO in the regenerator.

SO_x emission control. It has not been necessary to control FCCU SO_x emissions in most locations. In those situations where SO_x control is desirable, feed hydrodesulfurization, which is used on some FCCUs to control product sulfur level and which will also reduce SO_x emission, has been used. However, during the past few years, two new technologies have become available for controlling FCCU SO_x emissions: flue gas scrubbers and SO_x control cracking catalysts.

Flue gas scrubbers. Flue gas scrubbers control both SO_x and particulates. Exxon started the first commercial FCCU flue gas scrubber at its Baytown refinery in 1974. Three other Exxon units were subsequently built, and two more are being designed for other oil companies. The Exxon units were placed downstream of a CO boiler, and use a sodium-based buffer solution circulating at high flow rates and pressures through spray nozzles to educt the flue gas into venturi scrubbers where SO_x and particulates are removed. In FCCUs equipped with HTR, flue gas is available at sufficient pressure, about 1.5 psi (10.34 kPa) above atmospheric pressure, to allow high-energy venturi scrubbers to be used. This approach uses flue gas pressure drop to atomize and mix the scrubbing liquid with the flue gas. SO_x is removed by reaction with the buffer solution, which is then regenerated by addition of caustic or soda ash. Particulates are removed by inertial impaction between the liquid droplets and the particulate in the gas stream. Flue gas scrubbers have demonstrated over 95% removal of SO_x and over 90% removal of particulates in commercial operation.

SO_x control cracking catalysts. The second new technology for controlling FCCU SO_x emissions is the development reported by Amoco of FCCU cracking catalysts which collect SO_x in the regenerator and transfer it to the reactor where it is converted to hydrogen sulfide (H_2S). H_2S, thus formed, can be scrubbed out of the product gases with monoethanolamine or some other suitable gas treating solution, and ultimately converted to sulfur in a Claus plant. Amoco reported tests on two commercial FCCUs in which SO_x emissions were reduced 40–75%. Amoco also claimed that the high cracking severity typical of molecular sieve catalysts remained unaffected (by inclusion of the sulfur trapping material) so that the conversion of feedstock and yields of cracked products were not impaired. The extent to which this technology is currently in use is unknown.

Particulate emission control. The typical FCCU regenerator has primary and secondary cyclones capable of reducing particulate emissions to the order of 0.2 grain per standard cubic foot (450 mg/Nm^3) on a dry-gas basis. External particulate control devices include tertiary cyclones, which are capable of a 50–65% further reduction in particulate emissions; electrostatic precipitators, which are capable of over 90% reduction in particulate emissions; and the flue gas scrubber described above, also capable of over 90% reduction in particulate emissions.

Water pollution control. The NH_3 and H_2S in FCCU waste water can be removed by steam stripping in a multistaged sour water stripper. Simple sour water stripper designs use a single tower and produce an overhead vapor containing NH_3, H_2S, and H_2O. This overhead vapor can either be flared or be sent to a Claus plant where the H_2S is converted to sulfur. Combustion conditions in a Claus plant are such that little of the NH_3 is converted to NO_x. More elaborate sour water strippers use two towers, the first of which produces an overhead vapor containing the majority of the H_2S, the second a vapor containing the majority of the NH_3.

The H_2S stream is fed to the Claus plant, and the NH_3 stream is flared or fed to NH_3 recovery.

The bottoms water from a sour water stripper still contains a significant amount of organic material. This material can be removed by a biological oxidation (BIOX) unit, which is typically a pond equipped with aerators in which bacteria which consume organics are grown. BIOX units can also remove some residual NH_3 not removed by the sour water stripper.

In summary, while FCCUs have a potentially large environmental impact, control methods capable of removing both air and water pollutants exist.

For background information *see* PETROLEUM PROCESSING in the McGraw-Hill Encyclopedia of Science and Technology.

[LEONARD S. BERNSTEIN]

Bibliography: J. D. Cunic et al., *Oil Gas J.*, pp. 69–73, May 22, 1978; I. A. Vasalos et al., *Oil Gas J.*, pp. 141–147, June 27, 1977.

Physiological ecology

Scorpions and beetles, especially species belonging to the family Tenebrionidae, are well represented in the world's desert regions. Their success in these potentially stressful environments is due largely to the evolution of a complex of behavioral, morphological, and physiological adaptations. Paramount among these adaptations are the avoidance of climatic extremes by burrowing, nocturnal surface activity, increased heat resistance, an impermeable integument with active mechanisms for water retention, a low metabolic rate, efficient elimination of excreta, and increased desiccation resistance. Emphasis in recent studies has been on the cellular and molecular basis for some of these adaptations and the monitoring of rate functions in unrestrained animals in their natural environment. Data generated from these experimental approaches have greatly increased understanding of the complex interrelationships among these desert arthropods and their environment.

Water loss. Water loss rates for desert scorpions are among the lowest reported for a terrestrial organism. Tenebrionid beetles exhibit slightly higher water loss rates, but these are still significantly less than rates for beetles from more mesic habitats. Until recently, these rate determinations were based on laboratory experiments, which often were artificial and ecologically meaningless. Extention of radiotracer techniques to free-roaming individuals has enabled investigators to measure water flux in a species' natural habitat and to provide a basis for checking the validity of previously determined laboratory data. An example of this experimental approach is the measurement of water exchange in laboratory and free-roaming populations of the desert tenebrionid beetle, *Eleodes armata*, using tritiated water. Water loss rates determined isotopically (WLR_i) for beetles decreased as the summer progressed (0.079 ml/g per day in early summer, compared with 0.062 ml/g per day in late summer). Significantly lower WLR_i (0.032 ml/g per day) were obtained for noncaptive winter beetles. WLR_i for beetles confined to artificial burrows (no food or water) were not significantly different from values for free-roaming beetles during either season. WLR_i for fed and hydrated beetles maintained in the laboratory under simulated summer conditions were four times higher than WLR_i for food- and water-deprived beetles. Similar tests on winter beetles under simulated winter conditions produced no significant differences in WLR_i.

Metabolic rate. Low metabolic rates help keep water loss at a minimum by reducing evaporative transpiration. Measurement of metabolism, like water exchange, is a relatively straightforward laboratory procedure; however, data obtained in this manner probably do not accurately represent the metabolism of an animal in its natural environment, where temperature and humidity are variable. Because of this, scientists have developed techniques that permit measurement of metabolism in field populations. One of the most promising of these techniques uses doubly labeled water ($^3HH^{18}O$), which provides data on body water content and water flux, as well as oxygen consumption.

The use of doubly labeled water to measure energy metabolism in free-roaming desert scorpions *(Hadrurus arizonensis)* represents the first application of this technique in an invertebrate. The mean metabolic rates of two groups of free-roaming scorpions were 0.326 and 0.329 ml O_2/g per day; these values are approximately two to three times higher than previously reported laboratory-determined metabolic rates at similar temperatures. Because the doubly labeled water technique follows every metabolic fluctuation of the individual animal and integrates total metabolism over the entire period of measurement, and because field animals are free to exhibit all levels of activity, the higher rates in field scorpions are expected. The substantial difference between estimates of metabolism by the $^3HH^{18}O$ method and by standard laboratory procedures, however, casts some doubt on the validity of this technique in invertebrates. Additional experimentation is necessary to determine whether the apparent problems reflect the need for technique refinement or are due to an inherent error in the technique principle when applied to invertebrates.

Role of scorpion gut. Radioisotopes were also recently used to examine the role of the scorpion gut in maintaining positive water balance. The ileum (midgut) of *H. arizonensis* was isolated for perfusion outside the body by using a flow-through technique developed by G. Ahearn. The results indicated that net water transport from the gut to the blood occurred in the absence of osmotic or hydrostatic pressure gradients, and that this flux was inhibited by metabolic poisons (sodium cyanide), was dependent upon the concentration of Na^+ in the lumen, and was inhibited by luminal K^+. This mechanism appears to be adaptive, for dehydration and starvation lead to luminal sodium and potassium concentrations that favor the uptake of water from the gut into the blood and, hence, reduce further depletion of critical body water.

Integument. Perhaps the most important morphological adaptation of desert arthropods for restricting water loss under hot, dry conditions is their impermeable integument. Several layers of the exoskeleton may play an active role in the waterproofing process, but for scorpions and tenebrionid beetles the principal barrier appears to be lipids associated with the epicuticle. Evidence supporting the role of lipids is based on increased transpiration rates following lipid extraction with organic solvents or abrasion of surface lipids, and on the reestablishment of permeabilities comparable to those observed for intact insects when the extracted lipids are plated onto artificial membranes or insect wings. The advent of modern analytical instrumentation such as the gas chromatograph and mass spectrometer and the probing capability of the electron microscope have made it possible to thoroughly analyze the chemical nature of these lipids and to provide information on their location within the epicuticle.

Epicuticular lipids. The chemical composition of epicuticular lipids has been determined for several desert scorpions. Cuticular lipids accounted for 0.03 to 0.09% of the fresh weight of *Paruroctonus mesaensis*, which inhabits sandy areas adjacent to dry river beds. Hydrocarbons were the most abundant lipid class; cholesterol, free fatty acids, and alcohols were also detected. Gas-liquid chromatographic analysis indicated that the hydrocarbon fraction was composed of saturated *n*-alkanes (straight-chain) and branched molecules, ranging from 21 to over 39 carbon atoms in length. A subsequent study compared the xeric species, *H. arizonensis*, with a montane scorpion, *Uroctonus apacheanus*. The surface densities of total lipids and hydrocarbons were significantly higher in *H. arizonensis*, which exhibited cuticular permeabilities approximately an order of magnitude less than those measured for *U. apacheanus* (see the illustration). Furthermore, epicuticular lipids of the more xeric *H. arizonensis* were characterized by higher proportions of long-chain branched hydrocarbons and long-chain saturated free fatty acids. Studies on plasma membranes and artificial bilayers have shown that long-chain, fully saturated lipid molecules are most effective in reducing permeability. Thus, results indicate that cuticular permeabilities can be altered to meet environmental requirements and that in scorpions predictable changes in epicuticular lipid composition are in part responsible.

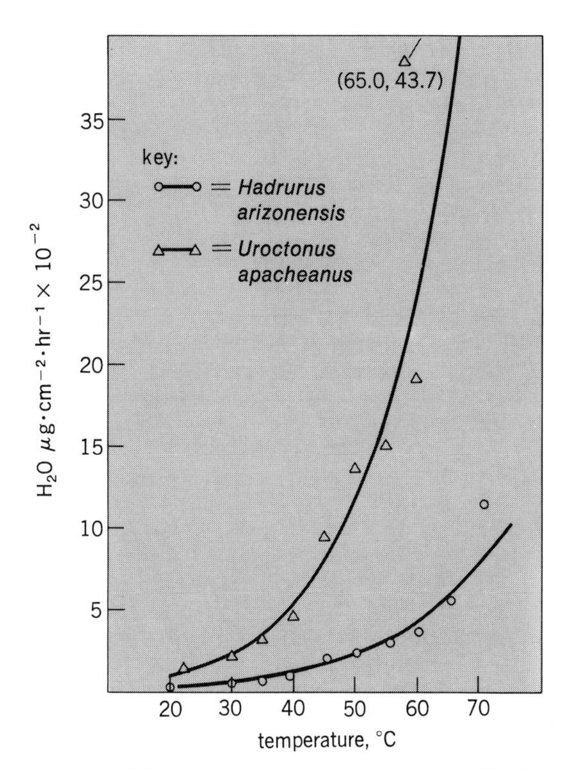

Effects of temperature on cuticular permeability to water in *Hadrurus arizonensis* and *Uroctonus apacheanus*. Plotted values represent sample means (minimum $N = 4$). *(From E. C. Toolson and N. F. Hadley, Cuticular permeability and epicuticular lipid composition in two Arizona vejovid scorpions, Physiol. Zool., 50(4):323–330, 1977)*

Comparative analysis of desert tenebrionid beetles*

Parameter	*Eleodes armata*†	*Crytoglossa verrucosa*	*Centrioptera muricata*	*Centrioptera variolosa*	*Pelecyphorus adversus*
Sample size	20	31	168	28	27
Sample weight, g	16.54	19.57	64.17	12.10	21.45
Total lipid, mg	29.73	24.99	18.98	1.90	5.07
Total hydrocarbon, mg	28.27	23.56	17.40	1.77	4.97
Percent hydrocarbon	95.1	94.2	91.7	93.2	94.3
Hydrocarbon/body weight, mg/g	1.71	1.20	0.27	0.15	0.23
Hydrocarbon size range (quantities >0.5%)	28 br–44 br‡	23–49 br	23 br–37 br	25–37 br	25–41 br
Percent *n*-alkanes	7.7	78.5	55.3	53.1	48.8
Percent branched	92.3	21.5	44.7	46.9	51.2
Cuticular water loss, mg/g/hr					
25°C	0.8	0.9	–	0.8	–
30°C	1.4	1.2	1.1	1.0	–
35°C	2.0	1.6	1.6	1.6	–
40°C	2.5	1.6	2.2	2.6	–
Transition temperature, °C	40.0	50.0	47.5	57.0	–

*All extractions with 100% hexane.
†Lipid data from *E. armata* group II summer beetles.
‡Numeral represents number of carbon atoms; br indicates branched molecule.
SOURCE: N. F. Hadley, Cuticular permeability of desert tenebrionid beetles: Correlations with epicuticular hydrocarbon composition, *Insect Biochem.*, 8:17–22, 1978.

Similar correlations were found in studies of cuticular permeability-lipid composition in desert tenebrionid beetles. Total surface lipids extracted from *E. armata* throughout the year ranged from 0.10 to 0.28% of their fresh weight. Hydrocarbons were most abundant, but percentage values varied depending upon the solvent used to extract the lipids (for example, chloroform:methanol removed more total lipids than did hexane). Hydrocarbons comprised over 25 components, all saturated, containing 26 to 44 carbon atoms. Branched alkanes accounted for over 92% of the total hydrocarbons, and were responsible for all components having chain lengths greater than 34 carbon atoms. Summer beetles and winter beetles acclimated to 35°C for periods of 5 and 10 weeks exhibited higher quantities of hydrocarbons and a higher percentage of long-chain components than did winter beetles in nature or appropriate controls.

The potential adaptive significance of this epicuticular lipid composition was examined further by comparing results for *E. armata* with similar analyses performed on four other sympatric species and correlating these data with the species' cuticular permeability (see the table). Cuticular transpiration increased linearly between 25 and 40°C in all species, but remained low in comparison to more mesic beetle species. Hydrocarbons accounted for over 90% of the total lipids in each species; they were completely saturated with odd-numbered chains containing from 25 to 33 carbon atoms predominating. The percentage of straight-chain hydrocarbons ranged from 7.7% in *E. armata* to 78.5% in *Cryptoglossa verrucosa*, the least permeable tenebrionid species. *Cryptoglossa verrucosa* was further characterized by having predominantly long-chain saturated *n*-alkanes, which should contribute to an effective cuticular water barrier.

Cuticle fine structure. Little information is available on the synthesis, transport, and deposition of these lipids in desert arthropods. A recently completed ultrastructural study of sclerite and intersegmental cuticle in *H. arizonensis* indicated that lipids are most likely distributed throughout the epicuticle and may be bound to protein or carbohydrate compounds. The latter belief is based on the failure of organic solvents to produce significant changes in the morphology of the epicuticular sublayers. *Cryptoglossa verrucosa* secretes waxes as amorphous blobs that reach the surface through narrow cylinders that open in the center of raised tubercles. At low humidities, the wax plug breaks into a fine basketlike network of filaments that completely covers the surface, giving the species a "blue" appearance. This morphological arrangement of surface waxes is potentially adaptive in that it both increases reflectance of solar radiation when the species is diurnally active and lengthens the diffusion pathway for water.

Summary. The majority of the adaptations exhibited by most desert scorpions and beetles are techniques for conserving rather than regaining water. Furthermore, most of these adaptations are not unique to desert arthropods, but are more highly developed and efficiently utilized in desert species than in their nondesert counterparts. Further studies on the adaptational biology of these groups will likely support these evolutionary trends.

For background information *see* ARACHNIDA; BEETLE; COLEOPTERA in the McGraw-Hill Encyclopedia of Science and Technology.

[NEIL F. HADLEY]

Bibliography: B. C. Bohm and N. F. Hadley, *Ecology*, 58(2):407–414, 1977; N. F. Hadley (ed.), *Environmental Physiology of Desert Organisms*, 1975; N. F. Hadley, *Insect Biochem.*, 8:17–22, 1978; E. C. Toolson and N. F. Hadley, *Physiol. Zool.*, 50(4):323–330, 1977.

Pipeline

Although past growth in pipeline transport in the Western world has been slow and steady, inflation and the recent concern about energy supplies have caused a surge of interest in hydraulic transport, especially for coal. In the United States the lack of Federal legislation for eminent domain has hampered progress, but 10 states now have such laws.

Plans for three new overland coal slurry pipelines in the United States and Europe were announced, and a large pipeline in Brazil for phosphate is scheduled to begin operation in 1978. Two new research pipeline facilities in England and West Germany were commissioned in 1978, one research pipeline facility is under construction, and another is planned, in the United States. One full-scale hydraulic haulage system in an underground coal mine has begun operations in West Germany, and another is under construction in the United States.

Three concepts for reducing or eliminating the problem of limited water supply received attention in 1978. Transporting coal in methanol is a new concept, and although coal in oil and in stabilized suspension are not new approaches, they are being considered.

Overland pipelines. The increase in construction of overland pipelines in the Western world has been steady over the past 4 decades, and the announced plans for additional construction indicates a spectacular increase in the next decade, as shown in Table 1. Nine of the pipelines planned for the 1980s are for transporting coal directly to electric generating plants, and five are for iron ore concentrates. The remaining two are for phosphate. The total capacity of the planned coal slurry pipelines is expected to be over 1.3×10^8 short tons (1.2×10^8 metric tons) per year and 2.1×10^7 tons 1.9×10^7 metric tons) per year for the iron ore concentrate pipelines. The coal pipeline figures reflect the effect of the increased concern about energy supply in the past few years. Table 2 shows the planned pipelines announced in 1978.

The seven coal slurry pipelines in the United States which are in various stages of planning are

Table 1. Overland pipelines in Western world

Decade	Number of pipelines	Annual capacity, 10^6 short tons (10^6 metric tons)	Total length, mi (km)
1940s	1	0.4 (0.36)	17 (27)
1950s	3	2.3 (2.1)	186 (298)
1960s	3	4.6 (4.2)	154 (246)
1970s	9	18.6 (16.9)	687 (1106)
1980s*	16	160.7 (145.8)	8235 (13,252)

*Planned

Table 2. Coal slurry pipelines planned in 1978

Location	Length, mi (km)	Diameter, in. (mm)	Annual capacity, 10^6 short tons (10^6 metric tons)
Kentucky to Florida	1500–1800 (2414–2897)	28–48 (711–1219)	15–45 (14–41)
Silesia, Poland, to Trieste, Italy	400 (644)	22 (559)	5 (4.5)
Energy Transportation Systems, Inc. (addition)	400 (644)	38 (965)	N.S.*

*N.S. = not specified.

being delayed by the need for uncontested water supplies, by the need for legislation granting the right of eminent domain, and by the lack of firm long-term coal contracts.

Federal legislation concerning eminent domain has been under consideration since 1974, but opposition by railroad companies and western states with limited water supply has delayed it. A study completed in 1978 by the Office of Technology Assessment concluded that there will be plenty of transport business for both railroads and pipelines and that delivery costs will be roughly comparable. Pipelines, however, were credited with being more reliable, less labor-intensive, and more acceptable esthetically and environmentally.

In July 1978 the House of Representatives voted to reject the most recent bill, H.R. 1609; however, other versions are certain to be introduced in the future. Presently, 10 states have legislation that either grants eminent domain to slurry pipelines or that can be interpreted as granting such rights. At least two other states are currently considering such legislation.

The considerations discussed above apply to overland pipelines in the United States for other transported solids as well, but none was announced in 1978. Overland pipelines in other countries have not experienced route acquisition problems because other national governments usually have greater control over property than in the United States.

Coarse-particle pipelines. Over the past few years experimental work in the study of coarse particles (especially coal) in large pipes has intensified, culminating in 1978 with the commissioning of two large research facilities in England and in West Germany. A large facility is nearing completion in the United States, and a moderate-size one is planned.

Also during 1978, one full-scale coal mine hoisting system began operation in West Germany, and a coal mine haulage and hoisting system is under construction in the United States. All of these facilities have pipe sizes of at least 6 in. (15 cm) and are designed to handle materials, primarily raw coal (uncleaned coal which contains rock and clay), with maximum particle sizes of 2 to 4 in. (5 to 10 cm). Table 3 lists these new facilities.

Aside from dredging, the great bulk of existing production systems that handle large particles are designed for the hoisting of raw coal from the underground workings to the surface. Research in the Eastern bloc nations has been in progress for many years, but reports of findings have been slow to reach the Western nations. The Soviet Union has several coarse coal haulage systems in operation, as does the People's Republic of China.

A number of articles on the transport of coarse-particle phosphates have been published, but a rough approach is used because of the lack of engineering data. The use of dredging is extensive, and sufficient work has been carried out to create a

Table 3. New coarse-particle facilities

Organization	Location	Pipe diameter, in. (mm)[a]	Circuit length, ft (m)	Vertical lift, ft (m)[b]
British Hydrodynamics Research Association Fluid Engineering	Cranfield, Bedford County, England	6 (150)	140 (43)	0
		8 (205)	373 (114)	43 (13)
		10 (255)	165 (50)	0
Steinkohlenbergbauverein	Essen, West Germany	10 (255)	690 (210)	13 (4)
		10 (255)[c]	150 (46)	13 (4)
		14 (353)[d]	N.S.[e]	N.S.[e]
U.S. Department of Energy	Pittsburgh, PA	6 (150)	720 (220)	90 (27)
		12 (305)	760 (232)	150 (46)
		18 (430)	790 (241)	150 (46)
State of Kentucky and University of Kentucky	Lexington, KY	6 (150)	600 (183)	–
Ruhrkohle Aktiengesellschaft, Hansa Mine	Dortmund, West Germany	10 (255)	10,000 (3050)	2790 (850)
		10 (255)§	12,300 (3750)	
Continental Oil Co./ Consolidation Coal Co. Loveridge Mine	West Virginia	8 (205)	*f*	0
		14 (333)	*f*	0
		12 (305)	13,500 (4115)	900 (274)

[a]Nominal pipe size. [b]Included in circuit length. [c]Alternate pipeline. [d]Future. [e]N.S. = not specified. [f]Variable.

large separate field of information. Deep-ocean mining is relatively new, and many of the problems of lifting minerals from depths as great as 15,000 ft (4575 m) have not been solved. Some academic studies have been published, but the bulk of the practical work has been performed by private industry and is considered proprietary.

Theoretical and experimental work. While the hydraulic transport of fine particles is fairly well understood, there have been no recent breakthroughs in the theory of coarse-particle transport. Until good reliable data are generated in test facilities, the approach to a theoretical solution will not be apparent. For many years researchers have unsuccessfully attempted to solve the problem by working on test facilities having pipelines of small diameter and length. Similarly, studies with low concentrations of solids (less than 20% by volume) have not been adequate to predict the performance of two-phase transport for higher concentrations. It is only when the experimental work is done at or near full-scale that results are considered to be reliable.

New concepts. Coal slurry is usually a mixture of coal and water; however, other fluids have been proposed as transport media. One proposal is to use part of the coal to produce methanol (CH_3OH); the rest could be mixed with the methanol and transported as a slurry. Water is not needed, and the mixture could be burned directly. In addition, the mixture is reported to have better transport characteristics than a water-based slurry due to its having a gellike consistency until pump pressure is applied, when it flows with normal fluidity.

A concept which has been mentioned in the past is coal in oil. This process would avoid the problems of obtaining water at the origin and of disposing of it at the destination. Research is being carried out to develop a burner which could efficiently burn the mixture, thereby avoiding the separation problem. If separation problems can be solved, the pipeline could be used for two-commodity transportation.

A third concept which is under more intensive review is that of using a transport medium composed of water containing particles fine enough to remain suspended at very low flow rates. At a sufficiently high concentration of particles, larger and heavier particles can be supported in the flow. Tests have indicated that concentrations of solids can be as high as 75% without major transport problems.

For background information *see* COAL; PIPELINE in the McGraw-Hill Encyclopedia of Science and Technology. [ANTHONY J. MISCOE]

Bibliography: A. J. Miscoe, *Hydraulic Transportation for Coal Mining: Workshop on Materials Handling for Tunnel Construction*, U.S. Department of Transportation, Transportation Systems Center, Colorado School of Mines, and ASCE and AIME Joint Materials Handling Task Committee, Aug. 3–5, 1977; C. A. Shook, *Can. J. Chem. Eng.*, 54:13–25, February–April 1976; Slurry pipeline support gains momentum, *Chem. Eng. News*, pp. 23–24, Apr. 17, 1978; E. J. Wasp, J. P. Kenny, and R. L. Gandhi, *Solid-Liquid Flow: Slurry Pipeline Transportation*, Trans Tech Publications Series on Bulk Materials Handling, vol. 1, no. 4, 1977.

Plant disease control

Plants can be protected against disease by using procedures that are essentially identical to those used to immunize animals. Recent work of this type has been carried out with cucurbits.

Immunization with *Colletotrichum lagenarium*. J. Kuć and S. Richmond reported that infection of a cotyledon or first true leaf (leaf one) of cucumber with the fungus *Colletotrichum lagenarium* systemically protected tissue above (developed or not yet developed) against disease caused by the pathogen. Physical damage or chemical injury did not elicit protection. Susceptibility in this interaction is characterized by the formation of large but defined lesions. Since many such lesions may form and coalesce on the leaves, stems, and fruit of cucurbits, the growth of plants and their productivity, as well as the quality of fruit, are adversely affected. Nevertheless, the restricted size of individual lesions suggests the presence of a mechanism for resistance in susceptible plants. It would appear, therefore, that the defense mechanism against the disease in the unprotected plant either is elicited with insufficient magnitude or is expressed too late.

This systemically induced protection is manifest as a delay in symptom expression and a reduction in the number and size of lesions. Infection of leaf one of cucumber when the second true leaf was one-fourth to one-third expanded systemically protected plants for 4–5 weeks, at which time plants had 8–12 large leaves. A second, booster inoculation 3 weeks after the first extended the time of protection into the fruiting period. Protection was elicited by and effective against six isolates of the fungus and was evident with 20 susceptible cultivars and 6 cultivars which expressed some resistance to the pathogen. Resistance in these cultivars is expressed, as in systemic induced resistance, by a delay in symptom appearance and a reduction in the number and size of lesions. A single lesion on leaf one produced significant protection. Protection was evident on the second leaf 72–96 hr after inoculating leaf one. Excising leaf one 72–96 hr after inoculating leaf one did not reduce protection of leaf two. Leaf two was protected if excised 96–120 hr after leaf one was inoculated. Thus it is evident that it is not necessary for the inducer to be present once protection has been initiated and that protection continues to be expressed in all developing leaves for 4–5 weeks. It is also evident that the protected leaf need not be attached to the plant to maintain protection.

Caruso and Kuć reported that the pattern of protection in watermelon and muskmelon resembles that in cucumber. Infection of the cotyledons or first true leaf of four cultivars of watermelon and four cultivars of muskmelon with *C. lagenarium* systemically protected the plants from disease caused by subsequent infection with the pathogen. Plants remained protected 4 weeks after the protecting inoculation. Race 1, 2, and 3 of the fungus and a single lesion on leaf one elicited significant protection.

Extending their studies to the field, Caruso and Kuć found that, in three separate trials, cucumber plants were systemically protected against *C. la-*

genarium by limited prior inoculation with the pathogen. In one trial with cucumbers, 90% of the drops of challenge inoculum applied to leaves of unprotected plants developed into lesions, whereas 18% developed into lesions on protected plants. Protection of watermelon was evident in two other trials, and there were indications that muskmelon could also be protected. In one trial with watermelons, 47 out of 69 unprotected plants died, as compared with only 1 out of 66 protected plants. Lesions on protected plants were reduced in number and size.

Cross-immunization. Although chemical or physical injury without the presence of an infectious agent did not elicit protection, A. Jenns and Kuć found that protection against *C. lagenarium* was elicited by the infection of cotyledons or leaf one with tobacco necrosis virus (TNV). Data also indicate that the infection of leaf one with the bacterial incitant of angular leaf spot, *Pseudomonas lachrymans*, protected against the fungus, and infection with the fungus protected against disease caused by the bacterium. Plant breeders have observed that cultivars are often either resistant or susceptible to both pathogens. Infection of cucumber with virus protected against disease caused by the fungus and bacterium. Jenns and Kuć have reported that systemic protection is graft-transmissible. The chemical signal responsible for protection originates with the pathogen but also appears to be produced or amplified by the host. The signal which protects cucumber also protects watermelon and muskmelon.

Protection mechanisms. Two or more distinct mechanisms may be responsible for the induced protection. There is no evidence that the systemic accumulation of classical phytoalexins is alone responsible for the protection of cucurbits. A chloroform-soluble inhibitor of the growth of *C. lagenarium* has been obtained from tissue surrounding lesions on protected and unprotected plants. The inhibitor may be important in restricting lesion development and may account for the normally restricted lesions characteristic of cucurbit anthracnose. Spore germination and appressorium formation are not inhibited on protected plants, but penetration from the appressoria appears to be reduced in protected plants. In a series of tests, penetration from appressoria was 20 to 40% in unprotected plants and from less than 1 to 5% in protected plants.

Conclusion. The above data support the hypothesis that susceptible plants have mechanisms for disease resistance which are effective if they are expressed early enough and with sufficient magnitude. Induced resistance can be systemic and of quite long duration. Systemic induced protection in plants is similar to immunization in animals at least in outward appearance. This phenomenon suggests a new approach for the practical control of disease in plants.

For background information *see* IMMUNOLOGY; PLANT DISEASE CONTROL in the McGraw-Hill Encyclopedia of Science and Technology.

[JOSEPH KUĆ]

Bibliography: F. L. Caruso and J. Kuć, *Phytopathology*, 67:1285–1292, 1977; A. E. Jenns and J. Kuć, *Physiol. Plant Pathol.*, 11:207–212, 1977; J. Kuć and F. L. Caruso, in P. A. Hedin (ed.), *Host Plant Resistance to Pests*, Amer. Chem. Soc. Symp. Ser., no. 62, pp. 78–89, 1977; J. Kuć and S. Richmond, *Phytopathology*, 67:533–536, 1977.

Plant evolution

Recent developments in plant evolution have focused on the evolution of higher land plants and the study of microfossil floras.

Evolution of land plants. The evolution of higher land plants (mosses and allies and vascular plants) is a critical benchmark in the history of life. Until recently, the beginnings of life on land had been shrouded in almost total obscurity. Assessment of the time, place, and environment of the evolution of terrestrial vegetation rested on a few scattered body fossils (megafossils) of vascular plants that straddle the Silurian-Devonian boundary. As recently as 1977, it was stated that there were no records of pre-Pridolian (latest Silurian) land plant life, either vascular or nonvascular.

Since about 1973, however, a new type of evidence has begun to accumulate, largely as the result of work by J. Gray and A. J. Boucot, that sheds more light on the course of pre-Devonian land plant evolution than has been discerned in the past 100–150 years. This evidence is based on dissociated organic microfossils of land plant type rather than on megafossils that are visible to the naked eye.

Plant organic microfossils. Plant microfossils are of three principal morphological types: trilete spores and spore tetrads; sheets of epidermal cells similar to cuticle of some land plants; and tubes with spiral thickenings that are similar to the tracheids of vascular plants. These microfossils are extracted from sedimentary rocks by rigorous chemical and mechanical means. They are found in a wide range of rock types, and are deposited in the continental environment as well as in the nearshore, shallow-marine environment where they were either air- or waterborne. Because the fossils are dissociated, no source plants have been identified, and taxonomists cannot attribute them to a specific group of plants, living or extinct. It is also not clear whether the three types of structures, which range in variability, come from the same plants. Thus, a wide spectrum of plants could be involved, representing various levels of phylogenetic and morphological evolution, and including both vascular and nonvascular plants at a level of development possibly similar to the mosses and their allies. Also worth considering are potentially transitional land plants—more advanced structurally than the algae from which land plants are believed to have derived, but not yet full-fledged land inhabitants—and enigmatic, extinct land plant groups for which there are fossil records but whose phylogenetic position is uncertain.

Although no specific taxonomic group(s) of plants can be selected as the definite source(s) of the microfossils, they can be regarded as diagnostic of plants living in a specific habitat. Among living plants, structures of similar morphology and durability that could withstand fossilization and the rigorous chemical extraction necessary to free them from rocks are found only in mosses and their allies and vascular plants. Similar or identical structures from body fossils are all from distinctly nonmarine groups of plants, including those of problematic taxonomic status which are now extinct. Thus, even though these microfossils are dissociated, their occurrence carries the same ecological constraints as if they were found connected to or within plant megafossils.

Trilete spores and spore tetrads. These structures are the most critical of the microscopic remains. Morphologically indistinguishable spores and spore tetrads are recovered from rocks bearing some of the earliest vascular plant fossils as well as from within the sporangia of the earliest vascular plants. Trilete spores and spore tetrads, indistinguishable from those produced by primitive vascular plants, are also found in fluviatile rocks containing remains of the earliest nonmarine, nonvascular plants.

The abundance and diversity of dissociated spores and spore tetrads far exceed the vascular and nonvascular plants known to have been land inhabitants prior to the Late Silurian, as well as in the Early Devonian, when body fossils of land plants first became common. This suggests that the megafossil record is not a reliable guide to the rate of evolution of higher land plants.

Modern morphological analogs of trilete spores and spore tetrads are reproductive bodies that are wind-dispersed, desiccation-resistant, and durable-walled, due to a chemically inert wall. The fossil spores presumably were also reproductive bodies. They are the first of the dissociated structures to appear as fossils. Thus a durable-walled spore may have been the first land plant–type structure that an emergent aquatic acquired in its transformation to a land plant, since ensuring reproductive capability via such a desiccation-resistant structure would have been a critical necessity in a nonaquatic environment.

Trilete spores and spore tetrads are the most common land plant–type microfossils recovered from pre-Devonian rocks. Their record now begins in the late Ashgillian (latest Ordovician), and they are found both in very nearshore, shallow-marine sediments, where spores and pollen of living land plants are most common in modern sediments, and in nonmarine rocks of Early Silurian age. As reported by Gray and Boucot, pre-Ashgillian rocks, including Late Cambrian–age rocks deposited in environmentally suitable shallow water or nearshore sites, have yet to yield land plant–type microfossils.

Trilete spores and spore tetrads have been found in pre-Devonian rocks from Canada, eastern and central United States, Great Britain, Norway, Sweden, Belgium, Spain, the Soviet Union, North Africa, and South America (Bolivia and Peru), indicating a geographically widespread group of spore-producing land plants. This record contrasts sharply with the pre-Devonian land plant megafossil record based on less than half a dozen occurrences, chiefly in eastern North America and Great Britain.

Cuticle- and tracheid-like remains. These two types of structures commonly accompany the trilete spores and spore tetrads. Cuticle-like sheets of cells are to be expected in land plants where water economy might be a problem, but not in a wholly aquatic organism, unless the aquatic adaptation was secondary.

The tracheid-like tubes have no exact counterpart in living groups of plants, unlike the trilete spores and spore tetrads. Similar or identical tubes have been identified, however, in an extinct group of land plants that flourished during the Silurian and Devonian. The presence of tracheid-like tubes only in land plants and their similarity to the tracheid of vascular plants, which functions in conduction of water and minerals in solution, suggest that these tubes may have had a similar function, that they may have evolved parallel to tracheids and that, like tracheids, they may have evolved in a variety of not too closely related land plants.

Conclusions. Several categories of land plant–type microfossils, extracted from pre-Devonian rocks but not yet recovered from pre-Ashgillian rocks, suggest that the uppermost Ordovician may mark the beginning of higher land plant life. The number, diversity, and geographic distribution of the microfossils support the presence of diverse land plant life, even though coeval megafossils are rare. There are good nonpaleontological reasons for suggesting that in the presence of habitable land some form of plant life escaped the confines of aquatic existence long before the appearance of undisputed vascular plants in the Late Silurian. Chief among them is the necessity for the formation of soil. Soils can be formed by the mechanical and chemical breakdown of bare rock, but soil nutrients are produced only by the decay of plants and plant parts. Bare rocks can be broken down chemically and rock minerals used directly only by a variety of algae, lichens, and some of the mosses and their allies. The decay of these plants in turn adds nutrients to developing soils in a form that can be utilized by vascular plants. There is thus reason to believe that a succession of plant life paved the way for vascular plants.

Many questions about early land plant evolution remain to be answered: How far back in time can land plant microfossils be traced? Do land plants have a single point of origin, or has their evolution involved multiple events occurring in widely separated places? Can the environment(s) of early land plant evolution be determined more specifically? The answers to these and other questions await further work on a global scale. [JANE GRAY]

Study of macrofossil floras. Recent developments in the study of macrofossil floras of Late Silurian–Early Devonian time include (1) the extension of the first demonstrable macrofossils back into late Ludlovian time; (2) the geographic extension of the range of zosterophylls to China, and their stratigraphic extension upward into Late Devonian (Frasnian) strata; (3) the discovery of an Early Devonian lycopod that demonstrates the oldest occurrence of paracytic stomata; (4) the still unproved suggestion that the first lycopods might have appeared in Silurian rather than Early Devonian (Siegenian) time; and (5) the elucidation of the anatomical structure of one trimerophyte, *Psilophyton dawsonii*.

Rhyniophytes. During the interval from approximately 405 to 380 million years (m.y.) ago, four subdivisions of vascular plants appeared on land and initiated a vigorous adaptive radiation.

Appearing first were the Rhyniophytina, characterized by leafless dichotomously branched stems, terminal sporangia containing trilete, resistant spores, and a solid centrarch vascular strand of helical elements. Only in Wales have all these characters been proved for Pridolian (Late Silurian, 405–395 m.y. ago) specimens. Similar-appearing fossils occur in New York State, Czechoslovakia, and Podolia in the Soviet Union. Recently D. Edwards and E. Davies demonstrated trilete spores in terminal sporangia and probable tracheids in late Ludlovian (the next stage older than the Pridolian) rocks in Wales. The specimens, although not assigned a generic name, presently represent the oldest known fragments of Rhyniophytina. All other accounts of supposed macrofossils of Ludlovian, or older, age are still completely unconvincing to date as evidence of vascular land plants. Resistant trilete spores are dispersed abundantly in older Silurian rocks, but they alone cannot prove the existence of macrofossils of vascular land plants. Rhyniophytina seem to have died out in Emsian (late Lower Devonian) time, although there are questionable reports of one genus in the Late Devonian.

Zosterophylls. A second group of macrofossils, Zosterophyllophytina, appeared in Early Devonian (Gedinnian) time. They too have leafless stems, but their sporangia, often reniform in shape, are borne laterally, often on short stalks, and dehisce distally. Their xylem strand is solid and exarch. Several new features are found in the group. Emergences range from hairlike to minute teeth to large teeth and large multicellular spines that may have terminated in glandular tips. Modified cells form a zone of dehiscence on the sporangia of some species. Stomata are present on the epidermal surface. Their restriction in some species to the aerial parts suggests that the plants grew on wet mud flats with their basal parts (rhizomes) immersed in the mud. Cortical cells in several species are thick-walled, suggesting a hypodermal tissue. Basal axes branch in a complex manner suggestive of a dense mat of subterranean stems. Axillary branches are present in the axil of some forkings of the aerial stems. It has been speculated that the axillary branches might be roots such as are found in modern *Selaginella* (small club moss). According to H. P. Banks, Zosterophyllophytina spread rapidly around the globe in the Northern and Southern hemispheres. The distribution of the group has been widened markedly by its recent discovery in Yunnan Province in China. Zosterophylls survived essentially unchanged into strata of Late Devonian age.

Early Devonian lycopods. Two more groups appeared in mid-Siegenian time, about 380 m.y. ago: the Lycophytina and Trimerophytina. The lycopods introduced additional new morphological characteristics, especially small leaves (microphylls). These may represent an elaboration of emergences such as those seen in zosterophylls (enation hypothesis). The small leaves are characterized by a single central vein and by stomata. S. Stubblefield and Banks recorded the oldest (Siegenian) occurrence of paracytic stomata in the genus *Drepanophycus*. Sporangia are apparently either axillary to the leaves and borne on short stalks or are adaxial on the leaves. Dehiscence is distal as in zosterophylls, and resistant trilete spores are produced. The xylem strand is solid and exarch. The early lycopods were plants with trailing rhizomes and erect, probably recurved, aerial branches. Both parts were clothed with leaves, arranged at first irregularly and subsequently in a spiral with a low helix so that cursory examination suggests a whorled condition.

Since the appearance of the lycopods, herbaceous representatives have been present continuously in the fossil record. Modern lycopods are very similar to early ones. The lycopod group appears to have been distinct from all other groups of vascular plants. Whether or not lycopods evolved from zosterophylls, the two groups share more characteristics than either does with any other group: lateral sporangia, distal dehiscence, and a solid exarch vascular strand. Only the Late Devonian and Carboniferous arborescent lycopods appear to represent evolutionary derivatives of the herbaceous lycopods, and these giants became extinct by the end of the Paleozoic Era.

Occurrence of lycopods in Silurian. There are recent reports of the occurrence of lycopods in rocks of Silurian age from both Libya and Australia. At present the evidence that the fossiliferous strata are in fact Silurian is too tenuous to be ac-

Fig. 1. Reconstruction of *Psilophyton dawsonii* showing sterile (below) and fertile (above) branches. *(From H. P. Banks, S. Leclercq, and F. M. Hueber, Anatomy and morphology of Psilophyton dawsonii, sp. n. from the late Lower Devonian of Quebec (Gaspé) and Ontario, Canada, Palaeontogr. Amer., 8:75–127, 1975)*

cepted. The Libyan specimens are poorly preserved and resemble similar poorly preserved lycopods from Middle and Late Devonian strata in the rest of the world. Those from Australia resemble collections from rocks of Siegenian-Emsian (Early Devonian) age in Australia.

Trimerophytes. Trimerophytina, in sharp contrast to Lycophytina, may have given rise to a number of more highly evolved groups. Trimerophytes share several characteristics with rhyniophytes: terminal sporangia, leafless axes, dichotomous branching (in part), and a solid centrarch xylem strand. Significant evolutionary changes are equally striking. The sporangia dehisce longitudinally and are borne in large clusters on much-branched lateral systems rather than singly on little-branched aerial stems. Branching of major axes in trimerophytes may be dichotomous or pseudomonopodial. It may be alternate and widely spaced or in a closely spaced spiral arrangement, or may even simulate a whorl.

One species *(Psilophyton dawsonii)* is anatomically preserved and has been studied by Banks, S. Leclercq, and F. M. Hueber. Its anatomical structure is advanced well beyond the slender xylem strand of rhyniophytes. The plant has alternate regions of vegetative and fertile branching (Fig. 1). The xylem strand enlarges markedly in the area of vegetative branching, gives off traces to vegetative branches in rapid succession, and then reverts to its previous size. Vegetative branches fork dichotomously and end in recurved tips that may be regarded as precursors of leaves. Above the vegetative region the xylem gives off alternately strands that simulate leaf traces and supply fertile branch systems. Each trace branches twice in rapid succession, producing externally an apparent trichotomy. In *P. dawsonii* the middle branch aborts, while in other genera the three may persist. Subsequent divisions are strictly dichotomous, and the ultimate segments terminate in a pair of sporangia. *Psilophyton dawsonii* had the potential for the evolution of leaves (by planation and webbing) from the ultimate branchlets of the vegetative branches and for the evolution of much larger leaves by the modification of the entire fertile branch system (Fig. 1). Xylem of *P. dawsonii* includes helical and more complex, scalariform-bordered pitted tracheids. The deposition of additional secondary wall outlines one or two rows of circular openings (Fig. 2). Banks, Leclercq, and Hueber suggest that these tracheids may represent an early stage in the evolution of circular-bordered from scalariform-bordered pits. Trimerophytina survived until early Middle Devonian (Eifelian) time but then seem to have been replaced by derivative groups, of which progymnosperms (plants whose anatomy is comparable with gymnosperms but whose reproduction is similar to pteridophytes) are the most abundant. Sphenophytina (horsetail group) and various primitive ferns are other groups that may have evolved from trimerophytes.

PLANT EVOLUTION

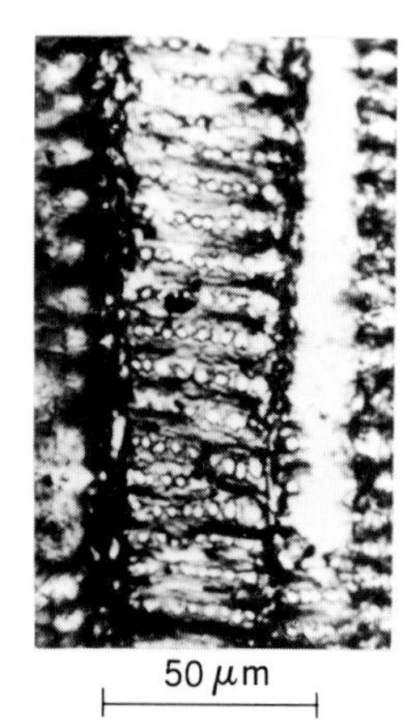

Fig. 2. Single tracheid of *P. dawsonii*, face view showing scalariform pitted tracheids and the added secondary wall material that outlines, between the scalariform bars, circular areas which may be precursors of circular-bordered pits. *(From H. P. Banks, S. Leclercq, and F. M. Hueber, Anatomy and morphology of Psilophyton dawsonii, sp. n. from the late Lower Devonian of Quebec (Gaspé) and Ontario, Canada, Palaeontogr. Amer., 8:75–127, 1975)*

Summary. Early Devonian (Gedinnian, Siegenian, and Emsian) time saw the establishment of the four groups discussed above as well as several enigmatic types of no known evolutionary significance. Terrestrial strata are continuous from Early into Middle Devonian time. Their precise stratigraphic position is difficult to correlate both with well-dated marine strata and between continents. Rapid evolution took place at that time (approximately 370–365 m.y. ago). Thus, conflicting accounts of the precise time of occurrence of other new subdivisions and new genera are to be expected, but they need not alter the major conclusions.

For background information *see* PALEOBOTANY; PLANT EVOLUTION in the McGraw-Hill Encyclopedia of Science and Technology.

[HARLAN P. BANKS]

Bibliography: H. P. Banks, in K. S. W. Campbell (ed.), *Gondwana Geology*, 1975; H. P. Banks, S. Leclercq, and F. M. Hueber, *Palaeontogr. Amer.*, 8:75–127, 1975; D. Edwards and E. C. W. Davies, *Nature*, 263:494–495, 1976; J. Gray and A. J. Boucot, *Geology*, August 1978; J. Gray and A. J. Boucot, *Lethaia*, 10:145–174, 1977; J. Gray and A. J. Boucot, *Science*, 173:918–921, 1971; S. Stubblefield and H. P. Banks, *Amer. J. Bot.*, 65:110–118, 1978.

Plant growth

Recent studies on plant growth have been concerned with cold hardiness and the possibility of genetically adapting crops to be more salt-tolerant.

Cold hardiness. Winter hardiness implies avoidance of or tolerance to all the cumulative effects of winter, including freezing, heaving, smothering, desiccation, and disease. Climatic, soil, plant, and cultural factors interact to determine the degree of injury incurred by a plant following the rigors of winter. Of all of the factors in the winter hardiness complex, cold hardiness, the ability to withstand low freezing temperatures, is of paramount importance. Resistance may take the form of certain physiological or morphological adaptations that allow the plants to either avoid or tolerate the imposed stresses. Avoidance and tolerance mechanisms may reside at either the whole-plant, tissue, or cellular level.

Characterization of freezing process. Freezing in plant tissues involves the redistribution of water with respect to both its physical state and location. When plants are subjected to temperatures below freezing, both the water in the cell and in the extracellular spaces initially supercools. Because the extracellular solution has a lower solute concentration and more effective ice nucleators, initial ice formation will occur extracellularly. As the vapor pressure of the extracellular ice-water mixture is less than that of the intracellular water, a vapor pressure gradient results. Vapor pressure equilibrium can be achieved either by the efflux of water out of the cell to the extracellular ice, resulting in cellular dehydration, or by intracellular ice formation. The manner in which equilibrium is achieved depends on the rate at which the cells are cooled in relation to the permeability of the cell membrane and the surface area/volume ratio. The temperature of the extracellular ice establishes the lower value of the vapor pressure gradient, and hence is responsible for water being removed from the cell to achieve vapor pressure equilibrium. However, the actual amount of water that has to be removed depends on the solute concentration (osmolality) of the intracellular solution. Whether this amount of water is removed depends on membrane permeability and the surface area available for efflux. If the flux is inadequate, the cells equilibrate by intracellular ice formation, which is generally lethal. However, under the freezing rates normally encountered in nature the cells equilibrate by extracellular ice formation. Two points should be emphasized: (1) even though ice formation occurs at only a few degrees below 0°C, water within the cell remains a liquid at temperatures as low as −20°C; and (2) individual cells contract rather than expand when extracellular ice formation occurs.

Effects of freezing on cells. The effects of the freezing process on the environment of the cell and its components include the obvious decrease in temperature, presence of ice crystals, and dehydration of the cell. There are several consequences of dehydration, including a reduction in cell volume and surface area, concentration of solutes, precipitation of some salts resulting in pH changes, and removal of water of hydration of macromolecules. P. Mazur refers to these as solution effects and notes that they all occur as a monotonic function of temperature. There is general agreement that under conditions of extracellular ice formation, decreases in temperature or the presence of ice crystals alone is not responsible for injury, and that the process of cellular dehydration is the most disruptive and injurious effect of the freezing process. However, dehydration has a number of effects on the cell, and it is within this array that there is a great divergence in hypotheses on the mechanism of freezing damage.

Freezing injury. The visual manifestations of freezing injury—a darkened, water-soaked, flaccid appearance—are apparent immediately following thawing. It is commonly believed that damage is the result of disruption of cellular membranes, especially the plasma membrane. Although there is general agreement that injury results in the loss of semipermeability, there are many ways in which this may occur, and the exact nature of plasma membrane damage at the molecular level is not known. Recent work by P. L. Steponkus and S. C. Wiest suggests that rupturing of the plasma membrane is the result of the freeze-thaw-induced contraction and expansion of the cell.

Cold acclimation. Temperate-zone plants exhibit an annual periodicity in their tolerance to freezing temperatures; in winter they are able to withstand temperatures of −30°C or lower, but in spring or summer they are susceptible to freezing temperatures and are easily killed by temperatures of −3 to −5°C. The increase in cold hardiness during the fall is referred to as cold acclimation, while the loss of hardiness in the spring is termed deacclimation. The cold hardiness of a given species is dependent on two factors: the inherent or genetic capacity of the species to acclimate and the conditioning or expression of this heritable capacity. Plants lacking the genetic capacity are considered unhardy or frost-sensitive species; those which possess the genetic capacity but have not experienced the proper cues for its expression are considered to be in an unhardy condition; and those which possess the genetic capacity and have received the proper environmental cues are considered to be in a hardy condition.

Not all plants necessarily respond similarly to the same environmental cues. What is considered to be an optimum or necessary cue for one species may vary considerably for different cultivars or ecotypes within that species. While environmental cues synchronize plant development with the environment, a species' responsiveness has taken centuries to evolve; thus, freezing injury in cultivated species can result from any factor which disrupts this synchrony. Such disruption may occur when individual cultivars are introduced into areas which are vastly or even slightly different from their natural habitat where centuries of selection pressures have evolved those individuals most closely synchronized with the prevailing environment.

Temperature. Temperature is the key environmental parameter for synchronizing a plant with the prevailing ambient temperatures. Low above-freezing temperatures (0 to 5°C) are conducive to cold acclimation in the fall, while warm temperatures are responsible for deacclimation in the spring. The progressive decline in temperatures—from the relatively high temperatures in early fall, through the low above-freezing temperatures in late fall and early winter, to freezing temperatures in winter—is extremely important in the acclimation process. Each stage has a distinct role in the overall process of acclimation.

Light and moisture. Light is the second major environmental factor affecting cold acclimation. Light may influence cold acclimation in either a photosynthetic or photoperiodic manner, depending on the species. A third parameter, moisture, can also significantly influence the hardiness of a given species. While light and temperature affect the hardiness of plants through their interaction with the acclimation process and the development of tolerance to the stresses of freezing, moisture greatly affects the freezing process itself.

Biochemical and physiological aspects of cold acclimation. Much attention has been given to the biochemical changes that occur during the period of cold acclimation. However, there is very little information detailing specific cause-and-effect relationships. Increases in cellular solutes such as sugars, amino acids, and organic acids may be important in mitigating the extent of ice formation and cellular dehydration. Alterations in membrane components, for example, lipids and proteins, may be instrumental in allowing the membranes to tolerate lower freezing temperatures. Thus, biochemical alterations may take place in the cellular environment so that either the freezing stresses are altered or there is direct protection of the sensitive membranes. Alternatively, cold acclimation may involve changes in the membrane itself so that its susceptibility to freezing stresses is decreased.

Summary. While an understanding of freezing injury and cold acclimation has steadily improved, the final answers have not yet been obtained. There is a significant amount of information on the physicochemical events associated with the freezing process, but the manner in which injury is caused is not fully understood. While the importance of several environmental cues in the cold acclimation process has been well established, the manner in which these cues are translated into increased resistance is unknown. In addition, the significance of most of the biochemical changes which occur during cold acclimation can only be

Fig. 1. Experimental genetic lines of barley growing in dune sand irrigated with undiluted sea water. *(From E. Epstein and J. D. Norlyn, Seawater-based crop production: A feasibility study, Science, 197:249–251, 1977)*

speculated upon. Insufficient information on what constitutes freezing injury at the molecular level precludes the final integration of the many known facts. [PETER L. STEPONKUS]

Genetic adaptation to salinity. Recent research has demonstrated the possibility that crops may be genetically adapted to tolerate much more saline soils and waters than present-day varieties. This development could be of considerable importance in the arid and semiarid regions of the world. Because of insufficient leaching by rain, many of the soils of these regions are saline, as is much of the irrigation water. This is a serious problem because of the sensitivity of virtually all presently used crop plants to salt. Genetic adaptation of crops to salinity may therefore become a useful stratagem in coping with this problem. It may even be possible to adapt crop species to sea water, which could then be used for irrigation and mineral nutrient supply for raising crops along some of the extensive coastal deserts of the world.

Fig. 2. Seeds of wheat from the world collection selected on a nutrient solution at 50% sea-water salinity. A few seeds have germinated and established seedlings. *(Courtesy of E. Epstein and the University of California)*

Plants and salt. The crops now used for food and fiber are almost without exception sensitive to salt to some degree. (Salt refers here to all types of sodium salts.) The most salt-sensitive crops (for example, the bean, *Phaseolus vulgaris*) may suffer a significant reduction in yield, by about 10%, at a salinity of the irrigation water of only 0.05%, or 500 parts per million (ppm). Such a salinity level is frequently encountered in the irrigation water available in arid and semiarid regions, and much higher salinities are by no means uncommon.

At the high temperatures characteristic of these regions, water rapidly evaporates from the soil and from leaf surfaces, that is, the rate of evapotranspiration is high. Salt is left behind, often resulting in its progressive buildup in the soil. These conditions may make it impossible to plant the more sensitive crops, forcing a shift to those able to tolerate relatively high salt concentrations, such as barley, *Hordeum vulgare*; but the salt tolerance of even these crops is limited.

However, there are wild plants, the halophytes, that grow in very saline environments, including sea water (35,000 ppm salt). Among these plants are those indigenous to salt marshes, such as *Spartina* (cordgrass) and to tropical shores, such as the mangroves, trees whose roots grow in soil saturated with sea water.

Possibility of salt-tolerant crops. Existing crop plants are generally intolerant of salt, but some plant life is capable of thriving in water with extremely high salinity. Recognition of this somewhat paradoxical situation prompted researchers to wonder whether it might be possible to combine within the same plant the desirable features of both groups: the economic utility of a plant crop and the salt tolerance of halophytic wild plants.

The possibility that this approach might be feasible was suggested by another fact. There are numerous collections or "banks" of seeds representing reservoirs of genetic variability. These collections contain seeds of important crops collected from all over the world, including commercial varieties, land races, local strains, and other genotypes; many thousands of these are seeds of important grains such as wheat and barley. Yet not a single species represented in any of these collections has ever been screened systematically or on a large scale for salt tolerance. Screening of this type for other desirable features such as disease resistance and cold hardiness has been highly successful. It therefore seemed likely that salt tolerance might also be uncovered in these and other collections of genetic variability if a systematic search for it were conducted.

Screening for salt tolerance. Scientists at the University of California, Davis, used a "composite cross" of barley as their initial stock from which to make selections for salt tolerance. It was synthesized earlier by other Davis plant scientists who intercrossed 6200 barley strains from all over the world, resulting in a mixture with great genetic variability. For screening, seeds and the plants they produced were exposed to synthetic nutrient solutions containing all the mineral nutrients that plants normally absorb from soil—a routine procedure of the plant nutritional laboratory and greenhouse. There was, however, this additional feature: the solutions were heavily salinized, either with ordinary table salt, NaCl, or with a synthetic sea salt mix.

Most of the thousands of seeds subjected to salt stress failed to germinate, or if they germinated and plants developed they failed to flower or set seed (grain). However, out of 7200 entries 22 seeds germinated, and the plants eventually matured and produced heads of grain. Grain of each successful entry was planted in soil to produce more plants of the same genotype for use in a field trial.

The site of this trial was a level area of dune sand at the University of California's Bodega Marine Laboratory, 80 km (50 mi) north of San Francisco. The grain was planted in conventional fashion in plots, some of which were irrigated with sea water, others with dilutions of sea water, and still others with fresh water, to serve as controls. Because sea water contains low concentrations of nitrogen and phosphorus—important plant nutrients—they were supplied in the form of commercially available fertilizers.

Figure 1 shows part of a plot irrigated with undiluted sea water. Each row of plants represents a genotype selected initially in Davis by the procedure described above. This and other experiments with barley indicate the biological feasibility of growing barley under a highly saline regime.

Similar work is under way with wheat, the world's foremost grain crop. Figure 2 shows wheat seeds from the world collection planted in a nutrient solution salinized with 50% sea water. Most of the seeds failed, but a few have germinated and established vigorous seedlings.

Crossing wild and commercial species. Screening genetically heterogeneous collections of seeds, as described above, is not always feasible. Various genetic lines of the tomato were found to have little difference in salt tolerance, which was not high. Another strategy was therefore used. Seeds of a wild, commercially useless species of tomato had been collected by a University of California, Davis, geneticist on Isla Isabella, one of the Galapagos Islands, where it grew just a few meters above high tide. The assumption that this tomato might be salt-tolerant was borne out when its seeds were germinated and plants grown in nutrient solutions. The plants survived even at full sea-water salinity, although their growth was impaired. This tomato was crossed with a commercial species, and progeny of the cross, grown at the Bodega test site on the coast in plastic greenhouse shelters, produced acceptable tomatoes the size of cherry tomatoes when irrigated with 70% sea water. Further improvement is expected.

Conclusions. It can be seen from these experiments that by appropriate techniques of screening and breeding genetic lines of crops may be obtained that can tolerate extremely saline environments. Such salt-tolerant crops may give growers in arid and semiarid regions new options for utilizing saline soils and water, and may even lead to the production of crops irrigated by sea water along sandy coasts.

For background information *see* PLANT GROWTH in the McGraw-Hill Encyclopedia of Science and Technology. [EMANUEL EPSTEIN]

Bibliography: E. Epstein, in M. J. Wright (ed.), *Plant Adaptation to Mineral Stress in Problem Soils*, 1977; E. Epstein and J. D. Norlyn, *Science*, 197:249–251, 1977; J. Levitt, *Responses of Plants to Environmental Stresses*, 1972; P. Mazur, *Annu. Rev. Plant Physiol.*, 20:419–448, 1969; A. Poljakoff-Mayber and J. Gale (eds.), *Plants in Saline Environments*, 1975; P. L. Steponkus, *Adv. Agron.*, in press; P. L. Steponkus and S. C. Wiest, in P. H. Li and A. Sakai (eds.), *Plant Cold Hardiness and Freezing Stress*, 1978.

Plate tectonics

Current hypotheses concerning the origin and evolution of oceanic crust are based on models for the generation of magmas from the Earth's mantle by fractional melting and on observations of the chemical, physical, and mineralogical properties of sea-floor samples. Data for these samples come mainly from rocks dredged from the sea floor and from samples collected by deep drilling of the *Glomar Challenger* (Deep-Sea Drilling Project). Rock exposures on land of material believed to be fragments of oceanic crust, rock series termed ophiolites, also play an important role in estimating the unexposed constituents of oceanic crust. The physicochemical processes involved in the formation of the new oceanic crust are believed to be closely related to the sea-floor spreading processes.

Oceanic crust origin. The main locus of generation of oceanic crust is the mid-ocean ridges such as the Mid-Atlantic Ridge, East Pacific Rise, and the Indian Ocean ridges. Less well-defined spreading centers are believed to exist in marginal basins adjacent to island arcs where oceanic crust is generated by processes similar to those which operate at mid-ocean-ridge spreading centers. Some of the marginal basins which appear to show evidence of this include the Lau Basin, the Marianas Trough, the North Fiji Plateau, and the Scotia Sea. At both the mid-ocean ridges and the marginal basins new ocean crust is formed by the leakage of basaltic composition magma from the Earth's mantle. This emplacement of new magma continually fills the rifts which form as the crust is distended and ruptured during the sea-floor spreading process.

The island arcs and volcanic arcs which rim the Pacific Ocean basin are considered to be an expression of magma generation in the mantle due to subduction of oceanic lithosphere. These subduction zones are also characterized by deep earthquakes (Benioff zones) and by deep oceanic trenches (such as the Tonga and Mariana trenches) which reach depths of 8 to 10 km. The island arcs are chains of volcanoes built on oceanic crust, and thus they represent a second locus of generation of oceanic crust by the transfer of melts derived from the mantle to the Earth's surface. Volcanic arcs, such as the Cascade Range of western North America and the Andes, are generally similar geologic features developed on continental crust and represent an addition of mantle-derived material to the continents. In both the island and volcanic (continental) arcs the new magmatic material leads to a thickening of the crust and to an increase in the volume of crustal rocks in the region of the arc.

In addition to the two major loci of oceanic crust generation described above, mantle-derived magmas are added to the oceanic crust at some transform fault boundaries and at isolated points within oceanic plates. Ideally, transform fault boundaries should be zones of slip and there should be neither dilation nor compression along the fault trace. Because the plates do not have simple boundaries and are not homogeneous lithosphere blocks, some transform fault plate boundaries have a dilational-stress component as well as a slip-stress component. Magma leaking out along these dilation areas forms volcanoes, usually submerged volcanoes or seamounts, which are superposed on the oceanic crust generated at the mid-ocean ridges.

Intraplate volcanism. Intraplate volcanism is formed by mantle-derived melts which punch their way up along zones of weakness within oceanic plates. These additions of crustal material are not necessarily aligned along the plate boundaries, and since they may erupt well away from the spreading centers, they form seamounts and islands much younger than the oceanic crust on which they are constructed. The Hawaiian Island chain is a linear array of volcanic islands formed by intraplate volcanism. The islands appear to increase in age northwesterly from the island of Hawaii, which includes the still-active volcanic centers of Mauna Loa and Kilauea. Other linear volcanic chains in the Pacific basin include the inactive Emperor Seamount chain which runs northward from the northwestern end of the Hawaiian chain, the Samoan volcanic chain which was last active in 1913, and the Tuamotu chain which has had recent volcanism at its southeastern end.

Tholeiitic basalt. The magma type which leaks out at the ridge crests to form the new oceanic crust is termed tholeiitic basalt. This rock type is mainly composed of olivine and calcic plagioclase, with clinopyroxene and iron-titanium oxides as less abundant constituents. The magma is quenched rapidly when it pours out on the sea floor, and the basalt is finely crystalline or glassy with a few large crystals (phenocrysts) of olivine, plagioclase, or both. The material formed at the ridge crests is remarkably uniform in composition for most of its chemical constituents. The mid-ocean-ridge tholeiites are Ca-, Mg-, Fe-, Cr-, and Ni-rich rocks and are characterized by low concentrations of incompatible elements, namely, alkali metals and elements such as Ba, Sr, U, Th, Pb, P, and rare earths. Their chemical characteristics suggest that ocean ridge basalts form by about 20–30% fractional melting of mantle material which was previously depleted in the incompatible elements.

Alkali basalt. Rocks typical of seamounts and many oceanic islands are also basalts but are enriched in alkali metals, Ba, Sr, U, Th, Pb, P, and rare-earth elements. These alkali basalts are believed to result from a small amount of fractional melting of undepleted mantle. From experimental studies of conditions for mineral stability, it is believed that the alkali basalts and similar rocks may be derived from depths on the order of 80–100 km. The mid-ocean-ridge basalts may represent melts derived from depths on the order of 30–50 km.

Crust modification. Oceanic crust, whether formed at mid-ocean ridges or seamounts, experiences complex modification as it moves away from the ocean ridge crests. During cooling, circulation of sea water through joints and fracture systems alters the original composition. The nature and extent of this alteration depend on the relative volumes of rock and sea water involved, but in general the basalt loses Si, Al, Ca, Mg, and Mn, it is enriched in alkalies and water, and the iron is oxidized. This chemical alteration is accompanied by mineralogical changes. Clay minerals, zeolites, and amphibole replace the original igneous minerals and glass. Carbonate and sulfates may fill cracks and gas bubble holes, and exposed rock surfaces may become coated with manganese oxide crusts. With time, these basaltic crustal rocks become buried beneath accumulations of sedimentary materials. The alteration processes described here affect all of the oceanic basaltic crustal rocks without regard to their site of origin. The end result of the aging process of oceanic crust is the development of a leached, oxidized, and hydrated basaltic crust overlain by sediments.

Island-arc volcanic rock. The volcanic rocks formed in island arcs differ in composition and mineralogy from the ocean ridge basalts and seamounts. The most common rock type, andesite, has a higher content of silica, alkalies, and alumina, and lesser abundances of Ca, Mg, Ti, and elements such as Cr and Ni than oceanic basalts. Magmas which form the island-arc rocks are believed to be derived deep in the mantle by fractional melting of the subducted oceanic lithosphere plates. Heating of the subducted plate, which consists in part of altered basaltic crust, leads to the formation of melts at depths on the order of 150–200 km. These melts are believed to be the source of the andesite magmas typical of island arcs. Support for this interpretation comes from laboratory experiments which attempt to duplicate the pressure-temperature conditions of the mantle and which use as experimental material samples having the presumed composition of the oceanic lithosphere. The chemistry and mineralogy of island-arc magmas and the explosive nature of their activity give further support for the hypothesis that melting of hydrated oceanic crust is involved in their origin, although there are good arguments for the contribution of mantle-derived material as well.

Linear volcanic chains. Linear volcanic chains present a problem of special significance to the plate tectonic hypothesis and to the origin of oceanic crust. As their name implies, they are long linear arrays of seamounts and oceanic islands. Some of the best-studied chains, for example, the Emperor and Hawaiian chains, show strong evidence that they have developed and evolved sequentially from the northernmost end to the southeastern end (Hawaii), which is still active today. This progression is characterized by a sequence of events which begins with the building of a shieldlike volcano, followed by collapse of the summit area and by less voluminous volcanic activity near the summit and on the flanks of the large volcanic edifice. A period of quiescence is accompanied by erosion and slow subsidence of the whole edifice. Quite commonly this is accompanied by the building up of fringing coral reefs. After an extensive period of erosion, that is, 1,000,000–2,000,000 years, another pulse of volcanism builds small cinder cones (such as Punchbowl Crater and Diamond Head in Oahu) which may be accompanied by outpourings of lava. This final outburst of volcanic activity is followed by further erosion, subsidence, and continued upward growth of coral reefs. Eventually only the reefs remain to form the irregularly circular atolls. If subsidence continues faster than reef growth, or if the reef dies, the reef and seamount may become totally submerged. This sequence of events is probably typical of most seamounts. The Hawaiian chain is of special interest because it illustrates different stages in this evolutionary cycle, from its active shield-building stage at Kilauea to the atoll stage on its northwestern end at Midway Island.

A widely accepted interpretation of the origin of the Hawaiian volcanic chain is that it formed as the Pacific Plate moved northwesterly over a spot more or less fixed with reference to the Earth's axis of rotation. This fixed spot is the location of a melting anomaly—which some researchers term a hot spot—in the deep mantle. This melting anomaly leaves its imprint on the moving lithosphere plate by the episodic generation of volcanoes as the plate passes over the melting anomaly. The orientation of the Hawaiian chain rather closely "fits" the path to be expected if a melting anomaly with a fixed position were leaving its trace on the moving Pacific Plate. The change in direction between the Hawaiian and Emperor chains is interpreted as being due to a change in the relative direction of plate motion over this melting anomaly some 40,000,000 years ago.

For background information *see* MAGMA; PLATE TECTONICS in the McGraw-Hill Encyclopedia of Science and Technology. [JAMES W. HAWKINS]

Bibliography: A. Ewart, W. Bryan, and J. Gill, *J. Petrol.* 14:429–466, 1973; J. Hawkins, *Earth Planet. Sci. Lett.*, 28:283–297, 1976; R. Kay, *Earth Planet. Sci. Lett.*, 38:95–116, 1978.

Population genetics

The genetic composition of natural populations is not static, but is subject to evolutionary change in response to mutation rate, the mating structure of populations, random genetic sampling, and the differential fitnesses of genotypes. Differential fitnesses are the unequal probabilities that describe genetic contributions to subsequent generations by individual genotypes. To describe the fate of a single protein mutant requires knowledge of a change in molecular structure and its consequence on molecular function. Altered gene functions interact with population size and structure to determine whether the mutant is lost from or fixed in the population. Since many critical variables, including population size and structure, cannot be directly estimated, various models have been formulated that attempt to predict the ultimate fate of mutant proteins, as well as levels of protein polymorphism in contemporary populations. Such models fall into two classes: those which incorporate natural selection (that is, fitness differences among genotypes), and neutral models whose predictions depend only upon mutation rate and population size, and not upon natural selection. For more than a decade, active debate has occurred between investigators explaining protein polymorphism by natural selection and those postulating neutral evolution.

It is now known that among closely related species, levels of protein polymorphism (that is, the number of allelic variants and their relative frequencies) are locus-specific: an enzyme that is highly polymorphic in one species tends strongly to be highly polymorphic in related species, and monomorphic (one-type) loci, exhibiting a single homozygous allele, also tend to be monomorphic in related species. This observation suggests either that natural selection operates in an identical fashion at loci in different species, or that polymorphism is principally due to locus-dependent mutation rates. Since polymorphism is not randomly distributed among loci, correlations between the degree of polymorphism and various aspects of molecular structure (such as quaternary structure and molecular size) have been recently investigated.

Subunit size and polymorphism. In *Drosophila*, heterozygosity (a measure of the equivalence of allele frequencies, averaged over all studied loci) at individual protein loci is positively correlated with enzyme subunit size (see illustration). Similar observations have been made in a variety of other organisms (see table) with the exception of humans, which appear to be a special case. In humans there is a significant increase in the number of electrophoretic alleles with subunit molecular weight, but no concomitant increase in heterozygosity. The increase in levels of genetic polymorphism with subunit size, which is in most cases a reflection of gene size, would suggest that the amount of extant polymorphism is a consequence of locus-dependent mutation rates. Larger genes would be larger targets for random mutation along the chromosome. Smaller genes, coding for

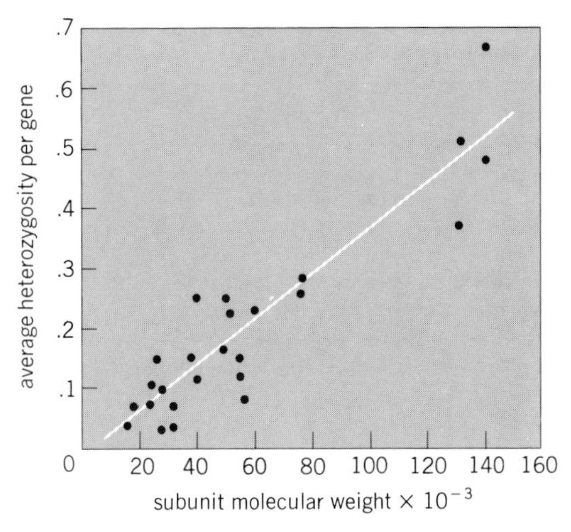

The relationship between average heterozygosity of enzymes and their respective subunit molecular weights in *Drosophila* species.

Summary of the results of correlation and regression tests of the independence of average heterozygosity of enzymes and their subunit molecular weight in various organisms

Organism	Correlation (r)	Explained variation (r^2)
Drosophila	.753	.567
Heliconiid butterflies	.545	.300
Primates (nonhuman)	.347	.121
Rodents	.653	.427
Reptiles	.649	.421
Salamanders	.369	.136
Fishes	.366	.134

smaller protein subunits, would experience relatively lower magnitudes of mutation and thus exhibit lower amounts of polymorphism.

Quaternary structure and polymorphism. Enzymes of differing quaternary structures may be expected to exhibit differing levels of heterozygosity, due to the loss of accessible atomic surface area during subunit association and the added structural constraints imposed by subunit contact sites. Several researchers have described the decreasing levels of polymorphism in enzymes of increasing structural complexity. H. Harris and colleagues have described such a relationship among quaternary classes of human enzymes and have attributed the observation to the very low incidence of polymorphism in multisubunit enzymes that form hybrid molecules between loci. It was postulated that multiple-locus multimeric enzymes that form molecular hybrids endow organisms with a greater degree of metabolic flexibility, thus alleviating the selective pressure to preserve polymorphism.

Although the data are suggestive, evidence for the influence of quaternary structure on heterozygosity is sparse and equivocal. Where comparisons of quaternary structures have been made, it is not possible to eliminate the potential influence of variations in subunit size from the test results. Where variations in size do not appear to be important, as for example in the human data, there is an apparent influence of quaternary structure, but this may be due more to gene duplication than the multimeric nature of the enzymes.

Neutral expectations. Neutral models have taken various forms. The "infinite allele" model of M. Kimura and J. Crow assumes that all mutations are novel and potentially detectable by the electrophoretic method. In 1973 T. Ohta and Kimura introduced the "stepwise mutation" model, where mutations are not electrophoretically unique but mutate between specific electrophoretic classes, sometimes called electromorphs. The stepwise mutation model is realistically suited to electrophoretic data, but both models assume that mutation rate is the same for all genes. The increase of levels of genetic polymorphism with increasing subunit molecular weight is an observation that seems consistent with a neutral model of varying mutation rate, wherein large genes experience a greater mutation rate than small genes. A neutral explanation for gene-size-related variation hinges upon the expectations of a neutral model that incorporates varying mutation rate.

Recently M. Nei and coworkers have begun to formulate a neutral theory which incorporates varying mutation rates among loci. In such a model, unlike both the infinite allele and stepwise mutation models, it is the distribution of mutation rates that determines the expectation of the model. This distribution has been estimated as a gamma function by Nei and colleagues, who have devised formulas to describe the expected relationship between average heterozygosity and the interlocus variance of heterozygosity. In all neutral models, the variance of heterozygosity among loci increases with the average heterozygosity among genes. However, in a model which incorporates varying mutation rate, the increase of the interlocus variance is more rapid. The larger the average among-locus heterozygosity, the larger will be the proportion of interlocus variance that can be attributed to a varying mutation rate. Hence, the higher the average heterozygosity, the larger the expected coefficient of determination (r^2) for a specific value of correlation. In principle, this is the reason why tests of the correlation in *Drosophila*, where the average among-locus heterozygosity is high, give considerably larger values of r^2 than comparable tests in other organisms (see table).

Observed correlations between subunit size and polymorphism are reasonably consistent with a neutral model incorporating a mutation rate that varies in a gamma distribution among genes. However, the amount of interlocus variance in heterozygosity explained by subunit size is somewhat lower than the neutral model would expect. Like any model, the varying neutral mutation model incorporates an assumption whose validity is difficult to evaluate. Congruence of observations with a neutral expectation does not prove neutral polymorphism. It does mean that if polymorphism is assumed to be neutral, it can be explained by a model with certain properties. This seems to be a general characteristic of all so-called tests of the neutral theory.

Cryptic variation. The extensive literature on genetic variation of enzymes that can be correlated with various aspects of molecular structure was collected over the past decade by extensive surveys that have utilized a single electrophoretic condition. Electrophoresis in any single condition of pH, ionic strength, and so forth, does not detect all existing genetic variation of enzymes. Recently several techniques, including rates of heat denaturation and electrophoresis in a variety of conditions of pH and acrylamide concentration, have revealed genetic variation of enzymes that had gone undetected by earlier electrophoretic methods. This variation is sometimes termed cryptic, and a single electrophoretic band composed of more than one allele has been termed an electromorph. Many additional gene variants can be discovered when new and diverse methodologies are employed. These recent findings emphasize the dependence of estimates of genetic variation on individual methodologies and make it difficult to assess the reliability of the data on polymorphism that has accumulated during the past several years. Very few enzymes have been studied by these newer techniques, but preliminary data suggest that there are greater numbers of undetected alleles at loci coding for enzymes with large subunit molecular weights. Since cryptic alleles are characteristically revealed within the most frequent electromorph, it is not yet possible to predict the effect these findings will have on overall levels of heterozygosity. This will depend on the population frequency of newly discovered alleles.

Population differentiation. Since the development of electrophoresis as a common tool for the analysis of population polymorphism, there has been a corresponding increase of interest in estimating genetic changes that accompany or promote speciation. Estimates of differentiation are expressed as genetic identity, which is a function of the number of alleles and their relative frequencies that are shared between populations (or species). The discovery that intrapopulation polymorphism is enzyme-specific has the consequence that intrapopulation and intraspecies estimates of genetic identity are also enzyme-specific. At highly polymorphic loci, that is, enzymes with large subunits, coefficients of genetic identity increase gradually with increasing taxonomic distance from 1.0 (total identity) to 0.0 (total nonidentity), and all intermediate values are possible. In contrast, loci that are monomorphic within populations exhibit only a single allele, and therefore genetic identity values between compared populations or species are restricted to 0.0 and 1.0, depending on whether they are shared or not.

Locus-specific levels of polymorphism within populations and locus-specific evolutionary divergence between populations suggest different forces are affecting the evolutionary behavior of individual genes. While this differing behavior can be correlated with molecular structure, this correlation does not in itself allow assessment of the adaptive significance of patterns of variation and differentiation. The adaptive nature of enzyme polymorphism remains an open question.

For background information *see* ALLELE; GENETICS; POPULATION GENETICS in the McGraw-Hill Encyclopedia of Science and Technology.

[RICHARD K. KOEHN]

Bibliography: M. Kimura (ed.), *Molecular Evolution and Polymorphism*, National Institute of Genetics, Mishima, Japan, 1977; R. K. Koehn and W. F. Eanes, Molecular structure and protein variation within and among populations, *Evol. Biol.*, 11:39–100, 1978; M. Nei, P. A. Fuerst, and R. Chakraborty, Subunit molecular weight and genetic variability of proteins in natural populations, *Proceedings of the National Academy of Science*, 75:3359–3362, 1978.

Porphyrin

Heme (Fig. 1), the iron complex of protoporphyrin, if bound to a number of proteins such as hemoglobin, myoglobin, and cytochrome P-450, reversibly binds molecular oxygen when the iron is in the ferrous oxidation state. When the heme is not bound to the protein, however, it undergoes irreversible oxidation to the ferric stage, which no longer binds oxygen. A number of approaches, by J. P. Collman, T. G. Traylor, and others, which involve modifications of porphyrin, or studies at low temperature, have been used to stabilize the iron-porphyrin oxygen adducts. More recently another approach, by N. Farrell, D. Dolphin, and B. R. James, using simple ruthenium porphyrins, has provided systems which reversibly bind molecular oxygen to give oxygen complexes which are stable at room temperature.

As shown in Fig. 2, complexes (where L = L′ = CH_3CN) were prepared from both octaethylporphyrin (OEP) and *meso*-tetraphenylporphyrin by photolyses of the corresponding complexes (where L = CO and L′ = CH_3CH_2OH) in acetonitrile. The course of a photolysis is shown in Fig. 3; the isosbestic points indicate the smoothness of the reaction. Photolysis in other solvents, including pyrrole, dimethylformamide (DMF), dimethylacetamide (DMA), and a variety of nitrogenous bases, gives rise to similar bis-solvated adducts.

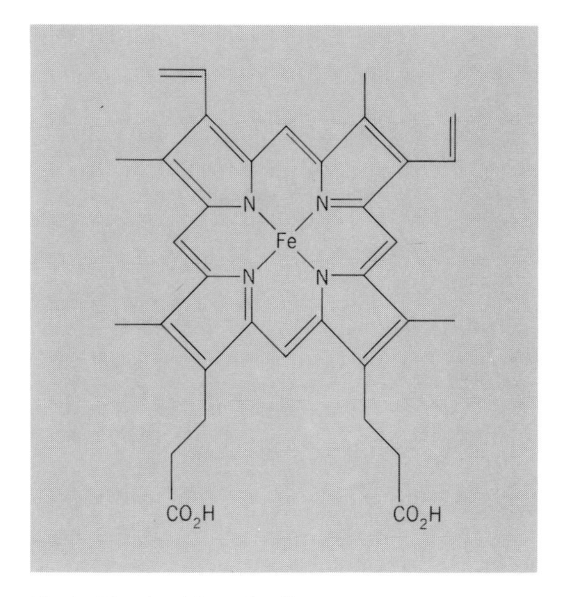

Fig. 1. Structural formula of heme.

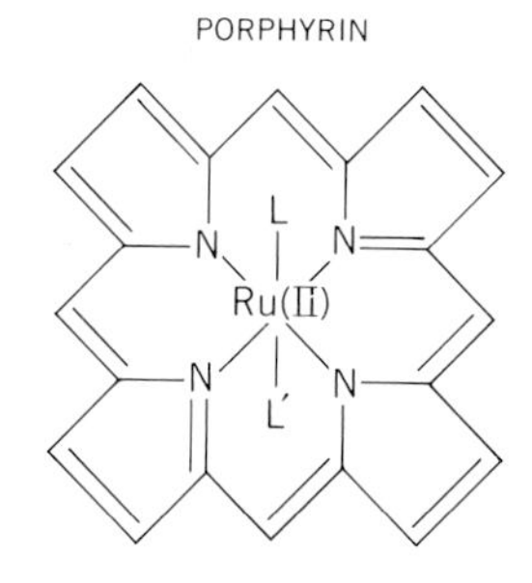

Fig. 2. Bis-ligated ruthenium(II) porphyrin.

When simple ferrous iron porphyrins are reacted with molecular oxygen, an initial 1:1 complex (Fig. 4) is formed. By itself the 1:1 complex does not undergo irreversible oxidation since oxygen is a two-electron oxidizing agent and ferrous iron a one-electron reductant. However, if a second iron porphyrin coordinates to the oxygen to give the complex in Fig. 5 (this can be prevented by sterically encumbering the porphyrin, which occurs when heme is bound to a protein), an irreversible reaction ensues to give the bridging peroxide (Fig. 6).

When $Ru(II)(OEP)(CH_3CN)_2$ in toluene reacts with molecular oxygen, a slow irreversible oxida-

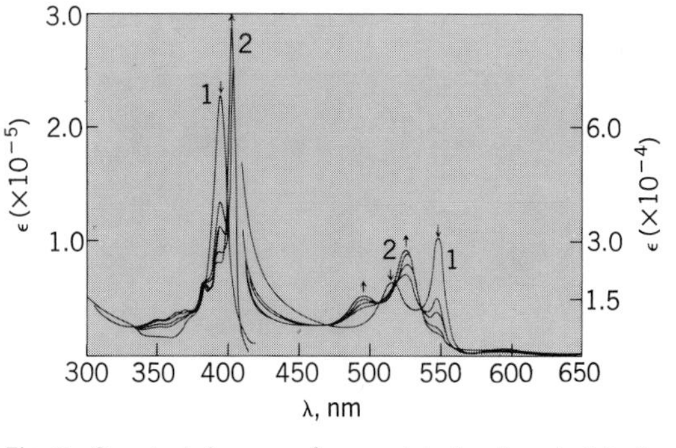

Fig. 3. Spectral changes observed during the photolysis of $Ru(II)(OEP)(CO)(CH_3CH_2OH)$, curve 1, in acetonitrile to give $Ru(II)(OEP)(CH_3CN)_2$, curve 2. Other curves represent intermediate steps in the reaction. Arrows indicate whether a particular peak is increasing or decreasing during the reaction. *(From N. Farrell, D. H. Dolphin, and B. R. James, Reversible binding of dioxygen to ruthenium (II) porphyrins, J. Amer. Chem. Soc., 100:324, 1978)*

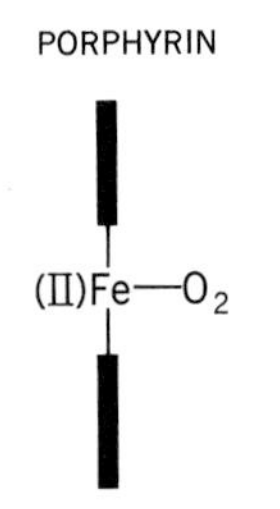

Fig. 4. The 1:1 ferrous porphyrin–dioxygen complex.

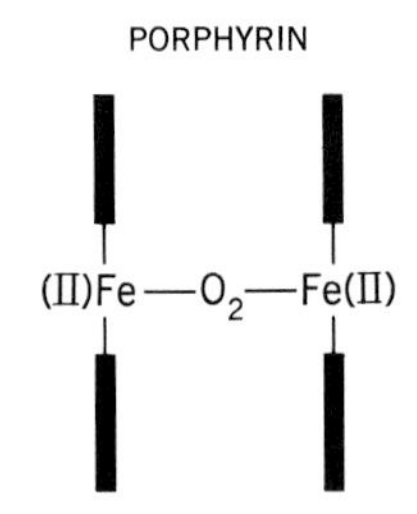

Fig. 5. Ferrous porphyrin–μ-dioxygen complex.

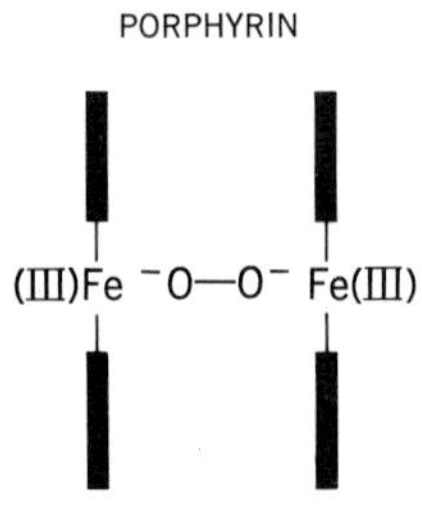

Fig. 6. Ferric porphyrin–μ-peroxide complex.

tion occurs which produces a ruthenium(III) complex. However, solutions of Ru(II)(OEP)$(CH_3CN)_2$ in DMF, DMA, or pyrrole reversibly absorb 1.0 mole of O_2 per ruthenium at room temperature. Figure 7 shows the spectral changes when the reaction is carried out in pyrrole. The rate of oxygenation depends upon the axially coordinated solvent; at 1 atm (10^5Pa) pressure of O_2 the reactions are pseudo–first order, with the reaction being almost instantaneous in DMA but having $t_{1/2} = 30$ min in DMF. The reverse deoxygenation is much slower and requires pumping for 24 hr. The addition of CO to the oxygenated complex generates Ru(II)(OEP)(CO)L with no change in the gaseous volume. All of the above chemistry is consistent with that shown in the scheme below. The stability

$$\mathrm{Ru(II)(OEP)(CO)L_2} \overset{-L}{\rightleftharpoons}$$

$$\mathrm{Ru(II)(OEP)L} \overset{O_2}{\rightleftharpoons} \mathrm{Ru(II)(OEP)(O_2)L}$$

$$\downarrow\uparrow \mathrm{CO}$$

$$\mathrm{Ru(II)(OEP(CO)L}$$

of the oxygen adducts is probably due to the slow off-rate for loss of oxygen, which is to be expected from substitution of inert ruthenium(II) complexes, and the correspondingly slow formation of the dimeric peroxide species is analogous to that in Fig. 6.

Apomyoglobin has been reconstituted with both Ru(II)- and Ru(III)-mesoporphyrin. Ru(II)-mesoporphyrin and reduction of the Ru(III)-containing complex produce ruthenomyoglobin, which reacts with carbon monoxide to give the CO-adduct similar to that of native myoglobin. At the present time addition of oxygen to ruthenomyoglobin appears to bring about an irreversible oxidation of the metal. In a manner analogous to complexes of the type shown in Fig. 2, ruthenomyoglobin also reversibly binds molecular nitrogen.

Oxidation of ruthenium(II)-containing porphyrins follows two paths. G. M. Brown and associates have shown that oxidation of the complex in Fig. 2 (where L = L′ = pyridine) results in oxidation of the metal to give the ruthenium(III) species. When CO is axially bonded, its strong π-accepting properties raise the potential for oxidation of the metal to a value higher than that of the porphyrin ring, which is instead oxidized to the porphyrin π-cation radical. In those cases where the axial ligands of the complex in Fig. 2 are substances other than CO (including CH_3CN, DMF, and DMA), oxidation initially occurs at the metal.

[DAVID DOLPHIN]

Bibliography: G. M. Brown et al., *J. Amer. Chem. Soc.*, 95:5939, 1973; J. P. Collman, *Account Chem. Res.*, 10:265, 1976; N. Farrell, D. H. Dolphin, and B. R. James, *J. Amer. Chem. Soc.*, 100: 324, 1978; T. G. Traylor, in E. E. van Tamelen (ed.), *Bioorganic Chemistry*, vol. 4: *Electron Transfer and Energy Conversion: Cofactors, Probed*, 1978.

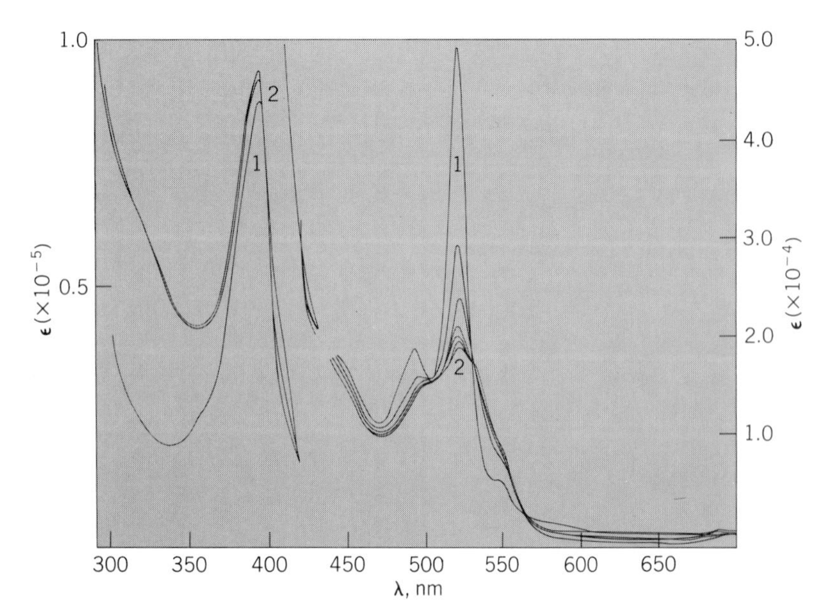

Fig. 7. Spectral changes during the oxygenation of Ru(II)(OEP)$(pyrrole)_2$ in pyrrole, curve 1, to Ru(II)(OEP)(O_2), curve 2. Other curves indicate intermediate steps in the reaction. *(From N. Farrell, D. H. Dolphin, and B. R. James, Reversible binding of dioxygen to ruthenium(II) porphyrins, J. Amer. Chem. Soc., 100:324, 1978)*

Printing

For the past 2 decades, banks, insurance companies, and similar institutions have employed magnetic ink character recognition (MICR) on all checks and certain other financial documents. MICR was chosen by the American Bankers Association (ABA) instead of optical character recognition (OCR) because signatures and other writing do not affect the magnetic sensing mechanism.

The specific type font of MICR selected by the ABA is known as E-13B because it was the fifth style (E) considered, is based on units of 0.013 in. (0.33 mm), and represents the second revision (B) of the original design. Its use speeds the processing of financial documents and results in fewer errors than manual handling. Another font, CMC-7, is used in Europe and South America.

In recent years the most noteworthy trends in this field have been the steadily increasing cost of handling checks rejected by computers as unreadable and the growing effort of financial institutions and printers to achieve quality control. Yet the problem remains formidable. Of about 32,000,000,000 checks printed in the United States each year, it is estimated that about 960,000,000, or 3%, prove defective. By 1980 the Bank Administration Institute predicts that the cost of repairing or replacing rejects will amount to \$435,000,000 annually, representing nearly 50% of total check-processing costs.

Causes of defects. Sometimes defects are caused by banks and other financial institutions; however, more often they are caused by printers. Most causes of the defects fall into one of four categories: the position of MICR symbols on checks, the magnetic strength of these symbols, their formation, and the application of ink to them. In this connection, all checks are divided into several so-called fields (see the illustration). All characters within these fields must be positioned in accordance with ABA standards, since even tiny deviations can sometimes cause rejects.

Improper positioning. One reason for improper positioning of characters is improper trimming of checks. When this occurs, characters do not appear in the correct horizontal or vertical positions. Excessive skewing so that characters do not appear totally upright also causes problems. Yet another difficulty lies in the abutment of characters

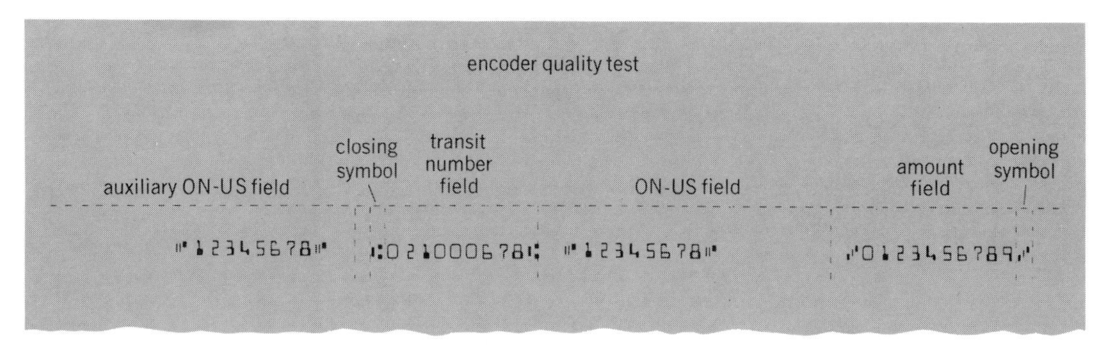

Encoder quality test used by the Irving Trust Company, New York.

in adjoining fields which can cause the computer to reject the check.

Magnetic strength. The second problem area involves the magnetic strength of symbols. Checks may be rejected by computers because this strength is excessive or insufficient, or because characters are overly embossed. The magnetic signal strength of MICR characters is rendered in relative terms. The first problem may arise because, when measuring the so-called signal strength with which each character is endowed, printers do not convert it into its proper relative strength. When this occurs, printers incorrectly adjust their presses and characters are not read correctly by computers. Relative signal strength may range from 50 to 200% of nominal strength (see the table).

Embossment, that is, penetration of the character into the paper, should not occur; rather, ink should lie flat on the paper surface. Yet embossment is an inherent problem in letterpress operations, which are usually used to print personal checks and bank deposit tickets, because letterpress penetrates MICR documents just as typewriter keys penetrate ordinary typing paper.

Character formation. Character formation is the third problem area in printing and processing MICR materials. Difficulties may arise because the dimensions or edges of characters are irregular or because the characters contain voids (that is, the absence of magnetic ink). The standards for character dimensions and edges are very precise, and allowable deviations are tiny. Voids can occur because ink is imperfectly laid down or because it does not adhere where required. Atmospheric conditions, dust, and poorly surfaced paper are also sometimes responsible for voids. Although deviations from proper placement of ink are allowed, they may not exceed 25% of a column or row of characters, and these columns or rows are nominally only 0.013 in. (0.33 mm) wide.

Relative signal strengths of MICR characters

Character	Nominal signal level	50% of nominal	200% of nominal
Dash	67	33.5	134
Transit	105	52.5	210
Amount	70	35.0	140
ON-US	100	50.0	200
0	130	65.0	260
1	85	42.5	170
2	105	52.5	210
3	85	42.5	170
4	105	52.5	210
5	105	52.5	210
6	105	52.5	210
7	75	37.5	150
8	105	52.5	210
9	165	82.5	330

Ink application. Possibly the most difficult aspect of check preparation is ink application. One reason for this is that magnetic inks are more viscous than other printing inks. This makes the art of printing more exacting when MICR is involved. Extraneous ink is perhaps the most common defect. Spots must not exceed 0.004 of an inch (0.10 mm) square on the front of a document and 0.006 of an inch (0.15 mm) square on the back. Failure to deposit ink uniformly is another frequent defect.

Prevention of errors. At present some banks and printers are taking more precautions to prevent errors. To this end, banks sometimes employ specially modified electronic sorters to screen sample incoming documents. These sorters print what computers can read, indicate what is unreadable, and sometimes analyze the problems in defective lines or sample runs as a whole. Sorters are effective within limits, meaning that they sometimes approve documents which are defective. As a result, approved documents must be checked further. Two instruments are widely used for this purpose: an MICR printing and layout gage and a 12× comparator with an E-13B reticle. The gage measures whether characters are correctly placed, while the comparator examines characters for size, shape, and defects such as voids.

Other equipment often used includes a magnetic signal-level tester, preferably with an oscilloscope, a 50× optical comparator, and a light-section microscope. The comparator measures the formation of characters and the application of ink to them. The microscope measures embossment.

To test documents for acceptability, random sampling is also used. The Irving Trust Company in New York, for example, uses the MIL-STD-105D sampling plan, which originated with the U.S. Defense Department.

Quality control. Printers are placing increasing emphasis on quality control, particularly with respect to paper, ink, and type font. These components are often tested before use.

Paper used on MICR documents must meet precise specifications. In particular, it must weigh at least 24 lb/basic box (90.2 g/m²) and be 0.004–0.0045 in. (0.10–0.11 mm) thick to provide proper density. In addition, paper should possess adequate resistance to tear and have smoothness of 80 to 130 Sheffield units, moisture content of 4.5 to 6%, wax pick of at least 11, and adequate brightness and printing opacity.

Because magnetic ink contains 50–60% iron

oxide, it is viscous. Thus printers should use it as is. Adjusting it with more than 3% of driers, oils, solvents, or toners creates problems. Printers should also try to use ink that is less than 6 months old because it tends to deteriorate with age. Magnetic inks dry more slowly than other types of ink. Thus, to avoid smudging, care must be taken not to trim documents until the ink has dried. For that reason cutting must be delayed at least 24 hr after printing.

Quality control of type font is equally important. Letterpress operators should use type about 0.001 in. (0.03 mm) smaller than would seem necessary because letterpress tends to increase the size of characters. Linotype operators must take great care to control heat, pump stroke, mouthpieces, and related factors, being especially alert to excessive pits and other irregularities in castings. Offset printers should try to avoid image distortions in their plates as well as pinholes in the opaquing process because pinholes cause extraneous ink to be laid down.

Today technically advanced printers have established formal quality-control plans indicating which components must be checked, as well as procedures and schedules for these checks. The plans also list the steps to be taken in case problems should arise.

For background information *see* PRINTING in the McGraw-Hill Encyclopedia of Science and Technology. [WILLIAM J. LATZKO]

Bibliography: American Bankers Association, *The Common Machine Language for Mechanized Check Handling: Final Specifications and Guides to Implement the Program*, 147 R3, 1967; American National Standards Institute, *American National Standard Bank Check Specifications for Magnetic Ink Character Recognition*, X3.3, 1970; American National Standards Institute, *American National Standard Print Specifications for Magnetic Ink Character Recognition*, X3.2, 1970; International Standardization Organization, *ISO Recommendation: Print Specifications for Magnetic Ink Character Recognition*, 1969.

Printing plate

Since the development of offset printing plates in the 1950s and electronic typesetting in the 1960s, it has become standard practice in the printing industry to make a full-size negative of copy. The negative is then used to expose the photosensitive surface of the printing plate. However, as more sensitive photographic systems became available for offset printing plates in the 1970s, the new technology made it possible to expose the printing plate directly without the use of the intermediate negative.

Direct plate exposure. In direct exposure systems, printing plates are directly exposed from copy in special cameras. These systems eliminate the requirement for silver-based film negatives in the preparation of printing plates. They also speed up the plate preparation process and reduce the amount of labor required. These camera-to-plate systems have generally been limited in usage to small printing plates (11 × 14 in., or 28 × 36 cm) and lower-quality printing (65-line screen). In the past 2 years four systems have been developed with a format size of 17 × 24 in. (43 × 61 cm) and improved quality (85- to 100-line screen). These systems have been used primarily for newspaper offset printing plates. Three of the systems use electrostatic coatings on aluminum of 6 to 12 mils (0.15 to 0.30 μm)—that is, coatings similar to those used on office copier drums—while the fourth system uses an electrostatic film, which after imaging is used to transfer a toner to a rubber cylinder, which in turn is used to transfer the toner to aluminum. The toner is fused to the aluminum to make the offset printing plate. The fourth system is known as an intermediate direct plate exposure (DPE) system, since it replaces the conventional negative with another material to transfer the copy image to the printing plate. The reduction in processing steps is shown in Fig. 1, which compares the processing steps required for conventional use of a film negative with the two forms of DPE systems.

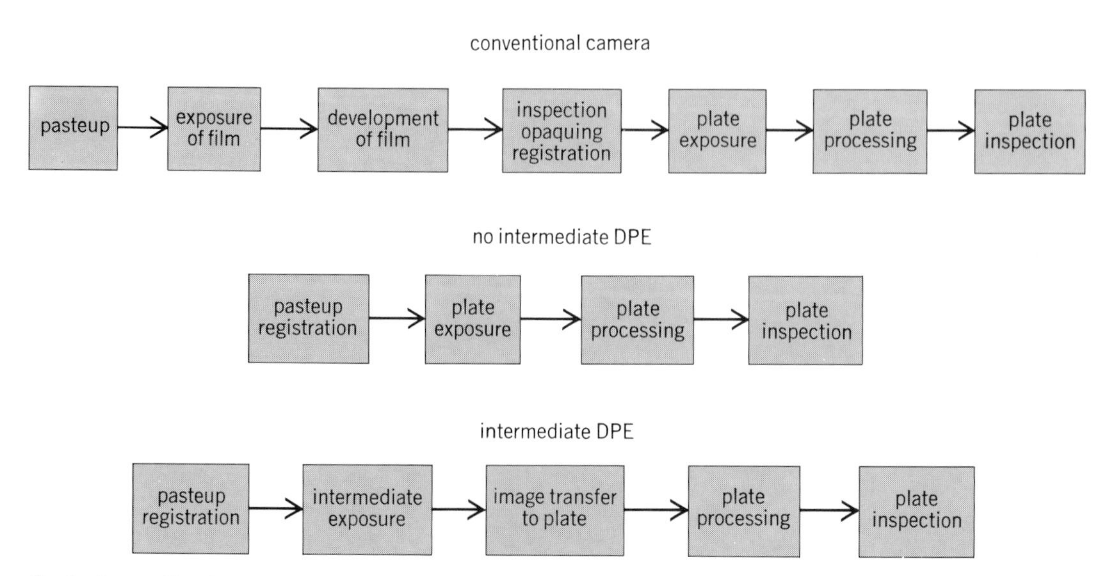

Fig. 1. Conventional camera system versus DPE system. *(Dunn Technology, Inc.)*

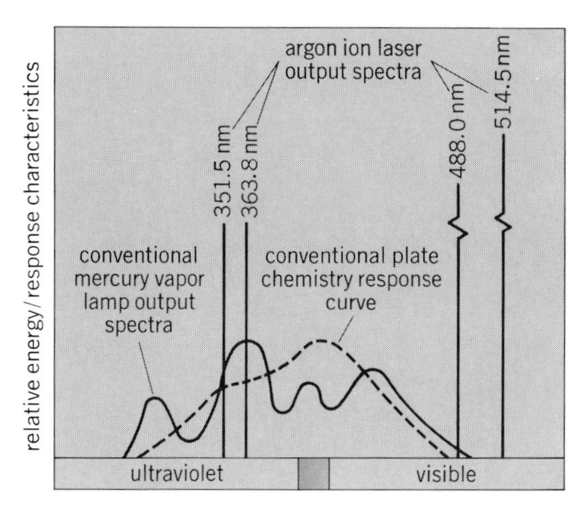

Fig. 2. Spectral comparison of exposure lasers, printing plate sensitivity, and ultraviolet lamp.

Table 1. Comparison of lasers currently used for exposing printing plates

Laser	Lines (position), A*	Power, W	Sensitivity required, mJ/cm²†
Ultraviolet argon	3511/3638	2	5–15
High-power visible argon	4880/5145	16	60–100
Medium-power visible argon	4880/5145	4	10–30
Low-power visible argon	4880	0.01–0.10	0.1–1.0
HeNe	6328	0.001–0.005	0.01–0.5
YAG	10,600	10	60–120

*1 A = 0.1 nm.

†Required sensitivity as a function of laser used for 3-ft²/min (2800 cm²/min) exposure (equivalent to a 1-min newspaper plate).

Laser scanning system. In the past few years a completely new technique has been developed that provides for direct exposure of conventional offset plates, intermediates, and electrostatic offset plates. This technique uses laser scanning systems to simultaneously read copy and expose the printing plate. These laser platemaking systems are expected to become the dominant means of plate preparation by the late 1980s. The principle underlying this type of system is the digital nature of the laser scanning technology, which allows for its interface with the electronic publishing systems of the not too distant future. That is to say, with electronic publishing systems it will be possible to expose printing plates with digital signals (from the computer) electronically controlling a laser scanner which images a printing plate. Thus, not only is the negative eliminated but also the copy, which is prepared on display screens attached to the computer. The capability of imaging printing plates from the computer was demonstrated in 1975 and two commercial systems of this type became available in 1978. In addition to their use with printing plates, lasers are being utilized in a variety of products to either input or output data from a computer, for example, optical character recognition devices, four computer laser printers announced in the past 2 years, and an all-electronic half-toning camera using a laser scanner.

Laser scanners for exposing printing plates are currently being used to scan the paste-up with a HeNe laser and to expose a printing plate (or intermediate) with a second laser. Some 70 systems have been ordered, mostly by newspapers, and growth in this field is expected to be dramatic throughout the 1980s. Currently there are two basic types of laser scanners available and several different lasers are being used to expose the printing plate.

Figure 2 illustrates the use of an argon laser in exposing conventional diazo photochemistry. The unique narrow spectral position of the laser lines available for exposure of ultraviolet photosensitive printing plate coatings has required the development of modified coatings to achieve the exposure speeds required by the printing industry (3 ft²/min, or 2800 cm²/min).

These developments proceeded rapidly in 1978, and several modified diazo coatings became available for use with the ultraviolet (3511/3638 A, or 351.1/363.8 nm) lines of the argon laser. In addition to conventional diazo coatings, photopolymer and electrostatic coatings are being used with the visible lines of the argon laser. Table 1 lists the lasers currently in use for exposing printing plates or intermediates. It also gives the required printing plate sensitivity for each of the lasers. Diazo coatings are used with the ultraviolet laser, photopolymer coatings with the medium-power visible laser, electrostatic coatings with the low-power visible laser, and intermediates with the yttrium-aluminum-garnet (YAG) laser.

Types of scanning. Two types of scanning are in use: the flatbed and the internal drum. Flatbed scanning is the most popular technique; the exposure portion of such a system is shown in Fig. 3. For the exposure section, the laser beam is optically directed to a modulator, which turns the laser on and off by deflection of the beam on or off an aperture. The modulator is controlled by a video signal from the read system or by digital signals from a computer. When the beam is on, it proceeds to a scanner which moves it across the width of the printing plate, exposing the appropriate image onto the plate. The plate is indexed in the length direction between each scan, and the full page is built up dot by dot and line by line until the full image is transferred to the printing plate. Several scanners are in use: vibrating- or oscillating-type scanners (Fig. 3), which use galvanometer principles; rotating multifaceted polygons that have mir-

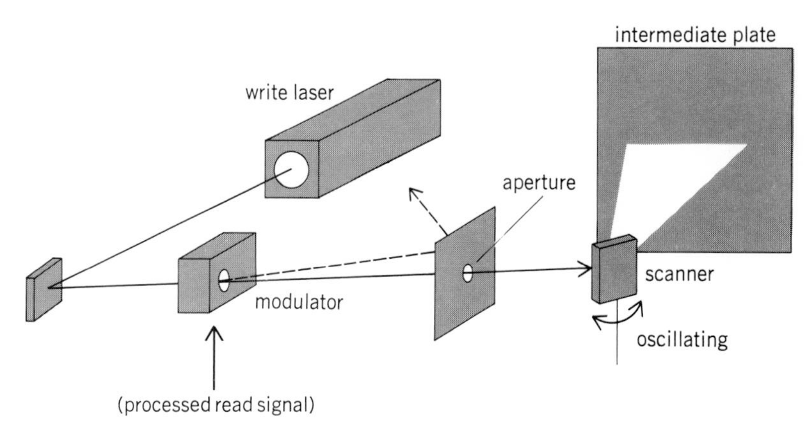

Fig. 3. Write laser modulation. *(Dunn Technology, Inc.)*

Table 2. Comparison of laser DPE and conventional camera

Features	Laser DPE	Conventional camera
Speed		
Starter page	4 min	12 min
Per page plate costs		
Negative associated costs	0	\$1.25
Wipe-on offset costs	\$0.85	\$0.85
Operating maintenance per page	0.38	(nominal)
Relative plate costs per page (one plate)	\$1.23	\$2.10
Annual savings		
Annual savings on page costs for plate (100 × 365 plates per year)	\$31,755	0
Annual labor savings (two people)	40,000	0
Total	\$71,755	0
Investment		
Number of units		
Cost	\$125,000	

rors on each face of the polygon; and rotating pyramids.

Not shown in Fig. 3 is a large flat-field lens that keeps the laser beam in focus across the width of the printing plate. The read system is similar to the exposing (writing) system in that a laser beam (typically HeNe) is directed toward the scanner (no modulator) and projected onto the copy in the same fashion as the exposure laser beam. The reflected laser light is measured by several different techniques, the most popular being a fiber optics bundle across the copy viewing the reflected light. For black areas the signal is zero, whereas for white areas it is one. This signal is electronically processed and then used to drive the modulator of the exposure scanner.

The internal drum scanner (Fig. 4) operates electronically in a fashion similar to the flatbed systems. However, the scanning technique is different in that the printing plate (and copy) is placed on the inside of stationary drums. Laser scanning is achieved by directing the laser down the center (axis) of the drum, where it strikes a 45° rotating mirror which scans one line around the drum and then indexes along the axis of the drum until the entire area is scanned (point by point and line by line). The diameter of the reading drum is larger than the diameter of the writing drum. This is to facilitate copy in the form of paste-ups, whose individually pasted-on pieces could fall off if placed on a drum as curved as the exposure drum. The differences in reading and exposing drums is accounted for electronically.

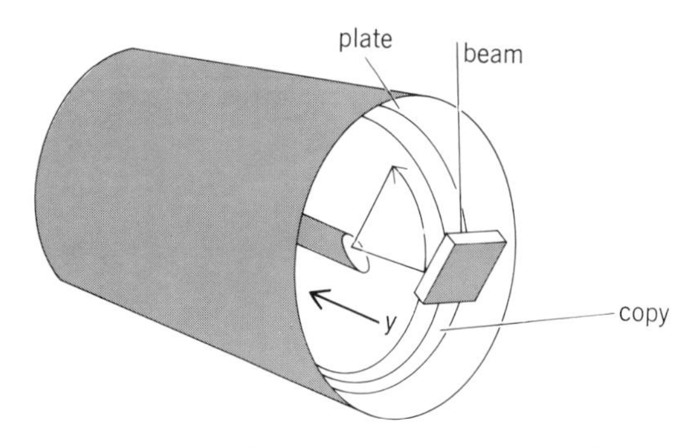

Fig. 4. Drawing of internal drum scanner; y = distance along the drum axis. *(Dunn Technology, Inc.)*

Time and quality advantages. Both the flatbed and internal drum techniques lead to high-quality 100-line screens with exposure speeds of approximately 3 ft²/min (2800 cm²/min). This quality is expected to be improved in the future to 150-line screens, allowing their use in a broader segment of the printing industry. It is likely that laser scanning systems are capable of producing better image quality than camera-based DPE systems, regardless of the exposure media used.

Table 2 shows the typical savings obtainable in time, consumables, and personnel for a typical United States newspaper (approximately 50,000 circulation) when using a laser DPE system, as compared with a conventional camera.

For background information *see* LASER; PRINTING PLATE in the McGraw-Hill Encyclopedia of Science and Technology.

[S. THOMAS DUNN]

Bibliography: S. T. Dunn, *Lasers and Newspaper Platemaking*, PIRA Seminar, Surrey, England, Oct. 10, 1975; S. T. Dunn, *The Possible Future of Printing Plates Production*, 1st International Foundation for Research in the Field of Advertising Symposium, Amsterdam, Jan. 24–25, 1978; Dunn Technology, Inc., Newport Beach, CA, *Market and Technology Report on Laser-Printing Plate Systems*, Rep. no. 2–55, May 1977.

Prochlorophyta

The prochlorophytes are a small group of microscopic algae whose existence was postulated by a Russian, C. Mereshkovsky, in 1905. They were briefly described as early as 1935 by H. G. Smith, but were fully recognized as a separate class of organisms only in 1976 by R. A. Lewin. They are green, prokaryotic algae, almost invariably associated with certain marine invertebrates. Thus far, only unicellular types have been found (see illustration). They grow either on the surfaces of colonies, or in the common cloacal chambers, of didemnid ascidians (colonial protochordates related to sea squirts). They are distinguished from other zoochlorellae (green algal symbionts of invertebrates) in that their cells are prokaryotic, that is, they have no membrane-limited nucleus, chloroplasts, or mitochondria. In 1975 these features were demonstrated independently by E. H. Newcomb and T. D. Pugh, working with cells from the Great Barrier Reef of Australia, and by Lewin and

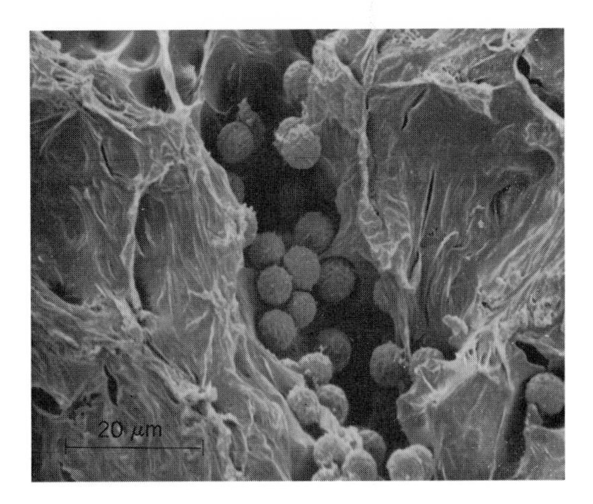

Prochloron cells in an oral groove of *Didemnum* at Isla San José, Baja California, Mexico.

L. Cheng, who studied similar cells from the Gulf of Mexico, Baja California. Though originally considered to be blue-green algae, prochlorophytes have more recently been considered sufficiently distinct to justify their assignment to a separate order, family, and genus, *Prochloron*. Prochlorophytes are distinguished from blue-green algae, the only other known class of prokaryotic oxygen-evolving photosynthetic organisms, by the absence of water-soluble blue or red bilin pigments and by the presence of chlorophylls *a* and *b*—features otherwise characteristic of eukaryotic green algae and higher plants.

If one accepts Mereshkovsky's hypothesis of "symbiogenesis," developed by L. Margulis in 1970, one could regard *Prochloron* as the only known descendant of green prokaryotes which may have existed 2 or 3×10^9 years ago, and which, by invading other, hitherto nonphotosynthetic cells, may have given rise to chloroplasts.

Although distinguished by special features of fine structure and a unique combination of biochemical constituents, the physiology of prochlorophytes is still unknown, since they have not yet been grown or studied in the laboratory. As soon as they can be cultured, other aspects of their biology will be investigated. Until then, it would be well to reserve judgment as to their possible evolutionary history and phylogeny, as well as to their role in the presumed symbioses in which they are now found.

For background information *see* ALGAE; BACTERIA, TAXONOMY OF in the McGraw-Hill Encyclopedia of Science and Technology.

[RALPH A. LEWIN]

Bibliography: R. A. Lewin, *Nature*, 261: 697–698, 1976; R. A. Lewin, *Phycologia*, 16:217, 1977; R. A. Lewin and L. Cheng, *Phycologia*, 14: 149–152, 1975; L. Margulis, *The Origin of Eukaryotic Cells*, 1970; C. Mereshkovsky, *Biol. Centralbl.*, 25:593–604, 1910; E. H. Newcomb and T. D. Pugh, *Nature*, 253:533–534, 1975; H. G. Smith, *Ann. Mag. Nat. Hist.* (10th ser.), 15:615–626, 1935.

Propeller, marine

Research on marine propellers has led to the development of a number of specialized propeller forms which are now being applied to the propulsion of major ocean-going ships. Among the most noteworthy are ducted, highly skewed, partially submerged, and contrarotating propellers.

Ducted propeller. This propeller is enclosed by a nonrotating coaxial nozzle formed so that either (1) the flow is decelerated inside the nozzle or (2) the flow is accelerated inside the nozzle. The first type is used primarily in military or special applications when it is desired to delay the inception of cavitation on the propeller blades. The second type, popularly known as a Kort nozzle, is used to improve the efficiency of the propulsion system where the load on the propeller is quite large in relation to propeller diameter. The interaction between the propeller and nozzle results in the nozzle producing some thrust in addition to that produced by the propeller. Kort nozzles have seen wide application in river tow boats, most notably those on the Mississippi River and its tributaries and the major rivers of Europe. More recently, with the growth in the size of oil tankers and the increasing cost of fuel oil, the Kort nozzle has been applied to the large oil, ore, and liquefied natural gas (LNG) carriers, particularly those built in Japanese and European shipyards (see Fig 1). The 215,000-ton (608,880-m^3) *Golar Nichu* built in Japan was one of the earliest of the very large oil tankers to be equipped with a ducted propeller. The improved propulsive performance (reduced fuel consumption) has led to numerous installations on other ships.

These ducts have a limitation which has not been completely resolved, namely, cavitation erosion on the inside wall of the ducts. This erosion is usually more severe than that on the propeller blade. The introduction of air inside the duct ahead of the propeller appears to reduce the severity of the erosion.

Highly skewed propeller. This type of propeller has blades which are in the form of scimitars, frequently with the tip of one blade aligning radially with the root of the following blade (see Fig 2). The objective of the highly skewed propeller is to reduce the propeller blade frequency forces induced into the ship's hull, thus reducing ship vibration. The increased power being installed in ships today

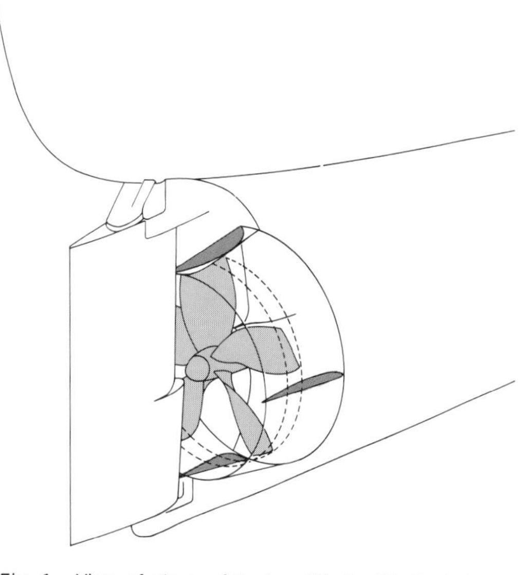

Fig. 1. View of stern of tanker fitted with ducted propeller.

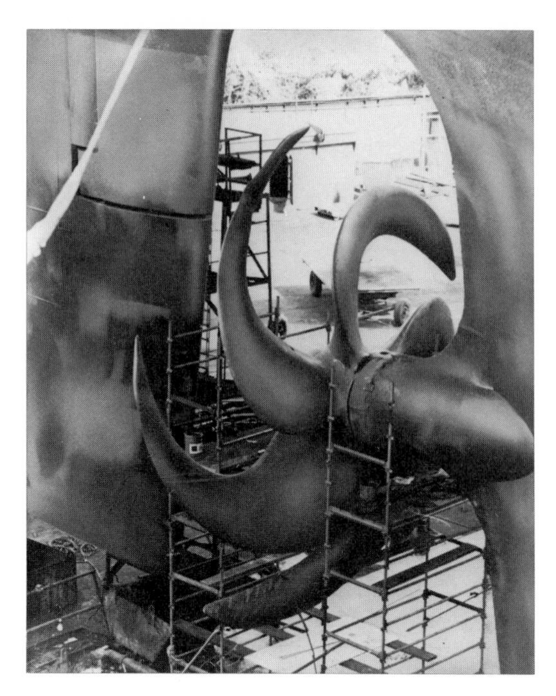

Fig. 2. Highly skewed propeller installed on SS *Ultra-Sea*.

and increased habitability requirements (reduced vibration and noise levels) have provided the incentive for this development. Advanced hydrodynamic propeller theory has made it possible to design these propellers with the same efficiency as more conventional marine propellers. The development of finite element techniques for structural design has allowed the prediction of the stress pattern in the blades.

Highly skewed propellers can reduce the blade frequency forces that are transmitted mechanically to the hull through the bearings as much as 70–80%. Those that are transmitted hydrodynamically through the pressure force on the hull can be reduced as much as 50% if there is little or no cavitation on the blade. If there is severe cavitation, hydrodynamic forces can be reduced about 10–20%. As a result of the reduced propeller blade frequency forces, this type of propeller is of greatest interest when ship vibration problems are anticipated. To date, highly skewed propeller installations have been made on three major commercial ships, one on an ore-bulk-oil (OBO) carrier and two on roll-on roll-off trailer ships with shaft horsepowers ranging from 24,000 to 37,000 (17.8 to 27.6 MW), to evaluate the overall performance of these propellers. The results of comparative tests with conventional propellers have indicated that overall propeller-induced vibratory forces are reduced about 50% and crew comfort has been increased, without increasing fuel costs and with only a minor increase in capital costs.

Partially submerged propeller. The partially submerged propeller is a by-product of the initial research work on supercavitating propellers. These propellers are intended for high-speed craft such as surface effect ships (SES), hydrofoils, and planing craft with high propeller-shaft rotational speeds, where it is not possible to eliminate cavitation on the blades. These propellers are designed to have fully developed blade cavities which spring from the leading edge and cover the entire back of the blade. The sections have sharp leading edges and are usually wedgelike in shape. The cavity can be filled either with vapor (supercavitating) or with air (ventilated). In the case of the partially submerged propeller, the air in the cavity is provided, naturally, at the air-water interface. These propellers can demonstrate high efficiency if the advance ratio, $J = V_{nD}$, is sufficiently high (over 1.4). V is the velocity of water into the propeller, n is revolutions per second, and D is the diameter of the propeller.

There are, however, specialized problems that have limited their application. High fatigue stresses occur due to the total loading and unloading of the propeller blade during each revolution. These cyclical loadings on the blades also transmit large blade frequency vibratory forces mechanically through the propeller shafts and bearings which have to be considered in using this type of thruster. They also build up a large hydrodynamic pressure force forward of the propeller which, when operating behind a large surface such as a planing hull, builds up a large lift force which, in turn, produces a large trimming moment by the bow. This is not true in the case of an SES with a narrow sidewall hull. Partially submerged propellers have been successfully applied to a 100-ton (283-m³) high-speed experimental SES, overcoming the problems noted above.

Contrarotating propeller. The search for improved propulsive efficiency of ships motivated by increased fuel costs has restimulated interest in the application of contrarotating propellers to major ships. Since contrarotating propellers are driven by coaxial contraturning shafts, it is possible, if they are properly designed, to recover the rotational energy that is lost in the wake behind a single propeller. The amount of energy that can be recovered is a function of the propeller loading and the advance ratio J of the propeller. The larger the two values, the higher the rotational losses; this provides a greater incentive for using contrarotating propellers. The larger loadings are associated with ships such as high-speed container ships. The reduction in power on container ships can be as great as 10%. Although the hydrodynamic aspect of such propellers is generally well known from model tests, the increased complication of the propulsion machinery and shafting design has inhibited application to major ships. In order to resolve some of these problems, the U.S. Maritime Administration and Sun Shipbuilding are conducting a joint program to investigate the application of a contrarotating propeller propulsion system to a high-speed trailer ship. If ship model tests and simulation studies establish technical and economic feasibility, application is expected in the near future.

For background information *see* PROPELLER, MARINE in the McGraw-Hill Encyclopedia of Science and Technology. [JACQUES B. HADLER]

Bibliography: R. A. Cumming, W. B. Morgan, and R. J. Boswell, *Soc. Nav. Architects Mar. Eng. Trans.*, vol. 80, 1972; J. B. Hadler and R. Hecker, in *7th Naval Hydrodynamics Symposium*, Office of Naval Research, DR-148, 1968; N. O. Hammer and R. F. McGinn, in *Ship Vibration Symposium*, 1978; Royal Institution of Naval Architects, London, *Proceedings of the Symposium on Ducted Propellers*, 1973.

Protective coloration

Examples of protective resemblance to inanimate objects among marine fishes are numerous, but mimicry between different marine species was considered, until recently, a comparatively rare phenomenon. Primarily as a result of an upsurge in direct underwater observation and study of fish behavior in the natural environment, a number of new discoveries have been made which significantly extend the number of known cases of mimicry in marine fishes and which increases the understanding of this phenomenon.

Cases of mimicry in fishes. Perhaps the earliest reported case of mimicry involving fishes was described by S. Schnee in 1905 between the ophichthid snake eel, *Myrichthys colubrinus*, and the poisonous banded sea snake, *Platurus colubrinus*, which it closely resembles. Another well-known example is the similarity of the black-marked pectoral fin of the common sole, *Solea vulgaris*, which when raised in a menacing manner is suggestive of the erected dorsal fin of the venomous weever fishes, *Trachinus draco* and *T. vipera*. Other early reported cases of mimicry in fishes were summarized by J. E. Randall and H. A. Randall.

Recent discoveries extend the number of known cases of mimicry among fishes. Known or suspected mimic species now include snake eels (Ophichthidae), groupers (Serranidae), dotty backs (Pseudochromidae), round heads (Plesiopidae), cardinal fishes (Apogonidae), snappers (Lutjanidae), coral bream (Nemipteridae), blennies (Blenniidae), flag blennies (Chaenopsidae), gobies (Gobiidae), surgeon fishes (Acanthuridae), soles (Soleidae), and leatherjackets (Monocanthidae). Among these widely diverse families, mimicry seems to have arisen independently, and in some groups, particularly the blennies, it is quite common. V. G. Springer and W. F. Smith-Vaniz reported nine different mimetic complexes involving 17 species of blenniid fishes and four other species. In addition, B. C. Russell, G. R. Allen, and H. R. Lubbock have reported 10 new cases of mimicry, 8 of which involve blenniid species.

Types of mimicry. Most cases of mimicry in fishes can be classified as either Batesian, Mullerian, or aggressive (Peckhammian) mimicry. Batesian mimicry, broadly, is the resemblance of a harmless or palatable species (the mimic) to a harmful or unpalatable species (the model). In Mullerian mimicry both mimic and model possess some undesirable qualities. Aggressive mimicry is the resemblance of a predatory species (the mimic) to a harmless or nonpredatory model. In some cases more than two species may be involved in a complex mimetic relationship, and elements of all three types of mimicry may be present.

Evolutionary origins. The evolutionary origins of mimetic resemblance can only be conjectured, but it seems likely that general convergent similarities in coloration, morphology, and behavior have preadapted certain species or groups of species for mimicry. The way in which even generally similar behavioral traits may become refined under selection pressure and lead to mimicry has been shown for the blenniid fish *Aspidontus taeniatus*, an aggressive mimic of the cleaner fish (a common harmless species), *Labroides dimidiatus*. By studying the behavior of other closely related blennies, W. Wickler has been able to reconstruct possible evolutionary pathways leading to the development of the un-blenny-like "dancing" movement of *Aspidontus*. This behavior seems to have originated from the vertical movement of the head, a motor pattern which occurs in all blennies during conflict between the tendencies to approach or to avoid an object. This may take the form of a simple nodding movement or, when superimposed on swimming, may resemble a sort of dance. In *Aspidontus* these movements seem to have become specialized as an interspecific signal important in mimicking the movements of the cleaner fish model.

Not all fish mimics, though, are as well adapted to their model as *Aspidontus*. Among the various species of the sabre-toothed blenny genus *Plagiotremus*, one finds varying degrees of mimetic specialization. Like *Aspidontus*, these blennies are aggressive mimics and feed almost exclusively by biting pieces of skin and body tissue from other unsuspecting fishes. Most species of *Plagiotremus* are not particularly deceiving to discern by the underwater observer. Some species, however, such as the eastern Pacific species, *P. azaleus*, regularly occur among aggregations of similarly colored fishes which they superficially resemble. This apparently confers a real mimetic advantage: prey fishes are less likely to notice the predatory blenny, thereby increasing its chances for successful attack. The widespread Indo-West Pacific *P. tapeinosoma* behaves in a similar way to *P. azaleus*, but is also interesting because it associates with different species over its range of distribution.

Other species of sabre-tooth blenny such as *P. rhynorhynchus* are slightly more specialized. *Plagiotremus rhynorhynchus* has colors similar to those of the cleaner fish *L. dimidiatus*, although the resemblance is convincing only between juveniles of the two species. Juvenile *L. dimidiatus* have a black body with a brilliant blue band running along each side of the back from head to tail. Juvenile *P. rhynorhynchus* are virtually identically colored and probably benefit from the resemblance in a manner similar to *A. taeniatus*. As the juveniles grow older, however, the color pattern of *P. rhynorhynchus* changes, and resemblance to *Labroides* is only superficial. The blenny is more elongate in form and also swims with an undisguised sinuous body movement. At this stage, *P. rhynorhynchus* usually is solitary and does not aggregate with other fishes. Nonetheless, its superficial resemblance to the cleaner fish seems to give it some mimetic advantage. Although small territorial reef fishes recognize the predatory blenny and avoid it, larger roving species such as snappers and parrot fishes—which, in effect, never learn to recognize their attacker—often approach unawares. At a distance these fishes may easily mistake the blenny for a cleaner fish. By employing a "hit-and-run" type of attack strategy and concentrating on roving fishes, *P. rhynorhynchus* is able to maintain a loose form of aggressive mimicry.

An even further degree of evolutionary specialization is shown by *P. laudandus*. On the Great Barrier Reef, this species mimics a poison-fanged blenny, *Meiacanthus atrodorsalis*. *Plagiotremus laudandus* has the same slate blue and yellow coloration as *M. atrodorsalis*, and also closely resembles it in shape and behavior. The impersonation is so convincing that *P. laudandus* is able to openly approach prey fishes. The relationship in this case appears to be a highly coevolved one; not only is *P. laudandus* an aggressive mimic of *M. atrodorsalis*, but because the latter is distasteful to predators it probably also benefits as a Batesian or possibly Mullerian mimic.

Multispecies mimic rings. The selection pressures involved in mimicry are such that many of the more highly specialized mimics resemble even local races or subspecies of the model. *Aspidontus taeniatus*, for example, shows geographic variation in color pattern which parallels that of its *Labroides* model. Such local variation may extend to all members of a mimetic ring. In the Fijian islands *P. laudandus* is a uniform bright yellow and mimics the local subspecies of *M. atrodorsalis*, which is also bright yellow. Also involved in this mimic ring are juveniles of the coral bream, *Scolopsis bilineatus*. In Fiji this species has also adapted to mimicking the yellow subspecies of *M. atrodorsalis*, the typical yellow and black banding having been replaced by a uniform yellow color.

Cases of multispecies mimic rings are not isolated discoveries. On the Great Barrier Reef the yellow and black banded poison-fanged blenny, *M. lineatus*, is mimicked by two other species, the blenny, *Petroscirtes fallax*, and the coral bream, *Scolopsis bilineatus*. In New Guinea another poison-fanged blenny, *M. vittatus*, is mimicked by at least two species, the apogonid, *Cheilodipterus zonatus*, and juveniles of the coral bream, *Scolopsis margaritifer*.

Mimicry of invertebrates. The phenomenon of mimicry in fishes is by no means limited to mimicking the other fishes. At least two species, juveniles of the bat fish, *Platax pinnatus*, and juvenile pomadasyids, *Plectorhynchus chaetodontoides*, are conspicuously colored and behave in a manner resembling soft-bodied noxious invertebrates such as a nudibranch or turbellarian flatworm. These seem to be cases of Batesian mimicry.

Perspectives. Early descriptions of mimicry were mainly of more spectacular specialized mimics, particularly those which were most striking to the human observer. It is evident from recent discoveries, however, that there are many cases where the relationship is less obvious. Even the superficial resemblance of one species to another may be sufficient to confer a real mimetic advantage under some circumstances, and many species probably derive some advantage as facultative mimics. These less specialized mimics may associate only casually with the model, sometimes mimicking only as juveniles and, compared with more obligative mimic-model relationships, the mimicry is much less coevolved. Facultative mimics may well represent early stages in the evolution of eventual obligative mimetic relationships. Clearly, there are likely to be many such intermediate cases, and recent discoveries suggest that mimicry, especially among tropical marine fishes, may well be as general and widespread a phenomenon as camouflage and protective resemblance.

For background information *see* PROTECTIVE COLORATION in the McGraw-Hill Encyclopedia of Science and Technology.

[BARRY C. RUSSELL]

Bibliography: J. E. Randall and H. A. Randall, *Bull. Mar. Sci. Gulf Caribbean*, 10:444–480, 1960: B. C. Russell, G. R. Allen, and H. R. Lubbock, *J. Zool. Lond.*, 180:407–423, 1976; V. G. Springer and W. F. Smith-Vaniz, *Smithsonian Contrib. Zool.*, 112:1–36, 1972; W. Wickler, *Mimicry in Plants and Animals*, 1968.

Quarks

Quarks have been postulated to be the fundamental building blocks of which all "elementary" particles are composed. These particles are believed to have electric charges equal to $\pm\frac{1}{3}$ or $\pm\frac{2}{3}$ that of an electron, unlike all known elementary particles, whose charges are intergral multiples of the electron charge. In the past few years the concept of quarks has come to play an increasingly central role in considerations of the structure of elementary particles. This has occurred even though quarks have not been observed in isolation with certainty. Indeed, the most widely accepted ideas about quark properties hold that quarks are always bound in pairs or triplets which constitute ordinary elementary particles, and that they cannot exist in the free, unbound state. The first section of this article discusses the search for free quarks and the first strong evidence for their existence. The second section discusses the evidence which is being gathered by many scientists for the number of different types or "flavors" of quarks, and summarizes the properties of the known quarks, the methods by which they have been found, and speculation about the properties of additional quarks.

Search for free quarks. The search for quarks was recently highlighted by the announcement by a group of Stanford University physicists of strong evidence for the existence of free quarks, with charge one-third that of the electron, on two niobium spheres, each of mass 9×10^{-5} g. This oil-drop-type experiment is sensitive to the presence of residual electric charge on freely suspended electrically isolated chunks of matter. If confirmed and validated, these results must be reconciled with the lower limits on quark mass and upper limits on quark production cross sections inferred from theoretical models based on the negative results of cosmic-ray and machine production experiments. Positive evidence for the existence of free quarks, furthermore, would force a reassessment of the commonly held belief that quarks are bound forever by forces which increase with increasing separation, and therefore that they cannot exist as fractionally charged entities in the free, unbound state.

Content-of-matter searches. The area of concern of the "quark searcher" covers fields ranging from high-energy physics to astrophysics, geophysics, spectroscopy, and chemistry. Cosmic-ray and machine production experiments assume that when nuclear interactions occur at sufficiently high energy, there is a reasonable probability that a meson or baryon might be dissociated into its hypothesized fundamental constituents with charge $\pm\frac{1}{3}$ or $\pm\frac{2}{3}$ that of an electron. Detection of these free quarks is achieved, for example, by searching for anomalously low charge-dependent ionization in particle detection chambers.

On the other hand, content-of-matter searches assume that there are free quarks already distributed at some low density throughout all matter. These free quarks are most likely attached to neutral atoms, thereby forming "quarked" atoms with fractional charge. Estimates of the expected density of quarked elements in material from the Earth's crust can be made by integrating the cosmic-ray flux over the age of the Earth and modeling the quark production cross section, taking into account mixing of the crustal material on a geologic time scale. There is also the possibility of a free-quark residue left over from the cosmological "big bang." *See* COSMOLOGY.

The most obvious way of detecting the presence of quarked elements in a content-of-matter search is to directly measure the residual electric charge on macroscopic matter. R. A. Millikan pioneered these techniques in his oil-drop experiments, in which he determined that the charge on liquid drops of mass 10^{-11} g appeared as integral multiples of the value e on the electron. (In one of his published reports Millikan noted that he had discarded one uncertain and unduplicated observation which gave a value of the charge on a drop some 30% lower than his final value of e. It is thus possible that he unknowingly detected a quark.)

Magnetic levitation. In order to increase the probability of finding quarked elements, which are presumably sparsely distributed throughout matter, it is desirable to measure the residual charge on as large a mass as is practicable. For this reason, a number of experimenters have chosen to use a magnetic levitation technique in which magnetic rather than electric fields are used to balance the force of gravity on a freely suspended mass. The record for mass has been recently obtained by G. Gallinaro, M. Marinelli, and G. Morpurgo at the University of Genoa. They used a feedback levitation scheme to measure residual electric charge on iron cylinders of mass 2×10^{-4} g, a factor of about 10^7 greater than the typical Millikan oil drop.

An attractive feature of the levitation experiments is that the test object is usually large enough to be easily recovered to repeat the measurements. The experimenter can therefore claim not only to have measured a quark, but also to have one firmly in possession.

The basic features of the levitation experiment of G. S. LaRue, W. M. Fairbank, and A. F. Hebard at Stanford are shown schematically in Fig. 1. With the exception of the magnetic detector, the same components in different configurations are common to all levitation experiments. In the Stanford experiment a superconducting niobium sphere, 0.28 mm in diameter, is supported in a vacuum at 4.2 K between two capacitor plates on the magnetic field produced by two superconducting coils. The suspension system is extremely stable because of the persistent supercurrents both on the surface of the sphere, which is a perfect diamagnet, and in the magnetic support coils. An alternating electric field is applied with the proper phase and at exactly the same frequency as the mechanical oscillations of the sphere. The final steady-state amplitude of motion is directly proportional to the excess charge on the sphere. The motion of the sphere can be monitored either optically or with a sensitive magnetic detector located nearby. The initial excess charge on a levitated sphere can be on the order of $10^4 e$, and therefore must be neutralized to within a few unit charges of zero by exposure to electron or positron radioactive sources. Many of the levitation experiments presently in progress use a feedback-stabilized suspension of a ferromagnetic mass at room temperature in order to overcome the difficulties of working at low temperatures.

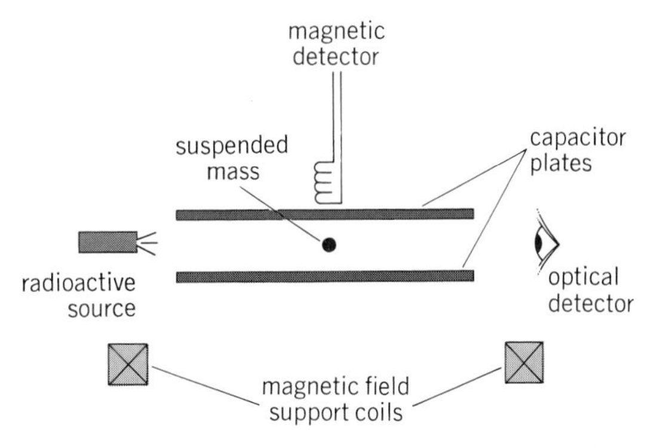

Fig. 1. Cross section of basic components in levitation experiment to detect quarks.

Experimental results. Unlike Millikan's experiment, the levitation experiments do not measure the absolute charge on the electron, but rather the residual charge calibrated with respect to unit electronic charge changes. This point is clarified in Fig. 2, which is a plot of idealized data revealing a residual charge of $+1/3$. The equally spaced set of Gaussian distributions represents an equivalent number of measurements taken for each charge state within $\pm 3e$ of zero total charge. The finite width of each distribution is due to noise. It is assumed, of course, that the smallest discrete displacement change is due to the addition or subtraction of single unit charges. The fact that the centroids of the Gaussian distributions (arrows of Fig. 2) are shifted from the origin by one-third of the unit-charge spacing implies that there is a residual fractional charge of $+1/3$ on the mass which cannot be neutralized by unit-charged electrons or positrons. Charge determinations are modulo one, and it is impossible to tell the difference between a charge of $+1/3$ and $-2/3$.

The Stanford group recently published data of the form displayed in Fig. 2 on eight niobium spheres. Two of the spheres with mass 9×10^{-5} g had apparent residual charges of $+(0.337 \pm 0.009)e$ and $-(0.331 \pm 0.070)e$, whereas the others were all near zero. There is a possibility that the presence

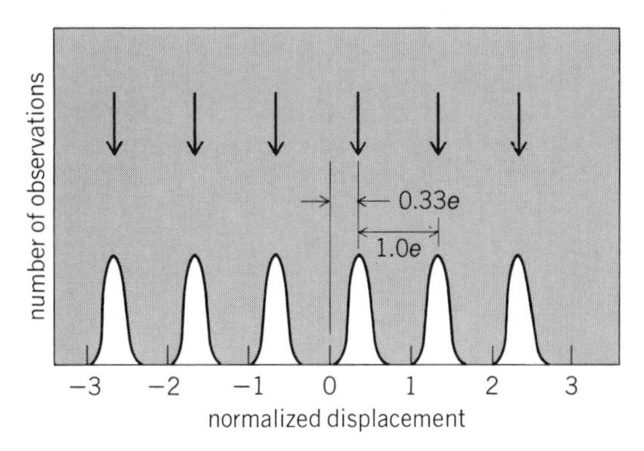

Fig. 2. Idealized data from levitation experiment showing an equal number of observations at each charge state and a residual charge of 0.33e.

Properties of quarks

Flavor	Electric charge	Baryon number	Strangeness	Charm	Contribution to R
u	$2/3$	$1/3$	0	0	$4/3$
d	$-1/3$	$1/3$	0	0	$1/3$
s	$-1/3$	0	1	0	$1/3$
c	$2/3$	0	0	1	$4/3$
b	$1/3$	0?	0?	0?	$1/3$

of electric and magnetic dipoles can give rise to spurious charge forces which masquerade as fractional charge. The Stanford physicists have argued that such effects are properly accounted for in their experiment. After the Stanford announcement, the Genoa group reported charge neutrality on three iron cylinders each of mass 2×10^{-4} g. This is in contrast to earlier measurements by E. D. Garris and K. O. H. Ziock at the University of Virginia which indicated a clustering of residual charges on 12 iron spheres about $+1/3e$ and $-1/3e$. Their conclusions were tempered, however, by systematic difficulties with spurious charge forces.

Outlook. In spite of the results of the Stanford group, no definite conclusions have been reached on the existence of unbound fractionally charged quarks. Independent confirmation or similar data on a greater number of spheres are needed. Attempts to enhance the sensitivity to quarks by increasing the mass in levitation experiments will most likely be unsuccessful because the displacement for a given electric field is inversely proportional to the mass, and many of the spurious charge forces also scale with the mass. There is also an increasing probability that radioactive contaminants would be included which could adversely affect the charge stability during the time it takes to make a measurement. The use of enrichment techniques to increase the concentration of quarked atoms on the levitated material is a more promising alternative.

The existence of free quarks in matter would have a profound impact on physicists' views of nature. Serious efforts to artificially produce or concentrate quarked atoms would follow in an attempt to discover the unique chemistry of entities which do not deal in the everyday currency of unit charge. [ARTHUR F. HEBARD]

Evidence for number of quarks. When the neutron was discovered by J. Chadwick in 1932 it provided the last element needed to construct a self-consistent theory of the structure of the nucleus built from protons and neutrons. For a short time, until many more "elementary" particles were found, it seemed that nature had used a minimum number of building blocks to constitute matter. The charmed quark, discovered by S. C. C. Ting and colleagues and B. Richter and colleagues in 1974, filled a similar need in clearing up some nagging problems in understanding the structure of hadrons. In 1977 another quark was found (and also another "unneeded" lepton), possibly indicating a quark and lepton spectrum as rich as the spectrum of hadrons.

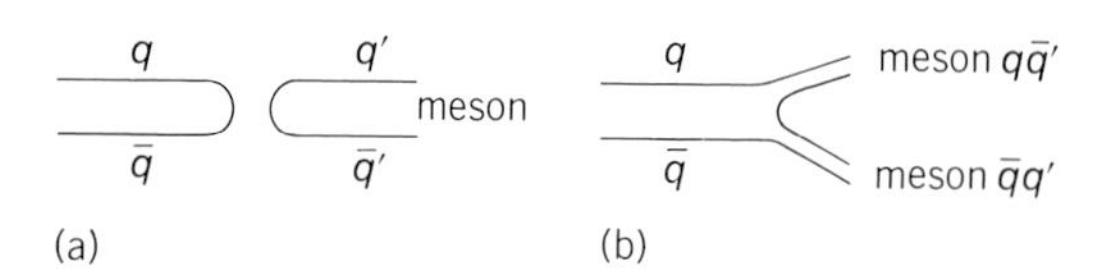

Fig. 3. Diagrams of quark reactions, illustrating Zweig rule. (*a*) Zweig-forbidden reaction. (*b*) Reaction which is Zweig-allowed, but with energy forbidden for low-lying bound states.

Leptons and hadrons. Elementary particles belong to one of two classes, the leptons and the hadrons, with the exception of the photon, to be discussed below, and hypothetical particles. Leptons behave like point particles without internal structure or spatial extent, and are subject to the electromagnetic interaction (if charged) and the weak interaction. Hadrons have a measurable size and are subject to the strong interaction in addition to the electromagnetic and weak interactions. While only five different leptons have been found experimentally (not counting their antiparticles), namely, the muon μ, the electron e, their corresponding neutrinos, ν_μ and ν_e, and the recently discovered tau lepton τ, there are several hundred known hadrons. Most hadrons are extremely unstable and decay rapidly. *See* LEPTON.

In 1964 M. Gell-Mann and G. Zweig recognized the order underlying this vast number of hadrons. The particles were found to form a representation of the SU_3 group. This group defines three basic items (quarks) and their antiquarks. The physical hadrons known at that time belong to those multiplets of the SU_3 representation formed by combining either a quark with an antiquark (mesons) or by combining three quarks (baryons). Many properties of the three basic quarks can be inferred once a number of hadrons are correctly assigned to their multiplets. The properties of the three original quarks and of the two found later are presented in the table.

Vector mesons. Of particular interest in the context of the search for new quarks is a class of mesons made of a quark q and an antiquark $\bar{q}$ of the same flavor. These mesons can have the quantum numbers of the photon, the field quantum of the electromagnetic interaction. Because of their spin angular momentum of 1 they are called vector mesons. There is at least one bound state, and sometimes several, of a quark with its own antiquark, differing by the radial quantum number of the quark system wave function.

Vector mesons are important for three reasons: (1) Having the quantum numbers of the photon, they can take the place of virtual photons in electromagnetic reactions, and exist as virtual photons part of the time. (2) Many vector mesons have long lifetimes. This is a consequence of the experimentally observed Zweig rule, which forbids reactions with unconnected quark lines, such as the one shown in Fig. 3*a*. If $q\bar{q}$ is bound so strongly that the

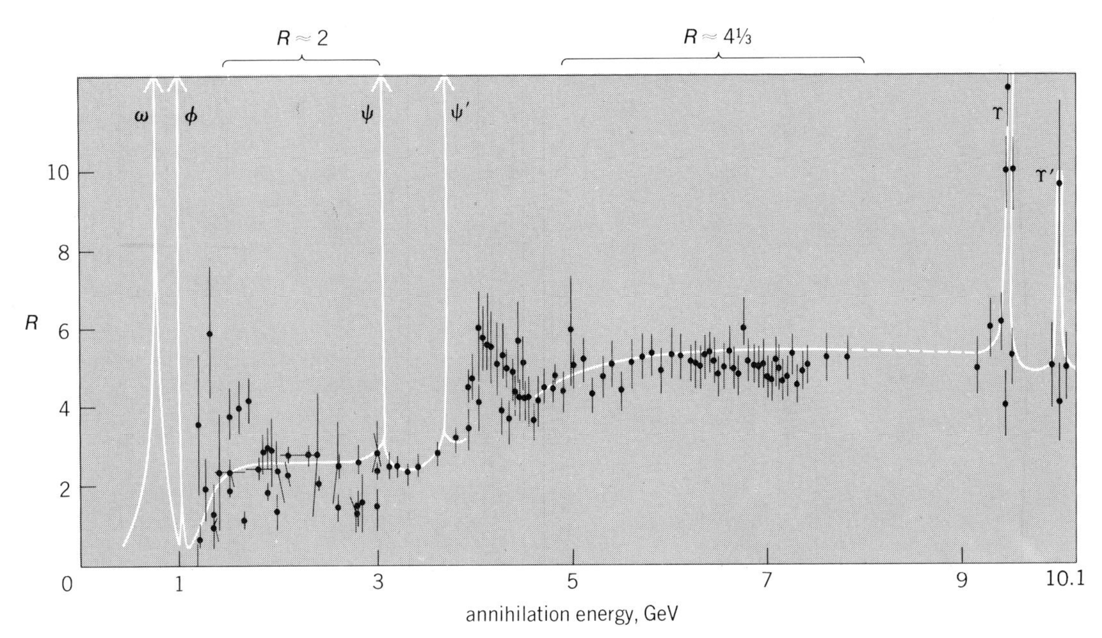

Fig. 4. Experimental data for the ratio R between e^+e^- annihilation into hadrons and e^+e^- annihilation into muon pairs. *(From S. D. Drell, When is a particle?, Phys. Today, 31(6):23–32, 1978; PLUTO Collaboration, Rep. DESY 78/21, 1978; DASP 2 Group and DESY-Heidelberg Collaboration, High-Energy Physics Conference, Tokyo, October 1978)*

mass of a pair of any mesons containing q or $\bar{q}$ is above the energy of the bound state, then the Zweig-allowed decay (Fig. 3*b*) is forbidden by energy conservation. Consequently, strong decays are greatly reduced in strength, leading to long lifetimes and narrow mass distributions for these mesons. (3) Vector mesons often decay into charged lepton pairs. With strong decays inhibited, electromagnetic decays become important. These electromagnetic decays proceed through emission of a virtual photon, which may materialize as an electron or muon pair, or a pair of lighter quarks which can decay into hadrons since they have a great deal of energy available in their rest system. Lepton pairs are important experimentally because they can provide a unique signature, making it possible to isolate the few events of interest from the great number of hadronic interactions.

Search for new quark flavors. Three phenomena have proved useful as the basis of methods designed to search for new kinds of quarks bound in hadrons: (1) resonances in the virtual photon spectrum in e^+e^- annihilations due to vector mesons; (2) the rise in the ratio $R = (e^+e^- \rightarrow \text{hadrons})/e^+e^- \rightarrow \mu^+\mu^-)$; and (3) resonances in the lepton-pair spectra observed in hadronic collisions. All three methods rely on the equality of quantum numbers between the virtual photon and vector mesons. The first two methods have been valuable in revealing the fine structure of resonances and understanding their decay properties. The third has been the most successful in finding new resonances.

Resonances in e^+e^- annihilation. When the energy available in the e^+e^- center-of-mass system coincides with the mass of a vector meson (which has the quantum numbers of the photon), the electron annihilation cross section gets much larger, as shown in Fig. 4. (The cross sections in Fig. 4 have been divided by the quantum electrodynamics part of the cross section $e^+e^- \rightarrow \mu^+\mu^-$, which falls as the inverse of the energy. The resulting ratio is called R.) In this case the virtual photon is indistinguishable from the vector meson, and as such is near its mass shell and lives longer. During its longer life span, the probability to materialize as a lepton or quark pair is enhanced, and the annihilation cross section rises.

Although huge resonance peaks at the vector meson masses have been seen for the first four quarks, the method becomes less powerful in the search for new flavors at higher masses. This is a consequence of three factors: (1) The energy width of the electron beams increases with rising electron energy due to the random emission of hard x-rays. (2) The continuum production cross section is almost constant with energy. (3) The resonance production cross sections fall rapidly with increasing mass. As an example, it took several weeks to confirm the existence of the Υ and Υ′ resonances at the DORIS storage ring (Fig. 4), although their mass was known to be about 1% accuracy. The signal-continuum ratio was only about 2:1 in this instance. All these factors are expected to deteriorate even further at higher energies, making this type of search a time-consuming procedure.

Rise in R. Even away from the vector meson resonances, the annihilation of electron pairs into hadrons contains information about the number of quarks with masses below one-half the electron-pair mass. The total annihilation cross section is a sum over the processes as shown in Fig. 5. The cross section is simply proportional to the sum over all final state pairs weighted with the square of the charge, since the coupling at the second vertex is again electromagnetic and the quarks are pointlike objects.

The ratio R between annihilation into hadrons and annihilation into muon pairs is shown in Eq. (1),

$$R = n_c \Sigma_i e_i^2 \qquad (1)$$

where e_i is the charge of the ith quark. The factor

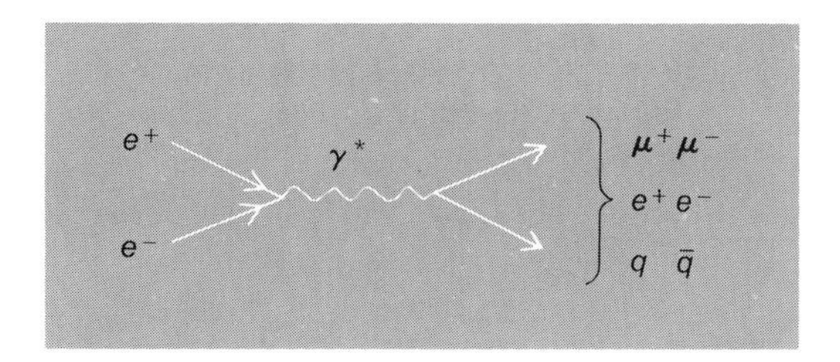

Fig. 5. Diagram of e⁺e⁻ annihilation processes.

n_c (equal to 3) takes into account the three possible colors for each quark, leading to three times as many different (yet indistinguishable) final states. The plot of R versus electron-pair mass (Fig. 4) indicates the contributions from the different quarks. (The τ lepton is treated in the same way as a quark here because it decays into hadrons and is not easily distinguished in experiments from quark-pair states.)

The contribution of the first three quarks (u, d, s) is accordingly two units of R (see the table), so that there is a plateau near $R = 2$ in Fig. 4. The charmed quark contributes $(2/3)^2 \cdot 3 = 4/3$ units, and the τ 1 unit, so that there is a second plateau near $R = 2 + 4/3 + 1 = 4\frac{1}{3}$. The b quark is expected to add only $1/3$ unit, small compared with the systematic errors which are estimated to be from $1/2$ to 1 unit in R, and its contribution to R is not expected to appear below 11 GeV.

The study of R as a function of electron-pair mass will be an important tool for obtaining a preliminary projection of how many more new quark flavors, or new leptons, might hide in the energy region (up to 30 GeV) which is accessible to the new generation of electron storage rings. The actual confirmation of such states is expected to be difficult and slow.

Lepton pair resonances in hadronic collisions. All vector mesons thus far have been discovered as products of collisions between hadrons (with the exception of the J/ψ, which was discovered simultaneously also in e^+e^- annihilations, and the ψ'). The detection makes use of the large probability of these mesons to decay into lepton pairs (e^+e^- or $\mu^+\mu^-$). Electrons and muons can be separated from hadrons quite easily. Electrons initiate electromagnetic showers, the total energy of which can be measured with high accuracy and greatly exceeds the energy deposited in such detectors by hadrons. Muons are not subject to the strong interaction and can therefore penetrate a large amount of matter (about 1 m of steel for every gigaelectronvolt of energy), making it practical to filter out all hadrons upsteam of the muon detectors. Either way, rejection ratios of $10^4$and more (depending on energy) can be achieved.

A typical experiment to detect lepton pairs made in hadronic collisions was performed by L. Lederman and collaborators at Brookhaven National Laboratories in 1969. A very intense beam of 10^{12} protons per accelerator cycle (3 s) was made to impinge on a solid uranium target (chosen for its high density and hence effective absorption of hadrons). The angles of the emerging muons were measured by a set of hodoscope counters, and their energy was estimated from their range in a thick iron absorber. From the angle and energy of the muons the mass of the muon pair $m_{\mu\mu}$ was calculated and the distribution of muon-pair masses (that is, the cross section $d\sigma/dm_{\mu\mu}$) was plotted (Fig. 6*a*). Multiple scattering and the coarse energy determination limited the mass resolution, so that the shoulder at about 3 GeV could not be proved to be a smeared mass peak (the J/ψ). In 1974 an electron-pair experiment by Ting and colleagues with good mass resolution found a pronounced peak at mass 3.1 GeV, the J/ψ particle (Fig. 6*b*). Starting in 1976, the Lederman group, working at Fermilab, performed a number of searches for vector mesons of yet higher mass, with both electron- and muon-pair detectors. The production cross section for these high-mass vector mesons falls rapidly with mass, and only a muon-pair experiment, in which all hadrons were blocked out with a 1-ton beryllium absorber, was sensitive enough to find the next state, the Υ. The data from the experiment, in which 400-GeV protons collided with a platinum or copper target, are shown in Fig. 7, where the cross section $d^2\sigma/dm_{\mu\mu}\,dy$ for producing muon pairs at an angle of 90° with the incident beam in the center-of-mass frame ($y = 0$) is plotted as a function of muon-pair mass. The ground state at 9.5 GeV was found, together with two excited states, the Υ′ at 10.0 GeV and the Υ″ at 10.4 GeV. The high mass resolution of this experiment was essential in first finding and then resolving these peaks. The Υ and Υ′ resonances have since been observed at the Deutsches Elektronen-Synchrotron Laboratory (DESY) in e^+e^- annihilation, where the cross section strongly supports a charge assignment of $1/3\,e$ to the new quark (b) from which the Υ is thought to be made ($\Upsilon = b\bar{b}$).

Significance. The charmed quark c had been predicted theoretically, and was in fact needed to explain the slow decay $K_L^0 \rightarrow \mu^+\mu^-$. In modern theories quarks as well as leptons are usually arranged in pairs, or doublets, as shown in notation (2). The c quark filled the empty spot in

$$\begin{pmatrix} u \\ d \end{pmatrix} \begin{pmatrix} s \\ c \end{pmatrix} \begin{pmatrix} b \\ ? \end{pmatrix} \quad \begin{pmatrix} e \\ \nu_e \end{pmatrix} \begin{pmatrix} \mu \\ \nu_\mu \end{pmatrix} \begin{pmatrix} \tau \\ ? \end{pmatrix} \qquad (2)$$

the second doublet, and its decays are as expected from that assignment. The b quark (as well as the τ), however, is "unneeded." Neither its partner in the doublet nor the manner in which it couples with the other quarks in weak decays is known. Earlier neutrino results (now believed wrong) led to theories predicting a quark of 5-GeV mass, which would exhibit an entirely new kind of coupling via "right-handed currents." [All neutrino interactions known thus far can be explained by a left-handed coupling only (V-A, where V and A are the vector and axial vector parts of the weak current).] It will be challenging to study both the decay channels and the coupling of the b quark, mostly by using the new e^+e^- storage rings.

Quark-quark strong potential. In analogy to the atom, there is for each quark flavor at least one bound state, and often several, between quark and antiquark below the "ionization limit," where it is energetically possible for the $q\bar{q}$ state to decay into two mesons ($q'\bar{q}$ and $\bar{q}'q$). Although energetically allowed, decays of the form $q\bar{q} \rightarrow q'\bar{q}'$ are

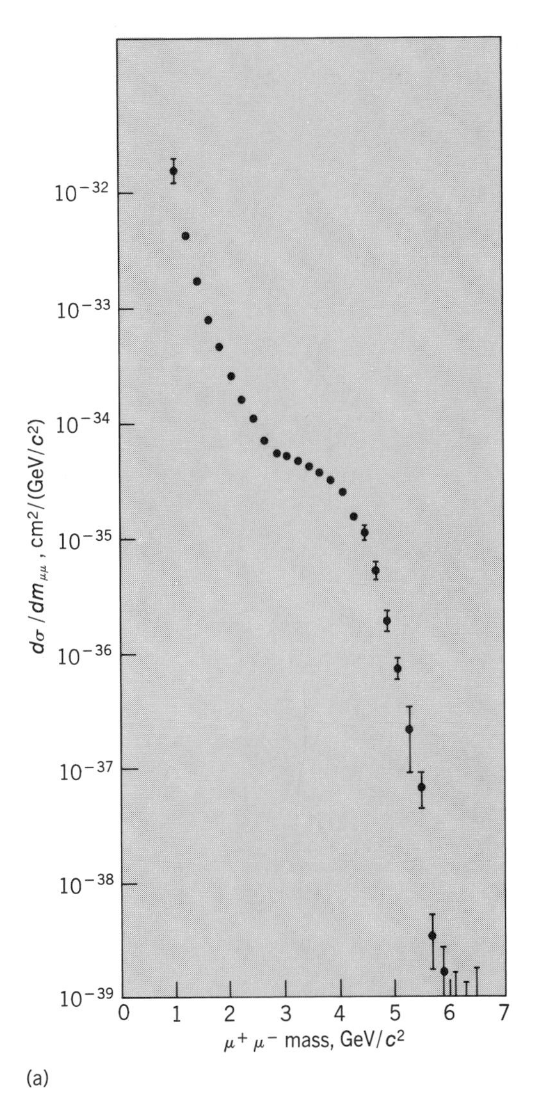

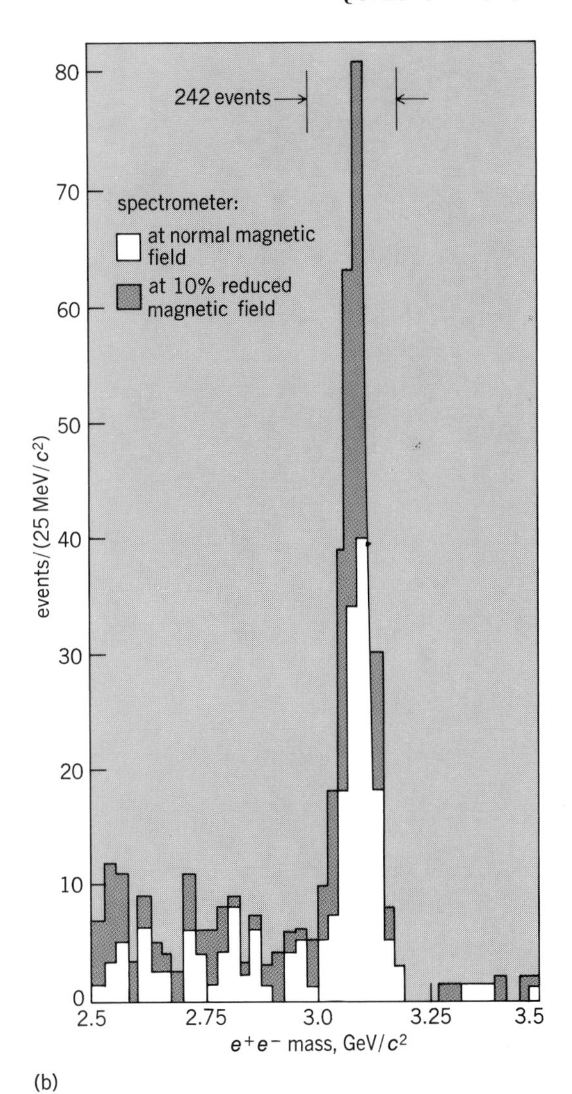

Fig. 6. Formation of lepton pairs in hadronic collisions. *(a)* Cross section for formation of muon pairs as function of muon-pair mass, observed in experiment with limited mass resolution *(from J. H. Christenson et al., Phys. Rev. Lett., 25:1523–1525, 1970)*. *(b)* Number of observed electron pairs as function of electron-pair mass. Data show existence of the J/ψ particle *(from J. J. Aubert et al., Experimental observation of a heavy particle J, Phys. Rev. Lett., 33:1404–1406, 1974)*.

forbidden by the Zweig rule, as discussed above.

The J/ψ has one excited state (the ψ' at 3.7 GeV), and the Υ has two. These states offer the first glimpse of the $q\bar{q}$ force as a function of distance. The strength of this force can be inferred from the excitation energy and the mean radius of the wave functions of the states involved. The analysis is made possible by the fact that both the c and the b quarks move at nonrelativistic velocities in their binding potential wells due to their large masses.

The knowledge of the force is important because recent advances in understanding the nature of matter were made possible by the radically new concept of a force increasing with separation and becoming negligible at small distance (asymptotic freedom). The verification and detailed determination of the force law is therefore of fundamental importance. From the excitation energies determined for the ψ and Υ systems, it is already known that a simple potential $V(r) = \alpha_S/r + \beta r$ predicts the right order of magnitude of the excitation, but this prediction needs refinement. A logarithmic potential can reproduce the splittings, but more information is needed to settle the question.

Speculation. The discoveries made thus far have aroused speculation about the mass of the next quark flavor (if any). Without stretching the data too much, the masses of the known quarks increase by roughly a factor of 3 each time (Fig. 4). No compelling reason for this has been put forth. If this rule is taken as an experimental indication, the next quark can be expected at about 14 GeV, and the next vector meson at 27 GeV. This energy range is presently accessible only at the Intersecting Storage Rings of the European Commission for Nuclear Research (CERN), where the low luminosity makes a search very time-consuming. Ting and colleagues have been looking for muon pairs at CERN for 2 years without having reported any results. The Tevatron, under development at the Fermilab, is the most promising tool to explore that mass region. Meanwhile, Lederman and colleagues de-

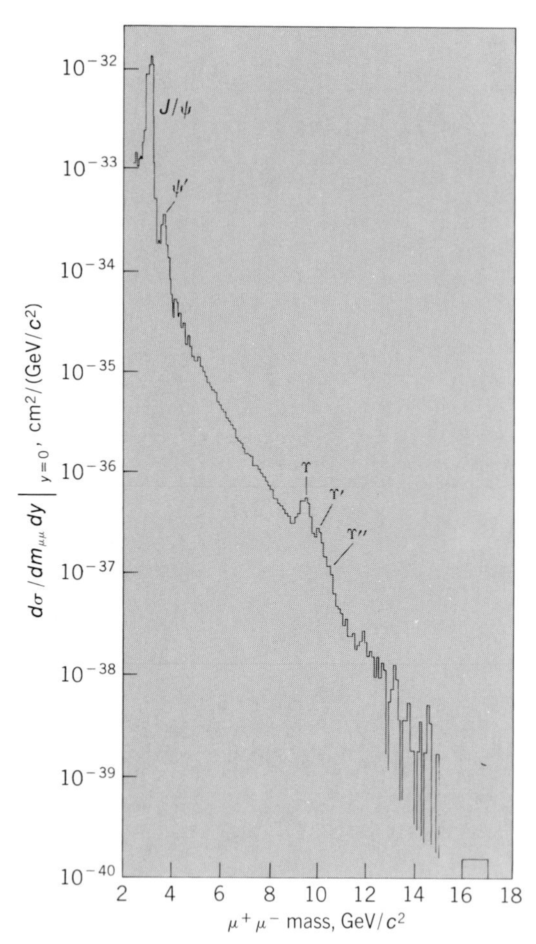

Fig. 7. Cross section for producing muon pairs at 90°, in collisions of 400-GeV protons with platinum or copper target, as function of muon-pair mass.

duced from Fig. 7 that there are no more $\mu^+\mu^-$ resonances of the same signal-continuum ratio as the Υ up to 15 GeV. A limit of 1/200 of the Υ cross section applies for any new such resonance with a mass between 14 GeV and the kinematic limit at 400 GeV, which corresponds to a mass of 27 GeV.

For background information *see* CHARM (QUANTUM MECHANICS); ELEMENTARY PARTICLE; J PARTICLE; QUARKS in the McGraw-Hill Encyclopedia of Science and Technology. [HANS JÖSTLEIN]

Bibliography: N. Calder, *The Key to the Universe*, 1977; S. Drell, *Phys. Today*, 31(6):23–32, June 1978; G. Gallinaro, M. Marinelli, and G. Morpurgo, *Phys. Rev. Lett.*, 38:1255–1258, 1977; E. D. Garris and K. O. H. Ziock, *Nucl. Instrum. Meth.*, 117:467, 1974; S. L. Glashow, *Sci. Amer.*, 233(4): 38–50, 1975; L. W. Jones, *Rev. Mod. Phys.*, 49: 717–752, 1977; D. M. Kaplan et al., *Phys. Rev. Lett.*, 40:435–438, 1978; G. S. LaRue, W. M. Fairbank, and A. F. Hebard, *Phys. Rev. Lett.*, 38:1011–1014, 1977.

Radio astronomy

Quasars and similar active galactic nuclei are still as enigmatic as when they were discovered 15 years ago. Among the most fascinating of the unexplained features of the observations is the phenomenon of "superluminal" expansion of some compact radio sources that are positionally associated with quasars and active galaxies. These sources are typically a few light-years across (1 light-year = 9.46×10^{15} m), and are frequently modeled by two components whose separation has in some cases been seen to increase over several years of observations. By using the red shift of the optical spectral lines from the quasar and an assumed value for the Hubble constant, a distance to the source can be inferred and, from the measured angular expansion rate, an apparent linear expansion rate is deduced. In a few sources there is good evidence that these speeds exceed the speed of light (c) by factors perhaps as large as 10. The problem is then to interpret this rapid expansion in the context of existing models of quasars.

Observation of superluminal expansion. Observations are made using the technique of very long base-line interferometry, whereby separate radio telescopes, often in different continents, simultaneously observe the same source for about half a day. By correlating the signals recorded at two telescopes, some information on the Fourier transform of the source brightness can be obtained. However, the Fourier transform is not as well sampled as in a conventional interferometer, and so in order to reconstruct a map of the source, approximate modeling techniques must be used. This is a very difficult procedure to carry out, and only in recent years has it become possible to have much confidence in even the gross features of the derived source structures.

Probably the best example of apparent superluminal expansion is furnished by the source 3C 345, which appeared to exhibit a double structure in observations made between 1970 and 1977. The measured angular separation of the two components is shown in Fig. 1. The slope of the line is 0.17 millisecond of arc per year. If one assumes a Hubble constant of 75 km s^{-1} Mpc^{-1} (1 megaparsec = 3.1×10^{22} m), the effective distance of the source is roughly 1200 Mpc and the observed expansion translates into a linear speed of about $5c$. The sources 3C 120 and 3C 273 also show good evidence for a similar expansion.

Explanation of observations. Although the theory of special relativity restricts physical velocities to less than c, there is, in fact, no objection to these apparent expansion velocities exceeding c. In particular, one is not forced to conclude that these

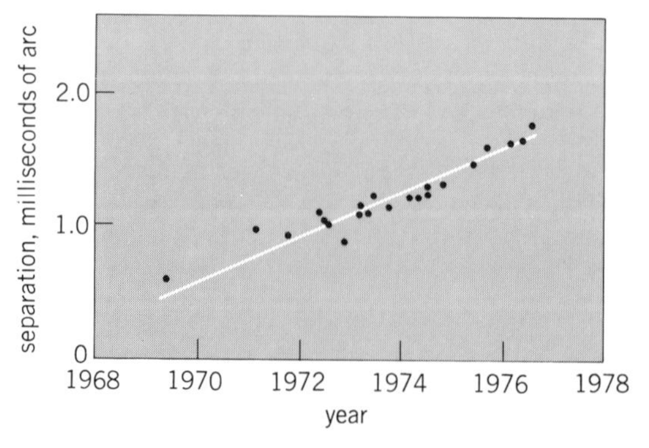

Fig. 1. Apparent separation of the two components of 3C 345 as a function of time. *(From M. H. Cohen et al., Radio sources with superluminal velocities, Nature, 268: 405–409, 1977)*

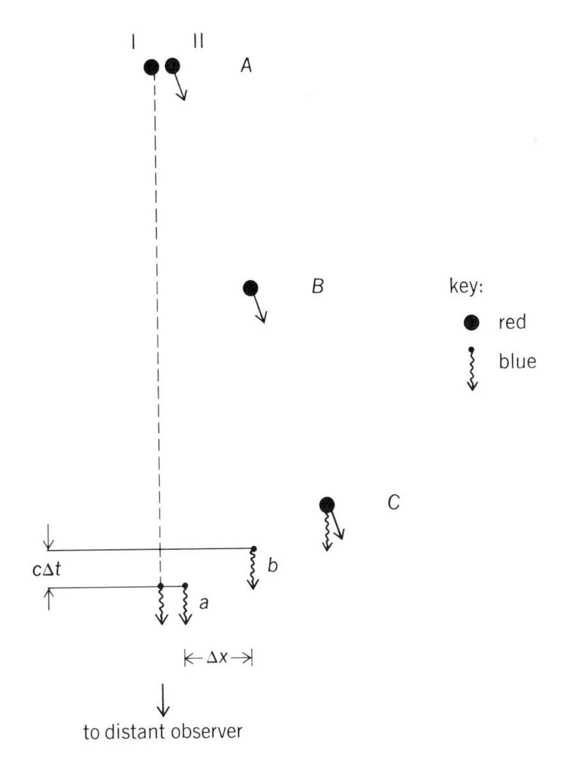

Fig. 2. Possible model for superluminal expansions.

sources are closer than the distances indicated by their red shifts. It is, in fact, likely that this superluminal expansion is an illusion produced by motions in the source with speeds close to c.

A simple example reinforces the point. If two self-luminous radio sources are created and one source moves at high speed toward a distant observer, then in computing what the observer sees, one must take into account the time it takes light to cross the region separating the two sources. The kinematics of this example are given in Fig. 2. Source I is stationary at A and source II moves with a speed of $0.9c$ in a direction making an angle of 20° with the observer. After a year, source II has moved 0.9 light-year to B. After 2 years it has moved 1.8 light-years to C, and at this time the radio waves that were emitted from A and B have traveled 2 light-years and 1 light-year respectively to a and b. A distant observer will see the waves at a and b separated transversely by a distance $\Delta x = (0.9 \sin 20°)$ light-years and separated longitudinally by a time interval $\Delta t = (1 - 0.9 \cos 20°)$ years. The apparent expansion speed is the quotient $\Delta x/\Delta t = 2.0c$. In this way, a source moving slower than c appears to be moving faster than the speed of light. Apparent superluminal expansion is thus possible.

This and many other similarly idealized models have been proposed to account for the observations, and in general it is not too difficult to account for motions as large as approximately $6c$. From the few well-studied cases, it seems that when apparent superluminal expansion does occur, it takes place along a preferred direction which is probably related to the spin axis of the innermost parts of the quasar. In many other sources where a nonexpanding compact radio source is seen, parallel radio structure is also observed on much larger angular scales.

Implications. The observations of these moving sources clearly indicate some very important facts about the nature of compact radio sources and their relationship to both the central power source of a quasar and to extended double sources. At present it is too early to choose among the many competing kinematical models and, in particular, to incorporate them within a self-consistent dynamical context. However, as the techniques and capability of very long base-line interferometry are constantly improving, it is likely that the basic geometry will be understood soon, and this should help considerably toward unraveling the basic mysteries of quasars. When the kinematics of these sources is well understood, it may become possible to use them to help measure the Hubble constant and deceleration parameter of the universe. *See* COSMOLOGY.

For background information *see* QUASARS; RADIO ASTRONOMY; RADIO TELESCOPE in the McGraw-Hill Encyclopedia of Science and Technology.

[R. D. BLANDFORD]

Bibliography: R. D. Blandford, C. F. McKee, and M. J. Rees, *Nature*, 267:211–216, 1977; M. H. Cohen et al., *Nature*, 268:405–409, 1977; A. C. S. Readhead, M. H. Cohen, and R. D. Blandford, *Nature*, 272:131–134, 1978.

Radio telescope

Electromagnetic radiation constantly bombards the Earth over a wide range of frequencies (or wavelengths). Within a certain frequency range, called the optical band, the radiation is visible. However, the picture of the sky that optical astronomers see is quite different from what is discerned at radio wavelengths. The Very Large Array (VLA), now under construction and in partial operation, is the latest and most powerful of all existing aperture synthesis interferometers, instruments that take radio pictures of the sky.

Figure 1 shows an aerial view of part of the array, with 17 of its eventual 27 elements visible, along with two arms of the Y-shaped track. Also visible are the antenna assembly building and other buildings for control and computing equipment, maintenance, cafeteria, and dormitory facilities. The site of the VLA, on the Plains of San Augustin near Socorro, NM, is remote, flat, high, and dry, factors which are all crucial to proper operation of the array. The VLA is being constructed by the National Radio Astronomy Observatory, part of Associated Universities, Inc., and is funded by the National Science Foundation.

Very Large Array operation. The power of the VLA lies in its resolution and sensitivity. When complete, the resolution of the VLA will be less than 0.1 second of arc, enough to read a newspaper 2 mi (3.2 km) away. It depends on the principle of interference between electromagnetic waves. Figure 2 illustrates the operation of a radio interferometer with two antennas. The amplified signals from a source in the sky are sent to the correlation receiver in the control building. First, though, the signal from antenna 1 must be delayed to make up for the extra distance B the signal traveled to antenna 2. The two waves received from a single source will then alternate between constructive and destructive interference, producing a sinu-

Fig. 1. Aerial view of Very Large Array under construction on Plains of San Augustin near Socorro, NM.

soidal pattern, called fringes, as shown in Fig. 2. From the amplitude and rate (or phase) of these interference fringes, the astronomer can measure the location and strength of a source in the sky. However, this information is limited to only a single line in the sky whose orientation is parallel to the line connecting the antennas. It also gives only information about sizes on the order of 10^5 (λ/D) seconds of arc, where λ is the observing wavelength and D is the separation of the antennas, or base line. To reconstruct a complicated picture, information must be compiled from a large number of base lines and orientations. In a technique developed in the late 1950s by astronomers in Cambridge, England, the rotation of the Earth is used to change the orientation of the antenna pair, as viewed from the source. By combining many such pairs and observing a source as it crosses the sky, one can simulate the operation of a single, very-large-aperture telescope. This is called Earth-rotation aperture synthesis.

VLA instrumentation. Each of the 27 VLA antennas is 82 ft (25 m) in diameter. Combining each antenna with every other produces 351 base-line pairs. The antennas are positioned along the Y-shaped railroad track, which has a total span of about 36 km. The antennas can be picked up and repositioned along the track by a specially built transporter; their placement depends on the size of the source which the astronomer is observing.

The parabolic surface of each antenna reflects and focuses the incoming waves to an asymmetric subreflector. This can then rotate to focus the radiation to one of four receiver systems, housed on the parabolic surface. The receivers are the heart of the system's sensitivity, and are maintained at a temperature of 18 K (−225°C). These receivers cover the four wavelength bands around 21, 6, 2, and 1.3 cm. The fundamental part of these receivers is a wide-band parametric amplifier, operating at 6 cm. The other wavelengths are first converted to 6 cm; the signals are then amplified about 10^6 times, before being sent to the control building along a buried waveguide. Here, a vast network of digital electronics and computers correlates and interprets the incoming data. It also sends out commands to keep the antennas and electronics operating in unison.

In order to produce the fringes with sufficient accuracy, all of the array electronics must be synchronized within about 3×10^{-12} s. A large amount of control and monitoring equipment is devoted to this task. It is also necessary to know the position of each antenna to a fraction of 1 cm. This involves astronomical calibrations each time an antenna is moved, and a series of corrections for the motion of the Earth, its precession, the solid tides in the Earth's surface, and even the gravitational deflection of waves passing near the Sun.

The computer network is divided into two systems. The synchronous system controls the antennas and electronics, monitors their performance, and receives the incoming data. It applies a number of corrections to the data, based on both instrumental behavior and external effects (such as atmospheric refraction). The data are then passed to the "asynchronous" system, where further editing, calibration, and processing lead the astronomer to the desired "picture" of the sky.

VLA use in radio astronomy. It is the combination of resolution and sensitivity which makes the VLA a spectacular new tool for radio astronomy. Detailed pictures of objects in the solar system, Galaxy, and "local neighborhood" of the universe have been studied for decades by optical astronomers. In many cases, however, the lack of comparative information at radio wavelengths has limited the understanding of the physical and chemical processes in these systems. There are also many areas where dust obscures the optical emission, but the radio picture emerges intact.

Work to date with the partially finished VLA includes detailed optical and radio comparisons of planets, planetary nebulae, clouds of ionized hydrogen, the central regions of the Galaxy, and other nearby galaxies. Also under study is the highly variable emission from stars and from objects discovered by x-ray satellites. Many astronomers believe these are collapsed objects that radiate as extremely hot, dense gas falls toward their surfaces. Much of the radiation in these systems is thermal; it originates in hot gases radiating over a broad spectrum.

However, most of the strong radio sources in the sky result from relativistic electrons spiraling around magnetic fields. This is nonthermal, or synchrotron (so called by analogy with Earth-based particle accelerators), radiation. The Galaxy also emits synchrotron radiation, but at a level 10^6 times less than the powerful radio galaxies and quasars. Details of these powerful radio sources are being examined at the VLA. One currently exciting area is the study of the thin streams, or "jets," that emanate from the nuclei of some radio sources. Study of these objects may help unravel the mechanism by which electrons are accelerated and maintained at relativistic energies hundreds of thousands of light-years from their origin.

During 1979 an additional spectroscopic capability will be installed at the VLA. One of its major uses will be to study the distribution of atomic hydrogen in the Galaxy and in other galaxies. In addition, the wealth of molecules now known in the Galaxy will be probed to learn about interstellar chemistry and the birth and evolution of stars.

Early VLA picture. Figure 3 is a picture of a radio galaxy, synthesized from VLA data at a wavelength of 6 cm, with a resolution of about 1.5 seconds of arc. In radio catalogs, this object is designated 3C 83.1B. In the optical, it is the elliptical galaxy NGC 1265, a member of the Perseus cluster of galaxies, which is about $2-3 \times 10^8$ light-years ($2-3 \times 10^{24}$ m) distant. The radio and optical pictures are quite different. The optical emission is from stars distributed over an elliptical region somewhat smaller than that shown in Fig. 3, and is centered on the burned-out spot, called the nucleus. The radio emission, however, is synchrotron radiation from electrons, presumably supplied to the thin streams by the nuclear source. From other studies of NGC 1265, it is known that the radio emission extends upward about 15 times further than shown here. This source is the prototype "head-tail" radio galaxy, so named for the long thin tails which stretch out behind the galaxy in its motion through the cluster gas.

The VLA picture is a close-up of details in the head. From detailed analysis of the emission, the magnetic fields in some of the bright spots are found to be 10^4 times weaker than on the surface of the Earth. This value is typical for many other extragalactic sources. However, it leads to problems with the theoretical models for NGC 1265, when combined with information on the shape of the streams. It is through such confrontation of observation with theory that knowledge is extended and refined.

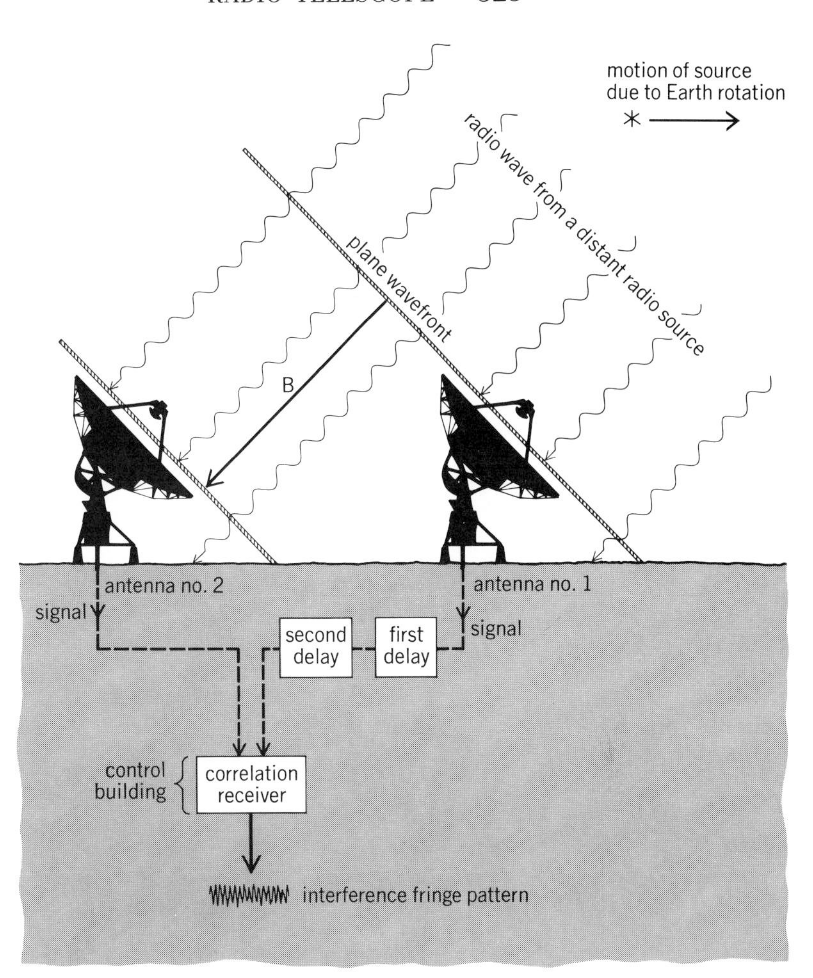

Fig. 2. Operation of a radio interferometer. Interference fringes shown here, from the intense radio source 3C 84, were the first ones produced at the VLA.

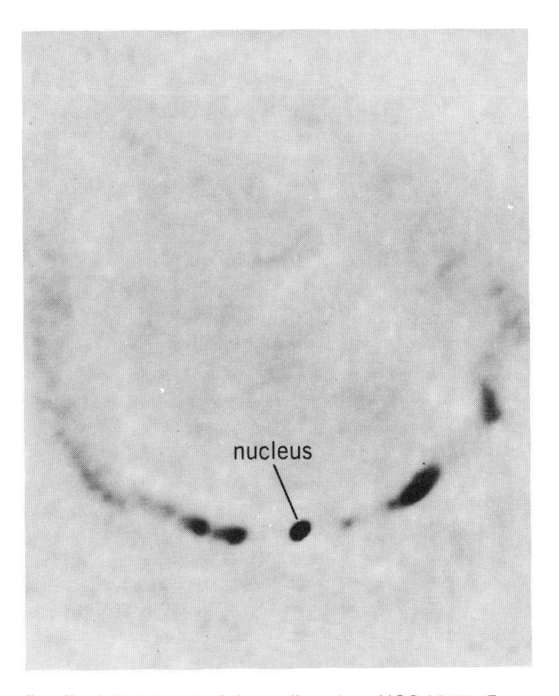

Fig. 3. A "picture" of the radio galaxy NGC 1265. *(From F. N. Owen, J. O. Burns, and L. Rudnick, VLA observations of NGC 1265 at 4886 MHz, Astrophys. J. (Lett.), 226:119; © 1978 by the University of Chicago Press; used with permission)*

Future use. With its completion in 1980, the VLA will be used to study astronomical objects from the Sun out perhaps to the edge of the visible universe. However, if one can be guided by past experience, its greatest discoveries may lie in directions which are yet unknown.

For background information *see* RADIO ASTRONOMY; RADIO TELESCOPE in the McGraw-Hill Encyclopedia of Science and Technology.

[LAWRENCE RUDNICK]

Bibliography: E. B. Fomalont and M. C. H. Wright, in G. Verschuur and K. Kellermann (eds.), *Galactic and Extra-Galactic Radio Astronomy*, pp. 256–290, 1974; F. N. Owen, J. O. Burns, and L. Rudnick, VLA observations of NGC 1265 at 4886 MHz, *Astrophys. J. (Lett.)*, 226:119, 1978; S. Weinreb et al., *IEEE Trans. Microwave Theory Techn.*, MTT-25:243–248, 1977.

Radioactive waste management

The decision by the Administration to defer indefinitely the reprocessing of spent nuclear reactor fuels was announced by President Jimmy Carter on Apr. 7, 1977. This decision forced a reconsideration of the plans and schedules that had been under development for handling spent nuclear fuel from reactors in the United States. *See* NUCLEAR POWER.

Nuclear power supplied 12% of the electric energy in the United States in 1977 (and more than 50% in several states). About one-third to one-fourth of the fuel in a nuclear power reactor is replaced annually. The fuel removed is highly radioactive and is called spent fuel. The spent fuel assemblies are placed in the storage pools at the plants. Utilities had anticipated that their spent fuel assemblies would routinely be shipped in specially designed shielded casks to reprocessing plants. Indeed, some had contracts with the Allied General Nuclear Services facility at Barnwell, SC, to accept such shipments, even if it meant temporary storage in pools at Barnwell until the plant achieved operation. Public hearings related to the Barnwell plant's permit to operate or even to store spent fuel were still in progress at the time of President Carter's announcement. The hearings have since been terminated without reaching any findings.

Therefore, utilities are expanding their on-site storage capability amidst uncertainty about if and when reprocessing will be permitted. At the same time, nuclear power itself is the subject of intense public debate, with one of the most visible and emotional issues being that of ultimate safe disposal of nuclear wastes.

Definitions. If spent fuel from a commercial nuclear power plant is reprocessed, the uranium and plutonium can be separated for recycle as fuel. The remainder from the reprocessing step is called high-level commercial waste. Similar waste from weapons program reprocessing is called military waste. Low-level waste includes contaminated towels, gloves, shoe covers, swabs, glassware, tools, resins, and containers. Some materials may actually be uncontaminated, but if they have been used in radiation areas it is simpler and safer to designate them low-level waste and package them for disposal in a shallow land burial site licensed for such use by the state or the Nuclear Regulatory Commission.

Methods of handling waste. Spent fuel can be stored in pools for decades if necessary. The water provides shielding and carries away the residual heat. The NRC has issued its draft Environmental Impact Statement on spent fuel storage pools and has found no significant impact or deterioration of fuel assemblies from this method of temporary storage.

The basic concept of reprocessing the spent fuel is to separate the unused uranium and the plutonium that was built up in the assemblies during the time the fuel was in the core. The uranium and plutonium would be recycled to produce electricity. The remaining high-level nuclear waste must be isolated from the environment.

High-level waste is held in liquid form at reprocessing plants. Next, the liquid is evaporated, and the solids are reduced to a granular powder by baking off all the moisture (calcining). This process has been demonstrated in laboratory- and pilot-scale operations and has been done on a routine basis for years in the reprocessing of naval reactor fuels. Several methods for preparing solidified high-level wastes have been extensively studied and demonstrated over the past decade. Calcined material can be mixed with glass and cast into thick glass rods and encased in steel cylinders. Finally, the cylinders are placed in concrete shielding blocks so that they can be shipped to the repository site and moved into it without causing significant radiation exposure to the operators.

Repository design. Isolation of the high-level wastes from the environment can be achieved through storage in salt beds or in caverns in granite, 1000 to 2000 ft (300 to 600 m) below the Earth's surface. The development of such a repository would use technology no different from that used in conventional mining; however, the repository sites must be carefully selected and studied extensively before they can be licensed to store radioactive wastes. A key characteristic is stability, and thus detailed studies need to be made of the geology, seismology, and hydrology. As examples, the existence of salt beds demonstrates that there has been no moisture at those locations for millions of years, and granite is one of the most stable underground formations known. The objective is to reduce as near to zero as practical the possibility of any significant quantities of radioactivity reaching the biosphere through any natural occurrences or through accidents. Therefore, all potential pathways need to be considered, and the risks of radioactive isotopes reaching groundwater or the Earth's surface need to be assessed.

Figure 1 is an idealized diagram of a geological repository. The actual design will depend on the detailed characteristics of the site. Specific design features need to be identified and developed in the engineering of each repository so as to provide the necessary protection for the public health and safety. For this reason, authorities such as the U.S. Geological Survey state that they have confidence that geologic storage of high-level wastes can be done acceptably, while at the same time itemizing those information gaps and uncertainties that need to be resolved before a specific design at a particular site can be licensed and the facility constructed.

Experiments and demonstrations. Actual experiments have shown that it is feasible and safe to

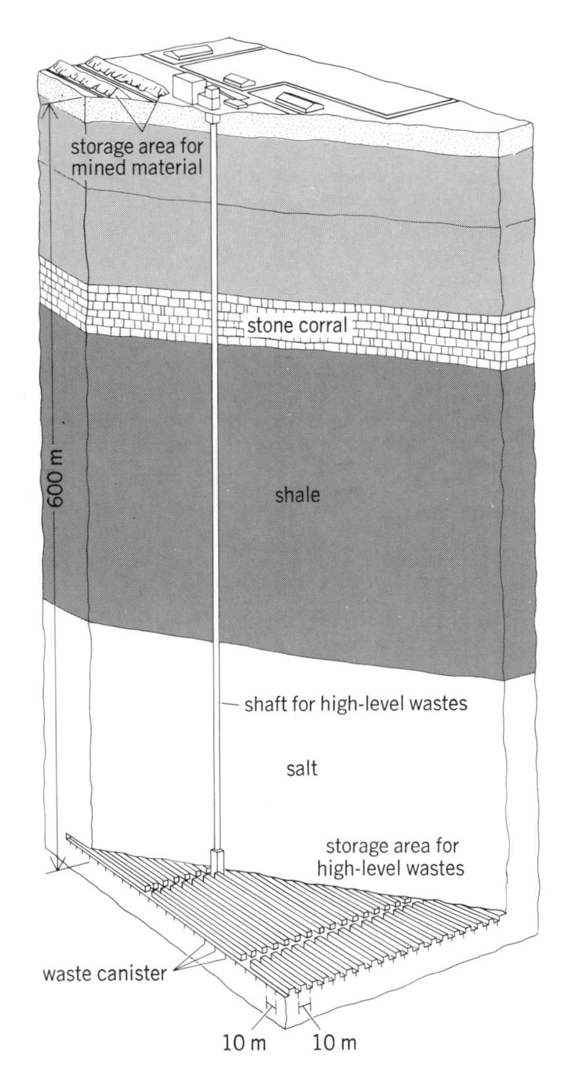

Fig. 1. Idealized diagram of a geological repository for storing radioactive wastes. *(From B. L. Cohen, The disposal of radioactive wastes from fission reactors, Sci. Amer., 236(6):21–31; copyright © 1977 by Scientific American, Inc.; all rights reserved)*

put shielding blocks containing canisters of radioactive waste in geologic formations. During an 18-month period in 1969–1970 a number of canisters simulating high-level radioactive wastes were placed in an underground salt vein and later removed. During that period the feasibility of remote handling of highly radioactive material and storing it underground without significant problems from heating or radiation was demonstrated. Longer-term demonstrations are planned at new sites.

In Germany both low-level and intermediate-level radioactive wastes have been put in permanent disposal locations in the Asse mine. No difficulties in handling, radiation exposure, or safety have been encountered. High-level wastes also will be tested at the Asse facility. A second permanent facility will be built for storing reprocessed commercial waste.

United States programs. The Congress gave the Atomic Energy Commission (now part of the Energy Research and Development Administration) authority to regulate the building of needed repositories and to accept commercial high-level wastes for permanent disposal. For several years the United States government has been engaged in the design, geologic studies, and preparation of environmental impact and safety analyses for an engineering verification of geologic storage. Processed and packaged wastes from military weapons production are to be emplaced and monitored in a facility known as the Waste Isolation Pilot Plant (WIPP), located in the desert outside Carlsbad, NM.

In June 1978 the Department of Energy released a task force report indicating that a commercial waste repository could be ready by 1988. This marks a slippage from the 1985 target date that had appeared in schedules in recent years. The 1985 target date had been considered reasonable despite the number of nuclear power plants in operation, because it was obvious that no reprocessing plants would be in operation until 1980 at the earliest. The time period between reprocessing and actual emplacement would be a minimum of 5 years. Thus a facility constructed earlier than 1985 would receive only military wastes. However, it would still be useful as concrete evidence that such a repository is practical.

Geologic storage of spent fuel. With indefinite deferral of reprocessing, some people are calling for geologic disposal of spent fuel assemblies. This concept has drawbacks on a resource basis because it would mean throwing away the remaining fuel value of the plutonium and of the uranium. Spent fuel assemblies are about the same size as the glass rods of processed waste and could also be placed in permanent geologic storage. The storage of spent fuel would be similar to that of processed waste, and the proper amount of material specified for safe storage in each location would be evaluated based upon heat dissipation rates.

Temporary storage in reactor pools. Until 1985 or later it appears inevitable that spent fuel will have to be stored in pools at reactors or in separate pools at facilities away from reactors. One such separate pool facility is operated by General Electric Company at Morris, IL (Fig. 2). The Congress is considering legislation to authorize the Federal government to build one or more such facilities. Meanwhile, pool capacity at reactor plants is being expanded by replacing standard storage racks with more closely spaced racks, some using boron as a neutron absorber, to permit even tighter packing of spent fuel assemblies.

Some older plants, whose owners had expected to be able to ship out spent fuel by this time, now find their storage pools nearing capacity. It is prudent to keep enough storage space available to unload an entire core. This has been required on numerous occasions in order to perform reactor modifications; in some cases it has been done in order to reduce worker exposure. Unless adequate spent fuel storage pools are constructed on a timely basis, there is the danger that a nuclear plant might be forced to shut down in the future due to lack of storage space. Since there is no significant risk or environmental impact from a spent fuel storage facility, its relatively small cost appears to be a justifiable expense.

Status in 1978. The Department of Energy task force recommended that an interagency group prepare a policy for nuclear waste management. This statement should identify the steps to be taken by each government agency involved and deter-

Fig. 2. Facility for storing spent reactor fuel at Morris, IL. *(General Electric Co.)*

mine whether congressional action is needed. Even if the entire commercial nuclear power program of the United States were to be stopped today and no more radioactive fission products ever generated, the Federal government would still bear the responsibility for implementing a complete program for permanent disposal of existing radioactive wastes. There are over 8×10^7 gal (3×10^5 m^3) of radioactive waste from 35 years of the United States nuclear weapons program in temporary storage today. There are already significant amounts of radioactive material in spent fuel and in operating commercial power reactors that also must be stored permanently. Since this must be done regardless, wastes from future nuclear power generation can be handled in the same way. In March 1978 the United States Court of Appeals decided that the NRC was acting properly in continuing to issue nuclear plant construction permits and operating licenses.

For background information *see* NUCLEAR POWER; RADIOACTIVE WASTE MANAGEMENT in the McGraw-Hill Encyclopedia of Science and Technology. [A. DAVID ROSSIN]

Bibliography: B. L. Cohen, *Sci. Amer.*, 236(6): 21–31, June 1977; G. Dau and R. Williams, *EPRI J.*, (6):6–14, July–August 1976; Energy Research and Development Administration, *Alternatives for Managing Wastes from Reactors and Post-Fission Operations in the LWR Fuel Cycle*, ERDA-76–43, May 1, 1976; U.S. Department of Energy, *Report of Task Force for Review of Nuclear Waste Manage-*

ment, February 1978; U.S. Department of the Interior, *Geologic Disposal of High-Level Radioactive Wastes: Earth Science Perspectives*, USGS Circ. 779, April 1978; U.S. Nuclear Regulatory Commission, *Draft Generic Environmental Statement on Handling and Storage of Spent LWR Fuel*, NUREG 0404, March 1978.

Radiocarbon dating

In mid-1977 a new radiocarbon dating method, based on ultrasensitive mass and charge spectroscopy, was developed. This method can detect less than three ^{14}C atoms in 10^{16} atoms of carbon. This concentration of ^{14}C atoms corresponds to a radiocarbon age of 70,000 years. In addition to its high sensitivity, the new method requires less than a milligram of sample, which is a thousand times less material than is ordinarily needed by the conventional β-decay counting method for radiocarbon dating. A further advantage is that the time required to measure the ^{14}C concentration in small samples is much less than that required by the conventional method. The great interest generated by the new method culminated in the convening of an international conference on the subject less than a year after the first demonstration of the feasibility of such methods for radiocarbon dating by a Rochester-Toronto-General Ionex collaboration.

Conventional radiocarbon dating. Assuming equilibrium, the production of ^{14}C by cosmic rays reacting with ^{14}N in the atmosphere is balanced by the loss of ^{14}C due to its β-decay, so that the concentration of ^{14}C in the atmosphere is maintained. After an organism dies and ceases replenishing its ^{14}C, the ^{14}C concentration in the carbon of its body decreases exponentially with a half-life of 5730 years. The ^{14}C concentration in a sample of formerly living carbon thus provides a measure of the time since the organism's death. In conventional carbon dating techniques the radioactivity of the carbon is measured, and by comparing this activity with that of samples of known ages, the age of the unknown may be determined, since radioactivity is proportional to ^{14}C concentration.

Early history. The use of accelerators as ultrasensitive mass spectrometers was originated by L. W. Alvarez and Cornog in 1939 to determine which isotope of mass 3, ^{3}He or ^{3}H, was radioactive. The application of mass spectroscopic methods to radiocarbon analysis was suggested later by various researchers, including H. Oeschger in 1970 and R. A. Muller in 1977. Muller was the first to suggest the use of high-energy accelerators for radioisotope dating.

Repeated attempts by M. Anbar and coworkers in 1974, 1975, and 1976 using conventional mass spectroscopy came increasingly close, but ultimately fell short by a factor of 10 of the goal of detecting ^{14}C at contemporary concentrations because of overwhelming background problems. In 1977 three groups almost simultaneously announced the ability to directly detect ^{14}C at contemporary concentrations or lower by using accelerators. The Berkeley group employed a cyclotron, whereas the Rochester and McMaster groups employed tandem electrostatic accelerators.

Principles of accelerator method. The accelerator method of radiocarbon dating does not make use of the radioactivity of ^{14}C for measuring its concentration, but utilizes techniques and devices originally developed for nuclear physics. The carbon atoms in a sample are first ionized, either positively or negatively, and subsequently accelerated with either a cyclotron or a tandem electrostatic accelerator. The use of acceleration is the main difference between the accelerator method and conventional mass spectroscopy, which cannot easily be utilized for radiocarbon dating because ^{14}C and ^{14}N differ in mass by only 1 part in 10^5, so that ^{14}N produces a ubiquitous overwhelming background. Once accelerated to about 5 to 10% of the velocity of light, however, ^{14}C and ^{14}N may be distinguished from each other, as well as from all other ion species, by the fact that they slow down at different rates as they pass through matter.

The details involved in the use of the two types of accelerator are somewhat different and will be described separately.

Cyclotrons. Positive ions produced in an ion source are injected into the center of the cyclotron and accelerated by a radio-frequency oscillating voltage in an increasingly larger spiral. After 100 or 200 cycles the fully accelerated beam of particles contains essentially only ^{14}N and ^{14}C, as each cycle is in effect a mass spectroscopic analysis. This beam of particles is then passed through a volume of absorbing gas, such as xenon, and since the ^{14}N slows down and stops sooner than does the ^{14}C, by using the correct pressure of gas the ^{14}C may readily be separated from the contaminating ^{14}N. A gas is used as the absorber since it is more uniform than a thin solid foil, and a heavy gas is used so that no nuclear reactions capable of forming background ^{14}C can take place in the gas.

Tandem electrostatic accelerator. Negative ions produced in a sputter ion source are injected into a tandem accelerator and accelerated by attraction to the central terminal, where the electrons are stripped from the ions, thus changing them to positive ions. They are then repelled from the terminal and further accelerated until they finish traversing the accelerator. Various ion species leave the accelerator, since the tandem electrostatic accelerator is not as selective as is the cyclotron in what it accelerates. These ions include ^{12}C, ^{13}C, ^{14}C, isotopes of oxygen and nitrogen, molecular compounds such as CH, CH_2, NH, and many others. After sufficient magnetic and electrostatic analysis, all of these ion species except ^{14}C may be removed.

Comparison of accelerators. The main advantage of the cyclotron is that its use results in a somewhat simpler system, since only ^{14}C and ^{14}N are in resonance with the cyclotron radio frequency and are thus accelerated. On the other hand, the tandem has the great advantage that it uses negative ions initially, and it was discovered that the negative nitrogen ion is so fragile that it does not survive acceleration. Also, by alternately injecting ions of mass 12, 13, and 14, the tandem method can sequentially and almost simultaneously measure the ^{12}C, ^{13}C, and ^{14}C atoms coming from the same sample. Thus the concentrations of all three isotopes may be measured almost continuously. By considering the $^{13}C/^{12}C$ ratio, the degree to which the organism preferred to absorb the heavier-mass carbon atoms may be inferred, and a correction may be applied to the measurement of the ^{14}C concentration.

Comparison of samples dated by conventional and accelerator methods*

Sample† and origin	Conventional age, years	Accelerator age, years
W-3629, Mount Hood, OR	220 ± 150	220 ± 300‡
W-3703, Mount Shasta, CA	4590 ± 250	5700 ± 400
W-3663, Lake Agassiz, ND	9150 ± 300	8800 ± 600
W-3823, Hillsdale, MI	39,500 ± 1000	41,000 ± 1100
Graphite	–	48,000 ± 1300

*From C. L. Bennett et al., Radiocarbon dating with electrostatic accelerators: Dating of milligram samples, *Science*, 201: 345, 1978.

†Samples dated by the USGS by the conventional method and labeled with USGS number were supplied by M. Rubin.

‡This case used as a calibration standard.

Results. As a check on the accelerator method, samples previously dated by the conventional method have also been dated by the direct detection method. Some examples of this comparison are given in the table. These samples were dated by the accelerator method using as little as 3.5 mg of carbon and taking a few hours of counting time. By contrast, the U.S. Geological Survey (USGS) measurements took a few days in some cases and at least 2 g of carbon.

At present no unknown items of historical or archeological interest have been dated by the new method.

Problems. The problems involved in using both cyclotrons and tandems are that they were designed for research in nuclear physics and are not stable enough to yield the high accuracy of conventional radiocarbon dating methods. The highest accuracy reported to date using the accelerator methods has been about 3 to 5%, in contrast with 0.18% obtainable with careful application of conventional techniques.

Another problem with using presently available accelerators is that they have built up a contamination of ^{14}C by exposure to a large amount of radiation over the years, and this ^{14}C background limits the maximum age measurable with a given accelerator. At present, plans are under way to design and build a dedicated dating accelerator having high stability and low contamination.

Potential applications. The high sensitivity and small sample sizes needed in the accelerator methods enable a wide variety of age measurements to be made that were hitherto impractical. For example, rare artifacts, of which only a few milligrams can be spared for analysis, may now be dated. Individual tree rings may be dated by the accelerator method to obtain a year-by-year calibration of the radiocarbon ages with the known tree ring dates.

Besides being useful for the dating of artifacts and geological specimens, the accelerator method has the great advantage that it is not restricted to radioactive elements, and it is in fact a method for ultrasensitive trace-element analysis. This aspect has been utilized already in the search for exotic states of matter, such as quarks and superheavy elements. However, the accelerator methods could also be used for conventional trace-element analysis. Perhaps one of the most exciting potential uses is in the area of medicine. The advantages of being able to perform radioisotope tracer analysis of human or animal metabolic systems with very small samples in very short times are clearly great.

The new accelerator methods promise to lead to an improved understanding of the history of humans and the Earth, and perhaps to enable the detailed tracing of the dynamics of human metabolism.

For background information *see* PARTICLE ACCELERATOR; RADIOCARBON DATING; TRACE ANALYSIS in the McGraw-Hill Encyclopedia of Science and Technology.

[CHARLES L. BENNETT]

Bibliography: C. L. Bennett et al., *Science*, 198: 508–509, 1977; H. E. Gove (ed.), *Proceedings of the Rochester Conference on Radiocarbon Dating with Accelerators*, Rochester, NY, 1978; R. A. Muller, *Science*, 196:489–494, 1977; K. H. Purser et al., *Rev. Phys. Appl.*, 12:1487–1492, 1977.

Reproductive behavior

Cichlid fish show strong parental behavior in protecting their broods of fry. Recently broods composed of two species were found in the care of parental fish. The occurrence of these mixed broods is proving to be more common than initially supposed. It appears that in the case of substrate spawners brood hybridization (presence of broods from two species) may be actively encouraged by the parents, but in the case of maternal mouth brooders of Lake Malawi such an association may be detrimental to the foster parent.

Cichlid fish occur in the fresh waters of Africa, America, and Asia, where they are important as a source of protein, and also as attractive aquarium fish which may be exported to provide foreign revenue for developing countries. The Cichlidae have become known among ichthyologists and aquarists for their well-defined territorial behavior, their elaborate behavioral sequences during courtship and spawning, and the remarkable manner in which they care for their young. Compared with most marine fish and many other groups of freshwater fish, the cichlids lay few eggs. It appears that the energy thus saved is spent in caring for the eggs and young to ensure a high percentage of survival of the progeny. The manner in which different cichlids protect their offspring has formed the basis of categorization.

Substrate spawners. Substrate spawners have a protracted courtship during which they form pairs. Both members of a pair may contribute to nest building. Courtship and nest preparation culminate in spawning, during which eggs are laid

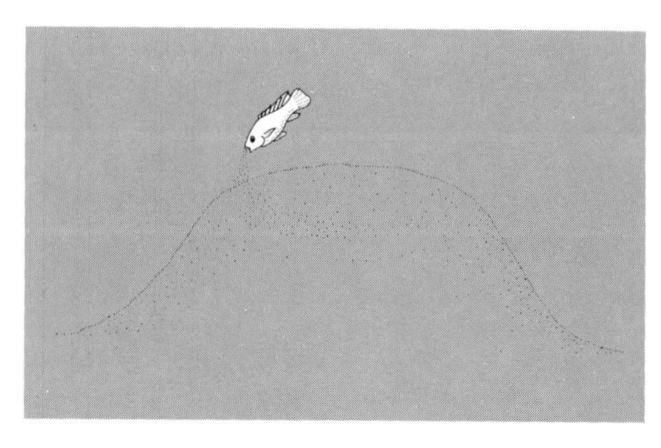

Fig. 1. Cichlid building a large turretlike nest by carrying sand onto the nest site.

on the substrate or a nearby plant and then fertilized. To ensure that eggs are not washed away, many species of substrate spawner glue their eggs to an object. Both parents guard the eggs and developing fry, and also aerate them by beating their fins so that a constant flow of water over the offspring is maintained. Furthermore, eggs and embryos are frequently cleansed by mouth, and any eggs or fry which die are removed. Occasionally parents may move their fry from one nest site to another by carrying them in their mouths. Parental care decreases as the fry become progressively independent.

Mouth brooders. The other major group of cichlids are the mouth brooders, which are considered to have evolved from substrate spawners. The mouth brooders lay fewer eggs than the latter, and all of these are held in the parents' mouth for the entire developmental period. In some species mouth brooding is practiced by males only, while in other species both parents participate. It is most common, however, for the females to bear the burden of parental duties alone. These maternal mouth brooders will be discussed in detail.

To attract gravid females, males become territorial, and some of those species which breed over sand may construct large, ornate nests or spawning sites by digging and carrying sand in their mouths (Fig. 1). Those species which hold territories over rock usually have a cleansed area on the rock to which females are led for spawning. After a brief courtship the female may lay eggs, which she then picks up in her mouth. In the African Great Lakes, predation on eggs is intense; consequently females lay very few eggs at a time, and immediately after each batch is deposited they swing around to snap them up. The males often do not have an opportunity to fertilize the eggs before the female collects them. Thus, to ensure fertilization, the male has marks resembling eggs on his anal fins; these egg dummies attract the spawning female, and while she attempts to collect these spots the male releases spermatozoa which flow over the anal fin to be taken into the female's mouth, fertilizing the eggs (Fig. 2). Once spawning is completed, females leave the breeding areas carrying a mouthful of eggs and join shoals of other incubating females in warm, well-oxygenated waters. The territorial males remain behind to court and spawn with other females.

Eggs and young in the mouth receive sufficient oxygen as the brooding female ventilates powerfully to pass a current of water over her progeny and her own gills. Once the young are large enough to swim and feed, the parent selects a safe site in which to release them. The fry remain together in a school and are guarded by the parent. They are responsive to the movements of their mother; if she moves suddenly, the fry scatter and plunge toward the bottom where they may hide. They then regroup, and if the parent angles herself and bobs slightly on her return, the young cluster around her mouth and enter (Fig. 3). The fry are also responsive to the others in the brood and show a strong schooling tendency. Fry are taken into the parent's mouth for protection when predators appear and, in Lake Malawi, they are also collected in the evenings when predatory pressure is greatest. In the morning they are released once again.

Fig. 2. Female collecting spermatozoa from the male while attempting to take "dummies" on the male's anal fin into her mouth.

Mixed broods. Recently A. J. Ribbink reported that a number of predatory species in Lake Malawi were found caring for mixed broods consisting of their own offspring, easily recognized because they have adult coloration and markings from the outset, and fry of another species differing emphatically in coloration and occasionally differing slightly in size. In those instances the foreign fry appeared to belong in all cases to a single species, *Haplochromis chrysonotus* Boulenger, which is a planktivorous fish found in surface waters. The foster parents *H. polystigma*, *H. macrostoma*, and *Serranochromis robustus* are predators which are usually found near the bottom in water less than 30 m deep. Since this initial discovery a number of other species have been found caring for mixed broods in Lake Malawi; several of them are predators such as *H. kiwinge* and *H. maculiceps*, but others are omnivorous or herbivorous such as *H. sphaerodon* and *H. fenestratus*. Furthermore, the foreign fry are not always *H. chrysonotus*, but other species which still need to be identified. The mixed brood phenomenon appears more common in Lake Malawi than intially supposed.

Cichlid cuckoos. It was suggested that *H. chrysonotus* might be "cichlid cuckoos" in the sense

Fig. 3. Cichlid retrieving fry which swim back into parent's mouth when recalled.

that foster parents guard the young. A complete parallel between *H. chrysonotus* and avian cuckoos is not tenable, however, as these cichlids have not been observed to deposit their young with foster parents, nor have "cuckoo fry" been observed to destroy the host's offspring. It is probable, however, that cuckoo fry are directly responsible for the loss of a number of host fry which they displace. A mouth brooder is limited in the number of eggs she lays by the capacity of her mouth, in which the eggs and fry are held. Consequently, if foreign fry are added to the brood and manage to scramble into the protective mouth before the native offspring, the chances of the parent losing some of her progeny to a predator are increased.

It is still not known how young *H. chrysonotus* join broods of other species, as their parents have not been observed to deposit them with the fry of a foster parent. Indeed, while a potentially surrogate parent is guarding its own young, it is unlikely to permit the approach of an adult *H. chrysonotus*. At this stage it appears as though *H. chrysonotus* release their fry among the rocks and the fry find their own way into the schools. Young cichlids tend to form schools, and when groups of fry of different species but similar size are placed together they shoal, benefitting from the protection afforded by the school. In Lake Malawi a large number of unattended groups of fry have been observed among the rocks; sometimes the shoals were homogeneous and on other occasions they were mixed. Thus, a behavior which results in *H. chrysonotus* young joining groups of fry of similar size, regardless of species, would enable them to gain the protection afforded by a foster parent. If this surmise is correct, those *H. chrysonotus* fry which do not find a surrogate parent would remain unprotected and their chances of survival would be reduced. Clearly, evolutionary selection pressures would favor those fry which sought foster parents and responded correctly to their signals. At present, evidence supporting these speculations is meager, and the mechanism by which fry of *H. chrysonotus* come to be in mixed broods still needs to be worked out.

As fry which join another brood may reduce the chances of the native fry receiving the protection intended for them, it is clearly not in the parents' interest to tolerate the intruders. By adding fry of different species to homogeneous broods artificially, it was found that the parents generally attempted to chase the new additions away and met with varying degrees of success.

Caring for predator's young. By contrast, Ken McKaye found that certain substrate spawners in Lake Jiloa, Nicaragua, actively kidnapped foreign fry and added them to their own broods. He considers that this strategy may reduce the probability of the fishes own young being taken, given an encounter with a predator. As substrate spawners usually have too many young to protect in their mouths, additions to their broods may be advantageous to the host species, but in mouth brooders additions would be detrimental as they might exclude the parent's own progeny from its mouth.

Also in Nicaragua, McKaye found a herbivorous fish, *Cichlasoma nicaraguense*, which cared for the progeny of one of its predators, *C. dovii*. It was suggested that this altruistic behavior may be repaid at a later date when *C. dovii* preyed upon another species of cichlid which competed with *C. nicaraguense* for the same natural resources.

For background information *see* REPRODUCTIVE BEHAVIOR in the McGraw-Hill Encyclopedia of Science and Technology. [A. J. RIBBINK]

Bibliography: K. R. McKaye, *Amer. Nat.*, 111: 301–315, 1977; K. R. McKaye and N. McKaye, *Evolution*, 31:674–681, 1978; A. J. Ribbink, *Nature*, 267:243–244, 1977.

Reptilia

Reptiles have been on Earth for at least 280,000,000 years. Even some present-day groups such as the turtles and crocodilians had ancestors 150,000,000–200,000,000 years ago, in the Triassic and Jurassic periods, that were structurally similar to modern representatives. Many intriguing mysteries surround the large reptiles, for example, the sudden disappearance of the dinosaurs, ichthyosaurs, and pteranosaurs at the end of the Mesozoic Era. Recently, some researchers have questioned how the control of body temperature was achieved by purportedly cold-blooded animals of enormous size and, in some instances, high metabolism and agility. A current theory proposes that dinosaurs were, in fact, warm-blooded animals, and maintained high body temperatures in a manner similar to birds and mammals! One aspect of testing this theory entails research on thermoregulation of living reptiles. Recent advances in electronic monitoring techniques may allow scientists to determine how large reptiles of today control body temperature (see illustration).

Alligators. All modern reptiles, including the closest living relatives of the dinosaurs, the crocodilians, are ectotherms, that is, they are dependent on external sources for control of body temperature. Thorough studies carried out with small lizards indicate that an individual can exercise control of its internal temperature by proportioning the amount of time spent in the shade or sun. However, some intriguing ecological and physiological findings about the ability of reptiles to control their body temperatures are being discovered in turtles and crocodilians.

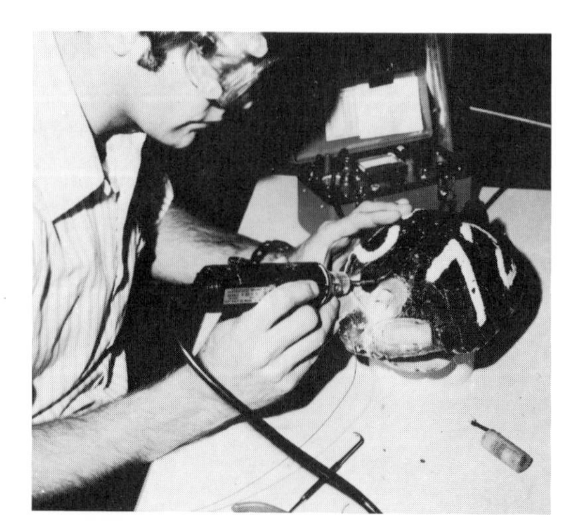

Electronic sensors are placed on the carapace of a turtle to monitor surface and internal temperatures. Radio signals transmit information to a receiving station.

In a large reservoir in the southeastern United States in which one arm is being warmed by effluent from a nuclear reactor, American alligators exhibit unusual behavior. During the winter, when the alligators normally hibernate, adult males in the population instead traveled to the heated portion of the lake. Thus, by taking advantage of an unusual situation, they were able to remain active and feed during the time of the year when they would normally be dormant. Why the adult males, rather than females or juveniles, display this type of behavior remains unknown at this time.

Subsequent studies have attempted to explain how heat transfer can occur efficiently in these large cold-blooded reptiles. Electronic sensors have been used to investigate how the alligator uses basking in the sun to control its temperature in a natural situation. The responses of a free-ranging alligator can be monitored with the use of special probes attached at selected regions of the body. Each sensor probe monitors a particular physiological or environmental variable. Information on rate of heartbeat, surface or internal body temperature, depth of the alligator in the water, and other variables can then be telemetered to researchers more than a mile (1.6 kilometers) away. This research has indicated that a large cold-blooded animal such as an alligator has remarkable control over its body temperature by using strictly external means.

Although ectotherms rely on thermal conditions in the environment to regulate their body temperatures, recent physiological experiments suggest that internal mechanisms involving the circulatory system may also play an important role in large individuals. Adult alligators, for example, can apparently redistribute stored heat throughout the body by means of the vascular system. This allows for the most energetically efficient use (or dispensation) of available (or excessive) heat. Smaller reptiles, including young alligators, are more responsive to their immediate thermal environment, and thus rely on behavioral, rather than physiological, thermoregulation in maintaining body temperatures within the critical range.

Turtles. Turtles are perhaps the prime example of sun-basking types among cold-blooded reptiles. The purpose of this behavior is most easily explained as a means of raising body temperature. However, scientists have also proposed other functions, such as drying up of external parasites and algae, maintenance of a vitamin requirement, or simply resting. The use of long-range radio telemetry techniques can now address the question of why turtles bask.

The mechanisms and complexities of thermoregulation in turtles are being studied in aquatic areas that receive thermally elevated effluent. If turtles merely bask to warm themselves on a cool day, then those living in warm-water areas should be able to obtain the heat they need without leaving the water. Research with turtles carrying temperature-sensitive probes that transmit radio signals suggests this to be true. Experimental animals in the warm portion of one particular cooling reservoir do not leave the water even on cool sunny days. However, those in the cool-water portion of the lake routinely bask under the same environmental conditions. In addition, surprising evidence was obtained to further support the hypothesis that turtles consciously thermoregulate by other behavioral adjustments to external thermal environments. For example, when conditions in the heated reservoir became too warm (several degrees warmer than a natural situation), the turtles burrowed into the mud where temperatures were much cooler. Thus they maintained their bodies at optimal temperature (around 28°C) even under artificial conditions.

Continuing research has also revealed that turtles can precisely control body temperature by adjusting for water temperature, amount and intensity of sunlight, wind, relative humidity, and other environmental factors. Hence, in natural environments turtles maintain their body temperatures at optimal levels during the warmer months by varying the amount of basking time.

Summary. The concept of warm-blooded dinosaurs is intriguing. However, work done to date with the large modern reptiles, such as turtles and alligators, which have finely tuned thermoregulatory control does not support the concept. This control is accomplished by proper utilization of microclimatological conditions in conjunction with certain physiological capabilities which are relatively unexplored as yet. Recent advances in biotelemetry have allowed the scientist to probe such biological mysteries more completely than in the past. Continued ecological and physiological research using biotelemetry should lead to other exciting discoveries.

For background information *see* CROCODILE; DINOSAUR; REPTILIA; THERMOREGULATION; TURTLE in the McGraw-Hill Encyclopedia of Science and Technology. [J. WHITFIELD GIBBONS]

Bibliography: J. A. Desmond, *The Hot-blooded Dinosaurs*, 1976; R. E. Gatten, Jr., *Copeia*, (4): 912–917, 1974; J. R. Spotila et al., *A Mathematical Model for Body Temperatures of Large Reptiles: Implications for Dinosaur Ecology*, 1973; E. A. Standora, in F. M. Long (ed.), *Proceedings of the 1st International Conference on Wildlife Biotelemetry*, Laramie, WY, 1977.

Respiratory system disorders

The respiratory distress syndrome (RDS) of the newborn infant (also known as hyaline membrane disease) is the most prevalent cause of death and disability in premature infants in the United States today. Over the years many theories have been proposed regarding the pathogenesis of this disorder, but only one has withstood the test of time. This theory holds that the disease is due to the deficiency of pulmonary surfactant, the phospholipid-protein complex that lines the alveoli, or small air sacs, of the lungs and prevents alveolar collapse due to surface tension. In recent years there have been tremendous advances in knowledge of the biochemical structure of surfactant, its development in the fetus, and the regulation of its synthesis and secretion by hormones. These advances will be discussed below.

Clinical features. RDS is seen almost exclusively in premature infants. These babies exhibit difficulty in breathing, grunting, and blue skin color (cyanosis) in the first few hours of life. Chest x-rays reveal a characteristic pattern consisting of generalized opacity of the lungs and consequent obscuring of the heart border. Because the lungs are radiopaque, the air in the bronchi stands out as radiolucent areas known as air bronchograms. The disease characteristically progresses for about 72 hr. For reasons not entirely clear, recovery then starts to occur and, in uncomplicated cases, is usually complete by 7 to 10 days of life. In babies who die, autopsy reveals a characteristic pattern of generalized atelectasis or collapse of the lungs, and the alveoli are found to be lined by proteinaceous material known as hyaline membranes. This clinical picture is characteristic of RDS, and the only other condition which exhibits some of these features is infection with the group B streptococcus. A related syndrome, termed shock lung, occurs in adults whose surfactant production may have been impaired by one of a number of pathological conditions.

Pathogenesis. The basic problem in these infants is believed to be deficiency of pulmonary surfactant on a developmental basis. In human fetuses, synthesis of pulmonary surfactant increases markedly at about 35 weeks of gestation, and infants born prior to that time are at risk for surfactant deficiency and subsequent development of RDS. As a result of this surfactant deficiency, there is increased surface tension at the air-alveolar interface and collapse of the alveoli of the lungs. This results in inadequate ventilation and oxygenation of the blood, so that the infants become hypoxic and acidemic. In turn, hypoxia increases the permeability of the pulmonary capillaries so that there is a seepage of fluid and protein from these blood vessels into the alveolar spaces. Once in the alveoli, this material coalesces to form the hyaline membranes characteristic of this disease. The presence of hyaline membranes is therefore a consequence and not a cause of this disorder. Another effect of the hypoxia and acidemia is constriction of the pulmonary blood vessels. This results in decreased blood flow to the lungs and shunting of blood from the right side of the heart to the left without passage through the lungs. A vicious cycle is then established as the synthesis of surfactant is inhibited by hypoxia and acidemia. As the baby's condition worsens, the synthesis of surfactant may be further inhibited, until in some cases death may eventually ensue.

Hypoxia or acidemia occurring during or immediately after delivery may also aggravate the situation and possibly precipitate the development of RDS in infants with a marginal ability to synthesize surfactant.

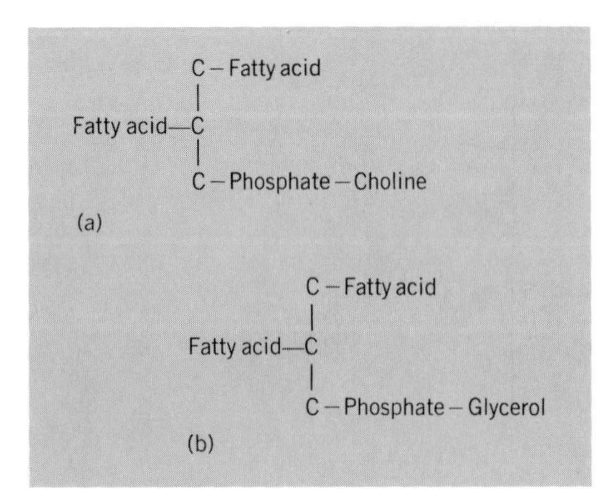

Structure of the major surfactant phospholipids. *(a)* Phosphatidylcholine. *(b)* Phosphatidylglycerol.

Therapy. Therapy focuses on maintaining oxygenation and keeping the airways open until the lungs begin to produce surfactant again. This is usually achieved by the administration of oxygen under continuous pressure. The application of pressure to the airways is believed to hold the alveoli open and thereby promote oxygenation. In some instances, mechanical ventilation is necessary to maintain gas exchange.

Surfactant system. Pulmonary surfactant is thought to be a lipid-protein complex. The function of the protein component of surfactant is not clear at present, but the lipid complex, which consists primarily of phospholipids, is believed to play a major role in lowering surface tension at the air-alveolar interface. The major phospholipid components of surfactant are phosphatidylcholine (lecithin) and phosphatidylglycerol (see illustration). About 80% of the phospholipid in pulmonary surfactant (obtained from animals by washing out the airways) is phosphatidylcholine and about 10% is phosphatidylglycerol. The phosphatidylcholine found in surfactant is unusual in that the two fatty acids which it contains tend to be saturated palmitic acid molecules. This disaturated species of phosphatidylcholine is uncommon in nature and is thought to be instrumental in lowering surface tension. Other phospholipids found in pulmonary surfactant include phosphatidylethanolamine, phosphatidylinositol, and phosphatidylserine.

Surfactant is synthesized by a large cuboidal alveolar cell, the alveolar type II cell. These cells are characterized by the presence of lamellar bodies in the cytoplasm. Lamellar bodies consist of concentric phospholipid layers wrapped around one another. The surfactant phospholipids are synthesized by enzymes located in the microsomal and soluble cytoplasmic fractions of the cell. The newly synthesized lipids are transported to the lamellar bodies, which are the storage sites of surfactant. The lamellar bodies are then secreted onto the cell surface by a process of exocytosis, they unravel to form tubular myelin, and this then lines the surface of the alveolar cells.

In humans and most animal species studied, surfactant production is present at a very low level until about 90% of gestation has elapsed. Then there is a marked surge in the production of surfactant, in the rabbit at about 27 days (term = 31 days), in the rat at 20 days (term = 22 days), and in humans at about 35 weeks (term = 40 weeks). The absence of large amounts of surfactant until this time is one of the reasons why animals born prematurely may not survive.

The regulation of surfactant production during pregnancy is thought to be under hormonal control. Hormones which have been shown thus far to stimulate surfactant production include corticosteroids, thyroxine, cyclic adenosinemonophosphate (cyclic AMP), prolactin, and estrogen. In addition, it has been shown that administration of thyrotropin-releasing hormone (TRH) to pregnant rabbits stimulates fetal surfactant production. TRH may act by stimulating thyroxine secretion. Administration of theophylline to pregnant rabbits stimulates fetal lung surfactant production because of its action as an inhibitor of cyclic AMP phosphodiesterase.

Maternal stress has also been shown to be an important stimulator of lung surfactant production. Thus, intrauterine infection, placental bleeding, insufficient placental blood flow, and prolonged rupture of the membranes surrounding the fetus have all been associated with increased surfactant production and a low incidence of RDS in the newborn infant. These stress factors may act by stimulating the release of hormones such as corticosteroids and thyroxine.

The secretion of surfactant appears to be under hormonal and nervous control. Agents which are related to autonomic nervous system function, such as epinephrine and pilocarpine, have been shown to be involved in this process.

RDS prevention. Recent developments have provided two approaches to the prevention of RDS. The first approach is the prediction of lung maturity prior to birth, so that infants with immature lungs are not delivered electively by obstetricians. The second approach involves the therapeutic acceleration of lung maturation by hormone administration.

L. Gluck and associates have shown that there is a good correlation between fetal lung maturity and the ratio of lecithin to sphingomyelin in the amniotic fluid. The fetal lung secretes surface-active phospholipids into the amniotic fluid via the trachea, and measurement of phospholipids in the amniotic fluid gives some indication of the state of phospholipid synthesis in the lung. It has been found that when the lecithin-to-sphingomyelin ratio is greater than two, the risk of the infant developing RDS is negligible. Thus, in situations where premature delivery is planned, for example, because of maternal disease, the lecithin-to-sphingomyelin ratio is examined. If the ratio is shown to be less than two, delivery is delayed until lung maturity is reached. Widespread use of this technique should decrease the incidence of RDS, as some studies have shown that up to 15% of cases occur after elective induction of labor or caesarian section.

As discussed earlier, many agents have been shown to accelerate lung maturation in animals. To date, only corticosteroids and thyroxine have been used clinically to prevent RDS in humans. There are now a number of studies which demonstrate that the prenatal administration of corticosteroids to mothers can result in a lower incidence of the syndrome in newborns. In order for corticosteroid administration to be effective, however, there must be at least a 24-hr delay between the time of administration of the hormone and the delivery of the baby. This is due to the fact that corticosteroids act by enzyme induction, and this process requires a certain amount of time. G. C. Liggins and associates in New Zealand were able to show in a large clinical trial that the incidence of RDS was 11% in infants born of pregnancies of less than 34 weeks' gestation and in which labor had been delayed for 24 hr to 7 days after corticosteroid administration, as opposed to a 40.2% incidence in infants not so treated.

Accordingly, there seems to be little doubt that corticosteroids are effective in accelerating lung maturation in humans. The major concern at this time is whether corticosteroid administration will adversely affect the structure or function of other organs of the body, particularly the brain. There is considerable evidence from animal studies that administration of relatively large doses of corticosteroids impairs brain cell division in the newborn and may result in a permanent reduction in brain cell number. Whether these data are applicable to humans has not yet been established, particularly since much lower dosages have been used in clinical trials. Studies are currently under way to evaluate the neurologic and developmental outcome in premature infants who have received prenatal corticosteroid treatments. If it can be shown that corticosteroids are nontoxic to the human fetus, this could be a major breakthrough in the prevention of RDS.

Physicians in Israel have recently demonstrated that instillation of thyroxine into the amniotic fluid also results in a decreased incidence of RDS in prematurely born infants. Less clinical information is currently available, however, on the efficacy or toxicity of thyroxine in the developing fetus.

Outlook. Recent developments in the understanding of the pathogenesis of RDS and of the influence of hormones on lung maturation make it likely that, in the not too distant future, a combination of the judicious use of the lecithin/sphingomyelin ratio and of hormone treatment will result in a significant lowering of the incidence of RDS in infants born to women who receive adequate prenatal care. The ultimate aim would be the prevention of premature labor itself, as RDS is essentially a complication of premature birth.

For background information *see* HORMONE, ADRENAL CORTEX; RESPIRATORY SYSTEM DISORDERS; THYROXINE in the McGraw-Hill Encyclopedia of Science and Technology. [IAN GROSS]

Bibliography: P. M. Farrell and M. E. Avery, *Amer. Rev. Resp. Dis.*, 111:657–688, 1975; I. Gross, *Fed. Proc.*, 36:2665–2669, 1977; L. M. G. van Golde, *Amer. Rev. Resp. Dis.*, 114:977–1000, 1976.

Semiconductor devices

In general, any electronic system has three functions: (1) data collection, (2) logic or signal processing, and (3) information output or power conversion and control. Since the 1940s the electronics industry has been under the revolutionary influence of semiconductor technology. The basis of this revolution was the invention of the junction transistor in 1948 and the subsequent development of single-crystal silicon technology used in device fabrication. These developments and those that followed paved the way for today's sensor and integrated circuits technology and power semiconductor technology. This article discusses the principles of operation of power semiconductor devices and the status and trends of their electronic capabilities.

Principles of operation. For the purpose of this discussion, power semiconductor devices are defined as devices which are applied to the conversion and control of power and which are capable of an absolute average current of at least 1A. These devices include the rectifier, transistor, and thyristor; the schematic structure and main terminal electrical characteristics of these devices are

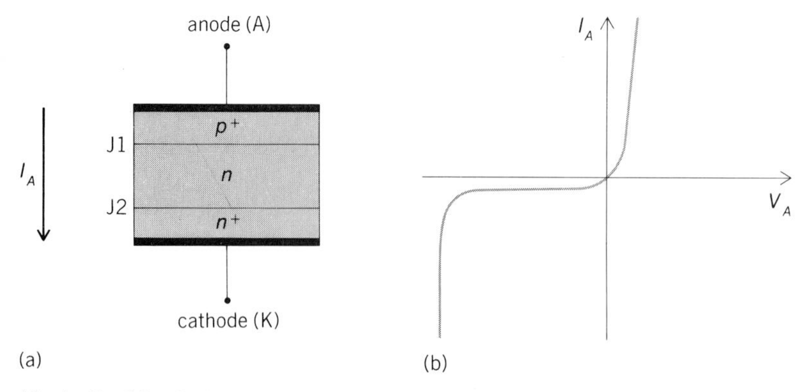

Fig. 1. Rectifier diode. *(a)* Structure. *(b)* Current-voltage characteristic.

shown in Figs. 1, 2, and 3. (Current is designated by the symbol I and its direction by an arrow.)

Rectifier. The rectifier, the simplest of the power semiconductor devices, is a two-terminal device having a high impedance to current flow in one direction and a low impedance in the opposite direction. This characteristic is obtained through the formation of a *pn* junction between opposite conductivity types of semiconductor, as shown in Fig. 1a. This *pn* junction (designated J1) acts as a barrier to current flow in one direction and injects current carriers across it in the opposite direction. In the latter case, the current carriers are collected by the n^+ contact layer.

The high-impedance state is characterized by a reverse blocking voltage capability, that is, the maximum voltage which can be safely applied without significant current flow, and is primarily determined by the physical parameters of the middle *n* region. This *n* region, by virtue of its charge storage capability, is also a major factor in determining the rectifier's power dissipation, which characterizes the low-impedance state. A rectifier's power limit is generally specified in terms of its average forward current capability as a function of maximum junction temperature and the applied heat sink.

Since the rectifier is a two-terminal device, it does not function as a power control element. However, its highly asymmetrical current-voltage characteristic (Fig. 1*b*) is ideally suited to ac to dc power conversion applications.

Transistor. This is a three-terminal device which is used as a discrete or integrated circuit in logic or signal-processing applications. However, bipolar transistors (both *npn* and *pnp* types) also find major use in the output stages of amplifiers, and are currently the focus of a major device development effort aimed at power-switching applications for motor control and power supplies.

The *npn* transistor shown in Fig. 2*a* consists of a collector junction J1 and an emitter junction J2. When a negative voltage is applied to the collector with respect to the emitter, J2 is reverse-biased and the transistor is always in a high-impedance state, similar to that described above for the rectifier. When a positive voltage is applied to the collector, the transistor has a variable impedance which is a function of the signal applied to the base terminal. Referring to Fig. 2*b*, point *a* on the current-voltage curve illustrates a high-impedance state when zero or negative voltage is applied to the base with respect to the emitter. With positive base voltage, however, a positive current I_{B1} will flow from base to emitter, resulting in a significantly larger increase in collector current. This collector current is a function of I_{B1} and the transistor's current gain, which in practice is greater than unity and can be greater than 1000. An important feature of the transistor is that it responds dynamically to the signal applied to the base, that is, removal of the base signal causes the transistor to revert to its high-impedance state. In fact, application of a negative base signal drives the transistor into high impedance in a shorter period of time. The low-impedance point *c* is a transistor-limited situation where the transistor is in "saturation" condition and the voltage drop across the device is composed of the sum of the collector and base drops and the difference in the voltage drops across J1 and J2. The current-voltage characteristic of the transistor can then swing along the load line *a-c* as determined by the base current. This type of operation is typical in the case where a transistor acts as a signal amplifier.

Of primary interest here, however, is the power-switching function of the transistor. In this case, the transistor is made to switch abruptly from the high-impedance (point *a*) to the low-impedance (point *c*) state and in reverse for the purpose of controlling power to a load. The saturation voltage (point *c*), switching speeds (both on and off), and blocking voltage capability are extremely important since they determine the power versus frequency limitations of the transistor. These characteristics are all functions of the three-dimensional physical parameters of the emitter, base, and collector regions of the transistor.

Thyristor. This is the generic term used to describe a class of devices which utilize a *pnpn* structure to switch from a high- to a low-impedance state. The simplest of the thyristors is the controlled rectifier, shown in Fig. 3*a*. The semiconductor material used in the fabrication of this device is predominantly silicon and results in a silicon controlled rectifier (SCR). The SCR and its companion silicon rectifier have the ability to handle large blocks of power at minimum cost per kilowatt.

The controlled rectifier exhibits a high impedance when either a negative or positive voltage is applied to the anode with zero gate voltage. However, with a positive anode voltage and sufficient

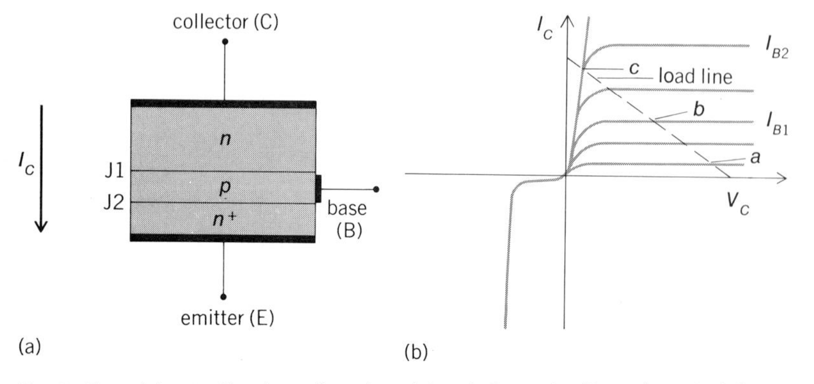

Fig. 2. Transistor. *(a)* Structure of *npn* transistor. *(b)* Current-voltage characteristic.

positive current applied to the gate, the device abruptly switches to a low-impedance state, that is, from point *a* to point *b* in Fig. 3*b*, to a current determined by the load of the circuit. Unlike the transistor, however, the controlled rectifier remains in the low-impedance state even if the gate current is removed. However, if the anode voltage is removed or switched to a negative value for a period of time equivalent to or greater than the turn-off time, the device reverts to the high-impedance state.

The operation of the controlled rectifier is best understood through the use of its two-transistor analog, shown in Fig. 4. In essence, the device consists of two transistors; a *pnp* and an *npn* are interconnected to form a regenerative feedback pair in which the base of each transistor is driven by the other transistor. In all bipolar transistors the current gains are a function of the emitter and emitter-collector current. The controlled rectifier utilizes this characteristic and is designed such that (1) with positive anode voltage and zero gate bias, the loop gain of the regenerative transistor pair is less than unity, resulting in the high-impedance (OFF) state; and (2) with positive anode voltage and a finite gate current, the loop gain goes to unity, each transistor is driven into saturation, and the controlled rectifier switches to a low-impedance (ON) state in a finite time called the turn-on time. In the ON state the forward voltage drop is the sum of the voltage drops across the *n* and *p* bases plus the algebraic sum of the junction drops. As in the case of the transistor, the limits of applicability of the controlled rectifier are determined by its turn-off time, blocking voltage capability, and ON-state voltage drop.

The thyristor family consists of other types of devices which can be triggered on by other means such as voltage, light temperature, and rate of rise of anode voltage. Functional integration has also occurred in power electronics: integrated pairs of antiparallel thyristors (triacs) have become industry standards for full-wave ac power control for lighting applications, and an integrated thyristor-rectifier pair has become accepted in traction drive applications.

For high-frequency applications the controlled rectifier and transistor technologies have been combined to produce a gate turn-off thyristor (GTO) which can be turned off as well as on by means of the gate.

Semiconductor materials and processes. Silicon is used almost exclusively today in the fabrication of power devices, a trend which is expected to continue in the foreseeable future. For the fabrication of one or many devices per wafer, 3-in. (7.62-cm) float zone silicon is available now in production quantities. Similarly, 4-in. (10.16-cm) silicon wafers will begin to be used in 1979 with 6-in. (15.24-cm) wafers expected to be in production by 1982. Neutron-transmutated silicon to achieve phosphorus-doped host wafers is being used for high-voltage (greater than 2000 V) devices and is expected to be used almost exclusively by 1980. This technology results in higher-performance power devices at lower cost due to improved process yields.

Diffusion and epitaxial deposition of silicon are the primary processes for achieving transistor and

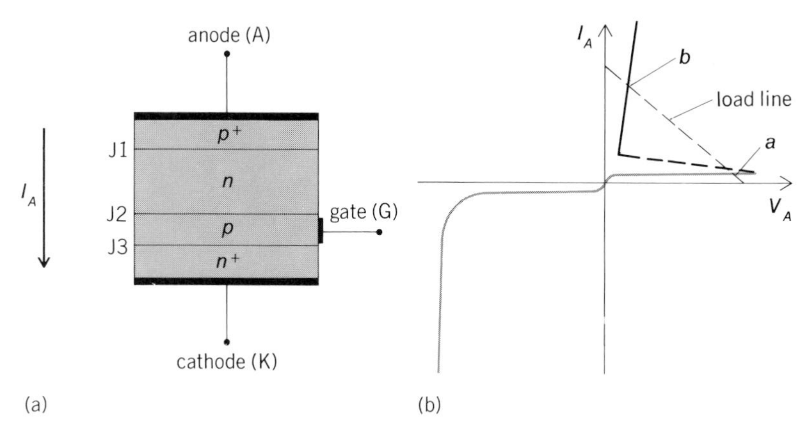

Fig. 3. Controlled rectifier. *(a)* Structure. *(b)* Current-voltage characteristic.

thyristor structures. Although high-performance rectifiers demand this processing, many of the low-voltage economy rectifiers will continue to be fabricated by alloy technology.

Packaging. Plastic encapsulation with soft-soldered assembly is the predominant form of packaging in low-current (less than 10 A) devices. In high-current (greater than 50 A) devices, glass or ceramic to metal seal technology is used for hermetic encapsulation. In these high-current devices, however, soft solders, high-temperature alloys, or pressure are used for contact to the silicon structure depending upon the cost/performance requirements.

The plastic and epoxy package has begun to be employed in high-current devices. This or some alternate form of low-cost packaging is expected to be a major technology change in the next 5 years. In addition, the cost pressures in specific medium-power applications such as motor control will continue present trends toward functional power hybrid circuits fabricated with multiple semiconductor chips in low-cost packages.

Electronic capabilities. In order to describe the electronic capabilities of power semiconductor devices, one must consider the maximum current and voltage capabilities, rather than simply power-handling capability. Since maximum frequency

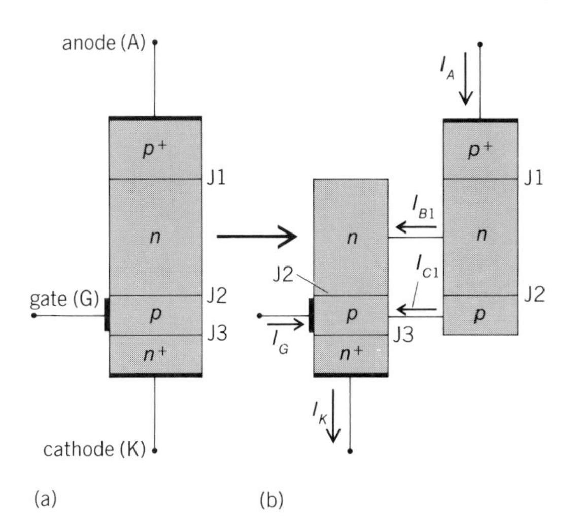

Fig. 4. Two-transistor analog of a controlled rectifier. *(a)* Structure of controlled rectifier. *(b)* Analog by means of *npn* and *pnp* transistors.

capability is the third key performance characteristic, the description of the power device capability in terms of a three-dimensional matrix appears necessary. Fortunately, the picture can be simplified greatly since the demand for power devices falls into three major frequency bands of operation: 50–60-Hz rectification and phase control, 1–2-kHz inverters, and above-20-kHz inverters or chopper applications. Inverter operation between 2 and 20 kHz is generally not considered since this results in audio noise, and major reduction in size, weight, and cost of reactive system components does not occur below 20 kHz. Since future device demands also fall primarily into these three frequency bands, the current and voltage capabilities of rectifiers, transistors, and thyristors will be described relevant to these frequency bands.

Rectifiers. Because of the fundamental differences between rectifier and thyristor or transistor operation, the current-voltage capability of rectifiers always exceeds that of thyristors or transistors. Today, rectifiers capable of handling an average current of 1500 A at a working voltage of 4500 V or a current of 6000 A at 600 V are available. Further penetration into higher powers are limited by application demands and availability of power control devices rather than by rectifier technology. A major demand, however, is for fast recovery rectifiers or Schottky diodes for power-switching applications above 20 kHz. Power Schottky diodes rated up to 50 V and 100 A are presently available but require major improvements in cost and reliability.

Transistors and thyristors. Since transistors and thyristors compete with each other in the current, voltage, and frequency spectrum, their electronic capabilities are best considered together. For 50–60-Hz applications the thyristor has a distinct advantage. SCRs capable of handling 1200 A at 5000 V or 1500 A at 3500 V are available primarily for high-voltage dc transmission applications. For lighting and motor control applications triacs are available with currents up to 100 A and voltages to 1200 V. These technologies are well developed, and future high-power improvements are demand-limited.

In 1–2-kHz inverter applications both thyristors and transistors are used. At 500 V and below, the transistor dominates due to its ability to be switched off by means of the gate rather than by anode commutation as in the case of the SCR. Today 500-A transistors are available for 250-V applications and are being extended to higher voltages. In the next 5 years such transistors will replace SCRs below 500 V. Above 1200 V, however, the SCR is expected to continue to dominate the 1–2-kHz inverter applications. SCRs capable of handling 2000 A at 2400 V are available today, and this capacity should be sufficient for the next few years.

Above 20 kHz the transistor has a distinct advantage over the SCR. In fact, no conventional SCR exists or is expected to be developed which can operate practically above 20 kHz. In contrast, 400-V 100-A switching transistors are available today, with 300-A transistors expected by 1980. The higher voltage applications are expected to be served by GTOs; 800–1200-V GTOs capable of handling 100–200 A have already been developed.

Summary. SCRs will dominate the low-frequency high-voltage applications; transistors will serve the high-frequency low-voltage applications; and GTOs will be developed to serve the high-voltage high-frequency end of the spectrum. Finally, integrated circuit technology is expected to impact power control and conversion devices only in the very-low-voltage, low-current, and very-high-frequency end of the spectrum where vertical metal oxide semiconductor (MOS) devices apply.

For background information *see* CONTROLLED RECTIFIER; RECTIFIER; SEMICONDUCTOR RECTIFIER; TRANSISTOR in the McGraw-Hill Encyclopedia of Science and Technology.

[RICHARD A. KOKOSA]

Bibliography: A. Blicher, *Thyristor Physics*, 1976; F. Gentry et al., *Semiconductor Controlled Rectifiers*, 1964; S. Ghandi, *Semiconductor Power Devices*, 1977; R. Kokosa and D. Muss, *IEEE Trans. Electron. Devices*, vol. ED-23, no. 8, August 1976.

Silicon

Crystalline silicon used in the semiconductor and solar cell industry is expensive to produce. Past attempts to solve this problem by substituting amorphous silicon (a-Si), an inherently less expensive, noncrystalline material, were unsuccessful; the deposition techniques (evaporation or sputtering) produced films with a high density of defects. A new deposition technique is now being used to produce better amorphous silicon. It involves the glow discharge of silane (SiH_4) and produces films with four orders of magnitude fewer defects than in previous films. As a result, amorphous silicon has been recently used to fabricate thin-film solar cells, with efficiencies up to 5.5%. In addition, this material has allowed measurements on transport and doping properties of tetrahedrally bonded amorphous semiconductors that had previously been impossible. The basic reason for this im-

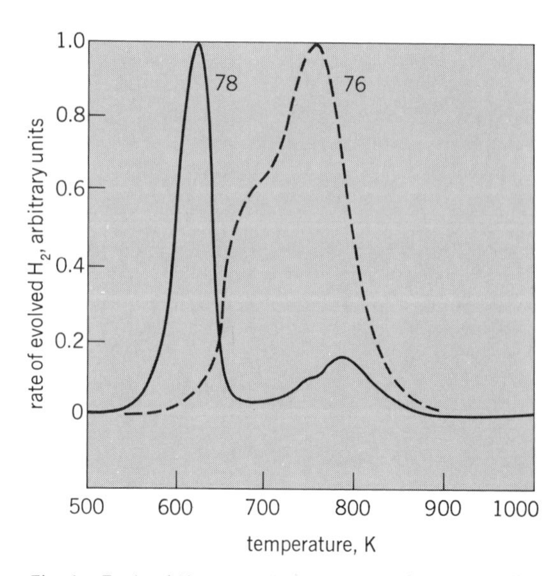

Fig. 1. Evolved H_2 versus temperature of two glow-discharge a-Si:H samples prepared at different substrate temperatures: sample 78 at room temperature and sample 76 at 250°C. *(From M. H. Brodsky et al., Quantitative analysis of hydrogen in glow discharge amorphous silicon, J. Appl. Phys., 30:561–563, 1977)*

provement is now believed to be due to hydrogen incorporated into the films during the deposition process. It satisfies dangling silicon bonds that would otherwise act as defects. The concentration of hydrogen is extremely high (up to 35 at. % in some cases). This is essentially a new material, silicon-hydrogen alloy (a-Si:H), which has extremely useful and interesting properties.

Fabrication. Films of a-Si:H are deposited on substrates placed in a glow-discharge plasma of SiH_4. The discharge can be sustained by direct-current or radio-frequency power. In either case, the applied energy must be sufficient to ionize the silane, allowing atoms of both silicon and hydrogen to collect on the substrate. The conditions of the discharge are important; variations in discharge power and pressure, gas flow rate, amount of argon dilution gas, and deposition rate can all change film properties and composition. This means that there is no single a-Si:H material, but a family of materials that depend on deposition parameters. This factor can be very important in preparing devices.

Substrate temperature is one of the most important deposition parameters. Films produced on room-temperature substrates always contain a high density of defects even under the best of conditions. Only at substrate temperatures above 200°C can high-quality films be made.

Doping is accomplished by adding appropriate gases to the silane. W. Spear and coworkers have carried out an extensive study of the two best dopants using this method, B_2H_6 (for *p* type) and PH_3 (for *n* type). J. C. Knights, T. M. Hayes, and J. C. Mikelsen, Jr., have recently studied *n*-type material doped with arsenic. They found from analysis of the extended x-ray absorption fine structure, the first direct evidence for substitutional doping in an amorphous semiconductor.

Role of hydrogen. One of the most important recent advances in this material during the last few years relates to the hydrogen in the films. Its presence in concentrations of 10 to 35 at. % has been positively established by mass spectroscopy, infrared absorption spectra, and evolution of H_2 during annealing. The infrared spectra is due to various silicon-hydrogen arrangements, and its absorption strength suggests that a large fraction, if not all, of the hydrogen is bonded to silicon. M. H. Brodsky and coworkers have identified at least two bonding types: a dihydride bond (SiH_2) and a monohydride bond (SiH). The former is primarily found in films made at room temperature, the latter in films made at higher temperatures. Hydrogen bound in either way can be evolved from the films by annealing at different rates. Figure 1 shows that the dihydride-bonded hydrogen (in room-temperature films) evolves first.

The hydrogen appears to function as a compensator for dangling bonds. H. Fritzsche found that the spin density increases and the photoluminescence efficiency decreases after hydrogen has been evolved. Both results are consistent with dangling bonds left on silicon atoms when the hydrogen is driven off. More recently, J. Pankove found that a dehydrogenated film can be rehydrogenated by exposure to an H_2 discharge, with the result that the dangling bonds become compensated again. Pankove measured the photoluminescence spectra of the a-Si:H at 79 K produced by 50 mW at 488 nm from an argon laser and detected by a cooled PbS photoconductor. Figure 2 shows his results; the photoluminescence decreases upon dehydrogenation and reappears upon rehydrogenation.

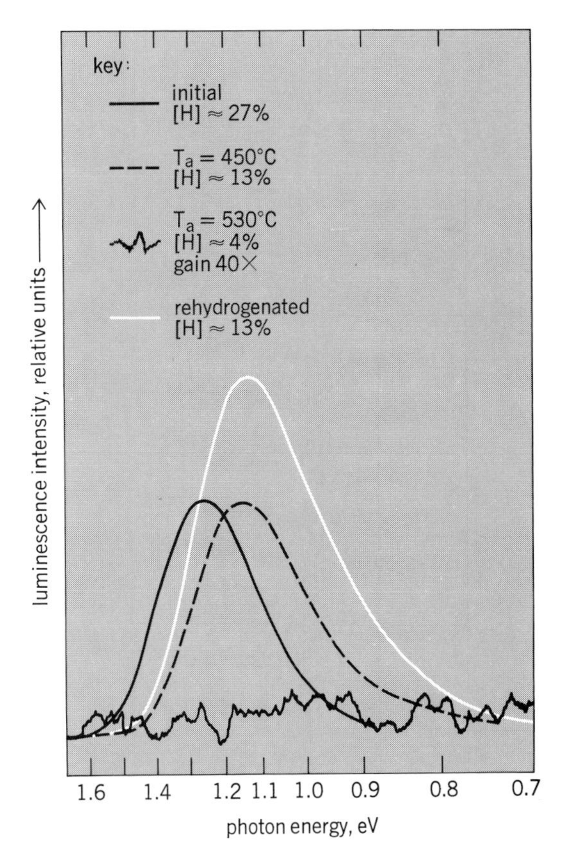

Fig. 2. Photoluminescence spectra of a-Si:H at 79 K. T_a is temperature of annealing treatment. Photoluminescence of curve 3 is plotted on a relative scale that is 40 times smaller than scale used for other curves. All hydrogen contents are given in atomic percent. *(From J. I. Pankove, Photoluminescence recovery in rehydrogenated amorphous silicon, J. Appl. Phys., 32:812–813, 1978)*

Dangling bonds are not the only defect type found in this material. Films prepared at room temperature have a high density of hydrogen, as well as a high density of defects. These films are composed of a silicon-hydrogen polymer rather than an interconnecting network of Si atoms. It is only at the higher substrate temperatures, where only the monohydride spectra are observed, that the benefits of hydrogen compensation are apparent.

Hydrogen also affects the optical properties of a material. In general, the greater the hydrogen concentration, the higher the optical gap. Values from 1.6 to 1.8 eV have been reported.

Conductivity. Recent conductivity studies on a-Si:H have been carried out primarily on doped samples by W. Spear and associates. Doping increases the conductivity and decreases its activation energy, which is consistent with the Fermi level being pulled toward the appropriate mobility edge. (The mobility edge is the energy level which separates localized from nonlocalized electron states in the conduction, or valence, band tail of

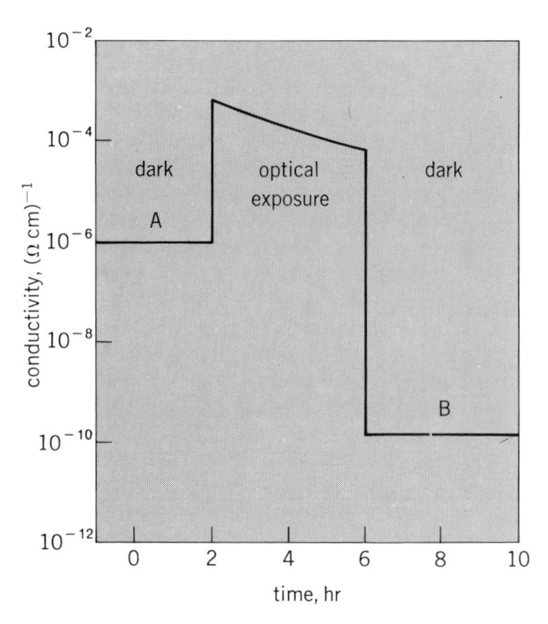

Fig. 3. Conductivity as a function of time of a-Si:H before, during, and after exposure to approximately 200 mW/cm² of light in the wavelength range of 600–900 nm at 25°C. *(From D. L. Staebler and C. R. Wronski, Reversible conductivity changes in discharge-produced amorphous Si, J. Appl. Phys., 31:292–294, 1977)*

states.) At room temperatures, transport is due to carriers above the mobility edges or hopping in shallow traps. Donors introduce a new transport path when their density exceeds about 10^{18} cm^{-3}.

A new conductivity effect in this material was recently reported by D. L. Staebler and C. R. Wronski. They found that the conductivity depends in a reversible way on the thermal and optical histories of the films. This result can explain much of the scatter in values seen in previous work, as shown in Fig. 3. A sample in state A is exposed to approximately 200 mW/cm² of red light. During exposure, photoconductivity slowly decreases. More spectacularly, after exposure the dark conductivity decreases by nearly four orders of magnitude. The new state B is stable at room temperature but can be reversed to the original state A by heating to temperatures above 150°C. The changes in dark conductivity are accompanied by changes in the activation energy. The results thus suggest a simple shift of the Fermi level, but the details of a specific model are not yet known. Only electron transport seems to be affected, and this may explain why such changes do not degrade the efficiency of solar cells during illumination with sunlight.

Solar cells. The first solar cell mode with a-Si:H was reported by D. E. Carlson and Wronski. It was a *pin* structure cell that gave a 2.4% conversion efficiency. More recently, Carlson and Wronski reported on a 5.5%-efficiency Schottky barrier cell. Although this efficiency is acceptable, it is still less than the theoretical maximum estimated for a-Si:H cells (15–17%). The problem is that the material still contains enough defects to limit the collection of holes. Thus the short-circuit current and fill factor are smaller than ideal. Increases in efficiency must come from improvements in the device configuration or improvements in the material, particularly in its hole transport properties. *See* SOLAR CELL.

For background information *see* CRYSTAL DEFECTS; SEMICONDUCTOR; SILICON; SOLAR BATTERY; SYNCHROTRON RADIATION in the McGraw-Hill Encyclopedia of Science and Technology.

[DAVID L. STAEBLER]

Bibliography: D. E. Carlson, *IEEE Trans. Electron Devices*, ED-24:449–453 (1977); H. Fritzsche, C. C. Tsai, and P. Persans, *Solid State Technol.*, 21:55–60, 1978; J. C. Knights, T. M. Hayes, and J. C. Mikelsen, Jr., *Phys. Rev. Lett.*, 39:712–715, 1977; W. E. Spear, *Adv. Phys.*, 26:811–845, 1977.

Social insects

There are both economic costs and benefits associated with sociality. In the aggregation of many individuals to one locality, the need for resources increases linearly with group size, while the availability of resources decreases asymptotically. As a result, there are economic limits on group size and sociality. However, these limits have been subjected to natural selection for improving the mechanisms of resource acquisition and processing.

Foraging mechanisms of social insects, particularly those of ants and bees, have recently been examined closely. These mechanisms have been found to be closely fitted for specific types of resources and their distributions; ants and bees have a variety of evolved behavior patterns or "strategies" that are suited to particular patchiness of resources in both space and time. Foraging can be perceived as optimization of individual foraging, or it can be viewed on a larger scale, where individuals are merely components of the colony foraging strategy.

Foraging optimization. Foraging of many types of animals has recently been analyzed to determine how closely it corresponds to predictions made originally by R. H. MacArthur, E. R. Pianka, and J. M. Emlen. These predictions are that if an animal forages "optimally" it will maximize the intake of some "currency," usually energy, by following simple rules in patterns of movement and choice of prey or food items. In practice, however, most animals do not have perfect knowledge of the potential food resources available so that they can maximize intake by following simple rules. And even if such knowledge were available, potential "optimal foraging" is usually difficult to document in nature because of many intervening variables, such as nutritional needs, satiation, predator avoidance, territoriality, and reproductive activities. For these reasons, optimal foraging has been explored primarily as a theoretical construct.

Studies on bumblebee foraging by B. Heinrich have recently examined the concept of optimal foraging empirically. The foraging worker bees, for all practical purposes, do not engage in predator avoidance, reproductive activity, and territorial defense. Their food intake on any one trip is unaffected by satiation or nutritional needs since food is exchanged in the nest.

Bumblebees are unable to communicate to hivemates the locations of rewarding food, and individuals of a colony act largely independently of each other in the field while foraging. Each attempts to maximize its returns, as individuals compete in a scramble competition where no colony controls or defends any feeding territory.

The major determinant of optimal foraging in the bees is that of assessing the resources, ranking them, and specializing in the most rewarding flower types. Each colony has specialists in different types of flowers, and the range of flower types utilized increases with competition. New foragers emerging from the colony find that the high-nectar flowers have already been depleted by specialists; relatively equal rewards are available in flowers of different rates with nectar or pollen production.

One deviation from optimal foraging is that the bees must sample and visit many unrewarding flowers before specializing. Furthermore, when resources have been depleted they continue to sample, usually not specializing strictly. A second deviation relates to foraging skill. It might be predicted from the MacArthur-Pianka model that when net food rewards available from the different flowers are approximately equal, as during competition, the bees would visit the different kinds of flowers as they come to them, thus minimizing traveling time. Instead, the bees pass over flowers of many kinds, maintaining "major" and "minor" specializations. For example, in an area with jewelweed, aster, and turtlehead flowers, one bee may visit aster primarily and jewelweed secondarily, bypassing all turtlehead. Another individual of the same species at the same time and place may visit turtlehead primarily and aster secondarily, bypassing all jewelweed.

A possible explanation for this behavior is that the real and perceived rewards are not the same due to differences in foraging skill. For example, skill in handling morphologically complex flowers such as jewelweed and turtlehead improves with experience, and inexperienced bees may visit many flowers without finding the rewards. Second, it takes inexperienced foragers longer to extract the nectar and pollen from complex flowers than experienced individuals. An inexperienced forager, given equal potential net rewards in the three types of flowers, would thus find more available in the aster, from which pollen and nectar are relatively easily harvested, than from the jewelweed and turtlehead.

There are other factors that may prevent the bees from harvesting the most rewarding flowers available. The placement of colonies may not allow them to minimize travel time to the most rewarding flowers. Also, both honeybees and bumblebees have a genetic predisposition to seek out certain flower types. For example, these bees generally learn to restrict themselves to blue flowers more rapidly than to white flowers. This may be adaptive in the evolutionary (though not always in the immediate) time scale because many flowers specifically adapted for bee pollination are blue and offer ample nectar and pollen.

The example of the bumblebees shows how the energy balance of the colony is achieved primarily by the combined individual initiative of the workers, where each attempts to find and manipulate appropriately the most rewarding flowers available. Competition for diffusely distributed resources does not involve aggression, being based on scramble competition by means of individual initiative. Energy balance of other social insects, which communicate the location of food resources among colony members, is seen as a community effort where strategy varies with the type and distribution of resources.

Clumped resources. Communication systems are of particular advantage to several insects harvesting clumped food resources, because the whole colony can potentially benefit from one forager's success in locating a rich food resource. The more diffusely the food is distributed, the more advantageous is independent foraging without communication among colony members.

The communication used for recruitment to rich food resources is usually also used in offense and defense. An example recently studied by L. K. Johnson and S. P. Hubbel involves *Trigona*, the stingless bees. Different species coexisting in the American tropics have different recruitment capabilities and aggressive tendencies. Fights lasting several days occurred at natural and artificial rich food resources. Sometimes up to 1800 dead were left at fights over the richest food resources. Johnson and Hubbel analyzed aggression in terms of costs and payoffs. Both the tendency to recruit hive-mates and attack as a group, and the tendency to remain at the food source and fight, depended on resource richness and compactness. Those less aggressive species which were always displaced from rich food resources presumably coexist in the same environment by their greater capacity to discover new food sources and to exploit them before being driven away.

The common imported (to America) honeybee *Apis mellifera* also recruits efficiently by using the symbolic dance language originally decoded by K. von Frisch and associates. There has been some controversy due to the claims of A. M. Wenner that the bees rely only on odor cues to locate food, and that their "dance" actually has no function in recruiting hive-mates. However, recently published experiments by J. L. Gould have put the controversy to rest. Gould has caused bees to "lie" by having them perform a dance indicating a different direction than that of the food source. Normally bees orient their dancing, and the dance interpretation, to gravity in the dark hive. Gravity symbolizes the azimuth direction toward the Sun. However, they will override the gravity stimulus and reorient dances if the Sun, or a light source acting as a "sun," is visible to them. Since both dancers and potential recruits are similarly reoriented by a light, no misdirection results. But Gould covered the ocelli of dancers, which made them six times less sensitive to light than untreated (recruit) bees, and adjusted the intensity of a light used to reorient the bees in the hive so that the dancing bees ignored the light and used gravity for orientation, while the recruits interpreted the dance with respect to the visible light. By moving the light to different angles Gould found that the recruits could be sent off in any direction indicated by the dancers, as would be predicted if they used dance language rather than relying on scent alone.

Gould suggested that the honeybees' behavior evolved as an adaptation in the Old World tropics to allow them to harvest from isolated, massively flowering trees scattered throughout the forest. The communication-recruitment system also allows them to raid other, weaker colonies.

A number of investigators have examined the foraging behavior of seed-eating ants of the deserts in the southwestern United States with the aim of discovering mechanisms of resource partitioning among species by means of different foraging strategies. B. Hölldobler found that *Pogonomyrmex rugosus* and *P. barbatus* recruit by pheromone trails to clumped seed resources. The recruitment trails may become more or less permanent trunk trails leading as far as 40 m from the nest. Foraging territories of these ants are nonoverlapping, being partitioned by the trunk trails. Trunk trails do not cross, and the trails function both to increase accessibility to distant food resources and to spatially partition the available foraging ground among neighboring colonies.

Diffusely distributed resources. Widely distributed resources can potentially require more energy to defend or harvest than they yield. Social insects usually do not defend diffusely distributed resources, and thus various strategies of efficient harvesting, generally involving much individual initiative, have evolved. Individual bumblebees, foraging from dispersed flowers, apparently initiate their search individually at random directions from the nest. Similarly, D. W. Davidson has found that most of the desert seed-eating ants that specialize on dispersed seeds forage individually.

The individually foraging ants tend to be large, perhaps allowing them to range widely about the nest. (Most of those that recruit to clumped resources by pheromone trails are small.) Apparently the foraging for seeds of low densities places a premium on individual initiative for their necessarily independent discovery. Group-foraging ants are more abundant than individually foraging ant species in desert areas having characteristically higher precipitation (associated with greater seed abundance). However, individual foragers continue to be active at times of the year when seed clumps have been depleted by group foragers. Specializations are thought to account for the ability of several seed-eating ant species to coexist.

A possibly unique strategy for harvesting dispersal seeds has been reported independently by R. Bernstein, and by S. W. Rissing and J. Wheeler for the seed-eating ant *Veromessor pergandei.* When confronted with low seed densities these ants leave the nest in columns, dispersing to forage individually from along or near the end of the column. However, rather than establishing quasipermanent trunk trails, successive *Veromessor* columns move in different directions. Successive columns (one or two per day) generally proceed in a roughly clockwise or counterclockwise direction about the nest entrance. Since the columns act to bring the ants to new foraging territory, it is thought that they function in reducing overlap of forager activity on already depleted ground.

Renewable resources. The potential conflict between immediate short-term foraging optimization and long-term energy balance can be great when the resources to be harvested are renewable and can be managed for maximum sustained yield. L. L. Rockwood has shown that *Atta* leaf-cutter ants apparently manage their resources for sustained yield. Colonies of these ants may each contain millions of individuals. The colonies are long-lived, remaining in place for 20 years or more. *Atta* colonies harvest fresh leaves, which are used to grow fungus, the actual food of the ants, in subterranean gardens. In agricultural areas of Central and South America with plant monocultures, a colony may strip a garden bare in a day or so. An ant colony is capable of harvesting several kilograms of leaves per day. But despite the ants' obvious potential to defoliate and kill trees near the nest, they seldom do so in their natural environment that contains many kinds of trees.

Rockwood found that trees were seldom fully defoliated, and the ants harvested preferentially some distance from the nest, with some ants foraging up to 140 m from the nest. Generally one trail continued to lead to a given harvesting area for a long period of time, but other trails switched daily from one source to another. The ants were attracted to novel sources; palatable leaves of any one type were harvested at a high rate only the first 2–3 days after discovery, at which time feeding rates leveled off and another source was attacked. The ants' behaviors function to spread the grazing pressure and thus to produce sustained yield by not killing host plants near the nest by overexploitation.

Controlling space. A food resource can be secured indirectly by controlling a foraging space. Many vertebrate and invertebrate animals are known to repel competitors from their feeding areas or territories. Hölldobler and E. O. Wilson have shown a complex pheromone communication system in the African weaver ants *Oecophylla longinoda* whereby these intensely aggressive ants acquire and control foraging space.

Weaver ants live in forest canopies where they make nests out of leaves bound together with the silk produced by their larvae. The ants of any one colony do not tolerate others in the section of canopy they control. Hölldobler and Wilson found that when scouts find unoccupied territory they recruit nest-mates to it by means of odor trails laid down by a newly discovered "rectal gland." By means of this long-range recruitment, nest-mates arrive at the new territory and claim it by marking it with fecal spots. The fecal material contains pheromone that tends to deter the invasion of alien workers. This pheromone persists at least 12 days.

If enemy ants are not deterred and are found invading an ant colony's territory, the defenders assemble nest-mates into clusters in the disputed area by dispensing another pheromone from another newly discovered pheromone source, the "sternal gland." Additional forces are recruited to the combat area by means of the long-range recruitment system of the rectal gland pheromone. As a result of these recruitment systems, the worker density at the combat area increases, and the intruders are eventually captured and killed.

Hölldobler observed a ritualized fighting behavior without physical combat during territorial border challenges of the desert honeypot ants, *Myrmecocystos mimicus.* When antagonists meet at disputed boundaries, they do not fight but rather engage in stereotyped display patterns with stilt-walking and head-to-head confrontation. Opposing colonies recruit workers to a "tournament area," and when one colony is considered the stronger the tournament ends and the weaker colony is raided and its ants annexed ("enslaved") to the stronger.

Wilson has explored a different defensive strategy at the nest in the territorial ant *Pheidole dentata*. *Pheidole* have small "minor" workers and large "majors" or "soldiers." The primary natural enemies of *Pheidole* are fire ants, principally *Solenopsis geminata*. Wilson has shown that when several *Pheidole* minors find fire ants near their nests, some individuals grapple with the intruders, while others run back to the nest dragging their abdomen and dispensing trail pheromone from the sting. Major workers quickly arrive along the recruitment trail and proceed with their massive mandibles to chop the fire ants, which are held by the minors, to pieces. The majors remain until the last of the *Solenopsis* has been dispatched.

A single fire ant can invoke a massive colony response, and the defensive strategy of *Pheidole* appears to be particularly adapted to destroy *Solenopsis* scouts. If *Solenopsis* scouts are successful and recruitment to a *Pheidole* nest can begin, the invaders have a good chance to kill the adult residents and eat the immatures. However, if large numbers of fire ants attack, the *Pheidole* reduce the fight to the nest periphery and, failing there, they abscond the nest. Other ants, which are not natural enemies of *Pheidole*, do not invoke the massive response that fire ants do.

Conclusions. Social insects have evolved various strategies to optimize long-term energy balance. Some, but not all, of these strategies involve the securing and defense of foraging territories. Mechanisms of harvesting are adapted to specific patterns of spatial and temporal distributions of resources, and mechanisms of offense and defense are sometimes adapted to specific enemies.

For background information *see* SOCIAL INSECTS in the McGraw-Hill Encyclopedia of Science and Technology. [BERND HEINRICH]

Bibliography: D. W. Davidson, *Ecology*, 58: 725–737, 1977; J. L. Gould, *Quart. Rev. Biol.*, 51: 211–244, 1976; B. Heinrich, *Ecol. Monogr.*, 46: 105–128, 1976; B. Hölldobler, *Behav. Ecol. Sociobiol.*, 1:3–44, 1976; B. Hölldobler and E. O. Wilson, *Proc. Nat. Acad. Sci. U.S.A.*, 74:2072–2075, 1977; L. K. Johnson and S. P. Hubbel, *Ecology*, 56:1398–1406, 1975; L. L. Rockwood, *Ecology*, 57:48–61, 1976; E. O. Wilson, *Behav. Ecol. Sociobiol.*, 1:63–81, 1976.

Soil chemistry

Recent research on soils has been concerned with high-gradient magnetic separation of soils, characterization of soil and clay minerals by electron spectroscopy for chemical analysis, and the application of far-infrared analysis to clay minerals.

High-gradient magnetic separation. Magnetic filters with high gradients permit fractionation of weakly magnetic minerals that formerly were considered to be nonmagnetic. Layer silicates, iron oxides, and other minerals in clay and silt fractions of soils can be concentrated by this technique without harsh chemical treatment. Trace minerals can also be concentrated for analysis and study.

Soils are complex mixtures of different minerals (mica, kaolinite, chlorite, goethite, quartz, and so on) inherited from parent rocks, altered or formed by weathering, and added to soils by biological processes and agents of transport (wind, water, ice, and gravity). These minerals contain silicon, aluminum, iron, magnesium, and other electropositive atoms bonded to negative oxygen atoms and hydroxyl (OH) groups. Minerals that contain iron (such as chlorite, mica, and goethite) are weakly magnetic and can be separated from nonmagnetic minerals (such as quartz) by high-gradient magnetic separation (HGMS).

Magnetic filter. To obtain magnetic separation of a particle, a gradient in the field strength must be produced across the particle (Fig. 1). For a particle to be trapped from suspension by the magnetic filter, the magnetic force acting on the particle must be greater than the drag force from the solution flowing through the filter. An electromagnet with a field strength of 1 to 2 tesla (10 to 20 kilogauss) fitted with a filter of fine stainless steel wool provides the necessary conditions. D. G. Schulze and J. B. Dixon employed a filter constructed of plastic tubes containing fine stainless steel wool. The steel wool focuses the magnetic lines of force to create the necessary gradient, as shown in Fig. 1*b*. The magnetic force acting on a particle is greatest when the radius of the ferromagnetic collector is about three times the radius of the particle to be captured. For clay particles, the stainless steel fibers are too large, but the sharp edges and barbs provide smaller radii with more effectiveness as filters than a cylindrical wire approximately equal in radius to the fiber. The magnetic separation is strongly particle-size-dependent. Coarse clay particles (2–0.2 μm) are separated more effectively than fine clay (<0.2 μm) particles. Increasing the time the particles are in the magnetic filter improves fractionation of the fine clay particles.

J. Neuzil and M. Kuzvart used a ridged plate magnetic filter to reduce the total iron content from 3.64 to 2.22% Fe_2O_3 and from 3.37 to 1.40% Fe_2O_3, respectively, in yellow and green kaolinitic deposits of <63-μm particle diameter. They reduced the total Fe_2O_3 content only from 1.11 to 1.04% in a <20-μm white kaolinitic deposit. Neuzil and Kuzvart's results demonstrated the applicability of HGMS to coarser particles where only a moderate magnetic gradient is required. Separation of clay particles requires a steel wool filter, which is more effective than the ridged plates. J.

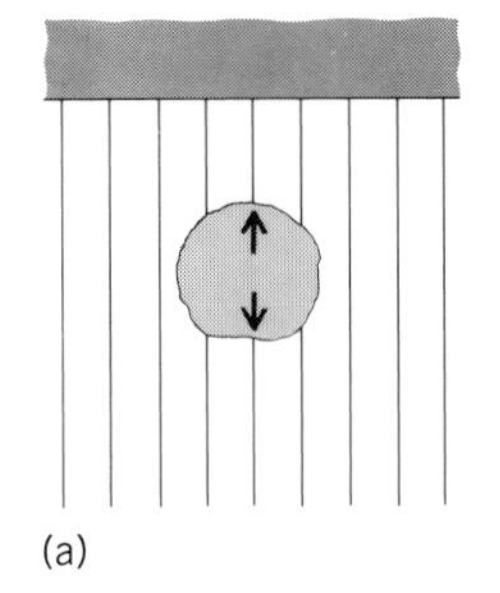

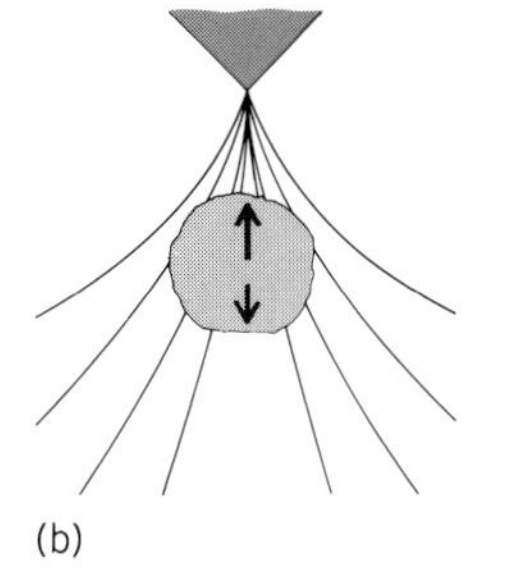

Fig. 1. Influence of *(a)* uniform magnetic field (no net force) and *(b)* nonuniform magnetic field (a net force) on weakly magnetic particles. *(Modified from H. Kolm, J. Oberteuffer, and D. Kelland, High gradient magnetic separation, Sci. Amer., 233:46–54; copyright © 1975 by Scientific American, Inc.; all rights reserved)*

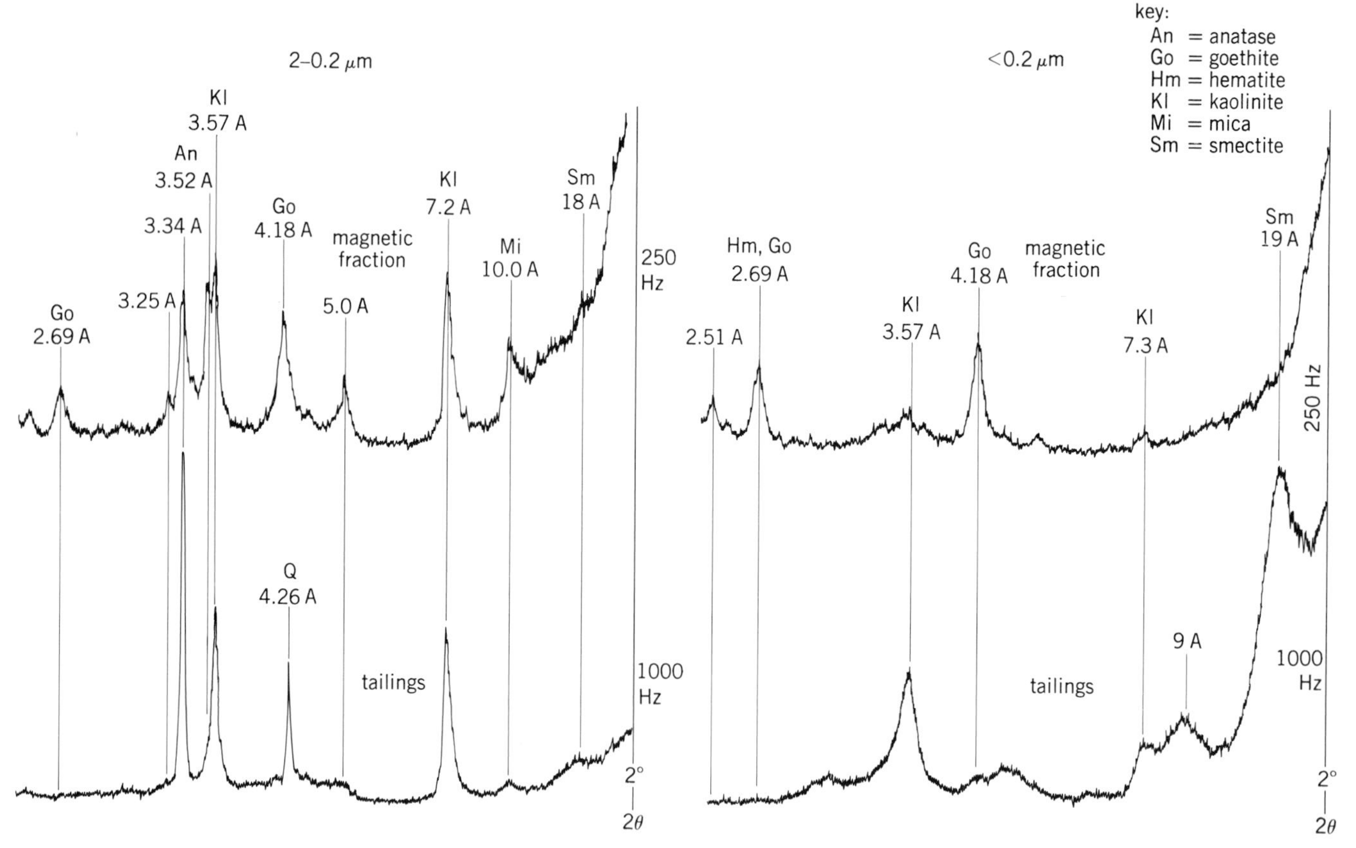

Fig. 2. X-ray diffraction patterns showing concentration of weakly magnetic minerals in the magnetic fraction and of nonmagnetic minerals in the tailings; θ values are converted through Bragg's law to angstrom values. 1 A = 0.1 nm. *(From D. G. Schulze and J. B. Dixon, High gradient magnetic separation of soils, Soil Sci. Soc. Amer. J., publication pending)*

Iannicelli used a stainless steel wool filter with the HGMS method to reduce titanium and iron content of commercial kaolinite of clay size (<2 μm).

Efficiency. The effectiveness of HGMS in fractionating Silawa soil clay is shown by the relative x-ray diffraction peak heights in Fig. 2. In the coarse clay (2–0.2 μm), mica, goethite, anatase, and rutile (or feldspar) are concentrated in the magnetic fraction. Quartz is concentrated in the tailings. Kaolinite is present in both the magnetic and tailing fractions of the coarse clay. In the fine clay (<0.2 μm), goethite, hematite, and anatase are concentrated in the magnetic fraction; smectite is concentrated in the tailings. Kaolinite occurs in the magnetic fraction, but is also present in the nonmagnetic fraction of the fine clay.

Table 1. Free iron oxide minerals in soil clay fractions concentrated by magnetic separation of suspended particles*

Soil	Free Fe_2O_3, % of sample: Original sample	Magnetic fraction	Tailings
	2–0.2-μm size fraction		
Aubrey	6.4	19.4	0.8
Beaumont	0.3	1.6	<0.1
Hiwassee	7.1	20.0	2.9
Ships	2.4	5.7	0.3
Silawa	7.7	27.7	0.6
Wockley	2.1	9.0	0.3
	<0.2-μm size fraction		
Aubrey	6.3	41.4	4.0
Beaumont	1.0	2.8	0.9
Hiwassee	18.2	53.1	12.7
Ships	2.8	13.6	1.9
Silawa	5.2	52.1	3.7
Wockley	2.0	14.6	1.6

*From D. G. Schulze and J. B. Dixon, High gradient magnetic separation of iron oxide minerals from soil clays, *Soil Sci. Soc. Amer. J.*, publication pending.

The yellowish-brown or red color of the magnetic fraction, as compared with the whiter color of the tailings, gives visible evidence of iron oxide removal. The separation of iron oxides from silicates is not complete, however, as shown by the free iron oxide percentages for clays from six soils (Table 1). The magnetic fractions of the coarse clays contain 1.6–27.7% free Fe_2O_3 compared with 0.1–2.9% Fe_2O_3 in the nonmagnetic (tailings) fraction. The fine clay (<0.2 μm) magnetic fractions were 2.8–53.1% free Fe_2O_3, compared with 0.9–12.7% free Fe_2O_3 in the tailings. These percentages indicate the presence of non–iron oxide minerals in the magnetic fraction and some residual iron oxides in the tailings. In spite of incomplete separation, magnetic fractionation enhances x-ray diffraction and other structural data on the magnetic minerals. Chemical studies of reactions of iron oxides with phosphates and nitrates will be more sensitive and easier to interpret when they are conducted on a sample much richer in iron oxides than the whole-soil clay.

Iron chlorite was recently separated by HGMS from a fine silt (5–2 μm) fraction of shale by first separating the magnetic minerals (largely chlorite and mica) and then progressively reducing the field strength for each successive split of the magnetic fraction. The final magnetic fraction was a chloritic sample with small amounts of mica and quartz impurities. Chlorite is a sensitive indicator of acid weathering of shale in mine spoils and in laboratory simulation of mineral weathering during soil formation. Characteristics of new soils formed in mine spoils can be forecast by investigating samples of chlorite and mica from the spoil material. Chlorite is a small but labile component of much overburden, and would be difficult to study without separation because its major x-ray diffraction peaks overlap those of kaolinite. Mica in the spoil may be an important source of potassium for plants, depending on the rate of potassium release.

[JOE B. DIXON]

Electron spectroscopy for chemical analysis. A recent development in soil chemistry is the use of electron spectroscopy for chemical analysis (ESCA) to characterize structural and adsorbed species in soil mineral systems. This nondestructive technique, also known as x-ray photoelectron spectroscopy (XPS), measures the binding energy of core-level electrons within atoms and molecules that compose the outermost surface (5-50 A, or 0.5–5.0 nm) of a sample. By analyzing the binding energy of the electrons, (1) the elemental composition of soil minerals can be determined; (2) the oxidation state, that is, the electronic charge, of an element or ion can be readily identified; and (3) in some cases, changes in coordination environment surrounding an element can be monitored.

Since its inception in 1967, ESCA has been used most widely in studies where the specific chemical composition of very thin surface layers is of critical importance, such as in metal catalysts, semiconductors, and electrodes. The necessity for high-vacuum conditions in the sample chamber makes the technique most effective with solids, although some progress has been made in the analysis of certain types of gaseous and liquid compounds. The first ESCA studies of aluminosilicate minerals were reported in 1971, and the technique subsequently was applied to clay and soil minerals by 1974.

Atomic ratio determination. The most recent advance in ESCA characterization of soil minerals is the quantitative determination of atomic ratios within the mineral structure. A typical ESCA spectrum of a clay mineral (Fig. 3) contains sharp peaks

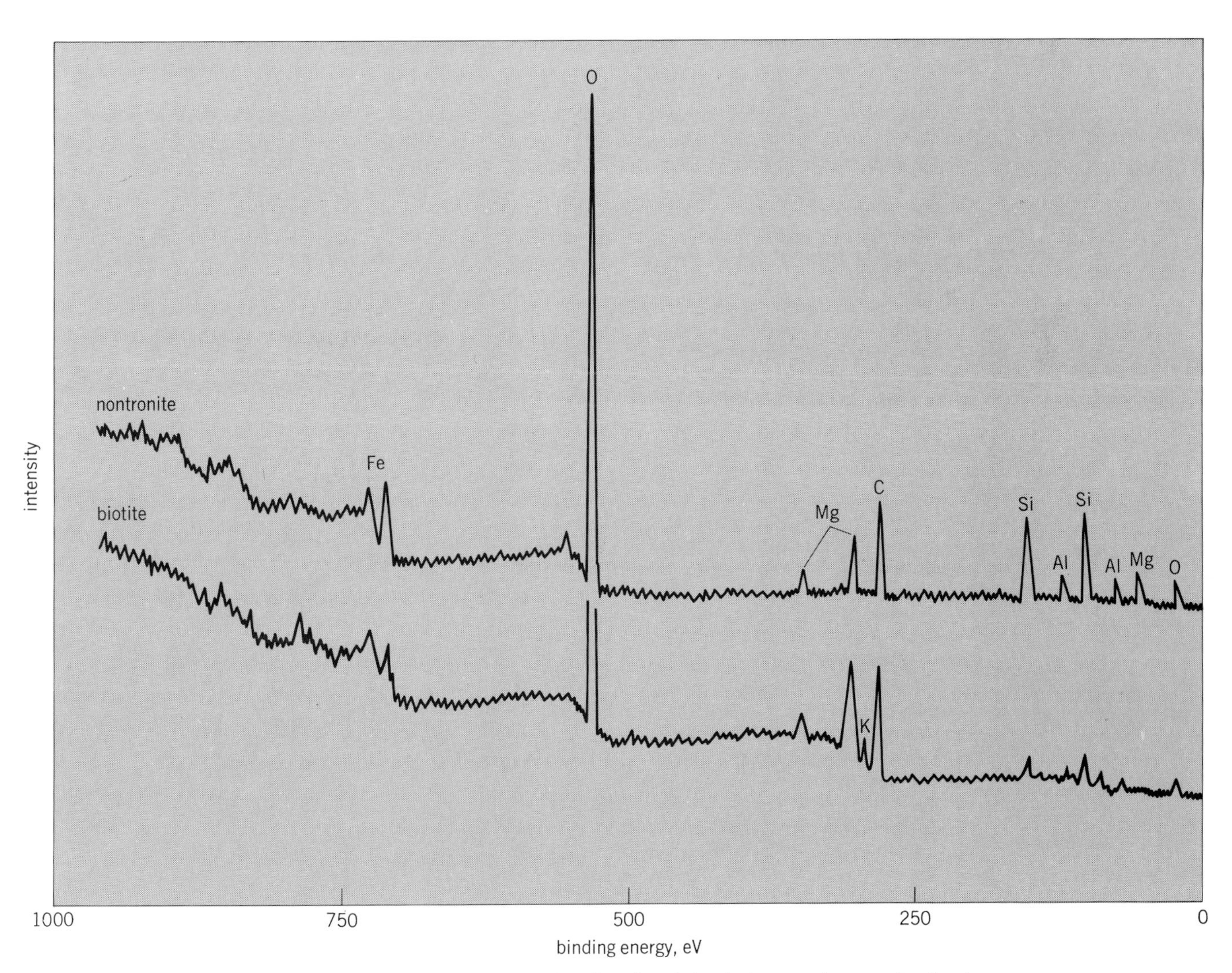

Fig. 3. ESCA spectra depicting electron binding energies from 0 to 1000 eV for nontronite and biotite. *(From J. W. Stucki, C. R. Roth, and W. E. Baitinger, Analysis of iron-bearing clay minerals by electron spectroscopy for chemical analysis (ESCA), Clays Clay Miner., 24:289–292; copyright 1976 by Pergamon Press, Ltd.)*

Table 2. Atomic ratios for several clay minerals

Mineral	Elements	Atomic ratio: Wet chemical method	Atomic ratio: ESCA method	Difference, %
Kaolinite (#865)	Si/Al	1.03	1.07	4
Kaolinite (Saint Austell)	Si/Al	1.04	1.09	5
Kaolinite (#7)	Si/Al	1.12	1.10	2
Talc	Si/Al	1.92	1.92	0
Montmorillonite (#23)	Si/Al	2.78	2.78	0
Montmorillonite (#22a)	Si/Al	2.63	2.70	3

corresponding to inner-shell electrons that have been ejected from the elements present in the surface layers of the mineral. Generally, every element present in the mineral can be identified qualitatively, and due to the quantum nature of electronic energy levels multiple peaks are often observed for a single element. With some effort one can use one or several of these peaks to determine accurately (within 3–5%) the relative atomic ratios of certain elements within a given sample, although at present ESCA is incapable of absolute quantitative analysis for elemental composition.

Among the most important factors that influence the accuracy of atomic ratio calculations from ESCA data of minerals are photoelectron cross sections, photoelectron escape depths, and the presence of surface contaminations. Some elements and electron states are more efficient than others in contributing electrons to the spectrum. Thus, since the intensity of a peak is a direct result of the number of electrons striking the detector, it is generally impossible to obtain accurate atomic ratios by simply comparing peak areas. One must first correct the peak intensity by using an efficiency factor, called the photoelectron cross section, for the specific element and electronic energy level associated with the peak. Predicted photoelectron cross sections are tabulated for most elements and levels, and values for several of the more important elements have been verified experimentally in a number of mineral samples. Experimental verification is essential since cross sections can be altered appreciably by placing the element in a different matrix, and few or no cross sections were calculated initially for mineral structures.

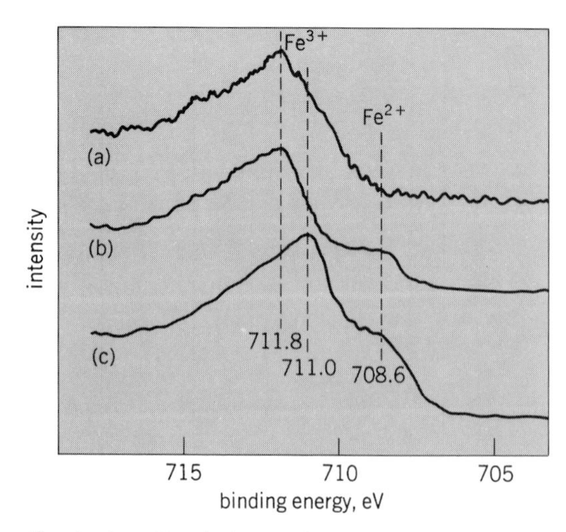

Fig. 4. Iron ($2p_{3/2}$) electron binding energies for nontronite (*a*) unaltered, (*b*) hydrazine-reduced, and (*c*) dithionite-reduced. *(From J. W. Stucki, C. R. Roth, and W. E. Baitinger, Analysis of iron-bearing clay minerals by electron spectroscopy for chemical analysis (ESCA), Clays Clay Miner., 24:289–292; copyright 1976 by Pergamon Press, Ltd.)*

Another source of error in using peak intensities directly for quantitative analysis is variation in sampling depth from one peak to the next. This arises from the fact that electrons with greater kinetic energy (lower binding energy) are able to escape from deeper within the sample than those of lesser kinetic energy (higher binding energy). It has been found that escape depth varies as the square root of kinetic energy and, as a first approximation, serves as an adequate correction factor. Deviations from this approximation increase as kinetic energy diminishes.

A third factor that must be taken into consideration is that quantitative analysis requires the surface measurement to be representative of the bulk material. ESCA is able to penetrate surface layers only, hence its usefulness as a quantitative tool is limited by surface contaminations and aberrations. Care must be taken to keep the outer surfaces of a sample free of obvious contaminants such as excess solutes deposited during drying or tenaciously adsorbed vapors.

Atomic ratios for silicon, aluminum, iron, magnesium, and oxygen have been calculated from ESCA spectra of biotite and kaolinite #865, and for silicon and aluminum in a number of other clay minerals (Table 2). When compared with ratios obtained from wet chemical analyses, the Si/Al ratios offer the best agreement (0–5%), and Si/Mg ratios are satisfactory; however, other permutations yield much greater differences. The close comparisons of Si/Al ratios are encouraging and indicate that the surface measurement of clay minerals does indeed reflect the overall composition of the mineral. As a result, the sampling depth must be significantly greater than the minimum estimate of 5 A (0.5 nm) and perhaps as great as 50 A (5.0 nm) or more, since the clay minerals consist of platelike sheets 7–10 A (0.7–1.0 nm) thick and when dried from solution readily orient themselves parallel to one another. Surface analysis of minerals therefore bears a distinct opportunity for bulk characterization.

Oxidation states. ESCA data have been used to study the effects of redox reactions on the structural and colloidal properties of nontronite (an iron-rich clay mineral). Chemical reduction of about 20% of the structural Fe^{3+} to Fe^{2+} in nontronite resulted in no structural changes other than the change in iron oxidation state (Fig. 4*a* and *b*). More extensive reduction (40%), however, produced distortions in the structure sufficient to shift the position of the original Fe^{3+} electron energy level by almost 1 eV (Fig. 4*c*). This shift is believed to be the result of a decrease in iron coordination number from 6 to 5.

Adsorbed species. Competitive adsorption phenomena and the charge states of various adsorbates have been studied by ESCA. By means of angle-resolved ESCA, information concerning spatial distribution of adsorbates on mineral surfaces may become available. One difficulty in interpretation that occurs with some adsorption data is that the charge states observed in dry samples under high vacuum may be altered when the mineral is placed in a moist environment that more closely resembles the natural soil condition. Interpretations of structural data are less susceptible to such differences. [J. W. STUCKI]

Far-infrared analysis. Recent work in the application of far-infrared analysis to clay minerals has been concerned with (1) the determination of far-infrared spectra for a wide range of clay minerals and (2) the effect of exchangeable cation on the far-infrared spectra of montmorillonite and rectorite. The far-infrared region of the spectrum has been used to investigate crystal vibrational modes. These vibrational modes are sensitive to the structure and composition of the mineral. Therefore, reactions involving clay minerals that result in either structural or compositional changes can be followed by using far-infrared spectroscopy.

There is very limited information on the far-infrared absorption spectra of clay minerals. Most interpretations of far-infrared spectra of clay minerals have been derived from observations and the application of group theoretical calculations of idealized mica structures. There has been no further information about far-infrared analysis of minerals since C. Karr, Jr.'s review article until the recent report by C. B. Roth concerning far-infrared analysis of clay minerals.

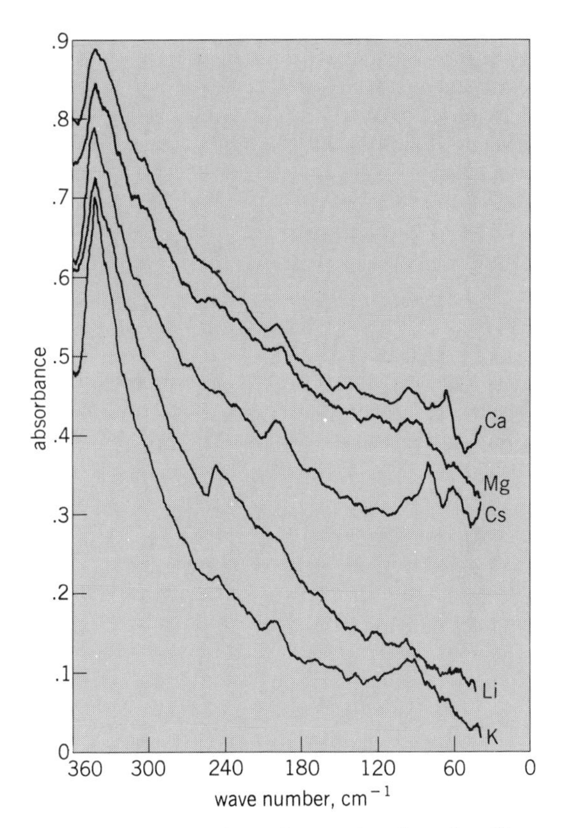

Fig. 6. Far-infrared absorption spectra of montmorillonite saturated with various cations.

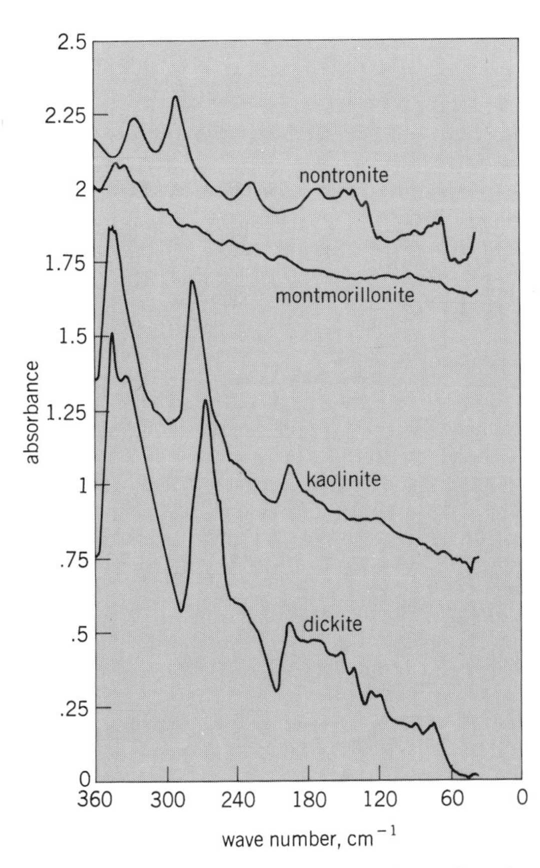

Fig. 5. Far-infrared absorption spectra of clay minerals.

The far-infrared absorption spectra of clay minerals is affected by the structure of the clay. Kaolinite, a clay mineral consisting of a silica tetrahedral sheet bonded to an alumina octahedral sheet, has absorption maxima at 368, 346, 278, and 195 wave numbers, in cm^{-1} (Fig. 5). On the other hand, montmorillonite, a clay mineral in which an aluminum octahedral sheet is sandwiched between two tetrahedral sheets composed of silica, exhibits strong absorption at 340 and 96 cm^{-1} with very weak absorption at 283, 246, and 202 cm^{-1}.

Differences in stacking between layers of similar clay minerals can also be studied with far-infrared spectroscopy. Dickite differs from kaolinite by only a slight change in the stacking of the layer composed of the silica tetrahedral and alumina octahedral sheets. The far-infrared spectra of dickite shows an additional absorption band at 334 cm^{-1}, with a shift in the band found in 278 cm^{-1} in kaolinite to 267 cm^{-1} in dickite (Fig. 5).

Compositional changes within the sheets of minerals having similar structures also affect the far-infrared spectra. Nontronite, a clay mineral structurally similar to montmorillonite except that iron instead of aluminum is found in the octahedral position, has absorption bands at 376, 292, 228, and 173 cm^{-1} that are not seen in the montmorillonite spectra (Fig. 5). The bands at 340 and 246

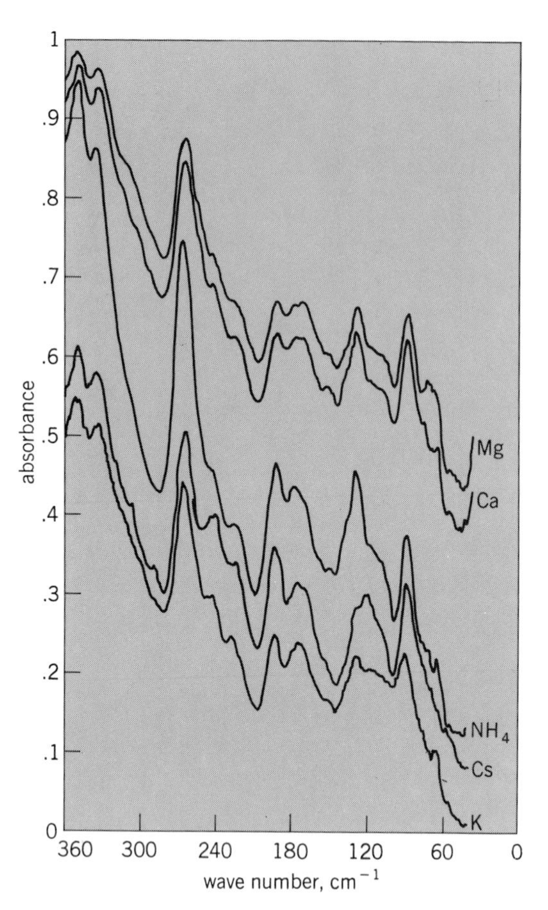

Fig. 7. Far-infrared absorption spectra of rectorite saturated with various exchange cations.

cm^{-1} in montmorillonite are shifted at lower wave numbers in nontronite, 328 and 228, respectively.

Recently it has been shown that the exchangeable cation can affect the far-infrared spectra of montmorillonite and rectorite. The far-infrared absorption band of montmorillonite, which is commonly located at 96 wave numbers when the clay is saturated with monovalent cations, is shifted downward by 4–6 wave numbers when divalent cations saturate the exchange complex (Fig. 6). This band is associated with the translational modes of the exchange cation. Cesium saturation of montmorillonite results in the appearance of far-infrared absorption bands at 80 and 60 wave numbers.

Rectorite, a clay mineral consisting of regularly interstratified mica and montmorillonite layers, has three far-infrared absorption bands at 286, 193, and 90 cm^{-1} that are not affected by the easily exchangeable cation (Fig. 7). Since the readily exchangeable cations are found only on the montmorillonite layers, bands that remain unaffected are thought to result from absorption by bonds associated with the mica layers. The shape and position of the absorption bands in the 129–124- and the 180–170-cm^{-1} regions are affected by the exchangeable cation in rectorite.

For background information *see* CLAY; CLAY MINERALS; SOIL; SPECTROSCOPY in the McGraw-Hill Encyclopedia of Science and Technology.

[CHARLES B. ROTH]

Bibliography: J. M. Adams et al., *Anal. Chem.*, 49:2001–2008, 1977; R. J. Baird et al., *Anal. Chem.*, 48:843–846, 1976; J. Iannicelli, *Clays Clay Miner.*, 24:64–68, 1976; M. Ishii, M. Nakahira, and H. Takeda, in *Proceedings of the International Clay Conference, 1969*, Tokyo, L. Heller (ed.), vol. 1, pp. 247–259, 1969; C. Karr, Jr., in H. A. Elton (ed.), *Progress in Nuclear Energy*, Analytical Chemistry Series 9, vol. 11, pp. 109–134, 1972; H. Kohm, J. Oberteuffer, and D. Kelland, *Sci. Amer.*, 233:46–54, 1975; M. H. Koppelman and J. G. Dillard, *Clays Clay Miner.*, 25:457–462, 1977; J. Neuzil and M. Kuzvart, in S. W. Bailey (ed.), *Proceedings of the International Clay Conference*, Mexico City, pp. 525–532, 1975; C. B. Roth, *Abstracts of the International Clay Conference, 1978*, Oxford, p. 255, 1978; C. B. Roth, *Agronomy Abstracts, 1977*, p. 189, 1977; D. G. Schulze and J. B. Dixon, *Soil Sci. Soc. Amer. J.*, publication pending, 1979; J. W. Stucki and C. B. Roth, *Soil Sci. Soc. Amer. J.*, 41:808–814, 1977.

Solar cell

The inherent simplicity of photovoltaic power systems provides one of the most conceptually attractive solar energy technologies. As with most alternative energy sources, the primary problem with photovoltaics is one of economics. For this reason, part of the U.S. Department of Energy's (DOE) solar energy program has been dedicated to producing high-purity silicon at low cost. A recently developed purification process using laser photochemistry promises to aid greatly in achieving this goal.

Cost of high-purity silicon. In a recent review of photovoltaic technology, H. Kelley of the Office of Technology Assessment stated that the cost of high-purity silicon production represents the single greatest technical barrier to meeting the overall cost goals for nonconcentrating solar arrays in the early 1980s. These goals require that the current price of purified silicon, $65/kg, be reduced to about $10/kg. In addition, the energy requirements for pure silicon production must be reduced by more than an order of magnitude. The basic silicon feedstock, metallurgical-grade silicon, is very inexpensive at $1/kg. The high price of solar-grade silicon represents the cost of producing silicon with the exceedingly high purity levels required by solar cells. Thus, the goals of the low-cost silicon project can be met only by devising low-cost, non-energy-intensive means of purification. Because laser photochemistry techniques can act selectively on a single component of a complex mixture, the application of laser-induced chemistry to materials purification promises to change that technology qualitatively, particularly for the ultrapure materials required by the semiconductor industry. *See* LASER CHEMISTRY.

Silane as a silicon source. Silane (SiH_4) gas is a silicon source used extensively throughout the semiconductor industry, so part of the low-cost silicon project is aimed directly at reducing the cost of SiH_4. In the DOE plan for the mid-1980s, SiH_4 is to become a primary intermediate in the production of silicon solar cells, mostly because it offers a less energy-intensive path to high-purity silicon than does the current intermediate, trichlorosilane ($SiHCl_3$). Amorphous silicon solar cells, which may supplant single-crystal cells because of their very low fabrication costs, have been made successfully only from SiH_4, primarily because a silicon-hydrogen "alloy" is necessary for good performance. An inexpensive means for purifying SiH_4 can thus have great benefits for a number of different silicon-based technologies. *See* SILICON.

Poor silicon resistivity can adversely affect the performance of silicon solar cells. The principal contaminants in SiH_4 that degrade the resistivity of silicon deposited from it are arsine (AsH_3), phosphine (PH_3), and diborane (B_2H_6). When SiH_4, AsH_3, PH_3, or B_2H_6 absorb ultraviolet light near 200 nm, the molecules are dissociated, and solid or polymeric materials are formed. Thus, if the impurities AsH_3, PH_3, or B_2H_6 present in SiH_4 could be preferentially dissociated, they could be removed from the gaseous SiH_4, and the SiH_4 would be purified.

Laser purification process. Experiments performed by J. Clark and R. Anderson at the Los Alamos Scientific Laboratory have shown that ultraviolet light at 193 nm can accomplish this purification. Light of this wavelength is absorbed 10,000 times more strongly by AsH_3 and PH_3 than by SiH_4. Experiments in which mixtures of SiH_4 with either PH_3, AsH_3, or B_2H_6 are irradiated with the 193-nm argon fluoride (ArF) laser show that the SiH_4 can be substantially purified by this technique. The illustration shows a schematic of the experimental apparatus used for laser purification of SiH_4. An ArF laser is directed into a cell containing SiH_4 gas. The laser excites impurities present in the SiH_4, causing the impurities to dissociate and precipitate out as solids or polymers, thus purifying the gas.

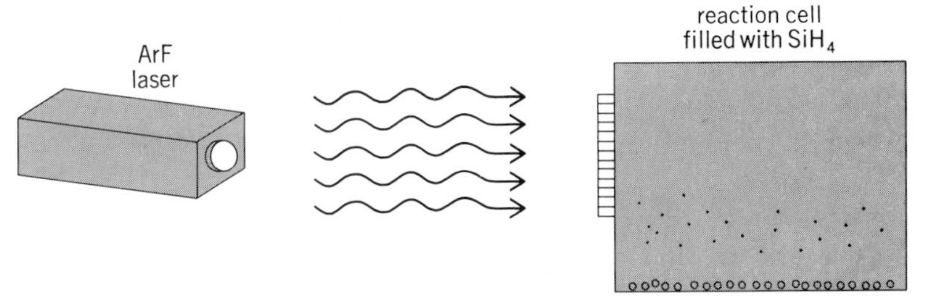

Schematic diagram of apparatus for laser purification of SiH_4.

Further experiments were conducted to investigate the chemistry involved when any of these compounds absorbs light from an ArF laser. The essential feature of this photochemistry is that upon absorption of an ultraviolet photon, SiH_4, AsH_3, PH_3, and B_2H_6 all lose an H atom. The remaining fragment, SiH_3, AsH_2, PH_2, or B_2H_5, precipitates out as a solid or polymer, and does not undergo any detrimental postphotolysis scrambling reaction. The H atom, which finds itself in essentially pure SiH_4, reacts with SiH_4 to form SiH_3 and H_2. The SiH_3 eventually forms a solid deposit. Thus, one SiH_4 molecule is lost each time an impurity molecule is photolyzed. For the impurity concentrations typical in SiH_4 (much less than 1 part per million), the loss of SiH_4 due to this postphotolysis reaction is negligible. In addition, it was found that the laser photons are used quite efficiently, with 30 to 40% of the impurity molecules which absorb a photon removed from the gas phase. Since it can be arranged to have each impurity molecule absorb a photon an arbitrarily large number of times by adjusting the amount of laser energy to which it is exposed, virtually all the impurities present in a sample can be removed.

Cost estimates. This knowledge, although by no means complete, makes possible preliminary cost and performance estimates for a laser process to produce solar-grade silicon. These cost predictions suggest that even in its present formative stage of development this process could purify SiH_4 for much less than $1/kg, making an almost insignificant contribution to the cost of SiH_4, even at the $10/kg level, which is the cost goal of the DOE program. Anticipated improvements in the process could drop this already modest projected cost substantially.

Removal of diborane. A second laser technique has been used to remove B_2H_6 from SiH_4. In this scheme the high peak power obtainable from a pulsed CO_2 laser is used to cause a dielectric breakdown arc in SiH_4, resulting in bond rupture for many of the molecules within the arc. After the arc has extinguished, some of the fragments produced recombine back to the parent molecules, while others polymerize and form solid deposits. Since SiH_4 is thermodynamically more stable than B_2H_6, it tends to recombine efficiently, whereas the B_2H_6 tends to be transformed into solid B_xH_y polymers. Experiments by S. Freund and W. Danen at Los Alamos have shown that more than 50% of the B_2H_6 can be removed with little loss of SiH_4. The high efficiency of CO_2 lasers suggests that this technique may also be economically practical.

Economic advantages. The low costs predicted for laser purification of SiH_4 may seem surprising, given the expensive technology normally associated with lasers. The explanation lies in the difference between the laser process and conventional purification techniques. Conventional methods exploit the small differences between the bulk physical properties (for example, boiling point and chemical affinity) of the contaminants and the desired species and therefore achieve rather small separation selectivities. The laser scheme, on the other hand, utilizes the often very large differences in the properties of individual molecules to obtain high selectivities. If, as is often the case in silicon purification, it is desired to remove 10 parts per billion of some impurity, the total amount of impurity in question is only 10 mg for each 1000 kg of material. In a conventional process this entire 1000 kg would be passed through a series of separation stages until the 10 mg of impurity was removed. In the laser process the laser light selectively "seeks out" the impurities and removes them. Because the laser light is used so efficiently, a small laser can process a large amount of material. Based on the economic predictions given above, a laser with an average output power of 250 W (much less than that of a typical street light) could purify more than 500 tons (4.5×10^5 kg) of silicon each year, the DOE estimate for 1985 SiH_4 demand. If envisioned improvements in the process prove feasible, these requirements for laser power, and hence the cost of the process, could decrease by a factor of 10 to 100.

Energy requirements. The amount of energy required for the laser technique appears to be especially promising. Current technology requires that the solar cells operate for about 4 years in an average climate before they begin to produce more energy than was used in their manufacture. Estimates for the laser process predict that if little of the SiH_4 is wasted during fabrication, energy payback time for the purification step would be about 6 min. Even if fabrication losses remain at their current levels, payback time would be less than 1 hr.

Additional applications. It is anticipated that development of photochemical purification techniques for photovoltaic materials will have important applications. One additional application is the attainment of higher yields in existing semiconductor manufacturing processes through increased process control. The laser technique should also be capable of producing SiH_4 of a purity which simply cannot be obtained with conventional techniques. The availability of such high-purity material could allow the fabrication of new, higher-performance solid-state devices.

For background information *see* LASER; SILICON; SOLAR BATTERY in the McGraw-Hill Encyclopedia of Science and Technology.

[JOHN H. CLARK]

Bibliography: J. H. Clark and R. G. Anderson, *Appl. Phys. Lett.*, 32:46–49, 1978; S. M. Freund and W. C. Danen, Purification of reagents using infrared lasers: Removal of diborane from Silane, *Appl. Phys. Lett.*, submitted for publication; H. Kelly, *Science*, 199:634–643, 1978; H. Macomber, in *Proceedings of the ERDA Semiannual Solar Photovoltaic Program Review Meeting, Silicon Technology Programs Branch, San Diego, CA, January 1977*, ERDA, pp. 68–72, 1977.

Solar energy

Solar energy has become the focus of much recent research and development since it is a constantly renewing energy source. Two recent developments are discussed below: (1) the development of a mechanical heat-storage device which would facilitate the use of solar energy in homes and (2) a method of fixing nitrogen into ammonia for fertilizers by using solar energy.

Solar heat storage in a rolling cylinder. The "rolling cylinder," a mechanical heat-storage device, invented at the General Electric Research and Development Center in Schenectady, NY, is under continuing development with the goal of creating a practical, cost-effective solution to the problem of storing solar heat for use when the Sun is hidden from view. Consisting of a cylindrical container lying on its side and rotating slowly about its axis, the rolling cylinder is filled with a chemical compound which stores latent heat of fusion as it melts from the solid to the liquid state; the compound later releases the heat as it reverts from liquid to solid. A short length of small-diameter tubing is attached to the cylinder at one end and rotates with it. Filled with seed crystals, the tube performs the vital nucleating function of inaugurating the phase change from liquid to solid without excessive supercooling. To date most research work has used Glauber's salt, sodium sulfate decahydrate ($Na_2SO_4 \cdot 10H_2O$), as the heat-storage material. However, many different phase-change storage materials are expected to be workable in the rolling cylinder, providing a choice of operating temperatures.

The rolling cylinder latent heat store is of great technical interest because it embodies simultaneous solutions to the three principal latent heat-storage problems: encapsulation and separation of undissolved salts; crystallization on heat-transfer surfaces; and excessive supercooling.

Preliminary work in laboratory-sized rolling cylinders has yielded results that are very encouraging for the prospects of a large store. These results include: 100% crystallization of Glauber's salt; no degradation of crystallization behavior visible after 200 freeze-thaw cycles; use of the cylindrical container as the heat-exchange surface; high heat-transfer rates because crystals do not adhere to or build up on the cylinder wall; repeatable nucleation; high volumetric efficiency since the cylinder is 90% filled; and automatic freeze-thaw cycling as the temperature of the surrounding air rises and falls. These results, taken together with expected good axial mixing, suggest that a large heat store could be built in a single large cylindrical tank, which probably would be long and narrow in shape and would fit into a single-family house as shown in the illustration.

Design of rolling cylinder system. The illustration shows one of many ways in which the rolling cylinder might be used in a solar home. Sun-heated air from rooftop collectors passes over and around the heat-storage cylinder in the basement which stores the heat collected in excess of present requirements. Later, when heat from the Sun is unavailable, dampers isolate the collectors, and heat from the storage is circulated throughout the house. The high energy-storage density of latent heat materials relative to other available storage media allows a large-capacity heat store to be packaged in a small volume. The entire system could fit inconspicuously in one end of the basement, leaving most of the floor space available for recreational or other uses.

Advantages of Glauber's salt. Preliminary work with the rolling cylinder heat store has concentrated primarily on Glauber's salt, since it may be the only heat-storage material which provides a store with a lower purchase price than either rock or water stores.

Presently available technology stores sensible heat in tanks of water or bins of rock. Heating systems of this kind are functional but have the disadvantage of being heavy, bulky, and costly. Solar energy experimenters have long believed that these disadvantages might be removed by storing energy as the latent heat of fusion of chemical compounds such as Glauber's salt, which has a melting point of 90.3°F (32.4°C) and a heat of fusion of 108 Btus/lb (0.070 kWh/kg). A Glauber's salt store theoretically would occupy one-eighth the volume of an equal-capacity water store and one-eighteenth the volume of a rock store. Since $Na_2SO_4 \cdot 10H_2O$, water, and rocks are all inexpensive materials, the main cost element of thermal storage is the storage container and insulation. A smaller store, therefore, has the potential to cost less. The rolling cylinder container may transform this potential into reality.

Encapsulation of undissolved salts. Glauber's salt has been recognized by solar energy workers as a desirable heat-storage material for more than 30 years. Unfortunately, it has a complicated melt-freeze behavior which has deterred its use. Glauber's salt melts incongruently; only about two-thirds of the sodium sulfate is soluble in the water of hydration at the melting temperature.

The undissolved portion is much denser than the liquid saturated solution and thus sinks to the bottom of the container. When freezing begins, the solid layer in the bottom of the container is promptly covered by a layer of newly formed decahydrate, which is also denser than the remaining liquid solution but less dense than the undissolved solid. The new layer of decahydrate acts as a diffusion barrier and slows down the rate of freezing to unusably low values. The rolling cylinder revolves at a rate of about 3 revolutions per minute, which is just enough motion to disperse the barrier layer and permit freezing to occur rapidly and completely.

Crystallization on heat-transfer surfaces. An important heat-transfer problem common to all liquid-solid phase-change latent heat systems is the formation of thermal-shock nuclei in the liquid boundary layer at the heat-transfer surface. These nuclei adhere to the surface, and quickly grow into a thick, dense layer of crystals as freezing proceeds. Since the thermal conductivity of solid salts is usually very low, this accumulating layer of crystals progressively decreases the rate of heat removal to such low values that it becomes a major design problem. The rolling cylinder permits control of boundary layer conditions so that thermal-shock nucleation does not occur, and no solid crystals adhere to the heat-transfer surface.

Nucleation. At the beginning of freezing it is imperative to introduce a near-replica of the desired crystal solid into the liquid. Otherwise the entire mass of liquid may supercool instead of freezing. Supercooling of 20 to 30°F (11 to 17°C) is quite possible in Glauber's salt, and a system in which this occurs would be undesirable for home heating. The nucleator tube attached to the rolling cylinder is long enough to extend outside the insulation surrounding the cylinder so that the crystals in the outer tip of the nucleator never become warm enough to melt. As the cylinder temperature falls below freezing, crystals grow along the nucleator tube and into the cylinder to inaugurate freezing behavior. Rapid crystal growth along the tube is ensured by purposely creating crystal defects having rapid growth velocities.

[CARLYLE S. HERRICK]

Nitrogen-reducing solar cells. Not all solar energy research is directed toward the production of fuels or electricity. A photochemical process has been developed in which ammonia is generated from water and nitrogen upon exposure to light. The catalyst for this unusual reaction, developed by G. N. Schrauzer, is a specially metal-doped oxide of titanium (TiO_2), which binds both water and nitrogen at room temperature. When this catalyst is irradiated with light of sufficient energy, reaction (1) occurs, in which the absorbed molecular

$$N_2 + 3H_2O \xrightarrow[\text{metal-doped } TiO_2]{\text{light quanta}} 2NH_3 + 1.5O_2 \qquad (1)$$

nitrogen is reduced to ammonia and oxygen gas evolves. Ideally, nitrogen-reducing solar cells using this reaction could be built in areas where conventional ammonia plants cannot be constructed and sunlight is abundant to produce fertilizer ammonia from water and atmospheric nitrogen.

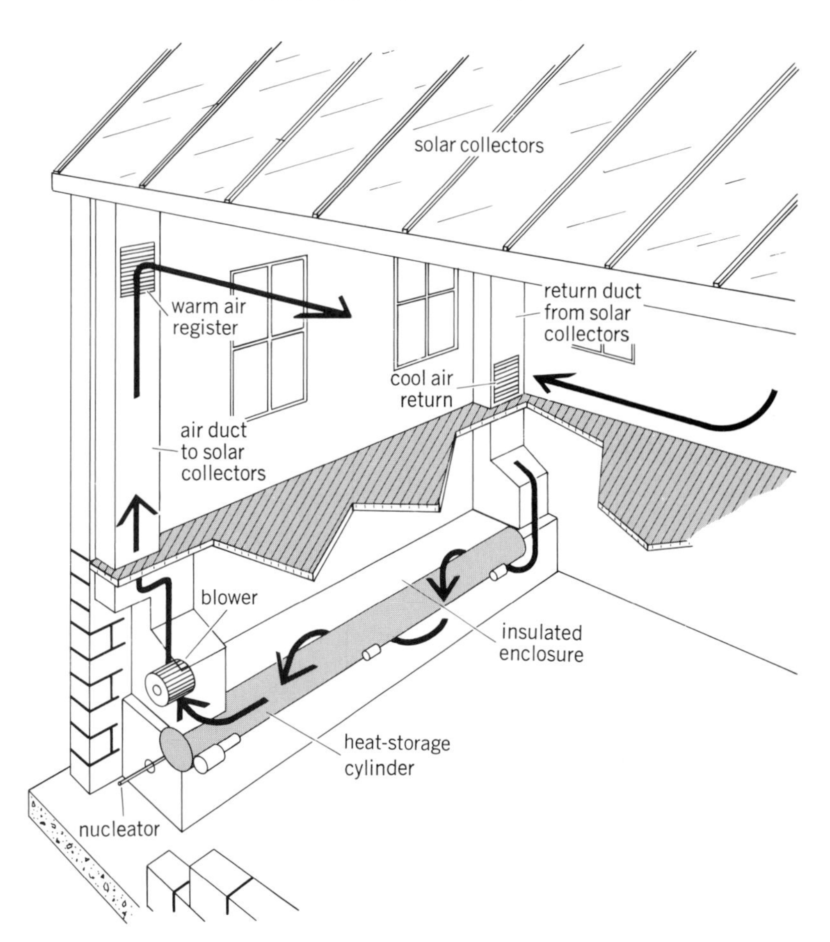

Design of solar home heating system using rolling cylinder heat storage.

Principles of the process. Reaction (1) combines the principles of photosynthesis and nitrogen fixation. In plant photosynthesis, solar energy is used to chemically split water into oxygen and chemical equivalents of hydrogen. The hydrogen equivalents provide the necessary reducing power for the reduction of carbon dioxide into carbohydrates. Reaction (1) utilizes solar energy (about 183 kcal/mole or 766 kJ/mole) for the same purpose but diverts the reducing power thus generated to convert molecular nitrogen into ammonia instead.

Nitrogen resists chemical reactions at room temperature. Only a few soil bacteria produce the enzyme nitrogenase, which can fix nitrogen from the air to yield ammonia through reaction (2). Ni-

$$N_2 + 6H^+ + 6e^- \xrightarrow[\text{nitrogenase enzyme}]{\text{adenosine-triphosphate}} 2NH_3 \qquad (2)$$

trogenase is an enzyme complex of two major components, an iron-molybdenum protein with a molecular weight of about 300,000 and an iron protein with a molecular weight of about 40,000. The reducing power required for reaction (2) is furnished by carbohydrates. Schrauzer and coworkers spent 8 years studying this process in the laboratory with the aim of developing simple working models of nitrogenase. Although they were successful in duplicating virtually all reactions of the enzyme in simpler artificial systems, they realized that any process which incorporates the chemistry of biological nitrogen fixation would be too complicated for use in the practical production of ammonia. However, this initial work laid the groundwork for the eventual discovery of the new process.

Schrauzer's laboratory also investigated several other systems that reduce molecular nitrogen under mild conditions. The titania-based photochemical system is the simplest, and appears to be the most promising of all those known to date.

Prospects. Presently, the new process is still too inefficient for immediate practical applications. An increase of the quantum efficiency of the process by at least a factor of 10 is necessary to make it practicable. However, with continuing research yields could be increased.

The main value of present findings lies in their potential to open the way toward the development of entirely new technologies. At present, the chief method of industrial ammonia synthesis is the Haber-Bosch process [reaction (3)], which yields

$$N_2 + 3H_2 \xrightarrow[\text{iron catalyst}]{500°C,\ 350\ atm} 2NH_3 \qquad (3)$$

an annual production of about 40,000,000 tons (36,000,000 metric tons) of ammonia. The new process would not replace this existing technology, but could be useful in underdeveloped countries. One advantage to this process is that it does not produce any harmful by-products. Also, unlike the Haber-Bosch process, which requires natural gas and coal for hydrogen production, it requires only sunlight as an energy source.

Natural occurrence. Schrauzer has suggested that reactions similar to reaction (1) may occur on certain planets, notably Mars, and that perhaps they may have occurred on Earth at one time. The rarefied atmosphere of Mars contains some water and nitrogen, and the surface of the planet is exposed to intense sunlight. These conditions would favor the photoreduction of chemisorbed nitrogen at the surface of certain min
um and iron). This possibility is being tested in the laboratory.

For background information *see* FERTILIZER; NITROGEN FIXATION; SOLAR ENERGY in the McGraw-Hill Encyclopedia of Science and Technology. [GERHARD N. SCHRAUZER]

Bibliography: J. W. Hodgins and T. W. Hoffman, *Can. J. Technol.*, 33:293, 1955; K. Kauffman and Y-C. Pan, *Thermal Energy Storage in Sodium Sulfate Decahydrate Mixtures*, University of Pennsylvania Rep. no. NSF/RANN/SE/GI/27976/TR72/11. Philadelphia, December 1972; G. N. Schrauzer and T. G. Guth, *J. Amer. Chem. Soc.*, 99:7189, 1977; M. Telkes, *Solar House Heating—A Problem of Heat Storage, Heating and Ventilating*, p. 68, May 1947.

Sound

Recently sound has been used for new studies of the properties of superfluid helium, and research has begun in several new areas of superfluid acoustics. Specifically, the existence of more than one type of sound wave in the superfluid has made it possible to use acoustic measurements to determine the fundamental thermodynamic properties of superfluid helium with greatly improved precision and consistency. Superfluid acoustics itself has been the subject of new studies which involve new superfluid acoustic waveguides and sound modes and second-order interactions between the different types of sound. Before discussing these topics in detail, it is useful to review the nature of sound in superfluid helium. *See* HELIUM, LIQUID.

Sound in superfluid helium. The model which accounts for the unusual properties of superfluid helium is the two-fluid model, in which superfluid helium is pictured as consisting of two independent, interpenetrating fluid components: a normal fluid component and a superfluid component. Each component has its own mass density (ρ_n and ρ_s, for normal and superfluid density, respectively) and its own velocity field ($\vec{V}_n$ and $\vec{V}_s$). The total fluid density ρ is given by the sum of the component densities: $\rho = \rho_n + \rho_s$. The normal component is an ordinary fluid; it has finite entropy (that is, it carries heat) and viscosity. The superfluid component has no entropy or viscosity; that is, it carries no heat and, below a critical velocity, moves without friction. The relative amounts of the normal and superfluid components (ρ_s/ρ or $\rho_n/\rho = 1 - \rho_s/\rho$) are thermodynamic functions of temperature and pressure.

First and second sound. With a thermohydrodynamic system consisting of two fluids, it is possible to have more than one type of propagating sound wave, as illustrated in Fig. 1. Figure 1*a* shows the spatial variation of a part of a sound wave in which both fluid components are moving together. In the center of the wave there is a buildup of fluid (or an increase in the total density) and

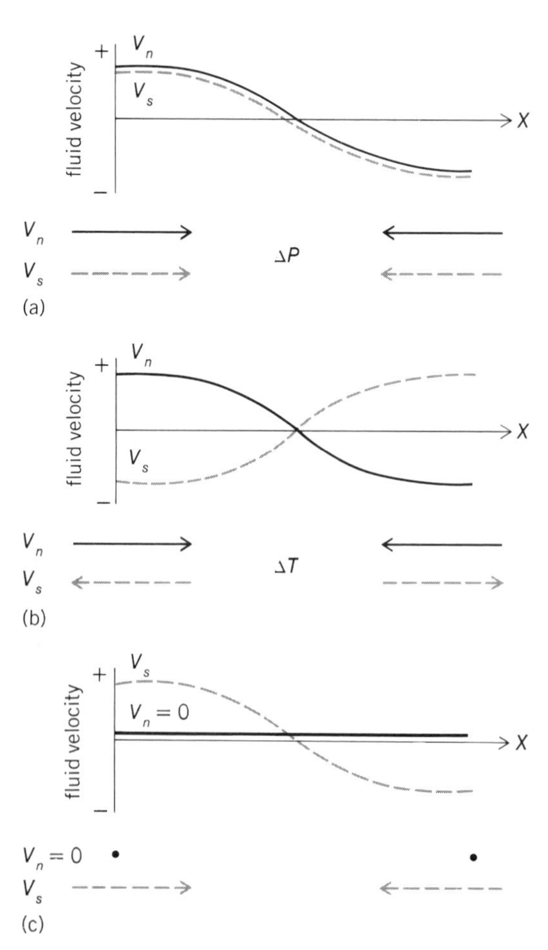

Fig. 1. Sound waves in superfluid helium. (*a*) Fluid components moving together. (*b*) Fluid components in counterflow. (*c*) Normal fluid clamped.

an increase in pressure Δp determined by the compressibility. The increase in pressure provides the restoring force for the wave motion just as in an ordinary sound wave; in superfluid helium this is called first sound and the velocity of propagation is denoted C_1. It is also possible to have a sound wave in which the two fluid components move in opposite directions, as shown in Fig. 1*b*. With this type of flow there would be negligible change in the density and pressure and hence no pressure restoring force. However, there would be a change in the relative amounts of the normal and superfluid components, corresponding to a temperature change ΔT. Furthermore, since the normal fluid carries entropy and the superfluid does not, there would also be a change in the specific entropy. Hence Fig. 1*b* represents a temperature or entropy wave; it is called second sound and is denoted C_2. Instead of having a mechanical (pressure) restoring force, this type of wave has a thermodynamic restoring force.

Fourth sound. Figure 1*c* shows another possible flow pattern for the two fluid components. In this case, there must be some environmental condition which clamps the normal fluid motionless through its viscosity. The superfluid component, having no viscosity, is still free to flow and form a sound wave. Several different types of sound are possible, depending on the dominant type of restoring force for the superfluid component. One normal-fluid-clamping environment, called a superleak, consists of a porous medium, such as a packed powder or Vycor glass, with pore sizes of 10 nm to 1 μm. In such small pores, any ordinary fluid with even the smallest known viscosity would be immobilized. If the walls of the superleak are rigid, the flow of the superfluid results in predominantly pressure waves, and a sound mode called fourth sound propagates with a velocity given approximately by $C_4 = [\rho_s/\rho]^{1/2}C_1$. Being a pressure wave, fourth sound is analogous to first sound, and ρ_s/ρ reflects the condition of purely superfluid flow. Theoretically there should also be a sound mode analogous to second sound with the normal fluid clamped. It should be a temperature wave with a predominantly thermal restoring force for the superfluid component and a velocity of propagation $[\rho_n/\rho]^{1/2}C_2$. The present experimental search for this sound wave is discussed below.

Third sound. Another normal-fluid-clamping environment consists of a substrate surface covered with a helium film with a thickness of 1–100 atomic layers. In such thin films all of the normal fluid is immobilized through its viscous interaction with the surface. As the superfluid component flows in a wave, it forms crests and troughs on the surface of the film similar to ordinary gravity waves on the surface of water. Unlike water waves, however, the restoring force is not gravity, but rather the dominating atomic-scale van der Waals force between the helium atoms in the film and the atoms in the substrate. This type of wave is called third sound. In ideal third sound only the superfluid component moves; hence there will be temperature changes but, since the superfluid carries no entropy, there will be no entropy changes in the wave. This particular wave is called adiabatic third sound. This wave is not generally observed experimentally because of the presence of helium gas above the free surface of the film. There are evaporation and condensation exchanges of atoms between the gas and film which cool and heat the film in such a way as to eliminate the temperature wave. Hence the customarily observed third sound is isothermal third sound. Present studies of adiabatic third sound are discussed below.

Probing helium properties with sound. Since ordinary sound or first sound is directly related to the compressibility of the fluid, a measurement of the sound velocity yields some information about the compressibility. However, the detailed expression for the ordinary sound velocity contains other thermodynamic quantities, so that a measurement of the sound velocity alone cannot result in a precise determination of the compressibility. However, in bulk superfluid helium (excluding the thin-film state) the three sound velocities C_1, C_2, and C_4 completely determine all the thermodynamic properties. Specifically, the pressure wave C_1 determines the compressibility, but requires the specific heat for a precise determination; second sound, being a thermal wave, determines the specific heat; and C_1 and C_4 determine ρ_s/ρ. Once these three fundamental thermodynamic quantities are known, all the other thermodynamic quantities can be found from their derivatives or integrals.

There are some definite experimental advantages in using sound velocities to determine thermodynamics. First, it is easy to make acoustic resonators for the different sounds which, because of the low acoustic attenuation of the superfluid, can determine the sound velocities to better than 0.1%, so that thermodynamics properties can be determined to within a few tenths of a percent. This is typically an order of magnitude improvement over existing calorimetric measurements. Furthermore, velocities can be measured simultaneously in one experiment with the resonators coupled so as to be at the same temperature and pressure. Hence there are no discrepancies resulting from inaccurate calibrations in different experiments, and the velocity and thermodynamic results are completely self-consistent. I. Rudnick and coworkers at the University of California, Los Angeles, have completed such sound velocity measurements and have compiled tables of the thermodynamic properties of superfluid helium.

Second-order interactions. The detailed two-fluid theory contains a set of thermohydrodynamic equations which, when solved to first order in deviations from equilibrium (the acoustic approximation), result in the various superfluid sound waves. When these equations are solved to second order, one obtains not only the usual nonlinear effects for classical fluids (such as streaming and shock formation) but, because of the already rich linear acoustics in the superfluid, one also obtains interactions between the different superfluid sound waves unique to this system.

A recently observed second-order interaction, called resonant mode conversion, is described as follows. When the amplitude of second sound is sufficiently large, second-order effects result in pressure waves accompanying the otherwise predominantly temperature waves of second sound. These pressure waves do not constitute a first-sound wave because they are traveling at the velocity of second sound, which is typically one-tenth

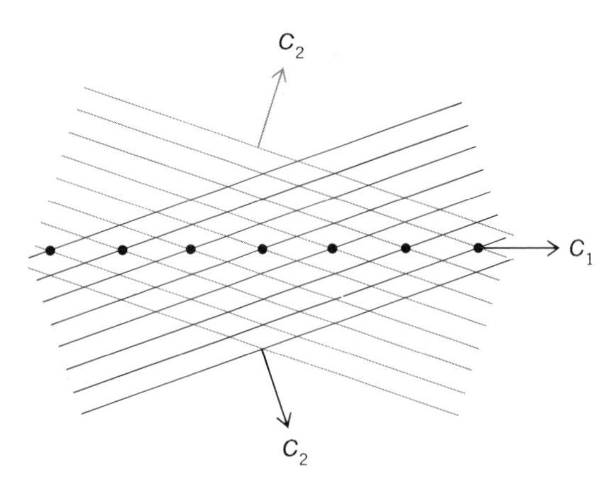

Fig. 2. Resonant mode conversion.

the velocity of first sound. However, the situation is different when one considers two intersecting plane-wave second-sound fields, as shown in Fig. 2. As the separate wavefronts progress, the points of intersection of the wavefronts (a few of them indicated by the dots in Fig. 2) undergo a scissor-like motion and travel along the line of dots with a velocity which can be much larger than C_2, and in fact equal to C_1, for a particular angle of intersection of the original two sound fields. In the second-order regime the dots indicate pressure wavefronts, and if they were made to move with velocity C_1, they would constitute a first-sound wave. Even if the region of interaction of the two intersecting second-sound waves were limited, a separate and distinct first-sound wave would propagate out of the interaction region. This process is termed mode conversion because two second-sound modes are converted into a first-sound mode; it is "resonant" because it occurs only at a precise angle of intersection between the two second-sound waves. This effect has recently been observed by S. L. Garrett and Rudnick. In their experiment the situation depicted in Fig. 2 was obtained by using a superfluid acoustic waveguide; other acoustic techniques were necessary to conclusively demonstrate the resonant mode conversion.

Superfluid acoustic waveguides. Acoustic waveguides for superfluid helium can determine the type of wave propagation not only through the wall boundary conditions but also through interior conditions. An open waveguide with rigid walls will propagate first or second sound, and a rigid waveguide filled with superleak will propagate fourth sound. A waveguide with a pressure-release surface (an open channel or canal) will propagate surface gravity waves, or if the helium depth is small, third sound will propagate. New types of superfluid waveguides have been studied recently which demonstrate some remarkable properties of the superfluid and which may lead to the discovery of a new fundamental sound mode.

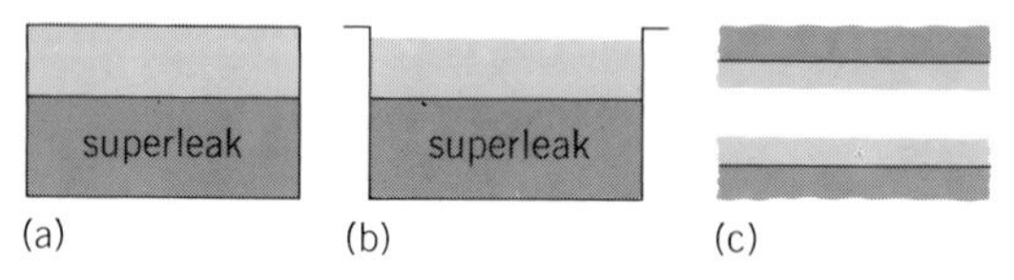

Fig. 3. Superfluid acoustic waveguides. *(a)* Closed and *(b)* open waveguides, both partially filled with superleak. *(c)* Parallel substrates covered with helium film.

Partially filled waveguides. Figure 3*a* shows a rigid waveguide partially filled with superleak; it is analogous to an electromagnetic waveguide partially filled with a dielectric. In the superfluid waveguide two modes may propagate. One mode is similar to second sound, with most of the fluid motion in the open part of the waveguide but with a velocity slightly different from C_2 depending on the amount of superleak. The other mode is a pressure wave which is a hybrid of first sound (from the open portion of the waveguide) and fourth sound (from the superleak portion of the waveguide); the velocity of propagation has a value intermediate between C_1 and C_4 depending on the amount of superleak in the waveguide. These sound modes have been used to study the characteristics of steady flows of superfluid down the waveguide; that is, a flow of superfluid in the open portion of the waveguide will Doppler-shift the C_2-like mode, but a flow of superfluid in the superleak will Doppler-shift only the C_1-C_4 mode. It has been found experimentally that when a persistent current flows down the waveguide, all the superfluid motion is contained in the superleak. This is remarkable because there is nothing to restrict the passage of superfluid into or out of the superleak.

Figure 3*b* shows a waveguide similar to the one in Fig. 3*a* except that it has a free, or pressure-release, surface so that the waveguide will not support a (low-order) pressure wave. This waveguide also has two propagating modes: a C_2-like mode similar to the one already discussed and a gravity surface wave with a velocity of propagation which is modified by the presence of the superleak in the channel.

Investigation of third sound. A third-sound waveguide, developed by J. M. Mochel and R. B. Hallock, is presently being used to investigate adiabatic third sound. Evaporation-condensation exchanges with the helium gas above the film modify the ideal superfluid wave. This may create the problem that energy is lost from the wave and the isothermal third sound is highly attenuated. The waveguide illustrated in Fig. 3*c* may eliminate this problem. The waveguide consists of two surfaces with adsorbed helium films separated by a small distance. This small distance keeps the gas volume and heat capacity small so that the gas will be less effective in modifying the superfluid wave. Experiments with a waveguide of this type, but without a parallel planar geometry, have shown an unattenuated sound mode which is probably adiabatic third sound. However, the nonplanar geometry introduced capillary forces which modified the velocity of propagation. Experiments with the planar geometry are presently under way to clarify the situation.

New sound mode. If the helium film thickness in this resonator is increased, then the van der Waals restoring force will cease to dominate. However, the thermal restoring force will remain, and if the film is still thin enough to keep the normal fluid clamped, the new sound mode discussed earlier, with velocity $[\rho_n/\rho]^{1/2}\, C_2$, will propagate. The present experimental search for this new sound mode is exciting, since it may be a new fundamental superfluid sound mode.

For background information *see* HELIUM, LIQUID; SECOND SOUND in the McGraw-Hill Encyclopedia of Science and Technology.

[JULIAN MAYNARD]

Bibliography: S. Garrett et al., Resonant non-linear mode conversion in He II, *Phys. Rev. Lett.*, publication pending; J. Maynard, *Phys. Rev. B*, 14: 3868–3891, 1976; S. Putterman, *Superfluid Hydrodynamics*, 1974; I. Rudnick, in *Proceedings of the International School of Physics, "Enrico Fermi" Course LXIII: New Directions in Physical Acoustics*, 1976.

Space flight

The year from September 1977 to September 1978 was a period of renewed Soviet emphasis on their crewed space program. Cosmonauts set two new records in succession for continuous space station (Salyut) occupation. The United States astronauts' cumulative total time in space was also surpassed by the Soviet cosmonauts. The Soviet Union continued to launch satellites at a rate of roughly eight per month, approximately four times that of the United States. In the United States, preparations were accelerated for the first shuttle orbital test. Of the 20 missions cited in the table, one-fourth involve spacecraft of the European Space Agency (ESA) and Japan. Crewed, scientific, and applications missions appear in approximately equal numbers.

Space shuttle development. As of September 1978, the National Aeronautics and Space Administration (NASA) was working toward its first orbital test flight *(OFT 1)* scheduled for the latter half of 1979. *See* SPACE SHUTTLE.

In July 1978 the eighth group of astronaut candidates joined the 27 active astronauts at the Johnson Space Center in Houston. The new group was composed of 15 pilot astronaut and 20 shuttle mission specialist candidates, including 6 women.

Soviet Union crewed program. On September 29, 1977, the space station *Salyut 6* was orbited successfully. In October *Soyuz 25* made an unsuccessful attempt to dock at the forward Salyut docking port. However, on December 11, cosmonauts Yuriy Romanenko and Georgiy Grechko latched *Soyuz 26* to the aft Salyut docking port. One month later they were joined by cosmonauts Vladimir

Major space missions from September 1977 to September 1978

Payload name	Launch date	Payload country or organization	Purpose and comments
Salyut 6	Sept. 29, 1977	Soviet Union	Occupied repeatedly by cosmonauts in 1977–1978
ISEE 1 and *2*	Oct. 22, 1977	United States *(ISEE 1)*; ESA *(ISEE 2)*	Magnetospheric and interplanetary research
Enterprise		United States	First operational-type landing of shuttle orbiter on concrete runway (Oct. 26, 1977)
Meteosat	Nov. 22, 1977	ESA	First European geostationary weather satellite
Soyuz 26	Dec. 10, 1977	Soviet Union	First successful use of second Salyut decking port; cosmonauts set new 96-day crewed flight record
Soyuz 27	Jan. 10, 1978	Soviet Union	First crewed resupply of an occupied orbiting space station
Progress 1	Jan. 22, 1978	Soviet Union	First Salyut refueling operation using automated tanker
IUE	Jan. 26, 1978	United States/ESA	Astronomical observatory satellite
FltSatCom A	Feb. 9, 1978	United States	Navy's first fleet communications satellite
NavStar 1	Feb. 22, 1978	United States	First prototype of Global Positioning System navigation satellite
Soyuz 28	Mar. 2, 1978	Soviet Union	First Soviet mission with non-Soviet (Czechoslovakian) cosmonaut
Landsat 3	Mar. 5, 1978	United States	Improved-resolution cameras
BSE	Apr. 7, 1978	Japan	First experimental Japanese broadcast satellite
OTS 2	May 11, 1978	ESA	Backup orbital test communications satellite (first OTS launch failed, Sept. 13, 1977)
Pioneer Venus 1	May 20, 1978	United States	Venus orbiter mission
Soyuz 29	June 15, 1978	Soviet Union	Cosmonauts set new crewed flight record on Sept. 20, 1978
Seasat 1	June 26, 1978	United States	First ocean resources monitoring mission
GEOS 2	July 14, 1978	ESA	Most complex ESA scientific spacecraft to date
Pioneer Venus 2	Aug. 8, 1978	United States	Venus multiprobe mission
ISEE 3	Aug. 12, 1978	United States	First mission to libration point between Earth and Sun

Dzhanibekov and Oleg Makarov in *Soyuz 27*, which used the forward docking port. Together, the three spacecraft formed a space configuration comparable in length to the United States *Skylab*/Apollo spacecraft combination, which is 36.1 m (118.5 ft) long. The three Soviet spacecraft weigh about 32,000 kg (70,000 lb), compared with the *Skylab*/Apollo weight of some 90,000 kg (200,000 lb). This twin Soyuz/Salyut mission accomplished the first four-person occupancy of a space station, the first crewed resupply of an occupied station, and the first multiple-transport spacecraft docking with a space station.

The *Soyuz 27* cosmonauts stayed aboard *Salyut 6* for only 5 days. They returned to Earth in *Soyuz 26*, thereby freeing the aft Salyut docking port. This port was used later in January by *Progress 1* in the first automated refueling of a space vehicle. During the 15 days it was docked with *Salyut 6*, the tanker transferred more than 2300 kg (5000 lb) of fuel and cargo. Its engines were used to change the orbit of the entire *Progress 1/Salyut 6/Soyuz 27* docked configuration. On February 6, 1978, *Progress 1* was separated from the space station. Two days later the tanker was commanded to reenter the atmosphere over the Pacific Ocean, thereby destroying itself in an uninhabited area. In March, Romanenko and Grechko returned to Earth uneventfully in *Soyuz 27* after 96 days in orbit, a new crewed flight duration record.

The Soviets began another similar sequence of related space flight operations on June 15, 1978. Alexander Ivanchenkov and Vladimir Kovalenok lifted off in *Soyuz 29* to dock with *Salyut 6*. In mid-August the *Progress 3* transport resupplied *Salyut 6* with consumables and nearly 450 kg (1000 lb) of oxygen regeneration equipment. On August 27 cosmonauts Valeriy Bykovskiy and Sigmund Jaehn, an East German air force officer, docked with *Salyut 6* in their *Soyuz 31*. After about a week in orbit, the *Soyuz 31* cosmonauts returned to Earth in the *Soyuz 29* spacecraft. This switch provided a fresher spacecraft for the return home. On September 20 the Soviets broke their own crewed flight duration record, and passed the 100-day mark 3 days later.

Planetary exploration. Two Viking landers have been on the Martian surface since the summer of 1976. After studying the data from the soil analysis experiments, biologists concluded that no unambiguous evidence had been obtained for the presence of life in the Martian soils at either landing site. The chemical reactions observed by Viking are thought to be due to the presence of powerful oxidizing agents (peroxides) on the surface of Mars which are produced by interactions between the Sun's powerful, virtually unshielded ultraviolet radiation and the tenuous but chemically complex atmosphere.

The composition of the Martian soil is interesting in other ways. According to N. W. Hinners, NASA Associate Administrator for Space Science, "An unusually high sulphur content in all the soil samples is suggestive of leaching by the upward migration of water with the subsequent formation of a salt-rich surface crust. . . . The soil composition is not what we would expect from the destruction of primary igneous rocks by various erosive forces, and we are still assessing the implications of this finding."

Excellent data concerning the meteorology of Mars have been obtained from the Viking landers. Surface pressure and wind data have been particularly illuminating. The summer circulation is dominated by atmospheric tidal and large-scale topographic effects. In the fall and winter the global circulation intensifies as heat is transferred poleward by huge wavelike cyclones and anticyclones similar to those encountered in Earth's temperate latitudes.

Unlike Earth, Mars exhibits a large seasonal global pressure cycle. When the frozen carbon dioxide polar caps grow, the barometric pressure decreases throughout the Martian atmosphere. Conversely, sublimation of the polar caps results in a rising barometer. The same phenomenon would be observed on Earth if its solar caps were cold enough to condense nitrogen and oxygen, its two predominant atmospheric gases.

Deimos and Phobos, the two tiny moons of Mars, have been found to be only about three-fifths as dense as the planet. These moons have dark surfaces whose color resembles that of the apparently carbonaceous asteroids much more than that of Mars.

Two United States missions were launched toward Venus during 1978. *Pioneer Venus 1* was to be inserted into orbit around Venus in December. Its planned orbital path was highly elliptical, with a distance of closest approach of 150 km (90 mi), so that the instruments could characterize the atmosphere's composition, motions, and cloud structure. A few days later, *Pioneer Venus 2* was to deliver four probes to the surface of Venus. The largest of these carried instruments to measure the winds, to determine the composition of the atmosphere and the clouds, and to observe the partial extinction of the Sun's radiation as the probe descended through the clouds. The three smaller probes were to make similar but less detailed measurements at widely separated locations around Venus. None of the probes was designed to survive on the planet's 430°C (800°F) surface, which is hot enough to melt lead.

Sun-Earth relations. Analysis of solar data gathered by astronauts using the Apollo telescope mount (ATM) on *Skylab* has produced more than 500 published articles and reports. An international ATM workshop on coronal holes studied the low-density, relatively cool regions of the Sun's outer atmosphere from which high-velocity ionized streams of the solar wind are emitted. A geomagnetic storm can now be predicted as long as 10 days in advance by observing a coronal hole as it comes into view around the eastern edge, or limb, of the Sun. The associated geomagnetic storm is caused by the impact of the solar wind stream on Earth's magnetosphere. Such storms are important because of their production of electrical surges in transmission lines and their interference with several forms of communications, particularly at high latitudes.

International Sun-Earth explorers (*ISEE 1* and *2*) were launched together in October 1977. They are in tandem flight in a highly elliptical Earth orbit to measure the response of the magnetosphere to the solar wind. Use of two tandem spacecraft provides the data necessary to separate magnetospheric variations in time from those with position. In August 1978 *ISEE 3* was launched to complete the triad.

For the first time a spacecraft was being maneuvered into a "halo" orbit around the libration point about 1,600,000 km (1,000,000 mi) toward the Sun from Earth. The concept is that of the libration point as a gravitational "anchor" to which *ISEE 3* can be "tethered" and from which position it can monitor the solar wind before the wind is distorted by its interaction with Earth's magnetosphere. However, if *ISEE 3* had been permitted to remain precisely at the libration point, its radioed data to Earth would have been drowned out by the Sun's powerful radio noise from the identical direction. The solution was to place *ISEE 3* into a loose orbit of about 150,000-km (90,000-mi) radius around the libration point in the plane perpendicular to the Earth-Sun line. The spacecraft was scheduled for final establishment in this halo orbit by late November 1978. Additional spacecraft hydrazine thruster burns are required about every 45 days to maintain the desired orbital characteristics.

Space astronomy. Measurement of the brightness of the nearest quasar in the ultraviolet was accomplished by the American scientist C. C. Wu by using data from the astronomical Netherlands satellite (*ANS*). More recently, Arthur Davidsen at the Johns Hopkins University acquired detailed ultraviolet "color" (spectral) information on this same quasar by using a precisely oriented sounding rocket. Analysis of these data demonstrates that the expansion rate of the universe at the present time is substantially slower than that long ago. In quantitative terms, the rate of slowing may be enough to result in a "closed" universe which will eventually contract. If so, there must exist tremendous quantities of matter in the universe which have not yet been detected or measured. This fundamental problem of cosmology is one which will undoubtedly require much more data and many more years to resolve.

Since its launch in January 1978 the international ultraviolet explorer satellite *(IUE)* has provided the most potent tool yet available to acquire high-resolution astronomical data in the far-ultraviolet. This satellite is another excellent example of international scientific teamwork: Great Britain provided the satellite's detectors and ESA the solar paddles. ESA operates *IUE* during one 8-hr shift per day, while NASA is responsible for the remaining 16 hr.

The first high-energy astronomical observatory *(HEAO 1)* launched in mid-1977 has been very productive. Many previously discovered x-ray sources have been located precisely enough to permit ground-based astronomers to locate the visible stars responsible for the x-ray emissions. Many new x-ray sources have also been discovered. One of these, named Nova Ophiuchi 1977, is of particular interest. It was discovered by *HEAO 1* as a sudden, vast increase in x-ray emission from a tiny region in the Ophiuchus constellation. The HEAO astronomers immediately passed on the source location to the Anglo-Australian Observatory in the Southern Hemisphere. There, astronomers discovered a new star, or nova, which was not present on older photographs of the same region of the sky. It is believed that this nova is part of a double (binary) star system of low mass.

Environmental applications. In November 1977 NASA launched *Meteosat*, the first European geostationary meteorological satellite. In addition to *Meteosat*, at the end of this reporting period there were six such satellites in orbit, five American and one Japanese. The Soviet Union had planned to launch its counterpart satellite by 1980, but because of delays in that program one of the United States "birds" will be used to complete the global network. The most important objective of this satellite network is to obtain wind velocities over the entire globe by tracking cloud positions on successive images every 30 min. The program goal is to build a precise computer model of the atmosphere so that more accurate weather forecasts can be made up to a week or more in advance based on such satellite data.

Since 1972, experimenters at several American laboratories have been using satellite-borne laser reflectors to measure the precise relative location of several sites in California on opposite sides of the San Andreas Fault. The precision of each set of measurements is 3 cm (1.2 in.). In July 1978 investigators at the NASA Goddard Space Flight Center announced that the tectonic plates on which these California sites were located had moved 9 cm (3.6 in.) closer to each other over the 6 years since the beginning of the measurements. The researchers observed that these results "seem to indicate that in the last 70 years the accumulated relative motion is of an order of 6–7 meters, the same magnitude of Earth motion which was associated with the 1906 San Francisco earthquake."

Earth resources detecting and monitoring. *Landsat 3*, launched in March 1978, differs from its predecessors in one important respect: the RCA return-beam vidicon (RBV) camera installation has been modified to provide one-half of the linear Earth surface coverage, at twice the resolution of the earlier Landsat RBV cameras. The *Landsat 3* RBV resolution is approximately 40 m, compared with the *Landsat 1* and *2* RBV and multispectral scanner (MSS) resolution of 80 m.

Ocean resources. In June 1978 *Landsat 3* was joined in orbit by its ocean-observing companion, *Seasat 1*. Economists at ECON, Inc., assessed the potential benefits obtainable through the use of data from an operational Seasat system to provide improved ocean condition and weather forecasts to be in the range of $859,000,000–2,700,000,000. This estimate is based on a postulated three to six spacecraft system, operating from 1985 to 2000. The benefits breakdown is as follows: one-third ocean fishing (ocean currents and temperatures data and safety improvements); one-fifth tanker operations (improved ocean and weather forecasts); one-fifth marine transportation (improved forecasts); one-sixth offshore oil and gas activities (production platform installation savings); one-tenth arctic operations (ice information for ice-breaking tankers). This important proof-of-concept mission was brought to a premature end on October 10, 1978, by a failure that occurred in the *Seasat*'s power system.

Navigation by satellite. The use of satellites for navigation started with *Sputnik 1* in 1958, when engineers tracking the satellite observed that the Doppler shift of the received telemetry signal could be used to determine the satellite's orbit. Conversely, they concluded that if the orbit was known, a receiver's position on Earth could be determined. After several years of work at the Johns Hopkins University's Applied Physics Laboratory, the Transit satellite series, and satellite navigation, were born. Latitude and longitude of a position are determined by receiving four successive satellite broadcasts, computing the Doppler hyperboloid of position for each one, and then solving four simultanous equations to give a point location. Actually, only three broadcasts are needed to solve for position, but the redundant fourth measurement improves accuracy, which is typically 0.1 km or better for this first-generation system.

The next-generation satellite navigation system is undergoing early testing in the field. Called Navstar or Global Positioning System (GPS), it provides three-dimensional position accuracies (latitude, longitude, and altitude) of 10 m, velocity of 3 cm/s, and time. The first prototype satellite, *NavStar 1,* was launched in February 1978. The receivers can be made small and lightweight and can be carried in ships, aircraft, missiles, ground vehicles, and even backpacks. The GPS system is scheduled for full 24-satellite operational capability by the mid-1980s.

Space communications. In 1978 there was rapid expansion of Earth's communications capability using satellites. The U.S. Navy's first fleet communications satellite, *FltSatCom A*, was launched in February. In April, *BSE*, the first Japanese experimental television broadcast satellite, was launched by NASA. In May, the backup *OTS 2* European communications test satellite was launched successfully. With this launch the ESA nations acquired the 14- and 11-GHz capability which had been impossible in 1977, due to the failure of *OTS 1* to achieve its desired geostationary orbit.

For background information *see* COMMUNICATIONS SATELLITE; MANNED SPACE FLIGHT; SATELLITES, APPLICATIONS; SATELLITES, SCIENTIFIC; SPACE FLIGHT; SPACE PROBE in the McGraw-Hill Encyclopedia of Science and Technology.

[JOHN F. CLARK]

Bibliography: *Aviat. Week Space Technol.*, pp. 40–166, Oct. 17, 1977, pp. 18–19, Jan. 23, 1978, p. 22, Feb. 13, 1978, p. 48, July 10, 1978, p. 17, Aug. 21, 1978, pp. 22–23, Oct. 16, 1978; A. Calio and N. Hinners, *Statements before the Subcommittee on Space Science and Applications, Committee on Science and Technology, United States House of Representatives*, Feb. 2, 1978.

Space shuttle

The United States commitment to a new era of space transportation in the 1980s with the advent of the reusable space shuttle will provide the ability to greatly reduce the cost of space flight and provide easy, rapid access to space for government-sponsored public services and for commercial, scientific, and defense purposes. The space shuttle will perform three basic categories of missions. The sortie missions will last from one to several weeks and use one or more of the Spacelab modules being developed by the European Space Agency; these missions will consist of a variety of experiments and observations. There will also be missions using additional propulsion stages to carry spacecraft to orbits beyond the capabilities of the space shuttle alone. In the third type of missions, satellites will be deployed directly into orbit or retrieved from orbit with the aid of a remote manipulator system.

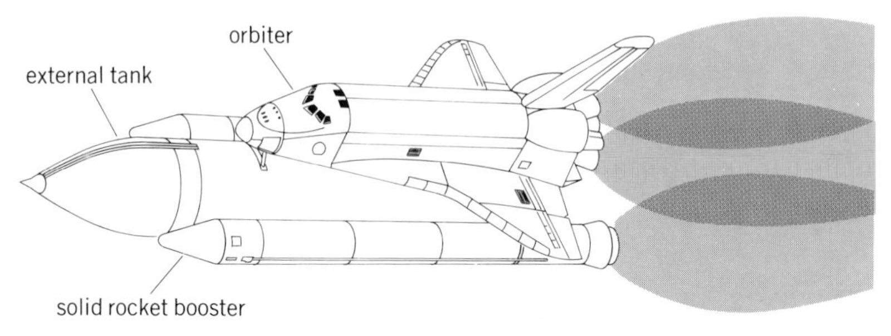

Fig. 1. Space shuttle system.

The capability of the space shuttle to perform these missions economically is the heart of the space transportation system. Reusability is the key factor in making these missions economically feasible. To place a satellite into orbit with an expendable launch vehicle would cost $25,400,000 in 1975 dollars; to place the same satellite into orbit with the space shuttle would cost only $16,300,000. Plans are being made to provide space shuttle operations with costs of (1) $26,300,000 in 1980 dollars for the United States government payload, and (2) $31,000,000 for the non–United States government payload.

Small self-contained payloads can be flown on the space shuttle for a maximum cost of $10,000 in 1975 dollars and a minimum cost of $3000. These payloads require no space shuttle services (power, deployment, and so forth) and will be flown on a space-available basis provided they weigh less than 200 lb (90.7 kg) and occupy less than 5 ft^3 (0.14 m^3) of volume.

Description. The space shuttle flight system (Fig. 1) consists of an orbiter, three space shuttle main engines (SSMEs), an external tank (ET), and a solid rocket booster (SRB) made up of two solid-fuel rocket motors. The orbiter, the SSME, and the SRB are all reusable; an ET is expended on each launch. Specifications of the space shuttle system and its components are given in Table 1.

The orbiter normally carries a crew of four, with provisions for a crew of as many as seven, and payloads. It can normally remain in orbit for 7 days, return to Earth with personnel and payload, land like an airplane, and be refurbished for a subsequent flight in 14 days.

Ascent and orbit insertion. The space shuttle is launched with the three SSMEs and the two SRBs burning in parallel (Fig. 2). A maximum dynamic pressure of approximately 650 lbf/ft^2 (31,122 N/m^2) is experienced at 62.5 s after launch at an altitude of 37,000 ft (11,278 m). The SRBs are separated at approximately 122 s after lift-off at an altitude of 142,200 feet (43,343 m), 25 nautical miles (46.3 km) down range from the launch site. After SRB separation, the orbiter continues to ascend, using the three SSMEs. The main engine cutoff (MECO) takes place 480 s after lift-off, when the orbiter has reached an altitude of approximately 380,000 ft (115,824 m). The ET separation occurs at MECO. The orbital maneuvering system (OMS) engines provide the additional velocity needed to insert the orbiter into an elliptical orbit having a minimum apogee of 150 n mi (277.8 km). The OMS engine cutoff occurs 600 s after launch at an altitude of 418,650 ft (127,604 m), when the orbiter is 1428.2 n mi (2645.0 km) from the launch site. At first apogee, the orbiter initiates the first of two maneuvers

Table 1. Specifications of the space shuttle system and its components

Component	Characteristic	Specifications
Overall system	Length	184.2 ft (56.1 m)
	Height	76.6 ft (23.3 m)
	System weight	
	Due east launch	4,490,800 lbm (2037.0 Mg)
	104° launch azimuth	4,449,000 lbm (2018.0 Mg)
	Payload weight	
	Due east launch	65,000 lbm (29.5 Mg)
	104° launch azimuth	32,000 lbm (14.5 Mg)
External tank	Diameter	27.8 ft (8.5 m)*
	Length	154.4 ft (47.1 m)
	Weight	
	Launch	1,648,000 lbm (747.6 Mg)
	Inert	70,990 lbm (32.2 Mg)
Solid rocket booster	Diameter	12.2 ft (3.7 m)
	Length	149.1 ft (45.4 m)
	Weight (each)	
	Launch	1,285,100 lbm (582.9 Mg)
	Inert	176,300 lbm (80.0 Mg)
	Launch thrust (each)	2,700,000 lbf (12.0 MN)
Separation motors (each SRB), four aft, four forward	Thrust (each)	22,000 lbf (97.9 kN)
Orbiter	Length	121.5 ft (37.0 m)
	Wingspan	78.1 ft (23.8 m)
	Taxi height	~57 ft (17.4 m)
	Weight	
	Inert	161,300 lbm (73.2 Mg)
	Landing	
	With payload	~203,000 lbm (92.1 Mg)
	Without payload	~173,000 lbm (78.5 Mg)
	Cross range	1,100 n mi (2037.2 km)
Payload bay	Length	60 ft (18.3 m)
	Diameter	15 ft (4.6 m)
Main engines (three)	Vacuum thrust (each)	470,000 lbf (2090.7 kN)
Orbital maneuvering system engines (two)	Vacuum thrust (each)	6,000 lbf (26.7 kN)
Reaction control system engines		
38 engines	Vacuum thrust (each)	870 lbf (3870.0 N)
6 vernier rockets	Vacuum thrust (each)	24 lbf (111.2 N)

*Includes spray-on foam insulation.

to circularize the orbit at an altitude of 150 n mi (277.8 km).

Return to Earth. Upon completion of orbital operations, the orbiter is oriented to a tail-first attitude. After the OMS provides the deceleration thrust necessary for deorbiting, the orbiter is reoriented with nose forward to the proper attitude for entry. The orientation of the orbiter is established and maintained by the reaction control system (RCS) down to the altitude at which the atmospheric density is sufficient for the pitch and roll aerodynamic control surfaces to be effective (at about 250,000 ft, or 76,200 m, altitude and 26,000 ft/s, or 7924.8 m/s, velocity). The yaw RCS remains active until the vehicle reaches an angle of attack of approximately 10° (at about 80,000 ft, or 24,384 m, altitude).

The orbiter entry trajectory provides energy management for an unpowered landing that includes lateral and downrange maneuvering. The trajectory, the lateral range, and heating are controlled through the vehicle angle of attack and bank angle. The angle of attack is established at 38° for the theoretical entry interface of 400,000 ft (121,920 m) altitude. The entry flight path angle is −1.19°. The 38° attitude is held until the speed is reduced to 21,700 ft/s (6614.2 m/s) at about 220,000 ft (67,056 m) altitude; is then reduced gradually to 28° until speed is reduced to 8500 ft/s (2590.8 m/s) at about 150,000 ft (45,720 m) altitude; and then is gradually reduced to 6° when the speed is approximately 1500 ft/s (457 m/s) at about 70,000 ft (21,336 m) altitude at the beginning of the terminal area energy management (TAEM) phase.

During the final phases of descent, flight path control is maintained by using aerodynamic surfaces. The TAEM phase is initiated to provide the proper vehicle approach to the runway with respect to position, energy, and heading. Final touchdown occurs at an angle of attack of approximately 16°. The maximum landing speed for a 32,000-lbm (14,515-kg) payload is approximately 210 knots (108 m/s).

Space shuttle main engines. Each of the three liquid-oxygen/liquid-hydrogen (LO_2/LH_2) SSMEs operates with a fixed nozzle area ratio (ratio of area

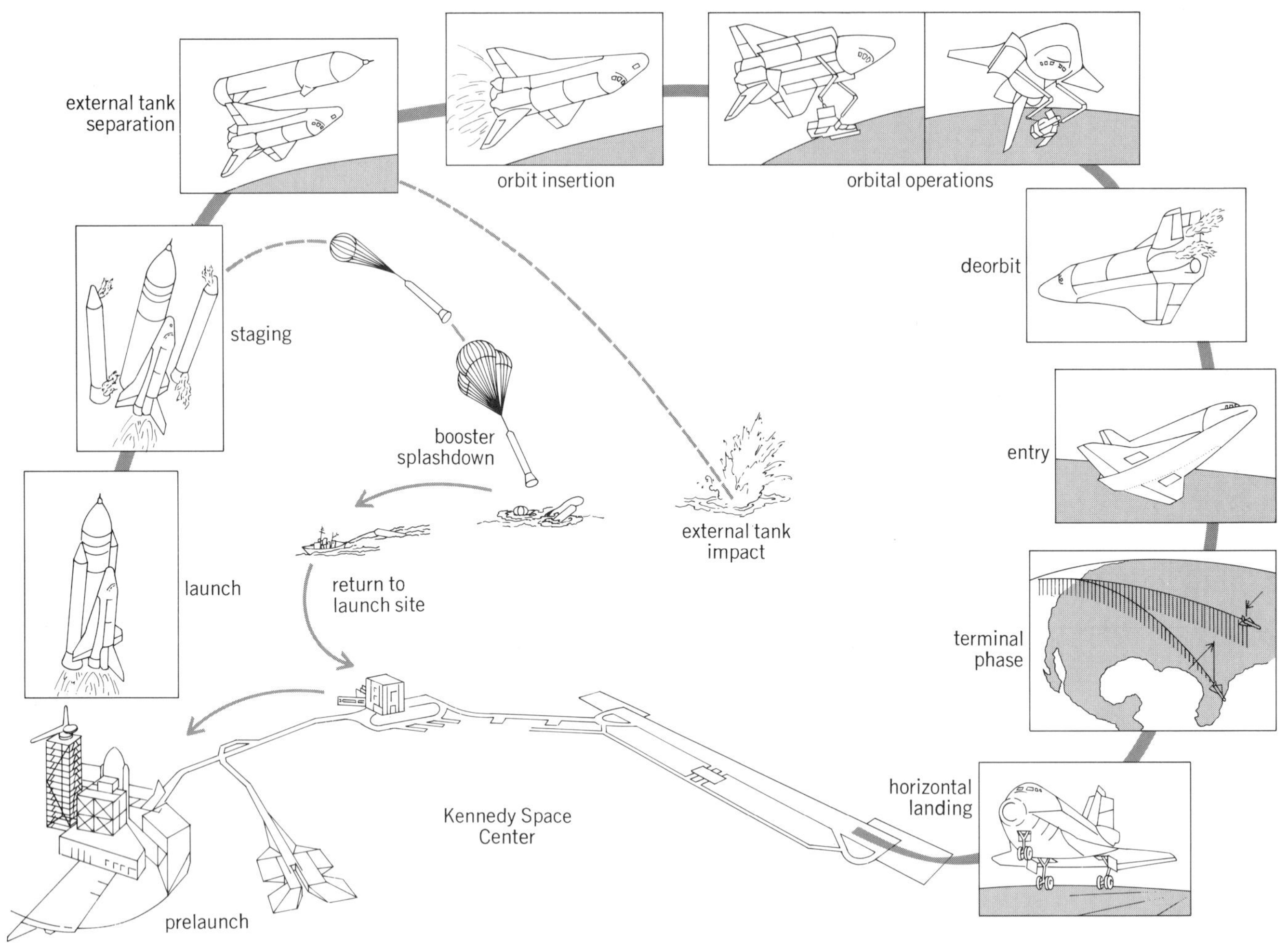

Fig. 2. Typical mission profile for space shuttle.

at the nozzle inlet to that at the throat) of 77.5:1, a mixture ratio (LO_2/LH_2) of 6:1, and a chamber pressure of 3000 psia (20,684 kN/m²) to produce a rated sea-level thrust of 375,000 lbf (1668 kN) and a vacuum thrust of 470,000 lbf (2091 kN). The engine can be throttled over a thrust range of 50 to 109% of rated thrust level to enable limitation of orbiter acceleration to 3 *g*. The engines are capable of being gimbaled ±10.5° in pitch and ±8.5° in yaw for flight control.

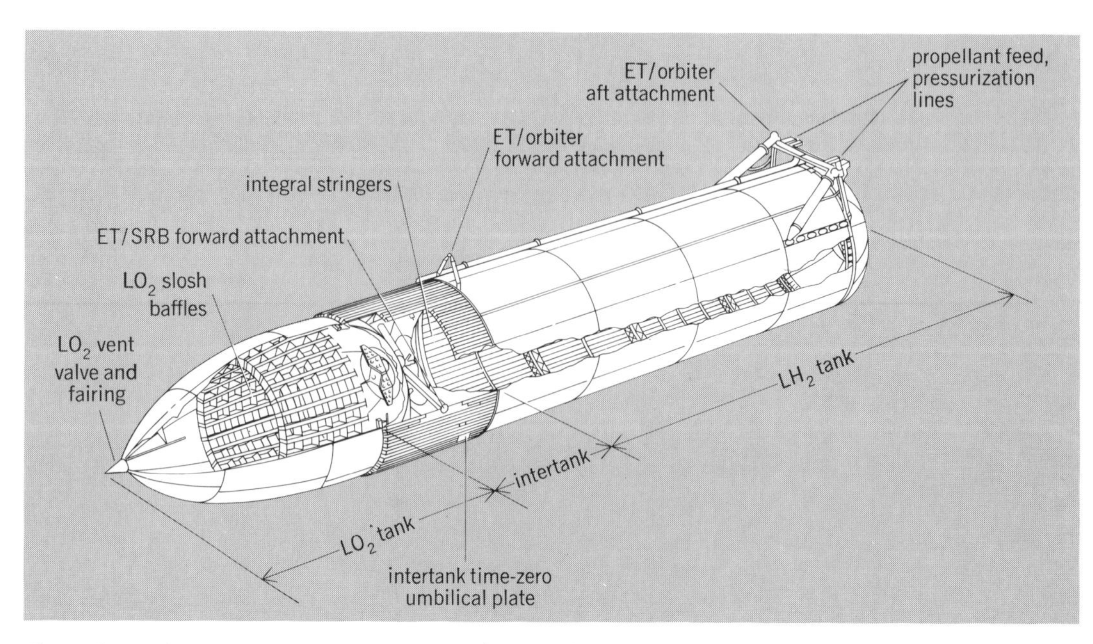

Fig. 3. Space shuttle external tank (tank 26 and subsequent).

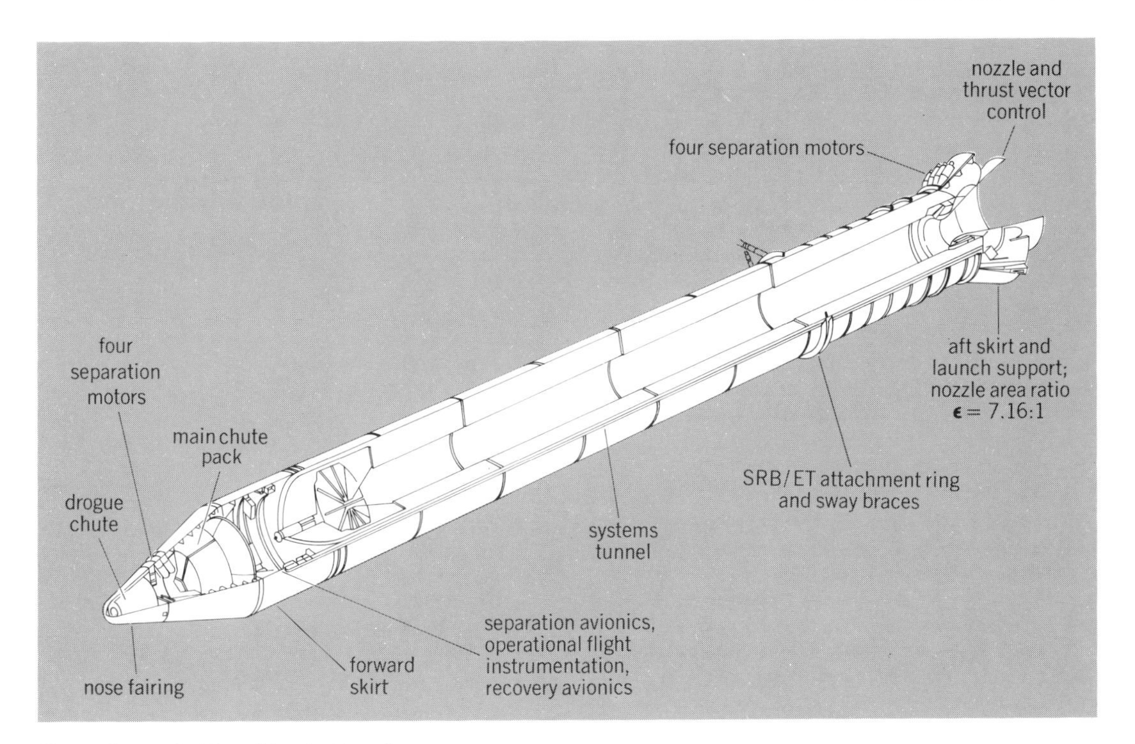

Fig. 4. Space shuttle solid rocket booster.

External tank. The ET (Fig. 3) provides 1,550,000 lbm (703,068 kg) of usable ascent propellant for the SSMEs. Five main propulsion fluid lines interface with the ET through disconnection fittings (disconnects) located at the bottom of the orbiter aft fuselage. The three hydrogen disconnects are mounted on a carrier plate on the left side of the orbiter (facing forward), and two oxygen disconnects are mounted on the right side.

The ET consists of a forward LO_2 tank, an unpressurized intertank, and an LH_2 tank. The LO_2 tank has a volume of 19,500 ft³ (552 m³) and is designed to operate at a nominal pressure range of 20 to 22 psia (137.9 to 151.7 kN/m²). The tank is an aluminum monocoque structure. The 55,552-ft³ (1573-m³) LH_2 tank is an aluminum semimonocoque structure, and operates at a nominal pressure of 32 to 34 psia (220.6 to 234.4 kN/m²). The ET is separated from the orbiter at MECO by means of pyrotechnic devices at the ET/orbiter forward and aft attachment points.

Solid rocket boosters. Each SRB (Fig. 4) weighs approximately 1,285,100 lbm (582,912 kg) and produces 2,700,000 lbf (12,010 kN) of thrust at sea level. The propellant grain is shaped to reduce thrust by approximately one-third at 55 s after lift-off to prevent overstressing the vehicle during the period of maximum dynamic pressure. The SRB thrust vector is controlled by a closed-loop hydraulic system having an omniaxial gimbal capability of 7.1°. The SRBs are released from the ET by pyrotechnic separation devices at the forward attachment and aft sway braces.

Orbiter dimensions and components. The orbiter (Figs. 5 and 6) contains the crew and the payload for the space shuttle vehicle system. The orbiter can deliver payloads of as many as 65,000 lbm (29,483 kg) with lengths as great as 60 ft (18.3 m) and diameters to 15 ft (4.6 m). The orbiter is comparable in size and weight to modern transport aircraft. It has a dry weight of approximately 161,300 lbm (73,164 kg), a length of 121.54 ft (37.04 m), and a wing span of 78 ft (23.8 m).

Most of the orbiter structure is conventional aluminum protected by reusable surface insulation. The forward fuselage structure is composed of aluminum alloy skin/stringer panels, frames, and bulkheads. The crew module, which is supported within the forward fuselage by four attachment points, is welded to create a pressure-tight vessel. The module has a side hatch for normal ingress and egress and a hatch from an airlock into the payload bay for extravehicular activity. The mid-fuselage is a 61-ft-long (18.6-m) section of primary load-carrying structure between the forward and aft fuselage sections. It includes the fin carry-through structures and the payload bay door. The aft fuselage includes a truss-type internal structure of diffusion-bonded titanium that transfers the main engine loads to the mid-fuselage and the ET. The wing is of conventional alloy construction. The elevons are aluminum honeycomb, and the body flap is of aluminum skin/stringer construction. The vertical stabilizer is a conventional aluminum structure. The rudder–speed brake assembly is divided into upper and lower sections. Each section is also split longitudinally and individually actuated to serve as both rudder and brake. Characteristics of the wing, vertical stabilizer, and control surfaces are given in Table 2. The payload bay door is a composite structure of graphite epoxy. The forward 30-ft-long (9.1-m) section of the door incorporates deployable radiators that are hinged and latched to the inner surface of the door. The forward radiators can be deployed in orbit. Fixed aft radiators are attached to the remaining inner surface of the door. The aft propulsion system (APS) pod structure is fabricated from graphite epoxy. It is 21.5 ft (6.6 m) long. The APS has two compartments, the OMS pod and RCS housing.

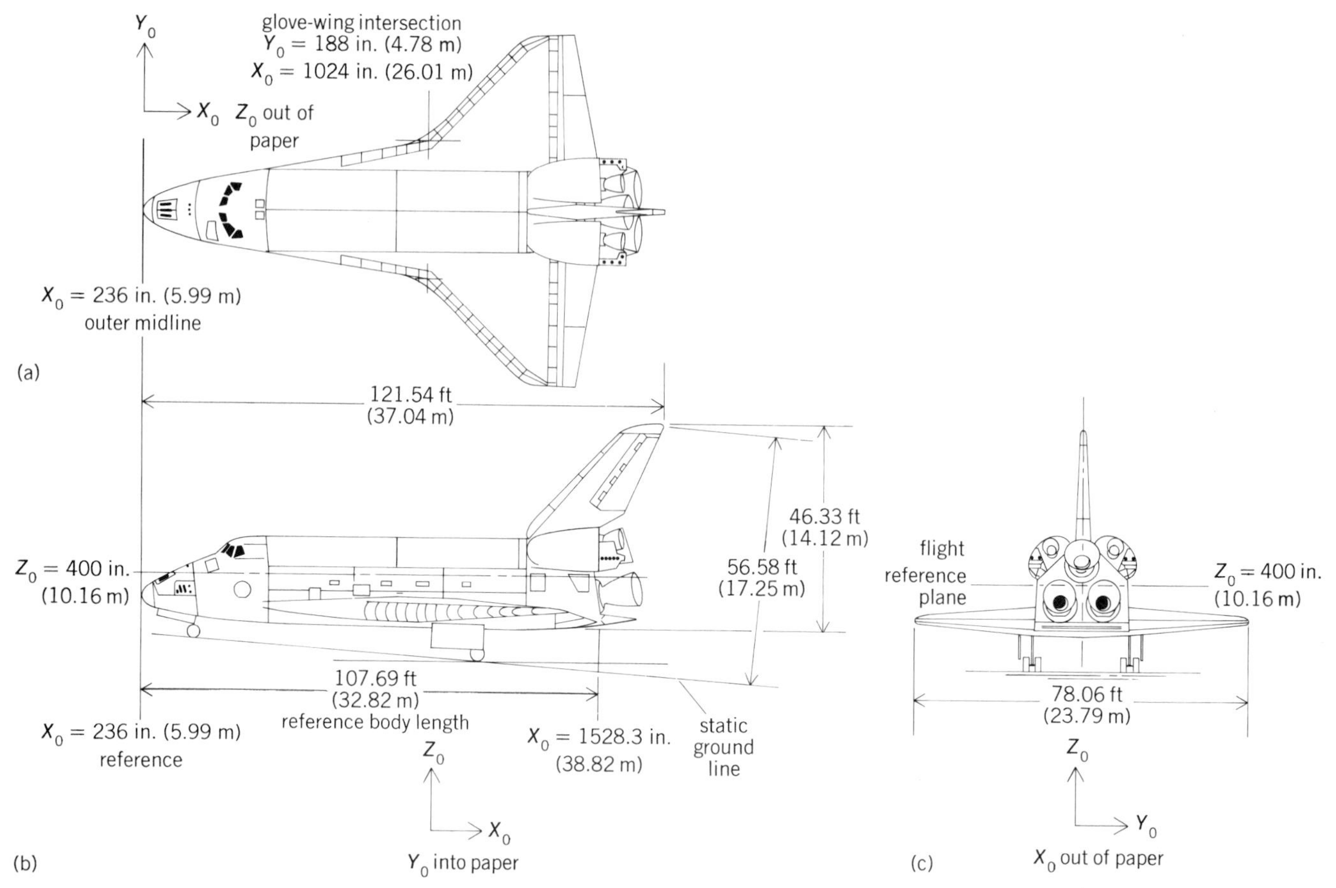

Fig. 5. Dimensions of the orbiter. *(a)* Top view. *(b)* Side view. *(c)* Rear view. The directions of axes of the reference coordinate system are indicated in each part of the figure.

Fig. 6. The orbiter at landing attitude. *(NASA)*

Table 2. Orbiter vehicle dimensions

Geometry	Wing	Vertical stabilizer
Area	2690 ft^2 (249.91 m^2)	413.25 ft^2 (38.39 m^2)
Aspect ratio	2.265	1.675
Airfoil, Y_0= 199 in. (5.05 m)*		Wedge
Sweep		
Leading edge	45°	45°
Wing glove	81°	–
Mean aerodynamic chord	474.81 in. (12.06 m)	199.81 in. (5.08 m)
Dihedral (trailing edge)	3°30′	–
Control surface	**Area**	**Maximum deflection**
Elevon (one side)	206.57 ft^2 (19.19 m^2)	−35 to +20°
Rudder	97.15 ft^2 (9.03 m^2)	±22.8°
Speed brake	97.15 ft^2 (9.03 m^2)	0 to 87.2° (total)
Body flap	135.75 ft^2 (12.61 m^2)	−11.7 to +22.55°

*Refers to coordinate system in Fig. 5.

Thermal protection system. The thermal protection system consists of materials applied externally to the primary structure of the orbiter that maintain the temperature of the airframe outer skin at less than 350°F (450 K). The materials are designed for 100 mission uses. The following are the types, temperature limits, and locations of the various materials used for thermal protection: (1) Coated high-temperature-resistant nylon felt is used in areas where temperatures are less than 750°F (672 K); that is, the upper cargo bay door, the mid- and aft fuselage sides, the upper wing surface, and the APS pod. (2) Low-temperature reusable surface insulation is used where temperatures are below 1200°F (922 K) and above 750°F (672 K). (3) High-temperature reusable surface insulation is used in those areas exposed to temperatures below 2600°F (1700 K) and above 1200°F (922 K), that is, the fuselage and the wing lower surfaces. (4) Reinforced carbon-carbon is used on areas, such as the wing leading edge and the nose cap, where predicted temperatures exceed 2600°F (1700 K).

Orbital maneuvering system. The OMS consists of two engines; each has a thrust of 6000 lbf (26.7 kN) and usable propellant [nitrogen tetroxide (N_2O_4) and monomethyl hydrazine (MMH)] of 23,876 lbm (10,830 kg). This system is used for orbit insertion, orbit transfer, rendezvous, and deorbit.

Reaction control system. The RCS employs thirty-eight 870-lbf (3.9 kN) thrusters (14 forward and 2 pods of 12 each aft) and six 25-lbf (111-N) thrusters (2 forward and 4 aft). The propellant for the RCS thrusters, as for the OMS engines, is N_2O_4 and MMH.

Mechanical subsystems. The orbiter mechanical subsystems, with the electrical and hydraulic actuators, operate the aerodynamic control surfaces, the landing/deceleration system, the payload bay doors, the deployable radiators, and the payload-retention and -handling subsystems. Orbiter/ET disconnects, vent doors, and a variety of other mechanical and pyrotechnic devices make up the balance of the mechanical subsystems.

The payload-handling subsystem consists primarily of the remotely controlled manipulator that can extract payloads from and insert payloads into the payload bay while the orbiter is in orbit.

The orbiter hydraulic system is driven by three 135-hp (100.7-kW) independent auxiliary power unit (APU) systems that derive their energy from the decomposition of hydrazine (N_2H_4). The APUs provide mechanical shaft power for the hydraulic pumps during prelaunch, ascent, entry, and landing phases. Each APU drives a hydraulic pump that supplies hydraulic power to actuate elevons, rudder–speed brakes, body flap, main engine thrust vector control, landing gear, brakes, and steering.

Electrical power system. The electrical power system (EPS) consists of three fuel cells together with distribution equipment and reactants required to supply electrical power to the electrical buses. The EPS can be functionally divided into two major elements: (1) The power generation subsystem, consisting of three fuel cells, each capable of delivering a minimum power of 2 kW, a continuous power of 7 kW, a peak power of 12 kW, and 27.5 to 37.5 V of direct current. (2) The power reactant storage and distribution subsystem, which can store hydrogen and oxygen capable of producing 2370 kW (8532 MJ) of delivered energy, and 168 lbm (76.2 kg) O_2 for life support.

Atmosphere revitalization system. The atmosphere revitalization system provides the flight crew and passengers with a conditioned environment that is both life-supporting and within crew comfort limitations. The cabin atmosphere, composed of 20% oxygen and 80% nitrogen, is maintained at a pressure of 14.7 psia (101.4 kN/m^2) and a temperature between 70 and 80°F (294 and 300 K).

Active thermal control subsystem. The active thermal control subsystem provides thermal control for several subsystems during mission phases. The active thermal control subsystem consists of a Freon heat-transport loop, a radiator system, an ammonia evaporation system, a flash evaporator system, and cold plates.

Other subsystems. The food, water, and waste management subsystems are designed to support hot and cold food preparation; to provide sterilization, storage, and dispensing of hot and cold potable water; and to collect and store condensates, disinfected urine, and feces for return to Earth. The smoke detection and fire suppression subsystem is designed to detect smoke and extinguish fires by using Freon 1301.

Avionics system. The space shuttle avionics system provides command functions and implementation; guidance, navigation, and control capability; communications; computations; displays and controls; instrumentation; and electrical power distribution and control for the orbiter, SSMEs, ET, and SRBs. The guidance, navigation, and control subsystem provides the following: (1) automatic and manual control capability for all mission phases; (2) guidance commands that drive control loops and provide steering displays for the crew; and (3) inertial guidance updated by star sensors or Doppler signals and by radio-frequency navigation aids for approach and landing.

The basic aerodynamic stability of the orbiter is augmented with body-mounted rate gyroscopes and accelerometers. Rotational hand controllers, rudder pedals, and trim controls enable manual control, and the guidance, navigation, and control computers provide commands for flight control functions.

Attitude information is obtained from redundant inertial measuring units (IMUs). Air data are provided by redundant probes deployed at lower altitudes. Gimbaled IMUs provide the navigation reference with star sensors for autonomous alignment and state-vector updating.

Landing is accomplished by means of a computer flight path generated in the guidance, navigation, and control computers. The inertial system is used for reference, with continuous updates from tactical air navigation (Tacan) equipment and from a microwave scanning beam landing system. Radar altimeter updates are used near touchdown.

The orbiter carries as many as 23 antennas for communication with ground stations, detached payloads, and crew undertaking extravehicular activity. Through these antennas, information is both transmitted and received at S-, K_U-, L-, C-, and P-band frequencies. Phase-modulated data can be transmitted to the spaceflight tracking and data network or to satellite control facility ground stations on two S-band carrier frequencies. Two additional S-band carrier frequencies are employed for ground-to-orbiter transmission of frequency-modulated data. The K_U-band link between ground stations and the orbiter by way of the tracking and data relay satellite system carries the same kinds of intelligence as the S-band system, but at wider bandwidths and higher data rates.

Testing. The major developments in the space shuttle system in the 12 months preceding August 1978 were the completion of the approach and landing test (ALT) program, the start of the main propulsion test article (MPTA) program, and the start of the mated vehicle ground vibration test (MVGVT) program.

Approach and landing test program. The ALT was started on February 18, 1977, and completed on October 26. The program began with flight of an inert (unpowered) orbiter on top of a Boeing 747 to calibrate the aerodynamic characteristics of the combined configuration. Once this objective was completed, a series of three flights was flown with the orbiter active, for the purpose of gaining confidence in the orbiter systems. The orbiter as described in the previous section was flown, except for the propulsion, the S- and K_U-band, the thermal protection, and a portion of the active thermal control equipment. In other words, the captive-active flights of the orbiter/747 enabled validation of such components as the APU, hydraulics, avionics, software, fuel cells, mechanical subsystems, and the environmental control system before separating the orbiter from the 747. During the captive-active flights, there were several occurrences of a leaking seal in the APU. This problem was corrected.

The next phase of the ALT program was free flight. In this phase, the 747 with the orbiter was flown to an altitude of approximately 28,000 ft (8534 m) at a speed of approximately 270 knots (139 m/s). At the appropriate separation flight condition, pyrotechnic bolts similar to ones used on the orbiter/ET were fired and the orbiter was separated from the 747. The net positive separation force between the vehicles under these conditions was 0.75 *g*. A total of five free flights were performed: three flights were made with an aft fairing on the orbiter to reduce buffeting on the 747, while the final two were made without the fairing.

Free flight 1. Free flight 1 was targeted for landing on a lakebed at the NASA Hugh L. Dryden Flight Research Center in California. The purpose of this test was to exercise all the active systems of the orbiter, and it was considered to be one of the best systems tests possible. Subsonic aerodynamic characteristics as well as handling qualities of the orbiter were also studied.

The state of the art is being taxed in two areas of the orbiter. One is the software required for the redundancy management system, especially the computers, and the other is the reusable thermal protection system.

From a software viewpoint, the most complex part of redundancy management is at the computation level. Here, all four computers in a redundant set are kept tightly synchronized by checking their operations 440 times per second. This checking is done at specified processing points, such as when a computer requests an input, initiates an internal operation, or generates an output. Failure at any checkpoint causes the computer to be dropped from the redundant set. Recognition of a failure is handled, as are all other operations, by all four computers operating in parallel. Each reports to the other three at every checkpoint, and each computer "votes" on itself as well as on the others.

Immediately after separation on free flight 1, the computer in charge failed and was voted out of the redundant set. The next computer assumed control; from then on, all systems were nominal. The orbiter made a perfect landing on the lakebed. This one test proved in flight that the most complex system worked as designed. After the flight, the reason for the computer being voted out of the redundant set was isolated and corrective action was taken.

Free flight 2. Free flight 2 was characterized by much the same profile as was free flight 1; the prime difference was that more vigorous exercising of the brakes was planned than was done in the previous flight. All systems of the orbiter operated nominally. The landing was nominal; however, upon brake application, a chattering occurred even though braking was normal. The rollout distance (distance from touchdown to halting of the orbiter) was 10,037 ft (3059 m).

Free flight 3. Free flight 3 was the final flight with the tail-cone fairing on the orbiter. The separation altitude was 29,500 ft (8992 m) at an airspeed of 250 knots (128.6 m/s). The free-flight phase was essentially the same as that of free flight 2. The flight plan called for a 1.8-*g* windup turn and application of both programmed test inputs and aerodynamic stick inputs for aerodynamic and flight control evaluation. After 23 s of coasting, following touchdown, gentle to moderate differential braking was performed commencing at a speed of approximately 150 knots (77 m/s). Moderate to hard braking was performed at a speed of approximately 115 to 120 knots (59 to 62 m/s). Chattering was again experienced during hard braking at about 110 knots (57 m/s). Nosewheel steering was engaged at a speed of 12 knots (6.2 m/s). The total rollout distance was 9184 ft (2799.3 m).

A postflight analysis indicated that the chattering of the orbiter during hard braking was due to an instability in the antiskid subsystem. The problem was analyzed, and a change was made in the antiskid lead-lag filter. The change was tested in the next free flight, and the chattering was eliminated.

Free flight 4. Free flight 4 was the first flight without the aerodynamic tail-cone fairing. Simulated main engines were exposed in the aft end of the orbiter. The free flight time for the orbiter with tail cone was approximately 5.5 min, whereas the free flight time without tail cone was approximately 2.5 min. The corresponding lift-to-drag ratios were approximately 8 and 4, respectively, whereas the corresponding flight path angles were approximately 14 and 24°, respectively. The main gear touchdown velocity with tail cone was approximately 190 knots (98 m/s); without tail cone, it was approximately 200 knots (103 m/s). The corresponding nose gear touchdown velocities were 140 knots (72 m/s) and 160 knots (82 m/s), respectively.

Free flight 4 was targeted for landing on the lakebed without the tail-cone fairing. The orbiter was separated from the 747 at an altitude of 28,200 ft (8595 m) and a speed of 248 knots (127.6 m/s). The subsequent 2-min 35-s free flight of the orbiter was performed as planned with application of angle of attack, sweep, and aerodynamic stick inputs for performance as well as stability and flight control test data. All systems operated nominally. The braking tests were nominal, and the chattering was eliminated. The vehicle stopped at a rollout distance of approximately 5725 ft (1745 m).

Free flight 5. Free flight 5 was the last flight to be made. This flight was performed without a tail cone and was targeted for landing on a concrete runway. The orbiter approach and landing were controlled manually in the control stick steering flight control mode through the entire free flight until touchdown. For the last 8 s before touchdown, there was a pitch oscillation caused by control stick inputs to control sink rate. These inputs kept the elevons rate saturated, and the flight control system did not respond to some roll inputs. The accumulated, very large roll commands apparently were triggered just at touchdown. The vehicle touched down softly with wings level, but skipped back into the air rolling right. A pilot-induced oscillation in roll then occurred for 4 s. The pilot ceased roll input momentarily, and the motion damped quickly just before a second touchdown, which occurred 6 s after the first. Total runway rollout distance from the first touchdown point was approximately 7930 ft (2417 m).

Postflight simulations and results. After this flight, a number of astronaut-in-the-loop simulations were run at the NASA Lyndon B. Johnson Space Center, at the prime contractor facility, and at the NASA Ames Research Center. The results of these simulations and of free flight 5 indicated the need for changes in the priority rate limiting between pitch and roll, for lower gains, for larger breakout forces in the hand controller, and for a reduction in the flight control transport lag (the delay involved in propagating signals from one point to another).

The basic results of the ALT program showed that the basic orbiter hardware and software systems were sound and that the subsonic aerodynamic characteristics were as predicted in the wind-tunnel test program. The Cooper-Harper rating for pitch axis control and landing was approximately 2.

Main propulsion test article program. The MPTA consists of the orbiter flight-type main propulsion system and an aft-end orbiter-type structure supported by a simulated orbiter airframe structure. Three main propulsion engines are installed in the orbiter. A flight-type ET is attached to the orbiter structure, and its lines are connected to the orbiter in a flight configuration.

To date, there have been four firings of the MPTA. These tests, performed at 90% of rated power level for durations of 2.5, 20, 40, and 100 s, have been extremely successful. The total system, that is, the orbiter, engines, and tank, has operated nominally. This testing is a major milestone in preparing for vertical flight.

Mated vehicle ground vibration test program. The MVGVT is composed of the orbiter flown in the ALT program, a flight ET, and flight SRBs. After completion of the ALT program, the orbiter was carried on top of the 747 to the George C. Marshall Space Flight Center (MSFC) in Huntsville, AL. The orbiter was mated to a flight ET in the test stand. The purpose of this program was to subject the total flight vehicle configuration to expected vibration levels and to obtain the resulting mode shapes and frequencies for verifying ascent flight control system design parameters.

In August 1978, testing of the orbiter and the external tank that simulates the time period when the SRBs have been separated was well under way. The SRBs were to be mated to this configuration in October 1978, and testing was to be completed in February 1979.

Development, building, and scheduling. On August 1, 1978, individual main engines were being tested, and the building of flight engines had begun. The ET was undergoing structural testing, and construction of the flight tanks was well under way. The SRBs were undergoing firing tests at contractor facilities in Utah and structural tests at MSFC. Construction of flight hardware was proceeding on schedule.

The building of Orbiter 102, which is the first vertical-flight orbiter, was well under way. Preliminary electrical checks had already been performed, and additional integrated electrical and hydraulic checks were scheduled to begin in August 1978. The thermal protection system was being installed. Subsystems, as previously described, were also being installed.

Construction of the launch facility at the John F. Kennedy Space Center was well under way. The orbiter processing facility, the mobile launcher, and all required facilities were being readied to accept the orbiter in the February 1979 time period. Thus the space shuttle could be ready to fly in the September 1979 time period.

The space shuttle transportation system is complicated, yet very well conceived. The design, testing, and building of components to date have shown that the purposes of the space transportation system can be met.

For background information *see* AIRFRAME; FUSELAGE; MANNED SPACE FLIGHT; SPACE STATION; SPACECRAFT STRUCTURE in the McGraw-Hill Encyclopedia of Science and Technology.

[AARON COHEN]

Spectroscopy

Auger electron spectroscopy (AES) has become, over the last 10–15 years, the most widely used of the available techniques for surface elemental analysis. The distribution of elemental species in the outermost layers of a solid surface represents, however, a rather low order of information. Other useful information would be how each atom is bound to the surface and in what chemical form, that is, what are its neighbors and what are their geometric and electronic arrangements. The determination of such chemical information is imperative if one is to unravel the details of even the simplest surface chemical reactions. Recent studies have shown that the energy distributions (lineshapes) of the characteristic lines observed in the Auger electron emission can be used to probe an atom's chemical environment.

Use of Auger lineshapes. Information about an atom's chemical environment is characterized by the spatial and energy distribution of its valence electrons. In AES, surface atoms are excited by removing one of the tightly bound core-level electrons (not directly involved in the chemical bonding) through electron or photon bombardment. In returning to the unexcited state, these atoms emit photons (x-ray fluorescence) and electrons (Auger electron emission) whose energies are characteristic of the excited atomic species because the distribution of core-level energies is characteristic. In the case of electron emission, which is more probable for low-energy core states (less than 2000 eV), transitions giving rise to the characteristic electron emission involve electrons residing in less tightly bound states relaxing to the empty (excited) core state while giving up the characteristic energy difference to one of the other less tightly bound electrons of the excited atom. The fact that one or both of the two electrons involved in the Auger transition can originate in the valence states implies that the energy distribution of the characteristic Auger lines should carry information about the energy distribution of valence levels and, thus, about the atom's chemical environment.

Other surface-sensitive techniques yield information on the energy distribution of valence levels, but the unique feature of the Auger process is that since the transition itself involves a very spatially localized core level, the distribution of Auger electrons should be sensitive to the valence-level configuration local to the particular atom being probed. Thus, in a molecule such as carbon monoxide, one can probe the distribution of valence electrons both in the neighborhood of carbon, by analyzing the carbon Auger electrons, and in the neighborhood of oxygen, by analyzing the oxygen Auger electrons. In a practical sense, the difference in 1*s* binding energies for carbon and oxygen separates the energies of the Auger electrons from these two atoms by about 250 eV.

Although changes in Auger lineshapes have often been observed during studies involving elemental analyses and occasionally have been taken advantage of, AES is seldom used as a probe of chemical state. The reasons for this are twofold: First, the valence-level distribution is not measured directly, but must be inferred from transitions between levels, and the details of these transitions are only poorly understood. Second, it is difficult to characterize special lineshapes in terms of known surface chemical makeup because there exist at present no reliable surface standards. The difficulty of producing surface standards is, of course, a reflection of the fact that the surface itself is a strong reactant. In the rest of this article, recent AES results using gas-phase molecules will be discussed. The goal of gas-phase studies is to develop a sufficient understanding of the chemical information contained in Auger spectra to be able to interpret spectral results from surface species. In terms of the use of AES as a "fingerprinting" probe of chemical state, these results demonstrate that atoms in similar chemical environments yield similar lineshapes, while those in different environments yield contrasting line-

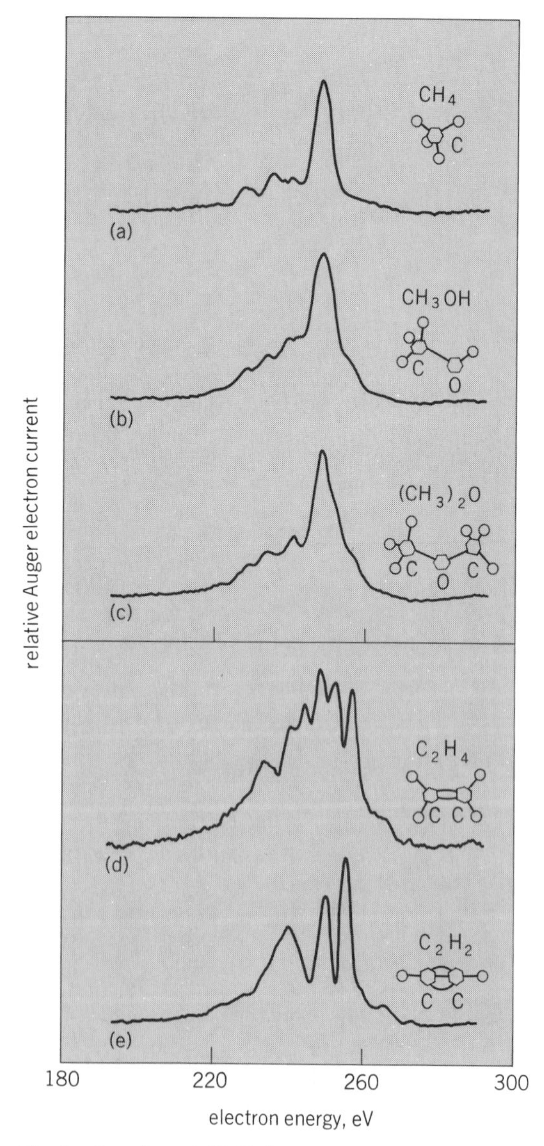

Fig. 1. Gas-phase C(KVV) Auger spectra and geometric configurations of molecules of *(a)* methane, *(b)* methyl alcohol, *(c)* methyl ether, *(d)* ethylene, and *(e)* acetylene.

shapes. In addition, results from a molecule having atoms of mixed environments will be shown to yield a superposition of spectral lineshapes.

Similar local environments. There are two advantages to using gas-phase molecules as absolute standards in characterizing the chemical information available from Auger spectroscopic results. First, the multitude of readily available molecular species makes it possible to vary the local environment about a specific atom in a completely systematic and controlled way. Second, a theoretical characterization of the spectroscopic results can be made using as a framework a wealth of existing calculations concerning detailed molecular orbital structures.

In characterizing the potential of AES as a fingerprinting probe of chemical environment, the first of these advantages makes it possible to choose a series of molecules, each containing the same local environment about a particular atom, to determine whether the spectroscopic results appear similar. In Fig. 1*a*–*c* the spectral results for carbon in the molecules methane (CH_4), methyl alcohol (CH_3OH), and methyl ether (CH_3OCH_3) are shown. These particular spectra are identified by convention as involving C(KVV) electrons, which means that the excited core state is a carbon K or 1*s* level and that both electrons involved in the Auger transition originate in valence states. In each of these molecules, the carbon atoms have a fairly consistent neighboring environment, that is, a tetrahedral bonding geometry which differs only in the replacement of a single hydrogen atom by oxygen. In each case the electron distribution about the carbon atom is consistent with sp^3 hybridization, and since the geometries of the bonding local to the methyl group carbon remain essentially unchanged, the molecular orbital distribution is expected to be quite similar.

The spectral lineshapes in Fig. 1*a*–*c* do, in fact, appear quite similar, consisting of one major peak at about 250 eV with three minor structures appearing at approximately 240, 235, and 230 eV. Subtle differences do, however, appear in the comparison. A high-energy shoulder appears in the methyl alcohol spectrum, while a double shoulder can be identified in the methyl ether results. Such differences in fine structure are certainly not unexpected since, although the overall electronic structure is similar, the symmetry present in methane has been broken in methyl alcohol and methyl ether. Such symmetry changes produce more subtle deviations in the local electronic structure and appear as additional features in the resulting Auger spectra.

The comparison of Fig. 1*a*–*c*, at least for this limited series of carbon-containing molecules, indicates that atoms with similar environments yield similar spectra. A necessary corollary is the requirement that contrasting carbon environments should produce contrasting spectra. For a molecular series containing such contrast methane (Fig. 1*a*) may be chosen and compared with ethylene (C_2H_4) and acetylene (C_2H_2). The overall molecular symmetry, as well as the local electronic structure, has now been changed. Ethylene (Fig. 1*d*) has a trigonal structure about carbon consistent with an sp^2 hybridization of the carbon atomic orbitals, while acetylene (Fig. 1*e*) is a linear molecule con-

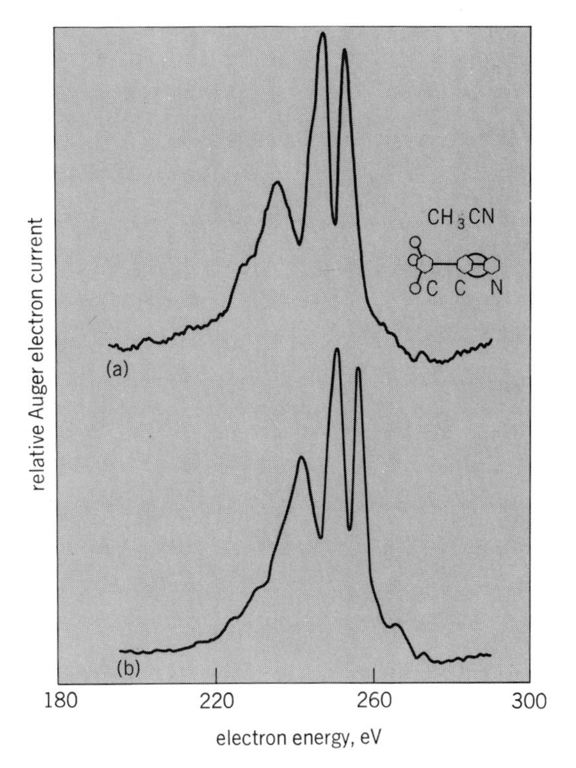

Fig. 2. Comparison of (*a*) the gas-phase C(KVV) Auger spectrum for the two carbon atoms in methyl cyanide with (*b*) the composite of the methyl alcohol and acetylene spectra of Fig. 1*b* and *e*.

sistent with *sp* hybridization. It is clear from a comparison of Fig. 1*a*, *d*, and *e* that the Auger results reflect these differences in local carbon environment.

Results for these series of simple molecules indicate the potential of AES as a probe of the local electronic state, and similar results have been obtained for numerous other molecules. Presumably, if one has a broad data base of spectroscopic results for, say, carbon in a variety of known environments, it should be possible to characterize the state of carbon in some unknown situation by a detailed analysis of the resulting carbon AES spectrum. Other results, however, are needed to demonstrate the full potential of AES as a probe of "local" environment.

Mixed local environment. There are, of course, virtually an endless number of molecules containing the same atoms in mixed chemical environments. For the present purpose, however, one should be picked for which spectral results corresponding to each of the individual environments already exist. An example of such a molecule is methyl cyanide (CH_3CN), shown in Fig. 2*a*. This molecule contains two carbon atoms: one methyl-group carbon (sp^3 hybridization) with a single bond to its carbon neighbor which itself is triply bonded (*sp* hybridization) to its nitrogen neighbor. The first of these carbon atoms has an environment very similar to the carbon atom of methyl alcohol shown in Fig. 1*b*. The cyanide-group carbon, on the other hand, has an environment quite similar to that found in acetylene (Fig. 1*e*).

The spectral results for this mixed-environment molecule are shown in Fig. 2*a*, which has an appearance remarkably like that of acetylene (Fig. 1*e*). In fact, the only major difference between these lineshapes is the relative prominence of a peak occurring at approximately 250 eV in methyl cyanide. However, the major structure in the methyl alcohol spectra is located at this same energy. If AES were truly a probe of local chemical environment, one might expect that the methyl cyanide lineshape would consist simply of the additive superposition of independent contributions from both methyl-like and acetylenic-like carbon. To verify this supposition, the methyl alcohol and acetylene spectra have been superimposed in Fig. 2*b*. The remarkable agreement in shape between the methyl cyanide structure and that for the composite spectra is certainly consistent with AES being a probe of the electronic structure local to each individual atom.

Conclusions. One of the principal advantages of the atomic specificity of AES is that information can be obtained for each distinct atomic species present. (Because Auger transitions involve two electrons above the core level, hydrogen, helium, and lithium do not show atomic Auger transitions.) In this discussion the lineshapes for carbon in a limited set of molecules have been emphasized. The results for oxygen in methyl alcohol and methyl ether and for nitrogen in methyl cyanide have not been discussed. These results, however, are in complete agreement with the conclusions reached from the carbon spectra that are shown in Figs. 1 and 2.

The overall conclusions to be reached from these rather preliminary results is that AES can give valuable information concerning an atom's local electronic environment. Recent theoretical calculations by D. R. Jennison have corroborated these conclusions and, in addition, have demonstrated that features in the spectroscopic results can be accurately identified as transitions between the various calculated molecular orbital states, and their intensities can be correlated with the atomic contributions to the various molecular orbitals. Such conclusions encourage the use of Auger analyses for molecular species interacting with solid surfaces. Preliminary studies involving molecular solids of several of the simple hydrocarbons shown here have been reported by R. R. Rye and colleagues. Again, systematic changes in spectral lineshapes show good correlation with detailed changes in local molecular bonding.

For background information *see* AUGER EFFECT; CHEMICAL STRUCTURES; SURFACE PHYSICS; VALENCE in the McGraw-Hill Encyclopedia of Science and Technology.

[ROBERT R. RYE; JACK E. HOUSTON]

Bibliography: D. R. Jennison, *Chem. Phys. Lett.*, submitted for publication; R. R. Rye et al., *J. Chem. Phys.*, 69:1504, 1978.

Spectroscopy, laser

Advances in laser technology and the exploration of novel coherent light techniques have created powerful new tools for spectroscopy. Recent progress in very-high-resolution spectroscopy is particularly noteworthy.

Lasers and Doppler-free spectroscopy. Several types of lasers, particularly dye lasers in the visible spectral region and color center lasers in the near-infrared, can be tuned continuously over wide wavelength ranges, while offering extreme spectral purity. J. Hall and associates, for instance, have reduced the bandwidth of a continuous-wave dye laser to a few kilohertz or less than 1 part in 10^{11}, using electronic feedback techniques for frequency stabilization.

Monochromatic tunable lasers can be powerful probes to investigate the structure of atoms and molecules. They can be used to study fine and hyperfine splittings, Zeeman and Stark splitting, light shifts, collision broadening, collision shifts, and other attributes of spectral lines. Perhaps more importantly, they can be used to measure wavelengths of spectral lines with unprecedented accuracy. The laser wavelength can also be locked to an atomic or molecular transition, providing an accurate standard of length or frequency. Such lasers have become important tools for precision metrology.

To take advantage of the very narrow instrumental line width of laser sources, however, it is necessary to overcome the Doppler broadening of spectral lines. Atoms which are relatively free and undisturbed are almost inevitably moving with high thermal velocity. Those atoms moving toward an observer appear to absorb light at higher frequencies than atoms at rest, and atoms moving away appear to absorb light at lower frequencies. In a gas, with atoms moving at random in all directions, the lines appear blurred, with typical Doppler widths on the order of 1 part in 10^6.

The oldest method to overcome this Doppler broadening is the transverse observation of a well-collimated molecular beam, so that the range of velocities along the line of sight is much restricted. However, it is difficult to observe rare species or molecules in short-lived excited states in this way. Fortunately, a number of clever schemes have been devised which permit Doppler-free spectroscopy of simple gas samples.

Saturation spectroscopy. Several of these methods use the laser light itself to "label" a group of slow-velocity molecules. The light intensity can be easily high enough to partly saturate the absorption of a spectral line. In other words, those atoms which have absorbed a light quantum are momentarily removed from the initial state. The absorption for that wavelength is thus reduced, at least until some relaxation process can replenish the supply of absorbing atoms.

In a widely used technique of saturation spectroscopy (Fig. 1), the output of a tunable laser is divided by a beam splitter into a stronger saturating beam and a weaker probe beam that traverse an absorbing gas sample along the same path but in opposite directions. When the saturating beam is on, it bleaches a path through the cell, that is, it depletes those atoms which are Doppler-shifted into resonance, and a stronger probe signal is received at the detector. As the saturating beam is alternately stopped and transmitted by a chopper, the probe signal is modulated. However, this happens only when both beams interact with the same atoms, that is, only atoms which are standing still or at most moving transversely. Thus the method selects only those atoms which have a near-zero velocity component along the laser beams. The application of this method to resolution of the fine structure of the Balmer-alpha line of atomic hydrogen is shown in Fig. 2. This technique can work only if the sample has measurable absorption. If the number of molecules is very small, it is sometimes possible to detect the absorption by looking at the fluorescence from the excited state.

Polarization spectroscopy. Recently, C. Wieman and T. Hänsch introduced another technique, laser polarization spectroscopy, which is capable of considerably more sensitivity than the saturated absorption method. A polarization spectrometer takes advantage of the fact that small changes in light polarization can be detected more easily than changes in light intensity. The laser output is again divided into a saturating beam and a probe. The probe "sees" the gas sample placed between nearly crossed linear polarizers so that only a very small amount of light arrives at the photodetector. Noise due to laser intensity fluctuations is greatly reduced in this way. The saturating beam is circularly polarized by a birefringent plate. Alternatively, a linearly polarized beam is used with its polarization axis rotated 45°.

Normally, in a gas, molecules have their rotation axes distributed at random in all directions. However, the probability for absorbing polarized light depends on the molecular orientation. Thus the saturating beam depletes molecules with a particular orientation preferentially, leaving the remaining ones polarized. These can then be detected with high sensitivity because they can change the polarization of the probe beam. A gyrotropic birefringence of the gas sample rotates the axis of the probe polarization, and an induced dichroism makes the probe light elliptically polarized. The probe thus acquires a component that can pass through the crossed polarizer into the detector. But again this happens only near the center of a Doppler-broadened line, where both beams are interacting with the same molecules.

Polarization spectroscopy makes it possible to detect fewer molecules with lower light intensity, so that external causes of line broadening and shifts are more easily avoided. By labeling molecules in a selected level, it can serve to unravel complex absorption spectra, as demonstrated by R. Teets and collaborators. In addition, it can provide information on the angular momenta of the levels involved. Although this method has been applied successfully in the infrared, near 10 μm, by V. Stert and R. Fischer, it works best near the visible spectral region for which good polarizers are available.

Saturated interference spectroscopy. F. Kowalski has recently demonstrated a promising related technique, saturated interference spectroscopy, which is in principle free from this limitation. Here, a gas sample is placed inside a Jamin interferometer. Consisting of two thick, parallel glass plates, this device splits the probe beam into two parallel, separate components which are later recombined. The two components cancel if the tilt angle of the plates is adjusted so that the phase difference is 180°. A saturating beam is sent in the opposite direction through the sample so that it overlaps with one of the probe components. Any light-induced change in absorption or refractive index will imbalance the system and will produce a detectable intensity change at the detector. Again,

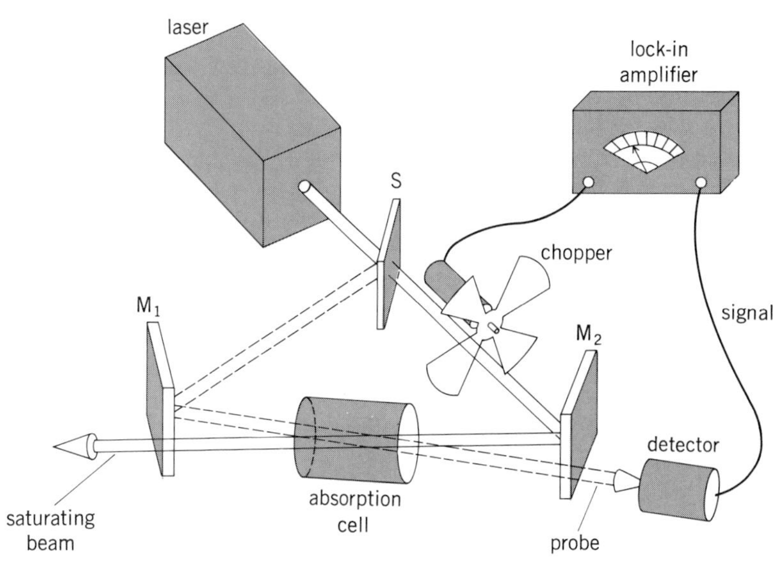

Fig. 1. Scheme of laser saturation spectrometer; S indicates the beam splitter and M_1 and M_2 are mirrors.

noise due to laser intensity fluctuations is greatly reduced by suppressing background light.

Resolution limits. The resolution of any of the methods described is of course limited by the natural line width of the transition. However, in practice, a large number of additional causes of line

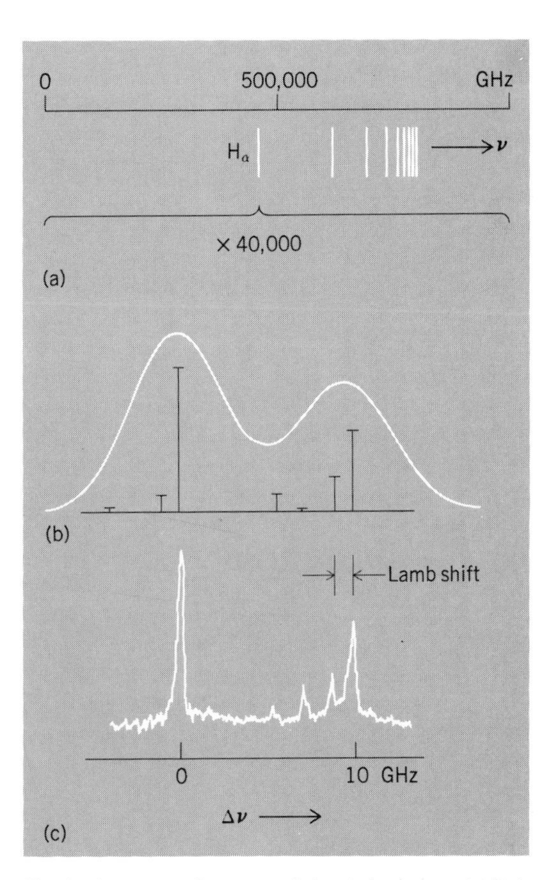

Fig. 2. Laser spectroscopy of atomic hydrogen. *(a)* Balmer series. *(b)* Doppler-broadened absorption profile (300 K) of the Balmer-alpha line with theoretical fine structure. *(c)* Saturation spectrum of Balmer-alpha line with resolved 2*S*-2*P* Lamb shift. *(Recorded by T. Hänsch and collaborators)*

broadening demand attention, including collision effects, power broadening and light shifts, relativistic "transverse" Doppler shifts, and imperfections in the laser wavefronts. Also, atoms change their velocity slightly when they emit or absorb a photon, in order to conserve momentum. This recoil causes a very small line splitting in saturation spectroscopy, which was first observed in a methane line at 3.39 μm by J. Hall and C. Borde and by V. Chebotaev and coworkers. Another and often more important effect is transit-time broadening: the number of field oscillations an atom can see, and hence the resolution, is limited by the finite time during which an atom traverses the laser beam.

Optical Ramsey fringes. Transit-time broadening has been overcome for a long time in radio-frequency spectroscopy of molecular beams by the use of two or more spatially separated excitation regions. Molecules leave the first field region in a coherent superposition of ground state and excited state and continue to oscillate at their resonance frequency until they enter the next field region. Their fate then depends on the relative phase between molecular and field oscillations. The molecules are either further excited or brought back to the ground state through stimulated emission. As first demonstrated by N. Ramsey, the spectrum thus acquires narrow-band interference fringes (Ramsey fringes) whose resolution is limited only by the total time of flight through the interaction regions.

However, there is an obvious difficulty in extending this method to the visible region. Because the wavelength of light is so small, even the minute spread of transverse velocities in a well-collimated molecular beam will tend to randomize any systematic phase relationships. Acting on a suggestion by Chebotaev and colleagues, J. Bergquist and coworkers demonstrated recently that optical Ramsey fringes are observable in saturation spectroscopy of molecular beams, if at least three spatially separated field regions are used. In this case, the preparation of the atoms in the first two fields depends on their exact transverse velocity and establishes the right phases to obtain interference effects in the third field.

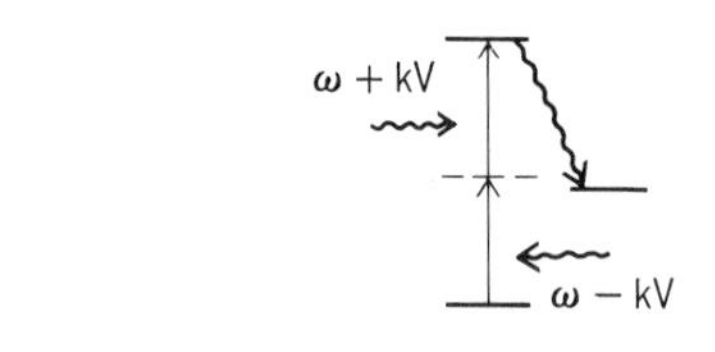

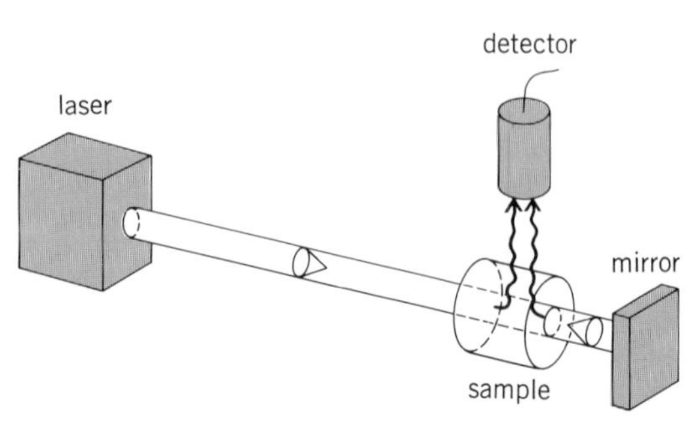

Fig. 3. Setup for Doppler-free two-photon spectroscopy.

Radiation cooling and trapping. Other techniques of high-resolution spectroscopy have been proposed which would utilize the laser light itself to slow down the atoms. In particular, A. Schawlow and Hänsch have pointed out that the resonant radiation pressure of intense laser light could be used to rapidly cool gases to very low temperatures. P. Toschek and collaborators and D. J. Wineland and colleagues have recently demonstrated radiation cooling of ions captured in a radio-frequency quadrupole trap or in a Penning trap. V. Letokhov and others have suggested spatially trapping slow atoms with laser light fields, taking advantage of dielectric forces.

Doppler-free two-photon spectroscopy. Doppler-free two-photon laser spectroscopy, however, is a completely different approach which does not rely on velocity selection. It has long been known that atoms can be excited to a state of the same parity as the ground state by simultaneously absorbing two photons which together provide the required energy. Excitation probability grows with the square of the intensity, and two-photon excitation with light has become observable only since the advent of strong laser sources.

Chebotaev and collaborators were the first to suggest that Doppler broadening in two-photon excitation can be eliminated to first order simply by reflecting the output of an intense monochromatic laser back onto itself and by placing a gas sample in the resulting standing wave field, so that the atoms can absorb two photons coming from opposite directions (Fig. 3). From a moving atom, one beam will appear Doppler-shifted toward the blue, and the other will appear shifted toward the red by an equal amount. The sum frequency is hence constant, independent of the atomic velocity. If the number of excited atoms is observed during a laser scan, a sharp resonance appears on a low, Doppler-broadened background, produced by each traveling wave separately. The Doppler-free signal is strongly enhanced because all atoms, regardless of their velocity, can contribute.

The practical feasibility of the method was first demonstrated in 1974 by B. Cagnac and colleagues, M. Levenson and colleagues, and Hänsch and colleagues. Recent applications include a measurement of the hydrogen ground-state Lamb shift by Wieman and Hänsch, who compared the two-photon spectrum of the $1S$-$2S$ transition with a polarization spectrum of the $n = 2-4$ Balmer beta line.

At wavelengths where only pulsed laser sources of relatively broad bandwidth are available, it is possible to increase the resolution of Doppler-free two-photon spectroscopy by excitation with two or more phase-coherent light pulses. This approach is related to Ramsey fringe spectroscopy with spatially separated light fields. Two-pulse spectra with sinusoidal fringe structure were first observed by M. Salour and C. Cohen Tannoudji. Comblike spectra with sharp, natural line width–limited fringes have recently been obtained by J. Eckstein and coworkers, exciting gas atoms with a train of picosecond pulses from an actively mode-locked continuous-wave dye laser. The well-controlled laser mode spectrum provides an accurate frequency calibration scale extending over hundreds of gigahertz.

For background information *see* DOPPLER EFFECT; HYPERFINE STRUCTURE; LASER; SPECTROSCOPY; STARK EFFECT; ZEEMAN EFFECT in the McGraw-Hill Encyclopedia of Science and Technology.

[THEO W. HÄNSCH]

Bibliography: N. Bloembergen (ed.), *Nonlinear Spectroscopy: Proceedings of the International School of Physics "Enrico Fermi,"* 1977; J. L. Hall and J. L. Carlsten (eds.), *Laser Spectroscopy III*, Springer Series in Optical Sciences, vol. 7, 1977; V. S. Letokhov and V. P. Chebotaev, *Nonlinear Laser Spectroscopy*, Springer Series in Optical Sciences, vol. 4, 1977; K. Shimoda (ed.), *High Resolution Laser Spectroscopy*, Topics in Applied Physics, vol. 13, 1976.

Steel

Improving the mechanical properties of steels so as to optimize and conserve raw materials is a great challenge to physical metallurgists, particularly in these times of energy crisis. An alloy design program with special emphasis on structural steels has been under way at the University of California, Berkeley, for the past dozen or so years. The research has been supported through the Materials and Molecular Research Division of the Lawrence Berkeley Laboratory with funds from the U.S. Department of Energy. The emphasis in this program is on control of properties through manipulation of microstructure, and involves sophisticated techniques of characterization such as transmission electron microscopy and diffraction. This research program has two main goals: (1) the development of strong, tough, economical, martensitic steels and (2) the development of duplex low-carbon ferrite-martensite steels for weight savings.

High-strength, tough structural steels. High-strength structural steels are used extensively for components such as aircraft landing gear, missiles, rocket casings, armor plate, and other defense applications. In addition, where such steels have high hardness and consequent abrasion resistance, they are used in mining operations (for example, as buckets and in comminution and mineral-processing operations). The limiting factor in the use of high-strength steels is their toughness. In practice, toughness and ductility are required to resist crack propagation and ensure sufficient formability for successful fabrication of the steel into engineering components. Many commercial high-strength steels in use today have been designed by experience, often by trial and error, and almost all

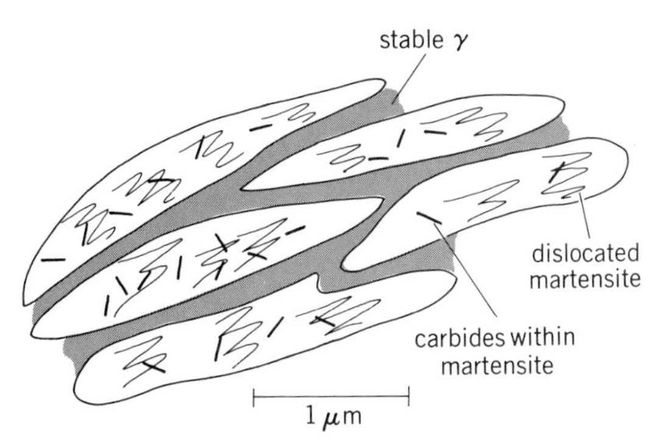

Fig. 1. Schematic of duplex austenite(γ)-martensite structure as resolved by transmission electron microscopy.

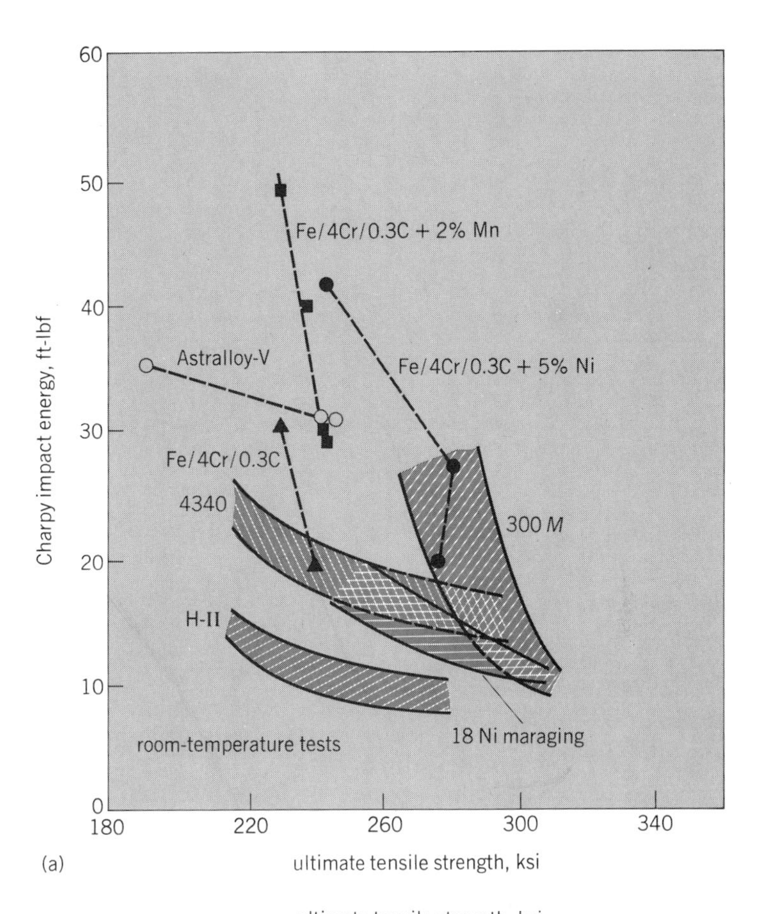

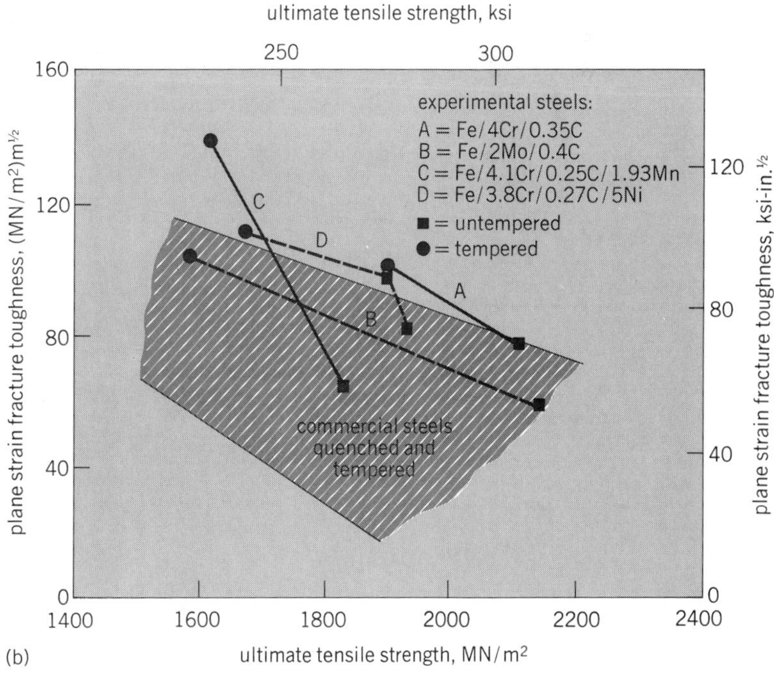

Fig. 2. Properties of experimental Fe/Cr/C/Mn and Fe/Cr/C/Ni steels compared with some commercial structural steels: *(a)* Charpy-strength and *(b)* K_{1c} toughness-strength relationships. Note the good strength-toughness properties even in untempered steels. 1 ft-lbf = 1.36 joules; kip per square inch = 690 newtons per square centimeter.

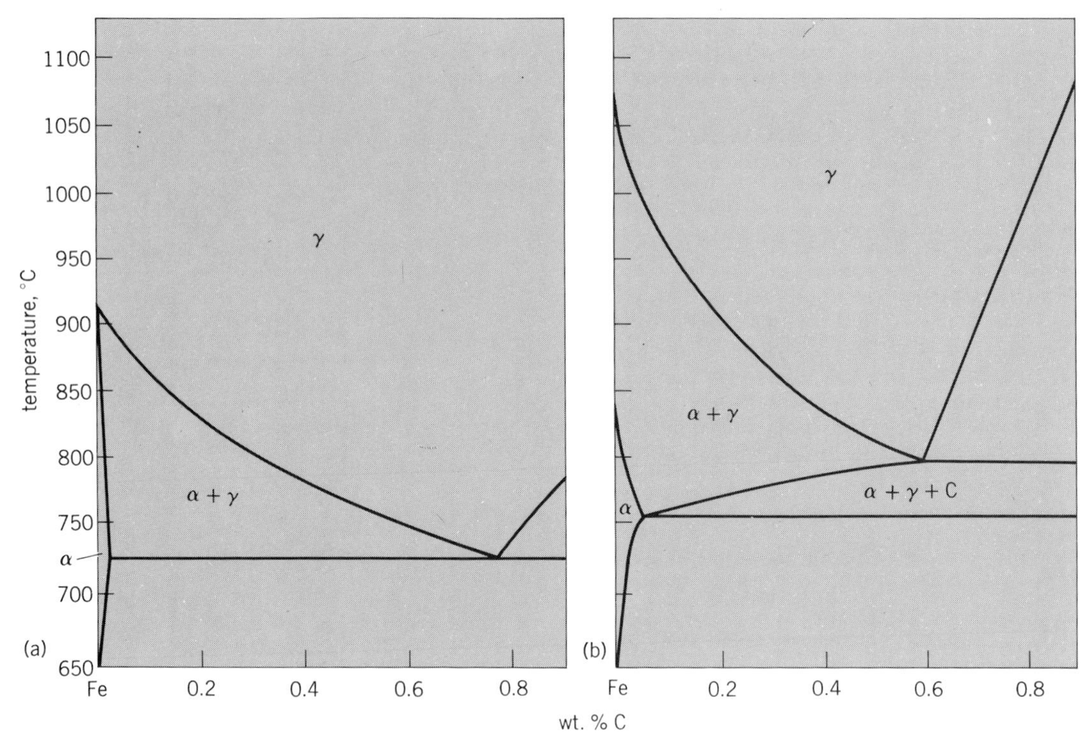

Fig. 3. Phase diagrams: *(a)* Fe-rich portion of Fe-C diagram; *(b)* Fe-rich portion of the 2.4 wt % Si diagram. The duplex structure is obtained by heating a normalized or quenched steel in the two-phase ($\alpha + \gamma$) field and quenching to transform γ to martensite.

of those at high-strength levels could benefit from improvements in toughness.

The University of California program has been under way since 1964 and involves a systematic study of the relation between martensitic microstructures and properties utilizing a series of Fe/C/X high-purity vacuum-melted experimental steels (where X is the substitutional solute). The martensite transformation, if controlled so that the inhomogeneous shear component occurs by slip and not by twinning, is the most efficient means of producing dense, uniformly dislocated fine-grained structures. The dislocations are an essential component for strength and toughness. The main factor controlling this aspect of the transformation is composition, especially percent carbon (affecting transformation temperature M_s and strength of the austenite), which must be regulated to maintain $M_s > 200°C$. As a result of detailed analysis by electron microscopy, an unexpected retained austenite was detected. In addition, it has been found that this retained austenite, if stable, promotes toughness. Consequently, the microstructure which corresponds to optimum properties is a duplex austenite-martensite structure, shown schematically in Fig. 1. The retained austenite has several beneficial characteristics: it is itself ductile and therefore enhances toughness, for example, by crack blunting; and it prevents brittle interlath carbide films from forming due to autotempering during the $\gamma \rightarrow$ martensite transformation. The steels are processed by vacuum induction melting, austenitizing at 1000–1100°C to dissolve all carbides, oil quenching, and tempering if higher toughness is required.

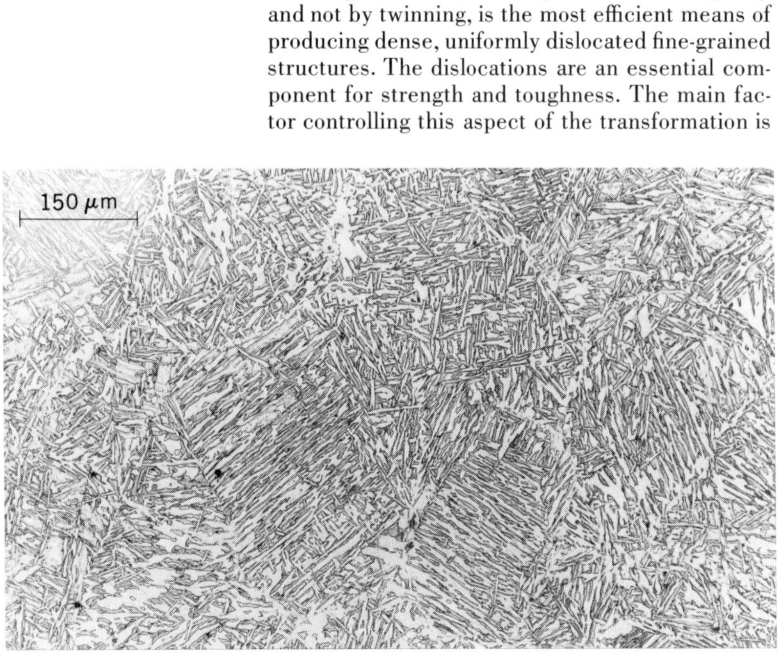

The current program's alloy design goals are: (1) the production of dislocated packet martensite; (2) surrounding of each martensite crystal in the packet by stable retained austenite (this involves optimizing chemical composition of the steel); and (3) hardening of the steel to avoid the creation of upper bainite (or other undesirable transformation products).

The program is aimed at optimizing these microstructural features by using minimal alloy contents. New steels based on Fe/Cr/C with Mn or Ni quaternary additions exhibit better combinations of strength and toughness than any existing commercial steel (Fig. 2). These new steels have strong potential for applications in the mining industry. Other properties of technological interest, including fatigue, corrosion behavior, and wear resis-

150 μm

Fig. 4. Optical micrograph showing the fibrous α-martensite microstructure within prior austenite grains.

tance, are also being investigated. Patents have been applied for.

Dual-phase steels. There is a need for higher-strength steels with good formability as dictated by current energy and resource conservation requirements. These requirements pose a challenge to the achievement of weight reduction and fuel savings as, for example, in the transportation industries (especially the automobile industry).

During the past several years duplex ferrite-martensite (DFM) steels have received increasing attention, since they contain characteristic microstructural features which combine high strength with good formability. The Japanese have been especially productive in this field. The strengthening principle of the DFM structure involves the incorporation of inherently strong martensite as a load-carrying constituent in a soft ferrite matrix, which supplies the system with the essential ductility. Thus these alloys can be considered equivalent to fiber composites.

Apart from these mechanical properties, however, there has been a general lack of fundamental understanding of the characteristic behavior of DFM steels. For instance, papers reporting the results of mechanical property tests generally made few direct correlations with duplex microstructures, except for some occasional optical metallography. The mechanical behavior of the two-phase materials, as well as the intricate interactions of parameters such as the size, shape, distribution, and volume fraction of martensite particles, must all be characterized and controlled in designing improved DFM alloys so that together they favorably contribute to the overall mechanical properties.

The main principles of the alloy development can be summarized as follows: (1) Obtain the (α + martensite) duplex structures by phase transformation alone; the optimum volume fraction is approximately 20% martensite. (2) Limit the percentage of carbon in martensite to approximately 0.3% to achieve strong, tough martensite (as in Fig. 1). (3) Make sure that alloys have a large slope in the $\gamma/\alpha+\gamma$ solvus line, to allow flexibility in heat treatment. (4) Optimize the properties of the constituent phases.

DFM structures can be produced in many different ways. In the Berkeley approach, only simple heat treatment by annealing is employed in the two-phase ($\alpha + \gamma$) field and quenching so as to transform the austenite phase (γ) to martensite without resorting to mechanical or thermomechanical treatments (Fig. 3). The processing method is as follows: conventional steel-making practice followed by heat treatment of up to 1/4-in. (0.635-cm) sections by either quenching or normalizing for 1000°C austenite condition, two-phase anneal (for example, at 950°C for Fe/2Si/0.1C steel), or rapid quenching to obtain ferrite-martensite composite.

From a practical point of view, it is desirable to have a large slope in the A_3 line (the $\alpha+\gamma/\gamma$ solvus line) so as to have flexibility in heat treatment for control of volume fraction of the two phases. This can be achieved by adding suitable ternary alloying elements (for example, Si) to the Fe-C base system, as is shown in Fig. 3*b*. In Fig. 3*a* any slight variation in the two-phase annealing temperature will shift composition as well as volume fraction of austenite (and hence martensite) to a great extent, whereas in the case of Fig. 3*b* less stringent control on the critical annealing temperature is required, thereby increasing the flexibility for heat treatment.

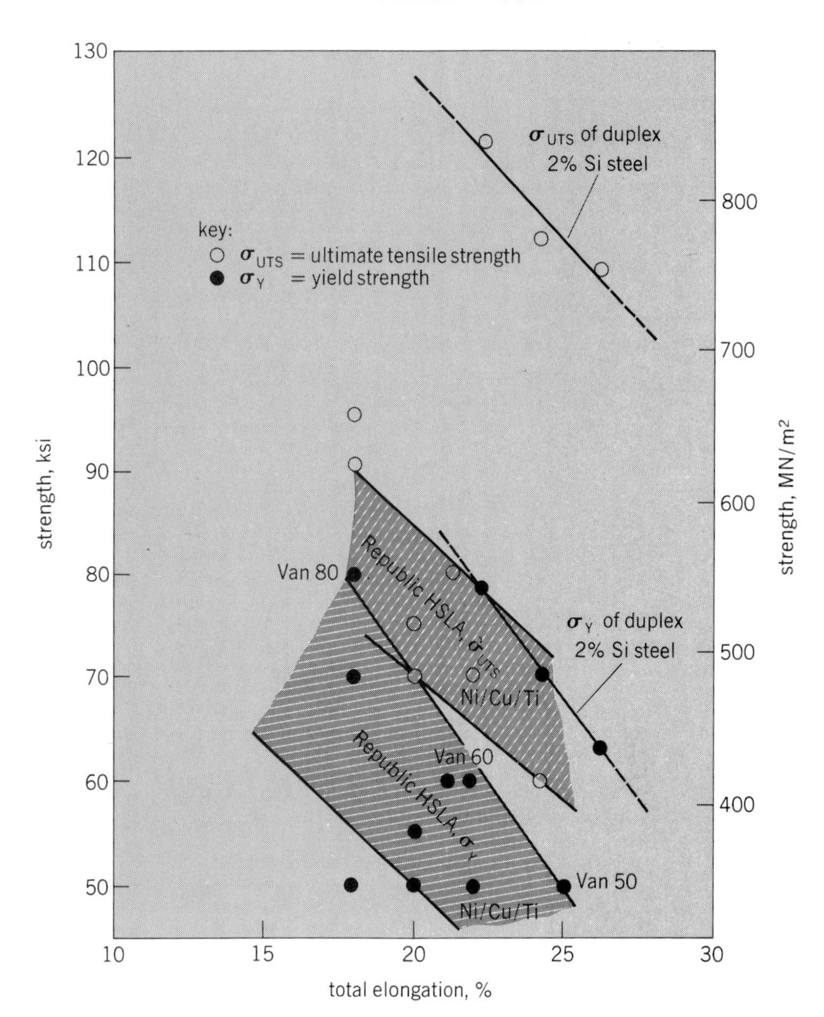

Fig. 5. Comparison of properties of Fe/2Si/0.1C duplex steel with some HSLA steels. The large ratio of ferrite to yield strength indicates good formability. *(From J. Y. Koo and G. Thomas, Design of duplex Fe/X/0.1C steels for improved mechanical properties, Met. Trans., 8A:525; copyright 1977 by American Society for Metals and Metallurgical Society of AIME)*

Utilizing these principles an Fe/2Si/0.1C DFM alloy has been developed which has a fibrous structure (Fig. 4) and excellent properties, as shown in Fig. 5 which compares the new steel data with those for some high-strength low-alloy (HSLA) steels. This steel has recently been patented by the U.S. Department of Energy.

It is difficult to accurately determine weight and

Table 1. Weight savings potential of HSLA steels*

Yield strength, psi†	Range of potential weight savings, %
50,000	22.5–40
60,000	29–50
70,000	34–57.1
80,000	38.8–62.5

*Data from D. G. Younger, Ford Motor Co.
†10^4 psi = 6.9×10^7 Pa.

Table 2. Effect of 100 lb (45 kg) weight reduction on automobile performance*

Size of automobile	Fuel economy effect, mpg (meters/liter)	0–10-second performance effect, ft (meters)
Small compact	+0.5 (211.7)	14 (4.2)
Intermediate luxury	+0.2 (84.7)	7 (2.1)

*Data from D. G. Younger, Ford Motor Co.

fuel savings if these steels are to replace existing ones, although as shown in Table 1 the ranges of weight savings gained by substituting HSLA steels for the current 30,000-psi (2.07×10^8 Pa) yield-strength steels have been estimated. Clearly, if there is no decrease in formability, any increase in strength means weight savings and thus fuel economy. Based on these estimates, the approximate direct worth of a 100-lb (45-kg) weight reduction on fuel economy and performance in automobiles can be approximated (see Table 2). Strength-critical parts offer excellent opportunities for weight savings which, on the average, can be 30% of the current weight if freedom to generate new designs is permitted.

In conclusion, from the standpoint of their superior properties and simplicity in composition and heat treatment, the new duplex steels show great promise, not only for the automobile industry but for other applications.

For background information *see* STEEL in the McGraw-Hill Encyclopedia of Science and Technology. [GARETH THOMAS]

Bibliography: R. L. Cairns and J. A. Charles, *JISI*, 205:1044, 1967; A. G. Grabe, U.S. Patent 2097878, 1937; R. A. Grange, *Proceedings of the 2d International Conference on the Strength of Metals and Alloys, ASM*, Pacific Grove, CA, p. 861, 1970; S. Hayami and T. Furukawa, Micro alloying 75, *Symposium on High Strength, Low-Alloy Steels, Products and Process*, Washington, DC, p. 56, 1975; G. Thomas, in R. I. Jaffee and B. Wilcox (eds.), *Proceedings of the Batelle Conference on Fundamental Aspects of Structural Alloy Design*, pp. 331–361, 1972.

Stem (botany)

Cacti (Cactaceae) are organisms of arid and semiarid regions, primarily in the Western Hemisphere, with highly succulent stems. In the course of evolution, cacti experienced reductions in leaf size and shifted photosynthetic processes to perennial stems. Large leaves may still be observed in certain taxa, but most cacti have small or vestigial leaves. Cacti have stem features ideally suited to their physiology and habitats; most are designed to withstand desiccation and overheating while retaining the capacity to produce and recycle energy year-round. Recently, many scientists have studied the structure and physiology of these perennial stems.

Crassulacean acid metabolism. All cacti studied to date exhibit crassulacean acid metabolism (CAM), which also characterizes perennial succulents in other plant families. For photosynthesis, most green vascular plants take up CO_2 through stomates in the light. CO_2 is immediately combined with ribulose 1,5-diphosphate (RuDP), the precursor of sugars. Stomates are closed in the dark, when photosynthesis cannot occur. In contrast, succulents with CAM open stomates in the dark to take up CO_2. CO_2 is combined with phosphoenolpyruvate in cytoplasm through carboxylation into oxaloacetate, which is converted to malate and, subsequently, other organic acids. These organic acids are then transported for storage to the vacuoles, where the chemical reaction reaches equilibrium at a very low pH. In the light, stomates normally close, organic acids are moved from the vacuoles back into the cytoplasm and are decarboxylated, and CO_2 is combined with RuDP for manufacture of sugars. Thus, during CAM, CO_2 uptake and utilization are temporally separated and the cell sap experiences vast diurnal changes in pH. An obvious advantage to the plant is that stomates are closed during the day, when evaporational loss of water would be greatest.

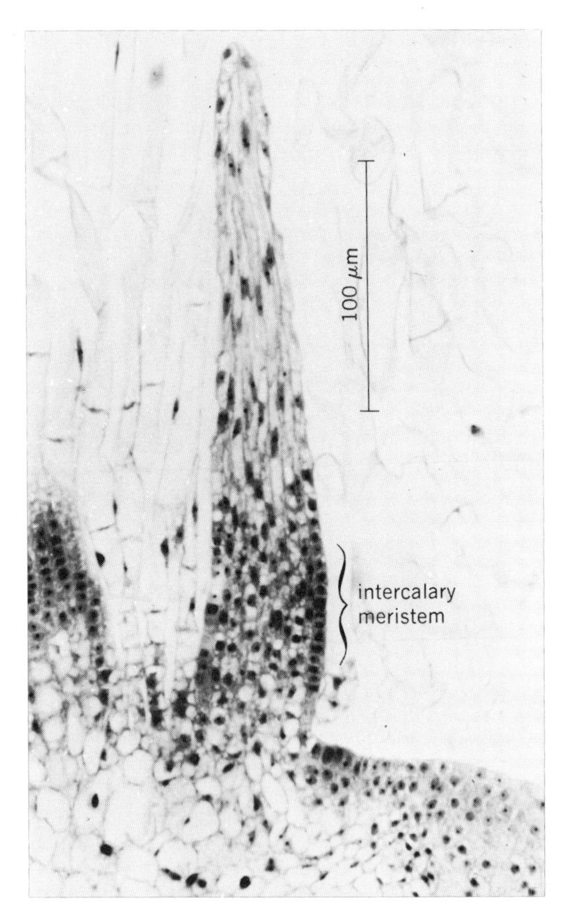

Fig. 1. Photomicrograph of a young cactus spine, showing intercalary meristem of the spine primordium. *(From J. D. Mauseth, Cytokinin and gibberellic acid–induced effects on the determination and morphogenesis of leaf primordia in Opuntia polyacantha (Cactaceae), Amer. J. Bot., 64:337–346, 1977)*

S. Szarek and I. Ting, among others, have shown that in cacti when stem water status is extremely high following heavy rains or irrigation, stomates open in the dark and remain open in the light; when water potential drops slightly, stomatal opening occurs only in the dark; and when plants are under water stress, stomates may open neither in the dark nor in the light. When stomates remain shut for long periods of time, the cactus functions as a closed system, recycling water, carbon, and energy with very low metabolic costs.

External features. Cactus areoles are highly modified branches (spur shoots) developed from axillary buds. These buds experience little elongation and produce spines as primordia, typically in helically alternate phyllotaxy, and trichomes. Once initiated, spines grow primarily from a basal intercalary meristem (Fig. 1). J. Mauseth has recently demonstrated that these primordia develop as normal, photosynthetic leaves under the influence of cytokinin, but develop as typical spines under the influence of gibberellic acid. A minute leaf generally subtends each areole, which is reduced in size to minimize water loss. Spines are so highly sclerified that water loss through areoles is very low, and they serve to protect plants against being eaten by animals and to promote dispersal of species reproducing in nature by rooting of stems. Some workers believe that spines may reflect intense insolation or decrease transpiration by decreasing airflow over the stem, thus increasing the boundary layer. These functions can be realized only where spines and hairs are very dense or wind velocity is extremely low. Several workers have discovered that certain small cacti have the ability to take up water through areoles; these species live in habitats where fog and dew are common.

In most species, areoles project above the stem on cylindrical tubercles (podaria) or elongate ribs, the latter formed by vertical fusion of areoles. Tubercles are greatly enlarged leaf bases with associated cortical tissue; because each tubercle is associated with a leaf primordium, they exhibit helically alternate phyllotaxy based on the Fibonacci summation series 2/5, 3/8, 5/13, 8/21, and so on. Species with ribs have helical phyllotaxy on which rib patterns are superimposed. In some ribbed cacti, for example, barrel cacti, rib number frequently equals denominators in the Fibonacci series, but most species show great variance in rib number among populations on a climatic gradient and also for individuals during a lifetime. In cold climates, where freezing temperatures are common, rib number frequently decreases as stem volume increases; thus, plants are selected with greater volume-to-surface area, presumably to withstand winter temperature minima.

Both ribs and tubercles appear to increase total photosynthetic surface; decrease full exposure to sun by partial shading of depressions, thus decreasing temperatures in valleys; and permit free seasonal expansion and contraction of the stem with changes in water status. Cactus stems diurnally increase in size in the light and decrease in the dark—phenomena related to CAM. Spherical growth forms minimize surface area with respect to volume and are ideal for decreasing water loss.

Platyopuntias have flattened stems or phylloclades called pads. Pads greatly increase the surface-to-volume ratio. No data explain why or how flattening evolved in tropical taxa, but several authors have claimed that pad orientation helps species survive in cold climates. Presumably, pads oriented in an east-west direction maximize winter and minimize summer temperatures in pads. Significant thermal differences in pads of various orientations have been demonstrated. The orientation of barrel cacti, for example, the compass cactus, may also function to minimize absorption of mid-day solar radiation; in contrast, the apex receives maximum radiation and heat, apparently to promote growth and reproduction without affecting plant water status.

Skin. The surface covering, or skin, of a cactus is designed to retain as much water as possible. As noted above, stomates open only when water status is favorable for photosynthesis and then mostly in the dark. In species with ribs or tubercles, stomates are concentrated in valleys, where transpiration rates are decreased by lower temperatures and wind movement. Cacti have extremely low densities of stomates in comparison with mesophytes, and stomatal resistances are high and increase markedly with increases in temperature. In hot, dry conditions, cactus stomates effectively seal off water loss by transpiration. Many xerophytes have sunken stomates to reduce transpiration; however, in most cacti, stomates are not sunken, probably allowing for maximum gaseous change in the dark for CAM.

Water loss is also limited by the presence on the skin of an effective, sometimes thick, cuticle, which resists transpiration from epidermal cells. Beneath the epidermis lies a hypodermis composed of cells with thick, hygroscopic walls (Fig. 2). The role of hypodermis in water retention is unknown, but certainly it permits expansion and contraction of the stem with changes in hydration. Hypodermal layers are interrupted by cavities from stomates to the internal tissues where photosynthesis occurs, facilitating gaseous exchange. Hypodermal cells are living but lack chloroplasts; thus, an important function of hypodermis is to allow photosynthetically active radiation to pass

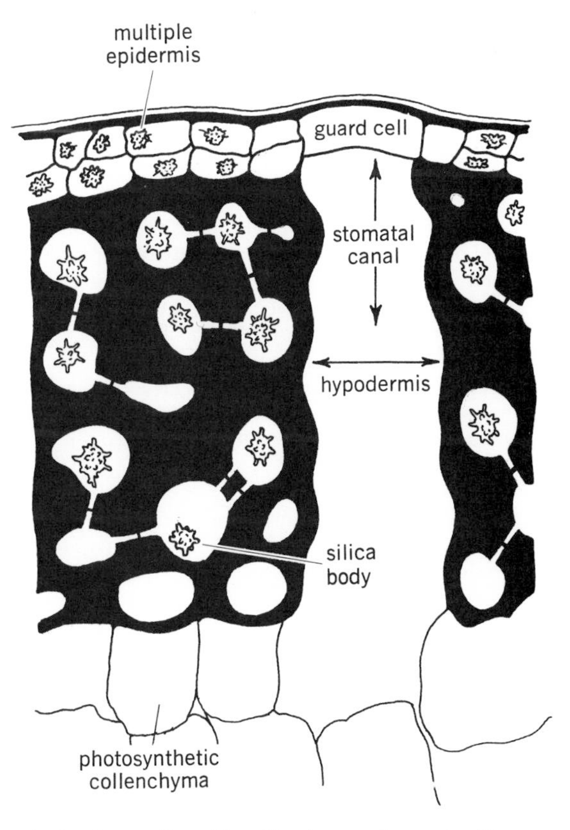

Fig. 2. Transection of the skin of organpipe cactus *Stenocereus thurberi*.

unimpeded to chloroplast-bearing cells beneath. Preliminary observations show that the skin is instrumental in stopping ultraviolet-β wavelengths from reaching chloroplasts. Crystals of calcium salts and silica found in certain cacti may influence light transmission as well as deter predation by insects with chewing mouthparts.

Internal tissues. In cactus stems, both pith and cortex are generally succulent, and are composed of large, highly vacuolate cells usually with thin, unlignified walls. Development of the much-enlarged outer cortex, including ribs and tubercles (and hypodermis), is principally by a subprotodermal meristem, the initials of which divide periclinally to produce cells in radial files. These cells also enlarge greatly, round up, and form intercellular spaces.

Although plastids are present in all cells of succulent tissues, chloroplasts are the largest and most abundant next to the hypodermis, where the light reaction of photosynthesis takes place. However, cells throughout succulent tissues may produce and store organic acids. Based on the presence of green tissues in very old stems, workers infer that cells may be photosynthetically active for over 100 years.

In many cacti, mucilage cells and vascular tissues differentiate late in the development of pith and cortex. Mucilage, a complex polysaccharide, helps to retain water in tissues while drawing water from the roots by increasing matrix forces. Even in periods of extreme water stress, these tissues have higher water content than many mesophytes. Vascular bundles pass through the enlarged cortex and pith of many species. Thus, succulent tissues are amply supplied with water and carbohydrates even though they may be isolated from growing regions. Both mucilage production and vascularization of cortex and pith show increases in many evolutionary lines.

Cacti suffering from water stress become fully rehydrated within 24 hr following a heavy rain. This quick recovery may be influenced by adaptations in the wood (secondary xylem). Cells of the wood in small cacti lack thick, well-lignified walls. Absence of secondary walls probably permits freer movement of water horizontally and vertically within the stem. Nonsucculent plants cannot tolerate woods that have mechanical weakness, but these small cacti have no mechanical problems because turgor of succulent tissues is sufficient to keep plants erect. Woods of most cacti have large regions of succulent tissues opposite areoles or ribs, which make it possible for water to diffuse between pith and cortex.

Unlike many desert plants, cacti cannot cool themselves through evaporation because stomates are closed during the day. Consequently, cacti develop internal temperatures significantly higher than the ambient. Highest temperatures and greatest diurnal thermal fluctuations occur in the skin, and lowest temperatures and smallest diurnal fluctuations in deep internal tissues. Temperatures are also highest on sides where mid-day radiation is greatest. Because internal temperatures are higher than in most plants, cacti probably have evolved organic molecules able to resist denaturation, for example, the proteins in hypodermis.

Because cacti are potentially excellent, moist habitats for pathogens, they have developed various protective mechanisms. Some cacti have alkaloids and triterpenes, which may discourage their being eaten by animals and may inhibit the growth of microorganisms. Great diurnal fluctuations in pH due to organic acid production may also inhibit growth of microorganisms, especially when pH is extremely low. Finally, when wounds are made, the cactus stem responds quickly with polyphenol production followed by callus and periderm formation, thereby sealing up the succulent tissues.

For background information *see* PHOTOSYNTHESIS; PLANT METABOLISM; STEM (BOTANY) in the McGraw-Hill Encyclopedia of Science and Technology.

[ARTHUR C. GIBSON]

Bibliography: A. C. Gibson, *Bot. Gaz.*, 138: 334–351, 1977; A. C. Gibson and K. E. Horak, *Ann. Missouri Bot. Gard.*, vol. 65, 1978; J. D. Mauseth, *Amer. J. Bot.*, 64:337–346, 1977; S. R. Szarek and I. P. Ting, *Amer. J. Bot.*, 62:602–609, 1975.

Stratigraphy

Marine sedimentary rock sequences (such as those defined by P. R. Vail and R. M. Mitchum, Jr.) are composed of successive aggradational cycles, on the order of 1 to 10 m thick, bounded by sharp transgressive surfaces. All cycles begin at a sharp surface of nondeposition or erosion which probably represents a basin-wide rapid subsidence event. Within each cycle sedimentary environments indicated by the rocks gradually become more shallow upwards. These punctuated aggradational cycles (PACs) characterize rocks deposited in all environments within a marine basin, from tidal flats to deep shelf, including bay and barrier complex deposits. This newly recognized phenomenon is significant for several reasons: (1) Possibly most important, it provides a new, genetically based approach for interpreting the depositional environments of stratigraphic sequences in the field. (2) It offers a new conceptual model for explaining the dynamics of thick accumulations of rock in sedimentary basins. (3) It suggests the potential for basin-wide correlation on a cycle-by-cycle basis if the transgressive surfaces are basin-wide events.

Historical perspective. Early explanations of the stratigraphic record were based on concepts of worldwide depositional and mountain-building events. The stratigraphic column was viewed as a "layer cake" of isochronous strata subdivided by synchronous breaks. By the late 19th century this doctrine of catastrophism had been replaced by a uniformitarian gradualism. Not only was the present considered to be the key to the past, but it was also implied that these processes operated continuously and uniformly over a vast span of time to produce the geologic record. It should be noted that most sedimentary processes projected into the past by using the doctrine of uniformitarianism are directly observable in their time scale. However, the rapid transgressive events of PACs are an order of magnitude too slow to be observed in the present. Because of their time scale these transgressive events have been overlooked in favor of directly observable smaller-scale features, which results in a gradualistic interpretation.

This evolution of stratigraphic concepts paralleled an analogous progression in evolutionary theory from special creation theories to the modern gradualistic Neo-Darwinian theory. In 1972 N. Eldredge and S. Gould proposed a new theory to explain aspects of biologic evolution in which gradualistic phyletic change is replaced by rapid speciation events (punctuations) separated by long periods of genetic and phenotypic stasis. Similarly, the hypothesis being advanced in this article states that most stratigraphic sequences are also produced by nongradualistic processes, that is, sediments accumulate during long periods of crustal stability which are punctuated by rapid subsidence events during which little or no deposition occurs.

Stratigraphic examples. Shallowing upward sedimentary cycles separated by sharp surfaces have been observed from most of the geologic periods. One of the best-documented cases was described by T. Ryer from Cretaceous nearshore stratigraphic sequences in western Colorado and Utah. Ryer concluded that these shallow marine strata record only periods of shoreline progradation. Shallow marine environments left no depositional record during periods of transgression. Therefore the deposits of large-scale transgressive sequences as well as those of progradational sequences consist totally of packages of prograding deposits.

A second recent example was described by N. Beukes from the base of the Transvaal Supergroup (Precambrian) of South Africa. Deposition takes place through a number of siliciclastic, mixed siliciclastic-carbonate, and carbonate cycles. According to Beukes, "the siliciclastic cycles represent progradational subtidal-tidal flat deposits, and the carbonate cycles progradational tidal flat deposits. These cycles are bounded by sharp erosional contacts which represent rapid transgressive ravinements." Beukes argued that the thickness of the subtidal plus intertidal part of a cycle equals the amount of rapid relative sea level rise (averaging about 2 m), and that this was caused by tectonically controlled basin subsidence produced by faulting.

Middle Ordovician and Lower Devonian limestones in the central Appalachians are currently being studied by E. J. Anderson and P. W. Goodwin. In rocks representing a wide spectrum of environments, PACs are ubiquitous. One example (see the illustration) is typical of the complete stratigraphic record of every locality studied. The Lower Devonian Roundout through Kalkberg formations are exposed in a continuous stratigraphic

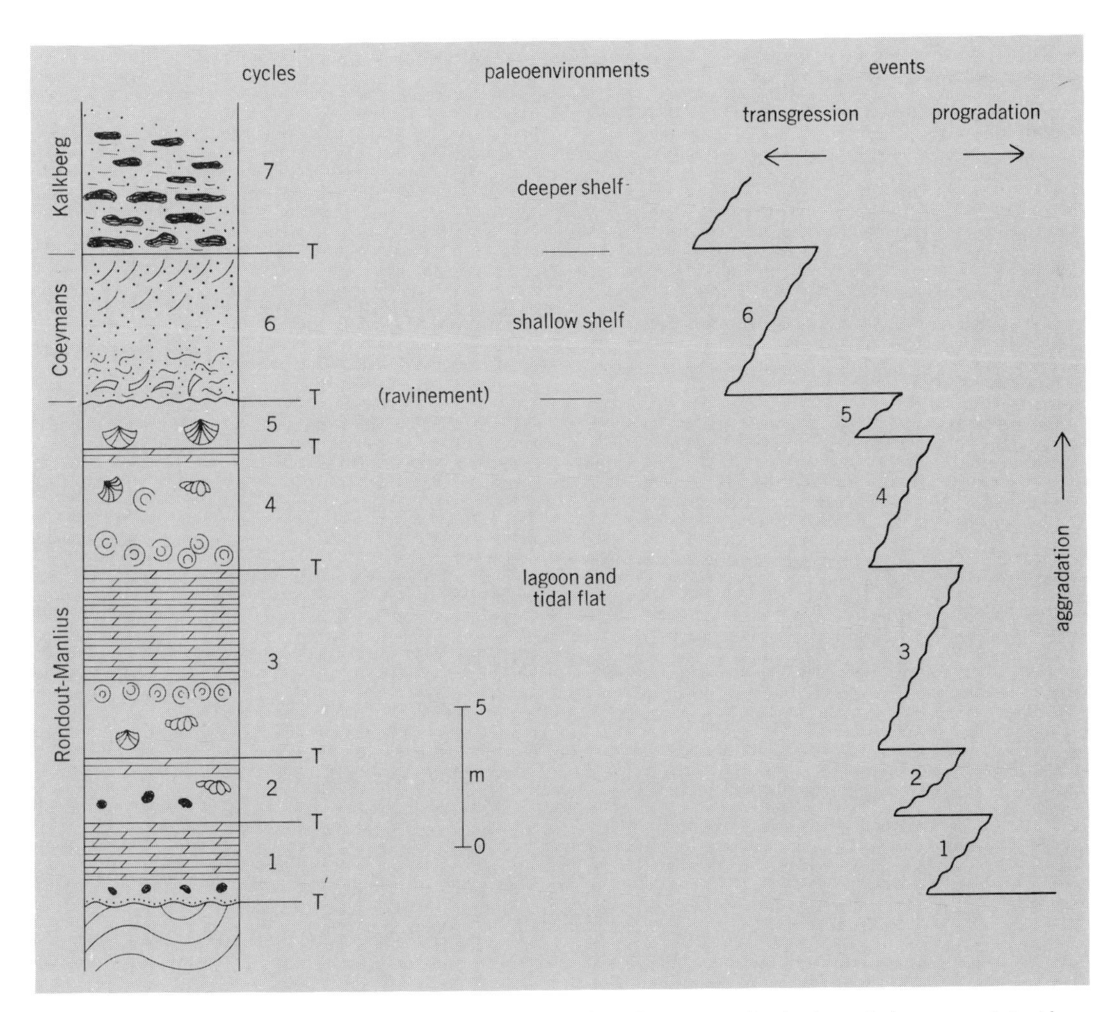

The Lower Devonian stratigraphic section at Catskill, NY, showing four formations divided into seven aggradational cycles separated by transgressive surfaces. Overall transgression is shown to be accomplished by a series of transgressive episodes separated by sedimentary aggradation. T = transgressive surface.

section at Catskill, NY. The section is divisible into seven cycles, each of which begins on a sharp transgressive surface and consists of deposits which gradationally shallow upward. In the first four cycles restricted subtidal fossil-bearing limestones grade vertically to tidal-flat dolomitic laminites. The fifth cycle has been truncated by ravinement, the sixth grades from bioturbated to current-washed calcarenite, and the seventh is a chert-bearing calcisiltite grading up into a bioturbated calcarenite.

Viewing the section as a succession of PACs leads to several significant stratigraphic observations. Formations (and members) may consist of one or more cycles, not necessarily the same number throughout their extent. For example, the Coeymans Formation is made up of one cycle at Catskill and three in central New York. However, within a formation, cycles are usually of a similar type. A change in cycle type denotes a major environmental change and therefore often coincides with established formational boundaries. When cycle and formational limits do not coincide or when two established formations (for example, Manlius and Roundout; see the illustration) consist of the same type of cycle, stratigraphic boundaries should be redefined in the interest of genetic integrity.

Careful analysis of cycle boundaries in sections of the Helderberg Group reveals two different types of sharply defined surfaces: erosional and nonerosional contacts. The top of the fifth cycle is truncated by a ravinement, a surface produced by shoreface erosion during landward migration of a barrier system. Other cycle boundaries, while sharp, are not erosional, in that bottoms of cycles do not contain lithologies representative of erosional flow conditions.

Perhaps of most interest is the fact that application of the cycle concept indicates that transgression is not a continuous process. Although the Helderberg section is generally transgressive in that each superjacent formation represents a deeper environment, the transgression is interrupted by numerous periods of shoaling. Transgression is renewed episodically, eventually prevailing over aggradation because subsidence is greater than supply. But the process is episodic rather than gradual and continuous.

Thus the PAC model offers a new conceptual approach to the study of stratigraphic sections. General paleoenvironments are characterized by a particular type of cycle, and specific environments are represented at predictable places in cycles. Stratigraphic units can now be defined and interpreted more genetically, and thick sequences can be analyzed as episodic changes in a single motif.

Mechanisms. Possible mechanisms to explain the empirically formulated hypothesis of PACs include: (1) tectonically produced basin subsidence; (2) eustatic rises of sea level, resulting from changes in sea-floor spreading rates or glacial melting; and (3) migration of environmental mosaics superimposed on gradual basin subsidence.

Changes in sea-floor spreading rates operate at the wrong periodicity to produce cycles (that is, they are too slow). Several environments such as the open shelf and lagoons do not contain a mechanism for causing lateral migration of environmental mosaics and for producing sharp cycle bases. Eustatic sea-level rises produced by melting of polar ice are difficult to apply to equable climatic geologic periods such as the Cretaceous, and cannot produce differential subsidence across a depositional basin. Tectonic mechanisms would appear to be good explanations for continental margins (geosynclinal areas), but are more difficult to apply to cycles in the stable craton. However, a tectonic mechanism of some type seems the best possibility for producing differential subsidence and thus developing a sedimentary basin containing thick stratigraphic sequences.

Conclusions. A new general hypothesis of deposition of epicontinental marine stratigraphic sequences has been formulated within the last year. It states that all deposition is aggradational and occurs during periods of crustal stability. These aggradational deposits are separated by nondepositional surfaces of rapid transgression. Thick rock sequences can thus be analyzed in the field as successions of cycles, all of which have the same basic motif. Stratigraphic units and paleoenvironments can be defined in terms of PACs, and problems of correlation and basin dynamics can be seen from a more genetic perspective.

Of the possible mechanisms to explain the existence of PACs, basin subsidence caused by abrupt crustal movements seems most likely. Current research efforts are aimed at integrating the concept of aggradational cycles with plate tectonic theory in an effort to better understand the processes controlling basin dynamics.

For background information *see* STRATIGRAPHIC NOMENCLATURE; STRATIGRAPHY in the McGraw-Hill Encyclopedia of Science and Technology.

[E. J. ANDERSON; P. W. GOODWIN]

Bibliography: N. J. Beukes, *Sediment. Geol.*, 18: 201–221, 1977; N. Eldredge and S. Gould, in T. Schopf (ed.), *Models in Paleobiology*, pp. 88–115, 1972; T. Ryer, *Geol. Soc. Amer. Bull.*, 88:177–188, 1977.

Streptomycetaceae

The Streptomycetaceae are bacteria which have a complex developmental cycle: they grow in the form of colonies that have a superficial resemblance to fungal colonies, with a mycelium that differentiates into spores. They produce nearly two-thirds of all known antibiotics, including some 70 compounds that are manufactured commercially, for example, such important therapeutic agents as the tetracyclines, erythromycin, streptomycin, chloramphenicol, and lincomycin. There have been some interesting recent developments in the genetics of these organisms.

Streptomyces plasmids. One finding is the identification of plasmids in streptomycetes and the discovery that they can be involved in the genetic determination of antibiotic production. Plasmids are circular deoxyribonucleic acid (DNA) molecules which can be regarded as supernumerary chromosomes. In various bacteria they are already known to carry genes controlling sexual reproduction, resistance to antibiotics or heavy metals, pathogenicity, or the ability to grow on unusual organic compounds. The characteristics which plasmids confer on their hosts are "dispensable," in that they are not required for normal growth.

Plasmids and antibiotic production. Among the streptomycetes, *Streptomyces coelicolor* has been studied in the greatest detail from a genetic point of view. It has two different plasmids, called SCP1 and SCP2, and produces two antibiotics, only one of which is determined by plasmid-borne genes. This is methylenomycin A, a compound which probably has no direct application in chemotherapy, although useful derivatives of it may perhaps be developed. In any event, it is of interest because it is the first example of an antibiotic with a biosynthetic pathway catalyzed by enzymes coded for by plasmid-borne genes. This was originally indicated by the discovery that only SCP1-carrying strains of *S. coelicolor* produce the antibiotic. Moreover, if they lose the plasmid, they not only cease to synthesize methylenomycin but become sensitive to it; evidently SCP1 carries a gene or genes determining methylenomycin resistance. The plasmid is also a sex factor, promoting conjugation (and thus its own transfer) between $SCP1^+$ and $SCP1^-$ strains in a manner comparable to the behavior of the F factor in *Escherichia coli.* SCP1 can be transferred to several other species of *Streptomyces* which produce antibiotics with totally different structures from methylenomycin; when they receive SCP1, they acquire resistance to methylenomycin and become producers of it.

Plasmids have recently been implicated in the synthesis of several other *Streptomyces* antibiotics, including at least two clinically important compounds, oxytetracycline and chloramphenicol. In these examples, however, the role of the plasmid appears to be rather indirect; the biosynthetic pathway of the antibiotic is probably catalyzed by enzymes coded by genes on the main chromosome of the organism—*S. rimosus* in the case of oxytetracycline and *S. venezuelae* in the case of chloramphenicol—while the plasmid in some way controls the expression of these genes.

Several further examples of a regulatory role for plasmids in the production of *Streptomyces* antibiotics have also been found. However, not all such antibiotics appear to have their synthesis controlled by plasmid-borne genes. The second antibiotic produced by *S. coelicolor*—a blue-red indicator pigment called actinorhodin—is synthesized under the control of a closely linked cluster of genes on the main chromosome of the organism.

Antibiotic resistance. The fact that the SCP1 plasmid of *S. coelicolor* determines methylenomycin resistance is of interest in relation to a recent theory of the origin of the antibiotic resistance genes found on plasmids in clinical isolates of disease-causing bacteria. Such "infectious" drug resistance is a major threat to the efficacy of important antibiotics in the treatment of human bacterial diseases. The resistance conferred on the bacteria by plasmids which they carry, or acquire by conjugation with other, harmless bacteria with which they come into contact, is often due to enzymes which inactivate antibiotics by adding phosphate, acetate, or adenylate groups to them. Enzymes which carry out the same reactions are found in the antibiotic-producing streptomycetes, and it has been suggested by Julian Davies and Raoul Benveniste that such streptomycetes are the primeval source of these antibiotic-inactivating enzymes; their role could be to protect the producing organism against its own antibiotic. The capacity of plasmids to transfer themselves between different species and even genera of bacteria, and to acquire genes from the various hosts through which they pass in the process, may have enabled antibiotic-resistance genes to spread from streptomycetes, probably via a series of other bacteria, to the pathogenic bacteria in which they are now widespread.

Plasmids and genetic manipulation. The second sex factor of *S. coelicolor*, SCP2, is apparently not involved in antibiotic synthesis. It can be isolated quite easily by methods developed for other bacteria. The procedure is to break open the cells by digesting their walls with the enzyme lysozyme, releasing the DNA which is then separated into chromosomal and plasmid components by differential centrifugation techniques. The SCP2 plasmid has a molecular weight of about 19,000,000, enough DNA for about 30 genes.

Recently it has been found by Mervyn Bibb and Judy Ward in D. A. Hopwood's laboratory that the SCP2 plasmid DNA can be reintroduced into the *S. coelicolor* organism (a process known as transformation) under certain special conditions involving treatment of protoplasts of the organism with the plasmid DNA. Protoplasts are obtained by treating the cells with lysozyme, this time in a sugar solution which balances the osmotic pressure inside the cells and thus protects them from bursting when the cell walls are removed. The resulting protoplasts are surrounded by the delicate plasma membrane which normally lies just inside the cell wall. The protoplasts take up plasmid DNA, and they can then be induced to resynthesize their cell walls and thus regenerate a normal culture of the organism.

The importance of this technique is that it provides a means for carrying out recombinant DNA procedures by inserting foreign genes into the plasmid "vector" before it is reintroduced into the organism. This should provide an important new tool for manipulating the genetic makeup of industrial *Streptomyces* strains, for example, in the development of new antibiotic structures combining some of the features of molecules produced by two different species.

Protoplast fusion. Another new technique of genetic manipulation in *Streptomyces* depends on the artificial fusion of protoplasts derived from two different strains. Hopwood and colleagues found that if mixtures of protoplasts are exposed briefly to a solution of polyethylene glycol—a substance already known to promote the fusion of plant protoplasts and animal cells—and then allowed to regenerate into normal cultures, these contain very high proportions of recombinant individuals carrying new combinations of the genes which differed in the two parental strains. The proportions of recombinants can be as high as 20%, in contrast to the frequency of about one in a million usually found when only the natural conjugation process is involved in gene transfer in a mixture of cells of two parental strains. The protoplast fusion technique is so simple and rapid that it promises to become a major tool in strain improvement when the objective is to produce new, desirable combinations of mutations which have occurred in divergent lines during the conventional type of strain selection program.

For background information *see* ANTIBIOTIC; BACTERIAL GENETICS; DRUG RESISTANCE; STREPTOMYCETACEAE in the McGraw-Hill Encyclopedia of Science and Technology. [DAVID A. HOPWOOD]

Bibliography: J. Davies and R. E. Benveniste, *Ann. N.Y. Acad. Sci.*, 235:130–136, 1974; D. A. Hopwood et al., *Nature*, 268:171–174, 1977; D. A. Hopwood and M. J. Merrick, *Bacteriol. Rev.*, 41: 595–635, 1977.

Superconducting devices

The phenomenon of superconducting quantum interference was discovered in 1962 at the Ford Scientific Laboratory in Michigan. The value of the phenomenon was immediately recognized, and the Ford group then developed a device based on the effect which has become known as the SQUID (superconducting quantum interference device). Since that time the SQUID has been steadily improved and is now commercially marketed by several small firms. It is widely used as an ultrasensitive detector of weak magnetic fields and is also the basis for a fast logic element.

SQUID configurations. The SQUID is realized in two configurations: (1) the direct-current (dc) SQUID, which contains two Josephson junctions in a superconducting loop; and (2) the radio-frequency (rf) SQUID, which has only one Josephson junction in the superconducting loop. The primary difference between the two configurations is that the state of the rf SQUID is determined from an rf measurement of the impedance of the ring, while the state of a dc SQUID is determined from dc measurements.

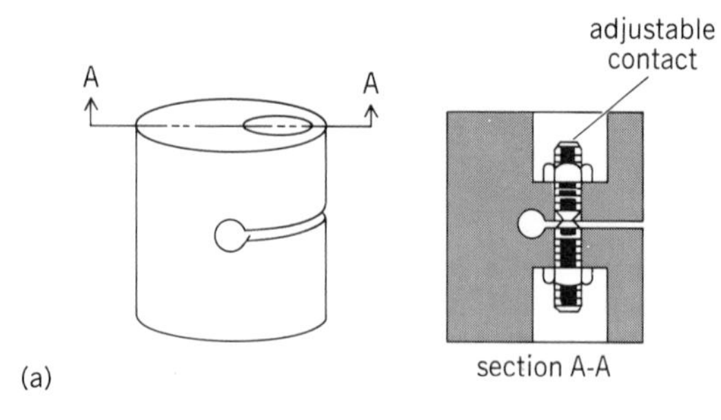

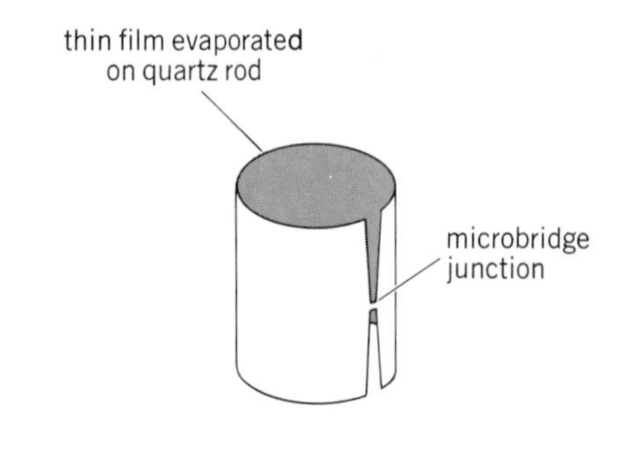

Fig. 1. Two types of rf SQUIDs: *(a)* point-contact SQUID; *(b)* microbridge SQUID.

The Josephson junction in a SQUID is realized as a weak connection between two pieces of superconductor. It can take the form of an insulating tunnel barrier between superconducting metal films (superconductor-insulator-superconductor, or SIS, junction), a narrow constriction in a superconducting metal film (microbridge), or a point contact between bulk superconducting metals. Figure 1 shows the point-contact and microbridge rf SQUIDs.

Because of their simplicity, rf SQUIDs of bulk niobium with point-contact junctions have been the most popular form to date. However, recent advances in the fabrication of thin-film microbridges and SIS junctions may cause a shift of popularity. Also, there has been renewed interest in thin-film versions of the double-junction dc SQUID in recent years.

Principle of operation. In a Josephson junction there is a current component I (the Josephson current) given by Eq. (1), where I_c is the maximum

$$I = I_c \sin \phi \qquad (1)$$

dissipationless current that the junction will sustain. An associated energy E (the Josephson coupling energy), which is the kinetic energy of the current I flowing through the junction, is given by Eq. (2). Here e is the electron charge, $\hbar$ is

$$E = -\frac{\hbar}{2e} I_c \cos \phi \qquad (2)$$

Planck's constant divided by 2π, and ϕ is the quantum-mechanical phase difference across the junction. The ϕ is related to the voltage V across the junction by Eq. (3). Finally, if a junction is con-

$$\phi = (2e/\hbar) \int V\, dt \qquad (3)$$

nected in a superconducting ring, as in an rf SQUID, then $\int V\, dt$ is just the magnetic flux Φ is the ring, so that $\phi = 2e\Phi/\hbar$. The fundamental constant $h/2e$ is generally called the flux quantum Φ_0. Its value is $h/2e \simeq 2.07 \times 10^{-15}$ weber, so that ϕ can be written as in Eq. (4).

$$\phi = 2\pi\Phi/\Phi_0 \qquad (4)$$

The inductance L associated with the kinetic energy E [Eq. (2)] can be written in the form of Eq. (5). From Eqs. (2), (4), and (5), it follows that the

$$1/L = d^2E/d\Phi^2 \qquad (5)$$

admittance, $1/2\pi fL$, is a periodic function of Φ. It is this admittance that is sensed in the readout scheme for the rf SQUID. When the coupling energy is at a maximum, the admittance is minimum (and vice versa).

The period of response, Φ_0, can be translated to a field periodicity B_0 by the relation $B_0 = \Phi_0/a$,

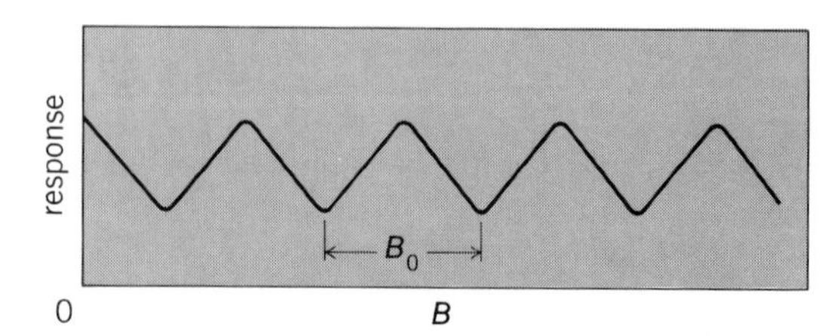

Fig. 2. Response of the rf SQUID to an applied magnetic field B.

where a is the area of the SQUID ring. For typical SQUIDs B_0 is of the order of 10^{-10} tesla. Figure 2 shows the SQUID output as a function of applied magnetic field. The output in its simplest form is digital, that is, one can electronically count the flux quanta as the applied magnetic field is changed. Each period of the response can be subdivided electronically into 10,000 or more parts, so that the demonstrated sensitivity of SQUID magnetometers is approximately 10^{-14} tesla (or 10^{-10} the strength of the Earth's magnetic field) in a bandwidth of 1 Hz. The SQUID sensitivity and bandwidth exceed those of all other magnetometers by a wide margin.

Magnetic field detection. One of the most straightforward applications of the SQUID is for measurements of fluctuations in the Earth's magnetic field. These fluctuations reflect activity in the Earth's magnetosphere, which is highly influenced by the solar wind. The variations in magnetic field, called micropulsations, not only are of direct scientific interest, but also provide a useful probe of the conductivity of the Earth. With the use of a SQUID and electric field probes, these pulsations provide conductivity information as a function of depth to many kilometers below the Earth's surface. High conductivity anomalies reflect geothermal activity which could be important as an energy source.

The SQUID has also been used in the laboratory to study magnetic properties of rock samples and other materials. The greatest advantage of SQUID systems for these measurements is their rapid access time. Measurements can often be accomplished in seconds, because samples need no special mounting. Competing techniques require special mounting of each sample since their operation depends on rapid vibration or rotation of the sample to produce an alternating magnetic field.

Biological systems. Perhaps the most intriguing SQUID application is in the detection of magnetic fields generated by biological systems. The human electrocardiogram and electroencephalogram are both manifestations of time-varying electric currents in the body. The magnetic fields associated with these currents extend beyond the body. The SQUID is the first magnetic sensor with sufficient sensitivity to provide a useful measurement of these very weak fields. Figure 3 shows both a magnetocardiogram and the instrument used to record it. An obvious advantage of this technique is the absence of any direct connections to the subject. There is, however, a concern about interfering magnetic fields. The typical environment inside most buildings includes powerline and other magnetic fields which are many times larger than the biomagnetic fields. However, the effect of these interfering fields can be reduced by the use of a special configuration of the SQUID called a gradiometer, by passive magnetic shielding, or by a combination of both of these methods.

The question that naturally arises with this application is whether there are any fundamental differences between the character of the electric and magnetic signals generated by the same source. The existence of such a difference is best illustrated by some recent brain research. In this study the magnetic response to external stimuli (light and sound) was recorded. The response to each stimulus was found to be highly localized in a different part of the brain. This result is in sharp contrast to electric recordings, which show a diffuse response over the entire head. The reason for this is related to the fact that the skull is a poor conductor of electricity. The currents that are driven by a local source in the brain reach the scalp through openings in the skull or by conduction through large areas of skull, and thus most spatial information is lost. The magnetic signal, on the other hand, is generated primarily by strong currents that flow very close to the source. This work could possibly have a large impact on the understanding of brain function.

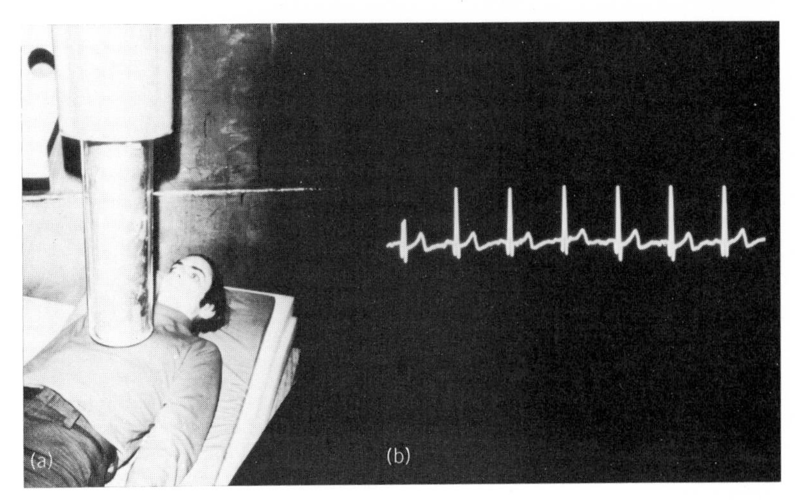

Fig. 3. Detection of magnetic fields generated by action of human heart. *(a)* Instrument being used to detect fields. *(b)* Magnetocardiogram generated by instrument.

Measurement of current and other quantities. Because of its sensitivity, the SQUID has found many other applications in metrology. A highly sensitive ammeter can be constructed by coupling a coil to the SQUID. The current to be sensed is thus converted to a magnetic field which is measured at the SQUID. Techniques have also been developed for measuring rf current, power, and attenuation. The most sensitive and precise instrument for measurement of dc ratios is likewise based on the SQUID. This instrument provides accurate methods for resistance ratio and voltage ratio measurement.

Fast switching applications. All the applications described thus far relate to the detection of weak magnetic fields. However, the dc SQUID has recently been developed for fast switching applications, that is, as an electronic logic or memory element. Present information indicates that computers or fast electronic instruments based on the SQUID switch could provide enormous improvements in speed and reductions in size. The SQUID circuits for this type of application are made of metallic (superconducting) thin films. The circuits are constructed by using photolithography, a process that is common to the fabrication of semiconductor circuits. The SQUID is induced to switch states by a weak magnetic field generated by a control line which runs over the SQUID. The time required for complete switching might ultimately be 10^{-12} s, but the best results thus far are about 10 times slower than that. While this rapid switching is highly useful, the prime advantage of the SQUID switch is the low level of heat power dissipated

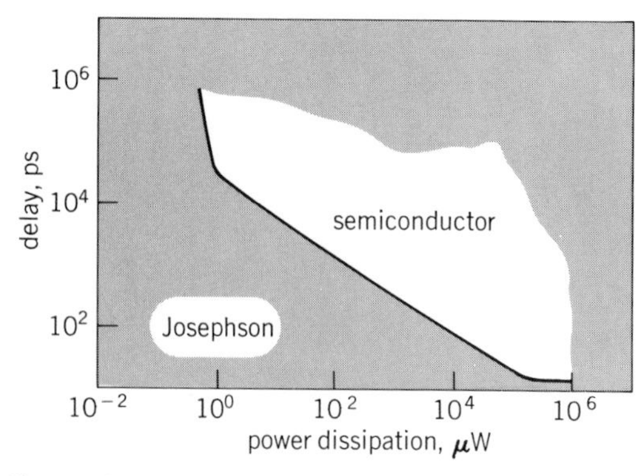

Fig. 4. Comparison of the SQUID (Josephson) technology with semiconductor technologies.

(Fig. 4). To understand why this is so important, it is necessary to consider the present state of computer development.

The controlling factor for fast computers is their size. The time required for signals to move from one part of the computer to another is limited by the velocity of light. Thus, to improve computer speed, the elements must be close to one another. However, they cannot be moved too close together, because it would be difficult to remove or dissipate the heat generated by each element. For example, the fastest large computer in existence today dissipates 115,000 W in a volume of 1.7 m^3. The major engineering feat represented by this computer is its cooling system, without which the computer would burn itself up. The SQUID switch dissipates 10,000 times less heat than its semiconductor rival and therefore permits much closer spacing of components. It has been estimated that the computer described above could occupy a volume 25,000 times smaller if SQUID circuit elements were used. The resulting computer would be more than 10 times faster.

The largest research program on this technology is being conducted by the IBM Corporation. Scientists at IBM are optimistic about their work, but much remains to be done. Smaller development programs, aimed at fast electronic instruments and computers, are being conducted at Bell Telephone Laboratories, the Aerospace Corporation, the University of California, Berkeley, and the National Bureau of Standards. It might be 10 years or more before the benefits of this research and development are realized.

For background information *see* JOSEPHSON EFFECT; SUPERCONDUCTING DEVICES in the McGraw-Hill Encyclopedia of Science and Technology. [JAMES E. ZIMMERMAN; D. B. SULLIVAN]

Bibliography: B. Deaver et al. (eds.), *Future Trends in Superconductive Electronics*, A.I.P. Conf. Proc. Ser. 44, 1978; H. D. Hahlbohm and H. Lubbig (eds.), *Superconducting Quantum Interference Devices and Their Applications*, 1977; V. L. Newhouse (ed.), *Applied Superconductivity*, chap. 1, 1975.

Superconductivity

Recent studies have shown that hydrogen can have important and varied effects on the superconducting properties of materials. First, in some cases hydrogen may be used as a bulk ingredient in order to raise the superconducting transition temperature T_c. Research in this area is motivated by the indications of an attractive pairing interaction resulting from the high-frequency vibrational modes of the proton lattice. For example, while palladium is not superconducting at any temperature, Pd-H has a T_c of 9 K and Pd-Cu-H has a $T_c =$ 16.6 K.

Second, hydrogen can be used as a dopant in the study of the nature of the transition to the superconducting state. Here the principal motivation results from the observation that some of the superconductors with the so-called A-15 crystal structure can be doped with hydrogen, resulting in a mild but systematic change in the crystal structure that provides a unique probe with which to try to understand the high T_c values of this important class of superconductors.

Third, it has recently been demonstrated that hydrogen can be present as an impurity in surprisingly large quantities, and it may have important influences on the superconducting properties of the material. Because hydrogen is usually the principal residual gas in high-vacuum systems, the problem of hydrogen impurities is especially important in materials made by vacuum deposition. Recent work on hydrogen as an impurity in Nb_3Ge superconductors will be discussed below.

Nb_3Ge superconductors. Nb_3Ge is the material with the highest known superconducting transition temperature, $T_c \sim 23$ K. While Nb_3Ge is of interest because of its high T_c, it is of additional special interest for at least two reasons. First, in spite of considerable study, it is not understood why this material has such a high T_c. Second, the high-T_c phase of Nb_3Ge is not the stable phase, but the metastable A-15 crystal structure. As a result, it is not possible to make Nb_3Ge in bulk, but it is quite easy to make it in thin films by evaporation, sputtering, or chemical vapor deposition methods. The reasons that the high-T_c metastable A-15 phase occurs under these deposition conditions are not known and are the focus of much experimental effort. Here the presence of hydrogen as an impurity is beginning to be explored, as is the more general question of the role of impurities in the stabilization of the metastable A-15 phase.

One approach to the study of why Nb_3Ge has such a high T_c is to measure complete sets of systematic data involving Nb_3Ge and closely related materials. One study has been to vary the stoichiometry away from the 3:1 ratio for pure Nb_3Ge. Somewhat surprisingly, R. C. Dynes, J. Poate, and collaborators at Bell Laboratories have demonstrated that T_c is not very sensitive to stoichiometry; even materials with a ratio of Nb to Ge atoms as low as 2:1 still have high T_c's. Another study has been to vary the temperature of the substrate onto which the Nb_3Ge films are deposited. Here a rather dramatic effect is found. The highest-T_c films are deposited at temperatures of about 700–750°C, and there is a rapid drop in T_c as the substrate temperature is lowered below this value. The generally accepted assumption is that this lowering of T_c is a result of the Nb-Ge phase diagram, and at lower temperatures phases other than the high-T_c A-15 phase are formed.

Hydrogen doping of Nb_3Ge. Studies have also been conducted to determine why, once the A-15 phase of Nb_3Ge has been successfully made, it has such a high T_c. Here there is much less freedom to vary parameters at will. One fruitful approach is to dope the Nb_3Ge with hydrogen. This has the effect of increasing the lattice constant without changing the crystal structure. Such a procedure provides a means for testing models of superconductivity which relate changes in T_c (ΔT_c) to the lattice constant and models which relate ΔT_c to electron/atom ratio.

This hydrogen doping procedure has recently been applied to two A-15 superconductors, Nb_3Ge and Nb_3Sn. The Nb_3Sn can be prepared in bulk samples by sintering Nb and Sn powders and, consequently, in the past more detailed studies were possible for Nb_3Sn than were possible for Nb_3Ge, which can be made only in thin films. The fact that Nb_3Ge is available only as thin films (typically 500 nm thick) has been a major experimental difficulty since it has severely limited the analytic measurements which could be applied to this important material. However, the recent development of nuclear back scattering and nuclear reactions to study the bulk and impurity composition and structure of thin films now allows many studies which were previously impractical. The quantitative measurement of the hydrogen content of the Nb_3Ge films doped with hydrogen is a good example of the application of such new technology.

Nuclear reaction analysis for hydrogen. The quantitative determination of the hydrogen content of solids traditionally has been a very difficult problem. Most of the newly developed analytic tools such as neutron activation analysis, Auger spectroscopy, back scattering, or fluorescence specifically do not work for hydrogen. More traditional methods suffer from hydrogen backgrounds from the walls of containing vessels or from water or hydrocarbons. In addition, these methods are difficult to apply on small samples. However, because of hydrogen's common occurrence as an impurity, its influences can be very important. Recently W. A. Lanford and collaborators at Yale developed a method for hydrogen analysis based on a resonant nuclear reaction between H and ^{15}N which now allows rapid, quantitative measurement of hydrogen content versus depth in any solid with a depth resolution on the order of 10 nm.

This procedure makes use of the $^{1}H + {}^{15}N \rightarrow {}^{12}C + {}^{4}He + 4.43$ MeV gamma-ray reaction as a probe for hydrogen. The sample to be analyzed is bombarded with ^{15}N and the number of characteristic 4.43-MeV gamma rays is used as a measure of hydrogen content. Because this is a resonant reaction, it has a large probability only when the ^{15}N is precisely at the resonance energy. For any given beam energy, this condition is fulfilled only within a narrow detection "window" at a certain depth beneath the surface. Hence, by measuring the yield of characteristic gamma rays versus ^{15}N-beam energy, one can deduce the hydrogen concentration profile, with the higher-energy ^{15}N probing deeper into the sample (Fig. 1).

Some of the results of using the ^{15}N hydrogen profiling method to study the effects of hydrogen doping on Nb_3Ge are shown in Figs. 2 and 3. Figure 2 shows the relationship between T_c and lattice parameter for Nb_3Ge films doped with varying amounts of hydrogen, while Fig. 3 relates T_c with

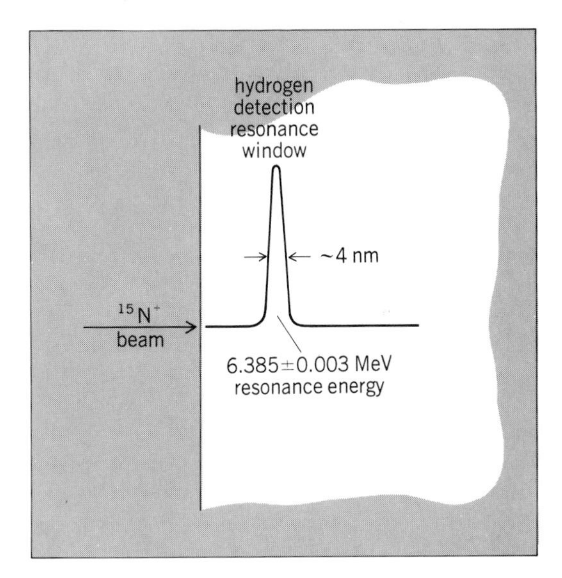

Fig. 1. Schematic representation of the ^{15}N nuclear resonance reaction method for measuring hydrogen concentration versus depth in any solid. *(From W. A. Lanford et al., A new precision technique for determining concentration vs depth of hydrogen in solids, Appl. Phys. Lett., 28: 566–568, 1976)*

hydrogen content for such films. The introduction of hydrogen expands the lattice constant enormously, up to 2.5%, while maintaining the crystal structure. Because of nonuniformities in thin films, the transition to the superconducting state does not occur at a single temperature, but rather over a range of temperatures. Thus, the vertical bars in Figs. 2 and 3 represent 97.5 to 2.5% transition temperatures, with the dots representing the 50%

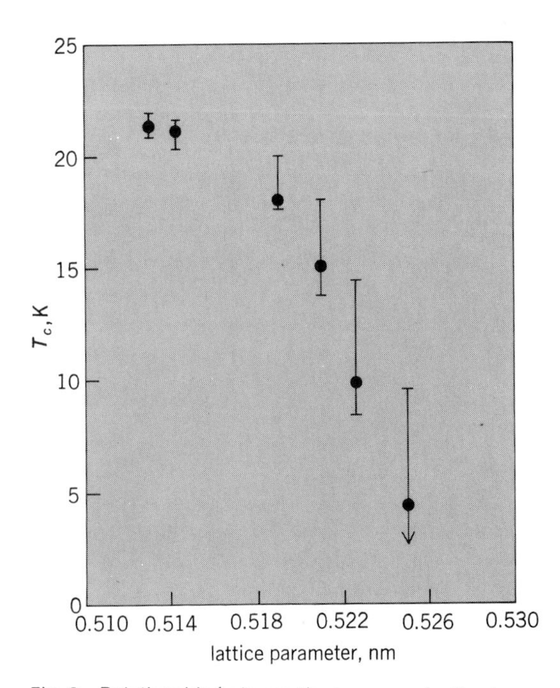

Fig. 2. Relationship between the superconducting transition temperature T_c and the lattice constant for Nb_3Ge films doped with hydrogen. *(From J. M. Rowell et al., Tunneling, x-ray and electron diffraction studies of the structure of Nb_3Ge films, IEEE Trans. Magnet., MAG-13: 644–647, 1977)*

transition temperature. No measurements of the superconducting properties of thin films were carried out below 4 K. Hence, transitions which occur below this temperature are given only as upper limits indicated by arrows in Figs. 2 and 3. The horizontal bars in Fig. 3 indicate the variations in hydrogen content.

A comparison of the results shown in Fig. 2 with those measured by L. J. Vieland and collaborators for Nb_3Sn is important in evaluating models which relate ΔT_c with changes in the lattice parameter. The results in Nb_3Sn show T_c decreasing much more rapidly with increasing lattice constant than in Fig. 2. On the other hand, the results shown in Fig. 3 relating ΔT_c with hydrogen atoms per Nb_3Ge are nearly identical to the results for Nb_3Sn. How general and how important this result is need further study, including similar measurement in other A-15 superconductors.

Impurity influence on phase stability. A question of continuing importance is what experimental parameters limit or control the T_c of Nb_3Ge films. Since it is the A-15 phase of Nb_3Ge that has the high T_c, a key part of this question is what influences the phase stability of the metastable A-15 phase. The influence of deposition temperature has already been mentioned; at low deposition temperatures, phases with low T_c become important.

While the question is certainly far from settled, there is evidence that impurities are important in stabilizing the A-15 phase. The most direct evidence for this comes from the work of J. R. Gavaler and other investigators who have found that they fail to make high-T_c Nb_3Ge if their vacuum system is too "clean," while if they allow some gas into the vacuum chamber during deposition, high T_c material is made. The contaminants most commonly discussed are hydrogen and oxygen. Both these elements have been found as impurities

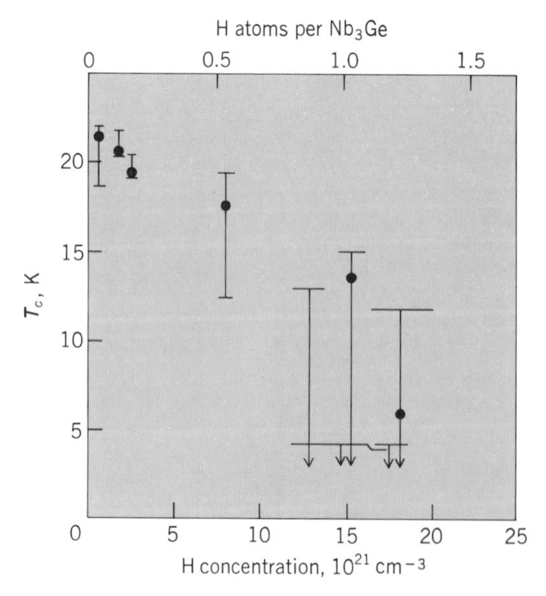

Fig. 3. Relationship between the superconducting transition temperature T_c and hydrogen content for Nb_3Ge films doped with hydrogen. *(From W. A. Lanford et al., Sensitivity of the T_c of Nb_3Ge to hydrogen content, Appl. Phys. Lett., 32:339–341, 1978)*

in Nb_3Ge films which were deposited in "standard" setups.

It is, however, difficult to assess the precise role that impurities play in these films because it has not generally been possible to vary the impurity concentration independently of other deposition conditions. For example, in the case of hydrogen impurities, a correlation between T_c and hydrogen content has been observed in a series of films deposited with the deposition temperature varied systematically. Films deposited at high temperatures have low hydrogen content and high T_c; those deposited at low temperature have high hydrogen content and low T_c. It is not yet known whether this should be interpreted as an effect where the presence of hydrogen is degrading the T_c of the Nb_3Ge or whether the lower deposition temperatures result in a different phase (or in defects in the A-15 phase) which can trap hydrogen effectively. In any case, it is now clear that hydrogen can be present in surprisingly large amounts in supposedly "clean" vacuum systems and can have important effects as an impurity.

For background information *see* SUPERCONDUCTIVITY in the McGraw-Hill Encyclopedia of Science and Technology. [WILLIAM A. LANFORD]

Bibliography: P. Duffer, D. M. Gualtieri, and V. U. S. Rao, *Phys. Rev. Lett.*, 37:1410–1413, 1976; J. R. Gavaler, in D. H. Douglass (ed.), *Superconductivity in d- and f-Band Metals*, pp. 421–427, 1976; W. A. Lanford et al., *Appl. Phys. Lett.*, 32: 339–341, 1978; L. J. Vieland, A. W. Wicklund, and J. C. White, *Phys. Rev.*, 11:3311–3316, 1975.

Superheavy quasiatoms

Since the beginning of the development of relativistic quantum mechanics, there has been a long-standing theoretical interest in fundamental questions associated with the behavior of the Dirac electron interacting with very strong electromagnetic fields. Recently, owing to the possibility of exploring such phenomena experimentally for the first time, such systems have been reexamined theoretically in the context of atomic physics. Extensive investigations by J. Rafelski, B. Müller, W. Greiner, and coworkers in Frankfurt and independent considerations by Ya. B. Zel'dovich and V. S. Papov in Moscow have shown that when the binding of electrons in atoms exceeds the energy equivalent of twice the electron mass, $2mc^2$ (where m is the electron mass and c is the speed of light), qualitatively new phenomena are expected to occur. In particular, if the electron charge is removed from the vicinity of nuclear charge Z, which exceeds the critical value Z_{crit} required to produce this overcritical binding, the vacated free space (that is, the so-called neutral vacuum of quantum electrodynamics) is predicted to be unstable to the spontaneous emission of positrons, unlike the well-documented stability to such decay for charges with $Z<Z_{crit}$.

In addition to this association with the electrodynamics of atomic systems, systems of particles in which the binding energy is a dominant fraction of the total energy are of great interest in connection with the physics of highly compressed nuclear matter and the strong gravitational fields surrounding black holes. High-Z atoms provide a convenient (and possibly the only) vehicle for exploring these strong binding phenomena in a situation where the interaction (electromagnetic in this case) is well understood, and therefore where precise theoretical predictions can be made.

Because of their fundamental relevance to several branches of physics, an experimental effort has been initiated during the past few years to test these theoretical ideas and to confirm understanding of the overcritical phenomena. In particular, attempts are being made to observe the unique positron emission process and, in general, to trace the energy-level structure of superheavy atoms into the region of overcritical binding in order to verify that overcritical binding does indeed exist in atoms. Superheavy quasiatoms are playing a pivotal role in this effort.

Overcritical binding and spontaneous positron emission. The Dirac equation is the basis for formulating the behavior of electrons interacting with electromagnetic fields. Its energy-level spectrum provides a convenient format for illustrating some of the physical content of the theory of overcritical binding in atoms and the related positron emission process.

As shown in Fig. 1, the solutions of the Dirac equation for an electron include (1) a continuum of positive energy states with total energy greater than mc^2, (2) a discrete set of energy levels whose bounds are mc^2 and $-mc^2$, and (3) a negative energy continuum of states which extends below $-mc^2$. (The rest-mass energy of the electron, mc^2, has been included in this diagram.) An essential feature of this spectrum is the symmetry between the positive and negative energy continuum states. It reflects the equal footing of particles (electrons) and their antiparticles (positrons) in the theory and in nature. This association of particles and antiparticles with the positive and negative energy continuum states, as well as some of the resulting physical consequences which follow from this connection, emerges from Dirac's hole hypothesis, which considers all the negative energy states occupied by electrons as the lowest energy state of free space (the neutral vacuum). A vacancy in this sea of electrons ("Dirac sea") materializes as a positron. The stability against the spontaneous transitions of positive energy electrons to negative energy states, allowed by energy conservation, is ensured in the theory by the complete occupancy of all the states in the lower continuum and by the Pauli principle which forbids the simultaneous occupancy of one state by more than one electron. Although Dirac's early hypothesis has been superseded by modern quantum field theory, the essential physical aspects of this model remain useful. For example, electron-positron pair production by photons can be viewed as an excitation of an electron by the photon from the Dirac sea into unoccupied positive energy states. The resulting vacancy escapes as a positron together with the ejected electron. The well-known threshold energy of $2mc^2$ for this process reflects the energy gap between the two continua shown in Fig. 1.

The pair production process is inherent in the Dirac theory of the electron and therefore plays a fundamental role in all atomic processes. For example, virtual electron-positron production (technically referred to as vacuum polarization), in which the pair cannot escape the atom due to the constraints of energy and momentum conservation, modifies the discrete set of energy levels of an electron bound in the electrostatic potential of a charged nucleus. For the experimentally accessible stable elements up to fermium ($Z = 100$), where the binding energy is only a small fraction of the energy gap between $\pm mc^2$, these effects are small. Nevertheless, precision measurements of vacuum polarization effects have provided the theory of quantum electrodynamics with some of its most exacting tests to date.

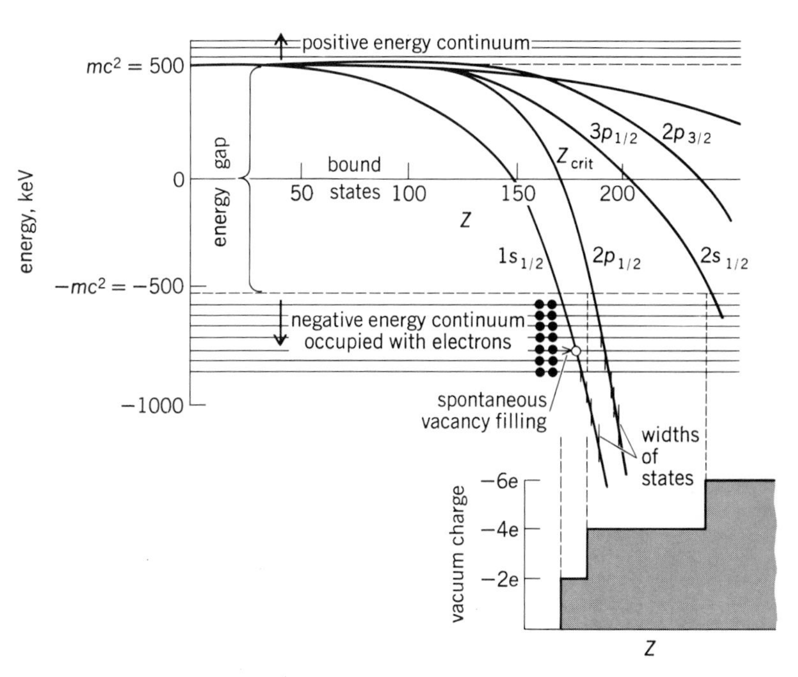

Fig. 1. Energies of the lowest-bound states of the Dirac equation plotted as a function of the nuclear charge Z for nuclei with a finite size. *(From B. Müller, J. Rafelski, and W. Greiner, Electron shells in overcritical external fields, Z. Phys., 257:62–77, 1972)*

However, as mentioned above, even more dramatic consequences of pair production are expected to occur beyond the known stable elements as the electronic energy becomes equal to $-mc^2$ for very large values of nuclear charge. Figure 1 shows that the deepest-bound $1s_{1/2}$ state achieves this binding energy for Z equal to about 173 in finite-size nuclei. This critical charge defines a boundary of fundamental significance and marks the threshold of the onset of a new phenomenon.

Below the critical charge, energy must be fed into the atomic system to create pairs. As is evident from energy considerations (Fig. 1), the required energy is a minimum and decreases with increasing Z if the electron of the pair remains bound to the nucleus in a vacant $1s_{1/2}$ state while the positron leaves the vicinity of the nucleus. It is equal to the energy difference between the $1s_{1/2}$ state and $-mc^2$, where the negative energy continuum begins. Eventually, beyond the critical charge, this energy difference changes sign and therefore becomes energetically favorable to produce pairs. Viewed in the context of the Dirac hole picture, as the vacant $1s_{1/2}$ state joins the negative energy continuum, it can be filled spontaneously with an electron from the Dirac sea with the simultaneous emission of a free positron (the ensuing hole in the Dirac sea) of a well-defined kinetic energy. The states that join the negative energy continuum also develop a width, schematically illustrated by the bars in Fig. 1, as they are shared by the continuum states, and the probability for spontaneous positron emission reflects this spreading width.

Thus the critical charge defines the boundary between endoergic and exoergic pair production in atoms. As a result of the spontaneous pair production in overcritical fields, an overcritically charged nucleus stripped of its surrounding electrons is inherently unstable. A positron is emitted, and an electron remains bound to the nucleus in an overcritically bound state. The lowest atomic state of the system surrounding the bare nucleus is no longer free of particles and neutral as originally prepared, but instead contains the bound electron cloud which is negatively charged. Thus as indicated schematically in Fig. 1, when the central nuclear charge is increased arbitrarily so that the bound states $1s_{1/2}$, $2p_{1/2}$, $2s_{1/2}$, and so forth, successively cross the energy of $-mc^2$, successive phase transitions can occur in which the neutral vacuum (no charged particles in the vicinity of the nucleus) increases its negative charge as an electron cloud surrounding the nucleus builds up. Since each of the curves in Fig. 1 actually represents two spin states, the vacuum charge increases by two units each time a curve crosses the critical energy. This new form of instability in overcritical fields can be viewed as a breakdown of the neutral vacuum. The resulting charged lowest-energy state, the so-called charge vacuum (a term adopted by Rafelski, Müller, and Greiner, who motivated and developed many of these ideas), differs inherently from the usual charged atomic state in undercritical systems. In the latter case, removing all the electrons from the nuclear charge leads to a stable neutral vacuum state. On the other hand, the neutral vacuum of an overcritical system always spontaneously decays to a charged vacuum and positrons, and therefore the space surrounding an overcritically charged nucleus can never be free of charges. Thus the charged vacuum is a necessary new concept for overcritical fields.

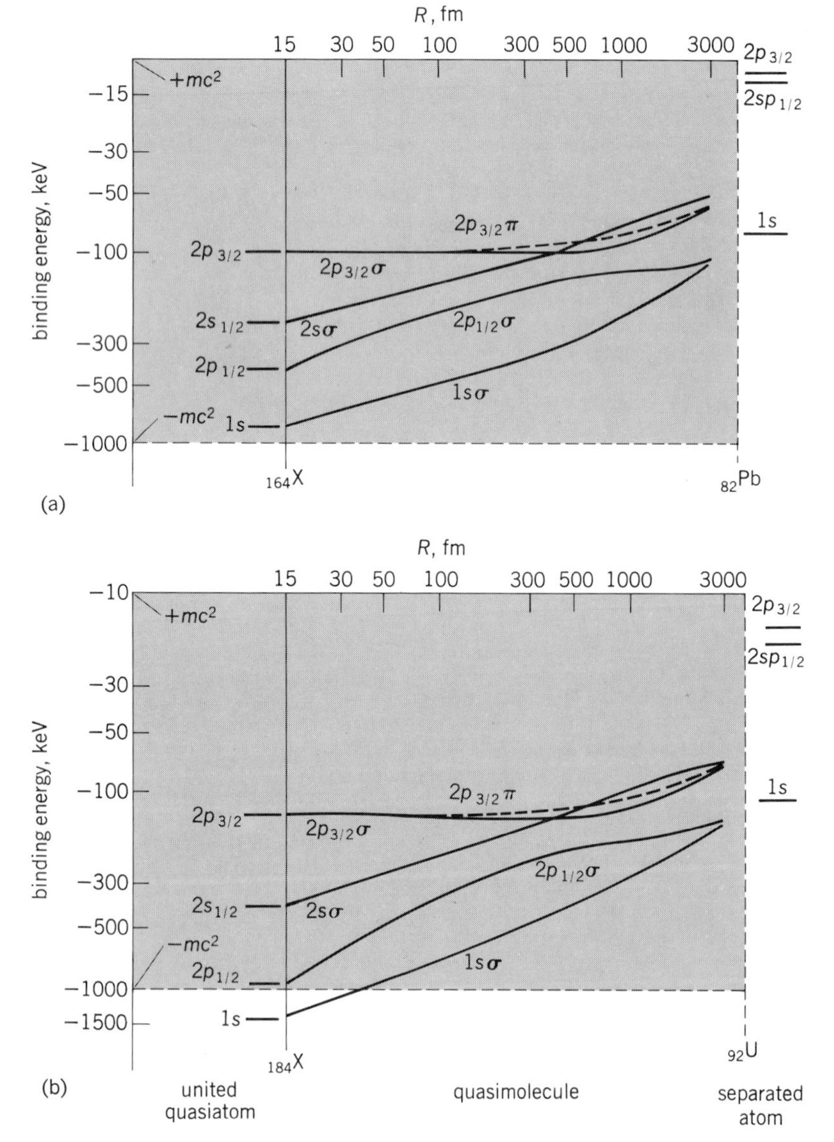

Fig. 2. Some of the lowest-energy electronic levels of quasimolecules as a function of internuclear separation *R*, showing development from the separated atom to the united quasiatom energy-level structures: *(a)* Pb + Pb system. *(b)* U + U system. *(From J. Kirsch et al., K-X-ray spectrum of the Pb + Pb quasimolecules, Phys. Lett., 72B: 298–302, 1978)*

The decisive experiment to verify these new aspects of quantum electrodynamics is to observe the spontaneously emitted positrons. For this purpose atoms of $Z \geq 173$ are required with the overcritically bound states unoccupied. With the region of overcritically charged nuclei being far beyond the present capability of producing superheavy nuclei, it is fortunate that the opportunity has developed recently to form short-lived superheavy atomic systems (quasiatoms) in heavy-ion collisions.

Overcritical fields in heavy-ion collisions. During the close collisions between two heavy ions, for the short time the nuclei are very close together, the electrons experience an electric field which is produced by the combined effect of the two nuclear charges.

For projectiles presently available at heavy-ion accelerations, the combined charge $(Z_1 + Z_2)$ can exceed the critical charge of approximately 173 and thus provide a source for overcritical fields. In fact, under stable bombarding conditions the collision system much resembles a quasiatom for a short period of time. If the electrons move fast compared with the nuclear motion so that they can adjust their distribution adiabatically to the changing nuclear potential fields, the whole electron cloud acts as if it were associated with both nuclear centers. The electrons evolve through a series of quasimolecular states as a function of the changing internuclear separation. At small internuclear separations, when the nuclei are well within the orbiting radii of the electrons, the system resembles a quasiatom of nuclear charge $(Z_1 + Z_2)$. For example, Fig. 2 shows how for Pb + Pb and U + U collisions the most-bound molecular orbitals evolve from the atomic states of the separated atoms to states of the united quasiatom (atomic number $Z_1 + Z_2$) at small internuclear separations. The Pb + Pb system never achieves overcritical binding. However, for U + U the potential binding of the tightest-bound $1s\sigma$ state exceeds $2mc^2$ for the internuclear separations less than a critical distance R_{crit} of about 34 femtometers. Therefore, as predicted in the case of stable atoms, in U + U collisions spontaneous positron emission can occur during the fraction of the collision time when the binding is overcritical. The needed $1s\sigma$ vacancy can be supplied early in the same collision by the ionization of $1s\sigma$ electrons in the time-dependent electric fields produced by the nuclear motion.

However, in contrast with the stable atom situation, in such dynamical systems other positron creation mechanisms, not involving overcritically bound states, complicate the observation of spontaneous positron emission. In addition to the latter process which proceeds without an external energy source, the time dependence of the electric field produced by the nuclei can also induce transitions of electrons from the negative energy continuum both to unbound states and to unoccupied bound states even above the critical binding energy. These two subcritical processes, induced emission into bound states and direct induced emission into continuum states, are shown in Fig. 3, along with spontaneous emission, in a plot of the evolution of the bound states with time during the collision. The energy required for each of these subcritical pair creation processes is supplied by the nuclear motion. Although they are undesirable backgrounds in the detection of spontaneous positron emission, the two effects are of considerable interest in themselves since they also reflect new aspects of the theory when strong electromagnetic fields are involved and when perturbation theory is not applicable.

Another important source of positron production to be considered in heavy-ion collisions is the excitation of internal degrees of freedom of the two nuclei through their internal Coulomb interactions and the subsequent decay of excited nuclear states by the well-known internal pair conversion process which competes with gamma radiation above 1.02 MeV. An understanding of all the subcritical pair production processes is required in order to carry out a meaningful study of the overcritical phenomena in heavy-ion collisions.

Search for positron emission. Beginning in the fall of 1976, a series of measurements have been carried out at the Gesellschaft für Schwerionenforschung laboratory in Darmstädt, West Germany, by an international collaboration of scientists to investigate positron creation in heavy-ion collisions. In order to selectively study both the subcritical and the overcritical positron production processes that are discussed above, the studies have included Pb + Pb, Pb + U, and U + U collision systems.

The measurements with the Pb + Pb systems, where only the subcritical phenomena contribute, have verified that the two quantum electrodynamic subcritical pair creation processes illustrated in Fig. 3 are fairly well understood. The extent of the agreement obtained between theory and experiment for both nuclear and quantum electrodynamic sources of positrons is demonstrated in Fig. 4. The differential cross section for positron production at a particular projectile scattering angle is plotted as a function of the distance of closest approach. (Measured contributions from the two sources were compared with calculations by J. Reinhart and colleagues.) In this case, the background from nuclear interactions could be readily determined experimentally and, as shown, it was

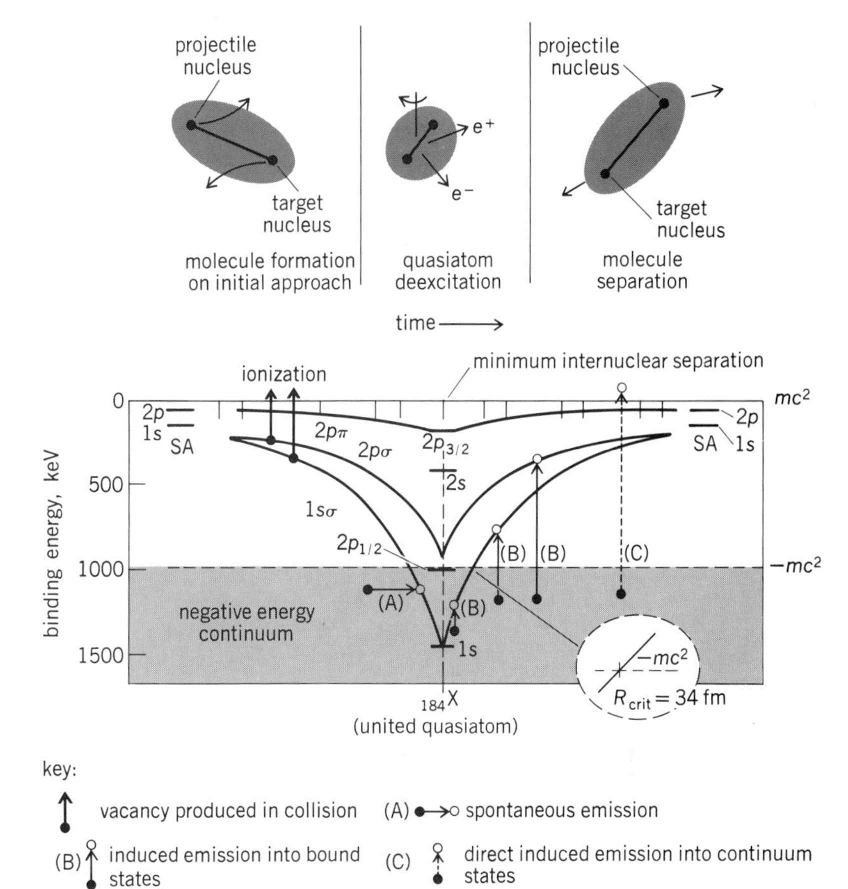

Fig. 3. Positron production mechanisms in heavy-ion collisions, shown on plot of energies of bound states as a function of time for U + U system.

found that the dominant contribution to the total positron yield was from nonnuclear processes. Further experiments also showed that the quantum electrodynamic effects again constituted the major contribution to the positron production in the higher $Z_1 + Z_2$ systems, and that in these cases this contribution could be separated effectively from the nuclear-related backgrounds. As shown in Fig. 5 for all these collision systems, the two quantum electrodynamic subcritical pair creation processes, induced by the superimposed Coulomb fields of projectile and target, can account successfully for the positron production probability per scattered particle as a function of the center-of-mass scattering angle. (Experimental results by C. Kozhuharov and colleagues were again compared with calculations by Reinhart and colleagues for these two processes.) There is some incompleteness in the calculation for the U + U system.

The most striking feature of the results obtained thus far is the almost exponential increase of the positron yield with increasing $Z_1 + Z_2$. Expressed as a power law proportional to $(Z_1 + Z_2)^n$, n has been determined to be approximately 17. This rapid increase in Z finds no other known analog in nature and singularly reflects the relativistic phenomena associated with strong electric fields and their strongly bound atomic states. However, although it has been possible to observe some of the dynamic features marking the effects of strong

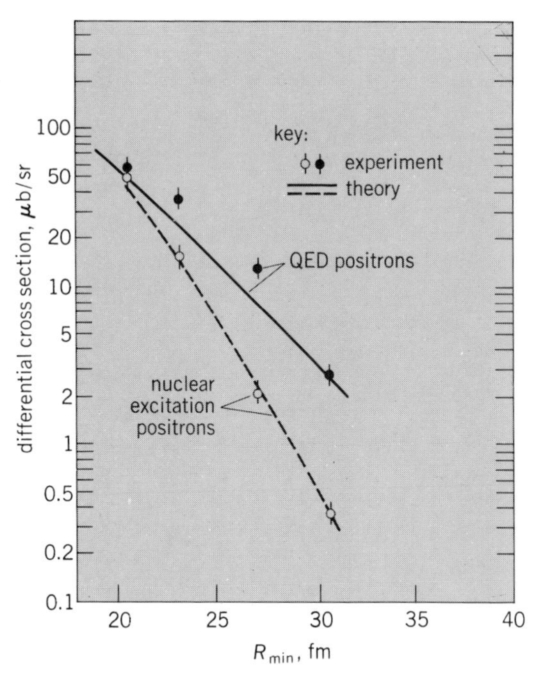

Fig. 4. Differential cross section with respect to the projectile ion-scattering angle (in the laboratory reference frame) for positron production in Pb + Pb collisions plotted against the distance of minimum approach, R_{min}, between nuclei during the collision. Data are shown for projectile ion-scattering angles in the range 45 ± 10°. *(From H. Backe et al., Observation of positron creation in superheavy ion-atom collision systems, Phys. Rev. Lett., 40:1443–1446, 1978)*

fields, a definitive signature for spontaneous positron creation has yet to be demonstrated experimentally in U + U collisions. One reason for this is that the sharp spontaneous positron spectrum

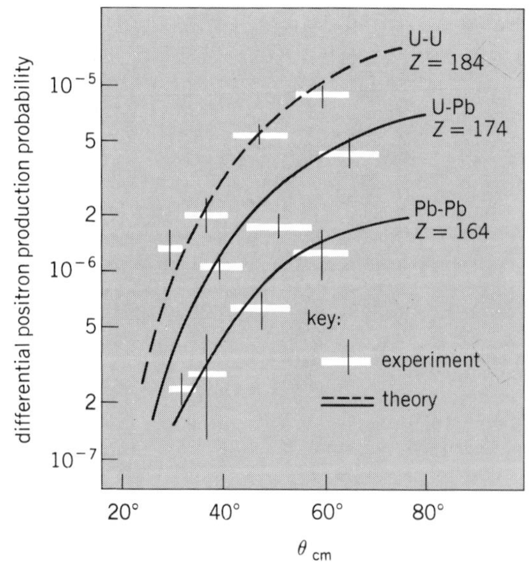

Fig. 5. Differential positron production probabilities per scattered particle as a function of the center-of-mass scattering angle θ_{cm} for 5.9-MeV-per-nucleon U + U, U + Pb, and Pb + Pb collisions. A positron energy interval of 100 keV is accepted centered about 490 keV, and nuclear-related backgrounds are subtracted.

expected from overcritically bound states in stable atoms loses this distinctive characteristic in the collision system due to collision broadening. Other possible identifying features are presently being studied theoretically. In the meantime, the experimental search continues, with more detailed experiments being carried out to examine the differential cross sections and positron spectra for subcritical and overcritical systems, as well as with explorations of higher-Z systems such as U + Cf, where the stronger binding is expected to enhance the spontaneous positron emission. The combined effort by theory and experiment has already revealed interesting new features of atomic physics, and it is expected that even more discoveries will follow.

For background information *see* ATOMIC STRUCTURE AND SPECTRA; PAIR PRODUCTION, ELECTRON-POSITRON; QUANTUM ELECTRODYNAMICS; QUANTUM THEORY, RELATIVISTIC; QUASIATOM in the McGraw-Hill Encyclopedia of Science and Technology.

[JACK S. GREENBERG]

Bibliography: H. Backe et al., *Phys. Rev. Lett.*, 40:1443–1446, 1978; J. D. Bjorken and S. D. Drell, *Relativistic Quantum Fields*, 1965; J. D. Bjorken and S. D. Drell, *Relativistic Quantum Mechanics*, 1964; J. Rafelski, L. P. Fulcher, and A. Klein, *Phys. Rep.*, 38C:227–361, 1978; J. Reinhardt and W. Greiner, *Rep. Prog. Phys.*, 40:219–295, 1977.

Surface physics

The past several years have seen the development of theoretical analyses and experimental techniques accurate enough to probe electrons confined to the surface of a solid or liquid. In part, these developments were prompted by important technical breakthroughs. Magnetic thin films are used to record information; printed circuitry, including surface-active field-effect transistors, metal oxide semiconductor field-effect transistors (MOSFETs), and other devices are used to process the information. All these applications require an understanding and control of charged carriers at surfaces or interfaces. Research has also been sparked by the hope of discovering high-temperature superconductors or, at the very least, a new generation of faster transistors and switching devices. Another motivation is application of this research to the important studies of corrosion and heterogeneous catalysis, both related to the quantum-mechanical properties of superficial electrons. Finally, there is the purely intellectual challenge. Due to topological considerations, particles in two dimensions are expected to behave quite differently than in three dimensions; interesting research has been concerned with identifying these differences.

Within inversion or accumulation layers at the surfaces and interfaces of semiconducting materials, there exist potential gradients steep enough to "freeze out" motion perpendicular to the surface. The carriers so confined constitute a two-dimensional fluid of electrons (or, possibly, holes) the density (n_s) of which may be varied in the range $n_s = 10^{11}-10^{13}/\text{cm}^2$ either by chemical means (such as dopants or ion implantation) or by physical means (such as applied electric, magnetic, or

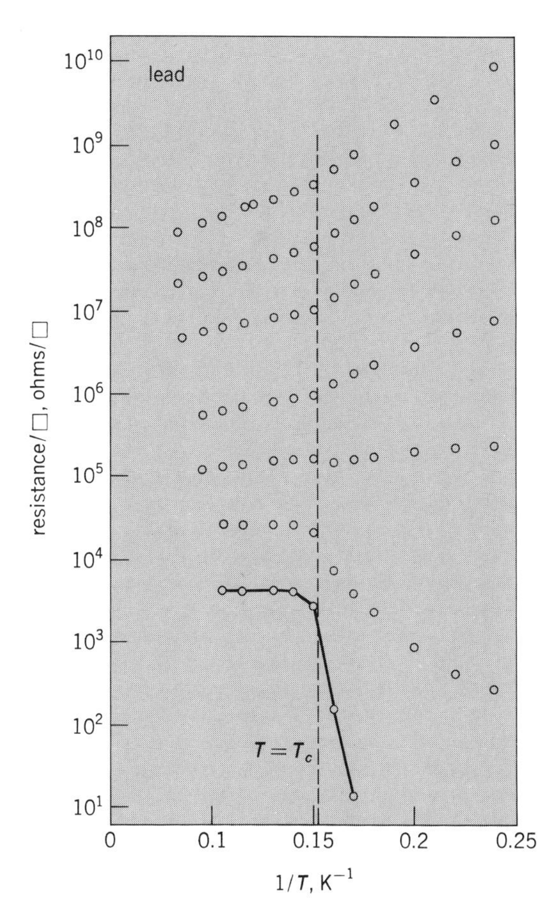

Fig. 1. Graph of resistance versus inverse temperature $1/T$ for ultrathin films of lead. *(From R. C. Dynes, J. P. Garno, and J. M. Rowell, Two-dimensional electrical conductivity in quench-condensed metal films, Phys. Rev. Lett., 40:479–482, 1978)*

stress fields). This gives enormous flexibility to the experimenter who, by extending the work to liquid helium, can further extend the range of densities to $n_s = 10^5 - 10^9/\text{cm}^2$.

Maximum metallic resistivity. One of the most remarkable concepts peculiar to two-dimensional electronics is that of a maximum metallic resistivity, introduced by J. T. Edwards, D. J. Thouless, and D. C. Licciardello. (Their work was motivated by some suggestions Sir Nevill Mott had made in 1968.) While the experiments referred to in their reports date back in some cases to the beginning of the 20th century, their analysis of maximum metallic resistivity is new and is based on the latest concepts in mathematical analysis (such as the renormalization group and scaling hypotheses). Briefly, when electrons in a two-dimensional conductor are increasingly scattered, for example, by thermal fluctuations, the resistivity can increase in the normal way only until the electronic mean free path equals the de Broglie wavelength of an electron at the Fermi energy. At this point, the resistivity changes from a maximum metallic value $0.12e^2/\hbar = 30{,}000\ \Omega/\square$ (where e is the electron charge, $\hbar$ is Planck's constant divided by 2π, and $\Omega/\square$ stands for "ohms per square," the resistivity in two dimensions being independent of the unit of length) to a different law of behavior, characteristic of an insulator or semiconductor rather than a metal. A recent experiment on various ultrathin lead films illustrates this behavior, as shown in Fig. 1. For resistances greater than 30,000 $\Omega/\square$, the dependence on temperature T is activated (that is, has the form $\exp(T/T_0)$, where T_0 is the activation energy divided by Boltzmann's constant) with an increase as the temperature is lowered through the critical temperature T_c, indicated by a vertical line in Fig. 1. Below 30,000 $\Omega/\square$, "metallic" behavior is observed, with superconductivity at temperatures below T_c. It is noteworthy that superconducting behavior can occur in the metallic phase. Indeed, superconductivity and two dimensions are not mutually exclusive, and thin films of various materials are known to be better superconductors (that is, to have a higher T_c) than their bulk counterparts.

Surface of liquid helium. It is possible to capture electrons on the surface of liquid helium. Whereas a potential barrier $V_0 \approx 1$ eV prevents penetration of the bulk by excess electrons, they

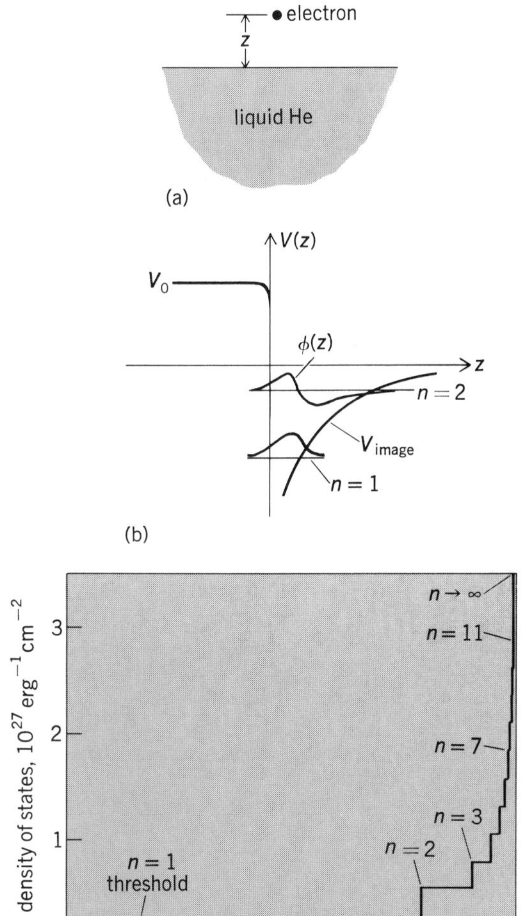

Fig. 2. Electron on the surface of liquid helium. *(a)* Position of electron relative to surface. *(b)* Potential energy near surface, two lowest bound-state energies (assuming no motion in x,y directions), and associated wave functions $\phi(z)$ in z direction. *(c)* Density of states, taking motion in x,y directions into account. *(From M. W. Cole, Electronic surface states of liquid helium, Rev. Mod. Phys., 46:451–464, 1974)*

are captured on the outer surface by the image potential $V(z) = -Qe^2/z$, where z is the perpendicular distance of the electron from the surface (Fig. 2*a*), $Q = \frac{1}{4}(\epsilon - 1)/(\epsilon + 1) \approx 7 \times 10^{-3}$; here, ϵ is the dielectric constant. The bound states are classified into subbands $n = 1, 2, \ldots$, according to the increasing number of nodes of the wave function $\phi(z)$ in the z direction (Fig. 2*b*). Within each subband the density of states (number of energy levels per unit energy) in the perpendicular directions x and y is a constant, a general feature of two dimensions. But this constant increases stepwise at each succeeding subband (Fig. 2*c*).

The electrons at the surface of liquid helium interact with the surface excitations of the host and thus constitute a microscopic probe of some of the most intricate aspects of this interesting superfluid. Recent research has touched upon such features as the scattering of the electrons by helium gas atoms in the vapor layer, (which is significant because the subband wave functions extend 10–50 nm or farther from the potential barrier at the surface), by capillary waves (known as riplons) just at the surface, or by each other through their Coulomb repulsion. The resulting electron-electron correlations and correlation energy, as well as the properties of the electron gas in applied magnetic fields of various orientations relative to the z axis, have been the subject of detailed study and delicate experimentation. The scattering times for electrons here are on the order of 10^{-7} s, about a million times longer than in the inversion layers of semiconductors. This is due in part to the lower density of electrons, and also to the greater perfection of the underlying stratum.

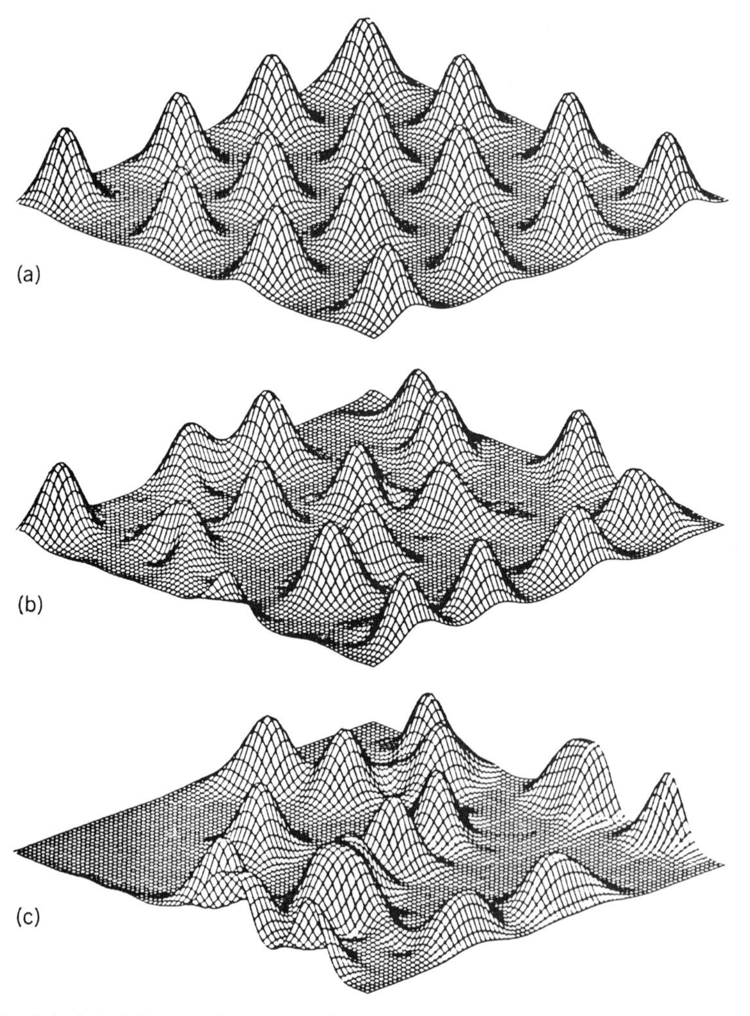

Fig. 3. Calculated electron density distributions in model two-dimensional system in strong magnetic field (*a*) in absence of random potential, (*b*) in presence of random potential of equal strength with electron-electron interactions, and (*c*) with dominant random potential. *(From H. Aoki, in G. Dorda and P. J. Stiles, eds., Electronic Properties of Two-Dimensional Systems, Proceedings of the International Conference on the Electronic Properties of Two-Dimensional Systems, Berchtesgaden, Germany, Aug. 19–22, 1977, North-Holland Publishing Company, 1978)*

Semiconductor surfaces and interfaces. The space charge regions on semiconducting materials are the most popular "laboratory" for the study of two-dimensional electrons. Layers on such diverse materials as Si, InSb, PbTe, and Te have been analyzed and studied as functions of doping and applied electric and magnetic fields. Special materials such as $Bi_2(Te_{1-x}S_x)_3$, with $0.33 < x < 0.5$, which spontaneously cleave into almost ideal two-dimensional structures have been studied for this purpose. Unconventional ideas are being explored. For example, the deposition of a grating on the surface will cause a periodic potential in the x direction and thus stimulate one-dimensional conductivity principally along the y direction. A magnetic field tends to localize electrons and hence promotes zero-dimensional "conductivity."

Much of the research has related to the collective effects such as the propagating charge-density waves known as plasmons. For a wavelength $\lambda = 2\pi/k$, where k is the wave number, in two dimensions the plasmon frequency is calculated to be $\omega = (2\pi n_s e^2 k/m)^{1/2}$, where m is the electron mass, which is to be contrasted with $\omega = (4\pi n e^2/m)^{1/2}$, where n is the volume density of electrons, independent of k, in three dimensions.

The equation of state of the two-dimensional Coulomb fluid is unusual. In the absence of a magnetic field the fluid is expected to be a fully ionized plasma at high temperature and an insulator, possibly a Wigner solid, at low temperatures. [In the late 1930s Eugene Wigner recognized that at low densities the electron gas might condense into a solid phase, due to the tendency of electrons to minimize their Coulomb repulsion energy. This is possible only if the cost in kinetic energy (the uncertainty principle mandates that the electrons' momentum be increased when they are spatially localized) is not too high. Thus it has been extraordinarily difficult to find experimental situations in three-dimensional physics where this can occur. This has sparked the search for this phenomenon in two dimensions.]

Several parameters can influence this behavior, including applied magnetic fields which render the condensed phase energetically more favorable, and the random potentials of thermal fluctuations and impurity atoms, which can distort the Wigner lattice, changing it into a Wigner glass at significant levels of disorder or into a purely random Anderson localization at extreme levels of disorder. Two new concepts are involved here. Electrons localized in a random pattern should form a glassy structure rather than the perfectly coordinated crystal that Wigner had in mind. Finally, P. W. Anderson conceived that waves could be localized about fluctuations when propagating through extremely disordered media, and this Anderson localization could affect the electrons even in the absence of the Coulomb repulsions responsible for the Wigner localization. The interplay among these various tendencies is demonstrated in Fig. 3, showing calculated electron density distributions in a model two-dimensional system in a strong magnetic field H, with $n_s = 0.16$ $eH/2\pi hc$ (where c is the speed of light). The Wigner lattice resulting from dominant electron-electron interactions (Fig. 3*a*) is changed to a Wigner glass by an equal-strength random potential (Fig. 3*b*) and to a random configuration by a dominant random potential (Fig. 3*c*).

It has been proposed that the uncertainty in the energy of a cyclotron orbit is proportional to the square root of the density of scatterers, rather than to their density as is usually the case. However, this is merely an example of the challenge of this new field of solid-state surface physics: every concept has to be reexamined, quantitative and sometimes qualitative reassessments of the basic concepts must be undertaken, and in some cases a new phenomenon is discovered.

For background information *see* HELIUM, LIQUID; SEMICONDUCTOR; SUPERCONDUCTIVITY; SURFACE PHYSICS; WAVES AND INSTABILITIES IN PLASMAS in the McGraw-Hill Encyclopedia of Science and Technology.

[DANIEL C. MATTIS]

Bibliography: H. Aoki, in J. F. Koch and G. Landwehr ((eds.), *Proceedings of the 2d International Conference: Electronic Properties of Two-dimensional Systems*, Physik-Department der Technischen Universität München, September 1977; M. W. Cole, *Rev. Mod. Phys.*, 46:451–464, 1974; A. Isihara and T. Toyoda, *Z. Phys. B*, 23:389, 1976; A. Isihara and T. Toyoda, *Ann. Phys.*, 106: 394, 1977; D. C. Licciardello and D. J. Thouless, *Phys. Rev. Lett.*, 35:1475–1479, 1975; A. Rajagopal and J. Kimball, *Phys. Rev. B*, 15:2819, 1977.

Tektite

The discovery of a new group of tektites, additional comparisons with lunar samples, and a new tektite find in Australia highlighted tektite research during the past year.

Tektites are natural terrestrial impact glass, generally found in small masses ranging in size from microscopic (microtektites) to tens of kilograms. The glass ranges from transparent to translucent, and is some shade of brown or green in transmitted light. However, some of the brown tektites are so dark that they appear black in reflected light.

Occurrence. Until very recently there have been only four known occurrences of tektites on the Earth's surface: (1) Occurrences in Czechoslovakia are the best-known tektites, with sites in both Bohemia and Moravia, where the tektites occur on the surface and in Miocene to Recent sedimentary rocks. The Bohemian tektites are especially well known because of their bright green color and widespread use in jewelry. (2) North American specimens are relatively numerous in Texas, rarer in Georgia, and only one has been found on Martha's Vineyard, MA. This is the oldest group of tektites, with radiometric ages of $34 \pm 1 \times 10^6$ years. (3) Ivory Coast tektites are found scattered along drainageways in densely vegetated areas, so that their total area of occurrence is not precisely known. (4) Australasian tektites occur over a vast area of Australia and Southeast Asia and are the most common of all the tektites. The largest known tektites occur in this group in Thailand and Indochina.

Ivory Coast tektites are important because they have been definitely linked to a terrestrial impact crater, the Lake Bosumtwi (sometimes called Ashanti) crater in Ghana, where it now seems certain that the tektites originated by fusion of the country rock in the impact event. Czechoslovakian tektites have the same potassium-argon and fission-track age as the Ries Crater located in southern Germany, but there is still some argument as to whether or not the tektites are produced by that specific impact.

New tektite group. A new occurrence of tektites has been reported in the Soviet Union by P. W. Florenski, and his report has been confirmed by workers in the United States and Germany. The Soviet tektites (see illustration), named irghisites by Florenski, occur within the weathered remnants of the Zhamanshin impact crater, which is located in the Northern Aral region of the Soviet Union approximately 200 km north of Aralsk. The crater structure is approximately 10 to 15 km in diameter. Florenski describes the impact structure and the tektites in detail, and provides convincing evidence that the tektites originated at the Zhamanshin site by fusion of the rocks that underlie the crater and that formerly occupied the volume of the crater. However, his analyses did not include determinations of water or ferrous/ferric iron contents. These determinations are important for tektites because the very low water contents and the high ferrous/ferric iron ratios found in previously known tektites have been used by some researchers as arguments against a terrestrial origin.

E. A. King and J. Arndt analyzed an irghisite for water, using an infrared spectrometric technique, and found that irghisites contain more water than tektites from other occurrences. This result is not surprising, as the irghisites were found within their source crater, and all other tektites were found several hundred or more kilometers from their source crater or possible source crater. Thus, the irghisites have received less heating due to the lesser kinetic energy input to their parent materials and have executed much shorter trajectories than the other tektites. Since the total water contents and the ferrous/ferric iron ratios are probably

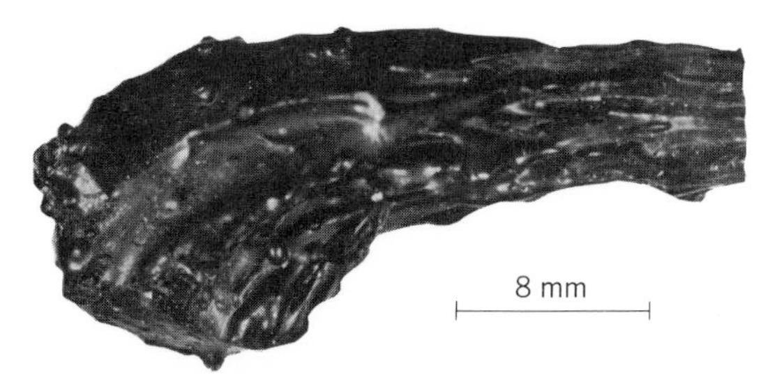

Soviet tektite (irghisite) from the Zhamanshin impact crater, Northern Aral region of the Soviet Union.

simple functions of the temperature and atmospheric partial pressure of water vapor and O_2 that the tektites have experienced while hot, the difference is easily explained.

This new group of tektites not only provides another example in which the tektites are genetically related to a known crater, but also is valuable in helping to understand more fully some of the variations in shock-melted rocks. The irghisites and associated impactite rocks will be the subjects of intensive research for the next few years.

Lunar sample comparisons. The great hope of some researchers who supported a lunar origin for tektites was that examinations of lunar samples returned by the Apollo and Luna missions would identify either tektites or materials which could be the parent materials of tektites. These examinations have continued for more than 8 years, and detailed studies of the small rock fragments from the lunar regolith and lunar breccias have identified some silica-rich rocks that are similar to granite in composition.

The implication is that somewhere on the surface of the Moon or within the lunar crust there may be substantial volumes of highly differentiated silica-rich rocks. The presence of silica-rich rocks on the Moon has provided momentary encouragement to the lunar-tektite-origin theorists. However, as the full chemistry of lunar rocks has become known, it is apparent that no lunar rocks have the trace- and minor-element distributions found in tektites, which closely resemble terrestrial rocks in their abundance patterns. Also, the lunar rocks are too old—ranging in age from approximately 4.6 to 3.1×10^9 years—to have the correct isotopic compositions to be the parent materials of tektites. Thus, the lunar samples have not in any way lent credence to a proposed lunar origin of tektites, but rather have added greatly to the already impressive arguments for a terrestrial origin.

Stratigraphic age of Australian tektites. There is a long-standing controversy regarding the age of the Australian tektites as determined by the stratigraphy of the surface rock units with which they were found. The geologic ages of many of these rock units are only a few thousand years, but the Australasian tektites almost uniformly yield potassium-argon and fission-track ages of 750,000 years. One group of researchers holds that the stratigraphic age of a few thousand years is the correct age of tektite formation and fall and that the older potassium-argon and fission-track ages are erroneous. Other workers claim that the tektites were reworked by erosion and redeposition from the stratigraphic units in which they originally fell and that the potassium-argon and fission-track ages correctly date the age of tektite fall and formation. The latter point of view has received some support from a tektite find at Hindmarsh Tiers in South Australia, where a tektite was found in a sample collected from a soil profile between 1 and 2 m below the present surface. However, the tektite-bearing soil unit is believed to range in age from the late Pliocene to late Pleistocene; thus, it does not completely resolve the dispute.

Remaining problems. One of the most difficult problems related to tektites is identification of the source of the Australasian strewnfield, although several candidate craters have been suggested. Because the Australasian tektites are very young geologically, one would expect the source crater to be still well preserved and a rather obvious feature of the Earth's surface. Although several source craters have been suggested for the North American tektites, the lack of a positively identified source is not considered serious as the North American tektites are the oldest known group of tektites and the source crater may have been covered by later sediments or destroyed by erosion. Also, there is the need for much additional work on the mechanics of large terrestrial impacts, so as to better understand the events in the cratering process that cause small molten bodies of silicate to be propelled to altitudes above the Earth's dense atmosphere, from which they can reenter the atmosphere and be scattered over thousands of miles.

For background information *see* TEKTITE in the McGraw-Hill Encyclopedia of Science and Technology.

[ELBERT A. KING]

Bibliography: P. W. Florenski, *Chem. Erde*, 36: 83–95, 1977; L. J. Furness, *Quart. Geol. Notes*, Geological Survey of South Australia, no. 63, p. 21, July 1977; E. A. King, *Amer. Sci.*, 65:212–218, 1977; E. A. King and J. Arndt, *Nature*, 269:48–49, 1977; R. D. Warner, G. J. Taylor, and K. Keil, *Proceedings of the 8th Lunar Science Conference*, vol. 2, pp. 1987–2006, 1977.

Television camera

Initially developed by Bell Telephone Laboratories in 1969, the charge-coupled device (CCD), a silicon chip organized as a solid-state television sensor, has significantly increased the application of television cameras beyond that obtained with vacuum-tube-based (image orthicon, vidicon, and so forth) television cameras. A typical CCD 525-line television solid-state array measures about 1 cm × 1 cm and, with appropriate optics, yields a 1-V video stream, essentially similar to video obtained from conventional television sensors after signal amplification.

Charge-coupled device. The basic structure of a CCD television sensor is a two-dimensional array of closely spaced metal oxide-silicon (MOS) capacitors which act as a matrix of analog shift registers designed to accept and transport information samples in the form of individual charge packets. These packets are electrons generated by the absorption of not only visible light (wavelengths of 0.4–0.7 μm) but also near-infrared radiation (0.7–1.1 μm). The absorption of such radiation by a silicon substrate results in the generation of electrons in a quantity directly proportional to the amount of incident radiation. It is this electrooptic property that enables the CCD to be used as an image sensor.

Charge packet integration and transport functions are controlled by varying the potential level of various electrodes located on the surface of the silicon substrate, a procedure known as clocking. Digital-type clock waveforms applied to the transport electrodes cause the stored charge packets to serially shift through the array and exit via an appropriate low-capacitance on-chip detector/amplifier. In effect, the CCD stores and transports samples of analog data without any appreciable loss of charge by using digital techniques.

Solid-state camera. The basic elements of a high-performance television camera based on CCDs are a conventional television lens, and a CCD sensor assembled in a small, hermetically sealed windowed enclosure complete with a thermoelectric device cooler (Fig. 1) that protects the sensor against ambient temperatures in the range from −50 to 50°C. For less stringent temperature requirements, the CCD may be assembled directly on a printed circuit card. This assembly is then complemented by driver, logic, and video processor electronic circuits (Fig. 2).

The driver electronics supplies the necessary clock-driving waveforms to the CCD array, plus direct-current (dc) bias levels needed to establish proper CCD operation. Typically, eight clocking drives and four dc bias levels are required.

The function of the logic electronics is to provide the necessary timing inputs to the CCD drivers, to control signals in the video processor, and to allow for gen-lock (external input synchronization). Typically a crystal oscillator is used as the stable time base of the logic system.

The most sophisticated electronic assembly of a CCD television camera is the video processor. Its basic function is to provide, as desired, automatic gain control, inputs to an external automatic light control mechanism (for example, an adjustable iris), blanking and synchronization insertion for the composite video signal, edge enhancement, gamma correction (CRT nonlinearity compensation), and black-level clamping.

Usually the electronics associated with a CCD television camera consists of two or three printed circuit cards (10 cm × 10 cm) which can be remotely located via a cable from the CCD lens/device assembly.

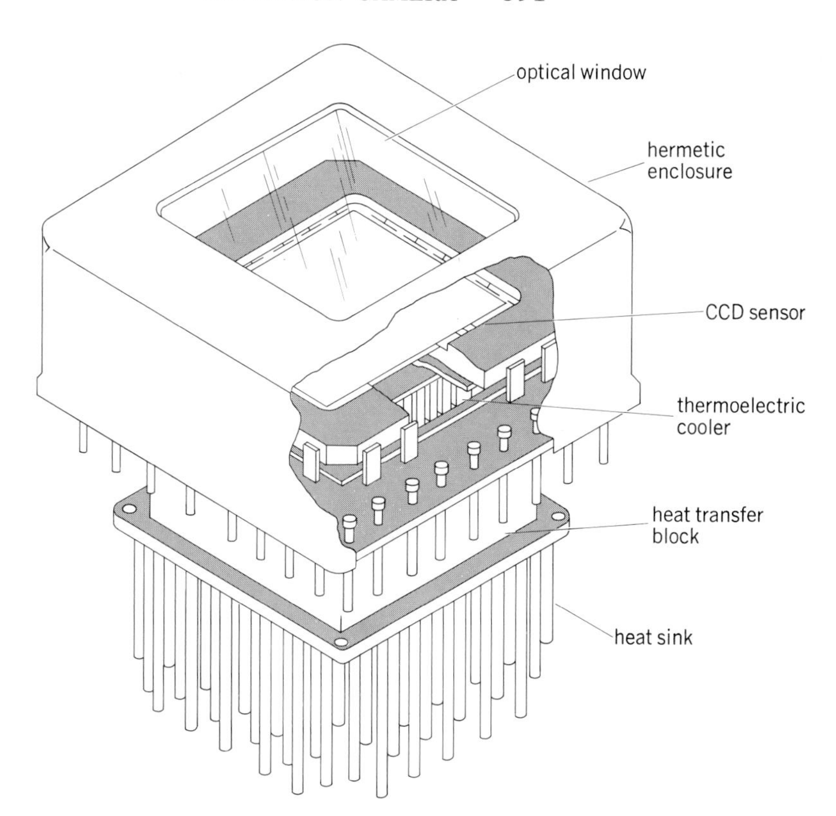

Fig. 1. A 525-line solid-state charge-coupled-device television camera head in thermoelectrically cooled, sealed assembly. CCD sensor consists of array, 488 elements long by 380 elements wide, for total of 186,000 elements.

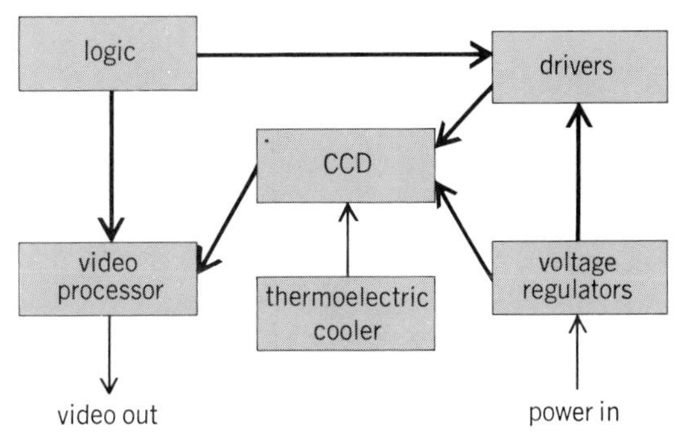

Fig. 2. Block diagram of solid-state television camera.

Applications. In an effort to achieve improved low-light-level performance, CCD television sensors have been successfully coupled to the output of image intensifier assemblies. The technique employed a fiber optic stub epoxied to the front (imaging) surface of the CCD array. The fiber optic acts as a coupler between the intensifier and the CCD array. This technique allows for gating (shuttering) of the system and yields intensifier-limited performance at light levels on the order of 10^{-6} footcandle (0.00001076 lux) on the input faceplace of the intensifier.

The fabrication of the CCD array results in intrinsically high metricity; that is, each individual photosite location (typically 200,000 photosites) is known to a precision of 1 part in 10^5. This property has made the CCD array useful for precise mensuration and automatic data processing.

The precise location of the photosites enables accurate registration of several arrays for the generation of the three (red, green, blue) spectral bands required for a solid-state color television camera. Several such cameras are now under development.

The inherent ruggedness of the solid-state CCD technology is an important advantage for television cameras subject to severe shock and acceleration inputs. An example of such an application is an artillery-launched CCD camera which yielded excellent imagery after repeated exposures to a 14,000-*g* shock load (137.2 km/s²).

For background information *see* INTEGRATED CIRCUITS; TELEVISION CAMERA TUBE in the McGraw-Hill Encyclopedia of Science and Technology.

[IRVING HIRSCHBERG]

Bibliography: G. F. Amelio, *Sci. Amer.*, 230(2): 22–31, February 1974; I. Hirschberg, *Opt. Spectra*, 9(11):28–33, November 1975; D. A. Hoagland and H. L. Balopole, in *Proceedings of the International Conference on the Application of Charge Coupled Devices*, San Diego, pp. 173–180, October 1975; C. H. Sequin and M. F. Tomsett, in *Charge Transfer Devices*, 1975.

Terrain sensing, remote

Significant advances have recently been made in the remote sensing of crop yields, with the completion of productive research in both the solar and thermal portions of the electromagnetic spectrum. Several independent techniques and combination approaches have been devised to estimate the time of harvest and the size of the ultimate grain yield of wheat. Work on forage crops such as alfalfa has also produced positive results. Experiments on other types of crops such as cotton and sugarbeets are in progress. In addition to predicting yields, some of the techniques can be used as management tools. The scheduling of irrigations from remotely acquired crop temperature data is perhaps the prime example. The fact that many of the new methods can be used from three different base levels of operation—satellite, aircraft, and ground vehicle—makes them quite versatile.

Solar techniques. Measurement of reflected solar radiation is the primary remote sensing technology in use today. Common aerial photography is the most basic form, followed by color infrared photography. The total spectrum of reflected solar radiation contains much more information than can be extracted by these means alone, however, and new research has begun to decipher its long-unfathomed code.

Albedo approach. The most simple of the new solar techniques applies to grain crops and involves the determination of crop albedo (the percentage of total incoming solar radiation reflected by the crop). Work in Arizona and California on Produra wheat, a durum used for spaghetti products, has shown that the final yield is inversely related to the minimum albedo, which occurs just prior to grain ripening, before the crop begins to turn a brown or yellow hue. That is, the lower the crop reflectance drops prior to ripening, the greater will be the grain production. For crops that completely cover the ground, the same relationship may well apply all over the world, for it is dependent only upon the spectral characteristics of the vegetation viewed. Thus, this approach indicates a possibility for global grain production estimates to be obtained from satellites.

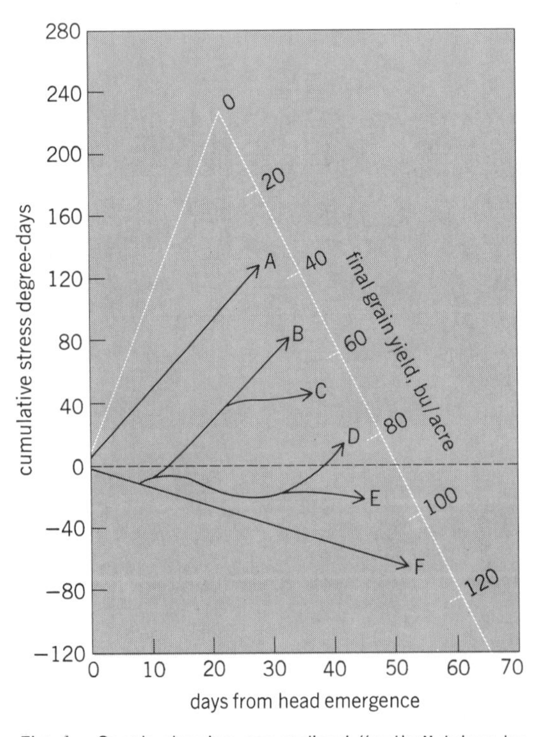

Fig. 1. Graph showing generalized "paths" taken by wheat crops through stress degree-days and irrigations. All but plot A were irrigated shortly before head emergence. Plot B had no additional irrigations, and plots C and D each had one more irrigation. Note the influence the time of irrigation had on final yield. Plot E had one additional irrigation, and plot F was irrigated almost every week. 20 bu/acre = 13.5 quintals/hectare. *(From R. D. Jackson et al., Crop temperature reveals stress, Crops Soils Mag., 29(8):10–13, 1977)*

Vegetation indices. Just like people, plant species have distinctive spectral "signatures" that can be used to identify them. The signatures are composed of the plants' specific reflectances for solar radiation in a variety of different small wavelength intervals. By combining these reflectances in a number of different ways (adding, subtracting, multiplying, dividing), various vegetation indices may be devised.

Historically, research on vegetation indices concentrated on deriving better and better combinations to relate to crop yields. In just the past year, however, a new approach has been taken. By obtaining data frequently over the entire crop growing season, it has been shown that integrations of several different indices over that period give good correlations with final grain production of wheat and barley. The present system of Landsat satellites does not acquire data frequently enough to make this approach currently feasible on a global scale. However, it can be used with aircraft or ground vehicle programs.

Another closely allied vegetation index technique developed this past year has to do with evaluating the rate of crop senescence. It was found that grain crops which produced the greatest yields remained green for a long time but then senesced very rapidly, so that grain yield was directly proportional to senescence rate. The technique thus requires data only over the period of grain ripening, but great frequency is again essential.

Thermal techniques. The basic parameter of the primary thermal technique is the stress degree day, defined as the crop-air temperature differential about an hour and a half past solar noon. It has been demonstrated that the integral of this parameter over the period of vegetative growth of alfalfa, from one cutting to the next, is linearly and inversely related to the amount of dry matter harvested. It has also been shown that the integral of the stress degree day over the reproductive growth period of grain crops—from the time of appearance of heads and awns to the initiation of senescence—is similarly related to the weight of grain produced. By plotting a time sequence of crop stress degree day progressions as in Fig. 1, it is also possible to determine when the available water in the crop's root zone is nearing depletion and irrigation will be needed. Thus, this technique can be used as a management, as well as a predictive, tool.

A complexity of this technique is that a graph

Fig. 2. Infrared thermometer used to take the temperature of a cotton crop to assess its degree of water stress.

such as Fig. 1 is both crop- and site-specific. The site specificity can be removed, however, by incorporating the classical growing degree day concept into the theory as a pseudo-time parameter. A more complicated graph then results, with the abscissa presented in growing degree days, defined as the difference between average daily air temperature and the minimum temperature below which the growth of the crop is inhibited.

A special utility of the thermal technique is that it allows field inhomogeneities to be detected and evaluated. Thermal scanners flown in an aircraft at altitudes of 200 to 400 m can produce a temperature map of each square foot of farmland below. Thus, hot spots where soil moisture is running low and evaporation has fallen below the potential rate can easily be spotted and used as an early warning signal for the field. Similarly, hand-held infrared thermometers like that of Fig. 2 can be used on the ground.

Combination techniques. Several different combination models have been constructed from the basic solar and thermal techniques. One is a multiplicative combination of a vegetation index and the classical growing degree day concept. Another is a more complex merger of the minimum albedo and stress degree day techniques. It has considerable potential, since it predicts both the time of senescence initiation and final grain yield of wheat from measurements of only reflected solar radiation and crop temperature. Because both of these measurements can be made remotely, this approach is appealing for applications from satellites and aircraft.

Current research. At the present time, research is in progress in a number of areas around the world on the remote sensing of crop yields. A major effort involves the Heat Capacity Mapping Mission (HCMM) satellite, launched by the United States in April 1978. There are 12 United States and 12 foreign investigators participating in the HCMM program. Although most of the ongoing studies are not directly testing the crop yield models described here, some are looking into complementary phenomena, such as plant water stress, soil water status, snow field extent, and rates of snowmelt. It is the ultimate combination of many of these techniques into a comprehensive surveillance program that may eventually provide the capability for the remote assessment of agricultural productivity on both regional and global scales.

For background information *see* ALFALFA; COTTON; PLANT, WATER RELATIONS OF; SUGARBEET; TERRAIN SENSING, REMOTE; WHEAT in the McGraw-Hill Encyclopedia of Science and Technology. [SHERWOOD B. IDSO]

Bibliography: S. B. Idso, R. D. Jackson, and R. J. Reginato, *Science*, 196:19–25, 1977; S. B. Idso, R. J. Reginato, and R. D. Jackson, *Nature*, 266: 625–628, 1977; J. P. Millard et al., *Photogrametric Engineering and Remote Sensing*, 44:77–85, 1978; R. J. Reginato, S. B. Idso, and R. D. Jackson, *Remote Sensing of Environment*, 7:77–80, 1978.

Thunderstorm

The increase in the number of commercial flights in recent years has often created heavy air traffic in and around major airports. During thunderstorms, in particular, planes take off and land while others are circling in holding patterns in stack-up positions. For the safety of aircraft operations it is important to understand the nature of air currents in and around thunderstorms.

Landing or takeoff beneath weak to moderate thunderstorms had been considered manageable until Eastern Airlines Flight 66 crashed in its final approach to John F. Kennedy (JFK) International

Fig. 1. Pine forest in northern Wisconsin blown down by a microburst, July 4, 1977. *(Courtesy of T. Theodore Fujita)*

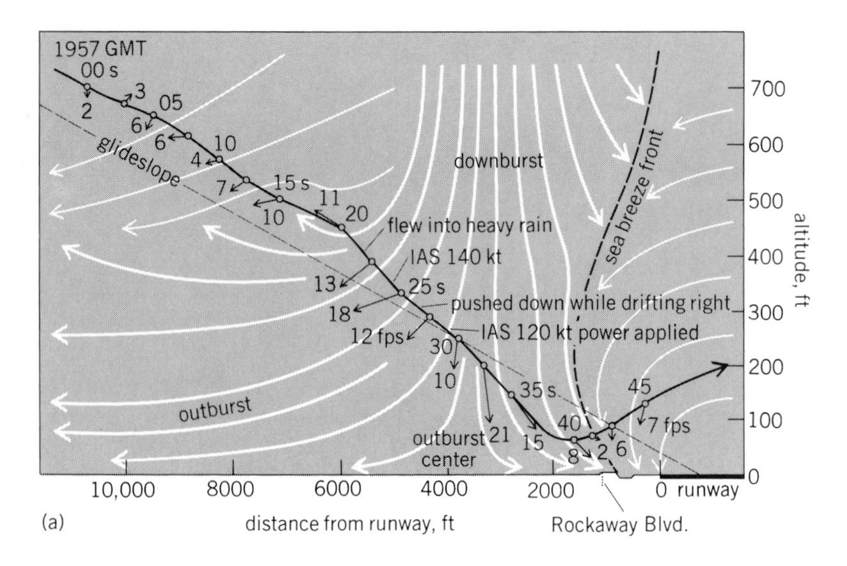

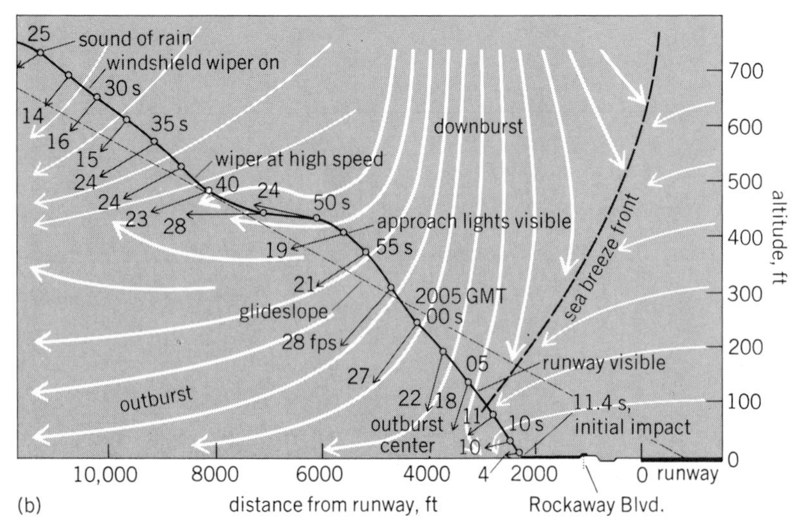

Fig. 2. Aircraft incidents at JFK International Airport caused by microbursts. *(a)* Path of an L-1011 through a microburst during approach to runway 22-L: the aircraft abandoned approach. *(b)* Path of a B-727 which crashed 7 min after the landing attempt by the L-1011. *(From T. T. Fujita and F. Caracena, Bull. Amer. Meteorol. Soc., 58:1164–1181, 1977)*

Airport, New York City. Meteorological investigation of this accident by T. T. Fujita identified the wind system which caused the accident as a downburst beneath the thunderstorm.

Downburst. A downburst is a strong downward current of air which induces an outward burst of damaging winds as it spreads over the ground. Wind speeds inside strong downbursts may reach F 2 intensity (113–157 mph or 182–253 km/hr). Some downbursts affect an area over 10 mi (16 km) wide and 20 mi (32 km) long, while others are only several miles across. These small downbursts are called microbursts (Fig. 1).

Comparison of four accidents caused by microbursts

Year	1975	1975	1976	1977
Date	June 24	Aug. 7	June 23	June 3
Local time	3:05 P.M.	4:11 P.M.	4:12 P.M.	12:59 P.M.
Airport	JFK	Denver	Philadelphia	Tucson
Flight phase	Landing	Takeoff	Landing	Takeoff
Number of fatalities	112	0	0	0
Number of injuries	112	15	106	0

Microburst. Because of the small horizontal extent, winds inside a microburst vary significantly. Frequently, wind direction reverses within only 1 to 2 mi (1.6 to 3.2 km).

Aircraft which land or take off in a microburst often encounter strong wind shear—a rapid change of wind direction, wind speed, or both, along the flight path. When an aircraft flies through a microburst, the head wind may change into a tail wind, resulting in a significant loss of airspeed. This loss of airspeed, in turn, induces a loss of the lift force. Consequently, the aircraft sinks or cannot gain altitude.

At JFK International Airport on June 24, 1975, an L-1011 aircraft caught in a microburst began sinking so rapidly that the pilot decided not to land. About 7 min later a B-727 aircraft (Eastern 66) flew into another microburst while approaching the same runway at JFK. This time, the wind shear in the microburst was so strong that the aircraft kept sinking until it hit the ground, 2400 ft (732 m) short of the runway (Fig. 2).

Aircraft accidents. During 1975–1977 four accidents in the United States were related to the penetration of a microburst during the critical phases of landing or takeoff (see the table). These accidents occurred during summer afternoons when thunderstorm activity is at its peak. A total of 112 persons were killed while 233 others were injured.

Flight safety. Microbursts with their horizontal extent comparable to the length of the runways could affect landing and takeoff operations. Furthermore, most microbursts develop rapidly, reaching maturity within only several minutes and weakening thereafter. This is the reason why one aircraft may encounter serious problems while the preceding or following aircraft experiences little or no difficulty.

A network of anemometers—devices which measure both wind direction and speed—are needed to detect microbursts in time for effective warnings. Such a network is extremely expensive, however. It is also known that most microbursts induce abrupt pressure rise, temperature drop, and surge of wind on the ground. Direct and indirect sensors to detect these parameters are either being developed or tested.

Prior to the completion of reliable microburst detectors, flight safety can be improved by the training of pilots via simulated flights in hypothetical microbursts of various sizes and intensities.

For background information *see* THUNDERSTORM in the McGraw-Hill Encyclopedia of Science and Technology. [T. THEODORE FUJITA]

Bibliography: T. T. Fujita, *Manual of Downburst Identification for Project NIMROD*, SMRP Pap. no. 156, 1978; T. T. Fujita and H. R. Byers, *Mon. Weather Rev.*, 105:129–146, 1977; T. T. Fujita and F. Caracena, *Bull. Amer. Meteorol. Soc.*, 58: 1164–1181, 1977.

Tornado

A recent development in meteorology is the possibility of using Doppler radar to operationally determine, in real-time, areas of potential tornado development and occurrence. Real-time means that a tornado can be identified in approximately the same time as is required for the data to be acquired. Tornadoes have been found to be related to mesocyclone circulations interior to severe convective activity. By accurately locating the region of circulation initiation, small areal warnings for tornadic activity can be made. Thus, smaller regions can be warned sooner than is presently possible, and appropriate actions can be taken to protect lives and property.

Mesocyclones. Flow fields inside convective storms have been examined for several years by using Doppler radar. A single Doppler radar measures the radial component of the motion of the water or ice particles along the radar beam. Mesocyclone vortex signatures have been detected in severe storms. A mesocyclone is a cyclonic circulation interior to a convective storm. A mesocyclone vortex appears on the radar display as a velocity couplet with adjacent velocity maxima of relatively different signs. It has dimensions of 1 to 10 km and a life cycle of up to 2 hours. If the mesocyclone has dimensions of less than 1 km and sufficiently high shear, it is called a tornado vortex signature (TVS) and is a signature of the tornado itself.

R. J. Donaldson proposed the initial interpretation of the mesocyclone vortex and rules for its identification in 1970. These recognition rules have been verified by dual Doppler measurements which can determine the complete velocity vector in two dimensions. Results of this analysis have shown an excellent agreement between severe weather occurrences and mesocyclone vortex development. Thirty-seven mesocyclones were identified from archived data, with all but two associated with severe weather, such as hail, damaging winds, or tornadoes; 62% were associated with reported tornadoes and 32% with hail or damaging winds. No tornadoes occurred unless they were preceded by a mesocyclone signature. The average lead time before tornado initiation was 36 min.

Real-time test. Based on the excellent results of the research data, an operational project was established by four governmental agencies of the U.S. Air Force and the National Oceanic and Atmospheric Administration (NOAA). The test was held for 3-month periods in the springs of 1977 and 1978 in Norman, OK, since this is the time of year when the most active severe weather occurs in the area. The results of the 1977 test were encouraging and consistent with the research data findings: Forty-five advisories were issued, based on Doppler data. Tornadoes were forecast correctly 94% of the time. The false alarm ratio (FAR), which is the proportion of the incorrect predictions to the total number of severe events, was .244, which is quite low for this type of meteorological prediction.

At the same time as these Doppler-based advisories were being issued, advisories were also being put out based on current operational criteria and methods. Conventional advisories include several counties in a state because of the inability of present radar systems to pinpoint tornadic storms. Doppler advisories were limited to only those counties directly threatened by the detected storms. Conventional-technique advisories correctly forecast tornado occurrence 69% of the time but had a high FAR of .73. These statistics must be viewed in the light of the fact that 67% of conventional advisories were issued based on public reports of tornado occurrence, not all of which were really tornadoes. As a result, these advisories had an average tornado lead time of only 1 min. Doppler-based advisories increased the lead time to 23 min.

Computer identification. In the spring of 1978, in conjunction with the real-time test, scientists at the Air Force Geophysics Laboratory in Massachusetts developed an automatic echo track and forecasting algorithm which analyzes data fed automatically from the radar processor to the computer. The program detects precipitation areas or severe storm cells based on variable parameters selected by the meteorologist. After cell detection and correlations are accomplished, forecasts are made of cell movement. Selected attributes of each cell are output for real-time use in forecasting severe storm development and motion.

At present, the complex nature of the mesocyclone structure and its variability has prevented its real-time automatic identification. However, an interactive feature of the computer-radar system permits the meteorologist to objectively pinpoint the mesocyclone on a color display in real-time and receive information on its location to the nearest 150 m and on the magnitude of the velocity maximum to the nearest meter per second. These features, combined with the automatically outputted associated severe storm attributes and motion, enable the forecaster to readily produce a storm track of the mesocyclone and determine warning areas.

Doppler radar advances. The use of Doppler radar has tripled the amount of raw data available, since not only reflectivity data but also radial velocity and velocity variance information are output. The operational use of Doppler radar thus poses the problem of data assimilation by the forecaster. This necessitates the mating of the radar with a computer to reduce the data to a manageable level. This fact, coupled with an increased need for automatic sensing of severe weather indicators, is the driving force behind current research efforts.

Results of recent and continuing efforts to transfer research technology in tornado identification to operational use have been encouraging. Significant increases in detection capability, decreases in FARs, and increases in severe storm warning lead time have provided the incentive for continued research in mesocyclone and TVS identifications. Mesocyclones and severe convective cells can be located to within 150 m in range by using a combination of interactive and automatic computer systems. A listing of significant attributes of the cells permits the easy assimilation of the Doppler radar data by forecasters. With continued efforts, an improved warning capability for severe storm and tornado occurrences is possible utilizing Doppler weather radars.

For background information *see* DOPPLER RADAR; TORNADO in the McGraw-Hill Encyclopedia of Science and Technology.

[CARLTON L. BJERKAAS]

Bibliography: C. Bjerkaas and R. Donaldson, *Preprints of the 18th Conference on Radar Meteorology*, no. 449, 1978; D. Burgess et al., *Preprints of the 18th Conference on Radar Meteorology*, no. 442, 1978; R. Donaldson, *J. Appl. Meteorol.*, 9:661, 1970.

Toxicology

The term "plastics" usually refers to synthetic insoluble polymeric materials but, in a more general sense, plastics can include synthetic elastomers (rubbers) and synthetic textiles. Since World War II the utilization of plastics has been extended to nearly every type of item used by people, including clothing, furnishings, construction materials, transportation vehicles, medical and biomedical devices, food and beverage containers, and various types of packaging materials. Plastics have also made possible technological breakthroughs such as the creation of artificial organs and numerous specialized medical devices. However, the rapid introduction of new plastics without proper assessment of their effects on living organisms and the environment has created certain problems. One type of problem is the potentially toxic or adverse effects of certain plastics on animals and humans when the plastic used is inappropriate to a specific application. Toxic problems have occurred in medical applications, in food applications, and in situations where fire and heat are applied to the plastic.

MEDICAL APPLICATIONS

Within the last quarter-century dramatic advances have been made in the use of plastics in biomedical and surgical applications.

Plastic devices and instruments. Each year hundreds of thousands of patients benefit from the use of plastic devices such as heart valves or dialysis units. Patients with kidney failure can be maintained by dialyzing the person several times a week through the use of an external artificial kidney. Within the past several years new plastic materials have been used to develop artificial kidneys which are sufficiently compact that the patient can be dialyzed at home or when traveling. Designs have already been made for an implantable artificial kidney, which will be available in the near future.

Much emphasis has been given to research relating to plastic heart assistance devices, including implantable artificial hearts. One such artificial heart was implanted in a man who was kept alive for several days before other medical complications caused his death. Artificial hearts have kept experimental animals alive for periods ranging from several weeks to several months.

The use of plastic implants and other products has made possible significant advances in surgical technique. In recent years older patients with cataracts have been restored to normal sight by the implantation of intraocular lenses. Plastics made into a thin resilient mesh are being used by surgeons as scaffolding to support abdominal tissue destroyed as a result of wounds such as those suffered in automobile accidents or industrial injuries. Plastic textiles knitted or woven into thin, flexible tubings are now used routinely as vascular grafts for diseased or injured blood vessels. Plastic "cement" is used in orthopedic surgery to anchor the hip prosthetic device to bone. Specially prepared liquid plastics which form into a pliable plastic when applied to tissue are being used to arrest aneurysms in the circulatory system. In animals and in selected humans, it has been possible to fashion a plastic tooth which can be implanted to replace an extracted tooth.

Each year over 100,000 persons undergo cosmetic surgery to improve facial and body structures. Women can elect to have breast implants, a procedure which can be performed in the surgeon's office; a variety of plastic breast prostheses are available. Men can receive surgical implants to simulate the appearance of genital organs.

Various plastics have been used to develop disposable items such as syringes, hypodermic needles, and catheter tubings and containers for the collection, storage, and administration of blood, other biological fluids, drugs, and nutritional products. Plastics have also made possible the development of intrauterine devices which release a birth control drug in a sustained and uniform manner for more than a year. A plastic-drug combination can be inserted into the eye and releases a potent medication over a period of several weeks, thereby eliminating the need to instill the eyedrops once or twice a day.

Concern for safe medical devices has existed for many years, but it was not until 1976 that Congress amended the Food, Drug and Cosmetic Act to exert greater control on the marketing of medical devices. All new medical devices must now meet certain standards of safety and efficacy. The safety aspect of the "Device Law" includes the requirement that toxicity of the material or device must be evaluated.

Toxicity. Toxic manifestation from a plastic material results primarily from the release of a chemical constituent(s) leading to: (1) local tissue response, (2) systemic toxicological response, (3) allergic response, or (4) carcinogenic, teratogenic, or mutagenic response. These responses may be caused by direct action, that is, an implant in contact with tissues or organs, or through indirect means, such as the administration of blood, blood components, and drugs which have been in contact with a plastic container or an administration device such as a tubing.

Local tissue responses have been noted in animal experiments and in humans when a polyvinyl chloride catheter containing a plasticizer and an organotin stabilizing agent was used. When in contact with tissue, the organotin-containing plasticizer migrated from the catheter to the tissue. In this case, the organotin compound was found to be highly irritating to tissue. Removal of this type of stabilizer from the catheter alleviated the problem.

The methyl methacrylate cement used in surgery of the hip comes as a two-component kit and, when ready for use, the components are mixed to form a doughlike substance. The surgeon then inserts the cement at the surgical site, and within a very short period of time the cement hardens into the final plastic. If not enough care is taken in mixing the components, monomer may be present which can be released to the circulatory system. The methyl methacrylate monomer causes a dramatic fall in blood pressure which can lead to death.

Silicone ball heart valves in patients have been found to sorb lipoidal constituents from the blood. This process, in turn, alters the shape of the valve and leads to cracks and chips in the material. If the valve is not removed and replaced, the patient is at risk and death is imminent.

In both animals and humans, breast prosthetic devices constructed from a cross-linked polyvinyl alcohol have been shown to sorb calcium and, within a relatively short implant period, the artificial breasts have become hard and brittle, with eventual fragmentation necessitating complicated surgical procedures to remove the particles. Breast implants made of silicone gel enclosed in a sheath of silicone film are extremely popular today. In certain females, scar tissue forms around the implant and the original soft, pliable implant becomes hard and unsightly.

Plastic intrauterine devices have caused complications in certain females, although the exact causative agent has not yet been determined. Certain dental adhesives and filling materials have produced necrotic responses in the pulp of teeth. Most of these problems can be resolved, however, if the dentist uses a plasticlike cavity liner.

Since the early 1950s flexible, clear polyvinyl chloride blood bags have been used to store, collect, and administer blood and blood products. In the late 1960s and early 1970s, however, it was found that the plasticizer, di-(2-ethylhexyl) phthalate, is released to the blood or product and administered to the patient. In animal experiments the phthalate esters have been found to be liver toxicants and, in large doses, to act as a teratogenic agent as well as possibly producing mutagenic effects in the animals. There is also speculation that the phthalate esters might aggregate blood components, which could cause occlusion of the pulmonary vessels leading to "shocked lungs" and death during surgery in which blood stored in the bags is used.

Carcinogenic and tumorogenic activity. A serious question has been posed as to the possible carcinogenic activity of plastics or their components. Vinyl chloride, the monomer used to manufacture polyvinyl chloride, is a potent carcinogen. All steps must be taken to reduce the free monomer to a level considered safe in a medical device. Other monomers or additives which have been demonstrated to be carcinogens in animal or other laboratory tests should be considered harmful in a medical device. However, if the monomer is completely reacted or if the various additives in the material do not leach from the device, the hazard to patients can be resolved.

Long-term implantation of plastic prosthetic devices may produce cancer in one of two ways: (1) The plastic may degrade in the body, releasing soluble chemicals which then act as carcinogenic agents. (2) The implant, through physical factors, may generate tumors at sites of implants. In fact, since the 1940s a number of reports have documented that solid materials which have sufficient surface area and uninterrupted surfaces cause cancer at sites of implants in rodents. If the same material is made into a powder, or made porous and into threads, its tumorogenic activity is reduced to levels found in control animals without implants. However, a long induction period is needed, ranging from 5 months to over a year, for these tumors to develop in the rats. If the implant is removed before this induction or latent period, the sites do not become cancerous. On the other hand, if the material is removed after the induction period, a tumor may still develop many months later. At times, it is difficult to determine whether a tumor has been caused by a degradation product or by the physical aspects of the plastic. Several case reports suggest that tumors have arisen from prosthetic devices implanted in humans. Since there is a very long induction period for tumors in humans, as compared with rodents, it may take from 20 to 30 years for a tumor to be revealed in a patient with an implant.

It is now well recognized that any new plastic considered for a medical or biomedical application should be subjected to a series of toxicity tests ranging from acute toxicity to long-term studies. One series of primary acute toxicity tests has been proposed by J. Autian. These same tests may also be adapted for food packaging materials.

FOOD APPLICATIONS

The food and beverage industry is a very large user of various types of packaging systems.

Packaging systems. A variety of plastics are employed in dispensing products. Many of these packaging systems are in the form of films or cartons. In the last several years the plastics industry has been able to develop a clear polyester bottle which is lightweight, break-resistant, and of low permeability. It is predicted that within a very short period of time these bottles will capture most of the soft drink and liquor market. The trend toward "fast foods" has led to the development of a variety of plastic-carton systems which can be heated in an ordinary or microwave oven.

Leachable constituents. The contact of foods and beverages with these various plastic systems suggests the possibility that leachable constituents from the plastic may migrate into the products and be consumed. If one or more of these leachable components are known carcinogenic agents such as vinyl chloride or are suspected carcinogens such as acrylonitrile, the population may be placed at risk. Other toxicological effects may also be present which may not be manifested for many years; particularly at risk are children who would be exposed to these leachable ingredients for decades.

Relatively little is known about the numerous leachable ingredients being used by the food and beverage industry. Up to the present, research on the toxicity of plastics used for foods has been minimal, and only a small data base is available to discern which plastics may present a toxic problem over years of use. The Food and Drug Administration, under the "food additive" clause of the Food, Drug and Cosmetic Act, is presently considering what steps should be taken to ensure that these leachable constituents do not present a health hazard.

PLASTICS EXPOSED TO FIRE AND HEAT

Plastics exposed to heat or fire will degrade, releasing a host of gases, many of which are extremely toxic to animals and humans.

Many of the synthetic fibers used for clothing and furnishings can be ignited by exposure to a flame. This combustibility can be retarded by the addition of chemicals referred to as flame retardants. However, these retardants may present their own toxic problems since they may generate other gases more toxic than the gases generated without them.

In addition, several of the flame retardants used for textiles have now been found to be carcinogens when tested in the laboratory. However, animal tests have not been completed to confirm carcinogenicity. It is clear, however, that any suspected carcinogen should not be added to wearing apparel since it may be released to the skin.

Teflon, when exposed to temperatures above 350°C, can release highly toxic substances. Workers smoking cigarettes while exposed to Teflon dust have developed asthmatic attacks. (The term "polymer fume fever" has been used to describe these toxic cases.) Absence from the work area, however, for 1 or 2 days has restored the general health of these workers.

Wrappers using polyvinyl chloride films to package meat have experienced toxic responses during the sealing of the film. In this case, the film is exposed to a hot wire which degrades the plastic and releases toxic gases. Proper ventilation or protective devices for the workers can prevent this problem.

For background information *see* FIRE TECHNOLOGY; FOOD ENGINEERING; ONCOLOGY; PLASTICS FABRICATION; POLYVINYL RESINS; TOXICOLOGY in the McGraw-Hill Encyclopedia of Science and Technology.

[JOHN AUTIAN]

Bibliography: J. Autian, *Artif. Organs*, 1:53–60, 1977; J. Autian, Toxicology of plastics, in L. J. Casarett and J. Doull (eds.), *Toxicology: The Basic Science of Poisons*, 1975; F. Bischoff, *Clin. Chem.*, 18:869–894, 1972.

Transplantation biology

In every species studied thus far, there exists a group of "strong" transplantation antigens, called alloantigens, which account for most of the rejection reaction to tissues transplanted within the same species. Although weaker antigenic systems may account for some rejections, especially when the graft recipient has been previously immunized with the same antigens, matching for the major antigens minimizes the intensity of immunological rejection. It has now been established that a single genetic region, encompassing a number of closely linked genes on a single chromosome, codes for several major histocompatibility antigens. Some of these antigens are identifiable by serological means, and others by their ability to induce a proliferative response by lymphocytes in tissue culture. The human major histocompatibility complex (MHC), analogous to H-2 in the mouse and RT1 in the rat, is called HLA, and consists of three serologically defined loci (sometimes abbreviated SD loci, old nomenclature) and one locus for the mixed lymphocyte culture (mixed lymphocyte reaction, abbreviated MLR, or LD locus, old nomenclature). In humans, the large number of possible alleles makes HLA the most polymorphic genetic system known. Phenotypic expression of antigens follows a simple mendelian codominant pattern. The antigens inherited from a given parent are referred to as a haplotype.

MHC antigens. There are two main classes of histocompatibility antigens, defined by their function and biochemical structure (see the table). Both classes are glycoproteins expressed predominantly on cell surfaces. Class I antigens are composed of two chains, a 44,000-dalton heavy chain and an 11,600-dalton light chain which is β_2-microglobulin. Class II antigens consist of two chains of 29,000 and 34,000 daltons, and have a more restricted tissue distribution.

Class I antigens are classically defined by serological methods, usually by employing a complement-dependent cytotoxic reaction to lyse cells bearing the relevant antigen. Antisera are obtained by immunization of appropriate inbred, congenic or recombinant strains of rodents, or by screening large numbers of sera from multiparous humans. Pregnancy represents an allogeneic exposure, and some mothers do make anti-HLA antibodies.

Class II antigens are serologically defined by similar techniques, by employing antibodies obtained from the same sources. A convenient distinction between class I and class II antigens is the fact that class II antigens are not expressed on normal thymus-derived (T) lymphocytes or platelets. Class II antigens are identical to, or closely associated with, antigens which initiate a proliferative response in lymphocytes from individuals bearing incompatible antigens of this class, and are identified by using the MLR culture technique.

Incompatibility for class I antigens usually produces no proliferative response, and rarely leads to more than weak proliferation. In the mouse, the antigen is called the lymphocyte activation determinant (LAD), and is coded for by genes in the I region (class II) of H-2, while the analogous region in humans is called HLA-D. Reference reagents for typing are lymphocytes homozygous for LAD or HLA-D, inactivated by x-irradiation or mitomycin C to make the reaction unidirectional. The responder cells' reaction to reference stimulating cells is a test for compatibility or incompatibility for the defined antigen. In the MLR, responder cells are T lymphocytes, while stimulating cells must express Ia-like determinants (B lymphocytes).

As shown in the table, there is no common nomenclature for the most thoroughly studied species. All show very close associations between LAD determinants and some of the class II antigens, called Ia in the mouse and D-related (DR) in humans. Their exact identity has not been established, but they may be separate products of closely linked genes, or even possibly separate determinants on the same molecule. In addition, there is recent evidence for the existence of class II–like antigens which are not related to MLR/LAD determinants.

Genetic organization. The MHC genetic region is on the short arm of the sixth chromosome in humans, and on the seventeenth chromosome in mice, encompassing 0.5 (mouse) to 1.5 (human) centimorgans map distance. Thus far, there have been identified two loci for class I antigens in the mouse (D, K), three in humans (A, B, C), and only one in the rat (Ag-B). The genetic organization of the MHC has been discerned by observations of recombination (crossing-over) in laboratory breedings, or by the study of large numbers of human families. The mouse I region maps between D and K, while the analogous D region is outside A, B,

Characteristics of gene products of the MHC

Antigen	Structure	Tissue	Identification technique	Nomenclature Mouse (H-2)	Rat (RT1)	Human (HLA)	Function
Class I	β_2-microglobulin plus heavy chain	Virtually all cells, except human red blood cells	Serology	D, K	Ag-B (A region)	A, B, C	Targets for effector antibodies and killer cells
Class II	α-chain plus β-chain	Absent on platelets, T cells, red blood cells, liver, and ?kidney	Mixed lymphocyte response (MLR) and serology (Ia)	I(LAD,Ia)	MLR/Ia (B region)	D, DR (Ia)	Key role in initiation of immune response (Ir genes)

and C, in the direction of the centromere. Less specific data are available for the rat.

In addition to the class I and II antigens, there are several other genes known to be within or close to the MHC region. Some of these determine enzyme polymorphisms and are useful in mapping studies. Others are genes for the complement components, C2, C4, and Bf. There is also evidence in humans that the cellular receptor for activated C3, but not C3 itself, is coded for by a sixth-chromosome gene. Finally, genes controlling the immune response to a variety of well-defined antigens map in the mouse I region and the rat B region. Clear evidence for such Ir genes in humans has not yet been obtained.

MHC function. It is apparent that transplantation is a modern practice, and that conservation through evolution of a closely linked genetic region in several species implies an important biological function. Since pregnancy involves maternal exposure to alloantigens, the question arises as to whether allogeneic incompatibility in some manner promotes survival, or whether additional mechanisms are necessary to circumvent rejection of the fetus. Present data favor the latter explanation. There is growing evidence that class I antigens function in normal immunobiology to focus effector mechanisms in dealing with cells which have been infected with viruses, and possibly also those which have undergone neoplastic transformation. Class II antigens are of the same region containing Ir genes, and these Ia antigens are thought to play a key role in cell-to-cell interactions during induction of the immune response.

T lymphocytes, B lymphocytes, and macrophages cooperate during the initial contact with antigen to develop a fully competent immune response. Resting B cells differentiate to become immunoglobulin-producing plasma cells, and some T cells differentiate into killer cells capable of inducing cytolysis of target cells bearing the relevant antigen. Another group of T cells provide the helper signal needed to effect these differentiative steps. Since T cells undergo proliferative responses in allogeneic mixtures without the addition of another antigen, it seems likely that normal immune reactions may involve recognition of Ia antigens which have been altered by a foreign antigen, and that T cells are concerned with focusing upon and reading out the Ia structures on surfaces of B cells and macrophages.

The association, and in some cases linkage, of some experimental and clinical disease entities to inheritance of certain MHC antigens points to the broad biological significance of genes of this region. Not all such observations are necessarily related to class II/Ir genes, however, since a variety of secondary and tertiary events occur in the evolution of the immune response as effector mechanisms are generated. The complement system is a case in point, and the existence of MHC-determined suppressor systems which modulate the immune response is now established. In the mouse, a subdivision of the I region, called I-J, determines a T-cell surface antigen marking a subpopulation of cells which play a suppressor role. Other genes unrelated to the immune system may be fortuitously present in the MHC region, an example being the gene for synthesis of the enzyme 21-hydroxylase in humans.

Tissue rejection. Both class I and II antigens are strong barriers to transplantation. Since the gene products of the MHC are defined by antibodies and cellular reagents produced as a result of grafting or exposure to incompatible tissue, these effectors (antibodies and killer cells) would be expected to play some role in destruction of a graft. Most early experimentation has been on skin grafts, and here cellular rejection mechanisms seem to predominate; however, vascularized organ grafts are readily damaged by antibodies as well, particularly when the recipient is already primed (sensitized) to the relevant antigens. Paradoxically, graft prolongation may occur under certain circumstances of passive transfer of antibodies to donor antigens, particularly those of class II, or after active immunization. There is preliminary evidence that such manipulations affect the balance of helper and suppressor T cells in favor of a blunted response, but more detailed immunogenetic study is necessary. Immunologically specific tolerance may be induced by fetal or neonatal exposure to donor antigens, but this is rarely achieved in adults. Engraftment of bone marrow cells provides additional complications of graft-versus-host reactions, since the graft contains immunocompetent cells.

Rejection of organ transplants performed for clinical purposes in conjunction with immunosuppressive drugs continues to be a major problem, with a 20% 1-year failure rate when haplotype-matched familial donors are used, and greater than 50% 1-year failure rate with cadaveric donors. Application of class II (HLA-D/DR) matching in the latter category is just beginning. Matching for class I (HLA-A, B, C) antigens only has shown marginal benefits.

For background information *see* ANTIGEN; CHROMOSOME; IMMUNOLOGY, CELLULAR; TRANSPLANTATION BIOLOGY in the McGraw-Hill Encyclopedia of Science and Technology.

[C. B. CARPENTER]

Bibliography: C. B. Carpenter, A. J. F. d'Apice, and A. K. Abbas, *Adv. Immunol.*, 22:1–65, 1976; D. Götze (ed.), *The Major Histocompatibility System in Man and Animals*, 1977; J. Klein, *Contemp. Topics Immunobiol.*, 5:297–336, 1976; G. D. Snell, J. Dausset, and S. Nathenson, *Histocompatibility*, 1976.

Tube worm

Although a number of quite different taxa can be included under the heading tube worms, the term is used in this article in its more common meaning for polychaetous annelids with calcareous tubes: the family Serpulidae of the Polychaeta. They can be economically important (second only to barnacles) in ship fouling. Serpulids are usually marine worms, a maximum of 10 cm long and 1 cm wide. The extremely osmoconforming (salinities of 1.6–55‰) genus *Ficopomatus* is typical for brackish environments, in which few other species may survive; one species, *Marifugia cavatica*, is known from fresh water in karst caves in Yugoslavia.

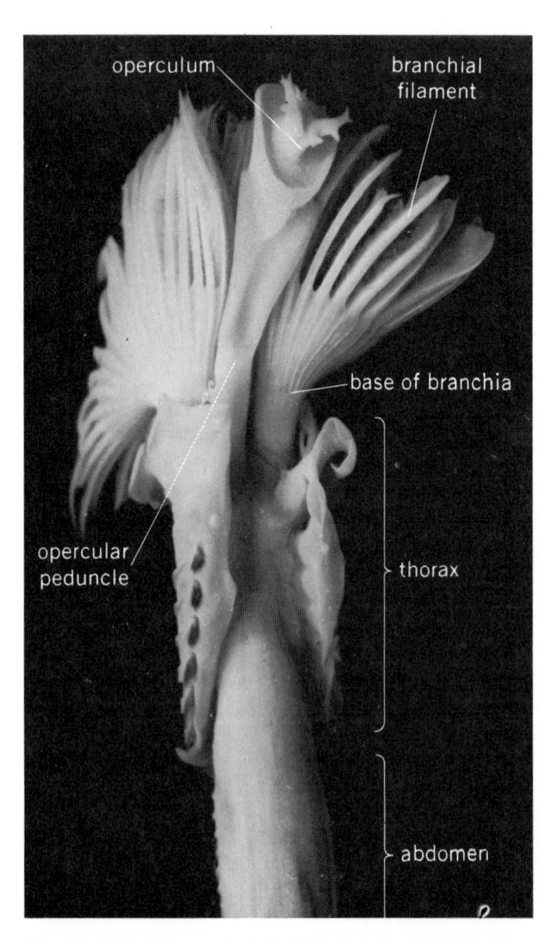

Fig. 1. Anterior part of *Spirobranchus giganteus*, removed from its tube, in dorsal view. *(Modified from H. A. ten Hove, Serpulinae from the Caribbean, I: The genus Spirobranchus, Stud. Fauna Curaçao, 32:1–57, 1970)*

After settling on a substrate, the worms secrete a tube, which they cannot leave, consisting of a fairly constant aragonite-calcite ratio for each taxon. The tubes are cemented firmly to any hard substrate, except for *Ditrupa*, which lives in a free tube in the sand. Anteriorly the worms have a number of featherlike tentacles (branchiae), used for respiration and feeding (Fig. 1). Rows of cilia on the pinnules (side branches) of the branchial filaments filter detritus and microscopical algae from the water. In feeding position, the tentacles extend beyond the tube opening; when disturbed, the animal withdraws them quickly into its protective tube. Generally, a modified branchial filament, the operculum, plugs the tube; this operculum may be shed periodically in some genera, notably *Hydroides*. Tubes are usually white, sometimes bluish or pink, and exceptionally transparent (*Placostegus* and various spirorbid species).

Taxonomy. Generally, three subfamilies are distinguished: Filograninae, Serpulinae, and Spirorbinae. In recent years family rank has been attributed to the Spirorbinae, with approximately 100 species, by specialists in this group. O. Hartman enumerates approximately 240 valid species of the Filograninae and Serpulinae; the same number is given in unpublished data of H. A. ten Hove. Since in the last decade about 70 new species have been described, as many names have been included in synonymic lists of other species. For *Spirobranchus tetraceros*, for example, 19 species names in eight genera have been used. Of 41 abyssal records, only 14 proved to be correct, including those where a nomenclatural change was necessary; the remaining 27 records are incorrect, doubtful, or unidentifiable. The taxonomy of tube worms is thus still incompletely known, although the use of scuba diving for collecting specimens has increased faunal knowledge, especially of cryptic species, in certain areas.

Distribution. Throughout the literature of the last decade one fact becomes evident: species previously thought to be "cosmopolitan" are nearly all complexes of more regionally limited species. The so-called circummundane *Ficopomatus enigmaticus* (formerly *Mercierella enigmatica*) in reality occurs only in temperate-subtropical areas; in the tropical region it is replaced by three other *Ficopomatus* species. As such, *F. enigmaticus* still has a rather large distribution, almost certainly due to spread of the species by transport on ship hulls (Fig. 2). As far as is known, most individual species are restricted to a single zoogeographical region. Exceptions to this rule are species distributed involuntarily by humans. *Hydroides dirampha*, *H. elegans*, and *H. dianthus* may belong to this category. Recently *H. ezoensis* has been imported to France with the Japanese oyster *Crassostrea gigas*; the Japanese seaweed *Sargassum muticum*, with its spirorbid epibiont *Pileolaria rosopigmentata*, may have reached the United Kingdom in a similar way or in ballast water of ships. Most species have a limited bathymetrical range; records of eurybathic species have been proved to be false.

Economic importance. Tube worms are a main constituent of the fouling communities of submerged surfaces such as jetties, canal walls, power

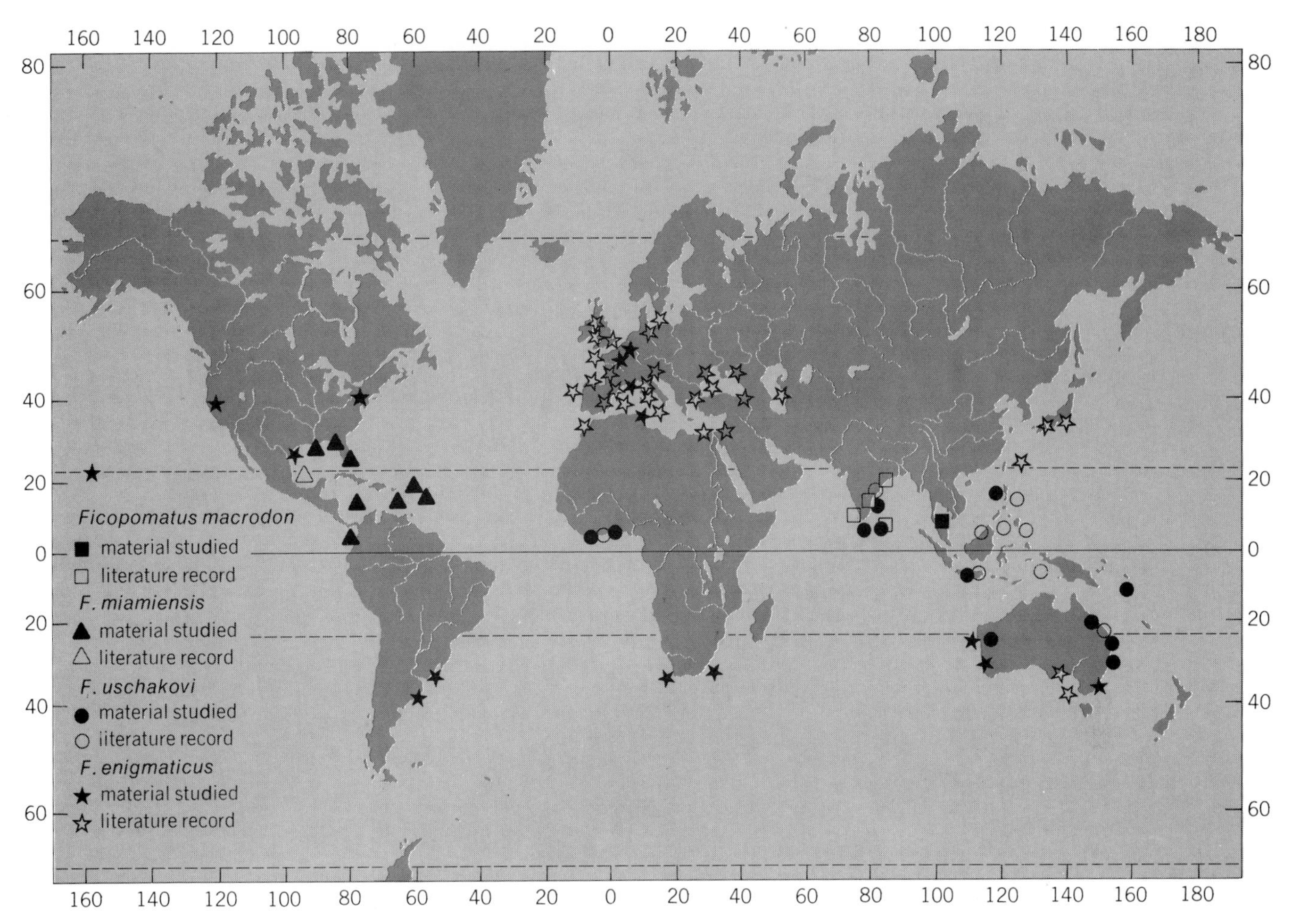

Fig. 2. Distribution of *Ficopomatus*. (*Adapted from H. A. ten Hove and J. C. A. Weerdenburg, A generic revision of the brackish-water serpulid Ficopomatus Southern 1921 (Polychaeta: Serpulinae), including Mercierella Fauvel 1923, Sphaeropomatus Treadwell 1934, Mercierellopsis Rioja 1945 and Neopomatus Pillai 1960, Biol. Bull., 154: 96–120, 1978*)

station pipes, and ship hulls, on which they can settle with water velocities of up to 2 km/hr (*H. elegans*). This species, often erroneously named *H. norvegica*, and *F. enigmaticus* can deposit a massive, 85-mm-thick layer of calcareous tubes within a year. Species of *Serpula* and the above-mentioned genera may form reeflike structures up to 50 cm thick and 300 cm in diameter in quiet lagoonal habitats. In the tidal zone of temperate regions such mass occurrence may lead to belt-forming by species of *Galeolaria*, *Pomatoceros*, *Pomatoleios*, and *Spirobranchus*. Clearly, this can be very damaging to marine installations. Moreover, mass settling can result in severe competition—possibly for food but certainly for space—with oysters, which has led to an estimated loss of 40% of the oyster crop in Hiroshima.

Reproduction and growth. Asexual reproduction has been reported for *Filograna* (including *Salmacina*), *Filogranula*, and *Josephella*. In *Filograna* the parent animal splits into two pieces. However, before actual fission occurs, a new cephalic region is formed in the middle of the parent, by transformation of abdominal to thoracic segments (morphallaxis). Through this scissiparity, chains of individuals, or a network of branching tubes, may develop. It certainly helps in exploiting a good settling site initially colonized by a single pelagic larva from another population.

Tube worms also reproduce sexually, the gametes being shed into the water. Hermaphroditism in not uncommon and is reported in the genera *Hydroides*, *Ficopomatus*, *Pomatoceros*, and *Filograna* and in the subfamily Spirorbinae; species of *Hydroides*, *Ficopomatus*, and *Pomatoceros* are protandrous, and species of *Filograna* and Spirorbinae are simultaneous hermaphrodites, leading to self-fertilization in some Spirorbinae. Whether the other species are gonochoristic is not yet known. Although ripe gametes may be found in many species throughout the year, reproduction may be seasonal, being influenced by temperature. In *Pomatoceros triqueter* settling of larvae is limited to the warm months in boreal areas, but is irregular year-round in temperature-subtropical areas. *Pomatoleios kraussii* also displays different types of seasonal breeding which are geographically determined.

Fertilization is external, in the water, except in species with brood care. Cleavage is spiral, holoblastic, and equal in *Ficopomatus*, *Galeolaria*, *Hydroides*, *Pomatoceros*, *Serpula*, and *Spirobranchus*, all Serpulinae with small eggs; it is unequal in *Filograna* and *Protula* (Filograninae), and in the Spirorbinae, which have large eggs with much yolk.

The trochophore larva of the Serpulinae is planktonic (except in *Chitinopoma serrula*, which has brood chambers lateral to its tube) and feeds for 1–8 weeks; the length of the larval stage depends upon season, salinity, and food availability. Larval incubation of yolky, nonfeeding embryos in the tube of some Filograninae and in the tube or operculum of Spirorbinae may result in a planktonic stage of only a few hours or in no pelagic stage at all; within a single spirorbid species the ratio of benthic to pelagic larvae may differ in different populations.

The larvae of most tube worms are dispersed by currents, their own swimming speed generally not exceeding 5 mm/s. Many species are gregarious during settlement, which partly explains mass occurrences. The larvae react to chemical and physical stimuli: in a single spirorbid species various strains may show genetically determined different preferences for algae as a substrate, or a locally determined photic behavior. Gregariousness may be density-dependent, only occurring in populations of 1 animal per square millimeter and above *(Ficopomatus)*, and results in its most extreme form in growth of tubes perpendicular to the substrate (Fig. 3). In spirorbid species, on the other hand, with a more rigid growth pattern, a density of 12 individuals per square centimeter may inhibit further crowding. Patterns of settling are the combined result of time of spawning, length of larval life, hydrographic factors, and settling behavior of the larvae; consequently, they vary geographically within a single species.

Fig. 3. Effect of crowding upon growth of tubes in *Spirobranchus polycerus* (*a*) from Barbados and (*b*) from Curaçao. (*From H. A. ten Hove, Serpulinae (Polychaeta) from the Caribbean, I: The genus Spirobranchus, Stud. Fauna Curaçao, 32:1–57, 1970*)

Initially the tubes increase in length very quickly (up to 1 cm/week in the first 6–8 weeks). Afterwards growth decreases; however, sometimes spasmodic bursts of growth can be observed. A new generation may thus overgrow and smother its slower-growing parents (or other animals) when settling is periodical and gregarious. The individual life-span of, for example, *Ficopomatus enigmaticus* may be 4–8 years; in Spirorbinae it is a few months at most.

Natural enemies. Little is known of the natural enemies of tube worms. Scattered records mention protozoan, trematode, gastropod, and copepod parasites. Whether they are specific is not known. Gastropoda, Echinodermata, Brachyura, and certain fishes are the main predators upon adult serpulids. Perhaps predator specialization upon a given conspicuous species (*Spirobranchus giganteus*) is prevented by the color variation. Nibbling of branchial crowns by fishes will not kill tube worms due to their regenerative powers.

Fossil tube worms. Paleontologists are confronted with the difficulty that in Recent literature the value of the tubes for identification has largely been underestimated. Consequently, the possible variation (which can be environmentally induced) in tubes on infraspecific, specific, and generic levels is little known. A further difficulty arises in differentiating serpulid and vermetid (Gastropoda) tubes. If the animal is not present, this distinction sometimes can hardly be made even in Recent tubes. Some fossil "serpulid" reefs have recently been demonstrated to be of vermetid origin.

Even within the tropics (Bonaire, Curaçao), the number of coexisting Recent species generally does not exceed 20; in a favorable habitat such as a coral reef (0–50 m depth) it may be 26; and in enclosed lagoons it is only about 12. The total number of species in this area is about 45. Not all species have characteristic tubes; for instance, in the largest Recent genus, *Hydroides* (approximately 60 species), most tubes are indistinguishable on the species level. Since 4–7 species of *Hydroides* may occur in one locality, with up to 14 in a given area, the number of coexisting tube types will be even less. In revisions of fossil serpulids the number of species may be two to three times higher; probably too many fossil taxa have been described.

For background information *see* ANNELIDA; POLYCHAETA in the McGraw-Hill Encyclopedia of Science and Technology. [HARRY A. TEN HOVE]

Bibliography: O. Hartman, *Catalogue of the Polychaetous Annelids of the World, Part II*, Allan Hancock Found. Publ. Occas. Pap., 23:354–628, 1959; O. Hartman, *Catalogue of the Polychaetous Annelids of the World, Supplement 1960–1965 and Index*, Allan Hancock Found. Publ. Occas. Pap., 23 (suppl.):1–197, 1965; H. A. ten Hove, in G. Larwood and B. R. Rosen (eds.), *Biology and Systematics of Colonial Organisms*, Systematics Association Special Volume, 11:281–298, 1978; P. Knight-Jones, E. W. Knight-Jones, and T. Kawahara, *Zool. J. Linn. Soc.*, 56(2):91–129, 1975; P. C. Schroeder and C. O. Hermans, Annelida: Polychaeta, in A. C. Geise and J. S. Pearse (eds.), *Reproduction of Marine Invertebrates*, vol. 3: *Annelids and Echiurans*, 1975.

Tunicata

The Larvacea (or Appendicularia) are a class of tadpole-shaped marine tunicates ranging in length from a few millimeters to 9 cm. Planktonic throughout their lives, they are one of the commonest members of marine zooplankton communities, particularly in coastal waters and over continental shelves. Larvaceans are herbivores exhibiting a unique feeding biology. They secrete and live within a balloon of mucus, the "house," equipped with minute filters which concentrate food particles as small as 0.1 μm. The nannoplankton (algae, and bacteria smaller than 20 μm) are their major food source. Recent research suggests that larvaceans are important in marine food webs as one of the few planktonic herbivores able to collect nannoplankton. In addition, their houses are of considerable ecological significance in open-ocean communities as food sources and habitats for other organisms.

House structure. Taxonomically, larvaceans are divided into three families, each with a different house structure. In the Oikopleuridae and Kowalevskiidae the house completely surrounds the animals, while in the Fritillaridae the house is deployed as a large bubble in front of the mouth (Fig. 1). Only the houses of the Oikopleuridae have been described in detail. In this family the house is a hollow sphere of gelatinous mucus, containing complex passageways and filters. Food-laden water, forced by the sinusoidal movement of the animal's tail, enters the house through two passages at the front, each of which is covered with fibers to form a screen (Fig. 2). These incurrent filters retain large particles, primarily diatoms and dinoflagellates, too large for the larvacean to eat. The pore size of the incurrent filters is dependent upon animal size and species.

Food small enough to pass through the incurrent filters travels down a passage containing the animal's tail and enters a complex filter within the house where particles become further concentrated. This feeding filter is composed of two porous membranes sandwiched together. As water and

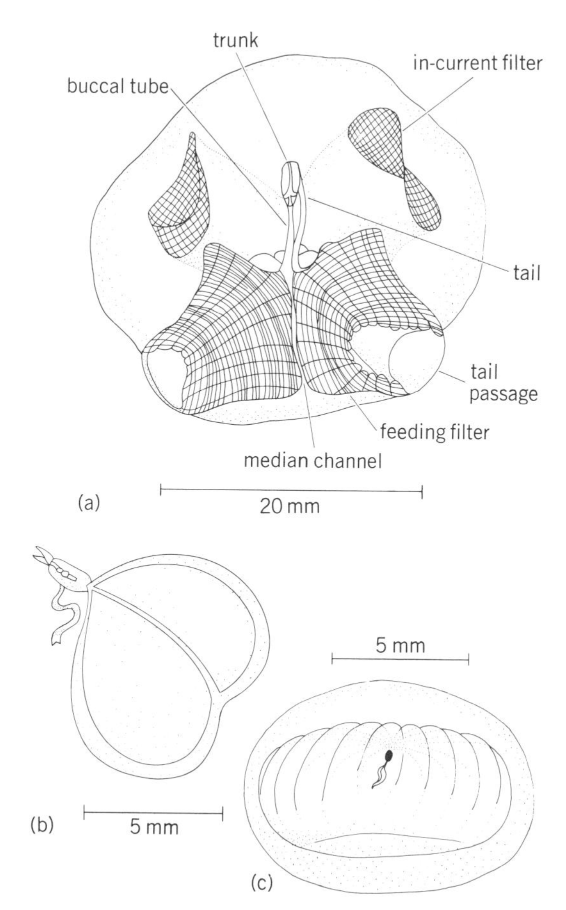

Fig. 1. Houses of the three larvacean families: (*a*) Oikopleuridae, (*b*) Fritillaridae, and (*c*) Kowalevskiidae.

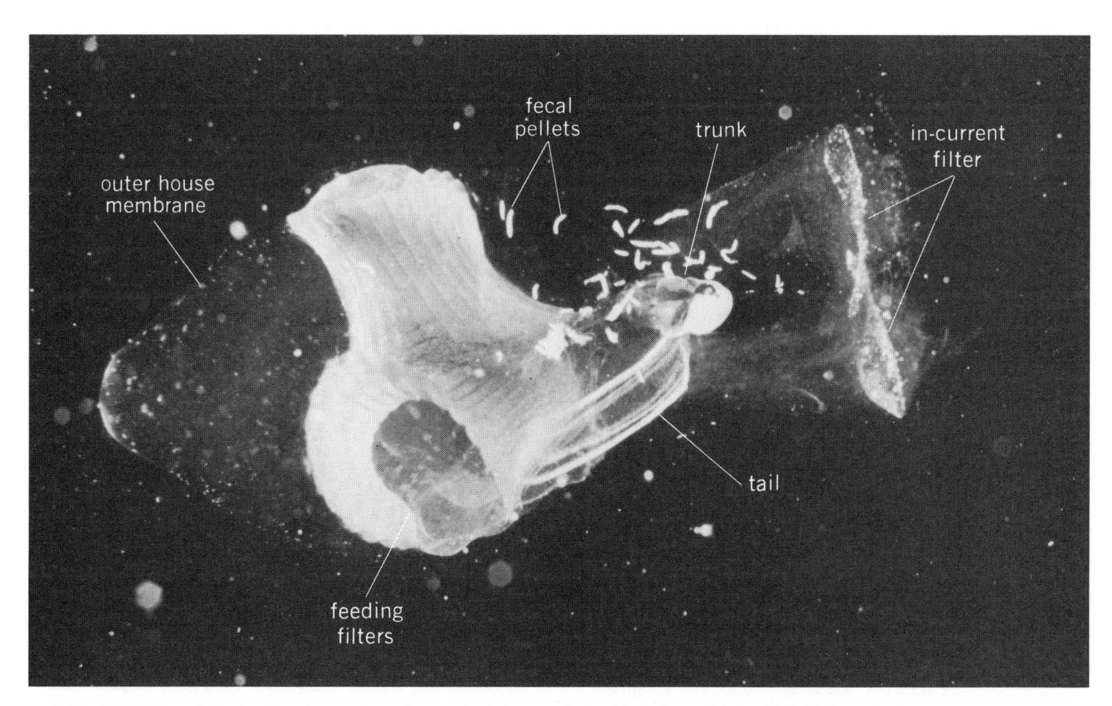

Fig. 2. *Megalocercus huxleyi*, a member of the family Oikopleuridae, within its torpedo-shaped house.

food enter between the membranes, hydrostatic pressure forces water out through the minute pores, leaving food particles trapped behind. Food particles are sucked up as a thick slurry through a hollow tube connecting the animal's mouth with the feeding filter. Filtered water, leaving the house through an exit passage, may propel the house slowly through the water.

The particles which enter the animal's mouth are collected by a method common to all tunicates, including sedentary ascidians. The endostyle, a gland near the mouth, produces a funnel of mucus which lines the pharynx and extends into the stomach. Water enters the mouth and passes through the funnel and out through the spiracles, structures homologous to vertebrate gill slits. Particles adhering to the mucus funnel are carried into the stomach and digested.

The house is a particularly efficient food-collecting device. G. A. Paffenhöfer found that one individual of the species *Oikopleura dioica* could filter up to 300 ml of sea water per day. Larger species may be able to remove the particles from several hundred milliliters of sea water per hour. Thus, larvaceans are significant grazers upon marine phytoplankton.

House building. The mucus that forms the walls and filters of the house consists of a mucopolysaccharide (sugar) that is secreted on the surface of the trunk by a layer of specialized glandular cells, the oikoplast epithelium. The pattern of this epithelium is characteristic for each species and determines the structure and shape of the house. Different groups of cells can manufacture mucus of different density and elasticity, so that on expansion the mucus takes on a variety of shapes that contribute to the complexity of the resulting structure. In the Oikopleuridae the oikoplast epithelium covers most of the trunk, while in the Fritillaria it is reduced to an area on the upper portion of the trunk and around the mouth. The oikoplast epithelium can be divided into regions based on the types of cells present and the parts of the house that are made there. One region of specialized cells, Eisen's oikoplast, is responsible for making the mesh covering over the incurrent openings of the house. Another region, Fol's oikoplast, produces the feeding filter.

The construction of a new house by a larvacean begins with the secretion of gelatinous material by the oikoplast cells. The mucus accumulates in a layer at the top of each cell. It is believed that a new membrane is fabricated directly under the mucus layer, moving the mucus to the outside of the cell. There, the mucus coalesces into the rudiments of a new house. Special structural fibers are manufactured by both Eisen's and Fol's oikoplasts. Although the secretion of the house may take several hours, some species can secrete houses in as little as 5 min.

After secretion the animal's trunk is covered with a thin gelatinous film which is carried in collapsed form until a new house is needed. The shape and final form of the house are predetermined at the time of secretion. The rudiment is ready to be expanded into a fully functional house.

Larvaceans expand new houses only after they have abandoned or discarded the old one. This occurs when the filters become clogged with particles and the house becomes less efficient at collecting food. Larvaceans may also abandon the house if threatened by numerous predators, including sardines, anchovies, jellyfish, and planktonic worms. Houses become clogged readily, and most members of the Oikopleuridae build about three to seven houses each day. Larvaceans can feed only while within the house. Thus, they are free-swimming only during house expansion.

Once the house is discarded, the free-swimming larvacean begins a complex behavioral sequence, first described by R. Fenaux, to expand the rudiment against its trunk into a fully functional house. The animal swims intermittently for several seconds. It then begins a series of violent movements to loosen the rudiment around the trunk and allow the animal to wriggle partially out of the collapsed house. These motions often cause the animal to somersault or cartwheel in the water. Rapid sinusoidal motions of the tail and nodding of the trunk expand the rudiment sufficiently to allow the animal to pull its tail within the house (Fig. 3), a motion requiring less than 1 s. The animal, now cramped within its tiny house, begins slow sinusoidal motions with its tail to force water into the house, which enlarges as water enters. The entire expansion process may take several minutes, during which time the opening at the front of the house, through which the animal has entered, gradually closes. Once the house reaches normal size, feeding begins again.

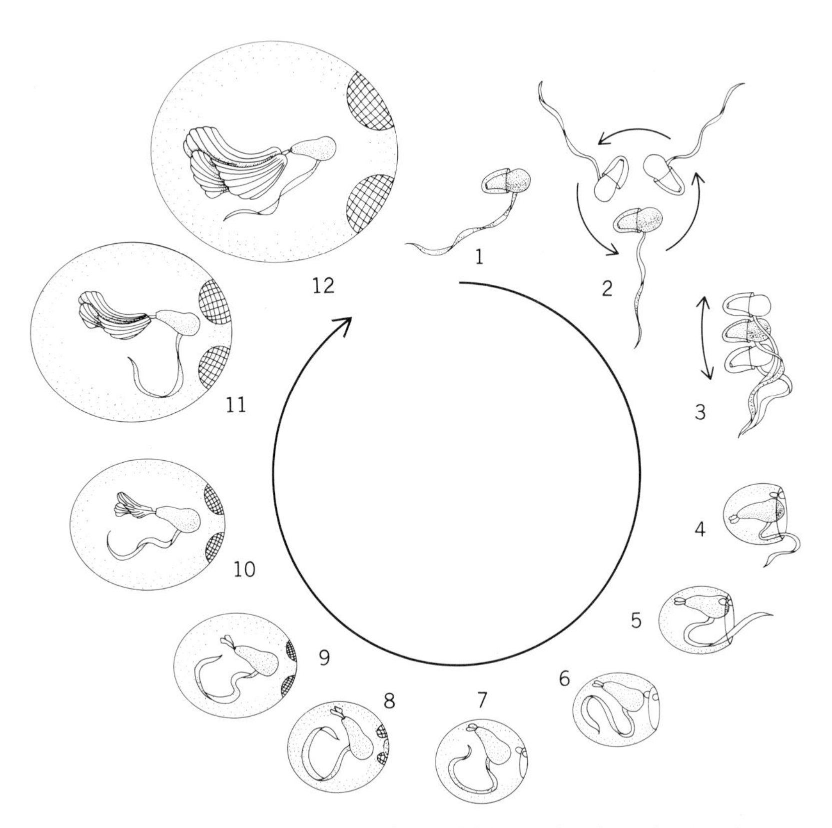

Fig. 3. A larvacean expanding its new house. (1) Free-swimming animal carries collapsed house rudiment against its trunk. (2–3) Cartwheel and nodding movements help lift the house away from the trunk. (4–8) The animal pulls its tail into the slightly expanded house. (9–11) Slow sinusoidal movements of the tail draw water into the house, (12) expanding it to normal size. (*From A. Alldredge, Appendicularians, Sci. Amer., 235:94–102; copyright © 1976 by Scientific American Inc.; all rights reserved*)

Ecological significance of house. Larvaceans themselves are important food for many schooling fish and planktonic predators and are major grazers upon marine algae. However, discarded larvacean houses also have considerable ecological significance in open-ocean communities. Several larvaceans, each building six houses per day, can quickly fill the surrounding water with discarded houses. Each discarded house is a concentrated package of phytoplankton, mucus, and detritus, a rich and concentrated food source in an environment where food is normally dilute, small, and scarce.

Discarded larvacean houses are significant to planktonic communities in three major ways. First, most filter-feeding adult marine zooplankton are unable to utilize nannoplankton since they have difficulty filtering particles smaller than 5 μm in diameter. However, many zooplankton, including copepoda, euphausiid larvae, and polychaete worms, have been observed feeding on the nannoplankton attached to larvacean houses, thus opening a new link in the planktonic food web. Second, discarded larvacean houses serve as sources of particulate carbon and other essential elements and aid in recycling and breakdown of organic matter. And third, houses provide habitats and surfaces upon which many planktonic animals can feed and live. Bacteria, protozoans, and other small organisms use larvacean houses as permanent habitats. The tiny solid surfaces and microhabitats provided by larvaceans introduce heterogeneity into an environment that is generally considered to be relatively homogeneous physically.

For background information *see* TUNICATA in the McGraw-Hill Encyclopedia of Science and Technology. [ALICE L. ALLDREDGE]

Bibliography: A. Alldredge, *Mar. Biol.*, 38: 29–39, 1976; A. Alldredge, *Sci. Amer.*, 235: 94–102, 1976; G. A. Paffenhöfer, *10th European Symposium on Marine Biology*, vol. 2, pp. 437–455, 1975.

Urban heat island

Radiometric measurements made aboard meteorological satellites have been widely used for studying diurnal temperature fluctuations at the Earth's surface and for mapping the ground temperature patterns. Recently, satellite infrared measurements have provided further insight into how centers of urban activity are able to modify the way in which the lower boundary of the atmosphere is heated, and therefore how cities can alter local weather conditions. It is well known that cities are usually warmer than their surroundings. The pattern of higher urban temperatures and cooler suburbs is called the urban heat island.

Land-use change. The causes of the urban heat island are complex; generally its appearance can be attributed to the effects of urbanization, particularly its effects on surface landscape. Although direct heat release to the atmosphere as a result of human activity constitutes an important contribution to local urban heating under some conditions, for example, in large northern cities such as New York or Montreal during winter (when the burning of fossil fuels for heat is appreciable), the urban heat island is more directly a result of the redistribution of the solar energy reaching the ground than a result of anthropogenic heating. Urbanized areas are generally characterized by a substitution of artificial materials such as concrete, brick, glass, and steel for natural vegetation. At ground level, the available radiant energy—that which is not reflected or emitted back to space—is divided into three forms of heat energy in a manner which is sensitively dependent upon the nature of the surface material and therefore upon the effects of urbanization. These forms of energy are evaporative (latent), sensible, and stored.

Changes in land use, occurring as a result of human activity, are reflected in surface temperature patterns, which are now resolvable in great detail by infrared radiometry from space. Conventional measurements of air temperatures, customarily made a meter or so above the surface, are relatively sparse and cannot adequately represent the detailed character of the surface temperature pattern. Moreover, the effective temperature at the surface-air interface responds much more sensitively to the nature of the ground surface than does the corresponding air temperature at the conventional measurement level slightly above ground level.

Surface temperature anomaly. Over Los Angeles and St. Louis—cities which have been studied extensively in regard to the urban heat island and the various other effects of urbanization—radiometric surface temperature measurements made from satellite and aircraft platforms show quite clearly that the surface temperature anomaly over the built-up areas can become very pronounced in summer. In these two cities, persistance of elevated surface temperatures into nighttime hours attests to the basic importance of solar energy as the principal mechanism for the formation of the heat island. Further, theoretical calculations show that the rate of heat flux from ground to atmosphere essentially governs the ground temperature fluctuation, and therefore both the heating and the temperature mutually depend upon the nature of the ground surface. Over a heterogeneous terrain, the ground temperatures (or the heat fluxes) are difficult to measure directly with any degree of spatial representativeness. However, the physics of heat transport can be mathematically modeled and reasonable temperatures or heat fluxes computed from the models, provided that one is able to determine numerical values for the surface parameters that govern heat flow at the ground. The practical importance of this idea will be explained later.

Surface character effect. The essence of how the surface character controls heating is illustrated by the fact that over moist terrain the available radiant energy is used mostly to evaporate water substance at the ground, and thus wet surfaces, or even those which merely evaporate efficiently, remain relatively cool during the day. Over urban areas the artificial canopy of buildings, roads, parks, houses, open lots, and so on is, on the whole, rather dry, and the ground canopy also restricts the transport of available moisture from substrate to surface. These artificial surfaces also store and retain heat during the day which is returned to the surface at night, thus maintaining a warm ground surface after sunset. Consequently,

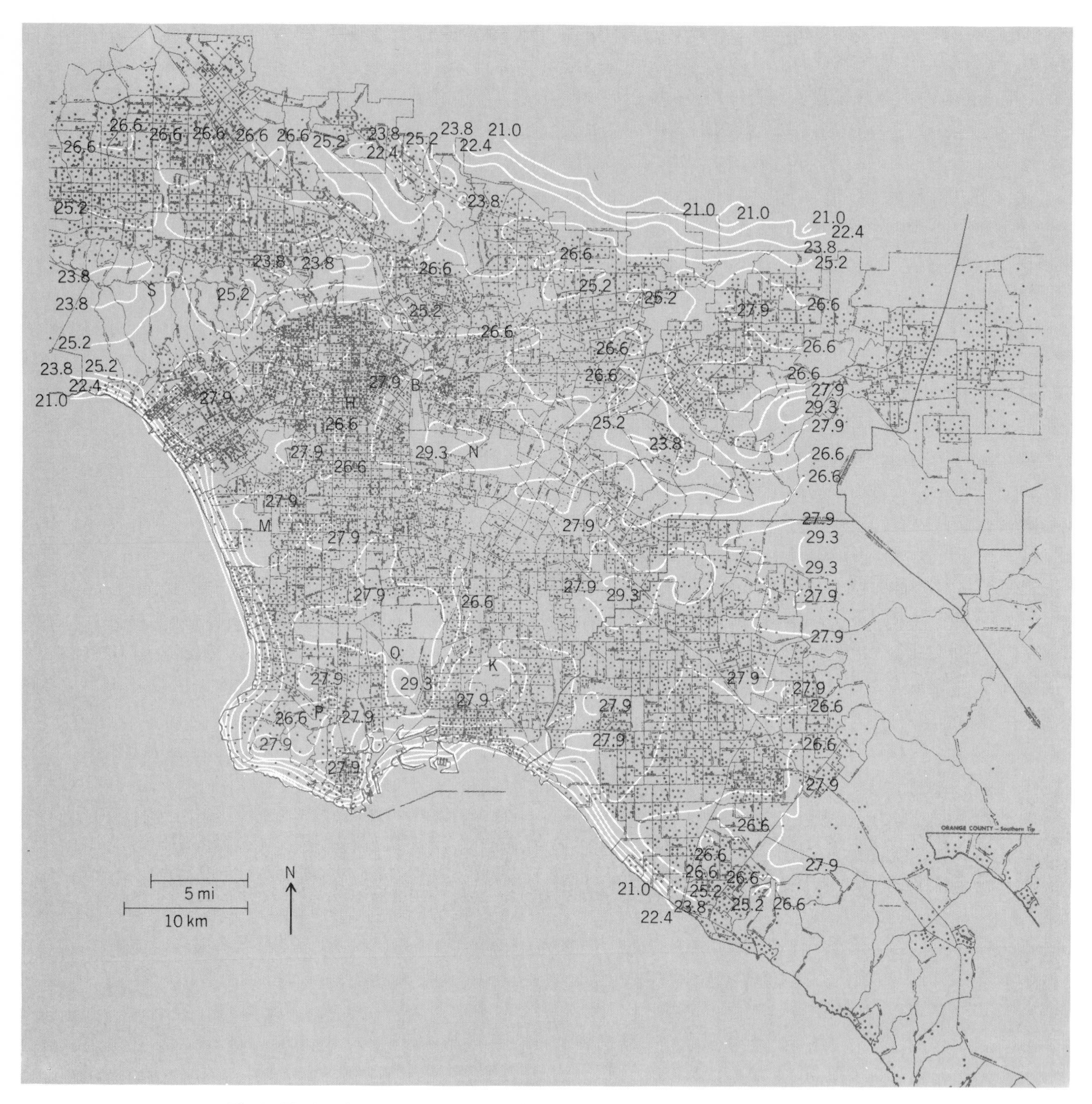

Fig. 1. The Los Angeles five-county area by 1970 census tracts. Surface infrared temperature analysis derived from *NOAA 5* VHRR measurements for orbit 879 at 0924 Pacific standard time, Oct. 8, 1976. The solar elevation and satellite nadir angles were, respectively, 39° and 16°. (*From T. N. Carlson and J. A. Augustine, Temperature mapping of land use in urban areas using satellite data, Earth Miner. Sci. Bull., 47(6):41–45, 1978*)

dry urban surfaces are not only warm during the day but remain warm at night. Other variations in surface temperature and therefore in surface heat flux are possible. For example, a dry, poorly conducting surface such as sandy soil is likely to be warm during the day in comparison with concrete, but the sand will cool off more rapidly than concrete at night, becoming cooler than concrete sometime during the night. In general, however, the landscape possesses a complete spectrum of surface types which produce an infinite distribution of surface temperature responses and thus a complex thermal map.

Los Angeles temperature observations. These concepts are illustrated by Figs. 1 and 2, showing a day-night pair of surface temperature maps derived from radiometric measurements made aboard one of the National Oceanic and Atmospheric Administration (NOAA) satellites. *NOAA 5*, which was launched in mid-1977, is a polar-orbiting sun-synchronous vehicle which contains a pair of very-high-resolution radiometers (VHRR), one operating in the visible part of the electromagnetic spectrum and the other in the thermal "window" portion of the infrared. The sun-synchronous schedule of these satellites affords the opportunity

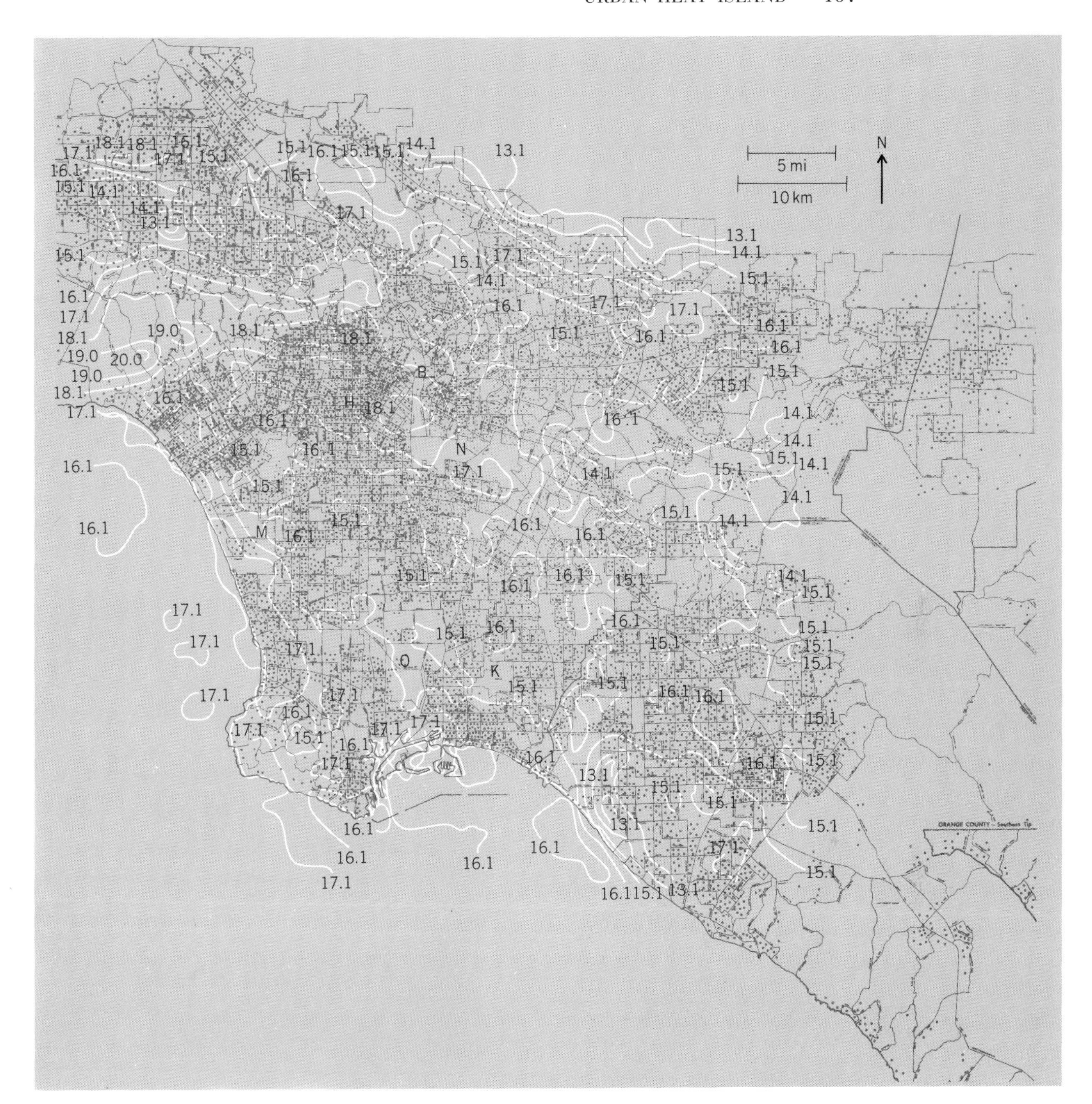

Fig. 2. The Los Angeles five-county area by 1970 census tracts. Surface infrared temperature analysis derived from *NOAA* 5 VHRR measurements for orbit 885 at 2026 Pacific standard time, Oct. 8, 1976. The satellite nadir angle was 23°. *(From T. N. Carlson and J. A. Augustine, Temperature mapping of land use in urban areas using satellite data, Earth Miner. Sci. Bull., 47(6):41–45, 1978)*

to view at the same point on the Earth's surface at 12-hr intervals, once during the day and once at night, approximately 9 A.M. and 9 P.M. local sun time, in the case of the NOAA satellites. At zero nadir angle (that is, with the satellite directly overhead) the resolution of these measurements is less than 1 km.

Figures 1 and 2 show two temperature patterns over Los Angeles for a 12-hr interval during October 1976. At first glance the patterns appear very complex, but the heat island emerges quite clearly as a pair of warm areas with surface temperatures of approximately 29°C located over a relatively unpopulated industrial zone (N) and also over a series of refineries farther to the south (O). The high-density suburbs (H) and central business district (B) are located close to the northern warm area. Lesser temperature maxima, approximately 28°C, appear over the airport region (M) and in the vicinity of a United States naval installation (K). The areas N, O, K, and M are all relatively unpopulated but have in common a general lack of vegetation and poorly conducting surfaces. Cooler temperatures prevail over the more vegetated Santa Monica Mountains (S) and over the Palos Verdes Peninsula (P).

On that October evening (Fig. 2) the high temperatures, approximately 18°C, are still centered near the inner city, but the maximum has now shifted to the high-density suburbs (H) and the downtown area (B). The oil fields (O) and naval base (K), which were relatively warm during the day, have cooled off more rapidly during the first several hours after sunset. The mountainous regions to the north (S) exhibit a far smaller temperature fluctuation than the downtown area, largely because of the vegetation canopy. Unlike the central business district, which is characterized by a strongly conducting but dry surface, the oil fields are associated with a dry, poorly conducting surface. One peculiarity in the early evening surface temperature pattern (Fig. 2) is the relatively high surface temperature, approximately 20°C, exhibited along the south-facing slopes of the Santa Monica Mountains (S). This feature is the result of differential heating of the slopes that face the afternoon sun; conversely, the north-facing slopes, which are shaded in the late afternoon, are relatively cool during the evening hours.

It is evident from inspection of such satellite-derived temperature analyses that the ground temperatures are strongly tied to the surface character, and to a considerable extent, day-night temperature fluctuation is explainable by the amount of available radiant energy converted into heat at the ground surface. The ground temperature thus emerges as a potential measure of land use, reflecting transformation of the landscape by people.

Climate modification by cities. Inherent in these temperature measurements is information concerning the ways in which the urban surface can modify the environment through input of heat to the atmosphere. Microclimatic features are closely tied to the pattern of surface heating, which influences wind flow and precipitation patterns over and downwind of a city. Results from the Metropolitan Meteorological Experiment (METROMEX) indicate that there is a 10–30% precipitation excess above background rainfall over portions of St. Louis and downwind of the city along the path of the urban heat plume. In summer, this precipitation often takes the form of thunderstorms which develop or pass over that city, with the maximum development of thunderstorms occurring over a 100-km-length swath to the northeast of the city.

Small-scale variations in wind flow and convergence of air near the surface associated with the urban heat island over St. Louis appear to be important factors in the enhancement of local precipitation and thunderstorms in the area. Precipitation increases occurred most notably when thunderstorm cells moved directly over the urban-industrial part of the city, where there was an increased mixing depth due to the enhanced heating at the surface. METROMEX clearly demonstrated that in some instances the urban heat island is associated with the initiation of local cumulus convection. Moreover, increasing evidence suggests that St. Louis is quite typical of urban areas in this regard.

Surface temperature observation. It is evident, therefore, that the observed changes in local weather resulting from urbanization are associated with factors governing the heat island. Because the ground surface is extensive, changing, and complex, the satellite, with its large field of view, high spatial resolution, and ability to make repeated observations of a particular surface feature, is ideal for probing the surface temperature response. The surface parameters necessary for calculating the surface heat budget in a mathematical model can, in principle, be derived from measurements of the ground temperature response. Although these parameters are not directly measurable, they can be considered in a physical model of the surface layers as effective terrain parameters which enable one to calculate the correct temperature (and therefore heating) response with the model, and which correspond to real characteristics of the terrain such as conductivity, wetness, and roughness.

The ground temperature response over large areas can be obtained from satellite radiometric measurements, such as those shown in Figs. 1 and 2. The terrain constants which govern surface heating may be inferrable from this type of measurement, and subsequently their values can be used to classify the land use in a numerical framework and to predict the heating response at some time in the future by using a model of the surface heat flux.

While climate modification on a global scale is still a debatable issue, small-scale but significant changes in climate have already occurred in and around urban areas. Satellite temperature measurements can be used to quantitatively define the extent and magnitude of these urbanized climate changes, and therefore to assess how future changes may affect the local microclimate.

For background information *see* METEOROLOGICAL SATELLITES in the McGraw-Hill Encyclopedia of Science and Technology. [TOBY N. CARLSON]

Weather forecasting and prediction

Research in weather forecasting has been concerned with the development and use of statistical methods to elucidate the complex interactions among the physical phenomena that determine weather, and with the further development of deterministic mathematical models. These and other factors have improved the accuracy of long-range forecasting.

Statistical weather prediction. In recent years electronic computers have made it possible to apply statistical prediction methods, notably multiple linear regression, to complex weather relationships more effectively than in the past. Computers have also contributed to advances in deterministic predictions from dynamical equations of the atmosphere. At the same time there has been a trend toward the unification of statistics and dynamics in weather prediction. Investigators have studied both the practical and theoretical problems associated with this unification.

Basic regression methods. The simplest regression equation is a linear relation between the weather element to be predicted (the predictand) and one or more other weather elements (the predictors). A sample of historical values of these elements is used to develop the equation by the least-squares method. A linear rather than nonlinear equation is preferred, largely for simplicity. Each equation is intended for a specific location or region under limited circumstances. For example, one predictand might be the nocturnal minimum temperature in January at a particular site, while the predictors might include the observed temperature and wind speed at that site at 8 P.M. on the previous evening.

A large sample of historical observations is desirable in developing the equation in order to ensure reliability when it is applied to independent data. Furthermore, practical experience has shown that the most valuable regression equations include predictors that have known physical relationships with predictands. A useful predictor could be a combination of two or more meteorological observations. Thus, formal knowledge and subjective experience are used to obtain equations that are objective forecast tools.

The use of a large number of predictors in one equation is not particularly advantageous, in part because correlations among the predictors may result in redundant predictive information. A method to overcome this problem is screening regression, first used extensively by R. Miller in the 1950s. In this method, predictors are selected according to their ability to account for the variability of the predictand and are rendered independent of each other by use of regression procedures that define new predictors in place of original ones. By this means, hundreds of possible predictors are usually reduced to a dozen or less.

If one wishes to forecast the probability of an event, such as the occurrence of precipitation or freezing temperatures, a technique called regression estimation of event probabilities (REEP) can be used. This technique was developed by Miller and has had wide application. Basically the procedure relies on the determination of the relative frequencies of the predictand as observed in predefined classes of one or more predictors in the developmental sample. A predictor may be continuous or binary. If it is continuous, the class containing its observed value is used to obtain the predicted probability.

In general, weather predictions from regression equations and other purely statistical methods are not useful beyond 24 hr. However, with the recent advent of numerical prediction methods that employ hydrodynamical and thermodynamical models to make large-scale weather predictions a day or more in advance, there has been a parallel development of methods that link statistics and dynamics to give more extended forecasts of weather elements than provided by dynamical models alone. Two important examples of such a combination are the perfect prognostication technique and model output statistics (MOS).

Perfect prognostication technique. From a preferably large sample of meteorological observations, a regression equation, for example, is obtained based essentially on concurrent values of predictors and the predictand. If there were perfect forecasts available for the predictors themselves, the regression equation would be maximally efficient in providing a value of the predictand. Lacking such perfect forecasts, however, forecast values from dynamical models are used to produce the final forecast.

An illustration of this technique is the prediction of maximum and minimum surface temperatures up to 60 hr in advance, as described by W. Klein and F. Lewis. They obtained multiple linear regression equations by the screening method from 18 years of daily data. Among the selected predictors were heights of, and thicknesses between, isobaric surfaces. (These variables are also predicted by dynamical models.) Equations were developed for more than 100 cities in North America and were used operationally for several years until replaced in 1973 by an improved method based on MOS.

An obvious disadvantage of the perfect prognostication technique is the disparity between forecast and observed values of predictors. However, as dynamical models improve, this disparity becomes smaller.

Model output statistics. This method resembles the perfect prognostication technique in that concurrent values of predictors and predictand are used to develop a forecast tool, most commonly a linear regression equation. However, MOS differs from the perfect prognostication technique in that the predictor values themselves are the output from dynamical model predictions.

Two disadvantages of this method are that developmental samples are not large because dynamical models have been in operation only for a short period of time and because the statistical equations must be rederived as the models improve. Nevertheless, at present MOS is the basis for many National Weather Service forecasts. One example is the probability of precipitation forecast first developed by H. Glahn and D. Lowry, which relies on multiple predictors and employs the REEP technique. Objective predictions are supplied to local forecasters and are valid up to 48 hr. The local forecaster can often improve on this prediction by using additional information and personal experience.

The MOS method is also used to predict, among other elements, maximum and minimum surface temperatures; surface wind speed and direction; and probabilities of frozen precipitation, precipitation amount, cloud amount, ceiling and visibility, thunderstorms, and severe local storms.

Initial-state errors. Because of the large gaps in distance between observing stations, uncertainties and errors arise in any representation of the initial state of a weather forecast. The effect of initial-state errors on the quality of predictions has been the subject of recent theoretical investigations.

In one of these studies T. Gleeson postulated a model of initial-error distributions that are functions of the observation network density, and used a standard propagation-of-error method to transform initial distributions into predicted distributions by means of dynamical model equations. The final distributions then yield theoretical probabilities of successful prediction by the dynamical model itself. These probabilities have been compared with observed relative frequencies with reasonably successful results.

A second approach to initial-state errors was developed by E. Epstein and is called stochastic dynamics. Whereas Gleeson's model is concerned with error distributions of individual weather variables, stochastic dynamics focuses on the composite multivariate-error distribution for all variables in a given dynamical model. Epstein,

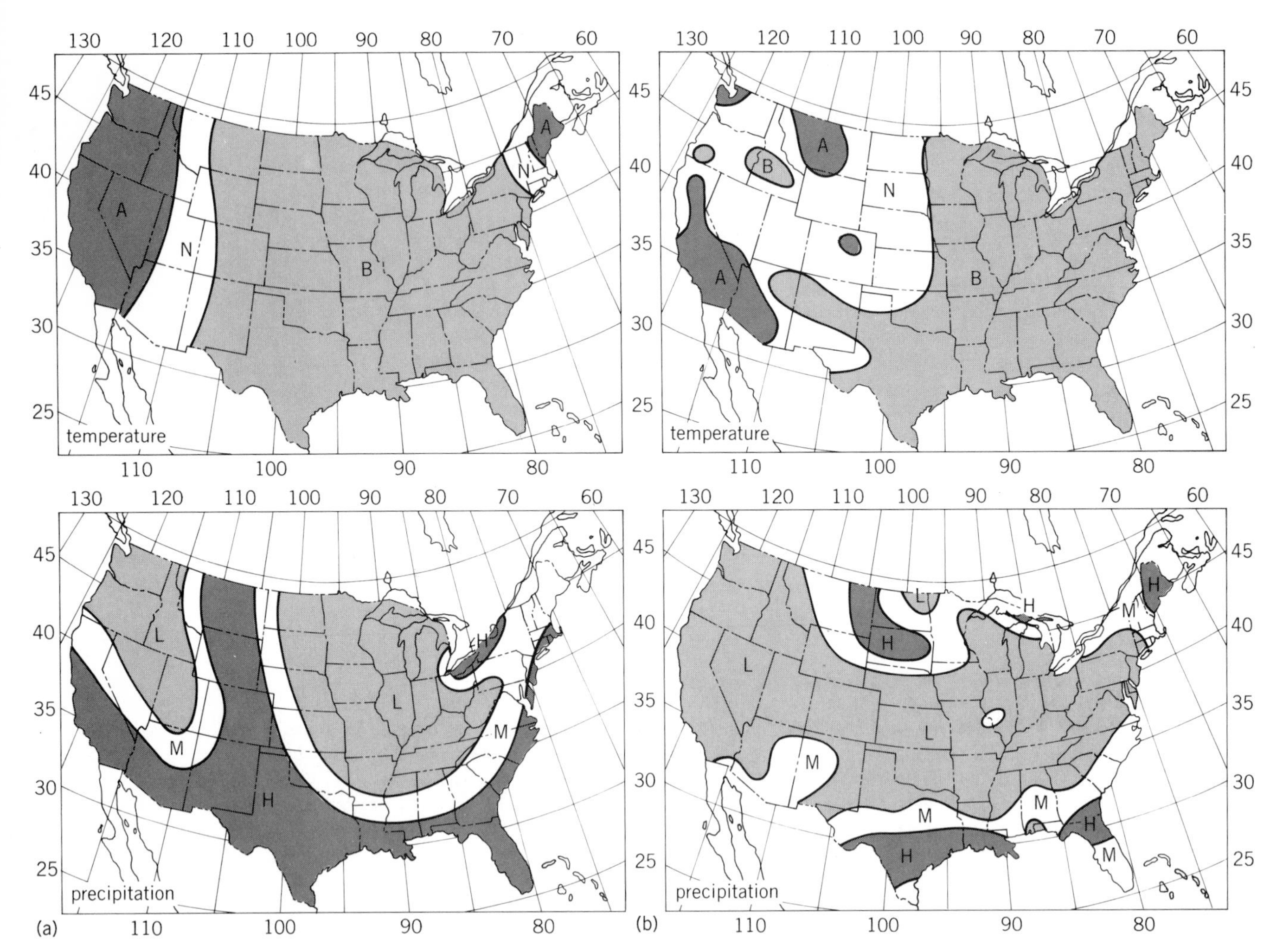

Maps of (*a*) predicted temperature and precipitation anomaly classes and (*b*) observed patterns for winter (December, January, February) 1976–1977. A stands for above-normal, N for near-normal, and B for below-normal temperatures. H stands for heavy, M for moderate, and L for light precipitation.

R. Fleming, and E. Pitcher have studied the evolution of such distributions with special attention given to the behavior of their means and variances. The results are of considerable theoretical interest, but there is a practical difficulty in applying stochastic dynamics to the real atmosphere. A large number of calculations is required to make predictions of multivariate distributions, and if a substantial volume of the atmosphere is under study the work is prohibitive. C. Leith has shown, however, that good approximations to predicted distributions can be achieved by moderate use of the Monte Carlo sampling method.

[THOMAS A. GLEESON]

Deterministic mathematical models. Recent progress in numerical weather prediction models used in routine operations has derived from improvements in the accuracy and efficiency of mathematical methods used to approximate the primitive (basic) equations of atmospheric motions. Operational forecasts over the past year that have incorporated these mathematical improvements show increased accuracy out to 3 days. Research has also focused on establishing more accurate initial conditions for numerical forecasts and on improving skill in inferring atmospheric behavior which is not directly predicted by the primitive equations. As an introduction, these developments will be placed in a brief historical and theoretical perspective.

History. The role of deterministic mathematical models of the atmosphere has become increasingly important since 1955 in producing daily weather forecasts. In the early phases of operational numerical weather prediction, models were necessarily simplistic mathematically, and were intended to simulate only the most basic physical behavior of the atmosphere that related to the propagation of deep weather systems in the troposphere. After 25 years of research and experience much has been discovered and, with the rapid development in computer technology, a considerable amount of the new information has been applied to operations.

In the United States, as in all other industrialized nations, numerical forecasting has evolved into a centralized process. Heavy emphasis is placed on reliable and fast communications equipment to disperse forecasts across the United States and throughout the world. Centralization has been necessitated by economics and by the need for large computers to process data and integrate the complex physical equations governing the behavior of the atmosphere.

Limiting factors. Limited computer power historically has restricted the degree of complexity meteorologists could introduce into their prediction schemes while still providing timely products to forecasters and managers. In the early phases, some important physical phenomena were completely ignored: solar and terrestrial radiation and the hydrologic cycle were considered too complex to handle. However, until recently a far more serious factor limiting the accuracy of predictions was the mathematical error associated with replacing the mathematically exact derivatives in the atmospheric equations with finite differences. This so-called truncation error manifests itself physically in terms of a general slowness in model weather disturbances and can produce incorrect answers when large disturbances interact with smaller ones. Both effects become serious as the motions approach the same scale (size) as the grid increment used in finite differencing.

As recently as 1975, most meteorologists viewed truncation error, and mathematical approximations in general, as the most important factor in limiting the forecast ability of operational models. Other less important factors include initial data error and the general inability to "specify" with sufficient accuracy the best initial state from which model time integrations are to be performed. The inability to specify the initial state is related to scales of motion one cannot "see" in the observational nets or on the grid mesh used for integrations. The interaction of these sub-grid-scale motions with motions that can be seen have been shown to have substantial impact on forecasts after a few days and are certainly responsible for minor problems within this period. Some success has been achieved by inferring the effect of sub-grid-scale behavior by relating it to evolutions on the numerical grid, but these approaches must necessarily be empirical.

Truncation error reduction. A number of approaches have been taken to reduce the truncation error associated with horizontal movement of weather systems. The most satisfactory approach from a theoretical point of view has been employed operationally in Australia and is called the spectral method. In this technique, once the harmonic functions describing meteorological variables have been calculated, the relationship between the variables and their derivatives is exact. Increased accuracy in the finite-differencing technique is attainable through the use of more than two points to approximate a first derivative (so-called higher-order finite difference). British meteorologists have incorporated this approach into their operational models.

The most expensive means of decreasing truncation error is through reduction of a grid increment. For each factor of 2 in increased resolution, the computational expense, everything else being equal, rises by a factor of 8. The United States has taken such a step recently with a valuable increase in forecast accuracy. The increase in cost was limited to a factor of only 3, however, with the improvement in efficient computer use. The United States experience has shown that while an increase of a factor of 2 has about the same advantage as the introduction of higher-order finite differencing as far as movement of high- and low-pressure systems is concerned, its contribution to increased accuracy in the movement of detailed moisture fields is considerably higher. Thus increased expense improves the forecast of elements of more immediate significance to most users. The move to higher-resolution grids has been made possible by the introduction of semi-implicit integration techniques that extend the time step two to four times beyond that conventionally used.

The ability of models with improved resolution to move weather systems in a manner more like the atmosphere is not the only reason for their success. The ability to react realistically with sources of energy in the simulated atmospheres is also crucial. In this respect spectral models have yet to be proved, but recent advances suggest that they may be equal to their finite counterparts.

Within the past year it has become apparent that models with grid spacing of less than 100 km offer little improvement in daily prediction at middle latitudes. The major hindrance to further improvement is associated with errors in the initial conditions. Even with the advent of satellite measurements, sparse data regions over the oceans remain a serious problem for downwind forecasters. In the United States the improvement of forecasts with newer models has been evident much more in the eastern states than in the western ones. Postevaluations of the least accurate forecasts verify that initial errors in the Pacific are most often the cause of poor performance.

Improvements in data input. Several developments should improve the input received from these data shadow areas. One involves improvements in indirect sensing techniques from satellites, along with more complete systems of quality control for recognition of cloud-contaminated measurements. Another advance is related to the use of completely automatic aircraft observational systems that employ satellites to transmit information rapidly back to forecast centers. An automated system using ocean buoys as platforms has been playing an increasingly important role in complementing surface reports from ships near continents. The observation of cloud movement from geostationary satellites over the Equator is allowing more accurate inferences of wind velocities, particularly near the Earth's surface and at levels in the upper troposphere. In cases where particularly serious weather events are likely, special observation plans are initiated employing reconnaissance aircraft and satellite imagery focused in more detail in time and space.

Model initialization systems. In order to fully utilize these new sources of weather information, more sophisticated model initialization systems are being developed. These systems differ from early approaches in their ability to logically account for variations in accuracy of observations from the varied data sources. Ideally, the observations also have to be adjusted to achieve a compatibility between model equations and atmospheric physics. This process of so-called data assimilation in models forces a close relationship between the models themselves and the means with which information is obtained from the atmosphere. The approach is far more expensive than that used in the past, but it is necessary to maximize the data.

Limited-area grids. Another approach aimed at

more effective use of information in dense data regions or during major storms is the use of limited-area grids. By reduction of the domain of integration with lateral boundaries determined by a larger model, perhaps global in extent, it is possible to dramatically increase grid resolution. While typically grids with increments of 50 km or so demonstrate no particular advantage over 100-km grids, they have been found to produce more useful guidance for winter blizzards, hurricanes, and heavy precipitation events.

Global Atmospheric Research Program. The developments discussed in this article are in part related to a worldwide program to improve weather forecasts. The Global Atmospheric Research Program (GARP) has as its objectives the solution to problems discussed above, and has organized global prediction experiments during 1978–1979 to determine whether recent advances in observational systems and modeling technology can be shown to provide a new level of predictability.

[JOHN B. HOVERMALE]

Long-range weather forecasting. The need for reliable long-range forecasts has been brought into sharp focus especially during the past few years when the United States has been plagued by drought in the Far West and severe cold and snowy winters in the East. The economic impact of these abnormalities has been felt in agricultural production, energy and water resources, recreational facilities (particularly skiing), and other industries. In spite of its increased importance, long-range forecasting over the past several years has progressed slowly. Nevertheless, there has been progress in the following areas: (1) improvement in teaching methods for weather forecasters; (2) the development of new generations of high-speed computers which can carry out many physical and statistical computations formerly impossible; (3) increased awareness of how conditions of the Earth's surface influence the overlying atmosphere—particularly sea-surface temperatures and the extent of ice and snow; and (4) the development of better means of translating prevailing atmospheric wind patterns into associated weather patterns.

Short-range or extended forecasting is now largely handled by machine computations modified by the human forecaster. These predictions permit detailed daily forecasts out to about 5 days, although their accuracy declines rapidly with time. Long-range forecasting, on the other hand, does not deal with details but only predicts statistical characteristics of the coming month's or season's weather—for example, whether it will be abnormally hot or cold or unusually wet or dry. Long-range forecasting methods, which have been improved over the past several years, include the following:

1. Statistical and kinematic techniques are being used to predict long-period trends in wind systems, particularly the prevailing wind patterns in the troposphere and up to the jet stream level. By using this procedure it is possible to track and extrapolate the prevailing waves in the upper westerlies, high- and low-pressure cells at the surface, and other features each 15 days.

2. Improved statistical techniques have been developed which predict pressure patterns at surface and aloft. These techniques are frequently based on linear regression methods based on a historical data sample of about 30 years. The net result is a series of prognostic equations.

3. New methods of selecting analogs have been developed. Analogs are cases which appear to be similar to the present situation and, therefore, can serve as a guide for future forecasts. These techniques of pattern recognition are now handled by computer, sometimes utilizing empirical orthogonal functions (eigenvectors). The forecaster may consult a few selected analogs from past years and then decide the best one to use. The sequence of weather developments leading up to the present along with the analog is also used as a selector.

4. New charts have been developed which show, by cross-correlations, the influence of the upper-level pressure deviation from normal in one area upon all other areas in the Northern Hemisphere. These so-called teleconnections have been determined for points 5° in latitude and 10° in longitude apart, and are stratified according to season. Teleconnections are the net expression of the macrophysics of the independence of systems in the atmosphere.

5. After wind and pressure patterns have been predicted, they must be interpreted in terms of desired weather elements. This procedure, called specification, is now done with the help of computers which, for example, can indicate quantitatively the temperatures and precipitation amounts for each city as a function of nearby and remote configurations of wind and pressure.

6. Perhaps some of the most important recent developments in long-range forecasting have come about from the interpretation of interactions between the ocean and atmosphere and between the cryosphere (ice and snow) and the atmosphere. These so-called boundary influences are very important in long-range forecasting because, unlike the quickly changing atmosphere, the sea and ice and snow retain vast amounts of heat for long intervals of time. In this manner they provide a "memory" for the atmosphere, continually forcing it to return to an abnormal pattern. These forces involve the distortion of the jet stream, prevailing storm tracks, and so on, which are immediately responsible for droughts, floods, variations in the frequency and areas of vulnerability to hurricanes, swaths of tornadoes, and other catastrophic events.

While the accuracy of long-range predictions leaves much to be desired, temperature forecasts in the United States for a month or season in advance are about 65% accurate (where accuracy due to chance is considered to be 50%), while precipitation forecasts are no more than 60% accurate. An example of a good forecast is shown in the illustration. Although considerable research is being undertaken, no one has been successful in predicting climatic abnormalities beyond a season. However, with the new emphasis being placed on climate and the impetus gained by the passage of the National Climate Act in the United States and the efforts of GARP, it is expected that general forecasts for 6 mo, a year, and perhaps a few years in advance will be possible before 1980.

For background information *see* WEATHER FORECASTING AND PREDICTION in the McGraw-Hill Encyclopedia of Science and Technology.

[JEROME NAMIAS]

Bibliography: W. A. Bourke, *Mon. Weather Rev.*, 102:687–701, 1974; J. A. Brown, Jr., and K. A. Campana, *Mon. Weather Rev.*, vol. 107, 1978; D. M. Burridge and A. J. Gadd, *The Meteorological Office Operational 10-Level Numerical Prediction Model*, Sci. Pap. no. 34, Her Majesty's Stationery Office, London, 1977; E. S. Epstein, *Tellus*, 21: 739–759, 1969; H. R. Glahn and D. A. Lowry, *J. Appl. Meteorol.*, 11:1203–1211, 1972; T. A. Gleeson, *J. Appl. Meteorol.*, 9:333–344, 1970; W. H. Klein and F. Lewis, *J. Appl. Meteorol.*, 9:350–359, 1970; J. Namias, *Geophysical Predictions*, chap. 9: Long-Range Weather and Climate Predictions, pp. 103–114, National Academy of Sciences, 1978; J. Namias, Short period climatic variations, *Collected Works of J. Namias, 1934 through 1974*, 1975; J. D. Stackpole et al., *How to Pick a New Operational Forecast Model*, Conference on Weather Forecasting and Analysis and Aviation Meteorology, Oct. 16–19, 1978.

Wind tunnel

A means for meeting the need for improved Reynolds number capability in wind tunnels has been provided at the NASA Langley Research Center in Hampton, VA, by the development of tunnels capable of operating at cryogenic temperatures, with minimum temperatures near 77 K (−321°F). Cooling the wind tunnel test gas to cryogenic temperatures results in a large increase in Reynolds number without raising dynamic pressure, while reducing the tunnel drive-power requirements. The ability to vary temperature independently of pressure and Mach number allows the independent determination of Reynolds number, Mach number, and aeroelastic effects on the aerodynamic characteristics of the model.

Increased Reynolds number. Interest in the development of commercial and military aircraft which operate efficiently at transonic speeds has resulted in a review of the problems of flow simulation in transonic tunnels. One of the more serious problems is the inability to test subscale models at Reynolds numbers sufficiently near full-scale values to ensure the usefulness of the wind tunnel data, particularly with respect to critical shock–boundary layer interactions. The problem is especially acute at transonic speeds where, because of the large drive-power requirements of transonic tunnels, economic forces have dictated the use of relatively small tunnels, and problems associated with high model loads have limited tunnel operating pressure. With ever-increasing aircraft size, existing transonic tunnels are becoming even more inadequate in test Reynolds number capability. A practical solution to the Reynolds number problem has been realized in the recent development of wind tunnels capable of operating at cryogenic temperatures.

Cryogenic concept. The use of low temperatures in wind tunnels was first proposed in England in 1945 as a means of reducing tunnel drive-power requirements at constant values of test Mach number, Reynolds number, and stagnation pressure. However, the need for increased Reynolds number was not as acute at that time, and no practical means of cooling was apparent. The concept lay essentially dormant until 1971, when a small group of researchers at the Langley Research Center recognized additional advantages to it and developed a practical approach for its application. The advantages of the cryogenic concept can be illustrated by means of the Reynolds number equation. Reynolds number *(R)*, which is the ratio of the inertia force to the viscous force, is given by Eq. (1), where ρ is the density of the gas, V is

$$R = \frac{\text{Inertia force}}{\text{Viscous force}} = \frac{\rho V^2 l^2}{\mu V l} \tag{1}$$

the velocity, μ is the viscosity, and l is a reference length. This equation reduces to the well-known Eq. (2), where M is Mach number and a is the

$$R = \frac{\rho V l}{\mu} = \frac{\rho M a l}{\mu} \tag{2}$$

speed of sound. As the temperature is decreased, ρ increases and μ decreases. As can be seen from Eqs. (1) and (2), both of these changes result in in-

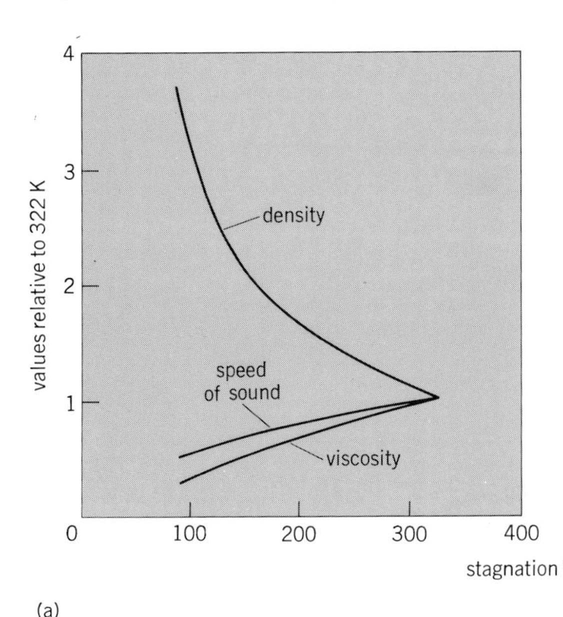

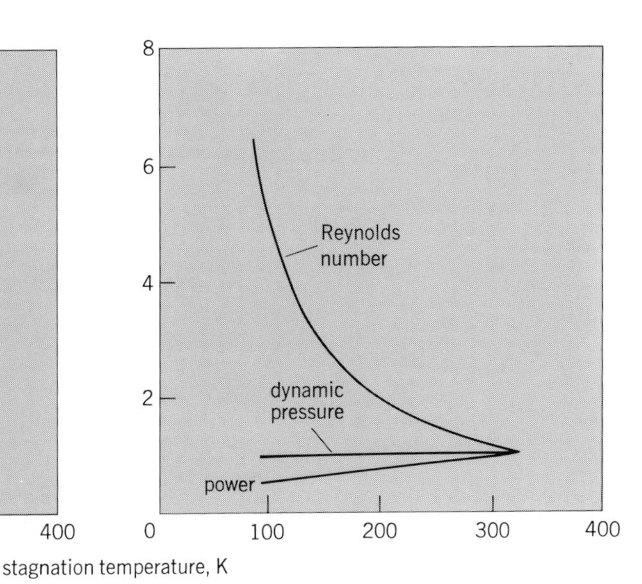

Fig. 1. Comparison of the effect of a reduction in temperature, at a Mach number of 1 and at constant stagnation pressure and tunnel size, on *(a)* gas properties and *(b)* test conditions and drive power.

creased Reynolds number. With decreasing temperature, a decreases. For a given Mach number, this reduction in the speed of sound results in a reduction in V which, while offsetting to some extent the Reynolds number increase due to the changes in ρ and μ, provides advantages with respect to dynamic pressure, drive power, and energy consumption.

It is useful to examine the underlying mechanism through which changes in pressure and temperature influence Reynolds number. To the first order, μ and a are not functions of pressure, while ρ is directly proportional to pressure. Thus, increasing pressure results in an increase in Reynolds number by raising the inertia force with a commensurate increase in model, balance, and model support loads. Also, to the first order, the dependence on temperature T is given by $\rho \propto T^{-1}$, $V \propto T^{0.5}$, and $\mu \propto T^{0.9}$. Thus, decreasing the temperature leaves the inertia force unchanged at a given Mach number due to the compensating effects of ρ and V^2. Increase in Reynolds number with decreasing temperature is thus due strictly to the large reduction in the viscous force term as a result of changes in μ and V with temperature.

The effects of a reduction in temperature on the gas properties, test conditions, and drive power are illustrated in Fig. 1. For comparison purposes, a stagnation temperature of 322 K (120°F) for normal tunnels is assumed. It can be seen that an increase in Reynolds number by more than a factor of 6 is obtained with no increase in dynamic pressure and with a large reduction in required drive power. To obtain such an increase in Reynolds number without increasing either the tunnel size or the operating pressure while actually reducing the drive power is extremely attractive and makes the cryogenic approach to a high-Reynolds-number transonic tunnel much more desirable than previous approaches.

Unique operating envelopes. The cryogenic concept also offers some unique and very desirable operating envelopes. Its wide range of stagnation temperatures, as an additional independent tunnel parameter, permits independent control of Reynolds number and dynamic pressure. Figure 2 indicates some of the implications of this unique capability for a constant-Mach-number case. It shows the envelope of dynamic pressure versus Reynolds number as bounded by the ambient-temperature line of 340 K (153°F) and the minimum-cryogenic-temperature line (based on liquefaction considerations) and minimum-pressure (1.0 atm, or 100 kPa) and maximum-pressure (8.8 atm, or 890 kPa) lines for a tunnel having a 2.5 × 2.5 m (8.2 × 8.2 ft) test section operating at a Mach number of 1. Since conventional, ambient-temperature pressure tunnels permit only minor temperature control—being essentially limited to operation along the ambient-temperature line—they encounter large changes in dynamic pressure, and therefore model deformation, with changes in Reynolds number.

In contrast, the cryogenic tunnel, with its large operating envelope, offers the following unique modes of operation: (1) a constant-dynamic-pressure mode in which the model shape can, for all practical purposes, be maintained constant while research is carried out on the effects of pure-Reynolds-number variations; and (2) a constant-Reynolds-number mode in which pure aeroelastic effects can be studied by variations in dynamic pressure.

The cryogenic tunnel can furnish many other combinations of Reynolds number and dynamic pressure as desired to match a particular aircraft flight condition. The ability to isolate these effects is extremely useful, since both aeroelasticity and Reynolds number can produce profound effects on critical aerodynamic phenomena, such as shock–boundary layer interactions.

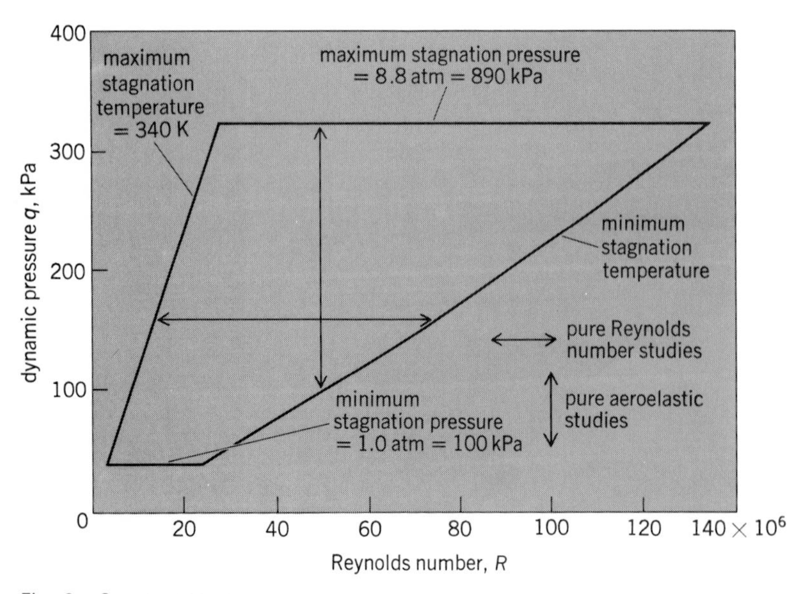

Fig. 2. Constant-Mach-number operating envelope for cryogenic nitrogen tunnel with 2.5 × 2.5 m test section, at a Mach number of 1. Reynolds number is based on a reference length of 0.25 m.

Application of cryogenic concept. Two relatively small fan-driven cryogenic wind tunnels were built during the development of the cryogenic wind tunnel concept at the Langley Research Center. The first, which was put into operation early in 1972, was a low-speed tunnel operated at atmospheric pressure. This tunnel, which had a test section of 18 × 28 cm (7 × 11 in.), was used to prove the validity of the concept as well as to develop the method of cooling the tunnel by direct injection of liquid nitrogen into the stream. Following the work with the low-speed tunnel, a larger cryogenic tunnel was built during 1973 which was capable of operating at pressures up to 5 atm (500 kPa) at transonic speeds. This tunnel had a test section of 34 × 34 cm (13.5 × 13.4 in.) and was used to verify the cryogenic concept at transonic speeds and provide design and operational experience necessary for the development of a large high-Reynolds-number cryogenic transonic tunnel.

Since 1972, several cryogenic wind tunnels have been built or are being planned in the United States and Europe. The largest of these is a new fan-driven tunnel under construction at the Langley Research Center. The tunnel, to be known as the National Transonic Facility (NTF), will have a 2.5 × 2.5 m (8.2 × 8.2 ft) test section and will be capable of operating from ambient to cryogenic temperatures at stagnation pressures up to 8.8 atm (890 kPa). By taking full advantage of the cryogenic concept, the NTF will provide Reynolds numbers of up to 120 × 10^6, which is an order of magnitude increase in Reynolds number capability over existing tunnels. As a national facility, the NTF will meet the high-Reynolds-number testing needs of NASA, the U.S. Department of Defense, other government agencies, the aerospace industry, and the scientific community.

For background information *see* REYNOLDS NUMBER; WIND TUNNEL in the McGraw-Hill Encyclopedia of Science and Technology.

[ROBERT A. KILGORE]

Bibliography: J. B. Adcock, R. A. Kilgore, and E. J. Ray, *Cryogenic Nitrogen as a Transonic Wind-Tunnel Test Gas*, AIAA Pap. no. 75-143, January 1975; R. A. Kilgore et al., *The Cryogenic Wind-Tunnel Concept for High Reynolds Number Testing*, NASA TN D-7762, 1974; O. W. Nicks and L. W. McKinney, *Status and Operational Characteristics of the National Transonic Facility*, AIAA Pap. no. 78-770, April 1978; E. C. Polhamus et al., *Astronaut. Aeronaut.*, 12(10):30-40, October 1974.

Wing

Recently a number of different aircraft have been developed with small, nearly vertical surfaces mounted at the wing tips (Fig. 1). These surfaces significantly decrease the drag or air resistance of the airplanes, which allows reductions in engine power and fuel consumption. These improvements provide increases in the range or payload of the airplanes. Since the shape and basic aerodynamic action of these surfaces are similar to those of wings, they are called winglets.

Fig. 1. Winglets on Gates Learjet model 28. *(Gates Learjet Corporation, Wichita)*

Function of vertical surfaces. The lift or upward force on a wing is produced by air pressures which are less than atmospheric above the upper surface and greater than atmospheric below the lower surface. At the wing tip the reduced pressures above the wing draw the air inward, while the increased pressures below the wing force it outward (Fig. 2). These cross flows persist behind the wing, where these effects, together with a downflow behind the wing and an upflow beyond the wing tip, combine to cause a vortex or swirling flow behind each wing tip. These vortexes are similar to small tornadoes. Substantial so-called induced drag is associated with these vortexes; at the usual airplane cruise flight conditions it is about 40 to 50% of total airplane drag. It has been recognized for many years that vertical surfaces at the wing tips should decrease the cross flows at the tips, lessen the strength of the resulting vortexes, and thus reduce the drag. As early as 1897 a patent was obtained by F. W. Lanchester for vertical surfaces at the wing tips. During the period from about 1925 to 1955 a number of wind tunnel experiments were conducted on various such surfaces, called end plates. However, all of these experiments indicated very little or no reduction in drag at airplane cruise conditions.

Design of winglets. Research on winglets was initiated at the Langley Research Center of the National Aeronautics and Space Administration in 1974. In contrast to the simple flat end plates tested earlier, winglets are designed with the same attention to detail as is used in designing wings. In particular, as for wings, the surfaces have specially designed airfoil shapes. (Airfoils are the cross sections of the wing or winglet parallel to the flight direction and perpendicular to the span or length of the surface.) For wings the airfoils are "cambered," that is, the curvature of the upper surface is greater than that of the lower. With such a shape, lift is produced more effectively. For winglets mounted above the wing tips (Fig. 1) the airfoils are cambered such that the inner sides have more curvature than the outer sides, as shown in Fig. 3. Further, as for wings, the winglets usually have relatively high aspect ratios, that is, the ratios of the length or span dimensions to streamwise or chord dimensions (Fig. 1).

Action of winglets. The action of the winglets is illustrated in Fig. 3. Because of the camber of the winglet airfoil, the winglet produces a significant side force directed inward. This inward force results in an outward redirection of the airflow behind the winglet in the same manner that the upward force on a wing causes a downward flow behind the wing. This redirection of the flow reduces the inflow behind the tip, resulting in a decrease in the strength of the associated vortex downstream of the tip and a reduction in the drag. Theoretical analyses and experiments indicate that the effectiveness of winglets in reducing induced drag is approximately proportional to the heights of the surfaces. However, increasing the height increases the aerodynamic loads imposed on the winglets and wing, which then require stronger, heavier structures to withstand these added loads. Analyses have indicated that a winglet height approxi-

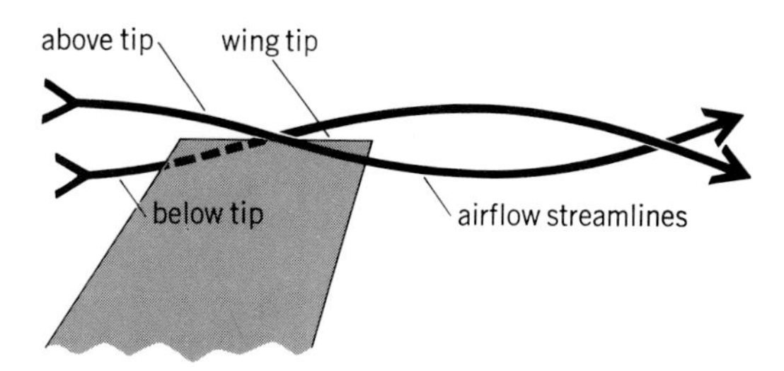

Fig. 2. Diagram of airflow causing vortex formation behind a wing tip. View is looking downward from above the wing tip.

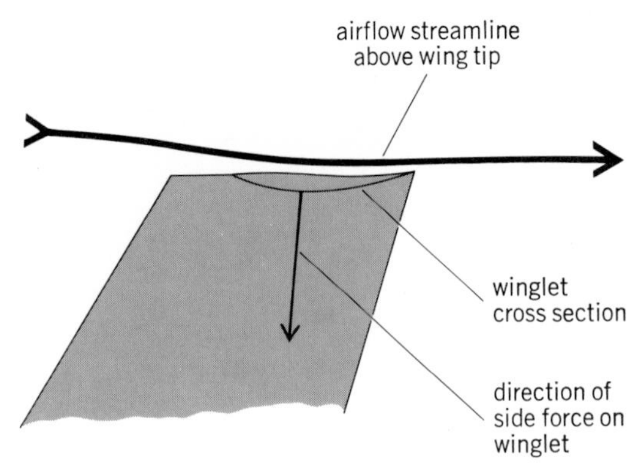

Fig. 3. Effect of upper surface winglet on cross flow at wing tip. View is looking downward from above the wing tip.

mately equal to the tip chord provides the most satisfactory compromise between the favorable aerodynamic effect and the adverse structural problem.

Positioning of winglets. For a given winglet height a winglet extending downward from the tip is almost as effective as the upward-extending version shown in Fig. 1. For such an arrangement the winglet airfoil is cambered with the greatest curvature on the outer surface, since the winglet in this position must reduce outflow behind the wing (Fig. 2). However, for an airplane with the wing mounted in a low position with respect to the fuselage, a lower winglet of the desired length would impact the ground during landing and takeoff. Therefore, for such airplanes the primary winglet must extend upward. For a wing mounted high on the fuselage, a downward-extending winglet is practical. Also, a winglet combination incorporating a rearward-located upper winglet and forwardly placed lower winglet may reduce the drag by a somewhat greater amount than a single upper or lower winglet. However, the aerodynamic gain may not justify the added cost and structural weight.

At the high subsonic flight conditions of jet transports and business jets, local regions of supersonic flow develop on the upper surface of the wing and the inner side of winglets mounted above the wing. When these regions are superimposed, a local shock wave which develops at the juncture of the two surfaces may cause a significant drag increase. To reduce or eliminate this problem, the upper winglets on such airplanes are placed rearward on the wing tip as shown in Fig. 1. Also, this problem is reduced by inclining the winglets outward from the vertical slightly and providing fairings at the junctures of the wing and winglets (Fig. 1).

Effectiveness of winglets. Extensive wind tunnel investigations of winglets on a number of different airplane configurations have indicated that these surfaces reduce induced drag by 10 to 20%, depending on the specific design of the airplane wing and the flight conditions. These improvements result in reductions of total airplane drag at cruise conditions of 4 to 8%. Further, these wind tunnel tests have indicated no adverse airplane stability or control problems associated with adding winglets. Winglets are now incorporated in the Rutan Varieze, the Israeli Aircraft Industries cargo aircraft, and the Gates Learjet model 28 (Fig. 1). They are part of the design of the forthcoming Grumman Gulfstream III. Also, winglets will be tested in flight on the U.S. Air Force KC-135 in 1979. If flight tests confirm the wind tunnel results, these surfaces will be retrofitted to most of the Air Force fleet of these aircraft. Further, wind tunnel tests of winglets on several commercial and military jet transport airplane configurations are continuing.

For background information *see* AIRFOIL; AIRPLANE; WING in the McGraw-Hill Encyclopedia of Science and Technology.

[RICHARD T. WHITCOMB]

Bibliography: S. G. Flechner and P. F. Jacobs, *Experimental Results of Winglets on First, Second, and Third Generation Jet Transports*, NASA TM-72674, May 1978; R. T. Whitcomb, *A Design Approach and Selected Wind-Tunnel Results at High Subsonic Speeds for Wing-Tip Mounted Winglets*, NASA TN D-8260, July 1976.

Wood preservation

Due to their toxic nature, all of the commercial wood preservatives presently used in the United States are effective in preventing attack by microorganisms. Most are classified as broad-spectrum preservatives, that is, they are effective against several different types of living systems. Because of environmental concerns and because prevention of wood decay is necessary to extend timber resources by increasing their service life, alternative methods of wood preservation which are not based on broad-spectrum toxicity are presently under investigation. Such nonconventional treatments are in the experimental stage. However, significant developments in the last year or two have shown the feasibility of these approaches.

Over the years, several nonconventional approaches for wood preservation have been reported. Research has been conducted on heat treatments, irradiation, polymer composites, and thiamine destruction, but none has been commercialized.

Chemical modification. One very promising approach is chemical modification of wood cell-wall components. Microorganisms, such as fungi, attack wood in the same way that the human stomach and intestine attack food. Both secrete enzymes which help break down the chemical structure of the food source into small, soluble units of nutrients. These enzyme reactions are very specific, so that if the food source or substrate is chemically changed, this digestive-type action can no longer take place.

The wood components that the fungi attack for food are cellulose and hemicelluloses contained in the cell wall. If these components have chemicals covalently bonded to them, the highly selective enzyme-substrate reaction is blocked. The chemicals used for substrate modification need not be toxic to the microorganism, because their effectiveness derives from rendering the substrate unrecognizable as a food source to support microbial growth.

By far the most abundant reactive chemical sites in wood are hydroxyl groups on the cellulose,

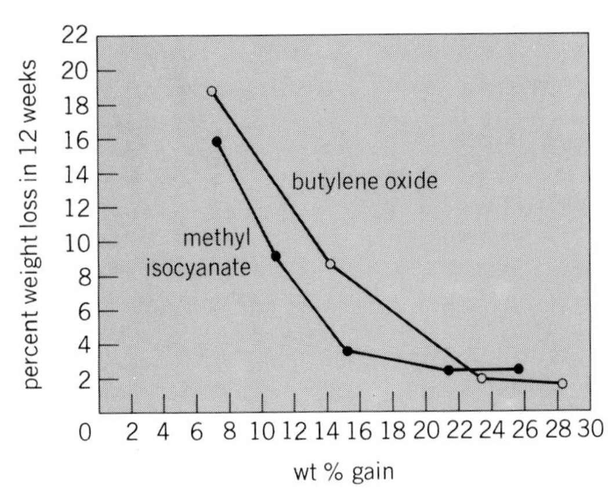

Resistance of modified southern pine to attack by *Gloeophyllum trabeum.*

hemicelluloses, and lignin. Chemicals being considered for wood modification must be capable of reacting with wood hydroxyls under neutral or mildly alkaline conditions at temperatures below 120°C. The chemical system should be simple, capable of swelling the wood structure to facilitate penetration, and volatile for easy removal of excess reagent after treatment. The complete molecule must react quickly with wood components, leaving no by-products, and must yield stable chemical bonds. The resulting product must still possess the desirable properties of untreated wood: it must have high strength, a good electrical insulator, and be safe to handle, gluable, and paintable. Two classes of reactive chemicals have been investigated that meet the above requirements—epoxides and isocyanates. Of the epoxides, butylene oxide shows the most promise at this time.

Epoxides. Butylene oxide reacts quickly with dry southern pine, with triethylamine as a catalyst, at 120°C and 150 psi (1 megapascal), and generates no by-products. At weight percent gains of 25–30, the modified wood shows good decay resistance in standard laboratory soil-block tests to *Gloeophyllum trabeum* (see illustration) and *Lentinus lepideus,* two widely distributed brown rot fungi well known for their wood-destroying activities. Untreated control blocks of southern pine lost 60–65 wt % in 12 weeks with *G. trabeum,* while butylene oxide–modified southern pine blocks lost 2–4%.

Because the epoxide becomes part of the wood cell wall, the volume of chemical added causes a proportional increase in wood volume. At about 25-wt % gain, the modified wood has swollen back to its original green volume. This bulking of the cell wall results in a decrease in swelling when the wood becomes wet. The reduction in swelling or antishrink efficiency resulting from butylene oxide modification is about 70% at weight percent gain levels of 25–30.

At these same levels of treatment, butylene oxide—modified southern pine was found to be resistant to attack by the subterranean termite *Reticulitermes flavipes* (Kollar). In a standard 2-week laboratory termite test, untreated control samples lost 31 wt % while the epoxide-modified samples lost 3 wt %. The modified wood was not toxic to the termites, nor did it repel them, but seemed to act as a feeding deterrent.

Mechanical properties of epoxide-modified hard maple as compared with untreated hard maple show a 10% reduction in crushing strength, 17% reduction in modulus of rupture, and 14% reduction in modulus of elasticity. Similar results were found for epoxide-modified basswood.

Isocyanates. The second class of chemicals that react quickly with wood hydroxyl groups are the isocyanates. Methyl isocyanate was found to react with southern pine uncatalyzed at 120°C and 150 psi to produce stable urethane bonds. There are no by-products with dry wood, and at weight percent gains of 25–30 an antishrink efficiency of 65–70 was achieved. At the same weight percent gains, isocyanate-modified southern pine blocks lost 2–3 wt % in the 12-week soil-block test, while untreated control blocks lost 50–60 wt % with *G. trabeum.* Weight losses due to fungal attack on butylene oxide–and methyl isocyanate–modified southern pine are given in the illustration.

Modification effect. Both modification treatments caused a slight darkening of light-colored woods and a slight bleaching of dark-colored woods. Water and organic solvent leaching tests showed that the chemicals bonded to the wood were stable and extremely weather-resistant.

Other applications. The concept of using bonded chemicals to change the basic properties of wood may have other applications besides rot and termite resistance and dimensional stability. These may include imparting such desirable traits as fire resistance, improved adhesion of glue or coatings such as paint, enhanced wood color, and resistance to corrosion, ultraviolet light, acids, or alkalies.

Although the modification treatment, if commercially feasible, will be more expensive than conventional preservation treatments, the nontoxic approach to wood preservation is an intriguing research concept. However, more research and development are necessary before these processes become commercially available.

Other nonconventional treatments are in preliminary research stages. Bound selective toxins, slow-release preservatives, repellents, feeding deterrents, and taking advantage of metabolic differences between microorganisms and humans are approaches presently under investigation.

For background information *see* WOOD, ANATOMY AND IDENTIFICATION OF; WOOD DETERIORATION; WOOD PRESERVATION in the McGraw-Hill Encyclopedia of Science and Technology.

[ROGER M. ROWELL]

Bibliography: R. M. Rowell, *Proceedings of the American Wood-Preservers Association,* vol. 71, pp. 41–51, 1975; R. M. Rowell, *Amer. Chem. Soc. Symp. Ser. No. 43,* pp. 47–56, 1977.

McGRAW-HILL YEARBOOK OF SCIENCE AND TECHNOLOGY

List of Contributors

List of Contributors

A

Abbott, Richard C. *Director of Marketing—Processable Laminations, Flexible Packaging Division, Continental Diversified Industries, Lombard, IL.* FOOD ENGINEERING (in part).

Alldredge, Dr. Alice L. *Department of Biological Sciences, University of California, Santa Barbara.* TUNICATA.

Anderson, Dr. E. J. *Associate Professor of Geology, Department of Geology, Temple University.* STRATIGRAPHY (coauthored).

Anderson, Dr. O. Roger. *Senior Research Associate, Biological Oceanography, Lamont-Doherty Geological Observatory, Columbia University.* FORAMINIFERA.

Andres, Dr. Klaus. *Bell Telephone Laboratories, Murray Hill, NJ.* ADIABATIC DEMAGNETIZATION.

Archibald, Dr. J. David. *Department of Geology and Geophysics, Yale University.* MAMMALIA.

Autian, Dr. John. *Materials Science Toxicology Laboratory, University of Tennessee Medical Units, Memphis.* TOXICOLOGY.

B

Banks, Dr. Harlan P. *Professor of Botany, Division of Biological Sciences, Cornell University.* PLANT EVOLUTION (in part).

Baron, Dr. Jonathan. *Department of Psychology, University of Pennsylvania.* PERCEPTION.

Bennett, Dr. Charles L. *Physics Department, Princeton University.* RADIOCARBON DATING.

Berkofsky, Prof. Louis. *Institute for Desert Research, Ben-Gurion University of the Negev, Beersheva, Israel.* DESERTIFICATION.

Bernstein, Leonard. *Petroleum Department, Exxon Research and Engineering Company, Florham Park, NJ.* PETROLEUM PROCESSING.

Billings, Dr. R. E. *Billings Energy Corporation, Provo, UT.* ENERGY SOURCES (in part).

Binkley, Dr. Sue A. *Biology Department, Temple University.* PERIODICITY IN ORGANISMS.

Bjerkaas, Capt. Carlton L. *Technical Consultant, Maynard, MA.* TORNADO.

Blandford, Prof. R. D. *Division of Physics and Astronomy, California Institute of Technology.* RADIO ASTRONOMY.

Breinan, Dr. Edward M. *Senior Materials Scientist, Materials Science Laboratory, United Technologies Research Center, East Hartford, CT.* LASER.

Briggs, Dr. D. E. G. *Department of Geology, Goldsmiths' College, University of London.* CRUSTACEA.

Brooks, Dr. Philip. *Department of Chemistry, Rice University.* MOLECULAR BEAMS.

Brynjolfsson, Dr. Ari. *Chief, Food Engineering Laboratory, Radiation Preservation of Food Division, U.S. Army Natick Research and Development Command, Natick, MA.* FOOD ENGINEERING (in part).

Bunn, Dr. Clive L. *Department of Genetics, National Institute for Medical Research, London.* CYTOPLASMIC INHERITANCE.

C

Carlson, Dr. T. N. *Department of Meteorology, Pennsylvania State University.* URBAN HEAT ISLAND.

Carpenter, Dr. C. B. *Peter Bent Brigham Hospital, Boston.* TRANSPLANTATION BIOLOGY.

Chu, Dr. T. Ming. *Professor and Director, Diagnostic Immunology Research and Biochemistry, Department of Health, Roswell Park Memorial Institute, Buffalo, NY.* CANCER (BIOLOGY).

Clague, Dr. David A. *Department of Geology, Middlebury College, Middlebury, VT.* OCEAN ISLANDS.

Clark, Dr. George R., II. *Department of Geology, Kansas State University.* ARCHEOLOGY.

Clark, Dr. John F. *RCA Laboratories, Princeton, NJ.* SPACE FLIGHT.

Clark, Dr. John H. *Los Alamos Scientific Laboratory, University of California, Los Alamos.* SOLAR CELL.

Clegg, Dr. John E. *Formerly, The Weapons Research Establishment, Adelaide, Australia.* HYDROGRAPHY.

Cocking, Prof. E. C. *Department of Botany, University of Nottingham, England.* BREEDING (PLANT).

Cohen, Aaron. *Manager, Orbiter Project, Lyndon B. Johnson Space Center, National Aeronautics and Space Administration.* SPACE SHUTTLE.

Cooper, Dr. Martin D. *Los Alamos Scientific Laboratory, University of California, Los Alamos.* NUCLEAR REACTION.

Corwin, Harold G., Jr. *Department of Astronomy, Royal Observatory, Edinburgh, Scotland.* GALAXY, EXTERNAL.

Cotton, Prof. F. Albert. *Department of Chemistry, Texas A & M University.* BOND ANGLE AND DISTANCE (coauthored).

Cracraft, Dr. Joel. *Associate Professor, Department of Anatomy, University of Illinois Medical Center.* ANIMAL SYSTEMATICS.

Creagan, Dr. Robert J. *Director of Technology Assessment, Westinghouse Electric Corporation, Pittsburgh.* ENERGY STORAGE.

Cummings, L. G. *Nuclear Insurance Consultants, Marsh and McLennan, Inc., New York.* NUCLEAR POWER (in part).

D

Denton, Prof. Bonner. *Associate Professor, Department of Chemistry, University of Arizona.* MICROCOMPUTERS.

Deutscher, Prof. Guy. *Physics Department, Tel Aviv University, Israel.* APPLICATIONS OF SUPERCONDUCTIVITY (feature).

Dixon, Dr. Joe B. *Professor of Soil Mineralogy, Department of Soil and Crop Science, Texas A & M University.* SOIL CHEMISTRY (in part).

Doering, Dr. Otto C., III. *Associate Professor, Department of Agricultural Economics, Purdue University.* AGRICULTURE.

Dolphin, Prof. David. *Department of Chemistry, University of British Columbia.* PORPHYRIN.

Doremus, Prof. R. H. *New York State Professor of Ceramics and Glass, Materials Engineering Department, Rensselaer Polytechnic Institute.* CHEMICAL DURABILITY OF GLASS (feature).

Dungan, Dr. Michael A. *Department of Geological Sciences, Southern Methodist University.* MAGMA (coauthored).

Dunn, Dr. S. Thomas. *President, Dunn Technology, Inc., Newport Beach, CA.* PRINTING PLATE.

E

Elder, Dr. Hugh Y. *Institute of Physiology, University of Glasgow.* CONNECTIVE TISSUE.

Ellis, Dr. David E. *Exploration Research Division, Conoco, Ponca City, OK.* HIGH-PRESSURE PHENOMENA.

Engelberg, Dr. Hyman. *Cedars-Sinai Medical Center, Los Angeles.* BLOOD.

Epstein, Prof. Emanuel. *Department of Land, Air and Water Resources, University of California, Davis.* PLANT GROWTH (in part).

F

Ferrante, Joseph, Jr. *Energy Systems Division, Wheelabrator-Frye Inc., Hampton, NH.* ENERGY SOURCES (in part; coauthored).

Fiory, Dr. Anthony T. *Bell Laboratories, Murray Hill, NJ.* MAGNETIC MATERIALS.

Frank, Dr. William M. *Research Associate, Department of Atmospheric Science, Colorado State University.* CYCLONE.

Frelinger, Prof. Jeffrey A. *School of Medicine, University of Southern California, Los Angeles.* IMMUNOLOGY.

Friedman, Dr. Herman. *Department of Microbiology and Immunology, Albert Einstein Medical Center, Philadelphia.* IMMUNITY.

Fujita, Prof. T. Theodore. *Department of Geophysical Sciences, University of Chicago.* THUNDERSTORM.

G

Ganotis, Chris G. *Energy Systems Division, Wheelabrator-Frye Inc., Hampton, NH.* ENERGY SOURCES (in part; coauthored).

Gibbons, Dr. J. Whitfield. *Savannah River Ecology Laboratory, Aiken, SC.* REPTILIA.

Gibson, Dr. Arthur C. *Department of Ecology and Evolutionary Biology, University of Arizona.* STEM (BOTANY).

Gleeson, Prof. Thomas A. *Department of Meteorology, Florida State University.* WEATHER FORECASTING AND PREDICTION (in part).

Godson, Dr. G. Nigel. *Associate Professor of Radiobiology, Yale University School of Medicine, Radiobiology Laboratories, New Haven, CT.* DEOXYRIBONUCLEIC ACID (DNA).

Gondhalekar, Dr. Awinash. *Francis Ritter National Magnet Laboratory, Massachusetts Institute of Technology.* FUSION, NUCLEAR (in part).

Goodwin, Dr. P. W. *Department of Geology, Temple University.* STRATIGRAPHY (coauthored).

Gordon, Dr. Donald T. *Department of Plant Pathology, Ohio Agricultural Research and Development Center, Wooster.* CORN.

Gordon, Prof. Robert J. *Associate Professor of Chemistry, Department of Chemistry, University of Illinois.* LASER CHEMISTRY.

Gray, Dr. Jane. *Department of Biology, College of Arts and Sciences, University of Oregon.* PLANT EVOLUTION (in part).

Greenberg, Prof. Jack S. *Physics Department, A. W. Wright Nuclear Structure Laboratory, Yale University.* SUPERHEAVY QUASIATOMS.

Griffin, Dr. Owen M. *Ocean Technology Division, Naval Research Laboratory, Washington, DC.* OCEANS AND SEAS.

Gross, Dr. Ian. *Assistant Professor of Pediatrics, Department of Pediatrics, Division of Perinatal Medicine, Yale University.* RESPIRATORY SYSTEM DISORDERS.

H

Hadler, Dr. J. B. *David W. Taylor Naval Ship Research and Development Center, Bethesda, MD.* PROPELLER, MARINE.

Hadley, Dr. Neil F. *Professor of Zoology, Department of Zoology, Arizona State University.* PHYSIOLOGICAL ECOLOGY.

Hänsch, Prof. Theo W. *Department of Physics, Stanford University.* SPECTROSCOPY, LASER.

Hardham, Dr. Adrienne R. *Department of Biological Sciences, Stanford University.* MERISTEM, APICAL.

Hawkins, Dr. James W. *Professor of Geology, Geological Research Division, Scripps Institution of Oceanography, La Jolla, CA.* PLATE TECTONICS.

Hayes, William C. *Editor in Chief, "Electrical World," McGraw-Hill Publications Company.* ELECTRICAL UTILITY INDUSTRY.

Hebard, Dr. Arthur F. *Bell Laboratories, Murray Hill, NJ.* QUARKS (in part).

Heinrich, Dr. Bernd. *Assistant Professor of Entomology, Department of Entomological Science, University of California, Berkeley.* SOCIAL INSECTS.

Henderson, Dr. Norman D. *Department of Psychology, Oberlin College.* BEHAVIORAL GENETICS.

Herrick, Dr. C. S. *Power Systems Laboratory, Research and Development Center, General Electric Company, Schenectady, NY.* SOLAR ENERGY (in part).

Herrmann, Prof. Robert B. *Associate Professor of Geophysics, Department of Earth and Atmospheric Sciences, St. Louis University.* EARTHQUAKE.

Hindman, Dr. James C. *Chemistry Division, Argonne National Laboratory.* CHLOROPHYLL.

Hirschberg, Irving. *Fairchild Camera & Instrument Corporation, Syosset, NY.* TELEVISION CAMERA.

Hopwood, Prof. David A. *John Innes Institute, Norwich, England.* STREPTOMYCETACEAE.

Houston, Jack E. *Supervisor, Sandia Laboratories.* SPECTROSCOPY (coauthored).

Hovermale, Dr. John B. *Chief, Atmospheric Modeling Branch, Development Division, National Meteorological Center, National Weather Service, Marlow Heights, MD.* WEATHER FORECASTING AND PREDICTION (in part).

Hunt, Ronald. *Southlands Experimental Forest, International Paper Company, Bainbridge, GA.* EUCALYPTUS.

I

Idso, Dr. Sherwood B. *U.S. Water Conservation Laboratory, Phoenix, AZ.* TERRAIN SENSING, REMOTE.

J

Jain, Prof. S. K. *Department of Agronomy and Range Science, University of California, Davis.* AMARANTH.

Jenkins, Dr. W. J. *Department of Chemistry, Woods Hole Oceanographic Institution, Woods Hole, MA.* DATING METHODS.

Jeong, Prof. Tung H. *Professor and Chairman, Physics Department, Lake Forest College, Lake Forest, IL.* HOLOGRAPHY.

Johnson, Dr. J. G. *Department of Geology, Oregon State University.* DEVONIAN.

Jöstlein, Prof. Hans. *Physics Department, State University of New York at Stony Brook.* QUARKS (in part).

Judd, Dr. L. R. *General Superintendent, Hidalgo Smelter, Phelps Dodge Corporation, Tyrone, NM.* FLASH SMELTING.

K

Kaufman, Peter. *Metal Tube Packaging Council of North America, New York.* FOOD ENGINEERING (in part).

Kehoe, John, Jr. *General Manager and Vice President, Energy Systems Division, Wheelabrator-Frye Inc., Hampton, NH.* ENERGY SOURCES (in part; coauthored).

Kiehl, Dr. Richard A. *Solid State Device Physics, Division 5133, Sandia Laboratories,* MICROWAVE SOLID-STATE DEVICES.

Kilgore, Dr. Robert A. *Langley Research Center, NASA.* WIND TUNNEL.

King, Prof. Elbert A. *Department of Geology, University of Houston.* TEKTITE.

Knapp, Dr. Gillian R. *Owens Valley Radio Observatory, California Institute of Technology.* STAR FORMATION (feature).

Knox, Prof. R. Bruce. *School of Botany, University of Melbourne.* FLOWER.

Kock, Dr. Winston E. *Herman Schneider Laboratory, University of Cincinnati.* HOLOGRAPHY, ACOUSTICAL.

Koehn, Dr. Richard K. *Associate Professor, Department of Ecology and Evolution, State University of New York at Stony Brook.* POPULATION GENETICS.

Kokosa, Richard A. *Manager—High Current Operations, Semiconductor Products Department, General Electric Company, Auburn, NY.* SEMICONDUCTOR DEVICES.

Kowal, Dr. Charles T. *Department of Astrophysics, California Institute of Technology.* ASTEROID.

Kuc, Prof. Joseph. *Department of Plant Pathology, University of Kentucky.* PLANT DISEASE CONTROL.

Kuhn, Dr. Peter M. *Atmospheric Physics and Chemistry Laboratory, U.S. Department of Commerce, National Oceanic and Atmospheric Administration, Boulder, CO.* CLEAR-AIR TURBULENCE (CAT).

Kuroda, Prof. Paul K. *Department of Chemistry, University of Arkansas.* NATURALLY OCCURRING NUCLEAR REACTOR (feature).

L

Landsberg, Prof. Helmut E. *Institute for Fluid Dynamics, University of Maryland.* CLIMATIC FLUCTUATIONS (feature).

Lanford, Prof. William A. *Physics Department, A. W. Wright Nuclear Structure Laboratory, Yale University.* SUPERCONDUCTIVITY.

Latzko, William J. *Assistant Vice President, Irving Trust Company, New York.* PRINTING.

Lenhard, Jerry R. *Department of Chemistry, University of North Carolina.* ELECTROCHEMISTRY.

Lewin, Dr. Ralph A. *Scripps Institution of Oceanography, University of California, San Diego.* PROCHLOROPHYTA.

Litman, Ira. *Flavor Fragrance Department, Stepan Chemical Company, Northbrook, IL.* COCOA POWDER AND CHOCOLATE (coauthored).

Lung, Iver A. *Air Force Geophysics Laboratory/LYD, Hanscom AFB, MA.* CLIMATOLOGY.

M

MacFarlane, Prof. R. D. *Department of Chemistry, Texas A & M University.* MASS SPECTROSCOPY.

Maneval, Dr. David. *Assistant Administrator for Technical Services and Research, Department of the Interior, Office of Surface Mining Reclamation and Enforcement.* MINING (in part).

Mark, Prof. Harry B. *Department of Chemistry, University of Cincinnati.* ELECTRODE.

Mattis, Prof. Daniel C. *Department of Physics, Polytechnic Institute of New York, Brooklyn.* SURFACE PHYSICS.

Maynard, Prof. Donald N. *Department of Plant and Soil Sciences, University of Massachusetts.* FERTILIZER.

Maynard, Prof. Julian. *Physics Department, Pennsylvania State University.* SOUND.

Mekolites, William E. *American Electric Power System, New York.* ELECTRIC POWER SYSTEMS.

Michaudon, Prof. André. *Service de Physique Nucléaire, Centre d'Etudes, Bruyères-le-Chatel, France.* FISSION, NUCLEAR.

Miscoe, Anthony. *Supervisory Mining Engineer, Hydraulic Transport Group, Department of Energy, Pittsburgh Mining Operations.* PIPELINE.

Mitchell, Donald. *Chief Mining Engineer, Gates Engineering Company, Berkeley, WV.* MINING (in part).

N

Namias, Dr. Jerome. *Scripp Institute of Oceanography, La Jolla, CA.* WEATHER FORECASTING AND PREDICTION (in part).

Nayfeh, Dr. Munir Hasan. *Department of Physics, Loomis Laboratory of Physics, University of Illinois at Urbana-Champaign.* ATOM.

Novotny, Prof. Milos V. *Department of Chemistry, Indiana University.* CHROMATOGRAPHY.

Numrych, Schelly. *Flavor Fragrance Department, Stepan Chemical Company, Northbrook, IL.* COCOA POWDER AND CHOCOLATE (coauthored).

O

Owen, Dr. Denis. *Department of Biology, Oxford Polytechnic, Oxford, England.* HUMAN ECOLOGY IN THE TROPICS (feature).

Oxner, Dr. Edwin S. *Siliconix Inc., Santa Clara, CA.* INTEGRATED CIRCUITS.

P

Pardue, Prof. Harry L. *Department of Chemistry, Purdue University.* ANALYTICAL CHEMISTRY.

Perl, Prof. Martin. *Stanford Linear Accelerator Center, Stanford University.* LEPTON.

Person, Dr. Philip. *Special Dental Research Program, Veterans Administration Hospital, Brooklyn, NY.* CARTILAGE.

Plantz, Dr. Philip E. *Applications Scientist, Leeds and Northrup Company, North Wales, PA.* FOOD ENGINEERING (in part).

Powers, Prof. Philip N. *Professor Emeritus of Nuclear Engineering, Purdue University.* NUCLEAR POWER (in part).

R

Ramanujam, Dr. P. S. *Institute of Physics, University of Aarhus, Denmark.* ATOMIC STRUCTURE AND SPECTRA.

Reichle, Dr. Alfred Douglas. *Exxon Research and Development Laboratories, Baton Rouge, LA.* CRACKING.

Reintjes, Dr. John. *Naval Research Laboratory, Washington, DC.* OPTICS, NONLINEAR.

Rhodes, Dr. J. M. *Johnson Space Center, NASA, Houston.* MAGMA (coauthored).

Ribbink, Dr. A. J. *Senior Fisheries Research Officer, Ministry of Agriculture and Natural Resources, Fisheries Department, Fisheries Research Station, Monkey Bay, Malawi.* REPRODUCTIVE BEHAVIOR.

Rossin, Dr. A. David. *System Nuclear Research Engineer, Commonwealth Edison Company, Chicago.* RADIOACTIVE WASTE MANAGEMENT.

Roth, Dr. Charles B. *Department of Agronomy, Purdue University.* SOIL CHEMISTRY (in part).

Rowell, Dr. Roger M. *Carbohydrate Chemistry Specialist, U.S. Department of Agriculture, Forest Products Laboratory, Madison, WI.* WOOD PRESERVATION.

Rudnick, Dr. Lawrence. *National Radio Astronomy Observatory, Charlottesville, VA.* RADIO TELESCOPE.

Russell, Dr. Barry C. *Biological Sciences, Macquarie University, New South Wales, Australia.* PROTECTIVE COLORATION.

Rye, Dr. Robert. *Sandia Laboratories, Albuquerque, NM.* SPECTROSCOPY (coauthored).

S

Salmon, Dr. Edward. *Assistant Professor, Department of Zoology, University of North Carolina.* CELL DIVISION.

Sancetta, Dr. Constance. *Department of Geology, Stanford University.* PALEOCEANOGRAPHY.

Schrauzer, Prof. Gerhard. *Department of Chemistry, University of California, San Diego.* SOLAR ENERGY (in part).

Sellin, Prof. Ivan A. *Department of Physics and Astronomy, University of Tennessee.* ION.

Sleper, Dr. David A. *Associate Professor, College of Agriculture, Department of Agronomy, University of Missouri—Columbia.* GRASS CROPS.

Staebler, Dr. David L. *RCA Laboratories, David Sarnoff Research Center, Princeton, NJ.* SILICON.

Stanley, George G. *Department of Chemistry, Texas A & M University.* BOND ANGLE AND DISTANCE (coauthored).

Steinberg, Dr. Meyer. *Process Sciences Division, Department of Energy and Environment, Brookhaven National Laboratory.* LINEAR ACCELERATOR.

Steponkus, Dr. Peter L. *Agronomy Department, Cornell University.* PLANT GROWTH (in part).

Stucki, Dr. J. W. *Assistant Professor of Soil Chemistry, Department of Agronomy, College of Agriculture, University of Illinois at Urbana-Champaign.* SOIL CHEMISTRY (in part).

Sullivan, D. B. *Electromagnetic Technology Division, National Bureau of Standards, Boulder, CO.* SUPERCONDUCTING DEVICES (coauthored).

Sumaya, Dr. Ciro V. *University of Texas Health Science Center at San Antonio.* EPSTEIN-BARR VIRUS.

T

Telschow, Prof. Kenneth L. *Department of Physics and Astronomy, Southern Illinois University.* HELIUM, LIQUID (in part).

Temple, Dr. Stanley A. *Department of Wildlife Ecology, University of Wisconsin.* ECOLOGICAL INTERACTIONS.

ten Hove, Dr. Harry A. *Laboratory for Zoological Ecology and Taxonomy, State University of Utrecht, Netherlands.* TUBE WORM.

Thies, Dr. Curt. *Professor of Chemical Engineering, Biological Transport Laboratory, Washington University, St. Louis.* MICROENCAPSULATION (feature).

Thomas, Prof. Gareth. *Department of Materials Science and Mineral Engineering, University of California, Berkeley.* STEEL.

Thouless, Prof. D. J. *Physics Department, Queen's University, Ontario, Canada.* HELIUM, LIQUID (in part).

Tinsley, Prof. Beatrice M. *Department of Astronomy, Yale University.* COSMOLOGY.

Topp, Dr. Michael. *Assistant Professor of Chemistry, Department of Chemistry, University of Pennsylvania.* FAST CHEMICAL REACTIONS.

Tuomi, Roger L. *Engineer, Forest Products Laboratory, U.S. Department of Agriculture, Madison, WI.* CONSTRUCTION METHODS.

Turcotte, Prof. Donald L. *Department of Geological Sciences, Cornell University.* EARTH, HEAT FLOW IN.

W

Ware, Dr. B. R. *Associate Professor, Department of Chemistry, Harvard University.* BIOPHYSICS.

Wells, Dr. Martin J. *Department of Zoology, University of Cambridge.* OCTOPUS.

Whitcomb, Dr. Richard T. *Aerospace Engineer, Langley Research Center, NASA, Hampton, VA.* WING.

Willis, Dr. Homer B. *Chief, Engineering Division, Directorate of Civil Works, Office of the Chief of Engineers, Department of the Army, Washington, DC.* DAM.

Wing, Dr. Elizabeth. *Associate Curator in Zooarcheology, Department of Natural Sciences, Florida State Museum, University of Florida.* DOG.

Yonas, Dr. Gerold. *Fusion Research Department, Sandia Laboratories, Albuquerque, NM.* FUSION, NUCLEAR (in part).

Z

Ziegler, Dr. A. M. *Associate Professor, Department of Geophysical Sciences, University of Chicago.* PALEOGEOGRAPHY.

Zimmerman, Dr. James E. *Cryoelectronic Metrology Program, Electromagnetic Technology Division, National Bureau of Standards, Boulder, CO.* SUPERCONDUCTING DEVICES (coauthored).

Zissis, Dr. George J. *Chief Scientist, Environmental Research Institute of Michigan.* INFRARED IMAGING DEVICES.

McGRAW-HILL YEARBOOK OF SCIENCE AND TECHNOLOGY

Index

Index

Asterisks indicate page references to article titles.

A

B

D

E

H

I

J

K

N

O

P

Q

R

S

T

U

V

W

X

Z